VERTEBRATE ICHNOLOGY

VERTEBRATE ICHNOLOGY

TETRAPOD TRACKS AND TRACKWAYS

Edited by

Spencer G. Lucas
New Mexico Museum of Natural History and Science, Albuquerque, NM, United States

Adrian P. Hunt
Flying Heritage and Combat Armor Museum, Everett, WA, United States

Hendrik Klein
Saurierwelt Paläontologisches Museum, Neumarkt, Germany

ELSEVIER

Elsevier
Radarweg 29, PO Box 211, 1000 AE Amsterdam, Netherlands
125 London Wall, London EC2Y 5AS, United Kingdom
50 Hampshire Street, 5th Floor, Cambridge, MA 02139, United States

ISBN: 978-0-443-13837-9

For information on all Elsevier publications visit our website at
https://www.elsevier.com/books-and-journals

Publisher: Candice Janco
Acquisitions Editor: Peter Llewellyn
Editorial Project Manager: Joshua Mearns
Production Project Manager: Sathyavani Deivasigamani
Cover Designer: Christian Bilbow

Typeset by TNQ Technologies

Contents

5. Triassic tetrapod footprints

Hendrik Klein and Spencer G. Lucas

6. Jurassic vertebrate tracks and traces

John R. Foster, Jerald D. Harris, Andrew R.C. Milner, Emese M. Bordy, Lara Sciscio, Diego Castanera, Matteo Belvedere, Lida Xing and Martin G. Lockley

7. Cretaceous vertebrate tracks

Lida Xing, Ignacio Díaz Martínez, Jens N. Lallensack, Kyung Soo Kim, Lisa G. Buckley, Anthony Romilio, Laura Piñuela, Donghao Wang, Qiyan Chen, Qing Yang, Chunyong Chou, Hang Yin and Yufei Jin

8. Ichnotaxonomy of Paleogene and Neogene tetrapod tracks

Nasrollah Abbassi, Ricardo Néstor Melchor and Ignacio Díaz-Martínez

Contributors

Nasrollah Abbassi Department of Geology, Faculty of Sciences, University of Zanjan, Zanjan, Iran

Matteo Belvedere Dipartimento di Scienze della Terra, Università di Firenze, Firenze, Italy

Emese M. Bordy Department of Geological Sciences, University of Cape Town, Rondebosch, South Africa

Michael Buchwitz Museum für Naturkunde Magdeburg, Magdeburg, Germany

Lisa G. Buckley Dinosaur Odyssey Science Museum, Guangxi, China

Diego Castanera Fundación Conjunto Paleontológico de Teruel-Dinópolis/Museo Aragonés de Paleontología, Teruel, Spain

Qiyan Chen School of the Earth Sciences and Resources, China University of Geosciences (Beijing), Beijing, China

Chunyong Chou School of the Earth Sciences and Resources, China University of Geosciences (Beijing), Beijing, China

Ignacio Díaz-Martínez Departamento de Ciencias de la Tierra y Física de la Materia Condensada, Facultad de Ciencias, Universidad de Cantabria, Cantabria, Spain

John R. Foster Utah Field House of Natural History State Park Museum, Vernal, UT, United States

Jerald D. Harris Utah Tech University, St. George, UT, United States; St. George Dinosaur Discovery Site at Johnson Farm, St. George, UT, United States

Adrian P. Hunt Flying Heritage Collection and Combat Armor Museum, Everett, WA, United States

Yufei Jin School of the Earth Sciences and Resources, China University of Geosciences (Beijing), Beijing, China

Kyung Soo Kim Department of Science Education, Chinju National University of Education, Gyeongnam, South Korea

Hendrik Klein Saurierwelt Paläontologisches Museum, Neumarkt, Germany

Jens N. Lallensack Departamento de Ciência da Computação, Universidade Federal de Minas Gerais, Belo Horizonte, Brazil

Martin G. Lockley Dinosaur Trackers Research Group, University of Colorado Denver, Denver, CO, United States

Antoine Logghe Centre de recherche en paléontologie—Paris, UMR 7207—CNRS-MNHN-Sorbonne Université, Muséum national d'Histoire naturelle, Paris, France

Spencer G. Lucas New Mexico Museum of Natural History and Science, Albuquerque, NM, United States

Lorenzo Marchetti Museum für Naturkunde Berlin, Leibniz-Institut für Evolutions- und Biodiversitätsforschung, Berlin, Germany

Ricardo Néstor Melchor Instituto de Ciencias de la Tierra y Ambientales de La Pampa (CONICET and Universidad Nacional de La Pampa), Santa Rosa, La Pampa, Argentina; Departamento de Geología, Universidad Nacional de La Pampa, Santa Rosa, La Pampa, Argentina

Andrew R.C. Milner St. George Dinosaur Discovery Site at Johnson Farm, St. George, UT, United States

Eudald Mujal Staatliches Museum für Naturkunde Stuttgart, Stuttgart, Germany; Institut Català de Paleontologia Miquel Crusafont (ICP-CERCA), Universitat Autònoma de Barcelona, Edifici ICTA-ICP, Catalonia, Spain

Laura Piñuela Museo del Jurásico de Asturias (Jurassic Museum of Asturias), Colunga, Spain

Anthony Romilio School of Biological Sciences, The University of Queensland, Brisbane, QLD, Australia

Lara Sciscio Jurassica Museum, Porrentruy, Switzerland; Department of Geosciences, University Fribourg, Fribourg, Switzerland

Matthew R. Stimson Steinhammer Paleontology Laboratories, New Brunswick Museum, St. John, NB, United States; Department of Geology, Saint Mary's University, Halifax, NS, Canada

Donghao Wang Frontiers Science Center for Deep-time Digital Earth, China University of Geosciences (Beijing), Beijing, China; School of the Earth Sciences and Resources, China University of Geosciences (Beijing), Beijing, China

Lida Xing Frontiers Science Center for Deep-time Digital Earth, China University of Geosciences (Beijing), Beijing, China; School of the Earth Sciences and Resources, China University of Geosciences (Beijing), Beijing, China; Dinosaur Odyssey Science Museum, Guangxi, China; State Key Laboratory of Biogeology and Environmental Geology, China University of Geosciences, Beijing, China

Qing Yang School of the Earth Sciences and Resources, China University of Geosciences (Beijing), Beijing, China

Hang Yin School of the Earth Sciences and Resources, China University of Geosciences (Beijing), Beijing, China

Vertebrate ichnology: Introduction and summary

Spencer G. Lucas[1], *Adrian P. Hunt*[2] *and Hendrik Klein*[3]

[1]New Mexico Museum of Natural History and Science, Albuquerque, NM, United States, [2]Flying Heritage Collection and Combat Armor Museum, Everett, WA, United States; [3]Saurierwelt Paläontologisches Museum, Neumarkt, Germany

1. Introduction

Ichnology (Greek ichnos, "trace") is the study of trace fossils, which are fossil evidence of the interaction of an organism with the substrate, and has been referred to as "fossil behavior." Trace fossils receive ichnotaxonomic names, and this parataxonomy is regulated by the International Code of Zoological Nomenclature (1999, article 1.2.1), which refers to ichnotaxa as "names based on the fossilized work of organisms (ichnotaxa)." Almost any organism could have produced a trace fossil, and, indeed, trace fossils of plants, invertebrate animals, and vertebrate animals have long been recognized and studied by paleontologists, many of whom call themselves ichnologists (Buatois and Mángano, 2011).

The study of vertebrate ichnology has largely been the study of fossil footprints, but also includes studies of fish trails and imprints, bromalites, dentalites, burrows, and nest structures. Study of vertebrate trace fossils began in the 1820s with classic work by British savant William Buckland, followed not long after in the 1830–50s by the work of American Edward Hitchcock (Fig. 1). Yet, despite a nearly 200-year history of research, vertebrate trace fossils were not regarded as important archives of vertebrate history until the last half-century or less. And, this has been a busy half-century of vigorous study, particularly of footprints and of coprolites. A book that presents a comprehensive review of vertebrate ichnology has long been overdue.

This is that book, the first volume to present a comprehensive review and extensive analysis of vertebrate trace fossils. It covers all aspects of the vertebrate

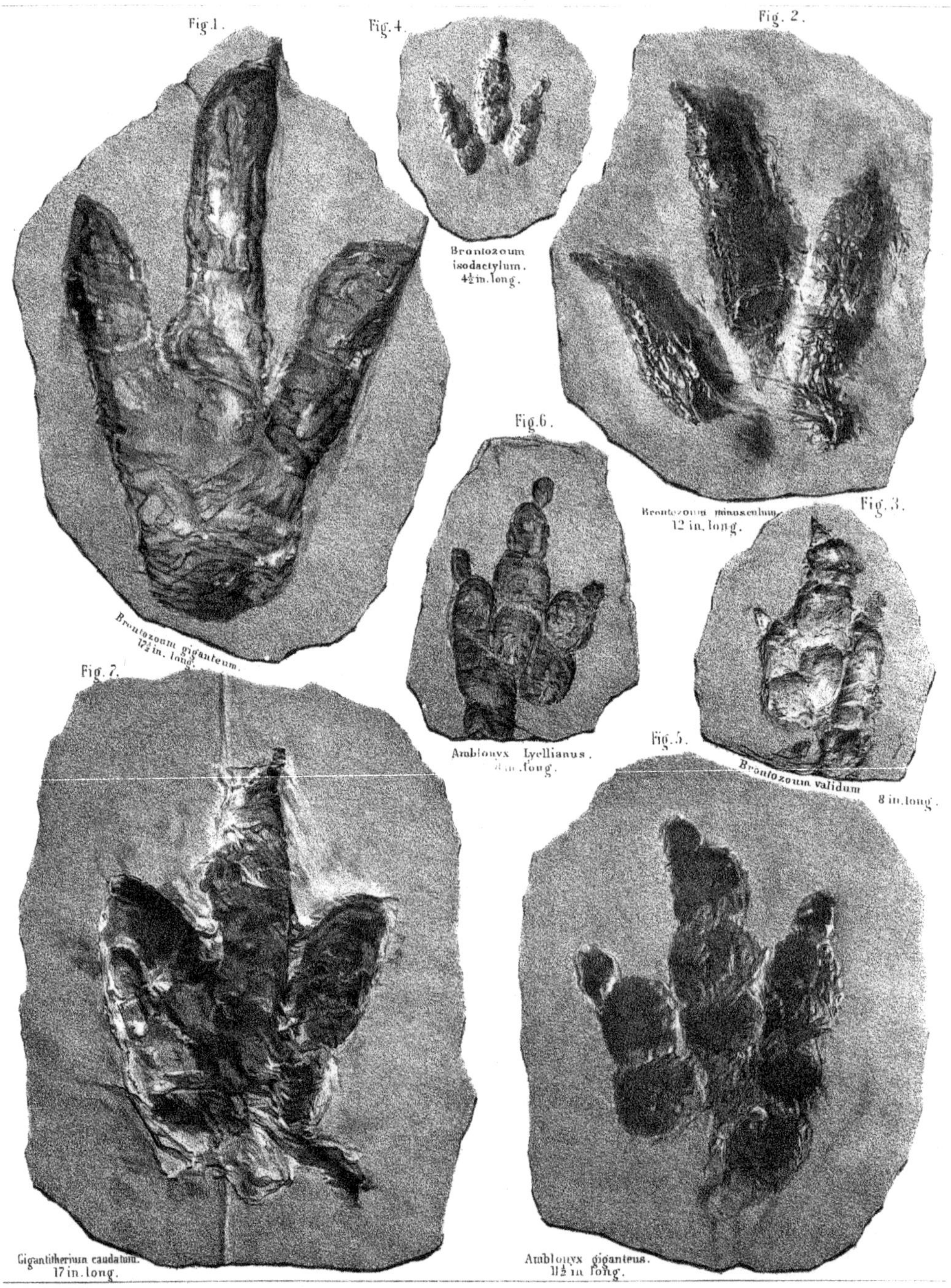

FIGURE 1 Some theropod dinosaur footprints from the Connecticut valley. *From Hitchcock (1858).*

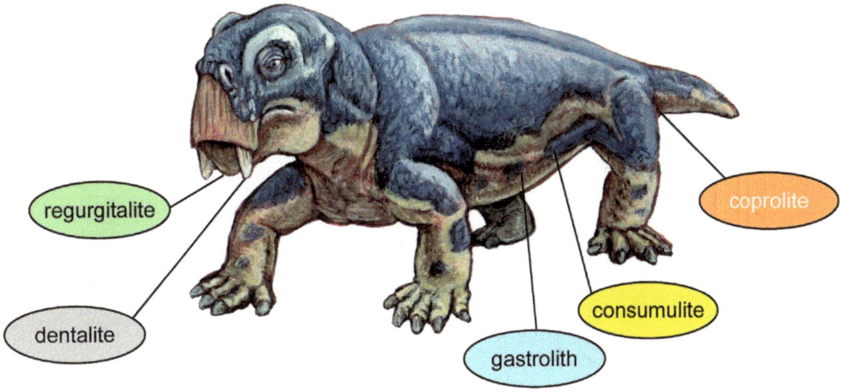

FIGURE 2 The trace fossils of vertebrates that are produced by consumption. *From Hunt and Lucas (2021). Artwork by Matt Celeskey.*

trace fossil record: tetrapod footprints, fish trails and imprints, dentalites, bromalites (regurgitalites, consumulites, coprolites, and others: Fig. 2), gastroliths, burrows, and much more. The record is reviewed here by prominent and current experts with extensive illustrations and up-to-date lists of the relevant scientific literature. Vertebrate trace fossils are used as a tool to solve problems of vertebrate biochronology and biogeography and to analyze and understand major events in the evolutionary history of vertebrates.

Here, we present an overview of this volume that summarizes its chapters. We also provide a broader context to understand the current status and hoped for future of vertebrate ichnology.

2. History

In this volume, Lucas et al. (2024) present an analytical history of the study of fossil vertebrate footprints and of the other major trace fossil groups reviewed in this volume. As already noted, most of the study of these trace fossils has focused on footprints and coprolites. It began with the classic work of William Buckland in the United Kingdom during the 1820s and 1830s, followed by research on vertebrate footprints in the eastern USA by Edward Hitchcock during the 1830–50s and the first scientific (binominal) name for a fossil footprint by Kaup in Germany during the 1830s. Together, by about 1850, their work and that of a few others provided a basis for a research program on fossil tetrapod footprints that included actualistic studies, identification of the tracemaker as critical to ichnotaxonomy and the use of footprints to identify the temporal and geographical distributions of fossil vertebrates in the absence of body fossils. Yet, footprint research languished until the 1980s, simply because vertebrate paleontologists saw little relevance of footprints to problems of vertebrate evolution, taphonomy, and paleoecology.

Coprolites were also long seen to be of little value to vertebrate paleontology, but in the 1950s research by archeologists on human coprolites developed to answer questions about diets, diseases, parasites, and climate. Research on dentalites, regurgitalites, consumulites, gastroliths, other bromalites, burrows, and nests was scattered and sporadic until the 1990s, when serious study of many of these traces began. At present, these traces are still in an early stage of research, with many records remaining to be documented and (except for dentalites and burrows) with ichnotaxobases and ichnotaxonomy in a very early stage of development.

The tetrapod footprint record extends from the Devonian through the Holocene and will continue to be expanded and applied to broad questions of tetrapod evolution, temporal and paleogeographic distribution, taphonomy, and paleoecology. Coprolites have a similar extensive and growing record and a developing ichnotaxonomy.

3. The vertebrate ichnology of locomotion

As already noted, most research in vertebrate ichnology has focused on studying tetrapod footprints, especially those of dinosaurs. Indeed, the footprint section makes up the bulk of this volume, in which most of the chapters review the tetrapod footprint record by geological time interval. Fishes also leave trace fossil evidence of motion, and study of their traces has developed much in the last half-century.

3.1 Methods and terminology

The methods of studying footprints were long based on two dimensional imaging by drawing or photography and with standard and long agreed on metrics (Fig. 3). In the 21st Century, new technology, especially photogrammetric imaging, some of which employs drones (Fig. 4), has improved the ability to capture three-dimensional data from traces and to more rigorously quantify their metrics.

Indeed, in this volume, Lallensack (2024) notes that both the methodology and terminology of vertebrate track ichnology are rapidly changing. Photogrammetry has improved the documentation of tracks, particularly allowing 3D morphology to be captured in great detail. Experiments and computer simulations have enhanced our understanding of the process of track formation. The recognition of penetrative tracks requires the re-evaluation of many existing ideas about trackmaker dynamics and undertracking. New methods are being used to evaluate quadrupedal gaits and gleno-acetabular distances from trackways. The application of artificial intelligence to tracks, which are

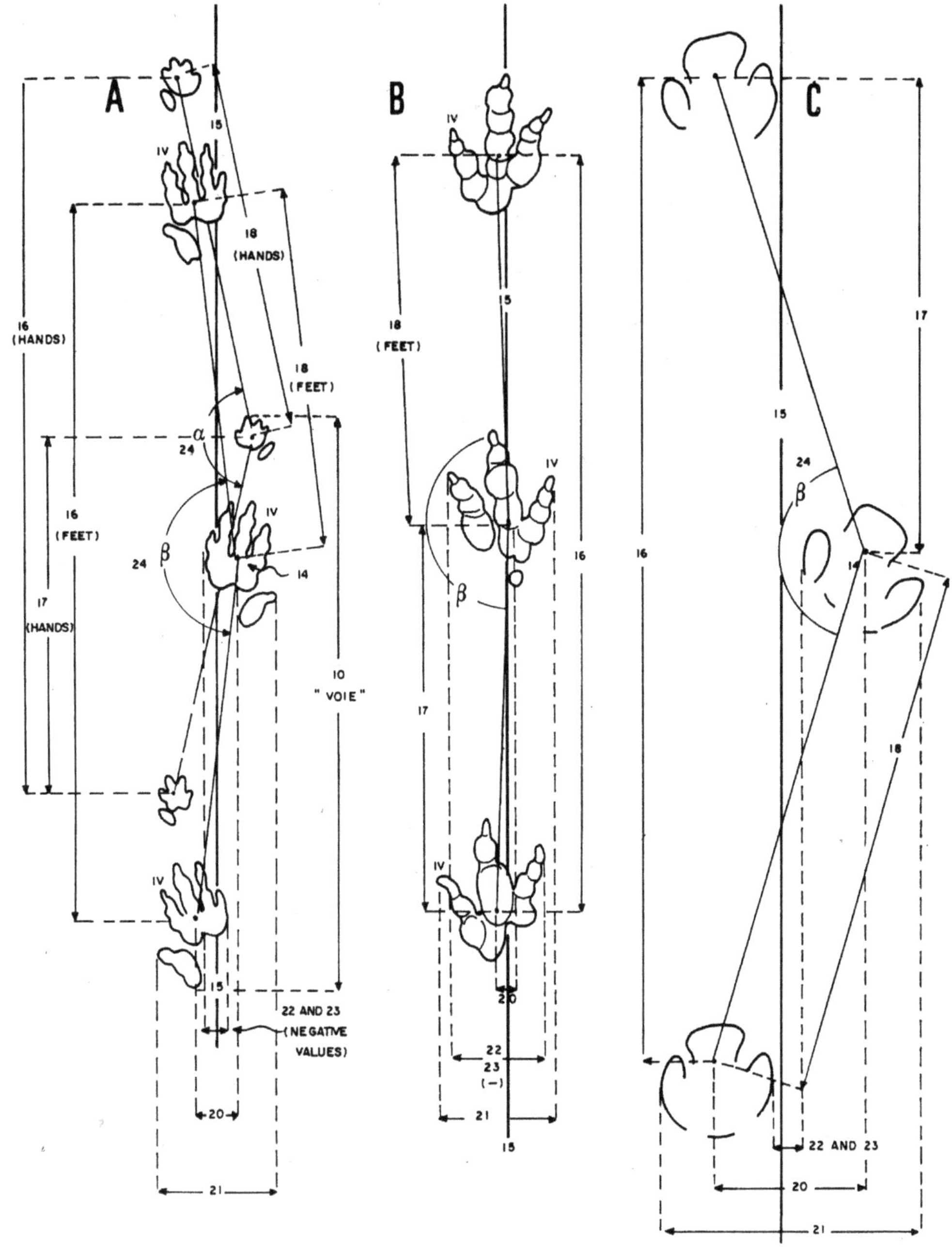

FIGURE 3 A standard measuring protocol for tetrapod trackways. (A) chirotherium, (B) bipedal theropod dinosaur, (C) overstepped trackway of a rhinoceros. *Modified from Leonardi (1987).*

FIGURE 4 A drone gathering photogrammetric data at the Lower Cretaceous Clayton Lake dinosaur tracksite in New Mexico, USA.

just beginning, may overcome important limitations of traditional quantitative analysis.

Lallensack begins with an overview of current challenges and debates in ichnological terminology. He then reviews both traditional and cutting-edge methodologies, starting with the documentation of tracks in the field and the definition of outlines and measurements. He follows by detailing qualitative methods and concepts for trackmaker identification, sources of variation and ichnotaxonomy, and, finally, reviews quantitative methods from simple averaging to more sophisticated statistical methods and the use of artificial intelligence.

3.2 Devonian tetrapod footprints

In this volume, Lucas (2024) reviews the Devonian record of tetrapod footprints (also see Lucas, 2015, 2019). Of the nine supposed Devonian tetrapod footprint records in the literature, only three can be verified as produced by a tetrapod trackmaker, and one of these may be younger than Devonian. These three are Genoa River, Australia (two trackways), Easter Ross, Scotland (one trackway) and Valentia Island, Ireland (eight trackways). The oldest tetrapod tracks are Middle Devonian (Givetian), and this is the oldest record of a tetrapod. The very sparse record of Devonian tetrapod tracks is of no other biostratigraphic and little paleobiogeographic significance. Bona fide Devonian tetrapod tracks are from nonmarine facies, so they do not support a marginal marine origin of tetrapods.

Devonian tetrapod trackways indicate lateral sequence walking and pelvic-limb-propelled, fully terrestrial (subaerial) locomotion in freshwater environments by at least some Devonian tetrapods. This raises the question of whether or not the tetrapodomorph fishes, which have larger forelimbs than hind limbs, are closely related to tetrapods or simply a group that converged on some tetrapod features. Body fossils of the Devonian tetrapod trackmakers are unknown, so a major clade, or clades, of early tetrapods have been sampled by the footprint record but not by the body-fossil record. This clade(s) is (are) more likely to have been the main line of tetrapod origins and evolution than the stem tetrapods that populate the Devonian tetrapod body-fossil record.

Thus, the entire understanding of tetrapod origins as now presented may be merely the understanding of the evolution of an early, dead-end radiation of aquatic stem tetrapods from a sarcopterygian ancestry. Other than the footprint record, the Devonian origin and subsequent early evolution of terrestrially adapted tetrapods remains largely undiscovered and unexamined. The Devonian footprint record impressively underscores the importance of footprints and other trace fossils for the understanding of the evolutionary history of vertebrates, not least because of the incompleteness of the body fossil record.

3.3 Carboniferous tetrapod footprints

In this volume, Lucas and Stimson (2024) review the Carboniferous record of tetrapod footprints. They note that Carboniferous tetrapod footprints are mostly of Euramerican origin and have been collected and studied since the 1840s. Very early in that study, Pennsylvanian footprints from North America refuted the idea that no terrestrial tetrapods lived during the Carboniferous, the first example of footprints demonstrating something not known from the tetrapod body-fossil record. During the 1850–1960s, many scattered reports of Carboniferous footprints from Europe and North America were published. The first syntheses were attempted during the 1970–80s, but the importance of the Carboniferous tetrapod footprint record to understanding tetrapod evolution has only been explored recently.

According to Lucas and Stimson, Carboniferous tetrapod footprint ichnotaxonomy is in need of much revision, and they recognize 14 well-known and likely valid ichnogenera: *Amphisauropus, Attenosaurus, Batrachichnus, Characichnos, Cincosaurus, Dimetropus, Dromopus, Hylopus, Ichniotherium, Limnopus, Matthewichnus, Notalacerta, Pseudobradypus,* and *Varanopus*. The temporal distribution of these ichnogenera provides the basis for a footprint biostratigraphy and biochronology of Carboniferous time that identifies four tetrapod footprint biochrons: (1) Valentian biochron of Middle Devonian-early Tournaisian age; (2) Bluebeachian biochron of middle-Tournaisian-early Bashkirian age; (3) Jogginsian biochron of early

Bashkirian-Kasimovian age; and (4) Robledoan biochron of Kasimovian-early Permian age.

Particularly significant to this biochronology is the Carboniferous tetrapod footprint record of the Maritimes Basin of eastern Canada (New Brunswick, Nova Scotia, and Prince Edward Island), which encompasses well-dated and stratigraphically superposed footprint assemblages of Early Mississippian-early Permian age. The Carboniferous tetrapod footprint record provides these important biostratigraphic datums: (1) oldest temnospondyls (middle Tournaisian); (2) oldest reptiliomorphs, likely anthracosaurs (middle Tournaisian); (3) oldest amniotes (early Bashkirian); and (4) oldest high fiber herbivores (Bashkirian). These footprint datums provide insight into some major events of the Carboniferous evolution of tetrapods.

3.4 Permian tetrapod footprints

In their review article in this book, Marchetti et al. (2024) note that Permian tetrapod footprints are abundant, relatively diverse, and widespread in Pangea. They consider 22 ichnogenera to be valid and attributed them to temnospondyls and lepospondyls (*Batrachichnus, Limnopus, Matthewichnus*), seymouriamorphs (*Amphisauropus*), diadectomorphs (*Ichniotherium*), nontherapsid synapsids (*Dimetropus, Tambachichnium*), nonmammalian therapsids (*Brontopus, Capitosauroides, Dicynodontipus, Dolomitipes, Karoopes*), captorhinomorphs (*Hyloidichnus, Notalacerta*), diapsids (*Dromopus, Paradoxichnium, Protochirotherium, Rhynchosauroides*) and parareptiles (*Erpetopus, Pachypes, Procolophonichnium, Varanopus*) (Fig. 5).

Marchetti et al. note that Permian footprints are more diverse than Carboniferous footprints, likely due to the ecological and morphological diversification of tetrapods. The Permian tracks indicate a generally raised, sprawling mode of locomotion, although some Permian ichnogenera are consistent with a semi-erect posture that clearly evolved before the Triassic. Permian tracks include examples of locomotion on inclined surfaces and penetrative traces, which markedly altered their morphological preservation. Permian footprints are sometimes associated with resting traces, scaly skin impressions, and swimming traces, and they have been found in various depositional environments, such as alluvial plains, lake margins, tidal flats, and deserts. The composition of the ichnoassemblages changes between different settings, especially the relative abundance of ichnotaxa, consistent with the ichnofacies concept. Nevertheless, a thorough review of Permian tetrapod ichnofacies is needed.

During the early Permian (Cisuralian), footprints are very well-represented at low paleolatitudes in western Europe, northern Africa, and North America. And, during the middle-late Permian (Guadalupian-Lopingian) they are well known as at mid-high paleolatitudes latitudes,

FIGURE 5 Three characteristic ichnotaxa of the early Permian and their trackmakers (not to scale). Left: *Batrachichnus* made by a small temnospondyl amphibian. Middle: *Dromopus* made by a neodiapsid. Right: *Dimetropus* made by a eupelycosaur. *From Lucas (2019). Artwork by Matt Celeskey.*

including South Africa and Russia. They are generally more diverse at low paleolatitudes. Tetrapod footprints are valuable for Permian chronostratigraphy; five tetrapod footprint biochrons are currently defined (ascending): *Dromopus, Erpetopus, Brontopus, Dicynodontipus,* and *Paradoxichnium*. Permian footprints are also key to the understanding of major faunal events such as the Artinskian Warming Event and the end-Guadalupian and end-Permian extinctions.

3.5 Triassic tetrapod footprints

In their review of Triassic tetrapod footprints, Klein and Lucas (2024) note that chirotheriid archosaur trackways from the famous Buntsandstein locality of Hildburghausen in Germany were the first scientifically described and named footprints (Fig. 6). Triassic tetrapod footprint assemblages are relatively diverse, and, besides chirotheriids, other archosaur ichnotaxa are often abundant, such as the small dinosauromorph *Rotodactylus* from Middle Triassic deposits or tridactyl dinosauriform—dinosaurian forms such as *Grallator* and *Atreipus*, which are common in the Late Triassic. Sauropodomorpha

FIGURE 6 *Chirotherium barthii* trackways on slab from type surface (Buntsandstein, Solling Formation, Middle Triassic) of Hildburghausen, Germany. Original at Naturhistorisches Museum Schleusingen. *Courtesy of Hartmut Haubold.*

are represented by the ichnogenera *Pseudotetrasauropus*, *Tetrasauropus*, *Eosauropus*, and *Evazoum*, the former two originally described from the Upper Triassic of southern Africa. Other common footprints, from some surfaces documented by mass occurrences, are the small lacertoid track *Rhychosauroides* (archosauromorph/lepidosauromorph). More rare are *Procolophonichnium* (procolophonid/therapsid), cynodont, and dicynodont therapsid footprints such as *Dicynodontipus*, *Pentasauropus*, and *Therapsipus*, and anamniote footprints such as *Batrachichnus*, tracks of turtles such as *Chelonipus* and of marine nothosaurs and placodonts (*Dikoposichnus*, *Anshunpes*).

Lower Triassic (Olenekian) footprint assemblages are known mostly from Europe (Volpriehausen, Detfurth, and Wióry formations, Germany, Poland), North America (Lower Moenkopi and Red Peak formations, Arizona, Wyoming), and North Africa (Timezgadiouine Formation, Argana Basin, Morocco). Middle Triassic (Anisian-Ladinian) footprint assemblages come from the classical Buntsandstein localities (Solling-Roet formations) in Germany, and equivalents of the Muschelkalk (Grafenwöhr-, Eschenbach-Vossenveld formations, Grés de Lyonnais) in Germany, The Netherlands and France, coeval deposits of northern Italy, from the Upper Moenkopi Formation/Group of Arizona and Utah, from the Timezgadiouine Formation of the Argana Basin, Morocco and from the Guanling Formation of China. Upper Triassic (Carnian-Rhaetian) footprint assemblages are found in Europe (Keuper and other deposits in Germany, France, Spain, Italy, Sweden, Poland), North America (Chinle Group, Newark Supergroup), South America (different units in Argentina, Bolivia, Peru, Brazil), East Greenland (Fleming Fjord Group), North Africa (Timezgadiouine Formation, Argana Basin, Morocco), southern Africa (Lower Elliot Formation, Lesotho), different continental deposits of China and in Australia (Sydney Basin, Queensland). Biostratigraphically, Triassic tetrapod footprints have been used to define five biochrons: (1) Dicynodont tracks (Induan-Olenekian), (2) *Protochirotherium* (Olenekian), (3) *Chirotherium barthii* (early Anisian), (4) *Atreipus-Grallator* (late Anisian-Ladinian), and (5) *Brachychirotherium* (Carnian–Rhaetian).

Tetrapod footprint assemblages from the Triassic give insight into the development of early Mesozoic terrestrial vertebrate communities. Innovations in the locomotor apparatus and gait, such as the evolution of the erect stance and bipedality, which partly began during the Permian, were perfected during the Triassic and were the likely key to the rise to dominance of the archosaurs.

3.6 Jurassic tetrapod footprints

Jurassic tetrapod footprints are abundant, diverse, and widespread in Asia, Europe, and North America. In this volume, Foster et al. (2024) review the extensive record of Jurassic tetrapod footprints and other vertebrate traces. In Australia and South America, their diversity is slightly lower, while in Africa they are both abundant and diverse, mainly concentrated in Morocco and in the Lower Jurassic of southern Africa. Jurassic tracks are as yet unreported from Antarctica.

Currently, there are more than 80 recognized Jurassic vertebrate ichnogenera, attributed to fishes, amphibians, turtles, lepidosaurs, crocodylomorphs, pterosaurs, mammaliamorphs, and various types of theropod, sauropodomorph, and ornithischian dinosaurs. Among Jurassic vertebrate traces are records of limping, running, and crouching theropod dinosaurs, a rare instance of a sauropod dragging its tail, swimming dinosaurs, burrowing mammaliamorphs, walking and swimming crocodylomorphs and turtles, walking pterosaurs, and animals turning—all examples of behavior preserved by the trace fossils. Jurassic footprints can be found in various depositional environments, including alluvial floodplains, river sand bars, lake margins, tidal flats, sand dunes, and desert oases, among others. The compositions of the ichnoassociations in these settings vary, particularly in the abundances of different ichnotaxa.

Jurassic vertebrate trace fossils document the emergence of dinosaur-dominated ecosystems during the early part of the Jurassic. They share some similarities with the track faunas of the first half of the Cretaceous, but they differ markedly from Triassic ichnofaunas.

3.7 Cretaceous tetrapod footprints

The abundance and widespread distribution of Cretaceous reptile-dominated tetrapods, including dinosaurs, can be attributed to the overall greenhouse climate, the highest sea level of the Phanerozoic, and the climax of Pangea rifting. This is also reflected in the extensive fossil record of Cretaceous tetrapod tracks, reviewed here by Xing et al. (2024).

They identify more than 112 tetrapod footprint ichnogenera in the Cretaceous, including 90 ichnogenera of nonavian and avian dinosaurs, 15 of other reptiles, and seven of other tetrapods. However, it is important to note that many of these ichnotaxa are of questionable validity due primarily to extramorphological variation, and extensive ichnotaxonomic revisions are needed.

The abundant Cretaceous tracks not only show kinematic features, but they also are records of behavior, including swimming, foraging, gregariousness, and crouching. Despite their almost global distribution (with the exception of Antarctica), Cretaceous footprint ichnospecies and ichnofaunas show a pronounced provincialism and vary across depositional environments. As the absolute ages of the stratigraphic units are ascertained, it becomes possible to establish comparable terrestrial ichnofacies and biostratigraphic standards based on tetrapod ichnotaxa. Meanwhile, they are used to interpret the distribution, migration, local endemism, and evolution of terrestrial faunas, especially in Europe-North America and East Asia, and provide additional data for analyzing the K–Pg extinctions.

3.8 Paleogene and Neogene tetrapod footprints

In this volume, Abbassi et al. (2024) review the ichnotaxonomy of Paleogene and Neogene tetrapod footprints, which were made by Amphibia, Reptilia, Aves, and Mammalia, and ichnotaxonomically include 2, 7, 25, and 63 valid ichnogenera, respectively. The review of Paleogene and Neogene ichnotaxa by Abbassi et al. is based on morphological comparisons, the analysis of the holotype material and its preservation quality (Fig. 7). Regulations proposed by the ICZN for ichnotaxa and well-known preservation scales were considered. The revision of the validity of Paleogene/Neogene tetrapod ichnotaxa leads Abbassi et al. to conclude that 12 named avian ichnogenera and eight named mammal ichnogenera are not valid. However, we note how few Paleogene-Neogene ichnotaxa of tetrapod tracks have been recognized, when compared to the Jurassic and the Cretaceous. Clearly, much more work is needed here.

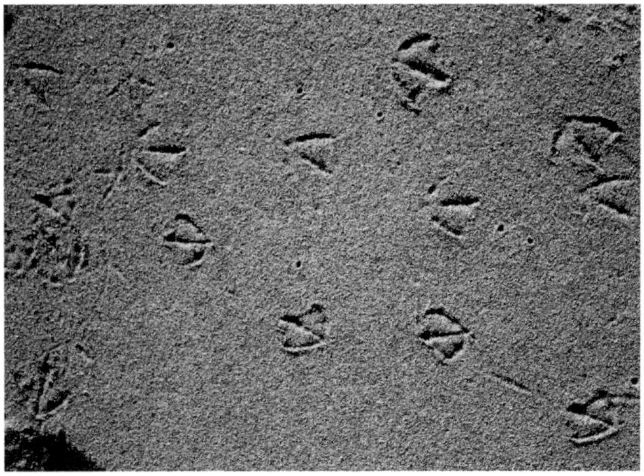

FIGURE 7 Photographs of some tracks made my living seagulls on a beach in Florida. Note the variation in shapes of the tracks. All of the gulls have webbed feet, but not all of the tracks preserve impressions of the interdigital webbing (and note that the presence of web impressions is an important ichnotaxobase for ichnologists who study bird footprints).

3.9 Quaternary tetrapod footprints

Neto de Carvalho et al. (2024a) here present a comprehensive assessment of the ichnological studies of Quaternary tetrapod footprint ichnotaxa and their possible producers. Except for Antarctica, the record of Quaternary vertebrate tracks is global. A total of 42 ichnogenera and 65 ichnospecies have been described from Africa, the Americas, Asia, and Europe; of these, 33 ichnotaxa were originally described from Quaternary deposits.

Europe and North America are the continents where the development of ichnology has the longest history and the most discoveries, and they dominate the fossil and subfossil records of Quaternary footprints. Despite great ichnological potential, ichnosites in Oceania have not received any formal ichnotaxonomic assessment, and such assessment only started to occur in Africa recently, mostly with regard to the large number of ichnosites that have been described in South Africa. Quaternary vertebrate ichnosites have been little reported in Asia, and in South America, the studies have concentrated mostly on the famous ichnosite of Pehuen Co in Argentina (Fig. 8), where 20 ichnotaxa have been described.

Neto de Carvalho et al. briefly discuss hominin tracks, but a separate chapter in the volume deals specifically with them.

FIGURE 8 The beach at Pehuen Co along the Atlantic coast of Argentina exposes sediments with numerous late Pleistocene mammal tracks. The oval depressions in the foreground of this photograph are the tracks of ground sloths.

3.10 Hominin footprints

In this volume, Neto de Carvalho et al. (2024b) review the hominin track record, referring to it as a study in ichnoarcheology. Although ichnoarcheology deals with any bioturbation, bioerosion, and biodepositional stuctures produced during the evolution of hominids, the paper focuses on a reappraisal of the record of limb dynamics interacting with natural and man-made substrates.

An updated review of the hominid footprint record recognizes 115 ichnosites worldwide, an unprecedented number resulting from the burst of interest during the last 2 decades and research in different fields, including ichnology, archeology, anthropology, and paleobiology. The review of Neto de Carvalho et al. is organized around major threshold events that definitely influenced hominin evolution, that is, bipedalism and dispersal worldwide; the use of limbs interacting with natural and artificial substrates to deliberately create symbolic meanings and artistic expressions; improving biomechanical capabilities through the development of footwear; and, finally, the development of aerospace technology producing a human ichnological record beyond Earth.

3.11 Inferring behavior from footprints

Vertebrate tracks are traces left by living animals, so they provide abundant evidence of behavior, locomotion, and paleoecology. In this volume, Hunt et al. (2024; also see Hunt and Lucas, 2021) review ichnological evidence of gregariousness and other behavior, determining speed, assessing gait and in recognizing pathologies. The majority of the work on these topics relates to dinosaurs for several reasons: (1) volume of literature, although there are arguably a similar number of papers on pre-Jurassic tracks; (2) the Taxophile Effect *sensu* Hunt et al. (2018), with elevated interest in dinosaurs relative to other vertebrates; and (3) the fact that nonavian dinosaurs were very successful organisms but have very limited modern analogs, so their paleobiology and behavior is poorly understood and of broad interest.

Hunt et al. distinguish two types of tracksites related to gregarious behavior: (1) transgressive sites where several individuals moved across an area in subparallel alignment; and (2) congregation sites where tracks of multiple individuals are concentrated at a location. They note that many factors must be considered in the evaluation of evidence of gregarious behavior, including: (1) ichnotaxonomy; (2) bedding plane; (3) time averaging; (4) subparallel orientation; (5) directional barriers; (6) ecological attractors; (7) intertrack spacing; (8) age structure; (9) speed and directional change; and (10) paleotopography. There are two scales of congregation sites; (1) microcongregation sites (up to tens or hundreds of square meters), which are concentrations of the same ichnotaxon focused on a discrete and often recognizable locus; and (2) mega-congregation sites are large tracksites (up to hundreds or thousands of square meters) with multiple trackways in which few, if any, demonstrate an obvious preferred orientation with no clear cause for the concentration. The first good track evidence for complex gregarious behavior in tetrapods is from the Early-Middle Triassic, but we expect that it will soon be recognized in Permian tracksites.

Paleopathologies have been recognized from fossil trackways on the basis of: (1) irregular gait; or (2) pathologic foot impressions. Consistent irregularities in pace and stride values are defined as irregular gaits, whereas limping gaits are the more extreme examples that are more likely to reflect cases of ichnopathology. The majority of reported ichnopathologies and putative ichnopathologies pertain to bipedal, tridactyl trackmakers, principally theropod dinosaurs. The narrow digits and long stride of theropods facilitate both the recognition of abnormal foot structure and of irregular pacing.

Characteristics of trackways reflect the gait of the track maker and thus have great potential in the study of the evolution of locomotion. However, gait analysis has received relatively little attention. Formulae for calculating speed from tetrapod trackways have been very useful, but they face challenges, notably those related to the nature of footprint preservation and of calculating trackmaker dimensions.

3.12 Fish trails and imprints

In this volume, Minter (2024) reviews the ichnotaxonomy, stratigraphic, geographic, and palaeoenvironmental distributions of fish trails and imprints. Ichnotaxa of fish trails and imprints are not diverse, and Minter reviews all currently named ichnogenera of fish locomotion trails (*Undichna* and *Parundichna*), resting imprints (*Broomichnium*, *Pisquiesichnus*, and *Raederichnus*), and feeding and nesting imprints (*Piscichnus*, *Osculichnus*, and *Daandavichnus*). He considers them to be morphologically distinctive and valid.

A comprehensive review of occurrences of fish swimming trails, resting imprints, and feeding and nesting imprints demonstrates that they range from the Silurian to Recent and they have a near-global distribution, having been reported from all of the continents except Antarctica. Fish trails and imprints, particularly *Undichna*, are most common from lacustrine environments, followed by marginal marine, alluvial, and tidal settings; with few occurrences from marine environments. Comparison with the body fossil record of fish suggests that these observations are the result of taphonomic controls on the preservation of fish trace fossils, and agrees with previous predictions that fish trace fossils are most likely to be preserved in low energy environments that were sites of relatively rapid burial, fine-grained sediment, and limited infaunal bioturbation.

Fish swimming trails have been attributed to jawless fish such as cephalaspids, as well as jawed fish such as cartilaginous fish (including acanthodians), and bony fish, including members of the Actinopterygii (ray-finned fish) and Sarcopterygii (lobe-finned fish). The more complex fish swimming trails, with greater numbers of features, may be attributed to producers with higher degrees of precision and certainty. Neoichnological observations have provided many insights into the producers of fish resting, feeding, and nesting imprints.

4. The Ichnology of Consumption

Besides footprints, the most common vertebrate trace fossils are bromalites (Fig. 9)—all trace fossils that represent food items that have entered the oral cavity and/or gastrointestinal tract of an animal and have been expelled (either anteriorly or posteriorly and either pre- or postmortem) from, or retained within, them (Hunt, 1992). However, if we view the entire process of consumption (Fig. 2), it begins with dentalites ("tooth marks"), traces produced on a substrate by the teeth or oral cavity of a vertebrate or invertebrate (Hunt et al., 2018). If food does not remain in the body to be excreted as feces, it may become a regurgitalite, a trace fossil that encompasses all manipulated or digested/partially digested food material egested via the oral cavity (Hunt, 1992). Sand and gravel may be swallowed and retained in the digestive tract to produce gastroliths to regulate buoyancy

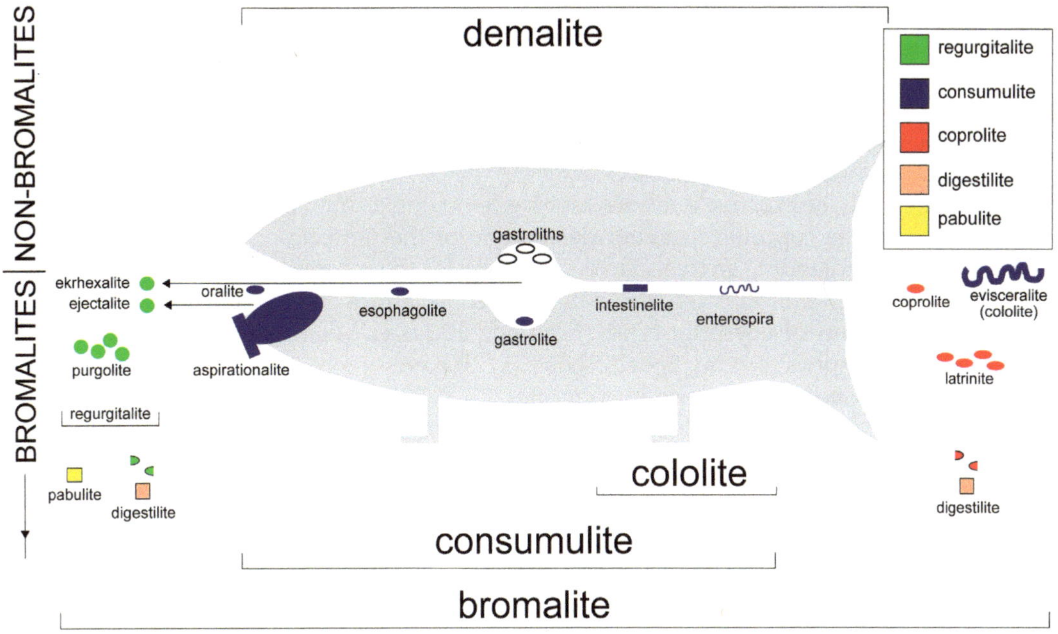

FIGURE 9 Selected classification terms for bromalites and associated trace fossils. *From Hunt and Lucas (2021).*

or as an aid to digestion. "Stomach contents" or "gut contents" are consumulites, and coprolites are fossil fecal material that has been ejected from the posterior end of the gastrointestinal tract (Buckland, 1829; Hunt and Lucas, 2021).

4.1 Methods

Coprolites represent a unique kind of trace fossil, that consists of a trace that has body fossil (bone, plant fragments, bacteria, fungi, parasites, among others) in it, and can include other traces too, such as dentalites and invertebrate burrows. This amount of information can be difficult to see in a small piece of fossilized feces. However, different methodologies are being used to decipher the identity of the coprolite producer and inclusions, digestive efficiency, and preservation. Here, Dentzien-Dias and Łaska (2024) present an overview of the different kinds of analyses of coprolites, separated into nondestructive, partially destructive, and destructive methods.

The simplest, nondestructive analysis involves overall morphological description. Generally, coprolites are divided into spiral or nonspiral. It is also important to examine inclusions (bone and plant fragments), surface marks (desiccation cracks), borings, dentalites, and adhesions (sand, plant, and shell impressions). The interior of coprolites can be examined by traditional thin sections, which cut the coprolite into slices. Coprolite thin sections can be analyzed under a Confocal microscope to survey preserved organic structures that are less evident in petrographic microscopy But, increasingly, noninvasive X-ray tomographic techniques are utilized to digitally examine them. One such technique is attenuation-based X-ray (micro-) computed tomography (CT).

Partially destructive analyses involve removing a small piece of one of the poles of the coprolite to avoid destruction of the morphology of the specimen. Petrographic analysis is very efficient for the recognition of inclusions and petrographic fabrics. The analyses of coprofabrics can be very useful for classification of coprolite types. Scanning Electron Microscope (SEM) and Energy-dispersive X-ray Spectroscopy (EDS) have shown exceptional details of soft tissues and bacteria in coprolites.

Destructive analyses include acid dissolution, which is a good method for studying parasite eggs and cysts from the Triassic to Holocene. Lipid analysis has been principally utilized for coprolites from Quaternary herbivores but also from the Mesozoic and Carboniferous. X-ray diffraction is widely used to determine the mineralogical composition of coprolites.

4.2 Vertebrate Dentalites

Dentalites ("tooth marks") are reviewed in this volume by Hunt and Lucas (2024a). Actualistic studies of them began with William Buckland in 1822, and an extensive fossil record of vertebrate dentalites exists, principally on bony substrates but also on invertebrate hard parts, coprolites, plants, lithic substrates, and others. There is a bias toward dentalites of large animals preserved on large pieces of substrate. The Taxophile Effect is also well seen in the marked disparity between the numbers of reports of dentalites on dinosaur bones relative to those on the bones of Cenozoic mammals.

Dentalites are nonrandomly distributed in time, with notable acmes in the Late Triassic, Late Cretaceous, and late Cenozoic (Fig. 10). Prior to the Quaternary, with the exception of the nonmarine Mesozoic, most reported dentalites are from marine facies. In the Quaternary, a large percentage of dentalites are from cave deposits. There is a clear seeker bias (seek and ye shall find), as careful investigation of large, existing collections, at least of vertebrate specimens, has always led to the identification of multiple examples of dentalites.

Dentalite ichnogenera of the Ichnofamily Machichnidae are *Machichnus*, *Knethichnus*, *Linichnus*, *Machichnus*, *Mandaodonites*, *Nihilichnus*, and *Heterodontichnites*. New ichnotaxa are introduced in this volume by Hunt and Lucas. Traces produced by mouth parts impacting the sediment surface are *Daandavichnus batoideum* and ichnospecies of *Osculichnus*—*O. labialis*, *O. tarnowskae*, *O. tepitsin*, and a new ichnospecies.

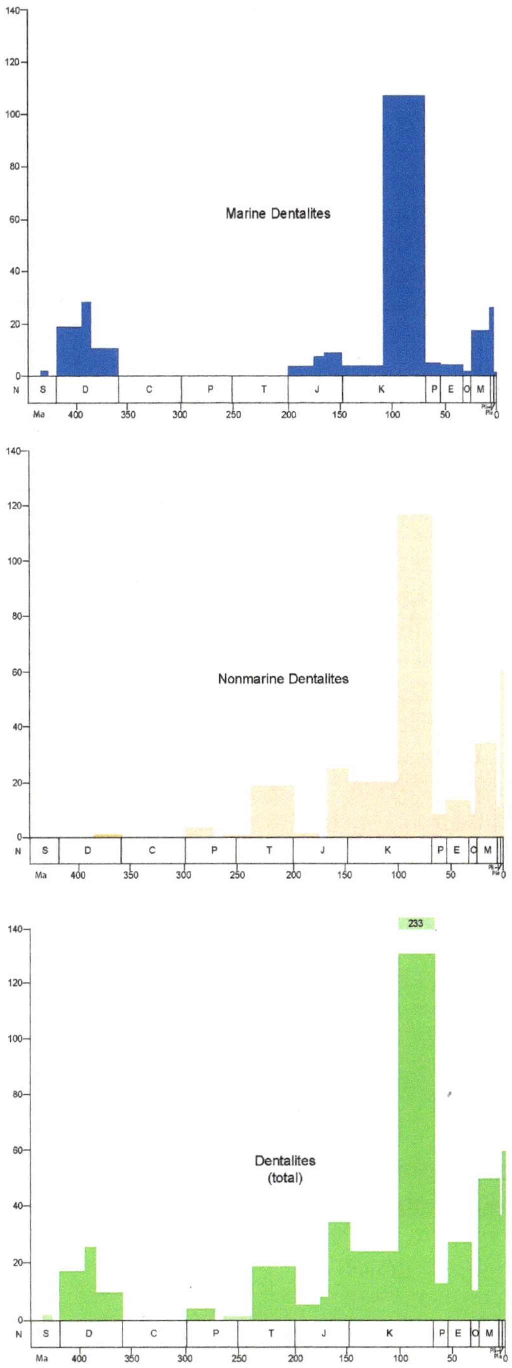

FIGURE 10 Temporal distribution of reports of dentalites. *From Hunt and Lucas (2021).*

4.3 Regurgitalites

Regurgitalites are the least studied bromalites with the smallest fossil record, and that record is reviewed here by Hunt and Lucas (2024b). Many identified vertebrate regurgitalites were produced by fish or birds and are preserved in a limited range of environmental settings (e.g., aquatic low energy). Thus, the fossil record of regurgitalites is strongly controlled by both taxonomic and taphonomic factors.

Regurgitalites can be recognized on the basis of a variety of features including geometry, paucity of matrix, biotic content (undigestible, size, evidence of digestion), dentalites, element representation and orientation. An ichnotaxobase for regurgitalites should rely on composition and morphology of the regurgitalites. Currently, there are only two named regurgitalite ichnotaxa—*Ostracobromus snowyensis* and *Conchobromus kinneyensis*—and both are from the Carboniferous of the United States.

4.4 Consumulites

Consumulites are material ingested and preserved within the gastrointestinal tract, and Hunt and Lucas (2024c) review their record here. Consumulites include aspirationalites, oralites, esophagolites, cropalites, proventralites, gastrolites, cololites, intesinilites, evisceralites, and enterospira (Fig. 9). The preservation of consumulites is favored by a relatively long gut retention time and low digestive acidity. Consumulites are principally associated with, or at least recognized in the context of, articulated skeletons, the majority of which are from aquatic environments. Articulated skeletons are common in Lagerstätten that preserve vertebrate fossils.

The vast majority of consumulites represent carnivorous animals. Large body size favors the recognition of consumulites. The combination of large body size and an aquatic lifestyle results in an extensive record of consumulites in large Mesozoic marine vertebrates, notably ichthyosaurs, plesiosaurs, and mosasaurs (Fig. 11). Hunt and Lucas (2024c) introduce the term matryoshkalite for a consumulite within a consumulite.

Consumulites should receive ichnotaxonic names. There are currently only four named consumulites, three preserved within the body cavity, *Werneburgichnus kinneyensis, W. varius,* and *Chondripilula zideki,* and the evisceralite *Hiabromus seilacheri.* Consumulites provide information on evolutionary and paleoecological issues, notably the evolution of digestive systems, including the function and location of organs, the evolution of diets within clades, and the ontogeny of feeding. Consumulites can also be Lagerstätten that preserve delicate organisms.

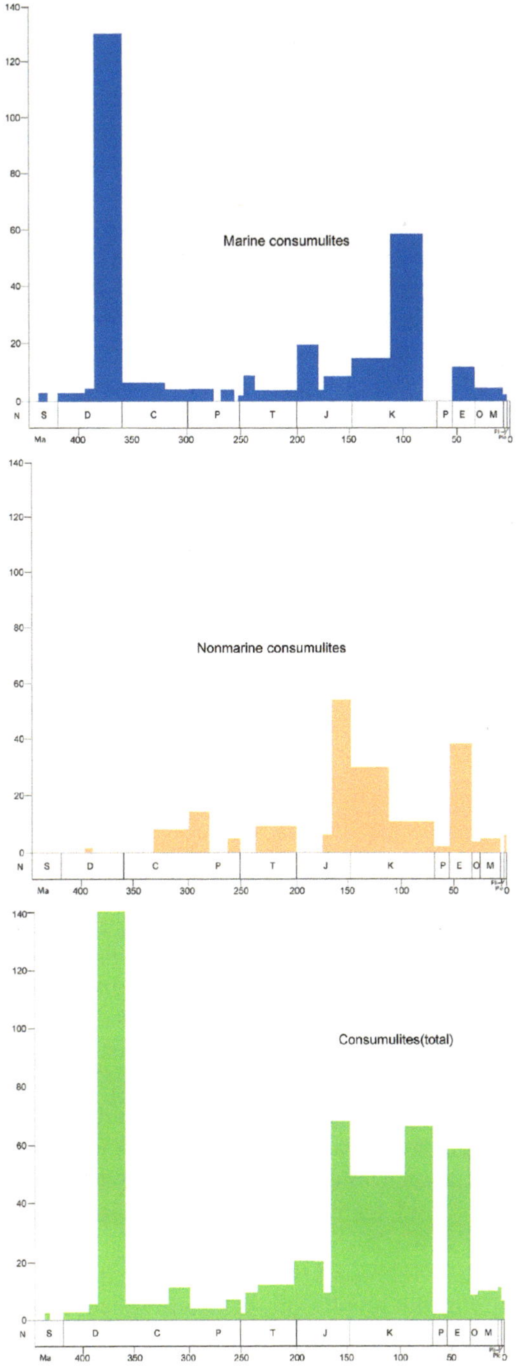

FIGURE 11 Temporal distribution of reports of consumulites. *From Hunt and Lucas (2021).*

4.5 Vertebrate coprolites

There is an extensive literature on coprolites, fossilized feces, large portions of which have been reviewed elsewhere (Hunt and Lucas, 2021) and is briefly reviewed in this volume (Hunt and Lucas, 2024d). After footprints, they are the most studied vertebrate trace fossils, and coprolites are the subject of an extensive and rapidly growing literature.

Coprolites can serve as proxies for biotaxa (the animals that produced them). However, the producer of individual vertebrate coprolites can usually only be identified at a high taxonomic level (order or higher) with very few exceptions before the Pleistocene, such as dicynodont therapsids from the Late Triassic of the United States and hyperodapedontid rhynchosaurs from the Middle Triassic of Brazil. Most Pleistocene vertebrate coprolites can be identified at the specific level on the basis of morphology and DNA (e.g., *Hyaenocoprus* derives from *Crocuta*). Regardless of the known taxonomic level of the producer, the distribution of a coprolite ichnotaxon is a proxy for a biological taxonomic unit. Vertebrate coprolites thus currently provide some biochronology in the Pennsylvanian to Paleogene and in the Quaternary. Although they are represented by fewer ichnotaxa, they do provide a biochronology through the entire Phanerozoic (Hunt et al., 1998; Hunt and Lucas, 2021).

Vertebrate coprolites can be used to delimit biogeographic regions. For example, there are two distinct biogeographic and taphonomic provinces for vertebrate coprolites in the Pleistocene. The *Castrocopros* province of the New World is characterized by a dominance of herbivore coprolites that are preserved almost exclusively in caves. The *Hyaenacoprus* province is in the Old World, and it is dominated by hyena coprolites that occur mainly in caves (Hunt and Lucas, 2020).

All trace fossils, including coprolites, are facies fossils. Spiral vertebrate coprolites in shallow marine environments define the *Crassocoprus* ichnofacies that ranges from the Mississippian to the Eocene. On a finer scale, vertebrate coprolites have allowed the discrimination of three ichnocoenoses in the nonmarine Late Triassic and four in an ecological transect from lacustrine to shallow marine conditions in the Late Pennsylvanian.

The first important inclusions to be recognized in coprolites were spores in arthropod coprolites and pollen in vertebrate herbivore coprolites. Subsequently, a wide range of organic materials, including soft tissues, have been identified in vertebrate coprolites, particularly muscle tissues, eggs, hair, parasites, bacteria, fungi and feathers. Thus, some coprolites represent Lagerstätten (Qvarnström et al., 2016).

Coprolites obviously represent the end point of digestion, so they can provide evidence about the evolution of feeding, diet and digestion. For example, vertebrate coprolites of the Silurian to Carboniferous demonstrate changing patterns of predation and food processing among carnivorous fishes as demonstrated by an increase in groundmass, increase in size and the diversification of spiral morphologies (Hunt and Lucas, 2021).

Coprolites provide ecological information about their producers, and Niedźwiedzki and co-workers have utilized them to reconstruct ecosystems and analyze diversity. Coprolites are also of utility in a number of other areas from the study of the evolution of human diseases to sedimentology to providing mineable resources (guanolite).

4.5 Gastroliths

Gastroliths are clasts of sand and/or gravel swallowed by an animal and retained in the digestive tract, and their record as trace fossils is reviewed here by Lucas and Hunt (2024). Their presence in various living vertebrates, especially crocodilians and birds, provides an actualistic basis for interpretation. Gastrolith distribution, however, is very irregular in extant and extinct vertebrates. Some crocodiles and marine mammals use/used gastroliths for buoyancy control, but all other gastrolith-bearing vertebrates appear to have used them in digestion.

Identification of fossil gastroliths is only certain when the sand/gravel are found as a concentrated mass in an anatomically plausible position within the abdominal region of a fossil skeleton. High polish does not characterize gastroliths. Among extinct vertebrates, plesiosaurs have the most extensive record of gastroliths, particularly the long necked, Cretaceous elasmosaurs. Some dinosaurs had gastroliths, but few if any sauropod dinosaurs did. Gastropods are also present in some fossil crocodilians, birds, and other tetrapods.

Gastroliths are clearly trace fossils as they are the work of animals that swallowed, concentrated, transported and, in some cases, altered them. Gastroliths thus are trace fossils in need of ichnotaxonomy that provide important insights into various behaviors, notably diet, digestion, buoyancy control, and habitat preferences.

4.7 Other bromalites

Dentalites, bromalites (regurgitalites, consumulites, coprolites), and gastroliths are the principal sources of ichnological evidence about vertebrate consumption, but less common bromalites (Fig. 9) and some other trace fossils also provide significant information reviewed here (Hunt and Lucas, 2024e). Pabulites represent fossilized food that never entered the digestive tract beyond the buccal cavity. They are very common in the fossil record, and the vast majority of specimens preserving dentalites are probably pabulites, Ichnotaxobases for pabulites include (1) anatomical representation; (2) degree of bone or shell breakage; (3) presence of dentalites (4) spatial arrangement; (5) incomplete preservation of soft tissue; and (6) evidence of particular methodology of predation and/or scavenging. Herein, Hunt and Lucas provide the first ichnotaxonomic name for a pabulite.

Digestilites are fossil specimens that preserve evidence of digestion. As food particles pass through the digestive tract they are subject to chemical and physical processes that result in characteristic damage. Digestilites can derive from regurgitated or defecated material. Micturalites are rare trace fossils produced by the interaction between liquid urine and a substrate.

There are several examples of trackways that purport to capture an act of predation, but none are convincing. Nests preserve abundant evidence of predation in terms of trace fossils such as dentalites and coprolites. Mololites are tooth wear features, including both mesowear and microwear.

5. The vertebrate ichnology of occupation and reproduction

Burrows are among the most common invertebrate trace fossils and have long been studied. However, vertebrate burrows have received much less attention until recently. Vertebrate nests are essentially unstudied ichnologically, but other trace fossil evidence of vertebrate reproduction has received some attention.

5.1 Vertebrate burrows

In this book, Melchor and Cardonato (2024) provide a revision of several aspects of vertebrate burrows and a critical assessment of the proposed ichnotaxa. The distinctive features of modern fish, amphibian, reptile, bird, and mammal burrows are summarized using selected examples from the literature, as an aid in the description and interpretation of fossil examples. Criteria for the identification of fossil vertebrate burrows and some keys to distinguish them from potentially similar structures are discussed. The methods to be applied in the field and laboratory for the study of these structures are outlined, and a set of ichnotaxobases to be applied in the recognition of ichnogenera and ichnospecies is also advocated.

There has been a substantial renewal of interest in the study of burrows in the last 20 years. Since 1970, approximately 180 papers dealing with vertebrate burrows were published, with about 80% of them in the last 20 years. New photogrammetric techniques (or other scanning techniques) have proved useful for documenting type material, especially when the field specimens are large and remain in the field. Three-dimensional models can be printed and deposited in a paleontological repository.

After a critical evaluation of all proposed ichnotaxa, Melchor and Cardonatto conclude that 15 ichnogenera and 23 ichnospecies of vertebrate burrows are available, and six ichnogenera and seven ichnospecies are unavailable. Available ichnogenera include *Alezichnos*, *Cornulatichnus*,

Daimonelix, *Katarrhedrites*, *Maneraichnus*, *Megaichnus*, *Megapermichnus*, *Nagtuichnus*, *Natatorichnus*, *Piscichnus*, *Redondarefugium*, *Reniformichnus*, *Tambia*, *Torridorefugium*, and *Yaviichnus*. Although the ichnotaxon is considered available, the vertebrate origin of *Tambia* is considered unlikely. The Oligocene-Miocene (especially the Miocene) is the interval with the most named vertebrate burrow ichnogenera, essentially related to mammals and, in many cases, to rodents.

5.2 Reproduction

Courtship-nesting behaviors are rarely reflected in body fossils but may be preserved as trace fossils. The study of trace fossils in conjunction with the reproductive behaviors of extant tetrapods is expanding our understanding of reproductive biology, filling gaps in the body fossil record and providing evidence that is independent of the latter. Trace fossils of tetrapods related to reproduction are reviewed here by Xing et al. (2024b). They are mainly known from dinosaurs, birds, crocodiles, turtles, and lizards, and include courtship scrapes (currently only known from nonavian theropods) and nesting traces. Most mammals do not produce eggs or recognizable nests, resulting in the relative rarity of their reproductive traces. Nesting traces other than fossilized eggs have been analyzed in detail only in the last ~50 years; such traces provide direct evidence for nest building and egg arrangement behaviors, and, possibly, allow for hypotheses regarding possible incubation behaviors and parental care.

6. The vertebrate ichnology of geoheritage

In 2021, the IUGS through IGCP Project 731 stated that "an IUGS Geological Heritage Site is a key place with geological elements and/or processes of scientific international relevance, used as a reference, and/or with a substantial contribution to the development of geological sciences through history." This would be a way to conserve the geoheritage, provide education to the public, and promote sustainable economic development based on geotourism (e.g., Gray 2013, 2019; Santangelo and Valente 2020). As Gray (2013, p. 239) put it, "the aim is to allow local communities to take ownership of their geological and other heritage by protecting it, promoting it and, by doing so, gaining some sustainable economic benefits from it." Various geoheritage sites have been, or could be, based on vertebrate trace fossils.

6.1 Geoconservation

For the last 3 decades, there has been a significant increase in interest among scientists and policy makers worldwide in the heritage significance of geology, particularly in the context of geoconservation, geotourism, and geopark initiatives. In this book, Neto de Carvalho et al. (2024c) provide a comprehensive overview of best practices for vertebrate ichnosite protection and geoconservation, and review the most important museums and visitor centers housing important or extensive collections of tetrapod trace fossils or their casts, and interpreted outdoor sites accessible to the public.

While the definition and conservation of dinosaur ichnosites has driven the geoconservation efforts in several countries and regions, the broader field of vertebrate ichnological geoheritage has seen varying levels of engagement. The late Martin G. Lockley played a pivotal role in promoting the scientific, educational, and geotourism value of dinosaur tracksites, and supported several projects all over the world through his Denver Dinosaur Trackers Research Group. This group's ongoing global presence underscores Lockley's enduring legacy and continues to inspire new geoconservation projects.

In contrast, non-dinosaur tetrapod ichnosites have inspired less enthusiasm for onsite conservation and public engagement. This disparity may stem from the less iconic nature of the trace producers, their smaller size and therefore less visually striking traces, or the challenges in preserving ichnosites in less consolidated rocks, such as the those containing Pliocene and Quaternary vertebrate tracks, including hominin footprints.

7. Current status and future of vertebrate ichnology

Vertebrate ichnology is now a vigorous and diverse subdiscipline of paleontology.

Each of the chapters in this volume documents an increasing degree of interest and sophistication of study in all areas of vertebrate ichnology. They also illustrate the widening range, both geographic and stratigraphic, of research on vertebrate trace fossils.

Indeed, in recent years vertebrate ichnology has become a stand-alone subdiscipline of paleontological research. Evolutionary developments of different vertebrate groups have partially been outlined based on data from the trace fossil record, for example, the evolution of the erect stance in basal archosaurs or the timing of the bipedal gait in dinosauromorphs. Vertebrate footprints are in many places and geologic time intervals much more abundant than skeletal fossils. Some characteristic footprint morphotypes have a restricted stratigraphic, but extensive geographical range. This makes them an excellent tool for biostratigraphy, paleobiogeography, and paleoecology. Thus, many vertebrate paleontologists now include ichnological data in their analyses and conclusions.

Lucas et al. (2024) in this volume discuss future tasks for vertebrate ichnology that we list and add to here:

1. Discovery and documentation of traces from time intervals (Late Paleozoic, Paleogene) and geographic regions (large parts of South America, Africa, Asia, Australia, and Antarctica) under-represented by the trace fossil record.
2. Ichnotaxonomic revisions of some largely oversplit footprint groups excluding extramorphological (substrate-related) factors.
3. Development of a workable ichnotaxonomy for nonfootprint traces, notably regurgitalites and consumulites.
4. Further use of photogrammetric and other metric data to evaluate, quantify, and document vertebrate trace-fossil morphologies.
5. Extend the use of databases of vertebrate ichnofossils including 3-D models.
6. Potentially utilize machine learning software and artificial intelligence (AI) to support ichnotaxonomic evaluation by recognizing different morphotypes and comparing them to known ichnotaxa.
7. Increase usage of nondestructive methodologies in the study of bromalites to exploit their potential as micro-Lagerstätten.

Despite its long history, vertebrate ichnology has great potential for future research. This is partly due to contemporary documentation techniques and computer software with 3D visualization that allow a more objective analysis of trace morphology. The shape of tetrapod footprints, for example, until recently was determined by the examining researcher who was drawing a simple outline of the footprint on superimposed transparency film, resulting from a more or less interpretive view. Furthermore, the researcher took measurements with a measuring tape. Most footprints studied in this way are still accessible in collections or in situ in the field. Thousands of them can be restudied now, and some can be reevaluated using photogrammetry or other methods. The results are important for ichnotaxonomic conclusions.

Indeed, a stable, widely accepted ichnotaxonomy is another challenging future task not only for the study of tetrapod footprints but for the other vertebrate traces such as bromalites, gastroliths, or burrows. Holotypes should not only be defined on original materials, but also based on a photogrammetric model or 3D-scan. Completely new insights and analyses can be based on the application of artificial intelligence (AI) software in ichnology, essentially in the identification of distinct morphotypes and ichnotaxa. However, this is still in an early stage and presently only being tested by different researchers. Finally, a comprehensive digital ichnology database with models of type specimens should be further established and made accessible for researchers, students, and other scientifically interested people.

Acknowledgments

First, and foremost, we thank the contributors to this volume. We asked them to produce landmark reviews of specific topics in vertebrate ichnology, and they did that, and all of that, in a timely manner. We also thank Jan Fischer and Lawrence Tanner for helpful reviews.

References

Abbassi, N., Melchor, R.N., Díaz-Martínez, I., 2024. Ichnotaxonomy of Paleogene and Neogene tetrapod tracks. In: Lucas, S.G., Hunt, A.P., Klein, H. (Eds.), Vertebrate Ichnology. Elsevier, Amsterdam.

Buatois, L.A., Mángano, M.G., 2011. Ichnology: Organism-Substrate Interactions in Space and Time. Cambridge University Press, Cambridge, p. 358.

Buckland, W., 1829. A Paper by Dr. Buckland. Proceedings of the Geological Society of London, pp. 142–143.

Dentzien-Dias, P., Łaska, W., 2024. Analyzing coprolites: Different methods for the study of coprolites. In: Lucas, S.G., Hunt, A.P., Klein, H. (Eds.), Vertebrate Ichnology. Elsevier, Amsterdam.

Foster, J.R., Harris, J.D., Milner, A.R.C., Bordy, E.M., Sciscio, L., Castanera, D., Belvedere, M., Xing, L., Lockley, M.G., 2024. Jurassic vertebrate tracks and traces. In: Lucas, S.G., Hunt, A.P., Klein, H. (Eds.), Vertebrate Ichnology. Elsevier, Amsterdam.

Gray, M., 2013. Geodiversity: Valuing and Conserving Abiotic Nature, Second edition. Wiley Blackwell, Chichester, UK.

Gray, M., 2019. Geodiversity, geoheritage and geoconservation for society. J. Geoheritage Parks 7, 226–236.

Hitchcock, E., 1858. Ichnology of New England. A Report on the Sandstone of the Connecticut Valley, Especially its Fossil Footmarks, Made to the Government of the Commonwealth of Massachusetts. William White, Boston, pp. 1–214. Printer.

Hunt, A.P., 1992. Late Pennsylvanian coprolites from the Kinney Brick Quarry, central New Mexico, with notes on the classification and utility of coprolites. N. M. Mus. Nat. Hist. Sci. Bull. 138, 221–229.

Hunt, A.P., Lucas, S.G., 2020. Hyena hegemony: biogeography and taphonomy of Pleistocene vertebrate coprolites with description of a new mammoth coprolite ichnotaxon. Ichnos 27, 111–121.

Hunt, A.P., Lucas, S.G., 2021. The ichnology of vertebrate consumption: Dentalites, gastroliths and bromalites. N. M. Mus. Nat. Hist. Sci. Bull. 87, 215.

Hunt, A.P., Lucas, S.G., 2024a. Vertebrate dentalites. In: Lucas, S.G., Hunt, A.P., Klein, H. (Eds.), Vertebrate Ichnology. Elsevier, Amsterdam.

Hunt, A.P., Lucas, S.G., 2024b. Regurgitalites. In: Lucas, S.G., Hunt, A.P., Klein, H. (Eds.), Vertebrate Ichnology. Elsevier, Amsterdam.

Hunt, A.P., Lucas, S.G., 2024c. Consumulites. In: Lucas, S.G., Hunt, A.P., Klein, H. (Eds.), Vertebrate Ichnology. Elsevier, Amsterdam.

Hunt, A.P., Lucas, S.G., 2024d. Vertebrate coprolites. In: Lucas, S.G., Hunt, A.P., Klein, H. (Eds.), Vertebrate Ichnology. Elsevier, Amsterdam.

Hunt, A.P., Lucas, S.G., 2024e. Pabulites, digestilites, micturalites and other trace fossil evidence of predation and consumption. In: Lucas, S.G., Hunt, A.P., Klein, H. (Eds.), Vertebrate Ichnology. Elsevier, Amsterdam.

Hunt, A.P., Lucas, S.G., Klein, H., 2018. Late Triassic nonmarine vertebrate and invertebrate trace fossils and the pattern of the Phanerozoic record of vertebrate trace fossils. In: Tanner, L.H. (Ed.), The Late Triassic World. Topics in Geobiology, 46. Springer Verlag, New York, pp. 447–543.

Hunt, A.P., Lucas, S.G., Klein, H., 2024. Evidence from fossil vertebrate tracks of gregarious behavior, ichnopathologies and locomotion. In: Lucas, S.G., Hunt, A.P., Klein, H. (Eds.), Vertebrate Ichnology. Elsevier, Amsterdam.

Hunt, A.P., Lucas, S.G., Lockley, M.G., 1998. Taxonomy and stratigraphic and facies significance of vertebrate coprolites of the Upper Triassic Chinle Group, Western United States. Ichnos 5, 225–234.

International Commisssion on Zoological Nomenclature, 1999. International Code of Zoological Nomenclature. London. International Trust for Zoological Nomenclature 306.

Klein, H., Lucas, S.G., 2024. Triassic tetrapod footprints. In: Lucas, S.G., Hunt, A.P., Klein, H. (Eds.), Vertebrate Ichnology. Elsevier, Amsterdam.

Lallensack, J.N., 2024. Methods and terminology in vertebrate track ichnology. In: Lucas, S.G., Hunt, A.P., Klein, H. (Eds.), Vertebrate Ichnology. Elsevier, Amsterdam.

Leonardi, G. (Ed.), 1987. Glossary and Manual of Tetrapod Footprint Palaeoichnology. Brasilia, Ministerio Minas Energie. Departemento Nacional Producão Mineral, p. 117.

Lucas, S.G., 2015. *Thinopus* and a critical review of Devonian tetrapod footprints. Ichnos 22, 136–154.

Lucas, S.G., 2019. An ichnological perspective on some major events of Paleozoic tetrapod evolution. Boll. Soc. Paleontol. Ital. 58, 223–266.

Lucas, S.G., 2024. Devonian tetrapod footprints. In: Lucas, S.G., Hunt, A.P., Klein, H. (Eds.), Vertebrate Ichnology. Elsevier, Amsterdam.

Lucas, S.G., Hunt, A.P., 2024. Gastroliths. In: Lucas, S.G., Hunt, A.P., Klein, H. (Eds.), Vertebrate Ichnology. Elsevier, Amsterdam.

Lucas, S.G., Stimson, M.R., 2024. Carboniferous tetrapod footprints. In: Lucas, S.G., Hunt, A.P., Klein, H. (Eds.), Vertebrate Ichnology. Elsevier, Amsterdam.

Lucas, S.G., Hunt, A.P., Klein, H., 2024. History of the study of vertebrate trace fossils. In: Lucas, S.G., Hunt, A.P., Klein, H. (Eds.), Vertebrate Ichnology. Elsevier, Amsterdam.

Marchetti, L., Mujal, E., Logghe, A., Buchwitz, M., Klein, H., Lucas, S.G., 2024. Permian vertebrate tracks. In: Lucas, S.G., Hunt, A.P., Klein, H. (Eds.), Vertebrate Ichnology. Elsevier, Amsterdam.

Melchor, R.N., Cardonatto, M.C., 2024. Vertebrate burrows. In: Lucas, S.G., Hunt, A.P., Klein, H. (Eds.), Vertebrate Ichnology. Elsevier, Amsterdam.

Minter, N.J., 2024. Fish trails and imprints. In: Lucas, S.G., Hunt, A.P., Klein, H. (Eds.), Vertebrate Ichnology. Elsevier, Amsterdam.

Neto de Carvalho, C., Helm, C., Melchor, R., Feola, S., Goldstein, D., Camens, A., 2024a. The Quaternary record of vertebrate tracks and trackways. In: Lucas, S.G., Hunt, A.P., Klein, H. (Eds.), Vertebrate Ichnology. Elsevier, Amsterdam.

Neto de Carvalho, C., Helm, C., Goldstein, D., Camens, A., Belaústegui, Z., Baucon, A., Muñiz, F., 2024b. Walk the world and beyond: Hominin tracks and trackways. In: Lucas, S.G., Hunt, A.P., Klein, H. (Eds.), Vertebrate Ichnology. Elsevier, Amsterdam.

Neto de Carvalho, C., Hunt, A.P., Alcala, L., Fischer, T., Helm, C., Page, K., Baucon, A., Kustascher, E., Lo Russo, G., Arena, L., Moreno, K., Xing, L., 2024c. Vertebrate ichnosites as Geological Heritage: a global synthesis of best practices in geoconservation. In: Lucas, S.G., Hunt, A.P., Klein, H. (Eds.), Vertebrate Ichnology. Elsevier, Amsterdam.

Qvarnström, M., Niedźwiedzki, G., Žigaitė, Ž., 2016. Vertebrate coprolites (fossil faeces): An underexplored Konservat-Lagerstätte. Earth-Sci. Rev. 162, 44–57.

Santangelo, N., Valente, E., 2020. Geoheritage and geotourism resources. Resources 9, 80. https://doi.org/10.3390/resources9070080.

Xing, L., Díaz Martínez, I., Lallensack, J.N., Kim, K.S., Buckley, L.G., Romilio, A., Piñuela, L., Wang, D., Chen, Q., Yang, Q., Chou, C., Yin, H., Jin, Y., 2024a. Cretaceous vertebrate tracks. In: Lucas, S.G., Hunt, A.P., Klein, H. (Eds.), Vertebrate Ichnology. Elsevier, Amsterdam.

Xing, L., Chen, Q., Wang, D., Lallensack, J.N., 2024b. The ichnology of reproduction in fossil tetrapods. In: Lucas, S.G., Hunt, A.P., Klein, H. (Eds.), Vertebrate Ichnology. Elsevier, Amsterdam.

History of the study of vertebrate trace fossils

Spencer G. Lucas[1], *Adrian P. Hunt*[2] *and Hendrik Klein*[3]

[1]New Mexico Museum of Natural History and Science, Albuquerque, NM, United States, [2]Flying Heritage Collection and Combat Armor Museum, Everett, WA, United States; [3]Saurierwelt Paläontologisches Museum, Neumarkt, Germany

1. Introduction

Though it is clear that ancient peoples observed and, in some cases, drew or collected trace fossils (Mayor and Sarjeant, 2001; Staker, 2006), the scientific study of vertebrate trace fossils began in Europe and North America in the 1820–30s. Most of the study of vertebrate traces during the last 200 years focused on footprints, but studies of other vertebrate trace fossils have increased significantly during the last 20 years. Nevertheless, there has been little coordination among different experts studying vertebrate traces until relatively recently, and there still is a sharp divide between those who study the trace fossils of vertebrate consumption (Hunt and Lucas, 2021c) and those who study footprints.

Here, we provide a history of the study of vertebrate trace fossils divided into the main categories of traces recognized in this book: footprints, coprolites, dentalites, regurgitalites, consumulites, gastroliths, other bromalites (the previous five are the ichnological products of consumption: Fig. 1), burrows, and nests. Our goal is to present an analytical history, with an overview of the narrative history, of the studies of these traces. At the end of this article, we identify what we believe should be the next steps in research on vertebrate ichnofossils.

2. Footprints

Most of the study of vertebrate trace fossils has been of footprints, extending back to the 1820s. A diverse literature has recounted most of the details of this history of study (Haubold, 1971, 2006; Sarjeant, 1974, 1983, 1987, 2000;

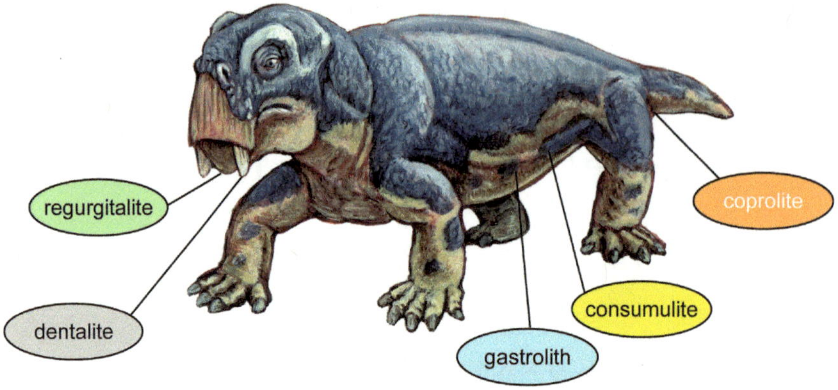

FIGURE 1 The trace fossils of vertebrates that are produced by consumption. *From Hunt and Lucas (2021c). Artwork by Matt Celeskey.*

Delair and Sarjeant, 1985; Sarjeant and Mossman, 1978; Cotton et al., 1995; Hunt et al., 1995; Lockley and Hunt, 1995; Tresise and Sarjeant, 1997; Lockley and Meyer, 2000; King et al., 2005; Rainforth, 2005; Díaz-Martínez and Pérez-García, 2012; Klein and Lucas, 2021; Klein et al., 2023; Bourgeois, 2024). To not become mired in the details, we have divided a fairly analytical history of the study of fossil vertebrate footprints into four phases: (1) first steps (1820–50s), (2) scattered studies (1860–1940s), (3) focused researchers (1940–70s), and (4) a renaissance of footprint studies (1980s-today).

2.1 First steps, 1820–50s

The scientific study of vertebrate (tetrapod) footprints began in the 1820–40s in Western Europe and North America. In about 1824, in Scotland, the Reverend Henry Duncan was presented with a slab bearing tetrapod footprints from the "New Red Sandstone" (Permian). Duncan collected more tracks and published on them in 1831. By then, he had sent information about them to William Buckland (Fig. 2), who famously conducted the first actualistic study of vertebrate trace fossils. Buckland thus had a tortoise walk on pie crust, soft clay, and wet sand to produce footprints very similar to those Duncan had acquired (Fig. 3). This prompted the later naming of the tracks *Chelichnus,* as Chelonia is a long used taxonomic name (based on a Greek root) for all turtles. It also indicates an important concept of vertebrate ichnology, in which the identity of the tracemaker is integral to ichnotaxonomy.

In Germany, Triassic footprints first came to scientific attention in a published letter written by F. K. L. Sickler (1834), a gymnasium director from the city of Hildburghausen, to the prominent anatomist Johann Friedrich Blumenbach. Sickler had discovered these footprints in 1833 while using slabs from a nearby Buntsandstein quarry for garden work. Kaup (1835a, b) soon named these tracks *Chirotherium barthii* and *C. sickleri,* which are the

FIGURE 2 Portrait of William Buckland in 1833. *Mezzotint by S. Cousins. From Hunt et al. (2012b).*

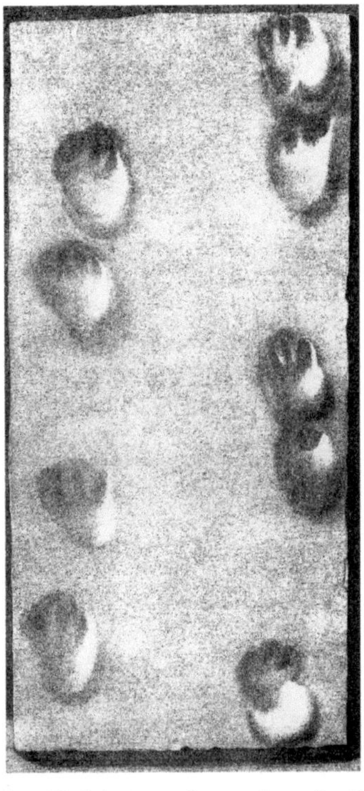

FIGURE 3 *Chelichnus* trackway. *From Buckland (1858).*

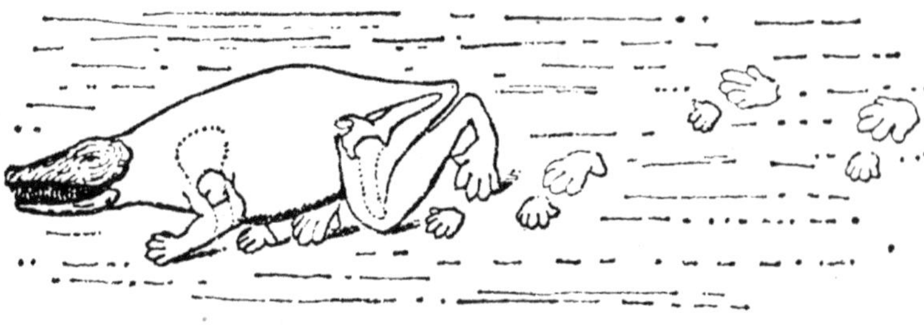

FIGURE 4 An early drawing showing a tetrapod walking cross legged to make *Chirotherium* footprints. *From Klein et al. (2023).*

first scientifically named vertebrate tracks. The mode of walking they indicated was soon (and famously) misinterpreted by various workers, who suggested that the amphibian producers crossed their legs while walking (Fig. 4). This mistake was based on the presence of a digit impression similar to a human thumb, which was positioned, however, on the outer side of the trackway. From the study of found skeletal fossils, it became evident later that this "thumb" is in fact the impression of the outer digit V, and *Chirotherium* obviously was made by a reptilian trackmaker (Klein et al., 2023).

In the Connecticut Valley of Massachusetts, USA, in 1836, Edward Hitchcock (Fig. 5) first published on footprints found in the Triassic-Jurassic strata, believing they were bird tracks (they are tracks of theropod dinosaurs) (Fig. 6). This work was culminated by a classic monograph (Hitchcock, 1858) and various other publications (see Rainforth, 2005; Pemberton et al.,

FIGURE 5 Edward Hitchcock. *From Tyler (1895).*

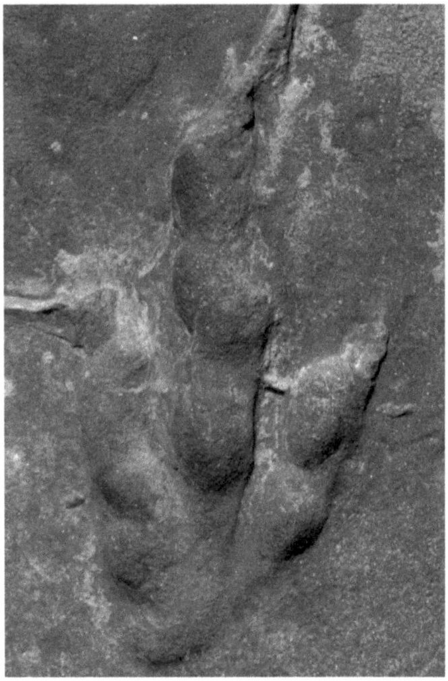

FIGURE 6 The first North American fossil footprints brought to scientific attention were discovered in Massachusetts and described by Edward Hitchcock. Hitchcock thought these tracks were made by birds, first referring to them as Ornithichnites, and later naming them *Grallator*. They have long been known to be the tracks of small, meat-eating dinosaurs. This track is approximately is 15 cm long.

2007; Bourgeois, 2024 for reviews of Hitchcock's work). Believing most of the tracks he studied were made by birds, Hitchcock initially referred to them as "Ornithichnites." Again, the ichnotaxonomy put emphasis on the identity of the trackmakers.

In 1841, William Edmund Logan discovered the first Carboniferous footprints at Horton Bluff in Nova Scotia, Canada, though his discovery was not published until 1863 by Dawson. Meanwhile, in Pennsylvania, USA, Alfred King found bona fide amphibian tracks (the type material of *Limnopus heterodactylus*: Lucas and Dalman, 2013; Lucas, 2015a), though he also described many Native American petroglyphs as fossil tracks. At the time, European geologists thought that no tetrapods lived during the "Coal Age" (Carboniferous). King's discoveries thus drew the attention of Charles Lyell, who, during his second trip to North America, examined King's finds, proclaiming most to be human-made petroglyphs but recognizing the authenticity of the Carboniferous amphibian footprints (Fig. 7). This is the first example of footprints providing important distributional information about tetrapods in the absence of a body-fossil record.

FIGURE 7 This is apparently the only authentic Carboniferous fossil track slab studied by Alfred King still available to scientists. It is in the collection of the Beneski Museum of Natural History, Amherst College, and is the type specimen of *Limnopus heterodactylus*.

An interesting example of the early sophistication of footprint research is provided by David Dale Owen, one of the first important American geologists. Thus, Owen (1842) resolved the identity and significance of supposed human footprints in a slab of limestone quarried on the western bank of the Mississippi River at St. Louis, Missouri, USA (Fig. 8). Brought to scientific attention in 1822, these "footprints" were much discussed by American and European savants with regard to their bearing on the antiquity of humankind. Owen's analysis of them was strikingly modern, not different from what an ichnologist would do today (Lucas, 2017).

First, he established the precise provenance of the "footprint"-bearing slab (after it was quarried, the slab had been sold and moved to Indiana). Owen then described the fossils of brachiopods and crinoids in the limestone. These fossils supported correlation of the limestone to the "mountain limestone" of Europe, which, in modern geological time terms means the limestone is of Mississippian age, on the order of 350 million years old. Owen then went on to evaluate the authenticity of the footprints, concluding that they were human carvings, not footprints. Evidence he marshalled for this included their isolation (only two tracks), their presence in a marine limestone (Owen found it difficult to envision submarine walking by humans followed

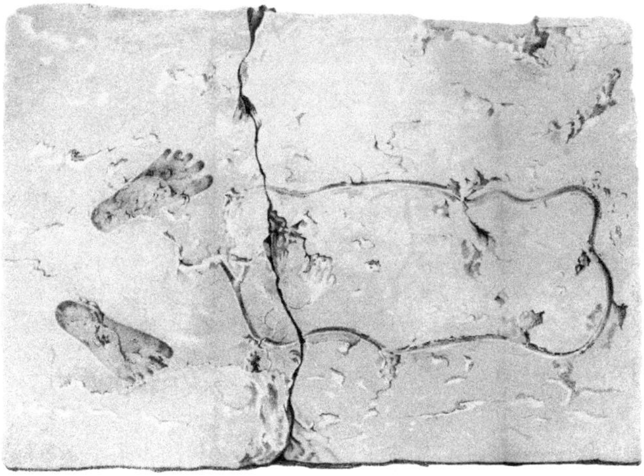

FIGURE 8 These supposed human footprints are in a piece of limestone quarried along the Mississippi River near St. Louis. First brought to scientific attention in the 1820s as proof of the great antiquity of humankind, they were judged to be carvings in Mississippian limestone by David Dale Owen in 1842, in a notably modern analysis. *From Owen (1842).*

by lithification of the tracks), and their great geological age. As Owen (1842, 27) well stated, "the impressions in question are not fossils, but an intaglio of artificial origin."

Other early footprint discoveries continued through the 1850s (see Sarjeant, 1987 for a detailed account), but we stress that the important conceptual and methodological aspects of footprint ichnology were already being explored/established by 1860: (1) actualistic research to understand the trackmakers and other aspects of fossil footprints; (2) the identity of the trackmaker as an important research objective and integral to footprint ichnotaxonomy; and (3) footprints providing important temporal and geographic distributional data in the absence of body fossils. It could be argued that a research program could have been based on these aspects of footprint ichnology, and such a program was almost fully developed by Edward Hitchcock. Yet, footprint ichnology would not have such a research program for more than a century.

2.2 Scattered studies, 1860–1940s

Although a research program for the study of vertebrate footprints had been crafted by the 1850s, the next 80 years encompassed a welter of diverse discoveries and studies of tetrapod footprints, but no extensive or unified research program developed. Thus, diverse authors reported tracks from the Devonian (Fig. 9), though none of these were actually tetrapod tracks (Lucas, 2015b), to Pleistocene from all of the continents except Antarctica.

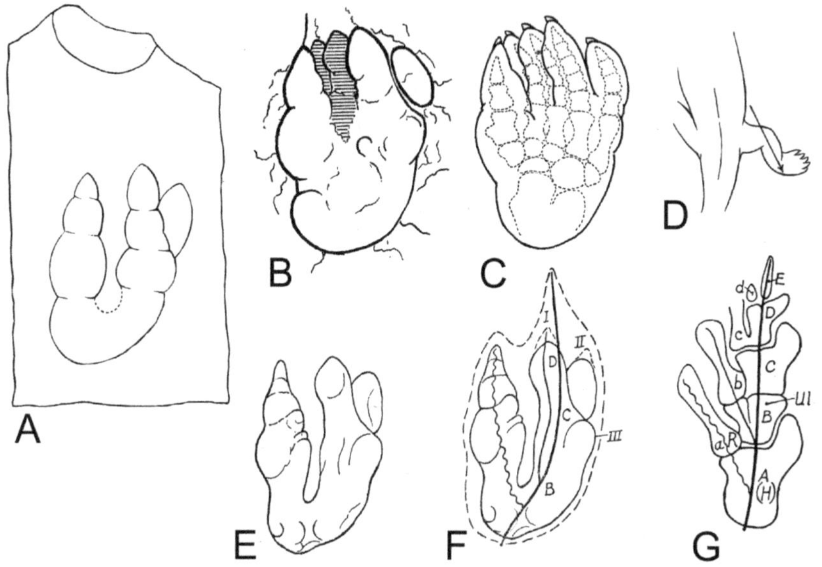

FIGURE 9 *Thinopus* was the first published Devonian tetrapod footprint, though it is not one. These are published drawings of the holotype of *Thinopus antiquus* (A, B, E), interpretation of that trace (C, D, F), and the proximal fin skeleton of the sarcopterygian fish (a supposed close relative) *Eusthenopteron* (G). For scale, length of "footprint" in A is ~84 mm. *From Lucas (2015b).*

Nevertheless, other than perhaps Winkler (1886), no syntheses were published, and no monographs or large studies were published, as had been done by Hitchcock. Some good work was published on Carboniferous tetrapod footprints from eastern Canada by William Dawson and George Matthew, but this work was of limited geographic and stratigraphic scope.

Little actualistic work was undertaken; taphonomy was only introduced as a broad concept in the 1940s, and sedimentology was in its infancy in the 1940s as a subdiscipline of geology. So, the understanding of footprint variation due to factors other than the shape of the trackmaker's foot did not exist. Hence, ichnotaxonomy was chaotic, with workers naming any unusual tracks found. Nevertheless, the first modern paleobiological analysis and detailed track-trackmaker correlation was undertaken by Soergel (1925) based on chirothere footprints and foot skeletons. His work influenced trackmaker reconstructions in all subsequent ichnological research.

A good example of footprint research in the early twentieth century is the classic work of Gilmore (1926,1927, 1928) on the tetrapod footprints from the Permian Coconino Sandstone in the Grand Canyon of Arizona, USA (Fig. 10). Gilmore named numerous footprint ichnospecies, but modern ichnotaxonomy (which is informed by taphonomic and sedimentological considerations) has reduced that number substantially (Marchetti et al., 2020).

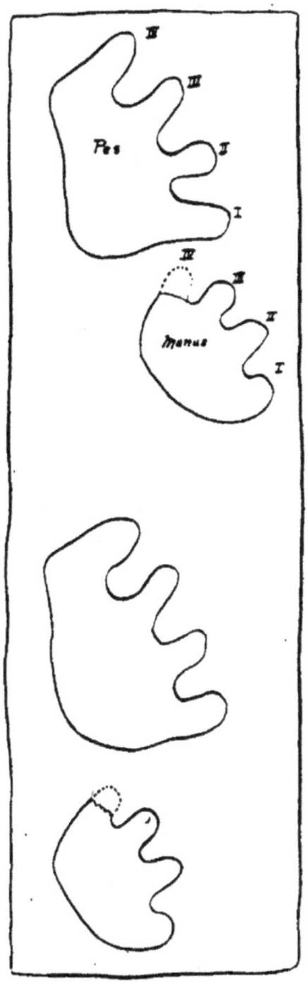

FIGURE 10 Gilmore (1927) coined the name *Baropus coconinoensis* (new ichnospecies) for these footprints from the Permian Coconino Sandstone in Arizona. In the latest ichnotaxonomic revision of the Coconino tracks, Marchetti et al. (2020) considered *Baropus coconinoensis* to be a junior subjective synonym of *Ichniotherium sphaerodactylum*. *Drawing from Gilmore (1927).*

2.3 Focused research, 1940–70s

After the Second World War, a major advance in footprint ichnology occurred with the first appearance of paleontologists whose primary research program was the study of tetrapod footprints (Figs. 11 and 12). These were Ukrainian Oleg Vialov (1904–88), Americans Frank Peabody (1914–58) and Donald Baird (1926–2011), French Georges Demathieu (1920–2010), English William A. S. Sarjeant (1935–2002), Italian Giuseppe Leonardi (1939) and German Hartmut Haubold (1941). These scientists not only published extensively on tetrapod footprints, but they introduced

FIGURE 11 Some important vertebrate ichnologists of the 20th century. (A) Oleg Vialov; (B) Frank Peabody; (C) Donald Baird; (D) George Demathieu. *(A) From Paliy (1990); (B) From Downs (1958); (C) From Sues et al. (2013); (D) Courtesy Pierre Demathieu.*

conceptual and methodological advances that have come to underpin all subsequent footprint research.

Peabody (1948) introduced the concept of extramorphology in the analysis of tetrapod footprints. Extramorphology refers to all of the factors that influence footprint shape other than the morphology of the trackmaker's foot—trackmaker behavior, substrate composition, taphonomy, etc. Peabody also published classic work on Triassic footprints from the American Southwest (cf. Klein and Lucas, 2010, 2021).

FIGURE 12 Some important vertebrate ichnologists of the 20th century. (A) William Sarjeant; (B) Giuseppe Leonardi; (C) Hartmut Haubold; (D) Martin Lockley. *(A) From Spalding et al. (2003); (B) Courtesy of Spencer G. Lucas; (C) Courtesy of Shinobu Ishigaki; (D) Courtesy of Gretchen Minney.*

Baird (1957) may have undertaken and published the first doctoral dissertation on tetrapod footprints, at Harvard University, on Early Jurassic footprints from the Connecticut Valley, following the work first undertaken by Hitchcock. Over the course of 40 years, Baird published numerous articles on footprints, primarily those of Permian and Triassic age.

Vialov (1965, 1966) published monographic studies of the late Cenozoic bird and mammal footprints from Miocene strata in the Carpathian Mountains of the western Ukraine. In these studies, he published the first extensive ichnotaxonomy of both bird and mammal footprints. Vialov (1972) also argued for the International Commission on Zoological Nomenclature to regulate names of trace fossils and introduced terminology for some trace fossil categories, including vertebratichnia for vertebrate trace fossils.

Sarjeant also favored more formal rules for ichnotaxonomic practice, even proposing his own code of ichnotaxonomic nomenclature (Sarjeant and Kennedy, 1973). Sarjeant was also an advocate of "one ichnology," meaning he favored footprint ichnotaxonomy based on the behavior of the track produced, not on its taxonomic affiliation. This is the method by which invertebrate ichnotaxonmy is undertaken (ethoichnotaxonomy of Hunt and Lucas, 2007) as opposed to the longstanding use of trackmaker identity in footprint identification (bioichnotaxonomy of Hunt and Lucas, 2007). Sarjeant also published prolifically on the history of footprint studies, and on diverse footprints ranging in age from Devonian to Pleistocene.

Demathieu published extensive work on French footprints of Mesozoic and Cenozoic ages. In particular, ichnofaunas from Middle Triassic deposits along the French Massif Central were the focus of much of his research. Although not a conceptual innovator, he was a methodological innovator, applying statistical methods to the discrimination of footprint morphologies and ichnotaxa. Demathieu was also one of the most prolific footprint ichnologists of the 20th century.

Leonardi was also extremely prolific and carried out much of his research on South American footprints, particularly in Brazil (Leonardi, 1994). His edited monograph titled "Glossary and manual of tetrapod footprint palaeoichnology" (Leonardi, 1987) is still an indispensable volume on methods used regularly by all students of fossil footprints.

Haubold stands out as the most important 20th century student of vertebrate footprints. He published the first grand syntheses of the tetrapod footprint record, in both monographic (Haubold, 1971) and book (Haubold, 1984) forms. He also published extensively on Permian and Triassic footprints from Europe, and, in the 1990s joined two of us (SGL and APH) to work on early Permian footprints from the American Southwest. This produced a wholescale revision of Permian footprint ichnotaxonomy based heavily on the application of the concept of extramorphology (Haubold, 1996, 2000). His great influence on students and collaborators is well seen in our work and in that of many other students of fossil footprints.

2.4 The vigorous subdiscipline of footprint ichnology, 1980s-present

In the 1980s, Martin Lockley (Fig. 12), who had primarily studied Ordovician brachiopods while a student in the UK, took a position at the University of Colorado at Denver in the USA. He soon encountered Cretaceous hadrosaur footprints, and his research shifted to a focus on footprint ichnology

(Lockley et al., 1983). Thus, Lockley rapidly realized that there was great opportunity in studying dinosaur footprints, especially to use them to interpret diverse aspects of dinosaur paleobiology and habitats, and he drew attention to the importance of dinosaur footprints (Lockley, 1984, 1986, 1987). As Lockley (1986, p. 45) well stated, dinosaur footprints represented "much-untapped data to be analyzed and integrated with current paleobiological and paleoenvironmental knowledge." Lockley saw this as the next step in the "dinosaur renaissance" that had begun in the 1970s based on the study of their body fossils.

Lockley had tremendous ambition and energy. He traveled widely and collaborated freely with many students and ichnologists around the world, studying dinosaur (and some other) tracks on all of the continents except Antarctica. Lockley published extensively, including a series of books authored, co-authored, and co-edited, on vertebrate footprints (Gillette and Lockley, 1989; Lockley and Hunt, 1994; Lockley and Meyer, 2000; Lockley, 1991, 1999; Lockley and Lucas, 2014). He worked primarily on Jurassic-Cretaceous dinosaur tracks in the American Southwest and eastern Asia (especially China and South Korea) and helped develop publicly protected parks for footprints, including a UNESCO World Heritage Site in South Korea and, perhaps, most famously, the Dinosaur Ridge park in Colorado (Fig. 13). It is fair to say that Lockley did more than any of his predecessors to create the current extent and vigor of footprint ichnology (Lucas et al., 2024).

FIGURE 13 The main tracksite at Dinosaur Ridge, Colorado; track outlines have been accentuated with charcoal. *Courtesy of Jeff Lamontagne.*

The last 40 years have thus seen the emergence of the research program that had its foundations in the work of Buckland, Hitchcock, and others by the 1850s to underpin a vigorous, global study of tetrapod footprints. Today, ichnologists around the world regularly undertake research on and publish on vertebrate footprints. The tetrapod footprint record is now known from the Middle Devonian (the oldest tetrapod fossils are footprints: Lucas, 2015b) to the Pleistocene. Diverse syntheses of this record have been published, and symposia on tetrapod footprint research are held regularly. The concept of extramorphology informs current footprint ichnotaxonomy, and the identity of the trackmaker is an integral part of footprint ichnotaxonomy (hopes for "one ichnotaxonomy" have been rejected). The footprint record provides important data on tetrapod distribution where body fossils are lacking, and has been used to evaluate biostratigraphy and paleobiogeography. The evolution of tetrapod locomotion and of paleoecology is also a focus of current research. New technology using drones and three-dimensional documentation such as photogrammetry and scanning are the latest technological innovations in footprint research (see the chapter by Lallensack in this volume).

3. Dentalites

Dentalites, generally called "tooth marks," have a diverse and long history of study, but have only been worked on in a more synthetic fashion very recently (Hunt and Lucas, 2021c, and see Hunt and Lucas in this volume). Drumheller-Horton (2012), Lucas (2016), Hunt et al. (2018), and Hunt and Lucas (2021c, and in this volume) reviewed the study of vertebrate dentalites, and that history is briefly summarized here.

3.1 Beginnings, 1820s

The study of dentalites began with William Buckland (1822, 1824; Fig. 1), who analyzed a Pleistocene vertebrate fauna from Kirkdale Cave in northern England (Fig. 14). He utilized element representation, bone damage and the dentalites to demonstrate that the bone deposit was not "Diluvial" (from the Flood) but actually represented an "Ante-diluvian" hyena den (Boylan, 1977). As part of his studies, Buckland fed cow bones to a captive hyena and compared the damage to that on bones from Kirkdale Cave.

3.2 A dearth of interest, 1830–1950s

Following Buckland's pioneering work, there was relatively little interest in dentalites for the remainder of the nineteenth century, with a few exceptions in Europe (Tournal, 1833; Dawkins, 1863). In the United States both

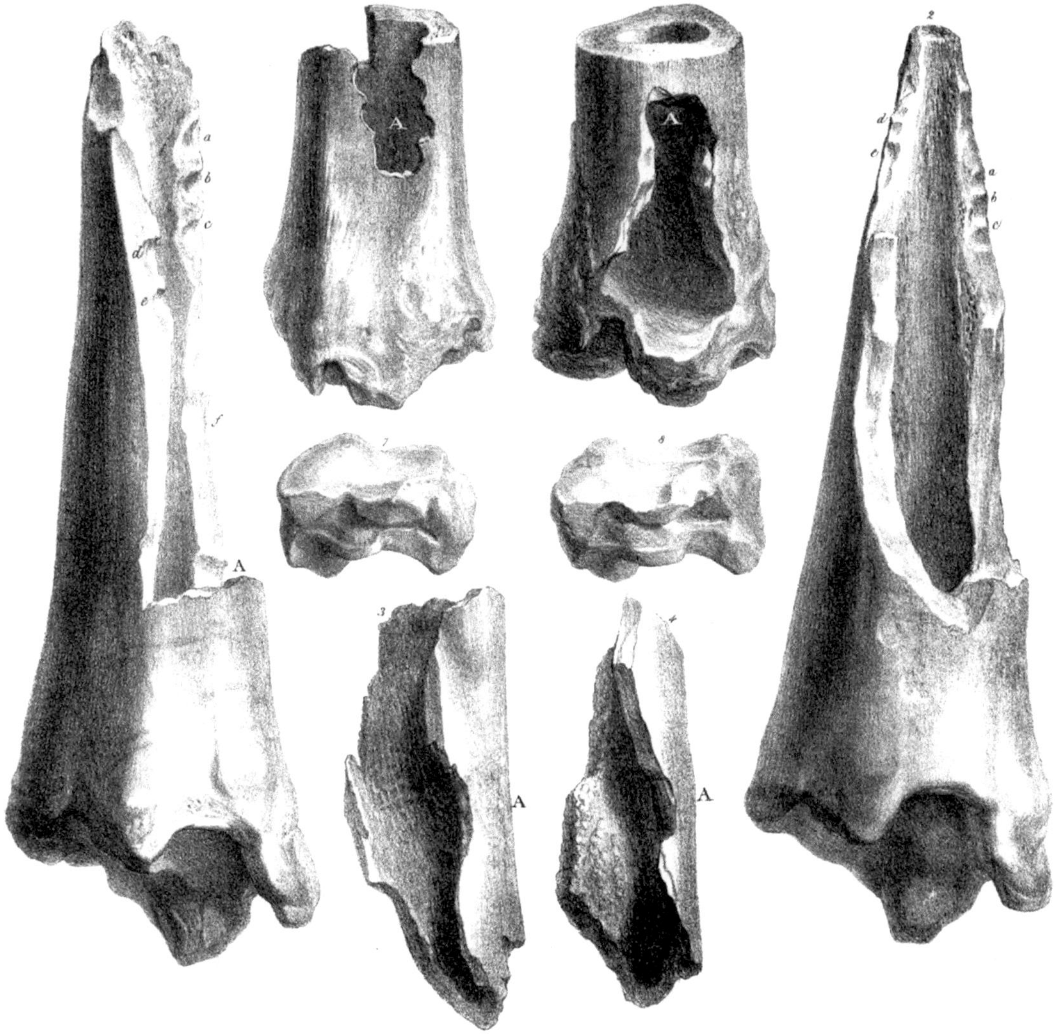

FIGURE 14 Comparison of Pleistocene and recent damage to bovid bones by hyena jaws (From Buckland, 1824, pl. 23). "1. Residuary part of the lower extremity of the tibia of an ox, which I saw given entire to a Cape hyaena in Mr. Wombell's traveling collection at Oxford, in December 1822: Marks of the teeth are definitely distinctly visible at a, b, c, d, e, f.... 2. Fragment of a similar tibia from Kirkdale cave, broken nearly in the same manner as No. 1, and bearing similar marks of teeth at a, b, c, d, e: In the recent, as in the antediluvian specimen, the lower condyle has, from its hardness, been left unbroken. 3. Splinter from another bome broken by the hyaena at Oxford: The cavity at A. was produced by the hyaena's tooth. 4. Similar splinter, bearing a similar cavity A., from the cave at Kirkdale, and partially incrusted with stalagmite. 5. Inside of the lower extremity of the recent specimen, No. 1, in which the hole A. was produced by the hyaena's bite. 6. Lower extremity of another tibia from Kirkdale, in which the form of the cavity A. resembles that in No. 5. 7. Scaphoid bone of the left carpus of an ox, which, with the other component bones of the carpus, lay all night untouched in the hyaena's cage at Oxford. 8. Similar bone from Kirkdale, equally untouched" (Buckland, 1824, p. 276–277).

Mudge (1877) and Williston (1898), noted mosasaur fossils in the Upper Cretaceous strata of the western United States with dentalites produced by sharks, but such observations were not developed further.

3.3 Developing interest, 1950–2000

During the 20th century, interest in dentalites grew in the context of a dichotomy between archaeological and paleontological studies (Hunt et al., 1994; Drumheller-Horton, 2012). Archeological studies of dentalites were driven by the advent of processual archaeology (Binford, 1981). A major motivation was to answer the question posed by the title of Brain's (1981) well-known book "The hunters or the hunted?" (also see Washburn, 1957). This was a reaction to Dart's (1957) famous hypothesis of the osteodontokeratic culture (i.e., tool kits made from bones and teeth) utilized by the early hominid *Australopithecus.*

Paleontological studies of dentalites were part of a broad resurgence of interest in vertebrate taphonomy (Behrensmeyer and Hill, 1980). The majority of this work, both actualistic and on fossils, related to fossil mammals, with a growing literature on fossil crocodilians. The ichnotaxonomy of dentalites began with Cruickshank (1986), who applied the first ichnotaxonomic name to them (but recently questioned by Zonneveld, 2022).

3.4 A maturing subdiscipline, 2000–present

Studies on dentalites have developed significantly during the past 20 years. The development of ichnotaxonomy was greatly stimulated by the publications of Mikuláš et al. (2006) and Jacobsen and Bromley (2009). There are now 25 named dentalite ichnotaxa (see Hunt and Lucas, 2021c, and in this volume).

Ongoing studies of dentalites are largely focused on dinosaurs, for which a diverse literature exists (see Hunt and Lucas, 2021c, and in this volume). These studies generally do not employ ichnotaxonomy and are focused on the implications of the dentalites for understanding feeding, scavenging, and food processing. Relevant to dentalites, which are a form of bioerosion, Pirrone et al (2014) proposed ichnotaxobases for trace fossils in bone. Lucas (2016) named two substrate-based nonmarine ichnofacies: the *Cubiculum* ichnofacies for modifications to bone, including dentalites, and the *Palaeoscolytus* ichnofacies for modifications to vegetation, including dentalites. Hunt et al. (2018) added the *Gaspeichnus* ichnofacies for modifications (including by teeth) to coprolites. Fernandez-Jalvo and Andrews (2016) recently published an extensive atlas documenting dentalites and other bioerosion on bones.

4. Regurgitalites

Regurgitalites are trace fossils of manipulated or partially/wholly digested food matter that was egested through the oral cavity (Hunt and Lucas, 2012, 2021c). They began to be studied sporadically about 100 years ago, and only recently have they been the subject of more detailed research.

4.1 First reports, 1930–1950

The description and discussion of regurgitalites started about a century after the first work on dentalites and bromalites. Götzinger and Becker (1932) described discrete accumulations of small fragments of *Inoceramus* shells from the Upper Cretaceous of Austria that they tentatively interpreted as coprolites. However, Häntzschel et al. (1968, p. 53) noted that they are "perhaps vomit balls, corresponding to the undigestible matter vomited by birds."

4.2 Growing interest, 1950–2000

Much of the work on regurgitalites in the second half of the 20th Century was conducted by German paleontologists, particularly with regard to the Jurassic Holzmaden and Solnhofen Lagerstätten, as well as some other localities. Italian paleontologists also conducted important studies.

The most prominent work in the United States during this time period was Zangerl and Richardson's (1963) meticulous study of the taphonomy of the Pennsylvanian Carbondale Formation of Indiana. They recognized numerous regurgitalites and introduced terminology to distinguish different types. However, subsequent actualistic taphonomic studies (Elder and Smith, 1988) have demonstrated that many of these specimens are not regurgitalites but instead represent decayed fish.

The first convincing strigilite (owl pellet) was described from the Miocene of New Mexico by Gawne (1975). Other strigilites were noted in the western United States. Regurgitalites were recognized from the Cretaceous chalk of Kansas and the Eocene lacustrine strata of the northwestern United States and British Columbia in Canada. Hunt (1992) introduced the term regurgitalite in a study of Pennsylvanian bromalites from New Mexico.

Starting in the 1970s there was an increased interest in the origin of microvertebrate fossil accumulations. Much work was focused on strigilites (and other ornithoregurgitalites) as an important source of such accumulations with Andrews' (1990) volume being particularly influential.

4.3 Accelerated interest, 2000–Present

During the first decade of this century, there was intermittent, but diverse, interest in regurgitalites. Aldridge et al. (2006) described the oldest putative regurgitalites from the Late Ordovician, and Sanz et al. (2001) published an influential paper in *Science* on a regurgitalite from the Lower Cretaceous of Spain. Burrow and Turner (2010) described the oldest definitive vertebrate regurgitalite from the Lower Devonian of Scotland. Multiple regurgitalites have been described from the marine Cretaceous of Kansas in the United States (Everhart, 2017). Avian regurgitalites have been described from the Paleogene of North and South America and, notably, the Eocene Messel Lagerstätte in Germany (Mayr and Schaal, 2016). The first purgolite (an accumulation of regurgitalites) composed of strigilites was described from the Oligocene of Wyoming in the United States (Lucas et al., 2012).

Most recognized vertebrate regurgitalites contain vertebrate remains, but angular shell debris can be used as a proxy for the prevalence of durophagous predators, including fish. Such regurgitalites were described from the Devonian, Mississippian, Triassic, and Jurassic. The older regurgitalites are prima facie evidence of durophagous feeding in the Paleozoic and run counter to the ill-conceived concept of a "Mesozoic marine revolution" (Lucas and Hunt, 2023).

Myhrvold (2011) and Hunt and Lucas (2012) reviewed terminology and introduced new terms, and several papers have reviewed criteria for the recognition of regurgitalites. Hunt et al. (2012c, d) named the first regurgitalite ichnotaxa, *Ostracobromus* and *Conchobromus,* from Carboniferous ichnofaunas in New Mexico and Montana in the United States.

In the last decade, there has been elevated interest in regurgitalites, which has increased rapidly during the past 5 years (see Hunt and Lucas in this volume). However, the study of regurgitalites is in an early phase during which many records need to be documented (see Hunt and Lucas in this volume).

5. Consumulites

Consumulites are food material ingested and preserved within the gastrointestinal tract, or elsewhere within the body cavity, of a fossil vertebrate (Hunt and Lucas, 2012). They have long been referred to in the vernacular as "stomach contents" or "gut contents" and largely reported incidentally as part of the description of a fossil vertebrate skeleton. Hunt and Lucas (2021c, and in this volume) discussed their history of study, which is reprised here.

5.1 A few reports from marine taxa, 1820s to 1900

In 1829, William Buckland had introduced the term coprolite for fossil feces. However, he utilized the term for both evacuated fecal material, and for what are now considered consumulites (Buckland, 1829, 1835, 1836; Hunt and Lucas, 2012). During the remainder of the 19th Century, there was sporadic interest in consumulites of Mesozoic marine reptiles and of Paleogene fish. Increased collection and study of mosasaurs later in that century, notably in western North America, resulted in more reports.

5.2 Sporadic reports, 1900–90

The earliest described nonmarine consumulite was Pleistocene plant debris of *Ephedra* sp. and *Salix* sp. in the oral cavity of a frozen woolly rhinoceros (*Coelodonta antiquitatis*) from Russia. In 1900, the first complete frozen mammoth with a preserved gastrointestinal tract was found. Subsequently, numerous frozen Pleistocene specimens of mammoths, bison, and horse have yielded consumulites, principally in Russia, but also in North America.

There were relatively few reports of consumulites for the first 9 decades of the 20th century, and the majority were just noted in the context of morphological and taxonomic descriptions. Boucot's (1990) influential compendium on the paleobiology of behavior and coevolution included numerous references to consumulites and two important review papers on consumulities in fish (Williams, 1990; Viohl, 1990).

5.3 Increasing interest, 1990–present

In the past 3 decades, there have been increasing reports of consumulites, in general, and publications in which the main focus is these bromalites, in particular. Everhart and others have described a large number of consumulites from the Late Cretaceous of the Western Interior of North America (cf. Everhart, 2017). Another principal area of focus in the 21st Century has been the consumulites from the Early Cretaceous Jehol Lagerstätte of China, principally in theropods and birds, as they relate to the development of the avian digestive system (crop, proventriculus, and ventriculus) (O'Connor, 2019). The Jehol Lagerstätte also yields consumulites in a mammal, a frog and in choristoderes.

Hunt and Lucas (2012, 2021c) introduced terminology for different types of consumulites, identified ichnotaxobases, and named two ichnogenera (Hunt and Lucas, 2021a). Nevertheless, the study of consumulites is still in a very early phase, and there is need for more documentation, ichnotaxonomy, and synthesis.

6. Gastroliths

Scientific observations and studies of fossil gastroliths extend back more than 300 years, but only in the last 20 years has comprehensive research on gastroliths been conducted, notably that of Wings (2004, 2007, 2015; Wings and Sander, 2007). That history is detailed in Hunt and Lucas (2021c) and briefly summarized here.

6.1 First steps, 1668–1907

According to Skoczylas (1978), in 1668 the German natural philosopher Henry Oldenburg (1619–77) described "stomach stones" in an extant crocodile. However, this report by Oldenburg (1668) was of observations of gastroliths in a caiman made by a Dr. Stubbes. Baker (1956) stated that Mayne (1854) first used the term *Gastrolithus* to refer to a "stone or calculus in the stomach." Lequat (1708) noted their presence in a recently extinct (within historical times) bird, the solitaire (*Pezophaps*), and this may be the oldest published reference to gastroliths in an extinct vertebrate.

Real discussion of gastroliths in extinct vertebrates began in the early 1900s. This was in a series of articles by Brown, Williston, Eastman, and Wieland that discussed plesiosaur and dinosaur gastroliths and debated the function of the former in buoyancy or grinding food. As part of that discussion, Wieland (1906, p. 820) "coined" the term gastrolith: "stomach stones, or *gastroliths*, as I shall conveniently call them." Wieland (1907, 1920) later advocated that dinosaur gastroliths have highly polished surfaces.

This discussion set up three ideas about gastroliths that have been much discussed in later studies: (1) plesiosaurs had gastroliths that were used either for diet or buoyancy control; (2) sauropod dinosaurs had gastroliths; and (3) high polish is characteristic of gastroliths.

6.2 Incidental reports, 1907–2004

What followed for about one century is a diffuse literature that documented gastroliths in various fossil vertebrates (particularly plesiosaurs) and a less extensive literature that debated their function. Particularly important was the work of Bryan (1931), who cogently argued that high polish does not character gastroliths, a conclusion ignored by many subsequent workers.

Much literature has focused on dinosaur gastroliths, in particular, identifying clasts without skeletal associations or with very loose skeletal associations as gastroliths, simply because they are highly polished. Archeologists have followed suit, identifying polished sand/pebbles at archaeological sites as bird gastroliths, although they lack any association with avian bones (Lucas and Hunt, 2021). And, some geologists have even identified dropstones and other exotic clasts as gastroliths.

6.3 Comprehensive research, 2004–Present

Wings (2004) undertook the first extensive study of gastroliths in extinct vertebrates. His work supports the concept that high polish does not characterize gastroliths and reinforces earlier conclusions that without a plausible skeletal association (i.e., mass of stones in the abdominal cavity), no sand/gravel can be verified as gastroliths. A large literature has grown up about plesiosaur gastroliths, and important research on Early Cretaceous birds from northeastern China has revealed important steps in the evolution of gastrolith use by both the avian crop and gizzard. Studies of crocodilian gastroliths have focused on debate over their use, either in buoyancy or digestion.

Hunt and Lucas (2021c) reviewed the entire record of gastroliths, proposing terminology and evaluating the significance of gastroliths to paleontological interpretation. An important point made by Hunt and Lucas (2021c) is that gastroliths are trace fossils in need of ichnotaxonomy. And, they provide important insights into various behaviors, notably diet, digestion, buoyancy control, and habitat preferences.

7. Coprolites

After footprints, coprolites are the most studied vertebrate trace fossils. Until this century, the study of coprolites was on the fringes of its science (paleontology or archeology) and has been disproportionately influenced by a small number of individuals (Hunt et al., 2012a). Several authors have reviewed the history of study of vertebrate coprolites (Reinhard and Bryant, 1992; Duffin, 2009, 2012a, b; Bryant and Reinhard, 2012; Hunt et al., 2012a; Pemberton, 2012; Shillito et al., 2020; Hunt and Lucas, 2021b, c, in this volume), and that history is briefly reviewed here.

7.1 Buckland and copromania, 1820–90s

Bromalites, later identified as coprolites, have been known since the late seventeenth century (Duffin, 2012a). However, it was only in the 1820s that William Buckland identified aggregates of apatitic subspheroidal objects from a Pleistocene deposit in Kirkdale Cave in northern England as fossil hyena feces (Buckland, 1822, 1824). This was a relatively straightforward conclusion, as the cave contained many hyena bones, and Buckland could demonstrate that the fossil specimens were comparable to modern hyena feces (Hunt and Lucas, 2021c).

Buckland made a much more important breakthrough in 1829, when he recognized spiral objects from the Early Jurassic of southern England as also being what he then called coprolites (Buckland, 1829; Hunt and Lucas, 2021c). This recognition, possibly aided by the famed fossil collector Mary

Anning (Duffin, 2012a), led him to identify these types of fossil from many stratigraphic intervals from the Carboniferous to the Neogene (Buckland, 1829, 1835). Buckland (1835) realized that modern sharks and rays have spiral valves, and he filled modern shark intestines with Roman cement to demonstrate that they might produce spiral feces (Fig. 15). Despite this innovative study, he still referred some spiral coprolites to ichthyosaurs.

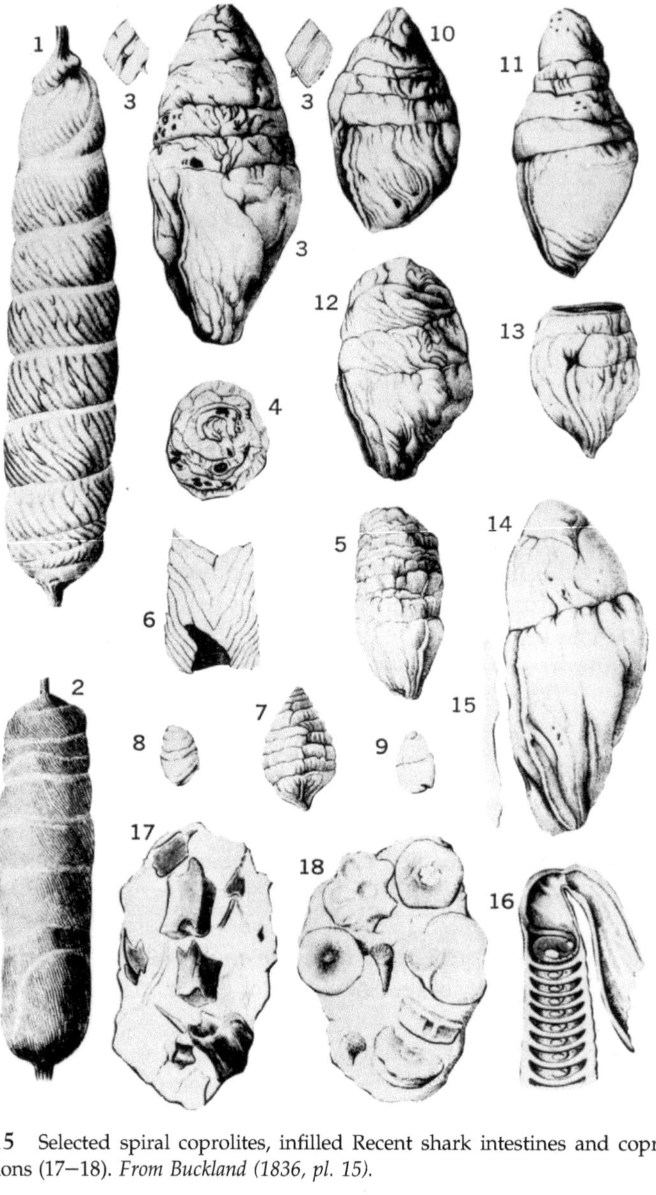

FIGURE 15 Selected spiral coprolites, infilled Recent shark intestines and coprolites with bone inclusions (17–18). *From Buckland (1836, pl. 15).*

Buckland's recognition of fossil feces sparked great interest and some levity (Duffin, 2012a; Pemberton, 2012). It inspired general interest in coprolites (but little serious research) that Duffin (2012a) termed "copromania," particularly in Victorian England, but also in Europe and in the United States, that lasted till about the middle of the 19th century.

7.2 New research, 1890–1910

At about the beginning of the 20th century, several workers explicitly addressed the classification of coprolites and other bromalites. During this period there were also a few significant descriptions of larger coprofaunas, the first since Buckland (1835). During this time period, Harshberger (1896) first drew attention to the potential importance of human coprolites.

7.3 The quiet years, 1910–50

For a period of 40 years, there were intermittent papers published on coprolites but none that were conceptually substantial or that described large samples. Two important papers on human coprolites were published during this time period (Loud and Harrington, 1929; Wakefield and Dellinger, 1936). They first used chemical and microscopic techniques to address issues of prehistoric human health.

7.4 Human and Pleistocene nonhuman coprolites, 1950–90

In the 1950s, Eric O. Callen first became interested in human coprolites (Callen and Cameron, 1955) and subsequently worked for several decades "with missionary zeal to convince archaeologists and others about the importance of [human] coprolite research" (Bryant and Reinhard, 2012, p. 380). One of his collaborators was Paul S. Martin. Martin was interested in Pleistocene coprolites and began to develop a large comparative collection of modern dung.

Callen's discovery of a method to reconstitute desiccated coprolites using trisodium phosphate ushered in the "Golden Age of [human] Coprolite Analysis" that lasted for just over 30 years, ending in the early 1990s (Reinhard, 2006; Shillito et al., 2020). This work focused largely on North America, principally on the western and southwestern United States. Other human coprolites were described from the southeastern United States, Mexico, and South America. Most studies were of the coprolites of *Homo sapiens,* but there were also studies of the coprolites of *Homo erectus* and of Neanderthals. Studies of both human and Pleistocene nonhuman coprolites addressed broad research questions during this time period, notably with regard to diet, climate, and parasites.

The decades of the 1960s through 1990s also saw an increased interest in paleontological coprolites. As the work of Callen had stimulated studies of human coprolites, Martin's research had a similar effect on the study of

Pleistocene vertebrate coprolites. Thus, Martin and his students and collaborators, undertook extensive research on the desiccated coprolites from caves in the American Southwest. Much of this work was focused on the palynology and paleobotany of the contents of coprolites, their geological ages and their relationship to late Pleistocene extinctions. Recent work saw the first application of DNA analysis to the study of ground sloth coprolites (Poinar et al., 1998, 2003).

The development of the study of geologically older coprolites was slower during this timeframe. The majority of this work was on Paleozoic coprolites.

7.5 A maturing subdiscipline, 1990–Present

While researchers in Quaternary studies and archeology were using coprolites as important research tools, most other paleontologists still treated them as not usually worthy of collection, let alone study, with the exception of some students of Paleozoic fish. This began to change in 1990s as there was a renewed interest in pre-Pleistocene coprolites. At that time, Karen Chin did much to popularize the paleontological study of coprolites as she and co-workers analyzed the coprolites of the most glamorous of fossil animals, dinosaurs. In a similar timeframe, two of us (APH and SGL) began to collect large samples of coprolites, especially of Triassic age, and to analyze them. Hunt et al. (1998) applied the first binomial ichotaxonomy to vertebrate coprolites, and various ichnotaxa have since been named. Ironically, as the study of human coprolites has declined, work on pre-Pleistocene coprolites has intensified and become more sophisticated.

Hunt et al. (2012b) edited the first scientific volume on vertebrate coprolites, which included diverse contributions that discussed the history of study, terminology, distribution, and importance of bromalites. This volume seems to have encouraged a subsequent increase in the quantity and diversity of coprolite studies. The last decade has been characterized by the use of new methodologies of study, notably micro CT scanning, first utilized by Jesper Milàn, and a focus on inclusions in coprolites of fossils of organisms, such as bacteria and beetles, and structures such as hair, that otherwise have a very limited and poor fossil record (see Dentzien-Dias and Łaska in this volume). Indeed, coprolites are now seen as time capsules or mini-Lagerstätten that contain a wealth of fossils not usually preserved in other contexts.

8. Other bromalites, burrows, and nests

There has been a relatively short history to the study of other vertebrate trace fossils, namely other bromalites (digestilites, pabulites, and micturalites), burrows, and nests. Hunt and Lucas (2021c) reviewed the limited

FIGURE 16 The corkscrew rodent burrow *Daemonelix*. By James St. John—*Daemonelix* burrows (Harrison Formation, Middle Miocene; Agate Fossil Beds National Monument, northwestern Nebraska, USA), Public Domain, https://commons.wikimedia.org/w/index.php?curid=37027739.

research on the other bromalites, most of which has taken place in the last decade. It is notable, though, that in the 1970s the idea that many microvetebrate fossil accumulations are composed of digestilites was advanced and has since gained much acceptance.

The first descriptions of fossil vertebrate burrows in the 19th century were of corkscrew-like structures from the Miocene of Nebraska that were named *Daimonelix* (Fig. 16). More comprehensive work on fossil vertebrate burrows began in the first half of the 20th century, including studies of aestivation burrows with lungfish remains from the Permian of the USA (Romer and Olson, 1954). Voorhies (1975) published a review of vertebrate burrows with fossil and extant examples. Other descriptions, partly introducing new ichnotaxa, followed (see the chapter by Melchor and Cardonatto in this volume).

Fossil nests have been relatively unstudied in an ichnological context. They are particularly well known for dinosaurs (Carpenter, 1999), but much documentation, identification of ichnotaxobases and an ichnotaxonomy remain to be undertaken.

9. Invertebrate versus vertebrate ichnology

Various authors (Osgood, 1970, 1975; Pemberton et al., 2007; Baucon et al., 2012) have reviewed the history of the development of invertebrate ichnology. The history of the two subdisciplines, vertebrate and invertebrate ichnology, is disparate.

By the 1850s, as noted above, key methodological aspects and conceptual foundations of vertebrate footprint ichnology were established, but research on vertebrate footprints would largely languish till the 1980s. Studies of most bromalites and gastroliths also have a similar trajectory—early comprehension, followed by more than a century of relative neglect.

In the case of invertebrate ichnology, an incorrect idea (that all invertebrate traces were fossil fucoids) hindered progress until at least the 1890s. In the case of vertebrate ichnology, progress was slowed by the judgment of ichnologists and vertebrate paleontologists that vertebrate trace fossils were not important to understanding broader questions of vertebrate evolution.

Invertebrate and vertebrate ichnology have long remained distinct subdisciplines, with few ichnologists working on both invertebrate and vertebrate traces. This in large part reflects the different traditions of both subdisciplines, as outlined by Hunt and Lucas (2007). Thus, the ethoichnotaxonomic tradition of invertebrate ichnology focuses on the trace as behavior and concerns itself relatively little with the identity of the tracemaker, especially because many invertebrate traces are homeomorphs that could have been made by very different trace makers. In contrast, the biotaxonomic tradition of vertebrate ichnology is very concerned with the identity of the tracemaker to establish the theoretical optimal morphology of the trace that can be distinguished from extramorphology. These different emphases are well embodied in current ichnological research and guarantee that there will never be "one ichnology."

10. Prospectus: The future of vertebrate ichnology

Based on this historical review, we believe that the future tasks for vertebrate ichnology should include:

1. Discovery and documentation of traces from time intervals and geographic regions under-represented by the trace fossil record. Ichnological field research in regions that have not been well studied will provide new trace fossil material that will help to answer questions about trackmaker affinities, migration and the dispersal of producers, as well as biostratigraphic and ichnotaxonomic problems. Large parts of South America, Africa, Asia, Australia, and Antarctica are under-represented in the trace fossil record because trace fossils have simply been overlooked or some regions have been completely unexamined. Important recent discoveries in Paleozoic-Mesozoic strata of North Africa, Greenland, and East Asia, but also in areas that are considered to have been well studied, such as in Arizona, support this.
2. Ichnotaxonomic revisions of some largely oversplit footprint groups, excluding extramorphological (substrate-related) factors. Thus, the

extensive splitting of footprint ichnotaxa (especially of dinosaurs), resulting largely from a failure to filter out extramorphology, needs to be reduced by defining strict, anatomically based ichnotaxa. In recent years, we have seen positive developments in this direction, but there are still numerous ichnotaxa in common use that can be considered invalid or nomina dubia.

3. Development of a workable ichnotaxonomy for non-footprint traces. For non-footprint vertebrate trace fossils, the identification of ichnotaxobases and the establishment of a reasonable classification and ichnotaxonomy will make it necessary to introduce further ichnotaxa, hopefully avoiding the mistakes made in much footprint ichnotaxonomy.
4. Databases of vertebrate traces should be extended and made available online, including 3D models. In particular, holotypes should be made available as photogrammetric images and, other than "plastotypes" and physical replicas, should be defined as valid type materials by ICZN. Machine learning software and artificial intelligence (AI) can also support ichnotaxonomic evaluation by recognizing different morphotypes and comparing them to known ichnotaxa.

Thus, an ambitious research program lies ahead for vertebrate ichnology, which is now a vigorous and diverse subdiscipline of paleontology.

Acknowledgments

We are grateful to the contributors to this volume and many other vertebrate ichnologists for their generosity and collaboration. We dedicate this article to the memory of Bill Sarjeant, who was the premier student of the history of footprint ichnology. Jesper Milàn and Sebastian Voigt provided helpful reviews that improved the content and clarity of this article.

References

Aldridge, R.J., Gabbott, S.E., Siveter, L.J., Theron, J.N., 2006. Bromalites from the Soom Shale Lagerstätte (Upper Ordovician) of South Africa. Palaeoecological and palaeobiological implications. Palaeontology 49, 857–871.

Andrews, P., 1990. Owls, caves and fossils: predation, preservation and accumulation of small mammal bones in caves, with an analysis of the Pleistocene cave faunas from Westbury-sub-Mendip, Somerset, UK. University of Chicago Press, Chicago, 239 p.

Baird, D., 1957. Triassic reptile footprint faunules from Milford, New Jersey. Bull. Mus. Comp. Zool. 117, 449–520.

Baker, A., 1956. The swallowing of stones by animals. Vic. Nat. 73, 2–95.

Baucon, A., Bordy, E., Brustur, T., Buatois, L.A., Cunningham, T., De, C., Duffin, C., Felletti, F., Gaillard, C., Hu, B., Hu, L., Jensen, S., Knaust, D., Lockley, M., Lowe, P., Mayor, A., Mayoral, E., Mikulas, R., Muttoni, G., Neto de Carvalho, C., Pemberton, S., Pollard, J., Rindsberg, A., Santos, A., Seike, K., Song, H., Turner, S., Uchman, A., Wang, Y., Yi-ming, G., Zhang, L., Zhang, W., 2012. A history of ideas in ichnology. In: Knaust, D., Bromley, R.G. (Eds.), Trace fossils as indicators of sedimentary environments. Developments in Sedimentology, 64. Elsevier, Amsterdam, pp. 3–43.

Behrensmeyer, A.K., Hill, A.P. (Eds.), 1980. Fossils in the Making: Vertebrate Taphonomy and Paleoecology. University of Chicago Press, Chicago, 338 p.

Binford, L.R., 1981. Bones: Ancient Men and Modern Myths. Academic Press, New York, 320 p.

Boucot, A.J., 1990. Evolutionary Paleobiology of Behavior and Coevolution. Elsevier, Amsterdam, 735 p.

Bourgeois, J., 2024. Trackways in the New Red Sandstone of the Connecticut River Valley, USA, the cradle of ichnology, 543. Geological Society London, Special Publications.

Boylan, P.J., 1997. William Buckland and the foundations of taphonomy and palaeoecology. Arch. Nat. Hist. 24, 361–372.

Brain, C.K., 1981. The Hunters or the Hunted? University of Chicago Press, Chicago, 384 p.

Bryan, K., 1931. Wind-worn stones or ventifacts—a discussion and bibliography. Reprint and Circular Series National Research Council [Report on Sedimentation 1929–1930], 98, pp. 29–50.

Bryant Jr., V.M., Reinhard, K.J., 2012. Coprolites and archaeology: The missing links in understanding human health. N. M. Mus. Nat. Hist. Sci. Bull. 57, 379–387.

Buckland, W., 1822. Account of an assemblage of fossil teeth and bones of elephant, rhinoceros, hippopotamus, bear, tiger and hyaena, and sixteen other animals; discovered in a cave at Kirkdale, Yorkshire, in the year 1821; with a comparative view of five similar caverns in various parts of England, and others on the continent. Philos. Trans. R. Soc. Lond. 112, 171–236.

Buckland, W., 1824. Reliquiae diluvianae; or, Observations on the Organic Remains Contained in Caves, fissures, and Diluvial Gravel, and on Other Geological Phenomena, Attesting the Action of an Universal Deluge. J. Murray, London, 303 p.

Buckland, W., 1829. A paper by Dr. Buckland. Proc. Geol. Soc. Lond. 1, 142–143.

Buckland, W., 1835. On the discovery of coprolites, or fossil faeces, in the Lias at Lyme Regis, and in other formations. Trans. Geol. Soc. Lond. 3 (series 2), 223–238.

Buckland, W., 1836. Geology and Mineralogy Considered With Reference to Natural Theology. In: The Bridgewater Treatises on the Power, Wisdom and Goodness of God as Manifested in the Creation. London, Pickering, 2 volumes, 599 p. and 128 p.

Buckland, W., 1858. The Bridgewater Treatises on the Power, Wisdom and Goodness of God, as Manifest in the Creation. Treatises VI, Geology and Mineralogy Considered with Reference to Natural Theology, 3rd ed. George Routledge, London.

Burrow, C.J., Turner, S., 2010. Reassessment of *"Protodus" scoticus* from the Early Devonian of Scotland. In: Elliott, K., Maisey, J.G., Yu, X., Miao, D. (Eds.), Morphology, Phylogeny and Paleobiogeography of Fossil Fishes. Verlag Dr. Friedrich Pfeil, Munich, pp. 123–144.

Callen, E.O., Cameron, T.W.M., 1955. The diet and parasites of prehistoric Huaca Prieta Indians as determined by dried coprolites. Proc. R. Soc. Canada 1955, 51.

Carpenter, K., 1999. Eggs, Nests, and Baby Dinosaurs: A Look at Dinosaur Reproduction. Indiana University Press, Bloomington, 336 p.

Cotton, W.D., Hunt, A.P., Cotton, J.E., 1995. Paleozoic vertebrate tracksites in eastern North America. N. M. Mus. Nat. Hist. Sci. Bull. 6, 189–211.

Cruickshank, A.R.I., 1986. Archosaur predation on an East African Middle Triassic dicynodont. Palaeontology 29, 415–422.

Dart, R., 1957. The osteodontokeratic culture of *Australopithecus promethius*: The total contents of the breccia hitherto. Transvaal Mus. Memoir. 10, 1–105.

Dawkins, W.B., 1863. Wookey Hole hyaena den. Proceedings of the Somersetshire Archaeological Natural History Society, 11, pp. 197–219.

Delair, J.B., Sarjeant, W.A.S., 1985. History and bibliography of the study of fossil vertebrate footprints in the British Isles: supplement 1973–1983. Palaeogeogr. Palaeoclimatol. Palaeoecol. 49, 123–160.

Díaz-Martínez, I., Pérez-García, A., 2012. Historical and comparative study of the first Spanish vertebrate paleoichnological record and bibliographic review of the Spanish chirotheriid footprints. Ichnos 19, 141–149.

Downs, T., 1958. Frank Elmer Peabody, PhD. 1914–1958. Bull. South. Calif. Acad. Sci. 57, 175–177.

Drumheller-Horton, S.K., 2012. An actualistic and phylogenetic approach to identifying and interpreting crocodylian bite marks [Ph.D. dissertation]. University of Iowa, Iowa City, 170 p.

Duffin, C.J., 2009. Records of warfare ... embalmed in the everlasting hills:" A history of early coprolite research, 2009. Mercian Geologist, pp. 101–111.

Duffin, C.J., 2012a. Coprolites and characters in Victorian Britain. N. M. Mus. Nat. Hist. Sci. Bull. 57, 45–59.

Duffin, C.J., 2012b. The earliest published records of coprolites. N. M. Mus. Nat. Hist. Sci. Bull. 57, 25–29.

Elder, R.L., Smith, G.R., 1988. Fish taphonomy and environmental inference in paleolimnology. Palaeogeogr. Palaeoclimatol. Palaeoecol. 62, 577–592.

Everhart, M.J., 2017. Oceans of Kansas—A Natural History of the Western Interior Sea, Second Edition. Indiana University Press, Bloomington. 427 p.

Fernandez-Jalvo, Y., Andrews, P., 2016. Atlas of Taphonomic Identifications: 1001+ Images of Fossil and Recent Mammal Bone Modification. Springer, New York, 359 p.

Gawne, C.E., 1975. Rodents from the Zia Sand Miocene of New Mexico. Am. Mus. Novit. 2586, 1–25.

Gillette, D.D., Lockley, M.G. (Eds.), 1989. Dinosaur Tracks and Traces. Cambridge University Press, Cambridge, 454 p.

Gilmore, C.W., 1926. Fossil footprints from the Grand Canyon. Smithson. Misc. Collect. 77 (9), 1–41.

Gilmore, C.W., 1927. Fossil footprints from the Grand Canyon II. Smithson. Misc. Collect. 80 (3), 1–78.

Gilmore, G.W., 1928. Fossil footprints from the Grand Canyon III. Smithson. Misc. Collect. 80 (8), 1–16.

Götzinger, G., Becker, H., 1932. Zur geologischen Gliederung des Wienerwaldflysches (Neue Fossilfunde). Jahr. Geol. Bundesanst. 82, 343–395.

Häntzschel, W., El-Baz, F., Amstutz, G.C., 1968. Coprolites: An annotated bibliography, 108. Geological Society of America Memoir, pp. 1–132.

Harshberger, J.W., 1896. The purposes of ethnobotany. Am. Antiq. 17, 73–81.

Haubold, H., 1971. Ichnia Amphibiorum et Reptiliorum fossilium. Encycl. Paleoherpetol. 18, 1–124.

Haubold, H., 1984. Saurierfährten. In: Wittenberg (Ziemsen), 2nd edition. Die Neue Brehm-Bücherei. 231 p.

Haubold, H., 1996. Ichnotaxonomie und Klassifikation von Tetrapodenfährten aus dem Perm. Hallesches Jahrbuch Geowissenschaften B, 18, pp. 23–88.

Haubold, H., 2000. Tetrapodenfährten aus dem Perm - Kenntnisstand und Progress 2000. Hallesches Jahrbuch Geowissenschaften B, 22, pp. 1–16.

Haubold, H., 2006. Die Saurierfährten *Chirotherium barthii* Kaup, 1835 – das Typusmaterial aus dem Buntsandstein bei Hildburghausen/Thüringen und das Chirotherium Monument. Veröffentlichungen des Naturhistorischen Museums Schleusingen, 21, pp. 3–31.

Hitchcock, E., 1858. Ichnology of New England. A Report on the Sandstone of the Connecticut Valley, Especially its Fossil Footmarks, Made to the Government of the Commonwealth of Massachusetts. William White, Printer, Boston, pp. 1–214.

Hunt, A.P., 1992. Late Pennsylvanian coprolites from the Kinney Brick Quarry, central New Mexico, with notes on the classification and utility of coprolites. N. M. Bur. Mines Miner. Res. Bull. 138, 221–229.

Hunt, A.P., Lucas, S.G., 2007. Tetrapod ichnofacies: A new paradigm. Ichnos 14, 59–68.

Hunt, A.P., Lucas, S.G., 2012. Classification of vertebrate coprolites and related trace fossils. N. M. Mus. Nat. Hist. Sci. Bull. 57, 137–146.

Hunt, A.P., Lucas, S.G., 2021a. A review of the bromalite ichnofauna from the Kinney Brick Quarry Lagerstätte (Late Pennsylvanian) of New Mexico, USA with descriptions of new consumulites and coprolites. N. M. Mus. Nat. Hist. Sci. Bull. 84, 217–236.

Hunt, A.P., Lucas, S.G., 2021b. Coprolites. In: Alderton, D., Elias, S. (Eds.), Encyclopedia of Geology, Second Edition), Volume 3. Elsevier, London, pp. 532–544.

Hunt, A.P., Lucas, S.G., 2021c. The ichnology of vertebrate consumption: Dentalites, gastroliths and bromalites. N. M. Mus. Nat. Hist. Sci. Bull. 87, 1–215.

Hunt, A.P., Lucas, S.G., Lockley, M.G., 1995. Paleozoic tracksites of the western United States. In: Lucas, S.G., Heckert, A.B. (Eds.), Early Permian Footprints and Facies. New Mexico Museum of Natural History and Science, Bulletin, 6, pp. 213–217.

Hunt, A.P., Lucas, S.G., Klein, H., 2018. Late Triassic nonmarine vertebrate and invertebrate trace fossils and the pattern of the Phanerozoic record of vertebrate trace fossils. In: Tanner, L.H. (Ed.), The Late Triassic World. Topics in Geobiology 46. Springer Verlag, New York, pp. 447–543.

Hunt, A.P., Lucas, S.G., Lockley, M.G., 1998. Taxonomy and stratigraphic and facies significance of vertebrate coprolites of the Upper Triassic Chinle Group, western United States. Ichnos 5, 225–234.

Hunt, A.P., Lucas, S.G., Milàn, J., Spielmann, J.A., 2012a. Vertebrate coprolite studies: Status and prospectus. N. M. Mus. Nat. Hist. Sci. Bull. 57, 5–24.

Hunt, A.P., Milàn, J., Lucas, S.G., Spielmann, J.A. (Eds.), 2012b. Vertebrate coprolites. N. M. Mus. Nat. Hist. Sci. Bull. 57, 387 p.

Hunt, A.P., Lucas, S.G., Spielmann, J.A., Lockley, M.G., 2012c. Bromalites from the Mississippian Bear Gulch Lagerstätte of central Montana, USA. N. M. Mus. Nat. Hist. Sci. Bull. 57, 171–174.

Hunt, A.P., Lucas, S.G., Spielmann, J.A., Suazo, T., Cantrell, A., 2012d. A re-evaluation of Late Pennsylvanian bromalites from the Kinney Brick Quarry Lagerstätte, New Mexico, USA. N. M. Mus. Nat. Hist. Sci. Bull. 57, 185–192.

Hunt, A.P., Meyer, C.A., Lockley, M.G., Lucas, S.G., 1994. Archaeology, toothmarks and sauropod taphonomy. In: Lockley, M.G., dos Santos, V.F., Meyer, C.A., Hunt, A.P. (Eds.), Aspects of Sauropod Palaeobiology, 10. Gaia, pp. 225–232.

Jacobsen, A.R., Bromley, R.G., 2009. New ichnotaxa based on tooth impressions on dinosaur and whale bones. Geol. Q. 53, 373–382.

Kaup, J.J., 1835a. Über Thierfährten bei Hildburghausen. Neues Jahrb. Mineral. Geol. Paläontol. 1835, 327–328.

Kaup, J.J., 1835b. Fährten von Beuteltieren. In: Das Tierreich, pp. 246–248.

King, M.J., Sarjeant, W.A.S., Thompson, D.B., Tresise, G., 2005. A revised systematic ichnotaxonomy and review of the vertebrate footprint ichnofamily Chirotheriidae from the British Triassic. Ichnos 12, 241–299.

Klein, H., Lucas, S.G., 2010. Tetrapod footprints – their use in biostratigraphy and biochronology of the Triassic, 334. Geological Society, London, Special Publications, pp. 419–446.

Klein, H., Lucas, S.G., 2021. The Triassic tetrapod footprint record. N. M. Mus. Nat. Hist. Sci. Bull. 83, 1–194.

Klein, H., Heckert, A.B., Celeskey, M., 2023. Chirotheres: Tracking the Ancestors of Dinosaurs and Crocodiles. Indiana University Press, Bloomington, Indiana, 360 p.

Leonardi, G. (Ed.), 1987. Glossary and Manual of Tetrapod Footprint Palaeoichnology. Brasilia,Ministerio Minas Energie. Departemento Nacional Producão Mineral, 117 p.

Leonardi, G., 1994. Annotated atlas of South America tetrapod footprints (Devonian to Holocene). CPRM, Brasilia, 247 p.

Lequat, F., 1708. Voyages et Avantures de Francois Lequat et de ses Compagnons en Duex Isles Desertes des Indes Orientales. London, David Mortier, 180 p.

Lockley, M.G., 1984. Dinosaur tracking. Sci. Teacher 51, 18–24.

Lockley, M.G., 1986. The paleobiological and paleoenvironmental importance of dinosaur footprints. Palaios 1, 37–47.

Lockley, M.G., 1987. Dinosaur tracks symposium signals a renaissance in vertebrate ichnology. Paleobiology 13, 246–252.

Lockley, M.G., 1991. Tracking Dinosaurs: A New Look at an Ancient World. Cambridge University Press, Cambridge, 238 p.
Lockley, M.G., 1999. The Eternal Trail: A Tracker Looks at Evolution. Perseus Books, New York, 334 p.
Lockley, M.G., Hunt, A.P., 1995. Dinosaur Tracks and Other Fossil Footprints of the Western United States. Columbia University Press, New York, 435 p.
Lockley, M.G., Lucas, S.G. (Eds.), 2014. Fossil footprints of western North America, 62. New Mexico Museum of Natural History and Science, Bulletin, pp. 1–508.
Lockley, M.G., Meyer, C.A., 2000. Dinosaur Tracks and Other Fossil Footprints of Europe. Columbia University Press, New York.
Lockley, M.G., Young, B.H., Carpenter, K., 1983. Hadrosaur locomotion and herding behavior: evidence from footprints in the Mesaverde Formation, Grand Mesa Coal Field, Colorado. Mt. Geol. 20, 5–14.
Loud, L.L., Harrington, M.R., 1929. Lovelock Cave. University of California Publications in American Archeology, and Ethnology, 25, 183 p.
Lucas, S.G., 2015a. Alfred King's forgotten fossil footprints. Penn. Herit. 41 (2), 16–25.
Lucas, S.G., 2015b. *Thinopus* and a critical review of Devonian tetrapod footprints. Ichnos 22, 136–154.
Lucas, S.G., 2016. Two new, substrate-controlled nonmarine ichnofacies. Ichnos 23, 248–261.
Lucas, S.G., 2017. Human footprints and the peopling of the Americas. Paleoamerica 3 (2), 97–100.
Lucas, S.G., Dalman, S.G., 2013. Alfred King's Pennsylvanian tetrapod footprints from western Pennsylvania. N. M. Mus. Nat. Hist. Sci. Bull. 60, 233–239.
Lucas, S.G., Hunt, A.P., 2021. Gastroliths in archaeology: A note of caution. Int. J. Osteoarchaeol. 32 (2), 541–543.
Lucas, S.G., Hunt, A.P., 2023. There was no Mesozoic marine revolution. Proc. 4th Int. Elect. Conf. Geosci. Proc. 87, 37. https://doi.org/10.3390/IECG2022-14819.
Lucas, S.G., Houck, K.J., Hunt, A.P., Lamontagne, J.J., Minney, G.K., Taylor, L.H., 2024. Martin G. Lockley (1950–2023): prime mover of vertebrate footprint studies. Ichnos 30, 173–178.
Lucas, S.G., Emry, R.J., Krainer, K., Hunt, A.P., Spielmann, J.A., 2012. Strigilites (fossilized owl pellets) from the Oligocene of Wyoming. N. M. Mus. Nat. Hist. Sci. Bull. 57, 325–335.
Marchetti, L., Francischini, H., Lucas, S.G., Voigt, S., Hunt, A.P., Santucci, V.L., 2020. Paleozoic vertebrate ichnology of Grand Canyon. In: Santucci, V.L. (Ed.), Grand Canyon National Park Paleontological Resource Inventory. National Park Service Nature Resource Technical Report, Washington, D.C., pp. 333–379
Mayne, R.G., 1854. An expository lexicon of the terms, ancient and modern. In: Medical and General Science. J. Churchill, London, 456 p.
Mayor, A., Sarjeant, W.A.S., 2001. The folklore of footprints in stone: From classical antiquity to the present. Hist. Biol. 8, 143–163.
Mayr, G., Schaal, S.F.K., 2016. Gastric pellets with bird remains from the early Eocene of Messel. Palaios 31, 447–451.
Mikuláš, R., Kadlecová, E., Fejfar, O., Dvořák, Z., 2006. Three new ichnogenera of biting and gnawing traces on reptilian and mammalian bones: A case study from the Miocene of the Czech Republic. Ichnos 13, 113–127.
Mudge, B., 1877. Notes on the Tertiary and Cretaceous periods of Kansas. Part I (Geology). U. S. Geological and Geographical Survey of the Territories [Hayden] for 1875, Ninth Annual Report, pp. 277–294.
Myhrvold, N.P., 2011. A call to search for fossilised gastric pellets. Hist. Biol. 24, 505–517.
O'Connor, J.K., 2019. The trophic habits of early birds. Palaeogeogr. Palaeoclimatol. Palaeoecol. 513, 178–195.
Oldenberg, H., 1688. Philosophical Transactions: Giving Some Account of the Present Undertakings, Studies and Labours of the Ingenious in Many Considerable Parts of the World, Volume 3. London, S. Smith and B. Walford, pp. 629–681.

Osgood Jr., R.G., 1970. Trace fossils of the Cincinnati area. Palaeontogr. Am. 6 (41), 276–439.

Osgood Jr., R.G., 1975. The history of invertebrate ichnology. In: Frey, R.W. (Ed.), The Study Of Trace Fossils: A Synthesis of Principles, Problems, And Procedures in Ichnology. Springer-Verlag, New York, pp. 3–10.

Owen, D.D., 1842. Regarding human foot-prints in solid limestone. Am. J. Sci. 43, 14–32.

Paliy, V.M., 1990. Oleg S. Vialov: 1904–1988. Ichnos 1, 52–53.

Peabody, F.E., 1948. Reptile and amphibian trackways from the Moenkopi Formation of Arizona and Utah, 27. University of California Publications, Bulletin of the Department of Geological Sciences, pp. 295–468.

Pemberton, S.G., 2012. William Buckland (1784-1856) and Henry De La Beche (1796–1855): The early history of coprolites. N. M. Mus. Nat. Hist. Sci. Bull. 57, 29–43.

Pemberton, S.G., Gingras, M.K., MacEachern, J.A., 2007. Edward Hitchcock and Roland Bird: two early titans of vertebrate ichnology in North America. In: Miller, W. (Ed.), Trace Fossils: Concepts, Problems, Prospects. Elsevier, Amsterdam, pp. 32–51.

Pirrone, C.A., Buatois, L.A., Bromley, R.G., 2014. Ichnotaxobases for bioerosion trace fossils in bones. J. Paleontol. 88, 195–203.

Poinar, H.N., Kuch, M., McDonald, G., Martin, P., Pääbo, S., 2003. Nuclear gene sequences from a late Pleistocene sloth coprolite. Curr. Biol. 13, 1150–1152.

Poinar, H.N., Hofreiter, M., Spaulding, W.G., Martin, P.S., Stankiewicz, B.A., Bland, H., Evershed, R.P., Possnert, G., Pääbo, S., 1998. Molecular coproscopy: dung and diet of the extinct ground sloth *Nothrotheriops shastensis*. Science 281, 402–406.

Rainforth, E.C., 2005. Ichnotaxonomy of the Fossil Footprints of the Connecticut Valley (Early Jurassic, Newark Supergroup, Connecticut and Massachusetts) [Ph. D. dissertation]. Columbia University, New York, pp. 1–1301.

Reinhard, K.J., 2006. A coprological view of ancestral Pueblo cannibalism: Debate over a single fecal fossil offers a cautionary tale of the interplay between science and culture. Am. Sci. 94, 254–261.

Reinhard, K.J., Bryant, V.M., 1992. Coprolite analysis: A biological perspective on archaeology. In: Schiffer, M.B. (Ed.), Advances in Archaeological Method and Theory, 4. University of Arizona Press, Tucson, pp. 245–288.

Romer, A.S., Olson, E.C., 1954. Aestivation in a Permian lungfish. Breviora 30, 1–8.

Sanz, J.L., Chiappe, L.M., Fernádez-Jalvo, Y., Ortega, F., Sánchez-Chillón, B., Poyatoariza, F.J., Pérez-Moreno, B.P., 2001. An Early Cretaceous pellet. Nature 409, 998–1000.

Sarjeant, W.A.S., 1974. A history and bibliography of the study of fossil vertebrate footprints in the British Isles. Palaeogeogr. Palaeoclimatol. Palaecol. 16, 265–375.

Sarjeant, W.A.S., 1983. Terrestrial Trace Fossils. Benchmark Papers in Geology 76. Stroudsbrug, Pa. Hutchinson Ross, 415 p.

Sarjeant, W.A.S., 1987. The study of fossil vertebrate footprints a short history and selective bibliography. In: Leonardi, G. (Ed.), Glossary and Manual of Tetrapod Footprint Paleoichnology. Brasília, República Federativa do Brasil, Ministerio das Minas e Energia. Deparatamento Nacional da Produçáo Mineral, pp. 1–19.

Sarjeant, W.A.S., 2000. The first Permian vertebrate footprints from the English Midlands: a tale of discovery and rediscovery. Paleontol. Soc. 4, 39–48.

Sarjeant, W.A.S., Kennedy, W.J., 1973. Proposal of a code for the nomenclature of trace fossils. Can. J. Earth Sci. 10, 460–475.

Sarjeant, W.A.S., Mossman, D.J., 1978. Vertebrate footprints from the Carboniferous sediments of Nova Scotia: A historical review and description of newly discovered forms. Palaeogeogr. Paleoclimatol. Palaeoecol. 23, 279–306.

Shillito, L.-M., Blong, J.C., Green, E.J., Van Asperen, E., 2020. The what, how and why of archaeological coprolite analysis. Earth Sci. Rev. 207, 103–196.

Sickler, F.K.L., 1834. Sendschreiben an Dr. Blumenbach über die höchst merkwürdigen Reliefs der Fährten urweltlicher Tiere in den Hessberger Steinbrüchen bei Hildburghausen. Programm des Herzoglichen Gymnasiums zu Hildburghausen, 16 p.

Skoczylas, R., 1978. Physiology of the digestive tract. In: Gans, C., Gans, K.A. (Eds.), Biology of the Reptilia, 8. Academic Press, London, pp. 589–717.

Soergel, W., 1925. Die Fährten der Chirotheria. Gustav Fischer Verlag, Jena, 92 p.

Spalding, D., Pemberton, S.G., McCrea, R., Lockley, M., 2003. William Antony Swithin Sarjeant (1935–2002): A celebration of his life and ichnological contributions. Ichnos 10 (2–4), 57–68.

Staker, A.R., 2006. The earliest known dinosaur trackers of Zion National Park, Utah. N. M. Mus. Nat. Hist. Sci. Bull. 37, 137–139.

Sues, H.-D., Hook, R.W., Olsen, P.E., 2013. Donald Baird and his discoveries of Carboniferous and early Mesozoic vertebrates in Nova Scotia. Atl. Geol. 49, 90–103.

Tournal, P., 1833. Considérations générales sur le phénomène des cavernes à ossemens. Ann. Chim. Phys. 52, 161–181.

Tresise, G., Sarjeant, W.A.S., 1997. The Tracks of Triassic Vertebrates: Fossil Evidence From North-West England. The Stationery Office, London, 204 p.

Tyler, W.S., 1895. A history of Amherst College during the administrations of its first five presidents, 2nd edition. Frederick H. Hitchcock, New York.

Vialov, O.S., 1965. Stratigrafiya Neogenovykh molass Predkarpatskovo progriba [Stratigraphy of the Neogene molasse of the PreCarpathian basin]. Naukova Dumka (Akademiya Nauk Ukrainskoy SSR Institut Geologii I Geokhimii Goryuchikh Iskopayemykh), Kiev, 165 p.

Vialov, O.S., 1966. Sledy zhiznedeyatelnosti organizmov i ikh paleontologicheskoe znacheniye [Traces of living organisms and their paleontological significance]. Naukova Dumka (Akademiya Nauk Ukrainskoy SSR Institut Geologii I Geokhimii Goryuchikh Iskopayemykh), Kiev, 219 p.

Vialov, O.S., 1972. The classification of the fossil traces of life. 24th International Geological Congress [Montreal, Canada], 7, pp. 639–644. Section.

Viohl, G., 1990. Piscivorous fishes of the Solnhofen lithographic limestone. In: Boucot, A.J. (Ed.), Evolutionary Paleobiology of Behavior and Coevolution. Elsevier, Amsterdam, pp. 285–303.

Voorhies, M.R., 1975. Vertebrate burrows. In: Frey, R.W. (Ed.), The Study of Trace Fossils. Springer-Verlag, New York, pp. 325–350.

Wakefield, E.F., Dellinger, S.C., 1936. Diet of the bluff dwellers of the Ozark Mountains, and its skeletal effects. Ann. Intern. Med. 9, 1412–1418.

Washburn, S.L., 1957. Australopithecines: the hunters or the hunted? Am. Anthropol. 59, 612–614.

Wieland, G., 1906. Dinosaurian gastroliths. Science 23, 819–821.

Wieland, G., 1907. Gastroliths. Science 25, 66–67.

Wieland, G., 1920. The longneck sauropod *Barosaurus*. Science 51, 528–530.

Williams, M.E., 1990. Feeding behavior in Cleveland Shale fishes. In: Boucot, A.J. (Ed.), Evolutionary Paleobiology of Behavior and Coevolution. Elsevier, Amsterdam, pp. p273–287.

Wings, O., 2004. Identification, Distribution, and Function of Gastroliths in Dinosaurs and Extant Birds With Emphasis on Ostriches (*Struthio camelus*) [Ph.D. dissertation]. Rheinische Friedrich-Wilhelms-Universität, Bonn, 187 p.

Wings, O., 2007. A review of gastrolith function with implications for fossil vertebrates and a revised classification. Acta Palaeontol. Pol. 52, 1–16.

Wings, O., 2015. The rarity of gastroliths in sauropod dinosaurs—a case study in the Late Jurassic Morrison Formation, western USA. Foss. Rec. 18, 1–16.

Wings, O., Sander, P.M., 2007. No gastric mill in sauropod dinosaurs: New evidence from analysis of gastrolith mass and function in ostriches. Proc. R. Soc. Lond. Ser. B 274, 635–640.

Winkler, T.C., 1886. Histoire de l'ichnologie. Haarlem. Les Héritiers Loosjes, 112 p.

Zangerl, R., Richardson, E.S., 1963. The paleoecological history of two Pennsylvanian black shales. Fieldiana, Geol. Memoirs 4, 352 p.

Zonneveld, J.P., 2022. On the status of the ichnotaxon *Mandaodonites coxii* Cruickshank, 1986. Ichnos 1–9. https://doi.org/10.1080/10420940.2022.2138378.

Further reading

Davis, J.W., 1887. The fossil fish of the Chalk of Mount Lebanon in Syria, 2. Scientific Transactions of the Royal Society of Dublin, Series, pp. 457–636.

Duncan, H., 1831. An account of tracks and footmarks of animals found impressed on Sandstone in the Quarry of Corncockle Muir. Dumfriesshire, 11. Transactions Royal Society, Edinburgh, pp. 194–209.

Haubold, H., 1990. Ein neuer Dinosaurier (Ornithischia, Thyreophora) aus dem unteren Jura des nördlichen Mitteleuropa. Rev. Paléobiol. 9, 149–177.

Hitchcock, E., 1836. Ornithichnology – description of the footmarks of birds (Ornithichnites) on New Red Sandstone in Massachusetts. Am. J. Sci. Ser. 1 (29), 307–340.

Pollard, J.E., 1990. Evidence for diet. In: Briggs, D.E.G., Crowther, P.R. (Eds.), Palaeobiology: A Synthesis. Oxford, Blackwell, pp. 362–367.

Richter, R., 1981. Untersuchungen zur Enährung von *Messelobunodon schaeferi* (Mammalia, Artiodactl. Senckenb. Lethaea 61, 355–370.

Seeley, H.G., 1877. On *Mauisaurus gardneri* Seeley, an elasmosaur from the base of the Gault of Folkestone. Q. J. Geol. Soc. Lond. 33, 541–546.

PART I

The ichnology of locomotion

CHAPTER

1

Methods and terminology in vertebrate track ichnology

Jens N. Lallensack

Departamento de Ciência da Computação, Universidade Federal de Minas Gerais, Belo Horizonte, Brazil

1. Introduction

Fossil tracks are an important data source that complements the body fossil record. Tracks allow conclusions to be drawn that cannot be drawn from body fossils alone. Because they represent a specific brief period in the animal's life, they provide a direct record of behavior, locomotion, and posture. Tracks also differ from bones in their preservation potential, as they are sedimentary structures and therefore occur in many rock units that are devoid of bones, filling gaps in the fossil record. As tracks are in situ structures unlikely to have been transported prior to burial, they can inform about biological communities and their habitat. Tracks also record the soft tissue anatomy of the feet, which is rarely preserved in the body fossil record.

Despite this potential, the analysis of tracks has proven to be methodologically challenging. Defining the boundary of a track (and therefore the measurements and landmarks) is problematic and often requires considerable interpolation and subjectivity. Although the anatomy of the trackmaker is often the focus of interest, substrate properties, foot movement, and postformational alterations will also influence the shape of the track. Determining which morphological feature conveys anatomical features and to what degree can be challenging and requires a high degree of human experience and therefore subjectivity. Because of the high degree of human subjectivity involved, palaeobiological hypotheses based on tracks have often been controversial and difficult to verify using rigorous objective methods.

In this chapter, I will discuss the many methodological challenges in vertebrate track ichnology and the methods that have been proposed to address them. Topics covered include terminology, field identification and documentation of tracks, their qualitative interpretation, quantitative shape analysis, standard statistical methods, analysis of trackway patterns, and

Vertebrate Ichnology
https://doi.org/10.1016/B978-0-443-13837-9.00009-3

ichnotaxonomy. The chapter concludes with a comment on the potential and limitations of artificial intelligence for the study of tracks.

2. Terminology

The terminology of vertebrate track ichnology is extensive but not uniform. The increasing interdisciplinarity of the field leads to collisions between traditional ichnological terminology and that of related fields such as biomechanics and zoology, and therefore requires constant adaptation. For example, gait terms such as "trot" and "pace gait" have long been confusing and contradictory, but consistent definitions of these terms have now emerged across disciplines in biology (Biknevicius and Reilly, 2006) and should be adapted by tetrapod ichnology. The same terms may also have different meanings in related disciplines. For example, "stride" often refers to the distance between two successive footfalls of the same foot, but in other fields of biology, the term refers to the entire step cycle. "Heel" often refers to the rear of a track, but in other fields, it is reserved strictly for the anatomical heel, which is only regularly impressed in plantigrade trackmakers. Such differences between related fields of research are a frequent source of confusion.

Even within the discipline of vertebrate track ichnology, terminology can be ambiguous. This is the case for the basic terms "track" and "footprint," for which different definitions have been proposed (e.g., Marchetti et al., 2019; Thulborn, 2012) which differ in important details from how these terms are generally used in the literature. Here, I define track as "the result of an interaction of the pes or manus with a substrate," which includes traces of fundamentally different formational modes such as swimming tracks and penetrative undertracks. The term "footprint" is often understood in a more narrow sense (the actual "print of the foot"), but remains vague.

Furthermore, terminology is not uniform between different subdisciplines. For example, the terminology proposed by Allen (1997) is often used when studying mammal tracks, whereas the terminology of Leonardi et al. (1987) has been most influential in studies of nonmammalian tracks. The terminology of hominid tracks is particularly different. Steps toward more standardized terminology within tetrapod ichnology may be fruitful, as some terms specific to subdisciplines may be useful in other subdisciplines (such as "pternion", the rear point of the heel, which is used to describe hominid tracks).

Similar discrepancies exist between vertebrate and invertebrate ichnology. For example, the term "elite track" (track of exceptional anatomical fidelity) is derived from the term "elite trace fossil" of invertebrate ichnology, where it adopts a different meaning (traces that dominate the fabric) (Bromley, 1990; Gatesy, 2003). The term "mark" is reserved for nonbiogenic structures in invertebrate ichnology, but in vertebrate ichnology, it also applies to traces of organism activity (e.g., "drag mark" and "claw mark"), which has led to much criticism (Ekdale et al., 1984; Jacobsen and Bromley, 2009; Vallon et al., 2015). However, differences in terminology may be justified by the different scopes of the two disciplines of ichnology and are not necessarily problematic if their use is unambiguous in the respective contexts (Zonneveld et al., 2022).

Terminology may need to be revised as new knowledge is gained. For example, the new concept of penetrative tracks required a revision of the term "undertrack" (Gatesy and Falkingham, 2020). As argued by Gatesy and Falkingham (2017), advances in understanding track formation may also require more far-reaching changes in terminology when terms relate to outdated assumptions. Traditionally, tracks have been viewed as imperfect copies of feet, modified to some extent by "extramorphological features", that is, features unrelated to foot anatomy (Peabody, 1948). This concept can be traced back to Hitchcock, who equated tracks with "petrified feet" that could be matched by comparative anatomy to the skeletons of their makers (Hitchcock, 1858). However, this view fails to appreciate the fact that variations during track formation typically have a greater influence on track morphology than modifications that occur after track formation (Gatesy and Falkingham, 2017). Track morphology must therefore be understood as the result of multiple sources of variation, rather than an imperfectly preserved copy of the foot (Gatesy and Falkingham, 2017).

In this context, Gatesy and Falkingham (2017) were particularly critical of the use of the term "preservation". According to the dictionary definition, "preservation" refers to the alteration of an original state. However, in tetrapod ichnotaxonomy, "preservation" is often understood as the fidelity of the track to the anatomy of the foot—even though this fidelity depends more on the processes during track formation than on actual preservation (i.e., post-formational alteration). Gatesy and Falkingham (2017) also argued that this second definition of "preservation" is illogical as features that never existed in a track cannot be preserved. These proposals sparked a controversial discussion (Falkingham and Gatesy, 2020; Gatesy and Falkingham, 2017; Marchetti et al., 2019, 2020), with Marchetti et al. maintaining the view that it is the foot itself that is preserved in a track (Marchetti et al., 2019, 2020).

Besides conceptual and logical problems, the two different definitions of "preservation" are in conflict as both are widespread within tetrapod track ichnology. For example, swimming tracks are regularly described as "well preserved" (e.g., Houck et al., 2010; Lockley and Gierlinski, 2014; Lockley and Milner, 2006; Milner et al., 2006; Milner and Lockley, 2016; Thomson and Droser, 2015; Vila et al., 2015; Xing et al., 2013), even though such tracks record only limited information on foot anatomy and would therefore have to be described as "poorly preserved" if the definition of Marchetti et al. (2019) was followed. Disciplines less concerned with ichnotaxonomy generally use the dictionary definition of "preservation", such as the study of hominid tracks (e.g., Bennett and Morse, 2014) and modern tracking (Elbroch, 2003; e.g., Halfpenny, 1986). The coexistence of different definitions of the same term leads to ambiguity, confusion, and misunderstanding, when instead we should aim for clear and intuitive terminology. Consequently, the term "anatomical fidelity" (Gatesy and Falkingham, 2017) (e.g., "high-fidelity track" and "low-fidelity track") is a more precise choice when referring to the correspondence with the trackmaker's foot.

3. Methodology and its challenges

3.1 Field identification

The first step in describing a tracksite is to verify that the structures observed are vertebrate tracks. This verification is not always straightforward, and structures such as weathering pits (see Breithaupt et al., 2021), fish-feeding traces (see Lucas, 2015), and arthropod trackways (see Caster, 1938) have been repeatedly mistaken for tracks (reviewed by Lallensack et al., 2022c). Four main criteria for the recognition of tetrapod tracks can be distinguished, the first being the least ambiguous and the last the most ambiguous (Lallensack et al., 2022c; Lucas, 2015):

(1) A regular trackway morphology with individual tracks in predictable positions, often in an alternating zigzag pattern and, in quadrupeds, with a differentiation into manus and pes;
(2) A consistent track morphology corresponding to known anatomy;
(3) The presence of deformation structures such as displacement rims and, in cross-section, transmitted undertracks;
(4) The temporal and environmental context.

3.2 The 3D revolution

Tracks differ from bones in that they are in situ structures that have not been transported from their place of origin. The context in which a track is found (e.g., as part of a trackway or tracksite) therefore often provides more information than the track itself. This raises the question of whether it is desirable to collect tracks and thus consequently remove them, at least in part, from their context, or whether they are better left in the field. Indeed, unlike bones, tracks are often not collected, especially when they are large (Bennett et al., 2013). This makes it even more important to properly document as much information as possible, as tracks, if left in the field, will degrade over time through weathering and erosion.

Traditionally, tracksites are documented by manual mapping, for example, by tracing with chalk and then transferring these tracings to transparent foil, and, for larger sites, by using grids laid over the surface. However, much information, such as the precise three-dimensional topography of the individual tracks, is lost in the process. The resulting sitemap is an interpretation that necessarily involves the subjectivity of the observer, but its independent validation may no longer be possible once the tracksite has degraded. To mitigate these problems, artificial casts (or moulds) made of silicone rubber, latex, or gypsum plaster are made of selected tracks or areas, especially of holotype specimens. However, for the majority of tracks found at a site, the production of such physical replicas is impractical.

Technological advances in laser scanning and photogrammetry now allow surfaces to be digitized accurately and at high resolution, securing enough information to "digitally preserve" entire tracksites (Bennett et al., 2013). Photogrammetry, the generation of 3D models from photographs has become the standard method due to its low cost and ease of use (Breithaupt et al., 2004; Falkingham, 2012; Matthews et al., 2016; Matthews and Breithaupt, 2001). Photographs are taken systematically to ensure coverage of the entire track surface, with each photograph overlapping other photographs (ideally 2/3 overlap). The camera should be approximately vertical (i.e., photographing tracks in top view), and strongly oblique angles should be avoided (note that moderate angling of the camera may be required for very small tracks). The camera must be moved in space before each shot is taken; simply panning the camera from a single position may result in a fish-eye effect. It is important to ensure that all photographs are sharp and in focus and have sufficient depth of field. A tripod may be necessary, especially for small tracks. As photogrammetry is scale independent, at least one physical scale must be included in the scene to allow for scaling during model processing. The resolution of the final model depends on the pixel resolution of the images, so greater detail requires smaller distances between the camera and the surface. It is important to avoid strong shadows and to ensure that the photographed surface is clean (as any remaining debris will become part of the model). Rotation and/or cropping of the images must be avoided. Detailed instructions on how to carry out photogrammetry of ichnological sites can be found in Matthews et al. (2016).

Photogrammetric documentation of larger tracksites can require hundreds of photographs, but can usually be completed in a few hours or less, drastically reducing the time spent in the field. Drones can be used to produce overview models of very large sites, or to document the spatial relationship between several smaller sites. Because photogrammetry does not necessarily require physical contact with the tracks, tracksites in inaccessible locations, such as under cliff overhangs, can be documented in great detail using zoom lenses (e.g., Lallensack et al., 2022a). Photogrammetry can even be applied to photographs that were not intended for this purpose, including historical photographs of sites that are now lost (Falkingham et al., 2014a; Lallensack et al., 2015).

3D models also allow effective two-dimensional visualization of the topography of a track surface, with different methods emphasizing different details, sometimes highlighting features that are difficult to see in the field (Fig. 1.1). Such visualization methods include height maps (where the image is color-coded according to depth; Fig. 1.1A), contour plots (which show a series of contour lines), shaded reliefs (where topography is enhanced by virtual low-angled light; Fig. 1.1B); ambient occlusion plots (where each point is shaded according to the surrounding topography; Fig. 1.1C), inclination maps (where each face is shaded according to its inclination, thus enhancing edges; Fig. 1.1D); and orthophotos (similar to photographs but without perspective distortion). Importantly, 3D models can be automatically fitted to a horizontal plane to provide a precise top view, at least when the track or trackway

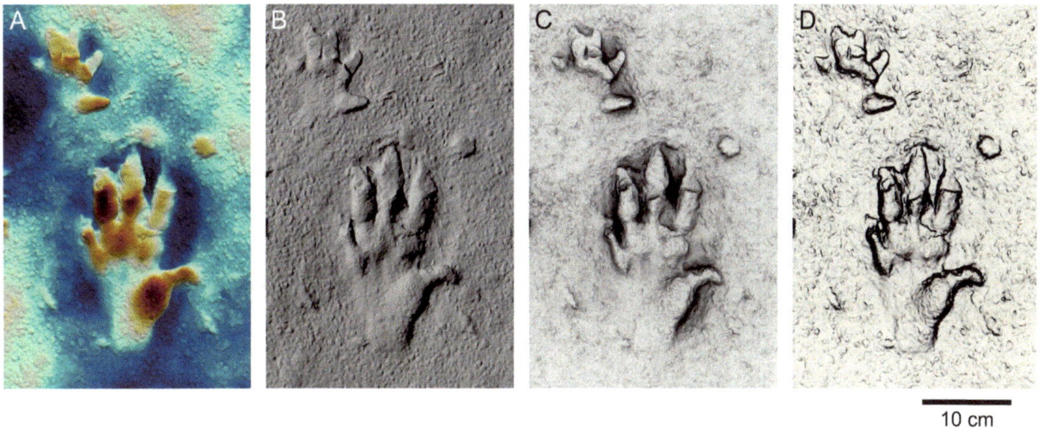

FIGURE 1.1 Different visualization methods applied to a 3D model of a *Chirotherium* manus-pes set from the Middle Triassic of Storeton, Cheshire, England (collection of the Oxford University Museum of Natural History). (A) Height map. (B) Shaded relief. (C) Ambient occlusion. (D) Inclination map.

is not impressed on a slope. 3D models can be visualized in orthographic projection (without perspective distortion), so the resulting 2D visualizations can be used as maps. Such maps can then be used to obtain measurements as well as interpretive outline drawings and site maps. A comprehensive methodology for processing and visualizing 3D models can be found in Lallensack et al. (2022b).

While 3D models do not degrade, they are highly susceptible to loss if not properly archived in public repositories, often more so than the original tracks. Data loss can occur due to hardware failure or damage, theft, software errors, or simply accidental deletion. Even with a robust backup strategy that includes multiple copies, the data are likely to be lost at some point, such as when the ichnologist retires. To ensure data availability for future generations of ichnologists, it is therefore essential to archive the 3D models and photogrammetric photographs in public data repositories such as Figshare (https://figshare.com) or Zenodo (https://zenodo.org/) (Falkingham et al., 2018).

A standard protocol (Falkingham et al., 2018) discusses best practice for recording, presenting, and archiving ichnological data, and has received widespread support. However, challenges and opportunities remain. Unlike body fossil data, which are collected in central databases such as the Palaeobiological Database (https://paleobiodb.org/) to enable broad data analysis, there is currently no central database that meets the requirements of ichnological data (Meek et al., 2020). A dedicated ichnological database including 3D models has been proposed (Belvedere et al., 2011), but remains dormant due to lack of funding (Meek et al., 2020). A further problem is that the International Commission of Zoological Nomenclature (ICZN), which governs trace fossil nomenclature, does not consider artificial replicas (digital or physical) as eligible holotypes, even though ichnotaxa have repeatedly been based on such "plastotypes" when the actual tracks could not be collected (Lucas and Harris, 2020). Finally, while the use of photogrammetry to document tracks has increased rapidly, proper archiving of such models is not yet common practice and should be actively encouraged.

Photogrammetric documentation of a tracksite should be accompanied by a thorough analysis of the sedimentology of the site, and in particular of the track layer, and only with such sedimentary analysis can a documentation be considered complete.

3.3 Track outlines: Where does a track start and end?

Unlike bones, tracks often do not have concrete margins but grade more or less continuously into the surrounding sediment. Nevertheless, such a margin is usually interpreted by the ichnologist, either by drawing an interpretive outline (Fig. 1.1A) or by taking linear measurements. Such interpretation requires a degree of subjectivity and outlines drawn by different observers tend to look different, sometimes drastically so, especially when there is disagreement over the responsible trackmaker taxon (Fig. 1.2B). When drawing outlines, presumed extramorphological features (features not related to the anatomy of the foot) are often disregarded, and anatomical features such as phalangeal pads are enhanced, sometimes even in cases where the evidence for the presence of such features is weak. Falkingham (2016) pointed out that the inherent subjectivity of such interpretations undermines any quantitative analysis, as the results will be equally subjective. (This is not to say that interpretive outlines should no longer be published—such outlines have great value in quickly conveying the key aspects of track shapes).

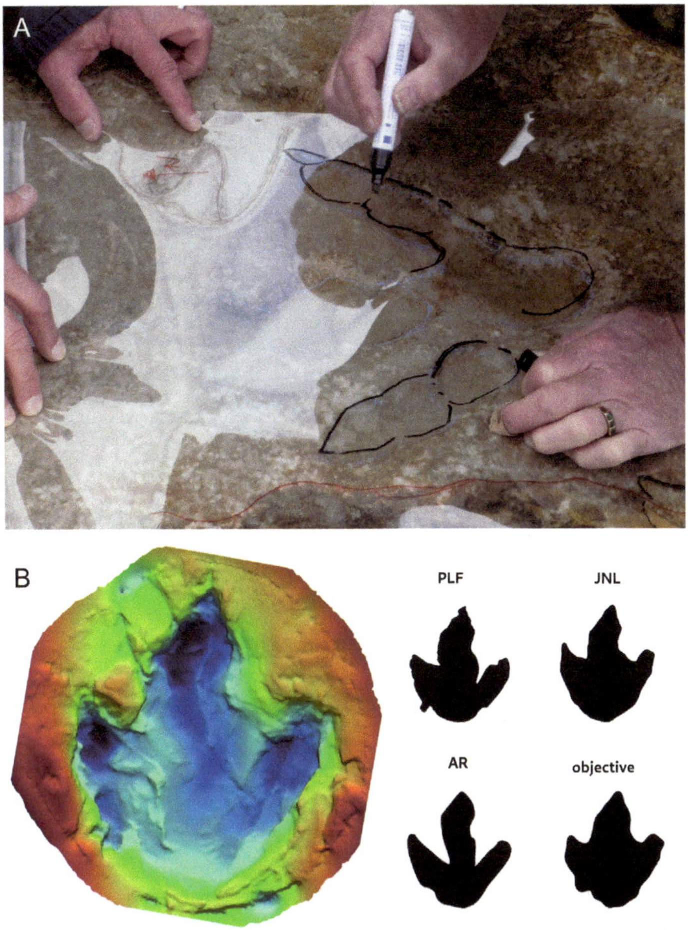

FIGURE 1.2 Definitions of track margins. (A) Martin Lockley tracing the outline of a theropod track in Obernkirchen, Germany (image licence: cc-by-sa). (B) Height map of track LQ1-3 of a large dinosaur from Lark Quarry, Australia, and interpretive outlines of the same track by three researchers (Peter L. Falkingham, Jens N. Lallensack, Anthony Romilio) and the objective outline generated by the algorithm of Lallensack (2019). The three researchers had different opinions on the responsible trackmaker taxon (PLF favored a theropod trackmaker, while AR assumed an ornithopod trackmaker), which influenced how the outlines were drawn.

This problem can be mitigated in a number of ways. First and foremost, 3D models of the tracks can be made available in addition to the interpretive outlines, allowing independent verification of the results (if a second, independent set of interpretive outlines drawn by another ichnologist provides the same quantitative results, confidence is increased). Furthermore, outlines can be defined in an objective manner, for example by selecting the contour line of a specific depth. Lallensack (2019) presented an algorithm that automatically interpolates the steepest slope of the impression based on a 3D model. Problematically, however, such approaches may work for most, but not all, tracks in a sample (e.g., a track may be too shallow to show meaningful contours in the first place). Furthermore, it is unclear how a single outline can be meaningful if it does not exist in the track in the first place.

Other approaches avoid reliance on single outlines. Falkingham (2016) suggested taking minimum and maximum outlines in addition to an interpretive "best guess" outline to obtain error values. Linear measurements can sometimes be taken between the centers of pad impressions (e.g., the "optical center method" of Mukhra et al., 2018, which is used in forensics), which tend to be less dependent on the definition of track margins, but this requires that such pads are preserved in the first place. Finally, the application of artificial intelligence methods may obviate the need to define track margins or landmarks altogether, as such methods can operate directly on images or 3D models (Lallensack et al., 2022d).

4. Qualitative methods

4.1 Trackmaker identification

An important and well-understood limitation of tracks as a palaeontological data source is their limited phylogenetic resolution; that is, trackmaker taxa are often identifiable only at the level of broader clades, and even then only with some degree of uncertainty. Species-level correlations are only possible in exceptional cases, and only when the fauna in question is well-known from both ichnofossils and skeletons (examples include Lockley and Hunt, 1994; Voigt et al., 2007). The ability to identify trackmaker taxa is often limited by the incomplete body fossil record, especially the rarity of complete foot skeletons, and the conservatism in foot anatomy found in some groups (e.g., Farlow et al., 2013). Ichnological methodology may still have potential to improve track-trackmaker correlations, for example by using methods to more accurately constrain gleno-acetabular length, and therefore body shape, based on trackways (Lallensack and Falkingham, 2022). In many cases, however, significant progress will depend on the discovery of better fossil material.

Three approaches to trackmaker identification are commonly used: Phenetic correlation, coincidence correlation, and synapomorphy-based correlation (Carrano and Wilson, 2001). Phenetic correlation uses the sum of shared characteristics between the tracks and the potential trackmaker, including morphology, posture, and behavior. Coincidence correlation is based on the geographic and stratigraphic distribution of tracks and their makers. Synapomorphy-based correlation, which is based on synapomorphies of body fossil taxa identified in tracks (Olsen, 1995), is considered the most rigorous of the three methods, but the identification of such synapomorphies is not always possible.

4.2 Sources of variation: What determines the shape of a track?

The majority of studies are concerned with track features that can be associated with taxa—both ichnotaxa and trackmaker taxa. However, freshly made tracks are always the result not only of foot anatomy but also of substrate properties at the time of track formation and the movements of the foot (Falkingham, 2014; Minter et al., 2007). For fossil tracks, two additional sources of variation have to be considered: Postformational alteration (the changes that occurred after the track was formed) and intravolumetric variation (the surface on which a track is exposed; e.g., a transmitted undertrack will differ in shape from a true track) (Gatesy and Falkingham, 2017). Furthermore, even when considering only features that inform about anatomy, these features may reflect individual or intraspecific variation rather than variation between taxa at the species level or above.

Because of this complexity, the interpretation of track shapes can be challenging and requires a considerable degree of human experience and knowledge. This hampers quantitative approaches, as several different sources of variation will often influence a given feature to varying degrees, so a given measure may not adequately capture the anatomy of interest in all tracks of a sample. More importantly, the human experience required to interpret anatomy from tracks necessarily introduces subjectivity that compromises quantitative analysis. Indeed, disagreement among ichnologists about how to interpret particular tracks and features is common, and many long-standing assumptions have recently been shown to be incorrect.

In the past, ichnologists have developed intuition about morphological variability primarily through the study of large numbers of fossil tracks, but have rarely used experiments or direct observations of modern animals to ground truth this intuition. This omission is now being rectified by an increasing number of ichnological studies of track formation using modern animals, experiments, and computer simulations (Falk et al., 2017; Falkingham et al., 2009, 2014b, 2020; Falkingham and Gatesy, 2014; Farlow et al., 2018; Gatesy and Falkingham, 2017; Hatala et al., 2023; Jackson et al., 2009; Milàn, 2006; Turner et al., 2020). The most significant recent advance is the recognition of penetrative tracks (Falkingham et al., 2020; Gatesy and Falkingham, 2020). Such tracks form when the foot sinks deeply into soft sediment and the track walls subsequently collapse, leaving a relatively shallow track on the tracking surface and possibly a series of penetrative undertracks on the underlying subsurface layers. So far, only penetrative tracks of tridactyl feet have been studied in detail. Such penetrative tracks tend to lack anatomical detail but may still be difficult to distinguish from regular surface tracks, while their shape and size, often with very narrow, curved digit impressions, often bear little resemblance to the foot that made them. A number of tridactyl ichnotaxa have been defined based on tracks that are probably penetrative (e.g., *Eoanomoepus*, *Magnoavipes*, *Saurexallopus*, *Trisauropodiscus*, and *Theroplantigrada*), and their revision must be the aim of future studies.

4.3 Ichnotaxonomy

Ichnotaxonomy is the naming and classification of trace fossils according to the rules of the ICZN (International Commission on Zoological Nomenclature, 1999), or "the code". Ichnotaxonomy mirrors, but is completely decoupled from, biological taxonomy—an ichnotaxon always applies to a trace fossil, not a biological species. Ichnotaxonomy is restricted to trace fossils, while modern traces are not to be named (Bertling, 2007). The approach of vertebrate ichnotaxonomy differs markedly from that of invertebrate ichnotaxonomy (Lockley, 2007). In the latter, trace fossils are based on the behavior of the trackmakers—consequently, traces of the same biological species showing different behaviors would be given different names. In vertebrate ichnotaxonomy, ichnotaxa are based on differences in foot anatomy as inferred from the tracks (Farlow et al., 2012), while behavior plays a minor role (Marchetti et al., 2019). Therefore, two tracks assigned to different ichnogenera are assumed to have been made by different species. It is also hoped that ichnotaxa correlate with biological taxa at a higher taxonomic level, although this is often not the case (Farlow et al., 2012). Higher-level ichnotaxa, such as ichnofamilies or ichnoorders, are rarely used, indicating limited practical utility—instead, the presumed trackmaker taxon is usually cited (but note the commonly used ichnofamily Chirotheriidae as an exception).

It follows from the above principles that only a fraction of all tracks discovered—those of high anatomical fidelity—would be suitable for naming. Tracks that record very different behaviors, such as swimming tracks, are typically excluded from ichnotaxonomy. There are exceptions, such as the scratch mark *Ostenichnus*, which has been interpreted as evidence for nest scrape display behavior (Lockley et al., 2016). However, such ichnotaxa are not comparable to regular track ichnotaxa that record walking locomotion, and must be considered as part of separate taxonomies. Tracks that show regular walking behavior but insufficient fidelity would also not qualify, but many ichnotaxa have been named on the basis of such material. This problem has been addressed with the introduction of the "preservation scale" (Belvedere and Farlow, 2017; Marchetti et al., 2019), here referred to as the "fidelity scale" for consistent terminology, which assesses tracks on four grades from low fidelity (0) to exceptional or "elite" fidelity (3).

A conceptual problem is that ichnotaxonomy relies on qualitative and more or less discrete features. In practice, however, features often grade continuously from one state into the next and separate ichnotaxa may form a continuum of morphologies (Lockley, 1998); such continua are often referred to as plexuses. This problem is particularly acute in tridactyl dinosaur tracks, for which very few discrete features are available but a large number of ichnotaxa are still in use. In Fig. 1.3, a canonical variate analysis (CVA) is performed to test how well popular theropod ichnotaxa (and theropod-like ornithischian ichnotaxa) can be discriminated based on a set of 23 landmarks (data from Lallensack et al., 2020). In Fig. 1.3A, 90% confidence ellipses are drawn for each included ichnotaxon, and the centers of these ellipses are marked with a star. Because CVA maximizes differences between groups, it is susceptible to overestimating these differences. Nevertheless, the CVA failed to find clear distinctions between the ichnotaxa, recognizing only four clusters: A cluster of highly elongate tracks (*Jialingpus, Anchisauripus, Grallator*); a cluster of larger theropod tracks of "average" morphology (*Eubrontes, Kayentapus, Changpeipus, Irenesauripus, Hispanosauropus, Asianopodus, Therangospodus, Megalosauripus*), and ornithischian tracks resembling theropod tracks (*Dinehichnus, Anomoepus*), and the theropod track *Minisauripus*. All these ichnotaxa, with the exception of the well-separated *Minisauripus*, form a single continuum, and the high degree of overlap of the ellipses may indicate over-splitting of tridactyl dinosaurian ichnotaxa, especially in the larger theropod (*Eubrontes*) cluster.

4.4 Neoichnology, experiments, and simulations

Observations of modern animals, experiments, and computer simulations form the empirical basis of interpretations of fossil tracks, and their importance cannot be underestimated (Fig. 1.4). Already the first scientific report of a fossil track, published by Duncan (1828) following observations by the Reverend William Buckland, was based on comparisons with modern tracks of a crocodile and three species of tortoises that Buckland had walked on different substrates (Pemberton et al., 2007). Fig. 1.4A shows the didactyl trackway of a

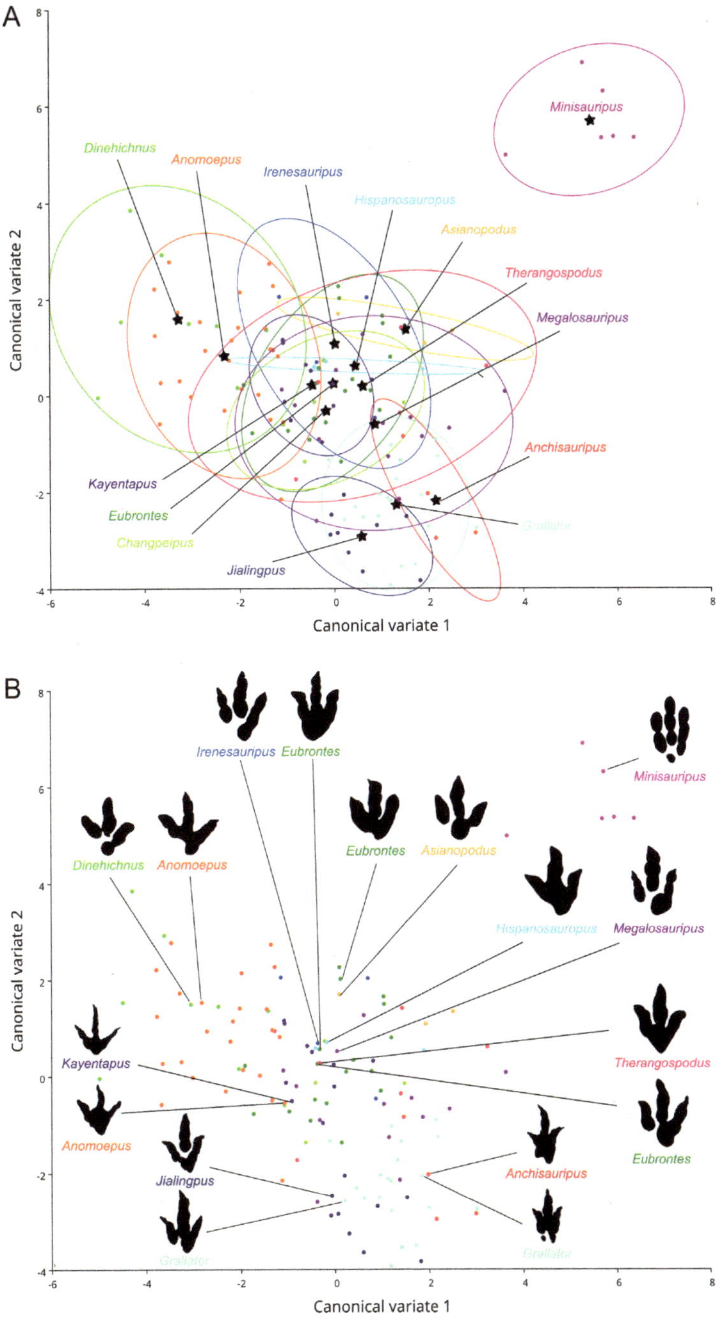

FIGURE 1.3 Discrimination between tridactyl ichnotaxa using canonical variate analysis, which maximizes distances between predefined groups, based on 23 landmarks. The data are taken from Lallensack et al. (2020) (A) 90% confidence ellipses are drawn to indicate the position of each ichnotaxon in shape space. The centers of ellipses (black stars) plot closely together especially for larger theropod tracks, indicating over-splitting of ichnotaxa. (B) Comparisons of tracks of different ichnotaxa that plot close to each other. Silhouettes from Lallensack et al. (2022d), and based on original tracings from the literature (*Grallator*: Olsen (1980); *Anchisauripus*: Niedźwiedzki and Pieńkowski (2004); *Jialingpus*: Xing et al. (2014); *Anomoepus*: Gierliński et al. (2004) and Ellenberger (1974); *Dinehichnus*: Lockley et al. (2009); *Irenesauripus*: Lockley et al. (2014); *Eubrontes*: Wagensommer et al. (2016), Ishigaki and Fujisaki (1989), and Lockley et al. (2013); *Asianopodus*: Li et al. (2015); *Minisauripus*: Lockley et al. (2008b); *Therangospodus*: Lockley et al. (1998); *Hispanosauropus*: Lockley et al. (2008a); *Megalosauripus*: Razzolini et al. (2017)).

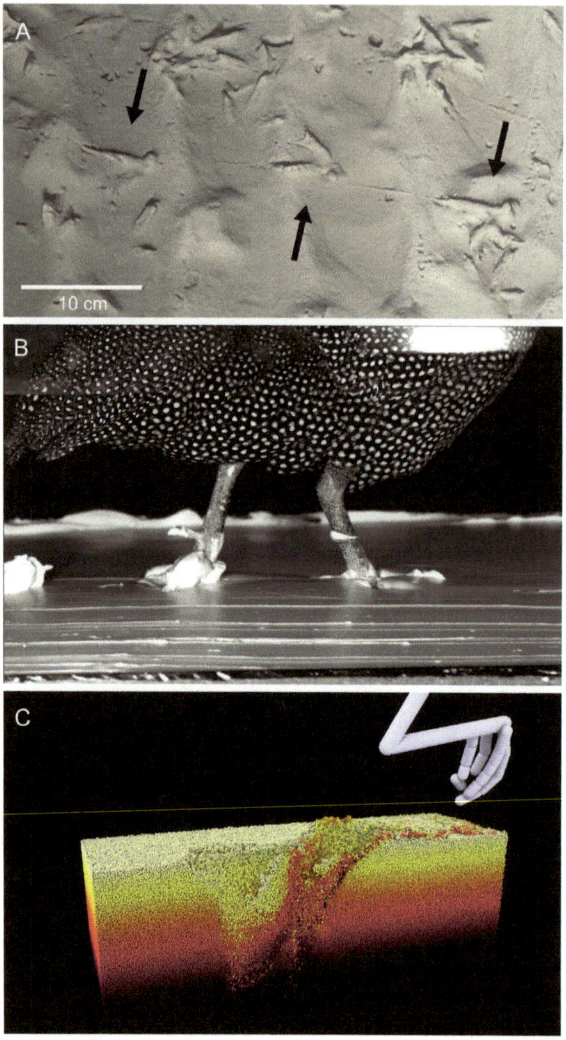

FIGURE 1.4 Observations of modern tracks (A), experiments (B) and computer simulations (C) form the basis for the interpretation of fossil tracks. (A) Tracks of the Eurasian oystercatcher (*Haematopus ostralegus*, here 3D model collected near Prestatyn, Wales) can occasionally be didactyl (as indicated by the arrows), which does not relate to obvious morphology in the oystercatcher foot, which may have implications for fossil didactyl tracks. (B) Helmeted guineafowl (*Numida meleagris*) walking over a prepared substrate of known consistency in the laboratory (Gatesy and Falkingham, 2017). (C) Computer simulation of a simplified foot moving through a digital substrate of individually modeled sediment grains (Falkingham and Gatesy, 2014a). *(B and C) Image courtesy of Peter L. Falkingham.*

Eurasian Oystercatcher (*Haematopus ostralegus*), although this apparent didactyly is not obvious from the foot anatomy of this wader. A similar didactyl trackway has been reported from a European herring gull (*Larus argentatus*) (Richter and Böhme, 2016). Although fossil didactyl dinosaur tracks are commonly assigned to deinonychosaurians (dromaeosaurids and troodontids), neoichnological observations like these call for caution in making such identifications.

Direct observation of the track formation process is hampered by the opacity of the substrate. In a groundbreaking study, Falkingham and Gatesy (2014) used X-ray reconstruction of moving morphology (XROMM) to image the legs of guineafowls as they moved through deep sediment (Fig. 1.4B). These leg movements were then modeled and incorporated into a sediment simulation (using the discrete element method, DEM) that traces the movement of each individual sediment grain (Fig. 1.4C). The digital tracks formed at different levels (i.e., penetrative undertracks) could then be revealed by horizontally splitting the virtual sediment (Falkingham and Gatesy, 2014).

Quantitative approaches using modern animals, experiments, or simulations have great unused potential, but so far are rarely used. Farlow et al. have done important work in this direction, measuring feet of modern birds and crocodiles to understand how much variability can be expected in the tracks of dinosaurs of different species (Farlow et al., 2022, 2018).

5. Quantitative methods

5.1 Mean or median, if any?

Ichnological studies often report averages of measurements rather than the individual measurements themselves. Averages are particularly useful when summarizing measurements from a single trackway because they cancel out the variation that occurs within the trackway (Demathieu, 1987). This intratrackway variation is caused by variations in substrate properties, foot movement, and preservation, whereas foot anatomy can be considered constant. Variation is also caused by measurement error, as the position of landmarks must be approximated when shape features are not distinct. Average values can therefore reflect the anatomy more reliably than single measurements. This concept has been extended to average track outlines (e.g., Lallensack et al., 2016; Wings et al., 2016) and average 3D models (Belvedere et al., 2018), and even to trackways (Buchwitz et al., 2021; Lallensack et al., 2019). However, averaging is not appropriate in all cases. Often, there may be few or even only one high-fidelity track, while the other tracks are of lower quality. In such cases, averages may not represent trackmaker anatomy as accurately as a selected representative track. Consequently, it is often necessary to select a subset of high-fidelity tracks on which to calculate the average.

A rarely considered issue is the choice of measure of average (central tendency). The arithmetic mean is almost always used in ichnological studies, but the mode (the most frequent value) and especially the median (which separates the upper half from the lower half of the values) are more appropriate in many cases, especially when outliers are present or the data are skewed. For example, when some of the tracks are affected by sediment collapse that shortens (but never lengthens) these tracks, the distribution of track length will be skewed, and the median or mode is appropriate.

While measures of central tendency such as the mean or median are useful, they necessarily remove information about how the values vary. However, this variation may potentially allow for valuable insights. The individual measurements can be summarized as an histogram to show the shape of their distribution. For example, if the distribution is bimodal (shows two distinct peaks), this could potentially indicate the presence of two separate taxa.

Another example is the variation of trackway measures that has been used to calculate gait parameters of quadrupedal trackmakers (Lallensack and Falkingham, 2022; Stevens et al., 2022). Therefore, it is preferable to publish all individual measurements rather than just their averages.

5.2 Logarithmic transformation, scaling, and statistical tests

Before testing for significant differences between samples or analyzing the relationship between two (bivariate) or more (multivariate) measures, three steps are often required: Logarithmic transformation, scaling to a common size, and tests for normal distribution:

- Logarithmic transformation can improve the statistical properties of the data (reduction of skewness and linearization), especially when allometry is present. The natural logarithm is usually used.
- Possible differences in the size of trackmakers must be taken into account when analyzing different trackways. Various approaches can be used, such as using ratios (e.g., track length/track width), or subtracting the mean of the log-transformed parameters from each individual value (Farlow et al., 2018). In landmark analysis, size is conveniently standardized during Generalized Procrustes Analysis (GPA) (see below). Note, however, that such scaling does not account for allometry, so the scaled values may still be correlated with size. To account for this, a second step may be necessary, where the parameter of interest is regressed against size and further analysis is performed on the regression residuals (e.g., Lallensack et al., 2020).
- Statistical tests may be parametric or nonparametric. Parametric tests (such as Student's *t*-test for equal means) assume that the parameters follow a normal distribution, whereas non-parametric tests (such as the Mann–Whitney test for equality of medians) do not make this assumption. To test whether a parameter is not normally distributed, the Shapiro–Wilk test can be used. Statistical tests commonly used in palaeontology are reviewed by Hammer and Harper (2006).

5.3 Regression analysis: Ordinary least squares or reduced major axis?

Linear regression is a standard method for analyzing the relationship between two (or more) variables. It derives a linear function, represented by a regression line, that describes the relationship between the two variables. The distances between the regression line and the individual data points are known as regression residuals. In biology and palaeontology, it is often necessary to linearize the data using logarithmic transformation (see above) before performing linear regression. An alternative is nonlinear regression (without logarithmic transformation), which can be more suitable in some cases (Xiao et al., 2011).

Two types of linear regression, ordinary least squares (OLS) and reduced major axis (RMA) regression, are common in palaeontology and biology, but are used in different scenarios: In OLS, a dependent variable (plotted on the y-axis) and an independent variable (plotted on the x-axis) must be defined, and the question is often how changes in the independent variable affect the dependent variable. OLS is asymmetric, which means that switching the axes results in different slopes of the regression line. RMA, on the contrary, is symmetric and is used when the choice of dependent and independent variables seems arbitrary (see Smith, 2009 for a detailed discussion of method choice).

Biological data, including ichnological data, are often affected by heteroscedasticity, which means that the variance changes along the x-axis. Strong heteroscedasticity affects regression analysis and must be accounted for (Rosopa et al., 2013). When heteroscedasticity is present, or when linearity is not given, Spearman's rank correlation coefficient can be useful to determine the strength and significance of the correlation between variables (Hammer and Harper, 2006).

5.4 Exploratory methods

To find new patterns in a multivariate data set, an exploratory method such as principal component analysis (PCA) can be used (Wold et al., 1987). PCA transforms the variables into a set of orthogonal axes, called principal components, which are arranged in order of the greatest variation they capture in the data. To develop intuition for this approach, it is useful to consider a simple example where only two variables, such as track length and track width, are plotted against each other. The coordinate system is then rotated so that the x-axis corresponds to the axis of greatest variation of the data points (e.g., if the data points form an ellipse, the x-axis would correspond to the long axis of that ellipse). The x-axis would then correspond to Principal Component 1 (PC1), and the y-axis, which is orthogonal to the x-axis, would correspond to Principal Component 2 (PC2). The orthogonality of the principal components is a useful feature, as it implies independence. If, for example, unscaled measurements of separate individuals are subjected to PCA, the first principal component can usually be assumed to capture size—and all other principal components can be assumed to be independent of size.

Only the first few principal components should be considered, as higher principal components capture less and less of the total variability of the data and become increasingly meaningless. Selected principal components (e.g., PC1 and PC2) can be plotted against each other in what is called a "biplot", showing both the scores (the position on the two axes) and the loadings (vectors indicating how much the individual original variables contribute to each of the two PCs) (Hammer and Harper, 2006). Such plots can show which variables are correlated with each other, and whether classes of data (e.g., separate ichnotaxa) plot separately. However, as an exploratory method, PCA does not indicate whether or not such correlations or differences between classes are statistically significant (Hammer and Harper, 2006). For this, appropriate tests must be used; for example, a multivariate analysis of variance (M-ANOVA) could be used to test for significant differences between classes.

Before performing a PCA, the data are often log-transformed and normalized with respect to variance (Hammer and Harper, 2006). Such normalization may be necessary when different units are to be analyzed (e.g., meters and angles) or when variables are on very different scales (e.g., track length and stride length) to prevent the dominance of variables with large variances.

Other useful exploratory methods include canonical variate analysis (CVA) and cluster analysis. CVA is similar to PCA, but maximizes the separation between predefined groups or classes (Darlington et al., 1973). It may also reveal which features are most appropriate for distinguishing these classes. It should be noted, however, that CVA will always find some separation of classes, even if these classes are arbitrary. Cluster analysis groups objects (such as specimens or taxa) in a dendrogram (tree), where objects in the same group (or cluster) are more similar to each other than to objects in other groups (Romesburg, 2004). Note that all these exploratory techniques can be applied to linear and angular measurements as well as landmark data.

5.5 Shape analysis

Track shape is traditionally captured by linear (e.g., track length and digit impression width) and angular (e.g., digit divarication) measurements, and such simple measures remain the main means of analyzing track shape (e.g., Farlow et al., 2018).

An alternative, more sophisticated method is geometric morphometrics—the analysis of landmarks (see Zelditch et al., 2004 for an introduction). Landmarks are single points (x,y or x,y,z coordinates) defined on the basis of relevant and homologous features that can be consistently identified in the tracks (e.g., the tip of digit impression II). Each track is described by a set of landmarks, and such sets of landmarks from different tracks must be aligned prior to analysis to make them comparable. Such alignment can be performed using GPA, which superimposes the track shapes using translation, rotation, and scaling. Absolute size is captured by a measure known as "centroid size", which is calculated from the sum of the squared distances between the center, or centroid, of the landmark configuration and each landmark. Multiplying this centroid size by the landmark coordinates gives the original size of the shape.

A key feature of geometric morphometrics is its ability to analyze how individual landmarks vary relative to all other landmarks. For example, measuring track length as a simple linear measure can tell us which tracks are particularly short, but unlike landmarks, it does not tell us whether this shortening is due to shortening of the digits or shortening of the posterior end of the track. On the contrary, this interdependence of landmarks can make it difficult to analyze certain shape features separately. Note, however, that linear and angular measures can be derived secondarily from landmarks, for example, by calculating the distances between landmarks (e.g., Lallensack et al., 2020).

By increasing the number of measures or landmarks, it is theoretically possible to capture every feature of a shape. However, this is not without problems. First, the more information that is captured, the more redundant and irrelevant information (e.g., information that does not relate to the shape of the foot) will be present in the data, which could compromise the analyses. Secondly, the inclusion of more measures or landmarks may require a reduction in the sample size to ensure that these measures/landmarks can be obtained from all the tracks to be analyzed. One solution to this second issue may be landmark-free approaches that do not rely on predefined landmarks (e.g., Bennett and Morse, 2014). For example, Pataky et al. (2008) developed a whole-foot analysis in which track shapes are aligned (registered) and rasterized, and each pixel location is then assumed to capture a corresponding area in the tracks. However, this approach requires that tracks can be precisely aligned and are relatively similar; if, for example, the interdigital angles vary, the same pixels could variably capture areas inside and outside of the impression, and can no longer be considered homologous.

5.6 Trackways

Trackway measures describe the positions and orientations of the tracks of a trackway relative to each other. These measures are taken between a reference point (landmark) that can be consistently and accurately identified in all tracks (when track rotation is to be measured, the long axis of the track must also be defined). The base of digit impression III is commonly used in reptile and amphibian tracks, while the tip of digit impression III is typically used in tridactyl dinosaur tracks. Obtaining the xy-coordinates of these reference points is the most efficient way to fully describe a trackway numerically, as most trackway measurements can be calculated from such coordinates (except for track rotation, heteropody, and internal and external trackway widths). A ready-to-use spreadsheet and R functions for calculating trackway measures from coordinates can be found in Lallensack et al. (2022b) and Lallensack and Falkingham (2022), respectively.

Tetrapods can increase their speed of locomotion by increasing stride length or stride frequency (or both). Sprawling tetrapods tend to modulate speed by changing stride frequency (Granatosky and McElroy, 2022), which cannot be measured directly in trackways, whereas changes in stride length are obvious. Alexander (1976) provided a formula for estimating locomotion speed from trackways, using stride length and hip height as inputs. Hip height must be estimated from tracks, usually by using pes track length as size proxy. Although this method "cannot claim to be accurate" (Alexander, 2006, p. 1850) and a large error has to be expected (Marmol-Guijarro et al., 2020), it has found wide application in vertebrate track ichnology. It should be noted that this method can only be used for trackmakers that are dynamically similar to mammals, and is therefore not applicable to sprawling tetrapods.

Recently, methods have been proposed to calculate symmetrical gaits of quadrupeds based on trackways (Lallensack and Falkingham, 2022; Polet and Hutchinson, 2022; Stevens et al., 2022), although this was previously thought to be impossible for many extinct trackmaker groups (Stevens et al., 2016). Quadrupedal gaits, such as the trot, the singlefoot walk, and the pace gait, can be described by a single parameter, limb phase, which describes the time lag between the footfall of a forelimb and that of the hindlimb on the same side of the body, expressed as a percentage of the step cycle duration (Fig. 1.5). In a pace gait (Fig. 1.5A),

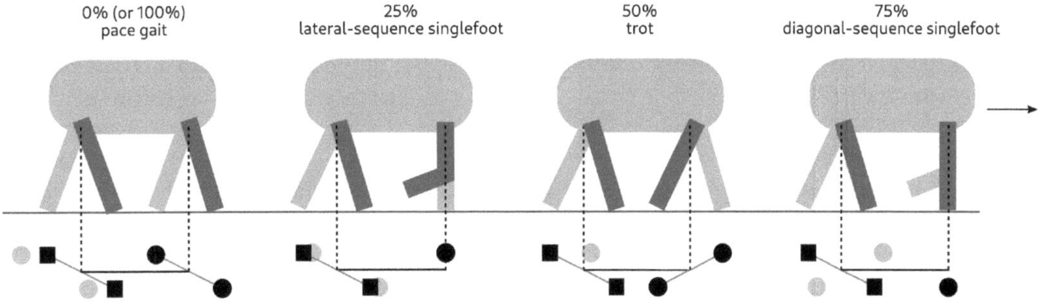

FIGURE 1.5 Quadrupedal symmetrical gaits, as determined by limb phase (i.e., the time lag between the footfall of a forefoot and that of the ipsilateral hind foot, expressed as a percentage of the step cycle duration). In a pace gait, the ipsilateral limbs move in unison, while in a trot, the diagonal limbs move in unison. In the two singlefoot walks, the time lag between all four footfalls is equal. The limb phase employed determines how the apparent gleno-acetabular distance has to be measured from the trackway. *Redrawn from Lallensack and Falkingham (2022).*

where the ipsilateral footfalls (e.g., right hind and right front) occur simultaneously, the limb phase is 0% (or 100%), whereas in a trot (Fig. 1.5C), where the diagonal footfalls (e.g., right hind and left front) occur simultaneously, the limb phase is 50%.

The choice of limb phase used by the trackmaker has an effect on the trackway pattern, particularly on distance measures between the manus and pes tracks: the manus-pes distance and the apparent gleno-acetabular distance (GAD). The GAD can be understood as a projection of the actual gleno-acetabular distance of the trackmaker onto the trackway and can therefore serve as a size proxy. However, there are different ways of measuring the GAD in a trackway, and the exact method of measurement depends on the limb phase employed. In two independent studies, Lallensack and Falkingham (2022) and Stevens et al. (2022) analyzed the variation of different GAD measures in long sauropod trackways and concluded that the analyzed sauropods used lateral-sequence diagonal-couplets (LSDC) gaits. These approaches require long and reasonably complete trackways that display changes in speed of locomotion, which are relatively rare in the fossil record, and produce erroneous results for trackways in which limb phase was tightly correlated with speed (Lallensack and Falkingham, 2022). Nevertheless, this new ability to calculate gaits from trackways and to constrain gleno-acetabular distances has great potential for future ichnological studies.

5.7 Tracksites

Tracksites can provide information about behavior and interactions between trackmakers. For example, multiple parallel trackways with consistent intertrackway spacings and estimated speeds may indicate possible gregarious behavior (Cotton et al., 1998; Currie, 1983; Getty et al., 2015). However, parallel trackways could also have been made by independent trackmakers traversing the surface at different times and coincidentally traveling in the same direction for other reasons, such as geographical barriers (e.g., a lake shore) (Myers and Fiorillo, 2009). Such accumulation of tracks on a surface over time is known as "time averaging". Extreme cases of time averaging include Holocene human tracks associated with tracks of Pleistocene megafauna as a result of re-exposure and subsequent re-burial of tracking surfaces (Rachal et al., 2021).

Little work has been done to constrain or estimate exposure durations based on modern tracking surfaces in different sedimentological settings, although such studies could provide valuable guidance for interpreting fossil tracks. Tracksites sometimes allow for the reconstruction of the succession of events; for example, overprinting of tracks can indicate the order in which trackmakers traversed the surface (e.g., Lockley et al., 2021). Furthermore, trackway orientations (i.e., the cardinal direction of travel) can be measured and plotted as a rose diagram to determine preferred directions of travel, which can also help to formulate hypotheses about the history of the tracksite.

Census studies count ichnotaxa and trackmaker individuals (assuming that one trackway represents one individual) across multiple tracksites at a regional or supra-regional scale, and can reveal the richness ("ichnodiversity", Buatois and Gabriela Mángano (2013)) and relative abundance ("ichnoabundance", Knaust et al. (2014)) of ichnotaxa. The results of such studies are likely to be affected by taphonomic bias. For example, theropod tracks tend to be much more common than ornithischian tracks, which may reflect the greater activity of the former rather than an actual overabundance of theropod trackmakers (Leonardi, 1989). However, the body fossil record is affected by different biases—integrative studies analyzing both the body fossil and track record therefore have great untapped potential for assessing the quality of the fossil record.

5.8 The rise of artificial intelligence

Applications of artificial intelligence (AI) in ichnology are still in their infancy, but, perhaps more than in other disciplines of palaeontology, have the potential to revolutionize quantitative analyses. In a pilot study, Lallensack et al. (2022d) trained an AI model to discriminate between outlines attributed to either theropod or ornithischian dinosaurs; the model consistently outperformed human experts on an independent test set. However, the potential of AI methods does not lie primarily in improved accuracy. More importantly, they can overcome the limitations of traditional approaches that have hampered quantitative track analysis:

- AI can be trained directly on raw data, including 3D models. This eliminates the step of subjectively measuring a track and defining its outline.
- Any possible track shape can be included in an analysis because there is no restriction that measurements must be taken consistently from all tracks.
- With 3D models as input, all potentially relevant shape information can be captured, rather than just a limited number of measurements.
- Multiple data formats (e.g., photographs, 3D models, and categorical and numerical data) can be combined in the same analysis.
- AI methods are highly generalizable and can be trained on modern animals to learn features relevant to a particular research question, and then applied to fossil tracks.

However, the application of AI also poses challenges. Large amounts of data are required to achieve best results, depending on the complexity of the input data (e.g., training from scratch on photographs requires many more training examples than training on a few linear measures). Currently, only a relatively small number of 3D models have been published, even though 2D visualizations of such models (e.g., height maps) may be an ideal data format for training such models. Even when sufficient data are available, it can be challenging to get the model to learn the features of interest (e.g., anatomically relevant features) and ignore irrelevant features (e.g., extramorphological features). It can also be challenging to formulate the exact problem to be solved by the AI. For example, it would be straightforward to train an AI to discriminate between ichnotaxa by providing it with training examples of such ichnotaxa. However, such an approach would be circular, as the quality of the model's output will be as subjective, and potentially flawed, as its input. Finally, because AI models are very complex, the results can be difficult to interpret; it is not always clear why a model produces a particular result. Progress is expected in this area of AI development; for example, gradcam (Selvaraju et al., 2020) can highlight areas on an image that were important for a model to make its decision (see Lallensack et al. (2022d) for an application on tracks).

6. Conclusion

The methodology and terminology of vertebrate ichnology are subject to continued change. This change is driven by new discoveries and insights, particularly into the process of track formation, and by the development of novel methods such as photogrammetry and, increasingly, artificial intelligence. It is also driven by the increasing interdisciplinarity of the field, where hypothesis testing using modern animals, experiments, and simulations is becoming increasingly important. Last but not least, there is a general tendency toward quantitative analyses in order to generate objective and repeatable results, following similar developments in other fields of palaeontology and biology. It is both a challenge and opportunity for future studies to further advance these developments, with the ultimate goal to unlock the full potential of tracks to further our understanding of ancient ecosystems.

Acknowledgements

This research was funded by the Alexander von Humboldt Foundation (Feodor-Lynen Research Fellowship).

References

Alexander, R.M., 2006. Dinosaur biomechanics. Proc. Biol. Sci. 273, 1849–1855.

Alexander, R.M.N., 1976. Estimates of speeds of dinosaurs. Nature 261, 129–130. https://doi.org/10.1038/261129a0.

Allen, J.R., 1997. Subfossil mammalian tracks (Flandrian) in the Severn Estuary, SW Britain: mechanics of formation, preservation and distribution. Philos. Trans. R. Soc. Lond. B Biol. Sci. 352, 481–518.

Belvedere, M., Baucon, A., Furin, S., Mietto, P., Felletti, F., Muttoni, G., 2011. The impact of the digital trend on ichnology: ICHNOBASE. In: Dinosaur Track Symposium Abstract Book, p. 15.

Belvedere, M., Bennett, M.R., Marty, D., Budka, M., Reynolds, S.C., Bakirov, R., 2018. Stat-tracks and mediotypes: powerful tools for modern ichnology based on 3D models. PeerJ 6, e4247.

Belvedere, M., Farlow, J.O., 2017. A numerical scale for qantifying the quality of preservation of vertebrate tracks. In: Falkingham, P.L., Marty, D., Richter, A. (Eds.), Dinosaur Tracks: The Next Steps. Indiana University Press, pp. 92–98.

Bennett, M.R., Falkingham, P., Morse, S.A., Bates, K., Crompton, R.H., 2013. Preserving the impossible: conservation of soft-sediment hominin footprint sites and strategies for three-dimensional digital data capture. PLoS One 8, e60755.

Bennett, M.R., Morse, S.A., 2014. Human Footprints: Fossilised Locomotion? Springer, Cham.

Bertling, M., 2007. What's in a name? Nomenclature, systematics, ichnotaxonomy. In: Miller, W. (Ed.), Trace Fossils. Elsevier, Amsterdam, pp. 81–91. https://doi.org/10.1016/B978-044452949-7/50131-5.

Biknevicius, A.R., Reilly, S.M., 2006. Correlation of symmetrical gaits and whole body mechanics: debunking myths in locomotor biodynamics. J. Exp. Zool. 305A, 923–934. https://doi.org/10.1002/jez.a.332.

Breithaupt, B.H., Chan, M.A., Seiler, W.M., Matthews, N.A., 2021. Weathering pits versus trample marks: a reinterpretation of the "dinosaur dance floor": a Jurassic Navajo Sandstone surface in the Vermilion Cliffs National Monument, Arizona. Palaios 36, 331–338. https://doi.org/10.2110/palo.2020.077.

Breithaupt, B.H., Matthews, N.A., Noble, T.A., 2004. An integrated approach to three-dimensional data collection at dinosaur tracksites in the rocky mountain west. Ichnos 11, 11–26. https://doi.org/10.1080/10420940490442296.

Bromley, R.G., 1990. Trace fossils, biology and taphonomy. Spec. Topics Palaeontol. 3, 280.

Buatois, L.A., Gabriela Mángano, M., 2013. Ichnodiversity and ichnodisparity: significance and caveats. Lethaia 46, 281–292.

Buchwitz, M., Jansen, M., Renaudie, J., Marchetti, L., Voigt, S., 2021. Evolutionary change in locomotion close to the origin of amniotes inferred from trackway data in an ancestral state reconstruction approach. Front. Ecol. Evol. 9, 266.

Carrano, M.T., Wilson, J.A., 2001. Taxon distributions and the tetrapod track record. Paleobiology 27, 564–582.

Caster, K.E., 1938. A restudy of the tracks of *Paramphibius*. J. Paleontol. 3–60.
Cotton, W.D., Cotton, J.E., Hunt, A.P., 1998. Evidence for social behavior in ornithopod dinosaurs from the Dakota group of northeastern New Mexico. U.S.A. Ichnos 6, 141–149. https://doi.org/10.1080/10420949809386445.
Currie, P.J., 1983. Hadrosaur trackways from the lower cretaceous of Canada. Acta Palaeontol. Pol. 28.
Darlington, R.B., Weinberg, S.L., Walberg, H.J., 1973. Canonical variate analysis and related techniques. Rev. Educ. Res. 43, 433–454. https://doi.org/10.3102/00346543043004433.
Demathieu, G.R., 1987. Use of statistical methods in palaeoichnology. In: Leonardi, G. (Ed.), Glossary and Manual of Tetrapod Footprint Palaeoichnology. Publicação do Departemento Nacional da Produção Mineral Brasil, Brasília, 117 pp.
Duncan, H., 1828. XI. Anaccount of the tracks and footmarks of animals found impressed on sandstone in the Quarry of corncockle muir, in Dumfriesshire. Earth Environ. Sci. Trans. Royal Soc. Edinburgh 11, 194–209.
Ekdale, A.A., Bromley, R.G., Pemberton, S.G., 1984. Ichnology: The Use of Trace Fossils in Sedimentology and Stratigraphy. https://doi.org/10.2110/scn.84.15.
Elbroch, M., 2003. Mammal Tracks and Sign – A Guide to North American Species, first ed. Stackpole Books, Mechanicsburg.
Ellenberger, P., 1974. Contribution à la classification des pistes de vertébrés du Trias: Les types du Stormberg d'Afrique du Sud (II éme partie: Le Stormberg Superieur - I. Le biome de la zona B/1 ou niveau de Moyeni: ses biocénoses). Palaeovertebrata, Memoire Extraordinaire 1–143.
Falk, A.R., Hasiotis, S.T., Gong, E., Lim, J.-D., Brewer, E.D., 2017. A new experimental setup for studying avian neoichnology and the effects of grain size and moisture content on tracks: trials using the domestic chicken (Gallus gallus). Palaios 32, 689–707. https://doi.org/10.2110/palo.2017.022.
Falkingham, P.L., 2016. Applying objective methods to subjective track outlines. In: Falkingham, P.L., Marty, D., Richter, A. (Eds.), Dinosaur Tracks: The Next Steps. Indiana University Press, Bloomington, pp. 72–81.
Falkingham, P.L., 2014. Interpreting ecology and behaviour from the vertebrate fossil track record. J. Zool. 292, 222–228.
Falkingham, P.L., 2012. Acquisition of high resolution three-dimensional models using free, open-source, photogrammetric software. Palaeontol. Electron. 15, 1–15.
Falkingham, P.L., Bates, K.T., Avanzini, M., Bennett, M., Bordy, E.M., Breithaupt, B.H., Castanera, D., Citton, P., Díaz-Martínez, I., Farlow, J.O., Fiorillo, A.R., Gatesy, S.M., Getty, P., Hatala, K.G., Hornung, J.J., Hyatt, J.A., Klein, H., Lallensack, J.N., Martin, A.J., Marty, D., Matthews, N.A., Meyer, C.A., Milàn, J., Minter, N.J., Razzolini, N.L., Romilio, A., Salisbury, S.W., Sciscio, L., Tanaka, I., Wiseman, A.L.A., Xing, L.D., Belvedere, M., 2018. A standard protocol for documenting modern and fossil ichnological data. Palaeontology 61, 469–480. https://doi.org/10.1111/pala.12373.
Falkingham, P.L., Bates, K.T., Farlow, J.O., 2014a. Historical photogrammetry: bird's Paluxy River dinosaur chase sequence digitally reconstructed as it was prior to excavation 70 years ago. PLoS One 9, e93247. https://doi.org/10.1371/journal.pone.0093247.
Falkingham, P.L., Gatesy, S.M., 2020. Discussion: Defining the morphological quality of fossil footprints. Problems and principles of preservation in tetrapod ichnology with examples from the Palaeozoic to the present by Lorenzo Marchetti et al. Earth Sci. Rev. 208, 103320. https://doi.org/10.1016/j.earscirev.2020.103320.
Falkingham, P.L., Gatesy, S.M., 2014. The birth of a dinosaur footprint: subsurface 3D motion reconstruction and discrete element simulation reveal track ontogeny. Proc. Natl. Acad. Sci. USA 111, 18279–18284. https://doi.org/10.1073/pnas.1416252111.
Falkingham, P.L., Hage, J., Bäker, M., 2014b. Mitigating the Goldilocks effect: the effects of different substrate models on track formation potential. R. Soc. Open Sci. 1, 140225. https://doi.org/10.1098/rsos.140225.
Falkingham, P.L., Margetts, L., Smith, I.M., Manning, P.L., 2009. Reinterpretation of palmate and semi-palmate (webbed) fossil tracks; insights from finite element modelling. Palaeogeogr. Palaeoclimatol. Palaeoecol. 271, 69–76.
Falkingham, P.L., Turner, M.L., Gatesy, S.M., 2020. Constructing and testing hypotheses of dinosaur foot motions from fossil tracks using digitization and simulation. Palaeontology 63, 865–880. https://doi.org/10.1111/pala.12502.
Farlow, J.O., Chapman, R.E., Breithaupt, B.H., Matthews, N.A., 2012. The scientific study of dinosaur footprints. In: Brett-Surman, M.K., Holtz, T.R., Farlow, J.O. (Eds.), The Complete Dinosaur. Indiana University Press, pp. 712–759.

Farlow, J.O., Coroian, D., Currie, P.J., 2018. Noah's Ravens: Interpreting the Makers of Tridactyl Dinosaur Footprints, Life of the Past. Indiana University Press, Bloomington, Indiana.

Farlow, J.O., Holtz Jr, T.R., Worthy, T.H., Chapman, R.E., 2013. Feet of the fierce (and not-so-fierce): pedal proportions in large theropods, other non-avian dinosaurs and large ground birds. In: Parrish, J.M., Molnar, R.E., Currie, P.J., Koppelhus, E.B. (Eds.), Tyrannosaurid Paleobiology. Indiana University Press, Bloomington, pp. 89–132.

Farlow, J.O., Lallensack, J.N., Müller, R.T., Hyatt, J.A., 2022. Pedal skeletal proportions of bipedal and potentially bipedal dinosaurs and other archosaurs: interpreting the makers of early Mesozoic footprints. Pbmb 63, 33–90. https://doi.org/10.3374/014.063.0201.

Gatesy, S., 2003. Direct and indirect track features: what sediment did a dinosaur touch? Ichnos 10, 91–98.

Gatesy, S.M., Falkingham, P., 2020. Hitchcock's leptodactyli, penetrative tracks, and dinosaur footprint diversity. J. Vertebr. Paleontol., e1781142 https://doi.org/10.1080/02724634.2020.1781142.

Gatesy, S.M., Falkingham, P.L., 2017. Neither bones nor feet: track morphological variation and 'preservation quality. J. Vertebr. Paleontol. 37, e1314298.

Getty, P.R., Hardy, L., Bush, A.M., 2015. Was the Eubrontes track maker gregarious? Testing the herding hypothesis at powder hill dinosaur park. Middlefield Connecticut.pbmb 56, 95–106. https://doi.org/10.3374/014.056.0109.

Gierliński, G.D., Pieńkowski, G., Niedźwiedzki, G., 2004. Tetrapod track assemblage in the Hettangian of Sołtyków, Poland, and its paleoenvironmental background. Ichnos 11, 195–213. https://doi.org/10.1080/10420940490444861.

Granatosky, M.C., McElroy, E.J., 2022. Stride frequency or length? A phylogenetic approach to understand how animals regulate locomotor speed. J. Exp. Biol. 225, jeb243231.

Halfpenny, J.C., 1986. A Field Guide to Mammal Tracking in North America. Big Earth Publishing.

Hammer, Ø., Harper, D.A.T., 2006. Paleontological Data Analysis. Blackwell Publishing, Malden, Oxfort, Carlton.

Hatala, K.G., Gatesy, S.M., Falkingham, P.L., 2023. Arched footprints preserve the motions of fossil hominin feet. Nat. Ecol. Evol. 7, 32–41. https://doi.org/10.1038/s41559-022-01929-2.

Hitchcock, E., 1858. Ichnology of New England: A Report on the Sandstone of the Connecticut Valley Especially its Fossil Footmarks, Made to the Government of the Commonwealth of Massachusetts. William White, Boston.

Houck, K.J., Lockley, M.G., Caldwell, M., Clark, B., Milàn, J., Lucas, S.G., Lockley, M.G., Spielmann, J.A., 2010. A well-preserved crocodylian trackway from the South Platte formation (lower cretaceous), golden, Colorado. New Mexico Museum Nat. History Sci. Bulletin 51, 115–120.

International Commission on Zoological Nomenclature, 1999. International Code of Zoological Nomenclature, fourth ed. International Trust for Zoological Nomenclature, London.

Ishigaki, S., Fujisaki, T., 1989. Three dimensional representation of *Eubrontes* by the method of Moiré topography. In: Gillette, D.D., Lockley, M.G. (Eds.), Dinosaur Tracks and Traces. Cambridge University Press, Cambridge, pp. 421–425.

Jackson, S.J., Whyte, M.A., Romano, M., 2009. Laboratory-controlled simulations of dinosaur footprints in sand: a key to understanding vertebrate track formation and preservation. Palaios 24, 222–238.

Jacobsen, A.R., Bromley, R.G., 2009. New ichnotaxa based on tooth impressions on dinosaur and whale bones. Geol. Q. 53, 373–382.

Knaust, D., Warchoł, M., Kane, I.A., 2014. Ichnodiversity and ichnoabundance: revealing depositional trends in a confined turbidite system. Sedimentology 61, 2218–2267. https://doi.org/10.1111/sed.12134.

Lallensack, J.N., 2019. Automatic generation of objective footprint outlines. PeerJ 7, e7203. https://doi.org/10.7717/peerj.7203.

Lallensack, J.N., Bordy, E.M., Lockley, M., Wings, O., 2022a. Relaunching the TY tracksite: tridactyl dinosaur footprints from the Lower Jurassic of southern Africa. Hist. Biol. 1–12. https://doi.org/10.1080/08912963.2022.2117042.

Lallensack, J.N., Buchwitz, M., Romilio, A., 2022b. Photogrammetry in ichnology: 3D model generation, visualisation, and data extraction. J. Paleontol. Tech. 20, 1–17.

Lallensack, J.N., Engler, T., Barthel, H.J., 2020. Shape variability in tridactyl dinosaur footprints: the significance of size and function. Palaeontology 63, 203–228. https://doi.org/10.1111/pala.12449.

Lallensack, J.N., Falkingham, P.L., 2022. A new method to calculate limb phase from trackways reveals gaits of sauropod dinosaurs. Curr. Biol. 32, 1635–1640. https://doi.org/10.1016/j.cub.2022.02.012.

Lallensack, J.N., Ishigaki, S., Lagnaoui, A., Buchwitz, M., Wings, O., 2019. Forelimb orientation and locomotion of sauropod dinosaurs: insights from the middle jurassic tafaytour tracksite (argana basin, Morocco). J. Vertebr. Paleontol. 5, 1–18. https://doi.org/10.1080/02724634.2018.1512501.

Lallensack, J.N., Owais, A., Falkingham, P.L., Breithaupt, B.H., Sander, P.M., 2022c. How to verify fossil tracks: the first record of dinosaurs from Palestine. Hist. Biol. 35, 924–934. https://doi.org/10.1080/08912963.2022.2069020.

Lallensack, J.N., Romilio, A., Falkingham, P.L., 2022d. A machine learning approach for the discrimination of theropod and ornithischian dinosaur tracks. J. R. Soc. Interface 19, 20220588. https://doi.org/10.1098/rsif.2022.0588.

Lallensack, J.N., Sander, P.M., Knötschke, N., Wings, O., 2015. Dinosaur tracks from the Langenberg Quarry (Late Jurassic, Germany) reconstructed with historical photogrammetry: evidence for large theropods soon after insular dwarfism. Palaeontol. Electron. 18 (2.24A), 1–34. https://doi.org/10.26879/529.

Lallensack, J.N., Wings, O., Heteren, A.H. van, 2016. Geometric morphometric analysis of intratrackway variability: a case study on theropod and ornithopod dinosaur trackways from Münchehagen (Lower Cretaceous, Germany). PeerJ 4, e2059. https://doi.org/10.7717/PEERJ.2059.

Leonardi, G., 1989. Inventory and statistics of the South American dinosaurian ichnofauna and its Paleobiological interpretation. In: Gillette, D.D., Lockley, M.G. (Eds.), Dinosaur Tracks and Traces. Cambridge University Press, pp. 165–178.

Leonardi, G., Casamiquela, R.M., Demathieu, G.R., Haubold, H., Sarjeant, W.A.S., 1987. Glossary and Manual of Tetrapod Footprint Palaeoichnology. Publicação do Departemento Nacional da Produção Mineral Brasil, Brasília.

Li, R., Lockley, M.G., Matsukawa, M., Liu, M., 2015. Important dinosaur-dominated footprint assemblages from the lower cretaceous tianjialou formation at the houzuoshan dinosaur park, junan county, Shandong province, China. Cretac. Res. 52, 83–100.

Lockley, M., Gierlinski, G., 2014. Jurassic tetrapod footprint ichnofaunas and ichnofacies of the Western Interior, USA. Volumina Jurassica 12, 133–150.

Lockley, M.G., 2007. A tale of two ichnologies: the different goals and potentials of invertebrate and vertebrate (tetrapod) ichnotaxonomy and how they relate to ichnofacies analysis. Ichnos 14, 39–57.

Lockley, M.G., 1998. Philosophical perspectives on theropod track morphology: blending qualities and quantities in the science of ichnology. GAIA 15, 279–300.

Lockley, M.G., Breithaupt, B.H., Matthews, N.A., Shibata, K., Hunt-Foster, R., 2021. A preliminary report on an early jurassic eubrontes-dominated tracksite in the Navajo sandstone formation at the mail station dinosaur tracksite, san juan county, Utah. Bull. NM Mus. Nat. Hist. Science 82, 195–208.

Lockley, M.G., Garcia-Ramos, J.C., Piñuela, L., Avanzini, M., 2008a. A review of vertebrate track assemblages from the Late Jurassic of Asturias, Spain with comparative notes on coeval ichnofaunas from the western USA: implications for faunal diversity in siliciclastic facies assemblages. ORYCTOS 8, 53–70.

Lockley, M.G., Gierliński, G.D., Dubicka, Z., Breithaupt, B.H., Matthews, N.A., Lockley, M.G., Lucas, S.G., 2014. A preliminary report on a new dinosaur tracksite in the Cedar Mountain Formation (Cretaceous) of Eastern Utah. In: Fossil Footprints of Western North America, pp. 279–285.

Lockley, M.G., Hunt, A.P., 1994. A track of the giant theropod dinosaur *Tyrannosaurus* from close to the Cretaceous/Tertiary boundary, northern New Mexico. Ichnos 3, 213–218.

Lockley, M.G., Kim, J.Y., Kim, K.S., Kim, S.H., Matsukawa, M., Rihui, L., Jianjun, L., Yang, S.-Y., 2008b. Minisauripus—the track of a diminutive dinosaur from the Cretaceous of China and South Korea: implications for stratigraphic correlation and theropod foot morphodynamics. Cretac. Res. 29, 115–130. https://doi.org/10.1016/j.cretres.2007.04.003.

Lockley, M.G., Li, J.-J., Rihui, L.I., Matsukawa, M., Harris, J.D., Xing, L.-D., 2013. A review of the tetrapod track record in China, with special reference to type ichnospecies: implications for ichnotaxonomy and paleobiology. Acta Geol. Sin. 87, 1–20.

Lockley, M.G., McCrea, R.T., Buckley, L.G., Deock Lim, J., Matthews, N.A., Breithaupt, B.H., Houck, K.J., Gierliński, G.D., Surmik, D., Soo Kim, K., Xing, L., Yong Kong, D., Cart, K., Martin, J., Hadden, G., 2016. Theropod courtship: large scale physical evidence of display arenas and avian-like scrape ceremony behaviour by Cretaceous dinosaurs. Sci. Rep. 6, 18952. https://doi.org/10.1038/srep18952.

Lockley, M.G., McCrea, R.T., Matsukawa, M., 2009. Ichnological Evidence for Small Quadrupedal Ornithischians from the Basal Cretaceous of SE Asia and North America, vol. 315. Geological Society, London, Special Publications, pp. 255–269. https://doi.org/10.1144/SP315.18.

Lockley, M.G., Meyer, C.A., Moratalla, J.J., 1998. *Therangospodus*: trackway evidence for the widespread distribution of a Late Jurassic theropod with well-padded feet. Gaia 15, 339–353.

Lockley, M.G., Milner, A.R., 2006. Tetrapod tracksites from the shinarump formation (chinle group, upper triassic) of southwestern Utah. New Mexico Museum Nat. History Sci. Bull. 37, 257–262.

Lucas, S.G., 2015. *Thinopus* and a critical review of Devonian tetrapod footprints. Ichnos 22, 136–154.

Lucas, S.G., Harris, J.D., 2020. The "plastotype problem" in ichnological taxonomy. Ichnos 27, 107–110. https://doi.org/10.1080/10420940.2019.1688802.

Marchetti, L., Belvedere, M., Voigt, S., Klein, H., Castanera, D., Díaz-Martínez, I., Marty, D., Xing, L., Feola, S., Melchor, R.N., 2020. Reply to discussion of "Defining the morphological quality of fossil footprints. Problems and principles of preservation in tetrapod ichnology with examples from the Palaeozoic to the present" by Marchetti et al.(2019). Earth Sci. Rev. 208, 103319.

Marchetti, L., Belvedere, M., Voigt, S., Klein, H., Castanera, D., Díaz-Martínez, I., Marty, D., Xing, L., Feola, S., Melchor, R.N., Farlow, J.O., 2019. Defining the morphological quality of fossil footprints. Problems and principles of preservation in tetrapod ichnology with examples from the Palaeozoic to the present. Earth Sci. Rev. 193, 109–145. https://doi.org/10.1016/j.earscirev.2019.04.008.

Marmol-Guijarro, A., Nudds, R., Folkow, L., Codd, J., 2020. Examining the accuracy of trackways for predicting gait selection and speed of locomotion. Front. Zool. 17, 17. https://doi.org/10.1186/s12983-020-00363-z.

Matthews, N.A., Breithaupt, B.H., 2001. Close-range photogrammetric experiments at dinosaur ridge. Mt. Geol. 38 (3), 147–153.

Matthews, N.A., Noble, T., Breithaupt, B.H., 2016. Close-Range photogrammetry for 3-D ichnology: the basics of photogrammetric ichnology. In: Falkingham, P.L., Marty, D., Richter, A. (Eds.), Dinosaur Tracks: The Next Steps. Indiana University Press, Bloomington, Indiana, pp. 28–55.

Meek, D.M., Eglington, B.M., Buatois, L.A., Mángano, M.G., 2020. IchnoDB: structure and importance of an ichnology database. Ichnos 28, 1–11. https://doi.org/10.1080/10420940.2020.1784157.

Milàn, J., 2006. Variations in the morphology of emu (*Dromaius novaehollandiae*) tracks reflecting differences in walking pattern and substrate consistency: ichnotaxonomic implications. Palaeontology 49, 405–420.

Milner, A.R.C., Lockley, G.M., 2016. Dinosaur swim track assemblages: characteristics, contexts, and ichnofacies implications. In: Falkingham, P.L., Marty, D., Richter, A. (Eds.), Dinosaur Tracks: The Next Steps. Indiana University Press, pp. 152–180.

Milner, A.R.C., Lockley, M.G., Kirkland, J.I., Harris, J.D., 2006. A large collection of well-preserved theropod dinosaur swim tracks from the Lower Jurassic Moenave Formation, St. George, Utah. New Mexico Museum Nat. History Sci. Bull. 37, 315–328.

Minter, N.J., Braddy, S.J., Davis, R.B., 2007. Between a rock and a hard place: arthropod trackways and ichnotaxonomy. Lethaia 40, 365–375.

Mukhra, R., Krishan, K., Kanchan, T., 2018. Bare footprint metric analysis methods for comparison and identification in forensic examinations: a review of literature. J. Forensic Legal Med. 58, 101–112. https://doi.org/10.1016/j.jflm.2018.05.006.

Myers, T.S., Fiorillo, A.R., 2009. Evidence for gregarious behavior and age segregation in sauropod dinosaurs. Palaeogeogr. Palaeoclimatol. Palaeoecol. 274, 96–104.

Niedźwiedzki, G., Pieńkowski, G., 2004. A dinosaur track association from the Early Jurassic deltaic deposits of Podole near Opatów, Poland. Geol. Q. 48, 333–338.

Olsen, P.E., 1995. A new approach for recognizing track makers. Geol. Soc. Am. Abstr. Progr. 72.

Olsen, P.E., 1980. Fossil great lakes of the newark supergroup in New Jersey. In: Field Studies in New Jersey Geology and Guide to Field Trips, pp. 352–398.

Pataky, T.C., Caravaggi, P., Savage, R., Parker, D., Goulermas, J.Y., Sellers, W.I., Crompton, R.H., 2008. New insights into the plantar pressure correlates of walking speed using pedobarographic statistical parametric mapping (pSPM). J. Biomech. 41, 1987–1994. https://doi.org/10.1016/j.jbiomech.2008.03.034.

Peabody, F.E., 1948. Reptile and Amphibian Trackways from the Lower Triassic Moenkopi Formation of Arizona and Utah, vol. 27. California University, Bulletin Department of Geological Sciences, pp. 295–468.

Pemberton, S.G., McCrea, R., Gingras, M.K., Sarjeant, W.A.S., MacEachern, J.A., 2007. History of ichnology: the correspondence between the Reverend Henry Duncan and the Reverend William Buckland and the discovery of the first vertebrate footprints. Ichnos 15, 5–18. https://doi.org/10.1080/10420940600864670.

Polet, D.T., Hutchinson, J.R., 2022. Estimating gaits of an ancient crocodile-line archosaur throughtrajectory optimization, with comparison to fossil trackways. Front. Bioeng. Biotechnol. 9.

Rachal, D.M., Zeigler, K., Dello-Russo, R., Solfisburg, C., 2021. Lake levels and trackways: an alternative model to explain the timing of human-megafauna trackway intersections, Tularosa Basin, New Mexico. Quater. Sci. Adv. 3, 100024. https://doi.org/10.1016/j.qsa.2021.100024.

Razzolini, N.L., Belvedere, M., Marty, D., Paratte, G., Lovis, C., Cattin, M., Meyer, C.A., 2017. *Megalosauripus transjuranicus* ichnosp.nov. A new Late Jurassic theropod ichnotaxon from NW Switzerland and implications for tridactyl dinosaur ichnology and ichnotaxomy. PLoS One 12, e0180289.

Richter, A., Böhme, A., 2016. Too many tracks: preliminary description and interpretation of the diverse and heavily dinoturbated Early Cretaceous 'Chicken Yard' ichnoassemblage (Obernkirchen Tracksite, northern Germany). In: Falkingham, P.L., Marty, D., Richter, A. (Eds.), Dinosaur Tracks: The Next Steps. Indiana University Press, pp. 334–357.

Romesburg, C., 2004. Cluster Analysis for Researchers. Lulu Press.

Rosopa, P.J., Schaffer, M.M., Schroeder, A.N., 2013. Managing heteroscedasticity in general linear models. Psychol. Methods 18, 335–351. https://doi.org/10.1037/a0032553.

Selvaraju, R.R., Cogswell, M., Das, A., Vedantam, R., Parikh, D., Batra, D., 2020. Grad-CAM: visual explanations from deep networks via gradient-based localization. Int. J. Comput. Vis. 128, 336–359. https://doi.org/10.1007/s11263-019-01228-7.

Smith, R.J., 2009. Use and misuse of the reduced major axis for line-fitting. Am. J. Phys. Anthropol. 140, 476–486. https://doi.org/10.1002/ajpa.21090.

Stevens, K.A., Ernst, S., Marty, D., 2022. Coupling length: a generalized gleno-acetabular distance measurement for interpreting the size and gait of quadrupedal trackmakers. Swiss J. Geosci. 115, 18. https://doi.org/10.1186/s00015-022-00418-9.

Stevens, K.A., Ernst, S., Marty, D., 2016. Uncertainty and ambiguity in the interpretation of sauropod trackways. In: Falkingham, P.L., Marty, D., Richter, A. (Eds.), Dinosaur Tracks: The Next Steps. Indiana University Press, pp. 226–243.

Thomson, T.J., Droser, M.L., 2015. Swimming reptiles make their mark in the Early Triassic: delayed ecologic recovery increased the preservation potential of vertebrate swim tracks. Geology 43, 215–218.

Thulborn, T., 2012. Impact of sauropod dinosaurs on lagoonal substrates in the broome sandstone (lower cretaceous), western Australia. PLoS One 7, e36208. https://doi.org/10.1371/journal.pone.0036208.

Turner, M.L., Falkingham, P.L., Gatesy, S.M., 2020. It's in the loop: shared sub-surface foot kinematics in birds and other dinosaurs shed light on a new dimension of fossil track diversity. Biol. Lett. 16, 20200309. https://doi.org/10.1098/rsbl.2020.0309.

Vallon, L.H., Rindsberg, A.K., Martin, A.J., 2015. The use of the terms trace, mark and structure. Ann. Soc. Geol. Pol. 85, 527–528. https://doi.org/10.14241/asgp.2015.014.

Vila, B., Castanera, D., Marmi, J., Canudo, J.I., Galobart, À., 2015. Crocodile swim tracks from the latest Cretaceous of Europe. Lethaia 48, 256–266.

Voigt, S., Berman, D.S., Henrici, A.C., 2007. First well-established track-trackmaker association of paleozoic tetrapods based on Ichniotherium trackways and diadectid skeletons from the Lower Permian of Germany. J. Vertebr. Paleontol. 27, 553–570. https://doi.org/10.1671/0272-4634(2007)27[553:FWTAOP]2.0.CO;2.

Wagensommer, A., Latiano, M., Mocke, H.B., D'Orazi, P., 2016. Dinosaur diversity in an early jurassic African desert: the significance of the Etjo Sandstone ichnofauna at the Otjihaenamaparero locality (Namibia). Neues Jahrbuch Geol. Palaontol. Abhand. 281, 155–182.

Wings, O., Lallensack, J.N., Mallison, H., 2016. The Early Cretaceous Dinosaur Trackways in Münchehagen (Lower Saxony, Germany) — 3D photogrammetry as basis for geometric morphometric analysis of shape variation and evaluation of material loss during excavation. In: Falkingham, P.L., Marty, D., Richter, A. (Eds.), Dinosaur Tracks: The Next Steps. Indiana University Press, Bloomington, pp. 56–71.

Wold, S., Esbensen, K., Geladi, P., 1987. Principal component analysis. In: Chemometrics and Intelligent Laboratory Systems, Proceedings of the Multivariate Statistical Workshop for Geologists and Geochemists, vol. 2, pp. 37–52. https://doi.org/10.1016/0169-7439(87)80084-9.

Xiao, X., White, E.P., Hooten, M.B., Durham, S.L., 2011. On the use of log-transformation vs. nonlinear regression for analyzing biological power laws. Ecology 92, 1887–1894. https://doi.org/10.1890/11-0538.1.

Xing, L., Lockley, M.G., Klein, H., Gierliński, G.D., Divay, J.D., Hu, S.M., Zhang, J.P., Ye, Y., He, Y.P., 2014. The non-avian theropod track *Jialingpus* from the Cretaceous of the Ordos Basin, China, with a revision of the type material: implications for ichnotaxonomy and trackmaker morphology. Palaeoworld 23, 187–199.

Xing, L., Lockley, M.G., Zhang, J., Milner, A.R., Klein, H., Li, D., Persons, W.S., Ebi, J., 2013. A new Early Cretaceous dinosaur track assemblage and the first definite non-avian theropod swim trackway from China. Chin. Sci. Bull. 58, 2370–2378.

Zelditch, M., Swiderski, D., Sheets, H.D., Fink, W.L., 2004. Geometric Morphometrics for Biologists: A Primer, first ed. Academic Press, Cambridge, Massachusetts.

Zonneveld, J.-P., Fiorillo, A.R., Hasiotis, S., Gingras, M.K., 2022. Tooth marks, gnaw marks, claw-marks, bite marks, scratch marks, etc: terminology in ichnology. Ichnos 1–9. https://doi.org/10.1080/10420940.2022.2058937.

CHAPTER

2

Devonian tetrapod footprints and tetrapod origins

Spencer G. Lucas

New Mexico Museum of Natural History and Science, Albuquerque, NM, United States

1. Introduction

Trace fossils often document the presence of organisms not known from the body-fossil record. This is the case with Devonian tetrapod footprints, which provide not only the oldest fossil record of tetrapods but also document tetrapods apparently not known from the skeletal record.

Lucas (2015) argued that Devonian tetrapod tracks and trackways can be recognized by three criteria: morphology of the manus and pes impressions that matches known Devonian tetrapod skeletal morphology; differentiation of the manus and pes impressions, in particular manus smaller than pes; and the alternating trackway pattern that results from lateral sequence walking in quadrupedal tetrapod locomotion. Using those criteria, a comprehensive review of the record of supposed Devonian tetrapod tracks/trackways by Lucas (2015, 2019) identified only three records that meet the criteria by which such trackways should be distinguished (Fig. 2.1). Here, that review is reprised, and the implications of the Devonian tetrapod trackway record for the understanding of tetrapod origins are discussed.

2. Supposed Devonian tetrapod footprints

Most reports of supposed Devonian tetrapod footprints have been of invertebrate traces or other structures that are not tetrapod footprints. These are as follows: (1) the holotype of *Thinopus antiquus* (Marsh, 1896), from the Upper Devonian of Pennsylvania, USA, likely the impression of coprolites, definitely not a footprint (Abel, 1935; Lucas, 2015); (2) a supposed tetrapod trackway from eastern Australia (Warren et al., 1986) that is either Late Silurian or Early Devonian in age, that has been reidentified as an arthropod trackway and assigned to *Diplichnites* isp. (Gourmanis et al., 2003); (3) the holotype of *Notopus petri* Leonardi (1983) from the Devonian of southern Brazil, reinterpreted as a trace similar to the starfish trace fossil *Asteriacites* (Roček and Rage, 1994); (4) two trackways from the Devonian of eastern Greenland (Friend et al., 1976) that are arthropod, likely *Diplichnites* (Lucas,

Vertebrate Ichnology
https://doi.org/10.1016/B978-0-443-13837-9.00014-7

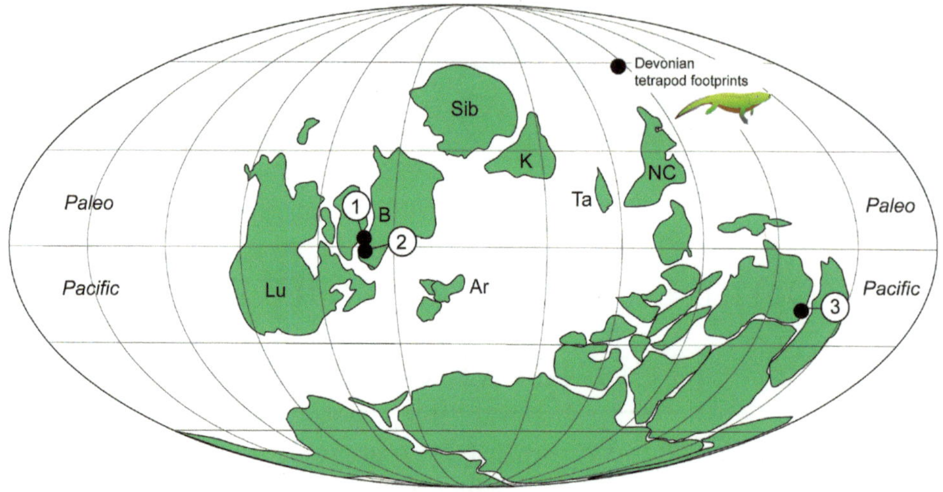

FIGURE 2.1 Devonian continental configuration showing tetrapod footprint localities. Abbreviations are *B*, baltica; *K*, Kazakstan; *Lu*, Laurentia; *NC*, North China; *Sib*, Siberia; *Ta*, Tarim. Localities are as follows: 1, Valentia Island, Ireland; 2, Easter Ross, Scotland; 3. Genoa River, Australia. *Clip art image of early amphibian by Arthur Escher.*

2015); (5) a supposed trackway from the Orkney Islands (Westoll, 1937; Sarjeant, 1974) that may be an invertebrate trackway or more likely is a plant impression (Lucas, 2015); and (6) supposed footprints and trackways from the Middle Devonian of Poland (Niedźwiedzki et al., 2010) are not tetrapod footprints or trackways and are more likely fish traces of the ichnogenus *Piscichnus* (Lucas, 2015, 2019). Thus, no structures (digit impressions, etc.) clearly identify these dish-shaped depressions as tetrapod tracks, and no alternating trackways are present. Indeed, some of the structures are larger than tracks made by any tetrapod before the Mesozoic. Niedźwiedzki and Szrek (2020) acknowledged that some of these depressions are fish traces but still maintained, without compelling evidence, that some are tetrapod tracks. Much has been made in the literature of this highly publicized report (see references cited in Lucas, 2015 and in Niedźwiedzki and Szrek, 2020), but it should disappear from the discussion of Devonian tetrapod footprints.

3. Devonian tetrapod footprints

Eliminating the six supposed records of Devonian tetrapod footprints leaves only three bona fide records: (1) Genoa River, Australia; (2) Easter Ross, Scotland; and (3) Valentia Island, Ireland.

3.1 Genoa River, Australia

Two tetrapod trackways from Upper Devonian (Frasnian) strata in New South Wales, Australia, are from nonmarine strata of fluvial origin of the Combyingbar Formation (Warren and Wakefield, 1972; Young, 2006). The trackways show differentiation of manus and pes sizes, alternating trackway patterns and, in some imprints, evident digit impressions (Fig. 2.2). Note that in Carboniferous-Permian tetrapod tracks, the manus impression is

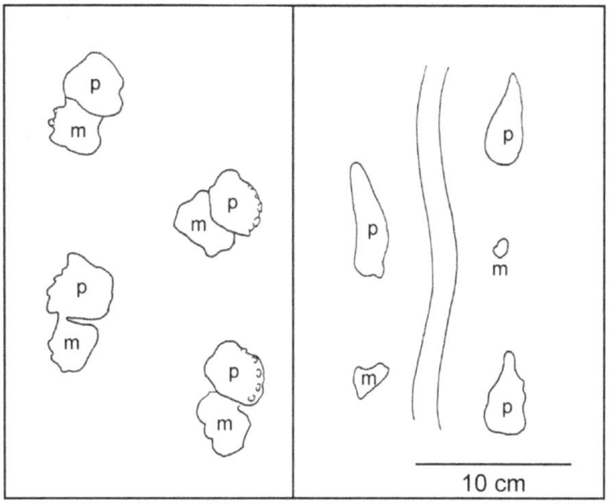

FIGURE 2.2 Drawings of parts of the Genoa River, Australia, tetrapod trackways (after Warren and Wakefield, 1972). The trackway on the left has overstepped manus and pes impressions and appears to show laterally oriented digits. The trackway on the right has a median drag and poorly preserved manus and pes impressions. Abbreviations are as follows: m = manus, p = pes. One scale bar for both images, and the direction of trackmaker motion is up the page.

smaller than the pes impression, so identification of Devonian manus versus pes impressions is based on this difference in size (Lucas, 2015). One trackway has a clear median drag impression, whereas the other lacks one. The estimated trackmaker gleno-acetabular length is 22 cm.

Both trackways show an alternating pattern and differentiation (by size) of the manus (smaller) and pes (larger) (Fig. 2.2). The trackway without a median drag impression preserves overstepped, broad, and short manus impressions smaller than pes impressions that are longer than wide. Some of the manus and pes impressions have at least five, short and blunt, laterally directed digit imprints. The other trackway shows a definite median drag mark and well-separated manus and pes impressions that lack definition. The overstepping in one trackway indicates body/limb proportions and locomotion speed of one of the trackmakers differed from that of the other trackmaker that is not overstepped.

Clack (1997) drew attention to how different the two trackways are, and argued that no know Devonian tetrapod could have made the tracks unless they were made subaqueously. However, Pridmore (1995) posited that the same kind of animal moving at different speeds produced the two different trackways. Associated mudcracks suggest the tracks were impressed subaerially (Young, 2006), and this would apparently exclude *Acanthostega* and *Ichthyostega*, so a more likely candidate for trackmaker is a tetrapod similar to *Tulerpeton* (Clack, 1997, p. 241). The Genoa River trackways were made subaerially, in a nonmarine setting by a tetrapod with a much larger pes than manus.

3.2 Easter Ross, Scotland

Rogers (1990) documented a tetrapod trackway from the Upper Old Red Sandstone of Easter Ross in northern Scotland. However, the Devonian age of this tracksite is not certain. According to Rogers (1990), it is ~900 m stratigraphically above a marine fossiliferous bed of

Givetian age and could range from Givetian to Tournasian in age. Marshall et al. (2018) raised the possibility that the Easter Ross footprints could be of Permian age. However, no biostratigraphic data tightly constrain the age of these footprints (Marshall et al., 2018), so I continue to tentatively consider them to be of Late Devonian age (Lucas, 2015, 2019).

According to Rogers, the track-bearing stratum is part of an eolian sabkha deposit in a sandstone bed with wind ripple and planar laminations, and therefore, the tracks were impressed subaerially. The tracks form an alternating trackway of three steps in which inferred manus and pes imprints can be differentiated on the basis of size (manus much smaller than pes). Rogers described the manus and pes as of similar size, but on the left side of the trackway, the inferred manus impressions are clearly smaller than the pes impressions. The glenoacetabular length of the tetrapod trackmaker is ~29 cm.

3.3 Valentia Island, Ireland

Stössel (1995) reported a trackway from the Devonian Valentia Slate Formation on Valentia Island in southwestern Ireland, and Stössel et al. (2016) provided much more detailed data on multiple trackways from this location (Fig. 2.3). A U–Pb age of ~385 Ma for an air-fall tuff is ~230 m stratigraphically above the trackway (Williams et al., 1997). This is a Givetian age, so the trackway is considered to be Middle Devonian, though the biostratigraphy (based on fishes and miospores) of the track-bearing stratigraphic interval is less precise and indicates a Mid-Devonian through Fammenian age (Stössel, 1995). A detailed sedimentological analysis by Stössel et al. (2016) indicates that the Valentia Island trackways were made along river margins in a freshwater setting.

At the Valentia Island tracksite, there are 150 tracks without median drag marks in a long, meandering trackway that extends for a course of about 10 m. This trackway shows both size differentiation of manus and pes tracks and an alternating trackway pattern of a quadruped (Fig. 2.4) with an estimated gleno-acetabular length of 38 cm and a coupling value of 1.2–1.4 (Stössel, 1995). No details of footprint morphology are preserved because of fracturing of the sediments, but Stössel drew attention to the tracks being wider than long, and suggested this indicated a polydactyl trackmaker (it also infers forward-oriented digits).

4. Discussion

4.1 Distribution

The Devonian tetrapod footprint record is too sparse to support definite conclusions about biostratigraphic, paleobiogeographic, and paleoenvironmental distributions of Devonian tetrapods. Thus, it lacks abundance, diversity, and index taxa that would be of use in biostratigraphy (Lucas, 2007, 2015, 2019). Nevertheless, the tetrapod trackways from Valentia Island in Ireland are no younger than Givetian, so this is the oldest record of tetrapods, and it provides an important datum that places tetrapod origins during the Middle Devonian, or earlier.

Tetrapod tracks of Middle Devonian age are known from Ireland, and Late Devonian tetrapod tracks are known from Scotland and Australia. The Irish record is the oldest tetrapod record, so it could indicate an origin of tetrapods in Euramerica. However, too

FIGURE 2.3 The oldest tetrapod fossils, Middle Devonian trackways from Valentia Island, Ireland. (A) Overview of trackway DO1, scale bar = 50 cm. (B) Western part of trackway surface showing trackways DO1-6, scale bar = 25 cm. (C) Trackway DO3, which is curved with a median groove, scale bar = 20 cm. *Modified from Stössel et al. (2016) from images provided by Iwan Stössel.*

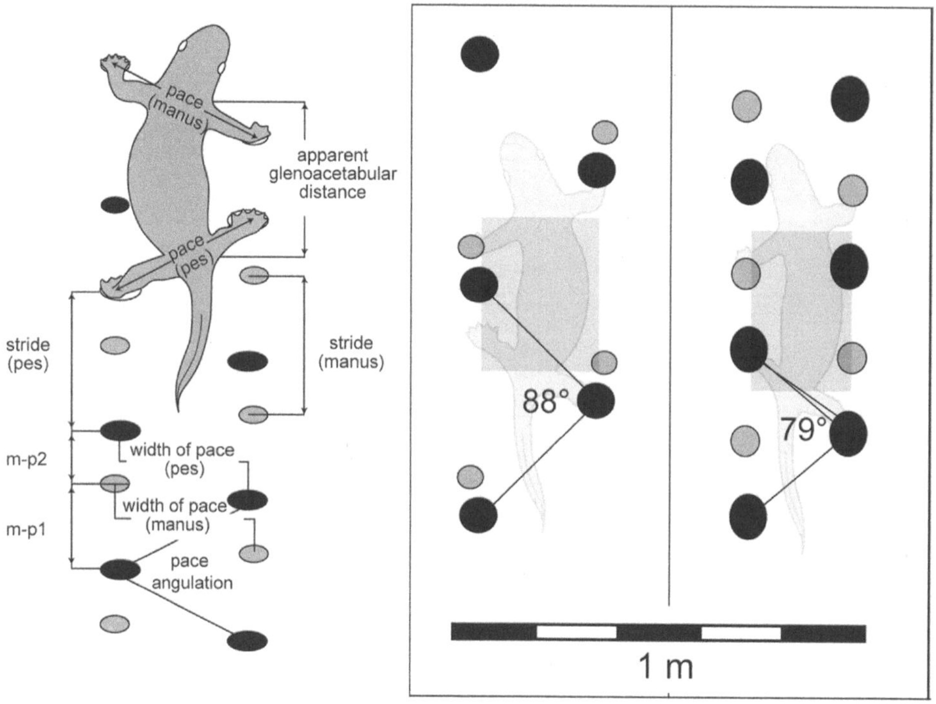

FIGURE 2.4 Interpretation of Middle Devonian trackways on Valentia Island, Ireland, showing lateral sequence walking by a tetrapod with a larger pes than manus. *Modified from Stössel et al. (2016) with images courtesy of Iwan Stössel.*

few Devonian tetrapod trackways are known to provide a reliable basis for paleobiogeographic interpretation.

The paleoenvironment of tetrapod origins has long been debated—marine or nonmarine, dry or wet (e.g., Retallack, 2011; George and Blieck, 2011; Clack, 2012; Schultze, 2013). The three bona fide Devonian tetrapod tracksites were formed in nonmarine environments. They thus provide no support for tetrapod origins in a marine or marginal marine setting, though that is not totally excluded by such a meager track record.

4.2 Footprints and tetrapod origins

The current understanding of tetrapod origins began to develop in 1930 with the discovery of tetrapod body fossils in the Devonian of eastern Greenland, which were the basis for naming *Ichthyostega*; later discoveries in eastern Greenland produced *Acanthostega* (Jarvik, 1996). These were long the only two genera of Devonian tetrapods known from body fossils, and they remain the only two such genera known from a significant sample of multiple specimens. In the last 20 years, the discovery of Devonian tetrapod fossils has accelerated, so that there are now 17 named genera, though most of these are based on very incomplete and fragmentary fossils (Clack and Milner, 2015; Olive et al., 2016; Lucas, 2021).

Devonian tetrapods are regarded as primarily aquatic (e.g., Coates and Ruta, 2007; Clack, 2012). Indeed, Coates and Clack (1995) argued that the limbs of *Ichthyostega* and *Acanthostega* were relatively inflexible paddles that were unlikely to have left the kinds of Devonian trackways attributed to tetrapods. This may in part explain the rarity of Devonian tetrapod footprints—most Devonian tetrapods spent little or no time walking on land.

The bona fide Devonian tetrapod footprints indicate an alternating pattern of limb-supported locomotion with a larger hind foot than forefoot (Figs. 2.2–2.4). This is the characteristic pattern of quadrupedal locomotion of post-Devonian tetrapods (e.g., Fillmore et al., 2012). The Devonian tetrapod track record thus indicates such locomotion was possible at least some of the time by some Devonian tetrapods. Body fossils of these Devonian tetrapods await discovery.

Current thinking identifies the "tetrapodomorph fishes" as the closest relatives, if not the ancestors, of tetrapods, pointing to *Panderichthys* from Latvia and Russia, *Livonania* from Latvia, *Elpistostega* from eastern Canada, and *Tiktaalik* from Arctic Canada (e.g., Daeschler et al., 2006; Carroll, 2009; Clack, 2012; Ahlberg, 2018). One hypothesis of tetrapod origins that has not received much discussion is that the sarcopterygian ancestors of tetrapods may yet remain undiscovered and may belong to a group other than the putative candidates, the tetrapodomorph fishes.

Thus, *Panderichthys* and *Tiktaalik* have robust pectoral fins, much larger than their hind fins, quite different from the larger pelvic limbs of tetrapods (Shubin et al., 2006; Swartz, 2012) (Fig. 2.5). There is thus a considerable gap between the locomotory apparatus of fully terrestrial tetrapods, driven by hind limb propulsion, and that of their supposed nearest relatives, driven by forelimb propulsion (e.g., Boisvert, 2005; Boisvert et al., 2008; Pierce et al.,

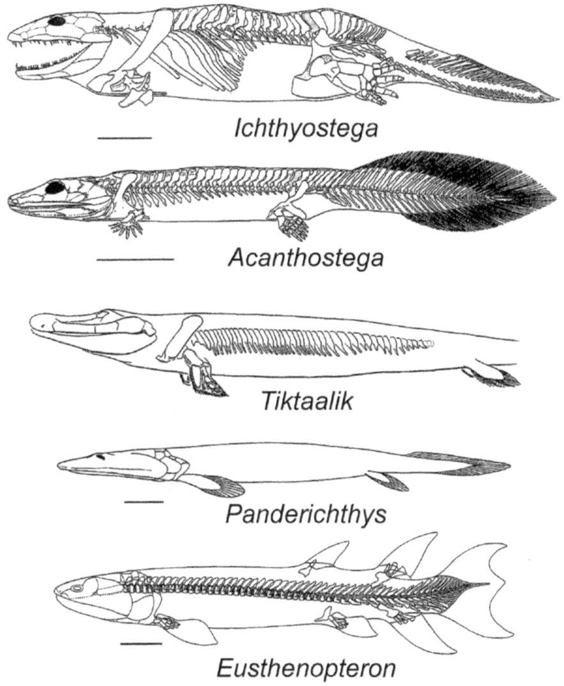

FIGURE 2.5 Lateral views of skeletons and body outlines of selected sarcopterygian and tetrapodomorph fishes and Devonian tetrapods. Scale bars = 10 cm; note that all of these animals have gleno-acetabular lengths between 19 and 54 cm. *After Carroll (2009), Fig. 3.1.*

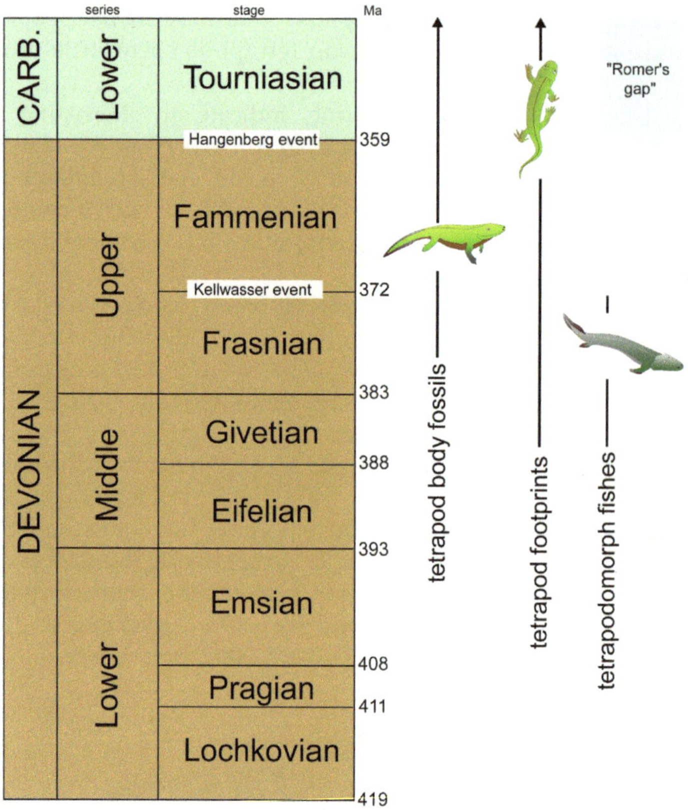

FIGURE 2.6 The Devonian-earliest Carboniferous timescale showing the stratigraphic ranges of tetrapod body fossils and footprints and some other key factors relevant to tetrapod origins. *Clip art images by Arthur Escher.*

2013). The Devonian tetrapod track record suggests this evolutionary gap had already been bridged by the Middle Devonian (Fig. 2.6).

This raises the question of whether or not the tetrapodomorph fishes are closely related to tetrapods or simply a group that converged on some tetrapod features. Bona fide Devonian tetrapod footprints indicate an alternating pattern of limb-supported locomotion with a larger hind foot than forefoot. The trackmakers of the Devonian footprints and trackways are essentially unknown, so a major clade, or clades, of early tetrapods have been sampled by the footprint record but not by the body-fossil record. This clade(s) is (are) more likely to have been the main line of tetrapod origins and evolution, not any lineage of known stem tetrapod. Thus, the entire understanding of tetrapod origins as now presented may be merely the understanding of the evolution of an early, dead-end radiation of aquatic stem tetrapods from a sarcopterygian ancestry. Other than the footprint record, the Devonian origin and subsequent early evolution of terrestrially adapted tetrapods remains largely undiscovered and unexamined.

Acknowledgments

I am grateful to John Marshall, Iwan Stössel, and John Warren for sharing images and information. Reviews by Adrian Hunt and Hendrik Klein improved the content and clarity of the manuscript.

Data availability

No data were used for the research described in this article.

Declaration of competing interest

The author declares that he has no known competing financial interests or personal relationships that could have appeared to influence the work reported in this paper.

References

Abel, O., 1935. Vorzeitliche Lebensspuren. Jena. Gustav Fisc. 64, 4.

Ahlberg, P., 2018. Follow the footprints and mind the gaps: a new look at the origin of tetrapods. Earth Environ. Sci. Trans. Royal Soc. Edinburgh 2018, 1–23.

Boisvert, C.A., 2005. The pelvic fin and girdle of *Panderichthys* and the origin of tetrapod locomotion. Nature 438, 1145–1147.

Boisvert, C.A., Mark-Kurik, E., Ahlberg, P.E., 2008. The pectoral fin of *Panderichthys* and the origin of digits. Nature 456, 636–638.

Carroll, R., 2009. The Rise of Amphibians: 365 Million Years of Evolution. The Johns Hopkins University Press, Baltimore, p. 360.

Clack, J.A., 1997. Devonian tetrapod trackways and trackmakers; a review of the fossils and footprints. Palaeogeogr. Palaeoclimatol. Palaeoecol. 130, 277.

Clack, J.A., 2012. Gaining Ground: The Origin and Evolution of Tetrapods, second ed. Indiana University Press, Bloomfield, Indiana, p. 523.

Clack, J.A., Milner, A.R., 2015. Basal tetrapoda. In: Handbook of Paleoherpetology 3A1, pp. 1–90.

Coates, M.I., Clack, J.A., 1995. Romer's gap: tetrapod origins and terrestriality. Bulletin du Muséum National d'Histoire Naturelle Series 4 (17), 373–388.

Coates, M.I., Ruta, M., 2007. Skeletal changes in the transition from fins to limbs. In: Hall, B.K. (Ed.), Fins into Limbs: Evolution, Development, and Transformation. University of Chicago Press, Chicago, pp. 15–37.

Daeschler, E.B., Shubin, N.H., Jenkins Jr., F.A., 2006. A Devonian tetrapod-like fish and the evolution of the tetrapod body plan. Nature 440, 757–763.

Fillmore, D.L., Lucas, S.G., Simpson, E.L., 2012. Ichnology of the Mississippian Mauch Chunk Formation, eastern Pennsylvania. New Mexico Museum Natural History Sci. Bulletin 54, 1–136.

Friend, P.F., Alexander-Marrack, P.D., Nicholson, J., Yeats, A.K., 1976. Devonian sediments of East Greenland II. Sedimentary structures and fossils. Meddelelser om Grønland 206 (2), 1–91.

George, D., Blieck, A., 2011. Rise of the earliest tetrapods: an Early Devonian origin from a marine environment. PLoS One 6 (7), e22136.

Gourmanis, C., Webb, J.A., Warren, A.A., 2003. Fluviodeltaic sedimentology and ichnology of part of the Silurian Grampians Group, western Victoria. Aust. J. Earth Sci. 50, 811–825.

Jarvik, E., 1996. The devonian tetrapod *Ichthyostega*. Fossils Strata 40, 1–213.

Leonardi, G., 1983. *Notopus petri* nov. gen. nov. sp.: un empreinte d'amphibien du Devonien au Parana (Bresil). Geobios 16, 233–239.

Lucas, S.G., 2007. Tetrapod footprint biostratigraphy and biochronology. Ichnos 14, 5–38.

Lucas, S.G., 2015. *Thinopus* and a critical review of Devonian tetrapod footprints. Ichnos 22, 136–154.

Lucas, S.G., 2019. An ichnological perspective on some major events of Paleozoic tetrapod evolution. Bolletino della Soc. Paleontol. Ital. 58, 223–266.

Lucas, S.G., 2021. Tetrapod origins. In: Alderton, D., Elias, S.A. (Eds.), Encyclopedia of Geology, second ed., vol. 3. Academic Press, United Kingdom, pp. 138–146. https://doi.org/10.1016/B978-0-08-102908-4.00090-4.

Marsh, O.C., 1896. Amphibian footprints from the Devonian. Am. J. Sci. 152, 374–375.

Marshall, J.A.E., Glennie, K.W., Astin, T.R., Hewitt, A.J., 2018. The Old Red Group (Devonian)-Rotliegend Group (Permian) unconformity in the Inner Moray Firth. Geol. Soc. London Special Publ. 471, 237–252.

Niedźwiedzki, G., Szrek, P., 2020. Non-tetrapod trace fossils from the Middle Devonian tetrapod tracksite at Zachełmie Quarry, Holy Cross Mountains, Poland. Palaeogeogr. Palaeoclimatol. Palaecol. 553, 109763.

Niedźwiedzki, G., Szrek, P., Narkiewicz, K., Narkiewicz, M., Ahlberg, P.E., 2010. Tetrapod trackways from the early Middle Devonian period of Poland. Nature 463, 43–48.

Olive, S., Ahlberg, P.E., Pernègre, V.N., Poty, E., Steurbaut, E., Clément, G., 2016. New discoveries of tetrapods (ichthyostegid-like and whatcheeriid-like) in the Fammenian (Late Devonian) localities of Strud and Becco (Belgium). Palaeontology 2016, 1–14.

Pierce, S.E., Hutchinson, J.R., Clack, J.A., 2013. Historical perspectives on the evolution of tetrapodomorph movement. Integr. Comp. Biol. 53, 209–223.

Pridmore, P.A., 1995. Submerged walking in the epaulette shark *Hemiscyllium ocellatum* (Hemiscyllidae) and its implications for locomotion in rhipidistian fishes and early tetrapods. Zool. Anal. Complex Syst. 98, 278–297.

Retallack, G.J., 2011. Woodland hypothesis for Devonian tetrapod evolution. J. Geol. 119, 235–258.

Roček, Z., Rage, J.-C., 1994. The presumed amphibian footprint *Notopus petri* from the Devonian: a probable starfish trace fossil. Lethaia 27, 241–244.

Rogers, D.A., 1990. Probable tetrapod tracks rediscovered in the Devonian of N Scotland. J. Geol. Soc. 147, 746–748.

Sarjeant, W.A.S., 1974. A history and bibliography of the study of fossil vertebrate footprints in the British Isles. Palaeogeogr. Palaeoclimatol. Palaecol. 16, 265–375.

Schultze, H.-P., 2013. The paleoenvironment at the transition from piscine to tetrapod sarcopterygians. New Mexico Museum Nat. History Sci. Bulletin 60, 373–397.

Shubin, N.H., Daeschler, E.B., Jenkins Jr., F.A., 2006. The pectoral fin of *Tiktaalik rosae* and the origin of the tetrapod limb. Nature 440, 764–771.

Stössel, I., 1995. The discovery of a new Devonian tetrapod trackway in SW Ireland. J. Geol. Soc. London 152, 407–413.

Stössel, I., Williams, E.A., Higgs, K.T., 2016. Ichnology and depositional environments of the Middle Devonian Valentia Island tetrapod trackways, south-west Ireland. Palaeogeogr. Palaeoclimatol. Palaeoecol. 462, 16–40.

Swartz, B., 2012. A marine stem-tetrapod from the Devonian of western North America. PLoS One 7 (3), e33683.

Warren, A.A., Jupp, R., Bolton, B., 1986. Earliest tetrapod trackway. Alcheringa 10, 183–186.

Warren, A.W., Wakefield, N.A., 1972. Trackways of tetrapod vertebrates from the upper devonian of victoria, Australia. Nature 238, 469–470.

Westoll, T.S., 1937. The Old Red Sandstone fishes of the north of Scotland, particularly of Orkney and Shetland. Proc. Geologists' Assoc. 48, 13–45.

Williams, E.A., Sergeev, S.A., Stössel, I., Ford, M., 1997. An Eifelian U-Pb zircon date for the Enagh tuff bed from the Old Red Sandstone of the Munster basin in NW Iveragh, SW Ireland. J. Geol. Soc. London 154, 189–193.

Young, G.C., 2006. Biostratigraphic and biogeographic context for tetrapod origins during the Devonian: Australian evidence. Alcheringa Special Issue 1, 409–428.

CHAPTER

3

Carboniferous tetrapod footprints

Spencer G. Lucas[1] *and Matthew R. Stimson*[2,3]

[1]New Mexico Museum of Natural History and Science, Albuquerque, NM, United States; [2]Steinhammer Paleontology Laboratories, New Brunswick Museum, St. John, NB, United States; [3]Department of Geology, Saint Mary's University, Halifax, NS, Canada

1. Introduction

Tetrapod footprints of Carboniferous age have been studied since the 1840s and are known primarily from North America and Western Europe at localities that were part of a Euramerican paleoprovince of equatorial Pangea (Fig. 3.1). These tetrapod footprints are a significant aspect of the fossil record of Carboniferous tetrapods that document temporal distributions and evolutionary events not known from the body-fossil record. Here, we review the Carboniferous record of tetrapod footprints and explain how it informs our understanding of some important biotic events in the evolution of Carboniferous tetrapods.

2. Some history

We offer a brief, analytical history of the scientific study of Carboniferous tetrapod footprints, which began in the 1840s. For a more detailed, narrative history, consult Sarjeant (1974, 1987), Sarjeant and Mossman (1978), Cotton et al. (1995), and Hunt et al. (1995). We divide the study of Carboniferous tetrapod footprints into three overlapping phases: (1) first discoveries and a significant conclusion (1840s), (2) diverse and mostly scattered discoveries (1840s–1970s), and (3) syntheses (1960s–Present).

2.1 First discoveries and a significant conclusion (1840s)

To the European geologists of the early 1800s, the "Coal Age" (=Carboniferous Period of current usage) was a time of an inhospitable atmosphere that made it impossible for tetrapods to live on land. The first discoveries of Carboniferous tetrapod footprints refuted that idea. The first Carboniferous footprints brought to scientific attention were discovered by

Vertebrate Ichnology
https://doi.org/10.1016/B978-0-443-13837-9.00006-8

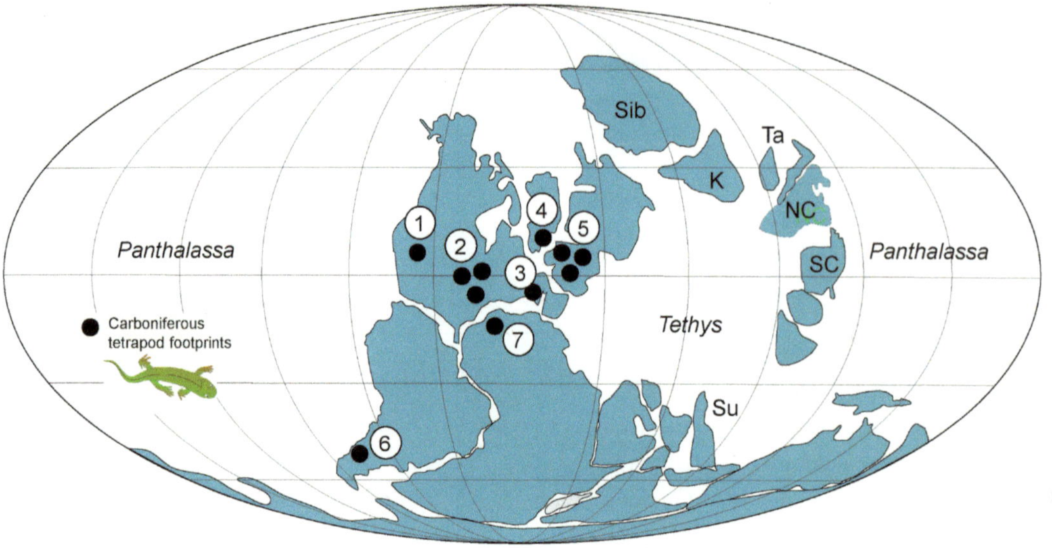

FIGURE 3.1 Carboniferous continental configuration showing tetrapod footprint localities. Abbreviations are as follows: K = Kazakstan, NC = North China, SC = South China, Sib = Siberia, Su = Subumasu, Ta = Tarim. Localities are as follows: (1) western USA; (2) eastern USA; (3) Nova Scotia, Canada; (4) east Greenland; (5) Europe; (6) Chile; and (7) Morocco. *Map lines delineate study areas and do not necessarily depict accepted national boundaries.*

William Logan in 1841 at Horton Bluff in Nova Scotia, Canada. They were presented to the Geological Society in London the very next year (Logan, 1842), and Lyell (1843, p. 185) described this discovery as "the first evidence of the reptilian class known in the Carboniferous rocks." But, although these footprints were later interpreted to be those of an amphibian, and named *Hylopus logani* by Dawson (1863a), this discovery had little impact until decades later.

Meanwhile, during the 1840s in Pennsylvania, USA, Alfred King found bona fide amphibian tracks (Fig. 3.2), the type material of *Limnopus heterodactylus* (King, 1845a; Lucas and Dalman, 2013; Lucas, 2015b), although he also described many Native American petroglyphs as fossil tracks (King, 1844, 1845a, 1845b, 1845c, 1846). Incidentally, *Limnopus* (=*Thenaropus*) *heterodactylus* King (1845a) was the first binomen applied to a Carboniferous footprint. King's discoveries drew the attention of Charles Lyell, who, during his second trip to North America, examined King's finds, proclaiming most to be human-made petroglyphs but recognizing the authenticity of the Carboniferous amphibian footprints (Fig. 3.2). The Carboniferous footprint discoveries of Logan and King demonstrated that terrestrial tetrapods lived during the Carboniferous. They are the first example of footprints providing important distributional information about tetrapods in the absence of a body-fossil record.

2.2 Diverse and mostly scattered discoveries (1840s–1970s)

Lea (1849, 1853, 1855; also see Fillmore et al., 2009, 2012) reported Mississippian age footprints from Pennsylvania, USA. Later important discoveries during the late 1800s in the USA

FIGURE 3.2 *Limnopus heterodactylus* holotype, from the Pennsylvanian Conemaugh group in eastern Pennsylvania, USA.

were of Pennsylvanian footprints in Pennsylvania (Mason, 1878; Leidy, 1879), Kansas (Mudge, 1873; Marsh, 1894), Indiana (Cox, 1874), and Missouri (Butts, 1891). Marsh's (1894) work was of particular significance as he introduced the ichnogenera *Dromopus* and *Limnopus*, which are common Late Pennsylvanian–early Permian footprint ichnotaxa (Fig. 3.3). In the Canadian Maritimes, important contributions to the Carboniferous tetrapod footprint record were made by Dawson (1872, 1878) and later by Matthew (1903a,b, 1904, 1905, 1910), who produced the first synthesis of the ichnotaxonomy of the then known Carboniferous footprints (Fig. 3.4).

In the United Kingdom, several Carboniferous footprint discoveries were made in the late 1800s: from the Millstone Grit at Corck, Ireland (Haines, 1852) and Cheshire, England (Binney, 1856). Barkas (1873, 1890) reported Carboniferous footprints from Northumberland. Geinitz reported the first Carboniferous footprints from Germany.

In the 20th century, most published work on Carboniferous footprints was sporadic reports of new records in the USA and in Europe. Thus, diverse records were variously documented in the USA from the states of Rhode Island, Massachusetts, Virginia, West Virginia, Ohio, Indiana, Oklahoma, and Colorado (e.g., Branson, 1910; Jillison, 1917; Lull, 1920; Henderson, 1924; Carman, 1927; Aldrich and Jones, 1930; Willard and Cleaves, 1930; Mitchell, 1931, 1933; Branson and Mehl, 1932; Toepelman and Rodeck, 1936; Colbert and Schaeffer, 1947; Dunkle, 1948). In eastern Canada, diverse records of Carboniferous footprints were also documented (e.g., Sternberg, 1933; Sarjeant and Mossman, 1978; van Allen et al., 2005; Wood and Miller, 2007; Falcon-Lang et al., 2007; Stimson et al., 2012, 2016a,b; Tanner, 2013; Allen et al., 2022).

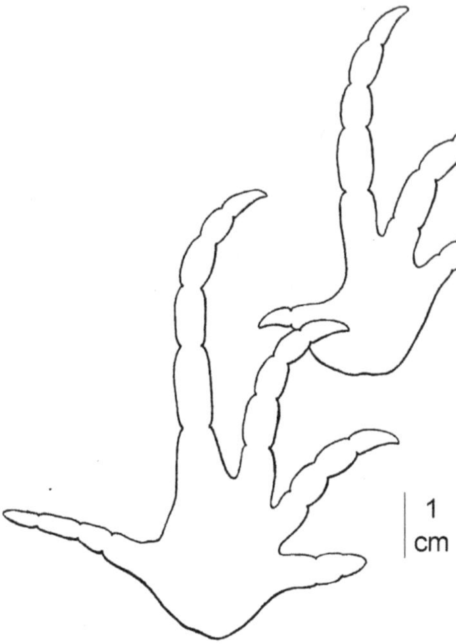

FIGURE 3.3 Marsh's (1894, pl. 3) illustration of one of the syntypes of *Dromopus agilis*.

In Europe, similar records were reported from Germany (Beck, 1915; Kukuk, 1926; Hahne and Wolansky, 1951; Wolansky, 1952; Weingardt, 1961; Haubold, 1970), France (Dollé et al., 1970; Gand, 1975; Langiaux, 1979, 1980, 1981; Langiaux and Sotty, 1975a, b, 1976, 1977) and Sardinia in Italy (Fondi, 1979, 1980). An important point is that almost all of these footprint records were from small outcrops with numerically and ichnotaxonomically limited assemblages. They represent more than a century of important discoveries, but, other than Matthew (1903a), no attempts were made to synthesize or analyze the Carboniferous tetrapod footprint record.

2.3 Syntheses (1960s—present)

As before, many diverse and scattered records of Carboniferous tetrapod footprints continued to be documented during the last 40 plus years, mostly from North America and Europe (e.g., Jake and Blake, 1982; Mietto et al., 1985; McClelland, 1988; Turek, 1989; Lane and Maples, 1990; Reisz, 1990; Martino, 1991; Schneider et al., 1992; Milner, 1994; Lockley and Hunt, 1995; Scarboro and Tucker, 1995; Mossman and Grantham, 1996, 2000; Blieck et al., 1997; Soler-Gijón and Moratalla, 2001; Calder et al., 2004; Falcon-Lang et al., 2004, 2010; Voigt et al., 2011; Fillmore et al., 2015; Niedźwiedzki, 2015; Schöllmann et al., 2015; Meade et al., 2016; Bird et al., 2019; Marchetti et al., 2018, 2020b; Weems and Lucas, 2020; Matamales-Andreu et al., 2021; Soler-Gijón and Díez Ruiz, 2023). Kuhn (1963) and Haubold (1971, 1984) published the first synthetic compilations of the tetrapod footprint record, which, of course, included the Carboniferous tetrapod footprints. Kuhn's (1963) was an

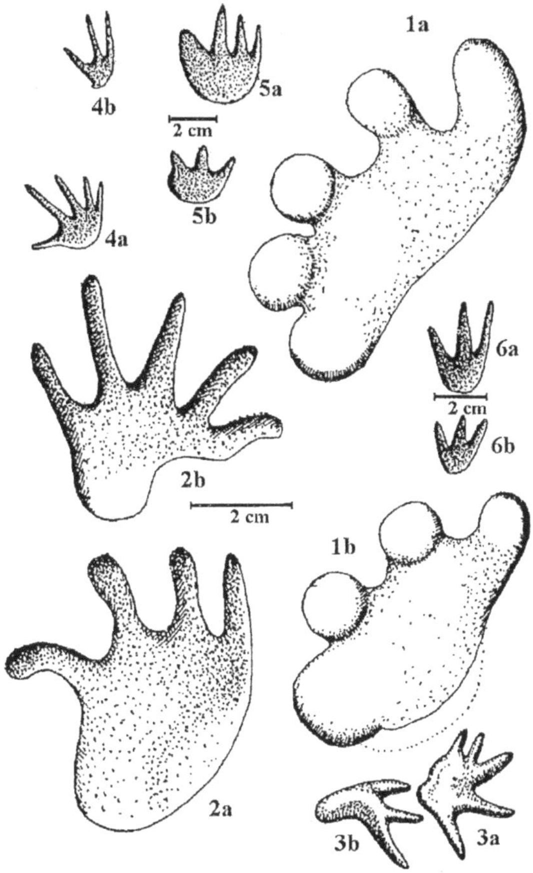

FIGURE 3.4 Matthew's (1903a) illustration of various Carboniferous amphibian footprints. He assigned them to *Baropezia* (1), *Megapezia* (2), *Asperipes* (3), *Cursipes* (4), *Barillopus* (5), and *Ornithoides* (6), all ichnogenera now considered invalid.

ichntaxonomic compilation, essentially a list of synonymies, whereas Haubold (1971) also evaluated some of the ichnotaxonomy and summarized the stratigraphic distributions of the footprints (also see Haubold, 1984).

Beginning in the 1990s—early 2000s, significant new Carboniferous footprint localities were discovered, collected, and/or studied: Early Mississipian ichnoassemblages at Blue Beach in Nova Scotia, Canada; the Late Mississippian Mauch Chunk localities near Pottsville, Pennsylvania, USA; the Early Pennsylvanian Union Chapel Mine in Alabama, USA (and, somewhat later, the Crescent Valley Mine ichnoassemblage, also in Alabama); and the Middle Pennsylvanian Keota ichnoassemblage in Oklahoma, USA (Lucas et al., 2004; Fillmore et al., 2012; Buta and Kopaska-Merkel, 2016; Lucas, 2019). Carboniferous footprint biochronology was developed by Lucas (2003, 2007), Fillmore et al. (2012), and Lucas et al. (2022), and recent syntheses have used the footprints to inform our understanding of Carboniferous tetrapod evolution (e.g., Lucas, 2019; Lucas et al., 2022; Marchetti et al., 2020c).

3. Carboniferous ichnotaxa

Carboniferous tetrapod footprints have had a long history of excessive ichnotaxonomic splitting that created a confused and confusing ichnotaxonomy (Table 3.1). Matthew (1903a) presented an early ichnotaxonomic revision (Fig. 3.4), and Haubold (1970, 1971), Sundberg et al. (1990), Haubold et al. (2005), Lucas et al. (2010a), Fillmore et al. (2012), and Marchetti et al. (2020c), among others, have clarified aspects of the ichnotaxonomy of some ichnogenera.

The need for a broad and detailed ichnotaxonomic revision is obvious (Hunt et al., 1995; Cotton et al., 1995; Lucas 2003,2007,2019). Nevertheless, to revise Carboniferous tetrapod footprint ichnotaxonomy is beyond the scope of this paper. Here, ichnogenera of Carboniferous footprints that we regard as valid (Figs. 3.5–3.13) are briefly characterized, and their

TABLE 3.1 List of most of the named Carboniferous tetrapod footprint ichnogenera. Names in boldface are considered valid here.

Allopus Marsh, 1894
Acripes Langiaux & Sotty, 1975 (preoccupied)
Ancylopus Carman, 1927
Anomoeichnus Carman, 1927
Anthracopus Leidy, 1880
Anticheiropus Sarjeant and Mossman, 1979 (preoccupied)
Asperipes Matthew, 1903
***Attenosaurus* Aldrich in Aldrich and Jones, 1930**
Barillopus Matthew, 1903
Baropezia Matthew, 1904
Baropus Marsh, 1894
***Batrachichnus* Woodworth, 1900**
Bipedes Aldrich, 1930
***Cincosaurus* Aldrich in Aldrich and Jones (1930)**
Collettosaurus Cox, 1874
Ctenerpeton Aldrich, 1930
Crucipes Butts, 1891
Cursipes Matthew, 1903
Dromillopus Matthew, 1905
***Dromopus* Marsh, 1894**
Hydromedichnus Kuhn, 1963 for *Hydromeda* Aldrich, 1930

TABLE 3.1 List of most of the named Carboniferous tetrapod footprint ichnogenera. Names in boldface are considered valid here.—cont'd

***Hylopus* Dawson, 1882**
Leptopus Langiaux and Sotty, 1975
***Limnopus* Marsh, 1894**
Limnosauripus Kuhn, 1959 for *Limnosaurus* Aldrich, 1930
***Matthewichnus* Haubold, 1970**
Megabaropus Baird, 1952
Megapaezia Matthew, 1903
Nanopus Marsh, 1894
***Notalacerta,* Butts (1891)**
Notamphibia Butts, 1891
Okypes Langiaux and Sotty, 1975
Onychopus Martin, 1922
Ornithoidipus Sternberg, 1933
Ornithoides Matthew, 1903
Palaeosauropus Hay, 1902 for *Sauropus* Lea, 1849
Parvives Willard and Cleaves, 1930
Peratodactylopus Sarjeant and Mossman, 1979
Pinguipes Langiaux and Sotty, 1975
Prolacertipes Dolle et al., 1970
***Pseudobradypus* Matthew, 1903**
Quadropedia Aldrich, 1930
Salichnium Muller, 1962
Schmidtopus Haubold, 1970
Sormiensipes Langiaux and Sotty, 1975
Sphaeropezium King, 1845 (replaced *Thenaropus* King, 1844)
Steganoposaurus Branson and Mehl, 1932
Stephanopus Gand, 1975
Tenuipes Langiaux and Sotty, 1975
Thenaropus King, 1844
Tridactylosaurus Barkas, 1883
Trisaurus Aldrich, 1930

Modified from Haubold et al. (2005).

FIGURE 3.5 Selected examples of Carboniferous tetrapod-footprint ichnogenera. (A) *Amphisauropus* (Middle Pennsylvanian, Oklahoma), (B) *Dromopus* (Late Pennsylvanian, Kansas), and (C) *Varanopus* (Late Pennsylvanian or early Permian, Nova Scotia).

inferred trackmakers are identified. We should note, though, that there are many ichnogeneric names, some still in use by various workers, applied to Carboniferous tetrapod footprint ichnogenera that we regard as either *nomina dubia* or as simply invalid (Table 3.1).

3.1 *Amphisauropus* Haubold, 1970 (Fig. 3.5A)

Type ichnospecies: *Amphisauropus latus* (Haubold, 1970).

Valid ichnospecies: *Amphisauropus kablikae* (Geinitz and Deichmuller, 1882).

Characteristics: Medium-sized quadrupedal trackway with plantigrade manus and pes. Manus and pes are pentadactyl. Pes is wider than long and oriented parallel to the midline. Digits are wide, short, and straight with rounded terminations. The digits diverge radially from the autopodium (124 degrees) with relative digit lengths of V < I < II < III < IV. With or without a wide, elliptical sole impression. The manus is wider than long and is slightly smaller than the pes and slightly oriented toward the midline. Digits are wide and

FIGURE 3.6 Examples of *Attenosaurus* from the Lower Pennsylvanian of Alabama.

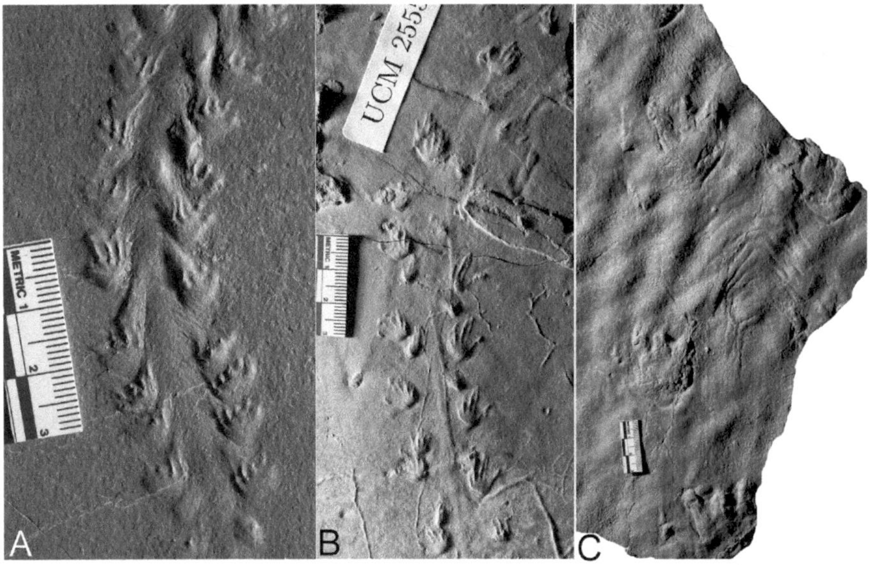

FIGURE 3.7 Selected examples of Carboniferous tetrapod-footprint ichnogenera. (A) *Batrachichnus* (Early Pennsylvanian, Alabama), (B) *Matthewichnus* (Early Pennsylvanian, Alabama), and (C) *Limnopus* (Mississippian, Nova Scotia).

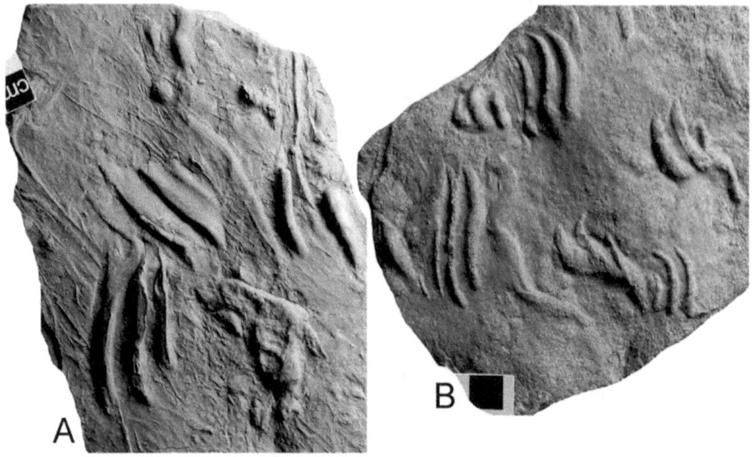

FIGURE 3.8 Two examples of *Characichnos* from the Lower Mississippian of Nova Scotia.

short with rounded terminations and with relative digit lengths of I < V < II < IV < III. Digits have a total divergence of 131 degrees. Sole impressions are wider then long, elliptical, and often deeply impressed. Some rare examples from Prince Edward Island in eastern Canada exhibit skin impressions on the autopodium and digits.

Temporal distribution: Middle Pennsylvanian—early Permian.

Trackmaker: The close resemblance of *Amphisauropus* tracks to the foot structure of seymouriamorph reptiliomorphs has long been used to identify them as the trackmakers (Voigt, 2005; Voigt and Lucas, 2018; Marchetti et al., 2017).

Comments: Most records of *Amphisauropus* are early Permian. However, its oldest record is from the Middle Pennsylvanian at Keota, Oklahoma (Lucas et al., 2004).

3.2 *Attenosaurus* Aldrich, Aldrich and Jones (1930) (Fig. 3.6)

Type ichnospecies: *Attenosaurus indistinctus* Aldrich (Aldrich and Jones, 1930).

Valid ichnospecies: *Attenosaurus indistinctus* Aldrich (Aldrich and Jones, 1930).

Characteristics: *Attenosaurus* is the footprint of a quadruped with long, thin, pointed, and often slightly curved digits (although at least some of the features of the digits likely reflect extramorphology). Both the manus and pes are pentadactyl, but this is only seen in a few specimens. Many tracks lack a sole impression and are little more than an arc representing the junction of the phalanges and the metapodials (metacarpals/metatarsals) behind the digit imprints. In the pes, digit III is longest, digits II and IV are slightly shorter and subequal in length, digit I is next in size, and digit V is shortest and often separated by a large angle of divarication from the remaining digits, which tend to be more closely packed and aligned. The manus appears to have the same relative digit lengths as the pes, with digit III longest. This is important, because it distinguishes the footprints of *Attenosaurus* from those of *Dimetropus* in which digit IV is always longest (Romer and Price, 1940; Voigt and Ganzelewski, 2010; Sacchi et al., 2014; Lucas et al., 2016). Trackways of *Attenosaurus* that preserve manus and pes impressions indicate that

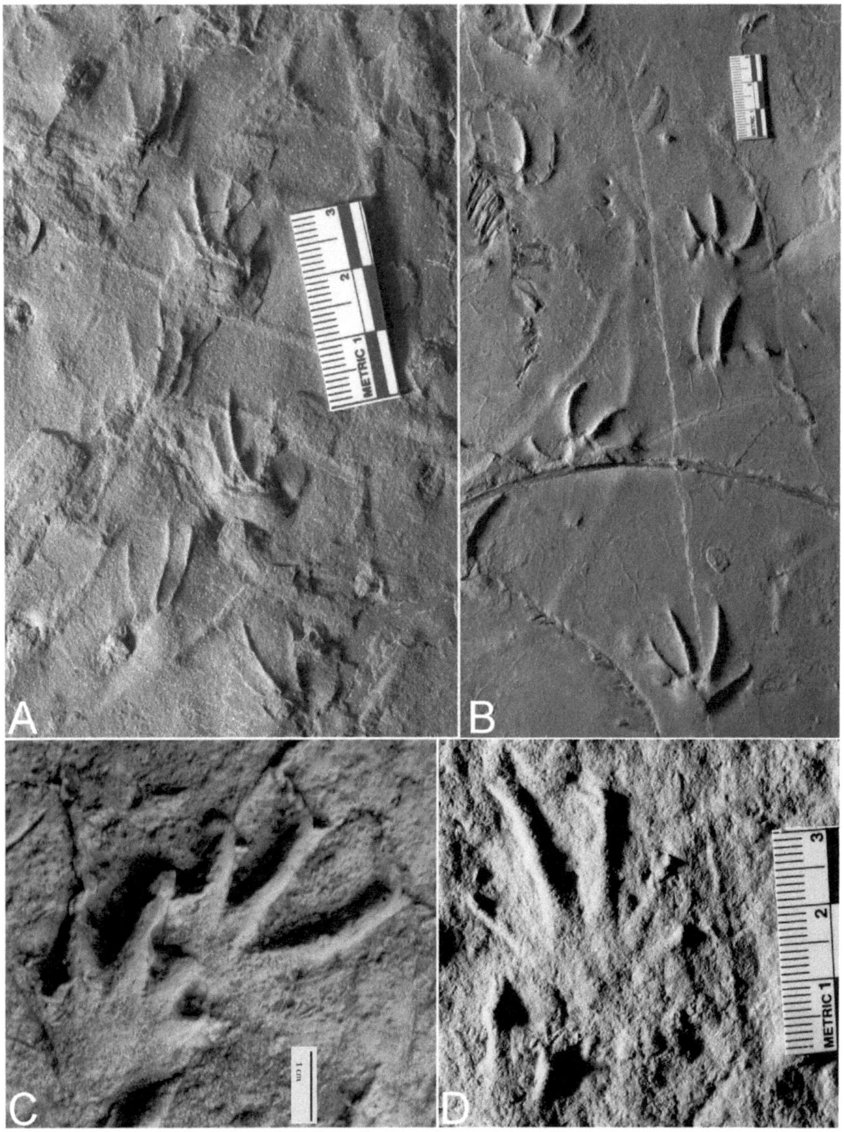

FIGURE 3.9 Selected examples of Carboniferous tetrapod-footprint ichnogenera. (A–B) *Cincosaurus* (Early Pennsylvanian, Alabama) and (C–D) *Notalacerta* (Middle Pennsylvanian, Indiana).

the pes is larger than the manus. Curvature of the manus and pes digits varies from concave toward or concave away from the midline. The most common morphology of *Attenosaurus* lacks a sole imprint. However, tracks with sole imprints have ovoid, round, or trapezoidal-shaped imprints, as is seen in *Dimetropus*. Footprints of *Attenosaurus* are among the largest tetrapod footprints of the Carboniferous, with pes lengths up to 25 cm.

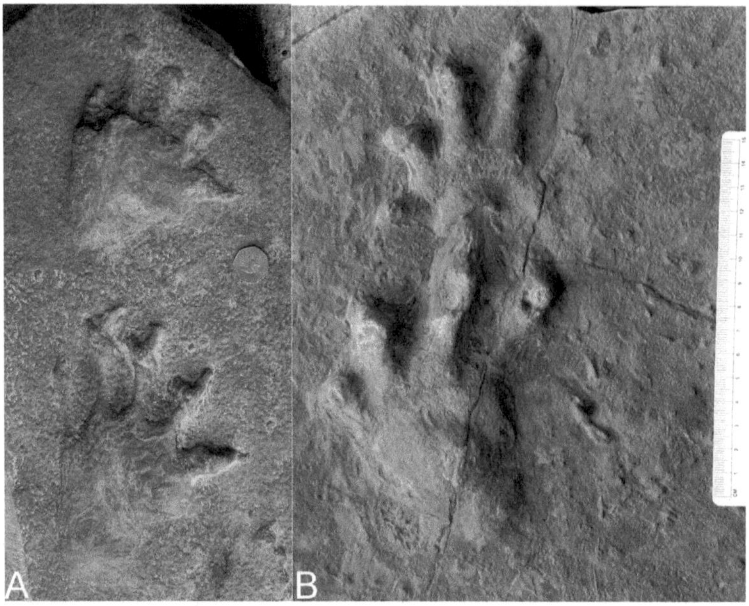

FIGURE 3.10 Selected examples of Carboniferous tetrapod-footprint ichnogenera. (A) *Dimetropus* (Late Pennsylvanian or early Permian, West Virginia) and (B) *Ichniotherium* (Late Pennsylvanian, UK).

FIGURE 3.11 *Hylopus hardingi* holotype (cast), Early Mississippian, Nova Scotia.

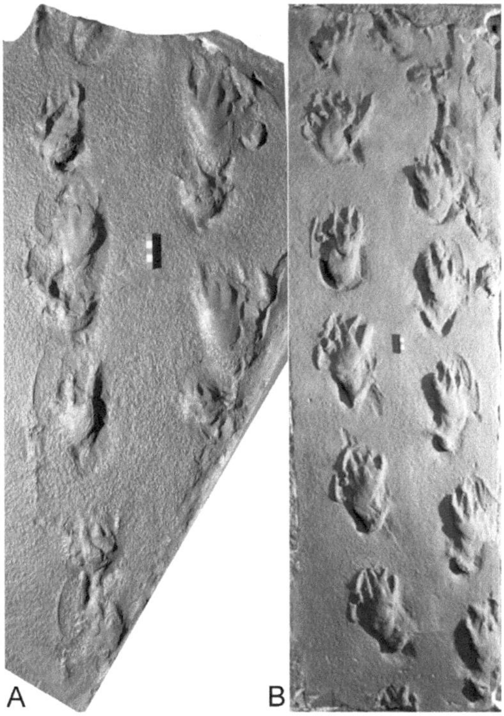

FIGURE 3.12 *Pseudobradypus unguifer*, holotype (A) and paratype (B).

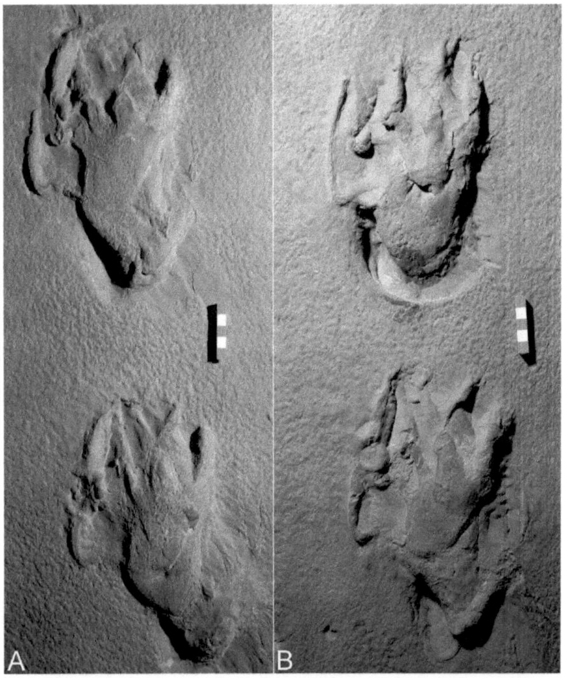

FIGURE 3.13 *Pseudobradypus unguifer*, paratype.

Temporal distribution: Early Mississippian-Middle Pennsylvanian.

Trackmaker: The trackmaker of *Attenosaurus* is generally considered to be a large reptiliomorph, likely an anthracosaur (Hunt et al., 2004; Haubold et al., 2005).

Comments: Aldrich (Aldrich and Jones, 1930, p. 13) named the ichnogenus *Attenosaurus* with two new ichnospecies, *A. indistinctus* and *A. subulensis*. The ichnospecies *A. indistinctus* is based on a convex, pentadactyl pes undertrack, whereas *A. subulensis* is based on a trackway. For that reason, Hunt et al. (2004) and Haubold et al. (2005) chose to consider *A. subulensis* as the type ichnospecies of *Attenosaurus* and to regard *A. indistinctus* as a junior subjective synonym based on an undertrack. However, Kuhn (1963, p. 9) had already designated *A. indistinctus* as the type ichnospecies, and we regard it as a senior synonym of *A. subulensis*.

Like Haubold et al. (2005), we recognize the problematic nature of the ichnotaxon *Attenosaurus*. The holotype is a single track, but it clearly represents a very large, pentadactyl track type that is easily recognized from other specimens. Thus, like Haubold et al. (2005), we believe that the range of variation in the sample of *Attenosaurus* from the Union Chapel Mine (UCM) in Alabama demonstrates a recognizable ichnogenus.

3.3 Batrachichnus Woodworth 1900 (Fig. 3.7A)

Type ichnospecies: *Batrachichnus plainvillensis* (Woodworth, 1900).

Valid ichnospecies: *Batrachichnus salamandroides* (Geinitz, 1861).

Characteristics: *Batrachichnus* is tracks of a quadruped with pentadactyl pes and tetradactyl manus imprints that are less than 40 mm long. Both imprints have indistinctly impressed proximal margins and distally rounded, straight digits. The manus imprints are as long as wide and have digits that are short, with digit III the longest, and digit IV shorter than digit II. Pes imprints are longer than wide, and digit length increases from I to IV, with digit V shorter than digit II. Pedal digit IV is about 1.5–2 times as long as the sole. Manus imprints are often slightly turned inward, and pes imprints are usually subparallel to the trackway midline. Tail/body traces may be present.

Temporal distribution: Mississippian-Middle Triassic.

Trackmaker: Tracks of this ichnogenus are attributed to temnospondyl or microsaur trackmakers and include the smallest known tetrapod footprints, with pes lengths of about 2–3 mm (Haubold, 1996; Voigt, 2005; Stimson et al., 2012; Allen et al., 2022).

Comments: The oldest record of *Batrachichnus* is in the Tournaisian strata at Blue Beach, Nova Scotia (Mansky and Lucas, 2013), and its youngest record is in the Triassic Moenkopi Formation of Arizona (Klein and Lucas, 2021). *Nanopus* clearly is a synonym (Hunt et al., 2005), and some salamander (lissamphibian) tracks are homeomorphic with *Batrachichnus* (Peabody, 1954,1959).

3.4 *Characichnos* Whyte and Romano 2001 (Fig. 3.8)

Type ichnospecies: *Characichnos tridactylus* (Whyte and Romano, 2001).

Valid ichnospecies: *Characichnos tridactylus* (Whyte and Romano, 2001).

Characteristics: Whyte and Romano (2001, p. 232) defined *Characichnos* as "two to four elongate, parallel hypichnial ridges (or epichnial grooves) which may be straight, gently curved or slightly sinuous. The termination of the ridges (or grooves) may be straight or slightly flexed."

Temporal distribution: The oldest records of *Characichnos* are from the Tournaisian strata at Blue Beach, Nova Scotia, and it has a long stratigraphic range through at least the entire Mesozoic.

Trackmaker: Originally attributed to a dinosaur, but subsequently attributed to diverse taxa, including temnospondyls (Lucas et al., 2010a).

Discussion: *Characichnos* is the most common ichnogenus of tetrapod "swim tracks," in being two to four, long, thin, well-spaced scratch marks (e.g., Lockley and Foster 2006). We treat it as a form ichnogenus applicable to the Carboniferous tetrapod swim tracks with which we are familiar. It is of behavioral and paleoecological significance, but of no value to biostratigraphy and biochronology.

3.5 *Cincosaurus* (Fig. 3.9A,B)

Type ichnospecies: *Cincosaurus cobbi* Aldrich (Aldrich and Jones, 1930).

Included ichnospecies: *Cincosaurus cobbi* Aldrich (Aldrich and Jones, 1930).

Characteristics: Quadrupedal trackway with semiplantigrade, pentadactyl manus, and pes. Digits are thin and pointed, and digit lengths increase from I to IV with digit V shorter and located backward and outward of the other digits. Pes slightly larger than manus (manus length = 15–40 mm). Pes digit III is longest, pes digits I–IV are usually curved inward, and digit V is separated to the side. Manus directed inward and pes parallel to the midline. Trackway narrow (width less than stride length) with pace angulation of manus imprints 100–120 degrees. This differs from *Notalacerata*, which has a relatively wide trackway and different digit proportions of the digits (Haubold, 1971; Schneck and Fritz, 1985; Haubold et al., 2005).

Temporal distribution: Early Pennsylvanian.

Trackmaker: Haubold et al. (2005) attributed *Cincosaurus* to an anthracosaur trackmaker.

Comments: Haubold et al. (2005, p. 96) regarded *Cincosaurus* from the Lower Pennsylvanian Union Chapel Mine locality in Alabama, USA, as a valid ichnogenus, but Lucas et al. (2022) tentatively suggested that it could not be distinguished from *Notalacerta*. However, here we treat *Cincosaurus* as a distinct ichnogenus noting that some material previously referred to it belongs to *Notalacerta*. Haubold et al. (2005) regarded many of the ichnotaxa named by Aldrich in Aldrich and Jones (1930) as junior synonyms of *Cincosaurus*, including the ichnogenera *Hydromeda, Limnosaurus, Quadropedia,* and *Triasaurus*.

3.6 *Dimetropus* Romer and Price 1940 (Fig. 3.10A)

Type ichnospecies: *Dimetropus berea* (Tilton, 1931).

Valid ichnospecies: *Dimetropus leisneranus* (Geinitz, 1863).

Characteristics: *Dimetropus* is footprints of a quadruped with pentadactyl manus and pes imprints that may be more than 200 mm long. The imprints have relatively short digits and proximally elongated palm and sole impressions. Digits increase in length from I to IV, and digit V is as long as digit III in the pes, whereas digit V is as long as digit II in the manus. The manus imprints are as long as wide, and the pes imprints are up to twice as long as wide. The imprints often preserve semispherical impressions of the metatarsal and metacarpal phalangeal pads. The digit tips are mostly pointed. The trackways have imprints either aligned parallel to the trackway midline or slightly turned outward. Tail/body traces are common.

Temporal distribution: Pennsylvanian-early Permian.

Trackmaker: *Dimetropus* was named for the inferred trackmaker *Dimetrodon*, but is more broadly regarded as the track of "pelycosaurs," excluding varanopids (Haubold, 1971; Voigt, 2005; Van Allen et al., 2005; Sacchi et al., 2014; Lucas et al., 2016).

Comment: We follow Voigt (2005) in recognizing only one ichnospecies of *Dimetropus*, but some other workers recognize at least two (e.g., Sacchi et al., 2014).

3.7 *Dromopus* Marsh 1894 (Fig. 3.5B)

Type ichnospecies: *Dromopus agilis* (Marsh, 1894) (Fig. 3.3).

Valid ichnospecies: *Dromopus lacertoides* (Geinitz, 1861).

Characteristics: *Dromopus* is the footprint of a digitigrade quadruped with pentadactyl manus and pes imprints as much as 80 mm long. The palm/sole is short, and the digits are long, slender, curved, and distally tapered. The footprint thus is described as lizard-like (lacertoid). Manus and pes imprints are very similar in shape. Both are about one third longer than wide, but the manus is about one fourth shorter than the pes imprint. Digit length increases from I to IV, with IV highly elongated, and digit V in a proximal position, laterally oriented and about as long as digit II. Marginal overstepping of the manus imprint by the pes imprint is common. Digit dragging and tail/body traces are seen only in tracks that were made on slippery ground.

Temporal distribution: Late Pennsylvanian—early Permian.

Trackmaker: *Dromopus* is interpreted as the footprint of various lizard-like parareptiles and eureptiles such as araeoscelids and bolosaurids (Haubold, 1971; Voigt, 2005; Voigt and Lucas, 2018). The FAD (first appearance datum) of *Dromopus* has long been used to define the beginning of the *Dromopus* biochron of Late Pennsylvanian—early Permian age (Lucas, 2003, 2007; Lucas et al., 2022).

3.8 *Hylopus* Dawson 1882 (Fig. 3.11)

Type ichnospecies: *Hylopus hardingi* (Dawson, 1882).

Valid ichnospecies: *Hylopus hardingi* (Dawson, 1882).

Characteristics: *Hylopus* is the footprints of a quadruped in which the manus and the pes are pentadactyl, though many referred specimens have a tetradactyl manus, likely due to the impression of digit V not being preserved. The pes is larger than the manus, both manus and pes are generally plantigrade, and the manus-pes may be overstepped. The manus digits are often relatively thin (but thicker than in *Notalacerta*) and curved medially, and digit IV is much longer than the others (digit IV > digit III > digit V > digit II). In the pes, the digits are straight to slightly curved medially and are more nearly of subequal length than in the manus (digit II > digit I = digit III > digit IV > digit V). In trackways, manus and pes prints are closely spaced, sometimes overstepped, divarication of the manus and pes to the midline is 15—30 degrees and body/tail drags are sometimes present (Sundberg et al., 1990; Fillmore et al., 2012; Marchetti et al., 2019).

Temporal distribution: Early Mississippian—Middle Pennsylvanian.

Trackmaker: The trackmaker of *Hylopus* was long considered to be an amphibian, likely a temnospondyl, primarily based on the idea that it had a tetradactyl manus (e.g., Haubold, 1971; Sundberg et al., 1990; Fillmore et al., 2012). However, specimens of *Hylopus* in which the manus is clearly pentadactyl are considered to be those of a reptiliomorph trackmaker (Marchetti et al., 2019; Lucas, 2019). The specimens with a tetradactyl manus are thus extramorphological variants in which the digit V imprints did not register.

3.9 *Ichniotherium* Pohlig 1892 (Fig. 3.10B)

Type ichnospecies: *Ichniotherium cottae.*

Valid ichnospecies: *I. cottae* and *I. sphaerodactylum* (Pabst, 1895).

Characteristics: *Ichniotherium* is footprints of a quadruped with pentadactyl manus and pes imprints that have enlarged, blunt digit tips. Pes imprints are up to 200 mm in length, about as long as wide, and the sole has a large, oval- to circular-shaped pad. The manus imprints are wider than long and about one-fifth shorter than the pes imprints. Digit length increases from I to IV, with digit V shorter than digit II or as long as digit III, depending on the ichnospecies. Trackways have pes imprints rotated outward and manus imprints subparallel to the midline or slightly to moderately rotated inward. Overstepping may be present, and tail/body traces are very rare and always discontinuous.

Temporal distribution: Early Pennsylvanian—early Permian.

Trackmaker: Based on the association of body and trace fossils in the early Permian Tambach Formation of central Germany, *Ichniotherium* is assigned with a high degree of certainty to diadectomorph trackmakers (Voigt, 2005; Voigt et al., 2007; Francischini et al., 2019; Mujal and Marchetti, 2020).

3.10 *Limnopus* Marsh, 1894 (Fig. 3.7C)

Type ichnospecies: *Limnopus vagus* (Marsh, 1894).

Valid ichnospecies: *Limnopus heterodactylus* (King, 1845a) and *L. primaevus* (Lea, 1849).

Characteristics: *Limnopus* is the footprints of quadrupeds with pentadactyl pes and tetradactyl manus imprints. The footprint morphology is essentially identical to the smaller footprint *Batrachichnus,* except for the fourth digit of the manus imprint, which is about as long as digit II in *Limnopus,* but about as long as digit I in *Batrachichnus*. The digits are also relatively thicker and shorter compared to the sole impression in *Limnopus* than in *Batrachichnus*. Total pes imprint length may reach up to 200 mm.

Temporal distribution: Early Mississippian—early Permian.

Trackmaker: The trackmakers of *Limnopus* are considered to be relatively large temnospondyls such as eryopids (Baird, 1952; Haubold, 1971; Voigt, 2005; Voigt and Lucas, 2015, 2018).

Discussion: Lucas et al. (2010a) argued for the distinctiveness of *Limnopus* and *Palaeosauropus*. But, having recently studied the large sample of such tracks from Blue Beach, Nova Scotia, we are convinced that the distinction of these ichnotaxa is subtle and should be made at the ichnospecific, not ichnogeneric level. Thus, we recognize two ichnospecies of *Limnopus, L. heterodactylus* (includes *L. vagus* and the other named ichnospecies, cf. Baird,

1952) and *L. primaevus*, the type ichnospecies of *Palaeosauropus*. Lucas et al. (2022) tentatively regarded *Baropezia* as a valid ichnotaxon, but it more likely is a synonym of *Limnopus*.

3.11 *Matthewichnus* (Haubold, 1970 (Fig. 3.7B))

Type ichnospecies: *Matthewichnus velox* (Matthew, 1905).

Valid ichnospecies: *Matthewichnus velox*, possibly *M. caudifer* (Kohl and Bryan, 1994).

Characteristics: *Matthewichnus* is the footprint of a quadruped in which the pes is much larger than the manus. Both manus and pes are ectaxonic and semiplantigrade, and the manus is tetradactyl and approximately as wide as long. The pes is pentadactyl and has digits II–IV longer than digit V. The manus imprint is medial to and sometimes overstepped by the pes imprint. Tail/body imprints are sometimes present. The manus is essentially identical to the manus of *Batrachichnus*: tetradactyl, short and relatively thick, blunt-tipped digits, and a rounded sole. However, the pes is very distinctive: pentadactyl with long and thin digits with pointed tips.

Temporal distribution: Early Mississippian–early Permian.

Trackmaker: A microsaur or a temnospondyl amphibian are generally identified as the trackmaker of *Matthewichnus* (e.g., Haubold, 1971; Haubold et al., 2005; Fillmore et al., 2012; Voigt and Lucas, 2015).

Discussion: All published records of *Matthewichnus* prior to Fillmore et al. (2012) were of Early Pennsylvanian age, and the Mauch Chunk Formation record extended the temporal range of *Matthewichnus* back to the Late Mississippian. Stimson et al. (2018, 2019) illustrated an example of *Matthewichnus* from the early Tournaisian Albert Formation of New Brunswick, further extending the ichnotaxon's range to the Early Mississippian. Voigt and Lucas (2015) first reported Permian records of *Matthewichnus* from the lower Permian Robledo Mountains Formation in New Mexico, and there are also new, unpublished records of the ichnogenus from the lower Permian Abo Formation in central New Mexico.

3.12 *Notalacerta* Butts 1891 (Fig. 3.9C,D)

Type ichnospecies: *Notalacerta missouriensis* (Butts, 1891).

Valid ichnospecies: *Notalacerta missouriensis* (Butts, 1891).

Characteristics: *Notalacerta* is the footprints of a quadruped with pentadactyl manus and pes imprints that are 15–30 mm long. The digit imprints are relatively long, very slender and terminate in pointed claw impressions. The digit tips may be bifurcated, and the digit imprints increase in length from digit I to digit IV, with pes digit V imprint about as long as digits II–III, and manus digit V imprint about as long as digit II. Digit I–IV imprints are usually distally bent inwards, and the digit V imprint is straight. The manus is about as long as wide, and the pes is about as long as wide or slightly wider than long. Palm and sole impressions are short and shallow, with a concave proximal margin in the manus and a convex proximal margin in the pes. Manus in front, lateral, or medial compared to the pes. In trackways, overstepping is rare, and the pes is parallel to the midline or turned outwards, and the manus is parallel to the midline or slightly turned inwards. Straight and continuous tail/body impressions are common (see Marchetti et al., 2020c).

Temporal distribution: Early Pennsylvanian—early Permian.

Trackmaker: *Notalacerta* is interpreted to be the footprints of early reptiles, such as protorothyridids (Marchetti et al., 2020c).

Discussion: The type specimen of the type ichnospecies of *Notalacerta* has apparently been lost, but Marchetti et al. (2020c) have designated a neotype and demonstrated the validity of *Notalacerta*.

3.13 *Pseudobradypus* (Matthew, 1903a (Figs. 3.12 and 3.13))

Type ichnospecies: *Pseudobradypus unguifer* (Dawson, 1872).

Valid ichnospecies: *Pseudobradypus unguifer* (Dawson, 1872).

Characteristics: Tracks of a quadruped in which the manus and pes are plantigrade and pentadactyl with relatively short, pointed digits. Manus with short and broadly rounded heel; pes heel narrowly round, and the sole is at least half the length of the pes impression; on the pes, digits II—V are anteriorly directed and not splayed, whereas digit I is posterior to the other digits and may point medially. Pes digit III > II > IV > V. The manus digits are more splayed than the pes digits and all point anteriorly. The manus and pes imprints are directed forward and sometimes overstepped. A body/tail drag is common. A long claw drag from pes digit V also is common (modified from Fillmore et al., 2012, p. 83).

The distinct differences between *Dimetropus* and *Pseudobradypus* are as follows: *Pseudobradypus* has relatively longer digits than *Dimetropus*; digit V is often divergent in *Pseudobradypus* and often seems only as a long, thin arcuate imprint; and *Dimetropus* tracks are relatively wide immediately behind the digits, so the tracks are triangular or trapezoidal with the most acute apex posterior, whereas those of *Pseudobradypus* have more parallel sides. *Dimetropus* characteristically has a pes digit IV longer than the others, digit I is not posteriorly located but instead is part of an arc that includes all five digits, the manus and pes are much wider than long and the sole impression is broadly triangular. Overstepping and tail drags are very rare in *Dimetropus* (e.g., Haubold and Sarjeant, 1973, Fig. 2; Gand and Haubold, 1984, Fig. 11; Hunt et al., 1995, Figs. 7 and 8; Voigt, 2005, Figs. 33 and 36; Sacchi et al., 2014, Figs. 9 and 12; Lucas et al., 2016, Figs. 3 and 5). Indeed, long trackways of *Dimetropus*, comparable to the holotype trackway of *P. unguifer*, are readily distinguished, particularly in lacking overstepping and tail drags.

Temporal distribution: Early Mississippian-Early Pennsylvanian.

Trackmaker: Haubold (1971) classified *Pseudobradypus* as a "captorhinomorph" track (also see Falcon-Lang et al., 2007), but its Mississippian records predate the lowest occurrences of amniote body fossils and footprints, so it is more likely a reptiliomorph footprint (Fillmore et al., 2012). Furthermore, Marchetti et al. reinterpreted the *Pseudobradypus* tracks described by Falcon-Lang et al. (2007) as *Hylopus*, an anamniote trackway.

Discussion: The holotype of *Pseudobradypus unguifer* is CMN (Canadian Museum of Nature) 4630 (Fig. 12; Sternberg 1933, pl. 35, Figs. 1 and 2; Harington et al., 2005, photograph on p. 56) and the paratype is CMN 4629 (Fig. 13; Sternberg, 1933, pl. 35, Fig. 3; Harington et al., 2005, photograph on p. 58). Both are from the Pennsylvanian strata at River Phillip, Nova Scotia.

Voigt and Ganzelewski (2010, p. 66) referred to "the ambiguity of the *Pseudobradypus* type material" because it "is lacking crucial details of the imprint morphology," largely due to the overstepping of pes and manus. However, we see these tracks as preserving many significant details that allow the ichnotaxon to be diagnosed. Nevertheless, Marchetti et al. (2019, p. 31) stated that *Pseudobradypus unguifer* was "defined based on poorly preserved material" and considered the taxon a nomen dubium. This is very questionable, as the holotype trackway of *P. unguifer* (CMN 4630: Fig. 3.12A) is a trackway of 11 footprints that include seven consecutive manus-pes couples (L132, R1-4) in convex hyporelief, the most extensive trackway that served as a holotype from the Pennsylvanian strata of Nova Scotia. Furthermore, the paratype of *P. unguifer* (Figs. 3.12 and 3.13) is an even longer and better-preserved trackway. If these specimens are "poorly preserved material" then almost all of the holotypes of late Paleozoic tetrapod footprints are more poorly preserved, as few are as extensive a trackway with as clearcut preservation as the holotype and paratype of *P. unguifer* (cf. Haubold, 1971, 1984; Fillmore et al., 2012 and many other references).

Thus, Marchetti et al. (2019, p. 33) referred to the holotype of *Pseudobradypus unguifer* as cf. *Dimetropus* isp., claiming that "the manus morphology is completely blurred, so it is impossible to assess with certainty the numbers and proportions of digit impressions." They evaluated this trackway as poorly preserved, grade 0.5–1.5 in their classification scheme, and claimed that "the manus does not show any features useful for ichnotaxonomic assignments" (p. 122). However, the paratype shows the clear morphology of the manus of *P. unguifer*, including a manus that is much smaller than the pes and has five digits, of which digit III is longest. The manus is much smaller than the pes, has a much shorter sole relative to digit length, and is wider than long. In both the manus and pes, digits III and IV are of similar length, though one or the other is often slightly longer. Digit V is slightly lateral, and overstepping is common. Divarication of the digits is relatively narrow, 45–60 degrees. In *Dimetropus*, digit IV is distinctly longest, the manus is usually longer than wide and both manus and pes are trapezoidal to triangular, with the most acute angle posterior. This is because the sole pad right behind the digits is widest. Divarication is high, up to 100 degrees.

Furthermore, Marchetti et al. (2019) supported assignment of the holotype of *P. unguifer* to cf. *Dimetropus* based primarily on the morphology of the pes, claiming close similarity to a specimen of *Dimetropus* from the lower Permian of Germany originally described by Voigt (2005) and illustrated by Marchetti et al. (2019, Fig. 3.5C). However, this specimen of *Dimetropus* shows some of the key features of *Dimetropus* identified here that distinguish it and *Pseudobradypus*, including pes digit IV longest, relatively long digits, digits directed outward and a broad, triangular sole impression.

3.14 *Varanopus* (Moodie, 1929 (Fig. 3.5C))

Type ichnospecies: *Varanopus curvidactylus* (Moodie, 1929).

Valid ichnospecies: *Varanopus curvidactylus* (Moodie, 1929) and *Varanopus microdactylus*.

Characteristics: Small (25–36 mm) quadrupedal trackways with semiplantigrade, pentadactyl manus, and pes with a manus that is slightly smaller than the pes arranged into manus-pes couplets with the pes impression oriented parallel to the midline and the manus oriented slightly toward the midline. The pes is as wide as long, while the manus is slightly wider than long. Digits are long and slender with tapered digit terminations sometimes with triangular claw impressions that curve toward the midline. Digits I–IV curve slightly toward the midline and are closely grouped while digit V is straight and directed distally lateral. Digits increase in length from digits I–IV with digit IV being longest in both manus and pes, and digit V being slightly shorter than digit II in the manus and slightly longer than digit II in the pes. Pace angulations vary from 74 to 103 degrees

Temporal distribution: Late Pennsylvanian–early Permian.

Trackmaker: Moodie (1929) originally identified an ophiacodontid synapsid as the trackmaker of *Varanopus*. Haubold (1971) favored a captorhinomorph tracemaker, and Haubold and Lucas (2003) suggested both captorhinomorphs and protorothyridids for *V. curvidactylus.* Marchetti et al. provided detailed analysis of skeletal remains for potential tetrapods and concluded that bolosaurians or acleistorhinids are among the most probable trackmakers.

Comments: Only two bolosaurs are known from Carboniferous body fossils, an unnamed taxon in the late Virgilian Bursum Formation of New Mexico (Harris et al., 2004; Falconnet, 2012), and the other (*Erpetonyx arsenaultorum*) at the Virgilian-Wolfcampian boundary of Prince Edward Island, Canada (Maritimes Basin) (Modesto et al., 2015). The oldest records of *Varanopus* are the same age (within resolution) as these body fossils, though most records of the ichnogenus are early Permian.

4. Distribution of Carboniferous tetrapod footprints

4.1 Introduction

Carboniferous tetrapod footprints have an almost strictly Euramerican distribution (Fig. 3.1), and Lucas et al. (2022) provided a detailed review of the distribution of Carboniferous footprints (Fig. 3.1). Here, we reprise that review.

Lucas (2019) listed five extensive footprint ichnoassemblages of Carboniferous age, and we add one more to that to make six, all from North America.

1. Blue Beach (also called Horton Bluff) in Nova Scotia, Canada (Fig. 3.14) encompasses thousands of footprints in the Early Mississippian (Tournaisian) Blue Beach and Hurd Creek members of the Horton Bluff Formation (Mansky and Lucas, 2013; Lucas and Mansky, in prep.). These strata were long considered to be strictly lacustrine in origin (i.e., a freshwater lake: Martel and Gibling, 1991), but later the depositional system was understood to have been at least periodically brackish; either a sizeable lagoon or a restricted marine embayment with periodic connections to the sea (Tibert and Scott, 1999). The tetrapod footprints at Blue Beach are associated with a diverse invertebrate ichnoassemblage of the *Cruziana* ichnofacies (Lucas and Mansky, in prep.). Tetrapod footprint ichnogenera from Blue Beach are *Attenosaurus, Batrachichnus, Characichnos, Hylopus, Limnopus,* and *Pseudobradypus.* This is the oldest extensive tetrapod footprint ichnoassemblage.

FIGURE 3.14 Overview of part of the Blue Beach (Horton Bluff) Lower Mississippian tracksite, Nova Scotia.

2. Mauch Chunk Formation outcrops in eastern Pennsylvania, USA (Fig. 3.15), are of late Visean age and are the oldest extensive, fluvial red-bed footprint ichnoassemblages (Fillmore et al., 2012). Tetrapod footprint ichnogenera are *Batrachichnus, Characichnos, Hylopus, Limnopus, Matthewichnus,* and *Pseudobradypus.*
3. In the sea cliffs at Joggins, Nova Scotia, Canada, an ~ 20 m thick interval of the Early Pennsylvanian Joggins Formation contains extensive footprints (Figs. 3.16 and 3.17). These were the subject of many early (and now invalid) ichnotaxonomic names and mostly belong to *Batrachichnus, Limnopus, Matthewichnus,* and *Notalacerta.*
4. At the Union Chapel Mine in Alabama (Fig. 3.18), the Early Pennsylvanian Pottsville Formation contains abundant tetrapod footprints and other traces (Haubold et al., 2005; Buta and Kopaska-Merkel, 2016). These belong to *Attenosaurus, Batrachichnus, Characichnos, Cincosaurus, Matthewichnus,* and *Notalacerta.*
5. The Crescent Valley Mine in Alabama has an ichnoassemblage very similar to that of the Union Chapel Mine and is also from the Pottsville Formation and of Early Pennsylvanian age (Buta et al., 2013).
6. At Keota in eastern Oklahoma, USA (Fig. 3.19), the Middle Pennsylvanian McAlester Formation contains an extensive, but low diversity footprint ichnoassemblage (Lucas et al., 2004). This is an ichnoasemblage of abundant *Batrachichnus* and *Notalacerta* and rare *Amphisauropus.*

4.2 Carboniferous tetrapod footprint record of the Maritimes Basin

The most extensive Carboniferous tetrapod footprint record comes from the Atlantic Canadian Maritimes Basin in Nova Scotia, Prince Edward Island, and New Brunswick that accumulated sediment from the Late Devonian to the early Permian. In the provinces of New Brunswick, Nova Scotia, and Prince Edward Island, tetrapod footprint assemblages

FIGURE 3.15 Two views of track-producing strata of the Upper Mississippian Mauch Chunk Formation in eastern Pennsylvania. (A) Lea's original type locality of *Palaeosauropus primaevus*. Note steeply dipping strata to left of person. (B) Steeply dipping Mauch Chunk strata at the Tumbling Run Reservoir.

of Mississippian (Horton and Mabou groups), Pennsylvanian (Cumberland, Morien, and Pictou groups), and early Permian (Pictou Group) ages can be superposed to provide a footprint record spanning the entire Carboniferous System (Figs. 3.20 and 3.21).

Although trackway assemblages are found throughout the Maritimes Basin, of particular interest is the nearly continuous stratigraphic succession in Nova Scotia. Most notable of these are the Lower Mississippian interval at Blue Beach, Nova Scotia, which sits within "Romer's gap" (Hunt et al., 2004; Mansky and Lucas, 2013); Lower Pennsylvanian trackway assemblages found at the UNESCO World Heritage Site at Joggins (e.g., Matthew 1903b; Calder, 1998, 2012; Stimson et al., 2012; Lucas et al., 2022); and the Carboniferous-Permian boundary tracksite at Brule (Van Allen et al., 2005).

Extensive work on the Maritimes Basin trackway record that began with Dawson (1844, 1845, 1863a, b, 1868, 1872, 1882, 1894, 1895), Matthew (1903a, b, c, 1904, 1905, 1910) and Sternberg (1933) produced the most complete single stratigraphic succession of Carboniferous footprints globally (Lucas, 2003,2007; Lucas et al., 2022). It encompasses footprint assemblages that range in age from earliest Mississippian to near the onset of the Permian, and we continue to consider it the standard for understanding the temporal succession of Carboniferous tetrapod footprint ichnotaxa (Lucas, 2003,2007; Lucas et al., 2022).

FIGURE 3.16 Overview of the principal track-producing interval of the Joggins Formation at the Joggins UNESCO World Heritage Site in Nova Scotia. Numbers at bottom are meter levels that correspond to those in Fig. 3.17.

4.3 Mississippian tetrapod footprints

Only two substantial tetrapod footprint assemblages of Mississippian age are known: from the Lower Mississippian (Tournaisian) Blue Beach and Hurd Creek members of the Horton Bluff Formation in Nova Scotia, Canada, and the Upper Mississippian (Serpukhovian) Mauch Chunk Formation in eastern Pennsylvania, USA. Other Mississippian-age footprint records are mostly single tracks or trackways and were listed by Lucas et al. (2022). Mississippian footprints are those of temnospondyls, "microsaurs" and reptiliomorphs. They belong to the ichnogenera *Attenosaurus, Batrachichnus, Characichnos, Hylopus, Limnopus,* and *Pseudobradypus.*

4.4 Pennsylvanian tetrapod footprints

Cotton et al. (1995), Hunt et al. (1995), Lucas (2003, 2007, 2019), and Lucas et al. (2022) reviewed the record of Pennsylvanian tetrapod footprints. Most of that record is in eastern and southwestern North America and in western Europe, and there are only a handful of sites elsewhere. As already noted, there are four extensive tetrapod footprint assemblages of Pennsylvanian age: Joggins, Nova Scotia, Canada; the Union Chapel Mine (UCM) and Crescent Valley Mine, Alabama, USA; and Keota, Oklahoma, USA (Figs. 3.16–3.19). The Joggins and UCM/Crescent Valley records are Early Pennsylvanian, whereas the Keota record is of Middle Pennsylvanian age (Lucas, 2019). Lucas et al. (2022) also outlined several, as yet undescribed, Pennsylvanian trackway assemblages from New Brunswick, Canada.

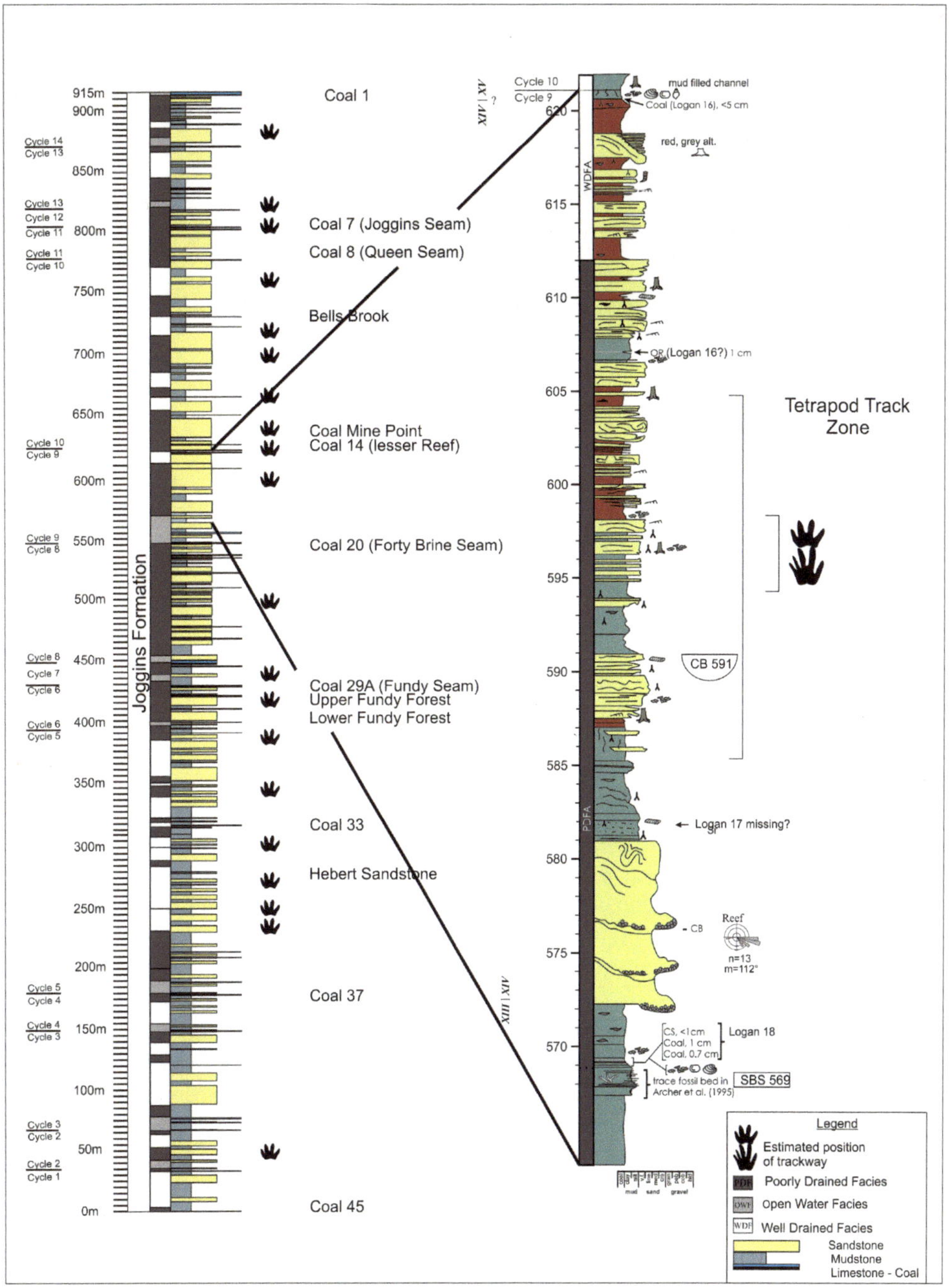

FIGURE 3.17 Measured stratigraphic section of the Joggins Formation at the Joggins UNESCO World Heritage Site, showing the tetrapod-footprint-bearing-interval.

FIGURE 3.18 Overview of the Union Chapel Mine, Alabama, developed in the Lower Pennsylvanian Pottsville Formation. (A) View of unexcavated highwall next to mined area. (B) Mined area, partially reclaimed; all the trace fossils were collected from the spoil piles of the mined area.

In Nova Scotia, the principal Pennsylvanian footprint-bearing strata are in the 932-m-thick Joggins Formation of Early Pennsylvanian (Morrowan/Bashkirian) age (Figs. 3.16 and 3.17). The record includes older, Lower Pennsylvanian strata of the upper Boss Point and Little River formations as well as younger, Lower Pennsylvanian strata of the overlying Springhill Mines and Ragged Reef formations. In total, these strata comprise more than ~4300 m of fossiliferous strata at the Joggins UNESCO World Heritage Site. Tetrapod tracks have been documented from a range of sedimentary facies, including well-drained red-bed and poorly drained gray, coal-bearing lithofacies assemblages. An ~ 20-m-thick interval (585–605 m of the Joggins Formation, see Calder, 1998, 2012; Davies et al., 2005; Stimson et al., 2012), is known to be one of the most footprint-productive intervals in the formation (Figs. 3.16 and 3.17) and has yielded examples of most of the known ichnotaxa from Joggins. Other Canadian Pennsylvanian tracksites from the Maritimes Basin were listed by Lucas et al. (2022).

In Alabama, Aldrich (in Aldrich and Jones, 1930) described several footprint ichnogenera, including their new ichnotaxa *Attenosaurus* and *Cincosaurus*, from the Lower Pennsylvanian Pottsville Formation of Alabama. Extensive footprint assemblages in the Pottsville Formation come from the Union Chapel Mine (Fig. 3.18) and the Crescent Valley Mine (Buta et al., 2013).

The Keota locality in eastern Oklahoma (Fig. 3.19) is the most extensive Middle Pennsylvanian footprint assemblage. The locality produces a prolific but low ichnodiversity footprint

FIGURE 3.19 Views of the Keota tracksite in the Middle Pennsylvanian McAlester Formation in eastern Oklahoma. (A,B) Overviews of the excavation for footprints. (C,D) Closeup views of the thin bedded, track-bearing strata.

assemblage that consists of abundant *Batrachichnus* and *Notalacerta* with rare *Amphisauropus* (Lucas et al., 2004; Marchetti et al., 2020c).

Other Pennsylvanian footprint records in the eastern United States are scattered, less extensive (often single footprints) and come from the states of Massachusetts, Rhode Island, Pennsylvania, Ohio, Indiana, Virginia, West Virginia, Kentucky, Tennessee, and Georgia (see Cotton et al., 1995; Lucas et al., 2022). Many ichnogeneric names have been applied to these footprints, but they mostly pertain to the ichnogenera *Batrachichnus, Dromopus, Hylopus, Limnopus, Matthewichnus, Notalacerta,* and *Pseudobradypus.* Scattered occurrences of Pennsylvanian tetrapod footprints are also known from the western United States (Wyoming, Colorado, Arizona, Oklahoma, Kansas, and Missouri). They are mostly assigned to *Batrachichnus, Limnopus,* and *Notalacerta* (see Hunt et al., 1995; Lucas et al., 2022).

From the Grand Canyon of northern Arizona, Rowland and Caputo (2018), and Rowland et al. (2020) reported tetrapod footprints from the Atokan (late Bashkirian—early Moscovian) Manakacha Formation. These footprints are so poorly preserved that any ichnogeneric assignment is questionable, but they are the oldest tetrapod footprints from an eolianite. Another important ichnoassemblage, from the Virgilian (Gzhelian) Wescogame Formation in the Grand Canyon of Arizona, includes cf. *Amphisauropus, Batrachichnus,* cf. *Limnopus* and *Varanopus* (Gilmore, 1927; Haubold, 1971; Marchetti et al., 2020a). It may also be one

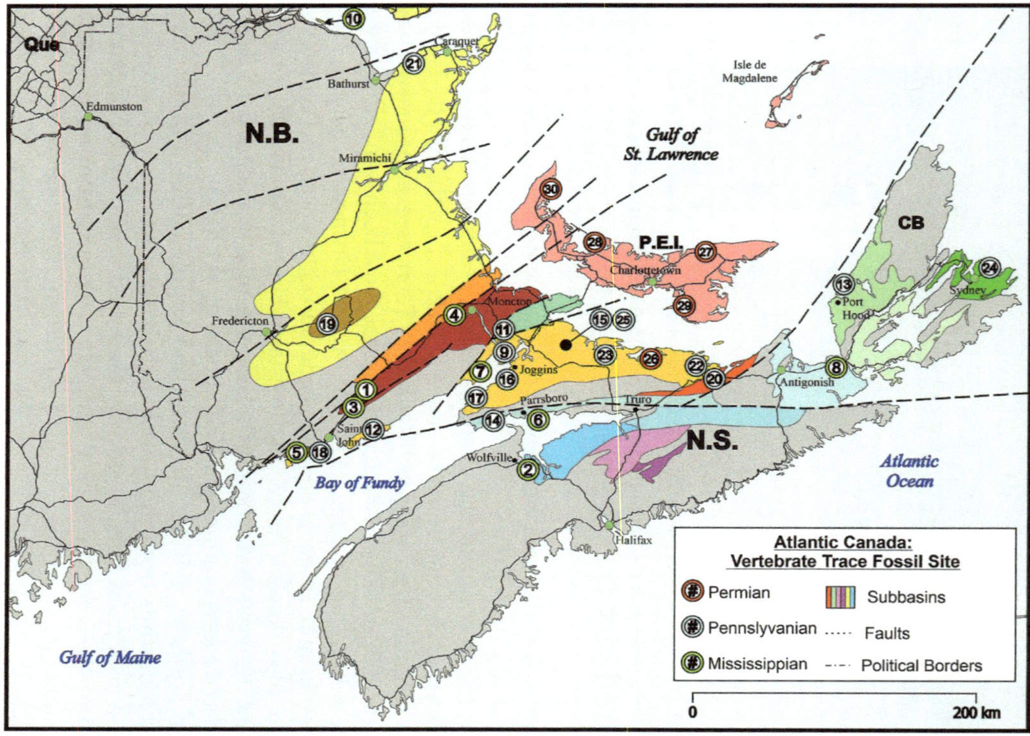

Map #	Location	Prov.	Age	Subbasin
			Permian	
30	Cavendish	PEI	Artinskian	PEI
29	Point Prim	PEI	Sakmarian	PEI
28	Malpeque	PEI	Asselian	PEI
27	Egmont Bay	PEI	Kasimovian - Gzhelian	PEI
26	Brule	NS	C/P Boundary	Cumberland
			Pennsylvanian	
25	Wentworth St.	NS	Westphalian D	Cumberland
24	Sydney Mines Fm.	NS	Westphalian D	Sydney
23	River Phillip	NS	Westphalian C-D	Cumberland
22	Abercrombie	NS	Westphalian C	Cumberland
21	Clifton	NB	Westphalian C-D	New Brunswick Platform
20	Stellarton	NS	Westphalian C	Stellarton
19	Grand Lake	NB	Westphalian C	Marysville
18	Fern Ledges	NB	Westphalian A-B	West Cumberland
17	Spicer's Cove	NS	Westphalian A-B	Cumberland
16	Springhill Mines, Joggins, Little River Fm.	NS	Westphalian A	Cumberland
15	Wentworth St.	NS	Westphalian A	Cumberland
14	Parrsboro Fm.	NS	Westphalian A	Debert-Kempton
13	Port Hood	NS	Westphalian A	Western Cape Breton
12	Tyemouth Creek	NB	Westphalian A	West Cumberland
11	Johnson's Mills	NB	Westphalian A	Cumberland
			Mississippian	
10	Heron Island	NB	? Visean to Serpukhovian ?	Restigouche Area
9	Marigouin Pennusula	NB	Serpukhovian	Cumberland
8	Riley's Beach	NS	Serpukhovian	Antigonish
7	Munson Pond	NB	Visean / Serpukhovian Boundary	Cumberland
6	West Bay	NS	Visean	Debert-Kempton
5	Lepreau Falls	NB	Visean	West Cumberland
4	Stilesville	NB	Tournaisian	Moncton
3	Bloomfield	NB	Tournaisian	Moncton
2	Horton Bluff	NS	Tournaisian	Windsor-Kennetcook
1	Norton	NB	Tournaisian	Moncton

FIGURE 3.20 Map of New Brunswick, Nova Scotia and Prince Edward Island showing Carboniferous tetrapod footprint localities in the Canadian Maritimes Basin. *Map lines delineate study areas and do not necessarily depict accepted national boundaries.*

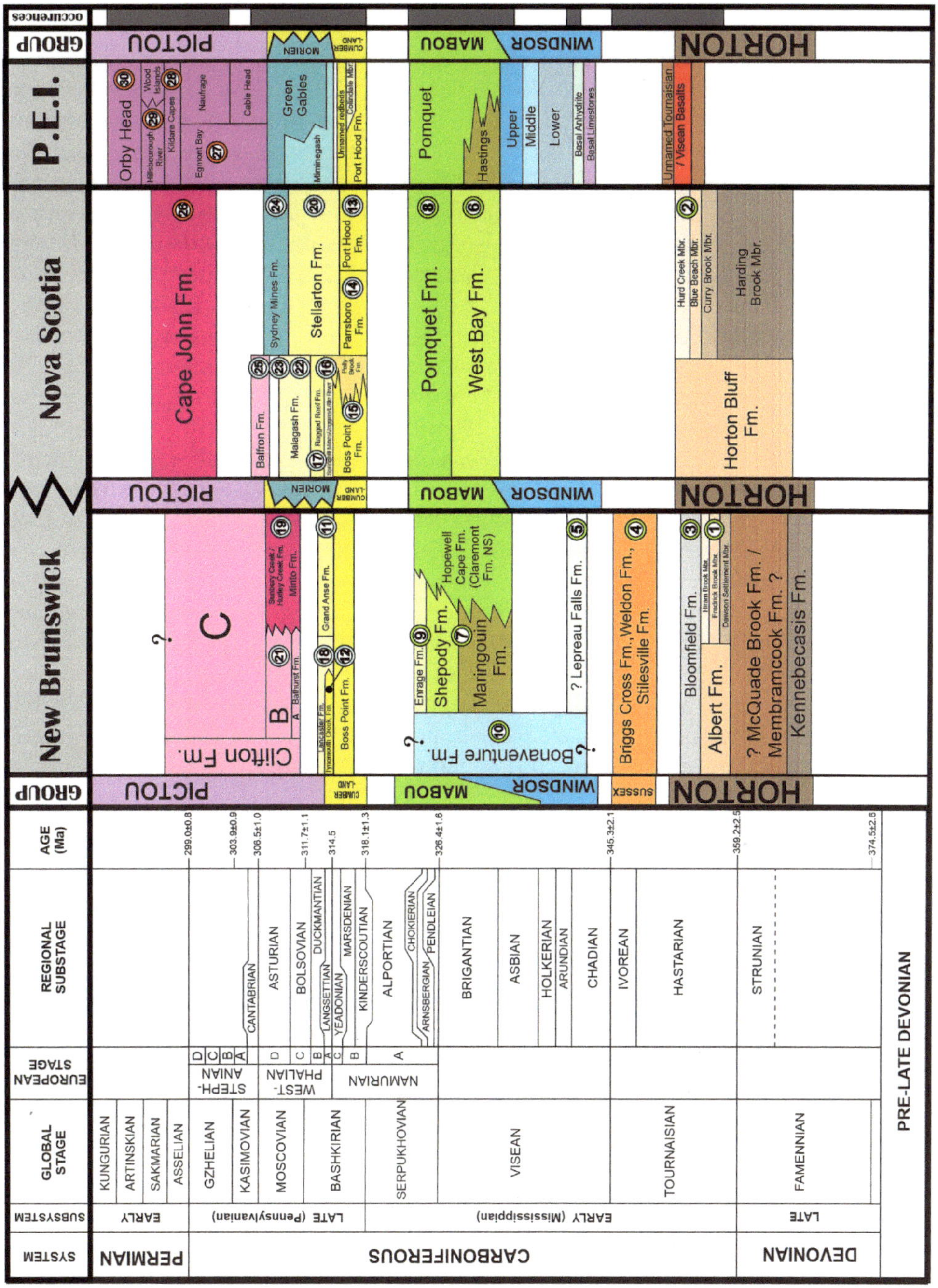

FIGURE 3.21 Correlation of Carboniferous tetrapod footprint localities in the Canadian Maritimes Basin. Locality numbers correspond to those in Fig. 3.20.

of the oldest tetrapod footprint assemblages from eolian strata. Footprints from the Virgilian (Gzhelian) Howard Limestone in Kansas include the type material of *Dromopus* (Fig. 3.3) and *Limnopus* (Marsh, 1894; Baird, 1952; Schoewe, 1956).

European Pennsylvanian tetrapod footprints come principally from Great Britain, France, and Germany, but there are also some records in Spain, Italy, Poland, and the Czech Republic (see Lucas et al., 2022). Lower-Middle Pennsylvanian tracksites are more common and are found in France, Germany (especially in the Ruhr Basin and in Sachsen) and the Czech Republic. These tracksites contain many of the same ichnogenera as their North American correlatives (e.g., *Batrachichnus, Limnopus, and Pseudobradypus*) and thus preserve a mixture of amphibian and reptiliomorph footprints. They also include the oldest records of the ichnogenera *Dimetropus* and *Ichniotherium* (Voigt and Ganzelewski, 2010). In Europe, Stephanian (approximately Late Pennsylvanian) tracksites are in England, Spain, France, Germany, and Italy. These are mostly records of *Batrachichnus, Dimetropus, Dromopus,* and *Limnopus.*

Outside of Euramerica, there have been only a few reports of Pennsylvanian tetrapod footprints, some of which are not tetrapod footprints or not of Pennsylvanian age: (1) northern Chile, temnospondyl footprints (Bell and Boyd, 1986) from a locality recently relocated that are being restudied; (2) the ichnogenus *Orchesteropus* from the Carboniferous of Argentina, originally considered tetrapod footprints (Frenguelli, 1950) but re-evaluated to be the footprints of arthropods (xiphosurans) (e.g., Aceñolaza and Buatois, 1993); (3) incomplete footprints identified as aff. *Dimetropus* from Shanxi, China (Lucas and Ataabadi, 2010) that were of questionable stratigraphic provenance but now appear to be of Permian age (Liu et al., 2020); (4) *Limnopus* from the Westphalian interval of the Vig Formation in Greenland (Milàn et al., 2016); (5) an ichnoassemblage from the Upper Pennsylvanian of the Souss basin of south-central Morocco consisting of footprints identified as *Batrachichnus, Dimetropus, Ichniotherium,* and *Limnopus* (Hmich et al., 2006; Lagnaoui et al., 2018); and (6) footprints assigned to cf. *Batrachichnus, Dimetropus,* cf. *Hylopus,* and *Notalacerta* from Westphalian strata in the Sidi Kassem basin of central Morocco (Lagnaoui et al., 2014).

5. Carboniferous footprint biochronology

5.1 Introduction

The use of fossil tetrapod footprints to subdivide Carboniferous time dates back to the late 1800s, when Marsh (1891, 1898) divided the North American Carboniferous into the "*Sauropus* beds" and "*Eosaurus* beds" (Fig. 3.22). For more than a century, no worker followed up on Marsh's use of a tetrapod footprint ichnotaxon to identify an interval of Carboniferous time. Thus, in his comprehensive reviews of the tetrapod footprint record, Haubold (1971, 1984) generally assigned Carboniferous tetrapod footprint assemblages to Mississippian and Pennsylvanian time intervals (Figs. 3.22 and 3.23). Lucas (2003, 2007) first presented a Carboniferous tetrapod footprint biochronology that recognized three intervals of time: Mississippian, Early-Middle Pennsylvanian (named the *Pseudobradypus* biochron), and Late Pennsylvanian (named the *Dromopus* biochron). Fillmore et al. (2012) renamed the *Pseudobradypus* biochron the *Notalacerta* biochron and named the Mississippian biochron the *Hylopus* biochron (Fig. 3.22).

Marsh (1891, 1898)	Haubold (1971, 1984)	Lucas (2003, 2007)	Fillmore et al. (2012)	Lucas et al. (2022)	this paper	AGE
Eosaurus beds	Pennsylvanian tracksites	*Dromopus* biochron	*Dromopus* biochron	*Dromopus* biochron	Robledoan biochron	PENNSYLVANIAN
		Pseudobradypus biochron	*Notalacerta* biochron	*Notalacerta-Dromopus* interval biochon	Jogginsian biochron	
Sauropus beds	Mississippian tracksites	Mississippian footprints	*Hylopus* biochron	*Hylopus* biochron	Bluebeachian biochron	MISSISSIPPIAN
				stem-tetrapod biochron	Valentian biochron	

FIGURE 3.22 Historical development of Carboniferous tetrapod footprint biochronology.

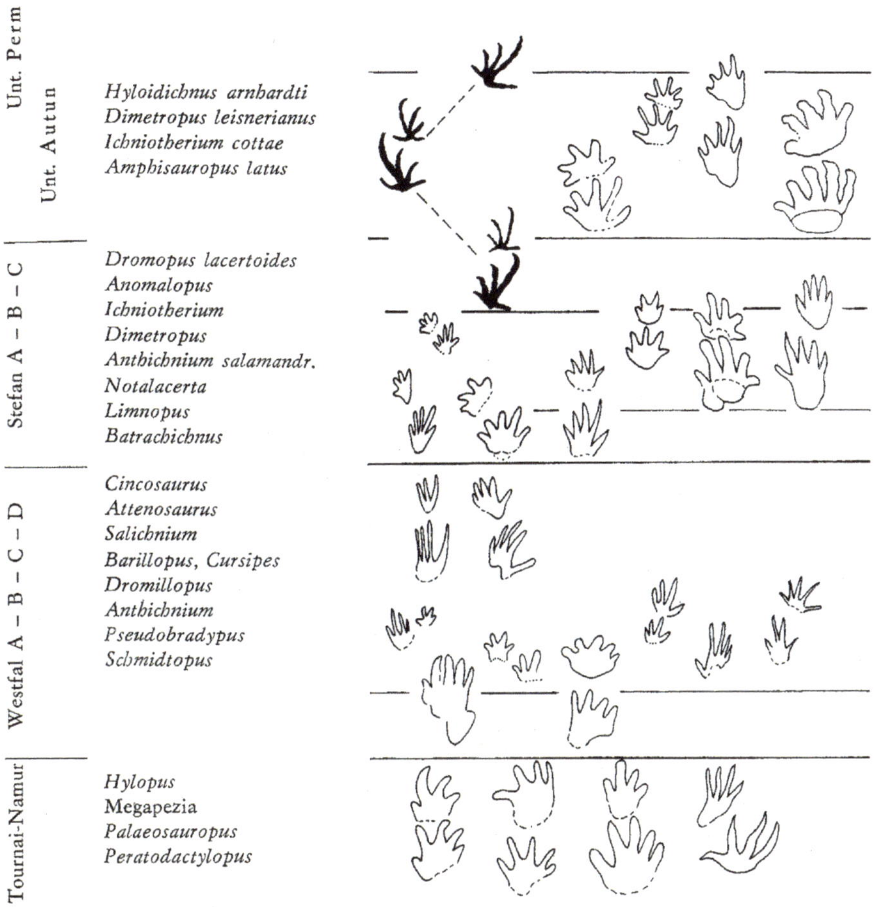

FIGURE 3.23 Haubold's (1984) depiction of the temporal distribution of Carboniferous footprints.

Lucas et al. (2022) refined this scheme, recognizing four tetrapod-footprint biochrons that encompass Carboniferous time (Fig. 3.22). Thus, they added a "stem-tetrapod biochron" for the Devonian and earliest Carboniferous footprint record before the first appearance datum (FAD) of *Hylopus* at Blue Beach, Nova Scotia. Given that *Notalacerta* is present in Permian strata (Voigt and Lucas, 2015; Marchetti et al., 2020c), they changed the name of the *Notalacerta* biochron to the *Notalacerta-Dromopus* interval biochron to identify an interval of time when *Notalacerta* is present before the FAD of *Dromopus*.

These changes underscore the problems of using ichnotaxonomic names to denote the biochrons. Thus, no Devonian ichnotaxa have been named, so a "stem-tetrapod biochron" is used and may change if an ichnotaxon is named from that time interval. Changes in the temporal range of an ichnotaxon, as happened with *Notalacerta*, force a name change. The solution to this instability is to name each footprint biochron after a locality, as then the name of the biochron need not change when the ichnotaxonomy or the temporal range of an ichnotaxon changes. The tetrapod body-fossil biochronology of the Carboniferous, Permian, and Triassic uses such locality-based names, as do other vertebrate biochronological constructs, such as the land-mammal "ages" of the Cretaceous and Cenozoic. Therefore, we take the step here of introducing new nomenclature for Carboniferous tetrapod-footprint biochronology and provide definitions and characterizations of each of the biochrons (Fig. 3.24). Correlation of these biochrons has been discussed in detail by Schneider et al. (2020) and Lucas et al. (2022).

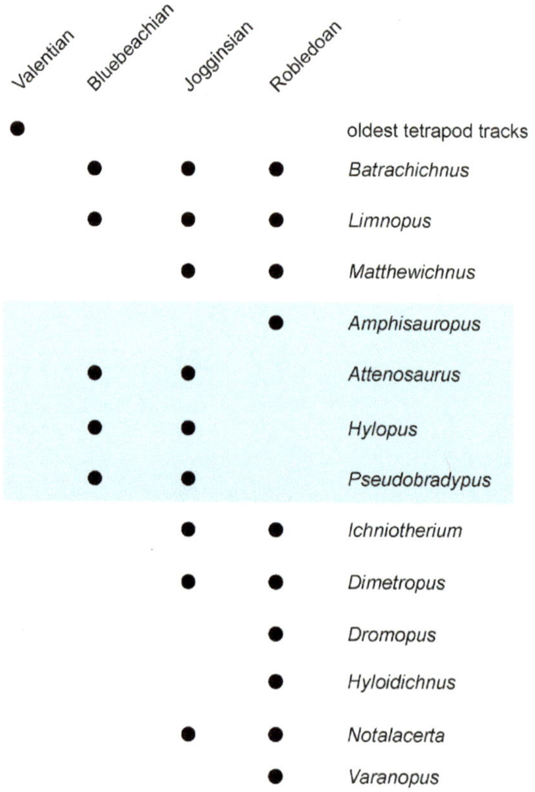

FIGURE 3.24 Carboniferous tetrapod footprint biochrons and the distribution of the ichnogenera.

5.2 Valentian biochron

Definition: The time between the first appearance datum (FAD) of tetrapod footprints and the FAD of *Hylopus*.

Characteristic footprint assemblage: Tetrapod footprints from the Valentia Slate Formation (Givetian) on Valentia Island, Ireland (Stössel, 1995; Stössel et al., 2016; Lucas, 2015a,2019).

Principal correlatives: The only other bona fide tetrapod footprint assemblages of Devonian age are in the upper Old Red Sandstone of Easter Ross in Scotland (Rogers, 1990) and the Combyingbar Formation in Australia (Warren and Wakefield, 1972). The Late Devonian age of the Scottish tracks is imprecise and open to question, and the Australian tracks have been assigned a Frasnian age (Lucas, 2015a, 2019).

Comments: This is a new name for the stem-tetrapod biochron of Lucas et al. (2022). As they noted, this biochron is very long (about 30 Ma), and there are no named ichnotaxa from a very sparse record. Much more discovery of Devonian tetrapod footprints needs to take place to improve this situation.

Note that Guo et al. (2023) recently claimed that the Valentia Island trackways resemble terrestrial trails of lungfishes, citing the descriptions of lungfish trackways by King et al. (2011) and Falkingham and Horner (2016). However, examination of the trackways in those articles indicates they bear little resemblance to the Valentia Island trackways.

5.3 Bluebeachian biochron

Definition: The time between the FAD of *Hylopus* and the FAD of *Notalacerta*.

Characteristic ichnoassemblage: The tetrapod footprint ichnoassemblages from the Tournaisian Blue Beach and Hurd Creek members of the Horton Bluff Formation at Blue Beach, Nova Scotia (see Lucas and Mansky, in prep.). Ichnogenera are *Attenosaurus, Batrachichnus, Characichnos, Hylopus, Limnopus,* and *Pseudobradypus*. These are "swim tracks" (*Characichnos*), temnospondyl or "microsaur" tracks (*Batrachichnus* and *Limnopus*) and the tracks of reptiliomorphs (*Attenosaurus, Hylopus* and *Pseudobradypus*).

Principal correlatives: The only other extensive Bluebeachian tetrapod footprint ichnoassemblage is from the Upper Mississippian (Serpukhovian) Mauch Chunk Formation in eastern Pennsylvania, USA (Fillmore et al., 2012). There are also a few other isolated records of tetrapod footprints/trackways from Mississippian strata in the USA, Canada, and the United Kingdom (Lucas et al., 2022).

Comments: Bluebeachian is a new name for the *Hylopus* biochron of earlier workers (e.g., Fillmore et al., 2012; Lucas et al., 2022). Like the Valentian, Bluebeachian is a long interval of time, about 30 my. It is the time of the oldest known temnospondyl and reptiliomorph tracks, which dominate the Bluebeachian footprint record. Also note that there are no trackways at Blue Beach that can be attributed to stem tetrapods (acanthostegids, ichthyostegids, etc.), though bones of stem tetrapods are present at Blue Beach (Anderson et al., 2015).

5.4 Jogginsian biochron

Definition: The time between the FAD of *Notalacerta* and the FAD of *Dromopus*.

Characteristic ichnoassemblage: Tetrapod footprints from the Early Pennsylvanian (Bashkirian) Joggins Formation in Nova Scotia. The ichnoassemblage is a mixture of amphibian (*Batrachichnus, Limnopus, Matthewichnus*), reptiliomorph (*Hylopus, Pseudobradypus*), and amniote (*Notalacerta*) tracks.

Principal correlatives: Outside of the characteristic ichnoassemblage, the most extensive tetrapod-footprint assemblages of Jogginsian age are from the Union Chapel Mine and Crescent Valley Mine in Alabama, USA, and the Desmoinesian McAlester Formation at the Keota locality in eastern Oklahoma, USA (Lucas et al., 2022). There are less extensive records in eastern Canada, the western and eastern USA and Europe (United Kingdom, France, Germany, Spain, Italy, Poland, and the Czech Republic), Greenland and Morocco (Lucas et al., 2022).

Comments: Jogginsian is a new name for the *Notalacerta* biochron or *Notalacerta-Dromopus* interval biochron of earlier workers (cf. Fillmore et al., 2012; Lucas et al., 2022). The FAD of amniote footprints marks the beginning of Jogginsian time. A mixture of amphibian, reptiliomorph, and amniote footprints is characteristic. The LOs of the reptiliomorph ichnogenus *Ichniotherium* and the amniote ichnogenus *Dimetropus*, ichnogenera well known from the later Robledoan, are in Jogginsian strata (Voigt and Ganzelewski, 2010).

5.5 Robledoan biochron

Definition: The time between the FAD of *Dromopus* and the FAD of *Erpetopus*.

Characteristic ichnoassemblage: Tetrapod footprints from the lower Permian (Sakmarian-Artinskian) Robledo Mountains Formation in the Robledo Mountains of New Mexico, USA (Voigt and Lucas, 2015). Footprints are those of amphibians (*Batrachichnus, Limnopus*), reptiliomorphs (*Amphisauropus*), and amniotes (*Dimetropus, Dromopus, Hyloidichnus, Robledopus,* and *Notalacerta*).

Principal correlatives: This is a new name for the *Dromopus* biochron of earlier workers (Lucas, 2003,2007; Fillmore et al., 2012; Voigt and Lucas, 2018; Lucas et al., 2022). Like the characteristic ichnoassemblage, most ichnoassemblages of the *Dromopus* biochron are of early Permian age (Voigt and Lucas, 2018). However, Late Pennsylvanian (Gzhelian) assemblages are known from the USA, eastern Canada, the United Kingdom, Germany, and Morocco (Lucas et al., 2022). Common amniote footprints associated with temnospondyl tracks are characteristic.

6. Carboniferous tetrapod footprints and biotic events

6.1 Introduction

The stratigraphic (temporal) distribution of tetrapod footprint ichnotaxa establishes biostratigraphic datums in the Carboniferous tetrapod fossil record, some different from those established by the body-fossil record (Fig. 3.25). These biostratigraphic datums are important to identifying and interpreting some key biotic events in Carboniferous tetrapod evolution and are briefly discussed here.

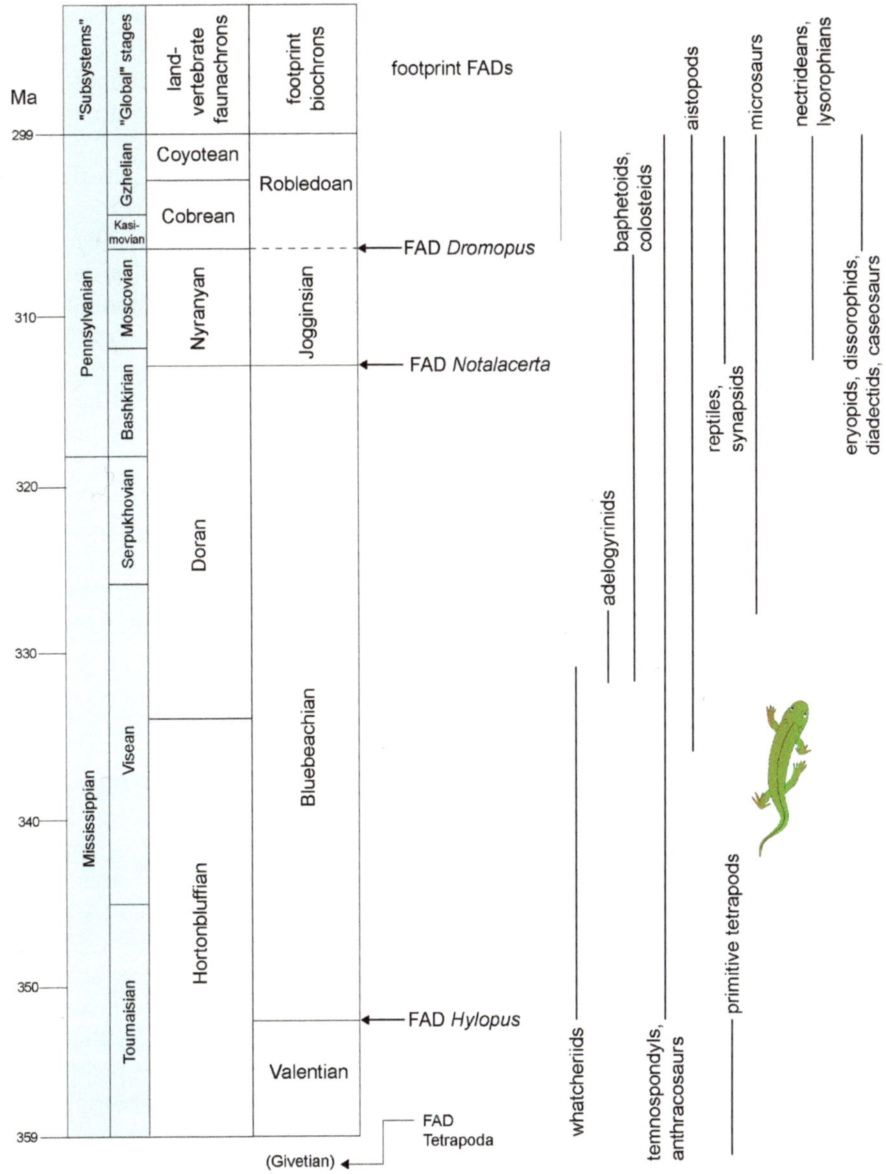

FIGURE 3.25 Summary of Carboniferus tetrapod-footprint biochronology and temporal distribution of twtrapod taxa. *Modified from Lucas et al. (2022).*

6.2 Pentadactyly and lateral sequence walking

The Late Devonian tetrapods known from body fossils were aquatic animals and appeared long before the oldest clearly terrestrial tetrapod faunas of the Carboniferous (Clack, 2002,2012). The interval between the Late Devonian aquatic tetrapods and the appearance of "terrestrial" tetrapods was long perceived to span much of the Lower Mississippian (the entire Tournaisian and first half of the Visean stages), at least 20 million years (e.g., Clack, 2002,2012). This interval was named "Romer's gap" and thought to conceal the poorly known/undocumented transition from stem-tetrapods that were not fully terrestrial to terrestrial tetrapods in crown-group lineages (Coates and Clack, 1995).

However, it is now clear that Romer's Gap was mostly an artifact of the fossil sample available that has been filled subsequently by the discovery of new fossils and the description of already known fossils (e.g., Mansky and Lucas, 2013; Clack and Milner, 2015; Clack et al., 2016; Lucas and Mansky, 2022a, b). Significantly, body fossils of ichthyostegids, acanthostegids, and tulerpetontids are now documented from Romer's gap in the Horton Bluff Formation at Blue Beach, Nova Scotia, confirming that these characteristic Devonian-grade tetrapods persisted well into the Lower Carboniferous (Mansky and Lucas, 2013; Anderson et al., 2015). They are found together with typical lower Carboniferous tetrapod taxa, suggesting the Devonian-Carboniferous transition in tetrapod evolution was less dramatic than once believed.

Devonian tetrapod footprints indicate lateral sequence walking by quadrupedal tetrapods with a smaller manus than pes (Lucas, 2015a,2019). However, they are too poorly preserved to allow estimates of digit numbers, though one trackway from Australia may indicate a trackmaker with more than five digits (polydactyl). The tetrapod footprints and trackways from Horton Bluff are numerous (in the thousands) and indicate hindlimb-propelled, lateral sequence walking by trackmakers that were not polydactyl.

6.3 Oldest temnospondyls and anthracosaurs

Footprints from the middle Tournaisian Horton Bluff Formation at Blue Beach, Nova Scotia, establish the presence of likely temnospondyl (*Batrachichnus, Limnopus*) and anthracosaur (*Attenosaurus, Pseudobradypus*) trackmakers. This is the oldest record of temnospondyls, which have a body-fossil record that begins during the late Visean (Lucas et al., 2022). Anthracosaur ("embolomere") bones have been reported from Blue Beach (Anderson et al., 2015), so the footprint and bone records coincide in establishing the oldest anthracosaur record. Both temnospondyls and anthracosaurs diversified significantly later in the Carboniferous (e.g., Milner, 1987,1990,1993; Carroll 2009), so these two evolutionary radiations have very deep roots early in Carboniferous time established/confirmed by their oldest footprints records. This supports Clack et al. (2016), who suggested that a deep split in the stem amphibians and stem amniotes occurred shortly after the Devonian-Mississippian boundary and was followed by rapid diversification.

Large Mississippian temnospondyl footprints from both the Horton Bluff and the Mauch Chunk formations indicate the presence of Mississippian temnospondyls much larger than any known from the coeval body-fossil record. Both the Horton Bluff Formation and the younger Mauch Chunk Formation also lack unambiguous amniote footprints, which supports a post-Visean origin of amniotes.

6.4 Oldest amniotes

The oldest amniote body fossils have long been considered to be those of *Hylonomus lyelli* from the Lower Pennsylvanian (Bashkirian) at Joggins, Nova Scotia (e.g., Dawson, 1895; Carroll, 1969; Steyer, 2012; Benton, 2015). *Hylonomus* is known from essentially complete skeletal material that reveals it to have been a small (up to 20 cm long), lightly built, lizard-like predator (Carroll 1969; Steyer 2012). The oldest unambiguous synapsid amniotes (*Asaphestera* sp.) have recently been recognized from the Joggins Formation as well, based on redescriptions of historical specimens by Mann et al. (2020). In the Pennsylvanian, the footprints of amniotes (eureptiles) are represented by *Notalacerta* (Marchetti et al., 2020c). Thus, the oldest definite amniote body fossils and footprints are of Early Pennsylvanian age, a notable and long-known correspondence of the body fossil and footprint records (e.g., Carroll 1969, 2009; Lucas 2019; Mann et al., 2020; Marchetti et al., 2020c).

6.5 Kasimovian bottleneck

The Late Pennsylvanian was a critical juncture in tetrapod evolution when many terrestrially adapted taxa first appeared (e.g., Sahney et al., 2008; Dunne et al., 2018; Lucas, 2023). The Middle Pennsylvanian (Moscovian) tetrapod record reflects a taphonomic megabias that favored preservation, discovery, and collection of aquatic tetrapods that lived in wetland paleoenvironments ("coal swamps"). The Kasimovian tetrapod body-fossil record is limited to seven localities, all but one in the USA, and two of which are singleton records, so it is less abundant, diverse, or widespread than earlier Moscovian and later Gzhelian tetrapod records. The same can be said of Kasimovian tetrapod footprint localities, which are few. This "Kasimovian bottleneck" hinders interpretation of tetrapod evolutionary events across the Middle–Late Pennsylvanian boundary (Lucas, 2023).

Significant changes did take place across that boundary, but they were spread out over Moscovian through Gzhelian time. Many of the perceived changes in tetrapods across the Middle–Late Pennsylvanian boundary are largely artifacts of facies changes and the Moscovian tetrapod taphonomic megabias and of the limited fossil record of Kasimovian tetrapods. Therefore, there is no simple link between Late Pennsylvanian tetrapod evolutionary events and changes in climate and vegetation (Lucas, 2023).

The change in the tetrapod footprint record that may represent part of the Kasimovian bottleneck is reflected in the Jogginsian-Robledoan tetrapod footprint biochron boundary. However, other than *Dromopus*, most of the tetrapod ichnogenera of the Robledoan biochron have older Pennsylvanian (and, in some cases, Mississippian) records (see above). A much older record of diadectomorph footprints (ichnogenus *Ichniotherium*) and of eupelycosaur (possibly limnoscelid?) footprints (ichnogenus *Dimetropus*) has been reported from the Early Pennsylvanian (Bashkirian) of Germany (Voigt and Ganzelewski, 2010). These records substantially predate the oldest body-fossil records of their inferred trackmakers, which are no older than Kasimovian. The Bashkirian *Ichniotherium* are particularly interesting as possibly representing the oldest fossils of a tetrapod high-fiber herbivore (Lucas et al., 2018).

These Bashkirian footprint records somewhat diminish the suddeness of the Kasimovian turnover in tetrapod evolutionary history by pushing back the record of two significant tetrapod groups (diadectomorphs and possibly eupelycosaurs minus varanopids) thought to have first appeared during the Kasimovian (Fig. 3.25). Future discoveries from Bashkirian and Moscovian strata may push back the supposed Kasimovian or Gzhelian first appearances of other tetrapod taxa. The tetrapod footprint record thus diminishes the apparent evolutionary turnover of tetrapods during the Kasimovian indicated by their body-fossil record.

6.6 Carboniferous–Permian transition

It has long been known that no substantial evolutionary turnover of tetrapods took place across the Carboniferous-Permian boundary (e.g., Romer, 1935; Olson, 1975; Lucas, 2004, 2006, 2017, 2018). Rather, a long-lasting succession of tetrapod body-fossil assemblages dominated by lepospondyl and temnospondyl amphibians, diadectomorphs, seymouriamorphs, and amniotes (notably, eupelycosaurs) appeared during the Late Pennsylvanian and persisted through much of the early Permian. This encompasses the Coyotean land-vertebrate faunachron of Lucas (2006), so he referred to it as the Coyotean chronofauna (Lucas, 2006,2017,2018).

As already noted, the oldest Permian tetrapod footprint assemblages are a continuation of those of the Upper Pennsylvanian. These are footprint assemblages of the Robledoan biochron characterized by the footprints of temnospondyls, seymouriamorphs, diadectomorphs, eupelycosaurs, and small parareptiles or early eureptiles. Of Kasimovian-middle Artinskian age, this footprint biochron parallels the Coyotean tetrapod chronofauna based on body fossils in showing that no substantial change of the tetrapod fauna took place across the Carboniferous-Permian boundary.

7. Conclusions

Based on the above, we offer these conclusions:

1. Carboniferous tetrapod footprints have been found primarily across Euramerica.
2. The Carboniferous record of tetrapod footprint assemblages in the Maritimes Basin of eastern Canada is the world's most extensive record of Carboniferous tetrapod footprints.
3. We recognize 14 valid tetrapod ichnogenera available for use at present, but note that a broad and detailed revision of Carboniferous footprint ichnotaxonomy is needed.
4. Carboniferous footprints identify a Valentian biochron of Middle Devonian-early Tournaisian age; Bluebeachian biochron of middle-Tournaisian-early Bashkirian age; Jogginsian interval biochron of early Bashkirian-Kasimovian age; and Robledoan biochron of Kasimovian-early Permian age.
5. "Romer's gap," a supposed paucity of Early Mississippian terrestrial fossils, includes the first substantial assemblage of tetrapod footprints at Blue Beach, Nova Scotia. These are the footprints of small and large temnospondyls and reptiliomorphs. This footprint assemblage establishes deep roots in the Tournaisian of the subsequent diversifications of temnospondyls and anthracosaurs.
6. No pre-Pennsylvanian amniote footprints are known, so the Early Pennsylvanian age of the oldest amniote footprints and body fossils is the same.

7. The Kasimovian encompasses a prolonged and complex change across the Middle–Late Pennsylvanian boundary with the appearance of new tetrapod taxa in the body-fossil record, notably the oldest high-fiber herbivores, the diadectomorphs and the edaphosaurid eupelycosaurs. However, the tetrapod footprint record documents much older records of diadectomorphs and eupelycosaurs (possible edaphosaurs) in the Bashkirian, thus diminishing the extent of tetrapod originations during the Kasimovian.
8. There is no substantial change in tetrapod footprint assemblages across the Carboniferous-Permian boundary, in agreement with the lack of substantial changes indicated by the body-fossil record across that boundary.

Acknowledgments

Many colleagues and collaborators made our studies of Carboniferous footprints in the field and museum possible, and enhanced our understanding of the subject. There are too many to name here, but we especially thank Luke Allen, Prescott Atkinson, Ron Buta, John Calder, David Fillmore, Brian Hebert, Danielle Horne, Adrian Hunt, Olivia King, Chris Mansky, Lorenzo Marchetti, Gloria Melanson, Rowan Norrad, Don Reid, Len Reid, Larry Rinehart, Edward Simpson, Sebastian Voigt, and Sonja Wood. We are grateful to Steven Hinds and Dr. Adrian Park, Geological Survey of New Brunswick, for stratigraphic information. Stimson also recognizes the assistance of Olivia King, Carlton University, for her field contributions in New Brunswick. Stimson acknowledges the support of the Natural Sciences and Engineering Research Council of Canada (NSERC, funding reference number 547631). Reviews by David Fillmore and Adrian Hunt improved the content and clarity of the manuscript.

References

Aceñolaza, F.G., Buatois, L.A., 1993. Nonmarine perigondwanic trace fossils from the late Paleozoic of Argentina. Ichnos 2, 183–201.

Aldrich, T.H., Jones, W.B., 1930. Footprints from the coal measures of Alabama. Alabama Museum Nat. Hist. Museum Pap. 9, 1–64.

Allen, L., Stimson, M.R., King, O.A., Norrad, R., Lucas, S.G., Mann, A., Hinds, S.J., Park, A.F., Calder, J.H., Maddin, H., Montplaisir, M., 2022. A *Batrachichnus salamandroides* trackway from the Minto Formation of central New Brunswick, Canada: implications for alternative tracemaker interpretations. Atl. Geosci. 58, 239–260.

Anderson, J.S., Smithson, T., Mansky, C.F., Meyer, T., Clack, J., 2015. A diverse tetrapod fauna at the base of 'Romer's gap. PLoS One 10 (4), e0125446.

Baird, D., 1952. Revision of the Pennsylvanian and Permian footprints *Limnopus, Allopus*, and *Baropus*. J. Paleontol. 25, 832–840.

Barkas, T.P., 1873. Illustrated Guide to the Fish, Amphibian, Reptilian and Supposed Mammalian Remains of the Northumberland Carboniferous Strata. Hutchings, London.

Barkas, T.P., 1890. Notes on the Numerous Newly Discovered Fossil Footprints in the Lower Calciferous Sandstones of Northumberland and Durham, 1889. Report of the British Association for the Advancement of Science, p. 565.

Beck, R., 1915. Über einen neueren Fund von Tierfährten innerhalb der sächsischen Steinkohlenformation. S. B. Abhandlungen Naturwissenschaft Gesellschaft Isis Dresden, pp. 49–52.

Bell, C.M., Boyd, M.J., 1986. A tetrapod trackway from the Carboniferous of northern Chile. Palaeontology 29, 519–526.

Benton, M.J., 2015. Vertebrate Palaeontology, fourth ed. John Wiley and Sons, New York.

Binney, E.W., 1856. On some supposed foot-marks in the Millstone Grit of Tintwhistle, Mottram-en-Longdendale, Cheshire. Q. J. Geol. Soc. Lond. 12, 350–354.

Bird, H.C., Milner, A.C., Shillito, A.P., Butler, R.J., 2019. A lower Carboniferous (Visean) tetrapod trackway represents the earliest record of an edopoid amphibian from the UK. J. Geol. Soc., Lond. https://doi.org/10.1144/jgs2019-149.

Blieck, A., Conti, M.A., Dalla Vecchia, F.M., Flügel, H.W., Gand, G., Hubmann, B., Lelièvre, H., Mariotti, N., Nicosia, U., Poplin, C., Schneider, J., Werneburg, R., 1997. Palaeozoic vertebrates of the Alps: a review. Bulletin de al Societe Géologique de France 168, 343–350.

Branson, E.B., 1910. Amphibian footprints from the Mississippian of Virginia. J. Geol. 18, 356–358.

Branson, E.B., Mehl, M.G., 1932. Footprint records from the Paleozoic and Mesozoic of Missouri, Kansas and Wyoming. Geol. Soc. Am. Bull. 43, 383–398.

Buta, R.J., Kopaska-Merkel, D.C., 2016. Footprints in Stone: Fossil Traces of Coal-Age Tetrapods. The University of Alabama Press, Tuscaloosa.

Buta, R.J., Pashin, J.C., Minter, N.J., Kopaska-Merklel, D.C., 2013. Ichnology and stratigraphy of the Crescent Valley Mine: evidence for a Carboniferous megatracksite in Walker County, Alabama. New Mexico Museum Nat. Hist. Sci. Bull. 60, 42–56.

Butts, E., 1891. Recently discovered footprints of the amphibian age in the upper coal-measure group of Kansas City, Mo. Kans. City Sci. 5, 17–19.

Calder, J.H., 1998. The Carboniferous Evolution of Nova Scotia, 43. Geological Society, London, Special Publications, pp. 261–302.

Calder, J.H., 2012. The Joggins Fossil Cliffs: Coal Age Galapagos. Nova Scotia Department of Natural Resources, Halifax.

Calder, J.H., Baird, D., Urdang, E.B., 2004. On the discovery of tetrapod trackways from Permo-Carboniferous redbeds of Prince Edward Island and their biostratigraphic significance. Atl. Geol. 40, 217–226.

Carman, J.E., 1927. Fossil footprints from the Pennsylvanian system in Ohio. Geol. Soc. Am. Bull. 38, 385–396.

Carroll, R.L., 1969. A Middle Pennsylvanian captorhinomorph and the interrelationships of primitive reptiles. J. Paleontol. 43, 151–170.

Carroll, R.L., 2009. The Rise of Amphibians: 365 Million Years of Evolution. The Johns Hopkins University Press, Baltimore.

Clack, J.A., 2002. Gaining Ground: The Origin and Evolution of Tetrapods, first ed. Indiana University Press, Bloomfield.

Clack, J.A., 2012. Gaining Ground: The Origin and Evolution of Tetrapods, second ed. Indiana University Press, Bloomfield.

Clack, J.A., Milner, A.R., 2015. Basal Tetrapoda. Handb. Palaeoherpetol. 3A1, 1–92.

Clack, J.A., Bennett, C.E., Carpenter, D.K., Davies, S.J., Fraser, N.C., Kearsey, T.I., Marshall, J.E., Millward, D., Otoo, B.K., Reeves, E.J., Ross, A.J., 2016. Phylogenetic and environmental context of a Tournaisian tetrapod fauna. Nat. Ecol. Evol. 1 (1), 1–11.

Coates, M.I., Clack, J.A., 1995. Romer's gap: tetrapod origins and terrestriality. Bull. Museum National Hist. Naturelle Paris 4 (17), 373–388.

Colbert, E.H., Schaeffer, B., 1947. Some Mississippian footprints from Indiana. Am. J. Sci. 245, 614–623.

Cotton, W.D., Hunt, A.P., Cotton, J.E., 1995. Paleozoic vertebrate tracksites in eastern North America. New Mexico Museum Nat. Hist. Sci. Bull. 6, 189–211.

Cox, E.T., 1874. *Colletosaurus indianaensis* Cox. Geological Survey of Indiana, Fifth Annual Report, pp. 247–248.

Davies, S.J., Gibling, M.R., Rygel, M.C., Calder, J.H., Skilliter, D.M., 2005. The Pennsylvanian Joggins Formation of Nova Scotia: sedimentological log and stratigraphic framework of the historic fossil cliffs. Atl. Geol. 41, 115–142.

Dawson, J.W., 1844. On the Lower Carboniferous rocks, or Gypsiferous Formation of Nova Scotia. Proc. Geol. Soc. Lond. 1843, 272–281.

Dawson, J.W., 1845. On the Newer Coal Formation of the Eastern Part of Nova Scotia. Proc. Geol. Soc. Lond. 1845, 504–512.

Dawson, J.W., 1863a. Air Breathers of the Coal Period. Balliere, New York.

Dawson, J.W., 1863b. Note on the footprints of a reptile from the coal Formation of Cape Breton. Can. Nat. Geol. 8, 430–431.

Dawson, J.W., 1868. Acadian Geology: An Account of the Geological Structure and Mineral Resources of Nova Scotia and Portions of the Neighbouring Provinces of British America, third ed. Oliver and Boyd, Edinburgh.

Dawson, J.W., 1872. Note on footprints from the Carboniferous of Nova Scotia in the collection of the Geological Survey of Canada. Geol. Mag. 9, 251–253.

Dawson, J.W., 1878. Acadian Geology: The Geological Structure, Organic Remains, and Mineral Resources of Nova Scotia, New Brunswick, and Prince Edward Island. Macmillan, London, Dawson Bros., Montreal.

Dawson, J.W., 1882. On the footprints of batrachians observed in the Carboniferous rocks of Nova Scotia. Trans. R. Soc. Lond. 173, 651–654.

Dawson, J.W., 1894. Some Salient Points in the Science of the Earth. Harper and Brothers, New York.

Dawson, J.W., 1895. Synopsis of the air-breathing animals of the Paleozoic in Canada, up to 1894. In: Trans. R. Soc. Canada, 12, pp. 71–88.

Dollé, P., de Lapparent, A.F., Montenat, C., 1970. Sur une dale à empreintes de pas lacertoides du houllier du basin du Nord-Pas-de-Calais. Ann. - Soc. Geol. Nord. 90, 63–68.

Dunkle, D.H., 1948. An interesting occurrence of fossil tracks in West Virginia. J. Washington Acad. Sci. 38, 130–132.

Dunne, E.M., Close, R.A., Button, D.J., Brocklehurst, N., Cashmore, D.D., Lloyd, G.T., Butler, R.J., 2018. Diversity change during the rise of tetrapods and the impact of the 'Carboniferous rainforest collapse. Proc. R. Soc. B 205, 20172730.

Falcon-Lang, H.J., Benton, M.J., Stimson, M., 2007. Ecology of earliest reptiles inferred from basal Pennsylvanian trackways. J. Geol. Soc. Lond. 164, 1113–1118.

Falcon-Lang, H.J., Gibling, M.R., Benton, M.J., Miller, R.F., Bashforth, A.R., 2010. Diverse tetrapod trackways in the lower Pennsylvanian Tynemouth Creek Formation, near St. Martin, southern New Brunswick, Canada. Palaeogeogr. Palaeoclimatol. Palaeoecol. 296, 1–13.

Falcon-Lang, H.J., Rygel, M.C., Calder, J.H., Gibling, M.R., 2004. An Early Pennsylvanian waterhole deposit and its fossil biota in a dryland alluvial plain setting, Joggins, Nova Scotia. J. Geol. Soc. 161, 209–222.

Falconnet, J., 2012. First evidence of a bolosaurid parareptile in France (latest Carboniferous earliest Permian of the Autun basin) and the spatiotemporal distribution of the Bolosauridae. Bulletin Societe géologique de France 183, 495–508.

Falkingham, P.L., Horner, A.M., 2016. Trackways produced by lungfish during terrestrial locomotion. Sci. Rep. 6, 33734.

Fillmore, D.L., Simpson, E.L., Lucas, S.G., 2009. Isaac Lea's *Palaeosauropus* (= "*Sauropus*") *primaevus*: a review of his discovery. Ichnos 16, 220–229.

Fillmore, D.L., Lucas, S.G., Simpson, E.L., 2012. Ichnology of the Mississippian Mauch Chunk Formation, eastern Pennsylvania. New Mexico Museum Nat. Hist. Sci. Bull. 54, 1–136.

Fillmore, D.L., Lucas, S.G., Simpson, E.L., 2015. Vertebrate ichnofossils from coal mine tailings of the Pennsylvanian Llewellyn Formation, eastern Pennsylvania. New Mexico Museum Nat. Hist. Sci. Bull. 67, 33–42.

Fondi, R., 1979. Orme di Microsauri nel Carbonifero Superiore della Sardegna. Memoir Societe Geologie Itali 20, 347–356.

Fondi, R., 1980. Le orme di tetrapodi plu antiche. In: I Vertebrati fossili italiani. Catalogo delta Mostra, Verona, Italy, pp. 41–45.

Francischini, H., Lucas, S.G., Voigt, S., Marchetti, L., Santucci, V.L., Knight, C.L., Wood, J.R., Dentzien-Dias, P., Schultz, C.L., 2019. On the presence of *Ichniotherium* in the Coconino Sandstone (Cisuralian) of Grand Canyon and remarks on the occupation of deserts by non-amniote tetrapods. PalZ. https://doi.org/10.1007/s12542-019-00450-5.

Frenguelli, J., 1950. Icnites en el Paleozoico superior del oeste Argentino. Rev. Asoc. Geol. Argent. 5, 136–148.

Gand, G., 1975. Sur quelques empreintes de pas de vertébrés récoltées dans le Stéphanien moyen de Montceau-les-Mines (Saône-et-Loire-France). Bulletin de la Société d'Historie Naturelle d'Autun 74, 9–31.

Gand, G., Haubold, H., 1984. Traces de vertébrés du Permien du bassin de Saint-Affrique (Description, datation, comparaison avec Celles du bassin de Lodève). Revue Géologie Meditérranéenne 11, 321–348.

Geinitz, H.B., 1861. Dyas oder die Zechsteinformation und das Rothliegende (Permische Formation zum Theil). Neues Jahrbuch für Mineralogie. Geologie und Paläontologie 1861, 65–67.

Geinitz, H.B., 1863. Ueber Thierfährten und Crustaceen-Reste in der unteren Dyas, oder dem unteren Rotliegendes, der Gegend von Hohenelbe. Beilage Natur Gesellschaft Isis Dresden 136–139.

Geinitz, H.B., Deichmüller, J.V., 1882. Die Saurier der unteren Dyas von Sachsen. Paläontographica 29, 1–46.

Gilmore, C.W., 1927. Fossil footprints from the Grand Canyon II. Smithsonian Misc. Collect. 80 (3), 1–78.

Guo, X., Retallack, G.J., Liu, J., 2023. Paleoenvironments of late Devonian tetrapods in China. Sci. Rep. 13, 20378.

Hahne, C., Wolansky, D., 1951. Eine neuere Fährtenfund eines Landwirbeltieres aus dem niederrheinisch-westfälischen Steinkohlengebirge. Gliickauf 87, 1025–1027.

Haines, O., 1852. Remarks to the Cork Cuvierian Society on a slab exhibiting fossil casts of footprints, from the Millstone Grit of Kilrush, Clare. Ann. Mag. Nat. Hist., Ser 2 (4), 433–434.

Harington, C.R., Foster, H., Holmes, R., Currie, P.J., 2005. Photographic catalogue of trackways in the Canadian Museum of Nature. Syllogeus 74, 1–151.

Harris, S.K., Lucas, S.G., Berman, D.S., Henrici, A.C., 2004. Vertebrate fossil assemblage from the Upper Pennsylvanian Red Tanks Member of the Bursum Formation, Lucero uplift, central New Mexico. New Mexico Museum Nat. Hist. Sci. Bull. 25, 267–283.

Haubold, H., 1970. Versuch einer Revision der Amphibien-Fährten des Karbon und Perm. Freiberger Forschungshefte Reihe C 260, 83–117.

Haubold, H., 1971. Ichnia amphibiorum et reptiliorum fossilium. Encycl. Paleoherpetol. 18, 1–124.

Haubold, H., 1984. Saurierfährten. A. Ziemsen Verlag, Wittenberg Lutherstadt.

Haubold, H., 1996. Ichnotaxonomie und Klassifikation von Tetrapodenfährten aus dem Perm. Hallesches Jahrb. Geowiss. B 18, 23–88.

Haubold, H., Lucas, S.G., 2003. Tetrapod footprints of the Lower Permian Choza Formation at Castle Peak, Texas. Paläontol. Z. 77, 247–261.

Haubold, H., Sarjeant, W.A.S., 1973. Tetrapodenfährten aus dem Keele und Enville Groups (Permokarbon) von Shropshire und South Staffordshire, Grossbrittannien. Zeitschrift der Geologische Wissenschaften Berlin 1, 895–933.

Haubold, H., Allen, A., Atkinson, T.P., Buta, R.J., Lacefield, J.A., Minkin, S.C., Relihan, B.A., 2005. Interpretation of the tetrapod footprints from the Early Pennsylvanian of Alabama. Alabama Paleontol. Soc. Monogr. 1, 75–111.

Henderson, J., 1924. Footprints in Pennsylvanian sandstones of Colorado. J. Geol. 32, 226–229.

Hmich, D., Schneider, J.W., Saber, H., Voigt, S., El Wartiti, M., 2006. New Continental Carboniferous and Permian faunas of Morocco – Implications for Biostratigraphy, Palaeobiogeography and Palaeoclimate, 265. Geological Society, London, Special Publications, pp. 297–324.

Hunt, A.P., Lucas, S.G., Lockley, M.G., 1995. Paleozoic tracksites of the western United States. New Mexico Museum Nat. Hist. Sci. Bull. 6, 213–217.

Hunt, A.P., Lucas, S.G., Pyenson, N.D., 2005. The significance of the Union Chapel Mine site: A Lower Pennsylvanian (Westphalian A) ichnological Konzentrat-Lagerstätte, Alabama, USA. Alabama Paleontol. Soc. Monogr. 1, 3–14.

Hunt, A.P., Lucas, S.G., Calder, J.H., Van Allen, H.E.K., George, E., Gibling, M.R., Hebert, B.L., Mansky, C., Reid, D.R., 2004. Tetrapod footprints from Nova Scotia: the Rosetta Stone for Carboniferous tetrapod ichnology. Geol. Soc. Am. Abstr. Progr. 36 (5), 66.

Jake, T.R., Blake Jr., B.M., 1982. 300 million year old footprints found in Tucker County. Mt. State Geol. 23–25.

Jillison, W.R., 1917. Preliminary note on the occurrence of vertebrate footprints in the Pennsylvanian of Oklahoma. Am. J. Sci. 194, 56–58.

King, A.T., 1844. Description of fossil footmarks supposed to be referable to the classes birds, *Reptilia* and *Mammalia* found in the Carboniferous strata in Westmoreland County, Pa. Proc. Acad. Nat. Sci. Phila. 2, 175–180.

King, A.T., 1845a. Description of fossil footmarks found in the Carboniferous series in Westmoreland County, Pa. Am. J. Sci. 48, 343–352.

King, A.T., 1845b. Description of fossil footprints. Proc. Acad. Nat. Sci. Phila. 2, 299–300.

King, A.T., 1845c. Footprints. Am. J. Sci. 49, 216–217.

King, A.T., 1846. Footprints in the coal rocks of Westmoreland County, Pa. Am. J. Sci., Ser. 2 (1), 268.

King, H.M., Shub in, N.H., Coates, M.I., Hale, M.E., 2011. Behavioral evidence for the evolution of walking and bounding before terrestriality in sarcopterygian fishes. Proc. Natl. Acad. Sci. USA 108, 21146–21151.

Klein, H., Lucas, S.G., 2021. The Triassic tetrapod footprint record. New Mexico Museum Nat. Hist. Sci. Bull. 83, 194 pp.

Kohl, M.S., Bryan, J.R., 1994. A new Middle Pennsylvanian (Westphalian) amphibian trackway from the Cross Mountain Formation, east Tennessee Cumberlands. J. Paleontol. 68, 655–663.

Kuhn, O., 1963. Ichnia tetrapodorum . Fossilium Catalogus l. Animalia, 101. s'Gravenhage. Junk, 176 pp.

Kukuk, P., 1926. Ober Fährten von Landwirbeltieren im Karbon des Ruhbezirkes. Zeitschrift Deutsche Geologische Gesellschaft 78, 600–613.

Lagnaoui, A., Voigt, S., Saber, H., Schneider, J.W., 2014. First occurrence of tetrapod footprints from Westphalian strata of the Sidi Kassem basin, central Morocco. Ichnos 21, 223–233.

Lagnaoui, A., Voigt, S., Belahmira, A., Saber, H., Klein, H., Hminna, A., Schneider, J.W., 2018. Late Carboniferous tetrapod footprints from the Souss basin, Western High Atlas Mountains, Morocco. Ichnos 25, 81–93.

Lane, N., Maples, C.G., 1990. Fossil footprints from Indiana. J. Vertebr. Paleontol. 9 (3), 32A.

Langiaux, J., 1979. Ichnologie: 5. *Baropezia polymorpha* n. sp. (Stéphanien terminal de Blanzy-Montceau). Revue Periodique de "La Physiophile," 90, 69–75.

Langiaux, J., 1980. Premières observations a Morteru (Stéphanien de Blanzy-Montceau) ichnites, fauna, flore. Revue Periodique de "La Physiophile," 93, 77–88.
Langiaux, J., 1981. Ichnologie, 6, Deux pistes nouvelles a Morteru. Revue Periodique de "La Physiophile," 94, 65–74.
Langiaux, J., Sotty, D., 1975a. Ichnologie, pistes de tetrapodes fossiles dans le bassin de houllier de Blanzy-Montceau. Revue Periodique de "La Physiophile," 82, 46–58.
Langiaux, J., Sotty, D., 1975b. Ichnologie, 2, pistes de tetrapodes fossiles dans bassin houiller de Blanzy-Montceau. Revue Periodique de "La Physiophile," 83, 57–68.
Langiaux, J., Sotty, D., 1976. Ichnologie, 3, trois nouvelles pistes de tétrapodes dans le Stéphanien de Blanzy-Montceau. Revue Periodique de "La Physiophile," 84, 46–54.
Langiaux, J., Sotty, D., 1977. Ichnologie, 4, pistes et empreintes dans le Stéphanien de Blanzy-Montceau. Revue Periodique de "La Physiophile," 86, 74–92.
Lea, I., 1849. On reptilian footmarks found in the gorge of Sharp Mountain near Pottsville, Pennsylvania. Proc. Am. Phil. Soc. 5, 91–94.
Lea, I., 1853. On the fossil foot-marks in the Red Sandstones of Pottsville, Schuylkill County, Pennsylvania. Trans. Am. Phil. Soc. 10, 307–315.
Lea, I., 1855. On the Fossil Foot-Marks in the Red Sandstone of Pottsville, pp. 1–16. Schuylkill County, Penna. T.K. and P.G. Collins, Philadelphia.
Leidy, J., 1879. Fossil footracks of the anthracite coal measures. Proc. Acad. Nat. Sci. Phila. 1879, 164–165.
Liu, J., Yi, J., Chen, J., 2020. Constraining assembly time of some blocks on eastern margin of Pangea using Permo-Triassic non-marine tetrapod records. Earth Sci. Rev. 207, 103215.
Lockley, M.G., Foster, J.R., 2006. Dinosaur and turtle tracks from the Morrison Formation (Upper Jurassic) of Colorado National Monument, with observations on the taxonomy of vertebrate swim tracks. New Mexico Museum Nat. Hist. Sci. Bull. 36, 193–198.
Lockley, M.G., Hunt, A.P., 1995. Dinosaur Tracks and Other Fossil Footprints of the Western United States. Columbia University Press, New York.
Logan, W.E., 1842. On the coal-fields of Pennsylvania and Nova Scotia. Proc. Geol. Soc. Lond. 3, 707–712.
Lucas, S.G., 2003. Carboniferous tetrapod footprint biostratigraphy and biochronology. Newsl. Carboniferous Stratigr. 21, 36–41.
Lucas, S.G., 2004. A global hiatus in the Middle Permian tetrapod fossil record. Stratigraphy 1, 47–64.
Lucas, S.G., 2006. Global Permian Tetrapod Biostratigraphy and Biochronology, 265. Geological Society, London, Special Publication, pp. 65–93.
Lucas, S.G., 2007. Tetrapod footprint biostratigraphy and biochronology. Ichnos 14, 5–38.
Lucas, S.G., 2015a. *Thinopus* and a critical review of Devonian tetrapod footprints. Ichnos 22, 136–154.
Lucas, S.G., 2015b. Alfred King's forgotten fossil footprints. Penn. Herit. 41 (2), 16–25.
Lucas, S.G., 2017. Permian tetrapod extinction events. Earth Sci. Rev. 170, 31–60.
Lucas, S.G., 2018. Permian Tetrapod Biochronology, Correlation and Evolutionary Events, 450. Geological Society, London, Special Publications, pp. 405–444.
Lucas, S.G., 2019. An ichnological perspective on some major events of Paleozoic tetrapod evolution. Bolletino della Soc. Paleontol. Ital. 58, 223–266.
Lucas, S.G., 2023. Middle to Late Pennsylvanian Tetrapod Evolution: the Kasimovian Bottleneck, 535. Geological Society, London, Special Publications, pp. 481–497.
Lucas, S.G., Ataabadi, M.M., 2010. Tetrapod footprints from the Upper Carboniferous of China. Ichnos 17, 163–165.
Lucas, S.G., Dalman, S.G., 2013. Alfred King's Pennsylvanian tetrapod footprints from western Pennsylvania. New Mexico Museum Nat. Hist. Sci. Bull. 60, 233–239.
Lucas, S.G., Mansky, C.F., 2022a. Tetrapod trace and body fossils from the Tournaisian Horton Bluff Formation at Blue Beach. In: Nova Scotia, Close Romer's Gap. Geoscience Canada (GAC-MAC-IAH-CNC-CSPG 2022 Halifax Meeting: Abstracts, 45), 49, pp. 150–151.
Lucas, S.G., Mansky, C.F., 2022b. Devonian-early Carboniferous ichnology and the origin and early evolution of tetrapods. Geosci. Can. 49, 150.
Lucas, S. G. and Mansky, C. F. in prep. Early Carboniferous ichnofossils from Blue Beach (Nova Scotia, Canada), and the Origin and Early Evolution of Tetrapods.
Lucas, S.G., Fillmore, D.L., Simpson, E.L., 2010a. The Mississippian tetrapod footprint ichnogenus Palaeosauropus: extramorphological variation and ichnotaxonomy. Ichnos 17, 177–186.

Lucas, S.G., Rinehart, L.F., Celeskey, M.D., 2018. The oldest specialized tetrapod herbivore: a new eupelycosaur from the Permian of New Mexico, USA. Palaeontol. Electron. 21 (3.39A), 1–42.
Lucas, S.G., Kollar, A.D., Berman, D.S., Henrici, A.C., 2016. Pelycosaurian-grade (Amniota: Synapsida) footprints from the lower Permian Dunkard Group of Pennsylvania and West Virginia. Ann. Carnegie Mus. 83, 287–294.
Lucas, S.G., Lerner, A.J., Bruner, M., Shipman, P., 2004. Middle Pennsylvanian ichnofauna from eastern Oklahoma, USA. Ichnos 11, 45–55.
Lucas, S.G., Stimson, M.R., King, O.A., Calder, J.H., Mansky, C.F., Brian, L., Hebert, B.L., Hunt, A.P., 2022. Carboniferous Tetrapod Footprint Biostratigraphy, Biochronology and Evolutionary Events, 512. Geological Society, London, Special Publications, pp. 933–963.
Lull, R.S., 1920. An upper Carboniferous footprint from Attleboro, Massachusetts. Am. J. Sci. 4 (50), 234–236.
Lyell, C., 1843. On the Coal Formation of Nova Scotia and on the age and relative position of the gypsum and accompanying marine limestone. Proc. Geol. Soc. Lond. 4, 184–186.
Mann, A., Gee, B.M., Pardo, J.D., Marjanović, D., Adams, G.R., Calthorpe, A.S., Maddin, H.C., Anderson, J.S., 2020. Reassessment of historic 'microsaurs' from Joggins, Nova Scotia, reveals hidden diversity in the earliest amniote ecosystem. Pap. Palaeontol. 1–21.
Mansky, C.F., Lucas, S.G., 2013. Romer's gap revisited: continental assemblages and ichno-assemblages from the basal Carboniferous of Blue Beach, Nova Scotia, Canada. New Mexico Museum Nat. Hist. Sci. Bull. 60, 244–273.
Marchetti, L., Mujal, E., Bernardi, M., 2017. An unusual *Amphisauropus* trackway and its implication for understanding seymouriamorph locomotion. Lethaia 50, 162–174.
Marchetti, L., Petti, F.M., Zoboli, D., Pillola, G.L., 2018. Vertebrate and invertebrate trace fossils in the Late Pennsylvanian (Carboniferous) fluvio-lacustrine San Giorgio Basin (South-West Sardinia): remarks on the oldest continental ichnoassociation of Italy. Ichnos 25, 94–105.
Marchetti, L., Belvedere, M., Voigt, S., Klein, H., Castanera, D., Díaz-Martínez, I., Marty, D., Xing, L., Feola, S., Melchor, R.N., Farlow, J.O., 2019. Defining the morphological quality of fossil footprints. Problems and principles of preservation in tetrapod ichnology with examples from the Palaeozoic to the present. Earth Sci. Rev. 193, 109–145.
Marchetti, L., Francischini, H., Lucas, S.G., Voigt, S., Hunt, A.P., Santucci, V.L., 2020a. Paleozoic vertebrate ichnology of Grand Canyon. In: Santucci, V.L. (Ed.), Grand Canyon National Park Paleontological Resource Inventory. National Park Service Nature Resource Technical Report, Washington, D.C., pp. 333–379
Marchetti, L., Muscio, G., Petti, F.M., Pillila, G.L., Zoboli, D., 2020b. The Carboniferous tetrapod ichnoassociation from Italy. J. Mediterr. Earth Sci. 12.
Marchetti, L., Voigt, S., Lucas, S.G., Stimson, M.R., King, O.A., Calder, J.H., 2020c. Footprints of the earliest reptiles. *Notalacerta missouriensis* – ichnotaxonomy; potential trackmakers, biostratigrapy, palaeobiogeography, and palaeoecology. Ann. Soc. Geol. Pol. 90. https://doi.org/10.14241/asgp.2020.13.
Marsh, O.C., 1891. Geological horizons as determined by vertebrate fossils. Am. J. Sci. 42, 336–338.
Marsh, O.C., 1894. Footprints of vertebrates in the coal measures of Kansas. Am. J. Sci. 348, 81–84.
Martel, A.T., Gibling, M.R., 1991. Wave-dominated Lacustrine Facies and Tectonically-Controlled Cyclicity in the Lower Carboniferous Horton Bluff Formation, 13. International Association of Sedimentologists, Special Publication, Nova Scotia, Canada, pp. 223–243.
Martino, R.L., 1991. *Limnopus* trackways from the Conemaugh Group (Late Pennsylvanian), southern West Virginia. J. Paleontol. 65, 957–972.
Mason, W.D.H., 1878. Letter on the batrachian foot tracks from the Ellengowan shaft in Schuyl Kill County. Proc. Am. Phil. Soc. 17, 716–719, 725.
Matamales-Andreu, R., Mujal, E., Galobart, A., Fortuny, J., 2021. Insights on the evolution of synapsid locomotion based on tetrapod tracks from the lower Permian of Mallorca (Balearic Islands, western Mediterranean). Palaeogeogr. Palaeoclimatol. Palaeoecol. 579, 110589.
Matthew, G.F., 1903a. An Attempt to Classify Paleozoic Batrachian Footprints, 9. Transactions of the Royal Society of Canada, pp. 109–121.
Matthew, G.F., 1903b. On batrachian and other footprints from the Coal Measures of Joggins. N.S. Bull. Nat. Hist. Soc. New Brunswick 5, 103–108.
Matthew, G.F., 1903c. New genera of batrachian footprints from the Carboniferous system in eastern Canada. Can. Rec. Sci. 9, 99–111.

Matthew, G.F., 1904. Note on the genus Hylopus of Dawson. Bull. Nat. Hist. Soc. N. B. 5, 247–252.
Matthew, G.F., 1905. New Species and Anew Genus of Batrachian Footprints of the Carboniferous System in Eastern Canada, 10. Transactions of the Royal Society of Canada, pp. 77–122.
Matthew, G.F., 1910. Remarkable Forms of the Little River Group, 3. Transactions of the Royal Society of Canada, pp. 115–133.
McClelland, S., 1988. Fossil footprints unearthed in eastern panhandle. Mt. State Geol. 88, 14–17.
Meade, L.E., Jones, A.S., Butler, R.J., 2016. A revision of tetrapod footprints from the late Carboniferous of the West Midlands, UK. PeerJ 4, e2718.
Mietto, P., Muscio, G., Venturini, C., 1985. Impronte di tetrapodi nei terreni Carboniferi delle Alpi Carniche. Gortania 7, 59–74.
Milàn, J., Klein, H., Voigt, S., Stemmerik, L., 2016. First record of tetrapod footprints from the Carboniferous Mesters Vig Formation in East Greenland. Bull. Geol. Soc. Den. 64, 69–76.
Milner, A.C., 1994. A Carboniferous reptile footprint from the Somerset Coalfield. PGA (Proc. Geol. Assoc.) 105, 313–315.
Milner, A.R., 1987. The Westphalian tetrapod fauna; some aspects of its geography and ecology. J. Geol. Soc. Lond. 144, 495–506.
Milner, A.R., 1990. The radiations of temnospondyl amphibians. In: Taylor, P.D., Larwood, G.P. (Eds.), Major Evolutionary Radiations. Systematics Association Special Volume, 42, pp. 321–349.
Milner, A.R., 1993. Biogeography of Palaeozoic tetrapods. In: Long, J.A. (Ed.), Palaeozoic Vertebrate Biostratigraphy and Biogeography. Belhaven Press, London, pp. 324–353.
Mitchell, H.R., 1931. Fossil footprints from the Pennsylvanian of Ohio. Ohio J. Sci. 31, 500–504.
Mitchell, H.R., 1933. Notes on another Pennsylvanian footprint from Ohio. Ohio J. Sci. 33, 48–49.
Modesto, S.P., Scott, D.M., MacDougall, M.J., Sues, H.-D., Evans, D.C., Reisz, R.R., 2015. The oldest parareptile and the diversification of parareptiles. Proc. R. Soc. B 282, 20121912.
Moodie, R.L., 1929. Vertebrate footprints from the red beds of Texas. Am. J. Sci. 97, 352–368.
Mossman, D.J., Grantham, R.G., 1996. A recently discovered amphibian trackway (*Dromillopus quadrifidus*) at Joggins, Nova Scoria. Can. J. Earth Sci. 33, 710–714.
Mossman, D.J., Grantham, R.G., 2000. Vertebrate trackways in the Parrsboro Formation (upper Carboniferous) at Rams Head, Cumberland County, Nova Scotia. Atl. Geol. 35, 186–196.
Mudge, B.F., 1873. Recent discoveries of fossil footprints in Kansas. Kansas Acad. Sci. Trans. 2, 7–9.
Mujal, E., Marchetti, L., 2020. *Ichniotherium* tracks from the Permian of France, and their implications for understanding the locomotion and palaeobiogeography of large diadectomorphs. Palaeogeogr. Palaeoclimatol. Palaeoecol. 547, 109698.
Niedźwiedzki, G., 2015. Carboniferous tetrapod footprints from the Lublin basin, SE Poland. GFF 137, 57–65.
Olson, E.C., 1975. Vertebrates and some problems of Permo-Carboniferous geochronology. In: Cys, J.M., Toomey, D.F. (Eds.), Permian Exploration, Boundaries, and Stratigraphy, vols. 75–65. West Texas Geological Society and Permian Basin Section SEPM Publication, pp. 98–103.
Pabst, W., 1895. Tierfahrten aus dem Rothliegenden von Friedrichroda, Tambach und Kabarz in Thüringen. Zeitschrift Deutsche Geologische Gesellschaft 47, 570–576.
Peabody, F.E., 1954. Trackways of an ambystomid salamander from the Paleocene of Montana. J. Paleontol. 28 (1), 79–83.
Peabody, F.E., 1959. Trackways of Living and Fossil Salamanders, 63. University of California Publications in Zoology, pp. 1–71.
Pohlig, H., 1892. Altpermische Saurierfährten, Fische und Medusen der Gegend von Friedrichroda i. Thüringen. In: Anonymous (Ed.), Festschrift 70. Geburtstag von Rudolf Leuckardt. Engelmann, Leipzig, 59–64. 1885.
Reisz, R.R., 1990. Geology and paleontology of the Garnett quarry. In: Cunningham, C., Maples, C.G. (Eds.), 1990 Society of Vertebrate Paleontology Upper Paleozoic of Eastern Kansas Excursion Guidebook, vols. 90–24. Kansas Geological Survey Open-file Report, pp. 43–48.
Rogers, D.A., 1990. Probable tetrapod tracks rediscovered in the Devonian of N Scotland. J. Geol. Soc. 147, 746–748. London.
Romer, A.S., 1935. Early history of Texas redbeds vertebrates. Geol. Soc. Am. Bull. 46, 1597–1658.
Romer, A.S., Price, L.I., 1940. Review of the Pelycosauria. Geol. Soc. Am. Spec. Pap. 28, 1–486.

Rowland, S.M., Caputo, M.V., 2018. Trackway of a sideways-walking basal tetrapod in the Pennsylvanian Manakacha Formation of Grand Canyon National Park. In: Society of Vertebrate Paleontology, Abstracts of the 78th Annual Meeting, 206.

Rowland, S.M., Caputo, M.V., Jensen, Z.A., 2020. Early adaptation to eolian sand dunes by basal amniotes is documented in two Pennsylvanian Grand Canyon trackways. PLoS One 15 (78), e0237636.

Sacchi, E., Cifelli, R., Citton, P., Nicosia, U., Romano, M., 2014. Dimetropus osageorum n. isp. from the early Permian of Oklahoma (USA): a trace and its trackmaker. Ichnos 21, 1–18.

Sahney, S., Benton, M.J., Falcon-Lang, H.J., 2008. Rainforest collapse triggered Carboniferous tetrapod diversification in Euramerica. Geology 38, 1079–1082.

Sarjeant, W.A.S., 1974. A history and bibliography of the study of fossil vertebrate footprints in the British Isles. Palaeogeogr., Palaeoclimatol., Palaecol. 16, 265–375.

Sarjeant, W.A.S., 1987. The study of fossil vertebrate footprints a short history and selective bibliography. In: Leonardi, G. (Ed.), Glossary and manual of tetrapod footprint paleoichnology. República Federativa do Brasil, Ministerio das Minas e Energia. De0ratamento Nacional da Produçáo Mineral, Brasília, pp. 1–19.

Sarjeant, W.A.S., Mossman, D.J., 1978. Vertebrate footprints from the Carboniferous sediments of Nova Scotia: a historical review and description of newly discovered forms. Palaeogeogr., Paleoclimatol., Palaeoecol. 23, 279–306.

Scarboro, D.D., Tucker, M.E., 1995. Amphibian footprints from the mid-Carboniferous of Northumberland, England: Sedimentological context, preservation and significance. Palaeogeogr. Palaeoclimatol. Palaeoecol. 113, 335–349.

Schneck, W.M., Fritz, W.J., 1985. An amphibian trackway (Cincosaurus cobbi) from the Lower Pennsylvanian ("Pottsville") of Lookout Mountain, Georgia: a first occurrence. J. Paleontol. 59, 1243–1250.

Schneider, J., Schamaev, M.I., Walter, H., 1992. Paläobiogeographie und Stratigraphie von Tetrapoden- und Arthropoden-Fährten aus dem paralischen Oberkarbon und Perm (Gzhel/Assel) des Donez Bassins. Freiberger Forschungshefte - A C445, 104–121.

Schneider, J.W., Lucas, S.G., Scholze, F., Voigt, S., Marchetti, L., Klein, H., Opluštil, S., Werneburg, R., Golubev, V.K., Barrick, J.E., Nemyrovska, T., Ronchi, A., Day, M.O., Silantiev, V.V., Rößler, R., Saber, H., Linnemann, U., Zharinov, V., Shen, S., 2020. Late Paleozoic-early Mesozoic continental biostratigraphy — links to the standard global Chronostratigraphic Scale. Paleoworld 29, 186–238.

Schöllmann, L., Ganzelewski, M., Piecha, M., Salamon, M., Voigt, S., Wrede, V., 2015. Auf großem Fuße - die älteste Tetrapodenfährte Deutschlands. Archäologie in Westfalen-Lippe 6, 23–25.

Schoewe, W.H., 1956. Geographic locations and stratigraphic horizon of Pennsylvanian fossil footprints discovered in Osage County, Kansas by B. F. Mudge in 1873. J. Paleontol. 30, 389–405.

Soler-Gijón, R., Moratalla, J.J., 2001. Fish and tetrapod trace fossils from the upper Carboniferous of Puertollano, Spain. Palaeogeogr. Palaeoclimatol. Palaeoecol. 171, 1–28.

Soler-Gijón, R., Díez Ruiz, A., 2023. "Carbonifero de Puertollano" Natural Monument (Puertollano basin, Spain): a window for the knowledge of early vertebrates. Spanish J. Paleontol. 38, 81–94.

Sternberg, C.M., 1933. Carboniferous tracks from Nova Scotia. Geol. Soc. Am. Bull. 44, 951–964.

Steyer, S., 2012. Earth before Dinosaurs. Indiana University Press, Bloomington and Indianapolis.

Stimson, M., Lucas, S.G., Melanson, G., 2012. The Smallest Known Tetrapod Footprints: Batrachichnus Salamandroides from the Carboniferous of Joggins, 19, pp. 127–140. Nova Scotia, Canada. Ichnos.

Stimson, M., Miller, R., Lucas, S., 2016b. Reassignment of vertebrate ichnotaxa from the Upper Carboniferous 'Fern Ledges,' Lancaster Formation, Saint John, New Brunswick, Canada. Atl. Geol. 52, 20–34.

Stimson, M., Miller, R., Lucas, S., Park, A., Hinds, S., 2016a. Redescription of tetrapod trackways from the Mississippian Mabou Group, Lepreau Falls, New Brunswick, Canada. Atl. Geol. 52, 1–19.

Stimson, M.R., King, O.A., MacRae, R.A., Miller, R.F., Hinds, S.J., Park, A.F., 2018. Ichnological, Palynological, and Biomarker studies on the Devonian–Carboniferous (Mississippian and Pennsylvanian) Strata of New Brunswick: implications for continental Biodiversity during Romer's gap. In: Energy, Mines, Petroleum Geoscience Project Summaries and Other Activities, pp. 90–102.

Stimson, M.R., King, O.A., Macrae, R.A., Gastaldo, R., Glasspool, I., Gensel, P.G., Basinger, J.F., Miller, R.F., Hinds, S.J., Park, A.F., Allan, L., 2019. The First Evidence of Terrestrial Vertebrates from the Lower Mississippian Albert Formation of New Brunswick. Implications for the Invasion of Continental Lacustrine Ecosystems and Biodiversity during Romer's Gap in Atlantic Canada in the Western Moncton Subbasin Energy, Mines, Petroleum Geoscience Project Summaries and Other Activities, pp. 70–71 (Abstract).

Stössel, I., 1995. The discovery of a new Devonian tetrapod trackway in SW Ireland. J. Geol. Soc. Lond. 152, 407–413.

Stössel, I., Williams, E.A., Higgs, K.T., 2016. Ichnology and depositional environments of the Middle Devonian Valentia Island tetrapod trackways, south-west Ireland. Palaeogeogr. Palaeoclimatol. Palaeoecol. 462, 16–40.

Sundberg, F.A., Bennington, J.B., Wizevich, M.C., Bambach, R.K., 1990. Upper Carboniferous (Namurian) amphibian trackways from the Bluefield Formation, West Virginia, USA. Ichnos 1, 111–124.

Tanner, L.H., 2013. A Pennsylvanian tetrapod trackway from Joggins, Nova Scotia: the size continuum of *Batrachichnus salamandroides*. New Mexico Museum Nat. Hist. Sci. Bull. 60, 427–432.

Tibert, N.E., Scott, D.B., 1999. Ostracodes and agglutinated Foraminifera as indicators of paleoenvironmental change in an early Carboniferous brackish bay, Atlantic Canada. Palaios 14, 246–260.

Tilton, E., 1931. Permian vertebrate tracks in West Virginia. Geol. Soc. Am. Bull. 42, 541–556.

Toepelman, W.C., Rodeck, H.G., 1936. Footprints in late Paleozoic red beds near Boulder, Colorado. J. Paleontol. 10, 660–662.

Turek, V., 1989. Fish and amphibian trace fossils from Westphalian sediments of Bohemia. Palaeontology 32, 623–643.

Van Allen, H.E.K., Calder, J.H., Hunt, A.P., 2005. The trackway record of a tetrapod community in a walchian conifer forest from the Permo-Carboniferous of Nova Scotia. New Mexico Museum Nat. Hist. Sci. Bull. 30, 322–332.

Voigt, S., 2005. Die Tetrapodenichnofauna des kontinentalen Oberkarbon und Perm im Thüringer Wald - Ichnotaxonomie, Paläoökologie und Biostratigraphie. Cuvillier Verlag, Göttingen, 305 pp.

Voigt, S., Ganzelewski, M., 2010. Toward the origin of amniotes: diadectomorph and synapsid footprints from the early late Carboniferous of Germany. Acta Palaeontol. Pol. 55, 57–72.

Voigt, S., Lucas, S.G., 2015. Permian tetrapod ichnodiversity of the Prehistoric Trackways National Monument (south-central New Mexico, U.S.A.). New Mexico Museum Nat. Hist. Sci. Bull. 65, 153–167.

Voigt, S., Lucas, S.G., 2018. Outline of a Permian Tetrapod Footprint Ichnostratigraphy, 450. Geological Society, London, Special Publications, pp. 387–404.

Voigt, S., Berman, D.S., Henrici, A.C., 2007. First well-established track-trackmaker association of Paleozoic tetrapods based on *Ichniotherium* trackways and diadectid skeletons from the lower Permian of Germany. J. Vertebr. Paleontol. 27, 553–570.

Voigt, S., Saber, H., Schneider, J.W., Hmich, D., Hminna, A., 2011. Late Carboniferous–early Permian tetrapod ichnofauna from the Khenifra Basin, central Morocco. Geobios 44, 399–407.

Warren, A.W., Wakefield, N.A., 1972. Trackways of tetrapod vertebrates from the Upper Devonian of Victoria, Australia. Nature 238, 469–470.

Weems, R.E., Lucas, S.G., 2020. The first record of vertebrate ichnofossils from the Norton Formation (Pennsylvanian, Middle Moscovian) in southwestern Virginia, USA. New Mexico Museum Nat. Hist. Sci. Bull. 83.

Weingardt, H.W., 1961. Tierfahrten-Funde aus der Steinkohlenzeit. Schacht und Heim, Werkzeitung der Saarbergwerke A.G., p. 7

Whyte, M.A., Romano, M., 2001. A dinosaur ichnocoenosis from the Middle Jurassic of Yorkshire, UK. Ichnos 8, 223–234.

Willard, B., Cleaves, A.B., 1930. Amphibian footprints from the Pennsylvanian of the Narrangasett basin. Geol. Soc. Am. Bull. 41, 321–327.

Wolansky, D., 1952. Ein neuer Fährtenfund im Karbon des Ruhrbezirks. Paläontol. Z. 30, 199–206.

Wood, N.J., Miller, R.F., 2007. A Mississippian trackway (Pseudobradypus ichnosp.) from the Enrage Formation, New Brunswick, Canada. Atl. Geol. 43, 180–186.

Woodworth, J.B., 1900. Vertebrate footprints on Carboniferous shales of Plainville, Masachusetts. Geol. Soc. Am. Bull. 11, 449–454.

Further reading

Chestnut Jr., D.R., Baird, D., Smith, J.H., Lewis, R.Q.,S., 1994. Reptile trackway from the Lee Formation (Lower Pennsylvanian) of south-central Kentucky. J. Paleontol. 68, 154–158.

Falcon-Lang, H.J., Benton, M.J., Braddy, S.J., Davies, S.J., 2006. The Pennsylvanian tropical biome reconstructed from the Joggins Formation of Nova Scotia, Canada. J. Geol. Soc. 163, 561–576.

Fichter, J., 1982. Tetrapodenfährten aus dem Oberkarbon (Westfalium A und C) west-und Südwestdeutschlands. Mainzer Geowissenschaft Mitteilungen 11, 33–37.

Fillmore, D.L., Simpson, E.L., Lucas, S.G., 2009a. Following Isaac Lea's footprints. Penn. Geol. 37, 5–11.
Fillmore, D.L., Lucas, S.G., Simpson, E.L., 2010. Invertebrate trace fossils in semi-arid to arid braided-ephemeral-river deposits of the Mississippian middle member of the Mauch Chunk Formation, eastern Pennsylvania, USA. Palaeogeogr. Palaeoclimatol. Palaeoecol. 292, 222–244.
Gastaldo, R.A., Bamford, M., Calder, J., DiMichele, W.A., Iannuzzi, R., Jasper, A., Kerp, H., McLoughlin, S., Oplustil, S., Pfefferkorn, H.W., Rossler, R., Wang, J., 2020. The coal farms of the late Paleozoic. In: Martinetto, E. (Ed.), Nature Through Time. Springer Textbooks in Earth Sciences, Geography and Environment.
Houck, K., Lockley, M.G., 1986. Pennsylvanian Biofacies of Central Colorado, 2. University of Colorado at Denver, Geology Department Magazine Special Issue, pp. 1–64.
Hunt, A.P., Lucas, S.G., 2005. Tetrapod ichnofacies and their utility in the Paleozoic. Alabama Paleontol. Soc. Monogr. 1, 113–119.
Hunt, A.P., Lucas, S.G., 2007. Tetrapod ichnofacies: a new paradigm. Ichnos 14, 59–68.
Hunt, A.P., Lucas, S.G., 2016. The case for archetypal vertebrate ichnofacies. Ichnos 23, 237–247.
Keighley, D.G., Calder, J.H., Park, A.F., Pickerill, R.K., Waldron, J.W., Falcon-Lang, H.J., Benton, M.J., 2008. Discussion on ecology of earliest reptiles inferred from basal Pennsylvanian trackways. J. Geol. Soc. 165, 983–987.
Lockley, M.G., Meyer, C.A., 2000. Dinosaur Tracks and Other Fossil Footprints of Europe. Columbia University Press, New York.
Lucas, S.G., Fillmore, D.L., Simpson, E.L., 2010b. Amphibian body impressions from the Mississippian of Pennsylvania, USA. Ichnos 17, 172–176.
Lucas, S.G., Mansky, C., Fillmore, D.L., Calder, J., Simpson, E.L., 2010c. Mississippian tetrapod footprint assemblages from Pennsylvania and Nova Scotia and the oldest record of amniotes: Geological Society of America. Abstr. Progr. 45 (5), 641–642.
Marsh, O.C., 1862. Description of the remains of a new enaliosaurian (Eosaurus acadianus) from the coal formation of Nova Scotia. Am. J. Sci. 34, 1–16.
Monks, J., 2002. A new occurrence of the reptile trackway *Notalacerta missouriensis* from western Indiana. J. Vertebr. Paleontol. 22 (3), 89A.
Moore, W.D., 1873. On footprints in the Carboniferous rocks of western Pennsylvania. Am. J. Sci. 3 (5), 292–293.
Patterson, R.P., 1971. Fossil trackways from the upper Pennsylvanian Monogahela Formation in southeastern Ohio. Earth Sci. 24, 181–185.
Pemberton, G., MacEachern, J.A., Gingras, M., 2009. Atlas of Trace Fossils. Elsevier Science, 300 pp.
Pickerill, R.K., 1994. Nomenclature and taxonomy of invertebrate trace fossils. In: Donovan, S.K. (Ed.), The Palaeobiology of Trace Fossils. The Johns Hopkins University Press, Baltimore, pp. 3–42.
Playford, G., 1963. Lower Carboniferous microfloras of Spitsbergen, part two. Palaeontology 5, 619–678.
Rowland, S.M., 2017. Trackway of a sideways-walking basal tetrapod in the Pennsylvanian Manakacha Formation in Grand Canyon National Park. Geol. Soc. Am., Abstr. Progr. 49, 6.
Rygel, M.C., Sheldon, E.P., Stimson, M.R., Calder, J.H., Ashley, K.T., Salg, J.L., 2014. Sedimentological framework of a portion of the Joggins Fossil Cliffs UNESCO World Heritage Site. Atl. Geol. 50, 249–289.
Schmidt, H., 1956. Die grosse Bochumer Oberkarbon-Fährte. Paläontol. Z. 30, 199–206.
Schmidt, H., 1963. *Herpetichnus erneri* n. sp., eine Reptilfährte aus dem westfälischen Oberkarbon. Paläontol. Z. 37, 179–184.

CHAPTER

4

Permian vertebrate tracks

Lorenzo Marchetti[1], *Eudald Mujal*[2,3], *Antoine Logghe*[4], *Michael Buchwitz*[5], *Hendrik Klein*[6] *and Spencer G. Lucas*[7]

[1]Museum für Naturkunde Berlin, Leibniz-Institut für Evolutions- und Biodiversitätsforschung, Berlin, Germany; [2]Staatliches Museum für Naturkunde Stuttgart, Stuttgart, Germany; [3]Institut Català de Paleontologia Miquel Crusafont (ICP-CERCA), Universitat Autònoma de Barcelona, Edifici ICTA-ICP, Catalonia, Spain; [4]Centre de recherche en paléontologie—Paris, UMR 7207—CNRS-MNHN-Sorbonne Université, Muséum national d'Histoire naturelle, Paris, France; [5]Museum für Naturkunde Magdeburg, Magdeburg, Germany; [6]Saurierwelt Paläontologisches Museum, Neumarkt, Germany; [7]New Mexico Museum of Natural History and Science, Albuquerque, NM, United States

1. Introduction

The Permian is a key time period in the evolution of terrestrial vertebrates (e.g., Lucas, 2017, 2018; Pardo et al., 2019). Important groups appeared and diversified, including the therapsid synapsids and the neodiapsids; others radiated, such as the temnospondyls, lepospondyls, diadectomorphs, captorhinomorphs, and parareptiles, and some went extinct, such as the lepospondyls and captorhinomorphs (e.g., Müller and Reisz, 2006; Ruta and Benton, 2008; Reisz et al., 2011; Ruta et al., 2011; Liu and Bever, 2015; Modesto et al., 2015; Brocklehurst et al., 2018). Key locomotory innovations, such as a semi-erect posture, were acquired for the first time (e.g., Bendel et al., 2023; Gônet et al., 2023). The major evolutionary radiations and extinctions had a remarkable impact on the tetrapod faunas (e.g., Lucas, 2017; Pardo et al., 2019). Nevertheless, the Permian skeletal record is often restricted to a few sites that do not overlap in time and paleolatitude, causing problems for the understanding of faunal distribution and evolution (e.g., Lucas, 2006, 2018).

In contrast, Permian tetrapod footprints have a remarkable record with regard to abundance, geographical and stratigraphical distribution, and morphological preservation. Therefore, they are also important to cover potential biases in the skeletal record (e.g., Voigt and Lucas, 2018). Moreover, although being the first vertebrate footprints ever described (Grierson, 1828), only in the last 50 years has their study thoroughly considered the taphonomic biases and pointed

Vertebrate Ichnology
https://doi.org/10.1016/B978-0-443-13837-9.00007-X

to an approach strictly relying on anatomy-consistent morphological features, which has a noteworthy higher potential than other approaches for stable ichnotaxonomy, biostratigraphy, paleobiogeography, paleoecology, and track-trackmaker correlation (e.g., Haubold, 1996; Voigt, 2005; Marchetti et al., 2019a). We herein review the Permian footprint record at the ichnogenus level, providing an overview of the taphonomy, ichnotaxonomy, ichnofaunal diversity, track-trackmaker correlation, posture, locomotion, paleoecology, and biostratigraphy.

2. History of study

The first tetrapod footprints described scientifically were from the Permian and more specifically from the Lopingian eolian sandstones of Scotland (Anonymous, 1828; Buckland, 1828; Grierson, 1828). After a few publications on the subject (e.g., Duncan, 1829, 1831; Pemberton, 1829; Blak, 1846), these footprints were officially described, being the first described ichnospecies in the Permian, "*Chelichnus duncani*" (Owen, 1842) and the first described ichnogenus in the Permian, "*Chelichnus*" Jardine (1850). During the 19th century, the first studies took place of Permian tetrapod footprints from fluvial and marginal lacustrine deposits, from Cisuralian localities now in Germany (Cotta, 1848; Pohlig, 1885, 1892; Pabst, 1895, 1896), the Czech Republic (Geinitz, 1861; Fritsch, 1895), Poland (Geinitz, 1863), and Italy (Geinitz, 1869; Curioni, 1870). Also, there was the first report of tetrapod footprints from the Lopingian red beds of Italy (Kittl, 1891).

In the first half of the 20th century, the first comprehensive studies of Permian tracks from the United States were carried out, from Cisuralian fluvial and eolian units of Arizona in the Grand Canyon (Lull, 1918; Gilmore, 1927, 1928), playa lake units of Texas (Williston, 1908; Moodie, 1929, 1930), and fluvial units of West Virginia (Tilton, 1931; Romer and Price, 1940). Tracks were also described from Kansas (Moodie, 1913). Tetrapod footprints were also first described from the Cisuralian of France (Fabre, 1872; Delage, 1912). Further tracks were described from the Cisuralian of Germany (Pabst, 1908; Schmidtgen, 1928; Korn, 1933), the Czech Republic (Fritsch, 1901), and Italy (Dozy, 1935), and from the Lopingian of Italy (Abel, 1929) and the Lopingian–Induan of South Africa (Nopcsa, 1923). The footprints from the Lopingian of South Africa were the first described records from Gondwana (Seeley, 1904, 1905).

In the second half of the 20th century, there was a noteworthy increase in studies on Permian tetrapod footprints, establishing the bases of modern-day Permian tetrapod ichnology. Extensive work has been done on the Cisuralian tetrapod ichnoassociations of Germany (Müller, 1954; Steiner and Schneider, 1963; Haubold, 1970, 1971a, 1973, 1998; Holub and Kozur, 1981; Martens, 1982; Fichter, 1976, 1983a,b, 1984, 1998; Fichter and Kowalczyk, 1983; Haubold and Stapf, 1998). At the same time, the first tracks from the Lopingian Zechstein of Germany—including trackways from eolian rocks—were described (Schmidt, 1952, 1959; Müller, 1959, 1970). Biostratigraphic schemes based on German footprints were proposed by Boy and Fichter (1982, 1988a,b). It is around that time that the first footprints from the Cisuralian of Austria were mentioned and described (Niedermayr and Scheriau-Niedermayr, 1979). Tetrapod footprints from Lopingian red beds of Italy were extensively studied and described, becoming the worldwide reference ichnoassociation for the Lopingian (Leonardi, 1951; Leonardi et al., 1975; Conti et al., 1975, 1977, 1987; Mietto, 1981; Ceoloni et al., 1988). More footprints were described from the Cisuralian fluvial to marginal lacustrine units of Italy (Berruti, 1969; Ceoloni et al., 1987; Conti et al., 1991; Santi and Krieger, 1999). These Italian records were also evaluated in terms of footprint biostratigraphy (Conti et al., 1997).

Extensive tetrapod ichnoassociations were described from Cisuralian fluvial to marginal lacustrine units of France (Ellenberger and Ellenberger, 1959; Heyler and Lessertisseur, 1962, 1963; Gand and Haubold, 1984; Gand, 1985, 1987, 1989, 1991, 1993a,b). More tetrapod ichnoassociations were described from Guadalupian units of France, constituting the best worldwide record from this time interval (Heyler and Lessettisseur, 1963; Heyler and Montenat, 1980; Ellenberger, 1983a; Demathieu et al., 1992; Gand et al., 1995). Because of the footprint diversity and overall stratigraphic continuity of the French track-bearing levels in the Cisuralian–Guadalupian time interval, biostratigraphic schemes of tetrapod footprints were proposed (Ellenberger, 1983b,c, 1984; Gand, 1987). Footprints from the Lopingian eolian units of Scotland were revised by McKeever and Haubold (1996). New footprints were described from Cisuralian units of Spain (Gand et al., 1997), Hungary (Majoros, 1965), Ukraine (Schneider et al., 1992), and Russia (Lucas et al., 1999), and from Guadalupian–Lopingian units of Spain (Robles and Llompart, 1987) and Russia (Tverdokhlebov et al., 1997). From northern Africa, Permian tracks were described for the first time from Guadalupian to Lopingian fluvial units of Tunisia (Newell et al., 1976) and from Cisuralian units of Morocco (Broutin et al., 1987). Important new tracksites were reported from the Guadalupian of South Africa (Smith, 1980, 1993; De Beer, 1986). Tracks from coal measures of Australia were alternatively dated as Lopingian or Early Triassic (Retallack, 1996; Warren, 1997); recent studies have attributed them to the latest Lopingian (McLoughlin et al., 2021). From Canada, tracks were reported for the first time from Cisuralian red beds of Prince Edward Island in eastern Canada (Mossmann and Place, 1989).

The most significant discovery in the United States was the Cisuralian megatracksite of the Robledo Mountains, in New Mexico (Hunt et al., 1993, 1995; MacDonald, 1994). The quantity and preservation of the material was a key to understanding the similarities between the North American and European track records in the Cisuralian, and the effect of taphonomy on tracks (Haubold et al., 1995; Haubold, 1996). Moreover, historic track-bearing sites from Texas (Sarjeant, 1971) and Arizona (Brand, 1979) were re-described. Regarding the first Permian footprint record from South America, tracks from the Cisuralian of Argentina were described during this time period (Cei and Gargiulo, 1977; Aramayo and Farinati, 1983; Melchor, 1997; Caselli and Arucci, 1999).

The 21st century has witnessed a further increase in studies of Permian footprints. The ichnotaxonomic concepts that had been developed in the previous 50 years were perfected (e.g., Haubold, 2000; Voigt, 2005; Marchetti et al., 2019a), the track-trackmaker correlation became more accurate (e.g., Voigt et al., 2007; Marchetti et al., 2017a; Logghe et al., 2021; Matamales-Andreu et al., 2023) and new methods such as 3D modeling were employed (e.g., de Klerk, 2002; Mujal et al., 2016, 2020). Footprints became an important tool in chronostratigraphy (e.g., Lucas, 2007; Voigt and Lucas, 2018), in paleoecology (e.g., Hunt et al., 2005; Hunt and Lucas, 2006, 2007; Marchetti et al., 2019b; Francischini et al., 2020), in the understanding of disruptive climatic events (e.g., Marchetti et al., 2022a), and in biomechanics, as regard locomotion and posture (e.g., Hunt and Lucas, 1998; Hopson, 2015; Nyakatura et al., 2019). Important revisional studies were carried out on the Cisuralian–Guadalupian tetrapod ichnoassociations of Germany (Voigt and Haubold, 2000; Voigt, 2005, 2007; Buchwitz et al., 2020), and a new footprint biostratigraphy was proposed (Voigt et al., 2012). Further studies on these ichnoassociations focused on paleoecology (Marchetti et al., 2018) and locomotion (Romano et al., 2016; Nyakatura et al., 2019; Buchwitz and Voigt, 2018; Buchwitz et al., 2021). New sites were reported from fluvial facies of the Lopingian of Germany (Buchwitz et al., 2017; Voigt and Fischer,

2020), and new revisions carried out, after a thorough ichnotaphonomic study, on eolian ichnoassociations from the same time interval (Marchetti et al., 2019c).

New, abundant and well-preserved material was described from the Cisuralian basins of Italy, the most diverse footprint association of the Kungurian (e.g., Nicosia et al., 2000; Santi and Krieger, 2001; Santi, 2003; Arduini et al., 2003; Avanzini et al., 2008; Marchetti et al., 2013a,b, 2014, 2015a,b; Petti et al., 2014; Citton et al., 2019a; Santi et al., 2020). Also, studies on biostratigraphy (Avanzini et al., 2001), locomotion (Bernardi and Avanzini, 2011; Petti et al., 2014; Marchetti et al., 2017a), paleoecology (Marchetti et al., 2015c, 2017b), and climatic events (Marchetti et al., 2022a) were carried out. Studies on Lopingian tracks from the red beds of the Dolomites in Italy were done by Nicosia et al. (2001), Valentini et al. (2007, 2008, 2009), and Citton et al. (2016, 2019b), and the ichnoassociation was comprehensively revised by Marchetti et al. (2019d, 2020a). Footprints have been described from the Cisuralian fluvial-lacustrine units of France (e.g., Barrier et al., 2009; Moreau and Gand, 2022; Moreau et al., 2022), and some important Guadalupian ichnoassociations have been revised (e.g., Gand et al., 2000; Marchetti et al., 2021a, 2022b) and biostratigraphic schemes proposed (Gand and Durand, 2006). Other studies focused on taphonomy and locomotion (Marchetti, 2019; Mujal and Marchetti, 2020; Logghe et al., 2021).

The footprint associations from the Lopingian eolian units of Scotland were thoroughly revised (Marchetti et al., 2019c). This was based on a strictly anatomy-based approach that considered "*Chelichnus*" as a taphotaxon *sensu* Lucas (2001). New material from Spain has been described and reported from the Cisuralian, Guadalupian and possibly Lopingian fluvial and lacustrine-like units of the Catalan Pyrenees (Voigt and Haubold, 2015; Mujal et al., 2016, 2017, 2018; Marchetti et al., 2021a; De Jaime-Soguero et al., 2023). Tetrapod footprints were also described from Cisuralian–Guadalupian fluvial units of the Balearic Islands (Matamales-Andreu et al., 2021a,b, 2022, 2023). New material was described from the Cisuralian of Austria (Voigt et al., 2015a,b), Poland (Voigt et al., 2012a,b) and the Czech Republic, the latter focusing on skin-fold traces (Calábková et al., 2022, 2023a). Possible Lopingian ichnoassociations have been described from Poland (Ptaszyński and Niedźwiedzki, 2004). However, these tracks are poorly preserved, so their ichnotaxonomy is dubious, as well as their age, which could be Early Triassic (Racki, 2005). New footprints have been described from the Cisuralian of Turkey (Gand et al., 2011) and from the Lopingian of Russia (Gubin et al., 2003; Surkov et al., 2007), including the longest known pareiasaur trackway (Kuznetsov, 2020). Studies on Permian tetrapod footprints from Africa witnessed a significant increase. New material has been described from Morocco, specifically from Cisuralian units (Hmich et al., 2006; Voigt et al., 2011a,b), Cisuralian-Guadalupian units (Zouicha et al., 2021) and also from Guadalupian-Lopingian units (Voigt et al., 2010; Hminna et al., 2012; Moreau et al., 2020). A new ichnoassociation has been reported from Guadalupian–Lopingian units of Niger (Smith et al., 2015), and previously known material from the Guadalupian–Lopingian units of Tunisia has been described (Contessi et al., 2018). Several new tracksites have been described from Guadalupian and Lopingian units of South Africa (de Klerk, 2002; Valentini et al., 2009; Marchetti et al., 2019e; Cisneros et al., 2020; Groenewald et al., 2023). The footprints from the Poortje Member of the Beaufort Formation were considered to have an age between Lopingian and Early Triassic (Smith and Botha-Brink, 2014; Marchetti et al., 2019e), Gastaldo et al. (2020) dated this unit as Lopingian.

Several new studies have detailed the Permian ichnoassociations of North America. New footprint material has been described from the Cisuralian red beds of Prince Edward Island, in Canada (Calder et al., 2004; Brink et al., 2012) and the Gzhelian–Asselian Cape John Formation of Nova Scotia, Canada (Van Allen et al., 2005). A large number of new footprint localities and material have been described from Cisuralian units of New Mexico (Lucas et al., 2001, 2005a,b, 2013; Voigt and Lucas, 2015a,b, 2016, 2017; Voigt et al., 2013a,b). The scientific significance of the Robledo Mountains megatracksite led to the establishment of the Prehistoric Trackways National Monument in New Mexico by the U.S. Government (Lucas et al., 2011a). Furthermore, nonamniote footprints were discovered in Cisuralian desert dune environments of Arizona (Francischini et al., 2020). Also from Arizona, important historical tracksites from Cisuralian red beds and eolian units have been re-evaluated (Marchetti et al., 2019b, 2021b) as have those from red beds of Texas (Haubold and Lucas, 2001, 2003). New tracksites were also described from different Cisuralian fluvial units of Texas (Lucas and Hunt, 2005; Lucas et al., 2011b), Oklahoma (Lucas and Suneson, 2002; Sacchi et al., 2014), Colorado (Voigt et al., 2005), Pennsylvania, West Virginia (Lucas et al., 2016) and eolian units of Utah (Chure et al., 2014). New ichnoassociations have also been described from South America, and more specifically from Cisuralian marginal lacustrine and eolian units of Argentina (Melchor, 2001; Melchor and Sarjeant, 2004; Krapovickas et al., 2015). New footprints have been described from the Permian/Triassic transition of Argentina (Citton et al., 2021), in a stratigraphic interval previously considered Late Triassic (Melchor and de Valais, 2006). Swimming traces have been reported from the Cisuralian of Brazil (Silva et al., 2009; Silva and Sedor, 2017), and poorly preserved tracks have been described from the Guadalupian–Lopingian of Brazil (Silva et al., 2012; Francischini et al., 2018) and Uruguay (Piñeiro et al., 2022).

After 200 years research on Permian tetrapod footprints, studies on this topic continue; and numerous actual projects initiated by different experts in the field will further increase our knowledge in the future.

3. Material and methods

For the purpose of this review, all of the valid ichnogenera occurring in the Permian have been discussed with a comparative diagnosis and precise reference to the type material, specifying whether the type material is Permian or not. Instead, the general description of the valid ichnogenera refers to Permian material only. Other, stratigraphically older and younger occurrences are mentioned, but not described in detail. Valid ichnospecies are mentioned but not discussed with a diagnosis or description. New assignments follow the provided diagnoses and descriptions. The term preservation refers to the morphological preservation of Marchetti et al. (2019a), when not otherwise specified. The terms ectaxony and mesaxony refer only to the relative digit length proportions.

3.1 Institutional abbreviations

DUMFM: Dumfries Museum and Camera Obscura, Dumfries, Scotland, UK.
FG: Geologisches Institut, TU Bergakademie Freiberg, Germany.

FMC: Field Museum of Natural History, Chicago, Illinois, USA.

GRCA: Grand Canyon National Park, Yavapai Geological Museum, Grand Canyon Village, Arizona, USA.

HF: Geologisches Institut, Martin-Luther-Universität Halle-Wittenberg, Germany.

IGPF: Museum of Paleontology and Prehistory "P. Leonardi," University of Ferrara, Italy.

L. U.; Art and Heritage Collections (Geology), University of Liverpool, England (UK).

MB: Museum für Naturkunde Berlin, Germany.

MCZ: Museum of Comparative Zoology, Harvard University, Cambridge, Massachusetts, USA.

MNG: Museum der Natur, Friedenstein Stiftung, Gotha, Germany.

MNHN: Muséum National d'Histoire Naturelle, Paris, France.

MSNM: Museo di Storia Naturale di Milano, Italy.

NHMMZ: Naturhistorisches Museum Mainz/Landessammlung für Naturkunde Rheinland-Pfalz, Mainz, Germany.

NHMS: Naturhistorisches Museum Schloss Bertholdsburg, Schleusingen, Germany.

NHMUK: British Museum of Natural History, London, England, UK.

NMMNH: New Mexico Museum of Natural History, Albuquerque, New Mexico, USA.

NML: Naturkundemuseum Leipzig, Germany.

NMS: Museum of Nature South Tyrol, Bozen/Bolzano, Italy.

PMJ: Phyletisches Museum Jena, Germany.

RAM: Raymond M. Alf Museum of Paleontology, Claremont, California, USA.

RGMW: Regionalmuseum Wolfhagen, Germany.

RM: Redpath Museum, McGill University, Montreal, Québec, Canada.

SNSB-BSPG: Bayerische Staatssammlung für Paläontologie und Geologie, Munich, Germany.

UCMP: University of California, Berkeley, California, USA.

UG: Faculty of Geoscience and Geography, Georg-August-Universität Göttingen, Germany.

UGKU: Urweltmuseum Geoskop, Thallichtenberg, Germany.

ULS: University La Salle, Beauvais, France.

UM: Université de Montpellier, France.

UR: University "La Sapienza" of Rome, Italy.

USNM: National Museum of Natural History, Smithsonian Institution, Washington, D.C., USA.

WVGES: West Virginia Geological and Economic Survey, Morgantown, West Virginia, USA.

YPM: Yale Peabody Museum, New Haven, Connecticut, USA.

3.2 Localities with specimens in situ

GF: Gansfontein farm (S 31°54.142′ E 021°28.854′), near Fraserburg, Northern Cape, South Africa.

KW: Klein Waterval (S 32°58.315′ E 022°24.939′), near Prince Albert, Western Cape, South Africa.

4. Systematic ichnology

4.1 *Batrachichnus* Woodworth, 1900 (Fig. 4.1A and B)

4.1.1 Type ichnospecies

Batrachichnus plainvillensis Woodworth (1900).

4.1.2 Holotype

MCZ 1052 (Woodworth, 1900: fig. 2, pl. 40.1). Trackway with tail trace, concave epirelief, and convex hyporelief. Coal Measures, Pennsylvanian, USA (Massachusetts).

4.1.3 Valid ichnospecies

Batrachichnus salamandroides (Geinitz, 1861).

4.1.4 Distribution

Mississippian–Middle Triassic. The Permian (mainly Cisuralian, but also Guadalupian–Lopingian) record of this ichnogenus includes localities from the Cisuralian of the USA (Arizona, New Mexico, Texas), Spain, France, Italy, Germany, the Czech Republic, Hungary, Morocco, and Argentina; the Guadalupian of France and South Africa; the Lopingian of ? Italy, Germany, and Russia (Table 4.1).

4.1.5 Diagnosis (after Woodworth, 1900, emended)

Small tracks (maximum size of 35 mm, that of *Limnopus* can be larger) of a four-footed animal, with fairly distinct impressions of fore and hind feet. Manus imprint generally about as

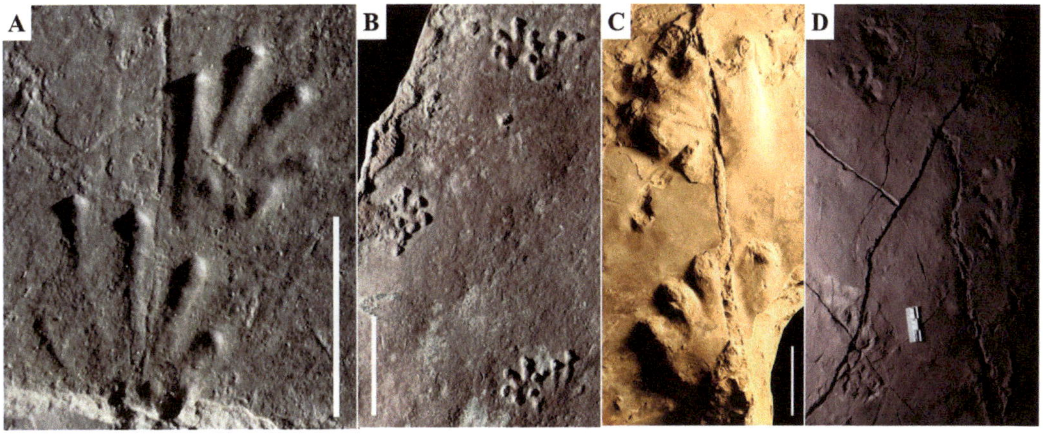

FIGURE 4.1 Temnospondyl nonamniote tracks. (A–B) *Batrachichnus salamandroides*. Oberhof Formation, Cisuralian, Germany. (A) MNG VF 4307-4314. Right pes-manus couple, convex hyporelief. (B) MNG 2364. Trackway, convex hyporelief. (C–D) *Limnopus heterodactylus*. (C) NHMMZ K 213. Left pes-manus couple, convex hyporelief. Meisenheim Formation, Cisuralian, Germany. (D) NMMNH P-24,603. Artificial cast of field specimen, trackway, concave epirelief. Abo Formation, Cisuralian, New Mexico, USA. Scale bars: A–B = 1 cm; C = 5 cm.

TABLE 4.1 Tetrapod footprint occurrences in Permian units. Uncertain occurrences with question mark and re-assignments in red.

Country	Basin	Formation/unit	Biochron	References	*Batrachichnus*	*Limnopus*	*Amphisauropus*	*Matthewichnus*	*Ichniotherium*	*Dimetropus*	*Tambachichnium*	*Brontopus*	*Dolomitipes*	*Karoopes*	*Capitosauroides*	*Dicynodontipus*	*Varanopus*	*Erpetopus*	*Procolophonichnium*	*Pachypes*	*Notalacerta*	*Hyloidichnus*	*Dromopus*	*Rhynchosauroides*	*Paradoxichnium*	*Protochirotherium*	new assignment	reference
Utah		Weber	*Dromopus*	Chure et al. 2014					X																		*Ichniotherium cottae*	Chure et al. 2014, fig. 4A-B
Arizona		Hermit	*Erpetopus*	Marchetti et al. 2021a, 2021b	X		X		X	X								X		X		X	X					
		De Chelly	*Erpetopus*	Marchetti et al. 2019b			X		X								X?	X					X?					
		Coconino	*Erpetopus*	Marchetti et al. 2019b			X?		X		X?						X	X					X?					
Colorado		Maroon	*Dromopus*	Voigt et al. 2005					X	X	X						X											
		Lyons	*Erpetopus*	Marchetti et al. 2019b													X?											
New Mexico		Abo	*Dromopus*	Voigt and Lucas 2017	X	X	X	X	X	X	X						X					X	X				*Matthewichnus* isp.	this work, Fig. 2D
		Abo (top)	*Erpetopus*	Voigt and Lucas 2017	X	X	X			X?								X					X					
		Arroyo de Alamillo	*Erpetopus*	Voigt and Lucas 2017	X		X											X					X					
		Sangre de Cristo	*Dromopus*	Voigt and Lucas 2015b	X	X		X?	X	X	X										X?	X?	X					
		Robledo Mountains	*Dromopus*	Voigt and Lucas 2015a	X		X	X		X							X				X	X	X				*Varanopus macdonaldi* n. comb.	Voigt et al. 2013, figs. 2-3
Oklahoma		Wellington	*Dromopus*	Sacchi et al. 2014						X																		
		Hennessey	*Erpetopus*	Lucas and Suneson 2002																		X	X				*Hyloidichnus* isp.	Lucas and Suneson 2002, figs. 5-6
Texas		Lueders	*Dromopus*	Lucas and Hunt 2005						X																		
		Clear Fork (base)	*Dromopus*	Lucas et al. 2011b	X		X?																X					
		Vale	*Erpetopus*	Lucas et al. 2011b						X																		
		Choza	*Erpetopus*	Haubold and Lucas 2003; Minter et al. 2007	X?												X	X					X				cf. *Batrachichnus* isp.	Minter et al. 2007, fig. 15B
		San Angelo	*Erpetopus*	Lucas and Hunt 2005, Marchetti et al. 2021a																X								
Ohio		Washington	*Dromopus*	Haubold, 1971a; Buchwitz and Voigt 2018		X			X														X					
West Virginia		Waynesburg	*Dromopus*	Tilton 1931; Baird 1952		X				X																		
		Washington	*Dromopus*	Lucas et al. 2016						X																		
Pennsylvania		Greene	*Dromopus*	Lucas et al. 2016						X																		
Prince Edward Island		Kildare Capes	*Dromopus*	Mossmann and Place 1989; Brink et al. 2012			X		X																			
		Hillsborough River	*Dromopus*	Calder et al. 2004													X										*Varanopus* isp.	Calder et al. 2004, fig. 5B
Nova Scotia		Cape John	*Dromopus*	Van Allen et al. 2005		X	X			X							X						X					
Scotland		Corncockle	*Dicynodontipus*	Marchetti et al. 2019c											X?	X			X?	X				X				
		Locharbriggs	*Dicynodontipus*	Marchetti et al. 2019c															X?	X								
England		Kenilworth	*Dromopus*	Marchetti et al. 2020b																	X?							
Spain		Sotres	*Erpetopus*	Gand et al. 1997		X?																X						
		Peranera	*Erpetopus*	Voigt and Haubold 2015, Mujal et al. 2016, Marchetti et al. 2021a	X	X			X?	X							X	X		X	X	X	X				*Erpetopus* isp.	Mujal et al. 2016, fig. 8

		Upper Red Unit (base)	*Erpetopus*	De Jaime-Soguero et al. 2023	X																	X	X					
		Upper Red Unit	*Brontopus*/ *Dicynodontipus*	Mujal et al. 2017, De Jaime-Soguero et al. 2023								X	X?							X		X	X				*cf. Dolomitipes* isp.	Mujal et al. 2017, fig. 6A-B
	Mallorca	Port des Canonge	*Erpetopus*/ *Brontopus*	Matamales-Andreu et al. 2021a, 2022						X										X?		X	X					
	Menorca	P3	*Brontopus*/ *Dicynodontipus*	Matamales-Andreu et al. 2021b														X?				X						
France	Brive	Grammont	*Dromopus*	Gand 1987	X	X																	X?					
		Meissac	*Dromopus*	Gand 1987						X																		
	Rodez	Serie II	*Dromopus*	Gand 1987	X	X	X			X													X					
	Saint-Affrique	Saint-Rome	*Dromopus*	Gand 1993a	X	X				X													X					
		Grès du Dourdou	*Dromopus*	Gand 1993a, Marchetti et al. 2020b	X	X				X											X		X					
		Saint Pierre	*Erpetopus*	Gand 1993a	X	X				X								X				X	X				*Erpetopus* isp.	Gand 1993a, fig. 9.1.C
	Lodève	Usclas Saint-Privat	*Dromopus*	Gand and Durand 2006	X	X?				X?													X					
		Tuilières-Loiras	*Dromopus*	Gand and Durand 2006, Mujal and Marchetti 2020	X	X			X	X													X					
		Viala	*Dromopus*	Gand and Durand 2006, Marchetti et al. 2020b	X	X	X			X							X						X					
		Rabejac	*Erpetopus*	Gand and Durand 2006, Marchetti et al. 2020b, 2021a	X	X				X	X							X		X	X?	X	X				*Erpetopus* isp.	Gand and Durand 2006, fig. 2.D.1
		Mérifons	*Erpetopus*	Gand and Durand 2006	X																	X						
		La Lieude	*Brontopus*	Gand and Durand 2006, Marchetti et al. 2019e, 2021a								X								X		X	X				*Hyloidichnus thaleri* n. comb.	Gand and Durand 2006, fig. 2.E.1
	Largentière	Lande	*Dromopus*	Gand 1993b	X	X				X																		
	Blanzy-Le Creusot	Grandmont	*Erpetopus*	Gand 1981														X				X					*Erpetopus* isp.	Gand 1981, pl. II
	Bas-Argens-Esterel	Bayonne	*Erpetopus*	Gand and Durand 2006, Marchetti et al. 2022b														X					X					
		Pradineaux	*Brontopus*	Gand and Durand 2006, Marchetti et al. 2022b	X?						X	X					X	X		X?		X	X					
		Le Mitan/Le Muy	*Brontopus*	Gand and Durand 2006, Marchetti et al. 2022b	X?															X			X					
		Pélitique	*Dicynodontipus*	Marchetti et al. 2022b	X										X	X	X					X		X				
	Barrot	Merveilles	*Erpetopus*	Barrier et al. 2009														X									*Erpetopus willistoni*	Barrier et al. 2009, figs. 5-8
Italy		Cala del Vino	*Erpetopus*	Citton et al. 2019a, 2023		X																X	X					
	Orobic	Pizzo del Diavolo	*Erpetopus*	Marchetti 2016; Marchetti et al. 2022a	X	X	X		X	X?							X	X				X	X				*Ichniotherium sphaerodactylum*	Marchetti 2016, fig. 2
	Collio	Collio	*Erpetopus*	Marchetti et al. 2022a	X?	X	X				X						X	X				X	X					
		Dosso dei Galli	*Erpetopus*	Marchetti et al. 2022a			X																X					
	Athesian District	Monte Luco	*Erpetopus*	Marchetti et al. 2022a		X												X		X?			X					
		Guncina	*Erpetopus*	Marchetti et al. 2022a																			X					
		Tregiovo	*Erpetopus*	Marchetti et al. 2022a	X													X?				X	X					

(Continued)

TABLE 4.1 Tetrapod footprint occurrences in Permian units. Uncertain occurrences with question mark and re-assignments in red.—cont'd

		Val Gardena Sandstone	*Paradoxichnium*	Marchetti et al. 2019d, 2020a	X?								X		X	X			X	X			X?	X	X	X?		
Austria		Laas	*Dromopus*	Voigt et al. 2015a			X		X														X					
Germany	Saar-Nahe	Altenglan	*Dromopus*	Voigt 2012						X?													X					
		Wahnwegen	*Dromopus*	Voigt 2012	X					X													X					
		Quirnbach	*Dromopus*	Voigt 2012	X				X?																			
		Lauterecken	*Dromopus*	Voigt 2012																			X					
		Meisenheim	*Dromopus*	Voigt 2012, Marchetti et al. 2020b	X	X			X?	X											X		X					
		Donnersberg	*Dromopus*	Voigt 2012	X?																		X					
		Wadern	*Dromopus*	Voigt 2012	X		X		X	X													X					
		Standenbuhl	*Dromopus*	Voigt 2012, Marchetti et al. 2020b	X?		X		X?	X?											X?		X					
	Zechstein	Rothenberg /Annweiler	*Paradoxichnium*	Voigt et al. 2023	X								X							X		X						
	Baden Baden	Michelbach	*Dromopus*	Voigt 2012					X	X																		
	Wetterau	Bleichenbach	*Dromopus*	Voigt 2012		X	X		X														X					
	Werra	Cornberg	*Dicynodontipus*	Marchetti et al. 2019c									X						X	X?				X				
	Thuringian Forest	Manebach	*Dromopus*	Voigt 2012																			X					
		Goldlauter	*Dromopus*	Martens, 1991; Voigt 2012	X	X	X		X	X	X						X						X				*Varanopus* isp.	Martens 1991, figs. 1-2
		Oberhof	*Dromopus*	Voigt 2012	X	X	X		X	X	X						X						X					
		Rotterode	*Dromopus*	Voigt 2012	X	X	X		X	X	X						X						X					
		Tambach	*Dromopus*	Voigt 2012, Marchetti et al. 2023	X?	X	X		X	X	X						X				X		X					
		Eisenach	*Brontopus*	Voigt 2005, Voigt 2012									X?		X?								X				cf. *Dolomitipes* isp. (1), cf. *Capitosauroides* isp. (2)	Voigt 2005, fig. 47A (1), fig. 47B-C (2)
	Doehlen-Briesnitz	Doehlen	*Dromopus*	Voigt 2012		X			X														X					
	Zechstein	Stassfurt	*Paradoxichnium*	Voigt 2012																					X			
	Ilfeld	Netzkater	*Dromopus*	Voigt 2012	X?																							
		Baumgarten	*Dromopus*	Voigt 2012					X?																			
	Saale	Hornburg	*Brontopus*	Buchwitz et al. 2020											X								X					
		Eisleben	*Dicynodontipus*	Buchwitz et al. 2023															X									
	Flechtingen	Foehrberg	*Dromopus*	Voigt 2012	X				X																			
		Dethlingen/ Hannover	*Dicynodontipus*	Buchwitz et et al. 2017, Buchwitz et al. 2023												X?		X?	X	X?					X?			
Poland	Intrasudetic	Słupiec	*Dromopus*	Voigt et al. 2012			X		X	X													X					
Czech Republic	Krkonose-Piedmont	Prosečné	*Dromopus*	Voigt et al. 2012	X		X																X					
	Boskovice	Padochov	*Dromopus*	Voigt 2005; Calabkova et al. 2022, 2023a, 2023b			X		X	X							X											

		Veverská Bitýška	*Dromopus*	Calabkova et al. 2023b					X																			
		Letovice	*Dromopus*	Mikulas and Martinek 2006, Calabkova et al. 2022, 2023a			X			X													X				*Dromopus lacertoides*	Mikulas and Martinek 2006, fig. 7A
Hungary		Turony	*Dromopus*	Barabásné Stuhl 1975	X																							
		Balatonfelvidék	*Paradoxichnium*	Majoros 1965, Kaszap 1968																		X					*Hyloidichnus* isp.	Majoros 1965, figs. 2-3
Ukraine	Donetsk	Kartamyshkaja	*Dromopus*	Schneider et al. 1992																			X					
Russia		Aksautskian	*Dromopus*	Lucas et al. 1999						X?																		
		Kinyrtchadskian	*Dromopus*	Lucas et al. 1999																			X?					
		Poldarsa, Isady Mb.	*Dicynodontipus*	Gubin et al. 2003, Kuznetsov et al. 2020																X							*Pachypes dolomiticus*	Kuznetsov et al. 2020, fig. 1
		Severodvinian horizon	*Dicynodontipus*	Tverdokhlebov et al. 1997	X																							
		Vyatkian Gorizont	*Paradoxichnium*	Surkov et al. 2007																X?							cf. *Pachypes* isp.	Surkov et al. 2007, fig. 4, fig. 5A
Turkey		Çakraz	*Erpetopus*	Gand et al. 2011													X										*Varanopus* isp.	Gand et al. 2011, fig. 4.1
Morocco	Kenifra		*Dromopus*	Voigt et al. 2011b	X	X			X	X													X					
	Tiddas		*Erpetopus*	Voigt et al. 2011a	X	X				X								X				X	X				*Erpetopus* isp.	Voigt et al. 2011, fig. 6B
	Koudiat El Hamra - Haiane	Koudiat El Hamra - Haiane	*Erpetopus/ Brontopus*	Zouicha et al. 2021	X?						X?											X	X					
		Cham-el-Houa	*Brontopus/ Dicynodontipus*	Moreau et al. 2020											X?			X				X					cf. *Capitosauroides* isp.	Moreau et al. 2020, fig. 4
		Ikakern	*Dicynodontipus*	Voigt et al. 2010; Hminna et al. 2012											X?			X		X		X	X				cf. *Capitosauroides* isp.	Hminna et al. 2012, fig. 5
Tunisia		Cheguimi	*Dicynodontipus*	Contessi et al. 2018												X						X					*Dicynodontipus geinitzi* (1), *Hyloidichnus thaleri* n. comb. (2)	Contessi et al. 2018, fig. 3A (2), fig. 3B (1)
Niger		Moradi	*Brontopus/ Dicynodontipus*	Smith et al. 2015																X?		X	X					
South Africa		Abrahamskraal	*Brontopus*	Marchetti et al. 2019e; Cisneros et al. 2020	X							X																
		Teekloof, Poortje Mb.	*Dicynodontipus*	Marchetti et al. 2019e									X	X	X?													
		Balfour, Oudeberg Mb.	*Dicynodontipus*	Marchetti et al. 2019e									X															
		Balfour, Poortje Mb.	*Paradoxichnium*	Marchetti et al. 2019e									X			X?			X					X				
Brazil		Rio do Rasto	*Dicynodontipus*	Silva et al. 2012												X			X					X				
		Pirambóia	*Paradoxichnium*	Francischini et al. 2018												X												
Uruguay		Yaguari	*Dicynodontipus*	Piñeiro et al., 2022										X?	X?					X?								
Argentina		Carapacha	*Erpetopus*	Melchor and Sarjeant 2004, Marchetti et al. 2020b	X													X			X	X					*Erpetopus* isp.	Melchor and Sarjeant 2004, fig. 6D
		Los Menucos	*Paradoxichnium*	Citton et al. 2021												X												
Australia		Coal measures	*Paradoxichnium*	Retallack 1996, Warren 1997									X			X											*Dolomitipes* isp. (1), *Dicynodontipus geinitzi* (2)	Retallack 1996, figs. 2-3 (1); Warren 1997, fig. 8 (2)

long as wide (different from *Amphisauropus*, which has a manus imprint wider than long) and about 1/3 shorter than pes imprint (different from *Matthewichnus*, which shows a relatively longer pes imprint). The manus imprint has four digit impressions (different from *Amphisauropus*, *Hylopus*, and *Ichniotherium*, which are pentadactyl); the pes imprint has five digit impressions. Relatively long and slender digit imprints (different from *Limnopus* and *Ichniotherium*, which have relatively shorter and/or thicker digit imprints). In the manus, digit IV is shorter than digit II (different from *Limnopus*, where digit IV is as long as digit II in the manus). Well-impressed circular basal pad of digit I (different from *Ichniotherium*, which lacks this feature). Broad trackways with alternated pes-manus couples (different from *Matthewichnus*, which shows secondary overstep and an irregular trackway pattern).

4.1.6 General description

Tracks of a quadruped. Manus tetradactyl and pes pentadactyl, both ectaxonic with well-splayed digit imprints and round, clawless digit tip imprints. Manus imprints (<20 mm long, usually around 10–15 mm) more commonly digitigrade to semiplantigrade, occasionally semidigitigrade to unguligrade, and more rarely plantigrade. Digit imprints of the manus are relatively slender, longer than the palm impression. Digits increase in length from I to III, which are straight to slightly curved inwards. Digit IV length between that of digits I and II, although it can be occasionally longer. Digit IV imprint separated from those of digits I–III (which form a group). Digit IV imprint is more proximally located with respect to the others and rotated outwards (occasionally also slightly curved outwards). Palm imprint wider than long. Round and deep imprint at the basal pad of digit I commonly present. The proximal margin of the palm is concave and shallower than the rest of the imprint. The round digit tip imprints are usually the deepest portions of the manus tracks. Pes tracks (up to ca. 35 mm long) larger than manus tracks, often longer than wide. They are semiplantigrade to plantigrade, although digitigrade and unguligrade imprints are also observed (usually missing digits IV and V). Digit imprints of the pes, like those of the manus, are relatively long and slender, longer than the sole impression. Digit imprints increase in length from I to IV, which are usually straight or slightly bent inwards. Digit V imprint commonly shorter than digit II separated from the digit I–IV group, more proximally located, and rotated and occasionally curved outwards. Sole imprint wider than long. Round and deep imprint at the basal pad of digit I commonly present. Pes imprints are usually shallower than manus imprints. Both manus and pes tracks are usually deeper medially, and the lateral portion (especially the region of pedal digits IV and V) is usually very shallowly impressed or not impressed at all. Footprints arranged in trackways of alternating manus-pes sets. Within each manus-pes set, pes imprints are more laterally positioned, and sometimes, they partially overlap the manus imprints. In trackways, manus tracks are commonly rotated inwards, whereas pes tracks are parallel to the midline or slightly rotated outwards. Pace angulation is usually <90° for both manus and pes tracks, with a clear alternating pattern. Stride length/pes length ratio variable, usually 2.0–3.0, but up to 7. Straight to sinuous tail traces are sometimes preserved.

4.1.7 Discussion

Batrachichnus was introduced by Woodworth (1900) based on a trackway from the Pennsylvanian of Massachusetts, with the ichnospecies *B. plainvillensis*. The most comprehensive recent review of this ichnogenus (see also reference works by Heyler and Lessertisseur, 1963; Haubold, 1970, 1971a; Gand, 1987; Haubold et al., 2005) was done by Voigt (2005), who considered *Batrachichnus salamandroides* (Geinitz, 1861) the only valid ichnospecies. Since this study, other authors have reported further records of *Batrachichnus*, extending its morphological, stratigraphic (lower Carboniferous–Middle Triassic), and paleogeographic ranges (e.g., Stimson et al., 2012; Mujal et al., 2016; Schneider et al., 2020; Cisneros et al., 2020; Klein and Lucas, 2021; Allen et al., 2022; Marchetti et al., 2022a,b). Additionally, we re-assign to cf. *Batrachichnus* isp. a specimen from the Choza Formation of Texas, because of its tetradactyl manus (Minter et al., 2007, fig. 15B).

Some *Batrachichnus* specimens from the Permian have been associated with trackways from subaquatic settings, sometimes even found in transition with other ichnogenera, such as *Lunichnium* (Petti et al., 2014) and *Characichnos* (Lucas et al., 2011b; Mujal et al., 2016; De Jaime-Soguero et al., 2023). In fact, *Batrachichnus* is mainly found in deposits related to environments with the presence of water. Neoichnological analyses of salamander trackways, which are similar to *Batrachichnus*, show that part of the morphological variability and trackway patterns depend on rheology and environmental conditions (including composition of the substrate and amount of water) (Brand, 1996; see also Melchor and Sarjeant, 2004). Such variability, including potential extramorphologies (e.g., Melchor and Sarjeant, 2004; Haubold et al., 1995; Marchetti et al., 2019a), supports a wide spectrum of habitats, but this alone cannot explain the wide range of morphologies that are more related to the anatomy of the autopodia and body plan of the trackmakers. Considering all these aspects, it is plausible that *Batrachichnus* contains several morphotypes that correspond to different ichnospecies (besides *B. salamandroides*), and/or even ichnogenera. However, an evaluation of this ichnotaxonomy is a large endeavor that needs a thorough review of all or most of the material currently attributed to this characteristic ichnogenus of the late Paleozoic.

4.1.8 *Trackmakers*

Potential trackmakers of *Batrachichnus* are generally interpreted to be small-sized temnospondyls such as branchiosaurids (e.g., Voigt, 2005), but also lepospondyls, including "microsaurs" (Petti et al., 2014; Allen et al., 2022). This is in agreement with the relative depth patterns of manus and pes tracks, suggesting a medial functional prevalence of the autopodia, a condition attributed to early tetrapods (Mujal et al., 2020; see also references therein). All in all, the relatively wide morphological range of this monospecific ichnogenus, together with the different locomotion styles that have been inferred, most probably reflect a wide range of trackmakers and environmental settings (linked to the presence of water), which is also in agreement with its long stratigraphic range (Schneider et al., 2020; Marchetti et al., 2022b).

4.2 *Limnopus* Marsh, 1894 (Fig. 4.1C and D)

4.2.1 *Type ichnospecies*

Limnopus vagus Marsh (1894).

4.2.2 Holotype

YPM 532 (Marsh, 1894: pl. 2.2, 3.2; Baird, 1952: pl. 122). Trackway, concave epirelief and convex hyporelief. Howard Limestone, Pennsylvanian, USA (Kansas).

4.2.3 Valid ichnospecies

Limnopus heterodactylus (King, 1845a,b).

4.2.4 Distribution

Pennsylvanian–Cisuralian. *Limnopus* is recorded from Cisuralian localities in the USA (New Mexico, Ohio, West Virginia), Canada (Nova Scotia), Spain, France, Italy, Germany, and Morocco (Table 4.1).

4.2.5 Diagnosis (after Marsh, 1894, emended)

Small to large tracks of a quadruped (maximum size of 200 mm, larger than *Batrachichnus* and *Matthewichnus*). Manus imprint about as long as wide (different from *Amphisauropus*, which has a manus imprint wider than long) and about 1/3 shorter than pes imprint (different from *Matthewichnus*, which shows a relatively longer pes imprint). The manus imprints have four digits (different from *Amphisauropus*, *Hylopus*, *Ichniotherium*, which are pentadactyl), while the pes imprints have five digits, all well developed. Relatively short and thick digit imprints, especially in the manus and digit IV of the manus, which is about as long as digit II (different from *Batrachichnus* and *Matthewichnus*, which show relatively longer and more slender digit imprints and a digit IV about as long as digit I in the manus). Well-impressed circular basal pad of digit I (different from *Ichniotherium*, which lacks this feature). Broad trackways with alternated pes-manus couples (different from *Matthewichnus*, which shows secondary overstep and an irregular trackway pattern).

4.2.6 General description

Footprints of a quadrupedal tetrapod. Manus tetradactyl and pes pentadactyl, both ectaxonic with well-splayed digits and round, clawless digit tip imprints. A characteristic relatively deep and round imprint is found as a basal pad of digit I of both pes and manus. Manus tracks are usually semiplantigrade, occasionally plantigrade. They are as long as wide with lengths generally >20 mm, often around 50–70 mm, but some can be >100 mm long. Digit imprints of stout aspect. Digit imprints increase in length from I to III. Length of digit IV imprint between those of digits II and III. Imprints of digits I to III often curved inwards, with the tips more markedly bent. Digit IV imprint separated from others, more proximally positioned and rotated outwards. Palm imprint relatively shallow, wider than long, with a concave proximal margin. Manus tracks are usually deeper medially, and the digit imprints increase in depth toward their tips. Pes tracks are plantigrade to semiplantigrade and larger than manus tracks. Digit imprints proportionally slightly longer than those of the manus imprints, but also with stout aspect. Digit imprints increase in length from I to IV, digit V about as long as digit II. Digit V imprint more proximally positioned than the others and rotated outwards. Digit imprints are generally straight to slightly curved inwards; their tips are also round and usually deeper, as in the manus digit imprints. Sole impression relatively round, with a convex heel, generally better impressed in the portion behind digits I to III, sometimes not imprinted behind digits IV and V. Pes tracks are deeper medially.

Generally, manus tracks are better preserved and more deeply impressed than pes tracks. In fact, the latter are sometimes poorly defined. Trackways composed of clearly alternating manus-pes sets, with manus tracks rotated toward the midline and pes tracks parallel to the midline or slightly rotated outwards. In manus-pes sets, pes tracks are usually slightly more laterally positioned, and sometimes partially overstep the manus tracks. Pace angulation usually <90° for both manus and pes tracks. Trackways generally wide, with broad space between couples of each side; stride length/pes length ratio very variable, from 2.4 to 5.5. A sinuous tail trace, usually shallow, is sometimes present.

4.2.7 Discussion

Limnopus was erected by Marsh (1894) based on a relatively small-sized trackway from the Pennsylvanian of Kansas, with the ichnospecies *Limnopus "vagus"*. *Limnopus* is a common component of lower Permian tetrapod ichnoassemblages. The typical morphologies of *Limnopus* are those reported by Baird (1952), from the Pennsylvanian–Cisuralian of Kansas, USA. Baird (1952) synonymized *"Allopus"* Marsh (1894) and *"Baropus"* Marsh (1894) with *Limnopus*. Haubold (2000) considered as valid the ichnospecies *L. vagus, L. "zeilleri"* and *L. "cutlerensis."* Voigt (2005) assigned all the *Limnopus* material from the Thuringian Forest Basin to L. *"vagus"* (see also Haubold, 1970, 1971a). Voigt (2005) highlighted the need for a thorough review of this ichnogenus, as in the case of *Batrachichnus* (see also discussion in Marchetti et al., 2013a; Stimson et al., 2016).

The unclear ichnotaxonomic status of *"Thenaropus" heterodactylus* King (1845a,b) generally hampered a comparison with *Limopus "vagus"* Marsh (1894). Lucas and Dalman (2013) revised the type material of *"T." heterodactylus*, assigned this ichnospecies to *Limnopus* and considered the ichnogenus *"Thenaropus"* invalid because it was introduced by King (1844) based on native American carvings mistaken for fossil footprints (cf. Lucas, 2015). Therefore, *Limnopus heterodactylus* (King, 1845a,b) is the first named *Limnopus* ichnospecies, and, lacking a clear differentiation from younger ichnospecies, it should be regarded as the only valid *Limnopus* ichnospecies (Lucas and Dalman, 2013; Marchetti et al., 2013a). Nevertheless, *Limnopus* includes a wide range of sizes and morphologies, the larger ones being represented by large tracks (>10 cm long) studied by DiMichele et al. (2012) and those of *L. "zeilleri"* (Gand, 1985, 1987). In this way, an ichnospecific differentiation might be possible, after a comprehensive review of the ichnogenus.

Tucker and Smith (2004) proposed to synonymize *Limnopus* and *Batrachichnus*, retaining the names as subichnogenera under the ichnogenus *Limnopus*. However, this was later rejected by Lucas et al. (2005a, 2011b), arguing that it created an unnecessary trinomial nomenclature. Indeed, the proposal by Tucker and Smith (2004) has not been followed in subsequent literature. In any case, the similarities between *Limnopus* and *Batrachichnus* are remarkable, and the two ichnogenera are mostly only (arbitrarily) differentiated by their size, at 20 mm of pes length (e.g., Stimson et al., 2016), although Voigt (2005) documented *Batrachichnus* pes tracks up to 35 mm long. Voigt (2005: pp. 78–79) identified some possible morphological differences between the two ichnogenera, including the width/length ratio of the manus tracks and the relative length of manus digit IV. He suggested applying the use of a plexus consisting of these two ichnogenera for material in which differentiation is equivocal (e.g., Voigt et al., 2011a,b). A further difference is the smaller pes length/manus length ratio in *Limnopus* with respect to *Batrachichnus*.

Limnopus is a very characteristic track in Cisuralian ichnoassociations (e.g., Voigt and Lucas, 2018). We herein re-assign to indeterminate tetrapod tracks a pes-manus couple from the Guadalupian/Lopingian of Niger, because the pes imprint is mesaxonic with median functional prevalence (Smith et al., 2015, Fig. 8A). Also, tracks of *Limnopus* have been associated with *Characichnos*, a swimming trace morphotype.

4.2.8 *Trackmaker*

A conspicuous feature of *Limnopus* material is that manus tracks are often better impressed than the pes tracks, which is related to the nature of the trackmakers, interpreted as medium- to large-sized terrestrial temnospondyls such as eryopoids (Marchetti, 2019). This is also in agreement with the relative depth patterns, suggesting a medial functional prevalence of the autopodia (Mujal et al., 2020). In any case, as mentioned above, a thorough review of this ichnogenus is also necessary to better clarify the ichnotaxonomy, as well as to better discriminate the potential trackmakers within Temnospondyli.

4.3 *Amphisauropus* Haubold, 1970 (Fig. 4.2A and B)

4.3.1 *Type ichnospecies*

Amphisauropus latus Haubold (1970).

4.3.2 *Holotype*

NHMS—AP 126-1a (Voigt, 2015: fig. 2). Trackway, convex hyporelief. Goldlauter Formation, Cisuralian, Germany.

4.3.3 *Valid ichnospecies*

Amphisauropus kablikae Geinitz and Deichmüller (1882).

4.3.4 *Distribution*

Pennsylvanian—Cisuralian. It is a relatively common component of the Cisuralian tetrapod ichnoassemblages from low latitudes of northern equatorial Pangaea, found in the present-day USA (Arizona, New Mexico, Texas), Canada (Prince Edward Island, Nova Scotia), France, Italy, Austria, Germany, Poland, and the Czech Republic (Table 4.1).

4.3.5 *Diagnosis (after Haubold, 1970, emended)*

Footprints of a quadruped, pace angulation usually around 90° (60—140°), imprints plantigrade to semiplantigrade, manus and pes pentadactyl (different from *Batrachichnus*, *Limnopus*, and *Matthewichnus*, which have a tetradactyl manus imprint), round digit tips, digit IV the longest, manus somewhat smaller than pes, divergence between manus and pes up to 50°, manus wider than long (different from *Ichniotherium* and *Hylopus*, which have manus imprints about as wide as long), mean divergence of pes digits I—V of 90°. Circular basal pad of digit I (different from *Ichniotherium*, which lacks this feature).

4.3.6 *General description*

Footprints of a quadrupedal tetrapod. Manus and pes plantigrade to semiplantigrade, pentadactyl, both ectaxonic with well-splayed digit imprints and round, clawless digit tips.

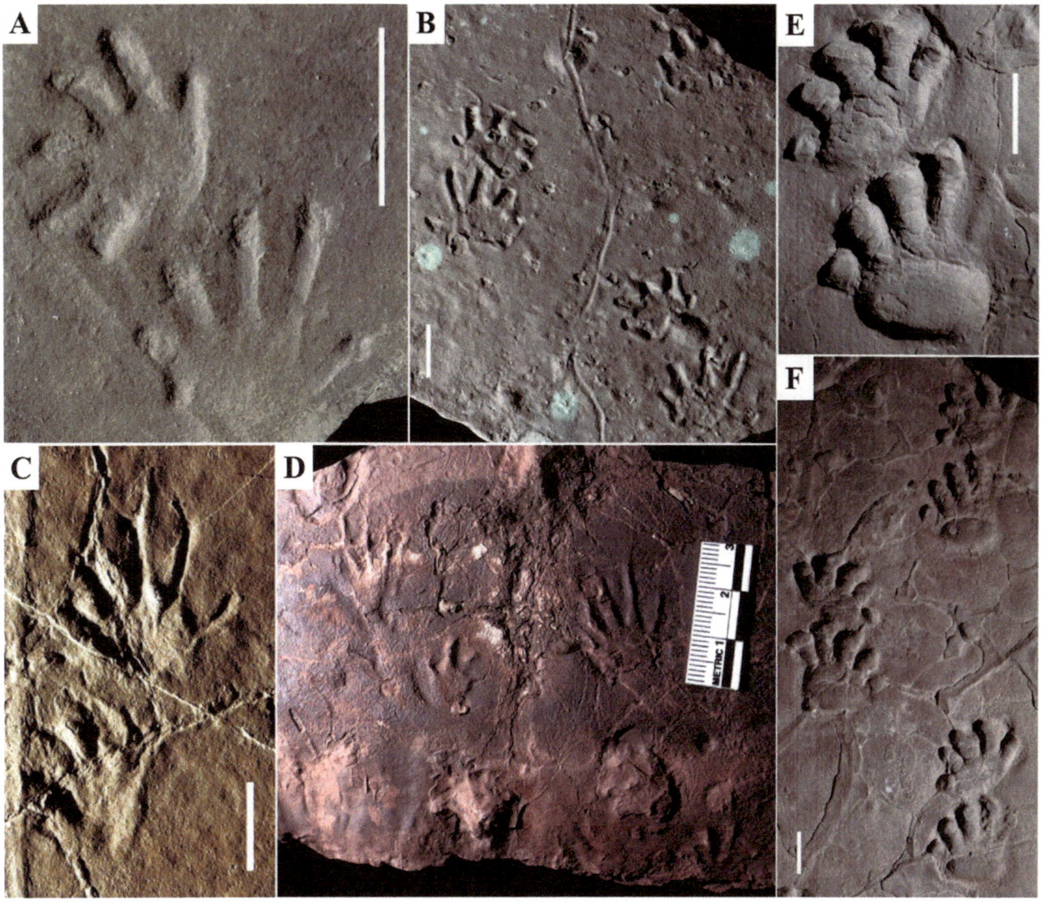

FIGURE 4.2 Non-temnospondyl nonamniote tracks. (A–B) *Amphisauropus kablikae*, track attributed to seymouriamorphs. (A) MNG VF 16. Right pes-manus couple, concave epirelief. Goldlauter Formation, Cisuralian, Germany. (B) MNG 7813-84. Trackway, concave epirelief. Rotterode Formation, Cisuralian, Germany. (C–D) *Matthewichnus* isp., track attributed to lepospondyls. (C) NMMNH P-23510-23511. Right pes-manus couple, complete secondary pes-manus overstep, concave epirelief. Robledo Mountains Formation, Cisuralian, New Mexico, USA. Courtesy of Sebastian Voigt. (D) NMMNH P-NN 1, partial trackway, convex hyporelief. Abo Formation, Cisuralian, New Mexico, USA. (E–F) *Ichniotherium sphaerodactylum*, track attributed to diadectids. Tambach Formation, Cisuralian, Germany. (E) MB.1969.54.258. Left pes-manus couple, convex hyporelief. (F) MB.1969.54.257. Trackway, convex hyporelief. Scale bars: A–C = 1 cm; E–F = 5 cm.

Manus tracks are wider than long, whereas pes tracks, which are larger (lengths from 10 mm up to 70 mm, or slightly more), are about as long as wide. In both manus and pes tracks digits increase in length from I to IV, with digit III almost as long as digit IV, and digit V length ranging from being the shortest to being slightly longer than digit II. Digits I to IV can be slightly curved inwards. Aspect of digit imprints relatively stout, usually they are bottle-shaped because of the wider and round proximal area connected to the palm and sole impressions. The digit tip imprints are sometimes markedly round, slightly deeper than the proximal portion of the digit imprints. Palm imprints markedly wider than long, with a general oval aspect, but sometimes with a slightly concave proximal margin. Sole imprints also wider than long, with a slightly convex heel. A round imprint usually defines the basal pad of digit I in both manus and pes tracks. In pes tracks, digit divergence is usually around 90° for I–V and around 70° for I–IV. Overall, manus tracks are more deeply impressed than pedal tracks and both manus and pes tracks are deeper medially, although in some cases, the tips of the manus digits III and IV are also relatively deeply impressed. In trackways, manus-pes sets are clearly alternating, with mean pace angulation of 90° in the manus and of 80° in the pes. Manus tracks are rotated inwards and pes tracks outwards to parallel to the midline. In manus-pes sets, pes tracks are slightly more laterally located. There is usually no overstepping of the pes on the manus, and only occasionally, the tips of the digits of the pes reach the proximal area of the palm impression. Stride length/pes length ratio ranges generally between 2.5 and 5.5. Sinuous tail impressions are sometimes present (Fig. 4.2B), as well as convex trails of digit drag traces.

4.3.7 *Discussion*

Haubold (1970) erected this ichnogenus based on material from the lower Permian of the Thuringian Forest Basin, Germany (see also Voigt, 2005, 2015), defining two ichnospecies, *A. "latus"* and *A. "imminutus,"* mostly distinguished by size. Later, Voigt (2005) synonymized the "*Saurichnites*" specimens referred to "*S.*" *kablikae*, originally described by Geinitz and Deichmüller (1882), with *Amphisauropus*, also synonymizing the two ichnospecies described by Haubold (1970). As a result, Voigt (2005) proposed the new combination *Amphisauropus kablikae,* including all the material referred to *Amphisauropus*, a conclusion accepted in the subsequent literature (e.g., Van Allen et al., 2005; Marchetti et al., 2017a; Calábková et al., 2022). This was justified by the description of additional well-preserved material from the type locality of *A. kablikae*, whereas the holotype is not well-preserved (Voigt, 2005, fig. 28A–B). Since this poorly preserved holotype does not show all the diagnostic features, *A. kablikae* could represent a *nomen dubium*, so further analyses should revise *Amphisauropus* at the ichnospecies level.

4.3.8 *Trackmaker*

The synapomorphy-based analyses carried out by Marchetti et al. (2017a), also including inferences of the functional prevalence of autopodia (e.g., Mujal et al., 2020), show that the most probable trackmakers of *Amphisauropus* are seymouriamorphs, a correlation that was already proposed in earlier studies of this ichnogenus (Haubold, 1970; Voigt, 2005).

4.4 *Matthewichnus* Haubold, 1970 (Fig. 4.2C and D)

4.4.1 *Type ichnospecies*

Matthewichnus velox (Matthew, 1905).

4.4.2 *Holotype*

RM 2.1139 (Matthew, 1905; pl. II.2; Haubold, 1970: fig. 8A). Right pes-manus couple, convex hyporelief. Joggins Formation, Pennsylvanian, Canada (Nova Scotia).

4.4.3 Valid ichnospecies

Matthewichnus velox (Matthew, 1905); *Matthewichnus caudifer* Kohl and Bryan (1994).

4.4.4 Distribution

Mississippian–Cisuralian. From the Permian, known from the Cisuralian of the USA (New Mexico) (Table 4.1).

4.4.5 Diagnosis (after Haubold, 1970, emended)

Small tracks of a quadruped (maximum size of 38 mm, smaller compared with *Limnopus*, which can be larger). Plantigrade to semiplantigrade impressions, foot pentadactyl, small sole surface rounded proximally, all digits arise on one line, II to IV the longest, rounded sole, manus with short digits, tetradactyl (different from *Amphisauropus, Hylopus, Ichniotherium*, which are pentadactyl). Manus imprint about as long as wide (different from *Amphisauropus*, which has a manus imprint wider than long) and much shorter than pes imprint (different from *Batrachichnus*, which shows a relatively shorter pes imprint). Relatively long and slender digit imprints (different from *Limnopus* and *Ichniotherium*, which have relatively shorter and/or thicker digit imprints). In the manus, digit IV is shorter than digit II (different from *Limnopus*, which shows digit IV as long as digit II in the manus). Well-impressed circular basal pad of digit I (different from *Ichniotherium*, which lacks this feature). Broad trackways with secondary overstep and irregular trackway pattern (different from *Batrachichnus* and *Limnopus*, which show regularly alternated pes-manus couples).

4.4.6 General description

Footprints of a quadruped in which the pes is much larger than the manus (1/3 longer), and the pes is up to 20 mm long. Both manus and pes are plantigrade to semiplantigrade and ectaxonic. The manus is tetradactyl and approximately as wide as long, with digit length increasing from I to III, digit IV is shorter than digit II. The pes is pentadactyl, slightly longer than wide, and has digit length increasing between digits I–IV, digit V is about as long as digit II. Digit imprints are generally well-splayed, straight, or slightly curved inward and have rounded tip impressions; sometimes, digit tip bifurcation can be observed. Digit imprints are generally longer than the palm/sole impressions. Palm/sole impressions are generally oval in shape and wider than long. Footprints are more deeply impressed medially than laterally. Trackways broad with low-pace angulation (usually about 60–90°) and irregular arrangement of pes and manus imprints. The manus can be turned inwards, and the pes is either parallel to the trackway midline or turned outwards. The manus imprint is medial to the pes imprint and sometimes overstepped by the latter (secondary overstep). Continuous tail/body drag traces are sometimes present.

4.4.7 Discussion

Matthew (1905) named "*Dromopus*" *velox* based on an isolated pes-manus couple from the Pennsylvanian of Joggins, Nova Scotia, Canada. Haubold (1970) made "*D.*" *velox* the type ichnospecies of a new ichnogenus, *Matthewichnus*, and created the new combination *Matthewichnus "woodworthi"* (Lull, 1920) from the Pennsylvanian of Massachusetts. Because of the poor preservation of the manual imprint in the type material, we consider *M. "woodworthi"* invalid. *M. caudifer* Kohl and Bryan (1994) from the Pennsylvanian of Tennessee, USA, has also been assigned to the ichnogenus, and there is disagreement over the validity of this ichnospecies (Haubold et al., 2005; Fillmore et al., 2012). Further work is needed to clarify the ichnotaxonomy of *Matthewichnus* at the ichnospecies level. Voigt and Lucas (2015a) published the first Permian records of *Matthewichnus*. *Matthewichnus* cannot be confidently identified without an associated manus-pes. This is because the isolated pes, in many extramorphological variants, could be mistaken for *Batrachichnus* or *Amphisauropus*, and the isolated manus, in many variants, could be identified as *Batrachichnus*. Only the very small, medially placed tetradactyl manus in association with the much larger, pentadactyl pes allows for a confident identification.

In the Cisuralian, *Matthewichnus* is known from the Robledo Mountains Formation and possibly Sangre de Cristo Formation of New Mexico (Voigt and Lucas, 2015a,b). We herein assign to *Matthewichnus* isp. a new specimen from the Abo Formation of New Mexico (Fig. 4.2D).

4.4.8 *Trackmaker*

A lepospondyl (including "microsaurs") or a temnospondyl amphibian are generally identified as the trackmaker of *Matthewichnus* (e.g., Haubold, 1971a; Haubold et al., 2005; Fillmore et al., 2012; Voigt and Lucas, 2015a). However, unlike the temnospondyl trackmakers of *Batrachichnus* and *Limnopus*, the trackmaker of *Matthewichnus* must have had a different body and limb structure, consistent with secondary overstep, an irregular trackway pattern and continuous body/tail drag traces, such as an elongated body and tail, common features in lepospondyls (e.g., Haubold, 1971a; Kohl and Bryan, 1994; Voigt, 2005).

4.5 *Ichniotherium* Pohlig 1892 (Fig. 4.2E and F)

4.5.1 *Type ichnospecies*

Ichniotherium cottae Pohlig (1885).

4.5.2 *Lectotype*

MNG 1781 (Buchwitz and Voigt, 2018: fig. 3G–H). Two consecutive pes-manus couples, convex hyporelief. Goldlauter Formation, Cisuralian, Germany.

4.5.3 *Valid ichnospecies*

Ichniotherium cottae (Pohlig, 1885), *Ichniotherium sphaerodactylum* (Pabst, 1895)

4.5.4 *Distribution*

Pennsylvanian–Cisuralian. In the Permian, it is known from the Cisuralian of the USA (Utah, Arizona, Colorado, New Mexico, Ohio), Canada (Prince Edward Island), Spain, France, Italy, Austria, Germany, Poland, the Czech Republic, and Morocco (Table 4.1).

4.5.5 *Diagnosis (after Voigt, 2005, emended)*

Pes and manus imprints plantigrade, pentadactyl (different from *Batrachichnus*, *Limnopus*, and *Matthewichnus*, which have a tetradactyl manus imprint). Manus imprint of *Ichniotherium* slightly wider than long, whereas in *Amphisauropus* it is distinctly wider than long. Relatively longer manual digit imprints compared with *Amphisauropus* and *I. "praesidentis." Ichniotherium* differs from *Hylopus* by the more ectaxonic pes and from *Amphisauropus* by the more ectaxonic manus. Different from *Amphisauropus*, *Hylopus* and *I. "praesidentis"* by lacking a prepollex imprint. *Ichniotherium* differs from *I. "praesidentis"* in the parallel orientation of its footprints with respect to the direction of movement, higher pace angulations, and lack of complete secondary overstep of the pes on the manus.

4.5.6 General description

Plantigrade, pentadactyl, ectaxonic footprints of intermediate size (60–150 mm). The pes is slightly larger than the manus (about 1/5 longer) and is as wide as long, and the manus is slightly wider than long. The pes often shows a medial-lateral decrease in relief, and the manus is usually more completely impressed. Thick, straight digits of moderate length. Digits increase in length between I and IV; manual digit V is shorter than digit II and pedal digit V either shorter than digit II or about as long as digit III. Pedal digit imprints relatively longer than manual digit imprints. Digits diverge at acute to obtuse angles (usually 70–130° between digits I and V). Digits II–III of the manus can be bent inward. All the digits have large, rounded "drumstick-like" terminations. The digits are more deeply impressed in the distal part. In well-preserved footprints, the digits are segmented by transverse, thin, and slightly curved grooves. The palm/sole is broad and wider than long, usually having elliptical shapes and is distinctly separated from the digits. Heel with a convex proximal end; equally long or somewhat shorter than the digits. Oval to round heel pad in pedal imprints, and in manus imprints, if present, less distinct. The trackway pattern is characterized by alternate locomotion with distinct close pes-manus couples. Trackways relatively narrow to moderately wide, pace angulation generally between 70 and 130°. Manus imprints positioned anterior to pes imprints, but in trackways of fast-moving individuals the posterior part of the manus can be partially overstepped by the latter (primary overstep). Manus imprints often slightly inward-oriented compared with pes imprints, which are either aligned parallel to the direction of movement or rotated slightly outward.

4.5.7 Discussion

Ichniotherium was introduced by Pohlig (1892), and the type ichnospecies is *I. cottae* (Pohlig, 1885). Voigt (2005) considered *"Sphaerodactylichnium"* Pabst (1895) as a younger subjective synonym of *Ichniotherium* and highlighted a possible synonymy of *Ichniotherium* with *"Megabaropus"* Baird (1952) and *"Cyclopus"* Heyler and Lessertisseur (1963), both later assigned to *Ichniotherium* by Buchwitz and Voigt (2018) and Mujal and Marchetti (2020), respectively. The material from the Cisuralian Thuringian Forest type locality of *Ichniotherium cottae*, the Gottlob quarry, includes only two specimens with partial step cycles, apart from individual tracks (Buchwitz and Voigt, 2018), and one of these (MNG 1781) is herein designated as lectotype. A more thorough description and definition of the ichnogenus *Ichniotherium* as an ichnotaxon is mostly based on the rich and well-preserved material from the Bromacker site of the Tambach Formation of Germany (Voigt and Haubold, 2000), which is likely of Sakmarian age (Lützner et al., 2021); the longest collected trackway includes more than 40 individual couples (Fichter, 1998). Footprints from the Bromacker site frequently feature transverse skin folds on the digit impressions and somewhat deeper and broadened, roundish toe tips, potentially indicating hoof-like claw sheaths. Pes length ranges from 45 to 140 mm in the large Tambach Formation sample, but larger tracks have been reported from some of the other occurrences (Buchwitz and Voigt, 2018; Mujal and Marchetti, 2020). Tail traces or body drag marks are rare.

As highlighted by Voigt (2005), the oldest ichnospecies probably assignable to *Ichniotherium* is "*Saurichnites rittlerianus*" Hochstetter (1868) from the lower Permian of the Czech Republic, but given that the type material was not figured and is apparently lost, this has to be considered a *nomen dubium*. Thus, the oldest valid ichnospecies is *I. cottae* (Pohlig, 1885) from the lower Permian Goldlauter Formation of Germany, considered the only valid ichnospecies from the Permian by Haubold (2000) and Voigt and Haubold (2000).

Voigt (2005), after a comprehensive review of the Thuringian Forest material, proposed the new combination *I. sphaerodactylum* (Pabst, 1895) from material of the lower Permian Tambach Formation of Germany and distinguished it from *I. cottae*, mostly based on the relative length of pedal digit V. This found support from discoveries in Colorado, Arizona, and Canada (Voigt et al., 2005; Brink et al., 2012; Marchetti et al., 2019b; Francischini et al., 2020). Voigt (2005) also synonymized "*Ichniotherium willsi*" Haubold and Sarjeant (1973) from the Pennsylvanian of England with *I. cottae*. This was followed by Buchwitz and Voigt (2018), who, however, recognized notable morphometric differences within material assigned to *I. cottae*. A revision of *Ichniotherium* at the ichnospecies level is needed to clarify whether these differences could be of ichnotaxonomic significance.

Voigt and Ganzelewski (2010) proposed the new combination "*I. praesidentis*" (Schmidt, 1956) from the Lower Pennsylvanian of Germany. However, according to the herein revised diagnosis of the ichnogenus, "*I. praesidentis*" (Schmidt, 1956) is not considered as part of *Ichniotherium*. This Carboniferous material, originally described by Schmidt (1956) as "*Megapezia praesidentis*" and later by Haubold (1970) as "*Schmidtopus praesidentis*" and by Voigt and Ganzelewski (2010) as "*Ichniotherium praesidentis*" is in need of revision.

Ichniotherium is a very typical track in the Cisuralian (e.g., Voigt and Lucas, 2018). This includes even footprints found in eolian paleoenvironments, clear evidence of nonamniotes in a desert (Francischini et al., 2020). In addition, we re-assign to *Ichniotherium cottae* a specimen from the eolian Weber Sandstone of Utah (Chure et al., 2014, fig. 4A–B), because the pes and manus imprint morphology and proportions are consistent with this ichnospecies. Also, we re-assign to *Ichniotherium sphaerodactylum* the footprints from the Cisuralian of Pizzo del Diavolo Formation of Italy originally described as *Amphisauropus* isp. (Marchetti, 2016, fig. 2). In fact, overall morphology and digit proportions are consistent with *I. sphaerodactylum*. The reduced length of digits, mentioned by Marchetti (2016) for an attribution to *Amphisauropus*, is only apparent, being a taphonomic effect due to tectonic stretching perpendicular to the footprint direction, and therefore is not relevant. Although the material described by Moreau and Gand (2022), fig. 6) shares some general similarities with *Ichniotherium*, the relatively short manual digit IV, the very wide manus imprint, and the morphology of the palm and sole impressions are rather consistent with *Amphisauropus*, therefore we re-assign this material to *Amphisauropus*.

4.5.8 Trackmaker

Since the discovery of diadectid skeletons at the Bromacker site, the assignment of *Ichniotherium* to diadectid producers, which display well-matched hand and foot proportions, quickly became popular (Fichter, 1998), replacing earlier correlations with pelycosaurgrade synapsids as likely trackmakers (see discussion in Voigt, 2005). The type ichnospecies *I. cottae* has been correlated with *Diadectes* and closely related diadectid genera based on pedal toe proportions (Voigt et al., 2007; Buchwitz and Voigt, 2018), but it was also correlated with other diadectomorph groups, such as limnoscelids (Mujal and Marchetti, 2020). Although *Ichniotherium sphaerodactylum* was originally described from the Bromacker site and correlated with the newly described *Orobates pabsti* as the best-fitting trackmaker (Voigt et al., 2007; see also Nyakatura et al., 2019), further occurrences in Canada, the United States, and Czech Republic from different stratigraphic levels suggest that this ichnotaxon also had a wider range of trackmakers. A slight difference in the trackway patterns of *I. cottae* and *I. sphaerodactylum* might be due to more elongate trunks in the producers of the latter (Voigt et al., 2007). An ancestral state reconstruction approach by Buchwitz et al. (2021), including trackway samples of "*I. praesidentis,*" *I. cottae* and *I. sphaerodactylum* as diadectomorph track types—in addition to various amniote and nonamniote Carboniferous to Permian trackways—suggests that the trackway patterns of *I. sphaerodactylum* specimens from the Tambach Formation are close to those predicted for the last common ancestor of amniotes.

4.6 *Dimetropus* Romer and Price, 1940 (Fig. 4.3A and B)

4.6.1 Type ichnospecies

Dimetropus berea (Tilton, 1931).

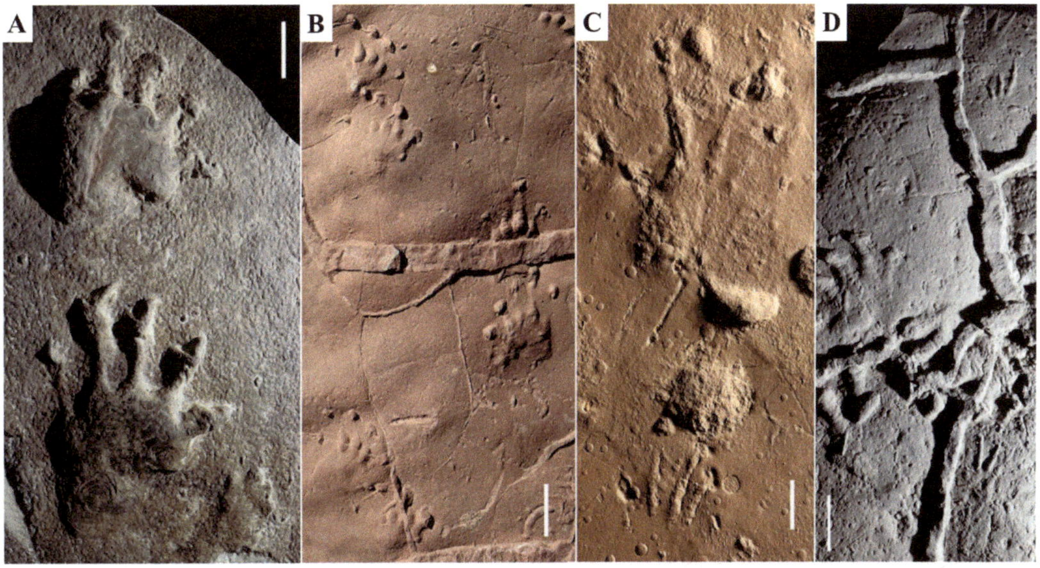

FIGURE 4.3 Non-therapsid synapsid tracks. (A—B) *Dimetropus leisnerianus*, track attributed to non-varanopid pelycosaurian-grade synapsids. (A) WVGES 126. Holotype of *Dimetropus "berea"*. Right pes-manus couple, convex hyporelief. Waynesburg Formation, Cisuralian, West Virginia, USA. (B) MNG 13490. Trackway, convex hyporelief. Tambach Formation, Cisuralian, Germany. (C—D) *Tambachichnium schmidti*, track attributed to varanopids. (C) NML 687. Right pes-manus couple, complete primary pes-manus overstep, convex hyporelief. Goldlauter Formation, Cisuralian, Germany. (D) PMJ-P-1321. Holotype. Trackway, complete primary pes-manus overstep, convex hyporelief. Tambach Formation, Cisuralian, Germany. Scale bars: A, B, D = 5 cm; C = 1 cm.

4.6.2 Holotype

WVGES 126 (Tilton, 1931: fig. 1A; Fig. 4.3A). Right pes-manus couple, convex hyporelief. Waynesburg Formation, Cisuralian, USA (W Virginia).

4.6.3 Valid ichnospecies

Dimetropus leisnerianus (Geinitz, 1863); *Dimetropus osageorum* Sacchi et al. (2014).

4.6.4 Distribution

Pennsylvanian–Cisuralian. In the Permian, known from the Cisuralian of the USA (Arizona, Colorado, New Mexico, Oklahoma, Texas, West Virginia, Pennsylvania), Canada (Nova Scotia), Spain, France, Italy, Germany, Poland, the Czech Republic, Russia, and Morocco; the Cisuralian-Guadalupian of Spain (Table 4.1).

4.6.5 Diagnosis (after Voigt, 2005, emended)

Plantigrade to semiplantigrade, pentadactyl (different from *Batrachichnus, Limnopus, Matthewichnus*), ectaxonic tracks of a quadruped, ranging from 50 to 200 mm in length (maximum size larger than any other pre-Guadalupian track, except *Limnopus*). Plantigrade pes significantly larger than the manus, with a fully impressed large sole extended proximally and laterally (different from other Carboniferous-Cisuralian tracks). Pedal digits relatively short compared with the palm/sole (unlike other Carboniferous-Cisuralian tracks) with thin and acuminate claw impressions, often bifurcated (contrasting nonamniote tracks). Medial parts of digits range from fully to nonimpressed, resulting in paw-like impressions (distinct from other nonsynapsid amniote tracks and *Tambachichnium*). Arcuate disposition of basal pads of digits large with evident hemispherical shape when visible (different from other anamniote, nonsynapsid amniote tracks and *Tambachichnium*). Manual digit proportions: $\mathrm{I} < \mathrm{II}\ \ \mathrm{V} < \mathrm{III} < \mathrm{IV}$. Pedal digit proportions: $\mathrm{I} < \mathrm{II} \leq \mathrm{V} \leq \mathrm{III} \leq \mathrm{IV}$ (proportions different from other synapsid tracks except *Tambachichnium*). Low total digit divergence in the pes, about 50–70° (contrasting nonamniote tracks and the amniote tracks *Dromopus, Hyloidichnus* and *Notalacerta*). Trackway pattern showing a simple alternating arrangement of pes-manus couples or, with lower pace angulation, a more indistinct pattern with the manus far from the pes and sometimes overstepped by the subsequent pedal impression (secondary overstep; different from all Paleozoic ichnotaxa except *Matthewichnus* and "*Ichniotherium praesidentis*"). Footprints aligned to midline; sometimes, pes slightly turned outwards (unlike other nonsynapsid amniote tracks).

4.6.6 General description

Tracks of quadrupedal tetrapods with pentadactyl, ectaxonic, plantigrade to semiplantigrade manus and pes imprints of relatively large size (50–200 mm). Manus imprints are as long as wide, and pes imprints are usually about one fourth longer than wide. Pes significantly larger than the manus. It typically has relatively short digit imprints compared with the palm/sole impressions. Digits increase in length from I to IV, and digit V is as long as digit III in the pes imprint, whereas digit V is as long as digit II in the manus imprint. Digit imprints closely grouped and more deeply impressed proximally and distally, forming typical paw-like impressions. Digit terminations enlarged and often showing sharp claw impressions; sometimes, the terminations are bifurcated. Digit imprints can show parallel grooves transverse to the digit length. Digit imprints may show semispherical impressions of the phalangeal pads which are circular, well-impressed, and arranged in an arcuate concave shape. Medial-lateral increase in relief of the pes and the manus imprints. Palm and sole commonly large, fully impressed, extended proximally and laterally, with a concave proximal margin. The sole impression is sometimes composed of two distinct parts, an elliptical metatarsal-digital part that is wider than long and a circular, smaller heel part. Scales may occur on heel imprints. Simple alternating arrangement of pes-manus couples, with manus in front of the pes, in some cases secondary overstep can be observed. Trackways with relatively low-pace angulation (75–120°). Pes usually parallel to the midline and manus usually directed medially. Tail drag trace is usually straight, but it is a relatively uncommon feature.

4.6.7 Discussion

Tilton (1931) identified a trackway from the lower Permian Waynesburg Formation of the Dunkard Group near Berea, West Virginia, USA, as the new species *Dimetrodon "berea."* The tracks Tilton (1931) assigned to *Dimetrodon* (a body-fossil genus) became the basis of a new ichnogenus, *Dimetropus* Romer and Price (1940) in the new combination *Dimetropus "berea."* As highlighted by Voigt (2005), *Dimetropus* has been consistently used for more than 60 years, whereas its older synonym "*Acrodactylichnium*" Pabst (1895) was never used subsequently and thus is considered a *nomen oblitum* (ICZN, 1999). Haubold (2000) considered as valid the ichnospecies *D. "berea," D. leisnerianus* and *D. "nicolasi."* The latter has been extensively used in French ichnoassociations (Gand and Haubold, 1984; Gand, 1987), although its type material is poorly preserved (Gand and Durand, 2006). In a revision of the ichnogenus *Dimetropus*, Voigt (2005) recognized only one valid ichnospecies, *D. leisnerianus* (Geinitz, 1863). He thus considered the Dunkard ichnospecies (and type ichnospecies of the ichnogenus) *D. "berea"* (Tilton, 1931) to be a junior subjective synonym of *D. leisnerianus*, as well as the ichnospecies *D. "acrodactylus"* (Pabst, 1897 as "*Ichnium acrodactylum*"; Haubold, 1998), *D. "salopensis"* (Haubold and Sarjeant, 1973), and *D. "nicolasi."* More recently, Sacchi et al. (2014) introduced *Dimetropus osageorum* from the lower Permian of Oklahoma, distinguishing it from *D. leisnerianus* and also from other *Dimetropus* ichnospecies previously considered synonyms by Voigt (2005). There is a wide range of extramorphological variation when trackways of *Dimetropus* are examined (e.g., Haubold et al., 1995; Hunt et al., 1995; Matamales-Andreu et al., 2021a). Thus, we follow Voigt (2005) in recognizing that this wide range of extramorphology in *Dimetropus* footprints makes discrimination of separate ichnospecies difficult and agree on the synonymy of the previously described *Dimetropus* ichnospecies with *D. leisnerianus*. Waiting for a comprehensive revision of the ichnogenus, we consider as valid only the two ichnospecies *D. leisnerianus* and *D. osageorum*.

4.6.8 Trackmaker

Dimetropus is considered to be the tracks of many Pennsylvanian and lower Permian non-therapsid synapsids ("pelycosaurs" in the traditional sense). Thus, it is assumed that the trackmakers include various eupelycosaurs, particularly edaphosaurids, ophiacodontids, and sphenacodontids, as well as caseasaurs (e.g., Gand, 1987; Haubold, 2000; Voigt, 2005). The Pennsylvanian to early Permian temporal range of *Dimetropus* largely corresponds to the stratigraphic distribution of the eupelycosaur and caseasaur potential trackmakers (Lucas, 2018, 2022).

4.7 *Tambachichnium* Müller, 1954 (Fig. 4.3C and D)

4.7.1 *Type ichnospecies*

Tambachichnium schmidti Müller (1954).

4.7.2 *Holotype*

PMJ-P-1321 (Müller, 1954: fig. 2, pl. 16.3, 17.1; Voigt, 2005: fig. 43B; Fig. 4.3D). Trackway, convex hyporelief. Tambach Formation, Cisuralian, Germany.

4.7.3 *Valid ichnospecies*

Tambachichnium schmidti Müller (1954).

4.7.4 *Distribution*

Cisuralian–Guadalupian. Cisuralian of USA (?Arizona, Colorado, New Mexico), France, Italy, and Germany; Cisuralian–Guadalupian of Morocco; Guadalupian of France (Table 4.1).

4.7.5 *Diagnosis (after Müller, 1954, emended)*

A lacertoid, semiplantigrade (pes) to digitigrade (manus) vertebrate track (different from *Rhynchosauroides* and *Paradoxichnium* which have a more digitigrade pes compared with the manus, and *Dromopus*, which has an identic pes and manus morphology). Digits I–IV of the forefoot markedly increasing in length, more than in the pes (different from *Rhynchosauroides* and *Paradoxichnium* which have a pes more ectaxonic than the manus and *Dromopus*, which has an identic pes and manus morphology). Toes with broader claw impressions than the manus (different from *Paradoxichnium*, which shows a manus with broader claw impressions). Digit V laterally oriented (different from *Paradoxichnium*, which generally shows a smaller digit IV–V divergence). Alternating gait, trackway width relatively low and step length relatively high (this differs from *Rhynchosauroides*, that has a relatively small step length), with complete primary pes-manus overstep (this is generally not observed in *Dromopus*).

4.7.6 *General description*

Quadruped track type, semiplantigrade to digitigrade, markedly ectaxonic, lacertoid imprint shape due to extended digit IV, usually between 70 and 90 mm imprint length, manus and pes imprints longer than wide. Elongate, slender, claw-bearing digits, digit length I < II < III < IV and V approximately as long as II. Digits II–IV more deeply impressed than digits I and V. Digits I–IV distally curved medially, more markedly in the manus than in the pes. Manus imprints morphologically similar to the pes imprints but smaller and with higher divergence of the middle digits, manus imprints sometimes *Dromopus*-like in appearance. In the pes imprints, the proximal parts of the middle toes II and III are parallel-oriented, and their claw imprints are inward- and parallel-oriented; digits I and V only rarely and partially preserved with digit V diverging at a 60–90° angle from IV, digit I only preserved as a claw imprint. Slightly convex proximal margin of the footprints; sometimes, skin folds on the digit imprints or scales on the sole impressions are present. Trackways characterized by an alternating arrangement of close pes-manus couples. Manus and pes imprints may overlap; sometimes, pes imprint placed in front (primary overstep) of or slightly behind the manus imprint; pace angulations usually between 100 and 130°. Pes imprints slightly outward oriented, manus imprints parallel to the trackway midline.

4.7.7 Discussion

Tambachichnium was introduced by Müller (1954) from the study of a trackway from the Tambach Formation, lower Permian of Germany. The validity of this ichnogenus was confirmed by Haubold and Stapf (1998) and Voigt (2005). This ichnogenus has been recognized also in the Cisuralian of the USA (e.g., Voigt and Lucas, 2017), France (Marchetti et al., 2021a), and Italy (Marchetti et al., 2022a); the Cisuralian-Lopingian of Morocco (Zouicha et al., 2021); and the Guadalupian of France (Gand et al., 1995). The ichnospecies "*Pseudosynaptichnium esterelense*" Gand et al., 1995 is herein considered a junior subjective synonym of *Tambachichnium schmidti*. Buchwitz et al. (2021) suggested that the trackway pattern of *Tambachichnium*, which is commonly marked by a relatively high-pace angulation and pes imprints placed in front of the manus imprints, might be indicative of a derived type of gait and a limb phase concordant with a lateral sequence walk instead of a walking trot. Such a functional difference or deviant body proportions could explain the notable difference from contemporaneous reptilian track types, such as *Dromopus*, *Notalacerta*, and *Varanopus*.

4.7.8 Trackmaker

In his study of tetrapod tracks from the Thuringian Forest succession, Voigt (2005) discussed araeoscelid diapsids, varanopids, and *Eudibamus*-like bolosaurids as possible producers of *Tambachichnium* tracks due to their lizard-like autopodia. Subsequently, most authors considered varanopids as the most likely trackmakers of *Tambachichnium* (e.g., Mujal et al., 2020; Zouicha et al., 2021). Marchetti et al. (2021c) discussed the similarity between *Dromopus* and the hand and foot skeletons of certain smaller varanopids, but also the overall similarity of *Dromopus* and *Tambachichnium*, suggesting that both were made by related producers. Noting that varanodontine varanopids such as *Aerosaurus*, *Varanops*, and *Tambacarnifex* have a more robust and less ectaxonic pes than manus—unlike smaller nonvaranodontine varanopids, araeoscelid diapsids, and typical *Dromopus* tracks—Marchetti et al. (2021c) considered varanodontines as the most likely producers of *Tambachichnium*.

4.8 *Brontopus* Heyler and Lessertisseur, 1963 (Fig. 4.4A and B)

4.8.1 Type ichnospecies

Brontopus giganteus Heyler and Lessertisseur (1963).

4.8.2 "Plastotype"

MNHN-LOD 70 (Heyler and Lessertisseur, 1963: pl. VIII.1). Right pes imprint, concave epirelief. La Lieude Formation, Guadalupian, France.

4.8.3 Valid ichnospecies

Brontopus giganteus Heyler and Lessertisseur (1963); *Brontopus antecursor* (Ellenberger, 1983a).

4.8.4 Distribution

Guadalupian. It occurs in Spain, France and South Africa (Table 4.1).

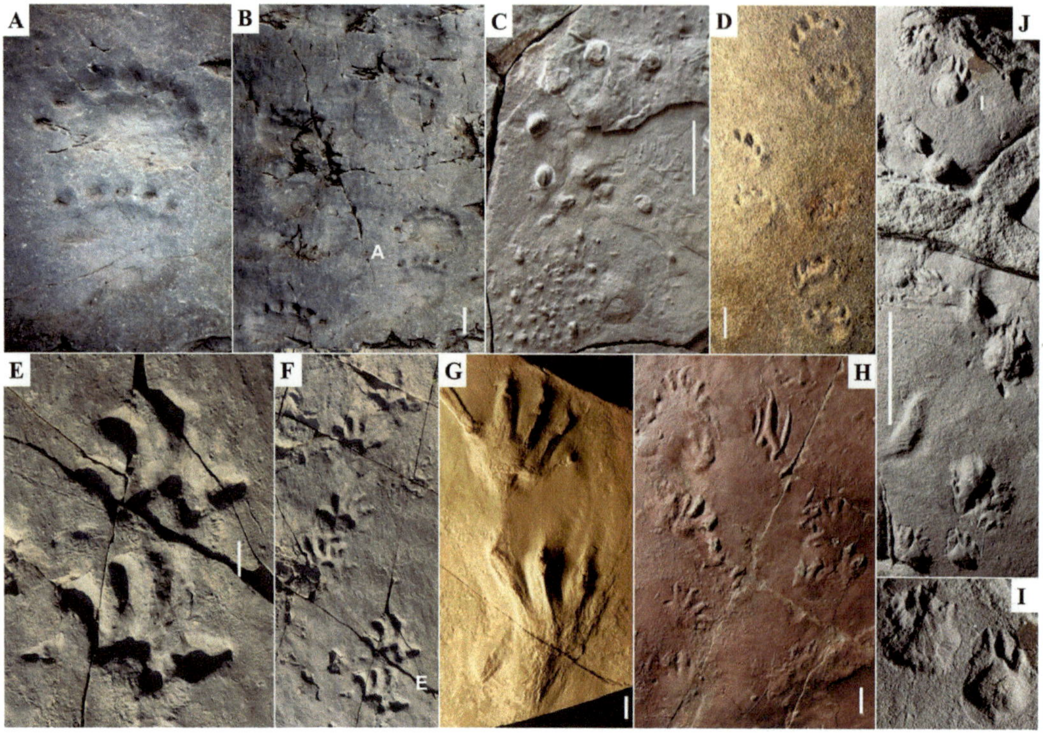

FIGURE 4.4 Non-mammalian therapsid synapsid tracks. (A–B) *Brontopus giganteus*, track attributed to dinocephalians. KW-TR 1, Abrahamskraal Formation, Guadalupian, South Africa. (A) Right pes-manus couple, concave epirelief. (B) Trackway, concave epirelief. (C–D) *Dolomitipes*, track attributed to dicynodonts. (C) UR/NS 34-19. *Dolomitipes accordii*, left pes-manus couple, convex hyporelief. Val Gardena Sandstone Formation, Lopingian, Italy. (D) UG-NN 1. *Dolomitipes* isp. Trackway, concave epirelief. Cornberg Formation, Lopingian, Germany. (E–F) *Karoopes gansfonteinensis*, track attributed to gorgonopsians. GF-TR 1, holotype, concave epirelief. Poortje member of the Teekloof Formation, Guadalupian, South Africa. (E) Right pes-manus couple. (F) Trackway. (G–H) *Capitosauroides talus*, track attributed to therocephalians. (G) ULS G 6, holotype. Right pes-manus couple, convex hyporelief. Pélitique Formation, Guadalupian, France. (H) NML-G 2020-0004 (TRO 6721). Trackway, convex hyporelief. Blätterton member, Hornburg formation, Guadalupian, Germany. (I–J) *Dicynodontipus geinitzi*, track attributed to cynodonts. DUMFM-NN 1, convex hyporelief. Corncockle Formation, Lopingian, Scotland, UK. (I) Right pes-manus couple, complete primary pes-manus overstep. (J) Trackway. Scale bars: B = 10 cm; C–E = 5 cm; G–H, J = 1 cm.

4.8.5 Diagnosis (after Marchetti et al., 2019e, emended)

Large (pes length of 230–455 mm), plantigrade, pentadactyl pes, and manus imprints of quadrupeds, with very short, distally rounded digit impressions of subequal length (unlike *Dimetropus*, which has relatively longer digit impressions and is ectaxonic). Manus imprint smaller than pes imprint or of the same size, wider than long (unlike *Dolomitipes*, which shows manual imprints of similar size or larger with respect to the pes). The posterior part of the palm impression is elongated laterally (unlike *Dicynodontipus* and *Dolomitipes*, which have a symmetrical palm impression). Low to very low-pace angulation (60–135°; unlike *Dicynodontipus* and *Dolomitipes*, which have higher pace angulation).

4.8.6 General description

Large (pes length of 230–455 mm), plantigrade, mesaxonic, pentadactyl manus, and pes imprints of quadrupeds, with broad palm/sole impression and very short, distally rounded digit impressions of subequal length, distinct from the palm/sole impression and separated from the latter by a groove. Manus imprint smaller than pes imprint or of the same size, wider than long, with similar structure, but asymmetrical, posterior part of the palm impression is elongated laterally. Relatively symmetrical foot imprint, as long as wide to longer than wide; shorter external digit impressions. Broad oval to hourglass-shaped sole impression, with an anterior metatarsal-digital part and a posterior elliptical heel impression.

Simple alternating arrangement of pes-manus couples, manus always in front of the pes, footprints generally aligned to the direction of locomotion, manual tracks can be inwardly rotated, low to very low-pace angulation (60–135°). Tail trace may be impressed.

4.8.7 Discussion

Brontopus giganteus was introduced by Heyler and Lessertisseur (1963) for an isolated but well-preserved right pedal imprint from the La Lieude Formation, Guadalupian of France. Partial in situ trackways were also described and used for the erection of this ichnogenus (Heyler and Lessertisseur, 1963). A second ichnospecies, *B. "circagiganteus,"* was introduced by Gand et al. (2000) from the same formation. Marchetti et al. (2019e) revised the ichnogenus and synonymized *B. "circagiganteus"* Gand et al. (2000) and *"Moschopus enormis"* Ellenberger (1983a) from the type locality, as well as a trackway from the Guadalupian Pradineaux Formation of France (Gand et al., 1995, fig. 7C), with *B. giganteus*. Marchetti et al. (2019e) assigned a well-preserved trackway from the Guadalupian Abrahamskraal Formation of South Africa to *B. giganteus* and more material from a different locality in South Africa to *Brontopus* isp. Marchetti et al. (2019e) introduced the new combination *Brontopus antecursor*, synonymizing *"Eocynodontipus" antecursor* Ellenberger (1983a), *"Pseudopithecopus recurvidigitus"* Ellenberger (1983a), *"Eocynodontipus sibleyrasi"* Ellenberger (1983a) and *"Planipes brachydactylus"* Gand et al. (2000) from the La Lieude Formation, and *"Planipes caudatus"* Gand et al. (1995) from the Pradineaux Formation. *Brontopus antecursor* was also reported from the Guadalupian Upper Red Unit of the Catalan Pyrenees (De Jaime-Soguero et al., 2023). Because this ichnogenus characterizes the Guadalupian, Marchetti et al. (2019c) established the *Brontopus* footprint sub-biochron of the *Erpetopus* biochron, which encompasses the time interval between the Roadian and the early Capitanian.

4.8.8 Trackmaker

Brontopus is attributed to dinocephalian therapsids, based on pedal and manual mesaxony, subequal digit length, broad palm/sole elongated laterally consistent with broad carpus/tarsus and thick metatacarpal/metatarsal V (Marchetti et al., 2019e). The ichnospecies *B. antecursor* has been attributed to anteosaurid dinocephalians, based on the shorter manus, the longer digits, the higher pace, and the occurrence of tail impressions (Marchetti et al., 2019d). The attribution of material assigned to *Brontopus* to other groups such as caseid synapsids and pareiasaurid parareptiles (Gand et al., 2000; Rubidge in Valentini et al., 2009) has to be discarded because they both show digit V reduction. Also, *Brontopus* forms paw-like impressions, typical of synapsids and not of parareptiles (Valentini et al., 2009).

4.9 *Dolomitipes* Marchetti et al., 2019d (Fig. 4.4C and D)

4.9.1 *Type ichnospecies*

Dolomitipes accordii (Ceoloni et al., 1988).

4.9.2 *Holotype*

NMS-NS 34/26 and UR-NS 34/26 (Ceoloni et al., 1988: figs. 5–7, 13; Marchetti et al., 2019d: fig. 5). Trackway, convex hyporelief. Val Gardena Sandstone Formation, Lopingian, Italy.

4.9.3 *Valid ichnospecies*

Dolomitipes accordii (Ceoloni et al., 1988); *Dolomitipes icelsi* (de Klerk, 2002).

4.9.4 *Distribution*

Guadalupian–Lopingian. It is known from the Guadalupian of Germany and South Africa, the Guadalupian-Lopingian of Spain; and the Lopingian of Germany, Italy, South Africa, and Australia (Table 4.1).

4.9.5 *Diagnosis (after Marchetti et al., 2019e, emended)*

Trackway of a quadrupedal tetrapod, relatively small to large (pes length of 40–250 mm) tracks with pentadactyl, semiplantigrade to plantigrade footprints. Manual imprint about one third wider than long and much more deeply impressed than the pes, contrasting with *Dimetropus* and *Dicynodontipus* which show a manus that is as long as wide, and *Brontopus*, *Dimetropus* and *Dicynodontipus*, which show a similar imprint depth of manus and pes. Pedal imprint of similar or slightly smaller size when compared to the manus, but different from *Dimetropus*, which shows a pes larger than the manus. Middle digit impressions II–IV of the pes distinctly longer than outer digit impressions and subequal in length (mesaxony), unlike *Dimetropus*, which is ectaxonic. Pedal digit impression V short, proximally and laterally positioned, different from *Dimetropus* and *Brontopus*, which show a relatively longer and nonproximal digit impression V.

4.9.6 *General description*

Trackway of a quadrupedal tetrapod, relatively small to large (pes length of 40–250 mm) tracks with pentadactyl, semiplantigrade (manus) to plantigrade (pes) footprints. Digit impressions relatively short, straight to bent inward distally (manus), terminating in enlarged tip impressions showing a subcircular shape and arcuate arrangement. Middle part of digits shallow or not impressed, often resulting in paw-like impressions. Manus mesaxonic to weakly ectaxonic. Well-impressed palm impression with concave proximal margin and large, rounded metacarpal-phalangeal pad impressions of all digit impressions. Manual imprint about one third wider than long and much more deeply impressed than the pes. Pedal imprint shallower and longer than wide or about as wide as long, of similar or slightly smaller size when compared to the manus. Middle digit impressions II–IV of the pes distinctly longer than outer digit impressions and subequal in length (mesaxony). Digit impressions I–IV are ~1/3 of the pes length, with rounded metatarsal-phalangeal pad impressions clearly separated from the postero-lateral, elliptical, and well-impressed heel impression. Digit impression V short, proximally and laterally positioned. Trackways consisting of alternating sets of manus and pes imprints with forward directed or inwardly rotated manus and a forward-directed pes. The manus is placed in line or slightly external to the pes, low (with possible partial secondary overstep) to relatively high (with possible partial primary overstep) pace angulation. No tail or body trace preserved.

4.9.7 Discussion

The ichnogenus *Dolomitipes* was erected by Marchetti et al. (2019d) in the new combination *Dolomitipes accordii* (Ceoloni et al., 1988) from the Lopingian Val Gardena Sandstone Formation of the Italian Dolomites. Part of this material had been previously assigned to *Ichniotherium* and *Limnopus* (Conti et al., 1977; Ceoloni et al., 1988). Nevertheless, these assignments were incorrect because of the pedal mesaxony and morphology and arrangement of basal digit pads, incompatible with these two ichnogenera, as well as the number of manual digits in the case of *Limnopus* (e.g., Voigt, 2005). *Dolomitipes accordii* was also described from the Palingkloof Member of the Balfour Formation in South Africa (Marchetti et al., 2019e), which is below the Permian–Triassic boundary according to Gastaldo et al. (2020). This material was previously figured by Watson (1960) and Smith and Botha-Brink (2014).

A second species of *Dolomitipes* was described from the Lopingian Oudeberg Member of the Balfour Formation, South Africa, by Marchetti et al. (2019e) with the new combination *Dolomitipes icelsi* (de Klerk, 2002). This ichnospecies has also been described from the latest Guadalupian Poortje Member of the Teekloof Formation, South Africa (Marchetti et al., 2019e). Similar footprint morphotypes, though relatively poorly preserved, were described from the Guadalupian–Lopingian of the Catalan Pyrenees (Mujal et al., 2017) and are herein assigned to cf. *Dolomitipes* isp.

Dolomitipes has also been described from the Lopingian eolian Cornberg Formation of Germany (Marchetti et al., 2019c) and reported from the Lopingian Rothenberg-Annweiler formations transition in the German Zechstein (Voigt et al., 2023). Also, we re-assign to *Dolomitipes* isp. specimens from the Lopingian of Australia (Retallack, 1996, figs. 2–3), because pes and manus imprint have a consistent morphology, proportions and arrangement. This material was originally described as "*Dicynodontipus bellambiensis*," herein considered a *nomen dubium* because of its relatively poor preservation. Two specimens from the Eisenach Formation of Germany are herein interpreted as isolated manus and pes imprints (Voigt, 2005, figs. 47A and 47D, respectively) and re-assigned to cf. *Dolomitipes* isp. This would be in agreement with a Guadalupian age for this unit.

4.9.8 Trackmaker

This ichnogenus has been attributed to dicynodont therapsids because of the heteropody, the mesaxony, the more deeply impressed and inward-directed manual imprints compared with the pedal imprints, which are instead parallel to the direction of progression, consistent with the postural dichotomy of dicynodonts (Marchetti et al., 2019d,e). This attribution was also proposed in previous studies (e.g., de Klerk, 2002; Smith and Botha-Brink, 2014). Other studies suggested a more conservative attribution to anomodont therapsids (Watson, 1960) or to edaphosaurid synapsids (Conti et al., 1977; Ceoloni et al., 1988). However, nondicynodont anomodonts may have been more ectaxonic, and edaphosaur synapsids have a different stratigraphic range (Marchetti et al., 2019d,e). Dicynodont trackmakers are thus retained for *Dolomitipes*.

4.10 *Karoopes* Marchetti et al., 2019e (Fig. 4.4E and F)

4.10.1 *Type ichnospecies*

Karoopes gansfonteinensis Marchetti et al. (2019e).

4.10.2 *Holotype*

GF-TR 1 (Marchetti et al., 2019e: fig. 6A–C; Fig. 4.4E and F). Trackway, concave epirelief. Poortjie Member of Teekloof Formation, Guadalupian, South Africa.

4.10.3 *Valid ichnospecies*

Karoopes gansfonteinensis Marchetti et al. (2019e).

4.10.4 *Distribution*

Guadalupian–Lopingian of South Africa and possibly Uruguay (Table 4.1).

4.10.5 *Diagnosis (after Marchetti et al., 2019e, emended)*

Medium- to large-sized (pes length of ~50–260 mm), semiplantigrade, pentadactyl, and weakly ectaxonic footprints (unlike *Brontopus, Dicynodontipus, Dolomitipes,* which show mesaxony) of a quadruped, with evident homopody (unlike *Capitosauroides,* which shows heteropody). The manus imprints are slightly more ectaxonic than the pes imprints (unlike *Capitosauroides,* which shows a slightly more ectaxonic pes imprint than the manus imprints). Digit imprints are relatively long compared with the palm/sole impressions (unlike *Dimetropus, Brontopus, Dicynodontipus,* and *Dolomitipes,* which have relatively shorter digit imprints). The palm/sole impression is very short (~75 mm in average) with a concave proximal margin (unlike *Hyloidichnus,* which does not have a concave proximal margin; and *Dimetropus, Brontopus, Dicynodontipus, Dolomitipes,* which have a large sole impression). Very large digit V (unlike *Dolomitipes* and *Hyloidichnus,* which show a relatively small pedal digit V). Footprints more deeply impressed laterally (unlike *Hyloidichnus,* which shows footprints more deeply impressed medially). Very high-pace angulation (110–170°) and possible pes-manus overstep (different from *Hyloidichnus, Dimetropus, Brontopus, Dolomitipes,* and *Capitosauroides,* which have a lower pace angulation and no complete primary pes-manus overstep).

4.10.6 *General description*

Medium- to large-sized (pes length of ~50–260 mm), semiplantigrade, pentadactyl, and weakly ectaxonic footprints of a quadruped, with evident homopody. The digit length increases between digits I–IV, and digit V is about as long as digit III. The relatively long tapering digit impressions have rounded terminations, with digit impressions I–IV curved inward distally, and a broad and proximal digit impression V about as long as digit impression II, outwardly directed and laterally curved distally. Middle part of digits shallower, digit I is usually not completely impressed. Relatively low digit I–IV divergence, higher digit IV–V divergence. The palm/sole impression is very short (~av. 75 mm) with a markedly concave proximal margin; the heel is rarely impressed. Footprints are more deeply impressed distally and laterally. Pes wider than long, manus wider than long and slightly smaller and slightly more ectaxonic than pes. In general, the trackway shows a simple alternating arrangement of pes-manus couples with common primary overstep of the pes on the manus (from partial to total), very high-pace angulation (110–170°), and no tail or body drag impressions preserved. The pes is directed slightly outward or parallel to the midline whereas the manus is directed slightly inward.

4.10.7 *Discussion*

The ichnogenus *Karoopes* was introduced by Marchetti et al. (2019e) with the ichnospecies *Karoopes gansfonteinensis* based on the study of abundant and well-preserved material from the Gansfontein locality in South Africa (uppermost Guadalupian Poortje Member of the Teekloof Formation). Some of these footprints had previously been figured and preliminarily described by De Beer (1986) and Smith (1993). More material from the Brandewyns Gat 214 locality of the same member, first figured by Smith (1980), was assigned to this ichnospecies by Marchetti et al. (2019e).

4.10.8 *Trackmaker*

Karoopes was attributed to gorgonopsian therapsids by Marchetti et al. (2019e) based on the semicircular disposition of basal digit pads, the lateral functional prevalence, the homopody, the slight ectaxony more pronounced in the manus than the pes and the high-pace angulation. Also, the semiplantigrady and digit imprints curved inward distally are consistent with the intratarsal joint of gorgonopsians. An attribution to other kinds of therapsids such as dinocephalians (e.g., De Beer, 1986; Smith, 1993) was discarded because of the semiplantigrady, the ectaxony, the high pace angulation, and the different stratigraphic range (Marchetti et al., 2019e). Also, an attribution of the small-sized *Karoopes* to dicynodonts or amphibians (Smith, 1980, 1993) was discarded because of the homopody and the lateral functional prevalence, respectively (Marchetti et al., 2019e; Mujal et al., 2020).

4.11 *Capitosauroides* Haubold, 1971b (Fig. 4.4G and H)

4.11.1 *Type ichnospecies*

Capitosauroides bernburgensis Haubold (1971b).

4.11.2 *Holotype*

HF 24 (Haubold, 1971b: fig. 5; Buchwitz et al., 2020: figs. 6A and 7A). Trackway, concave epirelief. Middle Buntsandstein, Early Triassic, Germany.

4.11.3 *Valid ichnospecies*

Capitosauroides bernburgensis Haubold (1971b); *Capitosauroides talus* (Heyler and Montenat, 1980).

4.11.4 *Distribution*

Guadalupian–Middle Triassic. In the Permian, it is known from the Guadalupian of France, Germany, and South Africa; the Guadalupian–Lopingian of Morocco and Uruguay; and the Lopingian of the UK (Scotland) and Italy (Table 4.1).

4.11.5 *Diagnosis (from Buchwitz et al., 2020, emended)*

Semiplantigrade pentadactyl tracks (different from *Dimetropus, Brontopus, Dolomitipes,* and *Dicynodontipus,* that are plantigrade). Pes larger than the manus (different from *Karoopes,* characterized by homopody). Relatively long and robust digit imprints that can be distally bent inwards or outwards (different from *Hyloidichnus* and *Varanopus,* that do not show outward bending). Digit IV imprints of the manus and pes are of equal length or slightly longer than digit III imprint, indicating a mesaxonic to slightly ectaxonic manus and slightly ectaxonic pes (different from *Amphisauropus, Dimetropus, Karoopes,* and *Varanopus,* that are ectaxonic). Shallow impression of digit I, which is often very short (different from *Amphisauropus, Hyloidichnus,* and *Varanopus,* that have a deep impression of digit I). Pedal digit V imprint is relatively long, set proximally and slightly separated from the digit I–IV group (different from *Amphisauropus, Varanopus,* and *Hyloidichnus,* in which the digit V imprint is connected with the digit I–IV group). Short to relatively long manual digit V imprint (different from *Amphisauropus, Hyloidichnus* and *Varanopus,* that have a relatively short manual digit V imprint). The manus is in line with the pes (different from *Amphisauropus, Hyloidichnus,* and *Varanopus,* in which the manus is more medial than the pes); manus oriented parallel to the midline, and pes is parallel to laterally oriented (different from *Hyloidichnus* that shows a more medially oriented pes).

4.11.6 *General description*

Semiplantigrade, pentadactyl tracks of intermediate size (pes length about 40–100 mm) of a quadruped. Relatively long, straight and robust digit imprints terminating in wide, triangular claw impressions or rounded tips. Relatively long and robust digit imprints that can be distally bent inwards or outwards. Digit length increases between digit I–IV imprints, with subequal digit II–IV imprints, these being more deeply impressed than digits I and V. Digit I imprint is the shortest, and digit V imprint is about as long as digit I in the pes and II in the manus. Digit V imprint in a proximal position but adjacent to digits I–IV and distally bent outwards. Concave to slightly convex proximal margin of palm and sole impressions. Palm relatively short, sole relatively longer and expanded in its proximal-lateral part. The sole impression is generally shorter than half foot length. Pes imprint slightly longer than wide, manus imprint slightly wider than long. Manus imprint about 1/3 shorter than pes imprint. Simple alternating arrangement of close pes-manus couples showing no overstep, with pes imprint parallel to trackway midline and manus imprint in line with pes imprint and parallel or turned slightly inward with respect to the midline. Step cycle relatively broad, and stride relatively low, with low-pace angulation (70–110°).

4.11.7 *Discussion*

The ichnogenus *Capitosauroides* was introduced from the Triassic of Germany (Haubold, 1971b), with the ichnospecies *C. bernburgensis*. Permian occurrences were first described by Marchetti et al. (2019d) from the Lopingian red beds of the Italian Dolomites, and subsequently by Buchwitz et al. (2020) from the Guadalupian Hornburg Formation of Germany. Marchetti et al. (2022b) introduced the new combination *Capitosauroides talus* (Heyler and Montenat, 1980) in their revision of the ichnofauna of the upper Guadalupian Pélitique Formation of France and re-assigned to this ichnospecies the mentioned material from Italy and

Germany. Also, material from the upper Guadalupian Poortje Member of the Teekloof Formation of South Africa and from the Lopingian Corncockle Formation of Scotland has been assigned to cf. *Capitosauroides* isp. (Marchetti et al., 2019c, 2019e). According to Marchetti et al. (2022b), morphologically similar material from Germany and Morocco is generally too incomplete and/or poorly preserved for a confident determination (e.g., Germany, Eisenach Formation, Voigt, 2005, fig. 47B–C; Morocco, Ikakern Formation, Hminna et al., 2012; fig. 5 and Cham-el-Houa Siltstone Formation, Moreau et al., 2020; fig. 4); these specimens are herein re-assigned to cf. *Capitosauroides* isp.

4.11.8 Trackmaker

This ichnogenus has been recently attributed to therocephalian therapsids (Marchetti et al., 2019e) because of the arcuate disposition of digit bases, the lateral functional prevalence of autopodia, the manual mesaxony, the heteropody, and reduction of digit V. The concave proximal margin of palm/sole and curved digit tips are consistent with the therocephalian intratarsal joint, which allowed for mobility between the astragalus and the calcaneum and an overall increase in lateral limb movement (Kemp, 2005). The attribution to anamniotes, as suggested in previous works (e.g., Haubold, 1971b; Marchetti et al., 2019d), is not in agreement with a lateral functional prevalence *sensu* Mujal et al. (2020), the digit base arrangement and the manual mesaxony.

4.12 Ichnogenus *Dicynodontipus* Rühle von Lilienstern 1944 (Fig. 4.4I and J)

4.12.1 Type ichnospecies

Dicynodontipus geinitzi (Hornstein, 1876).

4.12.2 Holotype

MNHN XXI (Rühle von Lilienstern, 1944: pls. 22–23). Trackway, convex hyporelief. Solling Formation, Middle Triassic, Germany.

4.12.3 Valid ichnospecies

Dicynodontipus geinitzi (Hornstein, 1876).

4.12.4 Distribution

Guadalupian–Middle Triassic. Guadalupian of France; Guadalupian–Lopingian of Tunisia and Germany; Lopingian of the UK (Scotland), Italy, South Africa, Australia, and Brazil. Lopingian-Early Triassic of Argentina (Table 4.1).

4.12.5 Diagnosis (after Haubold, 1971a, emended)

Pes and manus imprints of the same morphology (different from *Brontopus*, *Capitosauroides*, *Dimetropus*, *Dolomitipes*, and *Karoopes*). They are plantigrade (different from *Capitosauroides* and *Karoopes* that are semiplantigrade), pentadactyl with short, forward-pointing digits, digit IV longest (different from *Brontopus* and *Dolomitipes*, which show mesaxony), and digit V shifted slightly posteriorly. Relatively narrow trackways, pace angulation not below 100° at normal gait. Manus imprints positioned anterior to pes imprints, but in trackways of fast-moving individuals widely overstepped by the latter (different from *Brontopus*, *Capitosauroides*, *Dimetropus*, and *Dolomitipes*, which have broader trackways and no complete pes-manus overstep). No tail traces are observed.

4.12.6 General description

Pentadactyl, plantigrade imprints of small to intermediate size with short digit traces and large palm/sole impressions. Pes imprints have the same morphology of the manus imprints. The size of the pes imprints ranges between 20 and 95 mm; the manus is slightly smaller. Digit imprints are short, slender, straight, and distally tapering. Proportions of digit traces can generally be characterized as subequal in length, with digit IV longest, and digit V shifted slightly posteriorly. Interdigital angles are low, digits are almost parallel. The claw traces can be preserved as small, round impressions of the tip only, which are positioned at a short distance from the distal digit ends. Middle part of digits very shallow or not impressed, resulting in typical paw-like impressions. Digit bases arranged in a semicircular shape. Very broad and circular palm/sole impressions, longer than digit imprints. Relatively narrow trackways characterized by a high-pace angulation, up to 160°. Manus imprints are rotated inward relative to the pes imprints and are overstepped by the latter laterally (primary overstep).

4.12.7 Discussion

The ichnogenus was introduced based on Middle Triassic material from the German Buntsandstein (Solling Formation) by Rühle von Lilienstern (1944). The type ichnospecies *Dicynodontipus geinitzi* was originally assigned to a different ichnogenus, as "*Chelichnus*" *geinitzi* by Hornstein (1876). Other Buntsandstein material that has been synonymized with *Dicynodontipus* (Haubold, 1971a,b) was formerly published by different authors under *Chirotherium*, "*Agostropus*," *Dicynodontipus* and "*Onkichnium*" (Rühle von Lilienstern, 1939; 1944; Schmidt, 1959; Kuhn, 1963).

Other than the known Triassic material, Permian *Dicynodontipus* mostly consists of isolated tracks, but trackways have been recently described as well (e.g., Francischini et al., 2018; Marchetti et al., 2019c; Citton et al., 2021). In particular, the characteristic morphology with large round sole impressions and short digits being subequal in length and terminating in pointed claw traces, similar in both pes and manus imprints, justifies an assignment of the Permian material to the ichnogenus. From the Permian (Lopingian), *Dicynodontipus* is known from the Val Gardena Sandstone Formation of Northern Italy (Conti et al., 1977; Marchetti et al., 2019d) and the Rio do Rasto Formation of Brazil (Silva et al., 2012). Nevertheless, we consider "*Dicynodontipus penugnu*" as a *nomen dubium*, because it is based on poorly preserved material (Silva et al., 2012). We re-assign this material to *Dicynodontipus* isp. Only recently, new occurrences have been found in late Capitanian to Induan units of South America, Africa, and western Europe (e.g., Francischini et al., 2018; Marchetti et al., 2019c, 2019e, 2022b; Citton et al., 2021). We re-assign the material from the Guadalupian-Lopingian of Tunisia (Contessi et al., 2018, fig. 3A) to *Dicynodontipus geinitzi*, Also, we re-assign to *Dicynodontipus* a trackway from the Lopingian of Australia (Warren 1997, fig. 8). In fact, both records show morphology/proportions/trackway pattern consistent with this ichnospecies.

Some purported occurrences from the Lopingian of South Africa (de Klerk, 2002) were reassigned to the ichnogenus *Dolomitipes* (Marchetti et al., 2019e). Also, the ichnospecies *Dicynodontipus* "*bellambiensis*" Retallack (1996) from the Lopingian of Australia, is herein considered a *nomen dubium* and re-assigned to *Dolomitipes* isp.

4.12.8 Trackmaker

Dicynodontipus was originally attributed to dicynodont therapsids (Rühle von Lilienstern, 1944), probably because of the mesaxony to slight ectaxony, with reduction of lateral digit impressions. Nevertheless, the homopody and very high-pace angulation with complete primary pes-manus overstep during faster progression suggest a cynodont producer (e.g., Haubold, 1966). In fact, cynodont autopodial skeletons show mesaxony to slight ectaxony and homopody. They acquired a parasagittal gait, which is consistent with high-pace angulation and a complete primary pes-manus overstep (e.g., Broom, 1948; Jenkins, 1970; Blob, 2001; Fourie and Rubidge, 2007, 2009). The stratigraphic distribution of *Dicynodontipus* (Lopingian-Middle Triassic) is also consistent with the stratigraphic distribution of cynodonts (Lucas, 2006, 2010). The slightly more ectaxonic pes of *Dicynodontipus* is comparable with the phalangeal formula of 2-3-4-5-3 in the pes and 2-3-4-4-3 in the manus of some cynodonts such as *Thrinaxodon* (Hopson, 1995).

4.13 *Varanopus* Moodie, 1929 (Fig. 4.5A and B)

4.13.1 Type ichnospecies

Varanopus curvidactylus Moodie (1929).

4.13.2 Holotype

YPM 1106 (Moodie, 1929: fig. 8; Fig. 4.5A). Left pes-manus couple with lateral overstep, convex hyporelief. Choza Formation, Cisuralian, USA (Texas).

4.13.3 Valid ichnospecies

Varanopus curvidactylus Moodie (1929); *Varanopus microdactylus* (Pabst, 1896); *Varanopus macdonaldi* (Voigt et al., 2013a) n. comb.

4.13.4 Distribution

Cisuralian–Guadalupian. Cisuralian of the USA (Arizona, Colorado, New Mexico, Texas), Canada (Prince Edward Island, Nova Scotia), Spain, Italy, Germany, the Czech Republic, Turkey; Guadalupian of France (Table 4.1).

4.13.5 Diagnosis (after Moodie, 1929, emended)

Pes length 15–70 mm, larger than *Erpetopus* and *Notalacerta* and smaller than *Pachypes*. Distally curved digit imprints (different from *Hyloidichnus*, which show straight digit impressions). Relatively robust and elongated digit imprints (different from *Erpetopus* and *Notalacerta*, which show slim and frequently bent digit imprints, and *Pachypes*, which shows relatively shorter and wider digit imprints). Digit imprints terminating in sharp, triangular claw impressions (different from *Hyloidichnus* and *Pachypes*, which generally show enlarged bifurcated or rounded digit terminations, respectively). Marked increase in length of manual digit II–III imprints (different from *Pachypes*, which shows subequal manual digit II–IV impressions). Relatively long, forward-directed, and straight pes digit V impression (different from *Erpetopus*, *Pachypes*, and *Procolophonichnium*, which have a relatively shorter and

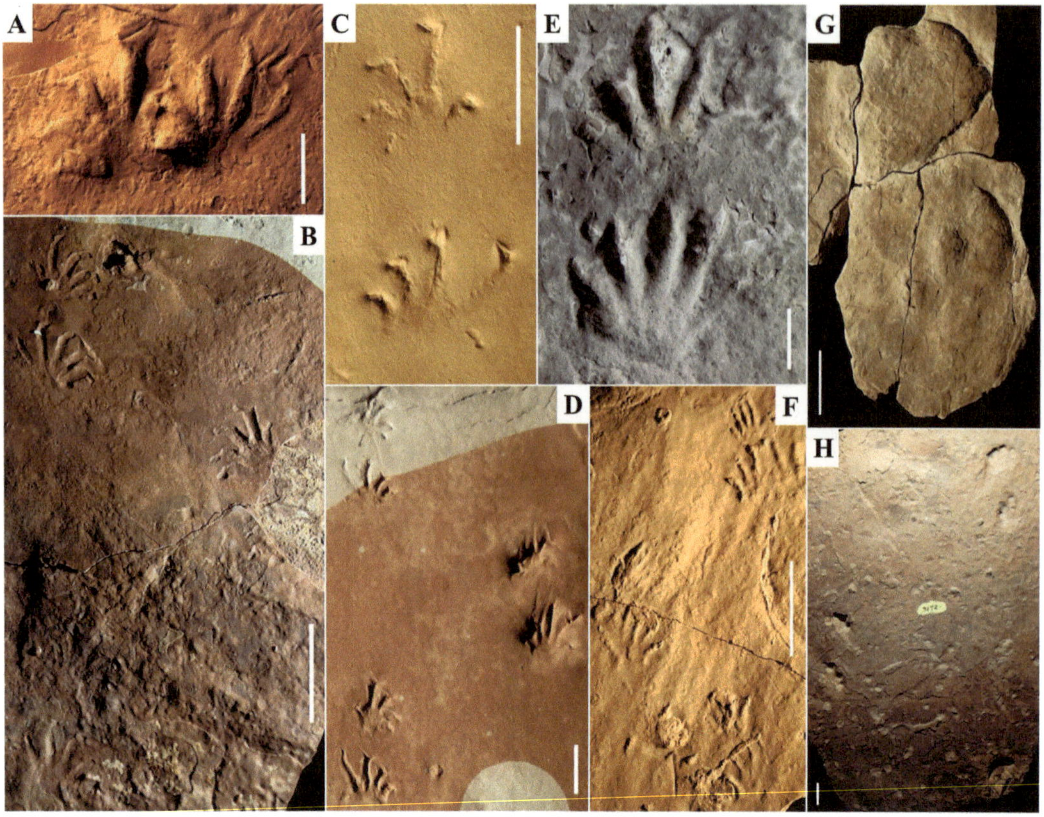

FIGURE 4.5 Parareptile tracks. (A–B) *Varanopus curvidactylus*, track attributed to bolosaurians. Choza Formation, Cisuralian, Texas. (A) YPM 1106, holotype. Left pes-manus couple with primary lateral pes-manus overstep, convex hyporelief. (B) NMMNH P-32392. Trackway, concave epirelief. (C–D) *Erpetopus willistoni*, track attributed to acleistorhinids, nyctiphruretids or millerettids. Choza Formation, Cisuralian, Texas. (C) HF 197. Right pes-manus couple, concave epirelief. (D) UGKU 1803. Trackway, concave epirelief. (E–F) *Procolophonichium*, track attributed to procolophonoids. (E) NHMUK-PV-R 3173. Plaster cast of holotype of *Procolophonichnium nopcsai*. Left pes-manus couple, convex hyporelief. Palingkloof member, Balfour Formation, Lopingian, South Africa. (F) NMS-NS 34/26. Holotype of *Procolophonichnium tirolensis*. Trackway, convex hyporelief. Val Gardena Sandstone Formation, Lopingian, Italy. (G–H) *Pachypes*, track attributed to pareiasauromorphs. (G) NMS-NS 34/27. *Pachypes dolomiticus*, left pes-manus couple, convex hyporelief. Val Gardena Sandstone Formation, Lopingian, Italy. (H) GRCA 3172. *Pachypes ollieri*, trackway, concave epirelief. Hermit Formation, Cisuralian, Arizona, USA. Scale bars: A, C–E, H = 1 cm; B, F = 5 cm; G = 10 cm.

laterally-bent pes digit V impression). Parallel/superimposed digit base impressions (different from *Notalacerta* and *Hyloidichnus*, which show radiating and not overlapping digit base impressions). Relatively short palm/sole impression (different from *Pachypes* and *Procolophonichnium*, which show a relatively longer palm/sole impression). Straight to concave proximal margin of the sole (different from *Notalacerta, Hyloidichnus* and *Pachypes*, which show a convex proximal margin).

4.13.6 General description

Semiplantigrade, ectaxonic, pentadactyl, and relatively small footprints (pes length of about 15–70 mm) of a quadruped. Long, straight, distally tapering digit traces terminating in triangular, thin, and sharp claw impressions. Manus about as long as wide and slightly to markedly smaller than pes. Manual digits I–IV with parallel bases and distally curved medially, digit V directed laterally. Manual digit trace V about as long as digit traces I–II. Pes wider than long or about as long as wide, with digit traces I–IV parallel at their bases, sometimes a medial-lateral digit base overlap is observed. Digit V slightly divergent from the digit I–IV group and forward-directed. Digit traces I–IV distally bent inward, digit trace V about as long as digit trace II or III. Very short palm/sole impressions, with straight to concave proximal margin. Manus more deeply impressed than the pes. Both pes and manus imprints show a medial-lateral decrease in relief, especially the pes imprint. Simple alternating arrangement of close pes-manus couples in broad to narrow trackways. Pes imprint generally parallel to the midline, and manus imprint oriented inward. Low to high-pace angulation (50–160°) with possible complete primary pes-manus overstep at higher gaits. Footprints can be associated with a tail drag trace.

4.13.7 Discussion

This ichnogenus was introduced by Moodie (1929) from the Cisuralian Choza Formation of Texas, with the ichnospecies *Varanopus curvidactylus*. Haubold and Lucas (2001, 2003) synonymized also the ichnospecies "*Laoporus wyldei*" Moodie (1930), from the same formation, with *V. curvidactylus*. This ichnospecies has also been described from the Coconino Formation of Arizona, and similar material comes from the De Chelly Formation of Arizona and the Lyons Formation of Colorado, all Cisuralian eolian units (Marchetti et al., 2019b).

A second *Varanopus* ichnospecies has been described by Haubold (1971a), who introduced the new combination *Varanopus microdactylus* (Pabst, 1896) from the Cisuralian Tambach Formation of Germany. Also, this ichnospecies has been described from the Cisuralian Oberhof and Rotterode formations of Germany (Voigt, 2005), the Peranera Formation of Spain (Voigt and Haubold, 2015), the Pizzo Del Diavolo and Collio formations of Italy (Marchetti, 2016; Marchetti et al., 2022a) and the Pennsylvanian-Cisuralian Cape John Formation of Canada (Van Allen et al., 2005). More material assigned to *Varanopus* is from the Cisuralian Maroon Formation of Colorado (Voigt et al., 2005), the Abo Formation of New Mexico (Lucas et al., 2001), the Padochova Formation of the Czech Republic (Voigt, 2005) and the Guadalupian Pélitique Formation of France (Marchetti et al., 2022b). Moreover, we consider "*Chelichnus incurvus*" Gand et al., 1995 a *nomen dubium* and re-assign this material to *Varanopus* isp. Also, we re-assign to *Varanopus* isp. a pes-manus couple from the Cisuralian of Canada (Calder et al., 2004, fig. 5B), a pes imprint associated with a body impression from the Cisuralian of Germany (Martens, 1991), a pes-manus couple from the Viala Formation of France (Gand, 1989, fig. 5B) and a pes-manus couple from the Cisuralian of Turkey (Gand et al., 2011, fig. 4.1). All these specimens show a relatively long, forward-directed impression of pedal digit V, digit base superimposition, medial functional prevalence and digit imprints distally curved inwards, all features of *Varanopus*.

The ichnogenus "*Robledopus*" was introduced with the ichnospecies "*Robledopus*" *macdonaldi* by Voigt et al. (2013a), from the Cisuralian Robledo Mountains Formation of New

Mexico, and is currently only known from this formation. This material has been previously variably attributed to "*Gilmoreichnus*," *Erpetopus*, and *Hyloidichnus* (Haubold et al., 1995; Hunt et al., 1995; Minter and Braddy, 2009). "*Gilmoreichnus*" was considered a *nomen dubium* by Marchetti et al. (2019a) and its holotype re-assigned to cf. *Hyloidichnus* isp. Nevertheless, the only possible character distinguishing "*Robledopus*" from *Varanopus* is the relatively long manual digit III, because the tail drag traces and outward-oriented pes imprints mentioned by Voigt et al. (2013a) as diagnostic features of "*Robledopus*" are also observed in *V. microdactylus* (e.g., Van Allen et al., 2005). Thus, a differentiation at the ichnospecies level within the first introduced ichnogenus *Varanopus* Moodie (1929) seems more appropriate, so we synonymize "*Robledopus*" and *Varanopus* and thus introduce the new combination *Varanopus macdonaldi* (Voigt et al., 2013a).

The ichnotaxon "*Varanopus rigidus*" was considered a *nomen dubium* by Marchetti et al. (2022b), and the holotype was re-assigned to cf. *Hyloidichnus* isp.

4.13.8 Trackmaker

Varanopus is currently attributed to relatively small-sized parareptiles such as bolosaurians (Marchetti et al., 2021c). This is in agreement with previous attributions of this material to small parareptiles (e.g., Müller, 1954; Haubold, 1971a; Haubold and Lucas, 2001). Some studies attributed this material to captorhinid or protorothyridid eureptiles (e.g., Haubold, 1971a; Fichter, 1983b; Haubold and Lucas, 2001; Voigt, 2005; Voigt et al., 2013a). For instance, *Varanopus macdonaldi* n. comb. has been previously attributed to protorothyridid reptiles because of its similarity with ichnogenera such as *Varanopus*, *Erpetopus*, *Notalacerta*, and *Hyloidichnus*, classically attributed to captorhinomorphs, and its more basal position suggested by the presence of a continuous belly/tail trace (Voigt et al., 2013a). However, the parallel and overlapping digit bases of *Varanopus* are incompatible with the well-splayed metatarsals of captorhinids and protorothyridids (Marchetti et al., 2021c).

4.14 *Erpetopus* Moodie, 1929 (Fig. 4.5C and D)

4.14.1 Type ichnospecies

Erpetopus willistoni Moodie (1929).

4.14.2 Holotype

FMC-UC 443 A (Moodie, 1929: fig. 4; Haubold and Lucas, 2001: fig. 3A). Trackway, convex hyporelief. Choza Formation, Cisuralian, USA (Texas).

4.14.3 Valid ichnospecies

Erpetopus willistoni Moodie (1929), *Erpetopus cassinisi* (Ceoloni et al., 1987)

4.14.4 Distribution

Cisuralian–Lopingian. From the Cisuralian, material assignable to *Erpetopus* has been described from the USA (Arizona, New Mexico, Texas), Spain, France, Italy, Morocco, and Argentina. It is also known from the Guadalupian of France and from the Guadalupian-Lopingian of ?Spain, ?Germany and Morocco (Table 4.1).

4.14.5 Diagnosis (after Moodie, 1929, emended)

Small, pentadactyl manus and pes, ectaxonic with a moderate increase in length of digit impressions I–IV (different from *Dromopus*, which shows a marked increase in length of digit impressions I–IV). Long, slender, and distally curved digit imprints (different from *Procolophonichnium*, *Pachypes*, and *Hyloidichnus*, which have thicker and/or straighter digit imprints). Digits parallel proximally and sometimes showing digit base overlap (different from *Notalacerta* and *Hyloidichnus*, which have well-splayed, radiating digits). Digit V shorter than digit II and bent laterally (different from *Varanopus*, *Notalacerta*, and *Hyloidichnus*, that have a relatively longer and/or not laterally bent pedal digit V). Digit impression V forward-directed (different from *Notalacerta* and *Hyloidichnus*, that have a laterally oriented digit V) and not separated from the digit impression I–IV group (different from *Dromopus*, which shows a separated and laterally oriented digit V impression). Very short palm/sole impression with concave to convex proximal margin (different from *Pachypes*, *Notalacerta*, and *Hyloidichnus*, which have a longer palm/sole impression and/or a convex proximal margin). Medial-lateral decrease in relief (different from *Dromopus*, which has a very shallow digit I impression).

4.14.6 General description

Footprints of a small quadruped (pes length 5–40 mm) with pentadactyl, ectaxonic, and semiplantigrade manus and pes imprints. The pes is about as long as wide, the manus is wider than long and about 1/3 shorter than pes. Footprints show a common medial–lateral decrease in relief, more marked in the pes imprints, which are less deeply impressed than the manus imprints. The digits are long, thin, tapering distally, and end in sharp claw impressions. Digit length proportions: I < V < II < III < IV. Digits I–IV are bent inward distally, and digit V is in a proximal position compared with digits I–IV and it is bent laterally distally. Digits rarely show transversal grooves. Digit I–IV bases parallel, sometimes overlapping. The sole/palm impression is short and rectangular. The sole/palm proximal margin is concave to convex. The trackways are constituted by a slightly irregular arrangement of pes–manus sets, with the pes close to the manus; complete primary overstep of the pes on the manus is possible at higher gaits. Pace angulation from low to high, on average around 90°. Manus placed more medially compared with the pes. The manus is usually oriented inward, the pes is variably oriented. In some cases, a sinuous to straight tail trace can be observed.

4.14.7 Discussion

Erpetopus was introduced by Moodie (1929) with the ichnospecies *Erpetopus willistoni* from the Cisuralian Choza Formation of Texas. From the same unit, the ichnospecies "*Microsauropus clarki*" Moodie (1929), "*M. acutipes*" Moodie (1929) and "*M. parvus*" Moodie (1930) were introduced, based on morphologically similar material. Haubold and Lucas (2001, 2003) synonymized all these ichnospecies under *Erpetopus willistoni*. Nevertheless, Gand and Durand (2006) supported the differentiation between the ichnogenera *Erpetopus* and "*Microsauropus*," based on the more slender digit imprints with claw traces of the latter. However, from the analysis of the type material and additional specimens from the type locality, these differences are clearly gait- and substrate-related. Therefore, "*Microsauropus*" is here considered a junior subjective synonym of *Erpetopus*. Some material from the Cisuralian of France is herein re-assigned to *Erpetopus willistoni* (Haubold and Lucas, 2001, fig. 12B; Barrier et al., 2009, figs. 5–8), because it shows all the diagnostic features of this ichnospecies.

A second *Erpetopus* ichnospecies was described by Marchetti et al. (2014) from the Cisuralian Collio Formation of Italy, with the new combination *Erpetopus cassinisi* (Ceoloni et al., 1987) and supported by the description of new material from the Pizzo del Diavolo Formation of Italy, following the preliminary hypothesis of Santi (2007). Nevertheless, this material should be revised, because its diagnostic features may be part of the *E. willistoni* range of morphologies. Recent studies highlighted the occurrence of a new track morphotype from Cisuralian units of Italy (Marchetti et al., 2015a,b) and New Mexico (Voigt and Lucas, 2017), provisionally assigned to *Varanopus* isp. The same morphotype, in France, was previously assigned to *Varanopus* isp. by Gand (1981), *V. curvidactylus* by Gand (1987), and cf. *V. curvidactylus* by Haubold and Lucas (2001). However, contrasting with this material, *Varanopus* is characterized by a forward-directed, straight, and relatively longer pedal digit V imprint (Moodie, 1929). Conversely, the occurrence of a pedal digit V distally bent outwards is a diagnostic feature of *Erpetopus*, so this new morphotype is herein renamed *Erpetopus* isp.

Erpetopus is a common ichnogenus in the time interval between the Artinskian and the Wuchiapingian, and its FAD defines the beginning of the *Erpetopus* footprint biochron (Voigt and Lucas, 2018). We herein re-assign to *Erpetopus* isp. a trackway from the Cisuralian of Spain (Mujal et al., 2016, fig. 8), and two isolated tracks from Morocco (Voigt et al., 2011a; fig. 6B) and Argentina (Melchor and Sarjeant, 2004, fig. 6D), because their morphology and digit arrangement and proportions are consistent with this ichnogenus.

4.14.8 Trackmaker

Erpetopus is currently attributed to small parareptile producers such as nyctiphruretids, acleistorhinids and millerettids (e.g., Marchetti et al., 2022a). The previous attribution to small reptiles such as protorothyridids and captorhinids (Haubold and Lucas, 2003; Gand and Durand, 2006) has to be discarded because of the parallel/overlapping digit bases of *Erpetopus*, consistent with metatarsal overlapping, which is observed in parareptiles and diapsids but not in protorothyridids and captorhinids (e.g., Marchetti et al., 2021c).

4.15 Procolophonichnium Nopcsa, 1923 (Fig. 4.5E,F)

4.15.1 Type ichnospecies

Procolophonichnium nopscai Kuhn (1963).

4.15.2 Holotype

Stored in SNSB-BSPG but lost (Seeley, 1904: fig. 1A–C). A plaster cast of the holotype is NHMUK-PV-R 3173 (Fig. 4.5E). Left pes-manus couple, convex hyporelief. Palingkloof Member, Balfour Formation, Lopingian, South Africa.

4.15.3 Valid ichnospecies

Procolophonichnium nopscai Kuhn (1963); *Procolophonichnium tirolensis* (Ceoloni et al., 1988).

4.15.4 Distribution

Guadalupian–Early Triassic. Guadalupian-Lopingian of Germany; Lopingian of ?UK (Scotland), Italy, Germany, South Africa, and Brazil (Table 4.1).

4.15.5 Diagnosis (after Klein et al., 2015, emended)

Pentadactyl and semiplantigrade (different from *Dromopus* and *Rhynchosauroides*, that can be semidigitigrade) ectaxonic footprints of small quadrupeds. Long, stout, and distally tapering digit impressions (different from *Varanopus*, *Erpetopus*, and *Notalacerta*, that have more slender digit imprints; and *P. lockleyi*" and "*P. nectouxi*, which do not taper), increasing in length from I–IV and curved inward distally (different from *Hyloidichnus*, which has straight digit imprints and *P. haarmuehlensis*" and "*P. polonicum*," which have digits bent outward distally). Digit impression IV subequal in length with digit impression III (different from *Dromopus* and *Rhynchosauroides*, which have a markedly longer digit IV). Digit impression V subequal in length with digit impression I or II (different from *Varanopus*, *Notalacerta*, "*P. lockleyi*" and "*P. nectouxi*," which have a longer pedal digit V imprint), forward-directed and straight or curved outward distally (different from *Notalacerta*, *Hyloidichnus*, *Dromopus*, and *Rhynchosauroides*, which have a laterally directed digit V; and *Varanopus*, which has a straight pedal digit V). Digit divarication about 70–100° (higher than "*P. lockleyi*" and "*P. nectouxi*," lower than "*P. haarmuehlensis*" and "*P. polonicum*"). Parallel and sometimes overlapping digit bases (different from *Notalacerta* and *Hyloidichnus*, which show well-splayed radiating digit imprints). Short rectangular palm/sole impression (different from *Pachypes*, "*P. lockleyi*" and "*P. nectouxi*, which show a longer palm/sole impression). Medial-lateral decrease in relief (different from "*P. haarmuehlensis*," "*P. polonicum*," *P. lockleyi*," and "*P. nectouxi*," which do not show this feature).

4.15.6 General description

Trackway of a quadrupedal tetrapod with pentadactyl semiplantigrade tracks; foot length ranges between 19 and 76 mm. Manus and pes imprints slightly differ in size and imprint proportion: the manus imprint is one fourth shorter than the pes imprint and slightly longer than wide or wider than long, whereas the pes is as wide as long. The compact, relatively long, robust, and straight digits increase in length between digits I–IV, manual digit V is shorter than digit I, pedal digit V is about as long as digit I or II. Digit tips are curved inward in digits I–IV, but parallel to the digit or rotated outward in V. Claws are triangular in shape and very thin or enlarged, depending on preservation. Digit V shows a characteristic deeply impressed and long basal pad in continuity with the digit and more proximal compared with the compact group of digits I–IV; this can generate a concave proximal palm/sole margin. Digits are parallel at their base and may show digit base overlap. Short palm/sole (less than half of the imprint length) with a concave to straight proximal margin. Trackways show an alternating arrangement of coupled manus-pes sets with parallel or variably rotated imprints (pes 13°/−27°; manus 21°/4°). Primary marginal overstepping of the pes on the manus has been observed. Pace angulation between 65 and 140°. The tail trace is shallow and sinuous, when preserved.

4.15.7 Discussion

Procolophonichnium was introduced by Nopcsa (1923) after the study of a pes-manus couple from the Palingkloof Member of the Balfour Formation of South Africa. The type ichnospecies *P. nopscai* was introduced later by Kuhn (1963). Further Permian material was described by Marchetti et al. (2019d) from the Val Gardena Sandstone Formation of Italy, with the new combination *Procolophonichnium tirolensis* (Ceoloni et al., 1988). The type species *P. nopscai* differs, among other features, in its imprint proportions from *P. tirolensis. Procolophonichnium* material has also been described from the Permian of Germany, Brazil and possibly Scotland (Silva et al., 2012; Buchwitz et al., 2017; Marchetti et al., 2019c). Other Triassic ichnospecies have been assigned to *Procolophonichnium* (see Klein et al., 2015 and references therein), including: "*P. haarmuehlensis*" (Holst et al., 1970), "*P. polonicum*" (Ptaszynski in Fuglewicz et al., 1990), "*P. nectouxi*" (Gand, 1977) and "*P. lockleyi*" Klein et al. (2015). Nevertheless, Klein et al. (2015) and Marchetti et al. (2019e) aimed at a morphological distinction of these Triassic *Procolophonichnium* ichnospecies from the Permian-Induan material, noting these different features: "*P. haarmuehlensis*" and "*P. polonicum*" have a concave proximal margin of the sole impression, a larger digit divarication, and digit impressions distally bent outwards; "*P. lockleyi*" and "*P. nectouxi*" do not show tapering digit impressions, have a digit impression V about as long as digit impression III and show lower digit divarication and an elongate heel impression. Based on the revised diagnosis of *Procolophonichnium*, we consider only *P. nopscai* and *P. tirolensis* as valid ichnospecies of *Procolophonichnium*. The ichnospecies "*P. haarmuehlensis*," "*P. polonicum*," "*P. nectouxi*," and "*P. lockleyi*" do not belong to *Procolophonichnium*, and their ichnotaxonomy has to be revised by further studies.

4.15.8 Trackmaker

In accordance with the first description and etymology of the ichnogenus *Procolophonichnium*, small parareptiles of the group Procolophonidae were initially considered as likely producers (Nopcsa, 1923). However, small reptilian producers cannot be ruled out for other specimens of *Procolophonichnium*, including the type material and the Permian *P. tirolensis* track sample (Marchetti et al., 2019d,e). Further studies, including a thorough revision of recently collected Permian *Procolophonichnium* specimens and their comparison with other similar small reptile track types, such as *Varanopus* and *Erpetopus*, might help to sort out distinct *Procolophonichnium* producer groups.

4.16 *Pachypes* Leonardi et al., 1975 (Fig. 4.5G and H)

4.16.1 Type ichnospecies

Pachypes dolomiticus Leonardi et al. (1975).

4.16.2 Holotype

IGPF 73-111 (Leonardi et al., 1975: figs. 2–4; Marchetti et al., 2020a: fig. 3). Right pes imprint, convex hyporelief. Val Gardena Sandstone Formation, Lopingian, Italy.

4.16.3 Valid ichnospecies

Pachypes dolomiticus Leonardi et al. (1975); *Pachypes loxodactylus* (Dudgeon, 1878), *Pachypes ollieri* (Ellenberger, 1983a)

4.16.4 Distribution

Cisuralian–Lopingian. It is known from the Cisuralian of the USA (Arizona, Texas), Spain, France, and Italy; the Cisuralian–Guadalupian of ?Spain, the Guadalupian of France; the Guadalupian-Lopingian of Spain, ?Germany, ?Niger and ?Uruguay; the Lopingian of the UK (Scotland), Italy, Germany, Russia and Morocco (Table 4.1).

4.16.5 Diagnosis (after Marchetti et al., 2021a, emended)

Pentadactyl and semiplantigrade footprints of a quadrupedal, heteropodal reptile, with tracks more deeply impressed medially. It differs from synapsid tracks such as *Brontopus, Capitosauroides, Dicynodontipus, Dimetropus, Dolomitipes,* and *Karoopes,* which have, instead, plantigrade footprints and/or tracks more deeply impressed laterally. Sturdy, short digits with rounded or flat terminations not separated from the palm/sole impression. This is generally different from reptile tracks such as *Dromopus, Erpetopus, Rhynchosauroides* and *Varanopus,* which have instead long and slender digits and sharp claw impressions and from synapsid tracks such as *Brontopus, Dicynodontipus, Dimetropus* and *Dolomitipes,* which have digit imprints separated from the palm and sole impressions. Digits I–IV closely grouped and proximally superimposed. This differs from tracks such as *Hyloidichnus* and *Notalacerta,* which show instead well-spread digit imprints. Pes digit imprints increase in length from I to IV, and digit V is about as long as digit I (this differs from *Varanopus* and *Notalacerta,* which show a longer pedal digit V). More symmetrical manus, imprints of digits II–IV being subequal in length; shorter external digits. Well-developed and deep sole and palm impressions, this differs from *Varanopus, Erpetopus, Procolophonichnium, Notalacerta,* and *Hyloidichnus,* which have shorter palm/sole impressions. Manus smaller in size and slightly turned inward with respect to the pes. Pes subparallel with respect to the trackway midline; this differs from *Amphisauropus,* which shows instead a markedly inward-turned manus and outward-turned pes. No tail-drag traces.

4.16.6 General description

Relatively small to large (pes length about 30–370 mm), semiplantigrade and pentadactyl footprints of a quadrupedal tetrapod with broad palm/sole impression and short and sturdy, distally rounded digits in continuity with the palm/sole impression. The manus imprint is wider than long and slightly smaller (about 1/3 shorter) and more deeply impressed than the pes imprint. Manus imprints are slightly ectaxonic to mesaxonic with external digit imprints shorter (especially digit V impression), but the lengths of digits II–IV are remarkably similar. The manual digit V imprint is in a proximal position. The palm impression is elliptical and wider than long, its proximal margin is straight to convex. Digit impressions are extremely short, tightly grouped, and superimposed for most of their length, and they have rounded terminations. The pes imprint it is about as long as wide, and clearly ectaxonic, because digit length increases between digits I–IV. Digit impressions are thick and end in rounded tips. The digit I–IV impressions are closely grouped, superimposed at their bases and may be distally bent inwards. The digit V impression is relatively short (about as long as I), thin, in a proximal position and may be distally bent outwards (laterally oriented). The sole impression is elliptical and wider than long, its proximal margin is convex. Digit terminations and the medial side of the footprint are more deeply impressed than the digit III–V bases, sometimes resulting in a crescent-moon shape in incompletely impressed tracks. The trackway pattern is characterized by a simple alternating arrangement of manus–pes couples with a short manus–pes distance and long stride length and width of pace compared with the pes length. Despite the long stride, the pace angulation is relatively low (about 70–110° for the pes and 80–115° for the manus). The pes imprint is aligned with the midline, and the manus imprint is in front of the pes imprint, slightly medially positioned and slightly rotated inward respect to the pes. No primary overstep and tail/body impressions have been reported for this ichnogenus.

4.16.7 Discussion

Pachypes was introduced from the Lopingian Val Gardena Sandstone Formation of Italy by Leonardi et al. (1975), with the ichnospecies *Pachypes dolomiticus*. The holotype consists of a large-sized, well-preserved, right pes imprint. Subsequent studies from the same units confirmed the validity of this ichnospecies by the description of a large quantity of well-preserved material, including pes-manus couples, incomplete step cycles, and a trackway (e.g., Conti et al., 1977; Valentini et al., 2008, 2009; Marchetti et al., 2019d, 2020a). Material assignable to *P. dolomiticus* has been described or reported from the Lopingian, and in particular from the Ikakern Formation of Morocco (Voigt et al., 2010), the Moradi Formation of Niger (Smith et al., 2015), two different Lopingian units in Germany (Voigt and Fischer, 2020; Buchwitz et al., 2017), and the Poldarsa Formation and possibly the Vyatkian horizon of Russia (Gubin et al., 2003; Surkov et al., 2007; Valentini et al., 2009; Kuznetsov, 2020). The ichnospecies "*Sukhonopus primus*" Gubin et al. (2003) is herein considered a junior subjective synonym of *Pachypes dolomiticus*.

Marchetti et al. (2019c) introduced the new combination *Pachypes loxodactylus* (Dudgeon, 1878), a more slender form of *Pachypes*, from the eolian Locharbriggs Formation of Scotland, and recognized it also from the eolian Corncockle Formation of Scotland and possibly the eolian Cornberg Formation of Germany. All these units are Lopingian.

A third, smaller, *Pachypes* ichnospecies was described by Marchetti et al. (2021a) under the new combination *Pachypes ollieri* (Ellenberger, 1983a) from the Guadalupian La Lieude Formation of France (Gand et al., 2000). This ichnospecies was also reported from the Guadalupian Le Muy Formation and a possible occurrence from the Pradineaux Formation of France, and from the Cisuralian Hermit Formation of Arizona, San Angelo Formation of Texas, Rabejac Formation of France, Peranera Formation (Lower Red Unit) of the Catalan Pyrenees (Spain), possibly the Port des Canonge Formation of Mallorca (Spain) and Monte Luco Formation of Italy (Marchetti et al., 2021a, 2022a; Matamales-Andreu et al., 2022).

Material from the Guadalupian-?Lopingian Upper Red Unit of Spain was assigned to *Pachypes* isp. (De Jaime-Soguero et al., 2023), and material from a different locality of the same unit can be assigned to cf. *Pachypes* isp. (Mujal et al., 2017).

4.16.8 Trackmaker

The material herein assigned to *Pachypes dolomiticus* has been consistently attributed to pareiasaurid parareptiles because of the short, thick, closely grouped digit imprints and the extreme reduction of digit V (Leonardi et al., 1975; Gubin et al., 2003; Valentini et al., 2009; Marchetti et al., 2019d; Kuznetsov, 2020). The attribution to therapsids of the material described by Surkov et al. (2007) is not supported by the track morphology, showing short and grouped digit imprints well impressed along their entire length, typical of pareiasaur tracks (Valentini et al., 2009). The ichnospecies *P. loxodactylus* is attributed to slender pareiasaur parareptiles (Marchetti et al., 2019c) and the ichnospecies *P. ollieri* to relatively smaller nycteroleter pareiasauromorphs (Marchetti et al., 2021a).

4.17 *Notalacerta* Butts, 1891 (Fig. 4.6A and B)

4.17.1 Type ichnospecies

Notalacerta missouriensis Butts (1891).

4.17.2 Neotype

USNM 7321 (Marchetti et al., 2020a, b: fig. 1B–D). Right pes-manus couple, convex hyporelief. Cement City Limestone, Chanute Formation, Pennsylvanian, USA (Missouri).

4.17.3 Valid ichnospecies

Notalacerta missouriensis Butts (1891).

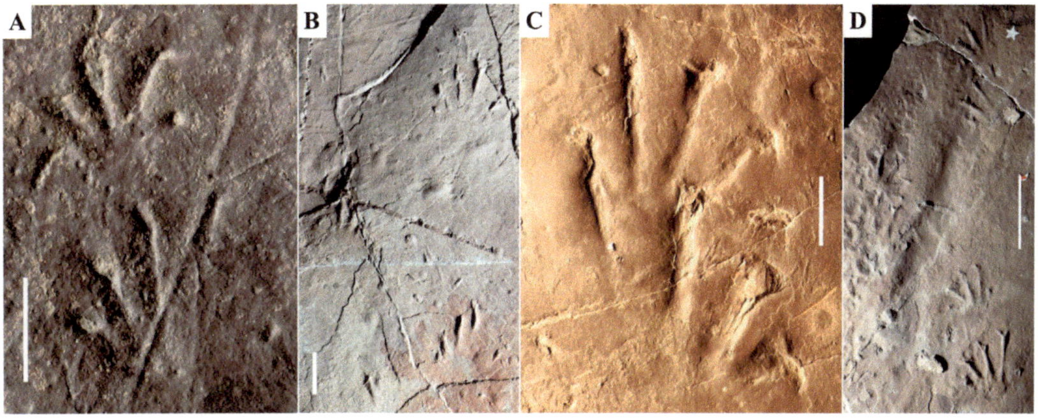

FIGURE 4.6 Nondiapsid eureptile tracks. (A–B) *Notalacerta missouriensis*, track attributed to basal eureptiles. (A) UGKU 1931. Left pes-manus couple, convex hyporelief. Peranera Formation, Cisuralian, Spain. (B) NMMNH P-62216. Trackway, convex hyporelief. Robledo Mountains Formation, Cisuralian. (C–D) *Hyloidichnus bifurcatus*, track attributed to captorhinids. (C) MNHN LOD 86. Holotype of *Hyloidichnus "major"*. Left pes-manus couple, concave epirelief. Rabejac Formation, Cisuralian, France. (D) USNM 11518. Holotype of *Hyloidichnus bifurcatus*. Trackway, convex hyporelief. Hermit Formation, Cisuralian, Arizona, USA. Scale bars: A–C = 1 cm; D = 5 cm.

4.17.4 Distribution

Pennsylvanian–Cisuralian. In the Permian, it is known from the USA (New Mexico), ?England, Spain, France, Germany and Argentina (Table 4.1).

4.17.5 Diagnosis (after Marchetti et al., 2020a, emended)

Semiplantigrade, pentadactyl footprints of small quadrupeds (pes length smaller than 25 mm, different from *Pachypes*, which has larger tracks). Each footprint has very slender digit imprints that differ from *Varanopus, Hylopus, Procolophonichnium, Pachypes,* and *Hyloidichnus*, which have robust digit imprints. Digit terminations may be acuminate, bifurcated, or pointed, with evidence of thin, sharp claw impressions (different from *Hylopus* and *Amphisauropus*, which show enlarged and/or rounded digit terminations). Manus with a different morphology compared with the pes (different from *Dromopus*, which shows the same morphology of manus and pes imprints). Pes digit IV clearly longer than pes digit III (different from *Varanopus* and *Procolophonichnium*, which show subequal pes digits III and IV). Pes digit V imprint as long as digits II–III (different from *Hyloidichnus, Erpetopus,* and *Pachypes*, which have a pes digit V imprint about as long as digit I). Well-spread digit imprints and digit V imprint in line with digits I–IV (different from *Dromopus, Erpetopus, Procolophonichnium,* and *Varanopus*, which show tightly grouped digit I–IV imprints superimposed at their base, and a proximal digit V imprint). Straight pes digit V imprint, different from *Erpetopus*, which has an outward-bent pes digit V imprint. Thin and usually medially directed pes digit I imprint (different from *Varanopus*, which shows a thick, well-impressed, and usually forward-directed digit I imprint). Variable orientation of digit imprints, especially digit III imprint, that can be parallel to digit IV in the manus and to digit II in the pes (different from *Hyloidichnus, Hylopus,* and *Varanopus*, which have digit imprints with consistent orientation). Medial-lateral decrease in relief of the pes (different from *Dromopus*, which is more deeply impressed in the distal part of digits III and IV). Very short and shallow sole and palm impressions, and absence of rounded basal pads (different from *Hylopus*, which may show a rounded basal pad of digit I and broad palm and sole impressions).

4.17.6 General description

Each footprint is semiplantigrade and has five slender digits, and each digit tapers from the base to the extremity. Digit termination acuminate or pointed, evidence of thin, sharp claws. Manus slightly wider than long, and pes about as long as wide, pes length of 10–25 mm, manus about 1/4 shorter. Marked increase in length between digits I–IV in both pes and manus but more marked in the pes, digit V as long as digit II (manus) or as long as or slightly shorter than III (pes). No digit superimposition, and digit V base in line with the bases of digits I–IV. The digits I–IV can be distally curved inward; digits long, compared with size of foot. Variably oriented digit imprints, especially digit III, which can be parallel to digit IV in the manus and to digit II in the pes. High total digit divergence, especially in the manus. Marked median-lateral decrease in relief in the pes. Very short and usually not impressed sole and palm traces, with straight to convex proximal margin. Simple alternating arrangement of close pes-manus couples, with a low to moderate pace angulation (60–120°), pes parallel to the midline or rotated outwards, manus slightly rotated inwards. Relatively common, thin, straight, and continuous tail drag trace.

4.17.7 Discussion

The ichnogenus *Notalacerta* has a morphology close to *Varanopus* and *Hyloidichnus*, which led to numerous misidentifications. It differs from *Varanopus* in the more slender and curved digit imprints, the less deeply impressed pes digit I, the nonsuperimposed digit bases, the nonconcave proximal margin of the sole, the relatively longer pes digit IV and the relatively shorter and more laterally directed pes digit V. It differs from *Hyloidichnus* based on the more slender, curved and mobile digit imprints and the relatively longer digit V. Marchetti et al. (2020a) provided further comparisons between *Notalacerta* and other similar ichnogenera (e.g., *Dromopus*, *Hylopus*). All the material of *Notalacerta* from the Cisuralian has been little discussed. Although revisions of the Carboniferous material have been undertaken (Marchetti et al., 2020b, 2021c), the Permian material is pending a thorough re-evaluation.

4.17.8 Trackmaker

Notalacerta has been attributed to reptiles since its introduction (Butts, 1891), although a temnospondyl trackmaker within the family Dissorophidae has also been proposed (Haubold, 1971a). Chesnut et al. (1994) performed a precise track-trackmaker correlation that ruled out the latter hypothesis and confirmed the clear reptilian morphology of *Notalacerta*. These authors attributed *Notalacerta* to "protorothyridid" nondiapsid reptiles (Chestnut et al., 1994; Marchetti et al., 2020b). They, indeed, share a small tarsal structure in agreement with the very short sole impression of *Notalacerta*, a relatively longer pes digit V, a carpal and tarsal structure consistent with a more marked medial-lateral decrease in relief, and a glenoacetabular length/pes length ratio similar to the same ratio measured from the *Notalacerta* trackways. Marchetti et al. (2020b) proposed the following nondiapsid eureptiles as potential trackmakers of *Notalacerta*: *Anthracodromeus*, *Hylonomus*, *Paleothyris*, and *Thuringothyris*, the last one known from the early Permian and thus consistent with Cisuralian *Notalacerta*.

4.18 *Hyloidichnus* Gilmore, 1927 (Fig. 4.6C and D)

4.18.1 Type ichnospecies

Hyloidichnus bifurcatus Gilmore (1927).

4.18.2 Holotype

USNM 11,518 (Gilmore, 1927: fig. 25; Fig. 4.6D). Trackway, convex hyporelief. Hermit Formation, Cisuralian, USA (Arizona).

4.18.3 Valid ichnospecies

Hyloidichnus bifurcatus Gilmore (1927), *Hyloidichnus thaleri* (Ellenberger, 1983a) n. comb.

4.18.4 Distribution

Cisuralian–Lopingian. Cisuralian of the USA (Arizona, New Mexico, Oklahoma), Spain, France, Italy, Morocco, and Argentina; Cisuralian–Guadalupian of Spain and Morocco; Guadalupian of France; Guadalupian-Lopingian of Spain, Morocco and Niger; Lopingian of Germany, Hungary and Morocco (Table 4.1).

4.18.5 Diagnosis (after Gilmore, 1927; emended)

Footprints of small to intermediate size (pes length 10–130 mm) (different from *Erpetopus* and *Notalacerta*, which can be smaller, and *Pachypes*, which can be larger). Long, very straight, bottle-shaped or tapering digit imprints and inward-directed, flat claw impressions, often bifurcated (different from *Erpetopus, Notalacerta, Pachypes, Procolophonichnium,* and *Varanopus*, which do not have bottle-shaped digit impressions or flat claw impressions). Radiating digit imprints not parallel/superimposed at their base (different from *Erpetopus, Pachypes, Procolophonichnium,* and *Varanopus*, which show parallel/overlapping digit bases). Digit divergence relatively high (different from *Erpetopus, Pachypes, Procolophonichnium,* and *Varanopus*, which have a low digit divergence). Digit proportions: I $\leq$ V < II < III < IV, digits III and IV of similar length (different from *Notalacerta* and *Varanopus*, which have relatively longer pes digit V imprints). Very short and rectangular palm/sole impression, with straight to convex proximal margin (different from *Pachypes*, which has a relatively broad palm/sole, and *Erpetopus, Procolophonichnium,* and *Varanopus*, which can show a concave proximal margin).

4.18.6 General description

Semiplantigrade, pentadactyl footprints of small to intermediate size (between 10 and 130 mm) of a quadruped. Ectaxonic footprints with digits I–IV increasing in length, digits III–IV of similar length and digit V relatively short (about as long as digits I–II). Digit imprints straight, elongated, and radiating, with high divarication angle of digit bases. Digit imprint termination in bifurcated or enlarged hourglass shape. Very short palm and sole impressions with straight to slightly convex proximal margins. Pes imprint about as wide as long. Manus imprint wider than long and shorter than the pes imprint. In both manus and pes imprints, the medial part of the footprints is more deeply impressed (more marked in the pes). Pes-manus couples in a simple and close alternating arrangement with low-pace angulation (around 100° on average). A slight primary overstepping of the pes imprint on the manus imprint can be observed. Manus imprint in front of the pes, more medially positioned and more inwardly turned along the broad trackway. Pes parallel to the midline. Curved and continuous digit drag traces and continuous straight to sinuous tail drag traces can be observed in association with the footprints.

4.18.7 Discussion

The type ichnospecies of *H. bifurcatus* Gilmore (1927) was described from the Permian Hermit Formation of the Grand Canyon (Arizona, USA). The ichnospecies *H. "whitei"* was also erected by Gilmore (1928) from the same site. Haubold (1971a) proposed the new combination *Hyloidichnus "major"* from the lower Permian of the Rabejac Formation in the Lodève Basin based on "*Auxipes* major" (Heyler and Lessertisseur, 1963). Later, Haubold (1973) proposed the new combination *H. "minor"* from the same locality based on "*Auxipes minor*" (Heyler and Lessertisseur, 1963) and erected a new ichnospecies, *H. "arnhardti"* Haubold (1973) from the lower Permian Oberhof Formation of the Thuringian Forest. Another ichnospecies, *H. "tirolensis"* Ceoloni et al. (1988), was described from the Lopingian of the Val Gardena Sandstone Formation but re-assigned to *Capitosauroides* isp. (Marchetti et al., 2019d). Haubold (2000) considered valid only *H. bifurcatus* and *H. "major."* Marchetti et al. (2022b) synonymized these two ichnospecies. Moreover, other ichnotaxa have been considered as extramorphological variants of *Hyloidichnus*. This is the case for "*Gilmoreichnus (Hylopus) hermitanus*" (Gilmore, 1927; Marchetti et al., 2013a) and *Varanopus "rigidus"* (Gand, 1989; Logghe et al., 2021; Marchetti et al., 2022b).

The ichnogenus "*Merifontichnus*" Gand et al., (2000), described from the Guadalupian La Lieude Formation of France, shares numerous morphological features with *Hyloidichnus*, as pointed out by Marchetti (2016). Potential differences include the similar size of the manus and pes, the short sole, the larger relative width of the pes, the similar length of the pedal digits III–IV, the pedal digit I straight and parallel to the proximal margin of the sole, the higher pace angulation and stride length. It has also shorter and thicker digits. Nevertheless, these features seem more suitable for an ichnospecific differentiation, so we consider "*Merifontichnus*" as a subjective junior synonym of *Hyloidichnus*, and synonymize it accordingly. "*Merifontichnus thalerius*" Gand et al. (2000) was first described from the La Lieude Formation of the Lodève Basin (Massif Central, France) as "*Macrochelichnus thaleri*" (Ellenberger, 1983a; Gand, 1987). Gand et al. (2000) justified the introduction of a new ichnotaxon because they considered "*Macrochelichnus thaleri*" not available because it did not fulfill the requirements of the ICZN (1999). However, "*Macrochelichnus thaleri*" Ellenberger (1983a), was indeed described, figured and reference was made to a type, as mandated by ICZN (1999), so it is an available name, is an objective senior synonym of "*Merifontichnus thalerius*," and it has priority. We synonymize "*Macrochelichnus*" with *Hyloidichnus* and introduce the new combination *Hyloidichnus thaleri* (Ellenberger, 1983a). Its morphology close to *Amphisauropus* has led to misidentifications of this ichnotaxon (Gand et al., 2000; Marchetti, 2016). *Hyloidichnus thaleri* n. comb. differs from *Amphisauropus* in: the similar size of the manus and pes, the short sole, the larger relative width of the pes, the similar length of the pedal digits III–IV, the pedal digit I straight and parallel to the proximal margin of the sole, the higher pace angulation and stride length, the nonmarked impression of the digit I basal pad, and the pes parallel to the trackway. Based on those differences, *Amphisauropus sensu* Marchetti et al. (2013b) from the Collio Basin is re-assigned to *Hyloidichnus thaleri* n. comb. Potential *Hyloidichnus thaleri* n. comb material has been described from the Cheguimi Sandstone of the Tebaga Formation (Newell et al., 1976; Contessi et al., 2018). Even though its attribution remains open with unnamed tracks (similar to *Karoopes*, see Marchetti et al., 2019e), we refer these tracks to *Hyloidichnus thaleri* n. comb., as this attribution is more in accordance with the track morphologies.

Hyloidichnus is a common track between the late Artinskian and the Wuchiapingian (Voigt and Lucas, 2018). Among tetrapod ichnoassociations including *Hyloidichnus*, it is relatively common to find two differentiated size classes of this ichnotaxon (Logghe et al., 2021; Matamales-Andreu et al., 2022, 2023). We herein re-assign material from the Cisuralian of Oklahoma (Lucas and Suneson 2002, figs. 5–6) and the Lopingian of Hungary (Majoros 1965, figs. 2–3) to *Hyloidichnus* isp., because it is in agreement with all the diagnostic features of the ichnogenus. Moreover, we re-assign to cf. *Capitosauroides* isp. the material assigned to "*Merifontichnus* isp." from the Lopingian of Italy (Marchetti et al., 2017c, fig. 4). In fact, the sole impression elongated proximally and the marked heteropody are rather consistent with *Capitosauroides*.

4.18.8 Trackmaker

Although first considered as belonging to nonamniote trackmakers such as seymouriamorphs (Haubold, 1971a), *Hyloidichnus* has been commonly attributed to captorhinomorph eureptiles (Voigt et al., 2010). The wide size range of *Hyloidichnus* tracks supports small captorhininid (e.g., *Captorhinus*), small-to medium-sized moradisaurine (e.g., *Tramuntanasaurus*), and large moradisaurine (e.g., *Moradisaurus*) trackmakers (Voigt et al., 2010; Matamales-Andreu et al., 2023; Rmich et al., 2023). A precise synapomorphy-based track-trackmaker correlation between *Hyloidichnus* from the late Guadalupian of Gonfaron and *Captorhinus aguti* Cope (1882) has been performed (Logghe et al., 2021). They share a pentadactyl manus and pes, a similar relative length of the digits, radiating and nonoverlapping metatarsals/proximal digit imprints, clawed digit tips, similar length ratios between (1) pes/manus length and metatarsus/metacarpus length and (2) pes/manus foot length and sole/palm impression length, similar length/width ratios of the pes (phalanges + tarsus) and of the manus (phalanges + carpus) (<1 for the manus and ~1 for the pes). *Labidosaurus* and *Captorhinikos* have also been proposed as potential trackmakers for larger *Hyloidichnus* tracks from the same site (Logghe et al., 2021). Matamales-Andreu et al. (2023) strengthened the correlation between small morphotypes of *Hyloidichnus* with captorhinids (both moradisaurines and nonmoradisaurines) and large morphotypes of *Hyloidichnus* with moradisaurine captorhinids based on the functional prevalence of the material analyzed by them. According to these authors, the gait and velocity of movement of the trackmakers (slow vs. fast locomotion) may also influence the functional prevalence inferred from differences in relative depth patterns of the manus and pes of *Hyloidichnus*.

Hyloidichnus thaleri n. comb. was first attributed to therapsid synapsids (Gand et al., 2000). However, this attribution was made on the supposed mesaxony of the studied tracks (a feature shared with therapsid tracks such as *Dicynodontipus* and *Brontopus*). The *Hyloidichnus thaleri* n. comb. tracks were since described as ectaxonic, ruling out this attribution. Various features, including (1) short palm/sole impressions, (2) straight digits with enlarged termination and median-lateral decrease in relief in the pes, in addition to (3) the ectaxony of the tracks, instead support an attribution to derived captorhinids (Marchetti, 2016). The relatively large size of *Hyloidichnus thaleri* n. comb., moreover, supports the heavily built moradisaurines as trackmakers (Sumida, 1989; Marchetti, 2016).

4.19 *Dromopus* Marsh, 1894 (Fig. 4.7A and B)

4.19.1 *Type ichnospecies*

Dromopus agilis Marsh (1894).

4.19.2 *Holotype*

YPM 519 (Marsh (1894): pls. II.3, III.3; Marchetti et al., 2021c: fig. 8A–C). Trackway, concave epirelief. Howard Limestone, Pennsylvanian, USA (Kansas).

4.19.3 *Valid ichnospecies*

Dromopus lacertoides (Geinitz, 1861).

4.19.4 *Distribution*

Cisuralian–Lopingian. It is known from the Cisuralian of the USA (Arizona, New Mexico, Oklahoma, Texas, Ohio), Canada (Nova Scotia), Spain, France, Italy, Austria, Germany,

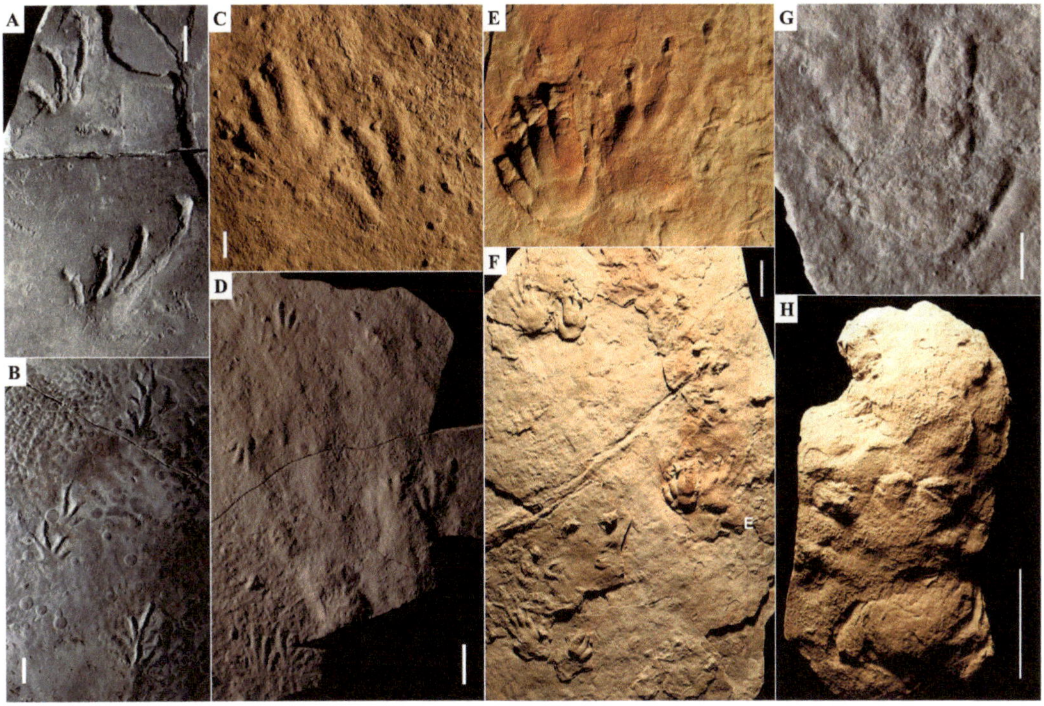

FIGURE 4.7 Diapsid eureptile tracks. (A–B) *Dromopus lacertoides*, track attributed to araeoscelids or varanopids. (A) MNG 13436. Right pes-manus couple, concave epirelief. Oberhof Formation, Cisuralian, Germany. (B) MNG JF 1. Trackway, convex hyporelief. Goldlauter Formation, Cisuralian, Germany. (C–D) *Rhynchosauroides pallinii*, track attributed to non-archosauromorph neodiapsids. IGPF 75-5. Holotype. Convex hyporelief. Val Gardena Sandstone Formation, Lopingian, Italy. (C) Left pes-manus couple, primary lateral pes-manus overstep. (D) Trackway. (E–F) *Paradoxichnium problematicum*, track attributed to non-archosauriform archosauromorphs. FG 20/1. Holotype. Convex hyporelief. Strassfurt Formation, Lopingian, Germany. (E) Left pes-manus couple, primary lateral pes-manus overstep. (F) Trackway. (G–H) cf. *Protochirotherium* isp., track attributed to early archosauriforms. Val Gardena Sandstone Formation, Lopingian, Italy. (G) UR NS 34/82. Left manus imprint, convex hyporelief. (H) UR 75/2. Right pes-manus couple, convex hyporelief. Scale bars: A–C, G = 1 cm; D, F, H = 5 cm.

Poland, the Czech Republic, Ukraine, ?Russia and Morocco; the Cisuralian–Guadalupian of Spain and Morocco; the Guadalupian of France and Germany, the Guadalupian-Lopingian of Spain and Morocco; the Lopingian of ?Italy and Morocco (Table 4.1).

4.19.5 Diagnosis (after Marsh 1894, emended)

A striking feature of fore and hind feet are the long, slender digit imprints, terminating in thin sharp claws (different from *Paradoxichnium*, terminating in wide triangular claw traces in the manus). Same morphology of pes and manus imprints (this differs from *Erpetopus*, *Varanopus*, *Tambachichnium*, *Paradoxichnium* and *Rhynchosauroides*). Manus as ectaxonic as the pes (different from *Tambachichium*, which has a manus more ectaxonic than the pes). Manus and pes semidigitgrade (different from *Tambachichnium*, which shows a semiplantigrade pes imprint, and from *Paradoxichnium* and *Rhynchosauroides*, which show a semiplantigrade manus imprint). Digit V imprint well-separated, proximal, straight, and oriented laterally (different from *Erpetopus*, which has a digit V with the base parallel to IV and distally bent, and from *Varanopus*, which has a digit V forward-directed). Strong overlap of digit I–IV imprint bases (different from *Notalacerta*, which shows radiating digits). Digit I shallow compared with digits II-V (different from *Erpetopus, Varanopus,* and *Notalacerta*, in which the medial part is more deeply impressed). Broad trackways that do not show a marked primary overstep of the pes on the manus (different from *Paradoxichnum, Rhynchosauroides,* and *Tambachichnium*, which show a marked primary overstep of the pes on the manus).

4.19.6 General description

Semidigitigrade, pentadactyl, ectaxonic footprints of a small- to medium-sized quadruped (pes length 8–80 mm). The pes is about one-third larger than the manus but shows identical morphology and proportions. Digit imprints are long, slender, distally tapering, ending in triangular claw marks that are directed medially. They may show elliptical digital pads. Digits I–IV are medially and distally curved and are tightly grouped, showing a marked proximal-lateral superimposition. Digit length markedly increases between digit I–IV imprints. Digit V is proximally located, straight, oriented laterally and about as long as digit II–III imprints. The palm/sole is very short, shallow, and rarely impressed. Footprints are semi-digitigrade, the most deeply impressed digit imprints are II–IV, and digit I and the bases of all other digits are very shallow and rarely impressed. Trackways are relatively broad and constituted by a simple alternating arrangement of pes-manus couples. The pace angulation is moderate (70–135°), the pes may partially overstep the manus proximally, and both pes and manus are oriented parallel to the midline or slightly laterally or medially. Tail and body traces are rarely observed. Continuous digit tip drag traces can occur, but they are generally rare.

4.19.7 Discussion

Marsh (1894) named *Dromopus "agilis"* for specimens from the Upper Pennsylvanian of Kansas, USA. An older subjective synonym is *"Protritonichnites"* Pohlig (1892), from the Cisuralian Goldlauter Formation, Germany. Nevertheless, while *Dromopus* had a wide usage in the literature from its introduction onwards, *"Protritonichnites"* was only used a second time by Holub and Kozur (1981), so almost 100 years after its introduction (Voigt, 2005). Given that *Dromopus* presents a case of prevalent usage as stated by the ICZN (1999), Voigt (2005) used the reversal of priority and considered *Dromopus* Marsh, 1894 as valid. Haubold (1996) and Voigt (2005) considered the type ichnospecies of the ichnogenus to be a junior subjective synonym of *Dromopus lacertoides* (Geinitz, 1861), and this has been followed in recent works (e.g., Lucas et al., 2011b; Marchetti et al., 2015a,b; Voigt and Lucas, 2015a, 2018). Some studies considered *Dromopus "didactylus"* (Moodie, 1930) from the upper Cisuralian Choza Formation of Texas to be a valid ichnospecies (Sarjeant, 1971; Gand, 1987; Gand and Durand, 2006). Haubold and Lucas (2001, 2003) considered *D. "didactylus"* a junior synonym of *Dromopus "palmatus"* (Moodie, 1929) from the same formation. The supposed differentiation between *D. "palmatus"* and *D. lacertoides* is based on a supposed larger distance between the bases of digit impressions IV and V and thus increased digitigrady in the former. However, Voigt (2005) questioned this argument, because it is based on a criterion difficult to verify due to the rarity of completely impressed footprints of *Dromopus*. For the time being, we follow

Voigt (2005) and consider *Dromopus lacertoides* as the only valid ichnospecies. Nevertheless, it is evident that a comprehensive revision of the ichnospecies of *Dromopus* is needed.

Dromopus is the most common track in the Cisuralian, and it is relatively abundant also in the Guadalupian. Lopingian records are instead rarer (e.g., Voigt and Lucas, 2018). We reassign to *Dromopus lacertoides* some material from the Cisuralian Letovice Formation of the Czech Republic (Mikuláš and Martínek, 2006; fig. 7A), because it shows the main diagnostic features of the ichnogenus.

4.19.8 Trackmaker

Dromopus has been considered the track of various lizard-like Permian parareptiles and eureptiles such as bolosaurids and araeoscelids (Haubold, 1971a, 1973, 1984, 2000; Fichter, 1979, 1983a; Voigt, 2005). Spindler et al. (2019) also proposed small varanopids as possible producers, based on digit proportions. Marchetti et al. (2021c), after an anatomy-based track-trackmaker correlation, restricted the potential producer of Upper Pennsylvanian-Guadalupian *Dromopus* to araeoscelid diapsids and nonvaranodontinae varanopids. Nevertheless, Lopingian occurrences were probably made by neodiapsids (Marchetti et al., 2019d).

4.20 *Rhynchosauroides* Maidwell 1911 (Fig. 4.7C and D)

4.20.1 Type ichnospecies

Rhynchosauroides rectipes Maidwell (1911).

4.20.2 Holotype

L. U. 8019 (Valentini et al., 2007: fig. 2). Right pes-manus couple with lateral overstep, convex hyporelief. Helsby Sandstone Formation, Early-Middle Triassic, UK (England).

4.20.3 Valid ichnospecies

R. pallinii Conti et al. (1977), *R. schochardti* Rühle v. Lilienstern, H., 1939, *R. tirolicus* Abel (1926), *R. hyperbates* Baird (1957).

4.20.4 Distribution

Guadalupian—Late Jurassic. Guadalupian of France; Lopingian of UK (Scotland), Italy, Germany, South Africa, and Brazil (Table 4.1).

4.20.5 Diagnosis (after Klein and Lucas, 2021; emended)

Lacertoid ectaxonic footprints of a quadruped. Pes slender and digitigrade; the smaller manus is more stout and more plantigrade (different from *Dromopus*, which is semidigitigrade, and from *Tambachichnium*, which has a more digitigrade pes imprint). Thin sharp digit termination (different from *Paradoxichnium* that has broad triangular claw marks in the manus). Length of digits increases from I to IV; digit V much shorter (different from *Dromopus*, *Paradoxichnium* and *Tambachichnium*, which have a relatively longer pedal digit V). Digits I—IV usually curved inward. Broad trackways with a marked lateral overstep of the manus by the pes (this is different from *Dromopus*, that does not show a marked lateral overstep).

4.20.6 General description

Pentadactyl, ectaxonic, longer than wide pes and manus imprints. Pes length of up to 89 mm, whereas the manus is up to 60 mm in length. Digits are slender with strong inward curvature along their complete distance or at distal ends only. They are moderately spread and increase in length from I to IV with digit IV being by far the longest. Small, acuminate claws form hook-like terminations. Digit V is small and postero-laterally positioned, consisting of a short impression that forms a right angle with the long axis of digit IV. Pes imprints are digitigrade or semiplantigrade and show a short metatarso-phalangeal area, whereas the smaller manus imprints are plantigrade with a distinct palmar surface. Well-preserved specimens display round phalangeal pads, their number suggesting a 2-3-4-5-?3 formula. Trackways have a relatively broad pattern, with pes imprints showing outward rotation, whereas manus imprints are rotated inward. The manus is laterally overstepped by the pes in all specimens. Stride lengths of trackways with pes imprints of 70 mm length were measured with 360 mm. The pace angulation is relatively low, between 70 and 125°. Manus medially positioned relative to the pes.

4.20.7 Discussion

Rhynchosauroides is an ichnotaxon originally described from Triassic deposits where it is often locally abundant. Maidwell (1911) introduced the ichnogenus with the type ichnospecies *R. rectipes* based on material from the Lower—Middle Triassic of Great Britain. Lopingian occurrences are known from the Val Gardena Sandstone Formation of northern Italy with *Rhynchosauroides pallinii* (Conti et al., 1977; Valentini et al., 2007; Marchetti et al., 2019d), which is the only ichnospecies from this time interval. Further Permian occurrences of the ichnogenus *Rhynchosauroides* are known from France, Germany, Scotland, and South Africa (Marchetti et al., 2019c,e, 2022b). Diagnostic features of *Rhynchosauroides* identified in the Permian material justify the assignment to this ichnogenus. Several morphologically identical footprints from the Val Gardena Sandstone Formation that were formerly described under different ichnogenera and ichnospecies have been synonymized more recently with *R. pallinii* (see Marchetti et al., 2019d). The ichnospecies *Rhynchosauroides "gangresci"* Silva et al. (2012) from the Lopingian of Brazil is herein considered invalid because the type material is poorly preserved. We reassign this material more conservatively to *Rhynchosauroides* isp.

4.20.8 Trackmaker

The ichnogenus *Rhynchosauroides* is widely recognized as a relatively small track of a non-archosauriform neodiapsid producer (e.g., Haubold, 1966, 1971b; Conti et al., 1977; Demathieu and Oosterink, 1983; Avanzini and Renesto, 2002; Valentini et al., 2007; Marchetti et al., 2019e; Klein and Lucas, 2021). This is mostly due to the long and slender digit impressions with marked ectaxony and sharp claw marks, the semidigitigrady to semiplantigrady with very short palm/sole impressions, and the proximal position and orientation of digit impression V (e.g., Romer, 1956; Evans and Haubold, 1987). The occurrence of *Rhynchosauroides* in the Palingkloof Member of the Balfour Formation (Marchetti et al., 2019e) is consistent with the record of small neodiapsid skeletal remains from the same member (Viglietti et al., 2016). Therefore, the attribution of *Rhynchosauroides* to small neodiapsids seems well-justified. Nevertheless, *Rhynchosauroides* contains numerous named ichnospecies and needs to be revised (cf. Klein and Lucas, 2021).

4.21 *Paradoxichnium* Müller, 1959 (Fig. 4.7E and F)

4.21.1 *Type ichnospecies*

Paradoxichnium problematicum Müller (1959).

4.21.2 *Holotype*

FG 20-1 (Müller, 1959: pls I–II, fig. 1; Fig. 4.7E and F). Trackway with lateral primary overstep, convex hyporelief. Stassfurt Formation, Lopingian, Germany.

4.21.3 *Valid ichnospecies*

Paradoxichnium problematicum Müller (1959).

4.21.4 *Distribution*

Guadalupian–Lopingian. Guadalupian–Lopingian of Germany; Lopingian of Italy (Table 4.1).

4.21.5 *Diagnosis (after Müller, 1959, emended)*

A lacertoid, semiplantigrade vertebrate track. Toes of the forefeet squat, with broad triangular claw traces and squat palm impression (different from *Dromopus*, *Tambachichnium* and *Rhynchosauroides*, which show very slender digits in manus imprints ending in thin claw imprints). Pes imprint more ectaxonic and digitigrade than the manus (different from *Tambachichnium*, which shows a more ectaxonic and digitigrade manus). Pedal digit V relatively long, about as digit III (different from *Rhynchosauroides*, which has a shorter pedal digit V). Trackways with an alternating gait, intermediate trackway width, and relatively large stride length. Hind foot with the foot axis turned outward (different from *Dromopus*, which has a pes imprint turned inward).

4.21.6 *General description*

Tracks of quadrupedal tetrapods with pentadactyl, semiplantigrade imprints of lacertoid character; pes more than 145 mm in length, manus about one third shorter. Manus imprints are more deeply impressed than pes imprints, and both are more deeply impressed in the digit II–IV group. Manus imprints semiplantigrade with thick, straight digits moderately increasing in length from I to IV, digit V about as long as II. Digits I and V slightly proximal compared with digits II–IV. Interdigital angles are strongly acute, sometimes with digit base overlap. Digits taper and can be slightly curved inward distally. Thin horizontal skin folds can be observed in the proximal part of digits II–IV. Distinct, broad, straight, or medially directed triangular claw traces. The palm imprint is oval, has a convex proximal margin, it is short and less deeply impressed than the digits. Pes imprints semiplantigrade with digit imprints markedly increasing in length between I–IV, digit V imprint is in a proximal position and about as long as digit III imprint. Digits II–IV are closely grouped, with small interdigital angles. Pes imprints with long and slender digits, straight to slightly curved inwards, tapering distally, oval phalangeal pads can be observed. Digits have triangular to rounded terminations. The sole imprint is oval, with a convex proximal margin. Trackways relatively broad (pace angulation pes around 80°, manus around 110°) and organized in an alternating arrangement of pes-manus couples, with primary lateral overstep of the pes compared with the manus. Manus medially positioned compared with the pes. Manus parallel to the midline to slightly turned inwards, pes rotated outwards.

4.21.7 Discussion

This ichnogenus was introduced by Müller (1959), with the ichnospecies *Paradoxichium problematicum*, from a well-preserved trackway from the Lopingian Stassfurt Formation, Zechstein of Germany. A further occurrence from a different locality of the German Zechstein was reported by Voigt et al. (2015b). *P. problematicum* was also described from the Val Gardena Sandstone Formation of Italy (Marchetti et al., 2019d). *Paradoxichium* has been used to establish the late Wuchiapingian-early Olenekian *Paradoxichium* footprint biochron (Voigt and Lucas, 2018).

4.21.8 Trackmaker

Paradoxichnium has been attributed to neodiapsids, and more specifically basal Archosauromorpha or Younginiformes, such as *Protorosaurus speneri*, which is present in the Kupferschiefer at the base of the Zechstein (e.g., Müller, 1959; Marchetti et al., 2019d).

4.22 Cf. *Protochirotherium* Fichter and Kunz, 2004 (Fig. 4.7G and H)

4.22.1 Type ichnospecies

Protochirotherium wolfhagense Fichter and Kunz (2004).

4.22.2 Holotype

RGMW 10.14.9.1 (Fichter and Kunz, 2004: fig. 3, pl. 2). Right pes-manus couple with scaly skin impressions, convex hyporelief. Detfurth Formation, Early Triassic, Germany.

4.22.3 Valid ichnospecies

Protochirotherium wolfhagense Fichter and Kunz (2004); *Protochirotherium hauboldi* (Ptaszyński in Fuglewicz et al., 1990).

4.22.4 Distribution

Lopingian–Early Triassic. In the Permian, it is known from the Lopingian of Italy (Table 4.1).

4.22.5 Diagnosis (after Klein and Lucas, 2021; emended)

Chirotheriids with plantigrade to semiplantigrade pes and manus imprints. Short, broad, straight digit I–IV imprints with rounded terminations (different from *Paradoxichnium* and *Rhynchosauroides*, which have more elongated and curved digit imprints with sharp claw impressions). Anterior digit group I–IV with digit III longest and digit IV being only slightly shorter (different from *Chirotherium*, which is more mesaxonic, *Paradoxichnium*, *Rhynchosauroides* and *Synaptichnium*, which have digit IV longest and *Isochirotherium*, which has a markedly shorter digit IV compared with III). Digit V in the pes with massive oval basal pad (different from *Brachychirotherium*, *Chirotherium*, and *Synaptichnium*, which have a smaller digit V basal pad), often strongly elongated proximally and with a separate, thinner phalangeal part that is nearly straight or slightly curved outward.

4.22.6 General description

The pentadactyl pes and possible manus imprints are up to 279 mm in length and 212 mm in width. Pes broad, semiplantigrade with relatively short and straight, robust digit traces, I–IV forming a compact anterior digit group. Thick and rounded phalangeal and metatarsophalangeal pad impressions, the latter arranged in a posteriorly concave arch. Claw traces appear to be robust but are indistinctly preserved in all specimens, sometimes laterally extended due to preservational peculiarities. Digit V, positioned posterolateral to the anterior digit group, has a massive oval to subtriangular basal pad with a thinner and recurved phalangeal portion. Digit proportions are characterized by longest digit III, and slightly shorter digits II, I, and IV. Digit IV is more faintly impressed and slightly laterally spread. The possible associated manus is smaller, short, and rounded, but relatively large if compared with the pes. Because of the incompleteness, exact manual digit proportions are uncertain, but digit I seems to be shortest. Digit V is missing. Small, triangular claw traces are preserved on digits I, II, and III. The possible manus is strongly rotated inward relative to the pes. Skin texture can be preserved on both pes and manus imprints.

4.22.7 Discussion

Originally, *Protochirotherium* was erected on Lower Triassic (Olenekian) tracks from Buntsandstein deposits of northern Hesse (Germany) (Fichter and Kunz, 2004, 2015). These are chirotheriids that differ from *Chirotherium* by relatively longer pedal digits IV and V. *Chirotherium* shows a strong tendency toward a functionally tridactyl (II, III, IV) mesaxonic pes, with digit III longest, II and IV shorter and subequal in length; I shortest and slightly shifted backward, manus rotated outward relative to the pes. *Protochirotherium* is functionally tetradactyl (I, II, III, IV) with digit IV being longer than II. In this feature and in the pes showing stronger outward rotation compared with the manus, it is similar to *Synaptichnium*. In some Triassic ichnoassemblages it can be difficult to distinguish between both ichnogenera (Klein et al., 2010, 2023; Klein and Niedźwiezki, 2012; Klein and Lucas, 2021). From upper Permian deposits (Val Gardena Sandstone Formation) of northern Italy, isolated pentadactyl pes and manus imprints have been tentatively assigned to cf. *Protochirotherium* isp. (Avanzini et al., 2011; Bernardi et al., 2015; Marchetti et al., 2019d). These would be the stratigraphically oldest occurrences of chirotheriids and *Protochirotherium* from the global record. Unfortunately, the material is incomplete and lacks a trackway showing a distinct locomotion pattern with preserved manual imprints. Therefore, we prefer a more tentative assignment for the Permian material with cf. *Protochirotherium*, keeping in mind that a new chirotheriid, or more general, archosaur ichnogenus could be present as well.

4.22.8 Trackmaker

Permian archosauriform footprints can be expected from known skeletal material of the stratigraphically oldest archosauriform *Archosaurus rossicus*, a proterosuchid from the upper Permian of Russia (Tatarinov, 1960). The early occurrence of archosauriform *Protochirotherium* trackmakers in the late Permian is likely and could possibly be documented in these footprints from Italy. More material is needed from the Permian to confirm this.

5. Ichnotaphonomy and ichnotaxonomy (Fig. 4.8)

Tetrapod footprints are affected by ichnotaphonomic processes that can alter the anatomy-consistent morphology during and after their registration in the sediment (Marchetti et al., 2019a). These processes generate morphological features that are not related to anatomy but rather to behavior, substrate conditions, impression on submerged sediment layers, bioturbation, tectonics, and weathering. This includes, for instance, sliding traces, digit tip bifurcation, incompletely impressed footprints, very shallow/deep footprints, penetrative tracks, overprinted footprints, undertracks, flattened footprints, stretched footprints, broken footprints, and eroded footprints, among others (Marchetti, 2019; Marchetti et al., 2019a; Lallensack et al., 2021). These morphological features not related to anatomy are commonly known as extramorphologies *sensu* Peabody (1948). Because extramorphological phenomena are extensive in the fossil footprint record of the Permian and generate a high morphological variability, their use in ichnotaxonomy brought an exceptional inflation of nomenclature with the erection of numerous phantom taxa *sensu* Haubold (1996) or taphotaxa *sensu* Lucas (2001). The total number of named Permian footprint ichnotaxa was estimated as up to 150 ichnogenera, 281 ichnospecies, and 370 binomina in the year 2000 (Haubold, 2000). This caused dramatic overestimation of the ichnofaunal diversity and created biostratigraphy without a solid base (Boy and Fichter, 1982, 1988a, 1988b; Ellenberger, 1983b, 1983c, 1984). Later on, the revision of the same ichnoassociations and new biostratigraphy based on fewer and anatomy-consistent ichnotaxa brought more reliable results (Gand and Durand, 2006; Voigt et al., 2012).

Moreover, the use of taphotaxa can lead to incorrect faunal interpretations. This is the case with the ichnogenus "*Gilmoreichnus*" Haubold (1970), which was based on poorly preserved material and therefore misinterpreted as a synapsid track (e.g., Voigt, 2005; Marchetti et al., 2019a). Material subsequently assigned to this ichnogenus (Haubold et al., 1995) was also poorly preserved, and this caused both incorrect ichnotaxonomic determinations and trackmaker attributions (see discussion in Marchetti et al., 2019a). In other cases, the use of taphotaxa can lead to an oversimplification of the ichnofaunal diversity, also with incorrect trackmaker attributions. This is the case with the ichnogenera "*Chelichnus*" and "*Laoporus*," which were considered the most abundant tetrapod ichnotaxa of Permian eolian ichnoassociations (e.g., Morales and Haubold, 1995; McKeever and Haubold, 1996), and were even used to establish different tetrapod ichnofacies (Lockley et al., 1994; Hunt and Lucas, 2006, 2007; Krapovickas et al., 2016). Marchetti et al. (2019b, 2019c) revised this material and demonstrated how "*Chelichnus*" and "*Laoporus*" are *nomina dubia*, their characteristic morphology actually being the result of the trackmakers sliding when walking on inclined dune surfaces. After selecting only anatomy-consistent material for analysis, the ichnoassociation that resulted is similar to, although less diverse than, contemporary ichnoassociations from alluvial and/or marginal marine environments (Marchetti et al., 2019b, 2019c).

These examples show distinctly how important it is, for a stable and meaningful ichnotaxonomy, to exclusively consider anatomy-consistent footprints and exclude any extramorphological features from track diagnoses. Tracks can be sorted by means of morphological

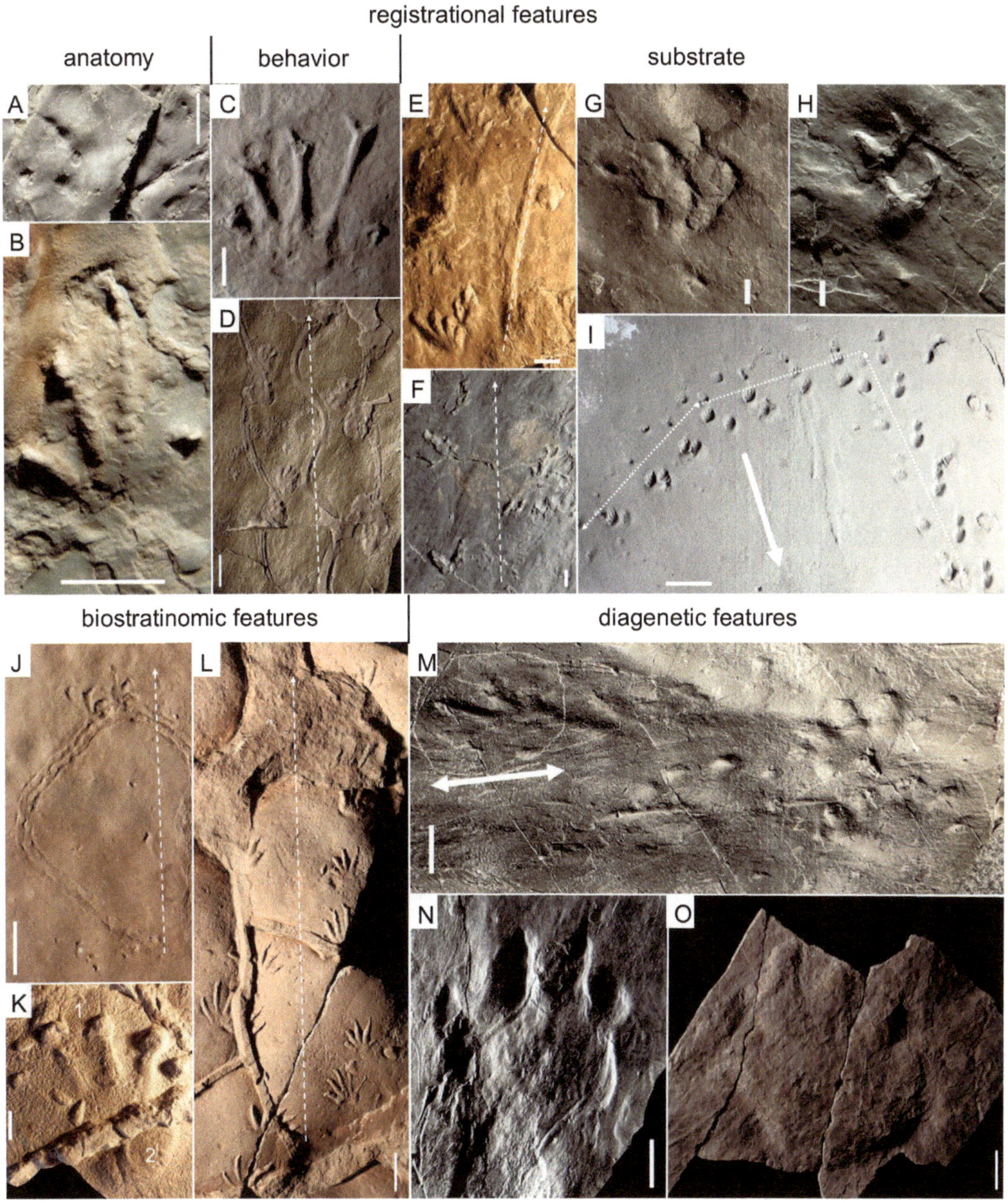

FIGURE 4.8 Permian examples of morphological features linked to registrational (A–I), biostratinomic (J–L) and diagenetic (M–O) processes. (A–B) Features derived from the trackmaker foot anatomy, scale and skin impression. Note the incomplete preservation of tracks. (A) Reptile footprint (*Dromopus*), concave epirelief. Cisuralian, Pizzo del Diavolo Formation, Italy. (B) Reptile footprint (*Erpetopus*), convex hyporelief, Cisuralian, Choza Formation, Texas. (C–D) features derived from the behavior of the trackmaker. (C) Digit tip bifurcation impression. Reptile pes

preservation, which is the preservation of anatomy-consistent footprint features, using a qualitative scale for each individual pes and manus (Belvedere and Farlow, 2016; Marchetti et al., 2019a). For the Permian record, an anatomy-consistent ichnotaxonomy was attempted, starting at the end of 20th century (e.g., Haubold, 1996; Voigt and Haubold, 2000; Voigt, 2005; Gand and Durand, 2006; Valentini et al., 2009). An additional tool to identify morphological/anatomical traits is the use of 3D visualization (mainly employing the photogrammetry technique: e.g., Mujal et al., 2020; Lallensack et al., 2022), which helps to elucidate relative depth patterns of footprints, which are consistent, and thereby helps to characterize ichnotaxa and potential trackmakers (Mujal et al., 2020).

As of now, there is relatively good consensus among Permian ichnologists about the number of tetrapod footprint ichnogenera that can be consistently distinguished based on morphological features. However, as noted above, various ichnospecies of Permian footprint ichnogenera merit re-evaluation. Compared with the Carboniferous tetrapod track record, the total number of ichnogenera and ichnospecies is relatively high, and we find it likely that the increase in ichnotaxonomic diversity is due to an increase in morphological disparity of the locomotor apparatus of tetrapods throughout the late Paleozoic. For the Mesozoic, research work aiming at a unified anatomy-consistent ichnotaxonomy and the removal of redundant ichnotaxa that lack suitable ichnotaxobases is still in progress and, thus, long term trends in ichnological diversity and disparity throughout the Paleozoic–Mesozoic ichnofaunal transition remain difficult to discern.

footprint (*Hyloidichnus*), convex hyporelief. Cisuralian, Hermit Formation, Arizona. (D) Continuous digit scratch traces and tail impression. Anamniote trackway (*Amphisauropus*), concave epirelief. Permian, Pizzo del Diavolo Formation, Italy. (E–I) Features conditioned by the substrate at time of impression. (E) Different digit width along the same trackway. Reptile pes-manus couples (*Rhynchosauroides*) and tail trace, Lopingian, convex hyporelief. Val Gardena Sandstone Formation, Italy. (F) Different digit morphology at the two sides of the trackway. Reptile trackway (*Dromopus*), convex hyporelief. Cisuralian, Pizzo del Diavolo Formation, Italy. (G–H) Different morphology on different stratigraphic layers. Reptile pes-manus couple (*Dromopus*), Cisuralian, Collio Formation, Italy. (G) True track, concave epirelief. (H) Undertrack, convex hyporelief. The image was mirrored to better compare the morphology. (I) Different trackway pattern and footprint morphology due to the substrate inclination. Reptile trackway on eolian foreset surface, Cisuralian, concave epirelief. Coconino Formation, Arizona. The large arrow indicates the original dip direction. (J–L) Features related to biostratinomy. (J) Invertebrate trace fossil superimposition on reptile pes-manus couples (*Erpetopus*), concave epirelief. Cisuralian, Choza Formation, Texas. (K) Tetrapod footprint superimposition (1), invertebrate trace fossil (2) and mud crack superimposition on nonamniote pes (*Ichniotherium*), convex hyporelief. Cisuralian, Tambach Formation, Germany. (L) Mud crack superimposition on reptile trackway (*Varanopus*), convex hyporelief. Cisuralian, Tambach Formation, Germany. (M–O) Features related to diagenesis. (M) Unidirectional stretch due to tectonics on nonamniote (*Amphisauropus*) and reptile (*Hyloidichnus*) footprints, concave epirelief (artificial cast). Cisuralian, Pizzo del Diavolo Formation, Italy. The arrow indicates the stretch direction. (N) Footprint flattened due to tectonics on nonamniote pes (*Limnopus*), concave epirelief (artificial cast). Cisuralian, Pizzo del Diavolo Formation, Italy. (O) Postexhumation footprint breakage on parareptile pes (*Pachypes*), convex hyporelief. Lopingian, Val Gardena Formation, Italy. Dashed arrows in D, E, F, I, J and L indicate the trackway midline and the direction of locomotion. Scale bars: A, C, F–H, J, N = 1 cm; B = 0.5 cm; D, I, O = 5 cm; E, K, L, M = 2 cm.

6. Potential trackmakers (Fig. 4.9)

A key question of vertebrate ichnology is inferring the track producer. This is a very difficult process, because of the high morphological variability of footprints due to ichnotaphonomic processes (e.g., Marchetti et al., 2019a) and the overall conservative nature of the appendicular skeleton of the producers, so that different species, genera, or even families can produce the same ichnospecies (e.g., Lucas, 2007). Moreover, appendicular skeletons are generally less studied due to a research bias (the cranial apparatus being generally more informative systematically) and are commonly highly disarticulated, with rare exceptions (e.g., Berman et al., 2000, 2004). Stratigraphic co-occurrence, size, and digit proportions were the most used criteria for track-trackmaker correlation in the Permian until the 21st century, and this, coupled with an inflated ichnotaxonomy, caused imprecise track-trackmaker correlations (e.g., Haubold, 2000). As regard ichnotaxa occurring in the Permian, this is the case, for instance, with *Capitosauroides*, *Dicynodontipus*, *Rhynchosauroides,* and *Varanopus*, all bearing the name of their supposed trackmaker groups, later found to be incorrect (e.g., Voigt and Lucas, 2018; Klein and Lucas, 2021).

In the 21st century, an anatomy-based ichnotaxonomy coupled with a morphofunctional correlation of footprints and autopodia brought more precise and solid track-trackmaker correlations in the Permian (e.g., Voigt et al., 2007; Marchetti et al., 2017a, 2019e, 2021a, 2021c; Buchwitz et al., 2020; Mujal and Marchetti, 2020; Logghe et al., 2021; Matamales-Andreu et al., 2023). So, in addition to the stratigraphic co-occurrence and the correspondence of autopodial digit proportions of tracks and foot skeletons, further features have been evaluated. Examples include gleno-acetabular length/pes length ratio in both skeletons and footprints (Marchetti et al., 2017a), size and shape of the carpus/tarsus compared with palm/sole (Voigt et al., 2007), occurrence of intratarsal articulations consistent with footprint semiplantigrady or digitigrady (Marchetti et al., 2019e, 2021c; Logghe et al., 2021; Matamales-Andreu et al., 2023), tarsal/metatarsal articulation consistent with digit V lateral orientation (Marchetti et al., 2021c), metatarsal overlap consistent with digit imprint base overlap (Marchetti et al., 2021a, 2021c) and phalangeal articulations forming a digital arcade consistent with paw-like impressions (Marchetti et al., 2019c), as well as relative depth patterns obtained from 3D models that aid in the determination of functional prevalence of autopodia (Mujal et al., 2020).

As a result, the 22 valid ichnogenera occurring in the Permian were correlated to all the main groups of Permian terrestrial tetrapods (e.g., Schneider et al., 2020). Aside from problems of correlation with multiple producer groups, as is the case in *Batrachichnus* and *Dromopus,* and the unclear phylogenetic placement of some producer groups, such as the varanopids (e.g., Ford and Benson, 2020), the Permian track-trackmaker correlation currently offers an unprecedented level of synthesis of the footprint and track record, increasing our understanding of locomotion, paleoecology, paleobiogeography, biostratigraphy and study of faunal evolution and events (e.g., Schneider et al., 2020; Buchwitz et al., 2021; Marchetti et al., 2022a, 2022b).

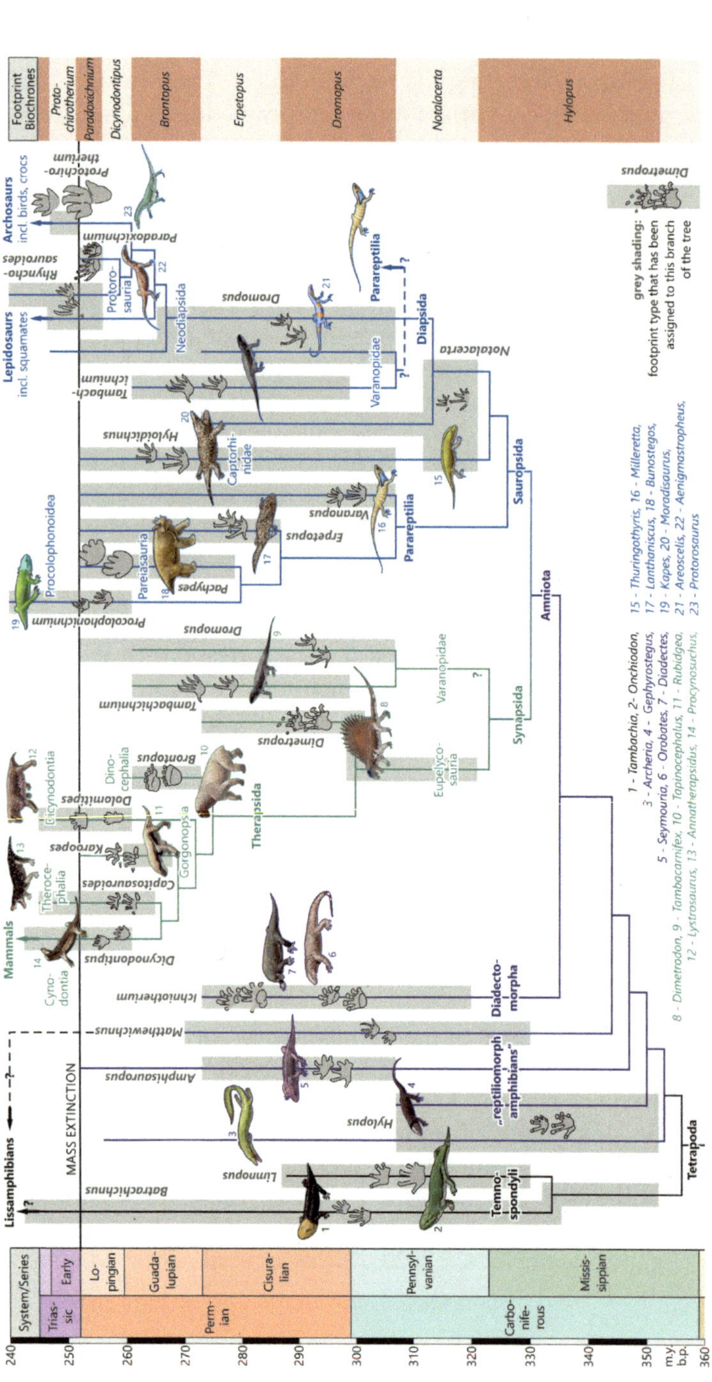

FIGURE 4.9 Permian tetrapod ichnogenera and inferred trackmaker groups, trackmaker phylogeny and stratigraphic ranges of both trackmakers and ichnogenera (in gray). Silhouettes drawn by J. Knüppe. Phylogeny based on Ruta and Coates (2007), Kemp (2011), Witzmann and Schoch (2018) and MacDougall et al. (2018). The alternative phylogenetic position of parareptiles and varanopids is based on Ford and Benson (2020).

7. Posture and locomotion (Fig. 4.10)

Tetrapod footprints are important for understanding the locomotion of extinct animals, because they are the direct evidence of their behavior during movement. Permian tetrapod trackways were all left by quadrupeds, and they are generally broad, with significant gauge width, typical of sprawling locomotion (e.g., Buchwitz et al., 2021). However, belly and tail drag traces are relatively rare. This suggests a raised sprawling locomotion, rather than a belly-dragging, sprawling locomotion (e.g., Hunt and Lucas, 1998; Nyakatura et al., 2014, 2019). Also, tracks are usually not turned outwards as expected by sprawlers, with a few exceptions, such as *Amphisauropus* (e.g., Voigt, 2005). The gait inferred from trackways was probably a walking trot, with clearly separated pes-manus couples and no or little overstep. In some trackways, complete secondary overstep of the pes on the manus is observed, and the stride is shorter, as in *Matthewichnus* (e.g., Voigt and Lucas, 2015a). This was probably caused by a relatively longer body of the trackmaker compared with the limbs, as explained in Leonardi (1987). Complete primary overstep of the pes on the manus is also observed. In this case, a lateral-sequence gait can be assumed (e.g., Buchwitz et al., 2021). Examples of this pattern are *Tambachichnium, Karoopes, Dicynodontipus, Varanopus, Erpetopus, Paradoxichnium,* and *Rhynchosauroides*. In general, this is associated with narrower gauge and relatively longer stride.

Some studies based on pace angulation suggested a postural transition from sprawling to more erect (parasagittal) between the Permian and the Triassic (Kubo and Benton, 2009). Nevertheless, trackways showing possible evidence of semi-erect posture are observed already in the Permian, in the ichnogenera *Dicynodontipus, Dolomitipes* and *Karoopes*, as well as tracks attributed to cynodont, dicynodont, and gorgonopsian therapsids, respectively (Marchetti et al., 2019e; Lucas, 2019). These footprints show relatively long stride, narrow gauge, and pes imprints forward-oriented, while the manus imprints are more inward-directed. This is indicative of a semi-erect posture, with the hindlimbs being more erected than the forelimbs. Therefore, the transition between sprawling and erect posture is likely to have started well within the Permian. Moreover, the occurrence of *Protochirotherium* is also potentially indicative of semi-erect archosauriform trackmakers, although trackways have not yet been found in Permian units (Bernardi et al., 2015). Also, stratigraphically older nontherapsid synapsid trackways generally show gauges narrower than expected, and this was interpreted as due to either higher lateral bending of the body (Hopson, 2015) or a less sprawling posture (Hunt and Lucas, 1998). A less sprawling posture can also potentially be inferred for some pareiasauromoph trackways (e.g., Marchetti et al., 2021a).

Other key anatomy-consistent features seen in footprints are the plantigrady/digitigrady of the trackmaker autopodia, and the different depth of distinct digital, plantar, and palmar portions of the footprint (relative depth patterns), which allow inference of functional prevalence of autopodia (Mujal et al., 2020). In general, during the Permian, synapsid tracks such as *Dimetropus* and nonamniote tracks such as *Ichniotherium* are more plantigrade, whereas reptile tracks, such as *Hyloidichnus* and *Varanopus,* tend to be semiplantigrade to semidigitigrade. More markedly digitigrade autopodial postures occur in diapsid and varanopid tracks, such as *Dromopus*. Regarding footprint functional prevalence, there is a clear distinction between medial-median footprint functional prevalence in nonamniote,

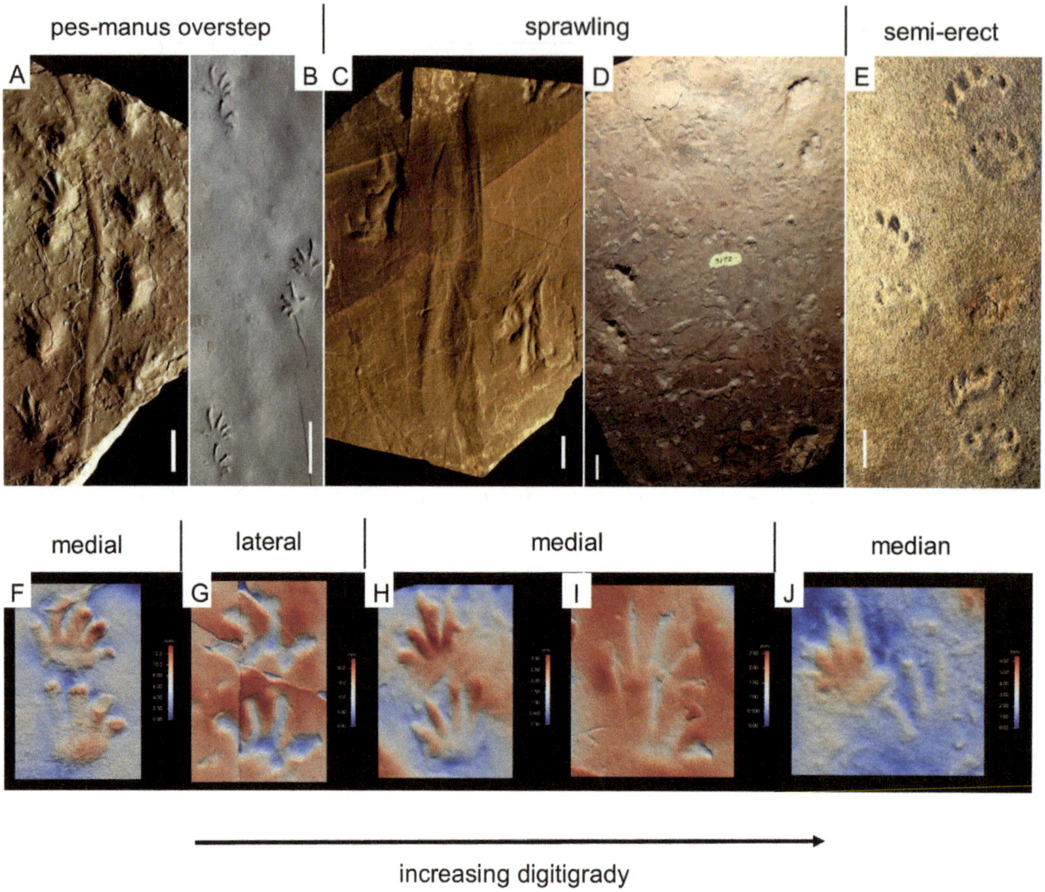

FIGURE 4.10 Posture and locomotion inferred from Permian tracks. (A–B) Pes-manus overstep. (A) NMMNH P-23312. *Matthewichnus* isp. Trackway, secondary overstep, convex hyporelief. Courtesy of Sebastian Voigt. (B) MCZ 107. *Erpetopus willistoni*. Trackway, primary overstep, concave epirelief. (C–E) Trackways and posture. (C) MSNM-V 7072. *Erpetopus willistoni*. Broad trackway with body, tail and digit drag traces, concave epirelief. Indicative of belly-dragging sprawling locomotion, rare in Permian. (D) GRCA 3172. *Pachypes ollieri*. Broad trackway without drag traces, concave epirelief. Indicative of lifted sprawling locomotion, the most common in the Permian. (E) UG-NN 1. *Dolomitipes* isp. Narrow trackway with mesaxonic pes imprints parallel to the direction of progression, concave epirelief. Indicative of semi-erect locomotion, relatively common from the late Capitanian onwards. (F–J) Footprint functional prevalence of Permian footprints, evidenced with false-color depth maps on 3D models of footprints. (F) Nonamniote tracks. MNG 1840. *Ichniotherium sphaerodactylum*. Right pes-manus couple, convex hyporelief. (G) Synapsid tracks. GF-TR 1. *Karoopes gansfonteinensis*. Right pes-manus couple, concave epirelief. (H) Parareptile tracks. NHMUK-PV-R 3173. *Procolophonichnium nopcsai*. Left pes-manus couple, convex hyporelief. (I) Captorhinomorph tracks. MNHN LOD 86. *Hyloidichnus bifurcatus*. Left pes-manus couple, concave epirelief. (J) Diapsid tracks. IGPF 75-5. *Rhynchosauroides pallinii*. Left pes-manus couple, convex hyporelief. Scale bars: A–D = 1 cm, E = 5 cm.

captorhinomorph, and parareptile footprints, median footprint functional prevalence in diapsid and varanopid tracks and median-lateral footprint functional prevalence in synapsid tracks (Mujal et al., 2020).

8. Substrate and locomotion (Fig. 4.11)

Locomotion can vary considerably in water-saturated substrates, and this is registered also by Permian footprints. The stride is significantly lower, and continuous digit and tail/belly drag traces are more frequently observed (e.g., Marchetti et al., 2017a; Logghe et al., 2021), as well as penetrative traces (Marchetti, 2019). The occurrence of continuous digit drag traces allows for a precise reconstruction of the entire swing phase of the trackmaker, as was done for *Amphisauropus* and *Hyloidichnus* (Marchetti et al., 2017a; Logghe et al., 2021). Also, the position of the body/tail during walking can be inferred from body/tail drag traces (Bernardi and Avanzini, 2011). Penetrative traces consist of footprints that collapsed right after the foot penetrated the upper substrate layer, whereas in the underlying layers the footprint morphology can be more distinct (Gatesy and Falkingham, 2020; Lallensack et al., 2021). In this particular case, in terms of morphology and foot anatomy, the tracks in the layers below can even be better preserved than the tracks on the original trampled surface (e.g., Marchetti, 2019). Locomotion can also vary significantly on inclined dune surfaces. This was observed for example in dune deposits from Arizona and Scotland (Marchetti et al., 2019b, 2019c). Trackways directed upslope generally have very short strides, whereas trackways running downslope show much longer strides with primary overstep, and trackways running transversely show marked asymmetries between right and left sides, as regard both track depth and arrangement. Moreover, the trackmaker digits were commonly sliding in opposition to the direction of the slope, resulting in relatively poorly preserved footprints from an anatomical perspective.

9. Associated traces (Fig. 4.12)

Trace fossils of tetrapods can also provide information on the external morphology of the producers through the impressions of the body, limbs, and tail, that sometimes show skin folds or scales. Considering the poor preservation of soft tissue during the Paleozoic (Eliason et al., 2017), this provides information of great value for paleoecological and morphological inferences.

Recently, Groenewald et al. (2023) described large body impressions and tail drag traces of putative Rhinesuchidae temnospondyls from the Guadalupian of South Africa, allowing the inference of the locomotion of large temnospondyls. Walter and Werneburg (1988) described the amphibian body impression *Hermundurichnus fornicatus*, from the Cisuralian of Thuringia, presenting scale-like impressions on the trunk and limbs and attributed to *Diplocaulus*. Ellenberger (1984) identified a supposed amphibian body impression from the lower Permian of Lodève (Mas d'Alary, France), "*Rachitomodepsa tuberose*," without any description. Heyler and Gand (2001, p. 46) figured a specimen with body and scale impressions associated with tracks from the Cisuralian of France. We re-assign these tracks to *Batrachichnus* isp. The most recovered structure in anamniotes are skin folds on palm and sole impressions. They are found in *Amphisauropus*, with recent work describing in detail transverse skin folds and flexion creases (e.g., Voigt and Lucas, 2017; Calábková et al., 2022) and more remarkably in *Ichniotherium*, especially from the lower Permian Bromacker locality in Germany, where

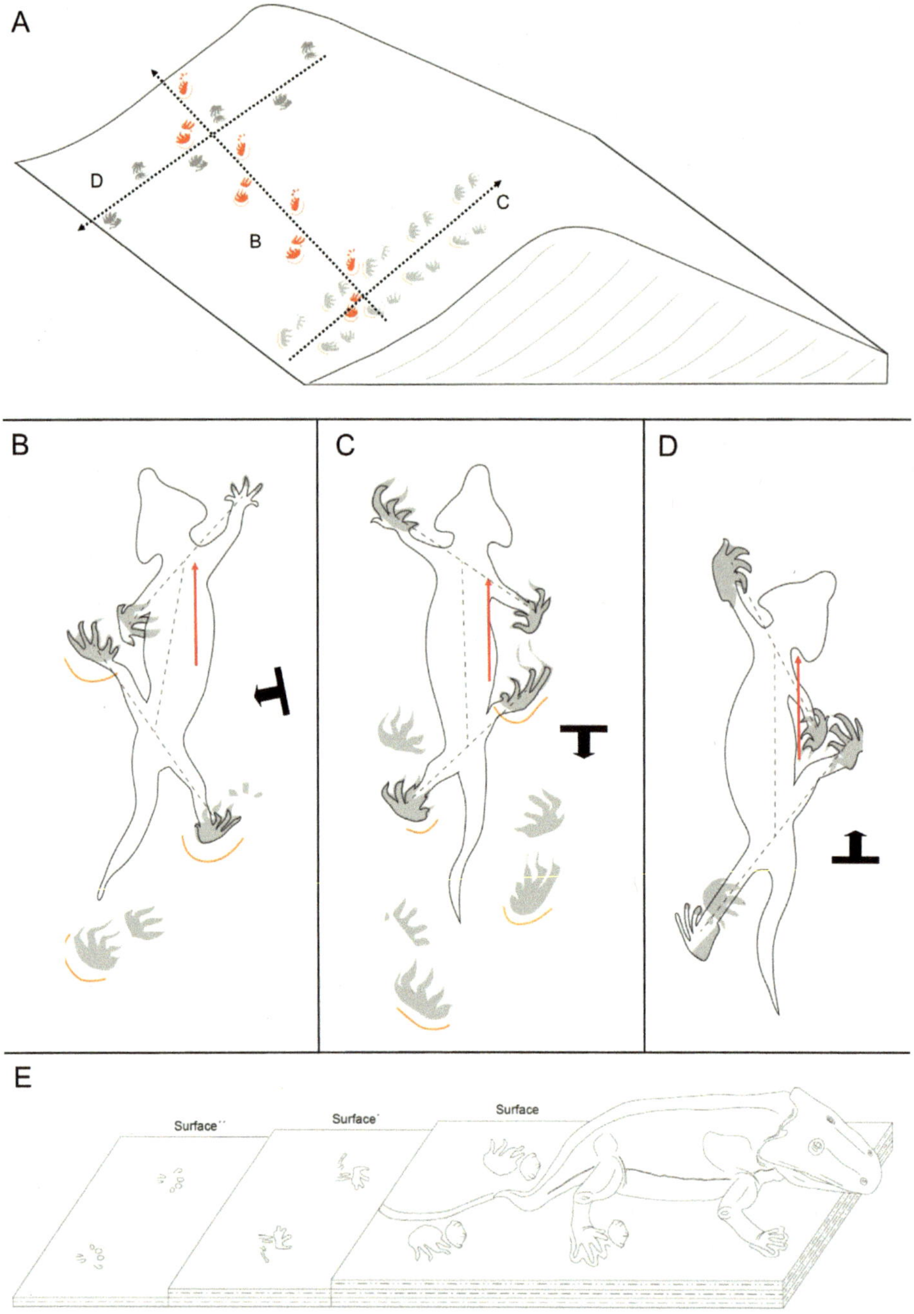

FIGURE 4.11 Models of locomotion and ichnotaphonomy based on the interaction between autopodia and substrate that generated the footprints. (A) Scheme of an eolian sand dune with trackways preserved on the foreset surface. (B–D) Different progression direction compared with the dip. Strike and dip symbols indicate the dip of the inclined surface, red arrows indicate the direction of progression. (B) Upslope progression with high lateral component and inferred position of the trackmaker, based on RAM-NN 1. Strongly asymmetric trackways, differential depth, upslope-rotation of the body. (C) Upslope progression with low lateral component and inferred trackmaker position, based on RAM-NN 1. Low pace angulation and very small asymmetry. (D) Downslope progression with low lateral component and inferred position of the trackmaker, based on UCMP 42944. High-pace angulation and pes-manus overstep. (E) Reconstruction of the formation of *Limnopus* tracks on different mudstone layers in water-saturated conditions (surface—true tracks, Surfacé—undertracks and Surfacé—deep undertracks) with the hypothetic trackmaker (eryopid temnospondyl). *Based on MNHN LOD 71/72.*

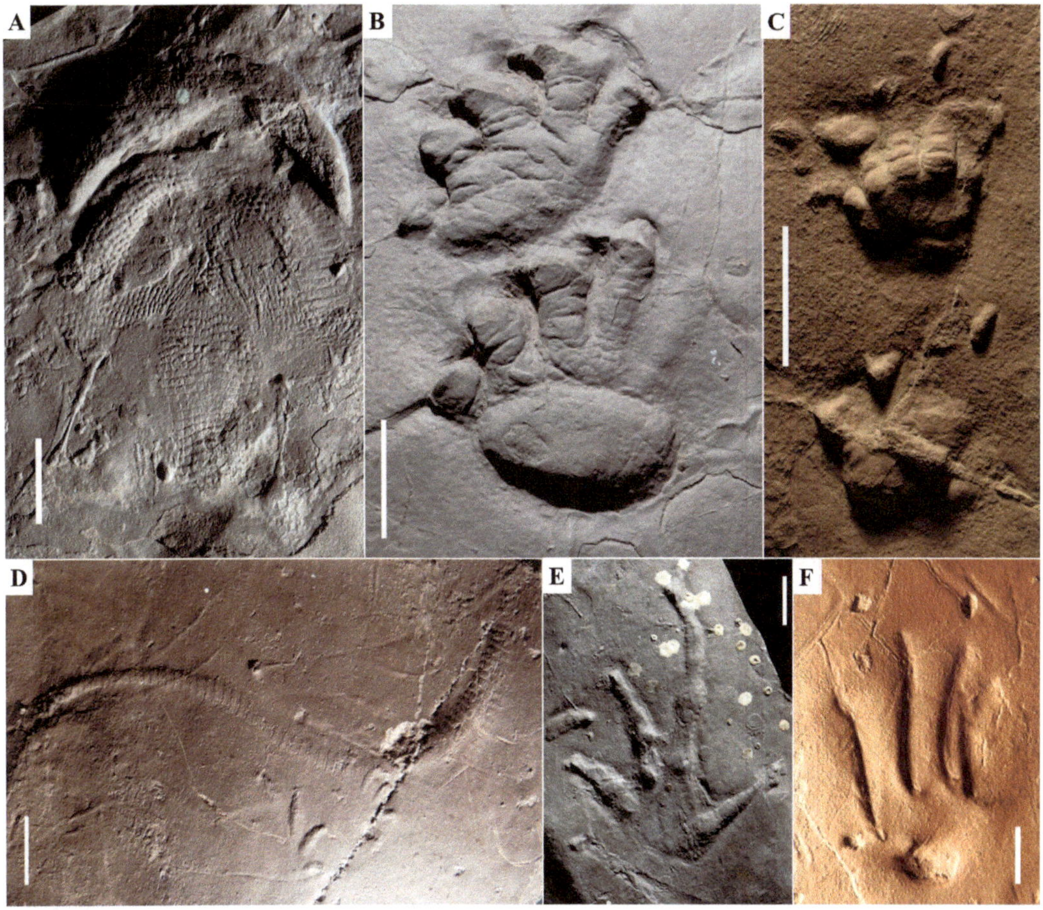

FIGURE 4.12 Associated traces. (A) NHMS FG 347/1. *Hermundurichnus fornicatus*, holotype. Lepospondyl body impression with scales, convex hyporelief. Rotterode Formation, Cisuralian, Germany. (B–C) Tambach Formation, Cisuralian, Germany. (B) MB 1969.54.258. *Ichniotherium sphaerodactylum*. Left pes-manus couple with skin folds on digit impressions, convex hyporelief. (C) MNG 1983. *Dimetropus leisnerianus*. Left pes-manus couple with skin folds on digit impressions, convex hyporelief. (D) UM LOD 180. Body and tail impressions with scales and *Erpetopus willistoni* right pes-manus couple, convex hyporelief. Rabejac Formation, Cisuralian, France. (E) MNG VF 2158. *Dromopus lacertoides*. Left pes imprint with scales on the sides of digit IV and skin folds on digit V. Goldlauter Formation, Cisuralian, Germany. (F) MNHN LOD 61. Swimming traces, set with four parallel traces, concave epirelief. Rabejac Formation, Cisuralian, France. Scale bars: A, D–F = 1 cm, B–C = 5 cm.

those skin folds are mostly perpendicular to the digit axis (e.g., Pabst, 1908; Voigt et al., 2007). Voigt et al. (2024) described a tail impression with scaly skin associated with *Ichniotherium cottae* tracks from the early Permian of Poland. No scale impressions have been recovered from temnospondyl tracks, although skin wrinkle traces are mentioned for *Limnopus* (Baird, 1952).

As regard traces attributed to synapsids, Pabst (1908) figured a body impression associated with *Dimetropus leisnerianus* footprints preserving the limbs and body traces with scaly skin impressions. This body impression was then preliminarily described by Haubold (1972). Niedźwiedzki and Bojanowski (2012) described a supposed sphenacodontid-body impression associated with *Dimetropus* tracks; however, no clear scale morphology distinguishes the impression from microbial mat-induced traces. *Dimetropus* preserves skin folds perpendicular to the digit axis (e.g., Voigt, 2005), as well as scale impressions, recovered on a *Dimetropus* pes imprint from the early Permian Rabéjac Formation of France (Heyler and Gand, 2001, p. 43). Calábková et al. (2023a) have described thin plantar and transverse creases on the sole of *Dimetropus* from the Czech Republic.

Reptile traces more often include body/scale impressions. Martens (1991) attributed footprints associated with resting traces preserving body, limb, and tail impression from the Cisuralian of Thuringia (Germany). Based on the new diagnoses established here, those footprints, originally assigned to "*Protritonichnites*" (*Dromopus*) *lacertoides*, are herein re-assigned to *Varanopus* isp. A body/tail impression with scales associated with "*Microphalangichnus vermivorus*," herein re-assigned to *Erpetopus* isp., has been figured by Ellenberger (1983b) from the lower Permian Rabéjac Formation of Lodève. Haubold (1972) described a body impression associated with *Dromopus* from the Cisuralian of Germany. A possible reptilian body trace has been figured from the Cisuralian of Morocco (Voigt et al., 2011a, fig. 2D). *Hyloidichnus* footprints also preserve oblique to straight although irregular skin folds transverse to the digit impression (e.g., Marchetti et al., 2022b), also interpreted as scales (Marchetti et al., 2015a). Perpendicular skin folds on the digits have also been recovered in *V. microdactylus* and *Erpetopus* footprints (e.g., Voigt and Haubold, 2015). Marchetti et al. (2019d) assigned a footprint to cf. *Pachypes* isp. presenting skin folds in the digits. Similar transverse skin folds have been recovered in *Tambachichnium* and *Dromopus*, the latter also presenting prominent triangular to ovoid scale impressions on the sides of the digit impressions (e.g., Marchetti et al., 2015a; Marchetti et al., 2022b). Scale impressions are also found in diapsid tracks such as *Paradoxichnium* (e.g., Marchetti et al., 2019d).

Most of the skin folds, scaly skin impressions and limb, body or tail impressions recovered in the Permian have thus been figured and/or described long ago and are in need of revision by recent imaging technique. Moreover, until recently, no focus has been made on the skin or scale impressions recovered in or associated with fossil tracks, that can be put in an evolutionary context to understand better the evolution of tetrapod integument. Moreover, recent collection investigations and fieldworks unveiled the presence of more well-preserved skin folds/scale impressions within tracks and body impressions that needs to be described to add to this knowledge of Permian tetrapod skin apparatus.

Another kind of traces often associated with Permian tracks are swimming traces, especially as regard nonamniote ichnotaxa such as *Amphisauropus* and *Batrachichnus*, but sometimes also parareptile ichnotaxa, such as *Erpetopus*, and synapsid ichnotaxa, such as *Dimetropus* and *Capitosauroides* (e.g., Voigt, 2005; Marchetti et al., 2013b, 2019e; Petti et al., 2014; Mujal et al., 2016; Buchwitz et al., 2020; De Jaime-Soguero et al., 2023).

10. Track paleoecology (Fig. 4.13)

Tetrapod footprints are a clear indication of the faunal composition of a locality, because these fossils are not affected by transport after their registration. Therefore, the ichnotaxa

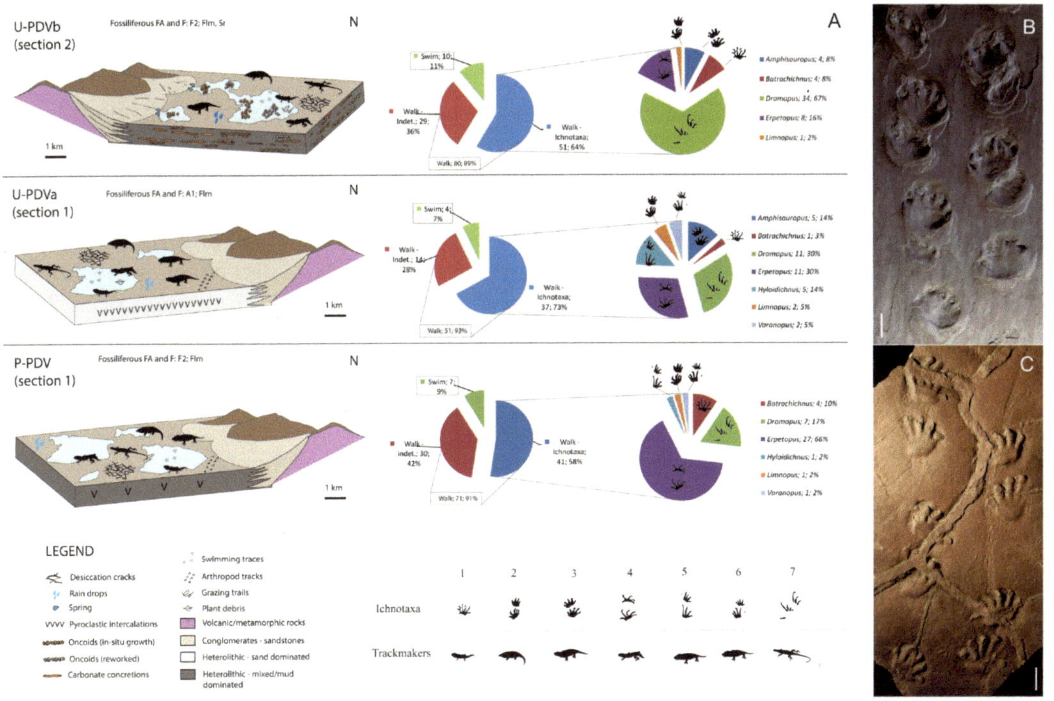

FIGURE 4.13 Tetrapod footprint paleoecology. (A) Paleoenvironmental reconstruction of the different lithozones of two study areas in the Cisuralin Pizzo del Diavolo Formation of the southern Alps (Lombardy, Italy) inferred from the facies analysis and paleoecology inferred from tetrapod footprints and other fossil remains. For each tetrapod track ichnocoenosis, behavior, taphonomy, and relative proportions between ichnotaxa inferred with the "track/slab"census method are indicated with pie charts. Ichnotaxa and putative trackmakers: (1) *Batrachichnus*, small temnospondyls; (2) *Limnopus*, medium-sized temnospondyls; (3) *Amphisauropus*, seymouriamorphs; (4) *Erpetopus*, small parareptiles; (5) *Hyloidichnus*, captorhinids; (6) *Varanopus*, small parareptiles; (7) *Dromopus*, aeroscelid eureptiles/varanopids. (B–C) facies-crossing ichnotaxa, *Ichniotherium sphaerodactylum*. (B) WICS-2 (NMMNH-P-80307). Trackway on a foreset dune surface, convex hyporelief. Desert paleoenvironment. Coconino Formation, Cisuralian, Arizona. (C) MNG 1840. Trackway on a tabular fluvial sandstone, convex hyporelief. Proximal floodplain paleoenvironment. Tambach Formation, Cisuralian, Germany. Scale bars: 5 cm.

found are a representation of the trackmakers walking in that area, and occurrences, nonoccurrences, co-occurrences, and relative proportions are important paleoecological signals (e.g., Hunt and Lucas, 2006; Mujal et al., 2016, 2018; Bernardi et al., 2017; Marchetti et al., 2017b; Voigt and Lucas, 2017; De Jaime-Soguero et al., 2021, 2023; Matamales-Andreu et al., 2022). Permian footprints are suitable for paleoecological studies because they are abundant, diverse and found in a wide array of depositional environments including lagoons, tidal flats, floodplains, lake margins, alluvial fans, and eolian dunes (e.g., Lucas, 2019). The ichnofaunal composition varies depending on the depositional environment (Mujal et al., 2016), as well as the relative proportions between ichnotaxa and footprint preservation (e.g., Marchetti et al., 2017b). Qualitative studies included an evaluation of the ichnofaunal composition and an approximate evaluation of the relative proportions between ichnotaxa among different facies (e.g., Voigt, 2005; Marchetti et al., 2015a, 2019b, 2019c; Mujal et al., 2016; Matamales-Andreu et al., 2022). These studies employed census methods on footprints in order to make simple statistics (e.g., Voigt and Lucas, 2017; Marchetti et al., 2017b). Other studies aimed to reconstruct the paleoenvironment through an evaluation of the entire fossil content of an area (e.g., Gand, 1989; Marchetti et al., 2015c). In the case of marginal marine ichnoassociations, the footprint relative abundances can be correlated with sequence stratigraphy (Bernardi et al., 2017). Since tetrapod footprints are clearly sensitive to facies changes, Hunt and Lucas (2006, 2007) introduced the tetrapod ichnofacies, which are specific ichnoassociations typical for different depositional environments. Facies-related changes in ichnofaunal compositions are known from Permian ichnoassociations as well (Hunt and Lucas, 2006). Nevertheless, the system of tetrapod ichnofacies has not been usually employed in subsequent studies and was also criticized, among other reasons, because of the relatively fast ichnofaunal replacement through time and the facies-crossing nature of tetrapod footprints (Santi and Nicosia, 2008). Moreover, some ichnofacies, such as the "*Chelichnus*" ichnofacies, need to be revised, as its reference ichnotaxon is not valid, and the ichnofaunal composition is very different from previous studies and more similar to the red beds ichnoassociations (Marchetti et al., 2019c). Further, quantitative studies are especially needed to redefine the tetrapod ichnofacies, particularly with regard to the Permian.

11. Track biostratigraphy (Fig. 4.14)

Anatomy-consistent ichnotaxa are a direct expression of the producer autopodia. Therefore they mirror the postcranial diversity and evolution through time and can be useful stratigraphic markers (e.g., Lucas, 2007). Tetrapod footprints were used in Permian local biostratigraphy since the works of Ellenberger (1983b, 1983c, 1984) in France and Boy and Fichter (1982) in Germany. The changes between the ichnoassociations were evident, but the unclear ichnotaxonomy provided unreliable results, as highlighted by subsequent work on the same ichnoassociations (Gand and Durand, 2006; Voigt et al., 2012). More Permian local footprint biostratigraphic schemes were proposed in Italy (Conti et al., 1997; Avanzini et al., 2001), the USA (Lucas et al., 2011b), and South Africa (Marchetti et al., 2019e). In the Permian, tetrapod footprints are abundant and widespread across Pangea, and the ichnofaunal changes are generally time-equivalent. As such, they were used to build chronostratigraphic schemes, through correlations with marine units and radiometric ages (Lucas, 2007; Voigt and Lucas, 2018; Schneider et al., 2020). In general, because of the low resolution, the ichnofaunal turnover is somehow slower than the faunal turnover, but it can be used nonetheless to build a meaningful chronostratigraphy, especially because of the low resolution of continental stratigraphy (Lucas, 2007).

Initially, Lucas (2007) introduced the *Dromopus* biochron for the Cisuralian and the *Rhynchosauroides* biochron for the Lopingian. Voigt and Lucas (2018) kept the *Dromopus* biochron for the Asselian–Artinskian time interval and introduced the *Erpetopus* biochron for the Kungurian–Guadalupian and the *Paradoxichnium* biochron for the Lopingian. Later, their boundaries slightly changed (Schneider et al., 2020). Marchetti et al. (2019c) introduced the *Brontopus* sub-biochron of the *Erpetopus* biochron for the Guadalupian. Marchetti et al.

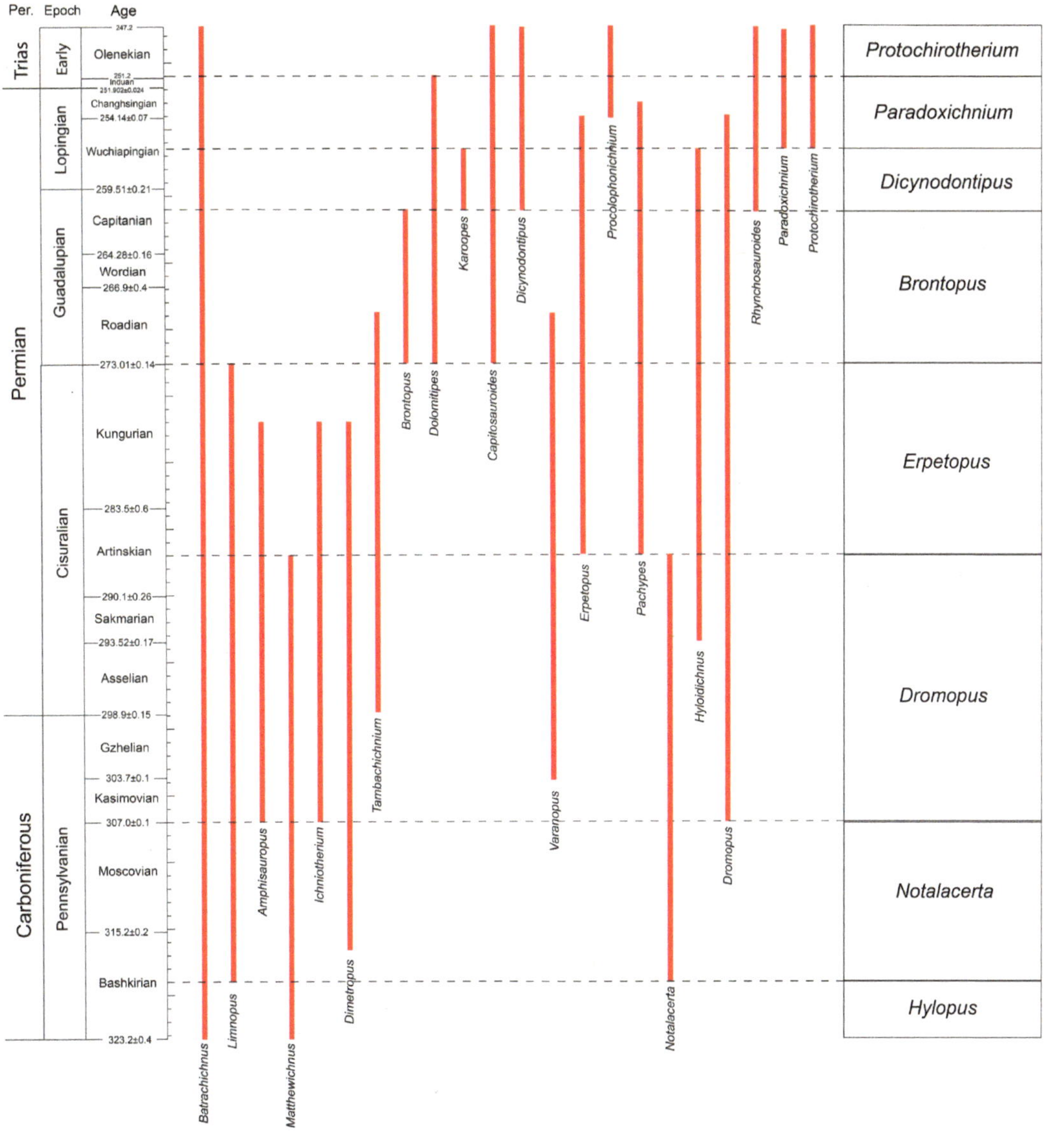

FIGURE 4.14 Permian tetrapod footprint chronostratigraphy.

(2022b) introduced the *Dicynodontipus* sub-biochron of the *Erpetopus* biochron for the late Capitanian–early Wuchiapingian. Here we revise the Permian footprint biochrons, and define them as follows:

(1) *Dromopus* footprint biochron (Late Pennsylvanian–early Artinskian). It is characterized by the FAD of *Dromopus* and includes the FADs of *Varanopus*, *Hyloidichnus*, and

Tambachichnium, and the LADs of *Matthewichnus* and *Notalacerta.* Other ichnotaxa: *Batrachichnus, Limnopus, Amphisauropus, Ichniotherium, Dimetropus.*

(2) *Erpetopus* footprint biochron (late Artinskian–Kungurian). It is characterized by the FAD of *Erpetopus* and includes the FADs of *Pachypes* and the LADs of *Limnopus, Amphisauropus, Ichniotherium,* and *Dimetropus.* Other ichnotaxa: *Batrachichnus, Tambachichnium, Varanopus, Hyloidichnus,* and *Dromopus.*

(3) *Brontopus* footprint biochron (Roadian–early Capitanian). It is characterized by the FAD of *Brontopus* and includes the FADs of *Dolomitipes* and *Capitosauroides,* and the LADs of *Brontopus* and *Tambachichnium.* Other ichnotaxa: *Batrachichnus, Varanopus, Hyloidichnus,* and *Dromopus.*

(4) *Dicynodontipus* footprint biochron (late Capitanian–early Wuchiapingian). It is characterized by the FAD of *Dicynodontipus* and includes the FADs of *Karoopes, Procolophonichium,* and *Rhynchosauroides,* and the LADs of *Erpetopus, Varanopus, Hyloidichnus,* and *Karoopes.* Other ichnotaxa: *Batrachichnus, Capitosauroides, Dolomitipes, Dromopus.*

(5) *Paradoxichnium* footprint biochron (late Wuchiapingian–Induan). It is characterized by the FAD of *Paradoxichnium* and includes the FAD of *Protochirotherium,* and the LADs of *Dolomitipes, Dromopus, Paradoxichnium.* Other ichnotaxa: *Batrachichnus, Capitosauroides, Dicynodontipus, Procolophonichnium,* and *Rhynchosauroides.*

12. Track paleobiogeography (Fig. 4.15)

During the Permian all continents were part of the supercontinent Pangea, so faunal exchanges were possible all over the supercontinent. This is reflected by similar tetrapod ichnoassociations in different localities across Pangea (e.g., Voigt and Lucas, 2018). However, during the Cisuralian, the great majority of footprint localities are situated in present-day North America and Western Europe. Only a small number of localities have been described from South America, Africa, and Asia (e.g., Nopcsa, 1923; Melchor and Sarjeant, 2004; Gand et al., 2011; Voigt et al., 2011a; Krapovickas et al., 2015). Nevertheless, with the exception of the western USA, the Cisuralian tetrapod footprint record of low paleolatitudes is by far less fragmentary and facies-related than the skeletal record and is characterized by a rather uniform paleobiogeographical distribution of the ichnotaxa within the same footprint biochrons (Marchetti et al., 2022a). During the Guadalupian and Lopingian, low-paleolatitude footprint localities are relatively common in western Europe and North Africa, and for the first time, also the mid-to high-latitude footprint record is well-represented, in both the southern hemisphere (Argentina, Brazil, South Africa, Australia; e.g., Retallack, 1996; Warren, 1997; Silva et al., 2012; Marchetti et al., 2019e; Cisneros et al., 2020; Citton et al., 2021; McLoughlin et al., 2021; Groenewald et al., 2023) and the northern hemisphere (Russia; Gubin et al., 2003; Surkov et al., 2007). During the Guadalupian-Lopingian, it can be observed that there was a higher abundance of reptilian tracks in lower latitudes, whereas skeletal and (to a certain degree) track records suggest a higher abundance of therapsids at higher latitudes (Bernardi et al., 2017; Marchetti et al., 2019c).

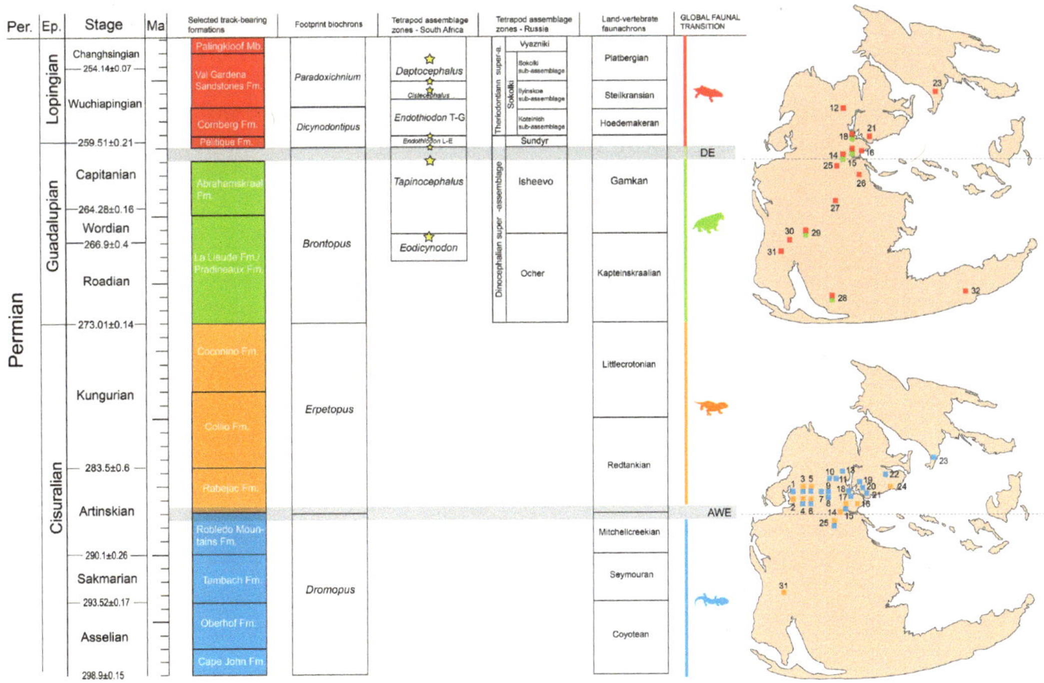

FIGURE 4.15 Permian tetrapod footprint record and faunal events. Age and effects of Artinskian Warming event (AWE) and late Capitanian extinction event (Dinocephalian Extinction event—DE) on tetrapod skeleton and footprint record. Localities: 1 = Utah, 2 = Arizona, 3 = Colorado, 4 = New Mexico, 5 = Oklahoma, 6 = Texas, 7 = Ohio, 8 = West Virginia, 9 = Pennsylvania, 10 = Prince Edward Island, 11 = Nova Scotia, 12 = Scotland, 13 = England, 14 = Spain, 15 = France, 16 = Italy, 17 = Austria, 18 = Germany, 19 = Poland, 20 = the Czech Republic, 21 = Hungary, 22 = Ukraine, 23 = Russia, 24 = Turkey, 25 = Morocco, 26 = Tunisia, 27 = Niger, 28 = South Africa, 29 = Brazil, 30 = Uruguay, 31 = Argentina, 32 = Australia.

13. Evolutionary events (Fig. 4.15)

Tetrapod footprints, and in particular anatomy-consistent ichnotaxa, are a reliable proxy for faunal associations, and can be integrated with the skeletal record (e.g., Lucas, 2019). During the Permian, the main faunal crises and radiations are clearly recorded. The Carboniferous–Permian transition, however, does not seem to affect significantly tetrapod ichnofaunas. In fact, the *Dromopus* biochron crosses this boundary (e.g., Voigt and Lucas, 2018; Lucas et al., 2022). Very significant for the footprint record is instead a sudden and consistent rise in the abundance and diversity of reptilian footprints (*Dromopus–Erpetopus* biochron transition), that happened by the mid-Artinskian (e.g., Marchetti et al., 2013a, 2019b, 2022a; Voigt and Lucas, 2017). This is coupled with a simultaneous decrease in non-amniote footprints, an increase in xeromorphic plants, and a decrease in hygromorphic plants. This has been related to a significant rise of carbon dioxide—probably of volcanic origin—that caused an increase in global temperature, an increase in seasonality at low latitudes and the melting of high-latitude ice sheets, named the Artinskian Warming Event of Marchetti et al. (2022a). Tetrapod footprints also testify to the first appearance and radiation of therapsid synapsids at low paleolatitudes, which happened in the early Guadalupian of France and Spain (e.g., Gand et al., 1995, 2000; Marchetti et al., 2019c; De Jaime-Soguero et al., 2023). Also, the end-Guadalupian extinction was recognized in the mid-high latitude footprint record of South Africa, in agreement with the skeletal record (Marchetti et al., 2019e) and in the low-latitude footprint record of France and Germany (although in Germany few middle Permian sites are known, see Buchwitz et al., 2020). This ichnofaunal event is older than the first low-latitude skeletal associations that show this transition and so is time-equivalent with the mid-high latitude extinction (Lucas, 2017; Marchetti et al., 2019c, 2022b). However, the tetrapod footprint record does not show significant changes at the end-Permian extinction (Lucas, 2017; Marchetti et al., 2019e), the most notable changes happened later, with the Olenekian recovery recorded by the appearance of the footprint associations of the *Protochirotherium* biochron (Klein and Lucas, 2021).

14. Conclusions and perspectives

- Selection of anatomy-consistent material is key for a meaningful ichnotaxonomy of Permian tracks;
- 22 Permian tetrapod footprint ichnogenera are considered valid;
- Tracks correlated with anamniotes include *Batrachichnus* and *Limnopus*, *Amphisauropus*, *Matthewichnus*, and *Ichniotherium*;
- Tracks correlated with synapsids are *Dimetropus*, *Tambachichnium*, *Brontopus*, *Capitosauroides*, *Dolomitipes*, *Dicynodontipus*;
- Tracks correlated with parareptiles are *Erpetopus*, *Pachypes*, *Procolophonichium*, and *Varanopus*. The ichnogenus "*Robledopus*" was herein considered to be a junior subjective synonym of *Varanopus*;
- Tracks correlated with eureptiles are *Notalacerta* and *Hyloidichnus*, *Dromopus*, *Paradoxichnium*, *Protochirotherium*, and *Rhynchosauroides*; The ichnogenus "*Merifontichnus*" was herein considered a junior subjective synonym of *Hyloidichnus*;
- In general, there is a need of revision of the valid ichnospecies;
- Permian footprints testify to a raised sprawling locomotion, but semi-erect postures can be inferred for *Dicynodontipus*, *Dolomitipes*, *Karoopes* (therapsids), and probably *Protochirotherium* (archosauriforms);
- Sedimentary facies have an influence on preservation, locomotion, and ichnofaunal composition, but tetrapod ichnofacies need to be revised;
- Permian tracks are well-represented at low-paleolatitudes; at mid-high paleolatitudes, there are only well-represented localities of Guadalupian to Lopingian age and are generally less diverse;
- five tetrapod footprint biochrons are defined: *Dromopus* (Kasimovian–early Artinskian), *Erpetopus* (late Artinskian–Kungurian), *Brontopus* (Roadian–early Capitanian), *Dicynodontipus* (late Capitanian–early Wuchiapingian), and *Paradoxichium* (late Wuchiapingian–Induan);
- Tetrapod footprints provide additional data supporting a better understanding of the effects of the Artinskian Warming Event, the end-Guadalupian extinction, and the end-Permian extinction of tetrapod faunas.

Acknowledgments

We sincerely thank all colleagues that worked with us on Permian tetrapod tracks. We thank S. Voigt for providing some photos of specimens stored at NMMNH and J. Knüppe for the drawings of Fig. 4.9. We thank all the curators and collection managers that helped our studies and all the institutions that funded our research. LM was funded by the Bundesministerium für Bildung und Forschung (BMBF) for the BROMACKER Project 2020–25. We are grateful to the editors and reviewers J.D. Moreau and A.P. Hunt for stimulating discussion.

References

Abel, O., 1926. Der erste Fund einer Tetrapodenfährte in der unteren alpinen Trias. PalZ 7, 22–24.

Abel, O., 1929. Eine Reptilienfährte aus dem oberen Perm (Grödener Sandstein) der Gegend von Bozen in Südtirol. Palaeobiol 2, 1–14.

Allen, L.F., Stimson, M.R., King, O.A., Norrad, R.E., Lucas, S.G., Mann, A., Hinds, S.J., Park, A.F., Calder, J.H., Maddin, H., Montplaisir, M., 2022. A *Batrachichnus salamandroides* trackway from the Minto Formation of central New Brunswick, Canada: implications for alternative tracemaker interpretations. Atl. Geosci. 58, 239–260.

Anonymous, 1828. Notice about Dumfriesshire footprints. Lond. Paris Obser. 141, 93–94.

Aramayo, S.A., Farinati, E., 1983. "*Paredichnus rodriguezi*" n. sp. Nuevas icnitas de tetrápodos de la Formación Los Reyunos (Grupo Cochicó), Pérmico inferior, Mendoza. Mundo Ameghiniana 4, 47–58.

Arduini, P., Krieger, C., Rossi, M., Santi, G., 2003. Early Permian vertebrates ichnoassociations in Scioc Valley-Orobic Basin – (Lombardy – Northern Italy). Neues Jahrb. Geol. P-M 7, 385–399.

Avanzini, M., Renesto, S., 2002. A review of *Rhynchosauroides tirolicus* Abel, 1926 ichnospecies (Middle Triassic: Anisian-Ladinian) and some inferences on *Rhynchosauroides* trackmaker. Riv. Ital. Paleontol. S. 108, 51–66.

Avanzini, M., Ceoloni, P., Conti, M.A., Leonardi, G., Manni, R., Mariotti, N., Mietto, P., Muraro, C., Nicosia, U., Sacchi, E., Santi, G., Spezzamonte, M., 2001. Permian and Triassic tetrapod ichnofaunal units of northern Italy: their potential contribution to continental biochronology. "Natura Bresciana" Ann. Mus. Civ. Sc. Nat. 25, 89–107.

Avanzini, M., Neri, C., Nicosia, U., Conti, A., 2008. A new Early Permian ichnocoenosis from the 'Gruppo vulcanico atesino' (Mt. Luco, Southern Alps, Italy). Studi Trentini di Scienze Naturali. Acta Geol. 83, 231–236.

Avanzini, M., Bernardi, M., Nicosia, U., 2011. The Permo-Triassic tetrapod faunal diversity in the Italian Southern Alps. In: Dar, I.A., Dar, M.A. (Eds.), Earth and Environmental Sciences. Tech, Rijeka, pp. 591–608.

Baird, D., 1952. Revision of the Pennsylvanian and permian footprints *Limnopus, Allopus* and *Baropus*. J. Paleontol. 26, 832–840.

Baird, D., 1957. Triassic reptile footprint faunules from Milford, New Jersey. Bull. Mus. Comp. Zool. 117, 449–520.

Barabásné Stuhl, A., 1975. Adatok a dunántúli újpaleozóos képződmények biosztratigráfiájához. Bull. Hungar. Geol. Soc. 105, 320–334.

Barrier, P., Montenat, C., de Lumley, H., 2009. Empreintes de pas de reptiles au Pic des Merveilles dans le Permien du Mont-Bego (Alpes-Maritimes). C.R. Palevol 8, 67–78.

Belvedere, M., Farlow, J.O., 2016. A numerical scale for quantifying the quality of preservation of vertebrate tracks. In: Falkingham, P.L., Marty, D., Richter, A. (Eds.), Dinosaur Tracks: The Next Steps. Indiana University Press, Bloomington, pp. 92–98.

Bendel, E.M., Kammerer, C.F., Smith, R.M., Fröbisch, J., 2023. The postcranial anatomy of *Gorgonops torvus* (Synapsida, Gorgonopsia) from the late Permian of South Africa. PeerJ 11, e15378.

Berman, D.S., Henrici, A.C., Sumida, S.S., Martens, T., 2000. Redescription of Seymouria sanjuanensis (Seymouriamorpha) from the Lower Permian of Germany based on complete, mature specimens with a discussion of paleoecology of the Bromacker locality assemblage. J. Vertebr. Paleontol. 20, 253–268.

Berman, D.S., Henrici, A.C., Kissel, R.A., Sumida, S.S., Martens, T., 2004. A new diadectid (Diadectomorpha), Orobates pabsti, from the early Permian of central Germany. Bull. Carnegie Mus. Nat. Hist. 35, 1–36.

Bernardi, M., Avanzini, M., 2011. Locomotor behavior in early reptiles: insights from an unusual *Erpetopus* trackway. J. Paleontol. 85, 925–929.

Bernardi, M., Klein, H., Petti, F.M., Ezcurra, M.D., 2015. The origin and early radiation of archosauriforms: Integrating the skeletal and footprint record. PLoS One 10, e0128449.

Bernardi, M., Petti, F.M., Kustatscher, E., Franz, M., Hartkopf-Fröder, C., Labandeira, C.C., Wappler, T., van Konijnenburg-van Citter, J.H.A., Peecock, B.R., Angielczyk, K.D., 2017. Late Permian (Lopingian) terrestrial ecosystems: a global comparison with new data from the low-latitude Bletterbach Biota. Earth Sci. Rev. 175, 18–43.

Berruti, G., 1969. Osservazioni biostratigrafiche sulle formazioni continentali pre-Quaternari delle valli Trompia e Sabbia. "Natura Bresciana" Ann. Mus. Civ. Sc. Nat. 6, 3–32.

Blak, J., 1846. Observations on a slab of new red sandstone from the Quarries at Weston, near Runcorn, Cheshire, containing the impressions of footsteps and other markings. Quart. J. Geol. Soc. 2, 65–68.

Blob, R.W., 2001. Evolution of hindlimb posture in nonmammalian therapsids: biomechanical tests of paleontological hypotheses. Paleobiology 27, 14–38.

Boy, J.A., Fichter, J., 1982. Zur Stratigraphie des saarpfälzischen Rotliegenden (? Ober-Karbon-Unter-Perm; SW-Deutschland). Z. dt. geol. Ges. 133, 607–642.

Boy, J.A., Fichter, F., 1988a. Ist die stratigraphische Verbeitung der Tetrapodenfährten im Rotliegend ökologisch beeinflußt ? Z. Geol. Wiss. Berlin 9, 877–883.

Boy, J.A., Fichter, J., 1988b. Zur Stratigraphie des höheren Rotliegend im Saar-Nahe-Becken (Unter-Perm); SW-Deutschland) und seiner Korrelation mit anderen Gebieten. N. Jb. Geol. Paläont. Abh. 176, 331–394.

Brand, L.R., 1979. Field and laboratory studies on the Coconino Sandstone (Permian) vertebrate footprints and their paleoecological implications. Palaeogeogr. Palaeoclimatol. Palaeocol. 28, 25–38.

Brand, L.R., 1996. Variations in salamander trackways resulting from substrate differences. J. Paleontol. 70, 1004–1010.

Brink, K.S., Hawthorn, J.R., Evans, D.C., 2012. New occurrences of *Ichniotherium* and *Striatichnium* from the lower Permian Kildare Capes formation, Prince Edward Island, Canada: palaeoenvironmental and biostratigraphic implications. Palaeontology 55, 1076–1090.

Brocklehurst, N., Dunne, E.M., Cashmore, D.D., Fröbisch, J., 2018. Physical and environmental drivers of Paleozoic tetrapod dispersal across Pangea. Nat. Commun. 9, 5216.

Broom, R., 1948. A contribution to our knowledge of the vertebrates of the Karroo Beds of South Africa. T. Roy. Soc. Edin-Earth. 61, 577–629.

Broutin, J., El Wartiti, M., Freytet, F., Heyler, D., Larhrib, M., Morel, J.L., 1987. Nouvelles découvertes paléontologiques dans le bassin détritique carbonate permien de Tiddas (Maroc central). Mécanique, Physique, Chimie, Sciences de l'univers, Sciences de la Terre C. r. hebd. Séances Acad. Sci. Série 2 305, 143–148.

Buchwitz, M., Marchetti, L., Voigt, S., Germann, S., Trostheide, F., Schneider, J.W., 2023. *Procolophonichnium* from the middle to late Permian of the North German Basin. Jahrestagung Paläontologische Gesellschaft, Jena, Germany. Abstract volume 52.

Buchwitz, M., Voigt, S., 2018. On the morphological variability of *Ichniotherium* tracks and evolution of locomotion in the sistergroup of amniotes. PeerJ 6, e4346.

Buchwitz, M., Luthardt, L., Marchetti, L., Trostheide, F., Voigt, S., 2017. A middle to late Permian assemblage of continental trace fossils from northern Germany. In: Helling, S., Hartenfels, S. (Eds.), Münstersche Forschungen zur Geologie und Paläontologie, 88th Annual Conference of the Paläontologische Gesellschaft, Program – Abstracts, p. 32. Münster, Paläontologische Gesellschaft.

Buchwitz, M., Marchetti, L., Jansen, M., Falk, D., Trostheide, F., Schneider, J.W., 2020. Ichnotaxonomy and trackmaker assignment of tetrapod tracks and swimming traces from the Middle Permian Hornburg Formation of Saxony-Anhalt (Germany). Ann. Soc. Geol. Pol. 90, 291–320.

Buchwitz, M., Jansen, M., Renaudie, J., Marchetti, L., Voigt, S., 2021. Evolutionary change in locomotion close to the origin of amniotes inferred from trackway data in an ancestral state reconstruction approach. Front. Ecol. Evol. 9, 674779.

Buckland, 1828. Note sur des traces de tortues observées dans les grès rouge. Ann. Sci. Nat. 13, 85–86.

Butts, E., 1891. Recently discovered footprints of the amphibian age in the upper coal measure group of Kansas city, Missouri. Kans. City Sci. 5, 17–19.

Calábková, G., Březina, J., Madzia, D., 2022. Evidence of large terrestrial seymouriamorphs in the lowermost Permian of the Czech Republic. Pap. Palaeontol. 8, e1428.

Calábková, G., Březina, J., Nosek, V., Madzia, D., 2023a. Synapsid tracks with skin impressions illuminate the terrestrial tetrapod diversity in the earliest Permian of equatorial Pangea. Sci. Rep. 13, 1130.

Calábková, G., Madzia, D., Nosek, V., Ivanov, M., 2023b. Tracking 'transitional'diadectomorphs in the earliest Permian of equatorial Pangea. PeerJ 11, e16603.

Calder, J.H., Baird, D., Urdang, E.B., 2004. On the discovery of tetrapod trackways from Permo-Carboniferous red-beds of Prince Edward Island and their biostratigraphic significance. Atl. Geol. 40, 217–226.

Caselli, A.T., Arucci, A.B., 1999. Huellas de tetrápodos en eolianitas de la Formación Patquía, Cuenza Paganzo, Provincia de La Rioja. I Simposio Argentino del Paleozoico Superior. Ameghiniana 36, 30.

Cei, R.L., Gargiulo, J., 1977. Icnitas de tetrápodos Pérmicos del sur de Mendoza. Ameghiniana 14, 127–132.

Ceoloni, P., Conti, M.A., Mariotti, N., Mietto, P., Nicosia, U., 1987. Tetrapod footprints from Collio Formation (Lombardy, northern Italy). Mem. Sci. Geol. 39, 213–233.

Ceoloni, P., Conti, M.A., Mariotti, N., Mietto, P., Nicosia, U., 1988. New late Permian tetrapod footprints from southern Alps. Mem. Soc. Geol. Ital. 34, 45–65.

Chestnut Jr., D.R., Baird, D., Smith, J.H., Lewis, R.Q.S., 1994. Reptile trackway from the Lee Formation (Lower Pennsylvanian) of south-central Kentucky. J. Paleontol. 68 (1), 154–158.

Chure, D.J., Engelmann, G.F., Good, T.R., Haymes, G., Hansen, R., 2014. The first record of vertebrate tracks from the eolian Weber Sandstone (Pennsylvanian-Permian), northeastern Utah: a preliminary report. Bull. N. M. Mus. Nat. Hist. Sci. 62, 95–102.

Cisneros, J.C., Day, O.M., Groenenwald, J., Rubidge, B., 2020. Small footprints expand Middle Permian amphibian diversity in the South African Karoo. Palaios 35, 1–11.

Citton, P., Nicolosi, I., Carluccio, R., Nicosia, U., 2016. Unveiling trampling history through trackway interferences and track preservational features: a case study from the Bletterbach gorge (Redagno, Western Dolomites, Italy). Palaeontol. Electron. 19, 1–20.

Citton, P., Ronchi, A., Maganuco, S., Caratelli, M., Nicosia, U., Sacchi, E., Romano, M., 2019a. First tetrapod footprints from the Permian of Sardinia and their palaeontological and stratigraphical significance. Geol. J. 54, 2084–2098.

Citton, P., Carluccio, R., Nicolosi, I., Nicosia, U., 2019b. Re-evaluation of *Chelichnus tazelwürmi*, a non mammalian therapsid-grade track from the Upper Permian Arenaria di Val Gardena. Hist. Biol. 31, 322–340.

Citton, P., de Valais, S., Díaz-Martínez, I., González, S.N., Greco, G.A., Cónsole-Gonella, C., Leonardi, G., 2021. Age-constrained therapsid tracks from a mid-latitude upland (Permian–Triassic transition, Los Menucos complex, Argentina). J. S. Am. Earth Sci. 110, 103367.

Contessi, M., Voigt, S., Bibonne, R., 2018. Permian tetrapod footprints from Tunisia. Ichnos 25, 119–127.

Conti, M.A., Leonardi, G., Mariotti, N., Nicosia, U., 1975. Tetrapod footprints, fishes and molluscs from the middle permian of the Dolomites (N. Italy). Mem. Geopal. Univ. Ferrara 3, 139–150.

Conti, M.A., Leonardi, G., Mariotti, N., Nicosia, U., 1977. Tetrapod footprints of the "Val Gardena Sandstone" (North Italy). Their paleontological, stratigraphic and paleoenvironmental meaning. Palaeontogr. Ital. 40, 1–91.

Conti, M.A., Leonardi, G., Mariotti, N., Nicosia, U., 1987. Una nuova località con icnofauna di rettili del Permiano superiore nelle arenarie di Val Gardena nei Pressi di Redagno (Bolzano, Italia) e una nuova lista ichnofaunistica (anfibi e rettili) della Stessa Formazione. In: Anais do X Congresso Brasileiro de Paleontologia, Rio de Janeiro, 19–25, pp. 319–331 de Julho.

Conti, M.A., Mariotti, N., Mietto, P., Nicosia, U., 1991. Nuove ricerche sugli icnofossili della Formazione di Collio in Val Trompia (Brescia). "Natura Bresciana" Ann. Mus. Civ. Sc. Nat. 26, 109–119.

Conti, M.A., Mariotti, N., Nicosia, U., Pittau, P., 1997. Selected bioevents succession in the continental Permian of the Southern Alps (Italy): improvements of intrabasinal and interregional correlations. In: Dickins, J.M., Zunyi, Y., Hongfu, Y., Lucas, S.G., Acharyya, S.E. (Eds.), Late Paleozoic and Early Mesozoic Circum-Pacific Events and their Global Correlation. Cambridge University Press, pp. 51–65.

Cope, E.D., 1882. Third contribution to the history of the vertebrata of the Permian formation of Texas. Proc. Am. Philos. Soc. 20, 447–461.

Cotta, B., 1848. Briefwechsel. Mittheilungen an den Geheimenrath VON LEONHARD gerichtet. N. Jb. Min. Geol. Paläont., Stuttgart 1848, 43–44.

Curioni, G., 1870. Osservazioni geologiche sulla Val Trompia. Memorie del R. Instituto Lombardo di Scienze e Lettere Serie III II, 1–60.

De Beer, C.H., 1986. Surface markings, reptilian footprints and trace fossils on a palaeosurface in the Beaufort Group near Fraserburg. C. P. Annale van die Geologiese Opname Republiek van Suid-Afrika 20, 129–140.

De Jaime-Soguero, C., Mujal, E., Dinarès-Turell, J., Oms, O., Bolet, A., Orlandi- Oliveras, G., Fortuny, J., 2021. Palaeoecology of Middle Triassic tetrapod ichnoassociations (middle Muschelkalk, NE Iberian Peninsula) and their implications for palaeobiogeography in the western Tethys region. Palaeogeogr. Palaeoclimatol. Palaeocol. 565, 110204.

De Jaime-Soguero, C., Mujal, E., Oms, O., Bolet, A., Dinarès-Turell, J., Ibáñez-Insa, J., Fortuny, J., 2023. Palaeoenvironmental reconstruction of a Cisuralian to Guadalupian terrestrial composite succession from the Catalan Pyrenees: implications for the evolution of tetrapod ecosystems in equatorial Pangaea. Palaeogeogr. Palaeoclimatol. Palaeocol. 632, 111837.

de Klerk, W.J., 2002. A dicynodont trackway from the *Cistecephalus* assemblage zone in the Karoo, east of Graaff-Reinet, South Africa. Palaeont. Afr. 38, 73–91.

Delage, A., 1912. Empreintes de pieds de grands quadrupèdes dans le Permien Inferieur de l'Hérault. Acad. Sci. Lett. Montpellier. Mém. Sect. Sci. 4, 221–267.

Demathieu, G., Oosterink, H.W., 1983. Die Wirbeltier-Ichnofauna aus dem Unteren Muschelkalk von Winterswijk (Die Reptilfährten aus der Mitteltrias der Niederlande). Staringia 7, 1–51.

Demathieu, G., Gand, G., Toutin-Morin, N., 1992. La palichnofaune des bassins permiens provencaux. Geobios 25, 19–54.

DiMichele, W.A., Lucas, S.G., Krainer, K., 2012. Vertebrate trackways among a stand of *Supaia* white plants on an early Permian floodplain, New Mexico. J. Paleontol. 86, 584–594.

Dozy, J.J., 1935. Einige Tierfährten aus dem Unteren Perm der Bergamasker Alpen. PalZ. 17, 45–55.

Dudgeon, P., 1878. Note on a new fossil footprint from the Permian sandstone of Dumfriesshire. Proc. Roy. Soc. Edinb. 9, 154–155.

Duncan, H., 1829. Account of the tracks of the foot-marks of animals found impressed in sandstone in the quarry of Corncockle Muir, Dumfries-shire. Am. J. Sci. Arts 15, 84–90.

Duncan, H., 1831. An account of tracks and footmarks of animals found impressed on Sandstone in the Quarry of Corncockle Muir. Dumfriesshire. Trans. Roy. Soc. Edinb. 11, 194–209.

Eliason, C.M., Hudson, L., Watts, T., Garza, H., Clarke, J.A., 2017. Exceptional preservation and the fossil record of tetrapod integument. Proc. Roy. Soc. B: Biological Sciences 284, 20170556.

Ellenberger, F., Ellenberger, P., 1959. Quelques pistes de vertébrés du Permien inférieur de Lodève. C. r. Acad. sci. Paris 248, 437.

Ellenberger, P., 1983a. Sur l'intérêt paléontologique de la dalle à pistes de la Lieude (commune de Mérifons, Hérault, France). Société de la Protection de la Nature du Languedoc Roussillon 1, 11–20.

Ellenberger, P., 1983b. Sur la zonation ichnologique du Permien moyen (Saxonien) du bassin de Lodève (Hérault). C. r. séances Acad. sci. Paris, Série 2, Mécanique-Physique, Chimie, Sciences de l'univers. Sci. Terre 297, 553–558.

Ellenberger, P., 1983c. Sur la zonation ichnologique du Permien inférieur (Autunien) du bassin de Lodève (Hérault). C. r. Acad. sci. Série 2, Mécanique-Physique Chimie, Sciences de l'univers, Sciences de la Terre 297, 631–633.

Ellenberger, P., 1984. Données complémentaires sur la zonation ichnologique du Permien du Midi de la France (Bassins de Lodève, Saint-Affrique et Rodez). C. r. Acad. sci. Paris, Série 2 (299), 581–586.

Evans, S.E., Haubold, H., 1987. A review of the upper permian genera *Coelurosauravus, Weigeltisaurus* and *Gracilisaurus* (Reptilia: Diapsida). Zool. J. Linn. Soc. 90, 275–303.

Fabre, G., 1872. Matériaux pour server à la description géologique du département de la Lozère. Bulletin de la Société d'Agriculture, Industrie Sciences et arts du département de la Lozère 23, 35–61.

Fichter, J., Kowalczyk, G., 1983. Tetrapodenfährten aus dem Rotliegenden der Wetterau und ihre stratigraphische Auswertung. Mainzer Geowiss. Mitt. 12, 123–158.

Fichter, J., Kunz, R., 2004. New genus and species of chirotheroid tracks in the Detfurth-formation (middle Bunter, lower Triassic) of Central Germany. Ichnos 11, 183–193.

Fichter, J., Kunz, R., 2015. Eine Tetrapoden-Fährtenvergesellschaftung im Mittleren Buntsandstein (frühe Trias, Olenekium) von Wolfhagen, Nordhessen. Z. Dtsch. Geol. Ges. 166, 253–273.

Fichter, J., 1976. Tetrapodenfährten aus dem Unterrotliegenden (Autun, Unter-Perm) von Odernheim/Glan. Mainzer geowiss. Mitt. 5, 87–109.

Fichter, J., 1979. Aktuopaläontologische Studien zur Lokomotion rezenter Urodelen und Lacertilier sowie paläontologische Untersuchungen an Tetrapodenfährten des Rotliegenden (Unter-Perm) SW-Deutschlands. Unpublished PhD thesis. Johannes-Gutenberg University, Mainz, 425 pp.

Fichter, J., 1983a. Tetrapodenfährten aus dem saarpfälzischen Rotliegenden (?Ober-Karbon – Unter-Perm; Südwest-Deutschland), Teil I: Fährten der Gattungen *Saurichnites, Limnopus, Ichniotherium, Amphisauroides, Protritonichnites, Gilmoreichnus, Hyloidichnus* und *Jacobiichnus*. Mainzer Geowiss. Mitt. 12, 9–121.

Fichter, J., 1983b. Tetrapodenfährten aus dem saarpfälzischen Rotliegenden (?Ober-Karbon-Unter-Perm; SW-Deutschland), Teil II: Fährten der Gattungen *Foliipes, Varanopus, Ichniotherium, Dimetropus, Palmichnus,* cf. *Chelichnus,* cf. *Laoporus* und *Anhomoiichnum*. Mainzer Naturw. Archiv. 21, 125–186.

Fichter, J., 1984. Neue Tetrapodenfährten aus den saarpfälzischen Standenbühl-Schichten (Unter-Perm; SW-Deutschland). Mainzer Naturw. Archiv. 22, 211–229.

Fichter, J., 1998. Bericht über die Bergung einer 20 t schweren Fährtenplatte aus dem Tambacher Sandstein (Unter Perm) des Thüringer Waldes und erste Ergebnisse ichnologischer Studien. Philippia 8, 147–208.

Fillmore, D.L., Lucas, S.G., Simpson, E.L., 2012. Ichnology of the Mississippian Mauch Chunk formation, eastern Pennsylvania. Bull. N. M. Mus. Nat. Hist. Sci. 54, 1–136.

Ford, D.P., Benson, R.B., 2020. The phylogeny of early amniotes and the affinities of Parareptilia and Varanopidae. Nat. Ecol. Evol. 4, 57–65.

Fourie, H., Rubidge, B.S., 2007. The postcranial skeletal anatomy of the therocephalian *Regisaurus* (Therapsida: Regisauridae) and its utilization for biostratigraphic correlation. Palaeont. Afr. 42, 1–16.

Fourie, H., Rubidge, B.S., 2009. The postcranial skeleton of the basal therocephalian *Glanosuchus macrops* (Scylacosauridae) and comparison of morphological and phylogenetic trends amongst the Theriodontia. Palaeont. Afr. 44, 27–39.

Francischini, H., Dentzien-Dias, P., Lucas, S.G., Schultz, C.L., 2018. Tetrapod tracks in Permo–Triassic eolian beds of southern Brazil (Parana Basin). PeerJ 6, e4764.

Francischini, H., Lucas, S.G., Voigt, S., Marchetti, L., Santucci, V.L., Knight, C.L., Wood, J.R., Dentzien-Dias, P., Schultz, C.L., 2020. On the presence of Ichniotherium in the Coconino Sandstone (Cisuralian) of the Grand Canyon and remarks on the occupation of deserts by non-amniote tetrapods. PalZ 94, 207–225.

Fritsch, A., 1895. Über neue Wirbeltiere aus der Permformation Böhmens nebst einer Übersicht der aus derselben bekannt gewordenen Arten. Sitzungsber. Königl. Böhm. Ges. Wiss. Math.-Naturwiss. Cl. 52, 1–17.

Fritsch, A., 1901. Fauna der Gaskohle und der Kalksteine der Permformation Böhmens. F. Řivnáč, Prag. 93–97.

Fuglewicz, R., Ptaszyński, T., Rdzanek, K., 1990. Lower Triassic footprints from the Świetokrzyskie (Holy Cross) Mountains, Poland. Acta Palaeontolo. Pol. 35, 109–164.

Gand, G., Durand, M., 2006. Tetrapod footprint ichno-associations from French Permian basins. Comparisons with other Euramerican ichnofaunas. Geol. Soc. Lond. Special Publ. 265, 157–177.

Gand, G., Haubold, H., 1984. Traces de Vertébrés du Permien du Bassin de Saint-Affrique (Description, datation, comparaison avec celle du bassin de Lodève). Géol. Méditerranéenne 11, 321–351.

Gand, G., Demathieu, G., Ballestra, F., 1995. La palichnofaune de vertébrés tétrapodes du Permien supérieur de l'Esterel (Provence, France). Palaeontogr. Abt. A 235, 97–139.

Gand, G., Kerp, H., Parsons, C., Martínez-García, E., 1997. Palaeoenvironnemental and stratigraphic aspects of animal traces and plant remains in Spanish Permian red beds (Peña Sagra, Cantabrian Mountains, Spain). Geobios 30, 295–318.

Gand, G., Garric, J., Demathieu, G., Ellenberger, P., 2000. La palichnofaune de vertébrés tétrapodes du Permien supérieur du bassin de Lodève (Languedoc-France). Palaeovertebrata 29, 1–82.

Gand, G., Tüysüz, O., Steyer, J.S., Allain, R., Sakinç, M., Sanchez, S., Sengör, A.M.C., Sen, S., 2011. New Permian tetrapod footprints and macroflora from Turkey (Çakraz Formation, northwestern Anatolia): biostratigraphic and palaeoenvironmental implications. C.R. Palevol. 10, 617–625.

Gand, G., 1977. Sur le matériel ichnologique récolté dans le Muschelkalk de Culles-les-Roches (Saône-et-Loire). Bull. Soc. Hist. Nat. Creusot. 35, 9–30.

Gand, G., 1981. Découverte de traces de Réptiles cotylosauriens dans le Permien du Bassin de Blanzy-Le Creusot (Saône-et-Loire, France). Une étape dans la succession des palichnofaunes de vertébrés tétrapodes. C. r. Acad. sci Paris, Série 3. Sciences de la Vie 292, 163–167.

Gand, G., 1985. Significations paléobiologique et stratigraphique de *Limnopus zeilleri* dans la partie nord du bassin permien de Saint-Affrique (Aveyron, France). Geobios 18, 215–227.

Gand, G., 1987. Les traces de vertébrés tétrapodes du Permien français. PhD Thesis. Université de Bourgogne. Édition Centre des Sciences de la Terre, Dijon, 341 pp.

Gand, G., 1989. *Varanopus rigidus*: Une nouvelle ichnoespèce de vertébré tétrapode du Permien Francais attribuable à des captorhinomorpha ou à des procolophonoidea. Geobios 22, 277–291.

Gand, G., 1991. Premières découvertes de traces de pas de vertébrés tétrapodes dans le bassin permien de Brive-la-Gaillarde (Corrèze). Géol. de la France 3, 43–51.

Gand, G., 1993a. Les empreintes animales et végétales du bassin permien de Largentière:(Ardèche, France). Bull. Soc. Hist. Nat. Autun 146, 8–41.

Gand, G., 1993b. La palichnofaune de Vertébrés tétrapodes du bassin permien de St-Affrique (Aveyron): comparaisons et conséquences stratigraphiques. Géol de la France 3–4, 41–56.
Gastaldo, R.A., Kamo, S.L., Neveling, J., Geissman, J.W., Looy, C.V., Martini, A.M., 2020. The base of the Lystrosaurus Assemblage Zone, Karoo Basin, predates the end-Permian marine extinction. Nat. Commun. 11, 1428.
Gatesy, S.M., Falkingham, P.L., 2020. Hitchcock's Leptodactyli, penetrative tracks, and dinosaur footprint diversity. J. Vertebr. Paleontol. 40, e1781142.
Geinitz, H.B., Deichmüller, J.V., 1882. Die Saurier der unteren Dyas von Sachsen. Palaeontographica 29, 1–46.
Geinitz, H.B., 1861. Dyas oder die Zechsteinformation und das Rothliegende (Permische Formation zum Theil). W. Engelmann, Leipzig, 130 pp.
Geinitz, H.B., 1863. Beiträge zur Kenntnis der organischen Überreste in der Dyas. Neues Jahrb. Miner. Geol. Paläont. 186, 385–398.
Geinitz, H.B., 1869. Über fossile Pflanzenreste aus dem Dyas von Val Trompia. N. Jb. Miner. Geol. Paläont. 1869, 456–461.
Gilmore, G.W., 1927. Fossil footprints from the Grand Canyon: Second contribution. Smithsonian Misc. Collect. 80 (3), 1–78.
Gilmore, G.W., 1928. Fossil footprints from the Grand Canyon III. Smithsonian Misc. Collect. 80 (8), 1–16.
Gônet, J., Bardin, J., Girondot, M., Hutchinson, J.R., Laurin, M., 2023. Locomotor and postural diversity among reptiles viewed through the prism of femoral microanatomy: Palaeobiological implications for some Permian and Mesozoic taxa. J. Anat. 242, 891–916.
Grierson, J., 1828. On footsteps before the flood, in a specimen of red sandstone. Edinb. J. Sci. 8, 130–134.
Groenewald, D.P., Krüger, A., Day, M.O., Penn-Clarke, C.R., Hancox, P.J., Rubidge, B.S., 2023. Unique trackway on Permian Karoo shoreline provides evidence of temnospondyl locomotory behaviour. PLoS One 18, e0282354.
Gubin, Y.M., Golubev, V.K., Bulanov, V.V., Petuchov, S.V., 2003. Pareiasaurian tracks from the upper permian of eastern Europe. Paleontol. J. 37, 514–523.
Haubold, H., Lucas, S.G., 2001. Die Tetrapodenfährten der Choza Formation (Texas) und das Artinsk-Alter der Redbed Ichnofaunen des Unteren Perm. Hallesches Jahrb. Geowiss. B 23, 79–108.
Haubold, H., Lucas, S.G., 2003. Tetrapod footprints of the Lower Permian Choza Formation at Castle Peak, Texas. PalZ 77, 247–261.
Haubold, H., Sarjeant, W.A.S., 1973. Tetrapodenfährten aus den Keele und Enville Groups (Permokarbon: Stefan und Autun) von Shropshire und South Staffordshire, Großbritannien. Z. Geol. Wiss. 1, 895–933.
Haubold, H., Stapf, H., 1998. The early Permian tetrapod track assemblage of Nierstein, Standenbühl Beds, Rotliegend, Saar-Nahe Basin, SW-Germany. Hallesches Jahrb. Geowiss. B 20, 17–32.
Haubold, H., Hunt, A.P., Lucas, S.G., Lockley, M.G., 1995. Wolfcampian (early Permian) vertebrate tracks from Arizona and New Mexico. Bull. N. M.Mus. Nat. Hist. Sci. 6, 135–165.
Haubold, H., Allen, A., Atkinson, T.P., Buta, R.J., Lacefield, J.A., Minkin, S.C., Relihan, B.A., 2005. Interpretation of the tetrapod footprints from the early Pennsylvanian of Alabama. Alabama Paleontol. Soc. Monogr. 1, 75–111.
Haubold, H., 1966. Therapsiden- und Rhynchocephalen-Fährten aus dem Buntsandstein Südthüringens. Hercynia 3, 147–183.
Haubold, H., 1970. Versuch der Revision der Amphibien-Fährten des Karbon und Perm. Freib. Forsch. C 260, 83–117.
Haubold, H., 1971a. Ichnia Amphibiorum et Reptiliorum fossilium. Encycl. Palaeoherpetol. 18, 1–124.
Haubold, H., 1971b. Die Tetrapodenfährten des Buntsandsteins. Paläonto. Abh. A 4, 395–548.
Haubold, H., 1972. Panzerabdrücke von Tetrapoden aus dem Rotliegenden (Unterperm) des Thüringer Waldes. Geologie 21, 110–115.
Haubold, H., 1973. Die Tetrapodenfährten aus dem Perm Europas. Freib. Forsch. C 285, 5–55.
Haubold, H., 1984. Saurierfährten. Wittenberg. Ziemsen-Verlag, 231 pp.
Haubold, H., 1996. Ichnotaxonomie und Klassifikation von Tetrapodenfährten aus dem Perm. Hallesches Jahrb. Geowiss. B 18, 23–88.
Haubold, H., 1998. The Early Permian tetrapod ichnofauna of Tambach, the changing concepts in ichnotaxonomy. Hallesches Jahrb. Geowiss. B 20, 1–16.
Haubold, H., 2000. Tetrapodenfährten aus dem Perm - Kenntnisstand und Progress 2000. Hallesches Jahrb. Geowiss. B 22, 1–16.
Heyler, D., Gand, G., 2001. Pistes et traces du Permien du Sud de la France dans les collections du Muséum National d'Histoire Naturelle (Paris). Bull. Soc. Hist. Nat. Autun 176, 7–50.

Heyler, D., Lessertisseur, J., 1962. Remarque sur les allures des Tétrapodes paléozoïques d'après les pistes du Permien de Lodève. Colloque du C.N.R.S. Paris 104, 123–134.
Heyler, D., Lessertisseur, J., 1963. Pistes de tétrapodes permiens dans la région de Lodève (Hérault). Mém. Mus. Natl. Hist. Nat. Série C 11, 125–221.
Heyler, D., Montenat, C., 1980. Traces de pas de vertébrés du Permien du Var. Intérêt biostratigraphique. Bull. Mus. Natl. Hist. Nat. 4, 407–435.
Hmich, D., Schneider, J.W., Saber, H., Voigt, S., El Wartiti, M., 2006. New continental Carboniferous and Permian faunas of Morocco – implications for biostratigraphy, palaeobiogeography and palaeoclimate. Geol. Soc. Lond. Special Publications 265, 297–324.
Hminna, A., Voigt, S., Saber, H., Schneider, J.W., Hmich, D., 2012. On a moderately diverse continental ichnofauna from the Permian Ikakern Formation (Argana Basin, Western High Atlas, Morocco). J. African Earth Sci. 68, 15–32.
Hochstetter, F.V., 1868. Saurier-Fährten im Rotliegend des Rossitz-Oslawaner Backens. Verhand. Geol. Bundes. 431–432.
Holst, H.K.H., Smit, J., Veenstra, E., 1970. Lacertoid footprints from the Early Middle Triassic at Haarmühle, near Altstätte, W. Germany. Proc. Sect. Sci. K. Ned. Akad. Wet. Amst. 73, 157–165.
Holub, V., Kozur, H., 1981. Revision einiger Tetrapodenfährten des Rotliegenden und biostratigraphische Auswertung der Tetrapodenfährten des obersten Karbon und Perm. Geol.-Paläontol. Mitt. Innsbruck 11, 149–193.
Hopson, J.A., 1995. Patterns of evolution in the manus and pes of non-mammalian therapsids. J. Vertebr. Paleont. 15, 615–639.
Hopson, J.A., 2015. Fossils, trackways, and transitions in locomotion. In: Dial, K.P., Shubin, N., Brainerd, E.L. (Eds.), Great Transformations in Vertebrate Evolution. University of Chicago Press, Chicago IL, pp. 125–141.
Hornstein, F.F., 1876. Entdeckung von Thierfährten im Buntsandstein bei Karlshafen. Brief an Geinitz. Neues Jahrb. Miner. Geol. Paläont. 1876, 923–924.
Hunt, A.P., Lucas, S.G., 1998. Vertebrate tracks and the myth of the belly-dragging tail-dragging tetrapods of the Late Paleozoic. Bull. N. M. Mus. Nat. Hist. Sci. 12, 67–69.
Hunt, A.P., Lucas, S.G., 2006. Permian tetrapod ichnofacies. Geol. Soc. Lond. Special Publications 265, 137–156.
Hunt, A.P., Lucas, S.G., 2007. Tetrapod ichnofacies: a new paradigm. Ichnos 14, 59–68.
Hunt, A.P., Lockley, M.G., Lucas, S.G., MacDonald, J.P., Hotton, N., Kramer, J., 1993. Early Permian tracksites in the Robledo Mountains, south-central New Mexico. Bull. N. M. Mus. Nat. Hist. Sci. 2, 23–31.
Hunt, A.P., Lucas, S.G., Haubold, H., Lockley, M.G., 1995. Early Permian (late Wolfcampian) tetrapod tracks from the Robledo Mountains, south-central New Mexico. Bull. N. M. Mus. Nat. Hist. Sci. 6, 167–180.
Hunt, A.P., Lucas, S.G., Spielmann, J.A., 2005. Paleoenvironmental transects and tetrapod biotaxonichnofacies in the Early Permian of the southwestern United States. Bull. N. M. Mus. Nat. Hist. Sci. 31, 49–51.
ICZN, 1999. International Code of Zoological Nomenclature, fourth ed. International Trust for Zoological Nomenclature, London.
Jardine, W., 1850. Note on Mr. Harkness's paper on the position of the impressions of footsteps in the Bunter Sandstone of Dumfriesshire. Ann. Mag. Nat. Hist. London 2, 208–209.
Jenkins Jr., F.A., 1970. Cynodont postcranial anatomy and the "prototherian" level of mammalian organization. Evolution 24, 230–252.
Kaszap, A., 1968. *Korynichnium sphaerodactylum* (Pabst) a balatonrendesi permben. Foldtani Kozlony 98, 429–433.
Kemp, T.S., 2005. The Origin and Evolution of Mammals. Oxford University Press, USA, 342 pp.
Kemp, T.S., 2011. The origin and radiation of therapsids. In: Chinsamy-Turan, A. (Ed.), Forerunners of Mammals. Indiana University Press, Bloomington, pp. 3–30.
King, T., 1844. Description of footmarks supposed to be referable to the classis Birds, Reptilia, and Mammalia found in the Carboniferous series in Westmoreland County, Pa. Proc. Acad. Nat. Sci. Philadelphia 2, 175–180.
King, T., 1845a. Description of fossil footmarks, found in the Carboniferous series in Westmoreland Country. Pa. Am. J. Sci. 48, 343–352.
King, T., 1845b. Footprints. Am. J. Sci. 49, 216–217.
Kittl, E., 1891. Saurierfährte von Bozen. Mitt. Sektion Naturk. Österreichischen Touristenklubs 3, 7.
Klein, H., Lucas, S.G., 2021. The Triassic tetrapod footprint record. Bull. N. M. Mus. Nat. Hist. Sci. 83, 1–194.
Klein, H., Niedźwiedzki, G., 2012. Revision of the Lower Triassic tetrapod ichnofauna from Wióry, Holy Cross Mountains, Poland. Bull. N. M. Mus. Nat. Hist. Sci. 56, 1–62.

Klein, H., Voigt, S., Hminna, A., Saber, H., Schneider, J.W., Hmich, D., 2010. Early Triassic archosaur-dominated footprint assemblage from the Argana Basin (Western High Atlas, Morocco). Ichnos 17, 215–227.

Klein, H., Lucas, S.G., Voigt, S., 2015. Revision of the ?Permian-Triassic tetrapod ichnogenus *Procolophonichnium* Nopcsa 1923 with description of the new ichnospecies *P. lockleyi*. Ichnos 22, 155–176.

Klein, H., Heckert, A.B., Celeskey, M., 2023. Chirotheres: Tracking the Ancestors of Dinosaurs and Crocodiles. Indiana University Press, Bloomington, Indiana, 360 pp.

Kohl, M.S., Bryan, J.R., 1994. A new Middle Pennsylvanian (Westphalian) amphibian trackway from the Cross Mountain Formation, east Tennessee Cumberlands. J. Paleontol. 68, 655–663.

Korn, H., 1933. Eine für die Kenntnis der Cotylosaurier des deutschen Perms bedeutsame Schwimmfährte von Tambach. Palaeobiologica 5, 169–201.

Krapovickas, V., Marsicano, C.A., Mancuso, A.C., de la Fuente, M.S., Ottone, E.G., 2015. Tetrapod and invertebrate trace fossils from aeolian deposits of the lower Permian of centralwestern Argentina. Hist. Biol. 27, 827–842.

Krapovickas, V., Mángano, M.G., Buatois, L.A., Marsicano, C.A., 2016. Integrated Ichnofacies models for deserts: Recurrent patterns and megatrends. Earth Sci. Rev. 157, 61–85.

Kubo, T., Benton, M.J., 2009. Tetrapod postural shift estimated from Permian and Triassic trackways. Palaeontology 52, 1029–1037.

Kuhn, O., 1963. Ichnia tetrapodorum. Fossilium Catalogus I 101, 176.

Kuznetsov, A.N., 2020. Passive-dynamic walkers of the late Paleozoic. Ameghiniana 57, 591–615.

Lallensack, J.N., Farlow, J.O., Falkingham, P.L., 2021. A new solution to an old riddle: elongate dinosaur tracks explained as deep penetration of the foot, not plantigrade locomotion. Palaeontology 65, e12584.

Lallensack, J.N., Buchwitz, M., Romilio, A., 2022. Photogrammetry in ichnology: 3D model generation, visualisation, and data extraction. J. Paleontol. Tech. 22, 1–18.

Leonardi, P., Conti, M.A., Leonardi, G., Mariotti, N., Nicosia, U., 1975. *Pachypes dolomiticus* n. gen. N. sp.; Pareiasaur footprint from the 'Val Gardena Sandstone' (Middle Permian) in the western Dolomites (N. Italy). Atti della Academia Nazionale dei Lincei 57, 221–232.

Leonardi, P., 1951. Orme di tetrapodi nelle Arenarie de Val Gardena (Permiano Medio-inferiore) dell'Alto Adige sud-orientale. Mem. Isti. Geol. Miner. Univ. Padova 17, 1–23.

Leonardi, G., 1987. Glossary and Manual of Tetrapod Footprint Palaeoichnology. Departamento Nacional de Produção Mineral, Brasilia, 75 pp.

Liu, J., Bever, G.S., 2015. The last diadectomorph sheds light on Late Palaeozoic tetrapod biogeography. Biol. Lett. 11, 20150100.

Lockley, M.G., Hunt, A.P., Meyer, C., 1994. Vertebrate tracks and the ichnofacies concept: implications for paleoecology and palichnostratigraphy. In: Donovan, S. (Ed.), The Paleobiology of Trace Fossils. John Wiley, London, pp. 241–268.

Logghe, A., Mujal, E., Marchetti, L., Nel, A., Pouillon, J.-M., Giner, S., Garrouste, R., Steyer, J.S., 2021. *Hyloidichnus* trackways with digit and tail drag traces from the Permian of Gonfaron (Var, France): new insights on the locomotion of captorhinomorph eureptiles. Palaeogeogr. Palaeoclimatol. Palaeoecol. 573, 110436.

Lucas, S.G., Dalman, S.G., 2013. Alfred King's Pennsylvanian tetrapod footprints from western Pennsylvania. In: Lucas, S.G., DiMichele, W.A., Barrick, J.E., Schneider, J.W., Spielmann, J.A. (Eds.), The Carboniferous-Permian Transition, Bull. N. M. Mus. Nat. Hist. Sci, 60, pp. 233–239.

Lucas, S.G., Hunt, A.P., 2005. Permian tetrapod tracks from Texas. Bull. N. M. Mus. Nat. Hist. Sci. 30, 202–206.

Lucas, S.G., Suneson, N.H., 2002. Amphibian and reptile tracks from the Hennessey Formation (Leonardian, Permian), Oklahoma County, Oklahoma. Oklahoma Geol. Notes 62, 56–62.

Lucas, S.G., Lozovsky, V.R., Shishkin, M.A., 1999. Tetrapod footprints from Early Permian Redbeds of the Northern Caucasus, Russia. Ichnos 6, 277–281.

Lucas, S.G., Lerner, A.J., Haubold, H., 2001. First record of *Amphisauropus* and *Varanopus* in the Lower Permian Abo Formation, central New Mexico. Hallesches Jahrb. Geowiss. B 23, 69–78.

Lucas, S.G., Minter, N.J., Spielmann, J.A., Smith, J.A., Braddy, S.J., 2005a. Early Permian ichnofossils from the northern Caballo Mountains, Sierra County, New Mexico. In: Lucas, S.G., Zeigler, K.E., Spielmann, J.A. (Eds.), The Permian of Central New Mexico, Bull. N. M. Mus. Nat. Hist. Sci, vol 31, pp. 151–162.

Lucas, S.G., Smith, J.A., Hunt, A.P., 2005b. Tetrapod tracks from the Lower Permian Yeso Group, central New Mexico. In: Lucas, S.G., Zeigler, K.E., Spielmann, J.A. (Eds.), The Permian of Central New Mexico, Bull. N. M. Mus. Nat. Hist. Sci, vol 31, pp. 121–124.

Lucas, S.G., Voigt, S., Lerner, A.J., MacDonald, J.P., Spielmann, J.A., Celeskey, M.D., 2011a. The prehistoric trackways national monument, Permian of southern New Mexico, USA. Ichnol. Newsl. 28, 10–14.

Lucas, S.G., Voigt, S., Lerner, A.J., Nelson, W.J., 2011b. Late Early Permian continental ichnofauna from Lake Kemp, north-central Texas, USA. Palaeogeogr. Palaeoclimatol. Palaeoecol. 308, 395–404.

Lucas, S.G., Krainer, K., Chaney, D.S., DiMichele, W.A., Voigt, S., Berman, D.S., Henrici, A.C., 2013. The Lower Permian Abo Formation in Central New Mexico. Bull. N. M. Mus. Nat. Hist. Sci. 59, 161–179.

Lucas, S.G., Kollar, A.D., Berman, D.S., Henrici, A.C., 2016. Pelycosaurian-grade (Amniota: Synapsida) footprints from the lower Permian Dunkard Group of Pennsylvania and West Virginia. Ann. Carnegie Mus. 83, 287–294.

Lucas, S.G., Stimson, M.R., King, O.A., Calder, J.H., Mansky, C.F., Hebert, B.L., Hunt, A.P., 2022. Carboniferous tetrapod footprint biostratigraphy, biochronology and evolutionary events. Geol. Soci. Lond. Special Public. 512, 933–963.

Lucas, S.G., 2001. Taphotaxon. Lethaia 34, 30.

Lucas, S.G., 2006. Global Permian tetrapod biostratigraphy and biochronology. Geol. Soc. Lond. Special Public. 265, 65–93.

Lucas, S.G., 2007. Tetrapod footprint biostratigraphy and biochronology. Ichnos 14, 5–38.

Lucas, S.G., 2010. The Triassic timescale based on nonmarine tetrapod biostratigraphy and biochronology. Geol. Soc. Lond. Special Public. 334, 447–500.

Lucas, S.G., 2015. Alfred King's forgotten fossil footprints. Pennsylvania Herit. 41, 16–25.

Lucas, S.G., 2017. Permian tetrapod extinction events. Earth Sci. Rev. 170, 31–60.

Lucas, S.G., 2018. Permian tetrapod biochronology, correlation and evolutionary events. Geol. Soc., Lond. Special Public. 450, 405–444.

Lucas, S.G., 2019. An ichnological perspective on some major events of Paleozoic tetrapod evolution. Boll. Soci. Paleontol. Ital. 58, 223–266.

Lucas, S.G., 2022. Carboniferous tetrapod biostratigraphy, biochronology and evolutionary events. Geol. Soci. Lond. Special Public. 512, 965–1001.

Lull, R.S., 1918. Fossil footprints from the Grand Canyon of the Colorado. Am. J. Sci. Ser. 4 (45), 337–346.

Lull, R.S., 1920. An upper carboniferous footprint from Attleboro, Massachusetts. Am. J. Sci., Ser. 4 (50), 234–236.

Lützner, H., Tichomirowa, M., Käßner, A., Gaupp, R., 2021. Latest carboniferous to early permian volcano-stratigraphic evolution in central Europe: U–Pb CA–ID–TIMS ages of volcanic rocks in the Thuringian Forest Basin (Germany). Int. J. Earth Sci. 110, 377–398.

MacDonald, J.P., 1994. Earths First Steps. Johnson Printing, Boulder, 290 pp.

MacDougall, M.J., Modesto, S.P., Brocklehurst, N., Verrière, A., Reisz, R.R., Fröbisch, J., 2018. Commentary: a reassessment of the taxonomic position of mesosaurs, and a surprising phylogeny of early amniotes. Front. Earth Sci. 6, 99.

Maidwell, F.T., 1911. Notes on footprints from the Keuper of Runcorn Hill. Proc. Liverpool Geol. Soci. 11, 140–152.

Majoros, G., 1965. Reptilian footprint from the permian of Balatonrendes. Földt. Közl 94, 243–245.

Marchetti, L., Avanzini, M., Conti, M.A., 2013a. *Hyloidichnus bifurcatus* Gilmore, 1927 and *Limnopus heterodactylus* (King, 1845) from the early Permian of Southern Alps (N Italy): a new equilibrium in the ichnofauna. Ichnos 20, 202–217.

Marchetti, L., Bernardi, M., Avanzini, M., 2013b. Some insights on well-preserved *Amphisauropus* and *Erpetopus* trackways from the eastern Collio Basin (Trentino– Alto Adige, NE Italy). Boll. Soci. Paleontol. Ital. 52, 55–62.

Marchetti, L., Santi, G., Avanzini, M., 2014. The problem of small footprints in paleoichnology: remarks on the Early Permian ichnotaxon *Erpetopus cassinisi*, a local species from Southern Alps (northern Italy). Rivista It. Paleontol. Stratigr. 120, 129–143.

Marchetti, L., Ronchi, A., Santi, G., Voigt, S., 2015a. The Gerola Valley site (Orobic Basin, Northern Italy): a key for understanding late early Permian tetrapod ichnofaunas. Palaeogeogr. Palaeoclimatol. Palaeoecol. 439, 97–116.

Marchetti, L., Ronchi, A., Santi, G., Schirolli, P., Conti, M.A., 2015b. Revision of a classic site for Permian tetrapod ichnology (Collio Formation, Trompia and Caffaro valleys, N. Italy), new evidences for the radiation of captorhinomorph footprints. Palaeogeogr. Palaeoclimatol. Palaeoecol. 433, 140–155.

Marchetti, L., Forte, G., Bernardi, M., Wappler, T., Hartkopf-Fröder, C., Krainer, K., Kustatscher, E., 2015c. Reconstruction of a late Cisuralian (Early Permian) floodplain lake environment: palaeontology and sedimentology of the Tregiovo Basin (Trentino-Alto Adige, Northern Italy). Palaeogeogr. Palaeoclimatol. Palaeoecol. 440, 180–200.

Marchetti, L., Mujal, E., Bernardi, M., 2017a. An unusual *Amphisauropus* trackway and its implication for understanding seymouriamorph locomotion. Lethaia 50, 162–174.

Marchetti, L., Tessarollo, A., Felletti, F., Ronchi, A., 2017b. Tetrapod footprint paleoecology: behavior, taphonomy and ichnofauna disentangled. A case study from the Lower Permian of the Southern Alps (Italy). Palaios 32, 506–527.

Marchetti, L., Belvedere, M., Mietto, P., 2017c. Lopingian tetrapod footprints from the Venetian Prealps, Italy: new discoveries in a largely incomplete panorama. Acta Palaeontol. Polon. 62, 801–817.

Marchetti, L., Voigt, S., Santi, G., 2018. A rare occurrence of Permian tetrapod footprints: *Ichniotherium cottae* and *Ichniotherium sphaerodactylum* on the same stratigraphic surface. Ichnos 25, 106–118.

Marchetti, L., Belvedere, M., Voigt, S., Klein, H., Castanera, D., Díaz-Martínez, I., Marty, D., Xing, L., Feola, S., Melchor, R.N., Farlow, J.O., 2019a. Defining the morphological quality of fossil footprints. Problems and principles of preservation in tetrapod ichnology with examples from the Palaeozoic to the present. Earth Sci. Rev. 193, 109–145.

Marchetti, L., Voigt, S., Lucas, S.G., Francischini, H., Dentzien-Dias, P., Sacchi, R., Mangiacotti, M., Scali, S., Gazzola, A., Ronchi, A., Millhouse, A., 2019b. Tetrapod ichnotaxonomy in eolian paleoenvironments (Coconino and De Chelly formations, Arizona) and late Cisuralian (Permian) sauropsid radiation. Earth Sci. Rev. 190, 148–170.

Marchetti, L., Voigt, S., Lucas, S.G., 2019c. An anatomy-consistent study of the Lopingian eolian tracks of Germany and Scotland reveals the first evidence of the end-Guadalupian mass extinction at low paleolatitudes of Pangea. Gondwana Res. 73, 32–53.

Marchetti, L., Voigt, S., Klein, H., 2019d. Revision of late Permian tetrapod tracks from the Dolomites (Trentino-Alto Adige, Italy). Hist. Biol. 31, 748–783.

Marchetti, L., Klein, H., Buchwitz, M., Ronchi, A., Smith, R.M., De Klerk, W.J., Sciscio, L., Groenewald, G.H., 2019e. Permian-Triassic vertebrate footprints from South Africa: ichnotaxonomy, producers and biostratigraphy through two major faunal crises. Gondwana Res. 72, 139–168.

Marchetti, L., Ceoloni, P., Leonardi, G., Massari, F., Mietto, P., Sacchi, E., Valentini, M., 2020a. The Lopingian tetrapod ichnoassociation from Italy, a key for the understanding of low-latitude faunas before the end-Permian crisis. Tetrapod ichnology in Italy: the state of the art. J. Mediterr. Earth Sci 12, 61–81.

Marchetti, L., Voigt, S., Lucas, S.G., Stimson, M.R., King, O.A., Calder, J.H., 2020b. Footprints of the earliest reptiles: *Notalacerta missouriensis*: ichnotaxonomy, potential trackmakers, biostratigraphy, palaeobiogeography and palaeoecology. Ann. Soc. Geol. Pol. 90, 271–290.

Marchetti, L., Voigt, S., Mujal, E., Lucas, S.G., Francischini, H., Fortuny, J., Santucci, V.L., 2021a. Extending the footprint record of Pareiasauromorpha to the Cisuralian: earlier appearance and wider palaeobiogeography of the group. Pap. Palaeontol. 7, 1297–1319.

Marchetti, L., Francischini, H., Lucas, S.G., Voigt, S., Hunt, A.P., Santucci, V.L., de Vertebrados, L.D.P., 2021b. Paleozoic vertebrate ichnology of Grand Canyon national Park. In: Santucci, V.L., Tweet, J.S. (Eds.), Grand Canyon National Park Centennial Paleontological Resources Inventory – A Century of Fossil Discovery and Research, vol. 1. Utah Geol. Ass. Special Publication, pp. 171–204.

Marchetti, L., Voigt, S., Buchwitz, M., MacDougall, M.J., Lucas, S.G., Fillmore, D.L., Stimson, M.R., King, O.A., Calder, J.H., Fröbisch, J., 2021c. Tracking the origin and early evolution of reptiles. Front. Ecol. Evol. 9, 696511.

Marchetti, L., Forte, G., Kustatscher, E., DiMichele, W.A., Lucas, S.G., Roghi, G., Juncal, M.A., Hartkopf-Fröder, C., Krainer, K., Morelli, C., Ronchi, A., 2022a. The Artinskian Warming Event: an Euramerican change in climate and the terrestrial biota during the early Permian. Earth Sci. Rev. 226, 103922.

Marchetti, L., Logghe, A., Mujal, E., Barrier, P., Montenat, C., Nel, A., Pouillon, J.-M., Garrouste, R., Steyer, J.S., 2022b. Vertebrate tracks from the Permian of Gonfaron (Provence, Southern France) and their implications for the late Capitanian terrestrial extinction event. Palaeogeogr. Palaeoclimatol. Palaeoecol. 599, 111043.

Marchetti, L., 2016. New occurrences of tetrapod ichnotaxa from the Permian Orobic Basin (Northern Italy) and critical discussion of the age of the ichnoassociation. Pap. Palaeontol. 2, 363–386.

Marchetti, L., 2019. Can undertracks show higher morphologic quality than surface tracks? Remarks on large amphibian tracks from the early Permian of France. J. Iber. Geol. 45, 353–363.

Marsh, O.C., 1894. Footprints of vertebrates in the coal measures of Kansas. Am. J. Sci. 48, 81–84.

Martens, T., 1982. Zur Stratigraphie, Taxonomie, Ökologie und Klimaentwicklung des Oberrotliegenden (Unteres Perm) im Thüringer Wald (DDR). Abh. Ber. Mus. Nat. Gotha 11, 33–57.

Martens, T., 1991. Ein besonderes fossil. PalZ 65, 225–226.
Matamales-Andreu, R., Mujal, E., Galobart, À., Fortuny, J., 2021a. Insights on the evolution of synapsid locomotion based on tetrapod tracks from the lower Permian of Mallorca (Balearic Islands, western Mediterranean). Palaeogeogr. Palaeoclimatol. Palaeoecol. 579, 110589.
Matamales-Andreu, R., Roig-Munar, F.X., Oms, O., Galobart, À., Fortuny, J., 2021b. A captorhinid-dominated assemblage from the palaeoequatorial permian of Menorca (Balearic Islands, western Mediterranean). Earth Environ. Sci. Trans. R. Soc. Edinb 112, 125–145.
Matamales-Andreu, R., Mujal, E., Dinar Esturell, J., Kustatscher, E., Roghi, G., Oms, O., Galobart, À., Fortuny, J., 2022. Early–middle Permian ecosystems of equatorial Pangaea: integrated multi-stratigraphic and palaeontological review of the Permian of Mallorca (Balearic Islands, western Mediterranean). Earth Sci. Rev. 228, 103948.
Matamales-Andreu, R., Mujal, E., Galobart, À., Fortuny, J., 2023. A new medium-sized moradisaurine captorhinid eureptile from the Permian of Mallorca (Balearic Islands, western Mediterranean) and correlation with the co-occurring ichnogenus *Hyloidichnus*. Pap. Palaeontol. 9, e1498.
Matthew, G.F., 1905. New species and a new genus of batrachian footprints of the carboniferous system in eastern Canada. Trans. R. Soc. Can 10, 77–122.
McKeever, P.J., Haubold, H., 1996. Reclassification of vertebrate trackways from the Permian of Scotland and related forms from Arizona and Germany. J. Paleontol. 70, 1011–1022.
McLoughlin, S., Nicoll, R.S., Crowley, J.L., Vajda, V., Mays, C., Fielding, C.R., Frank, T.D., Wheeler, A., Bocking, M., 2021. Age and paleoenvironmental significance of the Frazer Beach Member—a new lithostratigraphic unit overlying the end-Permian extinction horizon in the Sydney Basin, Australia. Front. Earth Sci. 8, 600976.
Melchor, R.N., De Valais, S., 2006. A review of Triassic tetrapod track assemblages from Argentina. Palaeontology 49, 355–379.
Melchor, R.N., Sarjeant, W.A.S., 2004. Small amphibian and reptile footprints from the Permian Carapacha Basin, Argentina. Ichnos 11, 57–78.
Melchor, R.N., 1997. Workshop "Ichnofacies and ichnotaxonomy of the terrestrial Permian". Ameghiniana 34, 168.
Melchor, R.N., 2001. Permian tetrapod footprints from Argentina. Hallesches Jahrb. Geowiss. B 23, 35–43.
Mietto, P., 1981. Una grande impronta di pareiasauro nel Permiano di Recoaro (Vicenza). Rendiconti Soc. Geol. It. 4, 363–364.
Mikuláš, R., Martínek, K., 2006. Ichnology of the non-marine deposits of the Boskovice basin (Carboniferous-Permian, Czech Republic). Bull. Geosci. 81, 81–91.
Minter, N.J., Braddy, S.J., 2009. Ichnology of an Early Permian intertidal flat: the Robledo Mountains Formation of southern New Mexico, USA. Spec. Pap. Palaeontol. 82, 5–107.
Minter, N.J., Krainer, K., Lucas, S.G., Braddy, S.J., Hunt, A.P., 2007. Palaeocology of an Early Permian playa lake trace fossil assemblage from Castle Peak, Texas, USA. Palaeogeogr. Palaeoclimatol. Palaeoecol. 246, 390–423.
Modesto, S.P., Scott, D.M., MacDougall, M.J., Sues, H.D., Evans, D.C., Reisz, R.R., 2015. The oldest parareptile and the early diversification of reptiles. Proc. R. Soc. B Biol. Sci. 282, 20141912.
Moodie, R.L., 1913. Some recent advances in vertebrate paleontology. Am. Nat. 47, 183–192.
Moodie, R.L., 1929. Vertebrate footprints from the red beds of Texas. Am. J. Sci. 100, 352–368.
Moodie, R.L., 1930. Vertebrate footprints from the red beds of Texas II. J. Geol. 38, 548–565.
Morales, M., Haubold, 1995. Tetrapod tracks from the Lower Permian DeChelly Sandstone of Arizona: systematic description. Bull. N. M. Mus. Nat. Hist. Sci. 6, 251–261.
Moreau, J.D., Gand, G., 2022. New data on the Permian ecosystem of the Rodez Basin: ichnofauna (traces of protostomians, tetrapods and fishes), jellyfishes and plants from Banassac-Canilhac (Lozère, southern France). Geodiversitas 44, 975–987.
Moreau, J.D., Benaouiss, N., Tourani, A., Steyer, J.S., Laurin, M., Peyer, K., Béthoux, O., Aouda, A., Jalil, N.E., 2020. A new ichnofauna from the Permian of the Zat valley in the Marrakech High Atlas of Morocco. J. Afr. Earth Sci. 172, 103973.
Moreau, J.D., Michelin, A., Fara, E., Gand, G., Galtier, J., Puech, G., Fouché, S., 2022. Ichnofossils and body fossils from the permian of the Sorgue valley (Saint-Affrique basin, southern France): palaeoenvironmental implications. Hist. Biol 1–14.
Mossmann, D.J., Place, C.H., 1989. Early Permian fossil vertebrate footprints and their stratigraphic setting in megacyclic sequence II red beds, Prim Point, Prince Edward Island. Can. J. Earth Sci. 26, 591–605.

Mujal, E., Marchetti, L., 2020. *Ichniotherium* tracks from the Permian of France, and their implications for understanding the locomotion and palaeobiogeography of large diadectomorphs. Palaeogeogr. Palaeoclimatol. Palaeoecol. 547, 109698.

Mujal, E., Fortuny, J., Oms, O., Bolet, A., Galobart, À., Anadón, P., 2016. Palaeoenvironmental reconstruction and early Permian ichnoassemblage from the NE Iberian Peninsula (Pyrenean Basin). Geol. Mag. 153, 578–600.

Mujal, E., Fortuny, J., Pérez-Cano, J., Dinarès-Turell, J., Ibáñez-Insa, J., Oms, O., Vila, I., Bolet, A., Anadón, P., 2017. Integrated multi-stratigraphic study of the Coll de Terrers late Permian-early Triassic continental succession from the Catalan Pyrenees (NE Iberian Peninsula): a geologic reference record for equatorial Pangaea. Glob. Planet. Change 159, 46–60.

Mujal, E., Fortuny, J., Marmi, J., Dinarès-Turell, J., Bolet, A., Oms, O., 2018. Aridification across the carboniferous–permian transition in central equatorial Pangea: the Catalan Pyrenean succession (NE Iberian Peninsula). Sediment. Geol. 363, 48–68.

Mujal, E., Marchetti, L., Schoch, R.R., Fortuny, J., 2020. Upper Paleozoic to lower Mesozoic tetrapod ichnology revisited: photogrammetry and relative depth pattern inferences on functional prevalence of autopodia. Front. Earth Sci. 8, 248.

Müller, J., Reisz, R.R., 2006. The phylogeny of early eureptiles: comparing parsimony and Bayesian approaches in the investigation of a basal fossil clade. Syst. Biol. 55, 503–511.

Müller, A.H., 1954. Zur Ichnologie und Stratonomie des Oberrotliegenden von Tambach (Thüringen). PalZ 28, 189–203.

Müller, A.H., 1959. Die erste Wirbeltierfährte (Paradoxichnium problematicum n.g. n.sp.) aus dem terrestrischen Zechstein von Thüringen. Monatsber. Dtsch. Akad. Wiss. 1, 613–623.

Müller, A.H., 1970. Neue Tetrapodenfährten aus dem terrestrischen Zechstein. Monatsber. Dtsch. Akad. Wiss. 12, 197–207.

Newell, N.D., Rigby, J.K., Driggs, A., Boyd, D.W., Stehli, F.G., 1976. Permian reef complex, Tunisia. Geol. Stud. 23, 75–112.

Nicosia, U., Ronchi, A., Santi, G., 2000. Permian tetrapod footprints fromW Orobic Basin (Northern Italy). Biochronological and evolutionary remarks. Geobios 33, 753–768.

Nicosia, U., Ronchi, A., Santi, G., 2001. Tetrapod footprints from the Lower Permian of western Orobic Basin (N. Italy). Permian continental deposits of Europe and other areas. Regional reports and correlations. "Natura Bresciana" Ann. Mus. Civ. Sc. Nat. 25, 45–50.

Niedermayr, G., Scheriau-Niedermayr, E., 1979. Eine Tetrapodenfährte aus dem Unter-Rotliegend von Kötschach in den westlichen Gailtaler Alpen, Kärnten—Österreich. Ann. Naturhist. Mus. Wien 83, 259–264.

Niedźwiedzki, G., Bojanowski, M., 2012. A supposed eupelycosaur body impression from the Early Permian oft he Intra-Sudetic Basin, Poland. Ichnos 19, 150–155.

Nopcsa, F., 1923. Die Familien der Reptilien. Fortsch. Geol. Paläontol. 2, 1–210.

Nyakatura, J.A., Andrada, E., Curth, S., Fischer, M.S., 2014. Bridging "Romer's Gap": limb mechanics of an extant belly-dragging lizard inform debate on tetrapod locomotion during the early carboniferous. Evol. Biol. 41, 175–190.

Nyakatura, J.A., Melo, K., Horvat, T., Karakasiliotis, K., Allen, V.R., Andikfar, A., Andrada, E., Arnold, P., Lauströer, J., Hutchinson, J.R., Fischer, M.S., Ijspeert, A.J., 2019. Reverse-engineering the locomotion of a stem amniote. Nature 565, 351–355.

Owen, R., 1842. Report on British fossil reptiles: part 2. Rep. Br. Ass. Advmt. Science (for 1841) 60–204.

Pabst, W., 1895. Thierfährten aus dem Rothliegenden von Friedrichroda, Tambach und Kabarz in Thüringen. Z. Dtsch. Geol. Ges. 47, 570–576.

Pabst, W., 1896. Thierfährten aus dem Oberrothliegenden von Tambach in Thüringen. Z. Dtsch. Geol. Ges. 48, 638–643.

Pabst, W., 1897. Die Thierfährten in dem Ober-Rotliegen den von Tambach in Thüringen. Z. Dtsch. Geol. Ges. 49, 701–712.

Pabst, W., 1908. Die Tierfährten in dem Rotliegenden Deutschlands. Nova Acta Leopold. 89, 316–481.

Pardo, J.D., Small, B.J., Milner, A.R., Huttenlocker, A.K., 2019. Carboniferous–Permian climate change constrained early land vertebrate radiations. Nat. Ecol. Evol. 3, 200–206.

Peabody, F.E., 1948. Reptile and amphibian trackways from the lower Triassic Moenkopi formation of Arizona and Utah. Univ. California Dep. Geol. Sci. Bull. 27, 295–468.

Pemberton, S.G., 1829. A short notice of the occurrence of footsteps in the Sandstone of Corn Cockle Muir. Dumfriesshire. Mag. Nat. Hist. J. Zool. Bot. Miner. Geol. Meteor. 1, 144–146.

Petti, F.M., Bernardi, M., Ashley-Ross, M.A., Berra, F., Tessarollo, A., Avanzini, M., 2014. Transition between terrestrial-submerged walking and swimming revealed by Early Permian amphibian trackways and a new proposal for the nomenclature of compound trace fossils. Palaeogeogr. Palaeoclimatol. Palaeoecol. 410, 278–289.

Piñeiro, G., Marchetti, L., Marmol, S., Celio, A., Xavier, P., Francia, M., Schultz, C., 2022. Enigmatic wood and first evidence of tetrapods in the Yaguarí Formation (Middle-Late Permian), Uruguay. Agrociencia Uruguay 26 (NE1), 504.

Pohlig, H., 1885. Saurierfährten in dem Unteren Rotliegenden von Friedrichroda. Verh. Natur. Ver. Preuss. Rheinl. Westf. 42, 285–286.

Pohlig, H., 1892. Altpermische Saurierfährten, Fische und Medusen der Gegend von Friedrichroda in Thüringen. In: Festschrift zum siebenzigsten Geburtstag von Rudolf Leuckardt. Engelmann, Leipzig, pp. 59–64.

Ptaszyński, T., Niedźwiedzki, G., 2004. Late Permian vertebrate tracks from the Tumlin Sandstone, Holy Cross Mountains, Poland. Acta Palaeontol. Polon. 49, 289–320.

Racki, G., 2005. "Late Permian" vertebrate tracks from the Tumlin sandstone of Poland: a commentary on some major implications. Acta Palaeontol. Polon. 50, 394–396.

Reisz, R.R., Liu, J., Li, J.L., Müller, J., 2011. A new captorhinid reptile, *Gansurhinus qingtoushanensis*, gen. et sp. nov., from the Permian of China. Naturwissenschaften 98, 435–441.

Retallack, G.J., 1996. Early Triassic therapsid footprints from the Sydney Basin, Australia. Alcheringa 20, 301–314.

Rmich, A., Lagnaoui, A., Hminna, A., Saber, H., Zouheir, T., Lallensack, J.N., 2023. Captorhinid trackways from mid- to late Permian red beds in Morocco: Implications for locomotion and the palaeobiogeography of northwest Gondwana. Palaeogeogr. Palaeoclimatol. Palaeoecol. 625, 111700.

Robles, S., Llompart, C., 1987. Análisis paleogeográfico y consideraciones paleoicnológicas del Pérmico superior y Triásico inferior en la transversal del río Segre (Alt Urgell, Pirineo de Lérido). Cuadernos Geol. Ibérica 11, 115–130.

Romano, M., Citton, P., Nicosia, U., 2016. Corroborating trackmaker identification through footprint functional analysis: the case study of Ichniotherium and Dimetropus. Lethaia 49, 102–116.

Romer, A.S., Price, L.I., 1940. Review of the Pelycosauria. Geol. Soc. Am. Spec. Pap. 28, 1–538.

Romer, A.S., 1956. The early evolution of land vertebrates. Proc. Am. Phil. Soc. 100, 157–167.

Rühle von Lilienstern, H., 1939. Fährten und Spuren im Chirotherium-Sandstein von Südthüringen. Fortsch. Geol. Paläontol. 12, 293–387.

Rühle von Lilienstern, H., 1944. Eine Dicynodontierfährte aus dem Chirotherium-Sandstein von Heßberg bei Hildburghausen. PalZ 23, 368–385.

Ruta, M., Benton, M.J., 2008. Calibrated diversity, tree tpology and the mother of mass extinctions: the lesson of temnospondyls. Palaeontology 51, 1261–1288.

Ruta, M., Coates, M.I., 2007. Dates, nodes and character conflict: addressing the lissamphibian origin problem. J. Syst. Palaeontol. 5, 69–122.

Ruta, M., Cisneros, J.C., Liebrecht, T., Tsuji, L.A., Mueller, J., 2011. Amniotes through major biological crises: faunal turnover among parareptiles and the end-Permian mass extinction. Palaeontology 54, 1117–1137.

Sacchi, E., Cifelli, R., Citton, P., Nicosia, U., Romano, M., 2014. *Dimetropus osageorum* n. isp. from the Early Permian of Oklahoma (USA): a trace and its trackmaker. Ichnos 21, 175–192.

Santi, G., Krieger, C., 1999. Nuove impronte di tetrapodi permiani della Valle dell'Inferno Lombardia-Italia settentrionale). Osservazioni sistematiche e biocronologiche. Geol. Insubrica 4, 27–33.

Santi, G., Krieger, C., 2001. Lower permian tetrapod footprints from Brembana valley – Orobic basin – (Lombardy, northern Italy). Rev. Paléobiol. 20, 45–68.

Santi, G., Nicosia, U., 2008. The ichnofacies concept in vertebrate ichnology. Studi Trentini sci nat. Acta Geol. 83, 223–229.

Santi, G., Marchetti, L., Schirolli, P., Ronchi, A., 2020. The Cisuralian tetrapod ichnoassociation from Italy: from historical findings to a standard reference status. In: Romano, M., Citton, P. (Eds.), Tetrapod Ichnology in Italy: The State of the Art, Journal of Mediterranean Earth Sciences, vol 12, pp. 39–59.

Santi, G., 2003. Early permian tetrapod footprints from the Orobic Basin (Southern Alps-Northern Italy): data, problems, hypotheses. Boll. Soc. Geol. It. spec. 2, 59–66.

Santi, G., 2007. A short critique of the ichnotaxonomic dualism *Camunipes-Erpetopus*, Lower Permian ichnogenera from Europe and North America. Ichnos 14, 185–191.

Sarjeant, W.A.S., 1971. Vertebrate tracks from the Permian of castle Peak, Texas. Texas J. Sci. 22, 343–366.

Schmidt, H., 1952. Neubeschreibung und Deutung einer Fährtenplatte aus dem Cornberger Sandstein (Perm). PalZ 25, 193–197.

Schmidt, H., 1956. Die große Bochumer Oberkarbon-Fährte. PalZ 30, 199–206.

Schmidt, H., 1959. Die Cornberger Fahrten im Rahmen der Vierfusler-Entwicklung [The Cornberger tracks in the context of quadruped development]. Abh. Hess. Landes. Bodenforsch. 28, 1–137.

Schmidtgen, H., 1928. Eine neue Fährtenplatte aus dem Rotliegenden von Nierstein am Rhein. Palaeobiol. 1, 245–252.

Schneider, J., Schamaev, M.I., Walter, H., 1992. Paläobiogeographie und Stratigraphie von Tetrapoden- und Arthropoden-Fährten aus dem paralischen Oberkarbon und Perm (Gzhel/Assel) des Donez Bassins. Freib. Forsch. C 445, 104–121.

Schneider, J.W., Lucas, S.G., Scholze, F., Voigt, S., Marchetti, L., Opluštil, S., Werneburg, R., Golubev, V., Barrick, J.E., Nemyrovska, T., Ronchi, A., Silantiev, V., Rößler, R., Saber, H., Linnemann, U., Zarinova, V., Shen, S., 2020. Late Paleozoic—early Mesozoic continental biostratigraphy — links to the standard global chronostratigraphic scale. Palaeoworld 29, 186–238.

Seeley, H.G., 1904. Footprints of small fossil reptiles from the Karroo rocks of Cape Colony. Ann. Mag. Nat. Hist. Lond. 14, 287–289.

Seeley, H.G., 1905. On footprints of small fossil reptiles from the Upper Karroo rocks of Cape Colony. Rep. Br. Ass. Advmt. Sci. 74, 549–550.

Silva, R.C., Sedor, F.A., 2017. Mesosaurid swim traces. Front. Ecol. Evol. 5, 22.

Silva, R.C., Sedor, F.A., Fernandes, A.C.S., 2009. Ichnotaxonomy, functional morphology and paleoenvironmental context of Mesosauridae tracks from Permian of Brazil. Rev. Bras. Geociencias 39, 705–716.

Silva, R.C., Sedor, F.A., Fernandes, A.C.S., 2012. Fossil footprints from the Late Permian of Brazil: an example of hidden biodiversity. J. South Am. Earth Sci. 38, 31–43.

Smith, R.M.H., Botha-Brink, J., 2014. Anatomy of a mass extinction: sedimentological and taphonomic evidence for drought-induced die-offs at the Permo-Triassic boundary in the main Karoo Basin, South Africa. Palaeogeogr. Palaeoclimatol. Palaeoecol. 396, 99–118.

Smith, R.M.H., Sidor, C.A., Tabor, N.J., Steyer, J.S., 2015. Sedimentology and vertebrate taphonomy of the Moradi Formation of northern Niger: a Permian wet desert in the tropics of Pangaea. Palaeogeogr. Palaeoclimatol. Palaeoecol. 440, 128–141.

Smith, R.M.H., 1980. The lithology, sedimentology and taphonomy of flood-plain deposits of the Lower Beaufort (Adelaide Subgroup) strata near Beaufort West. South Afr. J. Geol. 83, 399–413.

Smith, R.M.H., 1993. Sedimentology and ichnology of floodplain paleosurfaces in the Beaufort Group (Late Permian), Karoo sequence, South Africa. Palaios 8, 339–357.

Spindler, F., Werneburg, R., Schneider, J.W., 2019. A new mesenosaurine from the lower Permian of Germany and the postcrania of *Mesenosaurus*: implications for early amniote comparative osteology. PalZ 93, 303–344.

Steiner, W., Schneider, H.E., 1963. Eine neue Lauffährte mit Schwanzschleppspur aus dem Ober-Rotliegenden von Tambach. Geologie 11, 715–731.

Stimson, M., Lucas, S.G., Melanson, G., 2012. The smallest known tetrapod footprints: *Batrachichnus salamandroides* from the Carboniferous of Joggins, Nova Scotia, Canada. Ichnos 19, 127–140.

Stimson, M.R., Miller, R.F., Lucas, S.G., 2016. Reassignment of vertebrate ichnotaxa from the upper carboniferous 'fern Ledges', Lancaster formation, Saint John, new Brunswick, Canada. Atl. Geol. 52, 20–34.

Sumida, S.S., 1989. The appendicular skeleton of the Early Permian genus *Labidosaurus* (Reptilia, Captorhinomorpha, Captorhinidae) and the hind limb musculature of captorhinid reptiles. J. Vert. Paleontol. 9, 295–313.

Surkov, M.V., Benton, M.J., Twitchett, R.J., Tverdokhlebov, V.P., Newell, A.J., 2007. First occurrence of footprints of large therapsids from the Upper Permian of European Russia. Palaeontology 50, 641–652.

Tatarinov, L.P., 1960. Otkrytie pseudozhukhii v verkhnei permi SSSR. Paleontol. Zhurnal 4, 74–80.

Tilton, J.L., 1931. Permian vertebrate tracks in West Virginia. Geol. Soc. Am. Bull. 42, 547–556.

Tucker, L., Smith, M.P., 2004. A multivariate taxonomic analysis of the late Carboniferous vertebrate ichnofauna of Alveley, southern Shropshire, England. Palaeontology 47, 679–710.

Tverdokhlebov, V.P., Tverdokhlebova, G.I., Benton, M.J., Storrps, G.W., 1997. First record of footprints of terrestrial vertebrates from the Upper Permian of the Cis-Urals, Russia. Palaeontology 40, 157–166.

Valentini, M., Conti, M.A., Mariotti, N., 2007. Lacertoid footprints of the upper Permian Arenaria di Val Gardena Formation (Northern Italy). Ichnos 14, 193–218.

Valentini, M., Conti, M.A., Nicosia, U., 2008. Linking tetrapod tracks to the biodynamics, paleobiogeography, and paleobiology of their trackmakers: *Pachypes dolomiticus* Leonardi et al., 1975, a case study. Studi Trentini Sci. Nat. Acta Geol. 83, 237–246.

Valentini, M., Nicosia, U., Conti, M.A., 2009. A re-evaluation of *Pachypes*, a pareiasaurian track from the late permian. Neues Jahrb. Geol. Paläontol. Abh. 251, 71–94.

Van Allen, H.E.K., Calder, J.H., Hunt, A.P., 2005. The trackway record of a tetrapod community in a walchian conifer forest from the Permo-Carboniferous of Nova Scotia. Bull. N. M. Mus. Nat. Hist. Sci. 30, 322–332.

Viglietti, P.A., Smith, R.M., Angielczyk, K.D., Kammerer, C.F., Fröbisch, J., Rubidge, B.S., 2016. The *Daptocephalus* Assemblage Zone (Lopingian), South Africa: a proposed biostratigraphy based on a new compilation of stratigraphic ranges. J. Afr. Earth Sci. 113, 153–164.

Voigt, S., Calábková, G., Ploch, I., Nosek, V., Pawlak, W., Raczyński, P., Werneburg, R., 2024. A diadectid skin impression and its implications for the evolutionary origin of epidermal scales. Biol. Lett. 20 (5), 20240041.

Voigt, S., Fischer, J., 2020. GEOSKOP-Forschungsgrabung im südpfälzischen Zechstein. Pollichia-Kurier 36, 53.

Voigt, S., Ganzelewski, M., 2010. Toward the origin of amniotes: diadectomorph and synapsid footprints from the early Late Carboniferous of Germany. Acta Palaeontol. Polon. 55, 57–72.

Voigt, S., Haubold, H., 2000. Analyse zur Variabilität der Tetrapodenfährte Ichniotherium cottae aus dem Tambacher Sandstein (Rotliegend, Unterperm, Thüringen). Hallesches Jahrb. Geowiss. B 22, 17–58.

Voigt, S., Haubold, H., 2015. Permian tetrapod footprints from the Spanish Pyrenees. Palaeogeogr. Palaeoclimatol. Palaeoecol. 417, 112–120.

Voigt, S., Lucas, S.G., 2015a. Permian tetrapod ichnodiversity of the Prehistoric Trackways National Monument (south-central New Mexico, U. S. A.). Bull. N. M.Mus. Nat. Hist. Sci. 65, 153–167.

Voigt, S., Lucas, S.G., 2015b. On a diverse tetrapod ichnofauna from Early Permian red beds in San Miguel County, north-central New Mexico. New Mexico Geol. Soc. Guidebook 66, 241–252.

Voigt, S., Lucas, S.G., 2016. Permian tetrapod footprints from the Lucero uplift, Central New Mexico. New Mexico Geol. Soc. Guidebook 67, 387–395.

Voigt, S., Lucas, S.G., 2017. Early Permian tetrapod footprints from central New Mexico. Bull. N. M. Mus. Nat. Hist. Sci. 77, 333–352.

Voigt, S., Lucas, S.G., 2018. Outline of a Permian tetrapod footprint ichnostratigraphy. Geol. Soc. Lond. Special Public. 450, 387–404.

Voigt, S., Small, B., Sanders, F., 2005. A diverse terrestrial ichnofauna from the Maroon formation (Pennsylvanian–Permian), Colorado: biostratigraphic and paleoecological significance. Bull. N. M. Mus. Nat. Hist. Sci. 30, 342–435.

Voigt, S., Berman, D.S., Henrici, A.C., 2007. First well–established track–trackmaker association of Paleozoic tetrapods based on *Ichniotherium* trackways and diadectid skeletons from the Lower Permian of Germany. J. Vert. Paleontol. 27, 553–570.

Voigt, S., Hminna, A., Saber, H., Schneider, J.W., Klein, H., 2010. Tetrapod footprints from the uppermost level of the Permian Ikakern Formation (Argana basin, western High Atlas, Morocco). J.Afr. Earth Sci. 57, 470–478.

Voigt, S., Lagnaoui, A., Hminna, A., Saber, H., Schneider, J.W., 2011a. Revisional notes on the Permian tetrapod ichnofauna from the Tiddas Basin, central Morocco. Palaeogeogr. Palaeoclimatol. Palaeoecol. 302, 474–483.

Voigt, S., Saber, H., Schneider, J.W., Hmich, D., Hminna, A., 2011b. Late Carboniferous–Early Permian tetrapod ichnofauna from the Khenifra Basin, central Morocco. Geobios 44, 399–407.

Voigt, S., 2012a. Tetrapodenfährten im Rotliegend. In Lützner, H.G. Kowalczyk, G. In: Deutsche Stratigraphische Kommission, Stratigraphie von Deutschland X. Rotliegend. Teil I. Innervariscische Becken. Schriftenreihe Dtsch. Ges. Geowiss., vol 61, pp. 161–175.

Voigt, S., Niedźwiedzki, G., Raczyński, P., Mastalerz, K., Ptaszyński, T., 2012b. Early Permian tetrapod ichnofauna from the Intra-Sudetic Basin, SW Poland. Palaeogeogr. Palaeoclimatol. Palaeoecol. 313–314, 173–180.

Voigt, S., Lucas, S.G., Buchwitz, M., Celeskey, M.D., 2013a. *Robledopus macdonaldi*, a new kind of basal eureptile footprint from the early Permian of New Mexico. Bull. N. M. Mus. Nat. Hist. Sci. 60, 445–459.

Voigt, S., Lucas, S.G., Krainer, K., 2013b. Coastal-plain origin of trace-fossil bearing red beds in the Early Permian of Southern New Mexico, USA. Palaeogeogr. Palaeoclimatol. Palaeoecol. 369, 323–334.

Voigt, S., Klein, H., Bernardi, M., Kutatscher, E., Schneider, J.W., Saber, H., Marchetti, L., ietto, P., Scholze, F., Faath, V., 2015a. *Paradoxichnium problematicum* Müller, 1959 – a principal moprh of Late Permian tetrapod tracks and ist palaeoecological significance. In: 13th Annual Meeting Oft He European Association of Vertebrate Palaeontologists. Opole, Poland, 8-12 July 2015, p. 44. Abstracts.

Voigt, S., Marchetti, L., Schönlaub, H.P., Krawanja-Ortner, G., 2015b. Pre-dinosaur tetrapod footprints as a new asset of the Geopark Carnic Alps. Europ. Geoparks Mag. 12, 18.

Voigt, S., Fischer, J., Spindler, F., Marchetti, L., Klein, H., Glienke, S., Schindler, T., 2023. An exceptionally diverse tetrapod track assemblage from the late Permian of SW Germany. Abstract Jahr. Paläontol. Ges 94, 53–54.

Voigt, S., 2005. Die Tetrapodenichnofauna des kontinentalen Oberkarbon und Perm im Thüringer Wald – Ichnotaxonomie. Paläoökologie und Biostratigraphie. Cuvillier, Göttingen, 305 pp.

Voigt, S., 2007. Auf den Spuren der Saurier: Die Tetrapodenfährten aus dem Oberkarbon und Unterperm des Saar-Nahe-Beckens. In: Schindler, T., Heidtke, U.H.J. (Eds.), Kohlesümpfe, Seen und Halbwüsten. Dokumente einer rund 300 Millionen Jahre alten Lebewelt zwischen Saarbrücken und Mainz, Polichia Sonderveröffentlichung, vol 10, pp. 288–303.

Voigt, S., 2015. Der Holotypus von *Amphisauropus latus* Haubold, 1970 – ein besonderes Objekt permischer Tetrapodenfährten im Naturhistorischen Museum Schloss Berholdsburg Schleusingen. Semana 30, 39–49.

Walter, H., Werneburg, R., 1988. Über Liegespuren (Cubichnia) aquatischer Tetrapoden (?Diplocauliden, Nectridae) aus den Rotteröder Scichten (Rotliegendes, Thüringer Wald/DDR). Freib. Forsch. C. 419, 96–106.

Warren, A., 1997. A tetrapod fauna from the Permian of the Sydney Basin. Rec. Australian Mus 49, 25–33.

Watson, D.M.S., 1960. The anomodont skeleton. Trans. Zool. Soc. Lond. 29, 131–209.

Williston, S.W., 1908. Salamander-like footprints from the Texas Red Beds. Biol. Bull. 15, 237–239.

Witzmann, F., Schoch, R.R., 2018. Skull and postcranium of the bystrowianid *Bystrowiella schumanni* from the Middle Triassic of Germany, and the position of chroniosuchians within Tetrapoda. J. Syst. Palaeontol. 16, 711–739.

Woodworth, J.B., 1900. Vertebrate footprints on Carboniferous shales of Plainville, Massachusetts. Geol. Soc. Am. Bull. 11, 449–454.

Zouicha, A., Voigt, S., Saber, H., Marchetti, L., Hminna, A., El Attari, A., Ronchi, A., Schneider, J.W., 2021. First record of Permian continental trace fossils in the Jebilet massif, Morocco. J. Afr. Earth Sci. 173, 104015.

Further reading

Brocklehurst, N., 2021. The first age of reptiles? Comparing reptile and synapsid diversity, and the influence of Lagerstätten, during the Carboniferous and Early Permian. Front. Ecol. Evol. 9, 669765.

Durand, M., Gand, G., Châteauneuf, J.-J., 2002. Les bassins permiens de Provence. In: Livre-guide de la 16ème excursion annuelle de l'Association des Géologues du Permian, 9–11 mai 2002.

Falcon-Lang, H.J., Benton, M.J., Stimson, M., 2007. Ecology of earliest reptiles inferred from basal Pennsylvanian trackways. J. Geol. Soc. Lond. 164, 1113–1118.

Falcon-Lang, H.J., Gibling, M.R., Benton, M.J., Miller, R.F., Bashforth, A.R., 2010. Diverse tetrapod trackways in the Lower Pennsylvanian Tynemouth Creek Formation, near St. Martins, southern New Brunswick, Canada. Palaeogeogr. Palaeoclimatol. Palaeoecol. 296, 1–13.

Tüysüz, O., 2022. Geology of the Kurucaşile-Cide region, NW Türkiye. Bull. Miner. Res. Explo. 167, 149–178.

CHAPTER

5

Triassic tetrapod footprints

Hendrik Klein[1] *and Spencer G. Lucas*[2]

[1]Saurierwelt Paläontologisches Museum, Neumarkt, Germany; [2]New Mexico Museum of Natural History and Science, Albuquerque, NM, United States

1. Introduction

Fig. 5.1.

The Triassic footprint record is very diverse and presently comprises 35 ichnogenera and 67 ichnospecies considered here as valid. Of these, nine ichnogenera with nine ichnospecies were originally introduced based on non-Triassic material, six ichnogenera and ichnospecies on Jurassic-Cretaceous material, and three ichnogenera and ichnospecies on Permian material. In particular, dominant and abundant in the Triassic are the footprints of archosaurs, a group that underwent a significant rise and diversification after the Permian-Triassic extinction event (PTE), increasingly from the Early Triassic Olenekian. Chirotheres are characteristic Triassic archosaur footprints and present in nearly all continental deposits. From the Early Triassic, two ichnogenera, *Protochirotherium* and *Synaptichnium,* are known with very few ichnospecies that can be considered valid, whereas the highest ichnological diversity is known from the Middle Triassic with numerous ichnospecies of *Chirotherium, Isochirotherium,* and *Synaptichnium*. In the Late Triassic, only a few chirotheriid taxa are present, essentially *Brachychirotherium* and the ichnogenerically uncertain "*Chirotherium*" *lulli,* which can be considered the "last chirotheres." Non-chirotheriid archosaur tracks are known by small *Rotodactylus* ranging from the Early Triassic (Olenekian) to the end of the Middle Triassic (Anisian-early Ladinian) and by early indicators of the dinosaurian lineage with typical tridactyl mesaxonic forms appearing in the late Anisian-Ladinian and dominating many Late Triassic assemblages with the ichnogenera *Atreipus* and *Grallator*, that can be attributed to dinosauriform–dinosaurian trackmakers. Other archosaur and nonarchosaur tetrapod groups left their earliest footprints with *Apatopus* (phytosaur), *Evazoum* and *Eosauropus* (sauropodomorphs), and *Chelonipus* (stem-turtle). Survivors from the Permian are *Rhynchosauroides* (archosauromorph/lepidosauromorph), *Dicynodontipus*, *Capitosauroides* (cynodont therapsids), *Procolophonichnium* (parareptiles), and *Batrachichnus* (temnospondyl anamniotes). Based on the footprint record, we see a turnover from synapsid, parareptile, and basal eureptile communities of the Permian to those dominated by archosaurs in the Triassic, but

Vertebrate Ichnology
https://doi.org/10.1016/B978-0-443-13837-9.00013-5

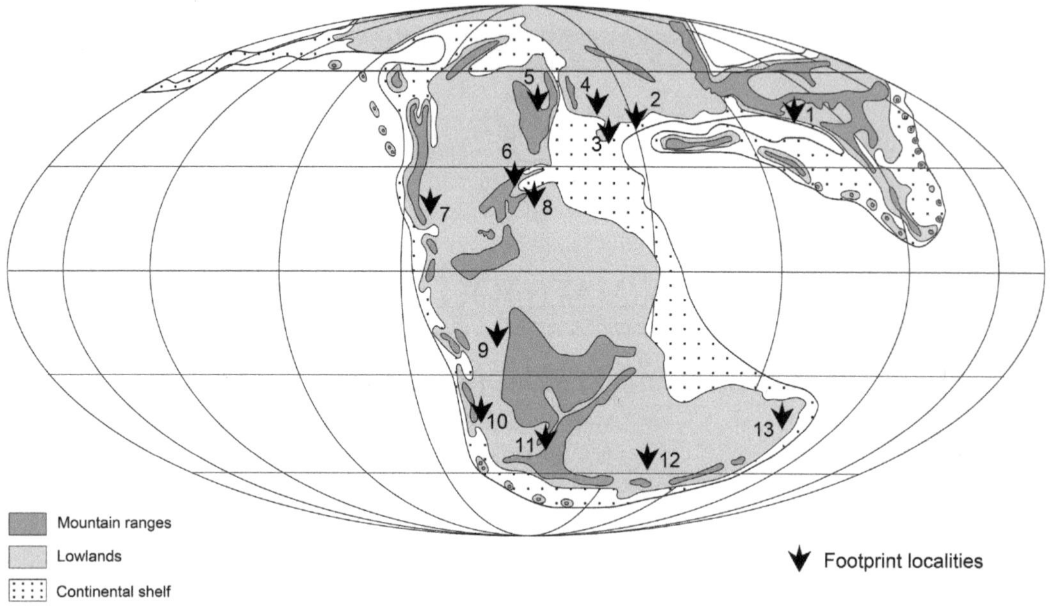

FIGURE 5.1 Paleogeography of the Triassic with position of tetrapod footprint localities. 1 = China; 2 = Italy; 3 = Spain; 4 = western Europe; 5 = Greenland; 6 = Newark Basin, eastern North America; 7 = Chinle Basin, western North America; 8 = Argana Basin, Morocco, northern Africa; 9 = Brazil; 10 = Argentina; 11 = southern Africa; 12 = Antarctica; 13 = Sydney Basin, Australia. *From Klein and Lucas (2021). Base map after Wing and Sues (1992). Map lines delineate study areas and do not necessarily depict accepted national boundaries.*

this is not a sudden change at the Permo-Triassic boundary, but instead spread out temporally from the Late Permian into the Middle Triassic. Footprint assemblages from the Triassic are a key to understanding the (1) recovery of terrestrial ecosystems after the Permian-Triassic extinction and (2) the early evolution of the crocodilian and dinosaur stem.

Here, we present a review of 200 years of Triassic footprint research and updated ichnotaxonomy. In our systematic part, we have essentially considered ichnogeneric levels and type ichnospecies as well as some common, widely accepted other ichnospecies. A more complete list of ichnotaxa including older names and synonyms has been listed together with authors in Table 5.1. In recent years Triassic footprints became increasingly important for biostratigraphy, paleobiogeography, ichnofacies and for understanding of archosaur and other tetrapod evolution. We therefore outline and discuss these issues in more detail in separate subchapters.

2. History of research

The first scientific description of Triassic tetrapod footprints is found in the work of Kaup (1835a,b) on the famous Buntsandstein assemblage with chirothere footprints from Hildburghausen, Germany. Furthermore, *Chirotherium barthii* and *C. sickleri* from this locality are the first binominally named tetrapod footprints. A note on this discovery was, however, given

TABLE 5.1 Tetrapod ichnogenera and ichnospecies of the Triassic. Left column: Ichnotaxa in bold are considered here as valid, others are either *nomina dubia* based on insufficiently preserved material or of uncertain ichnogeneric (quotation marks) and/or ichnospecific affinity. Middle column: Older names and synonyms corresponding to bold ichnotaxa of left column. Right Column: Authors and year of introduction referring to ichnotaxa listed opposite in left or middle column.

Ichnogenera and Ichnospecies	Older names/ synonyms	Authors
Chirotheriids		
		Beurlen, 1950
Brachychirotherium	Brachychirotherian	Baird, 1957
	Group	
		Beurlen, 1950
hassfurtense TI		(C.H. Hitchcock, 1889)
parvum	*eyermani*	Baird, 1957
	copei	Bock, 1952
		(Rühle v. Lilienstern,
thuringiacum		1938)
	coburgense	Aumann, 1960
		Kaup, 1835
Chirotherium	*Cheirotherion*	Sickler, 1836
	Cheirotherium	Sickler, 1836
	Chirosaurus	Kaup, 1835a
	Krokodilipus	Nopcsa, 1923
	Parachirotherium	Kuhn, 1958a
	Rigalites	von Huene, 1931
	Saurichnites	Kirchner, 1927
	Sphingopus	Demathieu, 1966
		Kaup, 1835
barthii TI	*Chirosaurus ibericus*	Navas, 1906
	Chiros. storetonensis	Lydekker, 1890
	Cheirotherium kaupii	Owen, 1842
	Saurichnites	Kirchner, 1927
	auraensis	Schuster, 1936
	S. gambachensis	
		Huene, 1941
	angustum	Reig, 1961
	bairdi	Rusconi, 1951
	higuerensis	Navas, 1906
	ibericum	Sickler, 1836
	majus	Demathieu and
	mediterraneum	Durand, 1991
		Morton, 1863
	storetonense	Kirchner, 1927
	vorbachi	Peabody, 1948
rex		Peabody, 1948
	moquinense	Kaup, 1835
sickleri		Nopcsa, 1923
	beasleyi	Berthold, 1835
	bipes	Willruth, 1917
	bornemanni	Sickler, 1836
	minus	Soergel, 1925
	pfeifferi	(Demathieu, 1966)

(*Continued*)

TABLE 5.1 Tetrapod ichnogenera and ichnospecies of the Triassic. Left column: Ichnotaxa in bold are considered here as valid, others are either *nomina dubia* based on insufficiently preserved material or of uncertain ichnogeneric (quotation marks) and/or ichnospecific affinity. Middle column: Older names and synonyms corresponding to bold ichnotaxa of left column. Right Column: Authors and year of introduction referring to ichnotaxa listed opposite in left or middle column.—cont'd

ferox		(von Huene, 1931)
ischigualastianum		(Avanzini & Wachtler,
ladinicum		2012)
		(Rehnelt, 1950)
postchirotherioides		
		Haubold, 1971b
Isochirotherium		
		(Peabody, 1948)
coltoni		(Demathieu, 1970)
coureli		Gand 1979
	circademathieui	Gand, 1979
	combelei	(Haubold, 1971a)
	demathieui	(Courel & Demathieu,
delicatum		1976)
		Courel & Demathieu,
felenci		1976
		(Egerton, 1838)
herculis		Abel, 1935
	?bipedale	Avanzini & Leonardi,
infernii		2002
		(Baird, 1954)
lomasi		(Peabody, 1948)
marshalli		(Haubold, 1967)
soergeli TI		(Haubold, 1971a)
	hessbergense	(Haubold, 1971a)
	jenense	
		Fichter & Kunz, 2004
Protochirotherium		(Fuglewicz et al., 1990)
hauboldi		Fuglewicz et al., 1990
	"Chirotherium",	Ptaszyński, 2000
	"Brachychirotherium	Ptaszyński, 2000
	","Isochirotherium"	Fuglewicz et al., 1990
	"Synaptichnium"	Ptaszyński, 2000
	B.wiorense	Ptaszyński, 2000
	I. gierlinskii	Fuglewicz et al., 1990
	I. sanctacrucense	Fuglewicz et al., 1990
	S.chirotherioides	
		Fichter & Kunz, 2004
wolfhagense TI		
		(Demathieu & Haubold,
	I. archaeum	1982)
		Nopcsa, 1923
Synaptichnium		(Peabody, 1948)
cameronense		(Peabody, 1948)
diabloense		(Rühle v. Lilienstern,
hildburghausense		1939)
		Ptaszyński, 2000
kotanskii		Demathieu & Haubold,
primum		1982
		Nopcsa, 1923
pseudosuchoides TI		Demathieu, 1970

TABLE 5.1 Tetrapod ichnogenera and ichnospecies of the Triassic. Left column: Ichnotaxa in bold are considered here as valid, others are either *nomina dubia* based on insufficiently preserved material or of uncertain ichnogeneric (quotation marks) and/or ichnospecific affinity. Middle column: Older names and synonyms corresponding to bold ichnotaxa of left column. Right Column: Authors and year of introduction referring to ichnotaxa listed opposite in left or middle column.—cont'd

	priscum	Demathieu & Gand, 1981
	argantobrivense	
Chirotheriid ichnospecies with uncertain ichnogeneric affinities		
		Beurlen, 1950
"Brachychirotherium"		Demathieu, 1971
circaparvum		Willruth, 1917
gallicum		(Haubold, 1967)
harrasense		(Soergel, 1925)
hessei		Demathieu & Leitz, 1982
	paeneparvum	
	paraparvum	Demathieu & Oosterink, 1988
kuhni		Demathieu & Haubold, 1982
lorteti		(Haubold, 1971a)
pachydactylum		Demathieu & Gand, 1973
praeparvum		(Haubold, 1967)
tintanti		Demathieu, 1971
"Chirotherium"		Kaup, 1835
atlensis		Biron & Dutuit, 1981
lulli		Bock, 1952
wondrai		Heller, 1952
Nomina dubia		
C. catalaunicum		Casanovas-Cladellas et al., 1979
C. culmbachense		Soergel, 1925
C. fucinii		Huene, 1941
C. gigas		Pohlig, 1893
C. huberi		Bock, 1952
?Otozoum swinnertoni		Sarjeant, 1970
Parasynaptichnium gracilis		Mietto, 1987
Ichnotaxa falsely assigned to chirotheriids		
"Chirotherium" geinitzi		Hornstein, 1876
"C." heterodactylum		King, 1845
"C." pabsti		Nopcsa, 1923
"C." peabodyi		Faber, 1958
"C." reiteri		Moore, 1873
"C." rubrum		Nopcsa, 1923
Prochirotherium permicum		Leonardi, 1951

(*Continued*)

TABLE 5.1 Tetrapod ichnogenera and ichnospecies of the Triassic. Left column: Ichnotaxa in bold are considered here as valid, others are either *nomina dubia* based on insufficiently preserved material or of uncertain ichnogeneric (quotation marks) and/or ichnospecific affinity. Middle column: Older names and synonyms corresponding to bold ichnotaxa of left column. Right Column: Authors and year of introduction referring to ichnotaxa listed opposite in left or middle column.—cont'd

Other		
Anchisauripus		Lull 1904
sillimani TI		E. Hitchcock 1843
Anshunpes		Xing et al. 2020
aquacursor TI		Xing et al. 2020
Apatopus		Baird 1957
lineatus TI		(Bock 1952)
Atreipus		Olsen and Baird 1986
milfordensis		(Bock 1952)
TI		(Heller 1952)
metzneri		(Baird 1957)
sulcatus		Olsen and Baird 1986
acadianus		Fraser and Olsen 1996
Banisterobates		Fraser and Olsen 1996
boisseaui TI		Woodworth 1900
Batrachichnus		(Geinitz 1861)
salamandroid		E. Hitchcock 1845
es		(Hitchcock 1843)
TI		Leonardi 1981
Batrachopus		Leonardi 1981
deweyi TI		Haubold 1971a
Brasilichnium		Haubold 1971a
elusivum TI		
Capitosauroides		Whyte and Romano
bernburgensis		2001
TI		R. v. Lilienstern 1939
Characichnos		R. v. Lilienstern 1939
tridactylus TI	*plieningeri*	Haubold 1971a
Chelonipus		Zhang et al. 2014
torquatus TI		Zhang et al. 2014
		R. v. Lilienstern 1944
Dikoposichnus		(Hornstein 1876)
luopingensis TI		Marchetti et al. 2019a
Dicynodontipus		(Ceoloni et al. 1988)
geinitzi TI		Lockley et al. 2018
Dolomitipes		Lockley et al. 2018
accordii TI		Lockley et al. 2006a
Eoanomoepus		Lockley et al. 2006a
latus TI		
Eosauropus		E. Hitchcock 1845
cimarronensis		E. Hitchcock 1845
TI		Nicosia and Loi 2003
Eubrontes		Nicosia and Loi 2003
giganteus TI		Lockley and Lucas 2013
Evazoum		E. Hitchcock 1858
sirigui TI		E. Hitchcock 1858
gatewayensis		Bock 1952
Grallator		Bock 1952
cursorius TI		Bock 1952
Gwyneddichnium	*minore*	Bock 1952
majore TI	*elongatum*	Yang and Yang 1987

TABLE 5.1 Tetrapod ichnogenera and ichnospecies of the Triassic. Left column: Ichnotaxa in bold are considered here as valid, others are either *nomina dubia* based on insufficiently preserved material or of uncertain ichnogeneric (quotation marks) and/or ichnospecific affinity. Middle column: Older names and synonyms corresponding to bold ichnotaxa of left column. Right Column: Authors and year of introduction referring to ichnotaxa listed opposite in left or middle column.—cont'd

		Yang and Yang 1987
		Ellenberger 1972
Pengxianpus		Ellenberger 1972
cifengensis TI		Lagnaoui et al. 2019
Pentasauropus		Nopcsa 1923
incredibilis TI		Kuhn 1963
argentinae		Ptaszyński 2000
Procolophonichnium		Ptaszyński 2000
nopcsai TI		Ellenberger 1972
Prorotodactylus		Ellenberger 1972
mirus TI		Ellenberger 1970
Pseudotetrasauropus	*lehmani*	(Gand et al. 2000)
bipedoida TI		Maidwell 1911
		Maidwell 1911
grandcomb.		(R. v. Lilienstern 1939)
Rhynchosauroides		Baird 1957
rectipes TI		Abel 1926
schochardti		Haubold 1966
hyperbates		Haubold 1966
tirolicus		Demathieu 1966
bornemanni		(Ryan and Willard
pusillus	*brunswickii*	1947)
petri	*moenkopiensis*	Haubold 1971a
	palmatus	(Lull 1942)
	majus	Demathieu 1970
	peabodyi	(Faber 1958)
	articeps	(Owen 1842)
		Peabody 1948
		Peabody 1948
Rotodactylus		Peabody 1948
cursorius TI		Haubold 1967
bradyi		Ellenberger 1972
matthesi		Ellenberger 1972
Tetrasauropus		Hunt et al. 1993b
unguiferus TI		Hunt et al. 1993b
Therapsipus		Ellenberger 1970
cumminsi TI		Ellenberger 1970
Trisauropodiscus		
aviforma TI		
Ichnospecies with uncertain ichnogeneric affinities		
		(Holst et al. 1970)
"*Procolophonichnium*"		(Gand 1977)
haarmuehlensis		Klein et al. 2015b
nectouxi		
lockleyi		

Bold = Ichnotaxa considered here as valid; TI = Type Ichnospecies

TABLE 5.2 Tetrapod ichnofacies and ichnocoenoses.

Archetypal tetrapod ichnofacies	Predominant trace fossil types	Constituent Triassic ichnocoenoses	Inferred environment
Chelichnus	Low diversity ichnofaunas (less than 4 ichnogenera) of tetrapod tracks that have equant shape with subequal manual and pedal impressions and short digit impressions	*Brasilichnium* (Late Triassic-Early Jurassic; Lockley et al., 1994)	Eolian crossbeds
Batrachichnus	Majority of tracks are of quadrupedal carnivores; medium-high diversity (4–8 ichnogenera)	*Chirotherium* (Early-Middle Triassic; Hunt and Lucas, 2007b); *Apatopus* (Late Triassic: Hunt and Lucas, 2006)	Tidal flat-fluvial plain
Brontopodus	Majority of tracks are terrestrial herbivores with small quantity (generally >10%) of terrestrial carnivore tracks; medium-high diversity (4–8 ichnogenera)	*Dicynodontipus* (Early Triassic: Hunt and Lucas, 2006); *Therapsipus* (Middle Triassic: Hunt and Lucas, 2006); *Procolophonichnium* (Middle Triassic: herein) *Brachychirotherium* (Late Triassic: Hunt and Lucas, 2006); *Evazoum* (Late Triassic: Hunt and Lucas, 2006)	Coastal plain, clastic or carbonate marine shoreline
Grallator	Medium-high diversity ichnofaunas (5–8 ichnogenera) with tracks (usually dominant) of tridactyl avian and nonavian theropods	*Grallator* (Late Triassic: Hunt and Lucas, 2007a,b)	Lacustrine margin
Characichnos	Parallel scratch marks and fish swimming trails (*Undichna*)	Unnamed	Shallow lacustrine/aquatic

From Hunt and Lucas (2007c).

earlier by Sickler (1834; see also 1835, 1836). Further pioneering studies of the Hildburghausen material and other Triassic material have been presented by Bornemann (1887), Walther (1917), and Willruth (1917). Soon after the German discoveries similar Triassic footprint assemblages with chirotheres were documented from Great Britain (Egerton, 1838; Morton, 1863, 1897; Woodward, 1902; Maidwell, 1911, 1914; Nopcsa 1923) and finally by Beasley, who proposed a classification of Triassic tetrapod footprints (Beasley, 1895, 1901, 1904, 1905, 1907, 1908, 1909). Ichnotaxonomic work on German Buntsandstein assemblages was also presented by Kirchner (1927) and Rühle v. Lilienstern (1939). Soergel (1925) published a detailed study of chirotheres and potential producers.

Beginning from the mid-20th century the work of Frank E. Peabody on footprints from the Moenkopi Formation of northern Arizona and Utah can be considered pioneering and has been influential on the ichnological community and the ichnotaxonomy of Triassic tetrapod footprints up to the present day (Peabody, 1948). Important studies were published soon after by Baird (1954, 1957) based on material from the Upper Triassic strata of the Newark Supergroup of the eastern USA. Modern views of Triassic footprint assemblages and an updated ichnotaxonomy were given by Haubold (1967, 1971a,b), Demathieu (1970, 1971, 1985, 1989), Demathieu and Haubold (1972, 1974) and Demathieu and Demathieu (2004), the core of their work being the assemblages from the Germanic Basin. Important documentational work on Triassic-Jurassic footprint assemblages from southern Africa, in particular from Lesotho, was done by Ellenberger (1970, 1972, 1974). This author introduced a large number of new ichnotaxa, many of them considered now as *nomina dubia* or synonyms of other ichnotaxa known from North America or Europe (Klein and Lucas, 2021). The main reason for the extreme ichnotaxonomic oversplitting of the southern African footprints was the presence of substrate-related extramorphological variations, a phenomenon not adequately considered and discussed by Ellenberger. Further ichnotaxonomic work was published by Beurlen (1950), Kuhn (1958a,b, 1963) and by Sarjeant (1967, 1970, 1996) based on central and western European material. From a global perspective, Triassic deposits from different countries and regions yielded extensive material largely increasing our knowledge of Triassic tetrapod communities based on studies of the footprint record from:

(1) North America (Olsen and Baird, 1986; Weems, 1987, 1992; Fraser and Olsen, 1996; Klein et al., 2006; Hunt and Lucas, 2007a; Lucas et al., 2010; Klein and Lucas, 2010b; Lockley and Lucas, 2013; Lockley et al., 2021; Milner et al., 2021; Klein et al., 2024), **(2) Greenland** (Klein et al., 2015a; Lallensack et al., 2017), **(3) South America** (Melchor and De Valais, 2006; Apesteguía et al., 2021), **(4) Europe,** Italy (Mietto, 1987; Leonardi and Lockley, 1995; Avanzini and Leonardi, 2002; Nicosia and Loi, 2003; Avanzini and Wachtler, 2012), Poland (Fuglewicz et al., 1990; Ptaszyński, 2000; Niedźwiedzki and Ptaszynski, 2007; Klein and Niedźwiedzki, 2012), Germany (Karl and Haubold, 1998; Haubold and Klein, 2000, 2002; Klein and Haubold, 2003, 2004; Fichter and Kunz, 2004, 2011; Klein and Lucas, 2018; Marchetti et al., 2020b), France (Courel and Demathieu, 2000; Demathieu and Demathieu, 2004; Gand and Demathieu, 2005; Gand et al., 2000, 2005), Switzerland (Cavin et al., 2012; Klein et al., 2016), and Great Britain (King et al., 2005; Clark and Corrance, 2009), Spain (Pérez-López, 1993; Gand et al., 2010; Mujal et al., 2017, 2018; Navarro and Moratalla, 2018; De Jaime Soguero et al., 2021; see Díaz-Martínez and Pérez-García, 2012; Díaz-Martínez et al., 2015; Klein and Lucas, 2021 for overview and literature therein); **(5) Southern Africa** (Olsen and Galton, 1984; D'Orazi Porchetti and Nicosia, 2007; Marchetti et al., 2019b, 2020; Abrahams et al., 2022), **(6) North Africa** (Klein et al., 2010, 2011; Lagnaoui et al., 2012, 2016; Zouheir et al., 2020; Moreau et al., 2024), and **(7) China** (Lockley et al., 2013; Xing et al., 2013a; Xing and Klein, 2019). Reviews of the ichnotaxonomy of global ichnotaxa have been given by Lockley et al. (2006a,b), Klein and Lucas (2013, 2021), Klein et al. (2015b), and Lucas et al. (2014). Studies on ichnofacies and biostratigraphy have been published by Lucas (2005, 2007), Hunt and Lucas (2007c), Klein and Haubold (2007), and Klein and Lucas (2010a, 2021).

In recent years numerous studies on the influence of the substrate on footprint shape and the extramorphological variation of footprints led to comprehensive ichnotaxonomic revisions of Triassic footprints (Milàn, 2006; Milàn and Bromley, 2006, 2008; Falkingham and Gatesy, 2014; Milàn and Falkingham, 2016; Marchetti et al., 2019c). Some authors proposed strict anatomically- (synapomorphy-) based approaches (e.g., Olsen and Baird, 1986; Haubold and Klein, 2000, 2002; Farlow et al., 2014; Lallensack et al., 2017). Some workers use statistical methods (e.g., Gand and Demathieu, 2005; Gand et al., 2005) or morphometric and landmark analyses for discrimination of footprints (Rasskin-Gutman et al., 1997; Karl and Haubold, 1998; Klein and Haubold, 2003; Lallensack et al., 2016). A large step toward more objective documentation and ichnotaxonomic evaluation was initiated by the application of 3D visualization by photogrammetry and laser scan methods (Manning, 2004; Petti et al., 2008; Matthews et al., 2016; Falkingham, 2012, 2016; Falkingham and Gatesy, 2014; Falkingham et al., 2018; Belvedere et al., 2017). Further methodological and terminological studies and discussions were presented by different authors (Klein and Haubold, 2004; Klein and Lucas, 2018, 2021; Belvedere and Farlow, 2016; Gatesy and Falkingham, 2017; Marchetti et al., 2019c, 2020b; Falkingham et al., 2020; Szewczyk et al., 2020; Turner et al., 2020).

3. Footprint preservation

Tetrapod footprints are the result of the interaction between the animal's feet and the substrate (Gatesy and Falkingham, 2017; Marchetti et al., 2019c). This means that the shape of imprints is not only a mere reflection of the foot anatomy, but also influenced by the dynamics of the foot and the substrate conditions such as grain size, water content, and plasticity of the sediment the animal walked on. When the foot is sinking deeply into the mud other effects can occur such as the registration of undertracks transmitted to a lower layer or penetrative tracks formed while cutting through different substrate layers accompanied by mud collapse. Often (not always) this causes a significant loss of anatomical details, for example, penetrative tracks mostly show very narrow digit traces that might lead to incorrect conclusions about the true foot anatomy and moreover lead to incorrect ichnotaxonomic assignments (Falkingham et al., 2020; Turner et al., 2020; Lallensack et al., 2022). Oversplit ichnotaxonomy and "endless" synonymy lists in subsequent revisions are often the results of substrate effects having been overlooked or simply ignored. In recent years, Belvedere and Farlow (2016) and Marchetti et al. (2019c) proposed different scales to define the quality of morphological preservation of tetrapod footprints.

The Triassic record has yielded numerous examples of ichnotaxonomic errors and footprints being introduced as new ichnotaxa that later turned out to be extramorphological variants and synonyms of formerly known ichnogenera or ichnospecies. Today, extensive materials from the global record for comparison and abundant specimens preserved in high-quality allow a more stable ichnotaxonomy, which, at least at the ichnogenus level, found a wide acceptance (Klein and Lucas, 2021). Also, numerous ichnospecies have such a characteristic imprint morphology, for example, some chirotheriids, such as the classical *Chirotherium barthii* and *C. sickleri*, that they can be easily identified, even from isolated imprints if being well-preserved. Moreover, the type surfaces with these ichnotaxa are considered a kind of reference for the understanding of Triassic archosaur footprints as a whole. Long trackways along the extensive type surfaces allow a more exact definition and diagnosis of the ichnospecies as well as the documentation of extramorphological variation (Haubold, 2006).

4. Ichnotaxa

Table 5.1.

4.1 Amphibian

4.1.1 Batrachichnus Woodworth (1900)

Fig. 5.2A.

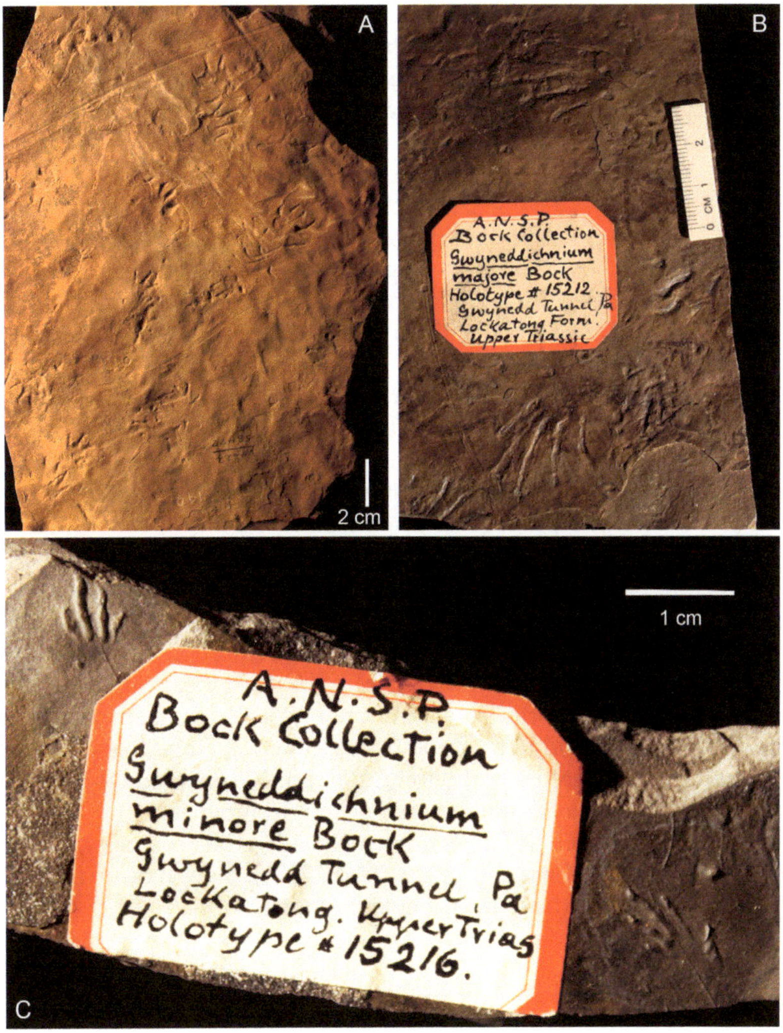

FIGURE 5.2 Rare tetrapod footprints from Triassic deposits. (A) *Batrachichnus* isp. (Temnospondyl) from Moenkopi Group of Colorado. (B–C). *Gwyneddichnium majore* (archosauromorph) from Lockatong Formation (Upper Triassic, Newark Supergroup) of Pennsylvania. (B) Holotype. (C) Holotype of G. *minore*, which is a synonym of *G. majore. (A) From Klein and Lucas (2010b). (C) From Lucas et al. (2014).*

Type ichnospecies: *Batrachichnus salamandroides* (Geinitz, 1861).

Other ichnospecies: Only the type ichnospecies is considered here as valid.

Distribution: Lower Carboniferous (Mississipian)–?Early or ?Middle Triassic. From the Triassic, only a single occurrence is known and comes from the Moenkopi Group of Colorado, USA (Klein and Lucas, 2010b).

Diagnosis (emended by Fillmore et al., 2012): Small (up to 20 mm long) footprints of a quadruped with four and five digits on manus and pes, respectively. Pes plantigrade to semiplantigrade and length less than 20 mm long (typically 10–15 mm). Pes digits I to III closely grouped together with increasing length, Digit IV longest and somewhat separate from the others; Digit V set somewhat posterior and laterally directed. Manus semiplantigrade, smaller than pes; increasing lengths of digits from I to III, Digit IV diverging outward from the grouping of I–III. Narrow quadrupedal trackways, common alternating manus-pes sets, small divergence of manus and pes axes, pace angulation rarely exceeds 90 degrees. Tail/body drag may be present (based primarily on Gilmore, 1928; Haubold 1971b, 1996; Melchor and Sarjeant, 2004).

Description: A single slab from the Moenkopi Group of Colorado showing a short trackway segment consisting of 3–5? pes and manus imprints as well as isolated imprints and scratch marks, represents the only Mesozoic specimen of *Batrachichnus*. The faintly impressed pes imprint is about 1.6 cm in length, pentadactyl and digitigrade with straight and slender digits that increase in length from I–IV; Digit IV is subequal in length with III, Digit V subequal with I and laterally spread. The smaller manus is 1.1 cm long and tetradactyl, digits either representing II–V or I–IV; the central digits are longest and subequal in length whereas outer digits are short. Ends of digits in both pes and manus are often bifurcated or knob-like. The trackway is broad with a pace angulation of 90–105 degrees and a stride length of 8.3 cm; the pes imprint is oriented parallel to the trackway midline and the manus imprint inwardly rotated. Associated may be thin traces of the tail.

Discussion: The overall morphology of the imprints, digit proportions, and the continuously tetradactyl, symmetrical manus trace suggest the presence of an amphibian trackway that can be assigned to the ichnogenus *Batrachichnus*. This is also supported by the presence of a tail trace that may be common in this ichnotaxon and other ones such as *Rhynchosauroides*. *Batrachichnus* is known from Carboniferous–Permian deposits in a wide distribution and had been introduced by Woodworth (1900). The first Mesozoic occurrence from the Moenkopi Group of North America is a rare ichnological document of amphibians from this unit that otherwise has abundant reptile tracks, whereas amphibians dominate the skeletal fauna. This can be considered a taphonomical effect due to different habitats and preservation potential in river or lake shore and channel environments.

Trackmaker: Small temnospondyls have been considered as the most likely trackmakers (Haubold, 1971b; Voigt, 2005; Klein and Lucas 2010b, 2021), though during the Carboniferous some *Batrachichnus* tracks may have been made by "microsaurs."

4.2 Lepidosauromorph–archosauromorph

4.2.1 Gwyneddichnium Bock (1952)

Fig. 5.2B and C.

Type ichnospecies: *G. majore* Bock (1952).

Other ichnospecies: Only known from the type ichnospecies.

Distribution: Eschenbach Formation (Middle Triassic) of Germany, Upper Triassic of Newark Supergroup, Pennsylvania and New Jersey, USA, Chinle Group of New Mexico, Colorado and Utah, USA.

Diagnosis (emended by Lucas et al., 2014): Footprints of a quadruped in which the manus and pes are pentadactyl and mesaxonic, the pes is substantially larger than the manus, digits are thin, long, straight to slightly curved, have nodular phalangeal pad impressions and terminate in claws. Differs from the most similar, lacertoid ichnogenus *Rhynchosauroides* in the digit proportions, with Digits III > II ≥ IV > I in *Gwyneddichnium*, whereas in *Rhynchosauroides* the proportions are IV > III > II > I. Also, the digits in *Rhynchosauroides* are thicker and display a greater curvature, whereas the digits of *Gwyneddichnium* are nearly straight or only slightly curved. Digit V in *Rhynchosauroides* is strongly recurved laterally, whereas in *Gwyneddichnium* Digit V is often nearly parallel to Digit IV.

Description (cf. Lucas et al., 2014): These are small (1–5 cm in length), mesaxonic pentadactyl pes and manus imprints with high heteropody (pes larger than manus). The digits increase in length from I–III, Digit III is longest, Digit IV is shorter or subequal in length with Digit II, and Digit I is shortest. Digit V is in a posterior-lateral position relative to other digits and often elongated, parallel, or slightly laterally abducted from Digit IV. Digits are long and thin and can be straight or slightly inward curved. In well-preserved specimens round to oval phalangeal pad impressions give digits a nodular shape. Claw traces are small and narrow, and can be elongated into scratches. Trackways of walking individuals are relatively broad, with the pes directed parallel to the midline and the manus, which is positioned anterior and close to the pes, pointing outward.

Discussion: *Gwyneddichnium* was originally introduced by Bock (1952) based on material from the Passaic Formation (Upper Triassic, Newark Supergroup) of New Jersey, USA. By the thin, slightly curved digits it might, in some examples, be confused with the ichnogenus *Rhynchosauroides*. However, while the latter is characterized by ectaxonic imprints with Digits I–IV increasing in length and Digit IV longest, *Gwyneddichnium* is rather mesaxonic with a longest Digit III and with only slightly curved or straight digit traces. A peculiarity of *Gwyneddichnium* is also the nodular shape of the phalangeal pads in well-preserved imprints. A detailed revision of the ichnogenus has been made by Lucas et al. (2014). Lockley et al. (1991) distinguished walking and swim trackways of *Gwyneddichnium*, which has been critiqued by Lucas et al. (2014) and by Klein and Lucas (2021).

Trackmaker: Originally small pseudosuchian archosaurs have been considered as possible trackmakers (Bock, 1952; Haubold, 1986). Based on congruence of the footprints with the feet of "*Tanytrachelos*" (Olsen and Flynn, 1989) the latter is presently the best match for *Gwyneddichnium* (see Lucas et al., 2014; Klein and Lucas, 2021 for more detailed discussion), and we adopt this interpretation here.

4.2.2 Prorotodactylus Ptaszyński (2000)

Fig. 5.3.

Type ichnospecies: *P. mirus* Ptaszyński (2000).

Other ichnospecies: *P. mesaxonichnus*.

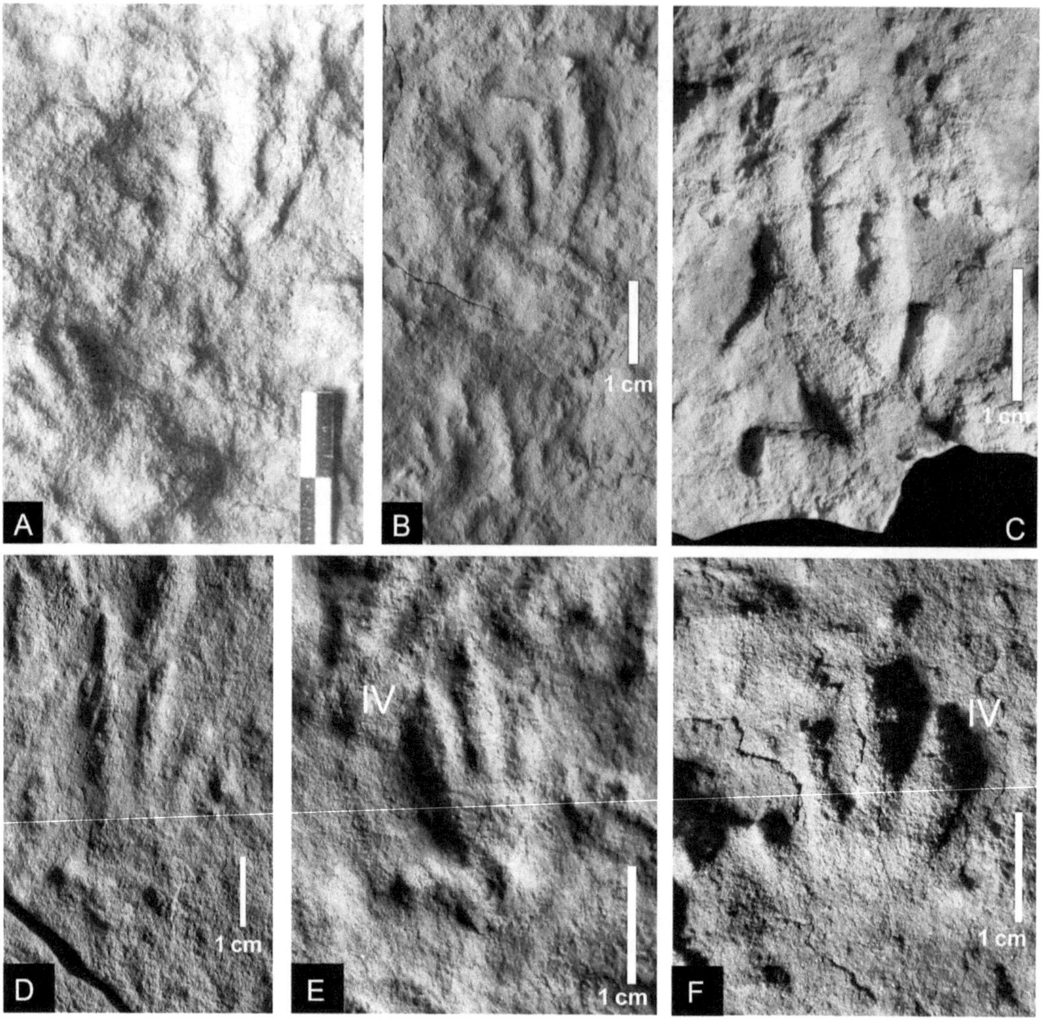

FIGURE 5.3 *Prorotodactylus mirus* from Wióry Formation (Lower Triassic, Olenekian) of the Holy Cross Mountains, Poland. (A–B) Pes-manus sets. (C–D) Pes imprints. (E–F) Manus imprints. *From Klein and Niedźwiedzki (2012).*

Distribution: Wióry Formation (Lower Triassic, Olenekian), Wióry, Poland, Detfurth Formation, Middle Buntsandstein, Olenekian) of Germany (Ptaszyński, 2000; Klein and Niedźwiedzki, 2012; Fichter and Kunz, 2013), Buntsandstein (Lower Triassic, Olenekian), Pyrenean Basin, Spain (Mujal et al., 2017).

Diagnosis (from Klein and Niedźwiedzki, 2012): Long, striding trackways with small, lacertoid, pentadactyl pes and manus imprints. Manus overstepped laterally by the pes. Pes rotated outward and manus rotated inward with respect to the midline. Digitigrade pes with Digits I–IV increasing in length, II–IV subparallel and tightly "bunched" with low divarication angle between Digits II and IV (≤10 degrees). Distinct, straight metatarsal-phalangeal axis forming posterior end of Digits II–IV. Digit I everted. Digit V rarely impressed, if present, in a posterolateral postion relative to Digits I–IV. Manus semiplantigrade, of chirotheroid shape, compact and rounded with postero-laterally positioned Digit V mostly impressed. Digit III always longest, followed by Digits IV, II, and I, which is shortest.

Description: Pentadactyl pes and manus imprints with straight, occasionally inward curved, long and slender digits that are parallel or slightly divaricating, increasing in length from I to IV. Especially in the pes Digits I–IV are tightly bunched, forming a narrow posterior end. Manus imprints have a Digit IV that is slightly shorter than Digit III, and, together with the relatively robust backward curved Digit V, this gives it a chirothere-like shape. In pes imprints a Digit V trace is commonly absent due to the more digitigrade appearance versus the plantigrade impression of the manus. The small claw traces are elongate triangular and acuminate in shape. Trackways broad with outward and inward rotation of the pes and manus, respectively, and with lateral overstep of the manus by the pes.

Discussion: Originally described from the Wiòry Formation (Lower Triassic, Olenekian) of Poland, it was later identified from other Olenekian deposits and localities in particular of the Buntsandstein in the Germanic Basin (Fichter and Kunz, 2013, 2015). An ongoing discussion about the validity of this ichnogenus concerns the similarity with *Rhynchosauroides* (see Klein and Niedźwiedzki, 2012; Klein and Lucas, 2021 for details). Ichnotaxonomically, the relatively short Digit IV < III and the tightly bunched posterior portion of Digits I–IV, similar to *Rotodactylus* (see below) distinguishes *Prorotodactylus* from *Rhynchosauroides* that has Digit IV longest and a less compact posterior margin, often lacking a metatarsophalangeal area due to the more digitigrade appearance of the pes. Therefore, we consider *Prorotodactylus* a distinct ichnogenus.

Trackmaker: There is no general consenseus about the attribution to either lepidosauromorph, archosauromorph, or dinosauromorph producers. Dinosauromorphs similar to the makers of *Rotodactylus* footprints (see below) have been proposed, in particular by Brusatte et al. (2011) and Niedźwiedzki et al. (2013). Alternative interpretations were discussed in Klein and Niedźwiedzki (2012) and Klein and Lucas (2021).

4.2.3 *Rhynchosauroides* Maidwell (1911)

Fig. 5.4.

Type ichnospecies: *R. rectipes* Maidwell (1911).

Other ichnospecies: *R. pallini, R. schochardti, R. tirolicus, R. hyperbates.*

Distribution: Long range in Upper Permian–Upper Jurassic age strata. Occurrences come from the Arenaria di Val Gardena Formation (upper Permian, Lopingian) of Northern Italy, Buntsandstein, Muschelkalk, Keuper and equivalents (Lower-Upper Triassic) in northern Spain, Great Britain, France, Germany, the Netherlands and Poland. Lower to Upper Triassic of North America (Moenkopi Group, Chinle Group, Newark Supergroup), South America (Argentina, Brazil), North Africa (Morocco), South Africa and China.

Diagnosis: Lacertoid ectaxonic footprints of a quadruped; broad trackways with lateral overstep of the manus by the pes. Pace angulation 70–125 degrees, depending on the velocity. Manus medially positioned relative to the pes. Pes slender and digitigrade; the smaller manus is stouter and more plantigrade. Length of digits increases from I to IV; Digit V much shorter. Digits I-IV usually curved inward.

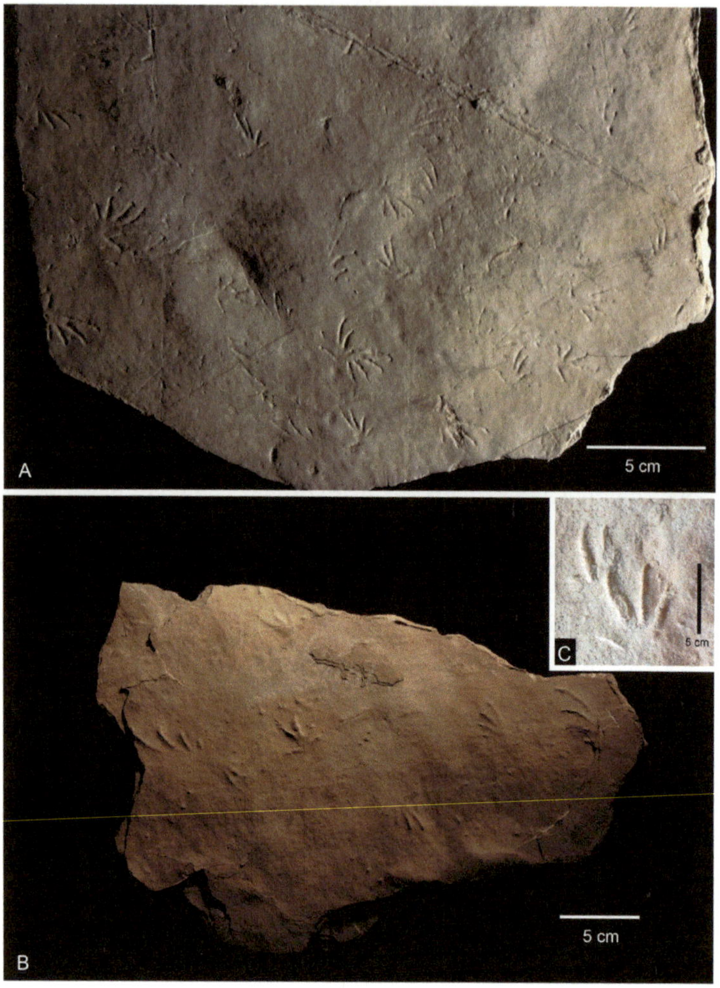

FIGURE 5.4 *Rhynchosauroides* (archosauromorph/lepidosauromorph) footprints from the Triassic. (A–B) *Rhynchosauroides tirolicus* from Middle Triassic deposits of the Southern Alps, Italy. (C) *Rhynchosauroides* isp. From Redonda Formation (Chinle Group, Norian-Rhaetian, Upper Triassic) of New Mexico, USA. *(A–B) from Klein and Lucas (2021). (C) From Lucas et al. (2010).*

Description: *Rhynchosauroides* are pentadactyl pes and manus tracks with thin digits that are inward curved or straight, often ending in hook-like terminations with small, acuminate claws. Digits increase in length from I to IV, and Digit IV is longest. Digit V is small and postero-laterally positioned, often consisting of a punctiform or short oval impression that can be curved backward. Pes imprints are digitigrade, whereas the smaller manus imprints are plantigrade with a triangular sole. Well-preserved specimens display round phalangeal pads, their number suggesting a 2-3-4-5-?3 formula. Trackways show a broad pattern, with pes imprints outwardly and manus imprints inwardly rotated. Characteristic is the lateral overstep of the manus by the pes in all *Rhynchosauroides* ichnospecies.

Discussion: *Rhynchosauroides* was introduced by Maidwell (1911) based on Middle Triassic footprint material from Great Britain. Subsequently, numerous ichnospecies have been described, of which only a few should be considered as valid. The ichnogenus is in need of revision, however, the overall similarity of tracks, and the variability of the trackway pattern due to changes of individual movement and velocity hampers the establishment of a stable ichnotaxonomy. Furthermore, morphological similarities with some ichnogenera such as *Gwyneddichnium*, *Procolophonichnium*, *Rotodactylus*, and *Prorotodactylus* can make an assignment difficult, in particular in isolated imprints lacking clear step cycles. Often *Rhynchosauroides* occurs in mass accumulations on trampled surfaces where no trackways can be identified. Nevertheless, distinct trackways have been documented, for example, from Middle Triassic deposits of northern Italy (Avanzini and Renesto (2002), from the Lower Triassic (Olenekian) of the Wiòry Formation of Poland (Klein and Niedźwiedzki, 2012) and from the upper Permian (Lopingian) of the Arenaria di Val Gardena Formation of northern Italy (see Permian chapter, this volume). Interestingly, the stratigraphically youngest occurrence is from the Upper Jurassic of Spain (Avanzini et al., 2010). The long range of the ichnogenus from the late Permian to the Late Jurassic suggests phylogenetically different trackmakers, perhaps various lepidosauromorph and archosauromorph diapsids (Klein and Lucas, 2021).

Trackmaker: Lepidosauromorph and/or archosauromorph trackmakers are convincing counterparts (Avanzini and Renestio, 2002). The long stratigraphic range suggests different members of these tetrapod groups with similar foot morphology as trackmakers (see Klein and Lucas, 2021 and discussion therein).

4.2.4 *Rotodactylus* Peabody (1948)

Fig. 5.5.

Type ichnospecies: *R. cursorius* Peabody (1948).

Other ichnospecies: *R. bradyi*, *R. matthesi*.

Distribution: Lower–Middle Triassic strata. Moenkopi Group (Lower–Middle Triassic, Olenekian–Anisian) of North America (Arizona, Utah), Volpriehausen, Detfurth, Solling, Eschenbach formations, Olenekian-Anisian) of Germany, Grés des Lyonnais (Middle Triassic, Anisian-Ladinian) of France, Wióry Formation (Olenekian) of Poland, Timezgadiouine Formation (Middle Triassic, Anisian-Ladinian) of Morocco, Haizer-Akouker unit (Middle Triassic) of Algeria.

Diagnosis (from Klein and Lucas, 2021, emended after Peabody, 1948; Haubold, 1971a; Klein and Lucas, 2010b; Klein and Niedźwiedzki, 2012): Moderately narrow trackways. Pes digitigrade with dominance of digit group II–IV and an extreme, posteriorly positioned, small, punctiform mark that constitutes the impression of Digit V. Manus similar to pes, but much smaller. The digit proportions are I < II < III < IV, and in the manus of some ichnospecies I < II < IV < III. Trackways preserve evidence of long strides and a primary, lateral overstep of the manus by the pes, though the values of overstepping and stride length show high variability.

Description: Tracks are pentadactyl, functionally tridactyl (II, III, IV), with straight or slightly inward curved digits increasing in length between II and IV. In the pes Digit IV is longest, in the manus IV can be longer, subequal, or slightly shorter than Digit III. Digits I and V traces are very small and punctiform only, mostly isolated from the rest of the imprint. The Digit V trace is in a far posterior position to the digit group I–IV, and in pes imprints this distance equals the length of Digit IV, in manus imprints being shorter. The posterior connection of Digits II–IV is narrow. Claw traces are very small, triangular, and acuminate. Well-

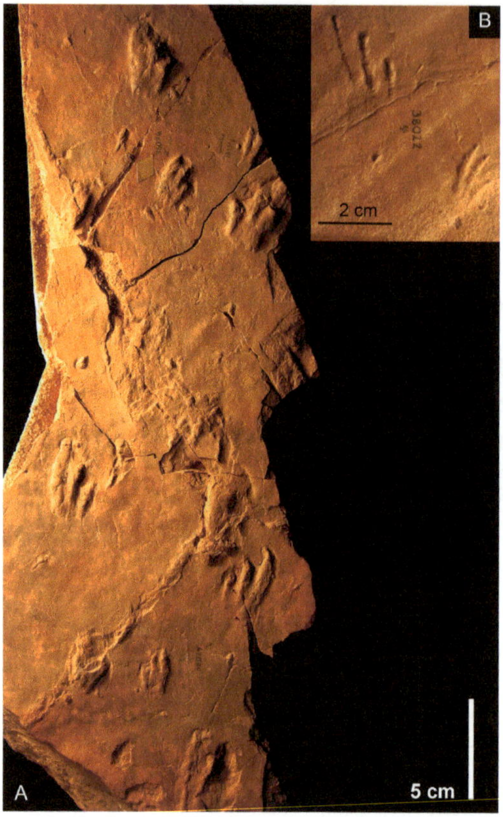

FIGURE 5.5 *Rotodactylus cursorius* (dinosauromorph) footprints from the Moenkopi Group (upper red formation) of Utah, USA. (A) Trackway (holotype of "*R. mckeei*"). (B) Pes-manus set. Note overstep of the manus by the pes and small, round impression of Digit V in a far posterior position.

preserved specimens often show rounded phalangeal pads, and, in deeper imprints, there are also metatarsophalangeal pad traces. The trackway pattern is characterized by very long strides with continuous lateral overstep of the manus by the pes. Pes imprints oriented parallel to the trackway midline, whereas manus imprints are rotated inward to various degrees. However, all parameters vary due to different speed of progression.

Discussion: Peabody (1948) introduced the ichnogenus when describing the famous material from the Moenkopi Formation of the western USA. He recognized peculiar features in the imprint morphology and the trackway pattern that distinguished it clearly from other lacertoid tracks such as *Rhynchosauroides*. This concerns: (1) Compact digit group II–IV with a tightly bunched posterior margin, (2) the punctiform Digit V impression, which, in particular, in the pes is largely separated from the rest of the imprint, (3) the long strides and (4) the continuous overstep of the manus by the pes. Peabody (1948) noted a dinosaur-like pattern and Haubold (1999) convincingly proposed dinosauromorphs similar to lagosuchids as the makers of *Rotodactylus* trackways, even if other groups such as basal archosauromorphs cannot be completely excluded. The discussion on this is still continuing. *Rotodactylus*, like

Rhynchosauroides, often occurs in mass accumulations on the surfaces indicating small agile tetrapods moving around quickly, thereby leaving abundant imprints in the fine-grained substrate along lake margins and river channels.

Trackmaker: Haubold (1999) convincingly proposed dinosauromorphs similar to lagerpetids as trackmakers based on similarities in the narrow pes and digit proportions with Digit IV longest, as well as leg proportions suggesting long strides and continuous overstep of the manus by the pes.

4.3 Turtle

4.3.1 Chelonipus *Rühle von Lilienstern (1939)*

Fig. 5.6.

Type ichnospecies: *C. torquatus* Rühle von Lilienstern (1939).

Other ichnospecies: Known only from the type ichnospecies.

Distribution: Solling Formation (Middle Triassic, Anisian), Stuttgart Formation (Upper Triassic, Carnian) of Germany, Upper Triassic of Spain, Lower Triassic of Wyoming and Utah (USA).

Diagnosis (after Haubold, 1971a,b): Tracks of a quadruped in which trackway width is almost as long as stride length; manus and pes tracks are in parallel rows and preserve three to five digits; the manus is broad and arched, with Digit III or IV longest; and the pes often has a rounded plantar surface and relatively longer digits compared with the manus.

Description: These are broad trackways of a quadrupedal tetrapod that show a typical understep where the hindfoot from the second step cycle can overstep the forefoot or alternatively can be placed just behind the forefoot of the same step series. The manus is broad, short, and arched, with digits III? or IV? being more distinct and longest. The pes shows four to five digit traces that are longer compared with those in the manus and occasionally has a well-developed plantar surface.

Discussion: The large variation in the trackway pattern can be observed in different specimens of *Chelonipus torquatus* from the Middle Buntsandstein (Anisian) and from the Schilfsandstein (Stuttgart Formation, Carnian) of Germany (Rühle von Lilienstern, 1939; Haubold, 1971a; Lichtig et al., 2018). The latter material was described as a distinct ichnospecies, *C. plieningeri*, by Haubold [1971a; see also original descriptions by Plieninger (1838) and Meyer and Plieninger (1844)] but synonymized with *C. torquatus* by Lichtig et al. (2018). A different interpretation of the Schilfsandstein trackway was given by Huene (1932), who considered a coelurosaurian dinosaur as the trackmaker, which is definitely incorrect. Coelurosaur tracks have a different, mostly tridactyl morphology and do not leave manus tracks with a broad arched shape. Lichtig et al. (2018) consider understep walking of the turtle trackmaker as the main cause of the variation in the trackway pattern. Interestingly, among ichnologists, there was no consensus about which imprint represents the manus and the pes, respectively. While Rühle von Lilienstern (1939) interpreted *C. torquatus* to show overstep with the manus imprint positioned behind the pes imprint, Haubold (1971a) considered the posterior impression to be the pes. Later, Avanzini et al. (2005), based on actualistic studies with turtles, demonstrated that the original interpretation by Rühle von Lilienstern (1939) might be correct. This was followed by Lichtig et al. (2018) and is adopted here.

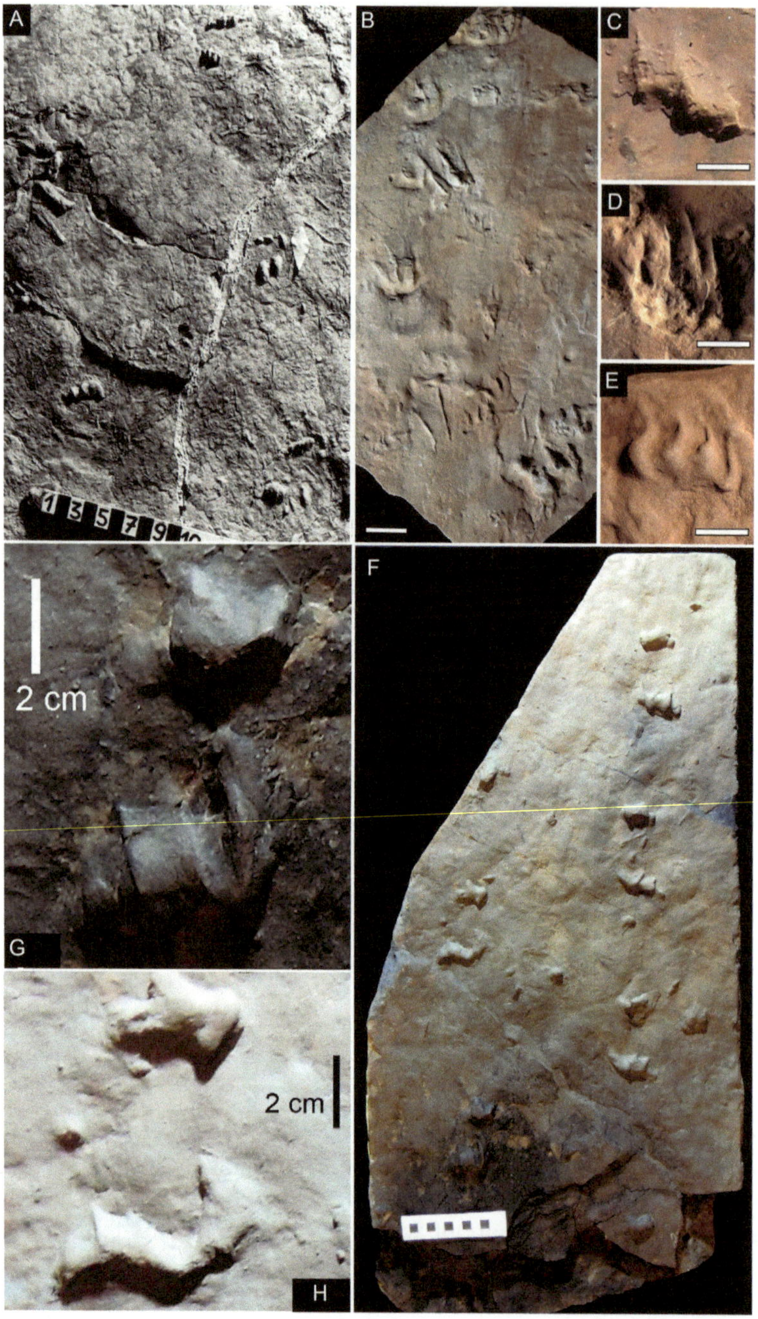

FIGURE 5.6 *Chelonipus* (prototurtle) footprints from the Triassic. (A) Holotype of *Chelonipus torquatus* from Solling Formation (Middle Triassic, Anisian) of Germany, scale bars in centimeters. (B–E) C. isp. From Red Peak Formation (Chugwater Group, Lower Triassic, Olenekian) of Wyoming. (B) Partial trackway. (C), (E) Manus imprints. (D) Pes imprint. (F–H) Trackway of *C. torquatus* from Stuttgart Formation (Upper Triassic, Carnian) of Germany (from Lichtig et al., 2018). (F) Overview. (G–H) Details with pes-manus sets.

Further possible *Chelonipus* material has been described from the Upper Triassic of Spain (Márquez-Aliaga et al., 1999; Reolid et al., 2018). In the USA, the Lower Triassic Red Peak Formation of Central Wyoming and the Virgin and Torrey formations (Thaynes and Moenkopi groups) of Utah yielded similar tracks that partly might represent *Chelonipus* (Lovelace and Lovelace, 2012; Thomson and Lovelace, 2014). Purported *Chelonipus* from the late Eocene of Texas (Sarjeant and Langston, 1994) is morphologically different from this ichnogenus, for example by the manus imprint that lacks the typical broad arched shape. It is obviously a turtle trackway but refers to a different ichnogenus.

Trackmaker: Turtles similar to *Proganochelys* are most likely the trackmakers. Upper Triassic deposits in Europe and North America have yielded body fossils of turtles, and some have preserved foot skeletons showing a convincing match.

4.4 Archosauriform

Phytosaur.

4.4.1 *Apatopus* Baird (1957)

Fig. 5.7.

Type ichnospecies: *A. lineatus* (Bock, 1952)

Other ichnospecies: Only known from the type ichnospecies.

Distribution: Carnian–Norian of North America (Newark Supergroup, eastern USA, Pekin, Stockton, Lockatong, Passaic formations; Chinle Group, western USA, Owl Rock Formation), Europe (Germany, Poland, northern Italy), North Africa (Argana and Ourika basins of Morocco) and Southeast Asia (Thailand).

Diagnosis (from Klein and Lucas, 2013, emended after Baird, 1957): Trackways of a quadruped with pace angulation of the pes ranging from 108 degrees to 120 degrees. Pes, but not manus, toed out. Pes long and narrow, semiplantigrade to plantigrade, pentadactyl with slender digits increasing in length from I to IV. Digit IV often very faintly impressed or missing; Digit V straight, antero-laterally oriented, and with posteriorly elongated "heel." Digits with well developed articular swellings and sharp claws. Manus pentadactyl, semiplantigrade, short, rounded, and symmetrical around Digit III, which is longest; manus positioned in the trackway anterior to or slightly medial to the pes.

Description: *Apatopus* is trackways of a quadruped with elongated semiplantigrade and ectaxonic pes imprints showing increasing lengths of Digits I–IV. Digits are long and straight, and in some cases can be curved inwardly or outwardly and moderately spread. Digit V is laterally to posterolaterally positioned, straight, anterolaterally oriented, and extends into a distinct posterior "heel." Manus imprints are short, rounded with Digit III being longest, II and IV shorter and subequal in length, I and V shortest. Claw traces are robust and sharp, sometimes extending into crocodile-like drag traces or scratches. Well-preserved imprints display rounded phalangeal and metatarsophalangeal pad traces. The distinct outward rotation of the imprints being strong in the pes and less distinct in the manus is characteristic of the trackway pattern. The manus is positioned anterior and close to the pes. Steps are relatively short. Sometimes a slight lateral overstep can be observed.

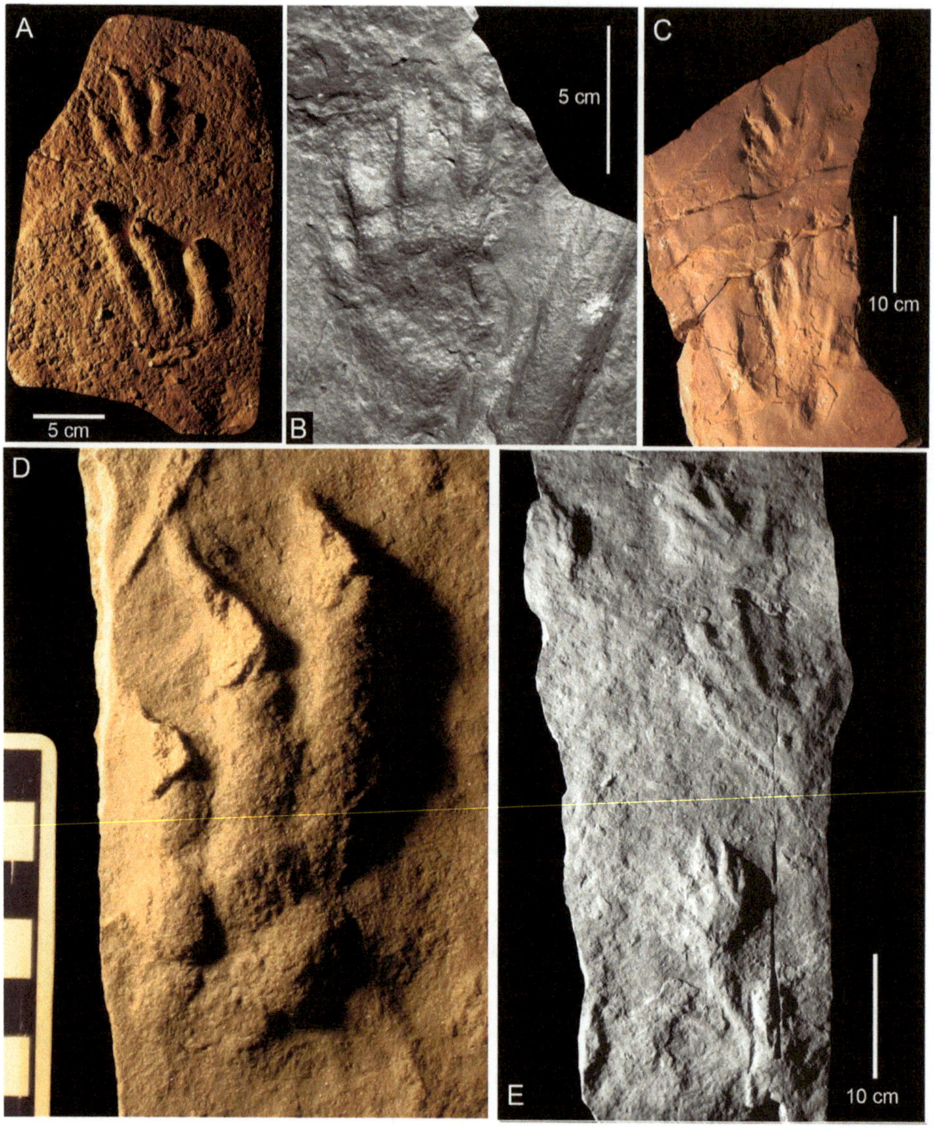

FIGURE 5.7 *Apatopus lineatus* (phytosaur) footprints. (A) Pes-manus set from holotype trackway, Passaic Formation (Norian), New Jersey, USA. (B) Pes from Passaic Formation (Norian) of Pennsylvania. (C) Pes-manus set from Timezgadiouine Formation (T5) of the Argana Basin, Morocco. (D–E) Pes and partial trackway from Hassberge Formation (Carnian) of Germany. *From Klein and Lucas (2021).*

Discussion: Baird (1957) established this ichnogenus with the type ichnospecies *Apatopus lineatus* when describing material from the Passaic Formation (Upper Triassic, Newark Supergroup) of New Jersey. The type was originally documented by Bock (1952) and assigned to *Otozoum lineatum*. *Otozoum* is an ichnogenus introduced by Hitchcock (1847). It is tetradactyl, was probably left by sauropodomorph producers, and well-known from the Lower Jurassic. *Apatopus* is generally accepted as a pentadactyl track of a quadrupedal phytosaur (Padian et al., 2010; Klein and Lucas, 2013). Compared with other Triassic archosaur tracks such as chirotheriid or grallatorid forms, *Apatopus* is relatively rare in the global record due to the semiaquatic lifestyle of the producers and the minor preservation potential in subaqueous environments. It has been described from North America, both the Newark Supergroup in the east and the Chinle Group in the west, from Europe (Germany, Poland) and from North Africa (Morocco) (see Klein and Lucas, 2013 for overview).

Trackmaker: Phytosaurs were the most likely producers of *Apatopus* tracks (Padian et al., 2010).

Chirotheriids.

4.4.2 *Brachychirotherium* Beurlen (1950)

Fig. 5.8.

Type ichnospecies: *B. hassfurtense* Beurlen (1950).

Other ichnospecies: *B. thuringiacum, B. parvum.*

Distribution: Upper Triassic. Hassberge and Löwenstein formations (Carnian-Norian) of Germany, Carnian of Italy, Timezgadiouine and Oued Oum Er Rbiaa formations (Carnian-Norian) of Morocco, Lower Elliot Formation (Norian–Rhaetian) of southern Africa, Ørsted Dal Formation (Fleming Fjord Group (Norian–Rhaetian) of Greenland, Chinle Group and Newark Supergroup (Carnian-Rhaetian) of USA and eastern Canada, Norian of Argentina.

Diagnosis (after Karl and Haubold, 1998): Chirotheriid trackways of a quadruped with plantigrade to semidigitigrade posture of the pes and manus. Pentadactyl pes imprint with rounded to oval shape and broad plantar surface. Pedal digit group I–IV nearly as wide as long and with an overall divarication of more than 40 degrees. Digits moderately spread, relatively short and broad with robust, rounded pads and narrow claws, Digits III and II longest, Digit IV always preserved. Pedal Digit V lacks a separated phalangeal segment or shows an indistinct trace of it, which is anteriorly oriented, but never recurved. In comparison with the pes, the manus imprint is much smaller, and manual digit group I–IV is mostly wider than long, but otherwise of a shape similar to the pes. Trackways with narrow pattern and relatively high pace angulation (160 degrees), moderate outward rotation of the pes, and stronger outward rotation of the manus.

Description: These are broad pes and manus imprints with short, thick digit traces and tiny claws. Digit proportions are characterized by Digit III being longest, followed by Digits II, IV, and I. So, Digit IV is shorter than Digit II. Digit V consists of an oval basal pad only, without a phalangeal portion. It can extend into an elongated heel, depending on the depth of the substrate. The manus is similar in shape and digit proportions, but much smaller. It is positioned anterior and close to the pes. Trackways are narrow and can show relatively long strides due to faster progression.

Discussion: *Brachychirotherium* was erected by Beurlen in 1950. The type material comes from the Coburger Sandstone unit (Hassberge Formation, Upper Triassic, Late Carnian) of northern Bavaria. Unfortunately, the lectotype designated by Karl and Haubold (1998; see also Karl and Haubold, 2000) in their revision of this material consists of an isolated pes imprint only, and other materials from this unit are at best preserved with an associated manus but never show a complete trackway sequence. Nevertheless, the characteristic and

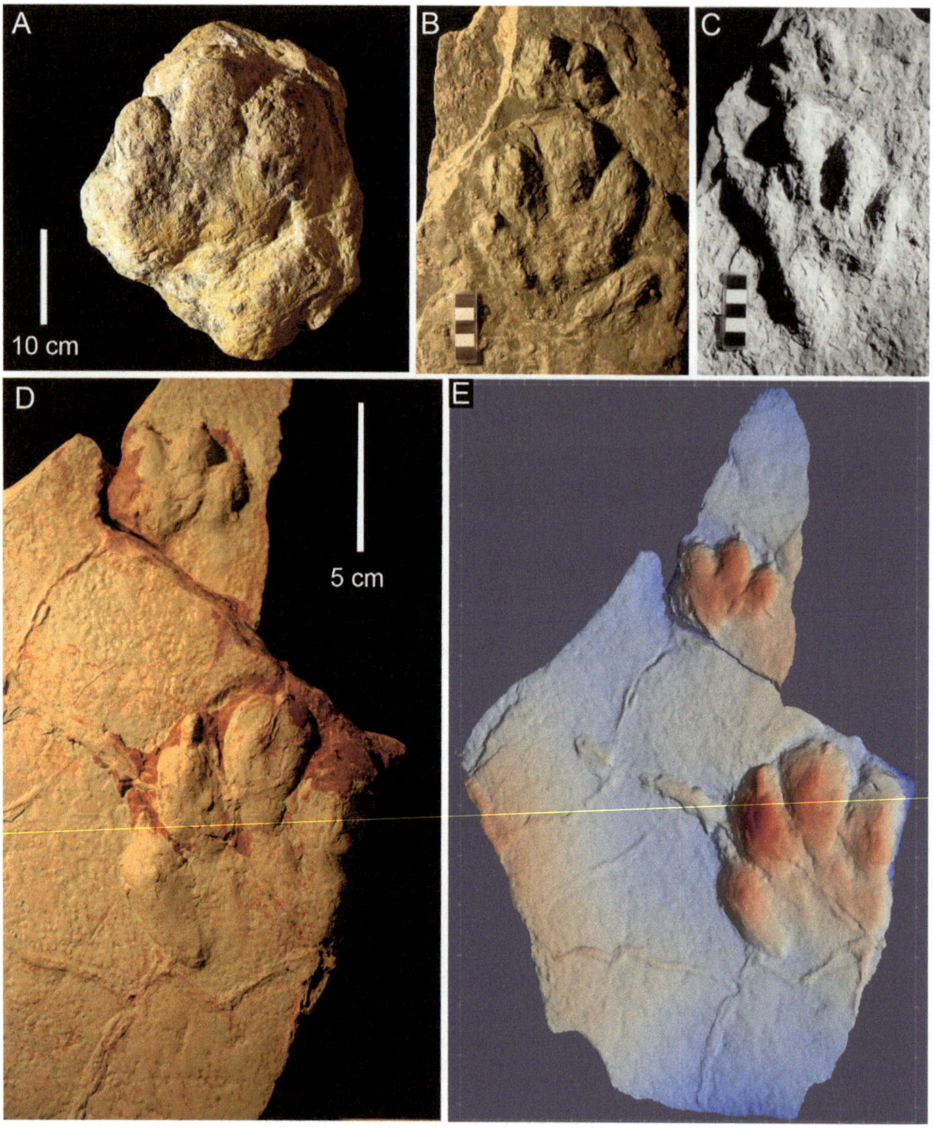

FIGURE 5.8 *Brachychirotherium* (archosaur) footprints from the Upper Triassic. (A) Lectotype of *B. hassfurtense*. (B) *B. thuringiacum*. (C) *B. parvum*. (D–E) *B. parvum* pes-manus set as photograph and false-color photogrammetric image. *(A–B) From Hassberge Formation (Upper Triassic, Carnian) of Germany. (C.) From Redonda Formation (Norian-Rhaetian) of New Mexico. (D–E) From Oued Oum Er Rbiaa Formation (Norian) of Morocco. From Klein and Lucas (2021), Hminna et al. (2021).*

diagnostic morphology with the broad pes imprint, the short, thick digit traces with tiny claws and the oval basal pad of Digit V lacking a phalangeal portion, are repeatedly observed in chirotheriids from coeval deposits and surfaces in Europe, North America, and North Africa (Karl and Haubold, 1998; Lucas et al., 2010; Klein et al., 2011; Hminna et al., 2013, 2021).

This makes *Brachychirotherium* a distinct morphotype and generally well accepted chirotheriid ichnogenus. Described occurrences in pre-Carnian deposits, especially from the Middle Triassic, but also from the Lower Triassic, differ from the Upper Triassic type and will be assigned to a new ichnogenus presented in a separate paper (see below).

Trackmaker: Crocodylian stem crown-group archosaurs can be considered. In particular, *Postosuchus*-like loricatans and aetosaurs, but also sphenosuchian crocodylomorphs have been proposed (Karl and Haubold, 1998; Klein et al., 2006, 2023b; Avanzini et al., 2010; Lucas et al., 2010; Lucas and Heckert, 2011). More complete foot skeletons of these groups for a comparative analysis are rare. Lucas and Heckert (2011) found a good match of preserved autopodia of the aetosaur *Typothorax* with *Brachychirotherium*.

4.4.3 *Chirotherium* Kaup (1835a)

Fig. 5.9.

Type ichnospecies: *C. barthii* Kaup (1835a).

Other ichnospecies: *C. sickleri, C. ferox, C. ladinicum, C. postchirotherioides, C. ischigualastianum.*

Distribution: Solling Formation (Middle-Upper Buntsandstein, Middle Triassic, Anisian) of Germany, various Middle-Upper Triassic (Anisian-Ladinian-Carnian) deposits of France, Germany, northern Italy, Spain, Great Britain, and Poland, Timezgadiouine Formation (T4-T5, Anisian-Ladinian-Carnian) of the Argana Basin, Morocco, Holbrook, and Anton Chico members of the Moenkopi Formation in Arizona and New Mexico, USA, Cerro de Las Cabras and Los Rastros formations (Middle-Upper Triassic) of Argentina, Guanling Formation (Middle Triassic) of Guizhou Province, China.

Diagnosis (from Klein and Lucas, 2021, emended after Haubold, 1971a, b): Trackways of quadrupeds or facultative bipeds with pentadactyl pes and manus imprints. Pes imprint with pronounced digit group II–IV; Digit III longest; Digit I reduced and thinner; pedal Digit V positioned slightly laterally behind the digit group I–IV; ratio of manus:pes area in completely preserved specimens = 1:2, up to 3.5. Pace angulation mostly 160–170 degrees.

Description: *Chirotherium* has a functionally tridactyl pes imprint, with Digit III being longest, followed by Digits IV, II, and I. Digit I is thinner than other digits and slightly posteriorly shifted. Therefore, the proximal margin of the digit group I–IV appears to form a posteriorly concave arc. Digit V consists of a distinct round to oval basal pad and a thinner phalangeal part, which is strongly backward curved, scarcely straight. The manus imprint is much smaller, with Digit III longest, followed by II, IV, and I. Digit IV is often laterally spread. Claw traces are triangular, acuminate, and robust. Trackways are narrow, with large pace angulation, and have variable stride lengths depending on the velocity of movement of the trackmaker. The orientation of the pes and manus imprints is rotated outward, slightly more in the manus compared with the pes.

Discussion: Older literature lists all chirotheriid footprints under the ichnogenus *Chirotherium* (see also Haubold, 1971a). It was the latter author (Haubold, 1971b) who proposed four different ichnogenera for chirotheres: (1) *Chirotherium*, (2) *Isochirotherium*, (3) *Brachychirotherium*, and (4) *Synaptichnium*. Later, Fichter and Kunz (2004) introduced *Protochirotherium* as a fifth chirotheriid ichnogenus based on material from the Lower Triassic Detfurth Formation (Middle Buntsandstein) of Germany. The ichnogenus *Chirotherium* was recently revised by Klein and Lucas (2021), to include all chirotheriids with a functionally tridactyl pes (II, III,

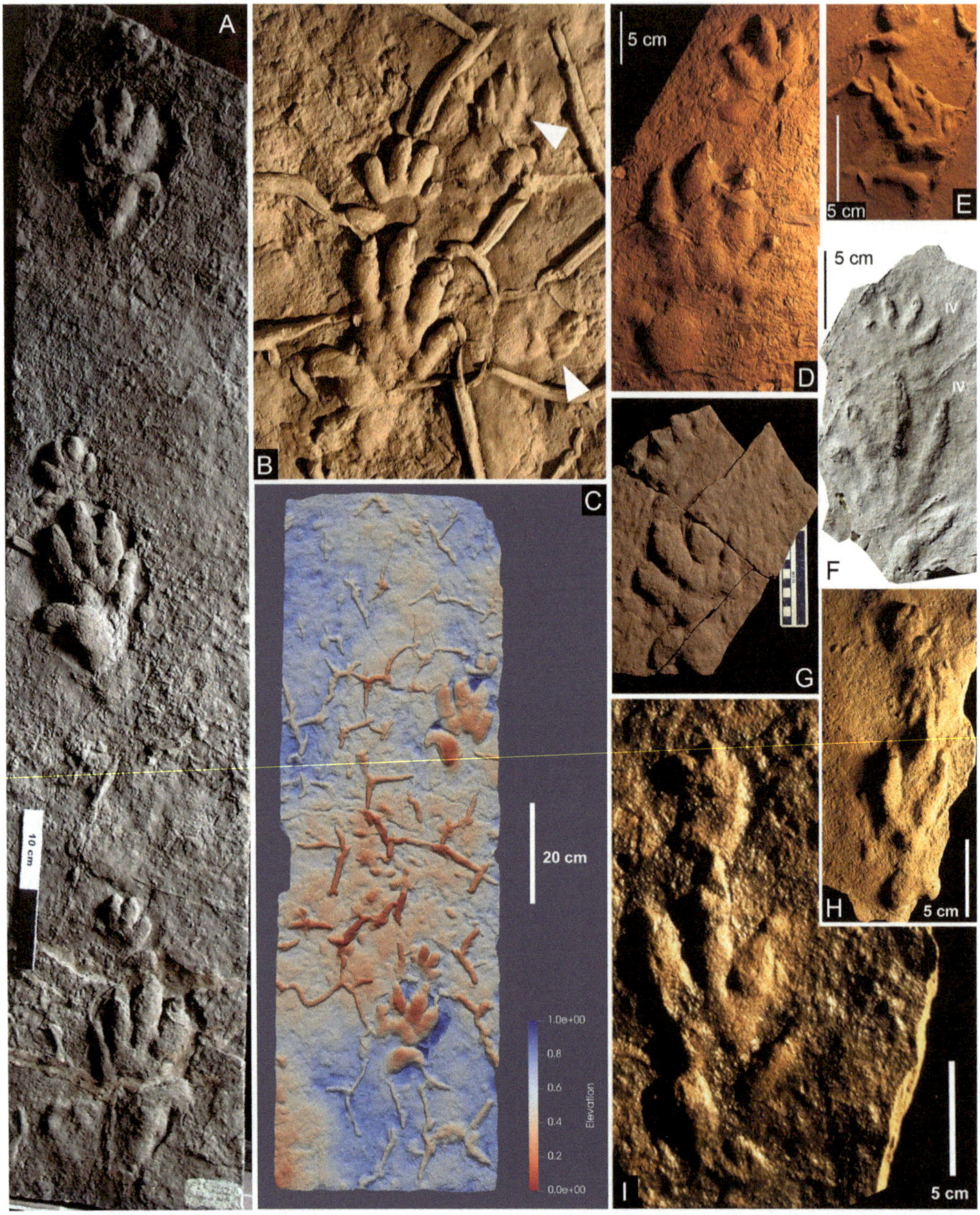

FIGURE 5.9 *Chirotherium* (archosaur) tracks and trackways. (A–C) *Chirotherium barthii* from type surface in the Solling Formation (Middle Triassic, Anisian) of Germany. (A) Trackway. (B) Pes-manus set associated with pes-manus sets of smaller *C. sickleri* (upper right) and *"Brachychirotherium" harrasense* (lower right). (C) Partial trackway as false-colour = r photogrammetric model. (D–E) *C. barthii* and *C. sickleri* pes-manus sets from Upper and Lower Moenkopi Formation (Olenekian-Anisian) of Arizona, USA. (F) *C. barthii* pes-manus set from Guanling Formation (Middle Triassic, Anisian) of Yunnan, China. (G) *C. barthii* pes manus set from Timezgadiouine Formation (T4) of the Argana Basin, Morocco. (H) *C. ferox* pes-manus set from Eschenbach Formation (Middle Triassic, Anisian) of Germany. (I) *C. postchirotherioides* from Benk Formation (Ladinian) of Germany. *From Klein and Lucas (2021) and Klein et al. (2023b).*

IV) and a shorter and posteriorly shifted pedal Digit I. These authors also re-assigned the ichnogenera *Parachirotherium, Sphingopus,* and *Rigalites* to *Chirotherium* with respective ichnospecies as new combinations. In chirotheriid footprints only *Chirotherium* shows a distinct tendency toward a mesaxonic, functionally tridactyl pes. All others can be characterized as either functionally tetradactyl (*Synaptichniium, Protochirotherium, Brachychirotherium*; Digits I–IV dominating) or showing a strong tendency toward functional didactyly (*Isochirotherium*; Digits II–III dominating).

Trackmaker: Crown-group archosaurs, both pseudosuchian (crocodile-line) and avemetatarsalian (dinosaur-bird-line) members with a functionally tridactyl, mesaxonic pes. Based on pes skeletons this morphology has been developed independently, for example in dinosauriforms and poposauroids (Farlow et al., 2014; Klein et al., 2023b).

4.4.4 *Isochirotherium* Haubold (1971b)

Fig. 5.10.

Type ichnospecies: *I. soergeli* (Haubold, 1967)

Other ichnospecies: *I. herculis, I. coltoni, I. lomasi, I. marshalli, I. coureli, I. felenci, I. delicatum.*

Distribution: Lower-?Upper Triassic. Solling Formation (Middle Buntsandstein, Anisian) and ?Löwenstein Formation (Norian) of Germany, Middle Triassic (Anisian-Ladinian) of France, Great Britain, Spain, and Italy, Timezgadiouine Formation (T4, Anisian-Ladinian) of the Argana Basin, Morocco, Moenkopi Group (Olenekian-Anisian, Lower-Middle Triassic) of Arizona, USA.

Diagnosis (from Klein and Lucas, 2021, emended after Haubold, 1971b): Narrow trackways of a quadruped with a pace angulation of ~165 degrees and relatively strong outward rotation of the pentadactyl pes and manus imprints (20–30 degrees on average). Pedal Digits II and III are longest, and Digit IV is mostly as long as Digit I and distinctly divergent from Digit III. Basal pad of Digit V close behind metatarsophalangeal area I–IV, and, in larger forms, progressively amalgamated with these. Distal phalangeal segment of Digit V only weakly loaded. Very small manus-ratio of manus:pes area = 1:4.5–6.8. Digit III in the manus is longest.

Description: Plantigrade to semiplantigrade imprints. *Isochirotherium* has peculiar digit proportions in the pes showing a tendency toward functional didactyly with Digits II–III being long, whereas Digits I and IV are very short. Digit III or, alternatively, in some ichnospecies Digit II is longest. In some smaller ichnospecies, for example, *I. delicatum,* these digit proportions are distinctly expressed. Digit V has a large oval basal pad, and, in well-preserved specimens, a separate, thinner phalangeal portion that can be recurved or straight. There is some similarity of *I. soergeli* pes imprints to those of *Chirotherium barthii,* in that both exactly show this latter feature, however, in *C. barthii* pedal Digit IV is longer. The pentadactyl manus is very small relative to the pes, short and rounded and positioned anterior or antero-medial to the pes. Digit III is longest. An interesting combination of features in the trackways is the narrow trackway width and strong outward rotation of the pes.

Discussion: Haubold (1971b) distinguished the ichnogenus *Isochirotherium* from other chirotheriids, noting the unique digit proportions and small size of the manus imprint as diagnostic characters. These features are continuously observed in all specimens, including the tracks of small and large individuals. *Isochirotherium* includes some of the smallest chirotheriid imprints as well as some of the largest ones. In small *I. delicatum* from the Middle Triassic

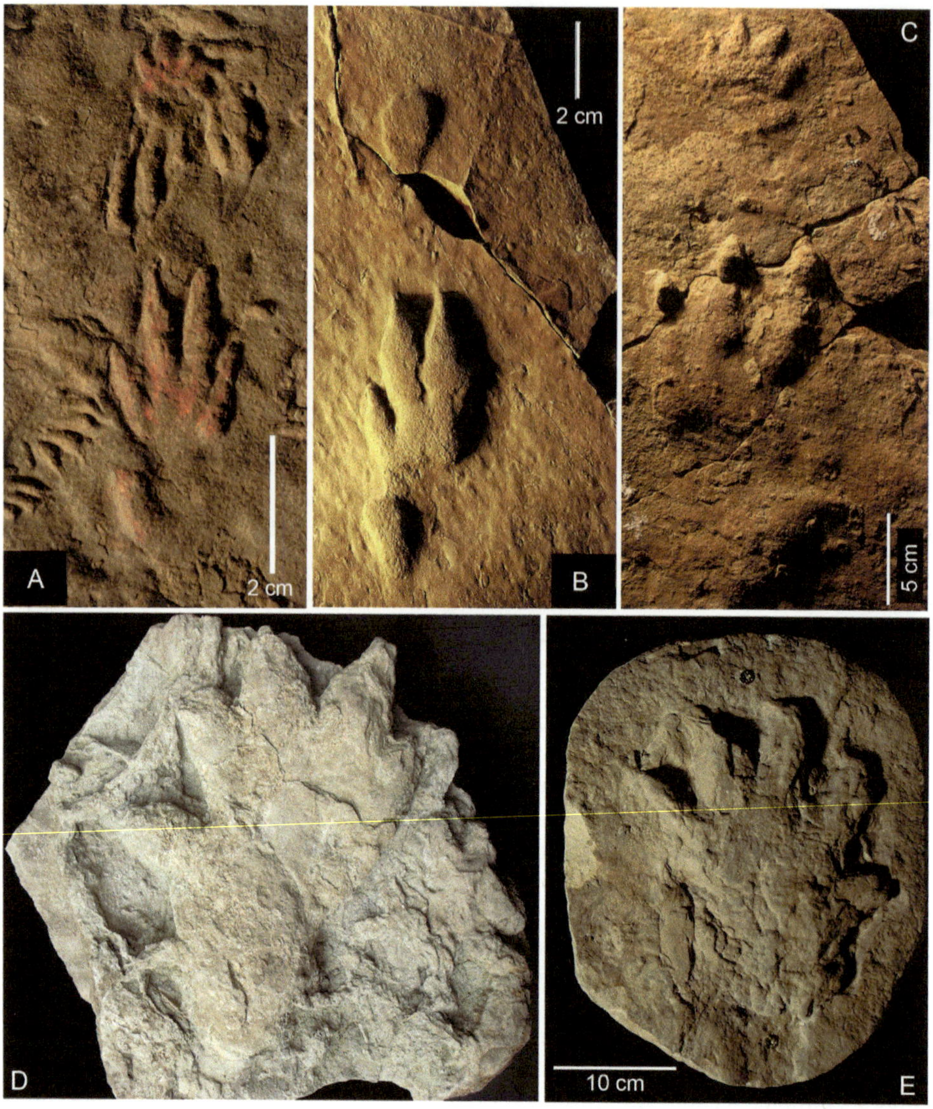

FIGURE 5.10 *Isochirotherium* (archosaur) tracks. (A) Small, possible juvenile *Isochirotherium soergeli* pes-manus set (colored reddish); note accompanying *Rotodactylus* footprints. (B). *I. delicatum* pes-manus set from Middle Triassic deposits of the Southern Alps, Italy. (C) *I. coureli* pes-manus set from Eschenbach Formation (Middle Triassic, Anisian) of Germany. (D–E) *I. herculis* pes imprints from the Solling Formation (Middle Triassic, Anisian) of Germany. Pes length of D ~30 cm. *From Klein and Lucas (2018, 2021), Klein et al. (2023b). (B) Courtesy Marco Avanzini,*

of northern Italy, Avanzini and Lockley (2002) documented ontogenetic growth and evidence for population structures of the trackmakers. Another ichnospecies, *I. herculis* from the Middle Triassic of Great Britain and Germany, belongs to the largest chirotheriids known from the global record (pes length >30 cm). There is some similarity with *Chirotherium rex* from the Moenkopi Formation of Arizona, and it cannot be completely excluded that the latter is a junior synonym of *I. herculis* (see also discussion below).

Trackmaker: Pseudosuchians and archosauriforms have been considered. Unfortunately, there is no optimal match with foot skeletons presently, and track-trackmaker correlations are still on weak grounds.

4.4.5 *Protochirotherium* Fichter and Kunz (2004)

Fig. 5.11.

Type ichnospecies: *P. wolfhagense* Fichter and Kunz (2004).

Other ichnospecies: *P. hauboldi.*

Distribution: Volpriehausen, Detfurth and Hardegsen formations (Middle Buntsandstein, Olenekian) of Germany, Wióry Formation (Olenekian) of Poland, Lower Triassic (Olenekian) of Austria, Timezgadiouine Formation (T3, Olenekian) of the Argana Basin, Morocco, Lower Triassic (Olenekian) of Niger, Moenkopi Group (Olenekian) of Utah, USA.

Diagnosis (from Klein and Lucas, 2021, emended after Fichter and Kunz, 2004): Chirotheriids with plantigrade to semiplantigrade pes and manus imprints. In the pes anterior digit group I—IV with Digit III longest and Digit IV being only slightly shorter; Digit V with massive, oval basal pad, often strongly elongated proximally and with a separate, thinner phalangeal part that is nearly straight or slightly curved outward. Manus rounded with Digit III longest, Digits II and IV shorter and subequal in length.

Description: These are plantigrade to semiplantigrade pes and manus imprints with relatively long Digits IV and V that can show similarities with *Synaptichnium* (see below). Compared with the latter they have a broader overall shape with robust digits, and the oval-round pedal Digit V ends in a mostly recurved phalangeal portion. Large, triangular claw traces are present. The type material from the Middle Buntsandstein (Lower Triassic, Olenekian) of Germany as well as the extensive material from the Wiòry Formation (Lower Triassic, Olenekian) of Poland show beautifully preserved skin impressions, not only along the digits but also covering the imprints of the sole. Trackways exhibit relatively short steps and low pace angulation. Pes imprints have a stronger outward rotation relative to the trackway midline, whereas the manus imprints are less outward rotated or parallel to the midline.

Discussion: Specimens more recently referred to *Protochirotherium* have formerly been known, for example, from isolated Buntsandstein tracks of Germany and from a diverse assemblage in the Wiòry Formation of Poland. They have been assigned to different ichnogenera such as *Chirotherium*, *Brachychirotherium* or *Isochirotherium* (Demathieu and Haubold, 1982; Fuglewicz et al., 1990; Fichter and Lepper, 1997; Ptaszyński, 2000; Niedźwiedzki and Ptaszyński, 2007). It was not until 2004 that Fichter and Kunz (2004) recognized features discriminating *Protochirotherium* as a distinct ichnogenus from all other chirotheriids. These authors introduced the new ichnotaxon based on well-preserved pes and manus imprints from the Middle Buntsandstein of the Detfurth Formation (Lower Triassic, Olenekian) of Germany. It was subsequently identified in the Wiòry Formation of Poland, in the Timezgadiouine Formation of Morocco, and in the Red Peak and Moenkopi formations of North America (Klein and Haubold, 2007; Klein and Lucas, 2010a, 2021; Klein et al., 2010, 2013; Klein and Niedźwiedzki, 2012), resulting in a reassignment of numerous specimens from these strata. *Protochirotherium* can be considered a key-ichnotaxon for the understanding of

FIGURE 5.11 *Protochirotherium* (archosaur) footprints. (A) *Protochirotherium wolfhagense* pes-manus set, artificial cast of holotype from the Detfurth Formation (Lower Triassic, Olenekian) of Germany. (B–C) *P. hauboldi* pes-manus sets from Wióry Formation (Lower Triassic, Olenekian) of the Holy Cross Mountains, Poland. (C) Artificial cast of syntype. (D) *Protochirotherium–Synaptichnium* plexus footprints and partial trackway from Timezgadiouine Formation (T3, Olenekian) of the Argana Basin, Morocco.

pre-Anisian chirotheriids and the evolution of their archosaurian producers. It is, furthermore, a bio-/ichnostratigraphically important index taxon for the Olenekian.

Trackmaker: Crocodylian-stem or noncrowngroup archosauriforms have been considered (Klein and Niedźwiedzki, 2012; Klein et al., 2013, 2023; Bernardi et al., 2015). This is based on the pes morphology with Digit III longest and the narrow trackway pattern. The latter feature also excludes more basal archosauriforms such as proterosuchids and erythrosuchids (Bernardi et al., 2015).

4.4.6 *Synaptichnium* Nopcsa (1923)

Fig. 5.12.

Type ichnospecies: *S. pseudosuchoides* Nopcsa (1923).

Other ichnospecies: *S. diabloense, S. cameronense, S. kotanskii.*

Distribution: Lower-Middle-?Upper Triassic. Wióry Formation (Olenekian) of Poland, Middle Triassic deposits of Germany, France, Great Britain, Spain and Italy, Timezgadiouine Formation (T3-T5, Olenekian-?Carnian) of the Argana Basin, Morocco, Moenkopi Group (Olenekian-Anisian) of the USA.

Diagnosis (from Klein and Lucas, 2021, emended after Haubold, 1971b; King et al., 2005): Moderately narrow trackways of a quadruped with a pace angulation that is only slightly larger than 160 degrees (140 degrees on average); imprints pentadactyl with strong outward rotation of the pes. Pes imprints with Digits I–IV increasing in length and Digit IV being longest or subequal with III; Digit V more laterally than proximally positioned behind front digits, slender, elongate, and straight or slightly curved outward and with oval basal pad. Manus more rounded, with Digit III being longest. Ratio of manus:pes area relatively large for chirotheriids = 1:1.8, up to 1:3.7 (1:2.5 on average). Sharp claws on Digits I–IV in both the pes and manus. Distinct, rounded phalangeal and metatarsophalangeal pad traces.

Description: *Synaptichnium* has a semiplantigrade, ectaxonic pes that is long and slender in overall shape. Pedal digits are increasing in length from I to IV with Digit IV being longest. Digit V is postero-laterally positioned, antero-laterally directed, and straight, with a round basal pad and a distinct, thinner phalangeal portion. Claw traces are elongated triangular and sharp. The manus is relatively large, round and with Digit III being longest. Trackways show a variable pattern with both high and relatively low pace angulation, long and short steps, obviously depending on different gaits and velocities of the producer. Pes imprints are strictly outwardly rotated, and manus imprints positioned parallel to the midline or slightly outwardly directed. The position of the latter is anterior to the pes or slightly antero-medial.

Discussion: *Synaptichnium* was introduced by Nopcsa (1923) based on isolated imprints from the Middle Triassic of Great Britain and later identified in Lower-Middle-?Upper Triassic deposits of Europe, North America, and North Africa (Haubold, 1971b; Klein and Lucas, 2010b, 2018, 2021 for overview; Klein et al., 2011; Lagnaoui et al., 2012). It differs from other chirotheriids by the pes imprint in which Digit IV is the longest.

Trackmaker: The ectaxonic pes and the relatively large manus reflect a more conservative ancestral foot constellation of the trackmaker, which can be seen in both proterosuchid noncrown-group archosauriforms and pseudosuchian archosaurs. Therefore, most likely different archosaur clades with a similar pes morphology were able to leave footprints of the *Synaptichnium* type, which is also supported by the long stratigraphic range of this ichnogenus.

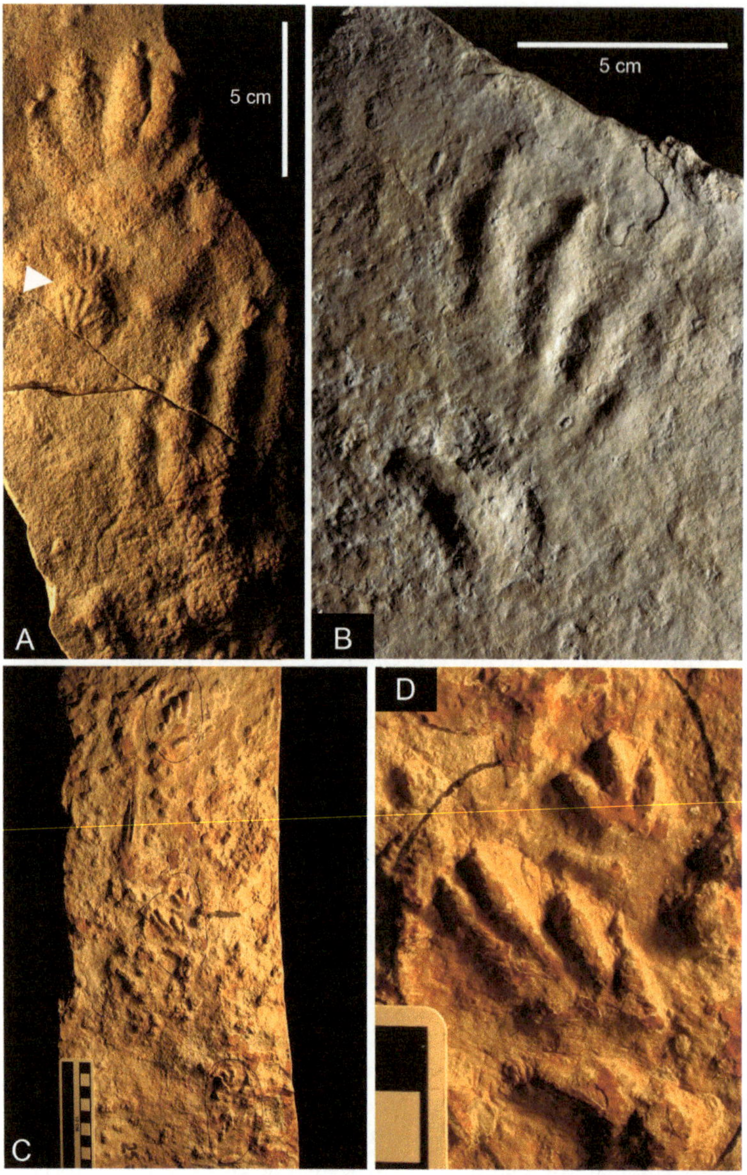

FIGURE 5.12 *Synaptichnium* (archosaur) footprints. (A) *Synaptichnium pseudosuchoides* from Eschenbach Formation (Middle Triassic, Anisian) of Germany. (B) *S. cameronense* from the Middle Triassic of the Southern Alps, Italy. (C–D) *S. diabloense* trackway and detail with pes-manus set from the Wupatki Member of the Moenkopi Formation (Lower Triassic, Olenekian) of northern Arizona.

4.4.7 *New ichnogenus (in prep.)*

Fig. 5.13.

Description: Specimens show pentadactyl pes and manus imprints. The pes has robust to moderately broad digits with rounded phalangeal and metatarso-phalangeal pads. In the pes Digits I–IV are subparallel in their orientation with Digit III being longest or subequal in length with IV, Digit II is shorter and I shortest. Digit V has a massive oval basal pad that can be elongated proximally into a larger "heel" area and distally into a thinner, often recurved phalangeal portion. The manus, which is positioned anteriorly or antero-medially to the pes, is relatively large compared with the pes, short and rounded with short digits. Manual Digit III is longest, II and IV are shorter and subequal in length, and I is shortest. Claw traces are small, narrow, and elongate, in the pes often separated from digits and in Digits II and III pointing antero-laterally. Trackways are narrow with high pace angulation with both pes and manus imprints being outwardly rotated.

Discussion: Problems arose with the identification of the ichnogenus *Brachychirotherium* in stratigraphically older (Lower-Middle Triassic) deposits. Various ichnospecies have been assigned to *Brachychirotherium* by different authors. Haubold (1971b) and Demathieu and Leitz (1982) listed *"B."praeparvum*, *"B."harrasense*, *"B."hessei*, and *"B."paeneparvum* from the Solling and Roet formations (Middle Triassic) of the German Buntsandstein. Others from the Middle Triassic of France and the Netherlands were introduced by Demathieu (1971), Demathieu and Gand (1973), and Demathieu and Oosterink (1983, 1988) as *"B."circaparvum*, *"B."pachydactylum* and *"B."paraparvum*. The *Brachychirotherium* diagnosis strictly applied to these older occurrences does not match the observed morphology of these materials. This concerns digit proportions with the relatively short pedal Digit IV and the lack of a distinct phalangeal portion in V, as well as the larger divarication of digits and the relatively small manus in *Brachychirotherium* versus a long Digit IV, a phalangeal part of V, subparallel orientation of digits and relatively large manus in the new ichnogenus. Furthermore, there is some morphological transition of purported Middle Triassic representatives to *Synaptichnium* (see Klein and Haubold, 2004; Klein and Lucas, 2018).

Trackmaker: Non-crown group archosauriform and/or pseudosuchian archosaurs.

4.4.8 *Chirotheriidae indet*

Fig. 5.14.

Further chirotheriid specimens have been described by different authors under "*Chirotherium*," "*Brachychirotherium*" or proposed new ichnogenera. We consider these assignments dubious and therefore list them under indeterminate forms. We give a short overview here. Some are based on incomplete materials, others are well-preserved but still wait for a detailed restudy.

4.4.8.1 *"Brachychirotherium tintanti"* Demathieu (1971)

Description: This is an isolated specimen being a pentadactyl manus imprint and associated fragmentary pes imprint showing terminal digit traces only.

Discussion: Demathieu (1971) illustrated this incomplete material as a new ichnospecies, however, the material is insufficiently preserved and cannot be determined more specifically than "indeterminate chirotheriid."

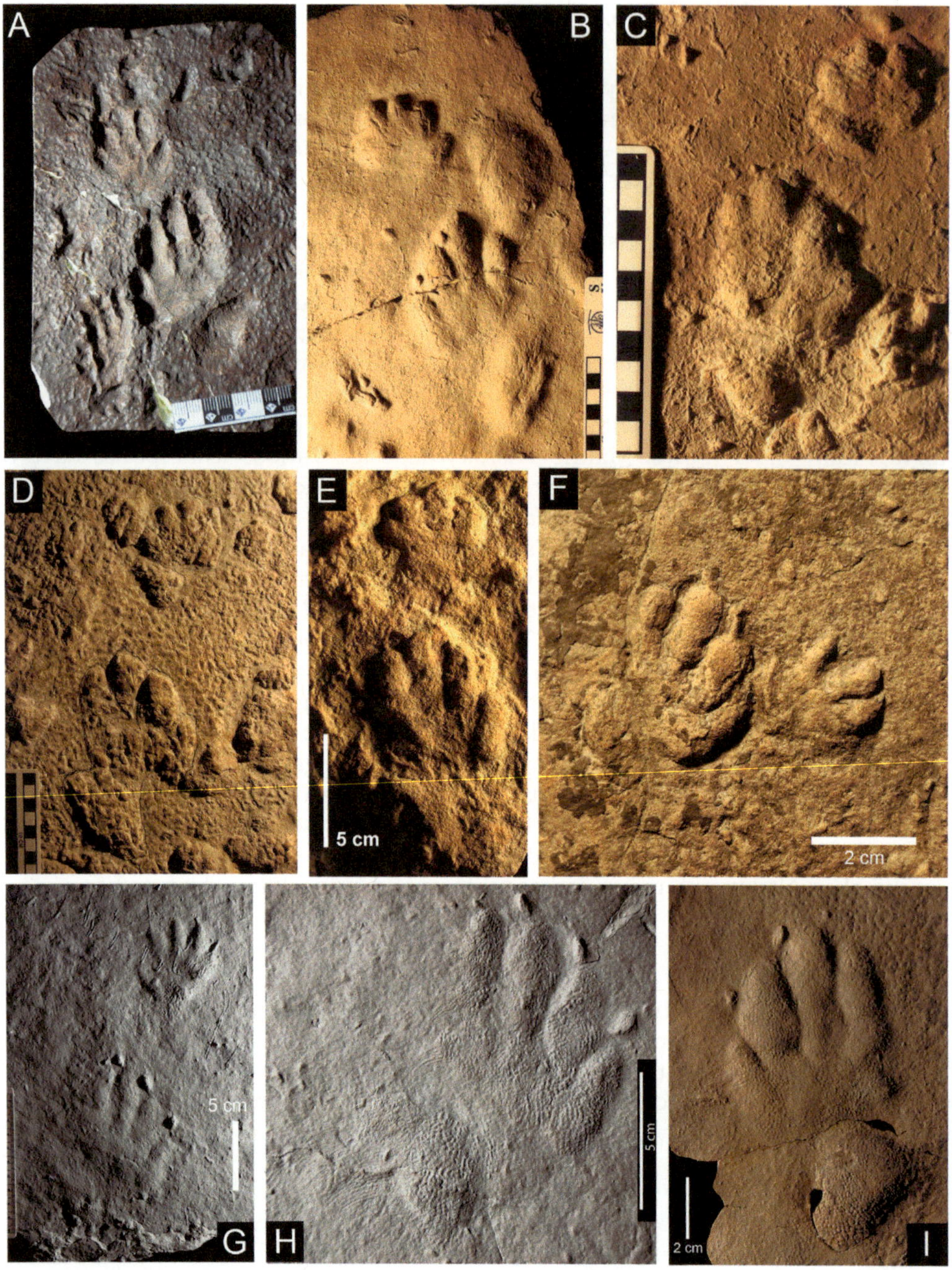

FIGURE 5.13 Footprints being considered to belong to a new ichnogenus. (A) "*Brachychirotherium*" *hessei* artificial cast of holotype pes-manus set from Upper Buntsandstein (Anisian) of Germany. (B–C) "*B.*"*circaparvum* pes-manus sets (Holotype) from the Middle Triassic of France. (D) "*B.*"*pachydactylum* from the Middle Triassic of France. (E) *Synaptichnium–Brachychirotherium*-like form pes-manus set from the Eschenbach Formation (Middle Triassic, Anisian) of Germany. (F) "*B.*"*praeparvum* small pes-manus set from Solling Formation (Middle Triassic, Anisian) of Germany. (G–H) "*B.*"*paraparvum* from Vossenveld Formation (Middle Triassic, Anisian) of The Netherlands. (I) "*B.*" aff. *parvum* from the Middle Triassic of the Southern Alps, Italy.

FIGURE 5.14 Chirotheriid archosaur footprints of uncertain ichnogeneric assignments. (A–C) "*Chirotherium*" *lulli* pes manus sets and isolated manus imprints from the Chinle Group (Upper Triassic, Norian–Rhaetian) of Utah, USA as photographs (A, B) and ambient occlusion photogrammetric image (C). (D) "*Chirotherium*" *wondrai* pes manus set (holotype) and pes from Steigerwald Formation (Upper Triassic, Carnian) of Germany; note tridactyl *Atreipus metzneri* associated on top. (E) Chirotheriid pes-manus set showing very large manus imprint at left. *(A–C) From Milner et al. (2021). (E) From Klein and Lucas (2018).*

4.4.8.2 *"Chirotherium atlensis"* Biron and Dutuit (1981)

Description: Several isolated chirotheriid pes and manus imprints (pes 15–20 cm in length) and two partial trackways on a larger slab from the Triassic (?Carnian) of the Ourika Basin in the Central High Atlas region of Morocco.

Discussion: Biron and Dutuit (1981) described these tracks under the new ichnospecies *Chirotherium atlensis*. However, the morphology of the imprints clearly excludes an assignment to the ichnogenus. Digit proportions with Digits II and III long and I and IV short as well as the stronger outward rotation of the pes suggest *Isochirotherium* as the probable ichnogenus being represented (see also Klein and Lucas, 2021). *Isochirotherium* is a characteristic Middle Triassic ichnogenus. Therefore, it cannot be excluded that the age of the track-bearing strata is Middle Triassic, not Upper Triassic (Carnian) as considered by Biron and Dutuit (1981). The material has recently been revised by Moreau et al. (2024), and these chirotheres were re-assigned to *Isochirotherium* isp. following suggestions of Klein and Lucas (2021).

4.4.8.3 *"Chirotherium catalaunicum"* Casanovas-Cladellas et al. (1979)

Description: Fragmentary pes track from the Buntsandstein (Middle Triassic) of the Catalonian Basin, Spain.

Discussion: This is incomplete, poor material and therefore assigned here to indeterminate chirotheriids.

4.4.8.4 *"Chirotherium (Brachychirotherium) gallicum"* Willruth (1917) (Bornemann ms.)

Description: Numerous chirotheriid footprints from the Middle Triassic of France, Spain, and Italy show a pentadactyl pes with proportions III > IV > II > I and a short and rounded manus with Digits II and III being longest. Digit V in the pes has a proximally elongated basal pad and a distinct thinner phalangeal portion distally.

Discussion: *Chirotherium* (*Brachychirotherium*) *gallicum* has been commonly identified from Middle Triassic deposits of central and southern Europe (Willruth, 1917; Leonardi, 1959; Demathieu, 1974; Pérez López, 1993; Demathieu and Demathieu, 2004; Gand et al., 2010; Marchetti et al., 2021). However, it appears that this ichnospecies may represent a wastebasket ichnotaxon for different chirotheriids that should better be reassigned to *Chirotherium barthii, Isochirotherium,* and others. Based on the overall shape of the imprints and digit proportions some of these footprints described under *Chirotherium* or *Brachychirotherium gallicum* show some similarities with the proposed new chirotheriid ichnogenus (see above). However, prior to a detailed restudy we prefer a tentative assignment to indeterminate chirotheriids.

4.4.8.5 *"Chirotherium (?) huberi"* Bock (1952)

Description: Incomplete isolated imprint from the Lower Lockatong Formation (Newark Supergroup, Carnian) of Pennsylvania.

Discussion: This material figured by Bock (1952) is too fragmentary and therefore cannot be determined with certainty.

4.4.8.6 "Chirotherium" lulli Bock (1952)

Fig. 5.14A.

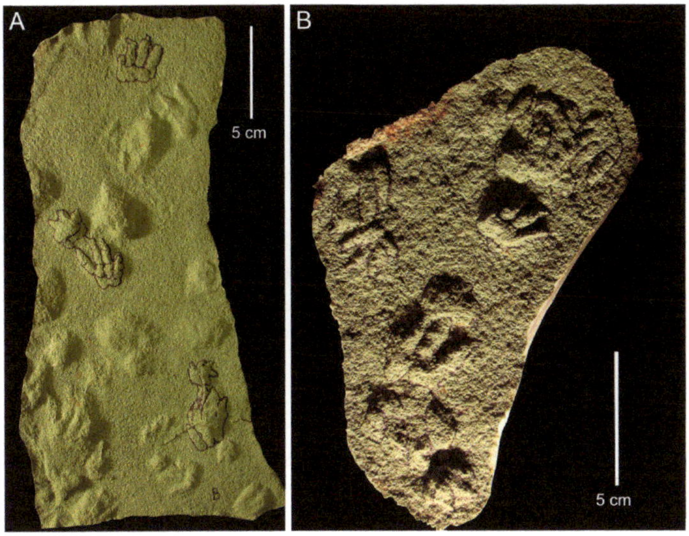

FIGURE 5.15 *Batrachopus* isp. (Crocodylomorph) partial trackway and isolated imprints from Lower Jurassic Moenave Formation of Arizona. Similar footprints are known from Triassic-Jurassic strata of the Newark Supergroup in eastern North America.

Description: These are small chirotheriid tracks (4.4 cm pes length) with pedal digit proportions III > IV > II > I. Digits are long and slender with sharp, elongated claw traces. The posterolaterally positioned Digit V has an oval basal pad and a distinct thinner and recurved phalangeal portion. Type trackway with 23.2 cm stride length and a relatively low pace angulation of 150 degrees. Pes and manus imprints strongly rotated outward by 29 degrees

Discussion: The ichnospecies *Chirotherium lulli* was introduced by Bock (1952) based on footprints from the Passaic Formation (Norian, Newark Supergroup) of New Jersey and studied in detail by Baird (1954). More recently, well-preserved *"C."lulli* pes and manus imprints have been discovered in the Chinle Group of Utah, which is the first evidence from western North America (Milner et al., 2021). An assignment to the ichnogenus *Chirotherium* is considered here problematic because diagnostic features of *Chirotherium* such as the functionally tridactyl mesaxonic pes cannot be observed in *"C."lulli*. Furthermore, the strong outward rotation of both pes and manus imprints and the relatively low pace angulation in *"C."lulli* is different from *Chirotherium*, which shows a minor outward rotation of the pes and manus. In overall appearance *"C."lulli* represents a more conservative, slightly ectaxonic pes morphology with a long Digit IV and distantly resembles *Synaptichnium*. The trackmaker was probably a pseudosuchian archosaur.

4.4.8.7 "Chirotherium" wondrai Heller (1952)

Fig. 5.14B.

Description: All footprints in the material are preserved as convex hyporeliefs on the lower surface of a sandstone bed. The pes imprints are up to 20 cm in length. Digit III is longest, followed by Digits II and IV that are subequal in length and Digit I that is shortest. The manus is short and rounded with Digit III longest, II and IV shorter, I shortest. Digit V in the pes with a massive oval basal pad and an elongated, thin phalangeal portion that is slightly recurved. Claw traces in the pes are thin and elongated, resulting from dragging movement; in the manus, these are short triangular, and acuminate. The type material is deeply impressed, showing lateral striations caused by the scales when sinking into the substrate.

Discussion: Based on the recent redefinition of *Chirotherium*, features of "*C.*"*wondrai* do not match the diagnosis of this ichnogenus. In particular, this concerns the lack of a functionally tridactyl, mesaxonic pes. "*C.*"*wondrai* has a rather broad appearance of the pes and relatively short, thick digits with rounded pads. There is some slight resemblance to *Brachychirotherium*, but the latter has no recurved phalangeal portion of Digit V. Because of the restricted number of specimens and isolated imprints from a single locality we are cautious with an ichnogeneric assignment, but keep the ichnospecies here considering the peculiar morphology of the imprints.

4.4.8.8 "*Dahutherium*" Montenat (1968)

Description: These are tetradactyl pes imprints (about 10 cm in length) with a longest Digit III, followed by IV and II. Digit I is shortest, thin, and posteriorly shifted. In some specimens, a small tri- to tetradactyl manus is associated anterior to the pes.

Discussion: Montenat (1968) gives no formal description and designation of a holotype. Therefore, "*Dahutherium*" can be considered a nomen nudum. There are similarities with the ichnogenus *Chirotherium*. Nevertheless, the incomplete and isolated specimens lead us to consider it an indeterminate chirotheriid (see also Klein and Lucas, 2021).

4.4.8.9 "*Parasynaptichnium*" Mietto (1987)

Description: Incomplete type material lacking the posterior portion of the pes. Pedal digits are slender and curved inward, showing small, triangular claws. Digit III is longest, followed by Digits IV and II, and Digit I that is shortest. The manus is pentadactyl with Digit III being slightly longer than Digits II and IV and Digits I and V being short.

Discussion: The type material is too fragmentary to justify a valid ichnogenus. It clearly represents a chirotheriid but does not permit a more specific assignment.

4.4.8.10 "*Pentichnus*" Biron and Dutuit (1981)

Description: Isolated, large, and broad pentadactyl imprints (28 cm in length and 31 cm in width) showing robust digit traces and triangular claws. Digits II and III are longest, I and IV shorter. Digit V is in a postero-lateral position to the anterior portion of the imprint.

Discussion: This is obviously a chirotheriid imprint, either pes or manus, but the material is too incomplete to be more specifically determined.

4.4.8.11 "*Pseudochirotherium*" Biron and Dutuit (1981)

Description: The ichnogenus *Pseudochirotherium* is based on isolated, incomplete, probably pentadactyl imprints.

Discussion: "*Pseudochirotherium*" is one of numerous new ichnotaxa introduced by Biron and Dutuit (1981) based on Triassic material from the Argana and Ourika basins of Morocco. Several have been re-assigned to other ichnogenera, or listed as indeterminate chirotheriids or *nomina dubia* by Lagnaoui et al. (2012, 2016). We consider "*Pseudochirotherium*" an indeterminate chirotheriid considering the fragmentary material for which this ichnogenus was introduced.

4.4.8.12 "*Swinnertonichnus*" Sarjeant (1967)

Description: Isolated, tridactyl pes imprint showing robust digit traces. The central Digit III is longest, while the lateral Digits II and IV are shorter. No claw traces are preserved.

Discussion: Even if being incomplete, from overall shape this material suggests a chirotheriid pes imprint. However, no distinct determination of an ichnogenus or ichnospecies is possible.

4.4.8.13 "Large manus chirotheriid"

Fig. 5.14C.

Description: Medium-sized (16 cm pes length) pes and manus imprints showing low heteropody. Digits are relatively short and broad. In the pes Digit III is longest, Digit IV and II are slightly shorter, I is shortest. Digit V has a massive basal pad that is proximally extended into a "heel," and with a thinner phalangeal portion distally. In the manus Digit III is longest, followed by II and IV, the latter being laterally spread. Digit V in the manus consists of an oval basal pad only. The manus is laterally overstepped by the pes.

Discussion: Klein and Lucas (2018) have published this material from the siliciclastic Lower Muschelkalk facies of Germany. The very low heteropody (1:1.4) is remarkable, and similar values have never been measured in chirotheriid footprints. For comparison, *Synaptichnium* is known thus far to have the lowest heteropody (1:1.8–3.7) of all chirotheriids. There is some morphological similarity with the proposed "new ichnogenus" (see above), however the latter has much higher heteropody. More material is needed to confirm that this feature is continuously occurring along complete trackways and larger track surfaces and therefore can confidently be considered anatomically based.

4.5 Crocodylomorph

4.5.1 Batrachopus *Hitchcock (1845)*

Fig. 5.15.

Type ichnospecies: *B. deweyi* (Hitchcock, 1843)

Other ichnospecies: Presently only the type ichnospecies is considered here as a valid ichnospecies. Further ichnospecies introduced by Hitchcock (1845, 1858) and Lull (1904, 1953) from Lower Jurassic deposits of the Newark Supergroup in eastern North America have been discussed among ichnologists without achieving a broader acceptance and could not be convincingly discriminated (see Rainforth, 2005; Klein and Lucas, 2021 for overview).

Distribution: Upper Triassic-Lower Jurassic-?Lower Cretaceous. Newark Supergroup and Glen Canyon Group (Lower Jurassic) of USA,? Lower Cretaceous of Thailand.

Diagnosis: Trackways of a quadruped with relatively low pace angulation. Pes imprints are functionally tetradactyl and digitigrade, with moderately spread digits. Digit III is longest, and Digit I shortest. Digit V is reduced to an oval pad posterior to and nearly in line with Digit III. The manus is pentadactyl and shows a wide digit divarication. It is strongly rotated outward relative to the pes, with Digit V pointing backward.

Description: Small tetradactyl pes imprints and pentadactyl manus imprints. In both, Digit III is longest, followed by IV and I. Digit V in the manus short if present, in the pes an oval pad at the posterior end might represent the reduced Digit V or simply a heel pad. Trackways show moderate outward rotation of the pes and extreme outward rotation of the manus.

Discussion: Hitchcock (1845) introduced this ichnotaxon based on material from Lower Jurassic deposits of the Newark Supergroup in eastern North America. Triassic representatives are scarcely documented, some are doubtful and might represent incomplete *Brachychirotherium* (Klein and Lucas, 2010c, 2021).

Trackmaker: Small terrestrial crocodylomorphs such as sphenosuchids have been considered (Klein and Lucas, 2010c).

4.6 Dinosauriform–dinosaur

4.6.1 Anchisauripus Lull (1904)

Fig. 5.16A.

Type ichnospecies: *A. sillimani* (Hitchcock, 1843)

Other ichnospecies: Only the type ichnospecies is considered here as valid. Thus far a convincing discrimination of Triassic grallatorid footprints, in particular at the ichnospecies level, has failed. Therefore, prior to a comprehensive revision, we prefer to use the type ichnospecies name only.

Distribution: Upper Triassic-Lower Jurassic. Upper Triassic to Lower Jurassic deposits of the Newark Supergroup in eastern North America, Ørsted Dal Formation of the Fleming Fjord Group of eastern Greenland, Upper Triassic of France, Sweden, and Slovakia, ?Upper Triassic-Lower Jurassic of the Stormberg Group, southern Africa.

Diagnosis (from Olsen et al., 1998): Medium-sized (~20 cm long), functionally tridactyl ichnite in which the Digit III projection ratio averages about 1.4, and the length-to-width ratio is about 2.0. Digit II tends to project more than IV along the axis of Digit III. Divarication of outer digits averages 27 degrees

Description: These are tridactyl, mesaxonic pes imprints of 15–25 cm length with slender digits and tapering claws. Digit III projects far beyond outer Digits II and IV that are subequal in length. This feature is the best criterion to avoid confusion with incomplete footprints of *Chirotherium*.

Discussion: *Anchisauripus* was originally described by Lull (1904) based on Lower Jurassic material from the Newark Supergroup of eastern North America. The ichnogenus was also identified from Upper Triassic deposits of North America (e.g., Weems, 1987) and from Middle-Upper Triassic deposits of Europe (Demathieu, 1989; Demathieu and Demathieu, 2004; Gand and Demathieu, 2005; Niedźwiedzki, 2011). Lower Jurassic *Anchisauripus* has been discussed as part of a continuum with different size classes: *Grallator* (<15 cm pes length)–*Anchisauripus* (15–25 cm pes length)–*Eubrontes* (>25 cm pes length) (Olsen et al., 1998). Triassic representatives of *Anchisauripus*-size grallatorids, in particular those from the Middle Triassic, have been included in the *Atreipus*–*Grallator* plexus, together with other tridactyl footprints of facultative bipeds formerly assigned to "*Coelurosaurichnus*" by Haubold and Klein (2000, 2002; see below).

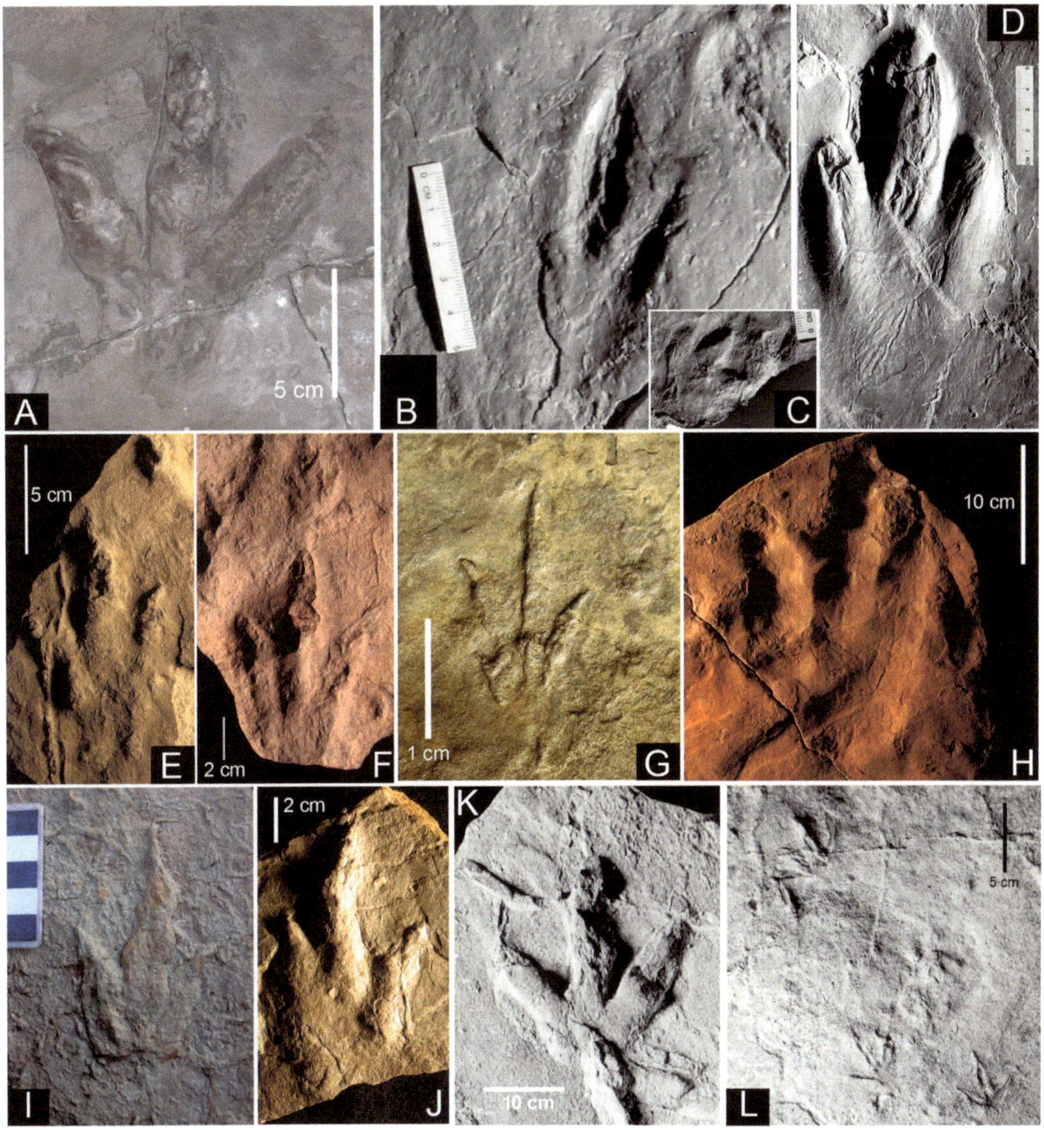

FIGURE 5.16 Tridactyl mesaxonic dinosauromorph–dinosaur footprints from the Upper Triassic. (A) Grallatorid footprint from the Ørsted Dal Formation (Norian–Rhaetian) of East Greenland. (B–D) *Atreipus milfordensis* pes-manus set, isolated manus and isolated pes from the Passaic Formation (Norian) of Pennsylvania. (E) *Atreipus* isp. pes-manus set from the Hassberge Formation (Carnian) of Germany. (F) *Atreipus* isp. from Bigoudine Formation (T6, ?Norian) of the Argana Basin, Morocco. (G) *Banisterobates boisseaui* pes-manus set from Dry Fork Formation (Carnian) of Virginia, USA; note presence of a hallux (pedal Digit I impression and very small round manus imprint. (H) *Eubrontes* isp. pes from Timezgadiouine Formation (T5) of the Argana Basin, Morocco. (I) *Grallator* isp. pes from Redonda Formation (Norian–Rhaetian) of New Mexico. (J) *Grallator* isp. From Hassberge Formation (Carnian) of Germany. (K) *Pengxianpus xifengensis* pes, holotype from Xujiahe Formation (Rhaetian) of Sichuan Province, China. (L) *Trisauropodiscus aviforma* ("bird-like") pes imprints from Lower Elliot Formation (Norian) of Lesotho, southern Africa. *(B–D) From Lucas and Sullivan (2006).* (F, H) *From Lagnaoui et al. (2012).* (K) *From Xing et al. (2013b).*

Trackmaker: Medium-sized theropod dinosaurs were the most probable producers of *Anchisauripus* tracks (Farlow and Lockley, 1993; Smith and Farlow, 2003; Lucas et al., 2006).

4.6.2 *Atreipus* Olsen and Baird (1986)

Fig. 5.16B.

Type ichnospecies: *A. milfordensis* (Bock, 1952).

Other ichnospecies: *A. metzneri* (Heller, 1952), *A. sulcatus* (Baird, 1957), *A. acadianus* (Olsen and Baird, 1986).

Distribution: Middle Triassic-Lower Jurassic. Lockatong and Passaic formations of the Newark Supergroup in eastern North America, Rock Point Formation of the Chinle Group of the western USA, Steigerwald and Hassberge formations (Keuper, Carnian-Norian) of northern Bavaria, Germany, Carnian of the Ardèche region, France, Travenanzes Formation (Carnian) of northern Italy, Timezgadiouine and Bigoudine formations (T5-T6, Carnian-Norian) of the Argana Basin, Morocco. Middle Triassic occurrences have been described and are included in an *Atreipus-Grallator* plexus by some authors (Haubold and Klein, 2002).

Diagnosis (after Olsen and Baird, 1986): Small (9–14 cm long), tulip-shaped pes impression with metatarsal-phalangeal pads of Digits II and IV that are oval to circular and often impressed. Distal phalangeal pads often more indistinct than more proximal pads. Hallux impression absent. Relative proportions of pes very similar to *Grallator*. Manus much smaller than pes, with Digit III longest, followed in length by II, IV, and I.

Description: Trackways of a quadrupedal producer with small (9–14 cm length) tridactyl mesaxonic pes imprints with Digit III longest and with tridactyl to pentadactyl manus imprints. The pes often shows distinct, round metatarsophalangeal pads in Digits II and IV. Manus imprints commonly lack digit traces I and V and have a short, laterally spread Digit IV.

Discussion: *Atreipus* has been erected based on Upper Triassic footprints from the Newark Supergroup of North America (Olsen and Baird, 1986; Lucas et al., 2006). The ichnogenus has also been documented from Middle-Upper Triassic deposits of Europe (Haubold and Klein, 2000, 2002; D'Orazi Porchetti et al., 2008) and North Africa (Lagnaoui et al., 2012, 2016). Haubold and Klein (2000, 2002) included material formerly described under the (now abandoned) name "*Coelurosaurichnus*" from Middle-Upper Triassic strata of Europe in the *Atreipus–Grallator* plexus, considering a facultative bipedal trackmaker leaving both *Atreipus* (with associated manus print) and *Grallator* (lacking a manus print) footprints.

Trackmaker: Different dinosauromorph-dinosaur groups have been considered as possible trackmakers. Olsen and Baird (1986) attributed *Atreipus* to ornithischian dinosaurs, Thulborn (1990) saw theropods as the probable trackmakers and Haubold and Klein (2000, 2002), recognizing evolutionary developments in the locomotor apparatus from quadrupedal to bipedal movement, interpreted Triassic representatives of this ichnogenus as dinosauromorphs.

4.6.3 *Banisterobates* Fraser and Olsen (1996)

Fig. 5.16C.

Type ichnospecies: *B. boisseaui* Fraser and Olsen (1996).

Other ichnospecies: Only known from the type ichnospecies.

Distribution: Only known from one locality in the Dry Fork Formation (Upper Triassic, Carnian) of the Dan River Group, Virginia, USA.

Diagnosis (after Fraser and Olsen, 1996): Ichnite of a small quadruped with tetradactyl pes but with Digit I very much reduced so that the pes is essentially mesaxonic. Digit III markedly longer than II and IV and Digit II slightly longer than IV. Manus poorly defined but with three very short digit impressions.

Description: This is a trackway consisting of three successive pes and manus imprints left by a very small individual. The pes is 1.8–2.5 cm in length, mesaxonic, functionally tridactyl (II, III, IV), occasionally with an impression of a short hallux (Digit I) and with sharp claw traces. The manus is very small with indistinct digit traces and positioned laterally to the pes. The posterior end of the pes track can be extended into a "heel" area. The trackway shows a stride of 12.2 cm, a relatively low pace angulation of 146 degrees and moderate outward rotation of the pes.

Discussion: *Banisterobates* is the trackway of a small dinosauriform and slightly resembles *Atreipus* by the presence of a grallatorid pes in combination with a small manus imprint. However, *Banisterobates* shows the trace of a hallux that is unknown in *Atreipus*. Because trackways of such small individuals of the *Atreipus* trackmaker are unknown there is still the possibility that *Banisterobates* could be a juvenile or "baby" *Atreipus*, and the presence of a hallux trace reflects a different gait and pes posture if compared to adults. This is supported by the plantigrade appearance of pes imprints, but this cannot be proved presently.

Trackmaker: A small quadrupedal dinosauromorph or dinosauriform is considered the most probable producer (see Klein and Lucas, 2021). Whether this was a juvenile or a small adult species cannot be decided.

4.6.4 *Eoanomoepus* Lockley et al. (2018)

Fig. 5.16D.

Type ichnospecies: *E. latus* Lockley et al. (2018).

Other ichnospecies: Only known from the type ichnospecies.

Distribution: Chinle Group (Late Triassic, Norian-Rhaetian) near Moab in southeastern Utah, USA.

Diagnosis (from Klein and Lucas, 2021; after Lockley et al., 2018): Transverse tridactyl tracks only two thirds as long as wide, hence more transverse than typical *Anomoepus* with digit traces little connected, and digit divarication wide but variable. Step short, trackway very wide, with low pace angulation, and outward rotation of pes. *Eoanomoepus* differs from *Anomoepus* in length/width ratio (0.60–0.70 vs. ~1.08) and lack of obvious inward rotation characteristic of the latter, well-known ichnogenus. The trackway width of *Eoanomoepus* differs from *Anomoepus* in being unusually wide, also with usually an irregular step and stride dimensions. *Eoanomoepus* also differs from *Anomoepus* in lacking evidence of a hallux (Digit I) trace and being more digitigrade.

Description: Small (9.0–9.6 cm long, 13.4–15.0 cm wide) and wider than long pes imprints. Digits are very slender with wide divarication (II–IV up to 109 degrees) and little connected. They lack pad impressions. Trackways with short steps, low pace angulation (<127 degrees), and pes imprints rotated outward.

Discussion: Differences from typical *Anomoepus,* such as the larger width of imprints, lack of a hallux trace, the larger trackway width, and outward rotation of imprints versus inward rotation in *Anomoepus*, differentiates this ichnogenus. The thin digit traces lacking pad impressions in the type material (Lockley et al., 2018, Fig. 9) resemble typical penetrative tracks (Falkingham et al., 2020; Turner et al., 2020; Lallensack et al., 2022) and therefore could be a preservational feature. This would be supported by the lack of a connection of digits in most pes imprints. There is also some similarity in imprint morphology with *Trisauropodiscus* (see below).

Trackmaker: A possible ornithischian dinosaur is proposed by Lockley et al. (2018).

4.6.5 Eubrontes *Hitchcock (1845)*

Fig. 5.16E.

Type ichnospecies: *E. giganteus* Hitchcock (1845).

Other ichnospecies: Prior to a comprehensive reanalysis of Triassic tridactyls we prefer to assign all representatives of the ichnogenus to the type ichnospecies to avoid unnessessary ichnotaxonomic splitting. It cannot be excluded that future studies will provide good arguments for discrimination of further ichnospecies.

Distribution: Upper Triassic (Carnian)–?Lower Cretaceous. Reports of Cretaceous *Eubrontes*, in particular from China, are difficult to confirm because of the unclear ichnotaxonomy of Chinese large tridactyl tracks from this time interval (see discussion and review in Xing et al., 2021). While Jurassic, in particular Early Jurassic, occurrences are abundantly known from a global record, Triassic *Eubrontes* is less abundant but commonly known from North America (USA), Greenland, North Africa (Morocco), southern Africa (Lesotho), and western Europe. In detail the material comes from the Newark Supergroup (Norian-Rhaetian) of Pennsylvania, New Jersey, and Virginia, the Blue Mesa Member of the Petrified Forest Formation (Upper Carnian) of Arizona, the Sloan Canyon Formation (Norian) of northeastern New Mexico, the Lower Elliot Formation (Norian-Rhaetian) of southern Africa, the Irohalene Member (T5, Carnian) of the Timezgadiouine Formation of the Argana Basin, Morocco, the Mercia Mudstone Group (Norian-Rhaetian) of Wales, Great Britain, the Hauptdolomit Group (Norian) of the Swiss Alps, Norian-Rhaetian deposits of southern, southwestern and possibly western France, Carnian deposits of northern Italy, the Tomanová Formation (Rhaetian) of the Tatra Mountains of Poland and Slovakia, the Bjuv Member of the Höganäs Formation (Rhaetian) of southern Sweden, and the Ørsted Dal Formation of the Fleming Fjord Group (Norian) of east Greenland. Previously reported Australian Triassic occurrences from the Sidney Basin have been reassigned recently to *Evazoum* (Romilio et al., 2021).

Diagnosis (from Olsen et al., 1998): Large (>25 cm long), functionally tridactyl ichnite in which the Digit III projection ratio is about 2.2, and the length-to-width ratio is about 1.4–1.5. Projection of Digits II and IV along the axis of Digit III about equal. Divarication of outer digits 30–40 degrees.

Description: These are large (>25 cm) grallatorid pes imprints known from surfaces with extensive trackways. Imprints are mesaxonic and relatively wide, with Digit III projecting less far beyond Digits II and IV when compared to conditions in typical *Grallator*. Digits are robust with well-developed rounded phalangeal pads and triangular acuminate claws. A large round metatarso-phalangeal pad IV is always and a small round metatarso-phalangeal pad II is occasionally preserved. Furthermore, in some specimens a hallux trace can be observed.

Discussion: *Eubrontes*, originally described from the Lower Jurassic of the Newark Supergroup in eastern North America, has been documented also from Upper Triassic deposits in North America, Europe, North Africa, and southern Africa. It has been discussed as part of a continuum with ontogenetic growth stages represented by the ichnogenera *Grallator* < *Anchisauripus* < *Eubrontes*. *Grallator*, which is the smallest (<15 cm pes length), shows the largest projection of Digit III beyond outer Digits II and IV, and *Eubrontes* (>25 cm) the smallest one, while *Anchisauripus* (15–25 cm pes length) shows an intermediate condition between *Grallator* and *Eubrontes*. Olsen et al. (1998) revised the type material of these three ichnogenera.

Trackmaker: The producer of *Eubrontes* tracks were relatively large theropods with a symmetrical mesaxonic pes. This has been widely accepted by most researchers. Nevertheless Weems (2003, 2019) argued for a sauropodomorph (e.g., *Plateosaurus*) origin of these tracks. Because of large differences in the morphology of the tracks and the plateosaur pes, we consider this interpretation as unlikely, following Lucas et al. (2006), Farlow et al. (2018), and several others.

4.6.6 Grallator Hitchcock (1858)

Fig. 5.16F.

Type ichnospecies: *G. cursorius* Hitchcock (1858).

Other ichnospecies: Only the type ichnospecies is considered here as valid prior to a comprehensive reanalysis of Triassic tridactyl grallatorid footprints.

Distribution: *Grallator* has a wide distribution in Upper Triassic deposits of North America (Newark Supergroup, Chinle Group, Carnian-Rhaetian), Greenland (Ørsted Dal Formation, Flemingfjord Group, Norian-Rhaetian), western Europe (Keuper, Carnian-Norian), North Africa (Morocco, Timezgadiouine Formation, T5-T6, Carnian-Norian), southern Africa (Lower Elliot Formation, Norian-Rhaetian) and Australia (Ipswich Coal Measures).

Diagnosis (from Olsen et al., 1998): Small (<15 cm long), bipedal, functionally tridactyl ichnite. Digit III projects relatively farther anteriorly, and the foot is narrower than in *Eubrontes* and *Anchisauripus* (length/width ratio close to or greater than 2). Hallux rarely impressed. Divarication of outer digits 10 degrees–30 degrees.

Description: This is the smallest (<15 cm) and most slender grallatorid of the proposed continuum *Grallator* < *Anchisauripus* < *Eubrontes*. Digit III is by far longest, while Digits II and IV are shorter and subequally projecting. The divarication between Digits II and IV is small. Well-rounded phalangeal pads are present, and a metatarso-phalangeal pad is mostly preserved in Digit IV only. Small, sharp triangular claw traces are impressed. Trackways show long strides (up to 117 cm) and a high pace angulation (up to 175 degrees). Orientation of pes imprints is parallel to the midline or rotated outward.

Discussion: Like *Anchisauripus* and *Eubrontes*, *Grallator* was first described from Lower Jurassic deposits of the Newark Supergroup in eastern North America (Hitchcock, 1858). Besides occurrences in Middle Jurassic up to Cretaceous strata it has been identified from Upper Triassic deposits in North America, South America Europe, North Africa, southern Africa, and East Asia (see Klein and Lucas, 2021, for overview). The long stratigraphic range suggests that different dinosauriform-dinosaurian producers with similar foot morphology left nearly identical pes prints assignable to the ichnogenus *Grallator*. Triassic representatives, in particular forms from the Upper Triassic, some formerly assigned to "*Coelurosaurichnus*," were placed in the *Atreipus–Grallator* plexus by Haubold and Klein (2000, 2002).

Trackmaker: Relatively small theropods are widely accepted as the trackmakers.

4.6.7 *Pengxianpus* Yang and Yang (1987)

Fig. 5.16G.

Type ichnospecies: *P. cifengensis*Yang and Yang (1987).

Other ichnospecies: Only known from the type ichnospecies.

Distribution: Only known from the Xujiahe Formation (Norian-Rhaetian) of Sichuan Province, China.

Diagnosis (after Xing et al., 2013): Large (>25 cm length), tridactyl, mesaxonic pes imprints of a biped showing a wide Digit II–IV divarication angle (up to 69 degrees). Digits of long and slender shape with rounded pads, Digit III the longest, followed by II and IV. Small, but pronounced circular metatarsal-phalangeal pad on Digit IV.

Description: The holotype material consists of a partial trackway with two successive mesaxonic, nearly symmetrical pes imprints (>25 cm pes length). Digit III is longest and Digit IV slightly shorter than II. Sharp, triangular claw traces are visible. Round metatarso-phalangeal pads are present, most distinct and separated by a notch from the phalangeal digit portion in Digit IV.

Discussion: The material from the Xuijahe Formation (Norian-Rhaetian) of Sichuan Province, China was first described by Yang and Yang (1987) and redescribed by Lockley and Matsukawa (2009) and finally by Xing et al. (2013b). There is some morphological similarity with *Eubrontes* or *Kayentapus*. The large digit divarication II–IV (up to 69 degrees) is different from typical *Eubrontes* and, together with the relatively slender digits, and the separated round metatarso-phalangeal pad IV, slightly resembles *Kayentapus*. The restricted sample, however, does not allow a reliable synonymization with one of these ichnotaxa. Therefore, we prefer to keep *Pengxianpus* presently as a distinct ichnogenus.

Trackmaker: Relatively large theropod similar to the *Eubrontes* trackmaker.

4.6.8 *Trisauropodiscus* Ellenberger (1970)

Fig. 5.16H.

Type ichnospecies: *T. aviforma* Ellenberger (1970).

Other ichnospecies: Only the type ichnospecies is considered here as valid.

Distribution: Lower-Upper Elliot Formation (Upper Triassic-Lower Jurassic) of southern Africa, Imilchil Formation (Middle Jurassic, Bajocian-Bathonian) of Morocco.

Diagnosis: Functionally tridactyl (with main digit group II–IV) footprints of small to medium-sized bipeds with occasional presence of a trace of the reversed semifunctional hallux. Footprints mesaxonic-symmetrical with Digit III being longest and lateral digits that are subequal in length. Tracks lack metatarsophalangeal pads of Digits II and IV, while that of Digit III is occasionally present, but not clearly separated from the phalangeal portion of Digit III. Digits are very narrow and widely divaricated (>90 degrees).

Description: Birdlike digitigrade, symmetrical mesaxonic pes imprints with slender digits, wide digit divarication II–IV, and sharp claw traces. Digits are functionally tridactyl and often show a reversed short hallux (Digit I) trace.

Discussion: The ichnogenus was introduced by Ellenberger (1970) based on footprints from the Lower Elliot Formation (Upper Triassic, Norian-Rhaetian) of Lesotho, southern Africa. Other material was described from the Upper Elliot Formation (Lower Jurassic) of this region (Ellenberger, 1970, 1972, 1974; Abrahams et al., 2022) and from the Middle Jurassic of Morocco (Gierliński et al., 2017; Klein et al., 2023a). There is some similarity of the imprints with those of the ornithischian ichnogenus *Anomoepus*, and, for example, Lockley and Harris (2010) considered *Trisauropodiscus* to be a junior synonym of the latter. Gierliński et al. (2017), however, confirmed the validity of the ichnogenus based on new material from the Middle Jurassic of Morocco and a detailed comparison with the type from southern Africa (see also Klein et al., 2023a). A further aspect in the discussion could be the preservation of some of the *Trisauropodiscus* materials that show very slender digit traces and could possibly indicate penetrative tracks (Gatesy and Falkingham, 2020; Falkingham et al., 2020; Turner et al., 2020; Lallensack et al., 2022). However, this has to be confirmed by further study.

Trackmaker: Most likely a nonavian theropod with birdlike feet (see discussions in Gierliński et al., 2017; Klein et al., 2023a).

4.7 Sauropodomorph

4.7.1 Eosauropus *Lockley et al. (2006a)*

Fig. 5.17A.

Type ichnospecies: *E. cimarronensis* Lockley et al., (2006a).

Other ichnospecies: Only known from the type ichnospecies.

Distribution: Revueltian-Apachean (Norian) of western USA (New Mexico, Arizona, Colorado, Utah), Greenland, China? western Europe.

Diagnosis: Narrow-gauge trackway of large quadruped with strong heteropody (pes larger than manus) and short step and stride. Pes elongate, oval, tetradactyl to pentadactyl, with long axis and prominent distal claw impressions rotated outward. Manus transverse, tetradactyl to pentadactyl with outwardly rotated digit impressions and concave posterior margins.

Description: These are trackways of quadrupeds with narrow gauge, short steps, low pace angulation, and strong outward rotation of pes and manus imprints. Tracks show strong heteropody, which means the manus is much smaller than the pes. The pes which is positioned anteriorly or antero-medially, close to the pes and can be slightly overstepped posteriorly by the latter, is plantigrade, elongate oval, and tetradactyl to pentadactyl, with short digit traces that show tapering ends and are pointing outwardly. Pes lengths with a mean of 21.7 cm and widths with a mean of 19.7 cm. The tetradactyl-pentadactyl manus is much smaller, short, and broad (7.7 cm long, 12.7 cm wide in holotype), showing a posteriorly concave margin.

Discussion: When discovered these trackways have been confused with those of chirotheres based on their overall appearance with a chirothere-like pes and the much smaller manus (Conrad et al., 1987). Later their distinct features and their sauropod nature were recognized (Lockley et al., 2001). Ichnotaxonomically they were referred to the ichnogenus *Tetrasauropus*, originally known from the Lower Elliot Formation (Upper Triassic, Norian-Rhaetian) of southern Africa (Ellenberger, 1970, 1972), and finally reassigned to a distinct ichnogenus *Eosauropus*, based on a comprehensive study of material from the Chinle Group of

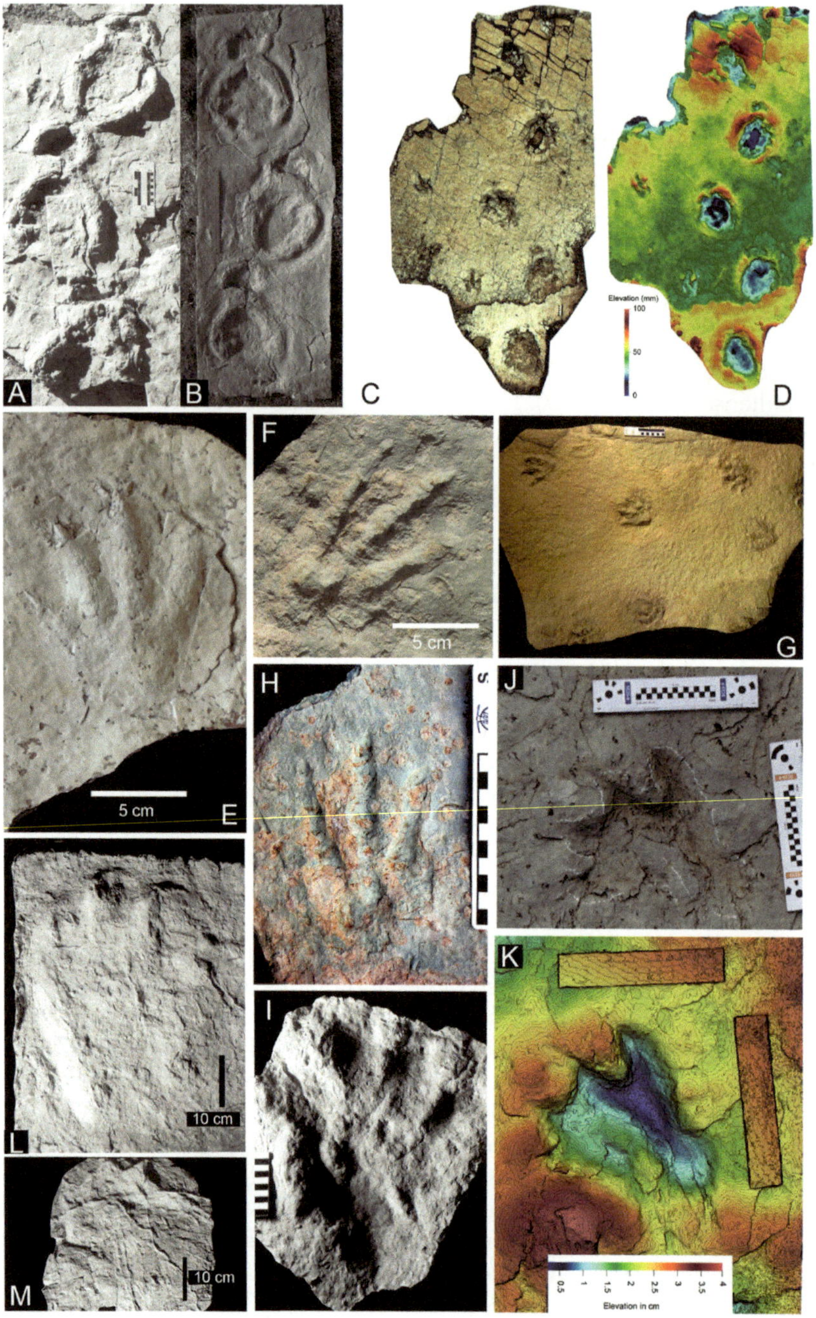

FIGURE 5.17 Upper Triassic sauropodomorph footprints. (A–B) *Eosauropus cimarronensis* trackway from Sloan Canyon Formation of New Mexico. (C–D) *Eosauropus* isp. trackway from Ørsted Dal Formation (Norian–Rhaetian) of East Greenland. (E) *Evazoum sirigui* holotype from Montemarcello Formation (Carnian) of Italy. (F–H) *E. sirigui* isolated pes imprints and trackway from Redonda Formation (Norian–Rhaetian) of New Mexico. (I) *Pseudotetrasauropus bipedoida*, artificial cast of holotype from Lower Elliot Formation of Lesotho, southern Africa. (J–K) *P. bipedoida* pes imprint from the Hauptdolomit Formation (Norian) of Graubünden, Swiss Alps, photograph and false-color photogrammetric image with contour lines. (L–M) *Tetrasauropus unguiferus* pes and manus imprint from Lower Elliot Formation of Lesotho, southern Africa, artificial casts from holotype trackway. *(A, B) from Lockley et al. (2006). (C–D) From Lallensack et al. (2017). (J–K) From Meyer et al. (2022).*

western North America (Lockley et al., 2006a). *Eosauropus* has been identified from several other North American localities (Lockley et al., 2011), Greenland (Lallensack et al., 2017), Europe (Lockley et al., 2006a) and China (Xing et al., 2018) (see Hunt et al., 2018 for overview).

Trackmaker: A sauropod interpretation of *Eosauropus* footprints is generally accepted (see discussions in Wilson, 2005; Wright, 2005; Lockley et al., 2006a; Lallensack et al., 2017; Xing et al., 2018). Lallensack et al. (2017) considered eusauropod trackmakers based on features such as the entaxonic semidigitigrade pes with laterally deflected unguals. The footprints support Triassic occurrences of skeletal remains of sauropods (Buffetaut et al., 2000, 2002).

4.7.2 Evazoum Nicosia and Loi (2003)

Fig. 5.17B.

Type ichnospecies: *E. sirigui* Nicosia and Loi (2003).

Other ichnospecies: *E. gatewayensis*.

Distribution: Montemarcello Formation (Carnian) of Italy, Chinle Group (Norian-Rhaetian) of western USA, Hassberge Formation (Carnian) of southern Germany, Ørsted Dal Formation of Fleming Fjord Group (Norian-Rhaetian) of East Greenland, Blackstone Formation, (Norian) of Queensland, Australia.

Diagnosis (after Klein and Lucas, 2021): Medium-sized bipedal tetradactyl footprints, ectaxonic to mesaxonic, functionally tridactyl, nearly as wide as long. First digit oriented forward. Digits evenly splayed, giving an overall fan-shaped impression. Rounded metapodial pad below Digits III and IV. Second and fourth digits are sub-equal in length, while the third is the longest. Relatively fleshy digits showing well-developed pads. Triangular, slightly smooth, long, claw marks on all digits. Trackways variable but with quite wide pace angulation ranging between 140 degrees and 170 degrees.

Description: Trackways of a biped. Small to medium-sized tridactyl-tetradactyl, functionally tridactyl (II, III, IV) pes imprints of elongate, slender shape. Digit proportions with Digit III longest, II and IV shorter and subequal in length, or IV slightly longer than II. Low mesaxony, Digit III projection beyond Digits II and IV less pronounced when compared to typical grallatorid footprints. Often a hallux impression (Digit I) is present, which can be curved inward. Claw traces are elongate triangular in shape and sharply pointed. A round to oval metatarso-phalangeal pad, slightly separated from Digit IV by a shallow crease, is commonly present, sometimes running into a broader "heel" posteriorly. Trackways are narrow with pes imprints turned inward relative to the midline.

Discussion: From the Montemarcello Formation (Upper Triassic, Carnian) of La Spezia, Italy, Nicosia and Loi (2003) described a tridactyl-tetradactyl morphotype that is clearly different from known grallatorid footprints. In particular, this concerns the low mesaxony, the configuration of pads, and the relatively long inward curved hallux. For the material, these authors introduced the new ichnogenus *Evazoum*. Later Lockley et al. (2006b) identified *Evazoum* on surfaces of the Chinle Group in North America, reassigning footprints formerly referred to the southern African ichnogenus *Pseudotetrasauropus* (see below) to the Italian ichnotaxon.

Trackmaker: A prosauropod dinosaur but also pseudosuchian archosaurs have been considered (Lockley et al., 2006b; D'Orazi-Porchetti et al., 2008; Lucas et al., 2010).

4.7.3 *Pseudotetrasauropus* Ellenberger (1972)

Fig. 5.17C.

Type ichnospecies: *P. bipedoida* Ellenberger (1972).

Other ichnospecies: *P. grandcombensis.*

Distribution: Lower Elliot Formation (Norian-Rhaetian) of southern Africa (Lesotho), ? Upper Triassic (Norian) of southwestern France, ?Upper Triassic Hauptdolomit and Kössen formations (Norian-Rhaetian) of Switzerland.

Diagnosis (after D'Orazi Porchetti and Nicosia, 2007): Trackway of a biped of large dimensions, digitigrade. Four straight digits impressions anteriorly oriented. A clear basal pad, related to the presence of a fifth digit, always present on the rear margin, slightly on the external border. Digits almost completely separated along their length. Foot axis orientation slightly variable, ranging from inward to outward.

Description: Trackways of a biped with relatively short steps and pes imprints being oriented nearly parallel to the trackway midline. Pes imprints large (~40 cm long), digitigrade-semiplantigrade and tetradactyl-pentadactyl. Digit III is longest, followed by II and IV, Digit I is shortest. If present Digit V positioned posterior to the digit group I–IV and consisting of an oval pad lacking a phalangeal portion. All digits with well-rounded phalangeal pads. Distal digit ends are round and lack claw traces, but this maybe a preservational feature.

Discussion: *Pseudotetrasauropus* was introduced by Ellenberger (1972) when describing footprint assemblages from the Lower Elliot Formation (Upper Triassic, Norian-Rhaetian) of Lesotho, southern Africa. Purported occurrences in the Chinle Group have later been reassigned to *Evazoum* (see Lockley et al., 2006b; Lucas et al., 2010 for ichnotaxonomic history of this material). Possible *Pseudotetrasauropus* is known from the Upper Triassic of southwestern France (Ellenberger, 1965; Ellenberger et al., 1970; Gand et al., 2000; see also Rainforth, 2003). In their revision of footprint assemblages from southern Africa, Olsen and Galton (1984) considered *Pseudotetrasauropus* a "bipedal *Brachychirotherium*." Indeed, there is some similarity of the pes imprints with chirotheriids, in particular with *Brachychirotherium*. However, because *Brachychirotherium* is the tracks of a quadruped, and, in the diagnosis, is defined to show an associated manus imprint, *Pseudotetrasauropus* cannot be assigned to this ichnogenus and should be distinct (see D'Orazi Porchetti and Nicosia, 2007). In recent years it has been generally accepted that *Pseudotetrasauropus* can be attributed to sauropodomorph dinosaurs, which is in accordance with known skeletal fossils from the Elliot Formation (Lockley and Hunt, 1995). Prosauropods, for example *Plateosaurus*, might be good candidates for the maker of *Pseudotetrasauropus* tracks (Lockley and Meyer, 2000; Lockley et al., 2006b; Meyer et al., 2018,2019). Lockley et al. (2006b) proposed to include similar sauropodomorph ichnogenera such as *Otozoum*, *Pseudotetrasauropus*, *Evazoum* and *Kalosauropus* in a plexus these authors called OPEK.

Trackmaker: A sauropodomorph (e.g., *Plateosaurus*) has been widely accepted as the trackmaker of *Pseudotetrasauropus* tracks (Lockley and Hunt, 1995; Lockley and Meyer, 2000).

4.7.4 *Tetrasauropus* Ellenberger (1972)

Fig. 5.17D.

Type ichnospecies: *T. unguiferus* Ellenberger (1972).

Other ichnospecies: Only the type ichnospecies is valid.

Distribution: Lower Elliot Formation (Norian-Rhaetian) of Lesotho, southern Africa, ? Hauptdolomit-Kössen formations (Norian-Rhaetian) of Switzerland.

Diagnosis (from D'Orazi Porchetti and Nicosia, 2007): Large-sized quadrupedal track, plantigrade, with a tetradactyl pes showing strong ectaxony, and with the foot axis almost parallel to the midline of the trackway. Strong claws in the pes bend inward; manus smaller than the pes (about 2/3) with four digits. Manus regularly placed in front of, and external to, the pes.

Description: Large tetradactyl, plantigrade pes, and manus imprints with sharp inward curved claw traces. The Digit I claw trace is longer than those of other digits in both pes and manus. Trackways are relatively broad with short steps and the manus positioned antero-lateral to the pes.

Discussion: *Tetrasauropus* is one of Ellenberger's (1970, 1972) numerous introduced ichnotaxa, but can be considered as a valid ichnogenus. The original material comes from the Lower Elliot Formation (Upper Triassic, Norian-Rhaetian) of Lesotho in southern Africa. Other units and locations where *Tetrasauropus* has been identified are Upper Triassic deposits of Switzerland (Lockley and Meyer, 2000; Meyer et al., 2013, 2018, 2019, 2022), Great Britain (Lockley et al., 1996; Lockley and Meyer, 2000), Greenland (Lockley and Meyer, 2000), the USA (Lockley and Hunt, 1995; Lockley et al., 2001) and Argentina (Melchor and de Valais, 2006). The material from Great Britain, Greenland, and the USA has recently been reassigned to the ichnogenus *Eosauropus* (Lockley et al., 2006a; Lallensack et al., 2017; see above), while the Argentine tracks were partly redated as Eocene (Melchor et al., 2013) and represent a non-dinosaurian tetrapod of unknown affinity. *Tetrasauropus* is different from *Eosauropus* in the parallel or slightly outward orientation of imprints which is strongly outward in the latter, and by the strong inward curvature of claw traces, which are outward directed in *Eosauropus*.

Trackmaker: Prosauropod sauropodomorphs were the most likely trackmakers (Lockley and Hunt, 1995; Lockley and Meyer, 2000; D'Orazi-Porchetti and Nicosia, 2007; and references therein).

4.8 Synapsid

4.8.1 Brasilichnium Leonardi (1981)

Fig. 5.18A and B.

Type ichnospecies: *B. elusivum* Leonardi (1981).

Other ichnospecies: *B. anaiti, B. saltatorum.*

Distribution: Chinle Group, Wingate, Moenave, Navajo formations of the USA, ?Elliot Formation of southern Africa, Botucatu Formation (Lower Cretaceous) of Brazil.

Diagnosis (from Klein and Lucas, 2021, emended after Fernandes and Carvalho, 2008; Buck et al., 2017): Trackway of a small quadruped with mean gleno-acetabular distance of 7.5 cm. Manus significantly smaller than pes; manus-pes distance increases when the walking speed decreases. Pace angulation relatively high for a quadruped. Heteropody is directly proportional to the positive rotation of the longitudinal axis of the pes (values $> 75°$). Longitudinal axis of the pes parallel to the trackway axis and with pace angulation $>125°$, and does not show heteropody. Pes impressions are elliptical with a nearly transverse major axis and anteroposterior axis slightly directed into the trackways; short digits, usually

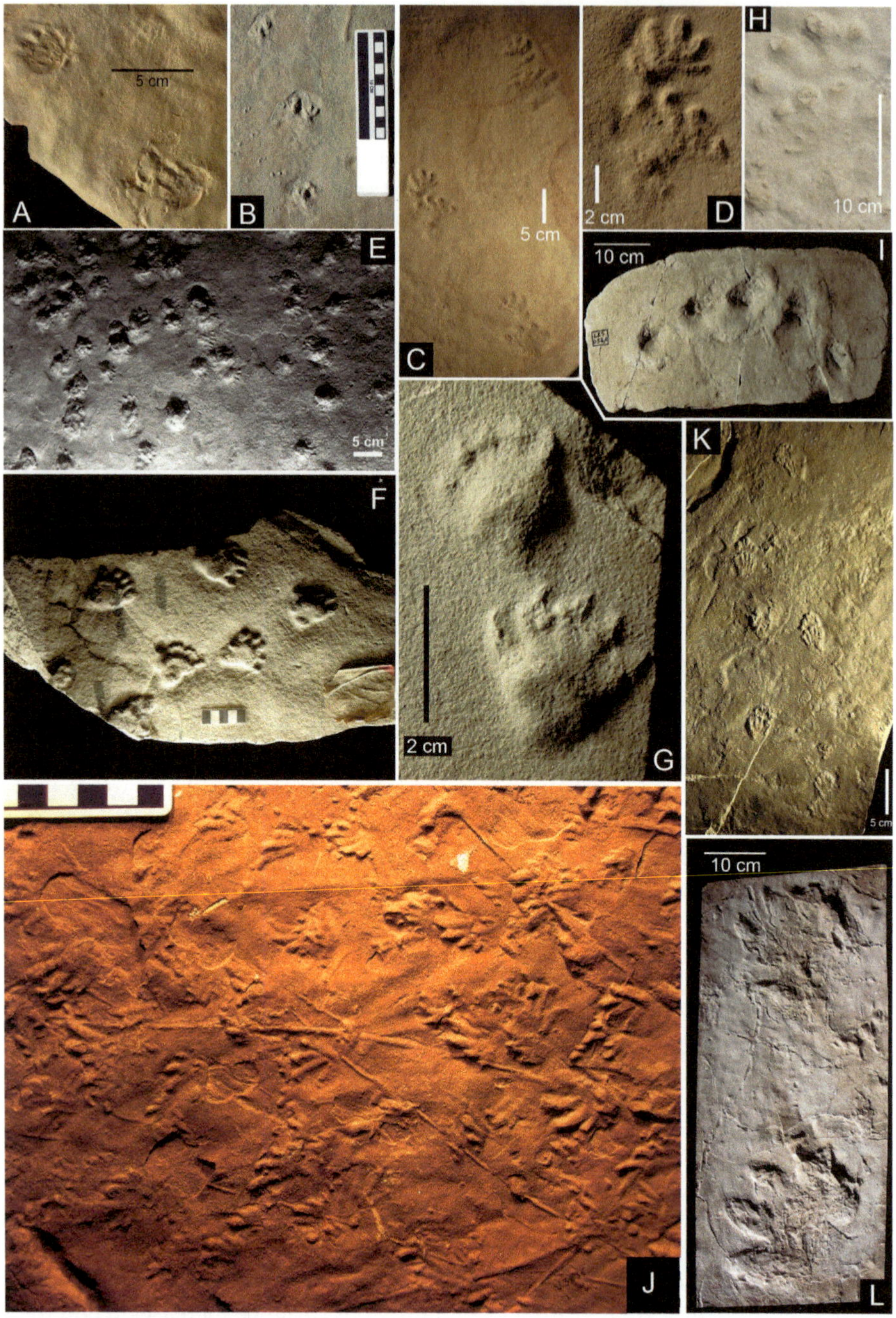

FIGURE 5.18 Therapsid footprints from Lower–Upper Triassic deposits. (A–B) *Brasilichnium elusivum* trackway from Redonda Formation (Norian-Rhaetian) of New Mexico, USA. (C–D) *Capitosauroides bernburgensis* trackway and detail from Solling Formation (Anisian). (E) *Dicynodontipus geinitzi* from Solling Formation (Anisian) of Germany. (F–G) *Dicynodontipus* isp. from Vera Formation (upper Permian–Lower Triassic) of Argentina. (H) *Dolomitipes accordii* from Balfour Formation (*Lystrosaurus* assemblage zone) of South Africa. (I) *Pentasauropus incredibilis* holotype from lower Elliot Formation of Lesotho, Southern Africa. (J) *Procolophonichnium haarmuehlensis* trackways from Wupatki Member of Moenkopi Formation (Olenekian) of Northern Arizona. (K) *P. lockleyi* trackway from Hassberge Formation (Carnian) of Germany. (L) *Therapsipus cumminsi* artificial cast of holotype trackway from Holbrook Member of Moenkopi Formation (Anisian) of northeastern Arizona. *(A, B) From Lucas et al. (2010). (E) From Haubold (1971a). (H) From Marchetti et al. (2019b). (I) Courtesy S. D'Orazi Porchetti. (J) From Klein and Lucas (2010b). (K) From Klein et al. (2015b). (L) Courtesy Martin Lockley.*

rounded, with possible phalangeal formula 2-3-3-3-3; the posterior autopodium is ectaxonic and tetradactyl (Digits II, III, IV and V) in a semiplantigrade condition. There is a greater hypex of Digit V in slight abduction. Anterior autopodium has at least four digits with claws; claws present or absent due to preservation. Tail impressions/drag marks absent.

Description: These are small, tetradactyl pes and manus imprints of a quadrupedal tetrapod with strong heteropody. The pes imprint is semiplantigrade, ectaxonic and wider than long. Digits are short and rounded. Digit V is laterally separated from other digits.

Discussion: Leonardi (1981) described *Brasilichnium* from the Lower Cretaceous? Botucatu Formation of Brazil. Fernandes and Carvalho (2008) revised the type material (see also D'Orazi Porchetti et al., 2018; Buck et al., 2017). Further well-known records are from the Navajo Sandstone (Lower Jurassic) of the western USA (Lockley, 2011). Triassic records are rare, and some of the best-known occurrences are from the Redonda Formation (Chinle Group) and the Wingate Formation of New Mexico and Colorado, respectively (Lockley et al., 2004; Klein et al., 2006; Lucas et al., 2010). Possibly *Brasilichnium* is also present in some material from the Stormberg Group of southern Africa (Ellenberger, 1972, 1974, 1975), however, this cannot be confirmed presently.

Trackmaker: Small nonmammalian or basal mammalian synapsid (Lockley et al., 2004; Fernandes and Carvalho, 2008; Lucas et al., 2010).

4.8.2 *Capitosauroides* Haubold (1971a)

Fig. 5.18C and D.

Type ichnospecies: *C. bernburgensis* Haubold (1971a).

Other ichnospecies: Only the type ichnospecies is valid.

Distribution: Middle Permian–Middle Triassic. Abrahamskraal-Teekloof formations, Beaufort Group (middle Permian) of South Africa, Hornburg Formation (Middle Permian) of Germany, Arenaria di Val Gardena Formation (upper Permian) of northern Italy, ?Wióry Formation (Lower Triassic, Olenekian) of Poland, Solling Formation (Buntsandstein, Anisian) of Germany.

Diagnosis (after Haubold, 1971a; Buchwitz et al., 2020, modified): Moderately wide trackways of a quadruped. Pace angulation of pes ~100 degrees. Pes and manus imprints pentadactyl, semiplantigrade, clumsy, without distinct proximal termination of the sole surface or specialized metatarso-/metacarpophalangeal area. Digits with broad, rounded distal ends lacking claw traces.

Description: Imprints with wide digit divarication and Digit V in a postero-lateral position relative to Digits I–IV. Manus slightly smaller than pes but with similar morphology. Imprints show strong outward rotation relative to the trackway midline.

Discussion: Haubold (1971a) introduced this ichnogenus with the type ichnospecies *C. bernburgensis* based on trackways from the Solling Formation (Buntsandstein, Anisian) of Bernburg, Sachsen-Anhalt, Germany. A possible occurrence with a distinct ichnospecies from the Wióry Formation (Olenekian) of the Holy Cross Mountains, Poland, refers to poorly preserved material (Ptaszyński, 2000; Klein and Niedźwiedzki, 2012) and has to be treated with caution, even if stratigraphically older material is known, for example from middle-upper Permian deposits of Germany (Buchwitz et al., 2020), northern Italy Marchetti et al., 2019a) and South Africa (Marchetti et al., 2019b).

Trackmaker: When described by Haubold (1971a), the original Buntsandstein material of *Capitosauroides bernburgensis* was attributed to temnospondyl amphibians based on the widely divaricated digits, the broad, round digit ends, and the indistinctly defined posterior palm/sole. This interpretation was questioned by Klein and Nieźwiedzki (2012) because temnospondyls and other amphibians should have a tetradactyl manus, whereas *Capitosauroides* clearly has five toe impressions. These authors proposed a nonamphibian tetrapod or reptiliomorph as the producer. Nevertheless, five-toed temnospondyl manus skeletons have been reported from the Middle Triassic of Australia (Watson, 1958) and from the Upper Triassic of Poland (Konietzko-Meier et al., 2020), inferred from preserved five metacarpalia. A different interpretation was given by Marchetti et al. (2019a, b), who considered therocephalian therapsids as trackmakers (see also Buchwitz et al., 2020).

4.8.3 Dicynodontipus *Rühle v. Lilienstern (1944)*

Fig. 5.18E–G.

Type ichnospecies: *D. geinitzi.*

Other ichnospecies: Only the type ichnospecies is considered here as valid.

Distribution: Upper Permian–Middle Triassic. Arenaria di Val Gardena Formation (upper Permian) of northern Italy, Palingkloof Member of Balfour Formation, (Lower Triassic, Induan) of South Africa, Solling Formation (Buntsandstein, Lower-Middle Triassic, Olenekian-Anisian) of Germany, Wióry Formation (Lower Triassic, Olenekian) of Poland, Middle Triassic (Anisian) of northern Italy, Vera Formation (upper Permian–Lower Triassic) of Argentina.

Diagnosis (after Haubold, 1971a,b): Relatively narrow trackways, pace angulation not below 100 degrees at normal gait. Manus imprints positioned anterior to pes imprints, but in trackways of fast moving individuals and with high pace angulation manus widely overstepped by the pes. Pes and manus imprints of the same morphology—plantigrade, pentadactyl with short, forward-pointing digits, Digit IV longest, and Digit V shifted slightly posteriorly.

Description: These are small pentadactyl pes and manus imprints (pes length up to 5.5 cm) with low heteropody (manus only slightly smaller than pes). The digits are straight and relatively short, subequal in length, showing the traces of small acuminate claws. A significant feature is the large round palmar/sole surface, which is often separated from the digits by a distinct groove. Trackways display a large pace angulation (160 degrees) and sometimes a lateral overstep of the manus by the pes due to faster progression speed.

Discussion: The original material of Rühle v. Lilienstern (1939, 1944) comes from the Solling Formation (Middle Buntsandstein, Anisian) of southern Thuringia, Germany. The type ichnospecies *D. geinitzi* was introduced in a different combination, *Chirotherium geinitzi,* by Hornstein (1876). Stratigraphically older occurrences come from upper Permian (Arenaria di Val Gardena Formation) deposits of northern Italy and from the Lower Triassic (Balfour Formation, Induan) of South Africa (Marchetti et al., 2019a, 2019b). In South America *Dicynodontipus* has been identified from the Vera Formation (upper Permian–Lower Triassic) of Argentina (Melchor and De Valais, 2006; Citton et al., 2018,2021). The latter material was originally described under different ichnogenera such *as Gallegosichnus, Calibarichnus, Palaciosichnus,* and *Stipanicichnus* (Casamiquela 1964,1975).

Trackmaker: Cynodont therapsids can be considered the most likely trackmaker. Dicynodonts, as inferred from the name, do not match these tracks (Marchetti et al., 2019a).

4.8.4 Dolomitipes Marchetti et al. (2019a)

Fig. 5.18H.

Type ichnospecies: *D. accordii* (Ceoloni et al., 1988)

Other ichnospecies: *D. icelsi.*

Distribution: Arenaria di Val Gardena Formation (upper Permian, Lopingian) of the Dolomites, northern Italy, Palingkloof Member of the Balfour Formation (?upper Permian, Lopingian–Lower Triassic, Induan) of South Africa.

Diagnosis (after Marchetti et al., 2019a): Trackway of a quadrupedal tetrapod with pentadactyl, semiplantigrade (manus) to plantigrade (pes) footprints. Pentadactyl manus different from *Limnopus*. Digit impressions straight to bent inwards distally (manus), terminating in enlarged tips showing a subcircular shape and an arcuate arrangement. Median digits of the pes distinctly longer than the outer digits and subequal in length (mesaxony), different from *Ichniotherium*, *Limnopus*, and *Dimetropus*. Manus weakly ectaxonic. Manual imprint about one third wider than long and more deeply impressed than the pes. Both features different from *Dimetropus* and *Dicynodontipus*. Well-impressed palm with concave proximal margin and large, rounded metacarpal-phalangeal pads of all digits. Pads of Digits II–V not present in *Ichniotherium* and *Limnopus*. Pedal imprint shallower and about as wide as long, of size similar to the manus. Digits I–IV about 1/3 of the pes length, with rounded metatarsophalangeal pads clearly separated from the proximal-lateral, elliptical heel. Sole structure different from *Ichniotherium*, *Limnopus* and *Dicynodontipus*. Digit V short and proximally and laterally positioned, different from *Dimetropus*. Trackways consisting of alternating sets of manus-pes imprints with inwardly rotated manus and forward-directed pes, relatively high gait, and no tail or body trace preserved.

Description: These are pentadactyl mesaxonic pes and slightly ectaxonic manus imprints that show very low heteropody (nearly equal size). The South African material has pes imprints of up to 14 cm length and 12.5 cm in width (Marchetti et al., 2019b). A distinct palm/sole is present. Digits often show pad impressions. The trackway pattern is characterized by a broad to moderately broad trackway width; the pes imprints are oriented parallel to the midline, and the manus imprints are either parallel to the midline or slightly inward rotated.

Discussion: The ichnogenus was introduced by Marchetti et al. (2019a). The type material comes from the Arenaria di Val Gardena Formation (upper Permian, Lopingian) of the famous Bletterbach section in the Dolomites of northern Italy. Another occurrence is in the Palingkloof Member of the Beaufort Formation of South Africa. The level is close to the Permian-Triassic boundary, and Gastaldo et al. (2020) consider it to be late Permian (Lopingian) in age (see also Marchetti et al. this volume). Because of the ongoing discussion on this and the close proximity, we consider it justified to mention *Dolomitipes* here in the Triassic chapter as well. *Dolomitipes* is here considered a valid ichnotaxon based on unique features of the morphology.

Trackmaker: Marchetti et al. (2019a,b) considered small-medium sized dicynodont therapsids, and we follow this here.

4.8.5 *Pentasauropus* Ellenberger (1972)

Fig. 5.18I.

Type ichnospecies: *P. incredibilis.*

Other ichnospecies: *P. argentinae.*

Distribution: Lower Elliot Formation (Upper Triassic) of Lesotho, southern Africa, Cerro de las Cabras Formation, Middle Triassic, and Los Menucos Group (Upper Permian–Lower Triassic) of Argentina.

Diagnosis (from D'Orazi Porchetti and Nicosia, 2007): *Pentasauropus* is the trackway of a quadruped, with very low heteropody. The track is usually wide, with a reduced pace angulation, and manus and pes axes face forward, or inward. When present the pes sole is rounded. No tail traces are present. The main characters shown by this ichnotaxon are five equally spaced traces of digits forming a broad, arcuate pattern that is convex anteriorly.

Description: Footprints of large quadrupeds with very low heteropody. Both pes and manus imprints preserved by round or oval impressions of the digit tips only, which are arranged in an arcuate pattern. Imprints of the pes sole are scarcely visible, if present at all. Trackways show short steps with the manus positioned anterior to the pes, often halfway off of the axis demarcating the stride.

Discussion: *Pentasauropus* is known from longer trackways in the Lower Elliot Formation (Upper Triassic, Norian-Rhaetian) of Lesotho in southern Africa. The pes and manus imprints with only the traces of the digit tips being preserved make it difficult to exclude the presence of incomplete undertracks. On the other hand, the combination of this feature with a characteristic trackway pattern repeatedly occurring in different trackways from various locations, suggests a distinct anatomical signal from a digitigrade to unguligrade producer. Therefore, we consider *Pentasauropus* a valid ichnogenus.

Trackmaker: Ellenberger (1970, 1972) and numerous authors subsequently following him considered prosauropod sauropodomorphs as the trackmakers. More recent analyses (D'Orazi Porchetti and Nicosia, 2007; Kammerer, 2018; Lagnaoui et al., 2019), however, showed that imprint morphology and trackway pattern closely match walking dicynodont therapsids, and, here, we agree with this view.

4.8.6 *Procolophonichnium* Nopcsa (1923)

Fig. 5.18J and K.

Type ichnospecies: *P. nopcsai* Kuhn (1963).

Other ichnospecies: *P. haarmuehlensis, P. nectouxi* and *P. lockleyi* differentiated by Klein et al. (2015b) and Klein and Lucas (2021) are presently being reanalyzed and will probably be assigned to other ichnogenera with new combinations. A comprehensive study is in preparation.

Distribution: Upper Permian–Lower Triassic. Palingkloof Member (Balfour Formation, Lower Triassic, Induan) of South Africa. Other occurrences are presently being reevaluated in a comprehensive study and ichnotaxonomic revision.

Diagnosis (from Klein and Lucas, emended after Haubold, 1971a,b, Klein et al., 2015b): Pentadactyl and semiplantigrade to plantigrade asymmetrical footprints of small quadrupeds with digits increasing in length from I–IV; Digit IV subequal in length to Digit III; Digit V subequal in length to Digit II. Manus similar in shape but smaller than pes and positioned anterior to or being slightly overstepped by the latter posteriorly. Differs from the most similar tracks of other ichnogenera in the following features: (1) from *Rhynchosauroides* and *Dromopus* by digit proportions, with minor differences of length and relatively longer Digit V, and (2) from *Rhynchosauroides, Dromopus,* and *Varanopus* by proportionately shorter and stouter digits, by (mostly) straight to outward curved digits versus the strong inward curvature of digits in the latter, and by semiplantigrade to plantigrade imprints versus digitigrade to semiplantigrade imprints observed in these latter ichnotaxa.

Description: These are the trackways of a small quadruped with pentadactyl pes and manus imprints. Digits are straight and relatively robust, terminating in small claws. Pes imprints show a short sole. Manus imprints are smaller than the pes imprints. Digit proportions with Digit IV and III subequal in length, II shorter and I shortest. Digit V is subequal in length with Digit II. Trackways are relatively narrow with the orientation of the imprints varying from parallel to the midline to outward or inward rotated.

Discussion: Nopcsa (1923) introduced the ichnogenus based on material from the Lower Triassic of South Africa. Later, the type ichnospecies *P. nopcsai* was erected by Kuhn (1963). The original holotype material stored in the Bavarian State Collection of Paleontology, Munich was destroyed in World War II. Therefore, a plaster cast in the Museum of Natural History, London and designated the "plasto-holotype" (Klein et al., 2015b) is the only available specimen and documentation. Klein et al. (2015b) revised the ichnogenus considering the following Triassic ichnospecies as valid: *P. nopcsai, P. haarmuehlensis, P. nectouxi, P. polonicum* and a new ichnospecies, *P. lockleyi*. They come from Lower-Upper Triassic deposits of South Africa, and from the Germanic Basin of Europe (Germany, France, The Netherlands, Poland). Other occurrences have been reported from the Arenaria di Val Gardena Formation (upper Permian, Lopingian) of northern Italy, the Timezgadiouine Formation (T4, Middle Triassic) of the Argana Basin, Morocco, from the Holbrook Member of the Moenkopi Formation (Middle Triassic, Anisian) of Arizona, USA, from the Passaic Formation (Upper Triassic, Norian) of New Jersey, and from the Santa Maria Formation (Carnian) of Brazil (Andrade-Silva et al., 2023). A restudy and detailed revision is on the way and will find some of these ichnospecies in combination with other ichnogenera.

Trackmaker: Procolophonid parareptiles and therapsids have been proposed as trackmakers (Klein et al., 2015b; Marchetti et al., 2019a, 2019b).

4.8.7 **Therapsipus** ***Hunt et al. (1993b)***

Fig. 5.18L.

Type ichnospecies: *T. cumminsi* Hunt et al. (1993b).

Other ichnospecies: Only known from the type ichnospecies.

Distribution: Moenkopi Formation (Middle Triassic, Anisian) of Arizona and New Mexico (USA), Lower Triassic (Wióry Formation) of the Holy Cross Mountains, Poland.

Diagnosis (after Hunt et al., 1993b): Large (pes length ~ 23–26 cm) footprints of a quadruped in which both pes and manus are pentadactyl, Digit I is poorly impressed and the digits have concave posterior margins; pace angulation ~ 85 degrees.

Description: Pes and manus imprints pentadactyl and relatively large. Pes imprints are 25 cm in length and 22 cm in width. Manus imprints are 21 cm in length and 22 cm in width. Pes imprints with Digit III longest and outer digits decreasing in length. Manus imprints with Digits III and IV longest and subequal in length and II and V shorter. Digit I is short and scarcely impressed. Broad trackways that show low pace angulation and short steps. The manus is positioned anterior to the pes and strongly rotated outward relative to the trackway midline by 60 degrees, while the pes is less rotated outward by 45 degrees.

Discussion: Compared with the therapsid ichnogenus *Dicynodontipus*, *Therapsipus* is much larger with imprints being relatively broad and shows a larger trackway width, whereas the former is more elongate in shape, and trackways are narrower. Furthermore, *Therapsipus* has a less defined, often posteriorly concave palmar/sole surface, which is distinctly round and separated from digit traces in *Dicynodontipus*. There is a slight similarity of imprint shape with some chirotheriid tracks in the short, robust digit impressions that form a distinct anterior group and in the posterior semiplantigrade or plantigrade sole. However, chirotheriids have a distinct posterolaterally positioned Digit V and a narrow trackway pattern with less outward rotated imprints. *Therapsipus* is documented with certainty only from uppermost Moenkopi strata in Arizona and New Mexico, USA (Hunt et al., 1993b; Klein and Lucas, 2021). A possible further occurrence from the Wióry Formation (Lower Triassic, Olenekian) of the Holy Cross Mountains, Poland (Klein and Nieźwiedzki, 2012) should be confirmed by more complete material and trackways.

Trackmaker: A large dicynodont as the trackmaker of *Therapsipus* has been widely accepted (Hunt et al., 1993b; Nesbitt and Angielczyk, 2002).

4.9 "Swim" traces left by buoyant or punting tetrapods

4.9.1 Anshunpes *Xing et al. (2020)*

Fig. 5.19A.

Type ichnospecies: *Anshunpes aquacursor*Xing et al. (2020).

Other ichnospecies: Only the type ichnospecies known.

Distribution: Only known from one locality in the Songzikan Member (Member I) of the Guanling Formation (Middle Triassic, Anisian-Ladinian) of Yunnan Province, southwestern China.

Diagnosis (from Xing et al., 2020): Plantigrade to semiplantigrade tetradactyl imprints, averaging 15 cm in length and 9 cm in width, with outwardly curved, blunt, or slightly tapering robust digits and broad sole area. Two medial Digits II and III longest, outer Digits IV and V shorter, with the lateral one being in a more posterior position relative to all others. Posterior margin of the sole often extended to a bi- or trilobed "heel." Trackways broad, with Tw = 45.8 cm on average. Imprints positioned opposite to each other and outwardly rotated relative to the midline. Different from: (1) *Characichnos* by complete semiplantigrade-plantigrade foot impressions with defined digits, versus scratch-like digit traces in the latter, (2) *Chelonipus* by longer foot and blunt digit impressions versus very short and broad foot imprints with sharp digit and claw traces in the latter, and (3) *Chirotherium* by the less compact anterior digit group II–IV and the relatively long Digit V.

Description: Plantigrade to semiplantigrade tetradactyl imprints, small to medium-sized (pes length 15 cm, pes width 9 cm). Trackways are broad with left and right imprints arranged opposite to each other, symmetrically relative to the trackway midline. Digits blunt and thick, outwardly curved. A broad sole area is impressed. Digit proportions with the

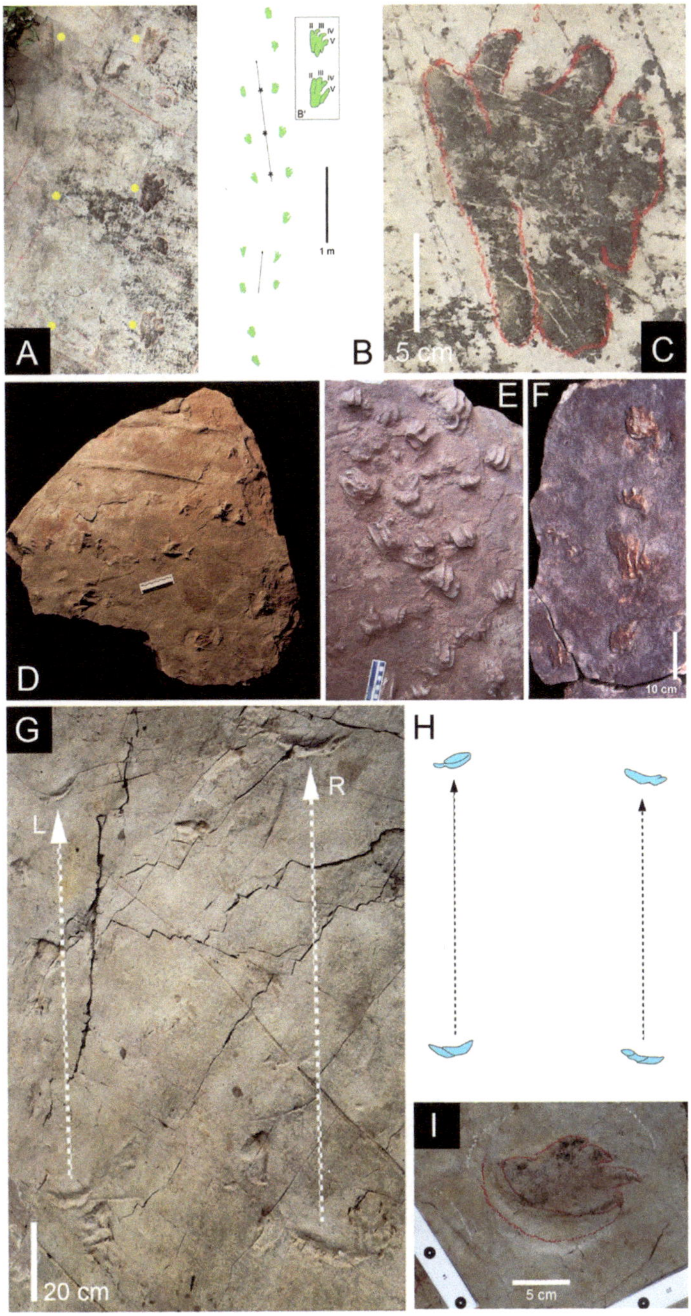

FIGURE 5.19 Subaqueous footprints and trackways of buoyant or swimming tetrapods. (A–C) *Anshunpes aquacursor* from Guanling Formation (Anisian) of Guizhou Province, China. (D) *"Barrancapus cresapi"* from Bull Canyon Formation (Norian) of New Mexico. (E–F) Unnamed swim tracks from the Volpriehausen Formation (Olenekian) and from the Eschenbach Formation (Anisian) of Germany. (G–I) *Dikoposichnus luopingensis* from the Guanling Formation of Guizhou Province, China. *(A–C), (G–I) from Xing et al. (2020). (F) From Klein and Lucas (2018).*

two middle digit traces longest and subequal in length, the medial and lateral digit traces shorter. The lateral digit trace is slightly posteriorly shifted relative to the other digits. Imprints sometimes posteriorly elongated by dragging traces and a bilobed "heel." Some imprints are preserved as bean-shaped extramorphological variants or elongated scratches only.

Discussion: *Anshunpes* with the ichnospecies *A. aquacursor* was introduced by Xing et al. (2020) based on abundant imprints and longer trackways from a large surface in the Guanling Formation near the city of Anshun, Guizhou Province. After first examination it became clear that they were left subaqueously at the bottom of shallow lagoons along the eastern Tethys coast by buoyant and swimming tetrapods, their feet touching the ground by typical punting movement. The track-bearing substrate is an argillaceous dolostone in a section comprising interbedded limestones and claystones that are rich in bivalves. *Anshunpes* reflects synchronous swimming movement of tetrapods leaving more complete foot imprints, with digits and sole impressions being similar in morphology to terrestrially walking animals. This is different from formerly described swim traces from the Guanling Formation, *Dikoposichnus luopingensis* (Zhang et al., 2014; see below) that show only crescent- or halfmoon-shaped imprints with indistinct digit traces. *Anshunpes* is considered here a valid ichnogenus based on characteristic features not found in any other swim traces and trackways.

Trackmaker: The documented broad trackway pattern with symmetrically arranged cycles, left and right foot positioned opposite to each other, suggests swimming or synchronous punting movements of tetrapods that were interpreted as marine placodonts or saurosphargids leaving impressions of their forefeet when foraging in shallow lagoons (Xing et al., 2020). Their skeletons are well-known from the Guanling Formation (Anisian-Ladinian) and from the overlying Falang Formation (Ladinian-Carnian). Preserved foot skeletons are almost complete and match the footprints very well.

4.9.2 Characichnos *Whyte and Romano (2001)*

Fig. 5.19B.

Type ichnospecies: *Characichnos tridactylus* Whyte and Romano (2001).

Other ichnospecies: No other ichnospecies known.

Distribution: *Characichnos* originally comes from the Saltwick Formation (Middle Jurassic) of Great Britain, but has been documented thus far from numerous Paleozoic, Mesozoic, and Cenozoic tracksites. In the Triassic it is common in the Moenkopi and Chinle groups Lower-Upper Triassic (Olenekian-Norian-Rhaetian) of North America, the Wióry Formation (Lower Triassic, Olenekian) of the Holy Cross Mountains, Poland, the Buntsandstein-Muschelkalk-Keuper (Olenekian-Norian) succession of the Germanic Basin and in numerous other units with a global distribution.

Diagnosis: Two to four elongate, parallel hypichnial ridges (or epichnial grooves) that may be straight, gently curved, or slightly sinuous. The termination of the ridges (or grooves) may be straight or slightly flexed.

Description: Series of (2–4) thin, long digital scratches that are arcuate, sinuous, or straight, with variable interdigital space. The anterior ends have acuminate terminations and the posterior margin occasionally has a sediment displacement mound or overhang.

Discussion: Whyte and Romano (2001) introduced this ichnotaxon for imprints consisting of sets of parallel furrows obviously attributed to buoyant or punting dinosaurs. Ichnotaxonomically the naming of so-called swim tracks or traces is problematical, because these scratch-like impressions characterize a certain behavior, and less distinctly the anatomy of the foot of the trackmaker. Moreover, a transition from walking tracks to "swim" tracks can be present. So, if a distinct trackmaker leaving both "swim" tracks and walking tracks can be convincingly demonstrated, the "swim" tracks might be given the name of the walking track version (e.g., *Grallator, Eubrontes* or *Gwyneddichnium;* see Milner et al., 2006; Lockley, 2006; Lockley et al., 2014; Klein and Lucas, 2021). Otherwise, they might be separately classified as has been done here.

Trackmaker: Buoyant or punting tetrapods, commonly dinosaurs, pterosaurs, crocodylomorphs, archosauromorphs, turtles, or lepidosauromorphs can leave *Characichnos* or *Characichnos*-like tracks and trackways.

4.9.3 Dikoposichnus *Zhang et al. (2014)*

Fig. 5.19G–I.

Type ichnospecies: *D. luopingensis.*

Other ichnospecies: No other ichnospecies known.

Distribution: Shizishan Member (Member II, Luoping Biota) of the Guanling Formation (Middle Triassic, Anisian) of Yunnan Province and Songzikan Member (Member I) of the Guanling Formation (Anisian-Ladinian) of Guizhou Province, China.

Diagnosis (after Zhang et al., 2014; Xing et al., 2020): Double row of imprints, preserved as concave epireliefs (bed) or convex hyporeliefs (sole). Each imprint is a narrow, V-shaped, slot-like depression, with a mound of sediment behind. Individual imprints may be elliptical to sigmoid-shaped in plan view, sometimes with an anterior sweep at the medial edge. Individual imprints are transverse to the direction of travel, spaced widely apart and with variable pace length. When preserved as trackways, prints generally occur in matching pairs, suggesting the limbs moved in concert, but they may be offset, produced by alternate left-right pacing. Overall trackway width is 30–70 cm, and individual prints are 5–20 cm along their longest axis, transverse to the direction of movement. Occasionally comprises a single trackway of prints.

Description: Half-moon or crescent-shaped broad and short imprints. Laterally, three to four digit traces with sharp ends, often elongated into scratches are visible. Trackways are broad with right and corresponding left side impressions positioned symmetrically and opposite to each other along the trackway midline.

Discussion: The ichnogenus *Dikoposichnus* was introduced by Zhang et al. (2014) based on extensive trackways from the Shizishan Member of the Guanling Formation (Middle Triassic, Anisian) of Yunnan Province, southwestern China. The track-bearing substrate is an argillaceous dolostone deposited in shallow marine marginal waters of the eastern Tethys. More recently, Xing et al. (2020) described a further locality with *Dikoposichnus* from the Songzikan Member of the Guanling Formation (Anisian-Ladinian) near the city of Anshun in Guizhou Province, southwestern China. At the latter locality these tracks are associated with abundant *Anshunpes* trackways that can be correlated with placodont or saurosphargid marine diapsids.

Trackmaker: Marine nothosaurs are considered the most likely trackmakers of *Dikoposichnus* trackways (Zhang et al., 2014).

4.10 Nomina dubia with different affinities

Here, we list Triassic ichnogenera that we consider to be nomina dubia and their likely affinities. See Klein and Lucas (2021) for further discussion of most of these ichnotaxa.

Agrestipus Weems, 1987 → Grallatorid.
Anatrisauropus Ellenberger, 1972 → Grallatorid.
Barrancapus Hunt et al. (1993a) → ?*Chelonipus*, ?*swim tracks*.
Batrachopodiscus Ellenberger, 1972 → Tetrapod footprint indet.
Bifidichnium Demathieu and Weidmann, 1982 → Tetrapod footprint indet.
Bosiutrisauropus Ellenberger, 1972 → Grallatorid.
Chelone? Morton (1897) → Tetrapod footprint indet.
Chelonichnium Schimper, 1850 → Tetrapod footprint indet.
Comptichnus Ellenberger, 1972 → Tetrapod footprint indet.
Cryptobranchichnus Huene, 1941 → Tetrapod footprint indet.
Cynodontipus Ellenberger, 1976 → ?Chirotheriid.
Delairichnus Haubold, 1971b → ?Therapsid footprint.
Deuterotetrapous Nopcsa 1923 → ?Chirotheriid.
Deuterotrisauropus Ellenberger 1972 → Grallatorid.
Dijaquesopus Ellenberger, 1972 → Tetrapod footprint indet.
Enigmatopus Biron and Dutuit, 1981 → Tetrapod footprint indet.
Furcapes Demathieu, 1977a → Tetrapod footprint indet.
Gregaripus Weems, 1987 → Grallatorid.
Huenepus Kuhn, 1958a → Tetrapod footprint indet.
Ichnites Hitchcock, 1837 → Tetrapod footprint indet.
Ingenierichnus Casamiquela, 1964 → Tetrapod footprint indet.
Lacertoidipus Ellenberger, 1972 → Tetrapod footprint indet.
Longipes Demathieu, 1977 → Tetrapod footprint indet.
Mafatrisauropus Ellenberger, 1972 → Grallatorid.
Marpurgichnium Schindewolf, 1928 → Tetrapod footprint indet.
Microtetrapodiscus Ellenberger, 1972 → Tetrapod footprint indet.
Minutipes Demathieu, 1970 → Tetrapod footprint indet.
Moltenotetrapodiscus Ellenberger, 1972 → Tetrapod footprint indet.
Nothosauripus Kuhn, 1958a → Tetrapod footprint indet.
Onkichnium Kuhn, 1963 → Tetrapod footprint indet.
Onychopoides Kuhn, 1958a → Tetrapod footprint indet.
Pachysaurichnium Demathieu and Weidmann, 1982 → Tetrapod footprint indet.
Paraophidichnium Demathieu, 1977 → Tetrapod footprint indet.
Paratrisauropus Ellenberger, 1972 → Grallatorid.
Plesiothornipos Harkness, 1850 → Tetrapod footprint indet.
Procolophonipus Rühle v. Lilienstern (1939) → *Procolophonichnium*, Tetrapod footprint indet.
Prototrisauropodiscus Ellenberger, 1972 → Grallatorid.
Prototrisauropus Ellenberger, 1972 → Grallatorid.
Pseudotrisauropus Ellenberger, 1972 → Grallatorid.
Psilotrisauropus Ellenberger, 1972 → Tetrapod footprint indet.
Quemetrisauropus Ellenberger, 1972 → Grallatorid.

Rogerbaletichnus Casamiquela, 1964→ Tetrapod footprint indet.
Rotodactylopus Ellenberger, 1970→ Tetrapod footprint indet.
Ruecklinichnium Kuhn, 1958a→ Tetrapod footprint indet.
Saurichnium Gürich, 1926→ Tetrapod footprint indet.
Saurischichnus Huene, 1941→ Tetrapod footprint indet.
Seakatrisauropus Ellenberger, 1972→ Grallatorid.
Senqutrisauropus Ellenberger, 1972→ Grallatorid.
Shimmelia Casamiquela, 1964→ ?Chirotheriid footprint.
Tetrapodichnus Haubold, 1971a→ Tetrapod footprint indet.
Thecodontichnus Huene, 1941→ Tetrapod footprint indet.
Tikoepentapodiscus Ellenberger, 1972→ Tetrapod footprint indet.
Trichristolophus Ellenberger, 1972→ Tetrapod footprint indet.
Tritotrisauropus Ellenberger, 1972→ Grallatorid.

4.11 Tetrapod footprints unnamed or given incorrect labels

Trackway of a quadruped: Klein and Lucas (2021), p. 103, fig. 72→ ?Therapsid trackway.

"*Coelurosaurichnus*": Demathieu and Oosterink (1988), p. 10–11, fig. 5A–D→ Chirotheriid or therapsid footprint.

"*Sustenodactylus*": Demathieu and Oosterink (1988), p. 15, fig. 8A→ Therapsid footprint.

"*Capitosauroides*": Demathieu and Oosterink (1988), p. 15, fig. 8B→ Tetrapod footprint indet.

"Capitosaurid footprints": Peabody (1948), fig. 9A–B, pl. 27A–B, Klein and Lucas (2010b), p. 60, fig. 58→ Tetrapod footprint indet.

"Vertebrate trace incertae sedis": Demathieu and Oosterink (1988), p. 15, fig. 8C→ Tetrapod footprint indet.

"*Dicynodontipus*": Retallack (1996), p. 303, fig. 2, p. 304, fig. 3, p. 306, fig. 4→ *Procolophonichnium*.

"*Dicynodontipus*": Silva et al. (2008a), figs. 3–6, 8C→ Tetrapod footprint indet.

"New genus 1": Olsen and Rainforth (2001), p. 139, fig. 51E→ ?*Evazoum*.

5. Biostratigraphy and paleobiogeography

Fig. 5.20.

Tetrapod footprints are useful for biostratigraphy and biochronology. If shown in a temporal-stratigraphic context, they document evolutionary developments of the locomotor apparatus of distinct tetrapod groups. Because of the restricted stratigraphic, and wide geographic distribution and abundance of some characteristic footprint morphotypes and taxa, a global biochronological succession for the Triassic has been established (Klein and Haubold, 2007; Klein and Lucas, 2010a, 2021). Earlier attempts were made by Lucas (2007) and Hunt and Lucas (2007b). Klein and Lucas (2010a) proposed five tetrapod-footprint-based biochrons for the Triassic Period, each demarcated by the first appearance datum (FAD; Lucas, 1998) of a distinct ichnotaxon or morphotype: (1) **Dicynodont tracks**, Induan–Early

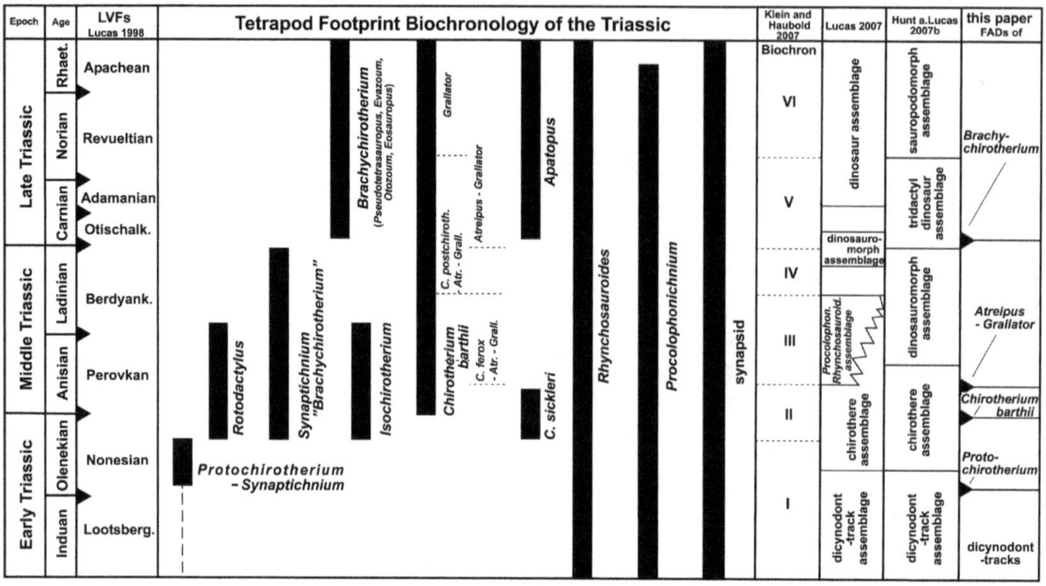

FIGURE 5.20 Biochronology and stratigraphic distribution of Triassic tetrapod footprints. *From Klein and Lucas (2010a).*

Olenekian, Lootsbergian–Nonesian; (2) ***Protochirotherium***, Early–Late Olenekian, Nonesian; (3) ***Chirotherium barthii***, Late Olenekian–Late Anisian, Nonesian–Perovkan; (4) ***Atreipus–Grallator***, Late Anisian–Ladinian, Perovkan–Berdyankanian); (5) ***Brachychirotherium***, Carnian–Rhaetian, Otischalkian–Apachian. More recently the *Chirotherium barthii* biochron was revised, and the FAD corrected to the Early Anisian, Perovkan (Klein et al., 2023b,2024).

6. Ichnofacies and ichnocoenoses

Table 5.2.

Hunt and Lucas (2006, 2007c) introduced terminology for the study of tetrapod footprint ichnofacies. An ichnocoenosis can be defined as a trace fossil assemblage produced by a biological community that can be characterized by morphological criteria (independent of depositional environment or biological affinities) (e.g., Bromley, 1996; McIlroy, 2004; Hunt and Lucas, 2007c). Seilacher (1964, p. 303) introduced the term ichnofacies for "general trace associations, or types of ichnocoenoses, representing certain facies with a long geologic range." Subsequently, Hunt and Lucas (2007c) defined five archetypal tetrapod footprint ichnofacies for nonmarine environments: *Chelichnus*, *Grallator*, *Brontopodus*, *Batrachichnus* and *Characichnos* ichnofacies and diverse ichnocoenoses (Table 5.2). The following brief review of Triassic tetrapod footprint ichnofacies and ichnocoenoses is based on Hunt and Lucas (2006, 2007a,b,c), Hunt et al. (2018) and Klein and Lucas (2021).

6.1 *Batrachichnus* ichnofacies

Hunt and Lucas (2007c) proposed the *Batrachichnus* ichnofacies for ichnofaunas in which the majority of tracks are of quadrupedal carnivores with a moderate-high diversity (four to eight ichnogenera). This ichnofacies represents tidal flat-fluvial plain environments from the Devonian to the Cretaceous. Hunt and Lucas (2006, 2007c) recognized two ichnocoenoses of this ichnofacies in the Triassic (Table 5.2).

Hunt and Lucas (2007c) named the *Chirotherium* ichnocoenosis for the well-studied Early-Middle Triassic ichnofaunas of Europe and North America that are dominated by *Chirotherium* tracks. Other common ichnotaxa are *Rotodactylus, Rhynchosauroides, Isochirotherium,* and *Synaptichnium*. Hunt and Lucas (2006) recognized a distinctive and pervasive ichnocoenosis throughout much of the Upper Triassic portion of the Newark Supergroup in eastern North America assigned to the *Apatopus* ichnocoenosis. Ichnofaunas of this ichnocoenosis lack *Evazoum* and *Eosauropus,* contain less than 50% *Brachychirotherium* and *Grallator* and are characterized by ichnotaxa that are rare or absent elsewhere, including *Apatopus* and *Gwyneddichnium*.

6.2 *Brontopodus* ichnofacies

Hunt and Lucas (2007c) proposed the *Brontopodus* ichnofacies for medium diversity ichnofaunas in which the majority of tracks are of terrestrial herbivores with a small quantity (generally >10%) of terrestrial carnivore tracks. This ichnofacies includes coastal plain/marine shoreline environments and some lacustrine shorelines, and it ranges from late Permian to Recent in age. Hunt and Lucas (2006, 2007c) recognized five ichnocoenoses within this ichnofacies in the Triassic (Table 5.2). The oldest ichnocoenosis within the *Brontopodus* ichnofacies occurs in the earliest Triassic (possibly restricted to the Induan) and is characterized by "dicynodont" footprints in southern Africa, Antarctica, and Australia. Hunt and Lucas (2006) termed this the *Dicynodontipus* ichnocoenosis.

Most of the Early Triassic to early Middle Triassic ichnofaunas represent the *Chirotherium* ichnocoenosis of the *Batrachichnus* ichnofacies. However, a small number of localities are dominated by therapsid tracks. Hunt and Lucas (2006) termed this the *Therapsipus* ichnocoenosis for the therapsid ichnotaxon from the Moenkopi Formation of Arizona. Hunt et al. (2018) identified the *Procolophonichnium* ichnocoenosis in the Anisian carbonate tidal flats of Germany and the Netherlands. It represents a temporal equivalent of the red-bed *Chirotherium* ichnocoenosis. This ichnocoenosis is dominated by tracks of *Procolophonichnium* and *Rhynchosauroides* with only rare chirothere tracks.

6.3 *Grallator* ichnofacies

Hunt and Lucas (2007c) proposed the *Grallator* ichnofacies for medium to high diversity ichnofaunas (five to eight ichnogenera) dominated by tracks of tridactyl avian and nonavian theropods (usually dominant) or of other habitual bipeds. Tracks of bipedal and quadrupedal ornithischians, sauropods, and herbivorous mammals are also locally common in this ichnofacies. Known from the Late Triassic to the Recent, this ichnofacies often characterizes lacustrine margin environments. There are four ichnocoenoses of the *Grallator* ichnofacies during the Triassic (Table 5.2).

As noted above, Hunt and Lucas (2007c) recognized a *Grallator* ichnocoenosis in the Late Triassic. There are many Late Triassic ichnofaunas in which the most abundant (>50%) ichnogenus is *Grallator*. Notable ichnofaunas occur in the uppermost strata of the Chinle Group or in the overlying Wingate Sandstone in Colorado; other prominent examples are in Wales, France, Germany, Italy, Switzerland, and Greenland. Hunt and Lucas (2006) noted that there is potential to subdivide the *Grallator* ichnocenosis, and it clearly includes several subichnocoenoses. Thus, for example, on the Colorado Plateau in the western USA, the upper and lower Wingate Sandstone have different subichnocoenoses: a lower *Eosauropus* subichnocoenosis includes *Brasilichnium*, *Brachychirotherium*, and *Eosauropus*, and an upper *Otozoum* subichnocoenosis includes *Eubrontes*, *Batrachopus* and *Otozoum*.

6.4 *Chelichnus* ichnofacies

Hunt and Lucas (2007c) proposed the *Chelichnus* ichnofacies for ichnofaunas that have a low diversity (fewer than five ichnogenera) of tetrapod tracks in which manus and pes tracks are equant in shape, of subequal size and have short digit impressions. This ichnofacies is recurrent on dune faces in eolian environments, and it extends from the early Permian to the Early Jurassic. Hunt and Lucas (2007c) redefined the *Brasilichnium* ichnofacies of Lockley et al. (1994) as an ichnocoenosis of this ichnofacies (Table 5.2). The *Brasilichnium* ichnocoenosis has a few Late Triassic records, in the Redonda Formation of eastern New Mexico and in the lower Wingate Sandstone in western Colorado.

6.5 *Characichnos* ichnofacies

Hunt and Lucas (2007c) proposed the *Characichnos* ichnofacies for medium diversity ichnofaunas in which the majority of tracks are tetrapod swimming traces (parallel scratch marks) and fish swimming trails (*Undichna*). This ichnofacies represents shallow lacustrine (and tidal) environments. Swimming traces are notable in various Triassic units in the western United States, including the Moenkopi Group (Lower-Middle Triassic) in Arizona, Utah, Wyoming, and New Mexico and the Chinle Group (Upper Triassic) in Arizona and New Mexico. There is no named Triassic ichnocoenosis of this ichnofacies.

7. Triassic tetrapod evolution based on the footprint record

Figs. 5.21 and 5.22.

Correlations of Triassic tetrapod footprints with complete or near-complete skeletal materials have been attempted by various authors (e.g., Olsen and Baird, 1986; Farlow and Lockley, 1993; Haubold, 1999; Haubold and Klein, 2002; Avanzini and Renesto, 2002; Weems, 2003, 2019; Padian et al., 2010; Brusatte et al., 2011; Lallensack et al., 2017; Lichtig et al., 2018; Farlow et al., 2014, 2022, 2023; Abrahams et al., 2022; Klein et al., 2023b). This concerns essentially archosauromorph and archosaur footprints and correlatives, but also synapsid

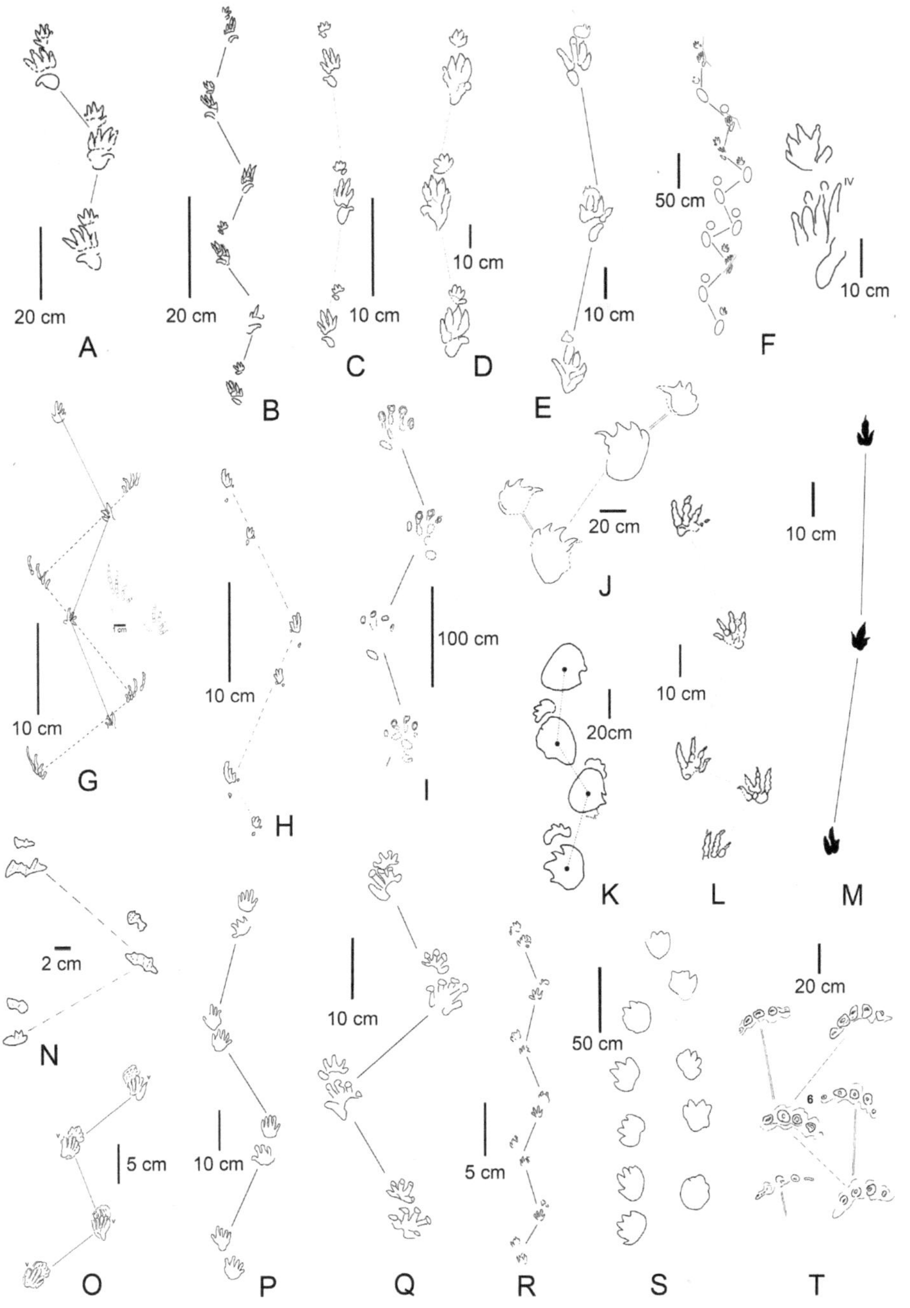

FIGURE 5.21 Trackways of Triassic tetrapods as interpretive outline drawings (from Klein and Lucas, 2021). **Archosauriform.** A. *Protochirotherium hauboldi*; Wióry Fm. (Olenekian), Poland. B. *Synaptichnium diabloense*; Wupatki Member (Moenkopi Fm., Olenekian), Arizona, USA. C. *Isochirotherium soergeli*; Solling Fm. (Buntsandstein, Anisian), Germany. D. *Chirotherium barthii*; Solling Fm. (Buntsandstein, Anisian), Germany. E. *Brachychirotherium parvum*;

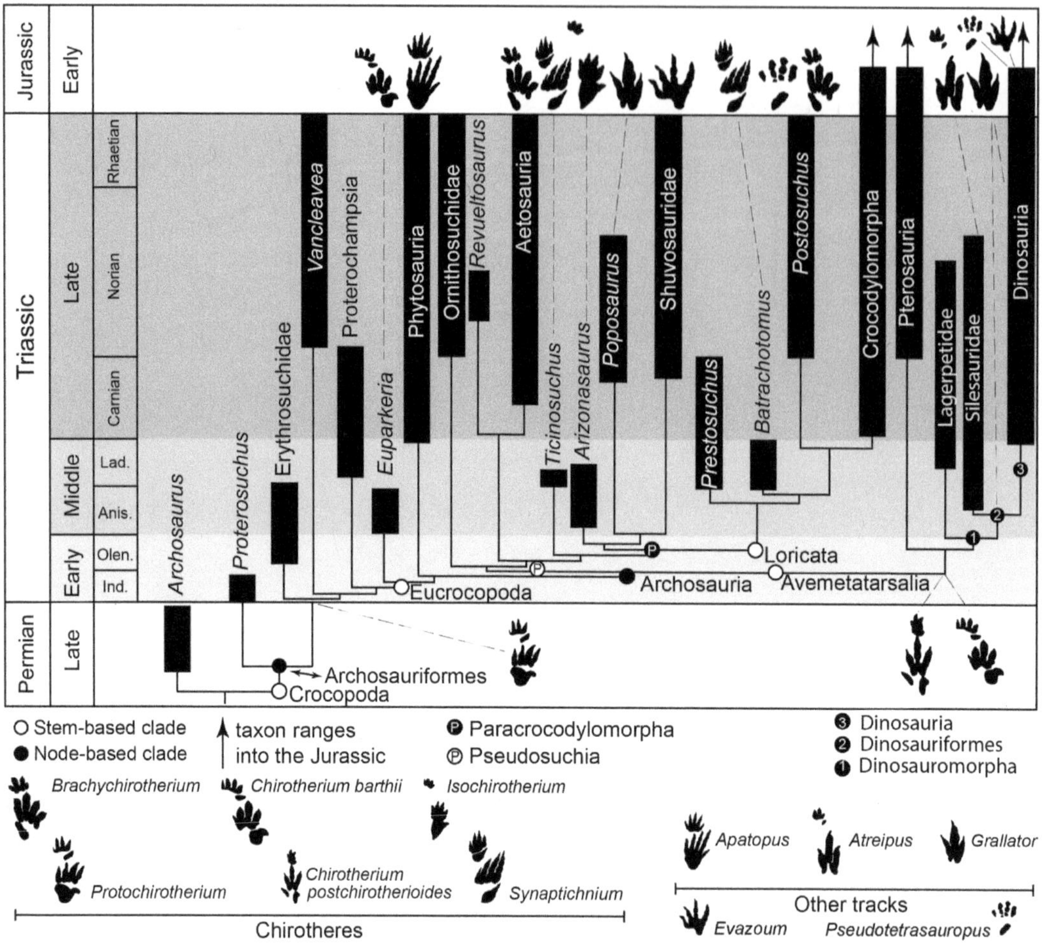

FIGURE 5.22 Phylogenetic relationships of archosaurs after Nesbitt (2011), with hypothetical position of archosaur footprints. Note alternative groups indicated for some ichnotaxa. *From Klein et al. (2023b).*

Redonda Fm. (Chinle group, Norian-Rhaetian), New Mexico, USA. F. *Apatopus lineatus*, trackway and detail with pes-manus; Timezgadiouine Fm. (T5, Carnian), Argana Basin, Morocco. G. **Archosauromorph–Lepidosauromorph.** *Rhynchosauroides tirolicus*; Middle Triassic, Southern Alps, Italy. **Dinosauromorph.** H. *Rotodactylus matthesi*; Solling Fm. (Buntsandstein, Anisian), Germany. **Sauropodomorph.** I. *Pseudotetrasauropus bipedoida*; lower Elliot Fm. (Norian), Lesotho, southern Africa. J. *Tetrasauropus unguiferus*; lower Elliot Fm. (Norian), Lesotho, southern Africa. K. *Eosauropus cimarronensis*; Sloan Canyon Fm. (Chinle group, Norian), New Mexico, USA. L. *Evazoum sirigui*; Redonda Fm. (Chinle group, Norian-Rhaetian), New Mexico, USA. **Dinosauriform–dinosaur.** M. *Grallator cursorius*; Redonda Fm. (Chinle group, Norian-Rhaetian), New Mexico, USA. **Turtle.** N. *Chelonipus torquatus*; Stuttgart Fm. (Schilfsandstein, Carnian), Germany. **Therapsid.** O. *Procolophonichnium lockleyi*; Hassberge Fm. (Coburger Sandstein, Carnian), Germany. P. *Dicynodontipus geinitzi*; Solling Fm. (Buntsandstein, Anisian), Germany. Q. *Capitosauroides bernburgensis*; Solling Fm. (Buntsandstein, Anisian), Germany. R. ?*Brasilichnium* isp. (Wingate Fm., Rhaetian), western Colorado, USA. S. *Therapsipus cumminsi*; Holbrook Member (Moenkopi Fm., Anisian), Arizona, USA. T. *Pentasauropus incredibilis*; lower Elliot Fm. (Norian), Lesotho, southern Africa.

footprints were analyzed based on Permian-Triassic materials (Marchetti et al., 2019a). Some are synapomorphy-based approaches based on diagnostic characters impressed and unequivocally documented in the footprints such as digit proportions, shape of sole impressions, or relative position and rotation of pes and manus imprints. However, because different tetrapod taxa can show similar foot morphology, footprints have a low taxonomic resolution if compared to skeletons, and may correspond to a family or even higher systematic unit. Nevertheless, distinct tendencies and improvements in the locomotor apparatus and autopodia are directly reflected in the footprints and trackways they leave. Examples are the development of the functionally tridactyl pes in different archosaur lines or the evolution of the erect stance convergently in archosaurs and synapsids (Fig. 5.21). This is important to have in mind when interpreting fossil footprints and their producers. After the Permian-Triassic extinction, recovery of terrestrial ecosystems included surviving tetrapod groups with temnospondyl nonamniotes, cynodont and dicynodont synapsids, parareptiles, and neodiapsids, the latter comprising lepidosauromorph and archosauromorph representatives. Footprints of all of these are known from Triassic deposits. With some probability and based on the known footprint record, the early evolution of the turtle-stem can be followed back to the Early Triassic, predating the oldest known skeletal fossils by more than 10 million years (Lichtig et al., 2018; see above).

The Triassic evolution of tetrapods and development of vertebrate communities on land were largely dominated by the further diversification and dispersal of archosaurs. This concerns the occupation of different ecological niches and habitats by well-adapted forms including archosauriform proterosuchids, erythrosuchids, proterochampsids, euparkeriids, phytosaurs, and more advanced members of the crown-group. The early evolution of crocodiles and dinosaurs falls into the Late Triassic as well as the first development of flight and the conquest of the skies by pterosaurs. Many of these are documented from both body- and ichnofossils, except in pterosaurs. In contrast to Jurassic and Cretaceous deposits that yielded abundant footprints, Triassic pterosaurs are only known from skeletal evidence. This phenomenon might be due to taphonomic instances, and early pterosaurs might have preferred environments that were not favorable for the preservation of footprints.

Convincing matches of Triassic archosaur footprints with known skeletal groups (Fig. 5.22) are *Rotodactylus*, which has been attributed to lagerpetid dinosauromorphs by Haubold (1999), *Apatopus*, which can be correlated with phytosaurs, *Synaptichnium*, which shows pedal digit proportions similar to the pseudosuchians, *Ticinosuchus* and *Batrachotomus* and *Brachychirotherium*, which have analogues in aetosaurs, but also in loricatans such as *Postosuchus*. Other archosaur footprints correspond to phylogenetically very distant groups such as *Chirotherium* that can be tentatively correlated with *Euparkeria*-like archosauriforms, but, based on the functionally tridactyl pes, also with basal avemetatarsalians and members of the dinosaur-bird line. Both dinosauromorph and crocodilian-stem producers can be considered for some typical tridactyl grallatorid footprints. This has been demonstrated more recently by Farlow et al. (2014), who showed that the paracrocodylomorph *Poposaurus* from the Upper Triassic of the western USA could potentially have left footprints generally considered to be those of theropod dinosaurs. A mostly complete pes is known from *Poposaurus* that shows a functionally tridactyl pattern that has obviously developed convergent with the typical dinosauriform pes. Nonrelated archosaurs with a similar pes morphology could have left the same type of footprints, being almost indistinguishable from the ichnofossil material. Different attributions and correlations with skeletons can be also considered for footprints such as *Pseudotetrasauropus* and *Evazoum* that can be correlated with sauropodomorphs but also with some pseudosuchian archosaurs.

Triassic tetrapod footprints largely support evolutionary developments and phylogenetic patterns derived from the known body fossils (Nesbitt, 2011), in some cases extending stratigraphic ranges as is the case with the first appearance of turtles. They do not identify mass extinctions of tetrapods across the Permo-Triassic or the Triassic-Jurassic boundaries.

8. Conclusions

Triassic tetrapod footprint assemblages are some of the most extensively studied ichnofossils. Their importance lies in their temporal-stratigraphic position demarcating recovery of terrestrial ecosystems after the Permian-Triassic extinction, including the rise and diversification of archosaurs with the early evolution of crocodiles and dinosaurs. Developments are reflected in the abundance of chirotheriid and other archosauriform–crocodylomorph and dinosauromorph ichnotaxa such as *Synaptichnium, Protochirotherium, Isochirotherium, Chirotherium, Brachychirotherium, Batrachopus* and *Rotodactylus*, and, beginning from the late Middle Triassic, the appearance of tridactyl forms such as *Atreipus, Grallator,* and *Eubrontes* documenting the transition to typical Jurassic dinosaur-dominated associations. Early sauropodomorph–sauropod dinosaurs are common at some Late Triassic localities and represented by the ichnotaxa *Pseudotetrasauropus, Tetrasauropus, Evazoum,* and *Eosauropus*. Nonarchosaur tetrapod groups are locally abundant such as more basal archosauromorphs and lepidosauromorphs represented by the ichnogenus *Rhynchosauroides*. Important is the appearance of stem-turtles beginning from the Early Triassic, as indicated by *Chelonipus* footprints. Less frequent components of Triassic assemblages are cynodont and dicynodont therapsids indicated, for example, by the ichnogenera *Dicynodontipus* and *Capitosauroides*. Scarce elements of ichnoassemblages are the footprints of nonamniotes such as *Batrachichnus*, which have been abundantly documented from Permian deposits.

Important tetrapod footprint assemblages are known from North America (Moenkopi and Chinle groups, Newark Supergroup), from South America (Lower–Upper Triassic deposits in Argentina, Bolivia, Peru, and Brazil), from Greenland (Fleming Fjord Group), from Europe (Buntsandstein, Muschelkalk, Keuper and equivalents), from North Africa (Timezgadiouine Formation in Morocco), from southern Africa (lower Elliot Formation in Lesotho and RSA), and finally from China (Guanling Formation in Guizhou and Yunnan provinces). Similar assemblages and ichnotaxa of archosaurs identified from different regions of Pangea suggest a global dispersal and distribution of their producers.

Biostratigraphically, five biochrons based on tetrapod footprints are distinguished: 1) Dicynodont tracks (Induan–Olenekian), 2) *Protochirotherium* (Olenekian), 3) *Chirotherium barthii* (early Anisian), 4) *Atreipus–Grallator* (late Anisian–Ladinian), and 5) *Brachychirotherium* (Carnian–Rhaetian).

Acknowledgments

We are grateful to many collection managers and curators for access to specimens, and to diverse colleagues for their collaboration and discussions. Ignacio Díaz-Martínez and Adrian P. Hunt provided helpful reviews of the manuscript.

References

Abel, O., 1926. Der erste Fund einer Tetrapodenfährte in der unteren alpinen Trias. Paläontol. Z. 7, 22–24.

Abel, O., 1935. Vorzeitliche Lebensspuren. G. Fischer, Jena.

Abrahams, M., Bordy, E.M., Knoll, F., Farlow, J.O., 2022. Theropod tridactyl tracks across the Triassic–Jurassic boundary in southern Africa: implications for pedal morphology evolution. Front. Ecol. Evol. Sec Paleontol. 10. https://doi.org/10.3389/fevo.2022.925313.

Andrade-Silva, M., Francischini, H., Ribeiro, A.M., 2023. On the presence of the ichnogenus Procolophonichnium Nopcsa, 1923 in the late triassic of Brazil. J. S. Am. Earth Sci. 132, 104689. https://doi.org/10.1016/j.jsames.2023.104689.

Avanzini, M., Leonardi, G., 2002. *Isochirotherium inferni* ichnosp. n. in the Illyrian (late Anisian, middle Triassic) of Adige valley. Bolletino della Soc. Paleontol. Ital. 41, 41–50.

Avanzini, M., Lockley, M., 2002. Middle Triassic archosaur population structure: interpretation based on *Isochirotherium delicatum* fossil footprints (Southern Alps, Italy). Palaeogeogr. Palaeoclimatol. Palaeoecol. 185, 391–402.

Avanzini, M., Renesto, S., 2002. A review of *Rhynchosauroides tirolicus* Abel, 1926 ichnospecies (Middle Triassic: Anisian-Ladinian) and some inferences on *Rhynchosauroides* trackmaker. Riv. Ital. Paleontol. Stratigr. 108, 51–66.

Avanzini, M., Wachtler, M., 2012. *Sphingopus ladinicus* isp. nov. From the Anisian of the Braies Dolomites (southern Alps, Italy). Bolletino della Soc. Paleontol. Ital. 51, 63–70.

Apesteguía, S., Riguetti, F., Citton, P., Veiga, G.D., Poiré, D.G., de Valais, S., Zacarías, G., 2021. The Ruditayoj-Tunasniyoj fossil area (Chuquisaca, Bolivia); A Triassic chirotheriid megatracksite and reinterpretation of purported thyreophoran tracks. Hist. Biol. 33 (11), 2883–2896. https://doi.org/10.1080/08912963.2020.1833328.

Aumann, G., 1960. *Brachychirotherium coburgense* n. sp.- eine neue Saurierfährte aus dem Mittleren Keuper von Löbelstein bei Coburg. Geologische Blätter für NO-Bayern 10, 58–65.

Avanzini, M., García-Ramos, J.C., Lires, J., Menegon, M., Piñuela, L., Fernández, L.A., 2005. Turtle tracks from the late Jurassic of Asturias, Spain. Acta Palaeontol. Pol. 50 (4), 743–755.

Avanzini, M., Piñuela, L., Garcia-Ramos, J.C., 2010a. First report of a late Jurassic lizard like footprint (Asturias, Spain). J. Iber. Geol. 36, 175–180.

Avanzini, M., Petti, F.M., Bernardi, M., Tomasoni, R., 2010b. Crocodile-like footprints from the upper Triassic (carnian) of the Italian southern Alps. New Mexico Museum of Natural History and Science, Bulletin 51, 55–60.

Baird, D., 1954. *Chirotherium lulli*, a pseudosuchian reptile from New Jersey. Bull. Mus. Comp. Zool. 111, 165–192.

Baird, D., 1957. Triassic reptile footprint faunules from Milford, New Jersey. Bull. Mus. Comp. Zool. 117, 449–520.

Beasley, H.C., 1895. An attempt to classify the footprints in the New Red Sandstone of this district. Proc. Liverp. Geol. Soc. 7 (4), 391–409.

Beasley, H.C., 1901. Notes on the type specimen of *Cheirotherium herculis* (Egerton). Proc. Liverp. Geol. Soc. 9, 81.

Beasley, H.C., 1904. Report on Footprints from the Trias: Part I. Report of the. British Association for the Advancement of Science, pp. 219–230.

Beasley, H.C., 1905. Report on Footprints from the Trias: Part II. Report of the British Association for the Advancement of Science, pp. 275–282.

Beasley, H.C., 1907. Report on Footprints from the Trias: Part IV. Report of the. British Association for the Advancement of Science, pp. 299–301.

Beasley, H.C., 1908. Report on Footprints from the Trias: Part V. Report of the. British Association for the Advancement of Science, pp. 300–304.

Beasley, H.C., 1909. Report on Footprints from the Trias: Part VI: Report of the. British Association for the Advancement of Science, pp. 150–163.

Belvedere, M., Farlow, J.O., 2016. A numerical scale for quantifying the quality of preservation of vertebrate tracks. In: Falkingham, P.L., Marty, D., Richter, A. (Eds.), Dinosaur Tracks: The Next Steps. Indiana University Press, Bloomington, pp. 92–98.

Belvedere, M., Bennett, M.R., Marty, D., Budka, M., Reynolds, S.C., Bakirov, R., 2017. Stat-tracks and mediotypes: Powerful tools for modern ichnology based on 3D models. PeerJ, e4247. https://doi.org/10.7717/peerj.4247.

Bernardi, M., Klein, H., Petti, F.M., Ezcurra, M.D., 2015. The origin and early radiation of archosauriforms: Integrating the skeletal and footprint record. PLoS One 10 (6), e0128449.

Berthold, C., 1835. Besprechung von Sickler's Sendschreiben an Blumenbach von 1834. Göttingische Gelehrte Anzeigen 52, 514–520.

Beurlen, K., 1950. Neue Fährtenfunde aus der fränkischen Trias. Neues Jahrbuch für Geologie und Paläontologie, Monatshefte, vol 1950, pp. 308–320.

Biron, P.E., Dutuit, J.-M., 1981. Figurations sedimentaires et traces d'activite au sol dans le Trias de la formation d'Argana et de l'Ourika (Maroc). Bulletin du Muséum National d'Histoire Naturelle. 4e, 3rd series, section C 4, 399–427.

Bock, W., 1952. Triassic reptilian tracks and trends of locomotive evolution. J. Paleontol. 26, 395–433.

Bornemann, J.G., 1887. Über fossile Tierspuren aus dem Buntsandstein Thüringens. Z. Dtsch. Geol. Ges. 39, 629–630 (Berlin).

Bromley, R.G., 1996. Trace Fossils: Biology and Taphonomy, second ed. Unwin Hyman, London. 361 pp.

Brusatte, S.L., Niedzwiedzki, G., Butler, R.J., 2011. Footprints pull origin and diversification of dinosaur stem-lineage deep into Early Triassic. Proc. R. Soc. Ser. B 278, 1107–1113.

Buchwitz, M., Marchetti, L., Jansen, M., Falk, D., Trostheide, F., Schneider, J.W., 2020. Ichnotaxonomy and trackmaker assignment of tetrapod tracks and swimming traces from the Middle Permian Hornburg Formation ofSaxony-Anhalt (Germany). Ann. Soc. Geol. Pol. 90, 291–320.

Buck, P.V., Ghilardi, A.M., Peixoto, B.C.P.M., Fernades, L.B., Fernades, M.A., 2017. A new tetrapod ichnotaxon from Botucatu Formation, Lower Cretaceous (Neocomian), Brazil, with comments on fossil track preservation on inclined planes and local paleoecology. Palaeogeography, Paleoclimatology, Paleoecology 466, 21–37.

Buffetaut, E., Suteethorn, V., Cuny, G., Tong, H., Le Loeuff, J., Khansubha, S., Jongautcharlyaku, S., 2000. The earliest known sauropod dinosaur. Nature 407, 72–74.

Buffetaut, E., Suteethorn, V., Le Loeuff, J., Cuny, G., Tong, H., Khansubha, S., 2002. The first giant sauropod from the Late Triassic of Thailand. C. R. Paleo 1, 203–258.

Casamiquela, R.M., 1964. Estudios icnológicos: Problemas y métodos de la icnologia con aplicación al estudio de pisadas Mesozoicos (Reptilia, Mammalia) de la Patagonia. Colegio Industrial Pio IX, Buenos Aires.

Casamiquela, R.M., 1975. Nuevo material y reinterpretacion de las icnitas Mesozoicas (Neotriasicas) de Los Menucos, Provincia de Rio Negro (Patagonia): 1° Congreso Argentino de Paleontología y Bioestratigrafía (Tucumán, 1974). Actas 1, 555580.

Casanovas-Cladellas, M.L., Santafé-Llopis, J.V., Gómez-Alba, J., 1979. Presencia de *Chirotherium* en el Triásico Catalán. Butll. Inf. Inst. Paleontol. Sabadell 9, 34–42.

Cavin, L., Avanzini, M., Bernardi, M., Piuz, A., Proz, P.-A., Meister, C., Boissonnas, J., Meyer, C.A., 2012. New vertebrate trackways from the autochthonous cover of the Aiguilles Rouges Massif and reevaluation of the dinosaur record in the Valais, SW Switzerland. Swiss J. Palaeontol. https://doi.org/10.1007/s13358-012-0040-0.

Ceoloni, P., Conti, M.A., Mariotti, N., Mietto, P., Nicosia, U., 1988. New late permian tetrapod footprints from southern Alps. Mem. Soc. Geol. Ital. 34, 45–65.

Citton, P., Díaz-Martínez, I., de Valais, S., Cónsole-Gonella, C., 2018. Triassic pentadactyl tracks from the Los Menucos group (Río Negro province, Patagonia Argentina): possible constraints on the autopodial posture of Gondwanan trackmakers. PeerJ 6, e5358. https://doi.org/10.7717/peerj.5358.

Citton, P., de Valais, S., Díaz-Martínez, I., González, S.N., Greco, G.A., Cónsole-Gonella, C., Leonardi, G., 2021. Age-constrained therapsid tracks from a mid-latitude upland (Permian-Triassic transition, Los Menucos Complex, Argentina). J. S. Am. Earth Sci. 110, 103367.

Clark, N.D.L., Corrance, H., 2009. New discoveries of *Isochirotherium herculis* (Egerton 1838) and a reassessment of chirotheriid footprints from the Triassic of the Isle of arran, Scotland. Scot. J. Geol. 45, 69–82.

Conrad, K., Lockley, M.G., Prince, N.K., 1987. Triassic and Jurassic vertebratedominated trace fossil assemblages of the Cimarron Valley region: implications for paleoecology and biostratigraphy. New Mexico Geological Society, Guidebook 38, 127–138.

Courel, L., Demathieu, G., 1976. Une ichnofaune reptilienne remarquable dans les grès Triasique de Largentière (Ardèche, France). Palaeontograph. Abteilung 151, 194–216.

Courel, L., Demathieu, G., 2000. Une nouvelle ichnoespèce *Coelurosaurichnus grancieri* du Trias supérieur de l' Ardèche, France. Geodiversitas 22, 35–46.

D'Orazi Porchetti, S., Bertini, R.J., Langer, M.C., 2018. Proposal for ichnotaxonomic allocation of therapsid footprints from the Botucatu Formation (Brazil). Ichnos 25, 192–207.

D'Orazi-Porchetti, S., Nicosia, U., Mietto, P., Petti, F.M., Avanzini, M., 2008. *Atreipus*like footprints and their co-occurrence with *Evazoum* from the upper Carnian (Tuvalian) of Trentino-Alto Adige: Studi Trentini di Scienze Naturali. Acta Geol. 83, 277–287.

De Jaime-Soguero, C., Mujal, E., Dinarès-Turell, J., Oms, O., Bolet, A., Orlandi-Oliveras, G., Fortuny, J., 2021. Palaeoecology of Middle Triassic tetrapod ichnoassociations (middle Muschelkalk, NE Iberian Peninsula) and their implications for palaeobiogeography in the western Tethys region. Palaeogeogr. Palaeoclimatol. Palaeoecol. 565, 110204.

Demathieu, G., 1966. *Rhynchosauroides petri* et *Sphingopus ferox*, nouvelles empreintes de reptiles de grès a bordure Nord-Est du Massif Central, D263. Comptes Rendus de l'Académie des Sciences, pp. 483–486.

Demathieu, G., 1970. Les empreintes de pas de vertébrés du Trias de la bordure Nord-Est du Massif Central. Cahiers de Paleontologie. CNRS, Paris, 211 pp.

Demathieu, G., 1971. Cinq nouvelles espèces d'empreintes de reptiles du Trias de la bordure nord-est du Massif Central. C. R. Acad. Sci. Paris D272, 812–814.

Demathieu, G., 1974. Les dalles a empreintes de pas de reptiles du Muséum d'Histoire Naturelle de Lyon. Nouvelles Archives du Muséum d'Histoire Naturelles Lyon 12, 5–12.

Demathieu, G., 1977. Des Microvertébrés dans le Trias Moyen du Lyonnais et du Máconnais révélé par leurs empreintes; signification paléoécologique. Geobios 10, 351–367.

Demathieu, G., 1985. Trace fossil assemblages in Middle Triassic marginal marine deposits, eastern border of the Massif Central, France. In: Curran, A.H. (Ed.), Biogenic Structures: Their Use in Interpreting Depositional Environments, vol 35. SEPM Special Publication, pp. 53–66.

Demathieu, 1989. Appearance of the first dinosaur tracks in the French Middle Triassic and their probable significance. In: Gillette, D.D., Lockley, M.G. (Eds.), Dinosaur Tracks and Traces. Cambridge University Press, Cambridge, pp. 201–207.

Demathieu, G., Demathieu, P., 2004. Chirotheria and other ichnotaxa of the European Triassic. Ichnos 11, 79–88.

Demathieu, G., Durand, M., 1991. Les traces de pas de Tétrapodes dans le Trias détritique du Var et des Alpes-Maritimes (France). Bulletin du Museum national d'Histoire naturelle, Paris, series 4 13C, 115–133.

Demathieu, G., Gand, G., 1973. Deux espèces ichnologiques nouvelles de grès à empreintes du Trias du Plateau d'Antully. Bulletin de la Société d'Histoire Naturelles d'Autun 67, 11–27.

Demathieu, G., Gand, G., 1981. Palichnologie: Interprétation paléoécologique de traces d'origine biologique et mécanique observées dans la carrière triasique de Pont d'Argent, part 1, vol 98. Bulletin de la Société d'Histoire Naturelles d'Autun, pp. 3–22.

Demathieu, G., Haubold, H., 1972. Stratigraphische Aussagen der Tetrapodenfährten aus der terrestrischen Trias Europas. Geologie 21, 802–836.

Demathieu, G., Haubold, H., 1974. Evolution und Lebensgemeinschaft terrestrischer Tetrapoden nach ihren Fährten in der Trias. Freiberger Forschungshefte - A C298, 51–72.

Demathieu, G., Haubold, H., 1982. Reptilfährten aus dem Mittleren Buntsandstein von Hessen. Hallesches Jahrb. Geowiss. 7, 97–110.

Demathieu, G., Leitz, F., 1982. Wirbeltier-Fährten aus dem Röt von Kronach (Trias, Nordost-Bayern). Mittl. Bayer. Staatssamml. Palaontol. Hist. Geol. 22, 63–89.

Demathieu, G., Oosterink, H.W., 1983. Die Wirbeltier-Ichnofauna aus dem Unteren Muschelkalk von Winterswijk (Die Reptilfährten aus der Mitteltrias der Niederlande). Staringia 7, 1–51.

Demathieu, G., Oosterink, H.W., 1988. New discoveries of ichnofossils from the middle Triassic of Winterswijk (The Netherlands). Geol. Mijnbouw 67, 3–17.

Demathieu, G., Weidmann, M., 1982. Les empreintes de pas de reptiles dans le Trias du Vieux Émosson (Finhaut, Valais, Suisse). Eclogae Geol. Helv. 75, 721–757.

Díaz-Martínez, I., Castanera, D., Gasca, J.M., Canudo, J.I., 2015. A reappraisal of the Middle Triassic chirotheriid *Chirotherium ibericus* Navás, 1906 (Iberian Range, NE Spain), with comments on the Triassic tetrapod track biochronology of the Iberian Peninsula. PeerJ 3, e1044. https://doi.org/10.7717/peerj.1044.

Díaz-Martínez, I., Pérez-García, A., 2012. Historical and comparative study of the first Spanish vertebrate paleoichnological record and bibliographic review of the Spanish chirotheriid footprints. Ichnos 19 (3), 141–149.

D'Orazi-Porchetti, S., Nicosia, U., 2007. Re-examination of some large early Mesozoic tetrapod footprints from the African collection of Paul Ellenberger. Ichnos 14, 219–245.

Egerton, P.G., 1838. On two casts in sandstone of the impressions of the hindfoot of a gigantic *Cheirotherium* from the New Red Sandstone of Cheshire. Proc. Geol. Soc. Lond. 3, 14–15.

Ellenberger, P., 1965. Découverte de pistes Vertébrés dans le Permien, le Trias et le Lias inférieur, aux abords de Toulon (Var) et d'Anduze (Gard). Comptes Rendus de l'Académie des Sciences Paris 260, 5856–5859.

Ellenberger, P., 1970. Les niveaux paléontologiques de première apparition des mammifères primordiaux en Afrique du sud et leur ichnologie. Establissement de zones stratigraphiques detaillées dans le Stormberg du Lesotho (Afrique du Sud) (Trias supérieur à jurassique). In: Second Gondwana Symposium, Proceedings and Papers. Council for Scientific and Industrial Research, Pretoria, pp. 343–370.

Ellenberger, P., 1972. Contribution à la classification des Pistes de Vertébrés du Trias: Les types du Stormberg d'Afrique du Sud (I): Palaeovertebrata, Memoire Extraordinaire, pp. 1–104.

Ellenberger, P., 1974. Contribution à la classification des Pistes de Vértebrés du Trias: Les types du Stormberg d' Afrique du Sud (II, Les Stormberg Superieur): Palaeovertebrata, Memoire Extraordinaire, 141 pp.

Ellenberger, P., 1975. L'explosion demographique des petities quadrupedes a l'alure de mammiferes dans les Stormberg Superieur (Trias) d'Afrique du sud apercu sur leur origine au Permien (France et Karoo). International Colloquium of the National Center for Scientific Research Bulletin 218, 409–432.

Ellenberger, P., 1976. Une piste avec traces de soies épaisses dans le Trias inférieur a moyen de Lodéve (Hérault, France): Cynodontipus poythrix nov. gen. nov. sp. les cynodontes in France. Géobios 9 (6), 769–787.

Ellenberger, F., Ellenberger, P., Ginsburg, L., 1970. Les dinosaures du Trias et du Lias en France et en Afrique du Sud, d'après les pistes qu'ils ont laissées. Bulletin de la Société Géologiques de France 7 (12), 151–159.

Faber, F.J., 1958. Fossiele voetstappen in de Muschelkalk van Winterswijk. Geol. Mijnbouw 20, 317–321.

Falkingham, P.L., 2012. Acquisition of high resolution three-dimensional models using free, open-source, photogrammetric software. Palaeontol. Electron. 15, 15.

Falkingham, P.L., 2016. Applying objective methods to subjective track outlines. In: Falkingham, P.L., Marty, D., Richter, A. (Eds.), Dinosaur Tracks, the Next Steps. Indiana University Press, Indianapolis and Bloomington, pp. 73–80.

Falkingham, P.L., Gatesy, S.M., 2014. The birth of a dinosaur footprint: Subsurface 3D motion reconstruction and discrete element simulation reveal track ontogeny. Proc. Natl. Acad. Sci. U.S.A. 111, 18279–18284.

Falkingham, P.L., Bates, K.T., Avanzini, M., Bennett, M.R., Bordy, E.M., Breithaupt, B.H., Castanera, D., Citton, P., Díaz-Martínez, I., Farlow, J.O., Fiorillo, A.R., Gatesy, S.M., Getty, P., Hatala, K.G., Hornung, J.J., Hyatt, J.A., Klein, H., Lallensack, J.N., Martin, A.J., Marty, D., Matthews, N.A., Meyer, C.A., Milàn, J., Minter, N.J., Razzolini, N.L., Romilio, A., Salisbury, S.W., Sciscio, L., Tanaka, I., Wiseman, A.L.A., Xing, L.D., Belvedere, M., 2018. A standard protocol for documenting modern and fossil ichnological data. Palaeontology 61, 469–480.

Falkingham, P.L., Turner, M.L., Gatesy, S.M., 2020. Constructing and testing hypotheses of dinosaur foot motions from fossil tracks using digitization and simulation. Palaeontology. https://doi.org/10.1111/pala.12502.

Farlow, J.O., Lockley, M.G., 1993. An osteometric aprroach to the identification of the makers of Early Mesozoic tridactyl dinosaur footprints. New Mexico Museum of Natural History and Science, Bulletin 3, 123–131.

Farlow, J.O., Schachner, E.R., Sarrazin, J.C., Klein, H.K., Currie, P.J., 2014. Pedal proportions of *Poposaurus gracilis*: Convergence and divergence in the feet of archosaurs. Anat. Rec. 297, 1022–1046.

Farlow, J., Coroian, D., Currie, P., 2018. Noah's Ravens: Interpreting the Makers of Tridactyl Dinosaur Footprints. Indiana University Press, Bloomington, Indiana, 750 pp.

Farlow, J.O., Lallensack, J.N., Müller, R.T., Hyatt, J.A., 2022. Pedal skeletal proportions of bipedal and potentially bipedal dinosaurs and other archosaurs: interpreting the makers of early Mesozoic footprints. Bull. - Peabody Mus. Nat. Hist. 63 (2), 33–90. https://doi.org/10.3374/014.063.0201.

Farlow, J.O., Kumagai, C.J., Klein, H., 2023. Alligators, crocodiles, and chirotheres: within-taxon and across-taxon variability in ichnologically relevant autopodial and other body proportions. In: Woodward, H.N., Farlow, J.O. (Eds.), Ruling Reptiles: Crocodylian Biology and Archosaur Paleobiology. Indiana University Press. https://doi.org/10.2307/jj.6047951.

Fernandes, M.A., Carvalho, I.S., 2008. Revisão diagnóstica para a icnospécie de tetrápode Mesozóico *Brasilichnium elusivuum* (Leonardi, 1981) (Mammalia) de Formação Botucato, Bacia do Paraná, Brasil: Ameghiniana, 45, 167–173.

Fichter, J., Kunz, R., 2004. New genus and species of chirotheroid tracks in the Detfurth-formation (middle Bunter, lower Triassic) of Central Germany. Ichnos 11, 183–193.

Fichter, J., Kunz, R., 2011. Neue Nachweise chirotheroider Fährten in der DetfurthFormation (Mittlerer Buntsandstein, Untere Trias) bei Wolfhagen. Geol. Jahrb. Hessen 137, 5–18.

Fichter, J., Kunz, R., 2013. "Dinosauromorph"Tracks from the Middle Buntsandstein (Early Triassic: Olenekian) of Wolfhagen, Northern Hesse, Germany: LNEG, vol 100. Comunicacoes Geologicas, pp. 81–88.

Fichter, J., Kunz, R., 2015. Eine Tetrapoden-Fährtenvergesellschaftung im Mittleren Buntsandstein (frühe Trias, Olenekium) von Wolfhagen, Nordhessen. Z. Dtsch. Geol. Ges. 166, 253–273.

Fichter, J., Lepper, J., 1997. Die Fährtenplatte vom Heuberg bei Gieselwerder: Philippia, vol 8, pp. 35–60.

Fillmore, D.L., Lucas, S.G., Simpson, E.L., 2012. Ichnology of the Mississippian Mauch Chunk formation, eastern Pennsylvania. New Mexico Museum of Natural History and Science, Bulletin 54, 136 pp.

Fraser, N.C., Olsen, P.E., 1996. A new dinosauromorph ichnogenus from the Triassic of Virginia: Jeffersonia. Contributions of the Virginia Museum of Natural History 7, 1–17.

Fuglewicz, R., Ptaszyński, T., Rdzanek, K., 1990. Lower Triassic footprints from the Swietokrzyskie (holy Cross) Mountains, Poland. Acta Palaeontol. Pol. 35, 109–164.

Gand, G., 1977. Sur le matèriel ichnologique recoltè dans le Muschelkalk de Culles-lesRoches (Saone-et-Loire). Bulletin de la Société d' Histoire Naturelle du Creusot 35, 21–44.

Gand, G., 1979. Description de deux nouvelles traces d'*Isochirotherium* observées dans les grès du Trias moyen de Bourgogne. Bulletin de la Société d'Histoire Naturelle du Creusot 37, 13–25.

Gand, G., De La Horra, R., Galán-Abellán, B., López-Gómez, J., Barrenechea, J.F., Arche, A., Benito, M.I., 2010. New ichnites from the middle Triassic of the Iberian ranges (Spain): paleoenvironmental and paleogeographical implications. Hist. Biol. 22, 40–56.

Gand, G., Demathieu, G., 2005. Les pistes dinosauroides du Trias moyen francais: Interprétation et réévaluation de la nomenclature. Geobios 38, 725–749.

Gand, G., Demathieu, G., Grancier, M., Sciau, J., 2005. Les traces dinosauroides du Trias supérieur francais: discrimination, interprétation et comparaison. Bull. Soc. Geol. Fr. 176, 69–79.

Gand, G., Vianey-Liaud, M., Demathieu, G., Garric, J., 2000. Deux nouvelles traces de pas de dinosaures du Trias supérieur de la bordure Cévenole (La Grand-Combe, SudEst de la France. Geobios 33, 599–624.

Gastaldo, R.A., Kamo, S.L., Neveling, J., Geissman, J.W., Looy, C.V., Martin, A.M., 2020. The base of the *Lystrosaurus* assemblage zone, Karoo basin, predates the end permian marine extinction. Nat. Commun. 11, 1428. https://doi.org/10.1038/s41467020-15243-7.

Gatesy, S.M., Falkingham, P.L., 2017. Neither bones nor feet: track morphology variation and preservation quality. J. Vertebr. Paleontol., e1314298 https://doi.org/10.1080/02724634.2017.1314298.

Gatesy, S.M., Falkingham, P.L., 2020. Hitchcock's Leptodactyli, penetrative tracks, and dinosaur footprint diversity. J. Vertebr. Paleontol. 40 (3), e1781142.

Geinitz, H.B., 1861. Dyas oder die Zechsteinformation und das Rothliegende (Permische Formation zum Theil). W. Engelmann, Leipzig, pp. 1–130.

Gierliński, G.D., Lagnaoui, A., Klein, K., Saber, H., Oukassou, M., Charrière, A., 2017. Bird-like tracks from the Imilchil Formation (middle Jurassic, Bajocian-Bathonian) of the central high Atlas, Morocco, in comparison with similar Mesozoic tridactylous ichnotaxa. Boll. Soc. Paleontol. Ital. 56, 207–215.

Gilmore, C.W., 1928. Fossil footprints from the Grand Canyon: second contribution: Smithsonian Miscellaneous collections vol 80, no. 3, 178 pp.

Gürich, G., 1926. Über Saurier-Fährten aus dem Etjo-Sandstein von Südwestafrika [On dinosaur tracks from the Etjo Sandstone of South-West Africa]. Paläontol. Z. 8, 112–120.

Harkness, R., 1850. Notice of a tridactylous footmark from the Bunter sandstone of Weston Point, Cheshire. Annals and Magazine of Natural History, series 6 (36), 440–442.

Haubold, H., 1966. Therapsiden- und Rhynchocephalen-Fährten aus dem Buntsandstein Südthüringens: Hercynia. NF 3, 147–183.

Haubold, H., 1967. Eine Pseudosuchier-Fährtenfauna aus dem Buntsandstein Südthüringens: Hallesches Jahrbuch für Mitteldeutsche Erdgeschichte, vol 8, pp. 12–48.

Haubold, H., 1971a. Die Tetrapodenfährten des Buntsandsteins: Paläontologische Abhandlungen A, vol 4, pp. 395–548.

Haubold, H., 1971b. Ichnia Amphibiorum et Reptiliorum fossilium. Encyclopedia of Paleoherpetology (18), 1–124.

Haubold, H., 1986. Archosaur footprints at the terrestrial Triassic-Jurassic transition. In: Padian, K. (Ed.), The Beginning of the Age of Dinosaurs. University Press, Cambridge, pp. 189–201.

Haubold, H., 1996. Ichnotaxonomie und Klassifikation von Tetrapodenfährten aus dem Perm: Hallesches Jahrbuch für Geowissenschaften B, vol 18, pp. 23–88.

Haubold, H., 1999. Tracks of the Dinosauromorpha from the early Triassic. In: Bachmann, G.H., Lerche, I. (Eds.), Triassic: Zentralblatt für Geologie und Paläontologie, part 1, vol 1998, pp. 783–795 (7–8).

Haubold, H., 2006. Die Saurierfährten *Chirotherium barthii* Kaup, 1835 – das Typusmaterial aus dem Buntsandstein bei Hildburghausen/Thüringen und das ChirotheriumMonument: Veröffentlichungen des Naturhistorischen Museums Schleusingen, vol 21, pp. 3–31.

Haubold, H., Klein, H., 2000. Die dinosauroiden Fährten *Parachirotherium–Atreipus–Grallator* aus dem unteren Mittelkeuper (Obere Trias: Ladin, Karn, ?Nor) in Franken. Hallesches Jahrbuch für Geowissenschaften, B22, pp. 59–85.

Haubold, H., Klein, H., 2002. Chirotherien und Grallatoriden aus der Unteren bis Oberen Trias Mitteleuropas und die Entstehung der Dinosauria: Hallesches Jahrbuch für Geowissenschaften, B24, pp. 1–22.

Heller, F., 1952. Reptilienfährten-Funde aus dem Ansbacher Sandstein des Mittleren Keupers von Franken: Geologische Blätter für NO-Bayern, vol 2, pp. 129–141.

Hitchcock, E., 1837. Fossil footsteps in sandstone and graywacke. American Journal of Science and Arts, Series 1 (32), 174–176.

Hitchcock, E., 1843. Description of five new species of fossil foot marks, from the Red Sandstone of the valley of the Connecticut River: Reports of the 1st, 2d and 3d. Meetings of the Association of American Gemologists and Naturalists 254–264.

Hitchcock, E., 1845. An attempt to name, classify, and describe the animals that made the fossil footmarks of New England. Proceedings of the 6th Annual Meeting of the Association of American Geologists and Naturalists 6, 23–25.

Hitchcock, E., 1847. Description of two new species of fossil footmarks found in Massachusetts and Connecticut, or, of the animals that made them. American Journal of Science, Series 2 (4), 46–57.

Hitchcock, E., 1858. Ichnology of New England. A Report on the Sandstone of the Connecticut Valley, Especially its Fossil Footmarks, Made to the Government of the Commonwealth of Massachusetts. William White, Printer. Boston, pp. 1–214.

Hitchcock, C.H., 1889. Recent progress in ichnology. Proc. Boston Soc. Nat. Hist. 24, 117–127.

Hminna, A., Voigt, S., Klein, H., Saber, H., Schneider, J.W., Hmich, D., 2013. First occurrence of tetrapod footprints from the continental Triassic of the Sidi Said Maachou area (western Meseta, Morocco). J. Afr. Earth Sci. 80, 1–7.

Hminna, A., Klein, H., Zouheir, T., Lagnaoui, A., Saber, H., Lallensack, J.N., Oukassou, M., 2021. The Late Triassic archosaur ichnogenus *Brachychirotherium*: first complete step cycles from Morocco, North Africa, with implications for trackmaker identification and ichnotaxonomy. Hist. Biol. 33 (5), 723–736.

Holst, H.K.H., Smit, J., Veenstra, E., 1970. Lacertoid Footprints from the Early Middle Triassic at Haarmühle Near Altstätte, W. Germany: Procedings Van Het Koninklijk Nederlandse Akademie Wetenschapen Sektion, B73, pp. 157–165.

Hornstein, F.F., 1876. Entdeckung von Thierfährten im Buntsandstein bei Karlshafen. Brief an Geinitz: Neues Jahrbuch für Mineralogie. Geologie und Paläontologie 1876, 923–924.

Huene, F. von, 1931. Die fossilen Fährten im Rhät von Ischigualasto in NordwestArgentinien: Palaeobiologica, vol 4, pp. 99–112.

Huene, F.v., 1932. Die fossile Reptil-Ordnung Saurischia, ihre Entwicklung und Geschichte. Monographien zur Geologie und Paläontologie 1, 1–361.

Huene, F.v., 1941. Die Tetrapoden-Fährten im toskanischen Verrucano und ihre Bedeutung. Neues Jahrbuch für Geologie und Paläontologie, Serie B 86, 1–34.

Hunt, A.P., Lucas, S.G., 2006. Triassic-Jurassic tetrapod ichnofacies. New Mexico Museum of Natural History and Science, Bulletin 37, 12–22.

Hunt, A.P., Lucas, S.G., 2007a. Late Triassic tetrapod tracks of western North America. New Mexico Museum of Natural History and Science, Bulletin 40, 215–230.

Hunt, A.P., Lucas, S.G., 2007b. The Triassic tetrapod track record: ichnofaunas, ichnofacies and biochronology. New Mexico Museum of Natural History and Science, Bulletin 41, 78–87.

Hunt, A.P., Lucas, S.G., 2007c. Tetrapod ichnofacies: a new paradigm. Ichnos 14, 59–68.

Hunt, A.P., Lockley, M.G., Lucas, S.G., 1993a. Vertebrate and invertebrate tracks and trackways from Upper Triassic strata of the Tucumcari basin, east-central New Mexico. New Mexico Museum of Natural History and Science, Bulletin 3, 199–201.

Hunt, A.P., Santucci, V.L., Lockley, M.G., Olson, T.J., 1993b. Dicynodont trackways from the Holbrook member of the Moenkopi Formation (middle Triassic: Anisian), Arizona, USA. New Mexico Museum of Natural History and Science, Bulletin 3, 213–218.

Hunt, A.P., Lucas, S.G., Klein, H., 2018. Late Triassic non-marine vertebrate and invertebrate trace fossils and the pattern of the Phanerozoic record of vertebrate trace fossils. In: Tanner, L.H. (Ed.), The Late Triassic World: Topics in Geobiology, vol 46, pp. 447–544.

Kammerer, C.F., 2018. The first skeletal evidence of a dicynodont from the lower Elliot Formation of South Africa. Palaeontol. Afr. 52, 102–128.

Karl, C., Haubold, H., 1998. *Brachychirotherium* aus dem Coburger Sandstein (Mittlerer Keuper, Karn/Nor) in Nordbayern: Hallesches Jahrbuch für Geowissenschaften, B20, pp. 33–58.

Karl, C., Haubold, H., 2000. Saurierfährten im Keuper (Obere Trias) Frankens, die Typen von *Brachychirotherium*: Berichte der Naturwissenschaftlichen Gesellschaft Bayreuth, vol 24, pp. 91–120.

Kaup, J.J., 1835a. Über Thierfährten bei Hildburghausen: Neues Jahrbuch für Mineralogie. Geologie und Paläontologie 1835, 327–328.

Kaup, J.J., 1835b. Fährten von Beuteltieren. In: Das Tierreich, pp. 246–248.

King, A.T., 1845. Description of fossil footmarks found in the Carboniferous series in Westmoreland County, Pa. Am. J. Sci. 48, 343–352.

King, M.J., Sarjeant, W.A.S., Thompson, D.B., Tresise, G., 2005. A Revised Systematic Ichnotaxonomy and Review of the Vertebrate Footprint Ichnofamily Chirotheriidae from the British Triassic: Ichnos, vol 12, pp. 241–299.

Kirchner, H., 1927. Über die Tierfährten im Oberen Buntsandstein Frankens. Paläontol. Z. 9, 112–121.

Klein, H., Haubold, H., 2003. Differenzierung von ausgewählten Chirotherien der Trias mittels Landmarkanalyse. Hallesches Jahrb. Geowiss. B 25, 21–36.

Klein, H., Haubold, H., 2004. Überlieferungsbedingte Variation bei Chirotherien und Hinweise zur Ichnotaxonomie nach Beispielen aus der Mittel- bis Ober-Trias (Anisium–Karnium) von Nordbayern. Hallesches Jahrb. Geowiss. B 26, 1–15.

Klein, H., Haubold, H., 2007. Archosaur footprints—potential for biochronology of Triassic continental sequences: New Mexico museum of natural history and science. Bulletin 41, 120–130.

Klein, H., Lucas, S.G., 2010a. In: Lucas, S.G. (Ed.), Tetrapod Footprints –Their Use in Biostratigraphy and Biochronology of the Triassic, The Triassic Timescale, vol 334. Geological Society, London, Special Publications, pp. 419–446.

Klein, H., Lucas, S.G., 2010b. Review of the tetrapod ichnofauna of the Moenkopi Formation/group (Early-Middle Triassic) of the American southwest. New Mexico Museum of Natural History and Science, Bulletin 50, 1–67.

Klein, H., Lucas, S.G., 2010c. The Triassic footprint record of crocodylomorphs – a critical re-evaluation. New Mexico Museum of Natural History and Science, Bulletin 51, 55–60.

Klein, H., Lucas, S.G., 2013. The late Triassic ichnotaxon *Apatopus lineatus* (Bock 1952) and its distribution. New Mexico Museum of Natural History and Science, Bulletin 61, 313–324.

Klein, H., Lucas, S.G., 2018. Diverse middle Triassic tetrapod footprint assemblage from the Muschelkalk of Germany. Ichnos 25 (2–3), 162–176.

Klein, H., Lucas, S.G., 2021. The Triassic tetrapod footprint record. New Mexico Museum of Natural History and Science Bulletin 83, 1–194.

Klein, H., Niedźwiedzki, G., 2012. Revision of the lower Triassic tetrapod ichnofauna from Wióry, holy Cross Mountains, Poland. New Mexico Museum of Natural History and Science, Bulletin 56, 1–62.

Klein, H., Lucas, S.G., Haubold, H., 2006. Tetrapod track assemblage of the Redonda Formation (Upper Triassic, Chinle Group) in east-central New Mexico – re-evaluation of ichnofaunal diversity from studies of new material. New Mexico Museum of Natural History and Science, Bulletin 37, 241–250.

Klein, H., Voigt, S., Hminna, A., Saber, H., Schneider, J.W., Hmich, D., 2010. Early Triassic archosaur-dominated footprint assemblage from the Argana Basin (western high Atlas, Morocco). Ichnos 17, 215–227.

Klein, H., Voigt, S., Saber, H., Schneider, J.W., Hminna, A., Fischer, J., Lagnaoui, A., Brosig, A., 2011. First occurrence of a middle Triassic tetrapod ichnofauna from the Argana Basin (western high Atlas, Morocco). Palaeogeogr. Palaeoclimatol. Palaeoecol. 307, 218–231.

Klein, H., Niedźwiedzki, G., Voigt, S., Lagnaoui, A., Hminna, A., Saber, H., Schneider, J.W., 2013. The tetrapod ichnogenus *Protochirotherium*Fichter and Kunz 2004, a characteristic early Triassic morphotype of central Pangea. Ichnos 20, 24–30.

Klein, H., Milán, J., Clemmensen, L.B., Frobøse, N., Mateus, O., Klein, N., Adolfssen, J.S., Estrup, E.J., Wings, O., 2015a. Archosaur Footprints (Cf. *Brachychirotherium*) with Unusual Morphology from the Upper Triassic Fleming Fjord Formation (NorianRhaetian) of East Greenland, vol 434. Geological Society, London, Special Publication, pp. 71–85.

Klein, H., Lucas, S.G., Voigt, S., 2015b. Revision of the ?Permian-Triassic tetrapod ichnogenus *Procolophonichnium* Nopcsa 1923 with description of the new ichnogenus *P. lockleyi*. Ichnos 22, 155–176.

Klein, H., Wizevich, M.C., Thüring, B., Marty, D., Thüring, S., Falkingham, P., Meyer, C.A., 2016. Triassic chirotheriid footprints from the Swiss Alps: ichnotaxonomy and depositional environment (Cantons Wallis &Glarus). Swiss Journal of Palaeontology 135, 295–314.

Klein, H., Gierliński, G.D., Oukassou, M., Saber, H., Lallensack, J.N., Lagnaoui, A., Hminna, A., Charrière, A., 2023a. Theropod and ornithischian dinosaur track assemblages from middle to ?late Jurassic deposits of the central high Atlas, Morocco. Hist. Biol. 35 (3), 320–346. https://doi.org/10.1080/08912963.2022.2042808.

Klein, H., Heckert, A.B., Celeskey, M., 2023b. Chirotheres: Tracking the Ancestors of Dinosaurs and Crocodiles. Indiana University Press, Bloomington, Indiana, 360 pp.

Klein, H., Lucas, S.G., Lallensack, J.N., Marchetti, L., 2024. Peabody's legacy: the Moenkopi Formation (Middle Triassic, Anisian) tetrapod ichnofauna– updates from an extensive new tracksite in NE Arizona, USA. Paläontol. Z. 98 (2), 357–389.

Konietzko-Meier, D., Teschner, E.M., Bodzioch, A., Sander, P.M., 2020. Pentadactyl manus of the *Metoposaurus krasiejowensis* from the Late Triassic of Poland, the first record of pentadactyly among Temnospondyli. J. Anat. 00, 1–11. https://doi.org/10.1111/joa.13276.

Kuhn, O., 1958a. Die Fährten der vorzeitlichen Amphibien und Reptilien. Bamberg, Meisenbach, 64 pp.

Kuhn, O., 1958b. Zwei neue Arten von *Coelurosaurichnus* aus dem Keuper Frankens. Neues Jahrbuch Geol. Palaontol. Monatsh. 437–440.

Kuhn, O., 1963. Ichnia Tetrapodorum. Fossilium Catalogus I, vol 101, 176 pp.

Lagnaoui, A., Klein, H., Voigt, S., Hminna, A., Saber, H., Schneider, J.W., Werneburg, R., 2012. Late Triassic tetrapod-dominated ichnoassemblages from the Argana Basin (western high Atlas, Morocco). Ichnos 19, 238–253.

Lagnaoui, A., Klein, H., Saber, H., Fekkak, A., Belahmira, A., Schneider, J.W., 2016. New discoveries of archosaur and other tetrapod footprints from the Timezgadiouine Formation (Irohalene member, upper Triassic) of the Argana Basin, western high Atlas, Morocco - ichnotaxonomic implications. Palaeogeogr. Palaeoclimatol. Palaeoecol. 453, 1–9.

Lagnaoui, A., Melchor, R.N., Bellosi, E.S., Villegas, P.M., Espinoza, N., Umazano, A.M., 2019. Middle Triassic *Pentasauropus*-dominated ichnofauna from western Gondwana: ichnotaxonomy, palaeoenvironment, biostratigraphy and palaeobiogeography. Palaeogeogr. Palaeoclimatol. Palaeoecol. 524, 41–61.

Lallensack, J.N., van Heteren, A.H., Wings, O., 2016. Geometric morphometric analysis of intratrackway variability: a case study on theropod and ornithopod dinosaur trackways from Münchehagen (Lower Cretaceous, Germany). PeerJ 4, e2059.

Lallensack, J.N., Farlow, J.O., Falkingham, P.L., 2022. A new solution to an old riddle: elongate dinosaur tracks explained as deep penetration of the foot, not plantigrade locomotion. Palaeontology 65 (Part 1), e12584. https://doi.org/10.1111/pala.12584.

Lallensack, J.N., Klein, H., Milán, J., Wings, O., Mateus, O., Clemmensen, L.B., 2017. Sauropodomorph Dinosaur Trackways from the Fleming Fjord Formation of East Greenland: Evidence for Late Triassic Sauropods: Palaeontologica Polonica, vol 62. https://doi.org/10.4202/app.00374.2017.

Leonardi, P., 1951. Orme di tetrapodi nelle Arenarie de Val Gardena (Permiano Medio-inferiore) dell'Alto Adige sud-orientale. Memorie dell'Istituto di Geologia e Mineralogia dell'Universita di Padova 17, 1–23.

Leonardi, P., 1959. Orme chirotheriane triassiche spagnole. Estud. Geol. 15, 235–245.

Leonardi, G., 1981. Novo icnogênero de tetrápode Mesozóico da Formação Botucatu. Araraquara, SP: Annales Academe do Brasil Ciências 54, 793–805.

Leonardi, G., Lockley, M.G., 1995. A proposal to abandon the ichnogenus *Coelurosaurichnus* Huene, 1941, a junior synonym of *Grallator* E. Hitchcock, 1858. J. Vertebr. Paleontol. 15 (Suppl. 3), 40A.

Lichtig, A.J., Lucas, S.G., Klein, H., Lovelace, M., 2018. Triassic turtle tracks and the origin of turtles: Historical Biology. https://doi.org/10.1080/08912963.2017.1339037 vol 30 (8), p. 1112–1122.

Lockley, 2006. Observations on the ichnogenus *Gwyneddichnium* and *Gwyneddichnium*-like footprints and trackways from the upper Triassic of the western United States. New Mexico Museum of Natural History and Science, Bulletin 37, 170–175.

Lockley, M.G., 2011. The ichnotaxonomic status of *Brasilichnium* with special reference to occurrences in the Navajo Sandstone (Lower Jurassic) in the western USA. New Mexico Museum of Natural History and Science, Bulletin 53, 306–315.

Lockley, M.G., Harris, J., 2010. On the trail of early birds: a review of the fossil footprint record of avian morphological evolution and behavior. Trends in Ornithological Research, Novapublishers, Hauppauge 1–63.

Lockley, M.G., Hunt, A.P., 1995. Dinosaur Tracks and Other Fossil Footprints of the Western United States. Columbia University Press, New York.

Lockley, M.G., Lucas, S.G., 2013. *Evazoum gatewayensis*, a new Late Triassic archosaurian ichnospecies from Colorado: implications for footprints in the ichnofamily Otozoidae. New Mexico Museum of Natural History and Science, Bulletin 61, 345–353.

Lockley, M.G., Matsukawa, M., 2009. A review of vertebrate track distributions in East and Southeast Asia. J. Paleontol. Soc. Korea 25, 17–42.

Lockley, M.G., Meyer, C.A., 2000. Dinosaur Tracks and Other Fossil Footprints of Europe. Columbia University Press, New York.

Lockley, M.G., Conrad, K., Paquette, M., 1991. In: Plumb, G. (Ed.), Distribution and Significance of Mesozoic Vertebrate Trace Fossils in Dinosaur National Monument. University of Wyoming National Park Service Research Center, pp. 85–90, 15th Annual Report.

Lockley, M.G., Hunt, A.P., Meyer, C., 1994. Vertebrate tracks and the ichnofacies concept: implications for paleoecology and palichnostratigraphy. In: Donovan, S. (Ed.), The Paleobiology of Trace Fossils. John Wiley, London, pp. 241–268.

Lockley, M.G., King, M., Howe, S., Sharp, T., 1996. Dinosaur tracks and other archosaur footprints from the Triassic of South Wales. Ichnos 5, 23–41.

Lockley, M.G., Wright, J.L., Hunt, A.P., Lucas, S.G., 2001. The Late Triassic Sauropod Track Record Comes into Focus: Old Legacies and New Paradigms, vol 52. New Mexico Geological Society, Guidebook, pp. 181–190.

Lockley, M.G., Lucas, S.G., Hunt, A.P., Gaston, R., 2004. Ichnofaunas from the Triassic-Jurassic Boundary Sequences of the Gateway Area, Western Colorado: Implications for Faunal Composition and Correlations with Other Areas: Ichnos, vol 11, pp. 89–102.

Lockley, M.G., Lucas, S.G., Hunt, A.P., 2006a. *Eosauropus*, a new name for a late Triassic track: further observations on the late Triassic ichnogenus *Tetrasauropus* and related forms, with notes on the limits of interpretation: New Mexico museum of natural history and science. Bulletin 37, 192–198.

Lockley, M.G., Lucas, S.G., Hunt, A.P., 2006b. *Evazoum* and the renaming of northern hemisphere "*Pseudotetrasauropus*:" Implications for tetrapod ichnotaxonomy at the Triassic-Jurassic boundary. New Mexico Museum of Natural History and Science, Bulletin 37, 199–206.

Lockley, M.G., Hups, K., Cart, K., Gerwe, S., 2011. A zone of sauropodomorph footprints in the basal Wingate Sandstone (latest Triassic) of western Colorado and eastern Utah: is *Eosauropus* a common ichnogenus in this region? New Mexico Museum of Natural History and Science, Bulletin 53, 337–343.

Lockley, M.G., Li, J.-j, Li, R-h., Matsukawa, M., Harris, J.D., Xing, L., 2013. A review of the tetrapod track record in China, with special reference to type ichnospecies: implications for ichnotaxonomy and paleobiology. Acta Geol. Sin. 87, 1–20 (English edition).

Lockley, M.G., Cart, K., Martin, J., Prunty, R., Houck, K., Hups, K., Lim, J.-D., Kim, K.S., Gierlinski, G., 2014. A bonanza of new tetrapod tracksites from the Cretaceous Dakota Group, western Colorado: implications for paleoecology. New Mexico Museum of Natural History and Science, Bulletin 62, 393–410.

Lockley, M.G., Foster, J.R., Gierlinski, G.D., Shibata, K., 2018 Upper. Triassic tetrapod track assemblages from the Chinle Group, Moab area, Utah. New Mexico Museum of Natural History and Science, Bulletin 79, 417–427.

Lockley, M.G., Klein, H., McHugh, J.B., Romilio, A., 2021. Fruita's first fossil footprint exhibit: the discovery of forgotten specimens in an historic former museum building. New Mexico Museum of Natural History and Science, Bulletin 82 (in press).

Lovelace, D.M., Lovelace, S.D., 2012. Paleoenvironments and paleoecology of a lower Triassic invertebrate and vertebrate ichnoassemblage from the red Peak Formation (Chugwater group), central Wyoming. Palaios 27, 636–657.

Lucas, S.G., 1998. Global Triassic tetrapod biostratigraphy and biochronology. Palaeogeogr. Palaeoclimatol. Palaeoecol. 143, 347–384.

Lucas, S.G., 2005. Tetrapod ichnofacies and ichnotaxonomy: Quo vadis? Ichnos 12, 157–162.

Lucas, S.G., 2007. Tetrapod footprint biostratigraphy and biochronology. Ichnos 14, 5–38.

Lucas, S.G., Heckert, A.B., 2011. Late Triassic aetosaurs as the trackmaker of the tetrapod footprint ichnotaxon Brachychirotherium. Ichnos 18, 197–208.

Lucas, S.G., Sullivan, R.M., 2006. Tetrapod footprints from the upper Triassic Passaic Formation near Graterford, Montgomery county, Pennsylvania. New Mexico Museum of Natural History and Science, Bulletin 37, 251–256.

Lucas, S.G., Klein, H., Lockley, M.G., Spielmann, J., Gierlinski, G.D., Hunt, A.P., Tanner, L.H., 2006. Triassic-Jurassic stratigraphic distribution of the theropod footprint ichnogenus *Eubrontes*. New Mexico Museum of Natural History and Science, Bulletin 37, 68–93.

Lucas, S.G., Spielmann, J.A., Klein, H., Lerner, A.J., 2010. Ichnology of the upper Triassic Redonda Formation (Apachean) in east-central New Mexico. New Mexico Museum of Natural History and Science, Bulletin 47, 75 pp.

Lucas, S.G., Szajna, M.J., Lockley, M.G., Fillmore, D.L., Simpson, E.L., Klein, H., Boyland, J., Hartline, B.W., 2014. The Middle-Late Triassic tetrapod footprint ichnogenus *Gwyneddichnium*. New Mexico Museum of Natural History and Science, Bulletin 62, 135–155.

Lull, R.S., 1904. Fossil Footprints of the Jura-Trias of North America: Memoirs of the Boston Society of Natural History, vol 5, pp. 461–557.

Lull, R.S., 1942. Chugwater footprints from Wyoming. Am. J. Sci. 240, 500–504.

Lull, R.S., 1953. Triassic life of the Connecticut valley: Connecticut state geological and natural history Survey. Bulletin 81, 1–331.

Lydekker, R., 1890. Catalogue of Fossil Reptilia and Amphibia in the British Museum, Part 4. Longmans, London, 296 pp.

Maidwell, F.T., 1911. Notes on footprints from the Keuper of Runcorn Hill. Proc. Liverp. Geol. Soc. 11, 140–152.

Maidwell, F.T., 1914. Notes on footprints from the Keuper. II (Int. Immunol.): Liverpool Geological Society Proceedings 12, 53–71.

Manning, P., 2004. A new approach to the analysis and interpretation of tracks: examples from the Dinosauria. In: McIlroy, D. (Ed.), The Application of Ichnology to Palaeoenvironmental and Stratigraphic Analysis, vol 228. Geological Society, London, Special Publications, pp. 93–123.

Marchetti, L., Belvedere, M., Voigt, S., Klein, H., Castanera, D., Díaz-Martínez, I., Marty, D., Xing, L., Feola, S., Melchor, R.N., Farlow, J.O., 2019c. Defining the morphological quality of fossil footprints. Problems and principles of preservation in tetrapod ichnology with examples from the Palaeozoic to the present. Earth Sci. Rev. 193, 109–145.

Marchetti, L., Collareta, A., Belvedere, M., Leonardi, G., 2021. Ichnotaxonomy, biostratigraphy and palaeoecology of the Monti Pisani tetrapod ichnoassociation (Tuscany, Italy) and new insights on middle Triassic Dinosauromorpha. Palaeogeogr. Palaeoclimatol. Palaeoecol. 567. https://doi.org/10.1016/j.palaeo.2021.110235.

Marchetti, L., Klein, H., Buchwitz, M., Ronchi, A., Smith, R.M.H., De Klerk, W.J., Sciscio, L., Groenewald, G.H., 2020a. Reply to discussion of "Permian–Triassic vertebrate footprints from South Africa: ichnotaxonomy, producers and biostratigraphy through two major faunal crises. In: Marchetti, L., Klein, H., Buchwitz, M., Ronchi, A., Smith, R.M.H., DeKlerk, W.J., Sciscio, L., Groenewald, G.H. (Eds.), Gondwana Res. 78, 379–383, 2019.

Marchetti, L., Klein, H., Buchwitz, M., Ronchi, A., Smith, R.M., De Klerk, W.J., Sciscio, L., Groenewald, G.H., 2019b. Permian-Triassic vertebrate footprints from South Africa: ichnotaxonomy, producers and biostratigraphy through two major faunal crises. Gondwana Res. 72, 139–168.

Marchetti, L., Klein, H., Falk, D., Wings, O., 2020b. *Synaptichnium* tracks from the middle Muschelkalk (Middle Triassic, Anisian) Bernburg site (Saxony-Anhalt, Germany). Ann. Soc. Geol. Pol. 90, 321–330.

Marchetti, L., Voigt, S., Klein, H., 2019a. Revision of Late Permian Tetrapod Tracks from the Dolomites (Trentino-Alto Adige, Italy): Historical Biology, vol 31, pp. 748–783.

Márquez-Aliaga, A., García-Forner, A., Martínez, C., Villena, J.A., 1999. Colección icnofósiles" del Museo de Geología de la Universidad de Valencia. Temas Geológico—Mineros. ITGE 26, 413—414.

Matthews, N., Noble, T., Breithaupt, B., 2016. Close-range photogrammetry for 3-D ichnology: the basics of photogrammetric ichnology. In: Falkingham, P.L., Marty, D., Richter, A. (Eds.), Dinosaur Tracks, the Next Steps. Indiana University Press, Indianapolis and Bloomington, pp. 29—55.

McIlroy, D., 2004. Some ichnological concepts, methodologies, applications and frontiers. In: McIlroy, D. (Ed.), The Application of Ichnology to Paleoenvironmental and Stratigraphic Analysis, vol 228. Geological Society of London, Special Publication, pp. 3—27.

Melchor, R.N., De Valais, S., 2006. A review of Triassic tetrapod track assemblages from Argentina. Palaeontology 49, 355—379.

Melchor, N.M., Sarjeant, W.A.S., 2004. Small amphibian and reptile footprints from the Permian Carapacha basin, Argentina. Ichnos 11, 57—77.

Melchor, R.N., Buchwaldt, R., Bowring, S., 2013. A late Eocene date for late Triassic bird tracks. Nature 495, E1.

Meyer, H.v., Plieningeer, T., 1844. Beiträge zur Paläontologie Württembergs, die fossilen Wirbeltierreste aus den Triasgebilden. Stuttgart, Schweizerbart, 132 pp.

Meyer, C.A., Marty, D., Thüring, B., Stecher, R., Thüring, S., 2013. Dinosaurierspuren aus der Trias der Bergüner Stöcke (Parc Ela, Kanton Graubünden, SE Schweiz): Mitteilungen der Naturforschenden Gesellschaften beider Basel, vol 14, pp. 135—144.

Meyer, C.A., Belvedere, M., Klein, H., Lallensack, J., Wings, O., Lucas, S.G., 2018. Tracking early sauropodomorphs. In: XVI Annual Meeting of the European Association of Vertebrate Palaeontologists (EAVP). Caparica June 29th—July 1st 2018.

Meyer, C.A., Klein, H., Wizevich, M.C., Stecher, R., 2019. Triassic sauropodomorph tracks with Gondwanan affinities from the central Austroalpine Nappes of Switzerland. In: 17th Swiss Geoscience Meeting, Fribourg, vol 107, 4.19.

Meyer, C.A., Klein, H., Wizevich, M.C., Stecher, R., 2022. Dinosaurierspuren aus der späten Trias der Val d'Uina (Piz S-chalambert, Scuol). Jahresber. Naturforschenden Ges. Graubuenden 122, 99—110.

Mietto, P., 1987. *Parasynaptichnium gracilis* nov. ichnogen., nov. isp., (Reptilia: Archosauria, Pseudosuchia) nell'Anisico inferiore di Recoaro, vol 39. (Prealpi vicentine — Italia): Memorie di Scienze Geologiche, pp. 37—47.

Milàn, J., 2006. Variations in the morphology of the emu (*Dromaius novahollandiae*) tracks reflecting differences in walking pattern and substrate consistency: ichnotaxonomic implications. Palaeontology 49, 405—420.

Milàn, J., Bromley, R.G., 2006. True tracks, undertracks and eroded tracks, experimental work with tetrapod tracks in laboratory and field. Palaeogeogr. Palaeoclimatol. Palaeoecol. 231, 253—264.

Milàn, J., Bromley, R.G., 2008. The impact of sediment consistency on track and undertrack morphology: Experiments with emu tracks in layered cement. Ichnos 15, 18—24.

Milàn, J., Falkingham, P.L., 2016. Experimental and comparative ichnology. In: Falkingham, P.L., Marty, D., Richter, A. (Eds.), Dinosaur Tracks, the Next Steps. Indiana University Press, pp. 15—27.

Milner, A.R.C., Lockley, M.G., Kirkland, J.I., 2006. A large collection of well-preserved theropod dinosaur swim tracks from the Lower Jurassic Moenave Formation, St. George, Utah. New Mexico Museum of Natural History and Science, Bulletin 37, 315—328.

Milner, A.R.C., Irmis, R.B., Lockley, M.G., Klein, H., Slauf, D.L., Romilio, A., 2021. First report of "*Chirotherium*" *lulli* from the upper Triassic Chinle Formation of San Juan county, Utah. In: Fossil Record 7, New Mexico Museum of Natural History and Science, Bulletin, vol 82, pp. 275—284.

Montenat, C., 1968. Empreintes de pas de reptiles dans le Trias moyen du plateau du Daüs près d' Aubenas (Ardèche). Bull. Sci. Bourgogne 25, 369—389.

Moore, W.D., 1873. On footprints in the Carboniferous rocks of Western Pennsylvania. Am. J. Sci. 3 (5), 292—293.

Moreau, J.-D., Benaouiss, N., Tourani, A., Jalil, N.-E., Bourquin, S., 2024. Tetrapod tracks from the Middle/Late Triassic of the Ourika and Yagour basins (Morocco): new ichnotaxonomical data based on the Biron collection. J. Afr. Earth Sci. 212, 105202.

Morton, G.H., 1863. Description of the footprints of *Cheirotherium* and *Equisetum* found at Storeton, Cheshire. Proc. Liverp. Geol. Soc. 1, 17.

Morton, G.H., 1897. The Geology of the Country Around Liverpool, Including the North of Flintshire, third ed. Philip, London. 319 pp.

Mujal, E., Fortuny, J., Bolet, A., Oms, O., López, J.Á., 2017. An archosauromorph dominated ichnoassemblage in fluvial settings from the late Early Triassic of the Catalan Pyrenees (NE Iberian Peninsula). PLoS One 12 (4), e0174693.

Mujal, E., Iglesias, G., Oms, O., Fortuny, J., Bolet, A., Méndez, J.M., 2018. *Rhynchosauroides* footprint variability in a Muschelkalk detrital interval (late Anisian–middle ladinian) from the Catalan basin (NE Iberian Peninsula). Ichnos 25 (2–3), 150–161.

Navarro, Ó., Moratalla, J.J., 2018. Swimming reptile prints from the Keuper facies (Carnian, Upper Triassic) of Los Gallegos new tracksite (Iberian range, Valencia province, Spain). J. Iber. Geol. 44 (3), 479–496.

Navas, L., 1906. *Chirosaurus ibericus* n. sp. Boletín de la Sociedad Aragonesa Ciencias Naturales 5, 208–213.

Nesbitt, S.J., 2011. The early evolution of archosaurs: relationships and the origin of major clades. Bull. Am. Mus. Nat. Hist. 352, 292.

Nesbitt, S.J., Angielczyk, K.D., 2002. New evidence of large dicynodonts in the upper Moenkopi Formation (middle Triassic) of northern Arizona: PaleoBios, 22 (2), 10–17.

Nicosia, U., Loi, M., 2003. Triassic footprints from Lerici (La Spezia, northern Italy). Ichnos 10, 127–140.

Niedźwiedzki, G., 2011. A late Triassic dinosaur-dominated ichnofauna from the Tomanóva Formation of the Tatra Mountains, central Europe. Acta Palaeontol. Pol. 56 (2), 291–300.

Niedźwiedzki, G., Ptaszynski, T., 2007. Large Chirotheriidae tracks in the early Triassic of Wióry, Holy Cross Mountains, Poland. Acta Geol. Pol. 57, 325–342.

Niedźwiedzki, G., Brusatte, S.L., Butler, R.J., 2013. *Prorotodactylus* and *Rotodactylus* tracks: an ichnological record of dinosauromorphs from the early–middle Triassic of Poland. In: Nesbitt, S.J., Desojo, J.B., Irmis, R.B. (Eds.), Anatomy, Phylogeny and Palaeobiology of Early Archosaurs and Their Kin, vol 379. Geological Society, London, Special Publications, pp. 319–351.

Nopcsa, F. von, 1923. Die Familien der Reptilien. Fortschritte der Geologie und Paläontologie 2, 1–210.

Olsen, P.E., Baird, D., 1986. The ichnogenus *Atreipus* and its significance for Triassic biostratigraphy. In: Padian, K. (Ed.), The Beginning of the Age of Dinosaurs. Cambridge University Press, Cambridge, pp. 61–87.

Olsen, P.E., Flynn, J., 1989. Field Guide to the Vertebrate Paleontology of Late Triassic Rocks in the Southwestern Newark Basin (Newark Supergroup, New Jersey and Pennsylvania): The Mosasaur, vol 4, pp. 1–35.

Olsen, P.E., Galton, P.M., 1984. A review of the reptile and amphibian assemblages from the Stormberg of southern Africa, with special emphasis on the footprints and the age of the Stormberg. Palaeontologia Africana, Haughton Memorial 92–116.

Olsen, P.E., Rainforth, E.C., 2001. The "Age of Dinosaurs" in the Newark Basin, with Special Reference to the Lower Hudson Valley. New York State Geological Association Guidebook, pp. 59–176.

Olsen, P.E., Smith, J.B., McDonald, N.G., 1998. Type material of the type species of the classic theropod footprint genera *Eubrontes, Anchisauripus,* and *Grallator* (Early Jurassic, Hartford and Deerfield Basins, Connecticut and Massachusetts, U.S.A.). J. Vertebr. Paleontol. 18, 586–601.

Owen, R., 1842. Report on British Fossil Reptiles: Report of the British Association of Advanced Sciences, vol 9, pp. 60–204.

Padian, K., Li, C., Pchelnikova, J., 2010. The trackmaker of *Apatopus* (Late Triassic, North America) implications for the evolution of archosaur stance and gait. Palaeontology 53, 175–189.

Peabody, F.E., 1948. Reptile and Amphibian Trackways from the Moenkopi Formation of Arizona and Utah, vol 27. University of California Publications, Bulletin of the Department of Geological Sciences, pp. 295–468.

Pérez-López, A., 1993. Estudio de las huellas de reptil, del icnogénero *Brachychirotherium*, encontradas en el Trias subbético de Cambil. Estud. Geol. 49, 77–86.

Petti, F.M., Avanzini, M., Belvedere, M., De Gasperi, M., Ferretti, P., Girardi, S., Remondino, F., Tomasoni, R., 2008. Digital 3D modelling of dinosaur footprints by photogrammetry and laser scanning techniques: Integrated approach at the Coste dell' Anglone tracksite (Lower Jurassic, Southern Alps, Northern Italy). In: Avanzini, M., Petti, F.M. (Eds.), Italian Ichnology: Studi Trentini di Scienze Naturali Acta Geologica, vol 83, pp. 303–315.

Plieninger, T., 1838. Tierfährten in der Keuperformation bei Stuttgart. Neues Jahrbuch für Mineralogie und Geologie 1838, 536–537.

Pohlig, H., 1893. Vorlage von Fließpapierabdrücken fossiler Wirbeltier-Fußstapfen: Abhandlungen des Naturhistorischen Vereins der Preußischen Rheinlande, vol 50, pp. 82–83.

Ptaszyński, T., 2000. Lower Triassic vertebrate footprints from Wióry, Holy Cross Mountains, Poland. Acta Palaeontol. Pol. 45, 151–194.

Rainforth, E.C., 2003. Revision and Re-evaluation of the Early Jurassic Dinosaurian Ichnogenus *Otozoum*: Palaeontology, vol 46, pp. 803–838.

Rainforth, E.C., 2005. Ichnotaxonomy of the Fossil Footprints of the Connecticut Valley (Early Jurassic, Newark Supergroup, Connecticut and Massachusetts) [Ph. D. Dissertation]. Columbia University, New York, pp. 1–1301.

Rasskin-Gutman, D., Hunt, G., Chapman, R.E., Sanz, J.L., Moratalla, J.J., 1997. The shapes of tridactyl dinosaur footprints: Procedures, problems and potentials. In: Wolberg, D.L., et al. (Eds.), Dinofest—International Proceedings, pp. 377–383.

Rehnelt, K., 1950. Ein Beitrag über Fährtenspuren im unteren Gipskeuper von Bayreuth: Berichte der Naturwissenschaftlichen Gesellschaft Bayreuth, pp. 27–36.

Reig, O.A., 1961. Acerca de la posición sistemática de la familia Rauisuchidae y del género *Saurosuchus* (Reptilia, Thecodontia). Publicaciones de la Museo Ciencia Naturale Mar del Plata 1, 73–114.

Reolid, M., Márquez-Aliaga, A., Belinchón, M., García-Forner, A., Villena, J., MartínezPérez, C., 2018. Ichnological evidence of semi-aquatic locomotion in early turtles from eastern Iberia during the Carnian humid episode (Late Triassic). Palaeogeogr. Palaeoclimatol. Palaeoecol. 490, 450–461.

Retallack, G.J., 1996. Early Triassic therapsid footprints from the Sydney basin, Australia. Alcheringa 20, 301–314.

Romilio, A., Klein, H., Jannel, A., Salisbury, S.W., 2021. Saurischian dinosaur tracks from the Upper Triassic of southern Queensland: possible evidence for Australia's earliest sauropodomorph trackmaker. Hist. Biol. https://doi.org/10.1080/08912963.2021.1984447.

Rühle v. Lilienstern, H., 1938. Fährten aus dem Blasensandstein (km 4) des Mittleren Keupers von Südthüringen. Neues Jahrbuch für Mineralogie, Geologie und Paläontologie (B) 80, 63–71.

Rühle v. Lilienstern, H., 1939. Fährten und Spuren im *Chirotherium*-Sandstein von Südthüringen. Fortschritte der Geologie und Paläontologie 12, 293–387.

Rühle v. Lilienstern, H., 1944. Eine Dicynodontierfährte aus dem *Chirotherium*-Sandstein von Heßberg bei Hildburghausen. Paläontologische Zeitschrift 23, 368–385.

Rusconi, C., 1951. Rastros de patas de reptiles pérmicos de Mendoza, vol 3. Revista de Historia y Geografia de Cuyo, pp. 42–54.

Ryan, J.D., Willard, B., 1947. Triassic Footprints from Bucks County, Pennsylvania: Proceedings of the Pennsylvania Academy of Sciences, vol 21, pp. 91–93.

Sarjeant, W.A.S., 1967. Fossil Footprints from the Middle Triassic of Nottinghamshire and Derbyshire: Mercian Geologist, vol 2, pp. 327–341.

Sarjeant, W.A.S., 1970. Fossil Footprints from the Middle Triassic of Nottingham and the Middle Jurassic of Yorkshire: Mercian Geologist, vol 3, pp. 269–282.

Sarjeant, W.A.S., 1996. A Re-appraisal of Some Supposed Dinosaur Footprints from the Triassic of the English Midlands: Mercian Geologist, vol 14, pp. 22–30.

Sarjeant, W.A.S., Langston, W., 1994. Vertebrate footprints and invertebrate traces from the Chadronian (Late Eocene) of Trans-Pecos Texas. Texas Memorial Museum Bulletin 36, 1–86.

Schimper, W.P., 1850. Palaeontologia Alsatica: Mémoires de Société d'Histoire naturelle. Strasbourg IV (1), 10.

Schindewolf, O.H., 1928. Studien aus dem Marburger Buntsandstein. Senckenbergiana 10 (3–7), 16–54.

Schuster, M., 1936. Die Gliederung des Unterfränkischen Buntsandsteins. II: Der Obere Buntsandstein oder das Röt. c. Das Obere Röt oder die Stufe der Röt-Tone (2. Die Oberen Röt-Tone mit den Myophorien-Schichten). Abhandlungen der Geologischen Landesuntersuchung am Bayerischen Oberbergamt, Munich 23, 53.

Seilacher, A., 1964, Biogenic Sedimentary Structures; inImbrie, J. and Newell, N., eds., Approaches to Paleoecology: Wiley, New York, p. 296-316.

Sickler, F.K.L., 1834. Sendschreiben an Dr. Blumenbach über die höchst merkwürdigen Reliefs der Fährten urweltlicher Tiere in den Hessberger Steinbrüchen bei Hildburghausen: Programm des Herzoglichen Gymnasiums zu Hildburghausen, 16 pp.

Sickler, F.K.L., 1835. Fährten bei Hildburghausen: Neues Jahrbuch für Mineralogie. Geologie und Paläontologie 1835, 230–232.

Sickler, F.K.L., 1836. Die vorzüglichen Fährtenabdrücke urweltlicher Tiere im bunten Sandstein: Hildburghausen (Kesselring), 8 pp.

Silva, R.C., Carvalho, I.S., Fernandes, A.C.S., Ferigolo, J., 2008. Pegadas teromorfóides do Triássico Superior (Formação Santa Maria) do Sul do Brasil. Rev. Bras. Geociencias 38, 98–113.

Smith, J.B., Farlow, J.O., 2003. Osteometric Approaches to Trackmaker Assignment for the Newark Supergroup Ichnogenera *Grallator, Anchisauripus*, and *Eubrontes*. Columbia University Press, New York, pp. 273–292.

Soergel, W., 1925. Die Fährten der Chirotheria: Jena. Gustav Fischer Verlag, 92 pp.

Szewczyk, L., Vennin, E., Moreau, J.-D., Gand, G., Verolet, M., Klee, N., Fara, E., 2020. Tracking dinosaurs in coarse-grained sediments from the Upper Triassic of Ardèche (southeastern France). Palaios 35, 447–460.

Thomson, T.J., Lovelace, D.M., 2014. Swim track morphotypes and new track localities from the Moenkopi and Red Peak formations (Lower-Middle Triassic) with preliminary interpretations of aquatic behaviors. New Mexico Museum of Natural History and Science, Bulletin 62, 103–128.

Thulborn, R.A., 1990. Dinosaur Tracks. Chapman and Hall, London, 410 pp.

Turner, M.L., Falkingham, P.L., Gatesy, S.M., 2020. It's in the loop: Shared sub-surface kinematics in birds and other dinosaurs shed light on a new dimension of fossil track diversity. Biol. Lett. 16, 20200309. https://doi.org/10.1098/rsbl.2020.0309.

Voigt, S., 2005. Die Tetrapodenichnofauna des kontinentalen Oberkarbon und Perm im Thüringer Wald-Ichnotaxonomie, Paläoökologie und Biostratigraphie. Cuvillier Verlag, Göttingen, 305 pp.

Walther, J., 1917. Über *Chirotherium*. Z. Dtsch. Geol. Ges. 69, 181–184.

Watson, D.M.S., 1958. A new labyrinthodont (*Paracyclotosaurus*) from the Upper Triassic of New South Wales. Bulletin of the British Museum (Natural History) Geology v 3 (7), 233–263.

Weems, R.E., 1987. A Late Triassic Footprint Fauna from the Culpeper Basin, Northern Virginia (U.S.A.), vol 77. Transactions of the American Philosophical Society, pp. 1–79.

Weems, R.E., 1992. A Re-evaluation of the Taxonomy of Newark Supergroup Saurischian Dinosaur Tracks, Using Extensive Statistical Data from Recently Exposed Tracksite Near Culpeper, vol 119. Virgina: Virginia Division of Mineral Resources Publication, 113127.

Weems, R.E., 2003. *Plateosaurus* foot structure suggests a single trackmaker for *Eubrontes* and *Gigandipus* footprints. In: LeTourneau, P.M., Olsen, P.E. (Eds.), The Great Rift Valleys of Pangea in Eastern North America. Columbia University Press, New York, pp. 293–313.

Weems, R.E., 2019. Evidence for bipedal prosauropods as the likely *Eubrontes* trackmakers. Ichnos 26 (3), 187–215. https://doi.org/10.1080/10420940.2018.1532902.

Whyte, M.A., Romano, M., 2001. A dinosaur ichnocoenosis from the Middle Jurassic of Yorkshire, UK. Ichnos 8, 233–234.

Willruth, K., 1917. Die Fährten von *Chirotherium*: Zeitschrift für. Naturwissenschaften 86, 395–433.

Wilson, J.A., 2005. Integrating ichnofossil and body fossil records to estimate locomotor posture and spatiotemporal distribution of early sauropod dinosaurs: A stratocladistic approach. Paleobiology 31, 400–423.

Wing, S.L., Sues, H.-D., 1992. Mesozoic and early Cenozoic terrestrial ecosystems. In: Behrensmeyer, A.K., Damuth, J.D., Dimichele, W.A., Potts, R., Sues, H.-D., Wing, S.L. (Eds.), Terrestrial Ecosystems through Time. University of Chicago Press, Chicago, pp. 327–416.

Woodward, A.S., 1902. Notes on footprints from the Keuper of South Staffordshire. Geological Magazine, new series 4, 215–217.

Woodworth, J.B., 1900. Vertebrate footprints on Carboniferous shales of Plainville, Massachusetts. Geol. Soc. Am. Bull. 11, 449–454.

Wright, J. L., 2005, Steps in Understanding Sauropod Biology; inCurry Rogers, K. A.andWilson, J. A., eds., The Sauropods: Evolution and Paleobiology: Berkeley, University of California Press, p. 252-284.

Xing, L.D., Klein, H., 2019. *Chirotherium* and first Asian *Rhynchosauroides* tetrapod trackways from the Middle Triassic of Yunnan, China. Hist. Biol. https://doi.org/10.1080/08912963.2019.1661409.

Xing, L.D., Klein, H., Lockley, M.G., Li, J., Zhang, J., Matsukawa, M., Xiao, J., 2013a. *Chirotherium* trackways from the Middle Triassic of Guizhou, China. Ichnos 20, 99–107.

Xing, L.D., Klein, H., Lockley, M.G., Chen, W., Ye, Y., Matsukawa, M., Zhang, J.P., 2013b. Earliest records of theropod and mammal-like tetrapod footprints in the Upper Triassic of Sichuan Basin, China. VertebrataPalAsiatica 51, 184–198.

Xing, L.D., Ba, J., Lockley, M.G., Klein, H., Yan, S.-W., Romilio, A., Chou, C.-Y., Scott Persons, W., 2018. Late Triassic sauropodomorph and Middle Jurassic theropod tracks from the Xichang Basin, Sichuan Province, southwestern China: First report of the ichnogenus *Carmelopodus*. J. Palaeogeogr. 7. https://doi.org/10.1016/j.jop.2017.11.004.

Xing, L., Klein, H., Wu, X., Lockley, M.G., Zeng, R., Romilio, A., 2020. Footprints of marine reptiles from the Middle Triassic (Anisian-Ladinian) Guanling Formation of Guizhou Province, southwestern China. Palaeogeogr. Palaeoclimatol. Palaeoecol. 558, 109943.

Xing, L.D., Lockley, M.G., Klein, H., Zhang, L.J., Romilio, A., Persons, W.S.I.V., Peng, G.Z., Ye, Y., Wang, M.Y., 2021. The new ichnotaxon *Eubrontes nobitai* ichnosp. nov. and other Saurischian tracks from the Lower Cretaceous Sichuan Province and a review of Chinese Eubrontes type tracks. J. Palaeogeogr. 10, 17.

Yang, X., Yang, D., 1987. Dinosaur Footprints from Mesozoic of Sichuan Basin. Science and Technology Publications, Chengdu City, pp. 1–30.

Zhang, Q., Wen, W., Hu, S., Benton, M.J., Zhou, C., Xie, T., Lu, T., Huang, J., Choo, B., Chen, Z.Q., Liu, J., Zhang, Q., 2014. Nothosaur foraging tracks from the Middle Triassic of southwestern China. Nat. Commun. 11, 3973. https://doi.org/10.1038/ncomms4973.

Zouheir, T., Hminna, A., Klein, H., Lagnaoui, A., Saber, H., Schneider, J.W., 2020. Unusual archosaur trackway and associated tetrapod ichnofauna from Irohalene Member (Timezgadiouine Formation, Late Triassic, Carnian) of the Argana Basin, Western High Atlas, Morocco. Hist. Biol. 32 (5), 589–601. https://doi.org/10.1080/08912963.2018.1513506.

CHAPTER

6

Jurassic vertebrate tracks and traces

John R. Foster[1], *Jerald D. Harris*[2,3], *Andrew R.C. Milner*[3], *Emese M. Bordy*[4], *Lara Sciscio*[5,6], *Diego Castanera*[7], *Matteo Belvedere*[8], *Lida Xing*[9,10] *and Martin G. Lockley*[11,a]

[1]Utah Field House of Natural History State Park Museum, Vernal, UT, United States; [2]Utah Tech University, St. George, UT, United States; [3]St. George Dinosaur Discovery Site at Johnson Farm, St. George, UT, United States; [4]Department of Geological Sciences, University of Cape Town, Rondebosch, South Africa; [5]Jurassica Museum, Porrentruy, Switzerland; [6]Department of Geosciences, University Fribourg, Fribourg, Switzerland; [7]Fundación Conjunto Paleontológico de Teruel-Dinópolis/Museo Aragonés de Paleontología, Teruel, Spain; [8]Dipartimento di Scienze della Terra, Università di Firenze, Firenze, Italy; [9]State Key Laboratory of Biogeology and Environmental Geology, China University of Geosciences, Beijing, China; [10]School of the Earth Sciences and Resources, China University of Geosciences (Beijing), Beijing, China; [11]Dinosaur Trackers Research Group, University of Colorado Denver, Denver, CO, United States

1. Introduction

The Jurassic was a crucial time in the development of continental ecosystems: dinosaurs rose to dominance on land in the early part of the period and many key dinosaur lineages, as well as new, nearly modern, vertebrate groups, established themselves during the Middle and Late Jurassic. The tetrapod tracks of the Jurassic comprise the first truly global footprint record, spanning most continents (Lucas, 2007).

In the Early Jurassic, the tetrapod record is dominated by ichnotaxa made apparently by theropod dinosaurs, although crocodylomorph, synapsid, ornithischian, and sauropodomorph tracks are also known. Some of the common track types from the Early Jurassic include *Grallator*, *Eubrontes*, *Batrachopus*, and *Brasilichnium* (Leonardi, 1987; Lockley and Hunt, 1995).

[a] Deceased 2023.

Vertebrate Ichnology
https://doi.org/10.1016/B978-0-443-13837-9.00015-9

The Middle Jurassic epoch is particularly interesting and significant because it marks the transition away from the dominance of Early Jurassic ichnotaxa, such as *Batrachopus*, *Eubrontes*, *Grallator*, and *Kayentapus*. Instead, more diverse ichnofaunas emerged, including new ichnotaxa associated with saurischian dinosaurs (theropods and sauropods), a greater diversity of thyreophoran ichnotaxa, and the first possible appearances of ornithopod ichnotaxa. In contrast, ichnotaxa attributed to nondinosaurian archosaurs (e.g., crocodylomorphs) and nonarchosaurian sauropsids (e.g., lepidosaurians) are relatively rare in the Middle Jurassic compared with the Triassic and Early Jurassic. The reasons for this are unknown, but it may be due in part to the scarcity of Middle Jurassic track sites compared with those from earlier time periods.

The Late Jurassic epoch represents the peak of ichnofaunas that originated in the Middle Jurassic. Tracks from theropods, sauropods, ornithopods, neornithischians, and some thyreophorans dominated this epoch. Some of the more common ichnogenera from these groups include *Megalosauripus* and *Hispanosauropus* (theropods); *Brontopodus* and *Parabrontopodus* (sauropods); *Anomoepus* and *Dinehichnus* (basal ornithischians and ornithopods); and *Deltapodus*, *Stegopodus*, and *Tetrapodosaurus* (thyreophorans). In addition, there were also several nondinosaurian ichnotaxa pertaining to pterosaurs, crocodylomorphs, turtles, and possibly lepidosaurs.

2. Regional summaries

2.1 Africa

Tracks are particularly abundant in the Lower Jurassic Stormberg Group, in the Roma Valley of western Lesotho, which is in the main Karoo Basin of southern Africa (Ellenberger et al., 2005; Abrahams, 2020; Abrahams et al., 2023). The track-rich Elliot Formation of the Stormberg Group (Rampersadh et al., 2018) is divided into lower and upper units, with the lower being exclusively Late Triassic (middle Norian–Rhaetian), and the upper being Hettangian, Sinemurian, and possibly lowermost Pliensbachian (Bordy et al., 2017). Vertebrate burrows from the upper Elliot Formation, Karoo Basin, South Africa have also been reported (Bordy et al., 2017).

Small, *Grallator*-like theropod tracks in a single trackway were reported by Bordy (2021) from the lower Clarens Formation (Pliensbachian) north of Maclear, Eastern Cape, South Africa. The site is on a path near an archaeological site with rock art called Storm Shelter. Bordy et al. (2022) described tracks from the lowermost Lesotho Formation of the lower Drakensberg Group (Pliensbachian–Toarcian) on sandstone interbeds between basalt flows at the Highlands Ichnosite to the south of Bethlehem, eastern Free State, South Africa. Bordy et al. (2020) also erected the new ichnofamily Delatorrichnopodidae, and the ichnogenus/ichnospecies *Afrodelatorrichnus ellenbergeri*. The Roma Valley, western Lesotho, part of the main Karoo Basin in southern Africa (Dornan, 1908; Abrahams et al., 2023), contains the famous Early Jurassic Moyeni Tracksite, about which much has been published (Ellenberger et al., 1963; Ellenberger, 1970; Smith et al., 2009; Wilson et al., 2009; Marsicano et al., 2014; Bordy et al., 2024). Tracksites from the Drakensberg Group of central Lesotho (e.g., Ralikohomo, Lekhalo-la-Maburu) are preserved on sandstones interbedded with flood basalt deposits

(Bordy et al., 2020). Tracks attributed to mammaliaforms include *Malutitetrapodiscus*, with ichnospecies *M. saltator*, *M. tenuis*, *M. minimus*, and *M. perlinax* (Ellenberger, 1970, 1972, 1975; Lockley et al., 2004c), and *Ralikhomopus aviator*, which has been attributed to a small, birdlike theropod (Ellenberger, 1974; Bordy et al., 2020). Wagensommer et al. (2016) described Early Jurassic African desert ichnofaunas from the Etjo Sandstone at the Otjihaenamaparero locality, Namibia.

Tracks have also been reported from the Early–Middle Jurassic in Tunisia (Contessi and Fanti, 2012; Niedźwiedzki et al., 2017), and a quadruped manus-pes trackway from the Early–Middle Jurassic of Zimbabwe preserves the paratype trackway of *Afrodelatorrichnus ellenbergeri* (Bordy et al., 2020). Early, Middle, and Late Jurassic tracks are known from Morocco and Niger (e.g., Jenny and Jossen, 1982; Ishigaki and Lockley, 2010; Ginsburg et al., 1966; Belvedere et al., 2010; Mudroch et al., 2011; Gierliński et al., 2017a,b; Klein et al., 2018, 2022; Oukassou et al., 2019; Oussou et al., 2023) as well.

2.1.1 *Madagascar*

The Middle Jurassic Bemaraha Formation contains a remarkable abundance of dinosaur tracks, representing the sole known occurrence of its kind in Madagascar (Wagensommer et al., 2010, 2012, 2023). These footprints, mainly from theropods and sauropods, contribute to our understanding of the Middle Jurassic, which has limited documentation globally. The Bemaraha Formation provides valuable insights into dinosaur behavior within a Jurassic marginal-marine environment. Currently, there are a total of 31 tracksites known in Bemaraha, mostly from the same stratigraphic level, and covering an area of $\sim 30\ km^2$ (Wagensommer et al., 2023). This makes it one of the largest occurrences of dinosaur tracks in the Southern Hemisphere. Theropod tracks greatly outnumber sauropod tracks indicating that sauropod tracks are less common and, possibly, that sauropods were rare in the Middle Jurassic Bemaraha mudflats. On the other hand, theropods were likely common, with little variation among track makers, mainly identified as *Kayentapus* isp. (Wagensommer et al., 2012, 2023).

2.2 Antarctica

Despite vertebrate traces from the Triassic and Eocene, no Jurassic tetrapod traces have yet been reported from Antarctica.

2.3 Asia

Tracks are particularly abundant in the Early Jurassic of Asia (e.g., Xing et al., 2013a), where ichnogenera such as *Batrachopus*, *Changpeipus*, *Eubrontes*, and *Jialingpus* are known. The Middle and Upper Jurassic are also well represented, with ichnotaxa such as *Carmelopodus*, *Crocodylopodus*, *Kuangyuanpus*, *Ravatichnus*, *Shensipus*, *Tuojiangpus*, and others known from the Middle Jurassic, and *Shirkentosauropus*, *Therangospodus*, *Jialingpus*, and others from the Late Jurassic. The Jurassic track record of China is particularly rich, including many cosmopolitan ichnotaxa such as *Anomoepus*, *Eubrontes*, *Grallator*, and *Kayentapus*, as well as endemic ones, such as *Shenmuichnus*; the record was reviewed by Lockley et al. (2013) and has been built on since by many new discoveries. Jurassic tracks and traces are also well known from Iran, Japan, Mongolia, and Tajikistan; tracks from the latter have been placed in several poorly known, apparently endemic ichnotaxa, such as *Gabirutosaurichnus*, *Regarosauropus*, and *Shirkentosauropus*, and are in need of review and thorough comparisons to other ichnotaxa worldwide.

2.4 Australia

The Jurassic track record of Australia is limited compared with the Triassic (Romilio et al., 2022) and Cretaceous records from the continent: Jurassic tracks have reported primarily from the Walloon Coal Measures (Long, 1998) and the Razorback Beds (Romilio et al., 2020, 2024). The Walloon Coal Measures include a track type possibly referable to *Changpeipus* (Haubold, 1971a), and the Razorback Beds include theropod and possibly basal ornithischian tracks (Romilio et al., 2020).

2.5 Europe

Early Jurassic ichnotaxa from Europe include *Batrachopus* from France and Romania, cf. *Wintonopus* from Poland, grallatorids and *Anomoepus*-like tracks from a number of sites, *Moyenisauropus* from Poland (Gierliński, 1991) and Portugal (Figueiredo et al., 2023), *Ameghinichnus* from Poland, and *Delatorrichnus*, also from Poland, among others.

Throughout Europe, there are abundant Middle and Late Jurassic vertebrate tracks, providing a diverse tableau of ancient life traversing the continent. Both siliciclastic and carbonate track-bearing strata dominate in both epochs. However, Middle Jurassic tracksites are relatively scarce compared with those of the Late Jurassic, so much so that Lockley and Meyer (2000) referred to this as the a "Middle Jurassic Gap." Nevertheless, several Middle Jurassic tracksites have now been identified in Portugal (e.g., Santos et al., 2009; Razzolini et al., 2016), the United Kingdom (e.g., Day et al., 2004, 2002; dePolo et al., 2020; Edgar et al., 2023; Romano et al., 2018; Romano and Whyte, 2003; Hudson et al., 2023), Denmark (e.g., Milàn and Bromley, 2005), and France (e.g., Gand et al., 2018; Moreau et al., 2020).

In Europe, the Late Jurassic is notable for its expansive track-bearing surfaces, colloquially termed "dinosaur freeways" (Lockley and Meyer, 2023). The Jura megatracksites, spanning Switzerland and France, are a prime example. The megatracksite phenomenon here is attributed partly to the emergence of large land areas during sea-level low-stands, connecting permanent landmasses such as the Massif Central and Iberian Massif, which facilitated faunal exchange across the Tethys region. The presence of interconnected pathways, influenced by fluctuations in sea level, explains the recurrence of shared ichnotaxa in these megatracksites from the Oxfordian to the Berriasian on the Jura carbonate platform (Meyer et al., 2006; Marty et al., 2018). In addition, the Iberian Peninsula has produced numerous vertebrate (mainly dinosaur) tracksites in transitional units from different basins, particularly the Lusitanian basin (e.g., Santos, 2008; Castanera et al., 2020, 2021) in Portugal and the Asturian (García-Ramos et al., 2006; Lockley et al., 2008b), Maestrazgo, and South Iberian (Aurell et al., 2016; Campos-Soto et al., 2019) basins in Spain. This wealth of tracksites has enabled comparisons between Europe and North America (Lockley et al., 2008b; Lockley, 2009), suggesting that both regions share some crocodylomorph, pterosaur, and dinosaur ichnotaxa. Overall, the abundance of localities and the recent advancements in ichnotaxonomy are laying the groundwork for future investigations into dispersal routes between Gondwana and Laurasia during the Middle and Late Jurassic. Collectively, this research has the potential to deepen our understanding of dinosaur evolution and faunal exchange between boreal and austral continents.

2.6 North America

The Early Jurassic is best represented in eastern North America by the Newark Supergroup, which spans the Triassic–Jurassic transition (Olsen and Rainforth, 2001), extending from North Carolina in the south to Nova Scotia in the north. In the southwestern United States, one of the most productive units is the Glen Canyon Group, composed of the Moenave Formation, Wingate Sandstone, Kayenta Formation, and the Navajo Sandstone in Utah, Arizona, and, in part, Colorado (Lockley and Hunt, 1995; Kirkland et al., 2014b). In northern and northwestern Utah, northwestern Colorado, and southern Idaho, the upper portion of the Nugget Sandstone is Lower Jurassic.

The Wingate Sandstone is a Late Triassic–Early Jurassic (Rhaetian–Hettangian), eolian unit spanning parts of Utah, Colorado, Arizona, and New Mexico. It may correlate laterally to the north with the Nugget Sandstone in northern Utah, and with the fluviolacustrine Moenave Formation in southwestern Utah and northwestern Arizona. Vertebrate traces are locally abundant and include the ichnogenera *Eubrontes, Grallator,* and *Anomoepus.*

The Moenave Formation is a Late Triassic to Early Jurassic (Rhaetian–Hettangian), laterally interfingering equivalent of the Wingate Sandstone of southwestern Utah and northwestern Arizona (e.g., Suarez et al., 2023). It is divided into two members: the lower, fluvial Dinosaur Canyon Member and the upper, lacustrine Whitmore Point Member. Vertebrate traces are locally abundant in the Dinosaur Canyon Member, and widespread in the Whitmore Point Member. The ichnolagerstätte at the St. George Dinosaur Discovery Site at Johnson Farm (SGDS) in southwestern Utah preserves a high diversity ichnofauna with *Eubrontes, Grallator, Kayentapus, Gigandipus, Anomoepus, Batrachopus,* indeterminate cynodont tracks, *Characichnos,* along with a high diversity of invertebrate traces (Milner et al., 2006a; Williams et al., 2006). Overlying or regionally adjacent formations that record similar ichnofaunas include the Kayenta Formation, Navajo Sandstone, Nugget Sandstone, and Aztec Sandstone.

The early and middle Middle Jurassic are particularly poorly represented by terrestrial formations in North America. Sauropod tracks from Mexico are Middle Jurassic in age (Ferrusquía-Villafranca et al., 1996), and a sauropod trackway from the Entrada Sandstone of Utah likely is as well (Foster et al., 2000; Milàn and Loope, 2007), but there are few other examples. But the Middle–Late Jurassic transition is well represented and rich in tracks in formations such as the Summerville, Curtis, Stump, and Sundance. The Late Jurassic is well represented with more than 80 tracksites in the Morrison Formation (Foster and Lockley, 2024). Tracks from these Upper Jurassic levels include forms such as *Chelonipus, Hatcherichnus, Parabrontopodus, Megalosauripus, Dinehichnus, Stegopodus,* and *Deltapodus,* among others.

2.7 South America

The Jurassic record of South America includes *Batrachopus* from the Saldaña Formation of Colombia (Early Jurassic) and *Grallator, Ameghinichnus, Delatorrichnus, Sarmientichnus,* and

Casamiquelichnus from the La Matilde Formation of Argentina (Middle Jurassic) (de Valais, 2011). The Upper Jurassic Baños del Flaco Formation in Chile produced a sauropod trackway and other footprints (Moreno and Pino, 2002, 2010; Moreno and Benton, 2005).

3. Systematic ichnology

Ichnogenera occurring in the Jurassic are discussed below. Each ichnogenus is accompanied by a diagnosis and reference to its type material. We explicitly state if the type material is not Jurassic in age. General descriptions of the valid ichnogenera and ichnospecies are restricted to the Jurassic material only. While valid ichnospecies are mentioned, we may not provide a diagnosis or description for each one.

Institutional abbreviations: ACM-ICH, "Appleton Cabinet," Beneski Museum of Natural History, Amherst, Massachusetts, USA; **AMNH**, American Museum of Natural History, New York, New York, USA; **ARSA**, see Leonardi (1981); **ARSB**, see Leonardi (1981); **BMNH**, British Museum of Natural History, London, UK; **BMNHC**, Beijing Museum of Natural History, Beijing, China; **BNHM**, Beijing Natural History Museum, Beijing, China; **BNSS**, Bournemouth Natural Science Society, Bournemouth, UK; **CEUM**, College of Eastern Utah Prehistoric Museum, Price, Utah, USA; **CFJW, CFNY, CFYH, CFZJ,** or **CFZW,** Chongqing Museum of Natural History, Chongqing, China; **CFZ-Zt**, Collection of the Facultad de Estudios Superiores Zaragoza, Universidad Nacional Autónoma de México, Iztapalapa, Mexico; **CM**, Carnegie Museum, Pittsburgh, Pennsylvania, USA; **CMN**, Canadian Museum of Nature, Ottawa, Ontario, Canada; **COLM**, Colorado National Monument, Fruita, Colorado, USA; **CPT**, refers to catalog at the Fundación Conjunto Paleontológico de Teruel–Dinópolis/Museo Aragonés de Paleontología, Teruel, Spain; **CRA**, Pterosaur Beach tracksite of Crayssac, France; **CU-MWC**, University of Colorado Denver-Museum of Western Colorado joint collection, now at University of Colorado Museum Boulder, Boulder, Colorado, USA; **DINO**, Dinosaur National Monument, Jensen, Utah, USA; **DP**, Dinosaur State Park, Rocky Hill, Connecticut, USA; **ESI**, Evolutionary Science Institute at the University of the Witwatersrand, Johannesburg, South Africa; **FGM**, Fundy Geological Museum, Parrsboro, Nova Scotia, Canada; **FHPR (also UFH)**, Utah Field House of Natural History State Park Museum, Vernal, Utah, USA; **GLCA**, Glen Canyon National Recreation Area, Page, Arizona, USA; **GPIT**, Geology Museum, University of Tübingen, Germany; **GSCA**, Geological Survey of Canada, Ottawa, Canada; **HYMVC**, Heyuan City Museum, Heyuan, Guangdong, China; **IGSP**, see Leonardi (1981); **IVPP**, Institute of Vertebrate Paleontology and Paleoanthropology, Beijing, China; **LuDM (now ZLJ)**, Lufeng Dinosaur Museum, Jingshan, Yunnan, China; **MACN-SC**, Museo Argentino de Ciencias Naturales, Buenos Aires, Argentina; **MÁFI**, Geological Institute of Hungary, Budapest, Hungary; **MGGC**, Collezione di Geologia "Museo Giovanni Cappellini," Bologna, Italy; **MHM**, Museum of Mount Holyoke College, South Hadley, Massachusetts, USA (specimens were destroyed by fire in early 2000s); **MJSN**, JURASSICA Museum (previously Musée Jurassien des Sciences Naturelles), Porrentruy, Switzerland; **MLP**, Museo de La Plata, La Plata, Argentina; **MM**, Montshire Museum of Science, Norwich, Vermont, USA (formerly from the collection of the Museum of Dartmouth College, Hanover, New Hampshire, USA); **MNA**, Museum of Northern Arizona, Flagstaff, Arizona, USA; **MNDPDSA**, Monumento Natural das Pegadas de

Dinossáurio da Serra de Aire (Serra de Aire Dinosaur Tracks Natural Monument), Bairro, Serra de Aire, Portugal; **MNHN**, Museu Nacional de História Natural da Universidade de Lisboa, Lisboa, Portugal; **MNRJ**, National Museum (Museu Nacional), Vertebrates Section, Quinta da Boa Vista, Tio de Janeiro, Brazil; **MNS**, Museo Numantino de Soria (Numantine Museum of Soria), Soria, Spain; **MPSUR**, Museo di Paleontologia, Sapienza Università di Roma, Rome, Italy; **MTM**, Hungarian Natural History Museum, Budapest, Hungary; **MUJA**, Museo del Jurásico de Asturias, Colunga, Asturias, Spain; **Muz. PIG OS**, Geological Museum of the Holy Cross Mountains Branch of the Polish Geological Institute, Kielce, Poland; **MWC**, Museum of Western Colorado, Fruita, Colorado, USA; **NGM**, Nanjing Geological Museum, Nanjing, China; **NHMUK**, The Natural History Museum, London, UK; **NJSM**, New Jersey State Museum, Trenton, New Jersey, USA; **NMMNH**, New Mexico Museum of Natural History and Science, Albuquerque, New Mexico, USA; **NMNHZ**, National Museum of Natural History of Zimbabwe, Bulawayo, Zimbabwe; **NWUV**, Institute of Cenozoic Geology and Environment, Department of Geology, Northwest University, Xi'an, China; **OTK**, Otari Folklore Museum, Otari Village, Nagano Prefecture, Japan; **PECMM&A**, Paul Ellenberger Collection at Morija Museum and Archives, Lesotho; **PECUoM**, Paul Ellenberger Collection at University of Montpellier, France; **PGN**, Dinosaur tracksite of Plagne, France; **QM**, Queensland Museum, Brisbane, Australia; **ROLM**, Rovereto Lavini de Marco Collections, Italy; **RU**, Rutgers University, New Brunswick New Jersey, USA; **SAM**, Iziko South African Museum, Cape Town, South Africa; **SGDS**, St. George Dinosaur Discovery Site at Johnson Farm, St. George, Utah, USA; **SGPV**, Museo Nacional de Historia Natural, Santiago, Chile; **SHN**, Sociedade de História Natural, Torres Vedras, Portugal; **SMNS**, Staatliches Museum für Naturkunde, Stuttgart, Germany; **SNHM**, Staatliches Naturhistorisches Museum Braunschweig, Germany; **SSM**, Springfield Science Museum, Springfield, Massachusetts, USA; **TGM**, Tashkent Geological Museum, Tashkent, Uzbekistan; **TMM**, Texas Memorial Museum, University of Texas, Austin, Texas, USA; **UC**, see **CU-MWC** and **UCM**; **UCM**, University of Colorado Natural History Museum, Boulder, Colorado, USA; **UCMP**, University of California Museum of Paleontology, Berkeley, California, USA; **UFH (also FHPR)**, Utah Field House of Natural History State Park Museum, Vernal, Utah, USA; **UMNH**, Natural History Museum of Utah, Salt Lake City, Utah, USA; **UW**, University of Wyoming, Laramie, Wyoming, USA; **USEGS**, University of Sheffield Environmental and Geological Sciences, Sheffield, UK; **YPM**, Yale Peabody Museum, New Haven, Connecticut, USA; **ZDM**, Zigong Dinosaur Museum, Sichuan, China; **ZION**, Zion National Park, Springdale, Utah, USA; **ZLJ (formerly LuDM)**, Lufeng Dinosaur Museum, Jingshan, Yunnan, China.

Accordiichnus
Manni, Nicosia, and Nobili 1999
Fig. 6.1
Type ichnospecies. *A. natans.*
Valid ichnospecies. *A. natans.*
Distribution. Lower Jurassic of Europe (Corniola Formation: Italy).
Type specimen(s). MPSUR NS 65, natural mold of a trackway of 22 apparent sets of imprints. Resin cast at the Museo di Cantiano.
Diagnosis. Traceway (terminology sensu Thomson and Droser, 2015) of a half-swimming tetrapod with pear-shaped forelimb imprints wider than small hind-limb imprints. Prints regularly alternate. Pes prints rarely overlap medial parts of manus prints.

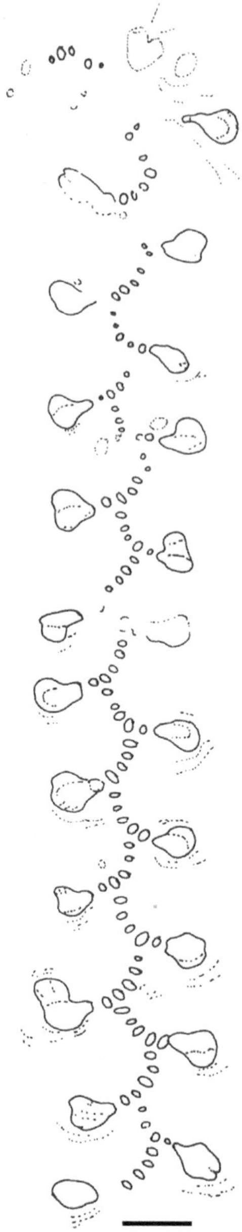

FIGURE 6.1 *Accordiichnus natans* holotype MPSUR NS 65 from the Lower Jurassic of central Italy. Scale = 10 cm. *From Manni et al. (1999).*

Description. See Fig. 6.1. The long axes of the manus footmarks (terminology sensu McAllister, 1989) lie nearly perpendicular to the traceway midline. Manus footmark pace angulation regularly roughly 45°. The pes footmarks are presumed to have been made only by the distal tips of the appendage and consist of ovate imprints, usually in sets of five and arranged in an anteriorly concave crescent.

Discussion. Traces similar to *Accordiichnus* found near the type locality were briefly described by Arduini (1996) and attributed to both "prosauropod" (basal sauropodomorph) and theropod dinosaurs. Manni et al. (1999) suggested that *Accordiichnus* was made by a marine reptile, possibly a nothosaur or basal plesiosaur, and further explored the possibility that the *Accordiichnus* trace maker was a hypothetical swimming reptile, one that was not as fully adapted to a "pelagic" lifestyle as an ichthyosaur or plesiosaur, but one that was more adapted to such a lifestyle than the extant marine iguana *Amblyrhynchus cristatus*. *Accordiichnus* traces differ markedly in footmark morphology and arrangement from *Dikosopichnus* traces from the Middle Triassic Guanling Formation of Guizhou, China, which were also ascribed to nothosaurs (Zhang et al., 2014), as well as from *Anshunpes* traces from the same unit that were ascribed to placodonts or saurosphargids (Xing et al., 2020). In particular, Zhang et al. (2014) described a marine-reptile interpretation for *Accordiichnus* as unconvincing.

Afrodelatorrichnus
Bordy, Rampersadh, Abrahams, Lockley, and Head 2020
Fig. 6.2
Type ichnospecies. *A. ellenbergeri.*
Valid ichnospecies. *A. ellenbergeri.*
Distribution. Lower Jurassic of Africa (Lesotho Formation, Drakensberg Group: South Africa) and Europe (Zagaje Formation: Poland); ?Lower–Middle Jurassic of Africa (Dande Sandstone Formation: Zimbabwe).

Type specimen(s). From Bordy et al. (2020): Holotype = Trackway A from the Highlands locality in South Africa. Silicone rubber cast (BP/6/744) housed at ESI. Paratype = unnamed Zimbabwean trackway cast from the ?Lower to Middle Jurassic, post-Karoo Dande

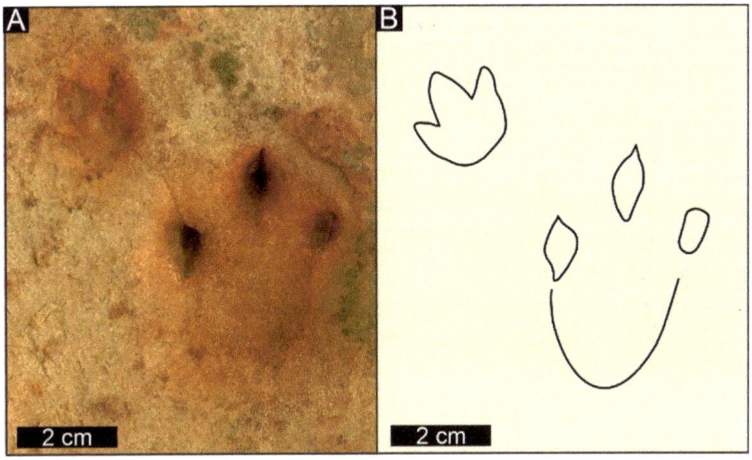

FIGURE 6.2 *Afrodelatorrichnus ellenbergeri*, quadruped ornithischian tracks from the Lesotho Formation, Highlands 1239 farm, South Africa. (A) Photograph of the fifth manus-pes set in the holotype trackway. Scale = 2 cm. (B) Interpretive outline of the tracks shown in A. Scale = 2 cm. For details, see Bordy et al. (2020, Figs. 6–8, which include details of the holotype trackway).

Sandstone Formation (Ntumbe River, Chewore South Safari Area, Lower Zambezi Valley, northern Zimbabwe) housed at NMNHZ, and replica GPIT1653.

Tracks preserved as natural impressions (concave epireliefs) on a sandstone surface (zone C/1, Lesotho Formation, Highlands 1239 farm, South Africa) figured in Bordy et al. (2020, Figs. 6–8). Recorded as silicon rubber cast BP/6/744 as well as digital models.

Diagnosis. Modified from Bordy et al. (2020): Trackway of a small, obligate quadruped with outwardly rotated tridactyl pes and smaller manus with indistinct digit traces (?tridactyl) recording a distinctive gait in which outwardly rotated pes registered outside, or "lateral" to, trackway midline and manus registered on or inside, or "medial" to, trackway midline. Moderate heteropody, with pes more than twice as long and wide as manus.

Pes tracks small tridactyl, mesaxonic, and slightly elongate: mean length of 3.8–4.3 cm and mean width of 3.1–3.4 cm. Digit III longest with an anterior triangle length-to-width ratio of 0.4–0.8. Pes outward rotation, although variable, is as high as 45° and averages 15–17.8°. Digit traces deep, fusiform, lacking differentiated digital pad traces and sometimes separate from shallow, broad, posteriorly convex "heel" trace in holotype trackway. Mean divarication of Digits II–IV is ~50°.

Manus tracks small, tridactyl, mesaxonic, with occasionally clear digit traces with a mean length of 2.0–1.7 cm and mean width of 1.6–1.9 cm. Manus trackway width smaller than pes trackway width due to manus registration medial to trackway midline: that is, typically inside pes tracks.

Larger than, and differing from, *Delatorrichnus* in pes trackway width and lack of pes-on-manus overstep. Also differing from *Delatorrichnus* in configuration of pes pad impressions, which are narrower and more widely separated in *Afrodelatorrichnus*.

Description. See Fig. 6.2. Marked contrast between moderately wide pes trackway width, more than twice the pes width, giving a pes pace angulation of about 150° compared with a narrow manus trackway width and high manus pace angulation (>180°). In the holotype trackway, consisting of seven manus-pes sets, the manus is situated anterior to the pes in two cases, but well "inside" (or medial to) the corresponding pes in five other cases. The four most distal manus tracks are rotated positively (inward) at an average of 21°. Likewise, the manus is situated anterior to, but well inside the pes in the paratype trackway. Rotation of the manus is only discernible in the last three manus-pes sets of the paratype trackway, with sequential values of 22° (out), 15° (in), and 8° (out). In the holotype trackway, the average left pes–right pes pace is 30.0 cm, and corresponding pes stride is 57.8 cm. The mean step and stride values for the paratype trackway, made by a smaller animal (based on pes length), are 20.6 and 38.9 cm, respectively. The longer step recorded in the holotype trackways is reflected by a higher pes pace angulation (154°) than in the paratype trackway (143°). However, the mean trackway width, 9.0 cm, is the same in both trackways, so the mean pes trackway width is more than twice pes width in both trackways. The mean manus pace and stride in the holotype trackway are 29.3 and 58.6 cm, respectively, with corresponding values of 18.3 and 38.0 cm for the paratype trackway. Thus, lengths of manus and pes steps and strides are similar. However, the mean manus trackway width is 2.6 and 3.4 cm, respectively, thus only slightly greater than actual manus track width. Such narrow manus trackway width values reflect the registration of the manus on or closer to the trackway midline than the pes, resulting in very high pace angulations up to 191° (means 181.2° and 181.3° for holotype and paratype, respectively).

Discussion. Reported from the surface of a fine-to very fine-grained sandstone interbed within the Karoo continental flood basalts (Pliensbachian–Toarcian), some 45 m above the top of the Clarens Formation at Highlands farm in South Africa (zone C/1, Lesotho Formation; Bordy et al., 2020). High pace angulation values in excess of 180° indicate that the track maker registered footprints "inside,""across," or medial to the trackway midline. The manus is also often situated inside the pes, and the hind feet are strongly rotated or "splayed" outward. This hind limb, "flat footed" straddle is accompanied by front limbs with little or no straddle and a strong tendency to register inwardly rotated manus prints medial to the trackway midline. The tridactyl pes tracks with outward rotation and without discrete digital pad traces, plus the smaller manus tracks with at least three digit traces, are indications that the track maker was an obligate quadruped, likely a small, ornithopod-like ornithischian, with fleshy, thickly padded feet. Gierliński and Niedźwiedzki (2002a) argued for the heterodontosaurid origin for Jurassic, *Atreipus*-like forms (i.e., *Delatorrichnus* and the more recently erected *Afrodelatorrichnus*). A manus-pes set ascribed to *Delatorrichnus* by Gierliński et al. (2004) as *Delatorrichnus* (specimen Muz. PIG OS-221/54) better conforms to the morphology described for *Afrodelatorrichnus* than *Delatorrichnus*.

Bordy et al. (2020) erected a new ichnofamily, Delatorrichnopodidae, to accommodate both *A. ellenbergeri* and *Delatorrichnus goyenechei*.

Ameghinichnus

Casamiquela 1961

Fig. 6.3

Type ichnospecies. *A. patagonicus.*

Valid ichnospecies. *A. patagonicus* (Casamiquela, 1961), *A. manantialensis* (de Valais, 2009), *A. mirabilis* (de Valais, 2009; Ellenberger, 1970).

Distribution. Lower Jurassic of North America (Towaco Formation, Newark Supergroup: New Jersey), Europe (Zagaje Formation: Poland), and Africa (Elliot Formation: Lesotho); Middle Jurassic of South America (La Matilde Formation: Santa Cruz, Argentina).

Type specimen(s). Natural mold of six manus–pes sets MLP 60-X-31-14A.

Diagnosis. Emended from Casamiquela (1961) by de Valais (2009): quadrupedal trackways with nearly homopod, symmetrical, and pentadactyl manus and pes imprints; manus–pes sets display alternate or opposite positions relative to the trackway midline. Inwardly rotated manus prints lie medial to outwardly rotated pes prints. Digit imprints subequal in both length and divarication angles. Tail drag impressions often present that are sinuous when tracks have an alternating arrangement and straight when tracks have an opposite arrangement.

Description. See Fig. 6.3. *Ameghinichnus patagonicus*, the most common ichnospecies, has manus–pes track sets with length/width ratios of 0.7 and manus and pes prints that lack claw imprints or phalangeal pads. *A. manantialensis* differs from *A. patagonicus* in relative manus–pes size proportions, digit length proportions and divarication angles, and the relative position of the manus print to the pes print; these may be ontogenetic differences, however (de Valais, 2009). Furthermore, *A. manantialensis* tracks are not known to exhibit a symmetrical, opposite arrangement, though this may be a consequence of small sample size (de Valais, 2009). de Valais (2009) accepted the synonymy of *Eopentapodiscus mirabilis*

FIGURE 6.3 *Ameghinichnus patagonicus* Casamiquela 1960 (A) MLP 60-X-31-14B, holotype showing trackways 1 and 2. Black arrows indicate travel directions. Scale = 5 cm. (B) MLP 60-X-31-14B, close-up first manus-pes set of trackway 2. Scale = 1 cm. *(A and B) Modified from de Valais (2009).*

(Ellenberger, 1970) with *Ameghinichnus* by Olsen and Galton (1984) because it demonstrates several characteristics of the ichnogenus but was unable to support ichnospecies-level synonymy and did not provide an ichnospecies-level differential diagnosis for the consequent *A. mirabilis*.

Discussion. Middle Jurassic *Ameghinichnus* tracks generally have been attributed to small mammaliaforms; older specimens generally have been attributed to either mammaliaforms or derived, nonmammaliaform eucynodonts. Specimens referred to *Ameghinichnus* from the Early Jurassic of New Jersey (Olsen, 1988, 1995) resemble *A. patagonicus* in having paired, roughly equal-sized manus and pes prints, but differ from *A. patagonicus* tracks in being larger, as well as by having longer, narrower, clawed digit imprints. de Valais (2009) recognized similarity between a lone track from the Early Jurassic of Poland (Gierliński et al., 2004) and specimens of *A. manantialensis* but could not attribute the former specifically to that ichnospecies. Several ichnotaxa erected by Ellenberger (1970, 1972, 1974, 1975) (including *Acropentapodiscus*, *Amphibiopodiscus* [sometimes misspelled as *Amphibiopus*], *Aristopentapodiscus*, *Dicynodontipus* [= *Calibarichnus*], *Dinopentapodiscus*, *Eoameghinichnus* [= *Ameghinichnus*], *Eopentapodiscus*, *Grypopentapodiscus*, and *Pseudameghinichnus* [= *Ameghinichnus*]) were synonymized with *Ameghinichnus* by Olsen and Galton (1984), but de Valais (2009) disagreed with some of these synonymies; she retained several of the ichnotaxa as provisionally valid and determined that others (e.g., *Eoameghinichnus* and *Pseudameghinichnus*) were based on material too incomplete or poorly preserved for adequate comparison.

Many *A. patagonicus* trackways demonstrate that the track maker progressed by saltation (hopping) as evidenced by the symmetrical positions of manus and pes prints on either side of the trackway axis. Some trackways transition from saltating to galloping (de Valais, 2009). Saltation has been documented for other Mesozoic mammaliaform tracks as well (e.g., Kim et al., 2017; Bordy et al., 2022), so it appears to have been a not uncommon mode of progression for such track makers.

Amphibiopodiscus
Ellenberger 1970
Fig. 6.4

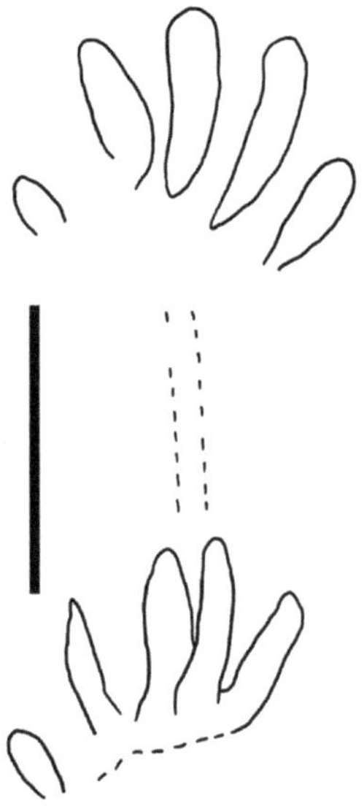

FIGURE 6.4 *Amphibiopodiscus saltator* Ellenberger (1974), two consecutive right pes impressions of a jumping temnospondyl amphibian at the Lower Moyeni tracksite in the lowermost upper Elliot Formation, Lesotho. Scale = 2 cm. *Retraced from Ellenberger (1970, Pl. IX, Fig. 71) and republished under a CC BY license, with permission from the Geological Society of South Africa, original copyright 1970.*

Type ichnospecies. *A. saltator.*

Valid ichnospecies. *A. saltator.*

Distribution. Lowermost Jurassic of southern Africa (lowermost upper Elliot Formation, zone B/1: Lower Moyeni/Quthing, Lesotho).

Type specimen(s). Multiple "light and bouncy" manus and pes tracks very irregularly distributed on a greenish-white, fine-grained, ripple-marked, algal-matted sandstone surface at the Lower Moyeni tracksite (zone B/1, lowermost upper Elliot Formation, Lesotho). Originally recorded as plaster casts 098 LES as well as 087 LES, 092 LES, 094 LES, and 095 LES, however, only plaster casts 087, 092, and 094 LES have survived at the PECUoM. Line drawing first illustrated as Plate IX, Fig. 71 in Ellenberger (1970).

Diagnosis. Modified from Ellenberger (1974): Pentadactyl pes impression consisting of digit impressions only, with Digit V preserved in the pes prints of stationary track makers. Track length and width nearly equal; length varies across four main sizes: 15 mm, 20–25 mm, 40–45 mm, and 75 mm. Predominance of Digit III over Digit IV. Slightly curved digits with distal gentle pads and without claw impressions. The digital pads are narrow, very subtle, separated by bends in the horizontal plane. Phalangeal formula is 3-4-5-5(-3?). Total divarication of Digits I–IV is up to 55° (I–V: up to 90°, II–III: 20°, III–IV: 20°). A rounded "heel" impression only appears in larger tracks. Sinusoidal skidding marks differ from the angular skidding in *Batrachopodiscus* (Upper Triassic of Lesotho), which was made by more a zig-zagging twister rather than a jumper. Prominence of Digit III differs from that of Digit IV in *Batrachichnus* (Haubold, 1971a).

Tetradactyl manus impression with Digit V print never preserved, similar or slightly smaller than the pes but similar in morphology with slightly more marked sliding/slipping marks than those of the pes. The impressions of the tetradactyl manus, outward curved digits, and the general lack of sole marks ("heel") set *Amphibiopodiscus* apart from *Ameghinichnus* (de Valais, 2009).

Description. See Fig. 6.4. Highly zigzagging (twirling) trackways of three–eight steps indicate a markedly irregular gait and a locomotion style somewhat influenced by size. Smaller tracks suggest mostly jumping progression with both feet, while larger tracks indicate gliding as well. The smooth, wavelike sliding or skidding marks associated with the tracks are specifically irregular and result from the frequent "sinusoidal" skidding of the pes and manus digits upon landing or, perhaps more precisely, at take-off before the wide "strikes." Some tracks occasionally are reminiscent of the softness and stretchiness of toad tracks. The tracks indicate a generally "agitated" activity throughout the tracksite that was heavily trampled/trodden by various vertebrates, including dinosaurs.

Discussion. Only reported from the Lower Moyeni tracksite (zone B/1, lowermost upper Elliot Formation) in Lesotho (Ellenberger, 1970, 1974), where the tracks are very irregularly distributed on a greenish-white sandstone surface, now covered by a protective shelter (Bordy et al., 2024). Amphibian tracks are uncommon in the Lower Jurassic of southern Africa. The pentadactyl pes and tetradactyl manus impressions of *Amphibiopodiscus saltator* were possibly made by a homogeneous group of small, jumping temnospondyl amphibians of variable sizes with digitigrade locomotion and a complex gait (Ellenberger, 1974).

Olsen and Galton (1984) made *A. saltator* a junior subjective synonym of *Ameghinichnus*, but de Valais (2009) disagreed, saying that while *Amphibiopodiscus* has pentadactyl manus and pes traces like *Ameghinichnus*, the digits curve outward and there is no sole mark even within deeply impressed tracks (Ellenberger, 1970, 1975). In addition, Ellenberger (1970) misspelled *Amphibiopodiscus* as *Amphibiopus*, a mistake repeated on a few occasions by subsequent authors (Olsen and Galton, 1984). Ellenberger (1970) attributed this ichnotaxon to amphibians, but some authors herein believe that *Amphibiopodiscus* may be a nomen dubium.

Anatopus
de Lapparent and Montenat 1967
Fig. 6.5

Type ichnospecies. *A. palmatus.*

Valid ichnospecies. *A. palmatus.*

Distribution. Lower Jurassic (Hettangian) of Europe, from a single site at Le Veillon, Vendée, France (de Lapparent and Montenat, 1967).

Type specimen(s). No catalog number given.

Diagnosis. Functionally tridactyl, small bipedal tracks with all digit traces separated from one another. Digit III is longer than Digits II and IV, and all digits have slender claw traces. Digit III appears to preserve three phalangeal pad traces, while Digits II and IV only show one. Divarication between Digits II–III is 30° while III–IV is 40°.

Description. Track length = 8.4 cm; track width = 9.6 cm. de Lapparent and Montenat (1967) claim that the type specimen possesses webbing between the digits (Fig. 6.5).

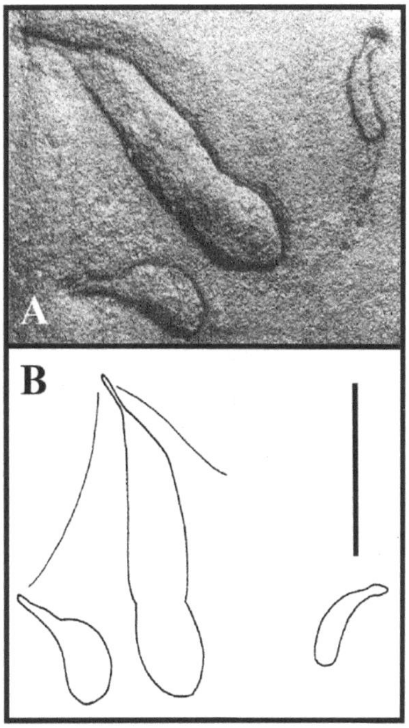

FIGURE 6.5 *Anatopus palmatus* de Lapparent and Montenat 1967. (A) black and white photo, and (B) interpretative sketch from de Lapparent and Montenat (1967). Scale = 5 cm.

Discussion. *Anatopus* is a problematic track type that cooccurs with large and small theropod and crocodylomorph tracks in a tidal flat paleoenvironment at Le Veillon, Vendée, France, which spans the Triassic–Jurassic boundary. *Anatopus* was described by de Lapparent and Montenat (1967) who tentatively suggested it was produced by an ornithischian; Lockley and Meyer (2000) later interpreted it as a poorly preserved *Grallator variabilis*, produced by a coelophysoid theropod. The authors herein strongly agree with Lockley (1998) and Lockley and Meyer (2000) that *Anatopus* was produced by a theropod and likely is synonymous with *Grallator*.

Anchisauripus
Lull1 1904
Fig. 6.6
Type ichnospecies. *A. sillimani.*
Valid ichnospecies. *A. sillimani, A. tuberosus.*

Distribution. Lower Jurassic (Hettangian–Toarcian) of eastern North America, Newark Supergroup (Hartford, Newark, and Fundy basins); Upper Triassic–Lower Jurassic of western North America (Glen Canyon Group, including the Moenave Formation); Upper Triassic–Lower Jurassic of Europe (Dolomia Principale of Italy; Woźniki Limestone, Zagaje Formation, Przysucha and Drzewica Formation of Poland; Tomanová Formation of Slovakia and Poland); Lower Jurassic of Africa (Etjo Formation of Namibia).

Type specimen(s). ACM-ICH 4/6, lectoholotype of *A. sillimani* according to Olsen et al. (1998) from Chicopee Falls, Chicopee, Massachusetts, upper part of the lower Portland

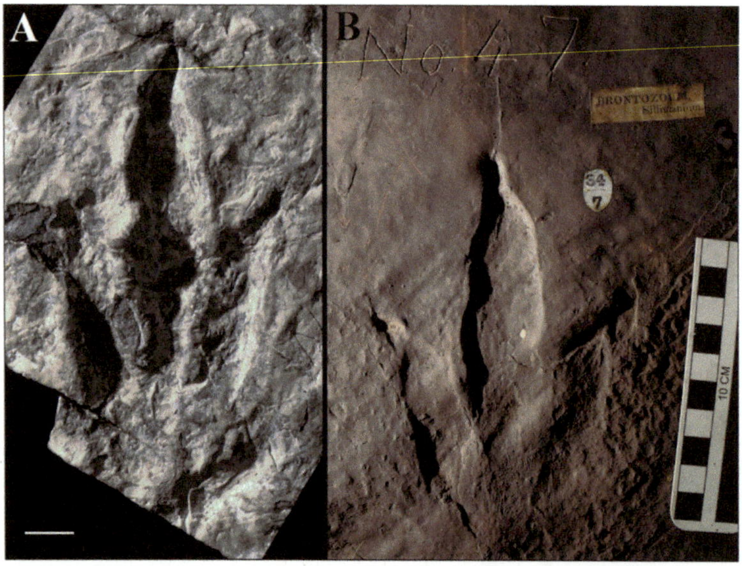

FIGURE 6.6 *Anchisauripus sillimani.* (A) Lectoholotype ACM-ICH 4/6 from the Lower Jurassic Portland Formation, Chicopee Falls, Massachusetts. Scale = 2 cm. (B) ACM-ICH 34/7 preserving skin impressions. *(A) Courtesy Paul Olsen. (B) Courtesy Patrick Getty.*

Formation. It is a natural mold transmitted track with indistinct pad traces. ACM-ICH 9/14 was figured by Lull (1904) and incorrectly referred as the holotype for his new ichnogenus, *Anchisauripus*. ACM-ICH 39/1 is the holotype of *A. tuberosus* (Hitchcock, 1841) from Turners Falls on the Montague shore, Massachusetts from the Turners Falls Formation. ACM-ICH 16/6 bears the lectotype of *Anchisauripus exsertus* (*Brontozoum exsertum* of Hitchcock, 1858, *A. exsertus* of Lull, 1904, and *Eubrontes exsertus* of Rainforth, 2005) from a gray micaceous sandstone of the Portland Formation from Smith's Ferry, Northampton, Massachusetts. In addition, Lull (1904) made ACM-ICH 56/1 the holotype for his new ichnospecies, *Anchisauripus hitchcocki* Lull (1904), from upper Portland Formation at Lily Pond, Massachusetts. ACM-ICH 16/1 is the holotype for *Anchisauripus minusculus* (Hitchcock, 1858), and it came from fine red shales of the upper Portland Formation at Lily Pond, Gill, Massachusetts.

Diagnosis. From Olsen et al. (1998): Bipedal, functional tridactyl, medium size (15–25 cm long) tracks. Digit III trace projected farther distally than in *Eubrontes*, but not as far as in *Grallator* (Digit III projection ratios >1.3 and <1.8, respectively). Tracks narrower than those of *Eubrontes*, but not as narrow as those of *Grallator* (length/width ratio near 2). Digit I trace (hallux) rarely registered. Divarication between Digits II–IV = 20–35°.

Description. *Anchisauripus* is diagnosed as tracks with an intermediate size range between *Grallator* and *Eubrontes*. See Fig. 6.6.

Discussion. Like many ichnotaxa described by Hitchcock from the Newark Supergroup of eastern North America, *Anchisauripus* has had a complicated and confusing history. Hitchcock (1836) named *Ornithichnites tuberosus*, but the specimens he figured have not been identified in the collection. However, three specimens figured, in 1841 are recognizable in the ACM collection (Olsen et al., 1998): ACM-ICH 39/1 (old number 54) from Turners Falls from the Turners Falls Formation, ACM-ICH 4/6 (or 23/13 of Hitchcock, 1865; old number 48) from the Portland Formation at Chicopee Falls, and ACM-ICH 35/31 (old number 52) also from Chicopee Falls, none of which were designated as type specimens (Olsen et al., 1998). ACM-ICH 4/6 and ACM-ICH 35/31 became the type series of *O. sillimani* (Olsen, 1980; Olsen et al., 1998). With no designated holotype, Olsen et al. (1998) designated ACM-ICH 4/6 as the lectoholotype of *O. sillimani* because of its clearer preservation than the other two specimens. In addition, Hitchcock (1845) placed *O. sillimani* tracks into *Eubrontes* and changed the ichnospecies name to *E. dananus* without any explanation. Then, Hitchcock (1847) again changed the name to *Brontozoum sillimani* without explanation (Olsen et al., 1998). Finally, Hitchcock (1848) once again inexplicably changed the ichnospecies name to *B. sillimanium* and continued to use that spelling thereafter (Olsen et al., 1998). Strangely enough, Hitchcock (1858) no longer listed ACM-ICH 46 and ACM-ICH 35/31 under *B. sillimanium*, but instead placed them under a new ichnospecies, *Grallator formosus* (Hitchcock, 1858, p. 75). Lull (1904) listed ACM-ICH 9/14 as the type specimen and type ichnospecies of the new ichnogenus *Anchisauripus*, but he used the name *A. dananus* rather than *A. sillimani*, with the latter having priority. Lull (1915, 1953) corrected this error (Olsen et al., 1998). ACM-ICH 4/6 is now established as the holotype specimen for *Anchisauripus sillimani*. In addition, Lull (1904, 1915, 1953) used ACM-ICH 31/73 as the type for *O. tuberosus*, but the specimen he figured is clearly ACM-ICH 31/72 (Olsen et al., 1998). In addition, ACM-ICH 39/1 was later determined to be the holotype of *O. tuberosus* (Olsen et al., 1998).

Anchisauripus parallelus is widely misused and is a senior synonym of *Grallator parallelus*. Lull (1904, 1915, 1953) specified ACM-ICH 54/8 as the holotype for *Anchisauripus parallelus* and subsumed the ichnospecies, *G. parallelus*, into *A. parallelus* (Olsen et al., 1998).

Anchisauripus has not been commonly used for Early Jurassic tracks in the southwestern United States, in contrast to its wide application in eastern North America. Lockley (2009) indicated that few workers have used the intermediate" ichnotaxon, *Anchisauripus* is now rarely mentioned, and is considered by many to be a synonym of *Grallator*. More broadly, few workers have followed Olsen et al. (1998) in applying *Anchisauripus* as a distinct ichnotaxon from *Grallator*.

Ancyropus nomen dubium
Hitchcock 1858
Fig. 6.7

Discussion. Avanzini et al. (2005) mention *Ancyropus* from the Lower Jurassic Portland Formation of Connecticut (Fig. 6.7), stating that they are not turtle tracks as suggested by Hitchcock (1858), but could be amphibian footprints instead. However, this ichnogenus and its two ichnospecies, *A. heteroclitus* and *A. jacksonii*, fall under Hitchcock's Leptodactyli, therefore thin-toed penetrative footprints with collapsed substrate that are likely extramorphological variants and display no identifiable characteristics (Rainforth, 2005a; Gatesy and Falkingham, 2020). *Ancyropus* likely is a nomen dubium.

Anomoepus
Hitchcock 1848
Fig. 6.8

Type ichnospecies. *A. scambus.*

Valid ichnospecies. *A. scambus.*

Distribution. Lower Jurassic of eastern North America, including rift valley deposits of the Newark Supergroup within the Deerfield (Hettangian-Turners Falls and Hettangian–Sinemurian Mount Toby formations), Hartford (Hettangian Portland Formation), and Newark (Hettangian Feltville and Towaco and Hettangian–Sinemurian Boonton formations) basins of the eastern United States, and the Fundy Basin of Nova Scotia, Canada (Hettangian–Sinemurian McCoy Brook Formation); western North America (Moenave [Hettangian], Kayenta [Sinemurian], and Navajo Sandstone [Sinemurian–Pliensbachian]

FIGURE 6.7 *Ancyropus heteroclitus* from the Lower Jurassic Portland Formation, Connecticut, USA. Scale = ~1 cm. *Redrawn from Hitchcock (1841), pl. 30.*

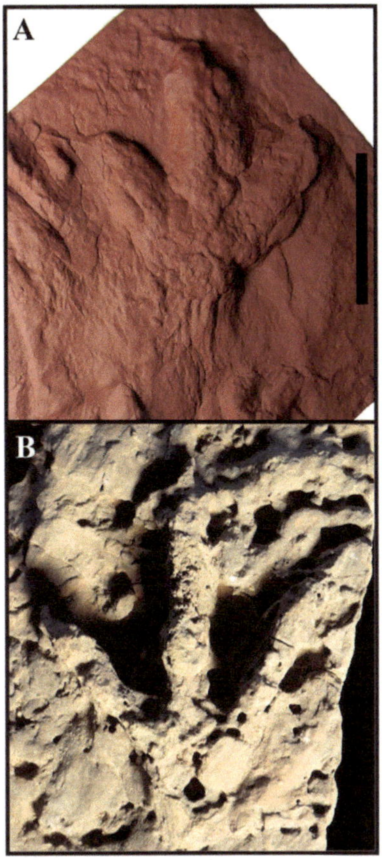

FIGURE 6.8 *Anomoepus*. (A) *Anomoepus* from Lower Jurassic Kayenta–Navajo transition, Red Hills Desert Garden, St. George, Utah (SGDS 1296; photo by Andrew Milner). Scale = 5 cm. (B) *Anomoepus-like track* SHN.(JJS).ICNO.56 from the Upper Jurassic of Portugal, classified as *Anomoepus*? isp. in Castanera et al. (2020). Track length 13.7 cm.

formations of the Glen Canyon Group). Lower Jurassic of Europe (Dolomitic Formation: France; Zagaje, Skloby, and Przysucha Ore-Bearing formations: Poland). Lower Jurassic of southern Africa (upper Elliot Formation in the Karoo Basin: Lesotho and South Africa). Lower Jurassic of South America (Marifil Volcanic Complex: Argentina). Upper Jurassic of Europe (Lastres Formation: Spain) and North America (Morrison Formation: Utah and Colorado). *Anomoepus*-like tracks have also been reported from the Upper Jurassic of Europe (Alcobaça and Lourinhã formations: Portugal).

Type specimen(s). ACM-ICH 34/40 (old number 140), lectotype of *Anomoepus scambus* is a natural mold with numerous metatarsal and overlapping manus impressions.

Diagnosis. Adapted from Olsen and Rainforth (2003): Small with pes <20 cm long, mostly bipedal and tetradactyl, although functional tridactyl. Metatarsophalangeal pad of Digit IV of pes positioned nearly, or directly in line with the axis of pedal Digit III in walking traces combined with pentadactyl manus. Pedal Digit I trace relatively long in walking tracks. Proportionally shorter pedal Digit III than in *Grallator* and other, contemporaneous theropod ichnotaxa. Manual Digit III trace is the longest, with all other digits decreasing in size symmetrically away from Digit II trace. They can appear subequal in poorer quality manus traces. Trackways can be bipedal and/or quadrupedal, and crouching traces are less common. Tail traces often present with crouching traces. Differs from Late Triassic *Atreipus* in the position of Digit IV metatarsophalangeal pad trace, and often missing the impression of the Digit II metatarsophalangeal pad, except when sitting. Consistently have wider divarication angles between digit traces II and IV than *Grallator* of similar size. Relatively larger manus traces with Digits I and V commonly registered.

Description. *Anomoepus* has a pentadactyl manus (when present) and a tetradactyl pes, but often the pes only produced tridactyl footprints (Fig. 6.8). Skin impressions are known from some specimens (Olsen and Rainforth, 2003; Lockley et al., 2018a,b).

Discussion. *Anomoepus* was produced by relatively small, gracile ornithischian dinosaurs (Olsen and Rainforth, 2003) that walked bipedally and quadrupedally. Tracks are recorded from marginal lacustrine and fluvial depositional paleoenvironments. The ichnotaxonomy of *Anomoepus*, like that of many other classic Connecticut Valley ichnotaxa first reported by Hitchcock and others, remains confusing to this day (Olsen et al., 1998; Olsen and Rainforth, 2003). Other proposed ichnospecies include: *Anomoepus crassus*, *Anomoepus cuneatus*, *Anomoepus curvatus*, *Anomoepus isodactylus*, *Anomoepus lacertoideus*, *Anomoepus major* and *Anomoepus minimus* (from the Newark Supergroup [Hitchcock, 1858; Dalman and Weems, 2013]), *Anomoepus intermedius* (from about 10 km west of Hailiutu, Wulatzhongqi, Nei Mongol, China [Li et al., 2010]), and *Anomoepus moabensis* (from Utah, Lockley and Gierliński, 2006). Extensive confusion still exists over homonyms and objective and subjective synonyms, although Olsen and Rainforth (2003) managed to sort out some of these problems.

Sauropus barrettii has been synonymized under *Anomoepus* (Olsen and Baird, 1986). Lockley and Hunt (1994) declared *Anomoepus* a senior subjective synonym of *Hopiichnus*. Lockley and Gierliński (2006) followed by declaring that *Hopiichnus shingi* is likely synonymous with *Anomoepus* and created the new combination *A. shingi* (with holotypes UCMP 83761 and UCMP 85387). *Anomoepus moabensis* is likely synonymous with "*Hopiichnus*" *shingi* (UCMP 83760; holotype cast of a trackway consisting of three footprints). See also Welles (1971), Irby (1993), and Farlow et al. (2018).

Anomoepus-like tracks have been discovered in Early Jurassic deposits worldwide (Díaz-Martínez et al., 2017 and references therein). Similar tracks also have been described in Late Jurassic formations in Spain, Portugal (Castanera et al., 2020), and the USA (Foster and Lockley, 2006; Lockley et al., 2008b, 2018; Piñuela Suárez, 2015). However, it is worth noting that some of the Late Jurassic specimens in Portugal "do not comply with one of the main features of the diagnosis," specifically that "pedal Digits II, III, and IV tend to be subequal in length" (Castanera et al., 2020b: 405). The oldest known occurrence of *Anomoepus* tracks in the American Southwest is found in association with abundant *Batrachopus* and *Grallator*, as well as a single in situ *Eubrontes*. These tracks are located within the post-End Triassic Extinction (ETE) portion of the upper Dinosaur Canyon Member (Rhaetian—Early Hettangian) of the Moenave Formation, Glen Canyon Group, at Olsen Canyon in southwestern Utah (Milner et al., 2011; Suarez et al., 2017). In St. George, in slightly younger, marginal lacustrine deposits at the SGDS, *Anomoepus* tracks are found on various stratigraphic horizons in the lower Whitmore Point Member of the Moenave Formation (Milner et al., 2006a, 2011; Lockley and Gierliński, 2006; Harris and Milner, 2015). Lockley and Gierliński (2006) considered these SGDS tracks to be the oldest prior to the discovery of the Olsen Canyon specimens.

Batrachopus
Hitchcock 1845
Fig. 6.9
Type ichnospecies. *B. deweyi.*
Valid ichnospecies. *B. deweyi, B. parvulus, B. dispar, B. bellus, B. gracilis, B. gilberti.*
Distribution. Upper Triassic–Lower Jurassic of North America (Newark Supergroup: eastern United States and Nova Scotia, Canada; upper Chinle Formation: Utah; Upper Triassic–Lower Jurassic Glen Canyon Group including the Moenave Formation (Upper Triassic and Lower Jurassic occurrences), Wingate Sandstone (Upper Triassic–Lower Jurassic occurrences), Kayenta Formation, and Navajo Sandstone of the southwestern United States).

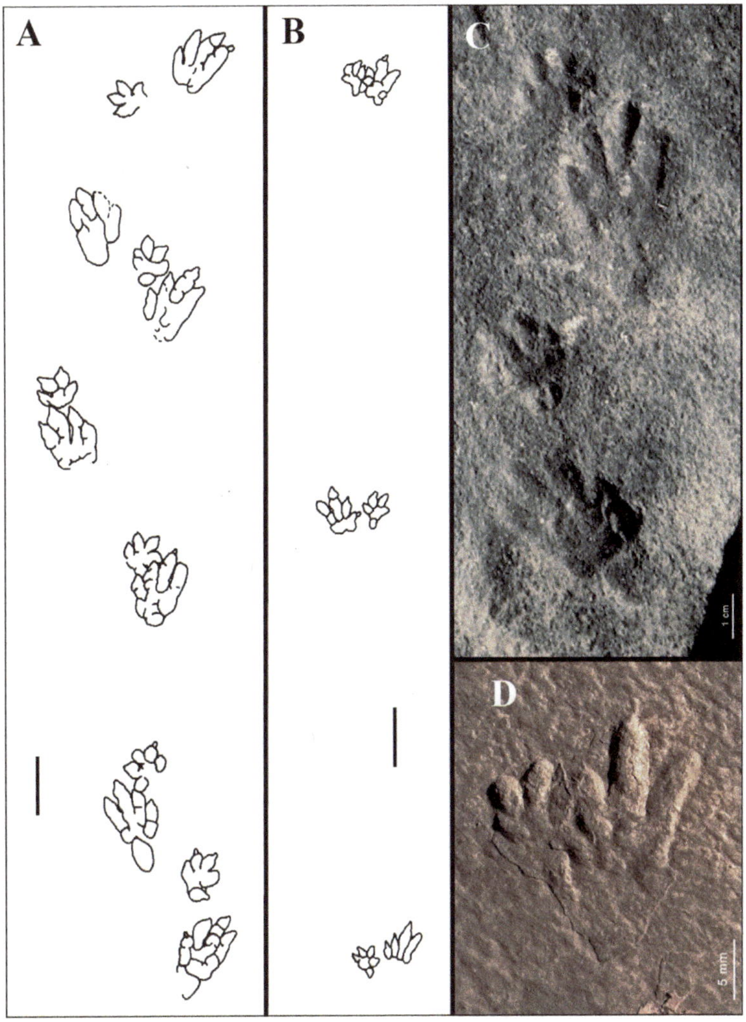

FIGURE 6.9 *Batrachopus* trackways. (A, C) *B. deweyi* neotype trackway (ACM-ICH 26/5) sketch on left, and (B, D) *B. "gracilis"* type (ACM-ICH 42/3) on right. Scale: A-B = 3 cm; C = 1 cm; D = 5 mm. *(B) Modified from Olsen and Padian (1986). (C and D) Courtesy Paul Olsen.*

Lower Jurassic of Europe (Dolomie Cubique Formation: France; Doman Member, Steierdorf Formation, Romania; Coimbra Formation: Portugal). Lower Jurassic of southern Africa (Stormberg Group: Karoo Basin of Lesotho and South Africa) and Middle?–Upper Jurassic of Morocco. Lower Jurassic of Asia (unnamed formation: Nei Mongol, China, and Fengjiahe Formation: Yunnan, China). Lower Cretaceous (Jinju Formation: South Korea), Lower Jurassic of South America (Saldaña Formation: Colombia).

Type Specimens. ACM-ICH 26/5 and ACM-ICH 26/6 (part and counterpart specimens) are the neotypes for *Batrachopus deweyi* Hitchcock 1845. These specimens are from an unknown locality, but Olsen and Padian (1986) indicated that the matrix resembled that from the East Berlin Formation, Hartford Basin, at Mount Tom, Massachusetts. ACM-ICH 20/4 is the holotype for *Batrachopus parvulus*, which was collected from sidewalks of Middletown (Hitchcock, 1865) or Middlefield (Lull, 1915, 1953), Connecticut, by Joseph Barratt around 1835 (Olsen and Padian, 1986). The lithology of this specimen resembles that of the Portland Formation sandstone, probably from flagstones from the old Portland Quarry. ACM-ICH 21/7 is the holotype of *Batrachopus dispar* Lull 1904 from the Turners Falls Formation, Lily Pond, Gill, Massachusetts (from Olsen and Padian, 1986).

Diagnosis. Small, quadrupedal archosaur tracks. Manus pentadactyl and rotated so that Digit II trace points forward, Digit IV trace points laterally, and Digit V points caudally. Pes is functionally tetradactyl and digitigrade, occasionally with impressions of Digit V which is reduced to an oval-shaped pad posterior to, and almost in line with Digit III trace. Digit III trace is the longest and Digit I is the shortest. Pes length ranges between 2 and 8 cm (from Olsen and Padian, 1986).

Description. Manus tracks are pentadactyl, although they can be tridactyl and tetradactyl, and they can display a variety of digit orientations (Fig. 6.9). Digit traces generally short and wide and sometimes display claw traces. Some *Batrachopus* examples from the Moenave Formation of southwestern Utah display skin impressions. Pes tracks can be ectaxonic and tetradactyl, but *Batrachopus* can show a variety of morphologies and extramorphological variations, which has resulted in several dubious ichnotaxa being named and misidentified. A major ichnotaxonomic overhaul of this ichnogenus is needed.

Discussion. *Batrachopus* is an abundant, Lower Jurassic ichnotaxon named originally based on specimens from the Newark Supergroup of New England in the eastern United States (Hitchcock, 1845). It is also becoming more abundant in the Glen Canyon Group of the American Southwest (Olsen and Padian, 1986; Lockley et al., 2004c; Milner et al., 2006a, 2011, 2023b) and from sites in Europe (de Lapparent and Montenat, 1967; Popa, 2000; Lockley and Meyer, 2000, 2004; Moreau et al., 2019) and the Middle–Late Jurassic of Morocco (Masrour et al., 2020).

Olsen and Padian (1986) synonymized *B. bellus* (ACH-ICH 26/21; Hitchcock, 1858), *B. gracilior* (ACM-ICH 46/3; Hitchcock 1865), and *B. gracilis* (ACM-ICH 42/3; Hitchcock, 1848) with the type ichnospecies, *B. deweyi*, but Rainforth (2005a) rejected these synonyms, arguing that *Batrachopus* has both gracile and robust forms (Rainforth, 2007). Gracile forms have longer, more slender digit traces that do not contact one another, while robust forms have shorter, wider digit traces that do have contact each other along their length. However, Rainforth (2007) synonymized *B. parvulus* with *B. deweyi*.

Batrachopus has many synonyms, or at least ichnotaxa placed in Batrachopodidae (Lull, 1904). For example, both *Comptichnus obesus* (ACM-ICH 55/5) and *Shepardia dehiscens*

(ACM-ICH 51/18) were synonymized with *B. bellus* by Rainforth (2007). Many *Batrachopus* ichnospecies are based on extramorphological variants (Lockley and Meyer, 2004). Haubold (1986) considered *Batrachopus* to be the only nondinosaurian ichnotaxon from the Lower Jurassic. Also, see discussion about *Antipus* from Europe in Lockley and Meyer (2004).

Batrachopus isp. is also known from about 10 km west of Hailiutu, Wulatzhongqi, Nei Mongol, China (Li et al., 2010).

Batrachopus tracks from Europe have been recorded in France (de Lapparent and Montenat, 1967; Moreau et al., 2019), and *Batrachopus gilberti* was an ichnospecies named by de Lapparent and Montenat (1967) from the Le Veillon sites in France. The ichnotaxon is known from the Lower Jurassic (Hettangian–Sinemurian) basal Valea Tereziei Member, Steierdorf Formation in the Anina area, Resita Basin, southern Carpathians, Romania (Popa, 2000). Note that Popa and Kedzior (2006) claim that the *Batrachopus* tracks described by Popa (2000) are from the lower Hettangian Doman Member, and not the overlying Valea Tereziei Member as originally indicated. These Romanian tracks (specimen P40/C2/22) were identified as *B.* cf. *deweyi* and described as a left manus track, a right pes track, and a tail trace alongside leaf impressions of *Podozamites paucinervis*. We agree with the identification as *Batrachopus*, however, photographs and illustrations by Popa (2000, Fig. 2, pl. 1) clearly show a partial trackway with the left track actually being a manus-pes set.

Batrachopus tracks occur in the Saldaña Formation near the town of Prado, Tolima, Colombia, which is suggested to be Upper Triassic–Lower Jurassic in age based on Late Triassic ammonites found in the lower part of the formation (Mojica and Macia, 1987).

Batrachopus isp. have been described from the Middle?–Upper Jurassic Imilchil megatracksite from the central High Atlas Mountains of Morocco (Masrour et al., 2020).

Behavioral variations in several in situ trackways from the lower Whitmore Point Member of the Moenave Formation (lower Hettangian) at SGDS in southwestern Utah display *Batrachopus* tracks transitioning from walk to swim in a marginal lacustrine setting (Milner et al., 2006b).

Selenichnus falcatus Hitchcock (1858) (types = ACM-ICH 42/6 and ACM-ICH 42/7) was originally attributed to small, bipedal theropod dinosaurs, but it is now regarded as crocodylomorph and synonymous with *Batrachopus* (Lockley et al., 2004c). It is from Turners Falls, Massachusetts (Hitchcock, 1858; Lull, 1953); Two specimens, SGDS 175 and SGDS 208, are from the Top Surface, Johnson Farm Sandstone Bed, lower Whitmore Point Member, Moenave Formation, Glen Canyon Group, SGDS, St. George, Washington County, Utah. Lower Jurassic, early Hettangian (Lockley et al., 2004c). As mentioned above, Lockley et al. (2004c) suggested that *Selenichnus* is likely an extramorphological variation of *Batrachopus* based on specimens from the Moenave Formation, but Rainforth (2007) rejected this.

Bipedopus
See *Brasilichnium* (also see Fig. 6.10).

Boutakioutichnium
Nouri, Díaz-Martínez, and Pérez-Lorente 2011
Fig. 6.11
Type ichnospecies. *B. atlasicus.*

FIGURE 6.10 *Bipedopus coloradensis* UCM 179-140 replica. Scale = 5 cm. *Photo by Andrew Milner.*

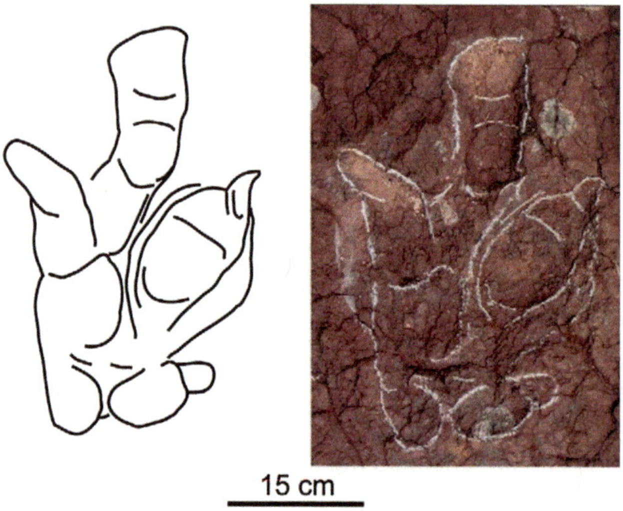

FIGURE 6.11 *Boutakioutichnium atlasicus* holotype 11IGR1-4. *Clipped from Nouri et al. (2011).*

Valid ichnospecies. *B. atlasicus.*

Distribution. Upper Jurassic of Morocco (Iouaridène Formation: Morocco).

Type specimen(s). Footprint 11IGR1.4. A plaster cast of the holotype is deposited at the Musée de Géologie d'Azilal (MGP, 1, 2011.7).

Diagnosis. Adapted from Nouri et al. (2011): Tetradactyl, digitigrade, tracks of a moderately large bipedal (theropod) dinosaur. Tracks are mesaxonic and digits terminate in pointed ends. Digit I (hallux) features two pads and the digit extends proximolaterally and appears nearly perpendicular to the axis of the foot. The first digital pad of Digit I aligns with the lateral end of Digit IV. Digit I is nearly equivalent in length to Digit II, with Digit III being the longest, and Digit II the widest. Metatarsal impressions are absent. The trackway width is narrow. The dimensions of the holotype are as follows: total footprint length is 45 cm (36 cm excluding the hallux), width is 30 cm, and the lengths of Digits I–IV are 18-22–26–23 cm respectively. The interdigital angles measure 44° for III, 25° for IIIII, and 29° for IIIIV.

Description See Fig. 6.11. Adapted from Nouri et al. (2011): Moderately large, tetradactyl tracks with pes lengths (hallux included) ranging between 38 and 48 cm, and widths between 26 and 31 cm, and a narrow (FL-FW)/FW ratio (0.1–0.5). All digits terminate in pointed margins and Digit III is the longest (20–25 cm), with Digit I (16–24 cm) longer than Digit II (18–19 cm). The toe extension (te) of the Digit III is relatively large (12–14 cm). *Boutakioutichnium* exhibits a broad Digit I, lacking a metatarsal impression, with the proximal area of the first pad positioned laterally. Divarication angle between Digits II–IV is low (33–67°), while I–II is high and variable (33–100°). In the well preserved footprints, digital pads are discernible, with Digit I often bearing the impression of two digital pads. Trackway can be described as narrow, with low pes rotation and high pace angulation (159–175°).

Discussion. In previous studies, *Boutakioutichnium atlasicus* was informally referred to as "*Eutynichnium atlasipodus.*" Only with the discovery of additional footprints, sharing characteristics of "*E. atlasipodus*" but distinct from those of the ichnogenus *Eutynichnium*, did it become necessary to revise and formally diagnose these tracks as conducted by Nouri et al. (2011). *B. atlasicus*, is known from several surfaces of the intermediate unit of the Iouaridène Formation (Upper Jurassic).

Most uniquely, *B. atlasicus* exhibits hallux features, currently unparalleled in the paleoichnological record, and prompting a discussion about the position and shape of Digit I (hallux) in theropod dinosaurs (Nouri et al., 2011). The hallux is characterized by its considerable robustness and length (17–24 cm), along with its lateral to medial orientation. Moreover, it possesses two digital pads, with the proximal area of the first digital pad positioned laterally. Importantly, *Boutakioutichnium* lacks a metatarsal traces. The presence of the hallux does not seem to be dependent on substrate conditions. Nouri et al. (2011) speculate that to leave an impression solely with the hallux (avoiding imprinting the metatarsus) it would be directed either medially or forward with the phalanges located relatively low and parallel to the ground.

The distinctive position and size of the hallux distinguish *Boutakioutichnium* from the typical osteological record of theropods. The track maker of *Boutakioutichnium* was likely a bipedal theropod dinosaur with relatively robust feet. Given the unique features of *Boutakioutichnium* and considering the known variations in the thickness and length of metatarsus I among different theropod groups, exceptions like therizinosaurids could potentially account for the observed characteristics of this ichnotaxon. Moreover, it is conceivable that

Boutakioutichnium footprints were made by a theropod species that has not yet been discovered or identified (Nouri et al., 2011).

Brasilichnium
Leonardi 1981
Fig. 6.12
Type ichnospecies. *B. elusivum.*
Valid ichnospecies. *B. elusivum, B. anaiti, B. saltatorium.*

Distribution. Lower Jurassic of the western North America (Navajo Sandstone: Utah; Aztec Sandstone: Nevada; Nugget Sandstone: Utah, Colorado, and Idaho—some of the Nugget Sandstone occurrences conceivably are Late Triassic in age), Europe (Hettangian Monte Zugna Formation: Lavini di Marco, Trento, Italy; Upper Pliensbachian, Drzewica Formation: Poland), and Africa (Pliensbachian–Toarcian Lesotho Formation: Lesotho). Also known from Upper Triassic of North America (Redonda Formation: New Mexico) and the Lower Cretaceous of Africa (Berriasian–Aptian Loia Formation: Kinshasa, Democratic Republic of the Congo; Twyfelfontein Formation: Damaraland, Namibia), Asia (Aptian Jinju Formation: Shaanxi, China; Sindong Group of Gyeongsang Supergroup: South Korea), and South America (Misiones Formation: Paraná Basin, Paraguay; ~Hauterivian–Barremian Botucatu Formation: Paraná Basin, Brazil). Upper Cretaceous of Africa (Cenomanian Kerker

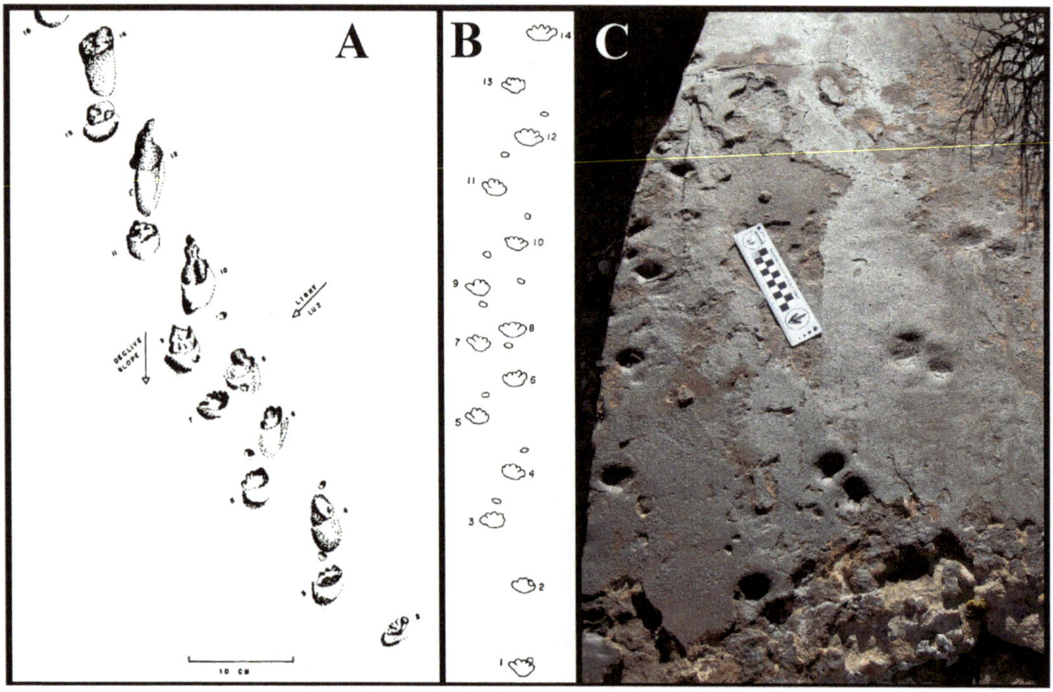

FIGURE 6.12 *Brasilichnium.* (A) *Brasilichnium elusivum* MRJ 3902-V (ARSB 78) sketch from Leonardi (1981). Scale = 10 cm. (B) Second interpretive sketch of MRJ 3902 V from Leonardi (1981), and (C) *Brasilichnium* isp. from Nugget Sandstone, Dinosaur National Monument, Utah. Scale in cm. *(C) Courtesy ReBecca Hunt–Foster.*

Member, Zebbag Formation: Jebel Boulouha West of Tataouine, Tunisia) and North America (Campanian Cerro del Pueblo Formation: El Pelillal, Coaluila, Mexico).

Type Specimens. For *B. elusivum*: MNRJ 3902-V (ARSB 78), holotype slab with 15 manus-pes sets in single trackway; MNRJ 3903-V (ARSB 127), paratype slab with single trackway with eight hind footprints; IGSP (ARSA 1), paratype slab with 13 manus-pes sets in single trackway.

Diagnosis. Quadrupedal trackway with footprints of small size, and pes frequently overprinting manus. The manus, when preserved, is located anterior to the pes and is very small in comparison. Manus-pes distances increase when speed of gait decreases. Relatively high pace angulation. Pes tracks are mediolaterally wider than anteroposteriorly long. Manus and pes are ectaxonic and tetradactyl, having short digit traces II–IV, and Digit I trace is always absent. Emended from Leonardi (1981).

Description. See Fig. 6.12. Digit traces are rounded proximally and taper distally with no distinct claw marks, but sometimes they can also be rounded distally (Leonardi, 1987). The longest digits are III–IV for *B. elusivum* (Buck et al., 2017b) or Digits II–III for *B. anaiti* (D'Orazi Porchetti et al., 2018). Tail drag traces are always absent. Tracks have rounded sole marks that are often wider than long, Sometimes *Brasilichnium* tracks have Digit II and III traces turned medially, while Digit IV and V traces curve laterally (D'Orazi Porchetti et al., 2018).

Discussion. Other *Brasilichnium* ichnospecies named are all from the Cretaceous and include *B. saltatorium* Buck, Ghilardi, Peixoto, Fernandes, and Fernandes 2017b and *B. anaiti* D'Orazi Porchetti, Bertini, and Langer 2017. *Brasilichnium* tracks have been reported from Upper Triassic to Upper Cretaceous formations globally (Lucas et al., 2010; Leonardi and de Souza Carvalho, 2020), and they have become somewhat of an ichnotaxonomic "wastebasket" due to extramorphological variation because of associated push-up rims, sediment collapse, and other deformation features as the result of the track makers walking on dune surfaces (Leonardi et al., 2007; Engelmann and Chure, 2017; Leonardi and de Souza Carvalho, 2020). *Brasilichnium*-like tracks have been reported from the Lower Triassic (Leonardi and de Souza Carvalho, 2020), but these have not been more precisely identified.

Brasilichnium is suggested to have been produced by tritylodonts and/or basal mammaliaforms (Lockley, 2011b; Leonardi and de Souza Carvalho, 2020). Lockley (2011b) placed *Brasilichnium* within the Ichnofamily Chelichnopodidae and in the *Brasilichnium* ichnofacies.

Several authors regarded *Semibipedopus* and *Bipedopus* as probable synonyms of *Brasilichnium* (Lockley and Tedrow, 2009; Lockley et al., 2011); Hunt and Lucas (2006) considered *Semibipedopus* a nomen dubium and likely synonymous with *Brasilichnium*, and Lockley (2011b) considered both *nomina dubia* that represent extramorphological variants of *Brasilichnium*.

The type ichnogenus and ichnospecies, *B. elusivum*, comes from the Lower Cretaceous (~Hauterivian–Barremian) Botucatu Formation of the Paraná Basin in Brazil (Pacheco, 1913; Huene, 1931; Leonardi, 1977, 1979, 1980, 1994; Leonardi and de Souza Carvalho, 2021). In addition, the first tetrapod tracks from Paraguay come from "paved flagstone squares" of the Lower Cretaceous Misiones Formation (a correlate of the Botucatu Formation) of the Paraná Basin (Leonardi, 1992, 1994; Leonardi et al., 2007; Leonardi and de Souza Carvalho, 2020). Other Cretaceous occurrences include the Lower Cretaceous (Berriasian–Aptian) Loia Formation, Kinshasa in the Democratic Republic of the Congo (Leonardi and de Souza Carvalho, 2020), the Lower Cretaceous Twyfelfontein Formation, Damaraland,

Namibia (D'Orazi Porchetti and Wagensommer, 2015), the Lower Cretaceous of Shaanxi, China (Xing et al., 2018b), the Aptian Jinju Formation, Sindong Group of Gyeongsang Supergroup, South Korea (Kim et al., 2017), the Upper Cretaceous (Campanian) Cerro del Pueblo Formation of El Pelillal, Coaluila, Mexico (Rodríguez-de la Rosa, 2003; Rodríguez-de la Rosa et al., 2004), and the Upper Cretaceous (Cenomanian) Kerker Member of the Zebbag Formation, Jebel Boulouha, west of Tataouine, Tunisia (Contessi, 2013).

Lower Jurassic occurrences of *Brasilichnium* are locally abundant in the eolian facies of the Navajo Sandstone of Utah (Stokes, 1978; Lockley, 2011b; Lockley et al., 1998f, 2014; Milner et al., 2011, 2023a,b), the Aztec Sandstone of Nevada (Rowland and Mercadante, 2014), and the Nugget Sandstone in Utah (Stokes, 1978; Engelmann, 2010, Chure et al., 2014; Engelmann and Chure, 2017), Colorado (Engelmann, 2010), and southern Idaho (Lockley et al., 2011). In Europe, a single footprint was reported from the Lower Jurassic (Hettangian) Monte Zugna Formation of Lavini di Marco, Trento, Italy (Piubelli, 2002; Petti et al., 2020), and cf. *Brasilichnium* was reported from the Holy Cross Mountains of Poland (Niedźwiedzki and Remin, 2008; Pacyna et al., 2022). *Brasilichnium*-like tracks also have been reported from the Lower Triassic Tumlin Sandstone of Poland (Sadlok, 2019) and the Lower Jurassic (Pliensbachian–Toarcian) Lesotho Formation of Lesotho (Bordy et al., 2020).

Breviparopus

Dutuit and Ouazzou 1980

Fig. 6.13

Type ichnospecies. *B. taghbaloutensis.*

Valid ichnospecies. *B. taghbaloutensis.*

Distribution. Middle Jurassic of Europe (Lealt Shale Formation: Scotland and Saltwick Formation: England). Upper Jurassic of Africa (Iouaridène Formation: Morocco).

Type specimen(s). In situ. No official labeling was provided by Dutuit and Ouazzou (1980); labeled "Trackway Bre" in Ishigaki and Matsumoto (2009a,b), trackway "4:00 A.m.1" in Nouri (2007), or "Deio-D" in Belvedere (2008) and Marty et al. (2010).

Diagnosis. Informal diagnosis translated and adapted from Dutuit and Ouazzou (1980):

FIGURE 6.13 *Breviparopus taghbaloutensis.* Deio D, *Breviparopus* type trackway, in October 2007. Tracks shown are from Deio D/15 to D/19 in the labeling of Belvedere (2008), corresponding to Brev. 16 to 20 in Ishigaki and Matsumoto (2009a). *Photo by Matteo Belvedere; also see Belvedere (2008).*

Narrow-gauge trackway with pes tracks intersecting the trackway midline and manus tracks located farther away from the trackway midline than pes tracks; manus tracks located in front of pes tracks and are (slightly) overprinted by the pes tracks; manus tracks have a slightly semicircular shape without evidence for digit impressions; pes tracks are oval shape with at least four digit impressions separated by three small interdigital displacement rims, and evidence for claw marks within the overall displacement rim; pronounced heteropody on the order of 1:3.6.

Description. See Fig. 6.13. The reference trackway (following the labeling of Belvedere, 2008) is a narrow-gauge (PTR: 50.2%, WAP/PL: 0.9; Marty et al., 2010), quadrupedal trackway composed of 33 tracks. The trackway is ~86 m long, including, however, a 22 m-long gap between tracks D/20 and D/21. After this long gap, the morphological preservation deteriorates, and more prints are missing (between D/22 and D/23, and between D/29 and D/30). Trackway direction changes twice along the trackway from 262°N to 250°N and then back to 269°N, while the stride length remains relatively constant.

The pes tracks are outwardly rotated and different between left and right tracks (15° and 30°, respectively), though no notable changes in rotation occur near the trackway turns. Pes tracks are deeper (up to twice as deep) than the manus tracks (Dutuit and Ouazzou, 1980). Manus tracks are outwardly rotated and slightly external to the pes. All manus tracks are at least partially overprinted, therefore the heteropody is quite marked and varies from 1:3 to 1:4.5 depending on the degree of overprinting (1:3.6 in Dutuit and Ouazzou, 1980, 1:4.5 in Nouri, 2007). However, for undeformed manus tracks not included in the type material, heteropody is suggested to be rather on the order of 1:3 (Marty et al., 2010). Pes tracks, as measured in Marty et al. (2010), are longer (110.7 cm on average) than wide (88.5 cm on average), the shape varies from elliptical to bell shaped; manus tracks are always wider (55.4 cm on average) than long (29.7 cm on average). Claw or digit impressions are absent on manus tracks and occasionally preserved in the pes tracks (Dutuit and Ouazzou, 1980; Ishigaki, 1985, 1989), although this has been questioned (Belvedere, 2008; Marty et al., 2010; Meyer and Monbaron, 2002).

Discussion. The ichnospecies has never been formally amended, although it is referenced in several papers as a trackway. A key unresolved matter revolves around the debate over the presence or absence of digit impressions, posing a point of contention among different authors. Notably, Ishigaki (1985) asserted the existence of digit impressions in some pes tracks, while Marty et al. (2010) argued the contrary.

Breviparopus taghbaloutensis has only been identified in the Upper Jurassic–?Lower Cretaceous Iouaridène Formation (Charriére et al., 2005) near the type locality. Tracks from the Aalenian–Bathonian of Yorkshire, UK, were assigned to *Breviparopus* isp. (Romano and Whyte, 2012); other large sauropod tracks from the Bathonian of Scotland were tentatively assigned to *Breviparopus* (dePolo et al., 2018), and *Breviparopus*-like tracks have been reported from Yorkshire (Romano et al., 1999; Romano and Whyte, 2003). Tracks from the Cerrito del Bote and El Chocón (Albian–Cenomanian) of Argentina were affiliated with *Breviparopus* isp., but later assimilated into *Sauropodichnus giganteus* (Calvo, 1991, 1999; Wright, 2005). The latter ichnotaxon was deemed a nomen dubium by Lockley et al. (1994a), a classification upheld by Salisbury et al. (2016) and Krapovickas et al. (2020).

Currently, no sauropod track maker group has been identified for *Breviparopus*. Traditionally, *Breviparopus* has been associated with diplodocoids based on their narrow-gauge trackways (Wright, 2005; see Marty et al., 2010). Body fossils found in North Africa within the study region range from the sauropod *Atlasaurus imelakei* (Middle Jurassic) to the basal rebbachisaurid *Rebbachisaurus garasbae* (Early Cretaceous; Belvedere, 2008; Marty et al., 2010), although none matches the size of the *Breviparopus* track maker.

Brontopodus
Farlow, Pittman, and Hawthorne 1989
Fig. 6.14
Type ichnospecies. *B. birdi* Farlow et al., 1989.
Valid ichnospecies. *B. plagnensis* (Upper Jurassic, France), *B. birdi* (Lower Cretaceous, USA), *B. pentadactylus* (Cretaceous, Korea), *B. changlingensis* (Upper Cretaceous, China).
Distribution. Middle Jurassic of North America (Entrada Sandstone, *Brontopodus* isp.: Utah); Upper Jurassic of North America (Morrison Formation, *Brontopodus* isp.: Utah and

FIGURE 6.14 *Brontopodus* isp. from the upper Entrada Sandstone at the Twentymile Wash tracksite in southern Utah is a six-step (manus-pes) trackway with a tail drag trace. It is the oldest evidence of eusauropods in the United States. Pes lengths approximately 60 cm each. See Foster et al. (2000). *Photo by John Foster.*

Colorado) and Europe (*B. plagnensis*—Couches du Chailley and Morillon formations: France; *Brontopodus* isp., Reuchenette Formation: Switzerland; Alcobaça Formation: Portugal; Villar del Arzobispo: Spain; Espichel unit: Portugal; Błaziny Oolite Limestones unit: Poland). Lower and mid-Cretaceous of North America (Arkansas and Utah) and Asia (*B. pentadactylus*, Korea; *B.* isp., China); Upper Cretaceous of Spain. Type ichnospecies from Lower Cretaceous of North America (Glen Rose Limestone: Texas).

Type specimen(s). *B. birdi*: AMNH 3065 and TMM 40,638-1; *B. plagnensis*: in situ trackway PGN 01.

Diagnosis. Wide-gauge, vertebrate trackways exhibiting a more pronounced positive rotation of the pes compared with the manus track relative to the trackway midline. The mesaxonic, pentadactyl manus is wider than it is long, with digit imprints arranged in an arc emphasized by a posterior indentation. Notably, the manus digit imprint III is the largest, while Digits I and V are generally smaller and slightly separated from the conjoined Digits II–IV. The pes tracks are typically elliptical in shape with the digits decreasing in size from I to V, with Digit V being the smallest and occasionally oriented posteriorly. Particularly well-preserved pes tracks may exhibit a unique elongated anterolateral claw trace pointing posterolaterally from Digit IV. Proportionally, the manus print is larger than the pes print (heteropody index HI = 1.86).

Differs from *Parabrontopodus* (Lockley et al., 1994a) in having a wider trackway gauge (the pes tracks medial edges do not overlap the trackway midline as in *Parabrontopodus*), manus tracks larger relative to the pes, and a less oval manus track shape. Differs from *Breviparopus* in having pes tracks farther laterally from the trackway midline than the manus tracks, pes track long axis closer to parallel with the direction of travel, and an overall wider gauge trackway. See Farlow et al. (1989) and Mazin et al. (2017).

Description. From the general ichnogenus diagnosis of Farlow et al. (1989): Sauropod tracks can vary in size, ranging from small to large. The length of the pes footprint is known to be between 50 and over 100 cm. In well-preserved tracks, the length and width of the manus footprint are about the same. The manus tracks are clawless and somewhat U-shaped, with slight separation between digit impressions I and V from the impression of conjoined Digits II–IV.

Pes tracks are longer than they are broad, and they have large, laterally directed claw marks at Digits I–III. These claw marks decrease in size from Digit I to Digit III. There is a small mark, either a claw, nail, or callosity, at Digit IV, and a small callosity or pad mark at Digit V. The marks at Digits IV and V are only visible in well-preserved footprints.

Manus tracks are often rotated outward in relation to the direction of travel. The manus track is located medially to a line through the long axis of the pes track, meaning that the centers of the manus tracks are somewhat closer to the trackway midline than the centers of the pes tracks.

Trackway is broad, with the left and right manus and pes footprints often positioned far away from the trackway midline. The trackway midline is approximately 1–1.5 times the length of the pes track. The outer boundaries of the trackway are defined by the pes tracks.

The distance between the manus and pes footprints is 0.5–1.2 times the length of the pes footprint. The stride length is roughly 2–5 times the length of the pes track. The step angle is generally between 100 and 120°. The glenoacetabular length is about 3–4 times the length of the pes track. Tail drag marks are rare or absent.

In Jurassic *Brontopodus* isp. specimens, the manus tracks are sometimes positioned laterally to the midline of the pes tracks, rather than medially. In addition, in at least one specimen found in the Entrada Sandstone of Utah, a tail drag trace is present (Foster et al., 2000; see Fig. 6.14).

Discussion. The ichnogeneric type material of *Brontopodus birdi* is from the Lower Cretaceous Glen Rose Limestone of Texas (Farlow et al., 1989). Similar tracks are also known from the same age in Arkansas and Asia (Lockley et al., 2002), and even in slightly younger strata in Utah (Lockley et al., 2014d). However, the majority of the occurrences of *Brontopodus* isp. are in the Early Cretaceous of North America, China, and Korea. Although *Brontopodus* is generally considered a wide-gauge sauropod trackway (Farlow, 1992), there is variation in the gauge. Some sauropod trackways are considerably wider (e.g., Middle Jurassic "*Polyonyx*").

The Upper Jurassic *Brontopodus plagnensis* from France (Mazin et al., 2017) shares characteristics with the three Cretaceous *Brontopodus* ichnospecies. These characteristics include the very similar pentadactyl clawless manus prints, the elongate pes prints, and trackway parameters. However, *B. plagnensis* can be clearly differentiated by the clawless pedal Digits I–III and the posterolateral clawlike mark on Digit IV. The absence of pes claw marks, particularly on the first three digits (which is diagnostic for *Brontopodus birdi*), may be questionable because claws could have been present but not recorded either due to their elevated position or due to the substrate competence. A careful observation of all the pes prints of *B. plagnensis* (PGN 01) leads to the conclusion that claw traces on the pes tracks are absent. This is also reflected in the manus tracks, which are also considered clawless.

Brontopodus has been reported from the Upper Jurassic Morrison Formation at the Hidden Canyon Overlook site (three parallel trackways; Barnes and Lockley, 1994) and the Copper Ridge (Valley City) site in Utah (Lockley and Hunt, 1995), and the Purgatoire River site in Colorado (Foster and Lockley, 2006). *Brontopodus*-like tracks have also been identified in the Upper Jurassic of Europe in the Reuchenette Formation, Switzerland (Marty et al., 2010; Meyer and Thüring, 2003; Sciscio et al., 2022), the Alcobaça Formation, Portugal (Meyer et al., 1994; Meyer and Pittman, 1994; Neto de Carvalho, 2016; Fürsich et al., 2022), the Pedra da Mua tracksite in Portugal (Lockley et al., 1994b), and the Cañada Paris, Las Cerradicas, and Miraflores I tracksites in Spain (Santisteban et al., 2009; Castanera et al., 2014). They have also been reported at the Błaziny site in Poland (Gierliński and Niedźwiedzki, 2002a).

A *Brontopodus* trackway with a tail drag trace was also reported from the Middle Jurassic Entrada Sandstone of southern Utah (Fig. 6.14; Foster et al., 2000). These Middle Jurassic tracks, along with similar but unattributed tracks from Oaxaca, Mexico (Ferrusquía-Villafranca et al., 1996), constitute the oldest records of eusauropod dinosaurs in North America.

Track makers of Jurassic *Brontopodus* appear to be large sauropod dinosaurs (Farlow et al., 1989), most likely macronarians (Lockley and Hunt, 1995) such as camarasaurids or titanosauriforms (Wright, 2005; Mazin et al., 2017).

Camptosauropus nomen dubium
Gabuniya and Kurbatov 1988
Fig. 6.15
Type ichnospecies. *C. vialovi.*
Valid ichnospecies. None.
Distribution. Middle Jurassic of Asia (Gabirut and/or Dzhirut formations: Tajikistan).

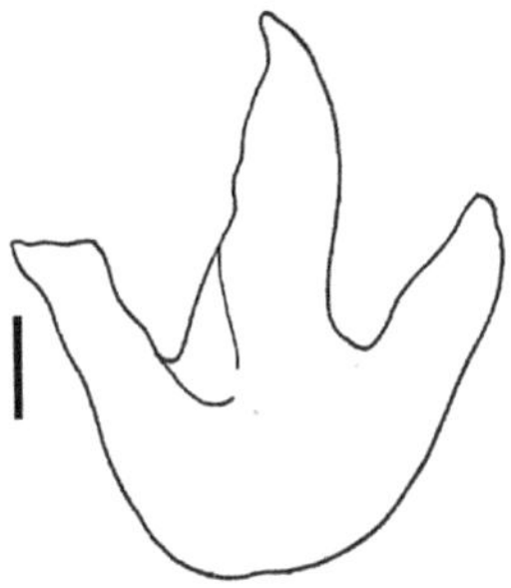

FIGURE 6.15 *Camptosauropus*. Scale = 6 cm. *From Gabuniya and Kurbatov (1988).*

Type specimen(s). Gabuniya and Kurbatov (1988) listed the holotype specimen as Ministry of Geology of the Academy of Sciences of the USSR No. 9, a right pes print. This ministry was apparently abolished in 1991, and its materials transferred to the Ministry of Natural Resources and Ecology of the Russian Federation; the current status and catalog number of the holotype is unknown. A cast of the holotype was reportedly reposited in the Museum of the Ministry of Geology of the Uzbek SSR (possibly now the Museum of Geology in Tashkent, Uzbekistan?), but no catalog number was provided. The holotype appears to be a natural mold.

Diagnosis. Medium-sized, tridactyl tracks of a bipedal dinosaur. Pes is wide (only slightly less than the track length) with a slightly separated and rounded "heel;" divarication angle between Digits II–IV is about 70°. The distal ends of the digit imprints generally are rounded, occasionally with a small, pointed claw imprint.

Description. See Fig. 6.15. Gabuniya and Kurbatov (1988) do not provide much description beyond the diagnosis. They described *Camptosauropus* tracks as "ornithine (bird)-like." Digit III is noted as being markedly longer than Digits II and IV, which is atypical of ornithopod dinosaur tracks. Gabuniya and Kurbatov (1988) specify the divarication angle between Digits II–IV as 70° in the diagnosis, but in the description mention that the angle between Digits II–III is 27° and between Digits III–IV is 25°, for a total of 52°. They mention several specimens present at the Ravat site, so one value set may pertain to the holotype and the other may be an average among tracks, but which is which is unclear. They further mention a variety of sizes of tracks of this morphology present at the site, which they attribute to various ontogenetic stages of the track maker(s).

Discussion. Romanovsky (1882) first mentioned dinosaur tracks in the "Sarvadan Lignite Formation" of the Ravat region of Tajikistan; however, no further references to this formation could be found. Gabuniya and Kurbatov (1988) cited the tracks as occurring in the "undivided Gabirut and Dzhirut formations." Rozhdestvensky (1964) cited the age of the strata in the area as being "within the limits of the [Early] Jurassic and the beginning of the Middle Jurassic," but Gabuniya and Kurbatov (1988) cited paleobotanical evidence that the strata were firmly Middle Jurassic in age. The initial track discoveries in the region, apparently those of theropod dinosaurs (Romanovsky, 1882; Rozhdestvensky, 1964), were destroyed during the construction of a highway, but more tracks were subsequently discovered nearby (Gabuniya and Kurbatov, 1988).

Camptosauropus prints were attributed to ornithopod dinosaurs, specifically to members of the "Camptosauridae." However, no such large-bodied ornithopod taxa are known from body fossils globally prior to the Late Jurassic. Nevertheless, a few pre-Late Jurassic tracks have been attributed to ornithopods, previously including *Wealdenichnites*, which is in fact Early Cretaceous (Kuhn, 1958a; see Milner et al., 2023b for a further review). Both larger (attributed to adults) and smaller (attributed to juveniles) tracks were reported at the type locality (Gabuniya and Kurbatov, 1988). *Camptosauropus* was considered a nomen dubium by Díaz-Martínez et al. (2015) due to its insufficient diagnosis and poorly figured holotype.

Camptosauropus tracks cooccur with *Coelurosaurichnus*, *Gabirutosaurichnus*, and *Ravatichnus* tracks (q.v. below).

Carmelopodus
Lockley, Hunt, Paquette, Bilbey, and Hamblin 1998
Fig. 6.16

Type ichnospecies. *C. untermannorum.*

Valid ichnospecies. *C. untermannorum.*

Distribution. Middle Jurassic of North America (Carmel Formation: Utah and Sundance Formation: Wyoming), Africa (Isli Formation: Morocco), Asia (Shaximiao Formation: Sichuan, China), and Europe (Forest Marble: England and Kościelisko Beds: Poland); Upper Jurassic of Europe (Morillon Formation: France); Lower Cretaceous of Asia (Phra Wihan Formation: Thailand). Possibly Lower Jurassic of Africa (unspecified formation: Morocco). Possibly Lower–Middle Jurassic of Africa (Mestaoua Formation: Tunisia). Possibly Middle Jurassic of Asia (Hojedk Formation [Shemshak Group]: Iran). Possibly Lower Cretaceous of North America (Cedar Mountain Formation: Utah).

FIGURE 6.16 *Carmelopodus untermannorum* (FHPR 4767) from the Middle Jurassic Carmel Formation of northeastern Utah, USA. Scale = 10 cm. *See Lockley et al. (1998a) Photo by John Foster.*

Type specimen(s). CU-MWC 184.12, replica of the first three footprints in a partial trackway of five consecutive, natural cast tracks. The fourth track in the sequence was collected and is reposited as CU-MWC 184.12a.

Diagnosis. Tridactyl, subsymmetrical tracks of a biped. Digital pad formula 2-3-3 for Digits II–IV; well-defined claw imprints are present. Tracks almost as wide as long, with wide digit divarication angles, especially between Digits II–III. Trackways narrow with the axis of Digit III parallel to the trackway axis.

Description. *Carmelopodus untermannorum* tracks characteristically lack any impression of a proximal pad on Digit IV, regardless of track size (and, presumably, the corresponding ontogenetic stage of the track maker). The distal phalangeal pad of Digit III tapers distally toward the claw imprint. Digit II is wider and shorter than Digit IV. The high length:width ratio of the tracks is a function of the high divarication angle between Digits II–III (upward of 67° and consistently 25%–75% greater than that between III–IV). See Fig. 6.16.

Discussion. Most authors have accepted that *Carmelopodus* tracks were made by small theropod dinosaurs (Lockley et al., 1998a). However, Diedrich (2011), without specifying reasons why, instead attributed the ichnotaxon to small ornithopod dinosaurs.

Tracks assigned to Carmelopodus isp. were mentioned from an unspecified, Lower Jurassic formation of Morocco by Molina-Pérez and Larramendi (2019). The holotype and associated specimens of *Carmelopodus* are from the Middle Jurassic Carmel Formation of northeastern Utah (Lockley et al., 1998a). Middle Jurassic specimens are also known from the lower Sundance Formation of Wyoming (Breithaupt et al., 2004), Forest Marble of England (Delair and Sarjeant, 1985; Lockley and Meyer, 2000), Kościelisko Beds of Poland (Gierliński and Nowacki, 2008), Shaximiao Formation of Sichuan, China (Xing et al., 2018a), and Isli Formation of Morocco (Gierliński et al., 2009b; Klein et al., 2022). Late Jurassic specimens have been reported from the Morillon Formation of France (Mazin et al., 2016) and Reuchenette Formation of Switzerland (Marty, 2008), though the latter have been likened more to *Kalohipus* (Castanera et al., 2018). The only Cretaceous instance of *Carmelopodus* isp. reported thus far is from the Phra Wihan Formation (Khorat Group) of Thailand (Kozu, 2017). Theropod tracks similar to *Carmelopodus* have also been reported from the Lower–Middle Jurassic Mestaoua Formation of Tunisia (Niedźwiedzki et al., 2017), the Middle Jurassic Hojedk Formation (Shemshak Group) of Iran (Abbassi et al., 2015), and the Lower Cretaceous Ruby Ranch Member of the Cedar Mountain Formation of Utah (Kirkland et al., 2014a).

Casamiquelichnus

Coria and Carabajal 2004

Fig. 6.17

Type ichnospecies. *C. navesorum.*

Valid ichnospecies. None.

Distribution. Middle Jurassic of South America (La Matilde Formation: Santa Cruz, Argentina).

Type specimen(s). MACN–SC–3380.3, set of seven slabs bearing natural casts of four pes tracks, associated drag impressions, and associated *Ameghinichnus* tracks.

Diagnosis. Tracks made by monodactyl or didactyl, bipedal track makers. Central digit imprint (presumably of Digit III) prominent; other digit imprints reduced to absent. Tracks associated with long, longitudinal grooves (ridges in the natural cast) oriented in the same

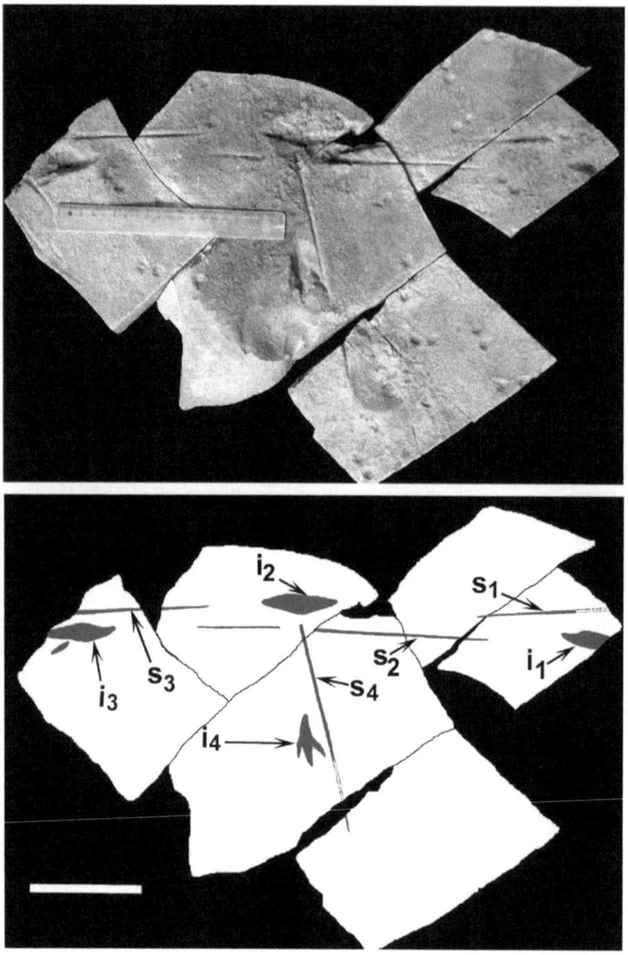

FIGURE 6.17 *Casamiquelichnus navesorum* MACN–SC–3380.3 holotype from the Upper Jurassic LaMatilde Formation, Argentina. Scale in lower = 5 cm. *From Coria and Carabajal (2004).*

direction as the tracks and lying close to and parallel to the internal side of each pes print. "Step angle" very wide.

Description. See Fig. 6.17. Coria and Carabajal (2004) contradictorily described the holotype *Casamiquelichnus* tracks as "low relief" but also "deep." They described the tracks as having clear contours that are curved near the midpoints of the tracks. No digital pads can be discerned. Despite the diagnosis of the tracks as monodactyl or didactyl, they inferred that the track maker was tridactyl based on small, shallow marks on two of the holotype prints that they interpreted as pertaining to digits lateral (or, presumably, medial) to the central digit. The associated groove impressions are shallower than their associated pes prints. Coria and Carabajal (2004) were uncertain about the origins of the grooves, discussing that they may be claw drags from a possibly hypertrophied claw on the foot opposite the one registering the associated pes print.

Discussion. Coria and Carabajal (2004) attributed *Casamiquelichnus* tracks to theropod dinosaurs. They differentiated it from the cooccurring *Sarmientichnus* by the following characteristics: its smaller size with respect to *Sarmientichnus*, the absence of a rough surface at the margins of *Sarmientichnus* tracks, and the presence of the associated, elongate, shallow grooves. de Valais (2011) synonymized *Casamiquelichnus* with *Sarmientichnus* (q.v.) because the holotype came from the same horizon that produced several *Sarmientichnus* traces, and its tracks exhibit no morphological features to distinguish them from *Sarmientichnus*. She discounted the report of possible lateral digit imprints reported on some prints by Coria and Carabajal (2004) and described the holotype prints as incomplete. She did not address the longitudinal grooves associated with the holotype *Casamiquelichnus* tracks.

Changpeipus
Young 1960
Fig. 6.18
Type ichnospecies. *C. carbonicus.*

Valid ichnospecies. *C. carbonicus* (Young, 1960), *C. longweimaoensis* (Li and Zhang, 2017) (see below).

Distribution. Lower Jurassic of Asia (Lufeng Formation: Yunnan, China); Middle Jurassic of Africa (Isli Formation: Morocco) and Asia (Sanjianfang Formation: Xinjiang, Yima

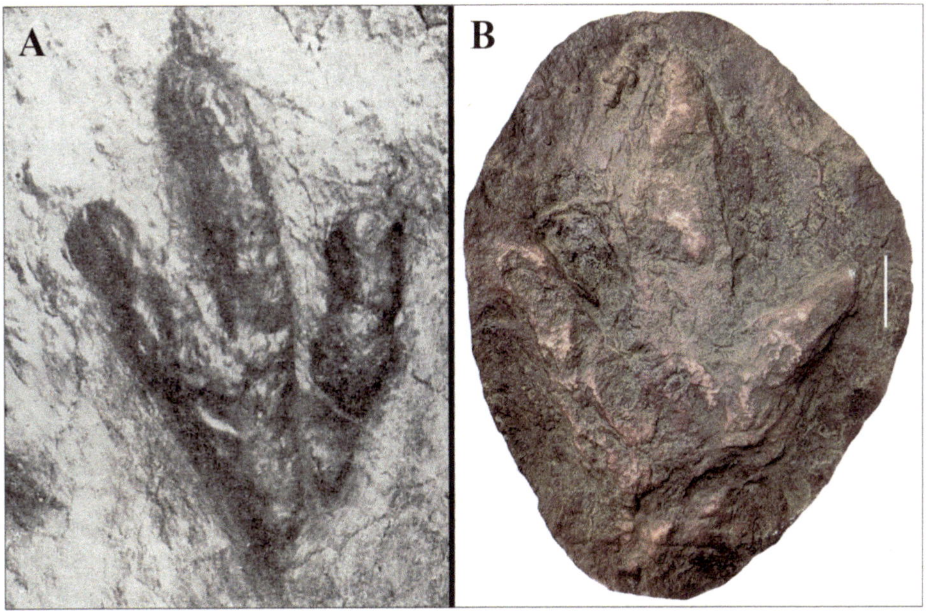

FIGURE 6.18 (A) *Changpeipus carbonicus*, IVPP V2472 holotype track 2, Liaoning, China. Track length ~27 cm. (B) *Changpeipus* cf. *carbonicus* (replica), UCM 185.16 from the Middle Jurassic Isli Formation, central high axis, Morocco; scale = 5 cm. *(A) From Young (1960). (B) Photo by Andrew Milner.*

Formation: Henan, and Yan'an Formation, Shaanxi, China); Lower Cretaceous (per Chen et al., 2006) of Asia (Fuxin Formation: Liaoning and ?Tuntainying Formation: Jilin, China). Possibly Middle–Upper Jurassic of Australia (Walloon Coal Measures).

Type specimen(s). Young (1960) confusingly reported the holotype as "Three footprints preserved in one slab, plates IV and V referred specimen: A single footprint plate VI Cat. No. V2472 and 2470, respectively." The locality information goes on to say "V2472 from Sungsankang coal mine, Huinan, Kirin [Jilin]. V2470 from the opening cutting coal mine, Hai-chou, Fusin, Liaoning." Then, in the description, he said "The footprints of Sungsankang have been discovered during the course of the operation of the coal mine ... There are seven complete footprints preserved in a [sic] area about two to three square meters and some fragmentary part of the same, thus footprints altogether as shown in Fig. 1... One of the better preserved footprints (marked in the sketch as 4) has been excavated by the Museum of Kirin [Jilin] provinces at Changchun and the other three marked as 2, 2a and 3 have been collected ... for our institute." However, in his Table 3, Young (1960) specified that his new ichnotaxon came from the "[b]ase of the Fusin coal-bearing beds Fusin, Liaoning," which appears to contradict the earlier specification of V2472 as coming from Jilin Province. Xing et al. (2009, 2014a) treated V2472, the slab containing three tracks, as the holotype. However, the fact that the holotype consists of two unassociated specimens from two different localities—V2472 from Jilin Province and V2470 from Liaoning Province—and two different stratigraphic units (see below) is problematic.

Diagnosis. From Young (1960): tridactyl, semiplantigrade tracks with a "deltoid" (triangular) general outline; Digit IV projects father anteriorly than Digit II. A small, associated track was interpreted by Young (1960) as a manus print and diagnosed as being tridactyl but with a rudimentary Digit III. The association of this print subsequently has been perceived as suspect (e.g., Zhen et al., 1989). Xing et al. (2009) added further diagnostic features: a faint claw imprint on Digit III, a metatarsophalangeal pad located essentially directly caudal to Digit III, and digit pad areas increase distally on all digits, most notably Digit III.

Description. In addition to the brief diagnosis by Young (1960), Zhen et al. (1989) noted that *Changpeipus* resembled *Grallator* except by being larger, having a well-developed heel imprint, and having greater digit divarication angles (Fig. 6.18). They similarly differentiated *Changpeipus* from *Eubrontes* by noting that it has a "more compact" heel and narrower digit imprints; the latter likely led to later comparisons between *Changpeipus* and *Kayentapus* (e.g., Lockley et al., 2013). Xing et al. (2009) described the clear heel area as triangular, though only in one preserved track, and therefore this feature is not diagnostic for the ichnotaxon. They also noted that the faint claw imprint on Digit III was likely extramorphological.

Discussion. *Changpeipus* has been consistently referred to as a theropod dinosaur ichnotaxon. The ichnogenus has a complex history, reviewed in detail by Lockley et al. (2013) and Xing et al. (2014a). It frequently has been likened to, even synonymized with, both *Grallator* (e.g., Gierliński, 1994) and *Eubrontes* (see below), and could even be a senior synonym of *Kayentapus* (Lockley et al., 2013). It also could be either synonymous with or part of a plexus of other, East Asian theropod ichnotaxa, such as *Jialingpus* and *Asianopodus* (Klein et al., 2022). Further, detailed work on all the specimens involved is required to resolve these issues.

The holotype specimens come from two different units: IVPP v2470 from the Fuxin Formation of Liaoning Province, and IVPP v2472 from the Tuntainying Formation (an equivalent of the Yixian Formation) of Jilin Province (Chen et al., 2006). Both units are Early Cretaceous in age, with the Fuxin Formation being substantially younger than the Yixian-equivalent Tuntainying Formation. Why Young (1960) referred to the source strata as Early–Middle Jurassic in age is unclear, but despite the correction by Chen et al. (2006), the Early–Middle Jurassic age attribution was repeated by Lü et al. (2007), Lockley et al. (2013), Xing et al. (2014a), and Li and Zhang (2017).

Xing et al. (2009) erected *C. pareschequier* for tracks from the Lower Jurassic Lufeng Formation of Yunnan Province; these were placed in *Eubrontes pareschequier* by Lockley et al. (2013), though no diagnostic characters of this ichnospecies that differentiate it from other *Eubrontes* ispp. were provided. *C. carbonicus* tracks have also been reported from the Middle Jurassic Sanjianfang Formation of Xinjiang (Xing et al., 2014a). Lü et al. (2007) erected *C. xuiana* for large tracks from the Middle Jurassic Yima Formation of Henan Province; these were synonymized with *C. carbonicus* by Xing et al. (2009) but listed in a table as *Eubrontes carbonicus* by Lockley et al. (2013). Whether the latter was intended to be *Eubrontes xuiana* is, however, unclear; no other mention in Lockley et al. (2013) firmly suggests that *C. carbonicus* should be sunk into *Eubrontes*. In any case, no diagnostic characters differentiating *E. carbonicus* (or *E. xuiana*) from other ichnospecies of *Eubrontes* were provided.

C. longweimaoensis was erected for tracks from the Middle Jurassic Yan'an Formation of Shaanxi, China by Li and Zhang (2017); this ichnospecies was differentiated from other *Changpeipus* ispp. by lacking a metatarsophalangeal pad imprint, having a lower average length:width ratio, having a different phalangeal pad formula, and different divarication angles between digits. *Changpeipus* cf. *carbonicus* tracks were reported from the Middle–Upper? Jurassic Isli and Imilchil formations of Morocco and cited as evidence of faunal exchange between northern Africa and East Asia during this time (Klein et al., 2022). Young (1979) erected another ichnospecies of *Changpeipus*, *C. luanpingeris*, based on tracks from an unspecified/unnamed unit in Hebei Province that also produced fossils of the fish *Lycoptera davidi*, a characteristic species of the Jehol Biota, which likely makes the source the Lower Cretaceous Yixian Formation. Young (1979) differentiated *C. luanpingeris* from *C. carbonicus* based on the former having a shorter Digit II impression. *C. luanpingeris* was synonymized with *C. carbonicus* by Xing et al. (2009) but declared a nomen dubium by Lockley et al. (2013); specimens of this ichnotaxon subsequently were placed in the new ichnogenus *Lockleypus luanpingeris* by Xing et al. (2018b). Haubold (1971a) referred to tracks from the Middle–Upper Jurassic Walloon Coal Measures of Australia as *C. bartholomaii*, but this is a nomen nudum because no holotype was established, and no formal diagnosis was provided.

Characichnos
Whyte and Romano 2001a
Fig. 6.19

Type ichnospecies. *C. tridactylus.*

Valid ichnospecies. *C. tridactylus.*

Distribution. Lower Jurassic of North America (Moenave and Kayenta formations: Utah) and Europe (Zagaje Formation: Poland); Middle Jurassic of Europe (Saltwick Formation: England); Upper Jurassic of Europe (Praia da Amoreira-Porto Novo: Portugal) and South America (San Salvador Formation: Chile); Upper Jurassic–Lower Cretaceous of Asia (Anning Formation: Yunnan, China). Possibly Middle Jurassic of Asia (Zhanghe Formation: Yunnan, China). Possibly Upper Jurassic of North America (Morrison Formation: Colorado).

FIGURE 6.19 *Characichnos.* (A) SHN.(JJS).ICNO.118 from Upper Jurassic of Portugal. Track width on left specimen = 7.5 cm. (B) From La Virgen del Campo tracksite, Lower Cretaceous Enciso Group, Spain. (C) SGDS 1447, Lower Jurassic Kayenta Formation, Washington County, Utah. (D) *Characichnos tridactylus* holotype slab from Middle Jurassic Saltwick Formation, UK. Top, photo of natural cast tracks; and bottom, interpretive sketch. (E and F) SGDS 167, *Characichnos* trackway likely made by a *Grallator* track-producer (i.e., coelophysid theropod) from Lower Jurassic Moenave Formation, St. George Dinosaur Discovery Site, Utah. (E) photo of natural cast trackways, and (F) inverted sketch (scale = 10 cm); (G) SGDS 361, details of cuticle and scale traces on natural cast digit trace from Lower Jurassic Moenave Formation, St. George Dinosaur Discovery Site, Utah (scale = 1 cm). *(A) From Castanera et al. (2021). (B) Courtesy Loic Costeur. (C) Courtesy David Slauf. (D) From Whyte and Romano (2001).*

Other occurrences: Devonian of Asia (Ust'-Chirka Formation: Russia); Carboniferous (Mississippian) of North America (Mauch Chunk Formation: Pennsylvania); Possibly Mississippian of North America (Mabou Group: New Brunswick, Canada); Carboniferous (Pennsylvanian) of North America (Rhode Island Formation: Massachusetts and Norton Formation: Virginia); Permian of North America (Robledo Mountains Formation: New Mexico) and Europe (Red Sandstone Group: France, Port des Canonge and Pedra de s'Ase formations: Mallorca, unnamed unit: Serbia, and Peranera Formation and Lower Red Unit: Spain); Lower Triassic of Europe (Middle Buntsandstein: Poland and "Buntsandstein Facies": Spain); Possibly Middle Triassic of Europe (Chiclana de Segura Formation ["Tabular Cover"]: Spain); Upper Triassic of North America (Redonda Formation, Chinle Group: New Mexico); Lower Cretaceous of Asia (Taoqihe and Feitianshan formations: Heilongjiang and Sichuan provinces, respectively, China) and Europe (Urbión Group: Spain); Possibly Lower Cretaceous of South America (Mulichinco and Agrio formations: Argentina); Possibly "mid"-Cretaceous of Australia (Winton Formation); Upper Cretaceous of North America (Ferron Sandstone: Utah) and Europe (Tremp Formation: Spain); Miocene of Europe (Zliv Formation: Czech Republic).

Note that this is not an exhaustive list of fossil swim traces. It is simply a list of specific attributions of swim tracks to *Characichnos*.

Type specimen(s). USEGS F00803, latex replica of a trackway consisting of at least six tracks preserved as positive hypichnia. Trackway depicted in Romano and Whyte (2015, 2003, Fig. 20)

Diagnosis. Elongate tracks consisting of two to four parallel hypichnial ridges or epichnial grooves that may be straight, gently curved, or slightly sinuous. Terminations of the ridges or grooves may be straight or sharply reflexed. Trackways consist of two rows of tracks (indicating a bipedal track maker) in which the long axes of the tracks are parallel to each other in straight trackways, and either parallel or oblique to the midline of the trackway.

Description. *Characichnos tridactylus* traces are, eponymously, tridactyl and longer than wide (length:width ratios 1.9–2.5). Individual ridges/grooves are variably developed in each tridactyl set, but they are always subparallel (Fig. 6.19). The central mark is always the longest and most pronounced (deepest), inferred to have been made by a tridactyl foot in which Digit III is the longest. The traces may be markedly asymmetrical in cross-section. The proximal ends of the traces can have well-defined, scalloped margins; others pass into rounded margins. In trackways, tracks on either side of the trackway axis are subparallel but may exhibit variable rotations with respect to the trackway axis, a function of the swimming track maker moving oblique to a current (Milner and Lockley, 2016; Milner et al., 2006b).

Discussion. The type *C. tridactylus* traces were attributed specifically to tridactyl dinosaurs, but Whyte and Romano (2001a) explicitly noted that tracks with the diagnostic *Characichnos* morphology could be made by a broad variety of tetrapods; this accords with a common practice in paleoichnology of recognizing traces as consequences of behavior, regardless of track maker identity, rather than attempting to restrict a trace type to a specific track-making taxon. Nevertheless, many subsequent reports of tridactyl *Characichnos* in the Mesozoic have been attributed specifically to theropod dinosaurs.

Characichnos is most commonly found in the Early Jurassic of North America, particularly in the Whitmore Point Member of the Moenave Formation in southwestern Utah. These traces have been linked to theropod dinosaurs (Milner et al., 2006b). In the contemporaneous

Zagaje Formation in Poland, *Characichnos* specimens were also discovered, but the specific dinosaurs responsible for these traces have not been identified (Pieńkowski et al., 2014). The type specimens of *Characichnos* come from the Middle Jurassic Saltwick Formation (Ravenscar Group) in England. While most of these specimens have been attributed to tridactyl dinosaurs, they have not been specifically identified as theropods (Romano and Whyte, 2015; Whyte and Romano, 2001a). Some of the specimens from this formation have an unclear origin (Romano and Whyte, 2010). Xing et al. (2023a) reported *Characichnos*-like traces from the Middle Jurassic Zhanghe Formation in Yunnan, China. However, they did not specify the exact dinosaur responsible for these traces. Late Jurassic *Characichnos* specimens have been discovered in the Praia da Amoreira-Porto Novo formation in Portugal, where they have been attributed to crocodylomorphs (Castanera et al., 2021). In the San Salvador Formation of Chile, Late Jurassic specimens have been found and linked to theropod dinosaurs (Soto-Acuña et al., 2015). In the Morrison Formation of Colorado, Late Jurassic *Characichnos*-type tracks were discovered as part of a trackway attributed to the tridactyl ornithopod ichnotaxon *Dinehichnus* (Lockley and Foster, 2006). At the Jurassic-Cretaceous boundary, specimens attributed to theropod dinosaurs have been reported from the Anning Formation in Yunnan, China (Xing et al., 2016a).

In the Paleozoic, the earliest report of a *Characichnos* trace is from the Devonian Ust'-Chirka Formation of northern Russia (Ahlberg and Beznosov, 2022), although this singular specimen was neither described in detail nor illustrated; presumably it would have been made by a basal tetrapodomorph. Carboniferous specimens are known from the Mississippian Mauch Chunk Formation of Pennsylvania, where they are associated with *Palaeosauropus* (temnospondyl) tracks (Fillmore et al., 2012). Stimson et al. (2016) also reported *Characichnos*-like traces from the Mississippian Mabou Group of New Brunswick, Canada. Pennsylvanian specimens are known from the Rhode Island Formation of Massachusetts (Getty et al., 2017a), where they were attributed to temnospondyl track makers, and Norton Formation of Virginia (Weems and Lucas, 2020), where they were attributed to an unspecified tetrapod. *Characichnos* traces become more common in the Permian, occurring in the Robledo Mountains Formation of New Mexico (Lerner and Lucas, 2015), where they were attributed to an unspecified amniote; the Peranera Formation of Spain (Mujal et al., 2015), where they were associated with *Batrachichnus* and *Limnopus* (temnospondyl) and *Amphisauropus* (seymouriamorph) tracks; the Lower Red unit of a Carboniferous–Triassic sequence in the Catalan Pyrenees, where they were associated with *Batrachichnus* (De Jaime-Soguero et al., 2023); the Port des Canonge and Pedra de S'Ase formations of Mallorca, where they were not attributed to any track maker (Matamales-Andreu et al., 2022); the Red Sandstone Group of France (Moreau and Gand, 2022), where they were attributed to unspecified, small tetrapods; and an unnamed red-bed unit in Serbia (Jovanović, 2013), where they were not specifically attributed to any taxon, but occur on the same surface as cf. *Batrachichnus* and cf. *Dromopus* tracks.

In the Mesozoic, Niedźwiedzki and Ptaszyński (2007) and Klein and Niedźwiedzki (2012) reported the presence of *Characichnos* in the Lower Triassic Middle Buntsandstein of Poland. These tracks were found to be associated with chirotheriid (basal archosauriform) tracks. In Spain, Mujal et al. (2017) reported the discovery of archosaurian *Characichnos* isp. in the Lower Triassic "Buntsandstein Facies." In addition, Reolid and Reolid (2017) reported tracks made by floating archosaurs from the Middle Triassic Chiclana de Segura Formation

("Tabular Cover") in Spain. Although they compared these tracks, they did not assign them to *Characichnos*. Late Triassic specimens were found in the Redonda Formation (Chinle Group) of eastern New Mexico, but the specific track maker was not identified (Lucas et al., 2010).

Tridactyl specimens have been reported from the Lower Cretaceous Feitianshan Formation of Sichuan, China (Xing et al., 2013a) and Taoqihe Formation of Heilongjiang, China (Xing et al., 2019a). The former were attributed to theropod dinosaurs, but the latter were not attributed to a specific type of track maker. Pazos (2009) reported possible *Characichnos* tracks from the Lower Cretaceous, marine Mulichinco, and Agrio formations of Neuquén, Argentina. Romilio et al. (2013) reported *Characichnos*-type swim tracks from the mid-Cretaceous Lark Quarry tracksite in the Winton Formation of Australia, which could be either theropod or ornithopod in origin. King et al. (2020) reported Late Cretaceous occurrences from the Ferron Sandstone of Utah, which they attributed to turtles, and Vila et al. (2015) reported Late Cretaceous occurrences from the Tremp Formation of Spain, which they attributed to crocodylians.

In the Cenozoic, *Characichnos* traces are far less common than in the Mesozoic. The only Cenozoic report thus far is from the Miocene Zliv Formation of the Czech Republic (Mikuláš, 2010), which was attributed to crocodylian trace makers.

Despite the fact that the type ichnospecies is specifically attributed to tridactyl track makers, the ichnogenus diagnosis is wide enough to include didactyl and tetradactyl traces as well (Whyte and Romano, 2001a). Conceivably, the diagnosis could be expanded to include pentadactyl traces, although among swim-track ichnotaxa, pentadactyl traces have only been reported for *Mesosaurichnium* (da Silva and Sedor, 2017). However, other swim-track-morphology ichnotaxa have been erected for nontridactyl traces (Lockley and Foster, 2006; see also Lockley et al., 2014a; Milner and Lockley, 2016). Furthermore, track makers with tetradactyl and pentadactyl feet are capable of creating tridactyl swim tracks, at least occasionally (e.g., Foster and Lockley, 1997; Lerner and Lucas, 2015; Lockley and Milner, 2006; Lockley et al., 2014a; Milner and Lockley, 2016; Mujal et al., 2015). A comprehensive review of tetrapod swim tracks and ichnotaxa, as well as broader or narrower diagnoses for the ichnotaxon, is needed to resolve this issue.

Hunt and Lucas (2007) recognized the *Characichnos* ichnofacies to encompass shallow lacustrine and tidal environments in which *Undichna* and tetrapod swim tracks, regardless of morphology or track maker, could be registered (q.v., Lockley et al., 2010a, 2020).

Chelonipus
Rühle v. Lilienstern, 1939
Fig. 6.20

Type ichnospecies. *Chelonipus torquatus* Rühle v. Lilienstern, 1939

Valid ichnospecies. *C. torquatus*, *C. plieninergi* (Triassic); *C. chadronicus*, *C. parvus* (Eocene)

Distribution. Triassic of Europe (Solling and Stuttgart formations: Germany; Keuper: Spain) and North America (Chugwater Group: Wyoming; Torrey and Moenkopi formations: Utah).

Upper Jurassic of Europe (Lastres Formation: Spain) and North America (Morrison Formation: Utah and Colorado). Cretaceous of North America (Dakota Group and "Mesaverde" Group: Colorado). Eocene of North America (Chuckanut Formation: Washington; Vieja Group: Texas).

FIGURE 6.20 *Chelonipus* tracks from the lower Morrison Formation (Upper Jurassic) of Utah. Plaster cast (CU-MWC 197.3) made from latex peel of natural casts. Scale = 10 cm. *Photo by John Foster. See Foster et al. (1999).*

Type specimen(s). SMNS 19107, slab with trackway.

Diagnosis. From Sarjeant and Langston (1994) and Mustoe (2019) (for family level, but *Chelonipus* the only ichnogenus): "Quadrupedal footprints with trackway very broad and stride short. Semiplantigrade to digitigrade; footprints broader than long and digits stiff and very short, bearing short, often blunt claws."

Description. See Fig. 6.20. Modified from Foster and Lockley (2006): Small symmetrical five-toed track, wider than long, with Digits I and V very short. Digits II, III, and IV subequal in length, tapering to a short sharp claw impression, with Digit III very slightly longer than II and IV. Trackway wide with short step and stride patterns. Some trackways may be composed of tracks that are variably preserved and may appear tri-, tetra-, or pentadactyl. The impressions of the feet are symmetrical with the central Digits II, III, and IV longer than the other digits, and Digit III the longest by only a very slight margin. Trackways wide with short steps and strides, but manus-pes differentiation may be difficult. This is likely a diagnostic characteristic of turtle trackways, as would be expected from their locomotor styles.

Discussion. *Chelonipus* appears to range from the Early Triassic to the Eocene–Oligocene boundary (Rühle v. Lilienstern, 1939; Sarjeant and Langston, 1994; Lichtig et al., 2018). None of the Jurassic-age referrals to *Chelonipus* have designated an ichnospecies for the tracks

(Foster et al., 1999; Lockley and Foster, 2006; Avanzini et al., 2005). The differences between *Chelonipus* and *Emydhipus* can be very subtle (Lockley et al., 2019).

The type material of *Chelonipus torquatus* (including *C. cuneiformis*) is from the Middle Triassic Solling Formation of Germany (Rühle v. Lilienstern, 1939), with *C. plieningeri* from the Upper Triassic Stuttgart Formation added by Haubold (1971a). Lichtig et al. (2018) considered *C. plieningeri* to be a synonym of *C. torquatus*. *Chelonipus* was later reported from the Upper Lower Triassic of Wyoming and Utah in the USA (Lovelace and Lovelace, 2012; Lichtig et al., 2018).

Chelonipus isp. was reported from the Upper Jurassic Morrison Formation in southern Utah and western Colorado (Foster et al., 1999; Lockley and Foster, 2006, 2024; Lockley et al., 2018a) and possibly the Upper Jurassic Lastres Formation of Spain (Lockley and Foster, 2006; see also Avanzini et al., 2005 on *Emydhipus*).

Chelonipus is generally attributed to turtles walking over wet substrate or poling over the bottom in shallow freshwater (Avanzini et al., 2005; Lockley and Foster, 2006).

Chonglongpus
Yang and Yang 1987
Fig. 6.21

Type ichnospecies. *C. hei.*

Valid ichnospecies. None.

Distribution. Middle Jurassic of Asia (Xintiangou Formation: Sichuan, China, per Chen et al., 2006).

Type specimen(s). CFZW46, natural mold of a single pes print from a trackway of 13 tracks, itself a portion of a longer trackway of 27 tracks.

Diagnosis. From Xing et al. (2021e): Tracks of a tetradactyl biped. Digit I imprint distinct, strong, and extending medially. Track length:width ratio 1.2; anterior triangle length:width ratio 0.39. Tracks subsymmetrical, with divarication angles II–III and III–IV subequal. Step length 2.4× track length.

Description. See Fig. 6.21. The imprints of Digits II and III are robust, while that of Digit IV is more slender. Digital pad formula 2-3-?3. The heel region is shallow, while the distal digit

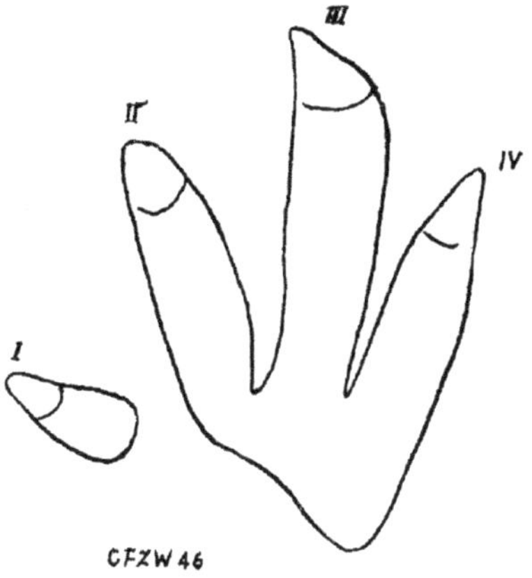

FIGURE 6.21 Sketch of *Chonglongpus hei* holotype mold. *Modified from Yang and Yang (1987), Fig. 13.*

imprints are deeper. The angle between the hallux imprint and the trackway axis is 58°; the divarication angles between Digits II–III and III–IV are about 29°.

Discussion. *Chonglongpus* is a theropod dinosaur ichnotaxon. Lockley et al. (2013) and Xing et al. (2021e) considered *Chonglongpus* synonymous with *Gigandipus*, but retained it as a distinct ichnospecies, *G. hei*.

Chongqingpus
Yang and Yang 1987
Fig. 6.22

Type ichnospecies. *C. nananensis.*

Valid ichnospecies. None.

Distribution. Middle Jurassic of Asia (Xintiangou Formation: Sichuan, China, per Chen et al., 2006).

Type specimen(s). CFZW CFNy1 from a trackway of 21 tracks.

Diagnosis. Tracks of a tridactyl–tetradactyl biped; Digit I sometimes present as a toe-tip-only imprint. Digit I imprint "half rotated" (oriented anteromedially rather than anteriorly). Digits II–IV curved inward, with straight digital pads and sharp claw imprints. Digital formula 2-2-3-4.

Description. See Fig. 6.22. For *C. nananensis*: narrow tracks that are longer than wide and with digit divarication angles I–II 55°, II–III 21–25°, and III–IV 28–30°. *C. microiscus* was differentiated from *C. nananensis* by being half the size and having narrower divarication angles (II–III 20°, III–IV 24°). *C. yemiaoxiensis* was acknowledged as similar to *C. nananensis* but differentiated from it by being smaller than *C. nananensis*, having reportedly slightly higher overall digit divarication angles (II–IV 55°), and having a different proportion of Digit IV with respect to Digit II.

Discussion. *Chongqingpus* tracks pertain to theropod dinosaurs. Lockley et al. (2003) considered *Chongqingpus* synonymous with *Grallator*; Lockley et al. (2013) considered it

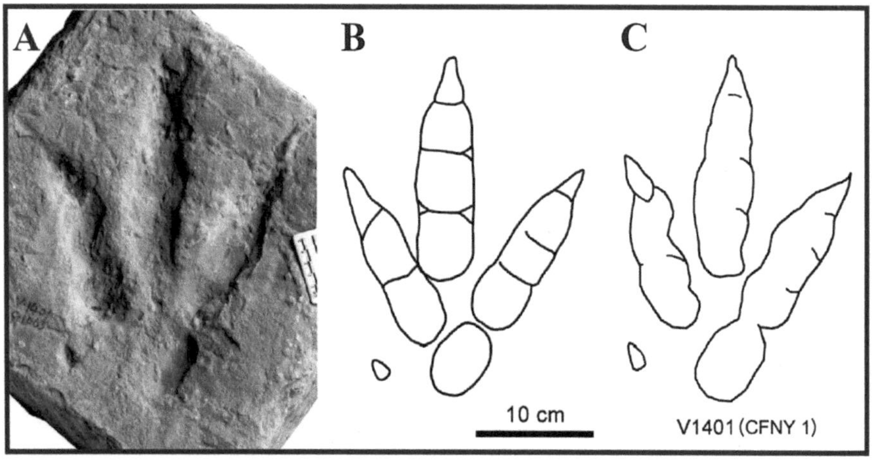

FIGURE 6.22 *Chongqingpus nananensis* V1401 (CFNY 1) holotype. (A) Photo. (B) Sketch from Yang and Yang (1987). (C) Xing et al. (2013) sketch of same specimen. *Modified from Xing et al. (2013d), Fig. 6.*

synonymous with both *Grallator* and *Kayentapus*, but retained the ichnospecies erected by Yang and Yang (1987) as distinct: *Kayentapus nananensis*, *Grallator microiscus*, and *Grallator yemiaoxiensis*. Xing et al. (2013d, 2021e) retained *K. nananensis*, but referred *C. microiscus* to *Grallator* isp.

Chuanchengpus
Yang and Yang 1987
Fig. 6.23

Type ichnospecies. *C. wuhuangensis.*

Valid ichnospecies. None.

Distribution. Middle Jurassic of Asia (Xintiangou Formation: Sichuan, China, per Chen et al., 2006).

Type specimen(s). CFZW101, a single pes print from a pair of trackways containing more than 10 tracks.

Diagnosis. Tracks of a small, tridactyl biped. Digit III projects farthest forward, followed by Digit IV and then Digit II. The imprint of Digit III is thick and long, that of Digit II is ovoid, and that of Digit IV is oblong.

Description. See Fig. 6.23. The imprint of Digit III is deeper than those of Digits II and IV. Digit divarication angles II–III 19° and III–IV 40°. All digits lack claw imprints.

Discussion. Yang and Yang (1987) considered *Chuanchengpus* dinosaurian but were uncertain as to more specific track makers. Lockley et al. (2003) considered *Chuanchengpus* too poorly preserved to be diagnostic, but Lockley et al. (2013) synonymized it with *Grallator*, a theropod dinosaur ichnotaxon, as *G. wuhuangensis*. Xing et al. (2021e) simply considered it *Grallator* isp.

Coelurosaurichnus
Huene 1941
Fig. 6.24

Type ichnospecies. *C. toscanus.*

Valid ichnospecies. None in the Jurassic. Triassic: *C. grancieri* (Courel and Demathieu, 2000), *C. kehli* (Beurlen, 1950), *C. kronbergeri*, *C. largentierensis* (Courel and Demathieu, 1976), *C. metzneri* (Heller, 1952), *C. moeni* (Beurlen, 1950), *C. palissyi* (Gand, 1976), *C. perriauxi* (Demathieu and Gand, 1972), *C. sabinensis* (Gand and Pellier, 1976),

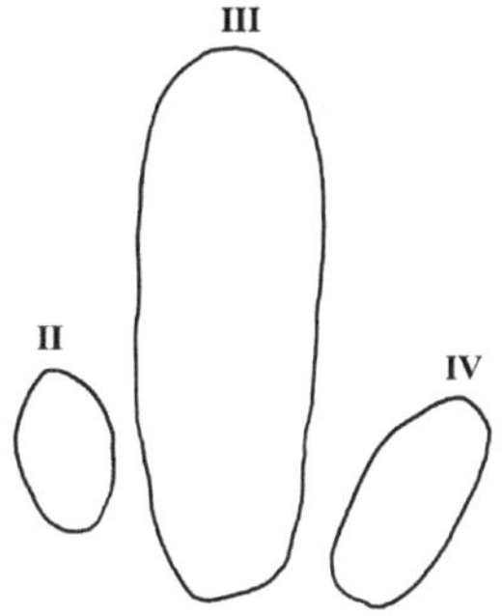

FIGURE 6.23 *Chuanchengpus wuhuangensis. Redrawn from Yang and Yang (1987, Fig. 21).*

FIGURE 6.24 Sketch of *Coelurosaurichnus romanovskyi*. Scale = 2.5 cm. *Modified from Gabuniya and Kurbatov (1988), Fig. 5.*

C. sassendorfensis (Kuhn, 1958b), *C. schlauersbachensis*, *C. ziegelangerensis* (Kuhn, 1958b). See Klein et al. (this volume).

Distribution. Questionably Middle Jurassic of Asia (Gabirut and/or Dzhirut formations: Tajikistan). Mainly Middle Triassic of Europe (Grès Inférieurs Formation: France and Quarziti del Monte Serra Formation: Italy) and Upper Triassic of Europe (unnamed unit: France and Hassberge and Löwenstein formations: Germany).

Discussion. The only non-Triassic report of *Coelurosaurichnus* is *C. romanovskyi* from the Middle Jurassic Gabirut and/or Dzhirut formations of Tajikistan (Fig. 6.24; see *Camptosauropus* for more details on this unit) (Gabuniya and Kurbatov, 1988). However, no holotype was designated, and the diagnosis is vague and insufficient, so the ichnotaxon is both a nomen dubium and a nomen nudum. Why the Tajikistani specimens were placed in *Coelurosaurichnus* at all is unclear.

Coelurosaurichnus tracks have a complex history of interpretation. They have been attributed alternately to theropod dinosaurs (e.g., Courel and Demathieu, 2000) or, more often, to "dinosauroids," presumably meaning nondinosaurian dinosauromorphs in modern parlance (e.g., Demathieu and Gand, 1972; Gand, 1976; Gand and Demathieu, 2005). *C. sabinensis* was compared with *Anomoepus* and *Moyenisauropus* (Gand and Pellier, 1976), ichnotaxa attributed to basal ornithischians and basal thyreophorans, respectively. Specimens attributed to *Coelurosaurichnus* often have both manus and pes prints, which would support a nondinosaurian interpretation. Yet, Leonardi and Lockley (1995) discussed reasons why *Coelurosaurichnus* should be abandoned as an ichnogenus and supported its synonymy with the typically theropodan ichnogenus *Grallator*, a position supported subsequently by several others (e.g., Hunt et al., 2018). This call notwithstanding, several new ichnospecies of the ichnogenus were erected subsequently (e.g., Courel and Demathieu, 2000; Gand and Demathieu, 2005). Thulborn (1993) suggested that *C. perriauxi*, possessing both manus and pes prints, was likely synonymous with *Atreipus*, but Gand and Demathieu (2005) supported its validity, as well as its distinction, as well as that of other C. ispp., from both *Anchisauripus* and *Grallator* based on a discriminant analysis. Many ichnospecies (e.g., *C. perriauxi*) were attributed to the ichnogenus without any specified reasoning or comparison to the holotype.

Most occurrences of *Coelurosaurichnus* are Middle–Late Triassic in age. The specimen constituting the type ichnospecies, *C. toscanus*, from the Middle Triassic Quarziti del Monte

Serra Formation of Italy, was reassigned to *Atreipus* by Marchetti et al. (2021). Middle Triassic ichnospecies from France—*C. largentierensis* (Courel and Demathieu, 1976), *C. palissyi* (Gand, 1976), *C. perriauxi* (Demathieu and Gand, 1972), and *C. sabinensis* (Gand and Pellier, 1976)—come from various localities in a unit currently called the Grès Inférieurs Formation; this unit has had its own convoluted nomenclatural history (Gand and Demathieu, 2005). However, *C. grancieri* comes from an unnamed, Upper Triassic unit near Ardèche, France (Courel and Demathieu, 2000). German ichnospecies come from Upper Triassic units: *C. kehli* and *C. moeni* (Beurlen, 1950) and *C. sassendorfensis* and *C. ziegelangernensis* (Kuhn, 1958b) are from the Löwenstein Formation (=Burgsandstein, = "*Semionotus*-sandstein"—q.v., Benton, 1993; Beutler, 1998; Moser, 2003); *C. metzneri* (Heller, 1952) is from the Hassberge Formation (=Kieselsandstein—q.v., Hunt et al., 2018).

Crocodylopodus
Fuentes Vidarte and Meijide Calvo 2001
Fig. 6.25

Type ichnospecies. *Crocodylopodus meijidei* Fuentes Vidarte and Meijide Calvo, 2001.

Valid ichnospecies. *C. meijidei.*

Distribution. Middle Jurassic of Asia (Hojedk Formation: Iran). Middle—?Upper Jurassic of Africa (Isli Formation: Morocco). Upper Jurassic and Jurassic/Cretaceous (Berriasian) transition of Europe (Alcobaça Formation: Portugal; Lastres and Huérteles formations: Spain). Lower Cretaceous of Asia (Jinju Formation: South Korea).

Type specimen(s). Lectotype MNS2002/96/2bis, trackway composed of four manus-pes sets from El Frontal tracksite.

Diagnosis. Revised by Lockley and Meyer (2004): "Ichnites of a tetrapod with pronounced heteropody. The manus track is pentadactyl, digitigrade with slender, sinuous digits, separated from the proximal base and terminating in claws; Digits I and V are opposed and are oriented antero-medially and postero-laterally, respectively with respect to the trackway axis; manus Digit II is therefore oriented almost parallel to the trackway axis, with Digits III

FIGURE 6.25 *Crocodylopodus*. Picture of track SHN.(JJS).ICNO.62 from the Upper Jurassic of Portugal classified as *Crocodylopodus* isp. Track length = 7.5 cm. *From Castanera et al. (2021).*

and IV oriented laterally. The pes impression is tetradactyl, plantigrade, with long slender digits which originate at the same level, ending in claws oriented toward the direction of progression of the track maker. Pes digit divarication, between I and IV high, up to about 70° in the holotype. Pes Digits II and IV subequal in length and slightly shorter than Digit II; Digit I shortest. The heel is oval, almost symmetrical with respect to the central axis [of the foot], and has a "little heel."

"Trackway narrow, about twice pes width. Inner margin of pes tracks falls on or near trackway axis: pes rotation slight or even inward, with Digit III (foot axis) typically more or less parallel to trackway axis. Placement and rotation of manus more variable but generally placed lateral to trackway and pes axes, with opposed Digits I and V typically aligned at about 40° to the trackway axis. Step 2–3 times pes length, pes pace angulation about 150°."

And, emended by Castanera et al. (2022): Star-shaped pentadactyl manual prints with slender digital impressions whose lengths vary as follows: dIII ≥ dIV = dII > dV = dI. Pedal track with digital impressions with the following length variations: dIII > dII ≥ dIV > dI. DIII is the longest and dI the shortest. Interdigital divarication varies from medium to high. Pronounced to medium heteropody (HI = 30%–40%). Manual tracks are laterally rotated, whereas the pedal tracks are slightly medially rotated. Intermediate-gauge trackway. Absence of tail, belly, or any other drag marks.

Description. See Fig. 6.25. Small-to medium-sized tracks that are characterized by pentadactyl manus prints (when all the digits are preserved) that are star-shaped and wider than long. The five digits have acute endings. Divarication between the digits is high, with dI and dV facing in opposite directions. Pes tracks are tetradactyl, subtriangular in shape, and longer than wide. The central digital impressions are longer than the lateral and medial ones. Generally, pes dIII is the longest, followed by dII–dIV, and dI is the shortest. The trackway of *Crocodylopodus* is narrow gauge (narrow-to intermediate-gauge), with high pace angulations, and typically no tail drag marks. The manus–pes impression distance is short. The heteropody varies from pronounced to moderate values.

Discussion. *Crocodylopodus* is known from the Upper Jurassic Lastres Formation of Asturias, Spain (Avanzini et al., 2007, 2010a). It has also been reported in several other Jurassic areas including the Middle Jurassic of Iran (Abbassi et al., 2015), the Middle–?Upper Jurassic of Morocco (Klein et al., 2018), and the Upper Jurassic of Portugal (Castanera et al., 2021). Recently, Lockley et al. (2020) reported *Crocodylopodus* trackways from the Lower Cretaceous of South Korea. *Crocodylopodus* tracks are thought to be produced by some groups of neosuchian crocodylomorphs, such as goniopholidids, atoposaurids, or bernissartids, that walked on land (as opposed to swimming or poling) (Avanzini et al., 2010a). These crocodylomorphs were common in the Late Jurassic–Early Cretaceous ecosystems. The trackway features of *Crocodylopodus*, and their differences from extant crocodylian trackways, indicate that they were possibly better adapted to terrestrial locomotion (see Castanera et al., 2022 and references therein).

Dahutherium
de Lapparent and Montenat 1967
Fig. 6.26
Type ichnospecies. *D. agilis.*
Valid ichnospecies. *D. agilis.*

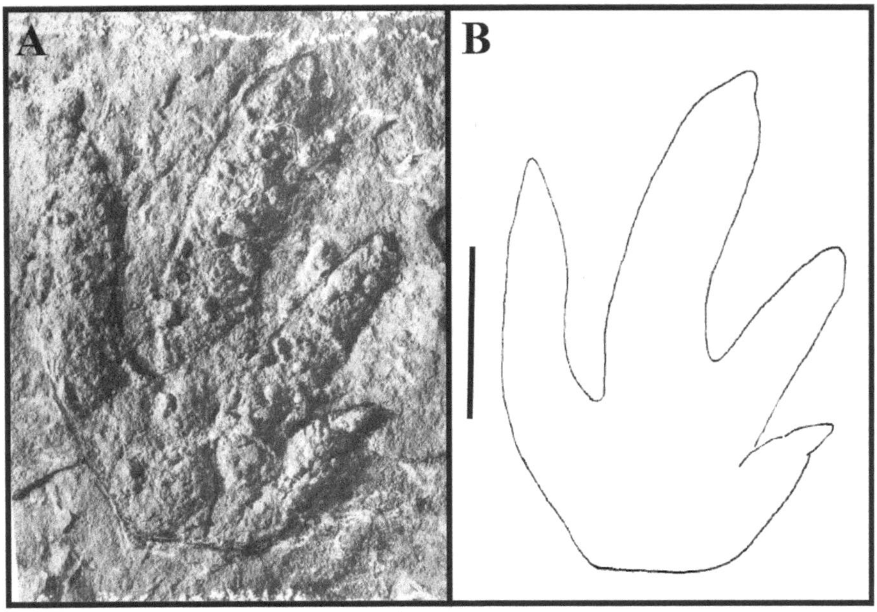

FIGURE 6.26 *Dahutherium agilis*. (A) Photo from de Lapparent and Montenat (1967, Pl. II). (B) Interpretive sketch from de Lapparent and Montenat (1967, Fig. 5). Scale = 5 cm.

Distribution. Lower Jurassic (Hettangian) of Europe, from a single site at Le Veillon, Vendée, France (de Lapparent and Montenat, 1967).

Type specimen(s). No catalog number given.

Diagnosis. From de Lapparent and Montenat (1967): Tetradactyl pes with Digit III being the longest and Digit IV slightly shorter. Digits II and I significantly decrease in length, respectively, compared with Digit III. All digits are inclined medially. Digit I displays a prominent claw trace, while all other digits do not show claw marks. All digits are elongate and separate from one another. The sole mark is moderately developed. The proposed right manus trace is partially obscured by the pes, and three digits are visible. No claw marks are evident.

Description. See Fig. 6.26. Tracks originally described from the Lower Jurassic of France (de Lapparent and Montenat, 1967) are based solely on two footprints; a right natural cast pes track, and an overprinted partial manus track. The pes track measures 16 cm long and 13 cm wide (de Lapparent and Montenat, 1967), which is quite large compared with *Batrachopus*. The manus measures 2.8 cm long and 3.5 cm wide (de Lapparent and Montenat, 1967), but the specimen is incomplete.

Discussion. *Dahutherium* was first described by de Lapparent and Montenat (1967) from the Lower Jurassic of Le Veillon, Vendée, France. Montenat (1968) assigned it to an ichnospecies, *D. agilis*. Klein and Lucas (2010) described *Dahutherium* as a questionable crocodylomorph ichnotaxon. It has been described from the Middle Triassic of France (Haubold, 1971a; Gand, 1974a,b); however, Klein and Lucas (2010, 2021) noted that the Middle Triassic record is most likely misidentified partial chirotherian tracks. Haubold (1968) attributed

Dahutherium to crocodylomorphs and placed the ichnogenus within the Family Batrachopodidae. More recently, Moreau et al. (2019) listed *Dahutherium* cooccurring with *Batrachopus* but did not comment on its validity.

Daimonelix

Barbour 1892

Type ichnospecies. *Daimonelix circumaxilis* Barbour, 1892; (Häntzschel, 1975) (Miocene)

Valid ichnospecies. *D. martini* Raisanen and Hasiotis, 2018 (Jurassic)

Distribution. Upper Jurassic of North America (Morrison Formation: Utah)

Type specimen(s). Holotype and paratypes are field specimens in the Salt Wash Member of the Morrison Formation near the Fremont River in Wayne County, Utah (Raisanen and Hasiotis, 2018).

Diagnosis. For Jurassic species only, from Raisanen and Hasiotis (2018): "Large, vertical to inclined, helical structure with narrow axis of coiling and steep angle of descent, with a horizontal to low-angle, lateral protrusion at or near the base, all without tunnels, chambers, or helices branching from the main structure or lateral protrusion."

Description. Carbonate-filled burrow casts of small diameter, oval, and slightly wider than high and ~3–14 cm. Comprising helical shafts that extend down from original surface to a horizontal tunnel that joins main tunnel at an L-shaped junction. Depth of tunnels ~34–156 cm, dip of helical whorls is 26–67°.

Discussion. The type specimens of *Daimonelix* were from the Miocene of Nebraska, but they have since been found in rocks as old as the late Paleozoic. The Jurassic ichnospecies *D. martini* was reported in the Salt Wash Member of the Morrison Formation in Utah by Raisanen and Hasiotis (2018). Although the type of *Daimonelix* is Cenozoic in age and may have been made specifically by beavers, the Mesozoic ichnospecies appears to have been made by indeterminate small vertebrates, possibly mammals (Hasiotis et al., 2004; Raisanen and Hasiotis, 2018). However, many animals, including modern varanids among reptiles, construct helical burrows (Doody et al., 2015).

Delatorrichnus

Casamiquela 1964

Fig. 6.27

Type ichnospecies. *D. goyenechei.*

Valid ichnospecies. *D. goyenechei.*

Distribution. Middle Jurassic of South America (La Matilde Formation: Santa Cruz, Argentina). Possibly Lower Jurassic of North America (Kayenta Formation: Utah) and Africa (Dande Sandstone Formation: Zimbabwe).

Type specimen(s). Natural cast trackway consisting of three manus–pes sets MLP 60-X-31–6.

Diagnosis. From de Valais (2011): quadrupedal trackways with symmetrical manus and pes impressions; manus prints lie medial to, and often in contact with, the pes prints. Manus and pes prints both longer than wide; the pes is nearly twice the size of the manus. Tridactyl–tetradactyl manus prints have subparallel digit impressions and are nearly rhomboidal in outline, with the long axis virtually parallel to the midline. Tridactyl pes prints are rotated outward from the trackway midline; robust digit imprints have low divarication angles. Digit

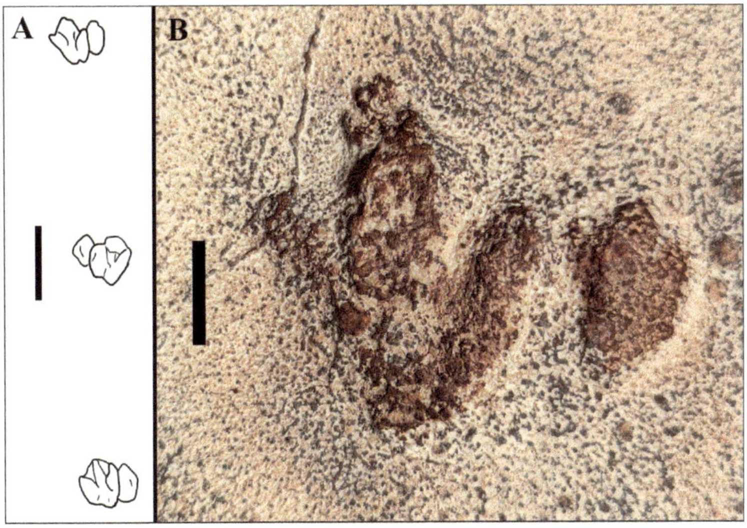

FIGURE 6.27 *Delatorrichnus goyenechei* from the Middle Jurassic LaMatilde Formation, Argentina. (A) Sketch of MLP 60-X-31–6, holotype (scale = 5 cm). (B) UCM 187.70 replica (scale = 1 cm). *Photo by Andrew Milner.*

III is the longest, and digital pads are absent. Trackways are relatively wide, with high pace angulation values for both the manus and pes.

Description. See Fig. 6.27. Without exception, *D. goyenechei* tracks at the type locality (occurring in four trackways containing a total of 11 manus and 12 pes prints) comprise manus–pes sets in which the pes has somewhat overstepped the manus so that the manus prints are partially overlapped by the pes prints. Manus prints are longer than wide (average length:width ratio of about 1.4); the subparallel digit imprints (I–IV) are wide and closely appressed, to the point that individual digit imprints are indistinct. Pedal digit imprints, despite their small sizes, are relatively thick, lacking any signs of digital pads. Divarication angles between pedal digit imprints are low and subequal (24–25°); the lengths of Digits II and IV likewise are subequal. Average pace angulation values for both the manus (172°) and pes (162°) are similar, as are stride lengths (298 and 294 mm, respectively).

Discussion. Casamiquela (1964, 1966) attributed *Delatorrichnus* tracks to small theropod dinosaurs, and this has occasionally been repeated (e.g., Bonaparte, 1979; Leonardi, 1989, 1994, although only referred to them as dinosaurian). However, the presence of palms-downward manus prints makes this attribution unlikely (Milner et al., 2009). Subsequently, most researchers have attributed *Delatorrichnus* to small ornithischian dinosaurs (Abrahams et al., 2017; de Valais, 2011; Gierliński and Niedźwiedzki, 2002b; Gierliński et al., 2004; Lockley et al., 2009; Otero and Reguero, 2013). Gierliński et al. (2004), Milner et al. (2009), and de Valais (2011) compared *Delatorrichnus* to *Atreipus*, a Late Triassic ichnotaxon that itself has been attributed to both theropods and ornithischians as well as to nondinosaurian dinosauromorphs (Olsen and Baird, 1986), plus other, similar, ichnotaxa.

Delatorrichnus tracks are essentially unknown outside the type locality. Small, Early Jurassic manus and pes tracks from the Zagaje Formation of Poland were assigned to *Delatorrichnus* isp. by Gierliński et al. (2004), but de Valais (2011) outlined how these traces did not match the diagnosis for the ichnotaxon. Instead, the Polish tracks correspond well with *Afrodelatorrichnus*. Similarly, manus-pes tracks from the Lower Jurassic Dande Sandstone Formation of Zimbabwe were compared, but not assigned, to *Delatorrichnus* by Lingham-Soliar and Broderick (2000). Small, Early Jurassic, pes-only tracks from the Kayenta Formation of Utah were likewise compared, but not assigned, to *Delatorrichnus* by Lockley (1986). Holotypic, Middle Jurassic specimens of *Delatorrichnus* are from the La Matilde Formation of Santa Cruz, Argentina (Casamiquela, 1964; de Valais, 2011).

Bordy et al. (2020) erected *Afrodelatorrichnus ellenbergeri* for tracks from the Lower Jurassic Lesotho Formation of South Africa; these differ from *Delatorrichnus* in being larger, having a greater pes trackway width, and lacking overlap between the pes and manus prints. In addition, the pes prints possess digital pad imprints that are narrow and widely spaced. Bordy et al. (2020) placed *Afrodelatorrichnus* and *Delatorrichnus* in the new ichnofamily Delatorrichnopodidae.

Deltapodus
Whyte and Romano 1994
Fig. 6.28

Type ichnospecies. *Deltapodus brodricki* Whyte and Romano, 1994.

Valid ichnospecies. *D. brodricki* (Whyte and Romano, 1994), *D. curriei* (Xing et al., 2013b), *D. ibericus* (Cobos et al., 2010).

FIGURE 6.28 *Deltapodus.* (A) Trackway image clipped from Slater et al. (2018); (B) USEGS F00768 holotype. (C) Orthophoto generated from 3D model of the manus-pes set 1CA23mp, part of the holotype trackway of *D. ibericus* (Cobos et al., 2010). Scale bar = 15 cm.

Distribution. Middle Jurassic of Europe (Saltwick Formation [Ravenscar Group]: England and Lealt Shale Formation [Great Estuarine Group]: Scotland) and South America (Lajas Formation [Cuyo Group]: Argentina); Upper Jurassic of North America (Morrison Formation: Utah) and Africa (Iouaridène Formation: Morocco); Upper Jurassic–Lower Cretaceous of Europe (Lourinhã Formation: Portugal; Villar del Arzobispo, Aguilar del Alfambra, Camarillas, Artoles, Tereñes, Lastres, Magaña, and Huérteles formations: Spain); Lower Cretaceous of Asia (Tugulu Group: Xinjiang, China) and Europe (Camarillas and Artoles formations: Spain).

Type specimen(s). USEGS F00768, latex mold and plaster cast of left manus–pes set.

Diagnosis. Adapted from Whyte and Romano (1994), by Cobos et al. (2010): The manus is entaxonic, broadly crescent-shaped in outline, and anteriorly convex. There is the occasional development of an inwardly (or medially) directed digit (pollex) impression. Pes is mesaxonic, oval to subtriangular in outline with the medial side slightly concave. In pes tracks, the three digit impressions appear as short projections or bluntly rounded points. Trackways are medium gauge (internal trackway width moderately wide: approx. 0.10–0.30 m).

Description. See Fig. 6.28. While still hypothesizing a sauropod origin for the ichnotaxon, Whyte and Romano (1994) noted that *Deltapodus* tracks were dissimilar to sauropod tracks by having manus prints with an inwardly directed pollex and more triangular, tridactyl, and mesaxonic pes prints with anteriorly, rather than anterolaterally, positioned digit imprints.

Deltapodus is characterized by mesaxonic subtriangular pes prints featuring three short, wide, and bluntly rounded digit impressions located anteriorly, providing evidence of hoof-like unguals. The manus prints exhibit shapes ranging from crescent to kidney-shaped. The main distinctions between the Jurassic ichnospecies (*D. brodricki* and *D. ibericus*) lie in the manus morphology and the proportions of the pes, *D. ibericus* being more elongate. *D. brodricki* displays a more crescentic manus shape with the presence of a pollex mark impression, while *D. ibericus* is more kidney-shaped, with a regular outline, and lacks this mark. In addition, notable diagnostic features of *D. ibericus* include the manus width equaling or being greater than that of the pes (Castanera et al., 2024).

Discussion. Originally reported from the Middle Jurassic of Europe (Saltwick Formation, Yorkshire, UK) by Whyte and Romano (1993, 1994), *Deltapodus* tracks were initially interpreted as of "probable sauropod" origin but were later reinterpreted as stegosaur by the same authors (Whyte and Romano, 2001a,b). However, McCrea et al. (2001) suggested that the quality of preservation of specimens from the type locality was not good enough to readily attribute them either to stegosaurs or ankylosaurs and instead referred to them simply as thyreophoran tracks. The tracks have subsequently been broadly accepted as the tracks of stegosaurian dinosaurs with, for instance, *D. ibericus* occurring in the same geological formation as body fossils of the stegosaurian *Dacentrurus* (Cobos et al., 2010).

Other Middle Jurassic specimens have been reported from the Lealt Shale Formation (Great Estuarine Group) of Isle of Skye, Scotland (dePolo et al., 2020); a Middle–Late Jurassic cf. *Deltapodus* specimen was reported from the Lajas Formation (Cuyo Group) of Argentina (Pazos et al., 2019). Late Jurassic specimens are known from the Morrison Formation of Utah (Milàn and Chiappe, 2009; Lockley et al., 2016, 2018a), the Iouaridène Formation of Morocco (Belvedere and Mietto, 2010), and the Lourinhã Formation of Portugal (Guillaume et al., 2022; Mateus et al., 2011). Despite being identified in the unit subsequent to the discovery of *Stegopodus* (Lockley and Hunt, 1998), *Deltapodus* has now been found at four localities in the Morrison Formation (in the Tidwell, Salt Wash, and Brushy Basin members) (Milàn and

Chiappe, 2009; Gierliński et al., 2010; Lockley et al., 2017, 2018a). These occurrences include the first *Deltapodus* trackway in North America, an 11 step trackway (Lockley et al., 2018a).

Additional Late Jurassic–Early Cretaceous specimens have been reported from the Teruel (Alcalá et al., 2012; Cobos et al., 2010; Gascón and Pérez-Lorente, 2017; Mampel et al., 2011; Castanera et al., 2024), Asturias (Lockley et al., 2008b; García-Ramos et al., 2011; Piñuela et al., 2014), and Soria (Pascual et al., 2012; Pascual-Arribas and Hernández-Medrano, 2015a) regions of Spain. Additional Upper Jurassic-Lower Cretaceous specimens have been reported from the Upper Jurassic–Lower Cretaceous Villar del Arzobispo and Lower Cretaceous Camarillas and Artoles formations of Spain, some of which are three-dimensional penetration tracks and include skin impressions (Cobos et al., 2010; Gascón and Pérez-Lorente, 2017).

Specimens from the Villar del Arzobispo Formation comprise the holotype of *D. ibericus* (Cobos et al., 2010), an ichnospecies diagnosed by: having a sauropod-like, kidney-shaped manus imprint that is about 1.3× wider than long and as wide as, or wider than, the pes prints, and that lies anterior or anterolateral to the pes print, with which it is sometimes in contact; and pes prints that are one-third longer than wide, 2× the manus length, and rotated strongly outward from the trackway axis. This ichnospecies has been recently identified in other Late Jurassic tracksites from the region (see Castanera et al., 2024).

Other Early Cretaceous *Deltapodus* specimens have also been reported from the Tugulu Group of Xinjiang, China (Xing et al., 2013b, 2021a); these were made the holotype of *D. curriei* by Xing et al. (2013b), which was differentiated from other *Deltapodus* ichnospecies by having entaxonic manus prints that are twice as wide as long, medially concave and laterally convex metacarpophalangeal region imprints, and distinct, robust digit I imprints; and pes prints that are more entaxonic than other *Deltapodus* ichnospecies. *D. curriei* tracks exhibit slight outward rotation with respect to the trackway axis. A *Deltapodus*-like trackway was identified from the "mid"-Cretaceous at the Clayton Lake dinosaur tracksite in northeastern New Mexico (Lucas and Dalman, 2016).

Dilophosauripus
Welles 1971
Fig. 6.29

Type ichnospecies. *D. williamsi.*

Valid ichnospecies. *D. williamsi.*

Distribution. Lower Jurassic Kayenta Formation, northeastern Arizona (Welles, 1971). Possibly Europe (Italy).

Type Specimen(s). UCMP 79690-4, holotype cast of single natural cast footprint. UCMP 79690-1,–2, and –3, hypodigm types are casts also from the type locality (loc. # UCMP V67239). Also has the number UCM 124-10 in Lockley collection at Boulder, Colorado.

Diagnosis. Functionally tridactyl and bipedal. Wide divarication between digits (II–IV = 54°) with broad digit traces having elongate claw marks.

Description. See Fig. 6.29. The inflated appearance of the digits could be the result of lithic compaction due to sediment load with denser sandstone compacting softer, finer-grained rocks (Lockley and Xing, 2015), and/or the deformation of bedding surfaces microbial mats below the actual track surface as transmitted footprints (Milner et al., 2023b).

Discussion. Several authors have suggested that *Dilophosauripus* is a morphological variant of *Eubrontes* and is likely synonymous. Others have considered it as valid.

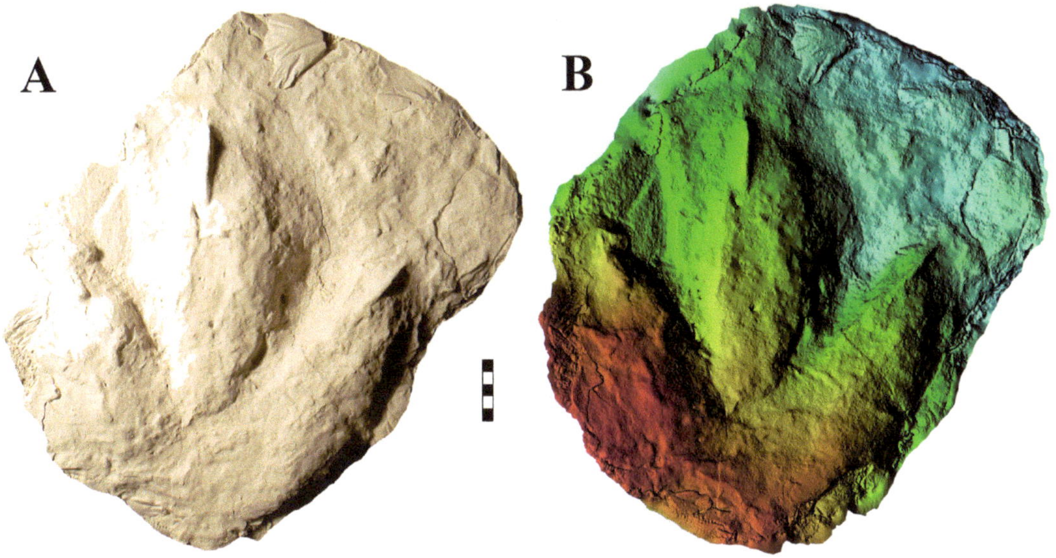

FIGURE 6.29 *Dilophosauripus williamsi*. (A) Holotype replica of UCMP 79690-4 (UCM 124-10). (B) Color contour image of UCMP 79690-4 (replica UCM 124-10). Scale = 5 cm.

Dinehichnus
Lockley, dos Santos, Meyer, and Hunt 1998
Fig. 6.30
Type ichnospecies. *D. socialis* Lockley et al., 1998b.
Valid ichnospecies. *D. socialis.*

Distribution. Upper Jurassic of North America (Morrison Formation: Colorado, Utah) and Europe (Błaziny Oolite Limestones, Wierzbica Oolite and Platy Limestones, Bałtów Coral Limestones: Poland; Alcobaça and Freixial formations: Portugal); Lower Cretaceous of Europe (Obernkirchen Sandstone [Bückeberg Formation]: Germany; Enciso Group: Spain). Possibly Upper Jurassic–Lower Cretaceous of Asia (Anning Formation: China). Possibly Upper Jurassic of Africa (Iouaridène Formation: Morocco). Lower Cretaceous of Asia (Jiaguan Formation: Sichuan, China).

Type specimen(s). Holotype CU-MWC 198.2, rubber molds and plaster and fiberglass casts of segment of three-step trackway five from the Boundary Butte locality in San Juan County, Utah. Topotypes CU-MWC 198.3, CU-MWC 198.4, and CU-MWC 196.1 all replicas also at the University of Colorado Museum. The originals were reported by Lockley et al. (1998b) as held by the Montezuma Creek Office of the Federal Bureau of Investigation.

Diagnosis. From Lockley et al. (1998b):"Trackway of small- to medium-sized biped with footprints about as wide as long. Tracks quadripartite, symmetric, and tridactyl with distinctive circular heel pad impression. Digit impressions consisting of single elongate, oval impressions sometimes with tapered claw impression. Digit divarication (II–IV) averaging about 90°. Trackways narrow with pace angulation of about 155°. Negative (inward) rotation

FIGURE 6.30 *Dinehichnus.* (A) Natural cast track, SHN.(JJS).ICNO.60, from the Upper Jurassic of Portugal. Track length = 18.7 cm. (B) CU-MWC 198.2, *D. socialis* holotype sketch. (C) UCM 198.2, *D. socialis* holotype replica. Scale = 10 cm. *(A) From Castanera et al. (2020). (B) Modified from Lockley et al. (1998). (C) Photo by Andrew Milner.*

of Digit III about 10–15° from trackway midline. Step length averaging a little more than three times foot length."

Description. See Fig. 6.30. The strong symmetry, approximate equidimensionality (as long as wide), and quadripartite nature of *Dinehichnus* tracks readily differentiate them from other small, tridactyl dinosaurian ichnotaxa. Although type *Dinehichnus* tracks have a pronounced, circular metatarsophalangeal pad imprint, separated from the digit imprints, the digit imprints lack signs of digital pads. Type *Dinehichnus* tracks occur in multiple, parallel trackways, suggesting gregarious behavior on the part of the track makers. The distances between steps in the trackways are relatively short, supporting an ornithopod interpretation (Lockley et al., 1998b).

Discussion. The type material was reported from the upper Salt Wash Member of the Morrison Formation at Boundary Butte in extreme southern Utah (Lockley et al., 1998b), and referred material came from the Salt Wash of Colorado National Monument in western Colorado (Lockley and Foster, 2006), which also includes possible swim tracks (see *Characichnos* above). Other Late Jurassic occurrences of *Dinehichnus* isp. were reported from the Holy Cross Mountains of Poland (Gierliński et al., 2009a) and the Lusitanian Basin of Portugal (Lockley et al., 1998b; Castanera et al., 2020). *Dinehichnus*-like tracks were reported from the Late Jurassic of Spain (Castanera et al., 2023a; García-Ramos et al., 2006), North Africa (Belvedere et al., 2010), and the Jurassic–Cretaceous boundary of China (Xing et al., 2014d, 2019b). Tracks referred to *Dinehichnus* also occur in the Early Cretaceous of Spain (Agirrezabala et al., 1985; Moratalla, 1993; Lockley et al., 1998b; Moratalla and Hernan, 2008; Viera and Torres, 1992). They further may occur in the Early Cretaceous of China (Xing et al., 2015a). Given that basal ornithopod dinosaurs persisted through the Cretaceous, future Early Cretaceous, as well as Late Cretaceous, occurrences would not be unexpected.

The track makers of *Dinehichnus* are thought to be dryosaurid ornithopods or relatively large basal neornithischians (Lockley et al., 1998b; Castanera et al., 2020). Despite the presence of basal ornithopods in the Middle Jurassic, particularly in China (e.g., Barrett et al., 2005), no definite Middle Jurassic *Dinehichnus* tracks are known there. A Middle–?Upper Jurassic occurrence was reported from the Imilchil and Isli formation complex of Morocco by Gierliński et al. (2009b), but this identification was refuted by Klein et al. (2022). Gierliński et al. (2001a) compared a Late Jurassic track from Poland to *Dinehichnus* and suggested it was camptosaurid in origin, but it lacks the characteristic, circular metatarsophalangeal pad.

Emydhipus
Fuentes Vidarte, Meijide Calvo, Meijide Fuentes, and Meijide Fuentes 2003
Fig. 6.31

Type ichnospecies. *E. cameroi.*

Valid ichnospecies. *E. cameroi, E. cerinensis.*

Distribution. Upper Jurassic of Europe (Lastres Formation: Spain; Lithographic Limestones: France); Lower Cretaceous of Europe (Oncala and Enciso groups Spain) and Asia (Lower Layer, Tugulu Group: China); "middle" Cretaceous of Africa (Midelt Formation: Morocco).

Type specimen(s). Holotype specimen *E. cameroi*: MNS2002/96.38 (Museo Numantino de Soria, Spain). Holotype of the Late Jurassic ichnospecies, *E. cerinensis*: n° 400.001 (original and cast) in the collections of the Département des Sciences de la Terre, Université Claude-Bernard Lyon, with the holotype manus and pes traces figured by Bernier et al. (1982, traces M22 and P24, plates 2–3).

Diagnosis. Translated from Fuentes Vidarte et al. (2003): Tracks from a quadrupedal tetrapod with different autopods. The anterior autopod leaves four long claws, straight and parallel oriented in the direction of travel. The posterior autopod is plantigrade, tetradactyl, with two central digits of the same size and two lateral, smaller and slightly divergent,

FIGURE 6.31 *Emydhipus cameroi* holotype, Museo Numantino de Soria 2002–96–38. (A) Natural cast trackway slab showing the positions of close-ups B and C. Scale = 10 cm. (B) Right manus-pes set close-up. Scale 2 Euros (diameter = 2.6 cm). (C) Left manus-pes set close-up. Scale 2 Euros (diameter = 2.6 cm). *(C) Modified from Pascual-Arribas and Hernandez-Medrano (2015b).*

with claws. Plantar area is short and rounded; the mean length of the track is 18 mm; digits are oriented in the direction of travel (Fig. 6.31).

Discussion. *Emydhipus* (*E. cameroi*) was described by Fuentes Vidarte et al. (2003) from a complete trackway found in the Huérteles Formation (Berriasian, Spain). Pascual-Arribas and Hernández-Medrano (2015b) documented new occurrences from other tracksites in the same formation. These authors summarized (their Table 3) the main features of *Emydhipus* in comparison with other turtle ichnotaxa, such as relative position of manus and pes (manus in front of the pes), size of the footprints (small), plantigrade pes, number of digits in both manus and pes (four) or distance of the autopods to the travel axis (similar distance). The largest Jurassic sample of *Emydhipus* comes from the Late Jurassic of Asturias (Avanzini et al., 2005). The authors described several trackways that were interpreted to have been produced by chelonians partially buoyed by water and also walking in wet or firm mud in a subaerial environment. In addition, *Chelonichnium cerinense* Demathieu and Gaillard 1982 (in

Bernier et al., 1982) was considered a nomen dubium and reassigned to *Emydhipus* under the new combination *Emydhipus cerinensis* by Avanzini et al. (2005). This is based on the shared features this trackway has with *Emydhipus* over *Chelonipus*. Bernier et al. (1982) proposed that *E. cerinensis* was created by "walking turtles" on a slope covered in moist mud, whereas the reassessment by Avanzini et al. (2005) suggested bottom walking in a shallow aquatic environment. *Emydhipus cerinensis* is found in association with *Saltosauropus latus* in the marine lithographic limestones. *S. latus* is proposed as a behavioral variant of *Chelonichnium* (*Emydhipus cerinensis*; Lockley and Foster, 2006; Lockley and Meyer, 2000).

Eotetrapodiscus

Fig. 6.32

Type ichnospecies. *E. cursor.*

Valid ichnospecies. *E. cursor, E. moyenensis.*

Distribution. Lowermost Jurassic of southern Africa (lowermost upper Elliot Formation, zone B/1: Lower Moyeni/Quthing, Lesotho).

Type specimen(s). Pes-manus impressions in red, very fine-grained sandstone ("muddy loess") immediately underlying the Lower Moyeni tracksite (zone B/1, lowermost upper Elliot Formation, Lesotho). Original specimen stored as sandstone slab* 101 LES (*E. moyenensis:* 102 LES, 131 LES) at the PECUoM (currently misplaced).

Diagnosis. Modified from Ellenberger (1974): Tetradactyl pes impression that is circular with short, rounded digital impressions. Pes appears more paraxonic than mesaxonic, with Digit I not impressed. Pes length is 8 mm (Digit I: 1 mm?; II: 2 mm; III: 3 mm; IV: 3 mm; V: 2 mm?); projection of Digit III forward from the tip of the lateral digits is 0.7 mm. Foot width is 8 mm. Impressions of Digits II–IV are distally distinct and located in front of a small sole. The digital pad impressions are forward directed and compact with blunt claw marks (? raised from the ground). Total divarication of pes Digits II–V is > 60°. The "heel" impression is hardly visible, defining a semidigitigrade pes.

Tetradactyl manus impression with a somewhat rudimentary and occasionally preserved Digit I impression. Manus length is 9 mm (Digit I: 2.5 mm?; II: 4 mm; III: 4 mm; IV: 3 mm; V: 2 mm?); projection of digit III forward from the tip of the lateral digits is 1 mm. Manus width is 12 mm. Total divarication of manus Digits I–V is > 70°. No "heel" impression, implying digitigrade movement. The broader manus prints are less pronounced than the pes prints.

Description. See Fig. 6.32. Poorly defined, small trackway of six consecutive pes-manus impressions, each with four digits. Trackway width is 18 mm, with a stride of 45 mm (step: 22–23 mm) and a pace angulation of 106°. The angularities of the pes and manus are 37° and 40°, respectively.

The distinct track features, notably the more pronounced pes prints compared with broader manus prints, suggest a minute, quadrupedal animal with cursorial progression. The track maker moved with the rear of the foot raised (forward-leaning legs) and with the manus lightly brushing the ground ahead of the foot, toward the outer side of the trackway.

Discussion. Only reported from red, very fine-grained sandstone forming the base of the Lower Moyeni tracksite that is characterized by greenish-white sandstone (zone B/1, lowermost upper Elliot Formation) in Lesotho (Ellenberger, 1970, 1974).

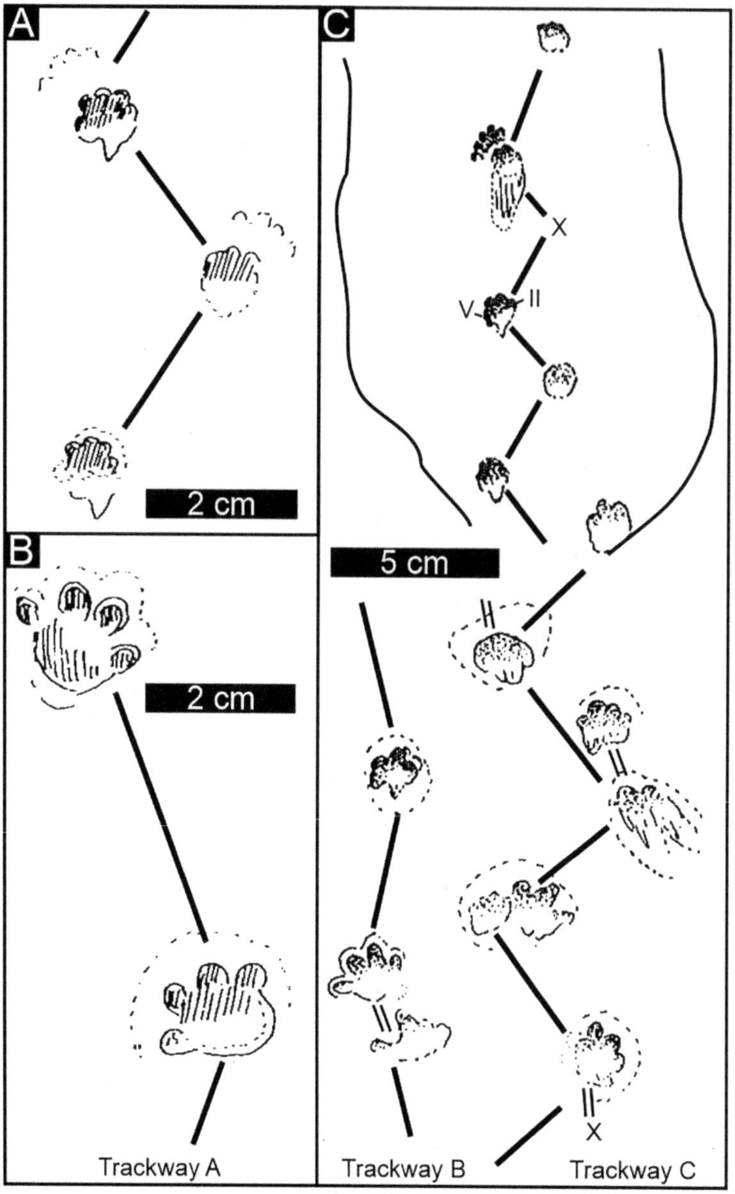

FIGURE 6.32 Tetradactyl tracks of possible mammal/protomammals with cursorial progression from the Lower Moyeni tracksite in the lowermost upper Elliot Formation, Lesotho. (A) *Eotetrapodiscus cursor*—three consecutive tetradactyl pes-manus impressions. (B) *E. moyenensis*—two consecutive tetradactyl pes impressions. (C) Trackway assemblage of a smaller *E. cursor* (top) and two larger *E. moyenensis* (bottom) on the same sandstone slab. *Retraced by EMB after Ellenberger (A, B: 1970, his figs. 72, 73; C: 1974, his plate H) and republished with permission from the Geological Society of South Africa and Palaeovertebrata.*

Ellenberger (1974) drew comparisons between *E. cursor* and *Ameghinichnus patagonicus* from Argentina, noting distinct locomotion variations (q.v., de Valais, 2009). *E. cursor*, unlike *Ameghinichnus*, was made by an animal trotting on its hind legs with a reduced, nonground-touching hallux. Ellenberger (1974, 1975) associated *Eotetrapodiscus* with basal, minuscule mammals (e.g., *Erythrotherium*); in a reevaluation of some Lesotho ichnotaxa, Lockley et al. (2004c) also categorized *E. cursor* as potential mammal/"protomammal" tracks. D'Orazi Porchetti et al. (2017, 2018) highlighted the similarity between *E. cursor* and *Brasilichnium*, both of which have been interpreted as tracks of small mammaliaforms.

Episcopopus
Ellenberger 1970
Fig. 6.33
Type Ichnospecies. *E. ventrosus.*
Valid ichnospecies. *E. ventrosus.*

Distribution. Lowermost Jurassic of southern Africa (lowermost upper Elliot Formation, zone B/1: Lower Moyeni/Quthing, Lesotho).

Type specimens. Pes-manus tracks on a greenish-white, fine-grained, ripple-marked, algal-matted sandstone surface at the Lower Moyeni tracksite (zone B/1, lowermost upper Elliot Formation, Lesotho). Also recorded as plaster casts SAM-PT-K10949 at Iziko, 087 LES, and 100 LES in the PECUoM, as well as digital models at UMMP.

Diagnosis. Modified from Ellenberger (1974) and Marsicano et al. (2014): Pentadactyl pes tracks comprise impressions of the clawless, rounded to somewhat laterally elongated tips of five digits only (no sole trace). The impressions of Digits II–V are arranged in a 20 × 20 cm, slightly inward-oriented arc. Digits II–IV make deeper impressions with the Digit I being very subtly imprinted just posterior to Digit II.

Tetradactyl manus tracks comprise shallow impressions of the gently rounded, clawless tips of Digits I to IV, of which Digits III and IV are associated with deeper, strongly curved drag marks that are convex anteriorly and that converge distally. The impressions of Digits I–IV are arranged in a 25 cm by 15 cm, inward-oriented, acute angle (i.e., an inverted "V" shape). Palm impressions are lacking.

The smooth, discontinuous, and irregular body impression ("heavy belly" drag) is 40 cm wide (Marsicano et al., 2014). This broad and slightly sinuous median body drag lacks evidence of regular undulation (Ellenberger, 1974). Occasional straight and narrow grooves run within the body drag, with a longer groove paralleling a shorter one.

Description. See Fig. 6.33. Originally recorded in a single, 23 m-long trackway of 82 consecutive impressions of pes-manus pairs and a straight body drag mark resulting from the track makers "heavy belly" flattening the ground below it. Trackway width is ~65 cm, with a remarkably regular stride of 110 cm and a pace angulation of 72°.

The digit only impressions are associated with the drag marks that, in the case of the manus, resemble a bishop's crook. The more tightly arched pedal digit impressions consistently overstep those of the manus of the previous stride and crosscut the manus drag impressions. The nearly parallel, trailing impressions of pedal Digits II and III deepen anterolaterally and are shorter and shallower than those of the manus. Marsicano et al. (2014) did not consider these latter features diagnostic of the ichnotaxon, but used them to interpret the locomotion (e.g., limited lateral motion of the pes away from the ground).

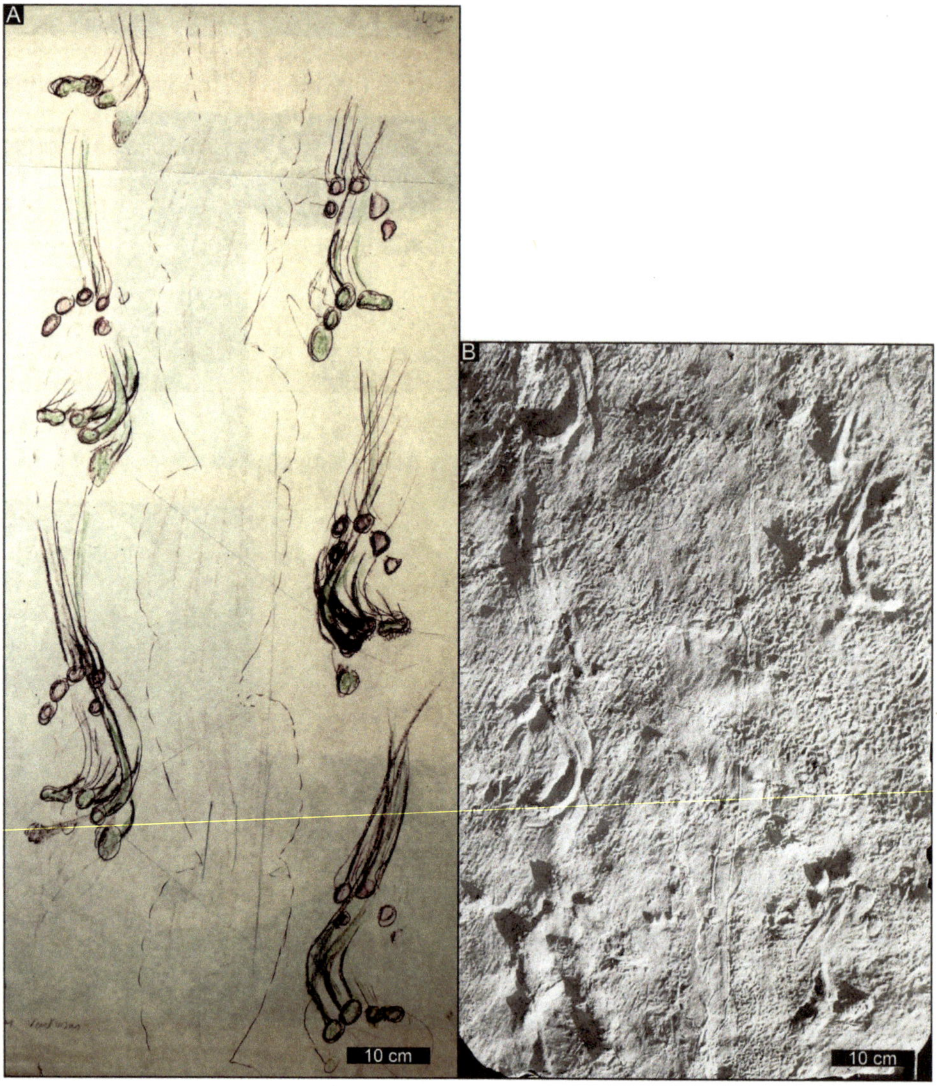

FIGURE 6.33 *Episcopopus ventrosus*—trackway segment from the Lower Moyeni tracksite in the lowermost upper Elliot Formation, Lesotho. (A) Hand-drawn illustration by P. Ellenberger, published as Plate I in Ellenberger (1974). (B) Unpublished oblique photo of the plaster cast trackway. Materials sourced from the PECUoM archives. Note that compared to Ellenberger's interpretation, Marsicano et al. (2014) considered the manus tetradactyl, the pes digital marks straight, and the belly drag to be narrower, discontinuous and more unevenly undulating. *(B) Photo by Emese Bordy.*

The fine, parallel streaks within the body drag are tail ridge impressions and/or tail drag marks.

Discussion. To date, *Episcopopus* has only been reported in the Lower Jurassic from greenish-white, fine-grained sandstone at the Lower Moyeni tracksite in Lesotho (zone

B/1, lowermost upper Elliot Formation; Ellenberger, 1970, 1974; Smith et al., 2009; Marsicano et al., 2014).

The consistent stride lengths suggest a steady pace of a short-legged, heavy-bellied, clawless, quadrupedal track maker. It slowly traversed the rippled sandbar with minimal trunk bending, using only the tips of its digits. Overstepping of the manus by the pes, distinctive digit prods, scrapes, intermittent belly, and straight tail drag marks indicate a movement pattern linked to stocky, large-headed temnospondyl amphibians, measuring up to 3.5 m in length (Smith et al., 2009; Marsicano et al., 2014).

Eubrontes
Hitchcock 1845
Fig. 6.34

Type ichnospecies. *E. giganteus.*

Valid ichnospecies. *E. giganteus, E. zigongensis, E. nobitai.*

Distribution. Upper Triassic of Europe (Tomanová Formation: Poland and Slovakia; upper "Buntsandstein" and Dolomías tableadas de Imón formations: Spain; Höganäs Formation: Sweden) and Africa (Timezgadiouine Formation: Morocco); Lower Jurassic of North America (Newark Supergroup in eastern North America and Glen Canyon Group (Moenave, Wingate Sandstone, Kayenta, and Navajo Sandstone formations and the laterally equivalent Nugget Sandstone in northern Utah) in western North America), Africa (Elliot Formation: main Karoo Basin of Lesotho and South Africa), and Asia (Fuxian and Zhenzhuchong formations, China); Late Jurassic of Europe (Cazals Formation: France); Lower Cretaceous of Asia (Jiaguan Formation, Sichuan, China).

Type specimen(s). ACM-ICH 15/3 (old number 38) is the holotype according to Olsen et al. (1998), a natural cast in gray fine sandstone from the Portland Formation, Dinosaur Footprint Reservation, Holyoke, Massachusetts; ACM-ICH 45/8 and ACM-ICH 45/1, *Eubrontes giganteus* Hitchcock 1836. ACM-ICH 54/10, *Eubrontes approximatum* (Hitchcock 1865), from gray shale at Turners Falls, Massachusetts. ACM-ICH 58/1, *Eubrontes divaricatus* (Hitchcock 1865), locality unrecorded. ACM-ICH 13/4, *Eubrontes platypus* Lull 1904, red shale from Turners Falls, Massachusetts. ACM-ICH 27/17, *Eubrontes tuberatus* (Hitchcock, 1858), red shale from Turners Falls, Massachusetts. ZDM 0032-F2, *Eubrontes zigongensis* (Gao, 2007) holotype, complete natural mold pes track from the Shaba Site, Weiyuan County, Sichuan Province, China. ZDM 0032-F1, F3–F6, *Eubrontes zigongensis* (Gao, 2007) paratypes, in the same trackway as the holotype from the Shaba Site, Weiyuan County, Sichuan Province, China.

Diagnosis. Large (>25 cm in length), bipedal, and functionally tridactyl tracks. Relatively short Digit III, broad pes, rarely preserving Digit I (hallux). Divarication between Digits II–IV averages 25–40° (from Olsen et al., 1998).

Description. See Fig. 6.34. *Eubrontes zigongensis* (Gao, 2007) as described by Xing et al. (2014e) is a large (mean TL and TW = 42.1 and 30.5 cm, respectively), tetradactyl but typically functionally tridactyl track with a short, thin anteromedially directed Digit I (hallux) trace and a pes L/W ratio of 1.4. The tridactyl portion of the track is nearly symmetrical with Digits II–III divarication angle (mean = 27°) about equal to Digits III–IV angle (mean = 28°). In the type trackway, the step is about three times the footprint length with a mean pace angulation as high as 165° (emended diagnosis from Xing et al., 2014e). Xing

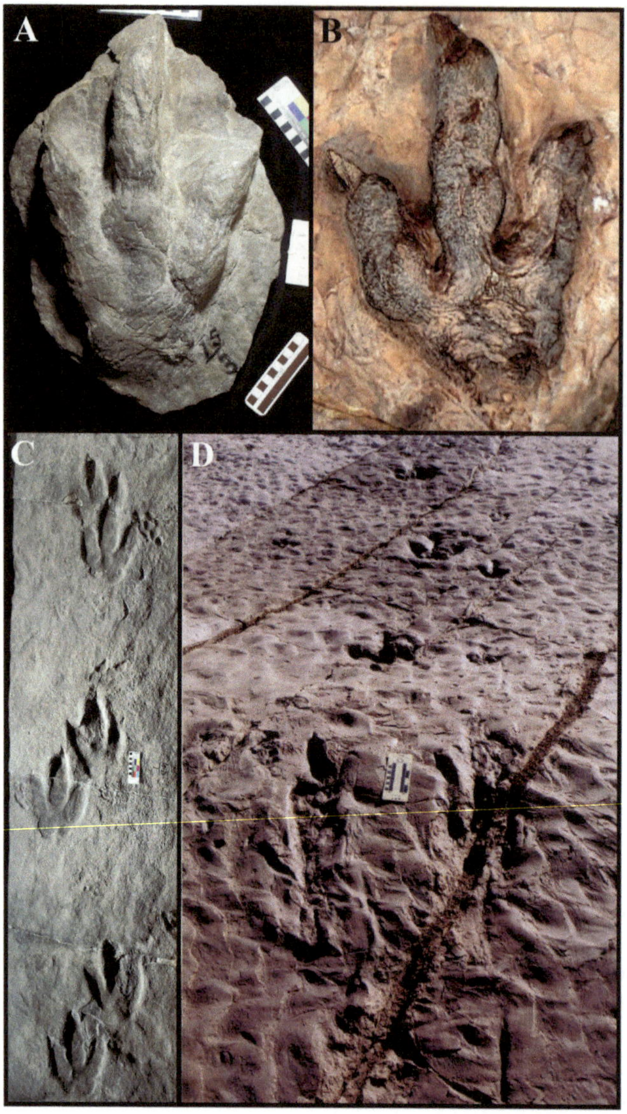

FIGURE 6.34 *Eubrontes.* (A) *E. giganteus* holotype ACM-ICH 15/3. Scale = 10 cm. (B) SGDS 9, Moenave Formation, Utah. (C) *E. giganteus* trackway. (D) SGDS 18-T1 crouching trace with hand traces, and trackway with associated tail drag marks. *(A and C) Courtesy James Farlow. (B) Courtesy Dan Whalen. (D) Courtesy David Slauf.*

et al. (2014e) estimate a speed for the type trackway of ~1.49 m/second or ~5.36 km/hour, with an estimated body length of the track maker to be ~5.4 m.

Eubrontes has other associated characteristics such as crouching traces and tail drag traces (Milner et al., 2009), running trackways, and skin impressions (Hendrickx et al., 2022).

Discussion. Hitchcock (1836) erected *Ornithichnites giganteus* for large, tridactyl tracks from Holyoke, Massachusetts. They were attributed to large birds (*Ornithichnites* = "stony bird tracks," although not used as an ichnogeneric name; Olsen et al., 1998). Hitchcock (1841) changed the name to *Ornithoidichnites*; Hitchcock (1845) then used the name *Eubrontes* ("true thunder") to reflect the name of the animal that produced the tracks (Olsen et al., 1998). Hitchcock never printed the name *Eubrontes* again, but instead substituted it with the name *Brontozoum* (Hitchcock, 1847; Olsen et al., 1998). Hay (1902) realized that *Eubrontes* had priority over *Brontozoum* and that *Ornithoidichnites* was invalid because it was not the name of an animal. All workers thereafter have followed Hay (1902; q.v., Lull, 1953). *Eubrontes* was considered valid because *Ornithoidichnites* was not used for 151 years, per ICZN Art. 23j(i) (Olsen et al., 1998). *Eubrontes giganteus* was finally established as the type ichnospecies by Baird (1957; q.v., Olsen et al., 1998).

Medium-sized theropods such as *Dilophosaurus* and *Liliensternus* are often used as model producers of *Eubrontes* and *Eubrontes*-like tracks, but Weems (2003, 2019) suggested *Eubrontes* was produced by "prosauropod" sauropodomorphs, and commonly used *Plateosaurus* as a model, though few other ichnologists have accepted this, in part because of the abundance of *Eubrontes* tracks in places where no skeletal evidence of "prosauropods" is known.

Eubrontes has been cited as an Early Jurassic indicator, at least in North America (e.g., Olsen et al., 2002), and indeed it is uncommon in the Late Triassic globally. Its Late Triassic occurrences are primarily European (Gierliński and Ahlberg, 1994; Pascual Arribas and Latorre Macarron, 2000; Niedźwiedzki, 2008, 2011); it has also been reported from Africa (Lagnaoui et al., 2012). The ichnotaxon becomes much more common in the Early Jurassic, particularly in the track-rich deposits of the Newark Supergroup in eastern North America (e.g., Olsen et al., 2002), the Glen Canyon Group and correlative strata in the American Southwest (e.g., Suarez et al., 2017), and southern Africa (Sciscio et al., 2016, 2017; Abrahams et al., 2020, 2021, 2023).

Eubrontes and *Eubrontes*-like tracks are also known from the Early Jurassic of Asia. *Weiyuanpus zigongensis* (Gao, 2007) was originally described based on a trackway with six footprints from the Lower Jurassic Zhenzhuchong Formation near Chongqing city, China (Gao, 2007; Chen et al., 2006). Xing et al. (2013e) considered *Weiyuanpus* a junior synonym of *Eubrontes*, but it had not been formally rediagnosed, so Xing et al. (2014e) considered it to be a valid ichnotaxon. *Eubrontes platypus* Lull 1904 was reported from the Lower Jurassic Fengjiahe Formation at the Xiyang Site, Yunnan Province, China (Zhen et al., 1986a,b; Chen et al., 2006); *Paracoelurosaurichnus monax* and *Youngichnus xiyangensis* from the same unit (Zhen et al., 1986a,b; Chen et al., 2006) were synonymized with *Eubrontes* because they are "indistinguishable" from one another (Lockley et al., 2013). *Eubrontes* has been reported and described from the Lower Jurassic Fuxian Formation in Shaanxi Province, China (Wang et al., 2016). "*Eubrontes*"*glenrosensis* Shuler (1935) was reported from about 10 km west of Hailiutu, Wulatzhongqi, Nei Mongol, China (Li et al., 2010). *Eubrontes nobitai* is from the Lower Cretaceous Jiaguan Formation of Sichuan Province, China (Xing et al., 2021d), demonstrating that the ichnotaxon persisted well beyond the Early Jurassic in Asia.

Eubrontes giganteus has been described from the Lower Jurassic Dolomitic Formation at the Mongisty Tracksite, Aveyron, Causses Basin, southern France (Moreau et al., 2021). *Eubrontes veillonensis* is from Le Veillon sites of France described by de Lapparent and Montenat (1967). *Eubrontes* is also known from the Middle Jurassic Lealt Shale Formation of the Isle of Skye, Scotland (dePolo et al., 2018). France is also home to a Late Jurassic occurrence of *Eubrontes* at the famous Crayssac Tracksite in the Cazals Formation (Mazin et al., 2000). See also Farlow et al. (2018).

Eutynichnium
Nopcsa 1923
Fig. 6.35
Type ichnospecies. *E. lusitanicum.*
Valid ichnospecies. *E. lusitanicum* (Nopcsa, 1923), amended Lockley et al. (1998c).
Distribution. Upper Jurassic of western Europe (Cabaços Formation: Portugal).
Type specimen(s). MNHN-MG-P261 (see Figs. 3 and 7 and Table III in Lockley et al., 1998c) with topotypes MNHN-MG-P262-264 and an unnumbered specimen from Cabo Mondego, Portugal.
Diagnosis. Adapted from Lockley et al. (1998c): Medium-sized, tetradactyl theropod footprint with large, functional Digits II–IV and a small, slender, anteromedially directed hallux (Digit I). Size range: pes length excluding hallux 37–40 cm (with hallux 46–58 cm); width 26–27 cm. Digits II–IV are relatively wide and nontapering, without clearly defined pad impressions. Short pace (<100 cm) and the trackway is narrow.

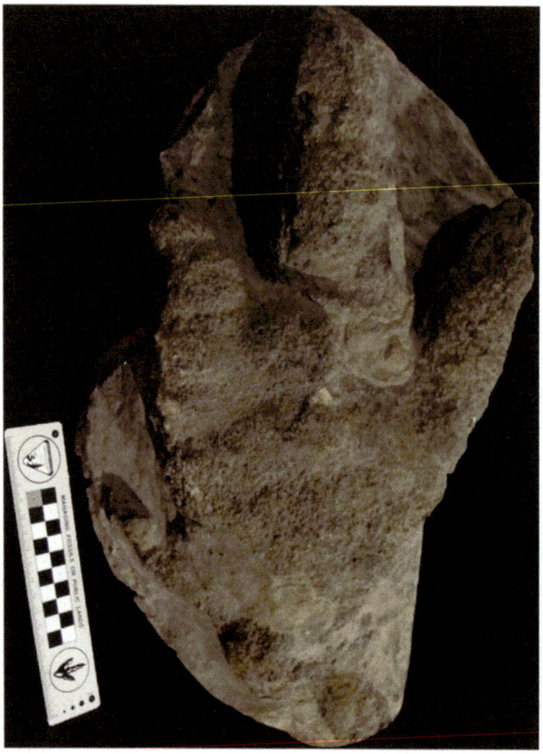

FIGURE 6.35 *Eutynichnium lusitanicum* type specimen, stored at the Museu Nacional de História Natural e da Ciência, Lisbon, Portugal. Scale = 18 cm (ruler). *Photo by Novella L. Razzolini.*

Description. *Eutynichnium lusitanicum* Nopcsa 1923 is a large, tetradactyl, gracile, mesaxonic, theropod track (Fig. 6.35). It has three robust, non tapering Digits (dII–dIV), no clear phalangeal pad impressions, and a broad heel region. It bears a slender, anteromedially facing hallux (dI) impression with a hallux:track length ratio of 0.1 (Xing et al., 2018d). Moreover, the angle between the axis of the hallux and the track axis is ~46° (Razzolini et al., 2017; Xing et al., 2018d).

Discussion. *E. lusitanicum* tracks from Cabo Mondego, housed in the Gomes collection at the MNHN, hold significant historical value as one of the earliest reported Jurassic track samples from Europe, dating back to the 1880s (Gomes, 1916). These tracks are natural casts preserved within a coarse sandy matrix. Their considerable depth, approximately 10–15 cm, accounts for the preservation of hallux and metatarsus traces, while the relatively short step (pace) suggests the track maker traversed a pliable substrate.

Thus far, *Eutynichnium* has only been employed to characterize material discovered in Portugal. Specimens present in the Gomes collections were utilized by Lockley et al. (1998c) to designate a holotype (strictly speaking, a lectotype) for formal description. Lockley et al. (1998c) suggested that the name should be retained to maintain historical consistency despite the original description of Nopcsa being deemed invalid due to its lack of reference to a type specimen.

In addition to the Gomes collection, other tracks from Cabo Mondego, including those in the collections of the Portuguese Geological Survey, are preserved as impressions, displaying differences in size and morphology from *Eutynichnium lusitanicum*. However, whether these differences stem from preservation or variations in track maker morphology is uncertain. Nonetheless, similar track impressions still exist in situ at Cabo Mondego (Lockley and Meyer, 2000: Fig. 7.3). These tracks measure approximately 60 cm in length and 50 cm in width, with a pace of 1.50 m, exhibiting wider digit divarication angles and more sharply tapering digits than in *Eutynichnium lusitanicum*. Morphologically, they resemble tracks found in other Portuguese sites labeled as *Megalosauripus* isp.

Furthermore, Belvedere et al. (2019) highlighted the morphological resemblance between *Megalosauripus* and *Eutynichnium*, particularly when excluding the Digit I trace from the comparison. Although this observation does not affect the validity of the ichnotaxon, it becomes crucial in identifying the track maker. The presence/absence, orientation, and morphology of the Digit I trace can play a role in the identification of the theropod track maker because these features can vary among theropods. Finally, it is worth noting that, in modern literature (e.g., Razzolini et al., 2017; Xing et al., 2018d), the ichnotaxon is often misspelled as "*Euthynichnium*".

Exocampe
Hitchcock 1858
Fig. 6.36
Type ichnospecies. *E. arcta* per Hay (1902).
Valid ichnospecies. None (see below).
Distribution. Lower Jurassic of North America (Turner Falls Formation [Newark Supergroup]: Massachusetts; and Whitmore Point Member, Moenave Formation [Glen Canyon Group]: Utah).

FIGURE 6.36 *Exocampe*. Two pairs of rare, cf. *Exocampe* tracks (SGDS 509) with associated *Batrachopus* and *Undichna* traces from the Lower Jurassic Whitmore Point Member, Moenave Formation, Utah, USA. Scale = 10 cm. *Photo by Andrew R.C. Milner.*

Type specimen(s). ACM-ICH 35/24, *Exocampe arcta* Hitchcock 1858. ACM-ICH 39/69 and ACM-ICH 41/39, *Exocampe ornata* Hitchcock 1858. ACM-ICH 55/4, *Exocampe minima* Hitchcock 1865.

Diagnosis. Tracks of a quadruped. Fore- and hind feet unequal in size. Most digit impressions curved and turned outward from the trackway axis. This diagnosis, by Hitchcock (1858), is, as was the case with many of his ichnotaxonomic diagnoses, depauperate in significant detail, and the ichnotaxon needs restudy.

Description. See Fig. 6.36. For *E. arcta*: When present, pentadactyl manus prints are roughly one-half the length of the pes prints and have digit imprints more strongly turned outward than those of the pes prints. Divarication angle between Digits I–V is 110°. Pes prints tetradactyl, with Digits II–IV oriented anteriorly but Digit I turned and curved outward from the track axis. Digit I imprint straight; all others curved. Divarication angle between Digits I–IV is 110°. Hitchcock (1858) referred to the pes prints as digitigrade, but went on to describe tapering heel imprints included with the tracks that are a little less than half the length of the entire print. Trackways are narrow gauge and have long step lengths; manus prints lie anterior to pes prints, but much farther from the trackway axis. For *E. ornata*: manus pentadactyl and digitigrade, although less so than the pes. Manus prints have four curved digit imprints directed anteriorly and a short one posteriorly; divarication between the four anteriorly oriented digits 130°. Manus prints rotated 40–70° outward from trackway axis. Pes tetradactyl and, as with *E. arcta*, reported as digitigrade but with a heel imprint. Some digit imprints terminate in claw imprints. Divarication angle between Digits I–IV is 100–130°. Pes prints rotated outward from trackway axis 15–25°. For *E. minima*: differs from *E. ornata* primarily in size and slenderness. Manus prints similar to those of *E. ornata* in having four anteriorly oriented digits and a short one oriented posteriorly; divarication of latter is 138°. Digit imprints are nearly straight.

Discussion. Hitchcock (1858) noted that the outward curvature of *Exocampe* digit imprints was turtle-like, but the narrowness of the body suggested a lacertilian track maker. Lull (1915) and Abel (1926) placed *Exocampe* under "amphibian" tracks, but Lull (1953) placed it under "Reptilia *incertae sedis*" tracks. More recently, *Exocampe* has been suggested to be sphenodontian in origin (Harris and Milner, 2015; Milner et al., 2011). Reexamination of the type and referred specimens is called for.

Rainforth (2005a) determined that *Exocampe* was indistinguishable from *Orthodactylus* Hitchcock 1858, which has page priority over *Exocampe* and therefore is a senior synonym, so she removed all ichnospecies of *Exocampe* to *Orthodactylus*. This has not received widespread support, although mentions of either *Exocampe* or *Orthodactylus* have been virtually nonexistent since 2005; certainly, no further analyses of specimens of either have been published.

Lull (1918) referred tracks from the Lower Permian Hermit Formation (per Gilmore, 1927) of Grand Canyon, Arizona to "*Exocampe* (?) *delicatula*," but Gilmore (1927) removed this ichnospecies to *Batrachichnus*. Numerous unpublished, online reports state that *Exocampe* is synonymous with the crocodyliform ichnotaxon *Batrachopus*, but these probably stem from mis-readings of the name *Batrachichnus* as applied by Gilmore (1927).

Fractisemita

Raisanen and Hasiotis 2018

Fig. 6.37

Type ichnospecies. *F. henrii.*

Valid ichnospecies. *F. henrii.*

Distribution. Upper Jurassic of North America (Morrison Formation: Utah).

Type specimen(s). Holotype and paratypes are field specimens in the Salt Wash Member of the Morrison Formation near Ticaboo in Garfield County, Utah (Raisanen and Hasiotis, 2018).

FIGURE 6.37 *Fractisemita* holotype (not collected), Morrison Formation, Utah, USA. *From Raisanen and Hasiotis (2018).*

Diagnosis. From Raisanen and Hasiotis (2018): "Unlined, sparse maze of Y-branched tunnels and shafts of varying diameter, angle from the horizontal, and length, with no distinguishable chambers."

Description. Unlined burrows filled with carbonate, containing no internal structures and comprising an interconnected network of cylindrical shafts and tunnels at varying angles and forming a maze pattern. Section lengths ~10–93 cm, oval cross-sections of ~5 cm average height and ~6 cm in width, inclined sections average 31° from horizontal (Fig. 6.37).

Discussion. *Fractisemita* has an architectural morphology, with its interconnected network of shafts at a range of angles, that appears to be unique among Mesozoic terrestrial burrows but is somewhat similar to other burrows attributed to tetrapods (Hasiotis, 2004; Raisanen and Hasiotis, 2018). The burrows appear to have been made by small vertebrates, possibly social mammals (Raisanen and Hasiotis, 2018). The structures may be burrows overprinted and distorted by carbonate cementation. Also see discussion of this ichnotaxon in the chapter on vertebrate burrows in this volume by Melchor and Cardonata, who consider *Fractisemita* to be a nomen dubium.

Gabirutosaurichnus
Gabuniya and Kurbatov 1988
Fig. 6.38

Type ichnospecies. *G. tianschanicus.*

Valid ichnospecies. *G. tianschanicus.*

Distribution. Middle Jurassic of Asia (Gabirut and/or Dzhirut formations: Tajikistan).

Type specimen(s). Gabuniya and Kurbatov (1988) listed the holotype specimen as No. 5, presumably of the Ministry of Geology of the Academy of Sciences of the USSR (because No. 1 at that institution is the holotype of *Ravatichnus*), a left pes print. This ministry was apparently abolished in 1991 and its materials transferred to the Ministry of Natural Resources and Ecology of the Russian Federation; the current status and catalog number of the holotype is unknown. A cast of the holotype was reportedly deposited in the Museum of the Ministry of Geology of the Uzbek SSR (possibly now the Museum of Geology in Tashkent, Uzbekistan?), but no catalog number was provided. The holotype appears to be a natural mold.

Diagnosis. Tridactyl to tetradactyl prints of a digitigrade biped. Digit imprints are long except for that of the sometimes-present Digit I, which is offset to the side; digit imprints

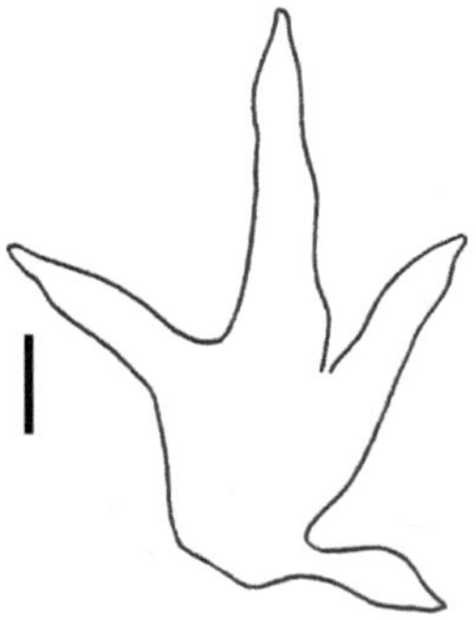

FIGURE 6.38 *Gabirutosaurichnus tianschanicus.* Scale = 6 cm. *From Gabuniya and Kurbatov (1988), Fig. 3.*

also often bear claw imprints. Digit III protrudes anteriorly; divarication angles between the primary digits are wide (up to 60° between Digits III and IV and 40° between Digits II and III).

Description. See Fig. 6.38. Gabuniya and Kurbatov (1988) likened *Gabirutosaurichnus* tracks to those of *Satapliasaurus*, but noted they were larger and had greater divarication angles. They also compared them to *Changpeipus* (Young, 1960) but dismissed synonymy because no *Changpeipus* tracks included a Digit I imprint and because Digit III does not protrude as "sharply anteriorly" (possibly meaning anterior prominence) as in *Gabirutosaurichnus*. Similarly, they dismissed synonymy with *Anchisauripus* because *Gabirutosaurichnus* tracks were longer and again had greater divarication angles between digits.

Discussion. *Gabirutosaurichnus* tracks are from the same locality as *Camptosauropus* (see above) and *Ravatichnus* (see below) tracks and have the same history. *Gabirutosaurichnus* tracks have been attributed to theropod dinosaurs. Romanovsky (1882) discussed tracks from the Ravat region of Tajikistan as *Brontozoum tianschanicum* (nomen nudum) based on similarities to *B. giganteum* (Hitchcock, 1847), which is a junior synonym of the theropod ichnospecies *Eubrontes giganteus* (Rainforth, 2005b). *Eubrontes* tracks are known to occasionally preserve Digit I imprints, so synonymy is possible. As illustrated, however, digit imprints of *Gabirutosaurichnus* appear to be thinner than those of typical *Eubrontes*, as well as *Gigandipus*. Further comparison to other Jurassic theropod ichnotaxa, such as *Carmelopodus*, *Eutynichnium*, *Megalosauripus*, and *Therangospodus*, is needed. *Gabirutosaurichnus* was mentioned by Sabirov (2008), but otherwise has received no scientific treatment and likely is in need of redescription and revision.

***Gigandipus* (=*Eubrontes*?)**
Hitchcock 1856
Fig. 6.39

Type ichnospecies. *G. caudatus*.

Valid ichnospecies. *G. caudatus*, *G. hei*.

Distribution. Lower Jurassic of eastern North America (Portland Formation, Newark Supergroup, Connecticut); Lower Jurassic of the southwestern United States (SGDS, lower Whitmore Point Member, Moenave Formation, Glen Canyon Group, Utah).

Type specimens. ACM-ICH 9/9 and ACM-ICH 9/10, *Gigandipus caudatus* Hitchcock 1858, from reddish shale at Lily Pond, Turners Falls, Massachusetts.

Diagnosis. From Milner et al. (2009): Like *Eubrontes* except it includes a medially or caudomedially oriented Digit I trace (Hitchcock, 1858; Lull, 1953; Bock, 1952); tail drag marks present in the holotype and other referred specimens that were suggested to have been diagnostic (Hitchcock, 1858; Weems, 2003).

Description. *Gigandipus* (Fig. 6.39) tends to have a more medially directed Digit I (hallux) (Hitchcock, 1856; Lull, 1953; Xing et al., 2014e), but it can also be anteromedially oriented (e.g., SGDS 50 (Milner et al., 2006a, 2009; Weems, 2019), which is more similar to *Eubrontes*. However, many examples exist of *Eubrontes* with Digit I traces oriented posteromedially instead (e.g., many of the left tracks in trackway SGDS 18-T1; Milner et al., 2009), more like the orientation of the hallux often illustrated for *Gigandipus caudatus* (Lull, 1953; Weems, 2019).

Discussion. Bock (1952) and Harris et al. (1996) regarded *Gigandipus* as an extramorphological variant of *Eubrontes* because the track maker sank deeper into the substrate, bringing

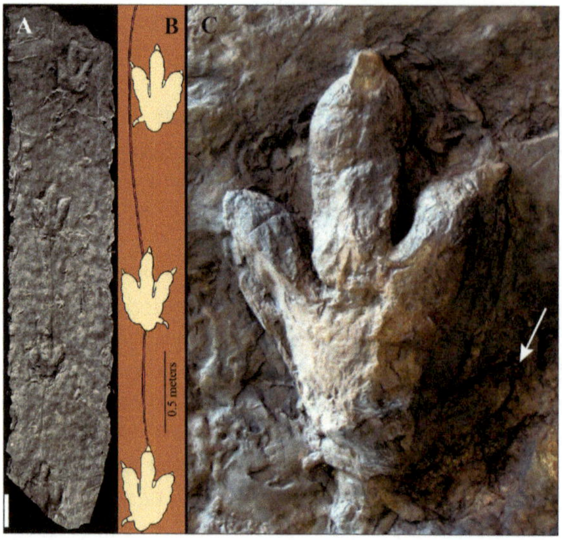

FIGURE 6.39 *Gigandipus.* (A) *G. caudatus* holotype, ACM-ICH 9/10 from the Portland Formation, Massachusetts (scale = 20 cm); (B) modified sketch of *G. caudatus* holotype, ACM-ICH 9/9 from the Portland Formation, Massachusetts; (C) SGDS 50, right natural cast track with white arrow point to end of distal Digit I trace oriented anteromedially. From the Moenave Formation, Utah. *(A) Courtesy Paul Olsen. (B) Modified from Lull (1953). (C) Photo by Andrew Milner.*

the more distally placed Digit I into contact with the substrate (Xing et al., 2014e). However, many *Eubrontes* tracks can lack Digit I trace impressions even though some tracks are deeper than some *Gigandipus* examples (Weems, 1992; Xing et al., 2014e). This could imply that different species of larger, Early Jurassic theropods that produced *Eubrontes* and *Gigandipus* footprints could have variations in Digit I (hallux) lengths and/or variation in placement and orientation of Digit I on the feet (Xing et al., 2014e). Therefore, it is considered that these two ichnogenera are likely to be synonymous in part (Weems, 2003; Rainforth, 2005a; Milner et al., 2009; Xing et al., 2014e). *Gigandipus hei* was described by Yang and Yang (1987) and Lockley et al. (2003).

Gigantosauropus nomen dubium
Mensink and Mertmann 1984
Fig. 6.40
Type ichnospecies. *G. asturiensis.*
Valid ichnospecies. None.
Distribution. Upper Jurassic of Europe (Tereñes Formation: Spain).
Type specimen(s). In-place trackway at La Griega Beach locality, Asturias, Spain.

Diagnosis. From Mensink and Mertmann (1984): Bipedal, tridactyl, plantigrade footprints with short, blunt toes and a clear heel extended from Digits III and IV; foot length 1.35 m, foot width 1.18 m, stride 2.35 m. Toe angle 20–25°, distinct foot pads.

Description. See Fig. 6.40. From Lockley et al. (2007b): "The shape of pes 1 seems to be typical of sauropods, while that of pes 2 is similar in size and it contains the possible

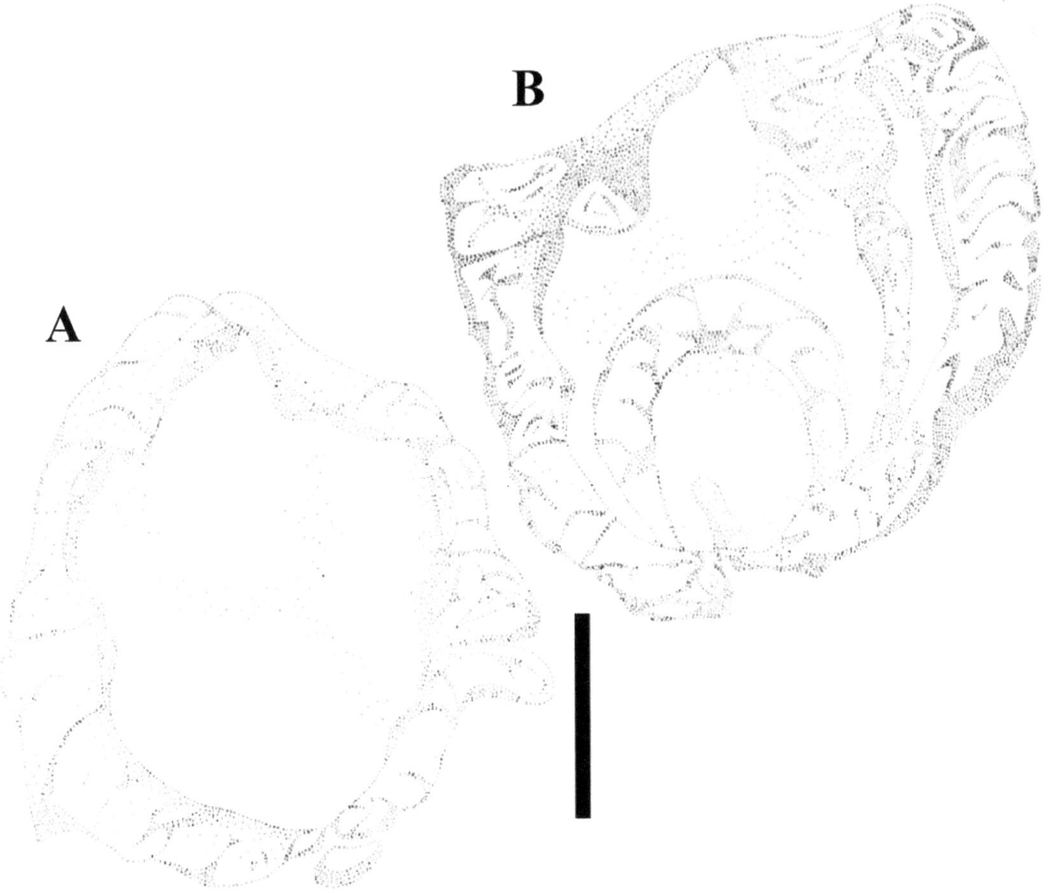

FIGURE 6.40 Sketches of *Gigantosauropus asturiensis* paratype (A) and holotype (B). Scale = 50 cm. *(B) Modified from Mensink and Mertmann (1984), Fig. 3.*

impression of one of the larger inner claws, perhaps Digit I. The ratio between the areas of manus 2 and pes 2 (and/or pes 1) may give an approximate indication of the manus/pes size ratio: i.e., the heteropody value is 1:6 between pes 2 and manus 2, or 1:5 for the ratio between manus 2 (and pes 1).The trackway also appears to have a relatively narrow gauge, with a maximum trackway width of only 16 cm between the inner margins of the pes. However, all these measurements are tentative given the sub-optimal preservation."

Discussion. The material assigned to *Gigantosauropus asturiensis* Mensink and Mertmann 1984 was reexamined by Thulborn (1990) and Lockley et al. (1994a) and shown to pertain to a sauropod, rather than a theropod, track maker. Lockley et al. (2007b) redescribed the trackway in detail and like Lockley et al. (1994a), established that it was not diagnostic. These authors regarded *Gigantosauropus* as a nomen dubium.

Grallator
Hitchcock 1858

FIGURE 6.41 *Grallator.* (A) *Grallator tenuis*, ACM-ICH 12/3, from the Portland Formation, Massachusetts. Scale = 10 cm. (B) *Grallator* isp. (SGDS 1289) from the lower Whitmore Point Member, Moenave Formation, SGDS, Utah. Scale = 14.5 cm. (C) *Grallator* slide trace (SGDS 74) from Moenave Formation, Utah. Scale = 18 cm (ruler). *(A) Courtesy Jim Farlow. (B) Photo by Andrew Milner. (C) Photo by Andrew Milner.*

Fig. 6.41

Type ichnospecies. *G. parallelus.*

Valid ichnospecies. *G. parallelus, G. cursorius.*

Distribution. Upper Triassic–Lower Jurassic of North America (Newark Supergroup: Connecticut, Massachusetts; Chinle, Moenave, Wingate, Kayenta, Navajo, Nugget, and other formations of the American Southwest), Africa (Clarens Formation: South Africa; Elliot Formation: Lesotho), Europe (Przysucha Formation: Poland; Mecsek Coal Formation: Hungary; Sandstones-Variegated Mudstones Formation and the Dolomitic Formation, France), and Asia (Fengjiahe and Tuchengzi formations: China). Middle Jurassic of South America (La Matilde Formation: Argentina). Upper Jurassic of Europe (Lastres Formation: Spain; Sand-Tonkomplex Member: Germany) and North America (Morrison Formation: Utah). Possibly Upper Jurassic of Europe (unspecified limestone unit: France).

Type Specimen(s). ACM-ICH 4/1a, *Grallator parallelus* Hitchcock 1847 is the natural cast holotype associated with *Anchisauripus sillimani* and the holotype of *Otozoum moodii* from the

Lower Jurassic Portland Formation at Moody's Corner (Moody homestead), South Hadley, Massachusetts (according to Olsen et al., 1998). The *G. parallelus* holotype is a trackway of four successive pes tracks. ACM-ICH 12/3 and ACM-ICH 12/4, *Grallator tenuis* Hitchcock 1858, from red shale at Turners Falls, Massachusetts.

Diagnosis. From Olsen et al. (1998): Small (<15 cm long), bipedal, functionally tridactyl ichnites, and Digit I (hallux) traces are rare. Digit III projects farther anteriorly, and the track is narrower than in *Eubrontes* and *Anchisauripus* with a length/width ratio near or greater than 2. Divarication angles between digit traces II–IV = 10–30°.

Description. See Fig. 6.41. *G. parallelus* is a small ichnospecies (<8 cm long) with a Digit III projection ratio of 1.3, and a length/width ratio of about 2.2 (Olsen et al., 1998). In addition, Digit II projects farther along the axis of Digit III and is more robust than the Digit IV trace, which is narrow. Divarication between Digits II–IV in *G. parallelus* averages 28°.

Discussion. *Grallator* tracks have been nearly universally acknowledged as theropod dinosaur tracks. Late Triassic and Early Jurassic track makers of *Grallator* were smaller coelophysoid theropods (Raath et al., 1990), such as *Coelophysis*, *Megapnosaurus*, and *Segisaurus*.

Hitchcock (1847) named the smallest theropod trackway on slab ACM-ICH 4/1a (now the holotype specimen for both *Otozoum moodii* and *Grallator parallelus*) *Brontozoum parallelum* (now *G. parallelus*), along with trackways on slab ACM-ICH 23/2 (old number 137) from the old ferry landing at Gill, Turners Falls, Massachusetts (Olsen et al., 1998). ACM-ICH 23/2 also preserves the type of *Antipus bifidus* (Hitchcock, 1858), now considered likely to be a synonym of *Batrachopus*.

Hitchcock used the name *Brontozoum parallelum* for small, tridactyl trackways on slabs ACM-ICH 4/1 and 23/2 until 1858, when he used these same specimens as the first ichnospecies for the new ichnogenus *Grallator*, but he placed them in a new ichnospecies, *G. cursorius* (Hitchcock, 1858, pl. 13, Fig. 3; pl. 33, Fig. 5; pl. 36, Fig. 8; pl. 58, Fig. 4). Clearly Hitchcock (1858) ignored the former name *B. parallelum* (Hitchcock, 1847), replacing it with *G. cursorius*, making this ichnospecies an objective synonym of *B. parallelum* (Olsen et al., 1998). Adding to the confusion, Hitchcock (1865) designated a new ichnospecies, *G. parallelus*, implying the same meaning as *B. parallelum*, making these two ichnospecies homonyms (Olsen et al., 1998). At the same time, Hitchcock (1865) did not list ACM-ICH 4/1 and 23/2 in the hypodigm of *G. parallelus*, nor did he mention *B. parallelum*. In any event, *G. parallelus* and *B. parallelum* are the same ichnospecies, and the correct ichnospecies for *Grallator* is *G. parallelus*, not *G. cursorius* (see Olsen et al., 1998 for further details).

Specimen ACM-ICH 54/8, listed as a syntype of *G. parallelus* by Hitchcock (1865), was called the holotype of the ichnogenus *Anchisauripus* by Lull (1904, 1915, 1953). To heighten the confusion, *Anchisauripus parallelus* has been used since by several authors (e.g., Baird, 1957; Haubold, 1971a; Smith, 1982). Olsen et al. (1998) formally recognized *A. parallelus* as a senior synonym of *G. parallelus*, with ACM-ICH 4/1 as the holotype specimen. They also stated that ACM-ICH 32/2 and 54/8 have no significance as type specimens (Olsen et al., 1998).

Sauropus barrattii (Lull, 1953) from the Lower Jurassic Portland Sandstone, Newark Supergroup at Gill, Massachusetts, is represented by specimen ACM-ICH 1/1, a coelophysoid theropod crouching trace now assigned to *Grallator tuberosus* (see Fig. 6.89; Gierliński, 1994; Lockley et al., 2003). Other crouching *Grallator* traces have been recorded from the Newark Supergroup, the Lower Jurassic of Poland, and from the Glen Canyon Group in the Kayenta and Navajo formations of Utah (Gierliński, 1996c; Milàn et al., 2008; Gierliński et al., 2009c; Lockley et al., 2014c; Milner et al., 2024). Running and limping trackways have also been recorded. Famous crouching *Grallator* traces, such as ACM-ICH 1/7, were assigned to *G. minusculus* by Gierliński (1994; q.v., Lockley et al., 2003).

Grallator has been reported from the Lower Jurassic Clarens and Elliot formations of South Africa and Lesotho, respectively (Bordy, 2021; Lallensack et al., 2022). It has also been reported from the Middle Jurassic La Matilde Formation of Argentina based on five isolated tracks and two associated in a partial trackway (de Valais, 2011). *Grallator lescurei* (Demathieu, 1990) from the Causses region of France was described based on specimens from Saint-Léons. *Grallator limnosus* is from the Lower Jurassic Fengjiahe Formation at the Xiyang Site of Yunnan Province, China and this ichnospecies is larger than the diagnosed size for *Grallator* (Zhen et al., 1986a,b; Chen et al., 2006). *Grallator madseni* is from the Cameron Dinosaur Tracksite, Dinosaur Canyon Member, Moenave Formation, Glen Canyon Group (Irby, 1995). *Grallator olonensis* (Lockley and Meyer, 2000) is from the Triassic—Jurassic-boundary-spanning Le Veillon sites of France described by de Lapparent and Montenat (1967). *Grallator sauclierensis* (Demathieu and Sciau, 1992; Lockley and Meyer, 2000) and other *Grallator* from the Causses region of France were redescribed by Demathieu et al. (2002). *Grallator tenuis* (Hitchcock, 1858) was suggested to occur in the Upper Hettangian Przysucha Formation, Gliniany Las near Mniów, Poland (Gierliński, 1996a). *Grallator tuberosus* (Hitchcock 1836) has been reported from the lower Sinemurian Mecsek Coal Formation, Pécsbánya Mine in Hungary (Gierliński, 1996b). *Grallator* isp. has been reported from the Upper Jurassic Lastres Formation of Spain (Castanera et al., 2016b), and cf. *Grallator* and grallatorid types are known from the Upper Jurassic Morrison Formation of the western USA (Lockley et al., 1998e, 2018a; Foster and Lockley, 1995; 2006). Grallatorid tracks are not necessarily rare in the Late Jurassic. Grallatorid tracks tentatively assigned to cf. *Kalohipus* (an otherwise Lower Cretaceous ichnotaxon) were identified in the Upper Jurassic Reuchenette Formation of Switzerland by Castanera et al. (2018). *Grallator* has also been reported from the Middle Jurassic of the Changdu Region of eastern Tibet (Li et al., 2022). In Europe, *Grallator* tracks have been reported from siliciclastic, biolaminate tidal flat and sand flat—sand bar facies of the Sand-Tonkomplex Member, Ornatenton Formation (Upper Kimmeridgian, Upper Jurassic) at Bergkirchen, Germany (Diedrich, 2011). Diedrich (2011) suggested dryosaurid ornithopods as a possible track-producers, although he also suggested theropods. Also from Europe, *Grallator*-like tracks are from the Middle Jurassic of the United Kingdom (Clark and Barco Rodriguez, 1998; Clark et al., 2005). See also Farlow et al. (2018) for more data and measurements.

Hatcherichnus
Foster and Lockley 1997
Fig. 6.42
Type ichnospecies. *H. sanjuanensis.*
Valid ichnospecies. *H. sanjuanensis.*

Distribution. Upper Jurassic of North America (Morrison Formation: Utah, Colorado, Wyoming), and Europe (Lastres Formation: Spain), and the mid-Cretaceous of Africa (Midelt Formation: Morocco). Possibly the Upper Jurassic of Europe (Lourinhã Formation: Portugal) and the mid-Cretaceous of North America (Dakota Group: Colorado).

Type specimen(s). Left manus and pes natural sandstone cast tracks, and associated tail drag, in roof of Gray Daun Mine (along Two Mile Creek, La Sal Mountains, US Department

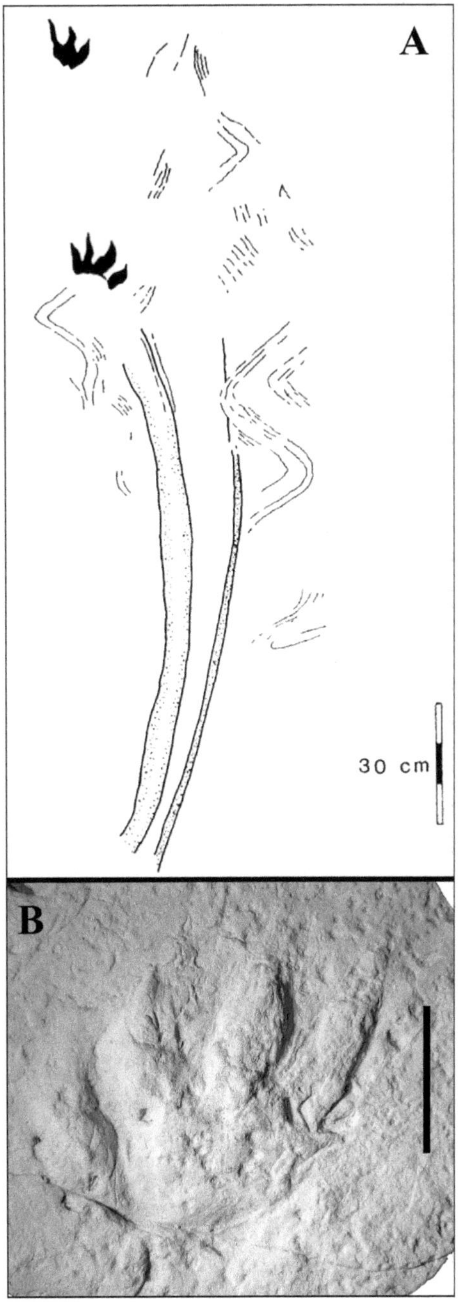

FIGURE 6.42 (A) *Hatcherichnus* holotype schematic based on tracing from roof of a uranium mine in southeastern Utah, USA (reversed to show view from above). Note left manus and pes, tail drag, and associated sedimentary "swish" marks, presumably from water movement. Salt Wash Member of the Upper Jurassic Morrison Formation. (B) *Hatcherichnus*, CM 924 (UCM 49167 replica), Four Mile Creek, Cañon City, Fremont County, Colorado, USA. Scale = 5 cm. *(A) From Foster and Lockley (1997). (B) Photo by Andrew Milner.*

of Agriculture Forest Service land, San Juan County, Utah); also preserved as plaster casts CU-MWC 197.1 (pes) and 197.2 (manus) at University of Colorado Museum.

Diagnosis. Modified from Foster and Lockley (1997): Tetradactyl pes impression consisting of digit impressions only, Digit III longest, Digit II second longest, Digits I and IV nearly equal in length and shorter than II and III by about 25%. Digit I and II impressions slightly recurved laterally. Claw impressions on digits. Total divarication between Digits I and IV about 70°. Manus impression with only Digits I–III preserved, increasing in length I through III. Digit impressions laterally recurved, and Digit II showing a claw impression.

Description. See Fig. 6.42. The type specimen appears to be a resting-to-swimming trace with a left manus and pes with only the digits impressed (I–III in the manus, I–IV and no heel imprint in the pes). A tail drag, consisting of parallel thicker and thinner grooves, is present behind and to the right of the pes. Associated with the feet and tail marks are sinuous sets of very small ripples that in at least one spot overprint the tail drag. The impression is of an animal in shallow water poling off the bottom with both left feet and swishing its tail to move forward.

Discussion. The track maker of *Hatcherichnus* appears to have been swimming or floating neosuchian crocodyliforms such as goniopholidids (Foster and Lockley, 1997; Avanzini et al., 2010b). Reported originally from the Salt Wash Member of the Morrison Formation (Upper Jurassic) in eastern Utah and undifferentiated Morrison near Garden Park, Colorado (Foster and Lockley, 1997), then from a second site in eastern Utah (Foster and Lockley, 2006; Lockley and Foster, 2010) and another from Wyoming (Foster and Lockley, 2006), and Asturias, Spain, from the Upper Jurassic Lastres Formation (Avanzini et al., 2007, 2010b; Lockley et al., 2010a). It has also been reported from the Cretaceous Dakota Group of Colorado (Kukihara et al., 2010) and Midelt Formation of Morocco (Klein et al., 2018). A probable crocodylian track from the Lourinhã Formation of Portugal (Mateus and Milàn, 2010) may be referable to *Hatcherichnus*, although it does appear to possess a heel impression.

Hatcherichnus is distinct from other crocodylian traces. Its pes tracks differ from those of *Kuangyuanpus* in having broadly similar length of Digits I and IV, and II and III, respectively, unlike the generally increasing length of I–IV in *Kuangyuanpus*; it also differs from that ichnogenus in having a tail drag. Its pes tracks likewise differ from those of *Batrachopus* in having Digits II and III of nearly equal length, a relatively shorter Digit IV, and in lacking both a heel impression and a Digit V impression. Manus prints differ from those of *Kuangyuanpus* in having only Digits I–III impressed, and those of *Batrachopus* in lacking impressions of Digits IV and V, in having less straight digits, and in having less divarication between Digits I and III.

Hispanosauropus
Mensink and Mertmann 1984
Fig. 6.43

Type ichnospecies. *H. hauboldi.*

Valid ichnospecies. *H. hauboldi.*

Distribution. Upper Jurassic of Europe (Lastres and Tereñes formations: Spain) and North America (Morrison Formation: Utah).

Type specimen(s). In situ tracks at Ribadesella, Spain, and photograph and illustration of holotype track in Mensink and Mertmann (1984: Figs. 4 and 5). According to Lockley et al.

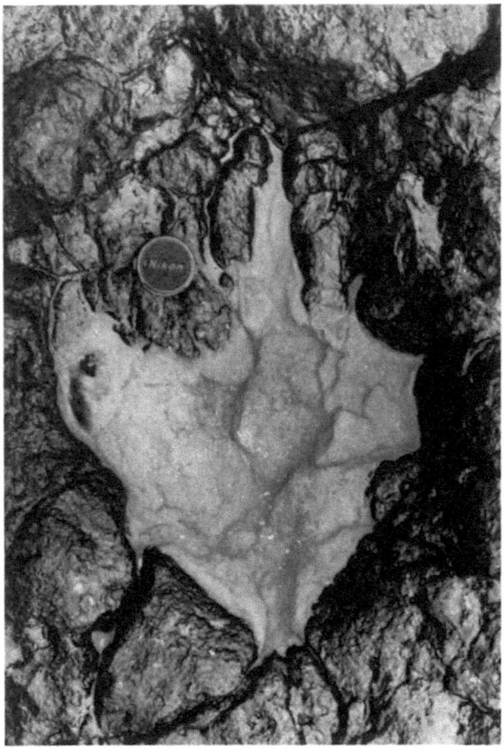

FIGURE 6.43 *Hispanosauropus* holotype, no specimen number. Lens cap for scale. *Clipped from Mensink and Mertmann (1984).*

(2007b), the in situ holotype may be lost to erosion because it could no longer be located in the field. They designated a paralectotype (their Fig. 5C) from the type locality.

Diagnosis. Moderately sized, tridactyl tracks (pes length ~ >52 cm, width ~ 36 cm) of a bipedal track maker. The heel trace is distinctly narrow and the pes length:width ratio is 1.4. The sole of the foot is fleshy.

Description. Moderately large, robust, tridactyl tracks with narrow "heel" and "sole" trace and short digits relative to the total pes length (Fig. 6.43). Track shape is elongated as defined by a high pes length-to-width ratio, and the total digit divarication is between 15° and 30°. The stride is 1.6 m.

Discussion. *Hispanosauropus* is a large theropod dinosaur ichnotaxon. Speculation of a megalosaurian (e.g., Avanzini et al., 2012) or allosauroid (Foster, 2015) affinity for these tracks generally has been acknowledged. Reported originally from the Upper Jurassic of Spain (Mensink and Mertmann, 1984; Lockley et al., 2007b), *Hispanosauropus* has since been identified in the Morrison Formation of western North America (Foster, 2015; Hunt-Foster et al., 2016). Comparisons were drawn by Mensink and Mertmann (1984) between *Hispanosauropus* and *Megalosauripus* and megalosaurid morphotypes. This was later explored by Lockley et al. (1998c, 2007b) who considered that *H. hauboldi*, while lacking a formal holotype, may be a junior synonym of *Megalosauripus* because it exhibits similarities to Late Jurassic cf.

Megalosauripus tracks found in Portugal and Germany. However, in the absence of a solid definition and diagnosis for *Megalosauripus*, Lockley et al. (2007b) asserted that *H. hauboldi* should remain a valid taxon until this is resolved. Subsequently, *Hispanosauropus* often has been considered a junior synonym of *Megalosauripus* (e.g., Piñuela Suárez, 2015; Razzolini et al., 2017; Rauhut et al., 2018). Avanzini et al. (2012) described large theropod tracks under the label "*Hispanosauropus* morphotype," stating that the relation with *Megalosauripus* is complex and "a revision is now needed, and we plan to undertake it in the near future." Foster (2015) maintained that Middle–Upper Jurassic *Megalosauripus* from the basal Morrison Formation and underlying units could be morphologically distinguished from stratigraphically higher occurrences of *Hispanosauropus*, also in the Morrison Formation, in the western USA. Further work is needed to properly resolve the *Hispanosauropus–Megalosauripus* relationship.

Hopiichnus
Welles 1971
See *Anomoepus* (also see Fig. 6.44).

Huanglongpus nomen dubium
Yang and Yang 1987
Fig. 6.45
Type ichnospecies. *H. shengouensis.*
Valid ichnospecies. None.

FIGURE 6.44 *Hopiichnus shingi*, UCM 124.7 (replica of UCMP 83760), Navajo sandstone, Arizona. Scale = 5 cm. *Photo by Andrew Milner.*

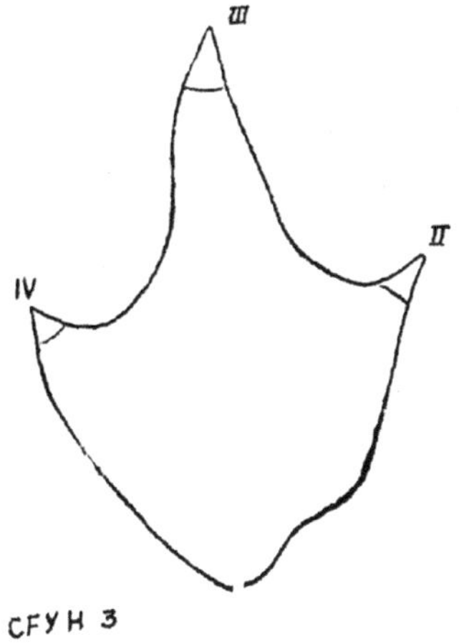

FIGURE 6.45 Sketch of *Huanglongpus shengouensis. Modified from Yang and Yang (1987), Fig. 19.*

Distribution. Upper Jurassic of Asia (Penglaizhen Formation: Sichuan, China, per Chen et al., 2006).

Type specimen(s). CFYH3, a single, natural mold, right pes print from a trackway of five tracks.

Diagnosis. Tracks of a tridactyl biped. Digit imprints lack segmented digital pads. All digits terminate in small, pointed claw imprints. Digit imprints possibly connected by webbing.

Description. See Fig. 6.45. Digit III is the longest; Digit II is the shortest. Digit divarication angles II–III = 32° and III–IV = 42°. Trackways narrow (high pace angulation).

Discussion. Yang and Yang (1987) were unable to pick a track-making taxon for *Huanglongpus*. Lockley et al. (2013) declared the material on which *Huanglongpus* is based to be too poorly preserved to be diagnostic and referred to it as a nomen dubium.

Iberosauripus

Cobos, Lockley, Gascó, Royo-Torres, and Alcalá 2014

Fig. 6.46

Type ichnospecies. *I. grandis* (Cobos et al., 2014).

Valid ichnospecies. *I. grandis.*

Distribution. Upper Jurassic and Jurassic–Cretaceous transition of Europe (Villar del Arzobispo and Huérteles formations: Spain).

Type specimen(s). In situ trackway 1CB at the El Castellar site (CT-1). The site is legally protected as a Paleontological Area by the Aragón Government. CPT-4055 is a cast with these three footprints (1CB1.4; 1CB1.5 and 1CB.6), on exhibit at the Museo Aragonés de Paleontología in Dinópolis–Teruel (Aragón, Spain).

Diagnosis. Adapted from Cobos et al. (2014): Tracks relatively wide with weak mesaxony and a wide, transverse, or poorly elongate, heel. Broad, slightly padded toes. These tracks show the following autapomorphies in relation to other large theropod tracks: (1) The anterior triangle subtended by the distal traces of Digits II, III, and IV is very obtuse, with a mean

FIGURE 6.46 *Iberosauripus grandis* orthophoto generated from 3D model of the track 1CB1.5, part of the holotype trackway (Cobos et al., 2014). Scale bar = 15 cm.

L/W ratio of about 0.3; (2) The distal impressions of Digits II and IV are almost equidistant from the perpendicular line that passes through the most proximal part of the "heel" in respect to the main axis of the track, along the midline of Digit III; (3) Digit III proportionally long in relation to the total track length; and (4) The distal anterior half of the track shows some parallelism among its digits (which occurs at several points in the anterior half of the track) due to a very low interdigital angle of less than 20° and a length/width (FL/FW) ratio between 1.20 and 1.30.

Description. The holotype consists of a 13 m-long trackway comprising eight consecutive footprints. These footprints are large, robust, tridactyl tracks with broad digits and weak mesaxony (Fig. 6.46). Digits show evidence of large, medially oriented claw marks. Pes prints are 54–58 cm in length and have widths between 42 and 48 cm, giving a length:width ratio of 1.28. The heel is rounded. The trackway has a pace angulation of 160°.

Discussion. *Iberosauripus grandis* was originally reported from the El Castellar site (CT-1, Villar del Arzobispo Formation) in Teruel province by Cobos et al. (2014), and the authors indicate that other previously described Berriasian tracks from Spain could be included in this ichnogenus at least at ichnogenus level, such as some specimens from the Oncala Group in Soria (Barco et al., 2005) and possibly La Rioja (Casanovas et al., 1995) provinces. Castanera et al. (2015) confirmed the presence of *Iberosauripus* in the Oncala Group (Huérteles Formation), restudying the specimen from Soria, and reported a new occurrence. Belvedere et al. (2019) noted the similarities between *Iberosauripus* and *Jurabrontes*, especially because both represent giant and robust theropod tracks but emphasized the difference in the outline and the digit proportions (dIII shorter in *Iberosauripus*), so the authors considered them two different ichnotaxa.

Iguanodonichnus

Casamiquela and Fasola 1968, emended by Moreno and Benton 2005

Fig. 6.47

Type ichnospecies. *I. frenki* (Casamiquela and Fasola, 1968).

Valid ichnospecies. *I. frenki.*

Distribution. Upper Jurassic of South America (Baños del Flaco Formation: Chile).

Type specimen(s). Trackway 3, cast of footprint 6, SGPV 1151; Termas del Flaco locality.

Diagnosis. From Moreno and Benton (2005): "Sauropod trackway of medium size (footprint length approximately 50–70 cm), characterized by small space between trackway midline and inside margin of pes tracks. Step angle less than 110°. Pes footprint longer than wide, with long axis rotated outward. Pes claw impression corresponding to Digit I is long and narrow and follows a straight line parallel to the footprint axis. Claw impressions in Digits II–IV are broad and much less developed than Digit I."

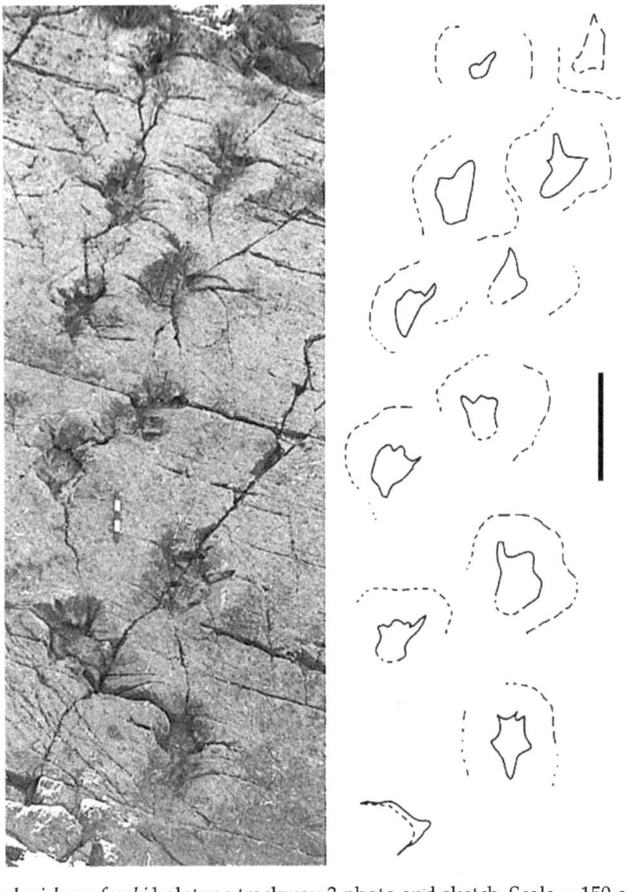

FIGURE 6.47 *Iguanodonichnus frenki* holotype trackway 3 photo and sketch. Scale = 150 cm. *Modified from Moreno and Benton (2005).*

Description. Eleven step "trackway 3" with deep but poorly preserved pes tracks and little indication of manus tracks (Fig. 6.47). Large mud rims. Pes tracks are long and narrower than many sauropod tracks, but discerning viable characters from preservational variation may be difficult. The 16 step "trackway 8" similarly preserves only pes tracks, also deep and with even more distinct mud rims.

Discussion. Originally described as an ornithopod trackway from the Early Cretaceous, this ichnogenus appears to represent the trackway of a Late Jurassic sauropod (Casamiquela and Fasola, 1968; Lockley et al., 2004b; Moreno and Benton, 2005). The ichnotaxon is generally similar in trackway features to *Parabrontopodus*, but with more anteriorly directed pes claw I and possibly a narrower pes. Preservational conditions make this uncertain, however.

Iguanodontipus
Sarjeant, Delair, and Lockley 1998
Fig. 6.48
Type ichnospecies. *I. burreyi.*
Valid ichnospecies. *I. burreyi.*
Distribution. Lower Cretaceous of Europe (Durlston Beds: UK; Bückeburg Formation: Germany; Salema Formation: Portugal; Oncala Group: Spain) according to Díaz-Martínez et al. (2015) and Figueiredo et al. (2022). Possibly Upper Jurassic of Europe (Lastres Formation: Spain). Possibly mid-Cretaceous of Africa (Midelt Formation: Morocco).
Type specimen(s). BNSS 33793b, slab containing a right and a left pes.

FIGURE 6.48 *Iguanodontipus*? track SHN.(JJS).ICNO.110 from the upper Jurassic of Portugal classified as Iguanodontipodidae indet in Castanera et al. (2020). Track length = 28 cm. *Photo by Diego Castanera.*

Diagnosis. Emended by Díaz-Martínez et al. (2015): Tracks belonging to Iguanodontipodidae featuring a small, rounded heel impression that is centered and narrow (as wide as the width of the proximal part of the Digit III impression); along with long, narrow digit impressions ending sharply at their distal ends.

Description. Medium- to large-sized tridactyl tracks, generally nearly as wide as long, exhibiting low mesaxony (Fig. 6.48). The tracks are relatively symmetrical, with robust digits, and are characterized by rounded to quadrangular metatarsophalangeal pad impressions that possess both lateral and medial notches. Digit III is longer than Digits II and IV, and all three have rounded distal ends. Moderate interdigital angles.

Discussion. The type ichnogenus was described from the Lower Cretaceous (Berriasian) Purbeck beds in Dorset, England (Delair, 1963) by Sarjeant et al. (1998), who also included other Lower Cretaceous iguanodontian footprints from Europe. Díaz-Martínez et al. (2015) reviewed iguanodontian tracks included in the ichnofamily Iguanodontipodidae, which is restricted to the Cretaceous. The authors concluded that *Iguanodontipus* tracks are restricted to the Purbeck and Wealden beds of England, the Bückeburg Formation in Germany, and the Oncala Group in Spain.

Piñuela Suárez (2015) mentioned Late Jurassic *Iguanodontipus* tracks from the Lastres Formation of Spain, but Piñuela et al. (2016) doubted the validity of the ichnotaxon. Thus, there are no definitive, Late Jurassic reports of this ichnogenus, but Jurassic tracks related to *Iguandontipus* have been described. Diedrich (2011) identified ?*Iguandontipus* from the Late Jurassic of Germany, and Cobos et al. (2015) described ornithopod tracks from the same epoch in Spain bearing similarities to *Iguanodontipus? oncalensis* (originally *Therangospodus oncalensis*, Lockley et al., 2000), an ichnotaxon documented in the Berriasian of Spain by Castanera et al. (2013) and later included in *I. burreyi* by Díaz-Martínez et al. (2015). Razzolini et al. (2017) described unnamed ornithopod-like tracks (their morphotype II) from the Late Jurassic of Switzerland that also share features with *I.? oncalensis*. Similarly, Piñuela et al. (2016) described Late Jurassic ornithopod tracks from the Late Jurassic not assignable to any known ichnogenus, suggesting they might represent a new ichnotaxon. Gierliński et al. (2017a,b) referred to tracks from the "mid"-Cretaceous of Morocco as strongly resembling *Iguanodontipus*.

Some authors (e.g., Piñuela et al., 2016) have raised doubts about the validity of *Iguanodontipus*, arguing that the type specimens might be theropod dinosaur undertracks. Consequently, this ichnotaxonomic issue has hindered the application of the name *Iguanodontipus* to new occurrences in the Late Jurassic of Portugal (tracks classified as Iguanodontipodidae indet.; Castanera et al., 2020) and in the Berriasian of Spain (tracks classified as cf. *Iguanodontipus* isp.; Castanera et al., 2023b). Moreover, in Portugal, some tracks from the Lower Cretaceous have been classified as either *Iguanodontipus* or *Caririchnium* (Santos et al., 2013; Díaz-Martínez et al., 2015; Figueiredo et al., 2022).

Jialingpus
Zhen, Li, and Zhen 1983
Fig. 6.49
Type ichnospecies. *J. yuechiensis.*
Valid ichnospecies. *J. yuechiensis.*

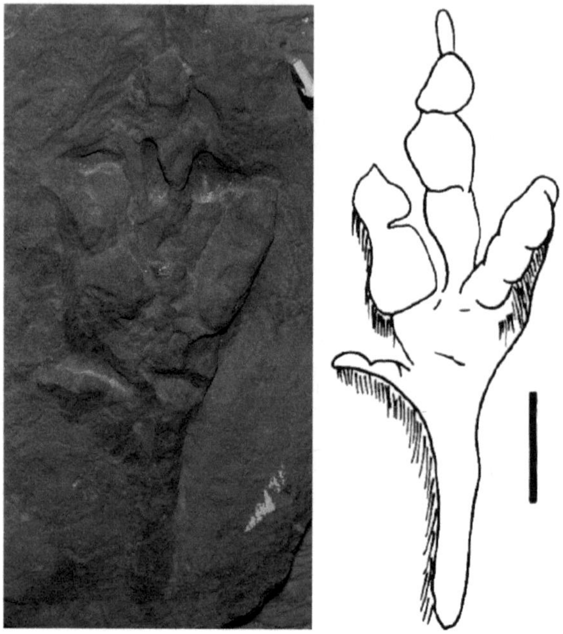

FIGURE 6.49 *Jialingpus yuechiensis*, BNHM SCFP24 holotype photo and schematic. Scale = 10 cm. *Photo by Jerry Harris at BNHM; schematic from Zhen et al. (1989).*

Distribution. Middle Jurassic of Africa (Isli Formation: Morocco)?; Upper Jurassic of Asia (Penglaizhen Formation: Sichuan, China) and Europe (Bałtów Platy Limestones, Bałtów Coral Limestones, and Wierzbica Oolite and Platy Limestones formations: Poland); Lower Cretaceous of Asia (Luohe Formation: Shaanxi, China, Dasheng Group: Shandong, China, and Tugulu Group: Xinjiang, China). Possibly Lower Jurassic of Asia (Ziliujing Formation: Sichuan, China). Possibly Upper Jurassic–Lower Cretaceous of Asia (Tuchengzi Formation: Liaoning, China).

Type specimen(s). BNHM SCFP24, a natural cast of a left pes. Replicas of the holotype and other specimens are in the UCM collections (Xing et al., 2014b).

Diagnosis. From Xing et al. (2014b): small-to medium-sized, functionally tridactyl tracks with hallux occasionally present; phalangeal pad formula 3-3-4-x (or 2-3-3-4-x if the hallux imprint is present). Length:width ratios range from 1.1 to 1.9, digit imprint divarication angles range from 44° to 70°, and the length:width ratio of the anterior triangle ranges from 0.70 to 0.93. The metatarsophalangeal area consists of two pad impressions: a larger one proximal to the imprint of Digit IV, and a smaller one proximal to the imprint of Digit II; collectively, the metatarsophalangeal area is larger than any individual digital pad imprint. In at least some specimens, Digit IV is shorter than Digit II. The first pad of Digit II is larger than that of Digit III. Also, in at least some specimens, claw imprints are largest on Digit II and smallest on Digit III. Divarication angles between Digits II–III are smaller than between Digits III–IV; the overall mean divarication between Digits II–IV is 48°.

Description. See Fig. 6.49. The figured holotype specimen of *Jialingpus* bears a Digit I and a narrow, elongate metatarsal imprint, but these features are neither diagnostic for or characteristic of other *Jialingpus* tracks from the same trackway as the holotype or other specimens (reposited at the IVPP) from the same locality (Xing et al., 2014b). Digit I imprints, when present, protrude medially from a position proximal to the metatarsophalangeal area; the metatarsophalangeal pad proximal to Digit II may be a coalesced pad for both Digits I and II (Xing et al., 2014b). Xing et al. (2014b) noted that Late Jurassic *Jialingpus* specimens tend to have smaller divarication angles between digits than do Early Cretaceous specimens.

Discussion. *Jialingpus* was initially compared with tracks of *Anomoepus intermedius* that bear similar elongate, narrow metatarsal imprints, and were consequently placed in the ichnofamily Anomoepodidae (Zhen et al., 1983, 1989). *Anomoepus* is widely regarded as an ornithischian dinosaur ichnotaxon, so transitively, *Jialingpus* was an ornithischian track. However, subsequent authors (e.g., Lockley et al., 2003) recognized its greater similarity to *Grallator*, a theropod dinosaur ichnotaxon, and considered *Jialingpus* theropodan. Some (e.g., Gierliński, 1994) synonymized *Jialingpus* with *Grallator*, but Lockley et al. (2013) retained it as a valid ichnotaxon pending a thorough review. Xing et al. (2014b) also considered it valid. Klein et al. (2022) noted that *Jialingpus*, *Changpeipus*, and *Asianopodus* could either by synonyms or at least part of a plexus of theropod track types.

Xing et al. (2017a) noted that some grallatorid tracks from the Lower Jurassic Ziliujing Formation of Sichuan, China resembled *Jialingpus*, but did not directly assign them to that ichnotaxon. Some Middle—?Upper Jurassic tracks from the Imilchil—Isli formation complex of Morocco were referred to *Jialingpus* isp. by Gierliński et al. (2009b), but Klein et al. (2022), while noting similarity, did not support explicit attribution to this ichnotaxon. Gierliński et al. (2009a) reported *Jialingpus* isp. from the Upper Jurassic Bałtów Platy Limestones, Bałtów Coral Limestones, and Wierzbica Oolite and Platy Limestones formations of Poland; they also reported *Jialingpus*-like tracks from the Upper Jurassic of Asturias, Spain. Xing et al. (2011a, 2013c) reported *Jialingpus* isp. from the Upper Jurassic and Lower Cretaceous strata of the Tugulu Group of Xinjiang, China. Xing et al. (2014b) likened several tracks assigned to *Grallator* by Fujita et al. (2007) from the Upper Jurassic—Lower Cretaceous Tuchengzi Formation of Liaoning, China to *Jialingpus*, but did not explicitly attribute them to that ichnotaxon. Xing et al. (2017b) referred some track specimens from the Lower Cretaceous Dasheng Group of Shandong, China to *Jialingpus*.

***Jinlijingpus* (=*Eubrontes*?)**
Yang and Yang 1987
Fig. 6.50

Type ichnospecies. *J. nianpanshanensis.*

Valid ichnospecies. None.

Distribution. Middle Jurassic of Asia (Xintiangou Formation: Sichuan, China, per Chen et al., 2006).

Type specimen(s). CFZJ6.

Diagnosis. Tracks of a tridactyl biped. Digit imprints II and III curved, with the tips curved inward, while the tip of Digit IV is curved outward. Digit IV slightly longer than Digit III; Digit II shortest. Digital pads round and somewhat discernible; formula 2-3-4. Tracks slightly wider than long.

FIGURE 6.50 *Jinlijingpus*, UCM 178.10 replica. Scale = 5 cm. *Courtesy Jacob Van Veldhuizen.*

Description. Digit divarication angles II–III = 22° and III–IV = 25° (Fig. 6.50).

Discussion. Yang and Yang (1987) considered *Jinlijingpus* a theropod dinosaur ichnotaxon similar to *Eubrontes*, but differentiated from the latter by having different digit imprint morphologies and divarication angles. Lockley et al. (2013) synonymized *Jinlijingpus* with *Eubrontes*, but retained it as a distinct ichnospecies, *E. nianpanshanensis*. Xing et al. (2021d) supported the synonymy with *Eubrontes*.

Jurabrontes

Marty, Belvedere, Razzolini, Lockley, Paratte, Cattin, Lovis, and Meyer 2018

Fig. 6.51

Type ichnospecies. *J. curtedulensis* Marty et al., 2018.

Valid ichnospecies. *J. curtedulensis, J. teutonicus, J. melphicticus.*

Distribution: Upper Jurassic of Europe (Reuchenette Formation: Switzerland; Süntel Formation: Germany), Lower Cretaceous of Europe (Calcare di Bari Formation: Italy).

Type specimen(s): MJSN-SCR011-553 (holotype); and paratypes: MJSN-SCR011-551, MJSN-SCR011-552, MJSN-BSY008-189, with associated paratype casts MJSN-SCR008-r134 and BSY008-r88.

Diagnosis: Adapted from Marty et al. (2018): Giant, tridactyl tracks with weak mesaxony and a small anterior triangle. Marked by the unique positioning of the proximal pad of Digit III, the presence of phalangeal pads on Digits II, III, and IV (2-3-4 configuration), and wide,

FIGURE 6.51 *Jurabrontes*. (A) Fiberglass cast *Jurabrontes curtedulensis* type trackway (SCR1500-T1-R4, L5) stored at Jurassica Museum, Porrentruy, Switzerland. Dr. Novella L. Razzolini for scale. Scale = 30 cm. (B) *Jurabrontes teutonicus*. Scale = 25 cm (from target centres). *(A) Photo by Matteo Belvedere; (B) modified from Meyer et al. (2021, Fig. 9).*

subtriangular claw traces. Interdigital divarication angles (II^III < III^IV) are narrow and slightly asymmetric. All digits display substantial width, and the heel region is asymmetrical. Trackways are classified as narrow and pace angulation varies between 151° and 172°.

Description: *Jurabrontes* is a large ("giant") tridactyl track with broad digits that taper and often end in subtriangular claw traces (Fig. 6.51). Claw traces are typically anteriorly directed, with those of Digits II and III often rotated inward and those of Digit IV outward relative to the trackway midline. A defining feature of the ichnogenus is the impression of the proximal pad of Digit III, which is not aligned with the elongation axis of the digit. The proximal pad trace of Digit IV creates an asymmetrical subtriangular heel shape. *Jurabrontes* has a weak mesaxony, which is typical of large tridactyl tracks (Lallensack et al., 2020). There are no hallux impressions present. The best-preserved pes impressions measure more than 50 cm in length and more than 40 cm in width. The largest track, including part of the metatarsal impression, measures 78.5 cm in pes length and represents one of the largest theropod tracks globally.

Discussion: *Jurabrontes* is a significant discovery, representing the first find of exceptionally large and robust theropod tracks at the Late Jurassic Jura megatracksites. It provides evidence of large theropods in the Late Jurassic, comparable in size to those from the Cretaceous Period.

Morphological similarities between *Jurabrontes* and several giant tridactyl tracks from Late Jurassic tracksites in Morocco, Spain, France, and Germany exist, but these have not yet been formally recognized, except for several tracks from the Iouaridène Formation that have been attributed to *J.* cf. *curtedulensis* by Belvedere et al. (2019). Tracks related to *Jurabrontes* have also been recognized in the Late Jurassic of the Lusitanian Basin, and have been assigned either to *Jurabrontes* isp (Belvedere et al., 2019). or cf. *Jurabrontes* isp. (Castanera et al., 2021).

In addition, after re-examining the Barkhausen material, Meyer et al. (2021) reclassified the former *Megalosauripus teutonicus* as *Jurabrontes teutonicus* comb. nov. The former association with *Megalosauripus* has been abandoned; the "historical" ichnospecies name has been kept to honor the formal ichnological intention of Kaever and de Lapparent (1974) and to allow the description of differences from the Swiss *Jurabrontes curtedulensis* (as also discussed in Marty et al. (2018) and Belvedere et al. (2019)). Barkhausen theropod tracks have a complex ichnotaxonomic history, stemming from the confusing use of the ichnogenus labels "*Megalosauripus*" and "*Megalosauropus*" during their initial definition (for further debate and discussion see Kaever and de Lapparent, 1974; Lockley and Meyer, 2000; Meyer et al., 2021).

All these co-occurrences suggest that there was a faunal exchange that could have taken place during sea-level lowstands in the Late Jurassic. This would have temporarily connected landmasses that were permanently exposed, such as the Iberian Meseta, the Massif Central, and the London Brabant Massif.

Recently, Antonelli et al. (2023) introduced a third ichnospecies, *J. melphicticus,* for some theropod tracks from the Lower Cretaceous of southern Italy. While these tracks have the same morphology and distinctive traits as *Jurabrontes*, they are generally smaller in size. More information about this taxon can be found in the Cretaceous chapter of this book.

Kalosauropus
Ellenberger 1970
Fig. 6.52
Type ichnospecies. *K. pollex.*
Valid ichnospecies. *K. pollex.*

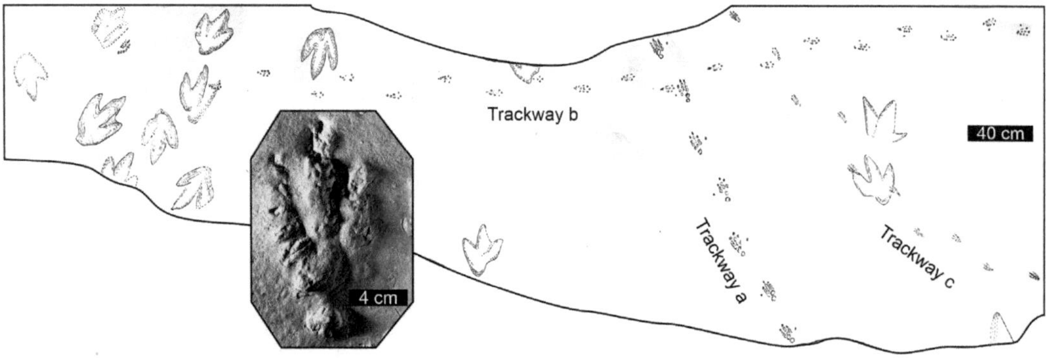

FIGURE 6.52 *Kalosauropus pollex* from the lower Clarens Formation, Leribe, Lesotho. Unpublished map showing trackways a, b, and c that are of *K. pollex var. victor, var. minor* and *var. minusculus*, respectively. For details, see Mukaddam et al. (2020). Inset is an unpublished photo of specimen LES 232.4, a track from Trackway a. *Materials sourced from the PECUoM archives; map drawn by P. Ellenberger.*

Distribution. Lower Jurassic of southern Africa (lower Clarens Formation, zone B/4: Leribe, Lesotho).

Type specimen(s). Pes only tracks on sandstone surface (zone B/4, lower Clarens Formation, Leribe, Lesotho). Preserved as plaster casts BM G053a–h (a total of eight specimens) at the PECMM&A; LES 232 (var. *victor*), 233 (var. *minor*), 234 (var. *miniscculus*) in the PECUoM as well as digital models.

Diagnosis. Modified from Ellenberger (1970) and Mukaddam et al. (2020): Pes only tracks of a small, consistently mesaxonic, tetradactyl biped. Track length range is ~10–15 cm with long gracile, slightly curving Digits III and especially Digit IV, preserving impressions of Digits II, III, IV and the claw mark of Digit I. Digit I is the shortest and diverges medially and greatly from Digit II. Digit I impression always present but only as a large distal, subtriangular claw mark. Digits III and IV have well-defined pad impressions and are often associated with pointed claw marks. Hindfoot impression ("heel") is distinctly circular and deep. Metatarsophalangeal pad (MTP) impressions of Digits III and IV appear fused, well-impressed, and MTP II is missing. There is a consistent, prominent posteromedial gap or embayment between the "heel" impression and the proximal pad impression of Digit II, which is typically inflated and larger than all other digital pads (i.e., the tracks possess a posteromedial notch behind Digit II, as in theropod tracks).

Description. See Fig. 6.52. Originally found on the same surface, tracks are recorded from three relatively narrow trackways (mean width of 12.7 cm) with a track:trackway width ratio of 1–1.25, a mean pace range of 28–45 cm, a mean stride range of 53–89 cm, and a pace angulation of 153°. The tracks are forwardly oriented with a very low inward pes rotation of ~6°. Digits II–IV are subparallel with Digit III being the longest and subequal in length to Digit IV. Digit II is the straightest, and Digits III and IV curve (convex laterally) with the curvature of Digit IV being the most prominent. Digits II–IV are never in contact with each other along their length. Digit I is the shortest and only displays a distal claw mark impression medially, and is well separated from the remainder of the track by a gap about as wide as the combined widths of Digits II–IV, that is, about half the track width. Differentiated from *Evazoum* by its high length–width ratio, and from *Otozoum* by relative size, lack of manus, slender digits, weakly and incompletely impressed Digit I but with a constant claw impression of this digit, lack of a metatarsophalangeal pad impression, an additional pad impression in Digit IV, and very sharp and acuminate claw traces (blunt claw traces in *Otozoum*). The fusing of metatarsophalangeal pads III and IV is similar to *Otozoum*, but a metatarsophalangeal pad trace on Digit II is not registered.

Discussion. Reported from the fine-grained sandstone surface at Leribe in Lesotho (zone B/4, lower Clarens Formation) (Ellenberger, 1970; Mukaddam et al., 2020), these otozoid-like trackways are attributed to small, gracile, tetradactyl, bipedal, basal sauropodomorph dinosaurs. *Kalosauropus* often is included in the OPEK plexus of track morphologies along with *Otozoum*, *Pseudotetrasauropus*, and *Evazoum*, and included in the ichnofamily Otozoidae. Mukaddam et al. (2020) comprehensively argued that *Kalosauropus* is distinct from *Otozoum* (q.v., Rainforth, 2003). Based on converging stratigraphical, geographical, and morphological evidence and coeval body fossil record in southern Africa, the track maker was likely an early, bipedal sauropodomorph, and tentatively a small massospondylid.

Kayentapus
Welles 1971
Fig. 6.53
Type ichnospecies. *K. hopii.*
Valid ichnospecies. *K. hopii, K. minor, K. ambrokholohali, K. wumanensis, K. nananensis, K. hailiutuensis, K. soltykovensis, K. ubacensis.*

Distribution. Lower Jurassic of North America (upper Portland Formation: Massachusetts and Connecticut; Wingate Sandstone and Moenave formations: Utah; Kayenta Formation: Arizona and Utah); Africa (upper Elliot Formation, Stormberg Group: Lesotho); Asia (uncertain formation: Nei Mongol, China and Fuxian Formation: Shaanxi Province, China); and Europe (Zagaje Formation: Poland; Höganäs Formation: Sweden; Mecsek Coal Formation: Hungary; Sacra Formation: Italy). Middle–?Upper Jurassic of Africa (Imilchil and Isli formations: Morocco).

Type specimen(s). UCMP 83668, holotype of trackway consisting of five footprints. UCMP 83669, hypodigm types from another trackway at the type locality (locality number UCMP V6898).

Diagnosis. Welles (1971) did not provide a separate diagnosis for the ichnotaxon from the description (see below). Lockley et al. (2011) provided the first thorough redescription of *Kayentapus*.

FIGURE 6.53 *Kayentapus hopei* Welles 1971, holotype replica, UCMP 83,668 from the Lower Jurassic Kayenta Formation, Tuba City area, Navajo Nation, Arizona (replica = UCM 124-11). Scale = 5 cm.

Description. See Fig. 6.53. Adapted from Welles (1971): Tridactyl, digitigrade, pes-only tracks. Pedal digit lengths II: 16.0–17.5 cm, III: 20.5–24.0 cm, IV: 18.0–19.2 cm, and overall pes length 34.0–35.5 cm. Width between tips of Digits II and IV 29.0–29.5 cm. Digit III projects 12.0–13.5 cm beyond Digits II and IV. Digit divarication angles: II–III: 26–42°, III–IV: 29–40°, and II–IV: 60–72°. Digit III parallel to track midline. Step lengths 40–100 cm. Metatarsophalangeal pad imprint isolated.

Lockley et al. (2011) supplemented the original description with: tridactyl tracks with sharp claw traces and strongly mesaxonic Digit III; claw traces on Digit III tracks are strongly oriented anteromedially; comparisons between tracks in the type trackway display significant variation in the orientation of the distal Digit II traces, yet Digit II proximal pad imprints are consistent in their positions; wide divarications between digits; and the proximal portions of the tracks are V-shaped. Trackway is narrow with long paces from 188 to 191 cm, and they report another trackway has an even greater pace, ranging from 234 to 239 cm.

Discussion. *Kayentapus* is a theropod dinosaur ichnogenus that some view with misgiving (P. Olsen, pers. comm., 2017). *Kayentapus hopii* Welles 1971, the type ichnospecies, is from the Sinemurian Kayenta Formation near Tuba City, Coconino County, Arizona (Welles, 1971). The ichnotaxon is common in Lower Jurassic strata of the American Southwest, including from lowermost Whitmore Point Member of the Moenave Formation, Washington County, Utah, and from the middle portion of the Wingate Sandstone within GLCA (Wood et al., 2021; Milner et al., 2024). These two sites represent the oldest occurrences of *Kayentapus* from the Glen Canyon Group. In addition, *Kayentapus*-like tracks were reported from the lowermost Navajo Sandstone at the John Wesley Powell Fossil Track Block within GLCA, Utah (Milner et al., 2023b).

Weems (1987) sank *Apatichnus minor* (*sensu* Lull, 1953, which was originally *Gigantitherium minus* from the upper Portland Formation of Massachusetts *sensu* Hitchcock, 1858) into *Kayentapus*, creating *K. minor*. Farlow et al. (2018) and Weems (2019) maintained *Kayentapus minor* as a valid ichnospecies.

Both *Zizhongpus wumanensis* and *Chongqingpus nananensis* (Yang and Yang, 1987) from the middle Shangshaximiao Formation, Chongqing, China were synonymized with *Kayentapus* by Lockley et al. (2013) and Xing et al. (2013c) but retained as distinct ichnospecies (*K. wumanensis* and *K. nananensis*, respectively). *Kayentapus hailiutuensis* Li et al. 2010, from about 10 km west of Hailiutu, Wulatzhongqi, Nei Mongol, China (Li et al., 2010), was described as small to medium size, tridactyl, and bipedal tracks with large digit divarication angles (II–III = 40°; III–IV = 43°) and no manus or tail traces (Li et al., 2010). In addition, *Kayentapus* was reported from the Lower Jurassic Fuxian Formation, Zizhou, Shaanxi Province, China (Wang et al., 2016). *Schizograllator* (Zhen et al., 1986a,b; Lockley and Matsukawa, 2009) was noted as being comparable to *Kayentapus* by Lockley and Hunt (1995); this comparison was supported by Piubelli et al. (2005) based on tracks from Italy, and by Xing et al. (2009) based on Asian footprints.

In Europe, *Kayentapus soltykovensis* Gierliński 1996b (formerly *Grallator* [*Eubrontes*] *soltykovensis;* Gierliński, 1991) is from the lower Hettangian Zagaje Formation, Poland (Gierliński and Pieńkowski, 1999; Gierliński et al., 2004), the Hettangian part of Höganäs Formation, Sweden (Gierliński and Ahlberg, 1994), and the lower Sinemurian Mecsek Coal Formation, Hungary (Gierliński, 1996b). *Kayentapus ubacensis* is from the Bathonian Calcaires a Stipites Formation of France (Gand et al., 2018). *Kayentapus*-like tracks have been also identified in the Late Jurassic of Spain (Avanzini et al., 2012; Guarido et al., 2024). Piubelli et al. (2005) described *Kayentapus* in detail from the famous Early Jurassic Lavini di Marco site near Rovereto, Trento, NE Italy from the upper Sarca Formation.

Ellenberger (1972) named *Deuterotrisauropus socialis*, *Neotrisauropus leribeensis*, *N. deambulator*, and *Kleitotrisauropus shososhoei* from the Lower Jurassic of Lesotho; all were suggested to be synonymous with *Kayentapus* by Lockley et al. (2011). *K. ambrokholohali* and several *Kayentapus*-like tracks in the Lower Jurassic upper Elliot Formation of the main Karoo Basin in southern Africa have been described (e.g., Sciscio et al., 2016, 2017; Abrahams et al., 2020, 2021, 2023). *Kayentapus* tracks have also been reported from the Middle–?Late Jurassic Imilchil and Isli formations of Morocco (Klein et al., 2022).

Kharkushosauropus nomen dubium
Dzhalilov and Novikov 1992
Fig. 6.54

Type ichnospecies. *K. kharkushensis.*

Valid ichnospecies. None (see below).

Distribution. Upper Jurassic of Asia (Karabil Formation: Tajikistan).

Type specimen(s). The specimens illustrated in Sabirov (2008: photographs 4–5) were apparently not collected or replicated (no such replicas are mentioned by Sabirov (2008)),

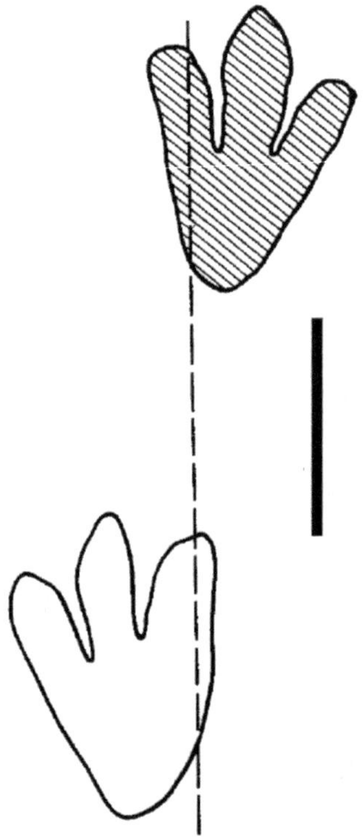

FIGURE 6.54 *Kharkushosauropus kharkushensis*. Scale = 50 cm. *Modified from Dzhalilov and Novikov (1992), Fig. 3.7.*

but Art. 72.5.6 and 73.1.4 of the ICZN (1999) permit specimens thus illustrated and accompanied by data (Recommendation 73C) to serve as holotypes even in the absence of collected, replicated, and/or reposited specimens. Though never desirable, and contentious (e.g., Santos et al. (2016), Shatalkin and Galinskaya (2017)), the practice is permissible.

Diagnosis. Tridactyl tracks of a large, digitigrade biped. Digit III longer than Digits II and IV. Low, subequal divarication angles between Digits II–III and III–IV. Digit imprints deeper distally than proximally and terminate in claw imprints. Rounded "heel" imprint well separated from the digit imprints.

Description. See Fig. 6.54. Twenty-nine tracks in five trackways were reported at the holotype locality. Digit imprints were described as being bounded distally by wide ridges of sediment, likely a complementary aspect to the distal depth of the digit imprints that indicate interplay between a giving substrate and the foot of the track maker as the track maker pushed off. The feature is thus extramorphological and should not be considered part of the diagnosis of the ichnotaxon. *Kharkushosauropus* tracks were compared with *Shirkentosauropus* from the same locality; Sabirov (2008) noted that *Kharkushosauropus* tracks differed by having greater divarication angles between digits and a wider, more separated "heel" area. *Kharkushosauropus* was also differentiated from *"Megalosauropus" uzbekistanicus* Gabuniya and Kurbatov 1988 by having a better developed "heel" area and thicker, more closely appressed digit imprints.

Discussion. Sabirov (2008) did not specifically attribute *Kharkushosauropus* tracks to a particular track-making taxon, but it was placed in the higher-level ichnotaxa Theropodina, Carnosauripodoidei, and Megalosauripodidae, suggesting that it should be attributed to theropod dinosaurs.

The diagnosis ("descriptions" of both Dzhalilov and Novikov (1992) and Sabirov (2008)) of *Kharkushosauropus* is vague and could as well describe numerous other theropod dinosaur ichnotaxa in broad terms; the ichnotaxon is probably best considered a nomen dubium pending further comparative work on the plethora of other Mesozoic ichnotaxa of tridactyl bipeds known globally.

Komlosaurus
Kordos 1983
Fig. 6.55
Type ichnospecies. *K carbonis.*
Valid ichnospecies. *K carbonis.*
Distribution. Lower Jurassic (Hettangian) of Europe (Mecsek Coal Formation: Hungary).
Type specimen(s). Holotype, MÁFI V12692 (Vt. 88), two natural mold footprints on a sandstone slab, from Mecsek Coal Formation, Komló area, Hungary.

Diagnosis. From Ősi et al. (2011): Functionally tridactyl and bipedal, narrow and long digit traces (compared with those of *Grallator* and *Kayentapus*) that often are curved. Pes length:width ratio <2.0. Digit divarication angles are high, averaging 72° between Digits II–IV. Metatarsal pad and Digit I (hallux) traces are common, and metatarsal traces merge with proximal digit traces.

Description. See Fig. 6.55. Track lengths range between 9.5–23 cm (mean 19 cm) long and 7–16 cm (mean 12.5 cm) wide. As is typical for theropod dinosaur pes tracks, Digit III is

FIGURE 6.55 *Komlosaurus* from the Lower Jurassic Mecsek Coal Formation, Hungary (UCM 176-2). Scale in cm. *Photo by Andrew Milner.*

longer than Digits II and IV. In addition, Digit III traces sometimes curve medially, reflecting rotation of the track axis toward the trackway midline. Claw imprints often project at different angles than their associated digits. The shallow metatarsal imprints, when present, typically are short and round but occasionally elongate anteroposteriorly. The position of the Digit I imprint varies with respect to the positions of the remaining digits, probably reflecting complex organism–substrate interactions (Ősi et al., 2011). Some footprints possess rather long Digit I traces that are anteromedially oriented (Ősi et al., 2011, Fig. 6A).

Discussion. The revised diagnosis by Ősi et al. (2011) was based on the type material plus an additional 27 tracks in seven trackways on slab PB1 (MTM V 2010.247.1), and 75 tracks in 14 trackways on slab PB2 (MTM V 2010.248.1-MTM V 2010.282.1). All specimens of this ichnogenus are restricted to the Mecsek Coal Formation of the Southern Transdanubia region (Mecsek Mountains), Hungary (Kordos, 1983; Ősi et al., 2011). Two different tracksites (PB1 and PB2) have produced more than 350 tracks in 48 trackways and were excavated in the Karolina Valley open pit mine at Pécsbánya, Hungary (Hips et al., 1988). Footprints at these two tracksites are mostly tridactyl and some preserve metatarsal traces, and different track depths into the substrate suggesting that they were produced at different times on the same track horizon (Ősi et al., 2011). Orientations of trackways and stride configurations at these tracksites suggest gregarious behavior among the theropod track makers. Ősi et al. (2011) estimated track maker speeds ranging from 6 to 14 km/hour. Gierliński (1996b) identified *Grallator tuberosus* at the Pécsbánya mine, as well.

In addition, the Pécs-Vasas open pit mine has produced 110 footprints on a single tracksite surface, which are now housed in the Hungarian Geological Museum (Ősi et al., 2011), but these specimens have yet to be described. Other isolated specimens come from the mine include one identified as *Kayentapus soltykovensis* by Gierliński (1996b).

Kordos (1983) suggested the tracks are ornithischian in origin, while others have suggested theropod track makers (Gierlinski, 1996b; Ősi et al., 2011). Gierliński (1996b) synonymized *Komlosaurus* with *Kayentapus*. We agree that *Komlosaurus* tracks were produced by small theropods and could be synonymous with *Grallator*. The digit traces are slender in *Komlosaurus* tracks, but this is likely due to sediment collapse into the tracks following registration (q.v., Milàn and Bromley, 2006).

From North America, Lockley and Hunt (1994) mentioned slender-toed, bipedal dinosaur tracks from the Moenave and Kayenta formations in the southwestern United States that they called cf. *Komlosaurus*. Also, Lockley and Hunt (1995) compared some unidentified, slender-toed footprints from the Kayenta Formation at the Lisbon Valley Oilfield Tracksite, San Juan County, Utah as being birdlike with wide divarication angles like *Komlosaurus* from Hungary. Two of us (ARCM and JDH) interpret these Lisbon Valley tracks as *Anomoepus*, which is one of the more common track types alongside *Grallator* at this tracksite (Lockley and Hunt, 1995, Fig. 4.9; Lockley and Gierliński, 2006).

Kuangyuanpus
Young 1943
Fig. 6.56
Type ichnospecies. *K. szechuanensis.*
Valid ichnospecies. *K. szechuanensis.*
Distribution. Middle Jurassic of Asia (Xiashaximiao Formation: Sichuan, China).
Type specimen(s). NGM C161, slab containing natural casts of four tracks.

Diagnosis. From Young (1943) and Lockley et al. (2010b): tracks of a quadruped. Pes tetradactyl with digit proportions I < II < III=IV. Manus tracks tetradactyl–pentadactyl, about one-third smaller than the pes prints, and have digit proportions I=V < II < III=IV. Digit imprints subparallel. Digits imprinted more strongly than metapodiophalangeal regions, which form faint arches posteriorly.

Description. The type *Kuangyuanpus* tracks (Fig. 6.56) exhibit classic swim-track morphology, preserving only the distal digit imprints and posterior mounds of pushed-up sediment (Lockley et al., 2010b).

Discussion. *Kuangyuanpus szechuanensis* was the first ichnotaxon formally named from China. Although Young (1943) was uncertain about the identity of the *Kuangyuanpus* track maker, it has been perceived as crocodylian in origin in recent years (Zhen et al., 1989, 1996; Lockley et al., 2010b, 2013). Without presenting morphological reasons why, Zhen et al. (1989, 1996) considered *Kuangyuanpus* synonymous with *Batrachopus*. Lockley et al. (2010b, 2013), however, treated it as a valid ichnotaxon of crocodylian swim tracks, similar to *Hatcherichnus*.

Young (1943) did not specify the formation from which the holotype *Kuangyuanpus* tracks originated, or even the age of the unit. Zhen et al. (1989, 1996) reported only that the unnamed unit was Middle Jurassic in age. Xing et al. (2021c) reported the track as coming from the Middle Jurassic Xiashaximiao Formation.

Lockley et al. (2010a) emended the *Hatcherichnus* ichnocoenosis of Hunt and Lucas (2007) as the *Kuangyuanpus–Hatcherichnus* ichnofacies (or ichnocoenosis).

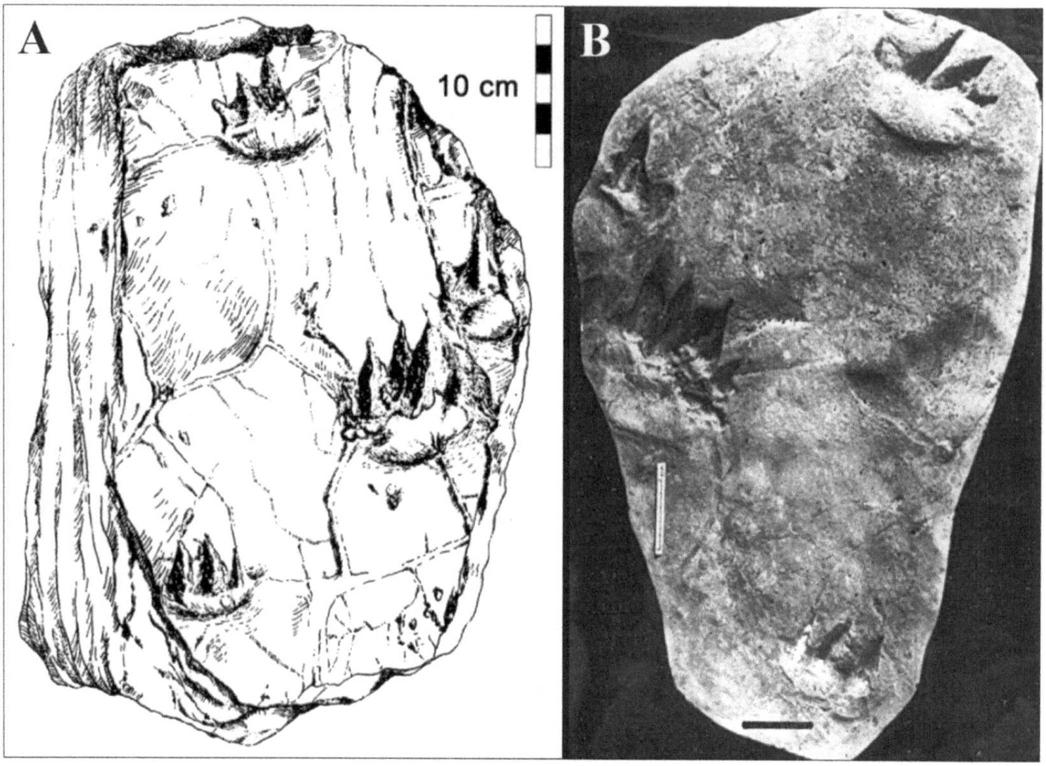

FIGURE 6.56 *Kuangyuanpus szechuanensis*. (A) Sketch of type specimen from Young (1943; cropped from Lockley et al., 2010a, Fig. 2); (B) photo of type specimen (reversed mold) as illustrated by Zhen et al. (1989, Fig. 19.4C; cropped from Lockley et al., 2010a, Fig. 3). Scale = 4 cm.

Lavinipes
Avanzini, Leonardi, and Mietto 2003
Fig. 6.57
Type ichnospecies. *L. cheminii.*
Valid ichnospecies. *L. cheminii.*

Distribution. Lower Jurassic (upper Hettangian—lower Sinemurian) of Europe (only recorded from the type locality, Calcari Grigi Formation: Lavini di Marci ichnosite, Rovereto, Italy).

Type specimen(s). Holotype is a rubber cast of manus-pes set 5 from ROLM 577 trackway and fiberglass replica of ROLM 577 2—5 manus-pes set. Paratypes are rubber cast of ROLM 577/10 manus print and rubber cast of ROLM 192/4 pes print. Topotype is trackway ROLM 577 that has been left in situ (Avanzini et al., 2003).

Diagnosis. Narrow, quadrupedal trackway with track length of ~43 cm and no space between the trackway midline and the inside margins of the pes tracks. Pes axis longer than wide, with the long axis parallel to the trackway midline or rotated inward slightly (up to ~12°). Pes tetradactyl with three short, anteriorly directed digits and a small Digit I. Digit

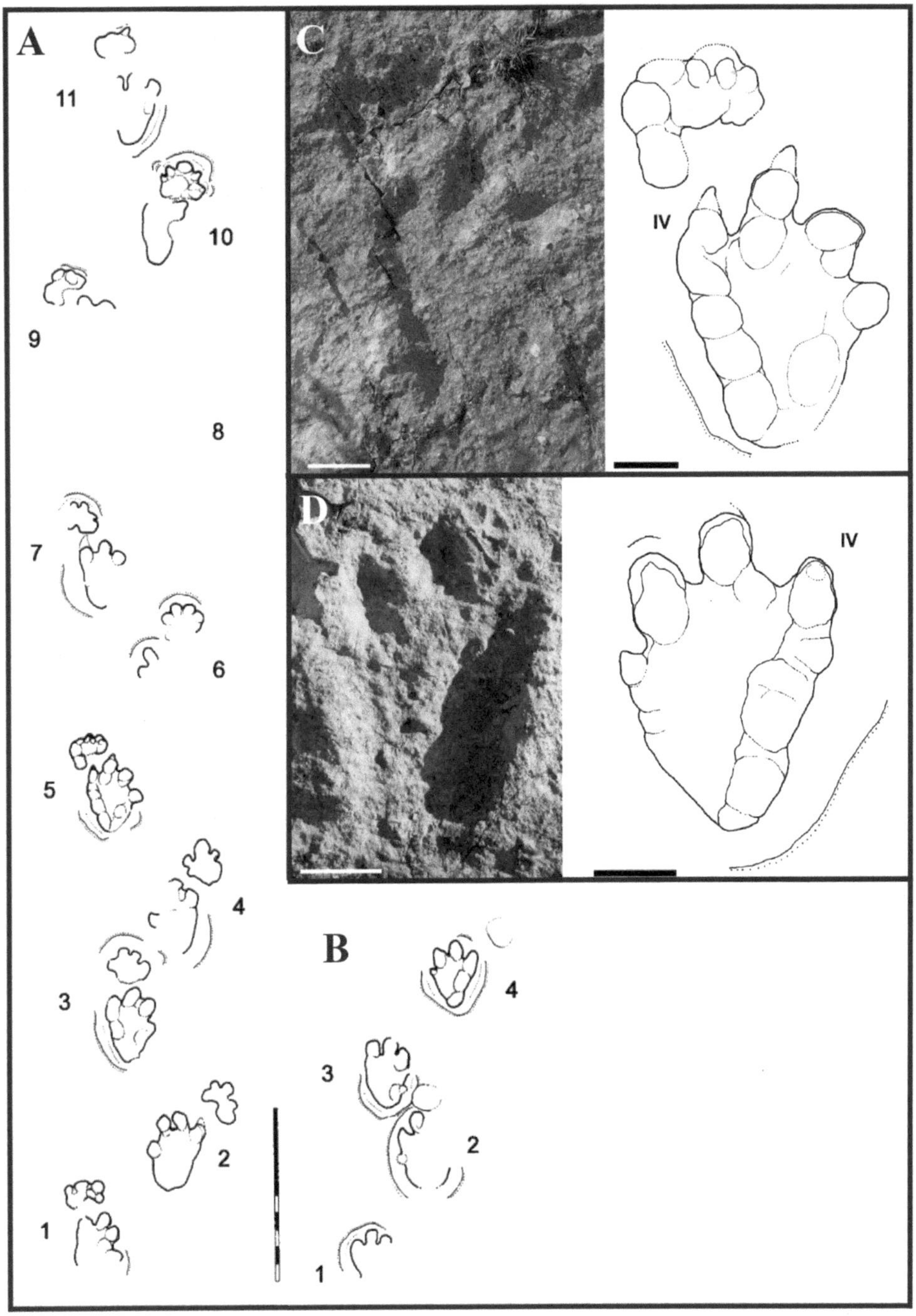

FIGURE 6.57 *Lavinipes cheminii*, Lower Jurassic Calcari Grigi Formation, Lavini di Marco tracksite, Italy. (A) Line drawing of trackway ROLM 577; and (B) line drawing of trackway ROLM 192 without manus prints. Scale = 1 m for A and B. (C) Holotype (MTSN ROLM577.5); scale = 10 cm. (D) Paratype (MTSN ROLM 192.4) pes track. Scale = 10 cm. *(B) From Avanzini et al. (2003), Fig. 2. (C) From Avanzini et al. (2003), Fig. 4. (D) From Avanzini et al. (2003), Fig. 6.*

traces rotated inward. Manus tracks pentadactyl or semicircular (manus/pes ratio = 0.44) and positioned lateral to the pes (from Avanzini et al., 2003).

Discussion. *Lavinipes cheminii* (Fig. 6.57) comes from the Lower Jurassic (upper Hettangian—lower Sinemurian) carbonate tidal flat deposits of the east-central Italian Alps (Avanzini et al., 2003). The Lavini di Marco dinosaur tracksite in the Calcari Grigi Formation is considered one of the richest tracksites in Europe, preserving theropod, sauropod (*Parabrontopodus*-like), and small ornithischian tracks and traces.

Sauropodomorph dinosaurs are the suggested producers of *L. cheminii* tracks (Avanzini et al., 2003). Some authors have considered this ichnogenus congeneric with *Otozoum* and *Tetrasauropus* (Lockley et al., 2022).

Liujianpus

Xing, Lockley, Zhang, Klein, Li, Miyashita, Li, and Kümmell 2016

Fig. 6.58

Type ichnospecies. *L. shunan.*

Valid ichnospecies. *L. shunan.*

Distribution. Lower Jurassic of Asia (Ziliujing Formation: Sichuan, China).

Type specimen(s). JYS1, a trackway comprising 14 manus (LM) and 15 pes (LP) tracks. A replica of tracks LM7 and LP7 are preserved as UCM 178.18.

Diagnosis. Trackway of a large quadruped. Manus prints pentadactyl to semicircular with blunt, discrete digit imprints. Manus prints strongly rotated outward (average 52°). Pes prints tetradactyl, with long, nearly equidimensional digit imprints, lengths III > IV=II > I, and outward rotation (average 38—40°). Pedal digit imprints subparallel to track axis. Sole area of the pes divided by a transverse crease located midway along the track length.

Description. See Fig. 6.58. *Liujianpus* tracks were described as having a combination of otozoid and sauropod (e.g., *Brontopodus*) track features (Xing et al., 2016a). Manus prints are wider than long, with length:width ratios ranging from 0.60 to 0.74. Pes prints have length:-width ratios ranging from 1.32 to 1.54. Prominent digit imprints (comprising 25%—30% of the track length) are slightly curved and parallel to the track axis. The transverse crease that divides the pes print about 40%—45% the length from the anterior margin of the track terminates in a distinct notch proximal to Digit I. The trackway is narrow, with the medial margins of the manus prints lying farther laterally from the trackway axis than those of the pes prints.

Discussion. *Liujianpus* tracks are attributed to basal sauropodomorph dinosaurs; indeed, skeletal remains of the basal sauropodomorph (sensu Apaldetti et al., 2011) *Gongxianosaurus* were recovered from the same unit (He et al., 1998). The only reported tracks similar to those of *Liujianpus* are unnamed tracks from the Early Jurassic of Morocco illustrated, but not described in detail, by Ishigaki (1988). *Liujianpus* differs from *Otozoum* by having larger manus prints, by lacking segmented (by digital pads) digit imprints and shorter digit imprints (25%—30% vs. 40%—50% track length), and by having greater outward rotation (Xing et al., 2016d). Similar differences differentiate *Liujianpus* from *Lavinipes*. *Liujianpus* tracks are similar to *Brontopodus* tracks in broad, general, nondiagnostic ways, such as having outwardly rotated manus prints and the "elephantine,""fleshy," and outwardly rotated pes prints with large, anterior claw imprints (Xing et al., 2016d). The pace angulation of *Liujianpus* is similar to those of sauropod trackways as well, but *Liujianpus* manus prints differ

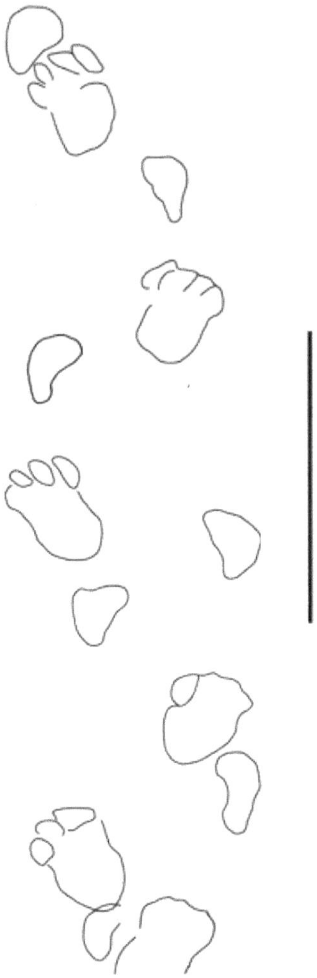

FIGURE 6.58 *Liujianpus shunan*, JYS1 holotype trackway sketch. Scale = 1 m. *Modified from Xing et al. (2016d), Fig. 5.*

in often exhibiting distinct traces of five digits; pes prints differ from those of sauropod tracks in having Digit I the shortest and Digit III the longest, as well as having a well-developed Digit IV imprint; typical sauropod tracks have a digit length formula I > II > III > IV > V.

Lufengopus (=Eubrontes?)
Lü, Azuma, Wang, Li, and Pan 2006
Fig. 6.59
Type ichnospecies. *L. dongi.*
Valid ichnospecies. None (see below).
Distribution. Middle Jurassic of Asia (Upper Lufeng Formation = Chuanjie Formation: Yunnan, China).

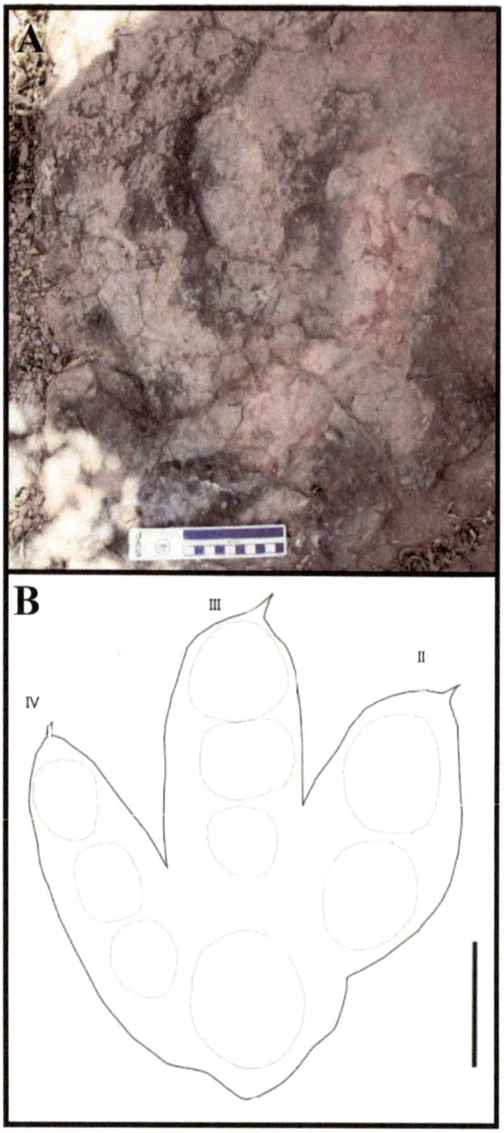

FIGURE 6.59 *Lufengopus* holotype, LuDM L 028 (now ZLJ BT1). (A) Photo; and (B) schematic. Scale = 10 cm. *(B) Modified from Lü et al. (2006).*

Type specimen(s). Complete natural mold of a left pes print LuDM L 028 (now ZLJ BT1); a cast of the specimen is stored at the Heyuan Museum as HYMVC-1.

Diagnosis. Large, bipedal theropod dinosaur track with separate phalangeal pads (formula 2-3-3) and a distinctive, circular heel pad. Digit impressions taper into claw imprints.

Description. The lone holotype track is wide and short with clearly visible phalangeal and heel pads (Fig. 6.59). Digits III and IV of equal length, though not of equal anterior prominence.

Discussion. The holotype specimen was reported by Lü et al. (2006) as the largest known Chinese theropod ichnotaxon (length 40 cm, width 35 cm). Xing et al. (2014c) redescribed the type specimen and two newly exposed, associated specimens (ZLJ BT2–3). They noted that the type specimen was somewhat smaller than initially reported, and followed Lockley et al. (2013) by sinking the ichnogenus into the broadly distributed theropod ichnotaxon *Eubrontes*.

"Malasaurus" nomen nudum
Malkani 2007
Type ichnospecies. *"M. mianwali."*
Valid ichnospecies. None.
Distribution. Middle Jurassic of Asia (Samana Suk Limestone: Pakistan).
Type specimen(s). None—three trackways totaling 10 tracks were declared the "holotype" (more correctly, these would constitute syntypes), but no specimen numbers or repository name were provided.
Diagnosis. None provided.
Description."*Malasaurus*" tracks were only described as large, pes-only prints; the manus prints were presumed to have been entirely overprinted by the pes prints. The trackways were described as narrow-gauge.
Discussion."*Malasaurus*" tracks were attributed to a large sauropod dinosaur. The ichnotaxon is a nomen nudum because of the absence of a diagnosis and the absence of a holotype. The mentioned trackway was described as revealing a confrontation between the sauropod track maker and a theropod track maker (see "*Samanadrinda*") moving at approximately 90° to the sauropod trackways.

Malutitetrapodiscus
Ellenberger 1970
Fig. 6.60

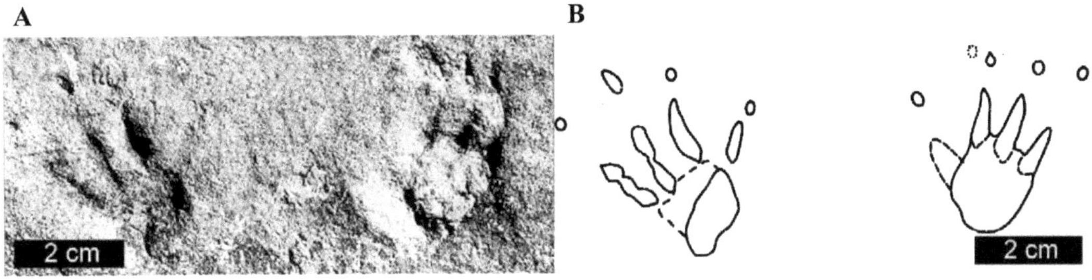

FIGURE 6.60 *Malutitetrapodiscus saltator*—side-by-side tetradactyl synapsid tracks in the Barkly East Formation, Ralikhomo, Lesotho. (A) Previously unpublished photograph of the conspicuous proximal "hop set" pair (newly relocated specimen LES 311.1 in the PECUoM). Scale = 2 cm. (B) Interpretive outline of the tracks shown in A. Scale = 2 cm. For details, see Bordy et al. (2022, Fig. 9, which includes details of the holotype trackway, including the distal "hop set", 2024).

Type ichnospecies. *M. saltator.*

Valid ichnospecies. *M. saltator.*

Distribution. Lower Jurassic of southern Africa (Barkly East Formation, zone C/3: Ralikhomo, Lesotho).

Type specimen(s). Paired tracks preserved as natural impressions (concave epireliefs) on a sandstone surface (zone C/3, Barkly East Formation, Ralikhomo, Lesotho) figured in Bordy et al. (2024, Fig. 9). Recorded as plaster cast LES 311.3, a newly relocated sandstone specimen LES 311.1 in the PECUoM as well as digital models at doi.org/10.6084/m9.figshare.19636593. Specimens LES 311.2, 312, and 313 appear to have been misplaced or lost.

Diagnosis. Modified from Ellenberger (1970, 1975) and Bordy et al. (2022): Small, tetradactyl pes tracks (~3.2 cm long and wide) with a consistent paired configuration and strong outward rotation. Digits II–IV show a total digit divarication of 60° and are deeply impressed proximally rather than distally. The same impression mode applies to Digit I, which is well separated from the other digits. The splayed digit imprints terminate in circular, very fine, pointy claw tip marks that are well in front of the distal phalangeal pad impressions. The slightly larger manus prints (~3.5 cm long and wide) have the same structure as the pes prints, but manual Digit II–IV imprints form a parallel bundle.

Description. See Fig. 6.60. In the type specimen, the paired tracks form a trackway that comprises two conspicuous "hop set" pairs with a hop (leap) distance of ~29 cm. The pairing of tracks across the trackway midline is symmetrical and implies a hopping gait. The hopping track maker progressed by jumping with both feet (in leaps of 29 cm), mostly on digits that ended in long, very fine, arching claws of which only the tips touched the ground. The manus claws were not only long and curved downward, but also curved inward, touching the ground far ahead of the distal phalangeal pads. Arrangement of the digit and claw marks implies not only raised toe tips but an opposable thumb capable of gripping. Differs from *Molapopentapodiscus supersaltator* (Van Dijk, 1978, Fig. 4) in digit trace curvature and hypex configurations.

Discussion. Reported from the surface of a very fine-grained sandstone interbed within the Karoo continental flood basalts (Pliensbachian–Toarcian), some 200–300 m above the top of Clarens Formation at Ralikhomo in Lesotho (zone C/3, Barkly East Formation (Ellenberger, 1970; Bordy et al., 2022). The paired tracks in the trackway are attributed to a small track maker with a jumping (saltating) gait. Based on converging stratigraphical, geographical, and morphological evidence and the coeval body fossil record of early mammaliaforms in southern Africa, the track maker was likely a small, hopping mammaliaform of rodent size.

***Megaichnites* (=*Eubrontes*?)**

Yang and Yang 1987

Fig. 6.61

Type ichnospecies. *M. jizhaoshiensis.*

Valid ichnospecies. None (see below).

Distribution. Middle Jurassic of Asia (Xintiangou Formation: Sichuan, China per Chen et al., 2006).

Type specimen(s). CFZW 161.

Diagnosis. Tracks of a digitigrade, tridactyl biped. Digit imprints wider distally than proximally, terminating in tapered claw imprints. Digit III longest and more anteriorly projecting

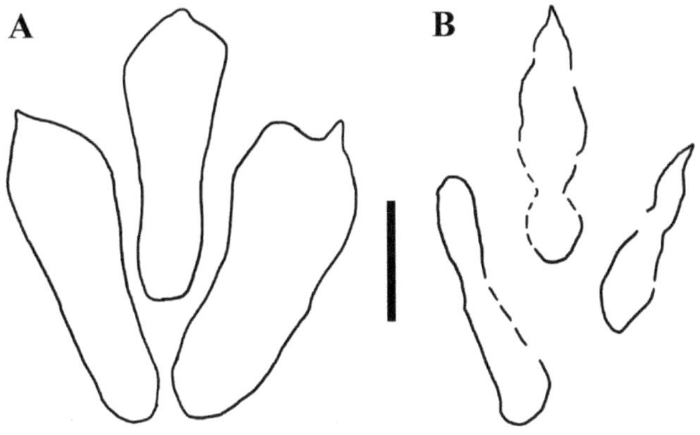

FIGURE 6.61 *Megaichnites jizhaoshiensis* holotype CFZW 161 from the Xintiangou formation, Sichuan, China. (A) Redrawn original interpretive schematic by Yang and Yang (1987). (B) Redrawn interpretive schematic by Xing et al. (2021e). Scale = 10 cm.

than Digits II and IV. Tracks "anchor"-shaped. Divarication angles between digit imprints moderate. Yang and Yang (1987) suggested the ichnogenus has a digital pad formula of 2-3-4, though no such pad imprints are discernible in their schematic outline.

Description. See Fig. 6.61. Divarication angles between digits moderate (II–III 25° and III–IV 28°). Digit IV is the longest, but Digit III projects the farthest anteriorly; Digit II is the shortest. Lockley et al. (2003) reexamined the type material, determined that the interpretation of Yang and Yang (1987) did not accurately reflect track morphology, and described the type specimen as having a more typically theropodan morphology, with narrower digits and discernible digital pads.

Discussion. *Megaichnites* is a theropod dinosaur ichnotaxon. Lockley et al. (2003) synonymized the ichnospecies as an indeterminate ichnospecies of *Kayentapus*, but Lockley et al. (2013) retained it as *Kayentapus jizhaoshiensis.* Xing et al. (2021d) referred the ichnotaxon to *Eubrontes* isp.

Megalosauripus
Lockley, Meyer, and Santos 1998c (first illustrated by Lessertisseur, 1955)
Fig. 6.62

Comment."Megalosaur" tracks have been the subject of a long-standing debate regarding their nomenclature. This debate has been fueled by old, incomplete, and informal descriptions (Lockley et al., 1996a, 1998c; Thulborn, 2001; Razzolini et al., 2017). Given this background, we support the proposal by Lockley et al. (1996, 1998c) to use the name *Megalosauripus* (with an "i") for the Late Jurassic to Late Cretaceous European, North American, and north African tracks. For the early–Late Cretaceous tracks found in Australia, the name *Megalosauropus*, with an "o.", has been recently used (e.g., Salisbury et al., 2016; Romilio and Godfrey, 2021).

Type ichnospecies. *Megalosauropus uzbekistanicus* Gabuniya and Kurbatov 1982, amended as *Megalosauripus uzbekistanicus* (Lockley et al., 1998c).

Valid ichnospecies. *M. uzbekistanicus, M. transjuranicus.*

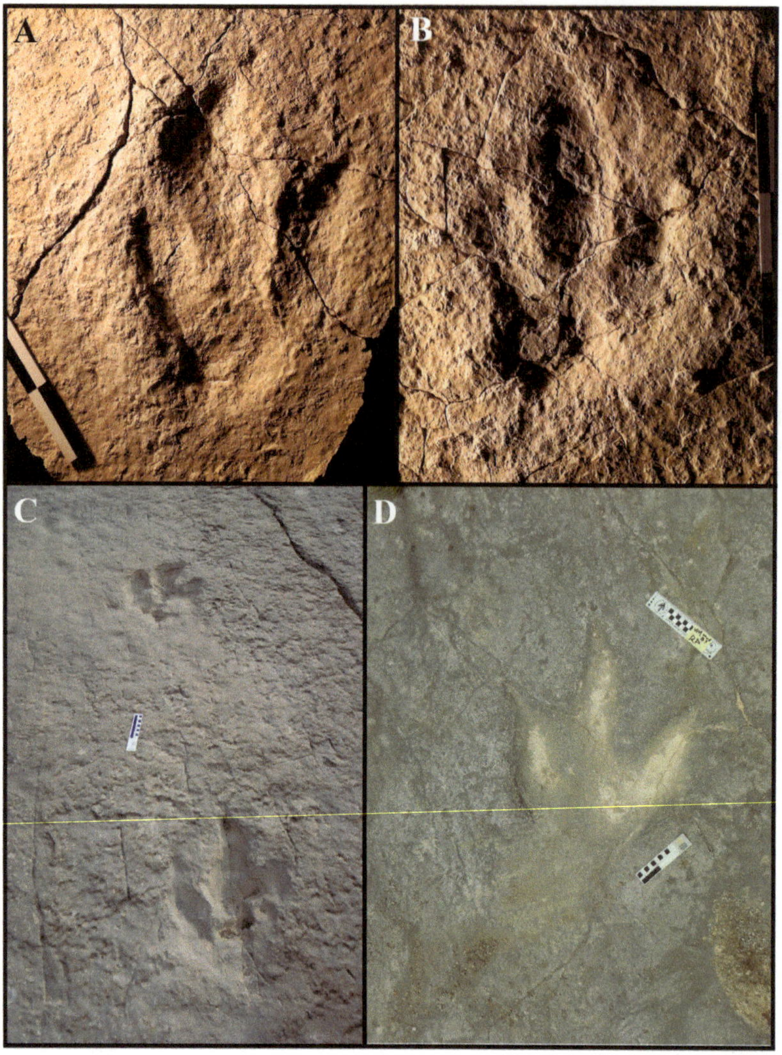

FIGURE 6.62 *Megalosauripus.* (A) *Megalosauripus transjuranicus* holotype track (TCH1030-T6-L1), stored at the Jurassica Museum, Porrentruy, Switzerland. Full scale bar = 20 cm. (B) *Megalosauripus transjuranicus* holotype track (TCH1030-T7-L2), stored at the Jurassica Museum, Porrentruy, Switzerland. Scale = 30 cm. (C) *Megalosauripus uzbekistanicus* paratype trackway (HA-1) from Turkmenistan. Scale = 10 cm. (D) *Megalosauripus* isp. tracks from the Vale de Mejos tracksite, Middle Jurassic, Portugal. Scale bars in cm. *(C) Photo courtesy Michela Contessi; from Fanti et al. (2013). (D) Photo courtesy Daniel Marty; from Razzolini et al. (2016).*

Type Specimen(s). Specimen 8/849, Tashkent Geological Museum, Turkmenistan. Paratypes from Turkmenistan redescribed in Fanti et al. (2013); cast and digital copies of these paratypes are stored in the Collezione di Geologia "Museo Giovanni Cappellini" (MGGC 21863, 21864, 21865).

Distribution. Middle Jurassic of North America (Curtis Formation: Utah), Europe (Serra de Aire Formation: Portugal; Scalby Formation and White Limestone Formation: UK), Africa (Isli Formation: Morocco; Bemaraha Formation: Madagascar). Upper Jurassic of North America (Morrison Formation: Colorado), Europe (Morillon Formation: France; Süntel Formation: Germany; Błaziny Oolite Limestones: Poland; Alcobaça, Cabaços, and Freixial formations, Portugal; Villar del Arzobispo Formation, Aguilar del Alfambra Formation: Spain; Reuchenette Formation: Switzerland), Africa (Djara Formation, Algeria; Iouaridène Formation: Morocco), and Asia (Tuchengzi Formation: China); Lower Cretaceous of Europe (Obernkirchen Sandstone: Germany; Oncala Group (Huérteles and Aguilar del Río Alhama formations) and Galve and Mirambe formations: Spain) and Africa (Tiloula and Tiout formations: Algeria); Upper Cretaceous of Africa (Rhelidha Formation: Algeria; Aoufous Formation: Morocco).

Diagnosis. Modified from Lockley et al. (1998c): Medium to large elongate tridactyl tracks, with holotype average footprint length of 53 cm and width of 36 cm, and exhibiting a phalangeal pad formula of 2, 3, and 4 corresponding to Digits II, III, and IV, respectively. The "heel" is predominantly formed by the impression of Digit IV, elongated and relatively longer than the impression of Digit III. Consequently, the back edge of the Digit III impression often aligns along the anterior-posterior axis with the penultimate digit pad of Digit IV. The trackway demonstrates significant variation in width, ranging from narrow to moderately wide, with pace angulation variable and as low as 120°. Pace shows alternation between short and long steps, with the type trackway exhibiting a pattern of 120–160–130–150–130–155–130 cm length steps (Lockley et al., 1996).

Description. Tridactyl tracks characterized by their large elongate, asymmetric morphology (Fig. 6.62), and with dimensions ranging from 39 to 72 cm in pes length and widths between 29 and 65 cm. Phalangeal pad impressions are well-defined only in adequately preserved specimens, conforming to the typical 2-3-4 theropod formula for Digits II, III, and IV, respectively. Trackways exhibit irregularities in pace and pace angulation, with step lengths alternating between ~120 and 130 and ~150–160 cm, and pace angulation between 120 and 175°.

Discussion. *Megalosauripus* exhibits notable variations in size, trackway width, pace, and pace angulation. For example, the trackway of *M. uzbekistanicus* showcases an irregular pattern consisting of alternating short and long steps, totaling 32 steps in all. This pattern is commonly observed in *M. uzbekistanicus* trackways, which are prevalent in a significant sample of Late Jurassic (Oxfordian–Kimmeridgian) carbonates across central Asia with the best-preserved from the Khodja-Pil Ata area, eastern Turkmenistan.

According to Lockley et al. (1998c), the majority of the available material from Europe and North America during that period could be attributed to *Megalosauripus*. Lockley et al. (1998c) highlighted the differences between *Megalosauripus* and *Eutynichnium*, a distinction subsequently confirmed by Razzolini et al. (2017) and Belvedere et al. (2019).

Razzolini et al. (2017) introduced the novel ichnospecies *Megalosauripus transjuranicus*. This ichnospecies is characterized by a unique impression of pad one in Digit IV. This pad exhibits a circular shape and serves as the widest and largest phalangeal pad within the track. It is seamlessly connected to the remainder of the Digit IV imprint, constituting the rounded heel of the track. Typically, this pad is approximately twice as wide as the rest of the Digit

IV impression. This distinguishing feature sets *M. transjuranicus* apart from other large theropod ichnotaxa, such as *M. uzbekistanicus* and *Jurabrontes*.

We note that the definition of *M. transjuranicus* is based on an extensive collection of paratypes and reference material (Razzolini et al., 2017). This comprehensive dataset, combined with its frequent occurrence in the Swiss Jura region, has facilitated the recognition of morphological variations within the tracks. These variations primarily stem from sedimentological or rheological alterations in the substrate and are not regarded as differences at the ichnospecies level. Nonetheless, they can contribute to the identification of the track maker and enhance the interpretation of poorly preserved tracks found in analogous environments elsewhere.

Megatrisauropus (=Kayentapus?)
Ellenberger 1972
Fig. 6.63

Type ichnospecies. *Megatrisauropus malutensis.*

Valid ichnospecies. None.

Distribution. Lower Jurassic Upper Stormberg "Transition beds" (Cave Sandstone, stage B/5), southern Africa (Ellenberger, 1972; Lockley et al., 2003).

Discussion. *Megatrisauropus malutensis* likely is a synonym of *Kayentapus ambrokholohali.*

FIGURE 6.63 *Megatrisauropus. Modified from Ellenberger (1970).*

Mirsosauripus
Dzhalilov and Novikov 1992
Fig. 6.64

Type ichnospecies. *M. tursunzadevi.*

Valid ichnospecies. *M. tursunzadevi.*

Distribution. Middle–Upper Jurassic of Asia (Kugitang Formation: Tajikistan).

Type specimen(s). The specimens illustrated in Sabirov (2008: photographs 6–7). The specimens illustrated were apparently not collected or replicated (no such replicas are mentioned by Sabirov, (2008)), but Art. 72.5.6 and 73.1.4 of the ICZN (1999) permit specimens thus illustrated and accompanied by data (Recommendation 73C) to serve as holotypes even in the absence of collected, replicated, and/or reposited specimens. Though never desirable, and contentious (e.g., Santos et al., 2016; Shatalkin and Galinskaya, 2017), the practice is permissible.

Diagnosis. From Dzhalilov and Novikov (1992) and Sabirov (2008): Tridactyl, ovoid tracks of a very large biped. Overall morphology indistinct, but wider anteriorly and possessing three short, wide, digit imprints that lack claw imprints. Dzhalilov and Novikov (1992)

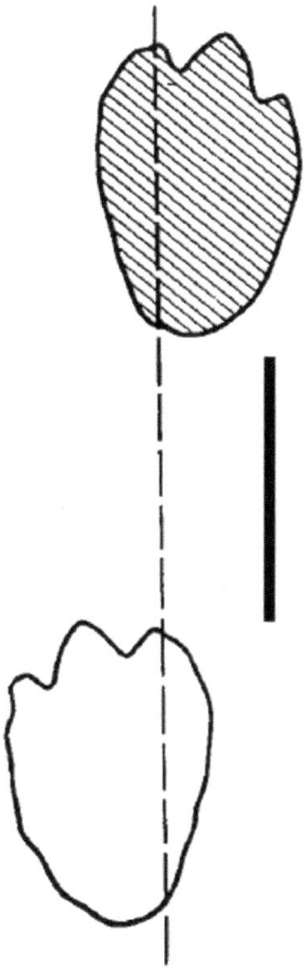

FIGURE 6.64 *Mirsosauripus tursunzadevi.* Scale = 50 cm. *Modified from Dzhalilov and Novikov (1992), Fig. 3.6.*

described the imprints of Digits II and IV as rounded, and Digit III as pointed, but Sabirov (2008) described all the digit imprints as "conical." Dzhalilov and Novikov (1992) described the imprint of Digit II as wider than Digit III, but Sabirov (2008) described the imprint of Digit III as wider than those of Digits II and IV. Divarication angles subequal between digits. The "heel" area is large and wide. In trackways, tracks overlap the trackway axis slightly, and prints are rotated slightly outward.

Description. See Fig. 6.64. Roughly 50 tracks in several trackways were reported at the holotype locality (Sabirov, 2008). *Mirsosauripus* tracks were described as being deep and surrounded by displacement rims of sediment—this was actually included in the diagnosis ("descriptions" of Dzhalilov and Novikov (1992) and Sabirov (2008)), but is certainly extramorphological.

Discussion. Sabirov (2008) noted that Novikov and Radililovsky (1984) attributed *Mirsosauripus* tracks to sauropod dinosaurs, but, based on the absence of manus prints associated with the pes prints, he followed Gabuniya and Kurbatov (1988) in considering the track maker instead a bipedal theropod dinosaur and that the tracks were perhaps related to those of "*Megalosauropus*". Lockley et al. (1994a) considered *Mirsosauripus* to be a nomen dubium.

Moyenisauropus
Ellenberger 1974
Fig. 6.65
Type ichnospecies. *M. natator*.

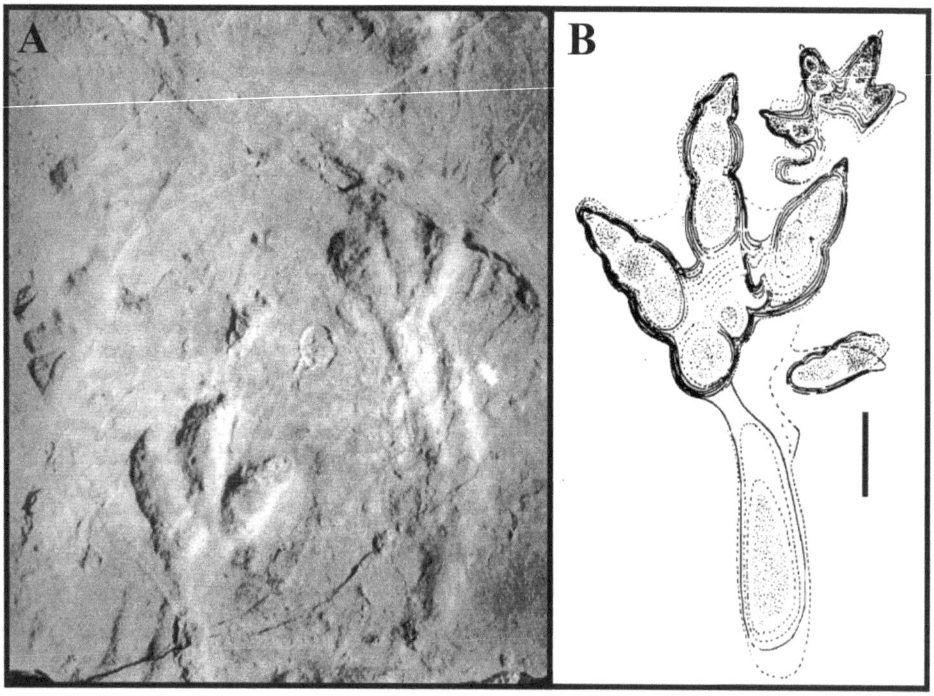

FIGURE 6.65 *Moyenisauropus natator* from Lower Jurassic Moyeni tracksite, Lesotho. (A) Photo, (B) sketch by P. Ellenberger. Scale = 5 cm. *(A) Photo by Emese Bordy. (B) Modified from Ellenberger (1974).*

Valid ichnospecies. *M. natator, M. karaszevskii, M. levicauda, M. lusitanicus.*

Distribution. Lower Jurassic of North America (Kayenta Formation, Kayenta–Navajo transition, and Navajo Sandstone: Utah), Europe (unspecified formation: Poland; Coimbra Formation: Portugal) and Africa (upper Elliot Formation, Stormberg Group: Lesotho).

Diagnosis. Emended diagnosis from Gierliński (1991): Typically bipedal, but sometimes quadrupedal. Pes functionally tridactyl and Digit I trace rarely registered. Tracks wider than long with robust pedal digits highly divaricated. Two phalangeal pads on Digit III. Pedal Digit IV is the longest. Pedal footprints directed inward in trackways. Crouching traces display metatarsal impressions. Divarication angles between Digits II–III always greater than Digits III–IV. Manus pentadactyl with short, robust digit traces of subequal lengths.

Description. See Fig. 6.65. *Moyenisauropus* generally has a digitigrade pes, but rare metatarsal traces are known from southern Africa (Ellenberger, 1972, 1974). Digits on both the manus and pes are far more robust than the more slender and gracile digits of *Anomoepus* (Gierliński, 1991a; Lockley and Gierliński, 2006). In addition, *Moyenisauropus* pes prints have subequal lengths and generally have greater divarication angles between digits, both unlike most examples of *Anomoepus* (Gierliński, 1991a).

Discussion. *Moyenisauropus* tracks have been attributed to basal, quadrupedal thyreophoran dinosaurs such as *Scelidosaurus* (Gierliński, 1999). The holotype tracks are from the Lower Jurassic upper Elliot Formation (Stormberg Group) at the famous Moyeni Site in Lesotho, southern Africa (Ellenberger, 1974). The ichnogenus has subsequently been reported from the Lower Jurassic Kayenta Formation, Kayenta–Navajo transition, and lower Navajo Sandstone of the Glen Canyon Group within GLCA in western North America (Lockley et al., 1998f, 2014c) as well as Portugal (Figueiredo et al., 2023).

Moyenisauropus karaszevskii Gierliński 1991a is from the upper Hettangian Przysucha Ore-bearing Series, Holy Cross Mountains, Poland. The holotype (MGIW 1560.II.9) is bipedal trackway with tridactyl tracks, sometimes with Digit I traces. Two phalangeal pads are present on Digit III; Digit IV is the longest, and trackways are wide, with a pace angulation of 103° (Gierliński, 1991a).

Olsen and Galton (1984) and Haubold (1984, 1986) regarded *Moyenisauropus* as junior synonyms of *Anomoepus*, but Gierliński and Potemska (1987) retained it as valid. Gierliński (1991) proposed moving *M. dodai, M. longicauda, M. minor, M. natatilis,* and *M. vermivorus*, all erected by Ellenberger (1974) to the ichnogenus *Anomoepus*. Gierliński (1991) retained the holotype *M. natator*, along with *M. levicauda* as valid, and placed his new ichnospecies *karaszewskii* into *Moyenisauropus*.

Several authors have continued to maintain *Moyenisauropus* as valid, and have erected new ichnospecies, such as *M. lusitanicus* from the Coimbra Formation of Portugal (Figueiredo et al., 2023). Other authors have considered *Moyenisauropus* to be a junior synonym of *Anomoepus* (Olsen and Galton, 1984; Thulborn, 1994; Wilson et al., 2009). Note that the correct spelling is *Moyenisauropus*; the ichnogenus has often been seen misspelled as "*Moyenosauripus*." In addition, Gand and Pellier (1976) misspelled the name as "*Moyenisauripus*".

Navahopus
Baird 1980
Fig. 6.66
Type ichnospecies. *Navahopus falcipollex* Baird 1980.

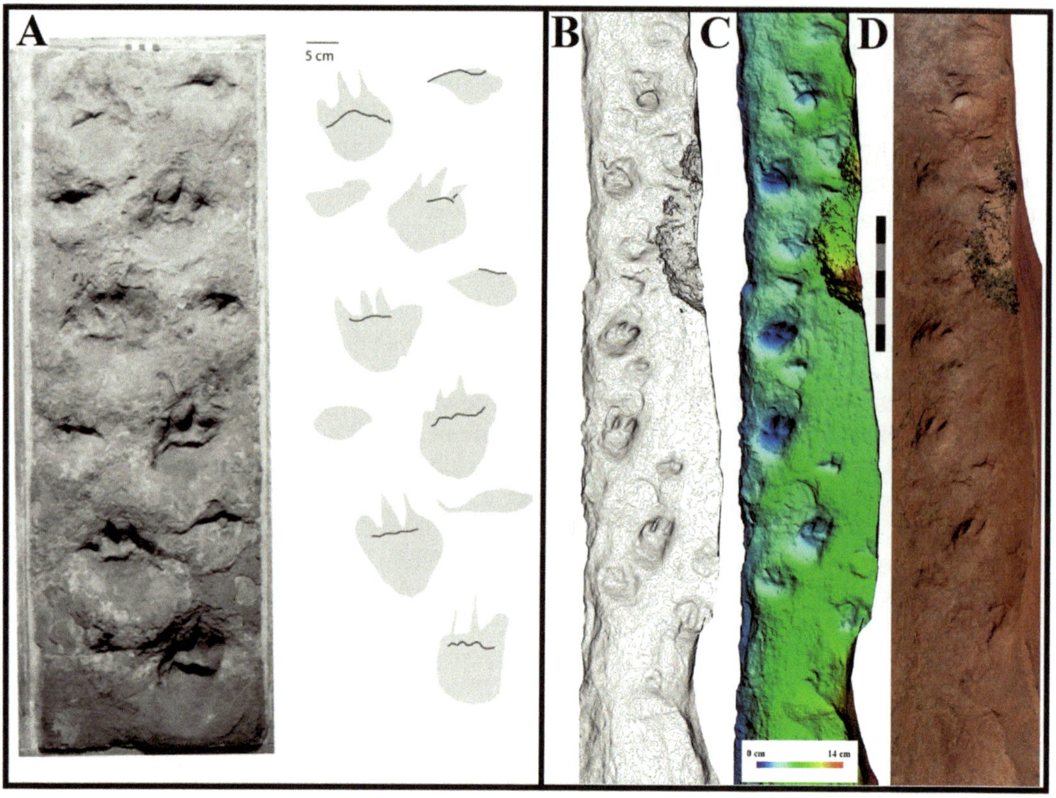

FIGURE 6.66 *Navahopus.* (A) Photo and sketch of *N. falcipollex* holotype replica (MNA V 3430) from the Navajo Sandstone, northern Arizona. (B–D) Photogrammetrically prepared images of *Navahopus coyoteensis* holotype from the Navajo Sandstone, Vermillion Cliffs National Monument, Utah. Scale = 50 cm. (B) Topographic contour map (1 mm contour intervals). (C) Color depth map (resolution 0.52 mm per pixel). (C) Digital orthoimage (resolution 0.26 mm per pixel). *(A and B) Modified from Hunt and Lucas (2006), Fig. 1. (B–D) Modified from Breithaupt et al. (2021), Fig. 3.*

Valid ichnospecies. *Navahopus falcipollex* Baird 1980; *Navahopus coyoteensis* Milàn, Loope, and Bromley 2008.

Distribution. Lower Jurassic, cross-stratified eolian dune facies of the Navajo Sandstone (Sinemurian–Pliensbachian), MNA Locality 226, 2 miles north of Copper Mine Trading Post on Kaibito Plateau, Navajo Nation, Coconino County, northeastern Arizona (Baird, 1980); near Vermillion Cliffs National Monument and "The Wave" in the Navajo Sandstone, Utah (Milàn et al., 2008).

Type Specimen(s). *Navahopus falcipollex* Baird 1980, holotype MNA V 3430 (old numbers MNA P3.339 and MNA G2.7092), a replica trackway with five manus/pes sets preserved in negative hyporelief (collected by Lionel F. Brady in 1958). *Navahopus coyoteensis* Milàn et al. 2008. According to Milàn et al. (2008, p. 198), a replica was to be made for the MNA, but a permit was pending at the time of publication and a replica likely does not exist. Breithaupt et al. (2021, Fig. 3) did create a photogrammetric model of the holotype, however.

Diagnosis. For *Navahopus falcipollex*: Quadrupedal trackways; pes tetradactyl, clawed, with digits increasing in length order from I, II, IV, and III; manus functionally tridactyl with short, clawed digit traces II and III directed forward, and horizontally recumbent, falciform pollex claw directed medially (from Baird, 1980). For *Navahopus coyoteensis*: having pedal digit impression IV of equal or greater length than Digit III. Pedal hypex of III–IV is much greater than hypices I–II and II–III (from Milàn et al., 2008).

Discussion. See Fig. 6.66. *Navahopus* is an uncommon, exclusively Lower Jurassic ichnotaxon represented by two ichnospecies, *N. falcipollex* (Baird, 1980) and *N. coyoteensis* (Milàn et al., 2008; Breithaupt et al., 2021), both from the Navajo Sandstone, Glen Canyon Group of the American Southwest. Tracks of *Navahopus* have been attributed to both sauropodomorphs (Baird, 1980; Milàn et al., 2008), and large therapsid synapsids (Lockley and Hunt, 1995; Shibata et al., 2006). Hunt and Lucas (2006) suggested *N. falcipollex* was likely produced by a large tritylodont. Milner et al. (2024) recorded new occurrences of *Navahopus* in the lowermost Navajo Sandstone in GLCA.

***Neotrisauropus* (=*Grallator*?)**
Ellenberger 1974
Fig. 6.67

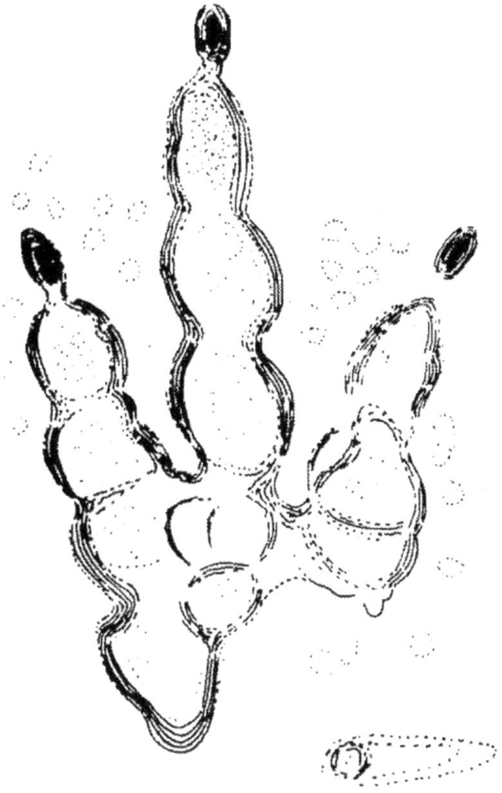

FIGURE 6.67 *Neotrisauropus deambulator* Ellenberger 1974 from the Lower Jurassic Elliot Formation, Moyeni tracksite, Lesotho. *Schematic modified from Ellenberger (1974).*

Type ichnospecies. *Neotrisauropus deambulator* Ellenberger 1974.

Valid ichnospecies. None.

Distribution. Lower Jurassic of southern Africa (Elliot Formation, Moyeni, Karoo Basin, Lesotho).

Diagnosis. See *Grallator*.

Description. See Fig. 6.67. Also see *Grallator*.

Discussion. Wilson et al. (2009) placed *Neotrisauropus* under *Grallator* following several authors (Olsen and Galton, 1984; Haubold, 1986; Gierliński, 1991a; Olsen and Rainforth, 2003). Therefore, both the type ichnospecies, *N. deambulator*, and others, including *N. mokanametsongensis* Ellenberger 1974, *N. lacunensis* Ellenberger 1970, and *N. leribeensis* Ellenberger 1970, are considered invalid.

Occitanopodus

Moreau et al., Trincal, Fara, Baret, Jacquet, Barbini, Flament, Wienin, Bourel, and Jean 2020 Fig. 6.68

Type ichnospecies. *O. gandi*.

Valid ichnospecies. *O. gandi*.

Distribution. Middle Jurassic of Europe (Calcaires à Stipites Formation: France).

Type specimen(s). Manus–pes track set LM1-LP1 of trackway CAS-1 preserved in situ on the roof of the Castelbouc No. 4 Cave (Lozère, southern France).

Diagnosis. Adapted from Moreau et al. (2020): This ichnotaxon exhibits a unique combination of characters that have not been observed in other wide-gauge sauropod tracks, such as *Brontopodus*, "*Polyonyx*"), and *Titanopodus*, as well as medium-to wide-gauge sauropod tracks like *Calorckosauripus* and *Oobardjidama*. The trackway is wide-gauge and displays pronounced heteropody, with the pedal track area being 4–6 times greater than the manual track area. The pes and manus tracks are aligned. The pes tracks are large (~1 m in length), pentadactyl, asymmetric, subcircular to oval in shape, and are positively rotated in relation to the trackway axis (at an angle of 0–27°). The tracks are generally as long as they are wide or slightly wider than they are long. The maximum width is located in the middle or slightly toward the anterior part of the track. The short digit impressions exhibit a strong outward rotation. The manus tracks are symmetrical and D-shaped, with a convex anterior shape. They lack distinct digit imprints and also display a positive rotation.

Description. See Fig. 6.68. *Occitanopodus* trackway CAS-1 is 18 m long and ~2.6 m wide. It consists of 12 pes tracks and 10 manus tracks, demonstrating a high level of heteropody. The pes tracks are asymmetrical, ranging from subcircular to oval in shape, and pentadactyl. They vary in length (56–94 cm) and width (82–116 cm). Pedal digit imprints are short and rotated outwardly. The pedal tracks in CAS-1 exhibit a slight positive rotation angle in relation to the trackway midline, measuring less than 27°. The manual prints have a symmetrical "D" shape, convex forward outlines, and are wider than they are long (~21–47 cm in length and 40–76 cm in width) with a small positive rotation angle (<27°). There are no distinct claw traces associated with the impression of manus Digit I, distinguishing them from other European Middle Jurassic tracks such as "*Polyonyx*" (Santos et al., 2009; in press; Moreau et al., 2020).

Stride lengths range between 3.1 and 3.5 m for both the manus and pes, and both have similar pace lengths (2.0–2.3 m). Moreau et al. (2020) considered *Occitanopodus* to be wide-

FIGURE 6.68 *Occitanopodus* LM1-LP1 holotype. *Modified from Moreau et al. (2020).*

gauge based on the tracks having width of the pes angulation pattern: pes length ratios (WAP/PL; sensu Marty, 2008) of 1.2–2.0 and a pes trackway ratio (PTR, sensu Romano et al., 2007) of 39%. However, they noted that the PTR value was close to those for "medium-gauge" sauropod trackways (sensu Marty, 2008).

Discussion. *Occitanopodus gandi* presents unique features not typically observed in other Late Jurassic and Early Cretaceous sauropod tracks. These features include wide-gauge tracks with significant heteropody, large-sized pentadactyl pedal tracks with outwardly rotated short digit impressions, and symmetrical "D"-shaped manual tracks devoid of digit or claw traces that form a small positive rotation angle relative to the trackway axis.

Occitanopodus tracks were granted ichnogeneric distinction from other sauropod dinosaur ichnotaxa based on having a unique combination of track characters unknown in other ichnotaxa.

Occitanopodus is a remarkable discovery of sauropod dinosaur tracks situated approximately 500 m below the surface of the Causse Méjean Plateau (within Castelbouc Cave) in southern France. *Occitanopodus* is currently known solely from its type locality, providing invaluable evidence of sauropods during the Middle Jurassic. The CAS tracks are preserved in convex hyporelief on the roof of a tunnel gallery in a natural cave at the interface between a (weathered away) marly limestone bed and a peloomicritic limestone bed. Two other sauropod trackways preserved on the same surface were not attributed to *Occitanopodus*: one (CAS-2) because it is poorly preserved and exhibits lower heteropody than *Occitanopodus*, and the other (CAS-3), a pes-print-only trackway, because of the absence of manus prints and narrow gauge trackway metrics.

Moreau et al. (2020) suggested that the maker of *Occitanopodus* tracks was most likely a titanosauriform sauropod. This aligns with interpretations of other Middle Jurassic sauropod trackways elsewhere in Europe. While it is important to note the limitations of interpreting wide-gauge sauropod trackways as belong to titanosauriforms, these trackways collectively provide evidence for the presence of titanosauriform sauropods in the Middle Jurassic, especially when the contemporaneous body-fossil record is depauperate (Bindellini and Dal Sasso, 2021; D'Emic, 2012). In the absence of skeletal remains, these trackways represent the first evidence of Middle Jurassic sauropods in France (Moreau et al., 2020).

Otozoum
Hitchcock 1847
Fig. 6.69

Type ichnospecies. *O. moodii.*

Valid ichnospecies. *O. moodii., O. pollex, O. grandcombensis.*

Distribution. Lower Jurassic, Portland Formation, Connecticut (Rainforth, 2003, 2005a); Lower Jurassic Kayenta Formation, Wingate Sandstone, and Navajo Sandstone (Sinemurian–Pliensbachian), Glen Canyon Group, Utah and Colorado (Lockley et al., 1998f, 2006, 2014; Lucas et al., 2006; Lockley and Gierliński, 2014; Milner et al., 2011). Many sites at GLCA (Lockley et al., 1998f, 2014); North Moccasin Mountain Tracksite near Kanab, Utah (Milner et al., 2011). Lower Jurassic Skłoby Formation (middle Hettangian), upper Gromadzice outcrop (Gierliński and Niedźwiedzki, 2005). Lower Jurassic Etjo Formation, Karoo Supergroup, Waterberg Plateau, Namibia (D'Orazi Porchetti et al., 2014). Lower Jurassic (Pliensbachian), Issil-n-Aït Arbi, central High Atlas, Morocco (Masrour and Pérez-Lorente, 2014; Masrour et al., 2015).

Type specimen(s). ACM-ICH 4/1a (old number 234), holotype (lectotype of Rainforth, 2003) natural cast trackway with four consecutive pes footprints in brown sandstone, collected by Pliny Moody in 1847. ACM-ICH 5/14, topotype natural cast in brown sandstone with two pes and two manus tracks. Specimens are from the Hettangian Portland Formation, Moody homestead, South Hadley, Massachusetts (Rainforth, 2003).

Diagnosis. From Lockley et al. (2006a): Pentadactyl pes with four digits traces directed anteriorly; Digit III is slightly longer than subequal Digits II and IV, which are longer than Digit I trace; Digit V trace is a subrounded pad posterior to the coalesced

FIGURE 6.69 *Otozoum.* (A and B) *Otozoum moodii* holotype ACM-ICH 4-1a from Lower Jurassic Portland Formation, Moody homestead, South Hadley, Massachusetts, (A) best preserved track, and (B) trackway slab which also contains the holotype trackway of *Grallator parallelus.* (C and D) Alternate images of best track from *Otozoum moodii* holotype slab. Note skin impressions preserved. Scales = 5 cm. (E) *Otozoum* from Navajo Sandstone at Antelope Island tracksite, GLCA, Utah. (F) *Otozoum* trackway discovered during March 2023 survey of Lake Powell (GLCA), Utah, because of drought. Scale = 10 cm. *(B) Photos courtesy James Farlow. (C and D) Modified from Rainforth (2003). (F) Photo by Andrew Milner.*

metatarsophalangeal pads of Digits III and IV; Digits II–IV subparallel, laterally convex, and blunt, anteriorly and medially directed claw impressions; Digit I impression less curved and slightly divergent medially with respect to Digit II; phalangeal pad formula, including claw marks = 2-3-4-5-1; pentadactyl manus much smaller than pes and rotated outwardly, with short digit traces; manual Digit I impression slightly longer and more prominent than digit traces II–V.

Description. See Fig. 6.69. The holotype trackway (ACM-ICH 4/1a; illustrated by Hitchcock, 1847) from the Lower Jurassic Portland Formation, Massachusetts, consists of four large, tetradactyl consecutive tracks of an animal walking bipedally, and preserves skin impressions (Lockley et al., 2022). *Otozoum* trackways have both wide and narrow gauges, as well as bipedal and quadrupedal gaits (Lockley, 2001, 2007; Lockley et al., 2022). ACM-ICH 5/14 preserves two manus tracks that belong to the same trackway as the pes prints, even though Farlow (1992) and Olsen et al. (1992) did not agree that these traces belonged together. However, an *Otozoum* trackway in the area of Moab, Utah (specimen UCM 184.41; Lockley et al., 2006a) demonstrated that all of the ACM-ICH 5/14 tracks are from the same trackway sequence.

Discussion. *Otozoum* is recorded from both Northern and Southern hemispheres. Sauropodomorphs are the likely producer of *Otozoum* tracks (Lockley et al., 2022). *Anchisaurus* occurs in the Hettangian Portland Formation in Connecticut and is a good candidate for the *O. moodii* track-producer (Rainforth, 2003), while *O. pollex* could have been produced by *Ammosaurus*, *Massospondylus*, and similar sauropodomorphs. Rainforth (2003) provides reasons why thyreophorans (e.g., *Scelidosaurus*) and other smaller ornithischians (e.g., *Lesothosaurus*) should be ruled out as track makers based on foot skeletal morphology. See Farlow et al. (2018) for good skeletal foot comparisons.

This ichnogenus is part of the *Otozoum–Pseudotetrasauropus–Evazoum–Kalosauropus* (OPEK) plexus (Lockley et al., 2006a) spanning the Upper Triassic (Carnian–Rhaetian) through Lower Jurassic (Hettangian–Pliensbachian?) of Pangaea. In addition, all four of these ichnogenera have been placed in the ichnofamily Otozoidae (Lull, 1904). The Late Triassic *Pseudotetrasauropus* had been suggested to be a synonym of *Otozoum* by some (Ellenberger, 1970; Haubold, 1971a), while others noted the similarities between the two but kept them separate (Rainforth, 2003; D'Orazi Porchetti and Nicosia, 2007). Also see *Otozoum grandcombensis* Gand et al. (2000).

Palamopus
Hitchcock 1845
Fig. 6.70
Type ichnospecies. *P. palmatus.*
Valid ichnospecies. *P. palmatus.*

Distribution. Lower Jurassic of North America (Turners Falls Formation [Newark Supergroup]: Massachusetts).

Type specimen(s). (lectotype) AC 32/6, shale? slab bearing natural casts of four tracks (past preparation has obliterated the manus prints such that only two damaged pes prints remain [Rainforth, 2005a]).

Diagnosis. From Hitchcock (1836 and 1848): manus and pes both tetradactyl, plantigrade (manus "imperfectly" so) with all digit imprints oriented anteriorly. Manus tetradactyl with

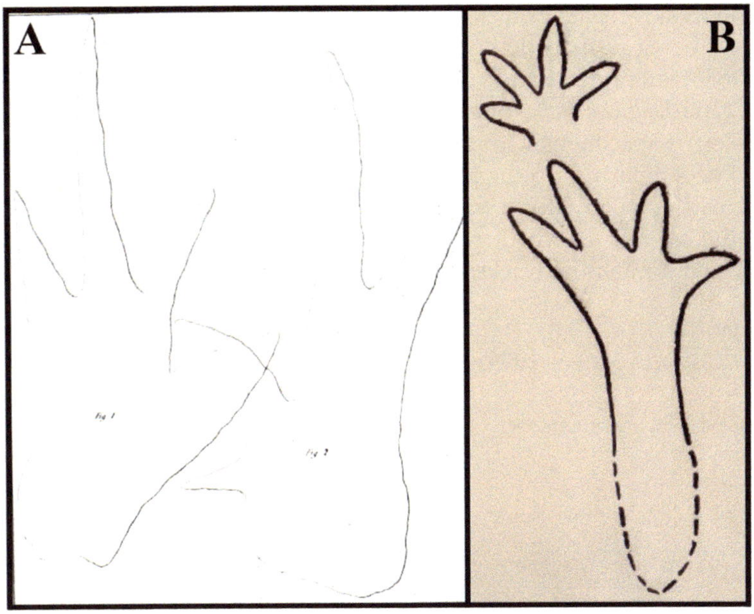

FIGURE 6.70 *Palamopus palmatus*. (A) Sketch from Hitchcock (1848). (B) *Palamopus* holotype, ACM-ICH 36/52, sketch from Lull (1915).

divarication angles I–II 30°, II–III 35°, III–IV 35°, and digit lengths III > IV > II > I. Heel wider than long. Pedal digits long and slender; Digits I and II lie close to one another (divarication angle 10°); other digits more equitably spaced (divarication angles II–III 30–35°; III–IV 35°; I–IV 75–80°). Digit lengths III > II > IV > I. Heel elongate (roughly 3× longer than wide). Axes of both manus and pes prints nearly parallel to trackway axis. Hitchcock (1836) proposed the tracks were attributable to a biped, but switched to proposing a plantigrade quadruped after subsequent preparation (Hitchcock, 1848).

Description. See Fig. 6.70. Hitchcock (1848) described *Palamopus* tracks as "leptodactylous" (slender-toed), a characteristic Rainforth (2005a) noted was prone to influence by substrate conditions. Manus prints lie anterior to pes prints. Hitchcock (1848) discussed a vague, posteriorly directed Digit V on the pes; this region of the specimen is now damaged (Rainforth, 2005a).

Discussion. *Palamopus* is an ichnogenus with a convoluted history. The type ichnospecies, *P. palmatus*, was initially erected by Hitchcock (1836) as *Ornithoidichnites palmatus*, then transferred without reason by Hitchcock (1837) to *Sauroidichnites palmatus* (*Sauroidichnites* is a suppressed name under the ICZN [Rainforth, 2005b]) before being placed without description or discussion in its own ichnogenus, *Palamopus* by Hitchcock (1845), in which the ichnospecies name was arbitrarily changed to *P. anomalus*. The ichnospecies *palmatus* was subsequently synonymized (incorrectly) with *Macropterna recta* because *Palamopus* was, ostensibly, a bipedal ichnotaxon (Hitchcock, 1848). The confusing situation has been reviewed thoroughly by Rainforth (2005a,b). Rainforth (2005b) petitioned the ICZN to conserve *Palamopus* *palmatus*. The petition received both support (e.g., Lucas, 2006) and dissent (e.g., Moser, 2006) before ultimately being ratified (ICZN, 2007).

Hitchcock (1836) attempted to describe the fossil tracks he had until that point as fossil bird tracks (hence "*Ornithoidichnites*"), but he had difficulty reconciling the pedal morphology of the type and referred specimens with those of birds. Lockley and Meyer (2004) noted that *Palamopus* bore some similarities to the ichnotaxon *Selenichnus*, which was considered synonymous with the early crocodyliform track *Batrachopus* (Lockley et al., 2004c). The damage to the holotype, the overall dearth of referred specimens, and the lack of subsequent study makes determining a potential track-making taxon difficult.

Parabrontopodus
Lockley, Farlow, and Meyer 1994a
Fig. 6.71
Type ichnospecies. *P. mcintoshi*.

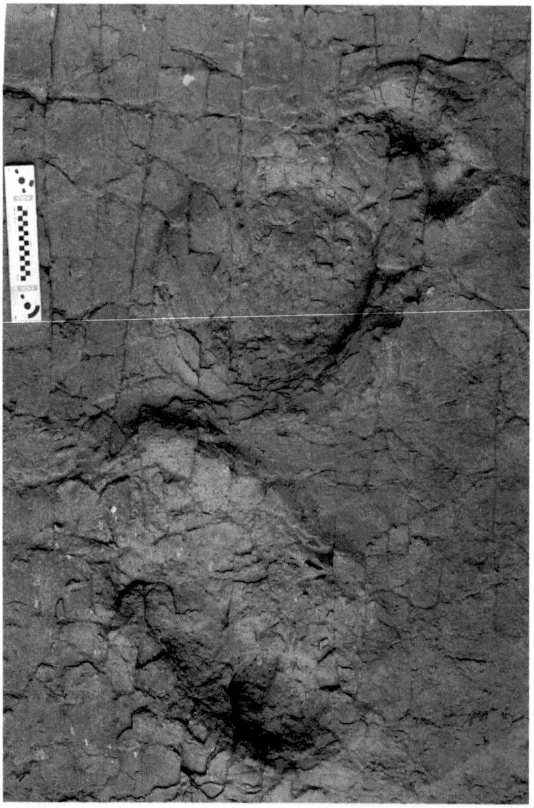

FIGURE 6.71 *Parabrontopodus barkhausensis* (prev. *Elephantopoides barkhausensis*). Orthophoto generated from 3D model of two manus pes couple of the trackway S1, referred as the type trackway of the *P. barkhausensis*. The upper set corresponds to the in situ holotype, and to the replica UCM 193.11 (Meyer et al., 2021). Scale = 25 cm, from target to target, or 15 cm along alternating black cm squares. *Photos and models by Christian Meyer.*

Valid ichnospecies. *P. mcintoshi*; *P. barkhausensis* (previously *Elephantopoides*) (Meyer et al., 2021).

Distribution. Lower Jurassic of Europe (Zagaje and Drzewica formations: Poland). Upper Jurassic of North America (Morrison Formation: Colorado) and Europe (Couches du Chailley Formation: France; Middle Süntel Formation: Germany; Poreč Formation: Croatia; Azóia Limestone: Portugal; Reuchenette Formation: Switzerland). Lower Cretaceous of Europe (Huérteles Formation and Enciso Group: Spain). Possibly Lower Jurassic of Europe (Monte Zugna Formation: Italy).

Type specimen(s). *P. mcintoshi*: Plaster cast of manus-pes set CU-MWC 190.5 now at University of Colorado Museum; original trackways at the Purgatoire River tracksite in southeastern Colorado. *P. barkhausensis*: In situ trackway "S1" (Meyer et al., 2021: Fig. 8) and manus-pes set cast (UCM 193.11) from the Barkhausen type locality. No holotype was designated by Kaever and de Lapparent (1974) for *Elephantopoides barkhausensis* (see also Friese, 1979). However, trackway S1 has been considered the most diagnostic by subsequent workers (e.g., Diedrich, 2011; Lockley et al., 2004a; Lockley and Meyer, 2000; Meyer et al., 2021). Several topotypes/paratypes are recognized from trackways S2–5 and S7–9 at the type locality.

Diagnosis. From Lockley et al. (1994a): "Narrow sauropod trackway of medium to large size (pes length ~50–90 cm), characterized by no space between trackway midline and inside margin of pes tracks. Pes footprint longer than wide with long axis rotated outward. Pes claw impressions, corresponding to Digits I, II, and III show strong outward rotation. Manus tracks semicircular and small in comparison with pes tracks (i.e., pronounced heteropody)."

For *P. barkhausensis*, emended by Meyer et al. (2021): Narrow to medium-gauge sauropod trackway of a small to medium sized sauropod (pes length ~ 30–47 cm). Oval pes with four digit traces and with the long axis of the track positively rotated relative to trackway midline. The manus tracks are small, semicircular. Trackway displays pronounced heteropody.

Description. *P. mcintoshi*: Narrow gauge trackways, with medial edges of pes tracks overlapping trackway midline. Manus tracks semicircular and wider than long; small relative to pes tracks. Pes tracks large (78 × 56 cm in holotype); similar to other sauropod pes tracks in having a large heel pad area, a relatively straight medial edge, convex lateral edge (Lockley et al., 1994a).

P. barkhausensis (Fig. 6.71): Adapted from Meyer et al. (2021): These narrow-to-medium gauge trackways (width from outer edges of pes: 67–108 cm) show pronounced heteropody. The pes are noticeably larger than the manus tracks, oval-shaped, with an average length of 36 cm and a width of 27 cm. No claw traces are present although four digit impressions are shallow. Semicircular manus tracks, outwardly rotated in relation to the pes (by 25–35°), with an average length of 13 cm and a width of 23 cm. Pace varies between 74 and 114 cm with stride lengths between 110 and 164 cm.

Discussion. Generally, *Parabrontopodus* trackways are held to have been made by diplodocid sauropod dinosaurs (Lockley et al., 1994a; Wright, 2005).

Parabrontopodus (Lockley et al., 1994a) was described from the lower Morrison Formation at the Purgatoire River tracksite in southeastern Colorado (Lockley et al., 1986) and was later identified at the Rancho Del Rio site in the Morrison Formation in central Colorado (Foster and Lockley, 2006). Since then, the ichnogenus has been used worldwide to classify "narrow-gauge" sauropod trackways throughout the Mesozoic (e.g., Marty, 2008; Marty et al., 2010). *Parabrontopodus* isp. has also been identified in European Late Jurassic tracksites in Portugal (Lockley and Santos, 1993), France (Le Lœuff et al., 2006; Mazin et al., 2016),

Switzerland (Marty et al., 2010; Sciscio et al., 2022) and Croatia (Mezga et al., 2007, 2017), making it the most common sauropod ichnotaxon in the Late Jurassic of Europe.

Elephantopoides barkhausensis holds a unique position in the annals of sauropod track research in Europe, being the first sauropod tracks to be described and named. However, it is noteworthy that the name "*Elephantopoides*" erroneously suggests an elephant as the track maker, and further adding to the confusion, the ichnospecies has often been misspelled as "*barkhausensis*" (Meyer et al., 2021). This misnomer, and the lack of a holotype, led to the classification of the ichnotaxon as a nomen dubium until the redescription and reexamination by Meyer et al. (2021): *Parabrontopodus barkhausensis* (Kaever and de Lapparent, 1974) comb. nov.

Meijide Fuentes et al. (2001) introduced a third ichnospecies, *Parabrontopodus distercii*; however, Moratalla (2009) deemed it a nomen nudum due to preservation issues. Castanera et al. (2012) emphasized that the gauge of the trackway varies along its length, deviating from the primary diagnostic characteristic of *Parabrontopodus*.

Paravipus

Mudroch et al., Richter, Joger, Kosma, Idé, and Maga 2011

Fig. 6.72

Type ichnospecies. *P. didactyloides.*

Valid ichnospecies. *P. didactyloides.*

Distribution. Middle Jurassic of Africa (Irhazer Group = Irhazer Shale: Niger).

Type specimen(s). SNHM NMB-1887-Sp, resin mold of a right footprint from a trackway preserved in a fine-grained sandstone. Track preservation suggests registration in moderately saturated sediment but after the formation of polygonal desiccation cracks preserved on the same surface.

Diagnosis. Didactyl tracks of a bipedal track maker. Digit III and IV impressions slender, subequal in length, and subparallel (divarication 5–15°); both taper into distinct claw imprints. Metapodial imprints absent. Digit II traces occasionally present as small round pad imprint slightly posterior to the proximal end of the Digit III impression.

Description. See Fig. 6.72. In addition to lacking metapodial traces, *Paravipus* tracks also lack digital pad imprints except for the occasionally present base of Digit II. Digit III imprints reportedly lie virtually parallel to the trackway axis. Roughly 120 tracks are preserved in five trackways (A–E); two of these (A and C) are oriented approximately 180° to another two (B and D), with tracks from A and C overlapping those of B and D. Trackways have high pace angulation values (150–180°).

Discussion. *Paravipus* tracks were attributed to walking, functionally didactyl, paravian ("deinonychosaurian") theropod track makers because of the consistent absence of elongate Digit II impressions across multiple trackways, as would be expected of a dromaeosaurid or troodontid track maker. If correct, these tracks would constitute some of the earliest evidence of such theropods globally, comparable in age to body-fossil occurrences in eastern China (e.g., Hu et al., 2018) and a report of two didactyl tracks from the Middle Jurassic Dansirit Formation of Iran (Xing et al., 2017c). They have subsequently been likened to theropod swim tracks (Lockley et al., 2016b; Milner and Lockley, 2016; Navarro-Lorbés et al., 2023), though they have not been synonymized with *Characichnos* or any other swim-track ichnotaxon. Poorly preserved, probably theropod swim tracks from the Jurassic–Cretaceous Tuchengzi Formation of Hebei, China reported by Xing et al. (2011b) bear some resemblance to *Paravipus*.

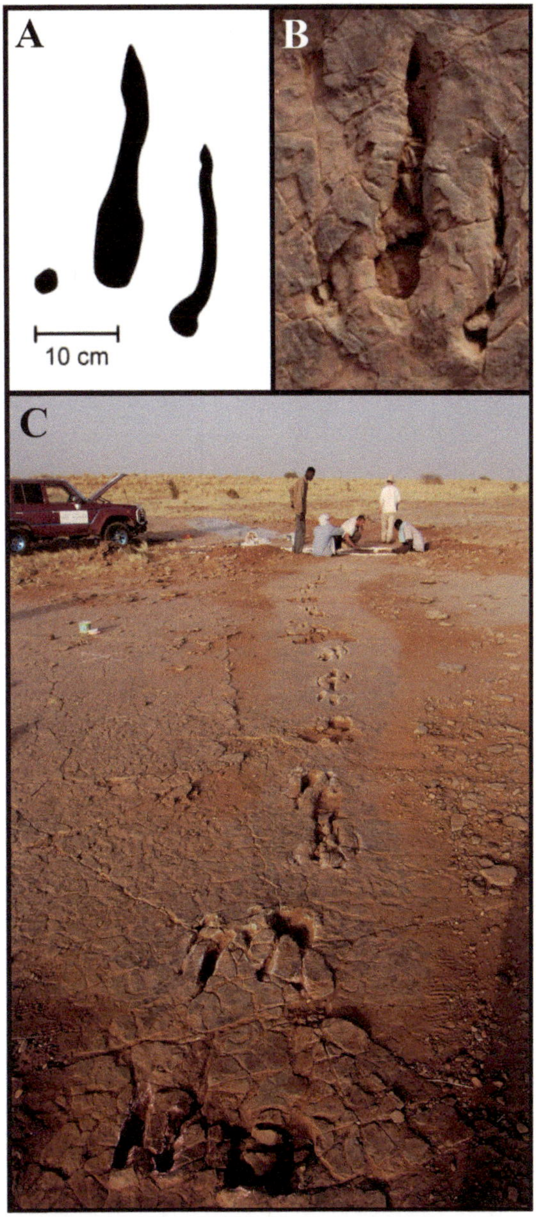

FIGURE 6.72 *Paravipus didactyloides*. (A and B) SNHM NMB-1887-Sp holotype, (A) sketch and (B) actual track photo at the same scale. Scale = 10 cm. (C) Trackway of type locality in Niger. *(B) Modified from Mudroch et al. (2011). (C) Courtesy Alexander Mudroch.*

Pengxianpus
Yang and Yang 1987
Fig. 6.73

Type ichnospecies. *P. cifengensis.*

Valid ichnospecies. *P. cifengensis, P. yulinensis.*

Distribution. Upper Triassic of Asia (Xujiahe Formation: Sichuan, China). Lower Jurassic of Asia (Fuxian Formation: Shaanxi, China).

Type specimen(s). *P. cifengensis*: CFZW1–2 (called CFPC1–2 in Yang and Yang (1987) and Xing et al. (2013e)), two consecutive, natural cast tracks on a mud-cracked slab. *P. yulinensis*: track ZJ-A1 on slab BNHM Ph001373, a natural mold footprint.

Diagnosis. From Xing et al. (2013e): Tracks of a large, tridactyl, mesaxonic biped. Digit imprints long and slender, with rounded digital pads. Digit length formula III > II > IV. Wide divarication angle between Digits II–IV (up to 69°). Small, circular metatarsophalangeal pad at the proximal end of the Digit IV imprint.

FIGURE 6.73 *Pengxianpus cifengensis,* holotype replica UCM 176-3. Scale = 5 cm. *Photo by Andrew Milner.*

Description. See Fig. 6.73. CFZW1, the better preserved of the two tracks on the holotype slab, has a possible digital pad formula of ?2-3-?3 (Xing et al., 2013e). The tapered distal ends of the imprints suggest the presence of claws. The metatarsophalangeal pad is distinct from the imprint of Digit IV, demarcated by a pronounced notch on the lateral margin of the track. Skin impressions are present on the metatarsophalangeal pad. Yang and Yang (1987) interpreted CFZW1 as being tetradactyl, but Lockley et al. (2013) and Xing et al. (2013e) denied the presence of a Digit I imprint; Xing et al. (2013e) described the supposed Digit I imprint as a sedimentary structure (probably a natural cast of a mud crack) adjacent to the track. Tracks apparently are not rotated with respect to the trackway axis. Wang et al. (2016) erected *P. yulinensis* for slightly younger tracks from Shaanxi; they based placement in *Pengxianpus* on the similarity of divarication angles, but differentiated the ichnospecies based on the younger tracks having more distinct digital pads, thicker digit imprints, and slightly narrower divarication angles.

Discussion. Yang and Yang (1987) referred *P. cifengensis* to a "prosauropod" (basal sauropodomorph) dinosaur track maker. Lockley and Matsukawa (2009) compared the ichnotaxon with *Atreipus*, an ichnotaxon that may be attributed to either basal dinosauromorphs or possibly basal ornithischian dinosaurs. Xing et al. (2013e) compared the ichnotaxon more favorably with *Kayentapus*, a theropod ichnotaxon, but retained the ichnotaxon as provisionally valid based on the small sample size available for *Pengxianpus*.

P. cifengensis is from the Upper Triassic Xujiahe Formation of Sichuan, China (Chen et al., 2006; Xing et al., 2013e; Yang and Yang, 1987). *P. yulinensis* is from the Lower Jurassic Fuxian Formation of Shaanxi, China (Wang et al., 2016).

***Plastisauropus* (=*Grallator*?) and *Platysauropus* (=*Eubrontes*?)**
Ellenberger 1970, 1974
Fig. 6.74
Type ichnospecies. *P. ingens*.
Valid ichnospecies. None.
Distribution. Lower Jurassic of Africa (Moyeni, Lesotho; Ellenberger, 1972).
Diagnosis. See *Grallator* and *Eubrontes*.
Description. See Fig. 6.74. Also see *Grallator* and *Eubrontes*.

Discussion. Likely equivalent to ichnotaxa in the *Grallator*—*Eubrontes* plexus. It seems that *Platysauropus ingens* (Ellenberger, 1970) was renamed to *Plastisauropus ingens* (Ellenberger, 1974; a track shown as Fig. 63 from Mokanametsong in zone B/1, so likely Hettangian). *Plastisauropus ingens* Ellenberger 1974 is slightly older and more *Grallator*-like than the more robust, larger *Platysauropus robustus* Ellenberger 1970, which is from zone B/4 at Phahameng (well into the Sinemurian of Lesotho). *Platysauropus robustus* was attributed to large, three-toed bipeds (Ellenberger, 1975, p. 423) and is similar to a large tridactyl southern Africa track called "*Ichnites euskelosauroides*" by Huene (1932) (q.v., Huene, 1941; Ellenberger and Ellenberger, 1958, p. 66; Ellenberger, 1970), and later synonymized with *Eubrontes* by Haubold (1971a). Crews with one of us (EMB) also found some large tridactyl tracks at the same Phahameng locality in Lesotho, but it seems *Platysauropus robustus* is slightly more elongate (TL/TW), has a higher Dp/TS (0.65) and stronger "heel" asymmetry than the unnamed

FIGURE 6.74 *Plastisauropus ingens*. Scale = ~20 cm. *Redrawn from Ellenberger (1970).*

Phahameng tracks (i.e., TL/TW of 1.35, Dp/TS between 0.47 and 0.5 and Dp/TL between 33% and 35%). Also, see Mukaddam et al. (2020) on *Platysauropus robustus*. In short, a preliminary hypothesis is that *Plastisauropus* is a possible *Grallator* synonym, whereas *Platysauropus* is a possible *Eubrontes* synonym.

Plesiornis
Hitchcock 1858
Fig. 6.75
Type ichnospecies. *P. fulicoides* (see below).
Valid ichnospecies. None (see below).

Distribution. Upper Triassic of North America (Manassas Sandstone: Virginia). Lower Jurassic of North America (Turners Falls Formation [Newark Supergroup]: Massachusetts) and Europe (Przysucha Formation: Poland).

Type specimen(s). AC 19/18, trackway of three natural mold pes prints.

Diagnosis. Hitchcock (1858) diagnosed *Plesiornis* only briefly in broad, vague terms: bird-like, tridactyl tracks with thick ("pachydactylous") digit imprints that terminate in blunt claws or "pellets." Pes prints rotated slightly inward toward the trackway axis. Gierliński (1996a) diagnosed *Plesiornis* (based on Polish specimens) as those of a tridactyl–tetradactyl biped with Digit I sometimes present. Digit III longest. Contradicting Hitchcock (1858),

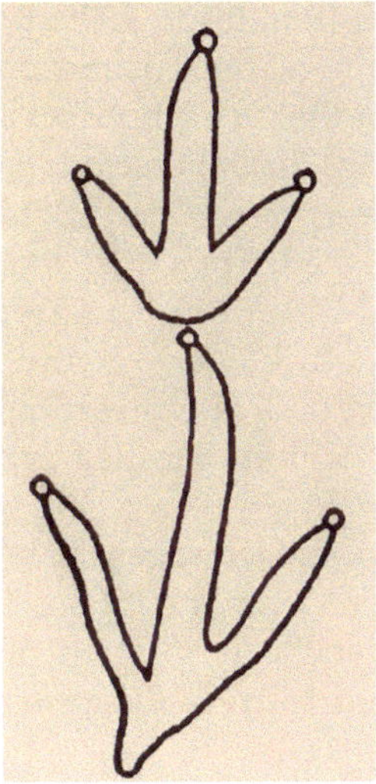

FIGURE 6.75 *Plesiornis pilulatus* Hitchcock, 1858 holotype ACM-ICH 13/1 from Lower Jurassic Turners Falls, Massachusetts. *From Lull (1915).*

Gierliński (1996a) further diagnosed *Plesiornis* as having thin digit imprints that lack discernible digital pads, though a metatarsophalangeal pad proximal to Digit III is well defined.

Description. See Fig. 6.75. Hitchcock (1858) described *P. quadrupes* (*fulicoides*; see below), his type ichnospecies, as pertaining to a quadruped, though Rainforth (2005a) does not report manus prints on the holotype specimen. According to Hitchcock (1858), manus prints tridactyl with divarication angles II–III 35° and III–IV 30°. Digital pad imprints absent. Axes of manus prints parallel to those of pes prints, to which they lie slightly anterior and sometimes somewhat laterally. Pes prints having divarication angles between Digits II–III 40° and III–IV 30°. Digit lengths III=IV > II and digital pad formula of 2-3-4. The absence of manus prints on holotype specimen AC 19/18 is puzzling, and suggests that Hitchcock (1858) based his description of *P. quadrupes* (*fulicoides*) on a different specimen (he cited five others referred to this ichnospecies, one of which [AC 16/14] was referred to *Anomoepus* by Rainforth (2005a)), or on multiple specimens collectively, some of which exhibit evidence of quadrupedal progression by the track maker.

Discussion. *Plesiornis* is another ichnotaxon with a confusing, convoluted history. Hitchcock (1843) initially erected the ichnospecies *Ornithoidichnites fulicoides* for specimen AC 19/18 (a lectotype designated by Olsen and Rainforth (2003)), from the Lower Jurassic Turners Falls Formation [Newark Supergroup] of Massachusetts. Hitchcock (1858) then renamed this ichnotaxon *Plesiornis quadrupes* and added two new ichnospecies to the ichnogenus: *P. pilulatus*, *P. aequalipes* (Rainforth (2005a) noted that the species name *quadrupes* is a junior objective synonym of *fulicoides*, and therefore invalid; likewise, *aequalipes* is a junior objective synonym of *Ornithoidichnites minimus*, and also therefore invalid.). Hitchcock (1865) added *P. mirabilis*, Hitchcock (1890) added *P. giganteus* (a nomen nudum [Rainforth, 2005a]), and Hay (1902) added *P. minor* for *Fulicopus minor* Hitchcock 1845. Lull (1904, 1915, 1953) sank *P. quadrupes* into *Anomoepus intermedius* (which was in error because the synonymized ichnospecies was, rather, *A. scambus* [Rainforth, 2005a]), *P. aequalipes* into *Argoides minimus*, and *P. mirabilis* and *P. giganteus* into various ichnospecies of *Anchisauripus*, leaving only *P. pilulatus* valid. *P. pilulatus* was not the type ichnospecies, however, and thus it cannot support the name *Plesiornis*, so Rainforth (2005a) erected "*Plesiornichnus*" to accommodate *pilulatus* as well as *giganteus* and *minimus* (*aequalipes*). Because this ichnogenus was proposed in a dissertation, it remains unpublished. Thus, *Plesiornis* remains in ichnotaxonomic limbo.

Hitchcock (1843, 1858) considered *Plesiornis* tracks avian; subsequently, different ichnospecies referred to *Plesiornis* have been attributed to both theropod and ornithischian dinosaur track makers. Those favoring a theropod interpretation for at least some of the specimens often have supported a "protoavian" affinity for the track maker (e.g., Gierliński, 1996a; Gierliński and Sabath, 1998).

Despite the questionable status of the ichnogenus, Weems and Kimmel (1993) and Gierliński (1996a) referred new specimens from the Upper Triassic Manassas Sandstone of Virginia and the Lower Jurassic Przysucha Formation of Poland, respectively, to *Plesiornis pilulatus*, and others have continued to treat *Plesiornis* as valid (e.g., Pacyna et al., 2022; Weems, 2019).

"Polyonyx"

Santos et al., Moratalla, and Royo-Torres 2009

Fig. 6.76

Comment. The sauropod ichnogenus name "*Polyonyx*" erected by Santos et al. (2009) is preoccupied by the decapod genus *Polyonyx* Stimpson, 1858, necessitating a new name for the ichnotaxon under International Commission on Zoological Nomenclature (ICZN) Art. 60. Although trace fossils are incorporated into the ICZN, conflicts may arise between biological taxonomy and ichnotaxonomy, prompting a review by Santos et al. (in press) and the establishment of a new ichnogeneric name.

Type ichnospecies. *P. gomesi.*

Valid ichnospecies. *P. gomesi.*

Distribution. Middle Jurassic of Europe (Serra de Aire Formation: Portugal). Possibly Middle–?Upper Jurassic of Africa (Isli Formation: Morocco).

Type specimen(s). A 142 m long trackway consisting of 94 consecutive manus–pes print sets in situ at MNDPDSA. Santos et al. (2009) reported that a cast (MNDPDSA-G5) consisting of a sequence of a left manus, two manus–pes prints sets, and a right pes print is reposited in the MNDPDSA center.

Diagnosis. From Santos et al. (2009): Wide-gauge quadrupedal trackway with low heteropody (manus-pes area ratio 1:2). Pes tracks exhibit four claw traces, with claw traces I–II positioned anteriorly and III–IV laterally oriented. Asymmetrical manus tracks with prominent Digit I traces oriented medially and associated with a posteriorly oriented, triangular claw trace. Manus Digits II–V display a slightly bent arrangement.

Description. See Fig. 6.76. Holotype trackway (G5) consists of 94 manus-pes sets and is ~142 m in length. It is a wide-gauge sauropod trackway with a unique, speech-bubble-shaped manus traces that possess clear digit impressions, including a large, medially oriented Digit I trace, and a substantial, posteriorly oriented triangular claw trace I. Pes prints show evidence of sharp claw traces in four digits. The ichnospecies demonstrates low heteropody.

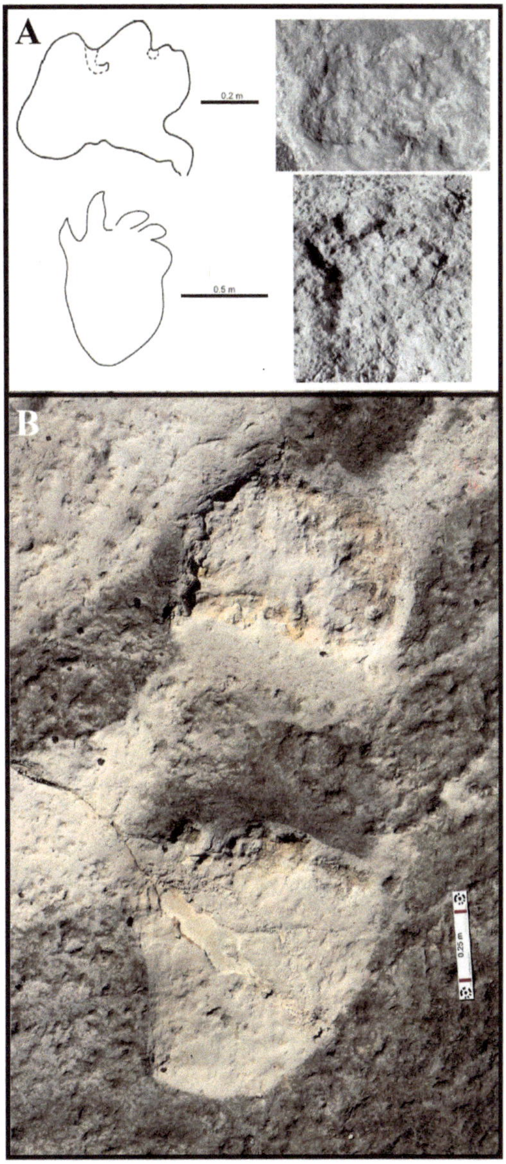

FIGURE 6.76 *"Polyonyx."* (A) MNDPDSA sample manus and pes printsx. (B) Right manus-pes couple from trackway G5 (Santos et al., 2009). Orthophoto from 3D models made in 2023. Scale = 25 cm from target to target. *(A) Modified from Santos et al. (2009). (B) Photo and models by Matteo Belvedere.*

"Polyonyx" trackways are wide gauge, having inner trackway widths 1.2× footprint width (medial edge of pes tracks separated by as much as 70 cm) and pace angulation values averaging about 95° (Santos et al., 2009). Gently arcuate manus prints are wider than long and have rounded, lateral Digit V imprints. They are unusual among sauropod tracks in exhibiting

imprints of Digits I–V, which supports their interpretation (see below) as basal sauropod tracks; more derived sauropods reduced or lost manual phalanges on Digits II–V. Pes prints are longer than wide and ovoid. Occasionally, the centers of the manus prints lie closer to the trackway midline than do the pes print centers (Santos et al., 2009). Pes prints occasionally overlap their associated manus prints. Both manus and pes prints are rotated outward relative to the trackway midline (from 25 to 50° for the manus and 25–42° for the pes).

Discussion. The holotype and associated specimens of "*Polyonyx*" are from the Middle Jurassic Serra de Aire Formation of Portugal (Santos et al., 2009, in press). "*Polyonyx*"-like tracks were also reported from the Middle–?Upper Jurassic Isli Formation of Morocco (Oukassou et al., 2019); these were subsequently more firmly assigned to the ichnogenus by Boutakiout et al. (2020) and Klein et al. (2022). "*Polyonyx*"-like tracks have been also identified in the Late Jurassic of Spain (Alcalá et al., 2014). Tracks referred to as aff. "*Polyonyx*" were reported from the Upper Jurassic–Lower Cretaceous Rupelo Formation of Spain (Fernández-Baldor et al., 2015) but have been since assigned to *Iniestapodus burgensis* (Torcida Fernández-Baldor et al., 2021).

Based on the characteristic speech-bubble-shaped manus prints with large pollex impressions, and despite being wide gauge (a presumed characteristic of titanosauriform sauropod dinosaurs), "*Polyonyx*" tracks are thought to have been produced by nonneosauropod eusauropod track makers (Castanera et al., 2016a; Santos et al., 2009) such as turiasaurs, as evidenced by the large relative sizes of the manus prints compared with the pes prints and the configurations of the manual digit imprints (Santos et al., 2009). Turiasaurs were present in the Middle and Late Jurassic of Europe and Africa (Royo-Torres et al., 2021).

Pooleyichnus
Sarjeant 1975
Fig. 6.77
Type ichnospecies. *P. burfordensis.*

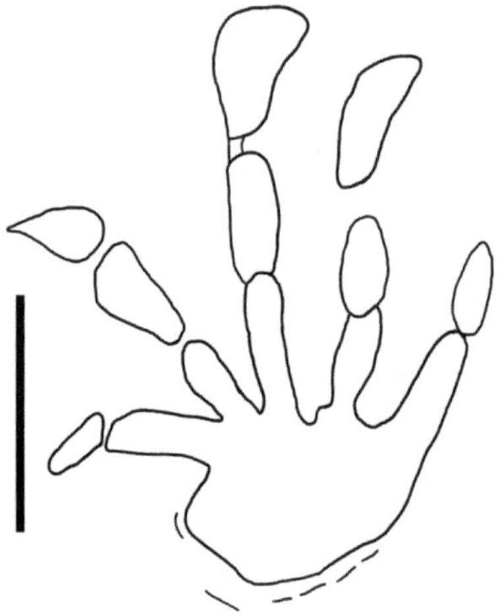

FIGURE 6.77 *Pooleyichnus burfordensis*, holotype, NHMUK R853. Scale = 2 cm. *Redrawn from Sarjeant (1975).*

Valid ichnospecies. *P. burfordensis.*

Distribution. Middle Jurassic of Europe (Stonesfield Slate, Inferior Oolite = Taynton Limestone Formation: England).

Type specimen(s). NHMUK/BMNH R853, natural cast of a single print (see below). Current catalogue number: PV R 893 that can be accessed online (NHMUK:ecatalogue:2272899) at: https://data.nhm.ac.uk/object/8556ca88-ca13-46e1-ad7b-4b894cb78467/1702512000000.

Diagnosis. Adapted from Sarjeant (1975): Pentadactyl, digitigrade ?pes/manus with digits spread radially. The total length of the tracks is 3.5 cm with a width of 3.3 cm and Digits I and V are comparatively short relative to the other digits with Digit III the longest (2.8 cm). Claw traces are noted on Digits II–V and the angle of claw trace direction varies from inwardly (in Digits II and III) to outward (Digits IV and V). Sarjeant (1975) provided a tentative phalangeal formula of 2- 3-3-3-3.

Description. The lone specimen of *P. burfordensis* is a natural cast of a right manus or pes (Fig. 6.77). The claw imprints on Digits II and III are directed inward, while those on Digits IV and V are directed outward. No claw imprint is visible on Digit I. Pentadactyl, digitigrade track with slim, radially arranged digit impressions; Digit III longest, Digits I and V relatively short; Digits II–V terminate in short claw imprints; apparent phalangeal formula 2-3-3-3-3. The longest length observed in Digit III, the notable separation between digits (with an interdigital divergence of 87°), and the presence of only three digit segments in Digits III and IV are characteristic (Panciroli and Romano, 2021).

Discussion. Discovered in 1886 near Burford, Oxfordshire, within what was then known as the Stonesfield Slate, a cast of a small vertebrate footprint found its way to the Natural History Museum, courtesy of the discoverer, Mr. C. Pooley. Despite being known for so long, the specimen was not formally studied for nearly 90 years (Sarjeant, 1975).

Sarjeant (1975) attributed *Pooleyichnus* to a mammalian (mammaliaform in modern parlance) track maker based on the small size of the track, having Digit III longest, and the radial arrangement of the digit impressions. This was the first such attribution of a track to a "mammalian" track maker and would be the first mammal trace from the Jurassic of the British Isles. Sarjeant (1975) further supported this attribution by noting that small mammaliaforms are known from the Stonesfield Slate (e.g., Goodrich, 1894), while comparably small reptiles are not.

Panciroli and Romano (2021) noted that despite the initial recognition of the track as a right print, it is depicted as a left in the illustrations. In addition, they discussed an alternative interpretation of the track, noting that a small imprint anterior to Digit II could be part of the digit, making it the longest digit imprint; they thus suggested that it is actually Digit IV, which would make the overall track morphology similar to that of *Rhynchosauroides* and conform better to a reptilian track maker. Potential Middle Jurassic reptilian track makers such as lepidosauromorphs or rhynchocephalians are known from elsewhere in the United Kingdom (e.g., Panciroli et al., 2020).

Pteraichnus

Stokes 1957

Fig. 6.78

Type ichnospecies. *Pteraichnus saltwashensis.*

Valid ichnospecies. *P. saltwashensis* (Stokes, 1957), *P. stokesi* (Lockley et al., 1995) (both Upper Jurassic). Lower Cretaceous forms: *P. parvus* (Meijide Calvo, 2001), *P. longipodus* (Fuentes

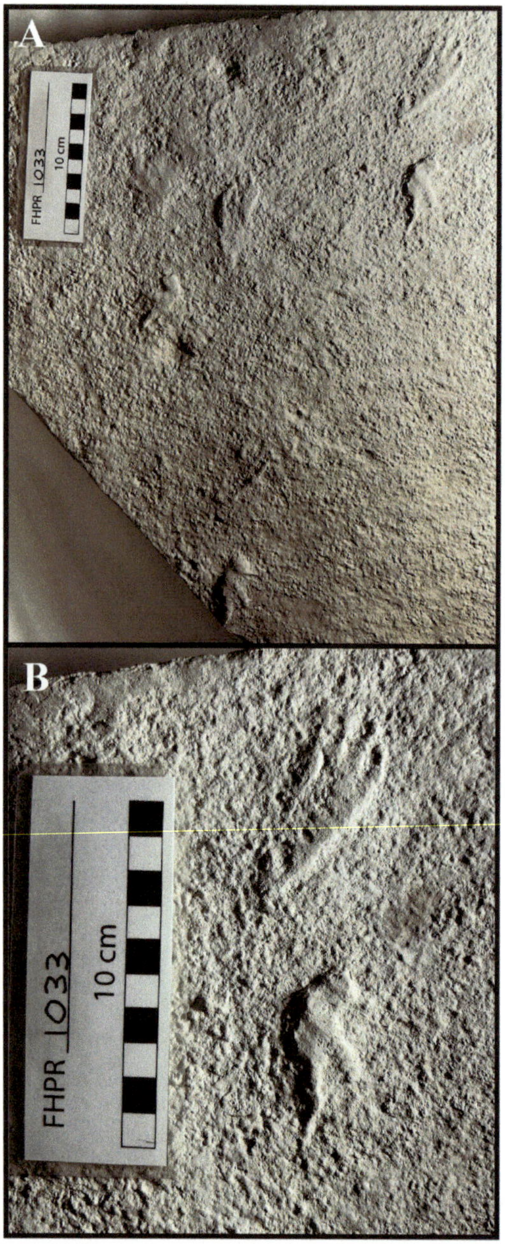

FIGURE 6.78 *Pteraichnus.* (A) Manus-pes sets of pterosaur trackway (FHPR 1033) from the Stump–Morrison transition (Upper Jurassic) at Flaming Gorge in northeastern Utah. Scale = 10 cm. (B) Close-up of one manus-pes set. Scale = 10 cm. *See Bilbey et al. (2005) for this locality. Photos by John Foster.*

Vidarte et al., 2004b), *P. palacieisaenzi* (Pascual Arribas and Sanz Pérez, 2000), *P. nipponensis* (Lee et al., 2010), *P. gracilis* (Ha et al., 2022), *P. koreanensis* (Lee et al., 2008).

Distribution. Middle Jurassic of North America (Summerville Formation: Colorado, Utah; Stump Formation: Utah; Sundance Formation: Wyoming) and Europe (Saltwick Formation: England). Upper Jurassic of North America (Morrison Formation: Arizona, Oklahoma, Wyoming) and Europe (Cazals Formation: Crayssac, France; Wierzbica Oolite and Platy Limestones: Poland; Tereñes and Lastres formations: Spain; Azoia Formation: Portugal). Lower Cretaceous of Europe (Oncala and Enciso groups: Spain), Asia (Hekou Formation: China; Haman Formation: Korea), and South America (Rio Limay Formation: Argentina). "Middle" Cretaceous of Africa (Midelt Formation: Morocco). Upper Cretaceous of Asia (Uhangri Formation: Korea), South America (Candeleros Formation: Argentina), and North America (Cerro del Pueblo Formation: Coahuila, Mexico; North Horn and Blackhawk formations: Utah). Possibly Lower Jurassic of North America (Aztec Sandstone: California; Navajo Sandstone: Arizona).

Type specimen(s). *P. saltwashensis*: UMNH VP580, nine-step (manus-pes) trackway.

Diagnosis. Modified from Stokes (1957) and Lockley et al. (1995): Relatively wide trackway of a quadruped. Asymmetric, tridactyl manus with digit length formula III > II > I. Impression of manus Digit III, elongate, curved, and posteriorly directed, parallel to the trackway axis. Pes is elongate, symmetrical, plantigrade, and tetradactyl with narrow digit imprints terminating in curved claw traces, and a V-shaped sole imprint. Outermost pes digit distinctly curved medially (except in *P. nipponensis* in which digits curve laterally). In trackways, the hindfoot usually overprints or oversteps the forefoot. Manus impressions often more deeply impressed than pes impressions. See also Billon-Bruyat and Mazin (2003): Quadruped track; manus prints on the same axis or more laterally to the pes prints; pes print anterior to the ipsilateral manus print; manus prints as or more deeply impressed than pes prints; elongate, subtriangular, plantigrade and tetradactyl pes print; toe II and III prints are slightly longer than I and IV, toe I–IV prints are clawed; elongate, asymmetrical, digitigrade and tridactyl manus print; digit I print anterior or anterolateral (generally with claw mark), digit II print anterolateral to posterolateral (rarely with claw mark), digit III print posterior (exceptionally with claw mark), digit prints increasing respectively in length; digit IV is rarely marked and limited to the impression of the proximal part of the fourth manus digit (oriented posteromedially); rounded impression in the medial margin of the manus print (impression of the fourth metacarpo-phalangeal joint); elongate penultimate phalanges can be observed in some exceptionally well-preserved manus and pes prints (inferred from digital pads when available).

Description. See Fig. 6.78. Tridactyl manus with two short digit impressions with short, blunt claw marks (usually directed laterally and posterolaterally) and third digit impression that is larger, longer, and directed most often well posteriorly. Tetradactyl pes with a narrow heel; outermost digit impression frequently curves medially. Pes tracks usually overstep manus impressions. Pace angulations vary but in the ichnogenus holotype are 105–115°. Ichnogenus holotype trackway consists of nine consecutive manus-pes sets.

Discussion. Valid ichnospecies include *Pteraichnus saltwashensis* (Stokes, 1957) and *P. stokesi* (Lockley et al., 1995) from the Upper Jurassic Morrison Formation, plus *P. parvus* (Meijide Calvo, 2001) and *P. longipodus* (Fuentes Vidarte et al., 2004b) from the Lower Cretaceous of Spain. Billon-Bruyat and Mazin (2003) considered only

P. saltwashensis valid. Most other occurrences around the globe are *Pteraichnus* isp. The Lower Cretaceous ichnospecies *Pteraichnus cidacoi* (Fuentes Vidarte, 2001), *P. vetustior* (Meijide Fuentes, 2001; Fuentes Vidarte et al., 2004a), *P. manueli* (Meijide Calvo, 2001; Fuentes Vidarte et al., 2004a), *P. palacieisaenzi* (Pascual-Arribas and Sanz Pérez, 2000), all from Spain, are *nomina dubia* according to Sánchez-Hernández et al. (2009), although Pascual-Arribas and Hernández-Medrano (2012) and Pascual-Arribas et al. (2015) demonstrated the validity of the latter. Ichnogenerically undesignated pterosaur tracks are known from a number of areas, including the Upper Cretaceous of Argentina (Díaz-Martínez et al., 2022) and the Upper Jurassic of Wyoming (Harris and Lacovara, 2004). See Xing et al. (this volume) for data on Cretaceous occurrences of *Pteraichnus*.

The track makers of *Pteraichnus* were proposed to be pterosaurs by Stokes (1957), but this was disputed by Padian and Olsen (1984), who argued for a crocodylian track maker. Most studies since then have determined that the track makers were indeed pterosaurs (Lockley et al., 1995; Mazin et al., 1995, 1997).

The *Pteraichnus saltwashensis* type trackway is from the Salt Wash Member of the Morrison Formation of Apache County, Arizona (Stokes, 1957). Referred Middle to Upper Jurassic specimens of the ichnogenus have come from many surrounding areas, including: the Summerville Formation in Colorado and Utah (Lockley and Wright, 2003; Mickelson et al., 2004; Lockley et al., 2008a; Barrett et al., 2008); the Tidwell Member of the Morrison Formation in the same part of northeastern Arizona as the type (Lockley and Mickelson, 1997); the Morrison Formation of Oklahoma (Lockley et al., 2008a); the Lake Como Member of the Morrison Formation in the Alcova region of central Wyoming, including the type material of *P. stokesi* (Lockley et al., 1995; Connely, 2006); the Windy Hill Member of the Sundance Formation of Wyoming (Meyers and Breithaupt, 2014); and the Windy Hill (Stump) or Tidwell (Morrison) members of the Flaming Gorge reservoir area in northeastern Utah (Bilbey et al., 2005). *Pteraichnus* has also been identified from the Lower Jurassic of the Aztec Sandstone of California (Reynolds, 2006) Navajo Sandstone of Utah (Stokes and Madsen, 1979); the Middle Jurassic of Yorkshire, England (Whyte and Romano, 2014); the Upper Jurassic of France (Mazin et al., 1995, 1997, 2003, 2009), Spain (Lastres and Tereñes formations; García-Ramos et al., 2006; Lockley et al., 2008a; Piñuela Suárez, 2015; Tera Group, Hernández-Medrano et al., 2017), Azóia Formation of Portugal (Mateus and Milàn, 2010), and Wierzbica Oolite and Platy Limestones of Poland (Pieńkowski and Niedźwiedzki, 2005; Elgh et al., 2019).

The ichnogenus also ranges into the Lower Cretaceous of Argentina, Spain, Asia, and South America (Calvo and Lockley, 2001; Fuentes Vidarte et al., 2004a,b; Lockley et al., 2008a; Sánchez-Hernández et al., 2009; Moratalla and Hernán, 2009; Hernández-Medrano et al., 2014; Pascual-Arribas et al., 2015c; Pascual-Arribas and Hernández-Medrano, 2012, 2016) and into the Upper Cretaceous of Asia and North and South America (Lockley et al., 2008a; Heredia et al., 2024).

The Lower Cretaceous *Purbeckopus* is rather similar to *Pteraichnus* (Wright et al., 1997), although Billon-Bruyat and Mazin (2003) questioned the former's pterosaurian origin at all.

Pullornipes
Lockley, Matsukawa, Ohira, Li, Wright, White, and Chen 2006
Fig. 6.79

FIGURE 6.79 *Pullornipes aureus,* UCM 214.21 holotype trackway A (courtesy Jacob Van Veldhuizen). Scale = 10 cm.

Type ichnospecies. *P. aureus.*

Valid ichnospecies. *P. aureus.*

Distribution. Upper Jurassic–Lower Cretaceous (Kimmeridgian–Valanginian) of Asia (Tuchengzi Formation, Kanjiatun, Liaoning Province, China).

Type specimen(s). UCM 212.21 and UCM 212.22 are plaster replicas of four consecutive footprints (1–4) and an additional seven consecutive footprints (30–36) respectively from the same trackway (from Lockley et al., 2006b).

Discussion. See Fig. 6.79. See Cretaceous ichnotaxonomy chapter for diagnosis, description, and discussion.

Ravatichnus
Gabuniya and Kurbatov 1988
Fig. 6.80

Type ichnospecies. *R. kotshnevi.*
Valid ichnospecies. *R. kotshnevi.*
Distribution. Middle Jurassic of Asia (Gabirut and/or Dzhirut formations: Tajikistan).

Type specimen(s). Gabuniya and Kurbatov (1988) listed the holotype specimen as Ministry of Geology of the Academy of Sciences of the USSR No. 1. This ministry was apparently abolished in 1991 and its materials transferred to the Ministry of Natural Resources and Ecology of the Russian Federation; the current status and catalog number of the holotype is unknown. The holotype likely is a natural mold, but whether an actual specimen was excavated and reposited or an artificial cast was reposited is unknown; an artificial cast is probable because that is the nature of the *Gabirutosaurichnus* holotype.

Diagnosis. Medium-sized, round tracks of a plantigrade, pentadactyl, parasagittal quadruped. The contour of the footprint covers the imprints of a cushion corresponding to the poorly defined phalangeal section of the five-toed foot/rounded tracks with the round prints of four or five digits and a "cushion." A somewhat elongate Digit III impression may indicate the presence of a claw.

Description. See Fig. 6.80. *Ravatichnus* tracks are illustrated by both Rozhdestvensky (1964) and Gabuniya and Kurbatov (1988) as possessing a larger, ovoid central pad flanked anteriorly by an arc of five smaller, ovoid to rectangular digit impressions, reminiscent of Triassic *Pentasauropus* tracks, but the illustration of the holotype by Gabuniya and Kurbatov (1988), as well as illustrations by Rozhdestvensky (1964) and Lockley et al. (1996b), suggest that the central pad is poorly defined and that the better defined digit impressions have axes that radiate outward from the central region in an upside-down U-shaped pattern.

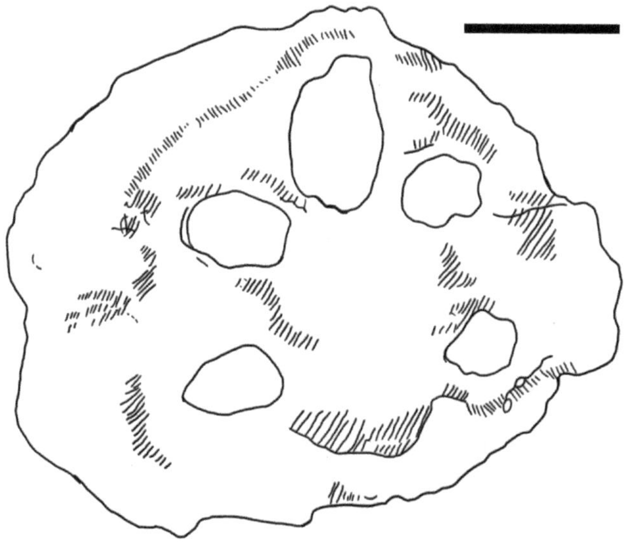

FIGURE 6.80 *Ravatichnus kotshnevi.* Scale = 20 cm. *Redrawn from Gabuniya and Kurbatov (1988).*

Discussion. *Ravatichnus* tracks are from the same locality as *Camptosauropus* and *Gabirutosaurichnus* tracks (see above) and have the same history. Rozhdestvensky (1964) attributed the tracks that would later become *Ravatichnus* to stegosaurian dinosaurs; Gabuniya and Kurbatov (1988) erected *Ravatichnus* and attributed them instead to sauropod dinosaurs, noting that stegosaurs were not pentadactyl. In either case, the tracks were described as those of a quadruped, but neither Rozhdestvensky (1964) nor Gabuniya and Kurbatov (1988) illustrated trackways as having both manus and pes prints, so the question is open whether *Ravatichnus* trackways are manus-only or pes-only. The ichnotaxon was subsequently mentioned in a review of central Asian track fossils by Lockley et al. (1996b), who cited the stegosaurian interpretation of Rozhdestvensky (1964). Lockley et al. (1996b) briefly mentioned and illustrated a track from the Middle Jurassic Saltwick Formation of Yorkshire, England, and referred it to *R. kotshnevi*, but this was refuted by Whyte and Romano (2001b), who interpreted that track as two superimposed *Deltapodus* tracks. *Ravatichnus*, if it is indeed a valid ichnotaxon, is thus restricted to Tajikistan at present. The ichnotaxon needs restudy.

"Ravatosauropus" nomen dubium
Dzhalilov and Novikov 1992
Fig. 6.81
Type ichnospecies. *"R. rozhdestvenskyi."*
Valid ichnospecies. None (see below).
Distribution. Middle Jurassic of Asia (Dzhizhukrut Formation: Tajikistan).

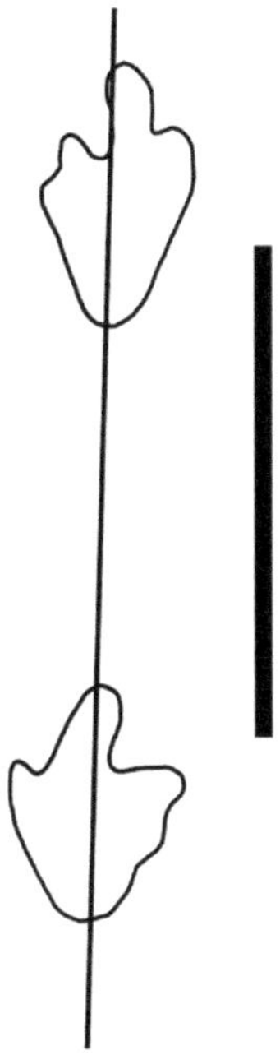

FIGURE 6.81 *"Ravatosauropus"*. Scale = 50 cm. *Redrawn from Dzhalilov and Novikov (1992), Fig. 3.3.*

Type specimen(s). None designated. Dzhalilov and Novikov (1992) designated the track shaded in their Fig. 3.3 as the holotype, but the figure in question is a vague line drawing that cannot serve as the depiction of a holotype because it lacks details, violating ICZN Recommendation 16F. The ichnotaxon thus appears to be a nomen nudum.

Diagnosis. Tridactyl tracks with vaguely defined lateral protrusions that may indicate a fourth digit. Digit III markedly larger than Digits II and IV. Low angles of divarication between digits.

Description. See Fig. 6.81. Only three tracks were reported at the type locality by Dzhalilov and Novikov (1992).

Discussion."*Ravatosauropus*" tracks were not ascribed to any particular track maker; indeed, Dzhalilov and Novikov (1992) considered the track "incertae sedis" in their classification scheme, although they noted that the tracks had previously been assigned to "*Carnosauropus thianschianicum*," ostensibly a theropod ichnotaxon. They also noted some similarity to *Babatagosauropus* and *Akmechetosauropus*, which they ascribed to hadrosaurs. Dzhalilov and Novikov (1992) noted that the quality of preservation of "*Ravatosauropus*" tracks was poor; coupled with the vague diagnosis ("description" in Dzhalilov and Novikov (1992)), the ichnotaxon is of dubious validity.

Regarosauropus nomen dubium
Dzhalilov and Novikov 1992
Fig. 6.82

Type ichnospecies. *R. manovi.*

Valid ichnospecies. None (see below).

Distribution. Middle–Upper Jurassic of Asia (Kugitang Formation: Tajikistan).

Type specimen(s). The specimens illustrated in Sabirov (2008: photographs 8–9). The specimens illustrated were apparently not collected or replicated (no such replicas are mentioned by Sabirov (2008)), but Art. 72.5.6 and 73.1.4 of the ICZN (1999) permit specimens thus illustrated and accompanied by data (Recommendation 73C) to serve as holotypes even in the absence of collected, replicated, and/or reposited specimens. Though never desirable, and contentious (e.g., Santos et al., 2016; Shatalkin and Galinskaya, 2017), the practice is permissible.

Diagnosis. From Dzhalilov and Novikov (1992) and Sabirov (2008): tracks of a small–medium, tridactyl, digitigrade biped. Narrow digit imprints do not contact one another but have modest divarication angles (25–35° per Dzhalilov and Novikov [1992] or 20–25° per Sabirov [2008]). Digit III longer and more anteriorly projecting than Digits II and IV. The metatarsophalangeal pad imprint is connected to the proximal end of the imprint of Digit II. Digit imprints narrow distally into claw imprints. Tracks are rotated slightly outward from the trackway axis.

Description. See Fig. 6.82. Roughly 45 tracks in several trackways were reported at the holotype locality (Sabirov, 2008). Dzhalilov and Novikov (1992) stated that they could not find described tracks with morphology comparable to *Regarosauropus*.

Discussion. Sabirov (2008) referred *Regarosauropus* tracks to theropod dinosaurs, specifically "coelurosaurs from the family Satapliosauridae," though the latter is an ichnofamily. Without naming them, Gabuniya and Kurbatov (1988) referred these tracks to their ichnotaxon *Gabirutosaurichnus*, but Sabirov (2008) disagreed, noting that *Regarosauropus* tracks

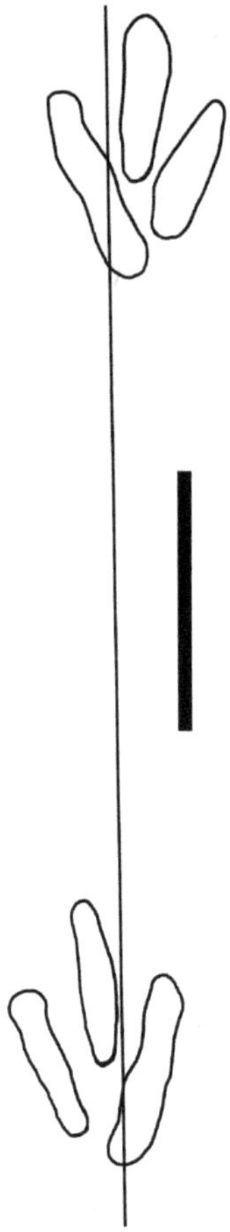

FIGURE 6.82 *Regarosauropus manovi*. Scale = 50 cm. *Redrawn from Dzhalilov and Novikov (1992), Fig. 3.5.*

lack Digit I imprints and have much smaller divarication angles. Sabirov also compared *Regarosauropus* to *Changpeipus carbonicus* Young 1960, which also differs in divarication angle values, and to *Coelurosaurichnus romanovskii* Gabuniya and Kurbatov 1988, which differs only in its greater size and narrower "heel" region.

The diagnosis ("descriptions" of both Dzhalilov and Novikov (1992) and Sabirov (2008)) of *Regarosauropus* is vague and could as well describe numerous other theropod dinosaur ichnotaxa in broad terms; the ichnotaxon is probably best considered a nomen dubium pending further comparative work on the plethora of other Mesozoic ichnotaxa of tridactyl bipeds known globally.

Reniformichnus
Krummeck and Bordy 2018
Fig. 6.83
Type ichnospecies. *R. katikatii.*

FIGURE 6.83 *Reniformichnus katikatii,* in situ from Lower Triassic Katberg Formation, Karoo basin, South Africa. *Clipped from Krummeck and Bordy (2018).*

Valid ichnospecies. *R. katikatii.*

Distribution. *Reniformichnus* burrows are reported from Triassic and Jurassic alluvial and fluvial floodplain deposits of Antarctica, Argentina, Poland, South Africa, and the Navajo Sandstone of the Glen Canyon Group in the western United States (e.g., Hasiotis et al., 2004, 2019; Bordy et al., 2011; Talanda et al., 2011; Krapovickas et al., 2013).

Type specimen(s). Holotype, PB/6/698, burrow casts from the Lower Triassic Katberg Formation of the Karoo Basin, Hobbs Hill locality near Katikati, Eastern Cape, South Africa.

Diagnosis. *Reniformichnus katikatii* from the Lower Triassic Katberg Formation, Karoo Basin, South Africa are subhorizontal burrows with a reniform (kidney-shaped) outline inclined at an angle of <30° and a straight to very gently sinusoidal form. Burrows have a slightly arcuate ceiling and an asymmetrical, bilobate floor when viewed in cross-section. The walls of burrow casts may be covered by digit scratch traces (from Krummeck and Bordy, 2018). See Fig. 6.83.

Discussion. Six burrow morphotypes have been reported from the Lower Jurassic Navajo Sandstone in the Moab region of Utah (Hasiotis et al., 2023). These include diagonal, vertical, spiral, prairie-dog-like complexes (Riese et al., 2011; Hasiotis et al., 2019, 2023).

From Hasiotis et al. (2023): "Their occurrence appears to be mostly mutually exclusive of each other and exact relationships are incomplete based on their limited exposures. Morphotypes are generally numbered and arranged in order of increasing size and complexity. Morphotypes (M) I, II, and III are subhorizontal or gently inclined (<15° angle) and gently curved burrows with elliptical cross-sections (length:width ratio [L:W] of ~2:1). Most well-preserved burrows exhibit a medial ridge along the base. The surficial morphology of these burrows contains thin and elongated, subparallel scratch marks along the floor and walls, with some exhibiting transverse scratch marks on the upper parts of the walls and ceiling. These burrows are tentatively assigned to *Reniformichnus*, with three size ranges of diameter and length that do not overlap in size: (1) M-I is < 7 cm in diameter; (2) M-II is < 15 cm in diameter; and (3) M-III is < 40 cm in diameter. Morphotype IV is inclined at < 30° and less gently curved compared with the other traces. The elliptical cross section has an L:W ratio of ~1.4:1

as well as a medial ridge along the base of the burrow that is more subtle compared with M-1—III burrows."

Reniformichnus clearly has a great range, from the Permian to at least the Early Jurassic. Likely producers of these tetrapod burrows within eolian, and possibly interdunal erg environments, are tritylodonts and potentially mammals (Hasiotis et al., 2023; Milner et al., 2023b). Hasiotis et al. (2019) identified *Reniformichnus* in the Lower Jurassic (Sinemurian—Pliensbachian) Navajo Sandstone in the Moab area of southwestern Utah. Many of the Navajo burrow sites are associated with bounding surfaces that were undergoing pedogenesis in a variety of stages (Hasiotis et al., 2019). They have recognized two burrow morphotypes from simpler burrow systems like *Reniformichnus* and very complex architectures such as *Fractisemita* (Hasiotis et al., 2019).

Rhamphichnus

Mazin and Pouech 2020

Fig. 6.84

Type ichnospecies. *R. crayssacensis.*

Valid ichnospecies. *R. crayssacensis, R. pereiraensis, R. lafaurii.*

Distribution. Upper Jurassic of Europe (Cazals Formation: France).

Type specimen(s). CRA13.08, trackway of 90 steps and ~4 m length, in place at The Pterosaur Beach Museum. Associated in situ paratypes: CRA13.07 and CRA14.14. *R. pereiraensis*: in situ trackway CRA13.09. *R. lafaurii*: in situ trackway CRA00.61.

Diagnosis. Adapted from Mazin and Pouech (2020): Pentadactyl pes displays a divergent Digit V oriented away from the midline and directed posterolaterally. The associated tridactyl manus tracks are oriented subparallel to the pes and anteriorly. All impressions are digitigrade. The width of the quadrupedal trackway is determined by the manus, which is wider than the distance between pes tracks (manus trackway width/pes trackway width ratio: 1.7).

Description. See Fig. 6.84. *Rhamphichnus crayssacensis* consists of a well-preserved trackway with successive paces of a quadrupedal individual. Manus prints are tridactyl and with an outward positive rotation; three moderately diverging manual digit impressions (without claw marks) are oriented anteriorly to anterolaterally, in contrast to *Pteraichnus*.

The type specimen (and its ichnospecies) is a ~3.9 m trackway comprising 75 consecutive steps of a quadrupedal track maker with a pes length of 2.9 cm. The pentadactyl pes Digits I—IV are long and anteriorly orientated (Digits II and III are the longest, occasionally with claw traces). In contrast, the fifth digit is laterally orientated and converges with the tarsometatarsal trace posteriorly. Overall, the pes are slightly inwardly rotated relative to the trackway midline. Manus traces show three, slightly parted, digits without claw traces, and display only slight outward rotation (~27°) relative to the trackway midline. A large trace of the metacarpophalangeal joint is marked by a depression at the base of digit traces.

Discussion. Mazin and Pouech (2020) described three ichnospecies from the same deposit in the Upper Jurassic Cazals Formation of Crayssac, France, and attributed the tracks to "rhamphorhynchoid" pterosaurs. Firstly, it should be noted that there is a spelling discrepancy for the ichnogenus in the type specimens of *R. crayssacensis*, *R. pereiraensis*, and *R. lafaurii*, with both spellings "*Ramphichnus*" or "*Rhamphichnus*" appearing in Mazin and Pouech (2020). Given priority, "*Rhamphichnus*" as nov. igen., is considered the formal spelling (Wroblewski, 2023). The three morphotypes of *Rhamphichnus* are considered to be the tracks

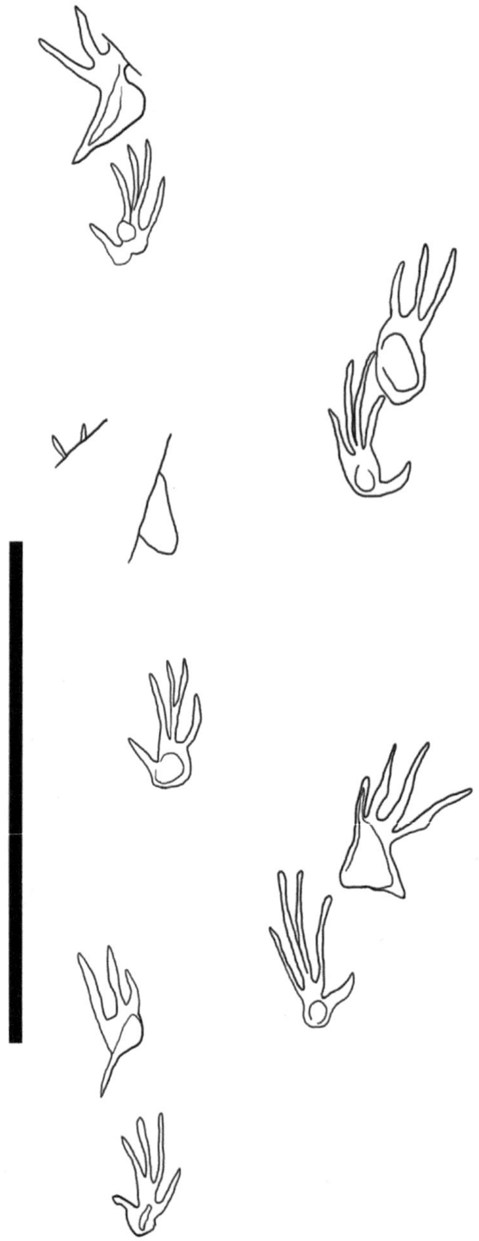

FIGURE 6.84 *Rhamphichnus* holotype CRA13-02. Scale = 10 cm. *Redrawn from Mazin and Pouech (2020).*

of nonpterodactyloid pterosaurs by Mazin and Pouech (2020). This was based on the pentadactyl pes-prints, a nonpterodactyloid characteristic when combined with tridactyl manus prints. Moreover, Mazin and Pouech (2020) proposed that the outward rotation of the manus is not a universally shared trait among pterosaurs. They suggested that nonpterodactyloid track makers, exemplified by *Rhamphichnus*, do not demonstrate the same outward rotation of the manus seen in pterodactyloid tracks during terrestrial locomotion. More recently, Wroblewski (2023) and Xing et al. (2023b) challenged the initial interpretation of *Rhamphichnus* as tracks made by nonpterodactyloid pterosaurians based on the poor resemblance to known pterodactyloid pterosaurs tracks and morphologies inconsistent with nonpterodactyloid taxa skeletal elements. Wroblewski (2023) proposed lepidosaurs and crocodylians as possible track makers, with *R. crayssacensis* considered made by crocodylian producers (and possibly related to *Crocodylopodus*), while *R. lafaurii* and *R. pereiraensis* are possibly rhynchocephalian in origin. As a result, Wroblewski (2023) and Xing et al. (2023b) concluded that no identified nonpterodactyloid pterosaurian tracks are currently known in the fossil record.

Rhynchosauroides
Maidwell 1911
Fig. 6.85
Type ichnospecies. *R. rectipes*.
Valid ichnospecies. *R. rectipes*. See Triassic chapter (this volume) for a complete list.
Distribution. Most occurrences of *Rhynchosauroides* are Triassic in age; see the Triassic chapter of this volume for details. The only Jurassic occurrences of the ichnogenus come from the Upper Jurassic of Europe (Lastres Formation: Spain) and possibly North America (Morrison Formation: Colorado).
Type specimen(s). LU 8019 (*R. rectipes*). (Also plaster cast of the specimen is housed at NHMUK as catalog number R3621.)
Diagnosis. Adapted from a revision by Haubold (1971b): Asymmetrical, ectaxonic track with both the manus and pes digit traces increase progressively from I to IV, with the Digit V shorter and outwardly projecting. Pes Digit IV is the longest and located on the external to

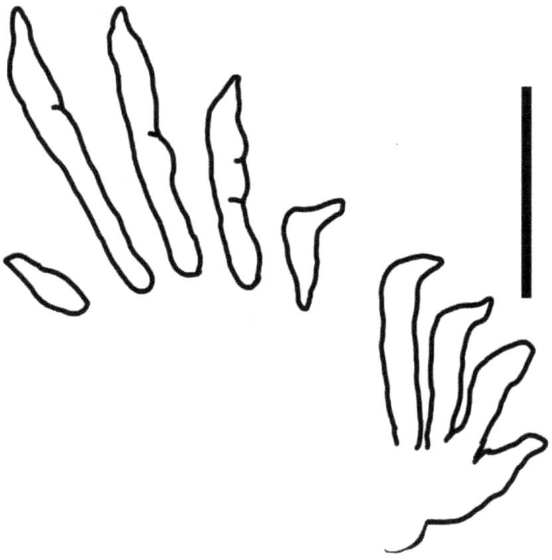

FIGURE 6.85 *Rhynchosauroides* holotype LU 8019. Scale = 2 cm. *Redrawn from Valentini et al. (2007).*

region of maximal load and Digits I–IV may curve inwardly. The pes is digitigrade, larger and more slender than the manus. The manus is plantigrade with a greater width and shorter digits than the pes. The manus is located closer to the trackway midline than the pes. According to Haubold (1971b), the pace, digit 2:3:4:5:3 phalangeal formula, length of pes Digit IV, collectively suggest "lacertoid" track makers. See Fig. 6.85.

Description. Avanzini et al. (2010c) described an in situ, Late Jurassic specimen of *Rhynchosauroides* from the Lastres Formation of Spain that is housed in the MUJA. The Lastres Formation trace is an isolated manus, ectaxonic, asymmetrical, and pentadactyl (length of 0.25 cm and a width of 0.24 cm). All digits are straight, curving inwards with discernible claw traces. Digits I–III run parallel to each other, while Digit IV is the longest. Digits I and II are of similar length, with Digit I featuring two phalangeal pads. Digit V is outwardly rotated, shorter than the other digits, and marked with a claw trace. Both the palm and digits are deeply impressed, with Digits III and IV displaying a higher degree of impression depth due to their external functional prevalence. The total divarication between Digits I–IV is 50°, and between Digits I–V, it is 105° (Avanzini et al., 2010c).

COLM 5172 is a field specimen from the Upper Jurassic Morrison Formation from western Colorado that has been referred to *Rhynchosauroides* (Shaffer et al., 2024).

Discussion: Although common in Upper Triassic rocks, *Rhynchosauroides* tracks are rare in the Late Jurassic, and to date, there are no novel ichnospecies identified from these time periods. The Morrison Formation tracks are from the lower part of the formation in the Tidwell Member at Colorado National Monument and from the Salt Wash Member in the Fruita Paleontological Area (Shaffer et al., 2024; Foster and Lockley, 2024).

Saltopoides (=*Grallator*?)
de Lapparent and Montenat 1967
Fig. 6.86

Type ichnospecies. *S. igalensis.*

Valid ichnospecies. None (see below).

Distribution. Lower Jurassic of Europe (France).

Type specimen(s). A plaster cast of the type has six consecutive footprints (de Lapparent and Montenat, 1967).

Diagnosis. Functionally tridactyl with only the distal portions of digits preserved. All digits display claw traces. Digits II and IV arranged asymmetrically to Digit III. Digit II is separated at a greater distance from Digit III and oriented more medially than Digit IV. The proximal pad of Digit IV is triangular in shape.

Description. See Fig. 6.86. Pace between successive holotype tracks has an average distance of 172 cm (de Lapparent and Montenat, 1967).

Discussion. Lockley (1998) and Lockley and Meyer (2000) stated that *Saltopoides* definitely falls within the size range of *Grallator*; they considered this ichnogenus unnecessary and probably synonymous with *Grallator*. The authors herein agree with this, although the issue has never been properly resolved. Clearly, small theropods were the track producers.

Saltosauropus
Demathieu and Gaillard 1984 (in Bernier et al., 1984)
Fig. 6.87

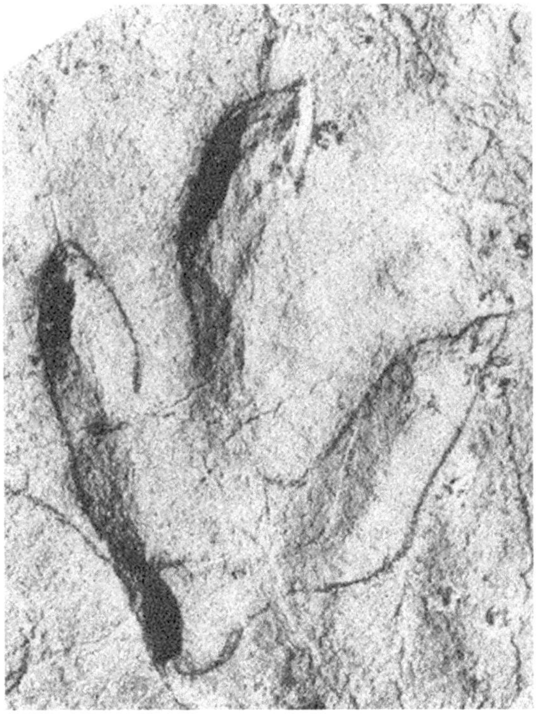

FIGURE 6.86 *Saltopoides igalensis* holotype from the Lower Jurassic of France. *From de Lapparent and Montenat (1967), pl. XII, Fig. 4.*

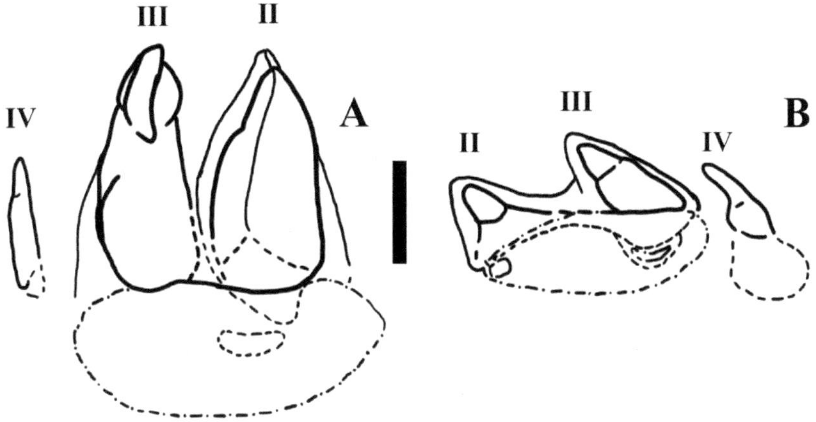

FIGURE 6.87 *Saltosauropus latus* from the Upper Jurassic of France, (A) left holotype track AG2-295A; (B) right track YD4-295A. Scale = 5 cm. *Redrawn from Bernier et al. (1984), Fig. 2.*

Type ichnospecies. *Saltosauropus latus.*

Valid ichnospecies. *Saltosauropus latus.*

Distribution. Upper Jurassic of Europe (Lithographic Limestone Formation of Cerin: France).

Type specimen(s). Specimen n° 400.002 (trace G2 from track A of bed 295 A; and cast), deposited in the collections of the Departément des Sciences de la Terre, Université Claude Bernard Lyon I.

Diagnosis. Translated and modified from Bernier et al. (1984): Paired, tridactyl tracks of a bipedal track maker where pes length is less than pes width. Digit III is the longest, Digits II and III are conjoined at their base and broad, whereas Digit IV is slender and separated from II–III. Each pes is rotated toward the trackway midline. Pedes are aligned with one another and are perpendicular to direct of movement with the ratio of the "jump length" to track width >1.

Description. See Fig. 6.87. Adapted from Demathieu and Gaillard (1984, in Bernier et al., 1984): *Saltosauropus latus* is distributed across several beds within the lithographic limestones at Cerin. These paired, tridactyl tracks show two footprints in the same locomotion phase separated by 5–9 times the average width of the track. The stride of the in-phase pairs is ~ 4–18 times the average length of the track. Tracks can be subdivided based on morphological and preservational characteristics: shorter traces that show only the ends of two digits ("short type"), or longer digit impressions showing more developed digital parts ("long type"). Both preservational types have a ridge that creates an angle with the presumed dIII directed toward the inside of the track at ~30–80°. Digits II and III are more deeply impressed than Digit IV. Claw sizes vary, with pads often indistinct in longer prints. The distance between the ends of Digits II–III is consistent regardless of whether the tracks fall into "short" or "long" types. Claw traces are present with longer, robust claws for Digits II and III, and shorter claw traces on Digit IV. Digital pads are not clearly visible.

The trackways also exhibit variable measurements corresponding to differences in the sizes of their makers, with correlations observed between stride and foot width, as well as between interior trackway width and foot width. In particular, the upper surface of bed 85 preserves five trackways (Demathieu and Gaillard, 1984 [in Bernier et al., 1984]). Among these trackways, two (A and B) likely were made by adults, while three (C, D, and E) were attributed to young animals. For an early statistical analyses of these tracks, see Demathieu (1970).

Discussion. Bernier et al. (1984) first proposed, as the ichnogenus name (*Saltosauropus*) suggests, that these tracks were made by a hopping dinosaur. Thulborn (1989, 1990), Lockley (1991), and Lockley and Meyer (2000) suggested that these were more likely to be made by large swimming turtles and reflected a behavioral variant. This suggestion was affirmed by Gaillard et al. (2003) with the description of new material (including manus tracks with five digits) from the same locality at Cerin. Lockley and Foster (2006) were the first to argue that the name *Saltosauropus* is misleading and propose it as a junior synonym and behavioral variant of *Chelonichnium* (*Emydhipus cerinensis*) or *Chelonipus*. However, considering historical priority and morphological similarity, they recommend referring *S. latus* to *Chelonipus latus*, aligning with recent discussions on *Chelonipus* tracks in Spain that are similar to those in the Morrison Formation.

"Samanadrinda" nomen nudum
Malkani 2007
Type ichnospecies. *"S. surghari."*
Valid ichnospecies. None.
Distribution. Middle Jurassic of Asia (Samana Suk Limestone: Pakistan).
Type specimen(s). None—a trackway of five tracks were declared the holotype, but no specimen numbers or repository name were provided.
Diagnosis. None provided.
Description.*"Samanadrinda"* tracks were only described as being tridactyl and slender-toed, about 61 cm long and 46 cm wide.
Discussion.*"Samanadrinda"* tracks were attributed to a large theropod dinosaur. The ichnotaxon is a nomen nudum because of the absence of a diagnosis and the absence of a holotype. The mentioned trackway was described as revealing a confrontation between the theropod track maker and a sauropod track maker (see *"Malasaurus"*) moving at approximately 90° to the theropod trackway.

Sarmientichnus
Casamiquela 1964
Fig. 6.88
Type ichnospecies. *S. scagliai.*
Valid ichnospecies. *S. scagliai.*
Distribution. Middle Jurassic of South America (La Matilde Formation: Santa Cruz, Argentina); Lower Cretaceous of Asia (Luohe Formation: Shaanxi, China).
Type specimen(s). MLP 60-X-31−1, natural mold of a right pes.
Diagnosis. From de Valais (2011): bipedal trackways composed of functionally didactyl tracks with the long axis nearly parallel to the midline. Digit I oriented posteriorly and Digit

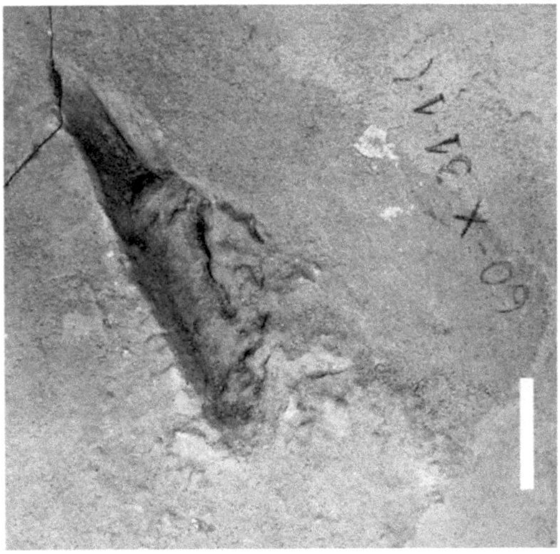

FIGURE 6.88 *Sarmientichnus scagliai*, MLP 60-X-31-1A, holotype from the Middle Jurassic LaMatilde Formation, Argentina. Scale = 3 cm *Clipped from Otero and Reguero (2013).*

III oriented anteriorly, both directly in line with one another and usually indistinguishable from each other; all other digits absent. Digit III bears a triangular claw impression. Straight marks parallel to the midline are often present. Trackways relatively narrow and have high pace angulation values.

Description. *Sarmientichnus* tracks from the type locality are typically deeply impressed and bounded by creases or sediment-collapse structures (Fig. 6.88). In addition, they often occur at the posterior end of the track as positive epichnia when the tracks are preserved as negative epichnia, and vice versa. Casamiquela (1964) originally attributed *Sarmientichnus* tracks to functionally monodactyl, running theropod dinosaurs progressing on only pedal Digit III, whereas Leonardi (1989) stated only that the individual tracks were narrow and made by a theropod that was a "highly specialized ... desert runner." de Valais (2011) agreed with the running theropod interpretation, but interpreted the tracks as being made by an anisodactyl, functionally didactyl track maker: the elongate, linear traces pertain to both an anteriorly oriented Digit III and an oppositely, posteriorly oriented Digit I, as indicated by the presence of a transverse, ridge-like structure (in some natural mold tracks) that irregularly divides the otherwise singular imprints into longer anterior and shorter posterior portions. *Sarmientichnus* tracks sometimes trail posteriorly into elongate, straight, narrow marks parallel to the track midline.

Discussion. de Valais (2011) discounted interpretations of *Sarmientichnus* tracks possessing lateral digit imprints (e.g., Bonaparte, 1983), describing said imprints as lateral deformations made by movement of the primary track-making Digit III. However, some specimens of *Sarmientichnus* isp. from the Early Cretaceous of China (Xing et al., 2018c) possess better-defined, albeit short lateral digit imprints, and were inferred to have been made by didactyl track makers.

Outside of its type locality in the Middle Jurassic of Argentina, *Sarmientichnus* has only been reported from the Lower Cretaceous Luohe Formation of Shaanxi, China (Xing et al., 2018c; Xing et al., 2021b). Xing et al. (2018c) declared *Sarmientichnus* a "form" ichnotaxon that is based on suboptimal preservation and that its apparent monodactyl (or didactyl, comprising Digits I and III) morphology is a consequence of imperfect preservation, supporting at least didactyly (of a more typical variety, consisting of functional Digits III and IV) for the track maker based on the Chinese specimens. The presence of larger, functionally didactyl theropods ("deinonychosaurians") by the Early Cretaceous is well understood based on both body fossils (e.g., Ostrom, 1969) and trace fossils (e.g., Li et al., 2007; Kim et al., 2008), and small representatives are known from the Middle Jurassic (e.g., Hu et al., 2018), but such taxa are not known from the Middle Jurassic of South America. If the interpretation of Xing et al. (2018c) is correct, the Argentinian occurrences of *Sarmientichnus* would indicate the presence of a lineage of theropods in South America far in advance of their known body fossil record.

Sauropus
See Fig. 6.89; also see *Grallator*.

Schizograllator
Zhen, Li, Rao, and Hu 1986
Fig. 6.90

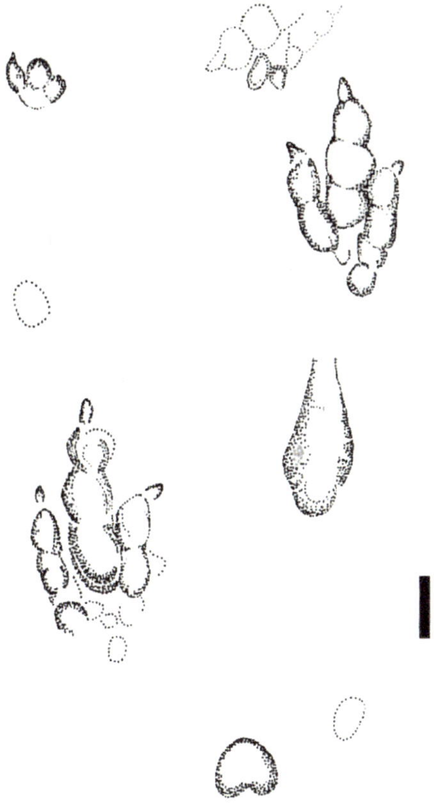

FIGURE 6.89 *Sauropus barratti* (ACM-ICH 1/1), now assigned to *Gralltor tuberosus*. Scale bar = 5 cm. *See Gierliński (1994) and Lull (1953).*

FIGURE 6.90 *Schizograllator xiaohebaensis* holotype BNHM BPV-FP4 (photo of replica UCM 178.3), Lower Jurassic Fengjiahe Formation: Yunnan, China. Scale bar = 5 cm. *From Zhen et al. (1989); photo by Andrew Milner.*

Type ichnospecies. *S. xiaohebaensis.*

Valid ichnospecies. None.

Distribution. Lower Jurassic of Asia (Fengjiahe Formation: Yunnan, China and KrIII Formation [Kuruma Group]: Japan).

Type specimen(s). *S. xiaohebaensis*: BNHM BPV-FP4, artificial casts of part of a trackway of 11 prints (Zhen et al., 1989). *S. otariensis*: OTK-001, six tracks on a slab; the ichnospecies holotype track is OTK = 001−3 (Matsukawa et al., 2005).

Diagnosis. Neither Zhen et al. (1986, 1989) nor Matsukawa et al. (2005) provided a diagnosis for *Schizograllator* separate from that of either *S. xiaohebaensis* or *S. otariensis*. The only characteristics common to the diagnoses of both ichnospecies are that they are tracks of a digitigrade, tridactyl biped, and that both have moderate digit divarication angles.

Description. See Fig. 6.90. For *S. xiaohebaensis*, tracks are slightly wider than long (length:-width ratio 0.93). Digit imprints terminate in claw imprints. Digital pads are elliptical and separated by wide creases. Digit divarication angles are moderate (II−III 30°; III−IV 45°). For *S. otariensis*: tracks are slightly longer than wide (length:width ratio 1.17). Divarication angles between digits are also moderate (II−III 33°; III−IV 32°). Digit III slightly longer than Digits II and IV.

Discussion. *Schizograllator* tracks were ascribed to theropod dinosaurs. Zhen et al. (1989: 191−192) stated "...the skeletal characters of BPV-FP4 are very similar to those of *Gigandipus caudatus*. However, *Gigandipus* is much larger than BPV-FP4 and has a hallux and tail trail. This specimen differs from *Anchisauripus* in that it has four digits, and from *Grallator* in that digit III is more than 50% longer than the lateral digits." However, Zhen et al. (1986a, b) gave the lengths of the digits as: II = 18 cm (including the 4.5 cm long claw imprint), III = 22 cm, and IV = 16.5 cm, dimensions that do not jibe with the relative proportions provided by Zhen et al. (1989). The source of the discrepancy is unclear. Zhen et al. (1989) noted similarity between *Schizograllator* and *Kayentapus*; Gierliński (1996a) followed the suggestion of Lockley and Hunt (1995) by synonymizing the former with the latter. Lockley et al. (2013) synonymized *S. xiaohebaensis* with *Kayentapus*, but retained the ichnospecies as *K. xiaohebaensis*; transitively, *S. otariensis* became *K. otariensis* as well. Xing et al. (2023c) noted that further analysis is necessary to determine whether or not *K. xiaohebaensis* is a valid ichnospecies.

S. xiaohebaensis is from the Lower Jurassic Fengjiahe Formation of Yunnan, China (Zhen et al., 1986a,b, 1989); *S. otariensis* is from the Lower Jurassic KrIII Formation of the Kuruma Group of Nagano, Japan (Matsukawa et al., 2005). Tracks similar to *S. otariensis* were reported from the Middle Jurassic Shemshak Group of Iran (Abbassi and Madanipour, 2014).

Semibipedopus

See *Brasilichnium* (Fig. 6.91).

Shanbeipus

Wang, Li, Bai, Gao, Dong, Hu, Zhao, and Chang 2016

Fig. 6.92

Type ichnospecies. *S. caudatus.*

Valid ichnospecies. *S. caudatus.*

Distribution. Lower Jurassic of Asia (Fuxian Formation: Shaanxi, China).

FIGURE 6.91 *Semibipedopus meekerensis* UCM 179-139 replica. Scale = 18 cm (ruler). *Photo by Andrew Milner.*

Type specimen(s). BNHM Ph001373, slab preserving one partial (ZJ-B0) and two complete (ZJ-B1 and -B2) natural cast tracks.

Diagnosis. Tracks of a small, tridactyl biped. Tracks longer than wide and possess clear digital pad imprints. Metatarsophalangeal pad IV larger than that of III. Tail drag impression often present.

Description. See Fig. 6.92. Tracks are rotated slightly inward toward the trackway axis. The holotype possesses two drag impressions: a thinner one that intersects all three tracks in the trackway, and a thicker, shorter one that only intersects one track (ZJ-B1). One of three trackways on paratype slab (BNHM Ph001374) also has a thin drag impression, but the other tracks on that slab, as well as on the other paratype (BNHM Ph 001372) lack this feature.

Discussion. *Shanbeipus* is a theropod dinosaur ichnotaxon. In addition to its unusual possession of tail drag impressions, the ichnotaxon was differentiated from *Grallator* based on having more slender digit imprints, from *Eubrontes* by its smaller size, and from *Kayentapus* by having greater divarication angles between Digits II–III than III–IV (the opposite of the condition in *Kayentapus*) (Wang et al., 2016). Similar differences in divarication angles

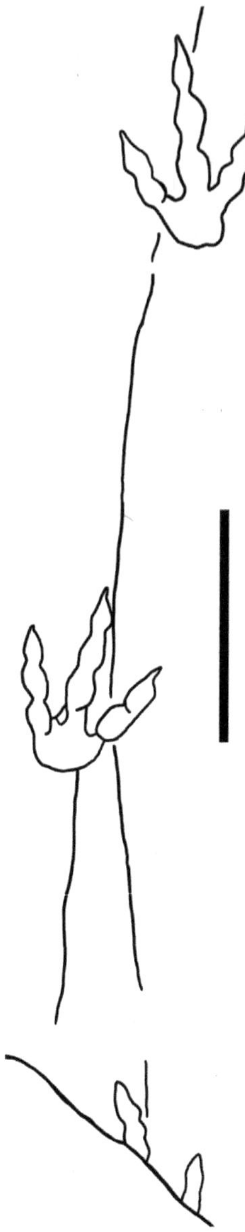

FIGURE 6.92 *Shanbeipus caudatus*, sketch of BMNHC Ph001373 holotype trackway showing tracks ZJ-B0, ZJ-B1, and ZJ-B2 and associated drag traces. Scale = 5 cm. *Redrawn from Wang et al. (2016), Fig. 5.*

were cited as differentiating *Shanbeipus* from *Shensipus*. The ichnotaxon was further differentiated from *Anomoepus* based on digit morphology, greater divarication angles, and track length:width ratios.

Shenmuichnus

Li, Lockley, Zhang, Hu, Matsukawa, and Bai 2012

Fig. 6.93

Type ichnospecies. *S. youngteilhardorum.*

Valid ichnospecies. *S. youngteilhardorum, S. wangi.*

Distribution. Lower Jurassic of Asia (Fuxian Formation: Shaanxi and Lufeng Formation: Yunnan, China) and Europe (Zagaje Formation: Poland).

Type specimen(s). Natural mold trackway C (15 manus–pes sets) at the Shenmu Museum; a replica of tracks C6 and C7 is cataloged as BNHM Ph-906.

Diagnosis. From Li et al. (2012): Tracks of a quadruped. Manus pentadactyl; pes tridactyl with blunt ungual traces. Similar to *Moyenisauropus*, but differs by having a larger manus

FIGURE 6.93 *Shenmuichnus youngteilhardorum*, replica (UCM 184-109) of track C6 from holotype trackway from the Lower Jurassic Lufeng Formation, Yunnan, China. Scale = 5 cm. *Photo by Andrew Milner. From Li et al. (2012).*

(reduced heteropody), lacking hallux traces on pes, having blunt, rather than sharp, ungual traces on the manus and pes, and lacking distinctive metapodial and tail traces.

Description. See Fig. 6.93. For *S. youngteilhardorum*: manus prints are wider than long (length:width ratio roughly 0.73) and sometimes crescentic, lacking clear digit imprints. Pes prints are nearly equidimensional (length:width ratios roughly 1.0) but slightly asymmetrical. Pedal digits were described as terminating in blunt ungual imprints. Digits II and III each have two pads; the count on Digit IV is unclear. The divarication angle between Digits II–IV ranges from 50° to 74°. Manus prints lie slightly anterior to the pes prints. Trackways are wide gauge with slight inward rotation of Digit III (Li et al., 2012). For *S. wangi*: pentadactyl manus prints symmetrical along Digit III and with length:width ratios of 0.6. Manus prints rotated outward 35°. Tridactyl pes prints have robust digit imprints, weak mesaxony, and digit divarication angles ranging from 67° to 73°. Pes prints seem to include a metatarsophalangeal pad imprint in line with Digit III, as indicated by an indentation on the posteromedial side of the track. Pes prints rotated inward 22°. Neither the manus nor the pes possess claw imprints. *S. wangi* has greater heteropody than *S. youngteilhardorum*. Trackways narrow gauge (average pace angulation 92° for the manus and 125° for the pes) (Xing et al., 2016e).

Discussion. *Shenmuichnus* tracks are attributed to basal thyreophoran dinosaurs. Li et al. (2012) differentiated *Shenmuichnus* from *Moyenisauropus*, noting that the latter has a tetradactyl, not tridactyl, pes, as well as sharp claw imprints. They also compared it to *Deltapodus*, noting marked similarities, but positing that various differences may be functions of substrate and preservation conditions. Xing et al. (2016e) differentiated *S. wangi* from *Anomoepus* and *Moyenisauropus* based on the presence of sharp claw imprints and the occasional presence of a hallux imprint in the latter two, as well as the possession of long, slender manual digit imprints in *Anomoepus*. They differentiated *S. wangi* from both *Deltapodus* and *Tetrapodosaurus* based on the shorter "heel" imprints in the pes of *Shenmuichnus* plus tetradactyly in the pes of *Tetrapodosaurus*.

Holotype *S. youngteilhardorum* tracks are from the Lower Jurassic Fuxian Formation of Shaanxi, China (Li et al., 2012); *S. wangi* tracks are from the Lower Jurassic Lufeng Formation of Yunnan, China (Xing et al., 2016e). Oukassou et al. (2022) referred tracks formerly labeled as *Moyenisauropus* cf. *natator* from the Lower Jurassic Zagaje Formation of Poland to *Shenmuichnus*.

Shensipus
Young 1966
Fig. 6.94

Type ichnospecies. *S. tungchuanensis.*

Valid ichnospecies. ?*S. tungchuanensis* (Young, 1966; see below), *S. xiaoliheensis* (Li and Zhang, 2017).

Distribution. Middle Jurassic of Asia (Chiloo = Zhiluo Group and Yan'an Formation: Shaanxi, China).

Type specimen(s). IVPP V 3229, sandstone slab bearing two natural cast tracks. Specimen apparently now lost (Xing et al., 2015b). Xing et al. (2015b) reported that an artificial mold of the holotype tracks is preserved at the Fukui Prefectural Dinosaur Museum in Japan, but did not provide a specimen number.

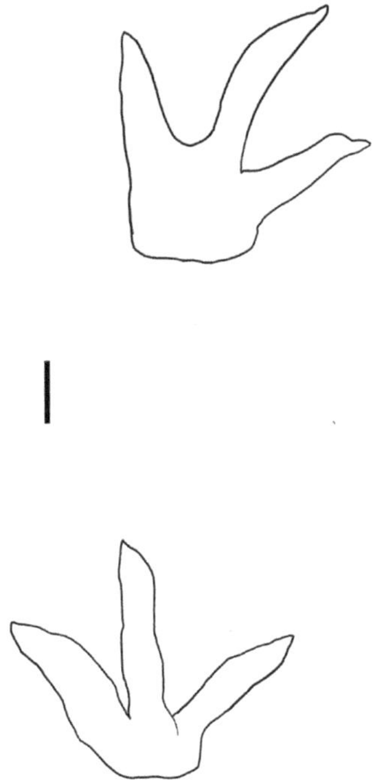

FIGURE 6.94 *Shensipus* holotype IVPP V 3229. Scale = 2 cm. *From Young (1966).*

Diagnosis. Tridactyl track with slender digits, wide divarication between Digits II and IV, small, "pellet-like" claw impressions, and a small heel imprint.

Description. Tracks relatively small and of equal length and width (Fig. 6.94). Metatarsophalangeal region in line with the axis of Digit III and expanded such that it may include a short Digit I imprint. Track details, such as metatarsophalangeal and digital pads, are difficult to discern because the casts are "strongly amalgamated with the surrounding rock" (Xing et al., 2015b: 6). If the two preserved tracks constitute a trackway, the tracks are inwardly rotated ~20–30°. Anterior triangle length–width ratio is 0.37, and average divarication angle between Digits II–IV is 86°, with greater divarication between Digits III and IV than between II and III.

Discussion. Young (1966) interpreted *Shensipus* as having been made by a small theropod dinosaur. Lockley et al. (2013) upheld the validity of the ichnogenus but declared it to be of uncertain track maker affinity. Xing et al. (2015b) suggested that *Shensipus* was a junior synonym of *Anomoepus*, an ornithischian track ichnogenus, though they retained the ichnospecies as *A. tungchuanensis* without providing diagnostic information to differentiate the ichnospecies from any other *Anomoepus* ichnospecies. Several small, basal neornithischians are known from the Middle Jurassic elsewhere in China (e.g., Barrett et al., 2005), so an ornithischian interpretation for these tracks is feasible. Li and Zhang (2017), while reporting new specimens of *S. tungchuanensis* and erecting *S. xiaoliheensis*, attributed the tracks to a theropod track maker.

The ichnogenoholotype of *Shensipus* is from the Middle Jurassic Zhiluo Group of Shaanxi, China. *S. xiaoliheensis* was reported from the Middle Jurassic Yan'an Formation, also in

Shaanxi (Li and Zhang, 2017). *S. xiaoliheensis* was differentiated from *S. tungchuanensis* by its size, having smaller overall angles of divarication between digits, and having an occasional Digit I imprint (discussed but not figured).

Shirkentosauropus
Dzhalilov and Novikov 1992
Fig. 6.95
Type ichnospecies. *S. shirkentensis.*
Valid ichnospecies. *S. shirkentensis* (see below).
Distribution. Middle–Upper Jurassic of Asia (Kugitang Formation: Tajikistan).

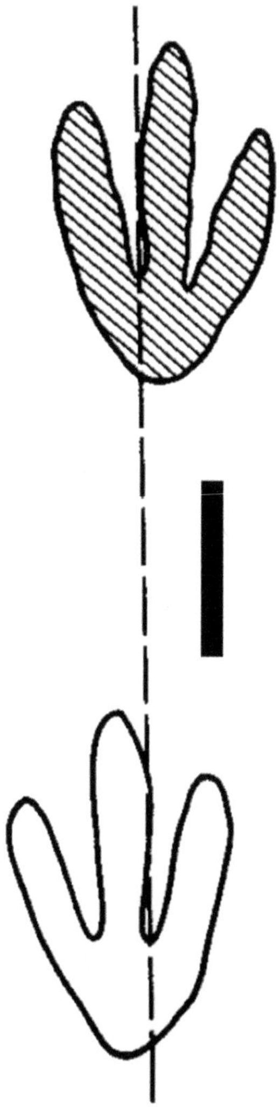

FIGURE 6.95 *Shirkentosauropus shirkentensis*. Scale = 50 cm. *Modified from Dzhalilov and Novikov (1992); Fig. 3.3.*

Type specimen(s). The specimens illustrated in Sabirov (2008: photographs 2–3). The specimens illustrated were apparently not collected or replicated (no such replicas are mentioned by Sabirov (2008)), but Art. 72.5.6 and 73.1.4 of the ICZN (1999) permit specimens thus illustrated and accompanied by data (Recommendation 73C) to serve as holotypes even in the absence of collected, replicated, and/or reposited specimens. Though never desirable, and contentious (e.g., Santos et al. (2016); Shatalkin and Galinskaya (2017); Santos et al. (2016); Shatalkin and Galinskaya (2017)), the practice is permissible.

Diagnosis. From Dzhalilov and Novikov (1992) and Sabirov (2008): tridactyl tracks of a medium–large, digitigrade biped. Pes print is wide with a detached, short, wide, and rounded "heel" imprint that accounts for no more than one-third the length of the track. Digit imprints narrow, subparallel, and distally rounded or slightly pointed, lacking clear claw imprints. Dzhalilov and Novikov (1992) described Digit III as longer than Digits II and IV, but Sabirov (2008) described Digit III as subequal in length to Digits II and IV. Digit imprint widths consistent along their lengths. Digit divarication angles subequal. Track axes almost parallel to trackway axis; trackway has high pace angulation.

Description. See Fig. 6.95. Roughly 200 tracks in 36 trackways were reported at the holotype locality (Sabirov, 2008). Sabirov (2008) compared *Shirkentosauropus* tracks only to Late Jurassic theropod tracks of *"Megalosauropus"uzbekistanicus* Gabuniya and Kurbatov 1988; Sabirov (2008) noted that *Shirkentosauropus* tracks differed by having greater width, a wider, rounder "heel," parallel digits lacking claws, and a less prominent Digit III. Some *Shirkentosauropus* tracks were mentioned as having more elongated "heel" imprints, which were attributed to the foot sliding during track registration (Sabirov, 2008).

Discussion. Despite being placed in the higher-level ichnotaxa Theropodina (Dzhalilov and Novikov, 1992; Sabirov, 2008), *Shirkentosauropus* tracks were attributed by Sabirov (2008) to ornithopod dinosaurs—specifically to "camptosaurs"—based on the wide and round "heel" area and absence of claw imprints, as well as the "configuration of the forefoot," even though no manus imprints were mentioned or described. Pedal digit imprint morphology conforms to those of typical ornithopod tracks, but possessing subparallel digit impressions is not typically ornithopodan.

Further work is needed to better diagnose *Shirkentosauropus*, and comparisons to the plethora of other Mesozoic ichnotaxa of tridactyl bipeds known globally are needed to better establish whether or not it is a valid ichnotaxon.

Stegopodus
Lockley and Hunt 1998
Fig. 6.96

Type ichnospecies. *Stegopodus czerkasi* Lockley and Hunt 1998.

Valid ichnospecies. *S. czerkasi.*

Distribution. Upper Jurassic of North America (Morrison Formation: Utah) and Europe (Bałtów Coral Limestone: Poland). Possibly Middle Jurassic of Europe (Bagå Formation: Denmark). Possibly Middle Jurassic and Jurassic–Cretaceous boundary of Africa (Isli and Oued El Atchane formations, respectively: Morocco). Possibly Middle Jurassic of Asia (Chuanjie Formation: China).

Type specimen(s). UCM 195.1, cast of right manus track, and referred UCM 195.2, cast of right(?) pes track.

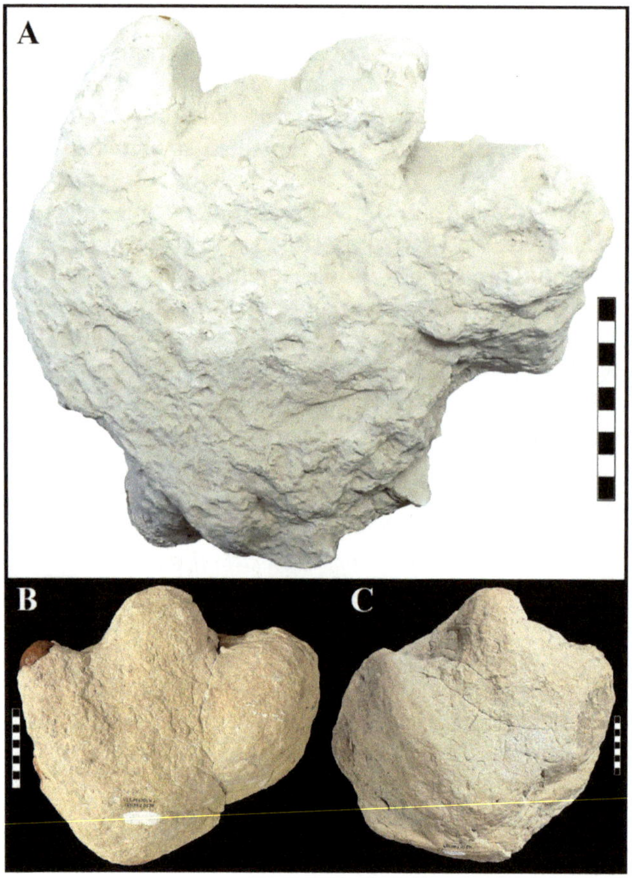

FIGURE 6.96 *Stegopodus czerkasi.* (A) Replica of holotype manus (UCM 195.1) from the Morrison Formation near Moab, Utah. (B) CEUM 20570, natural cast pes from the vicinity of Cleveland-Lloyd Quarry, Utah. (C) CEUM 20551, natural cast pes from the vicinity of Cleveland-Lloyd Quarry, Utah. Scale = 10 cm for all. *Photo A courtesy Jacob Van Veldhuizen (UCM), and B and C courtesy Josh Lively (CEUM).*

Diagnosis. Modified from Lockley and Hunt (1998) and Gierliński and Sabath (2008): Manus track is entaxonic, functionally tetradactyl, wider than long, and subtriangular in shape with the apex forming the posterior heel impression. The manual digit impressions are short and blunt. Medium to large, tridactyl blunt-toed pedal prints with blunt toe impressions; full pes track approximately twice as long as the manus track. The pedal digits are very short, broad, and moderately divaricated, more so than *Deltapodus*. Pedal digits barely project beyond the hypex. Pes is asymmetrical, with the proximal pad much shorter than in *Deltapodus* and located posterolaterally.

Discussion. See Fig. 6.96. *Stegopodus* is hypothesized to represent the tracks of stegosaurian dinosaurs (Lockley and Hunt, 1998; Oukassou et al., 2022). Lockley et al. (2017) noted that the apparent difference in lengths of the heel pad between *Deltapodus* (long) and *Stegopodus* (short) may be due to preservational variation. Still, given the differences between the heel pads and the digit divarication of these two ichnotaxa, we also note the possibility that some isolated pes tracks referred to *Stegopodus* may in fact be tracks of large ornithopods (e.g., Lockley and Hunt, 1998), which are rare but have been reported from the Morrison Formation (Harris, 1998).

The type material of *Stegopodus* is from the base of the Salt Wash Member of the Morrison Formation (Upper Jurassic) north of Moab, Utah (Lockley and Hunt, 1998). *Stegopodus* has since been reported from several other sites in the Morrison Formation of Utah (Gierliński and Sabath, 2008; Gierliński et al., 2010), as well as the Upper Jurassic of Poland (Gierliński and Sabath, 2002). Late Jurassic tracks from Spain referred to *Stegopodus* (Gierliński and Sabath, 2008) were considered ornithopodan by Piñuela et al. (2016). Possible referrals also have been reported from the Middle Jurassic of Morocco and Denmark (Milàn, 2011; Oukassou et al., 2022). *Stegopodus* may also occur near the Jurassic–Cretaceous boundary in the Oued El Atchane Formation of the Middle Atlas of Morocco (Oukassou et al., 2022).

Stenonyx
Lull1 1904
Fig. 6.97
Type ichnospecies. *S. lateralis.*
Valid ichnospecies. *S. lateralis.*
Distribution. Lower Jurassic of eastern North America (Turner Falls Formation [Newark Supergroup]: Massachusetts) and Europe (Zagaje and Drzewica formations: Poland). Possibly Lower Jurassic of western North America (Navajo Sandstone: Utah) and Europe (Hasle Formation: Denmark).
Type specimen(s). (Lectotype) AC 47/40, shallow natural cast of a single pes print.

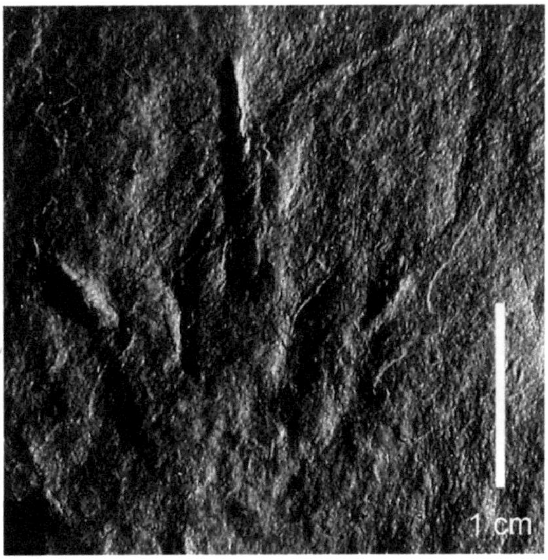

FIGURE 6.97 *Stenonyx lateralis* holotype ACM-ICH 47-40 from the Lower Jurassic Turners Falls Formation, Massachusetts. Scale = 1 cm. *From Klein et al. (2022).*

Diagnosis. From Hitchcock (1865) and Lull (1904): tridactyl tracks of a digitigrade biped. Digit imprints wide despite the minute size of the track. Divarication angle between Digits II–IV 60°. Long, slender claws and distinct phalangeal pad imprints are present; claw imprints on the outer digits originate on the outsides of the digit imprints.

Description. See Fig. 6.97. For *S. lateralis*, Lull (1904) described the following digit divarication angles: II–III 32° and III–IV 28°. He kept *Stenonyx* separate from *Grallator* based on the more slender claw imprints and the locations of the claw imprints with respect to the digit imprints, in which he contradicted Hitchcock (1865) by noting that the imprints on Digits II and III originate on the median sides of the digit, but on the outside of Digit IV. Dalman (2012) further described *Stenonyx* as differing from *Grallator* in having a proportionally shorter Digit III, a wider overall divarication angle between Digits II–IV, and more widely spaced metatarsophalangeal pads on Digits II and IV.

Discussion. *Stenonyx* was originally named *Leptonyx* Hitchcock 1865, but this name was preoccupied simultaneously by a passeriform, a mustelid, and a gastropod (Rainforth, 2005b). Lull (1904) provisionally included *Stenonyx* along with *Grallator* in the ichnofamily Grallatoridae (Melchor et al., 2019), which was and is widely accepted as pertaining to theropod dinosaurs. Dalman (2012) noted that at least some differences between *Stenonyx* and *Grallator*, such as greater divarication angles in the former, may be size-dependent ontogenetic differences. Weems (1992) suggested that *Stenonyx* represented a juvenile form of *"Apatichnus"minor*, and that *"A."minor* belonged in *Kayentapus minor*. Subsequently, Gierliński (1996b) accepted *Stenonyx* as a synonym of *Kayentapus*, even though *Stenonyx* would have nomenclatural priority. Rainforth (2005a) described the lengths of the claw imprints as likely extramorphological, caused by dragging of the claws when the foot was extracted from the substrate; she further noted that the lack of detail of the heel area in the holotype made it difficult to ascertain if the track pertains to either *Anomoepus*, an ornithischian dinosaur ichnotaxon, or *Eubrontes*, another theropod ichnotaxon. Klein et al. (2022) discussed *Stenonyx* in relation to its morphological similarity to the theropod ichnotaxon *Wildeichnus*, to which some ?Middle–Upper Jurassic tracks from the Isli Formation of Morocco were referred.

Stenonyx is represented by only two tracks from the Lower Jurassic Turners Falls Formation of Massachusetts (Hitchcock, 1865). Tracks referred to cf. *Stenonyx* from the Lower Jurassic Zagaje and Drzewica formations of Poland and the Navajo Sandstone of Utah were reported by Gierliński and Niedźwiedzki (2002b) and by Milàn and Surlyk (2015), but differences exist between those and holotype *Stenonyx* (Gierliński et al., 2004). Nevertheless, Pieńkowski et al. (2014) more firmly assigned the Polish specimens to *Stenonyx*. An isolated track from the Lower Jurassic Hasle Formation of Denmark was compared favorably to the Polish specimens (Milàn and Surlyk, 2015).

Synaptichnium
Nopcsa 1923
Fig. 6.98
Type ichnospecies. *S. pseudosuchoides.*
Valid ichnospecies. *S. pseudosuchoides, S. cameronense, S. diabloense.*
Distribution. Middle Jurassic of England (Scalby Formation: Yorkshire, United Kingdom).

FIGURE 6.98 The Enigmatic Burniston Trackway manus track (EBT1) of *Synaptichnium* from the UK. Scale = 3 cm. *Clipped from Panciroli and Romano (2021).*

Material. *Synaptichnium* is a Triassic chirotheriid ichnogenus. Nevertheless, an in situ, Middle Jurassic specimen from England labeled Enigmatic Burniston Trackway "EBT" was compared favorably, but not explicitly assigned, to this ichnogenus by Panciroli and Romano (2021), who noted the marked temporal gap between the specimen and all other *Synaptichnium* occurrences. The specimen currently appears lost due to erosion.

Diagnosis: From Klein and Lucas (2021): "Moderately narrow trackways of a quadruped with a pace angulation that is only slightly larger than 160° (140° on average); imprints pentadactyl with strong outward rotation of the pes. Pes imprints with Digits I–IV increasing in length and Digit IV being longest or subequal with III; Digit V more laterally than proximally positioned behind front digits, slender, elongate, and straight or slightly curved outward and with oval basal pad. Manus more rounded, with Digit III being longest. Ratio of manus:pes area relatively large for chirotheriids = 1:1.8, up to 1:3.7. (1:2.5 on average). Sharp claws on Digits I–IV in both the pes and manus. Distinct rounded phalangeal and metatarsophalangeal pad traces."

Description. See Fig. 6.98. For the British specimen, adapted from Panciroli and Romano (2021): Trackway EBT consists of four to six tracks. The well-preserved manus (EBT1) consists of five subparallel digits with Digit III the longest in length (III > II > IV > V > I) and a divergent Digit V with the total digit divarication of ~40°. A total manus length of ~64 mm and width of ~38 mm is recorded. Pes tracks are laterally rotated relative to the manus and consist of at least three digits, but this is marred by their relatively poor morphological preservation.

Discussion. Panciroli and Romano (2021) showed that, at least, the Enigmatic Burniston Trackway manus (EBT1) exhibits similarities with the Triassic ichnogenus *Synaptichnium* (Nopcsa, 1923; King et al., 2005). Despite the resemblance between the Middle Jurassic EBT and the Triassic *S. pseudosuchoides*, the significant chronological gap between these tracks suggests shared morphological similarities in the track maker(s) (Panciroli and Romano, 2021). Currently, the block that preserved the Enigmatic Burniston Trackway is lost due to coastal erosion. However, Panciroli and Romano (2021) hoped that the reporting of this specimen from the Long Nab Member (Scalby Formation) will inspire future exploration and work within the region. H. Klein (pers. comm., 2024) noted that the figured specimen is nonchirotheriid-like and should be referred to as "tetrapod footprint indet."

Synaptichnium has been attributed to basal archosaur, crocodylomorph, or synapsid (possibly mammalian) track makers (Panciroli and Romano, 2021).

***Talmontopus* (=*Grallator*?)**
de Lapparent and Montenat 1967
Fig. 6.99
Type ichnospecies. *T. tersi.*
Valid ichnospecies. *T. tersi.*
Distribution. Lower Jurassic of Europe (France).

FIGURE 6.99 *Talmontopus tersi* holotype from the Lower Jurassic of France. Scale = 10 cm. *From de Lapparent and Montenat (1967), Plate XII, Fig. 1.*

Type specimen(s). de Lapparent and Montenat (1967) cited and figured a type specimen for *Talmontopus*, but did not provide a specimen number or repository information for it; technically, the ichnotaxon therefore is a nomen nudum.

Diagnosis. From de Lapparent and Montenat (1967): Functionally tridactyl, large track with all digits tapering distally. Claws cannot be differentiated from distal toe pads. The proximal "heel" is rounded.

Description. See Fig. 6.99. Divarication angles between digit traces II–III = 45° and III–IV = 30° (de Lapparent and Montenat, 1967). In addition, de Lapparent and Montenat (1967) misinterpreted webbing between distal ends of the digits.

Discussion. de Lapparent and Montenat (1967) named this ichnogenus from rocks that span the Triassic–Jurassic boundary at Le Veillon, France. Lockley (1998) stated that *Talmontopus* is likely synonymous with *Grallator* and clearly represents tracks of small theropod dinosaurs. We agree that *Talmontopus* is probably synonymous with an ichnotaxon within the *Grallator–Anchisauripus–Eubrontes* plexus. Montenat and Bessonnat (2003), however, referred *Talmontopus* to an ornithopod track maker.

Tetrapodosaurus
Sternberg 1932
Fig. 6.100

Type ichnospecies. *Tetrapodosaurus borealis* Sternberg 1932.

Valid ichnospecies. *T. borealis.*

Distribution. Middle Jurassic of North America (Zorrillo-Taberna Indiferenciadas Formation: Mexico). Upper Jurassic of North America (Morrison Formation: Colorado). Plus, Lower Cretaceous of North America (Gething, Gates, and Gorman Creek formations: British Columbia; Naturita [= Dakota] Formation: Colorado, Utah, and Patuxent Formation: Virginia), Europe (Ashdown, Vectis, and Wessex formations: England; Camarillas and Artoles formations: Spain), Asia (Jiaguan Formation: Sichuan, China); Upper Cretaceous of North

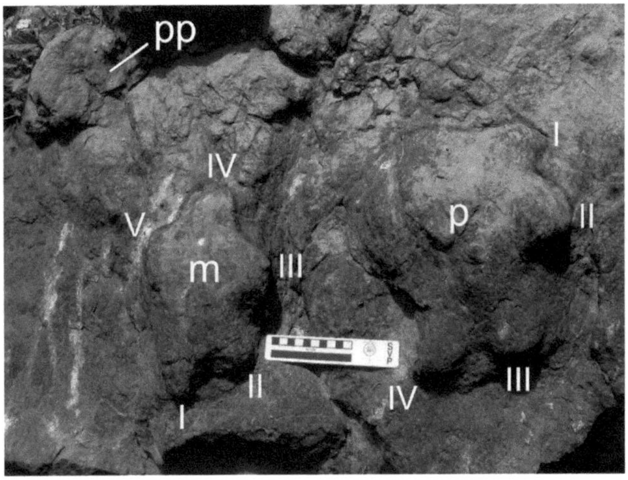

FIGURE 6.100 *Tetrapodosaurus* isp. from the Middle Jurassic of Zorrillo-Taberna Indiferenciadas Formation, Mexico. *From Rodriguez-de la Rosa et al. (2018).*

America (Chignik and Nanushuk formations: Alaska, Kaskapau Formation: British Columbia, and an unnamed formation: Yukon). Possibly Lower Cretaceous of Australia (Broome Sandstone). Possibly Upper Cretaceous of South America (El Molino Formation: Bolivia).

Type specimen(s). GSCA 8556, cast of trackway consisting of six tracks, two left manus and two left pes plus one right manus and one right pes from the Lower Cretaceous Gething Formation of British Columbia, Canada. The cataloged cast is part of a trackway consisting of 14 tracks (seven manus, seven pes).

Diagnosis. From Sternberg (1932) with modifications by McCrea (2000) and McCrea et al. (2001): nonwebbed tracks of a medium–large quadruped; short, broad (wider than long), posteriorly concave, tridactyl–pentadactyl, digitigrade manus prints situated anterior to semiplantigrade, tetradactyl pes prints. Manus approximately two-thirds the size of the pes. Manus Digits I and V often oriented almost opposite the direction of travel. Pedal digit imprints short and distally rounded.

Description. See Fig. 6.100. Sternberg (1932) postulated that the main bodies of *Tetrapodosaurus* tracks demonstrated that the appendages of the track makers were largely encased in large, singular pads from which only the distal tips of the digits protruded; these protrude farther than do the digit imprints in ceratopsid tracks (McCrea et al., 2001).

Discussion. *Tetrapodosaurus* was originally described by Sternberg (1932) from the Lower Cretaceous Gething Formation of British Columbia, Canada. It has been subsequently reported as cf. *Tetrapodosaurus* from the Morrison Formation of two sites in western Colorado (Lockley et al., 2014b; Foster and Lockley, 2014) and *Tetrapodosaurus* isp. from the Middle Jurassic (Bajocian) Zorrillo-Taberna Indiferenciadas Formation of Oaxaca, Mexico (Rodríguez-de la Rosa et al., 2018). This latter occurrence was reported as coming from a fallen block of fine-grained sandstone that preserved natural casts (convex hyporeliefs) of a manus–pes set (CFZ-Zt 305) and a partial pes (CFZ-Zt 306). The counterpart natural mold tracks were in a carbonaceous shale horizon in the same formation. If correctly assigned, these tracks would be the earliest occurrence of *Tetrapodosaurus*, or indeed of ankylosaurian tracks of any kind, yet reported globally.

All other definitive reports of *Tetrapodosaurus* tracks are Cretaceous in age, including the holotypes from the Gething Formation (Sternberg, 1932; McCrea et al., 2001) and related specimens from the underlying Gorman Creek Formation (McCrea et al., 2014) and overlying Gates Formation (McCrea, 2000; McCrea and Currie, 1998) of British Columbia, Canada. Early Cretaceous specimens also have been reported from the Naturita Formation of Colorado and Utah (Lockley et al., 2018b, 2021), the Patuxent Formation of Virginia (Weems and Bachman, 2015), the Ashdown, Vectis, and Wessex formations of England (Pond et al., 2014; Shillito and Davies, 2018), the Camarillas and Artoles formations of Spain (Herrero et al., 2017), and the Jiaguan Formation of China (Xing et al., 2016b). Late Cretaceous specimens have been reported from the Chignik and Nanushuk formations of Alaska (Fiorillo et al., 2019; May and Druckenmiller, 2009), the Dunvegan and Kaskapau formations of British Columbia (Arbour et al., 2020; McCrea et al., 2014), and an unnamed unit from Yukon (Gangloff et al., 2004). Specimens referred to cf. *Tetrapodosaurus* have been reported from the El Molino Formation of Bolivia (Riguetti et al., 2021).

Sternberg (1932) suggested that *Tetrapodosaurus* tracks were made by basal ceratopsids, a position supported by Gierliński and Sabath (2008), but Carpenter (1984) proposed that they were made by ankylosaurians, instead, and this has largely been followed since (Lockley and Hunt, 1995; McCrea et al., 2001; Rodríguez-de la Rosa et al., 2018).

Ichnogenerically undesignated ankylosaur tracks were also reported from the Upper Jurassic Guará Formation of southern Brazil, and similarities between these and the ankylosaurian tracks *Metatetrapous* (e.g., Hornung and Reich, 2014) were noted without formal assignment by Francischini et al. (2017).

Therangospodus
Lockley, Meyer, and Moratalla 1998
Fig. 6.101

Type ichnospecies. *T. pandemicus.*

Valid ichnospecies. *T. pandemicus.*

Distribution. Middle–?Upper Jurassic of Africa (Isli Formation: Morocco); Upper Jurassic of North America (Entrada Formation–Summerville Formation transition zone: Utah, Summerville Formation–Bluff Sandstone transition zone: Utah, Summerville Formation: Colorado and Utah, and ?Bell Ranch Formation: Oklahoma) and Asia ("Middle Complex": Turkmenistan); Upper Jurassic–Lower Cretaceous of Asia (Tuchengzi Formation: Hebei, China). Possibly Upper Jurassic of Asia (Shangshaximiao Formation: Chongqing, China) and Europe (Reuchenette Formation: Switzerland). Possibly Lower Cretaceous of South America (Agrio Formation: Argentina).

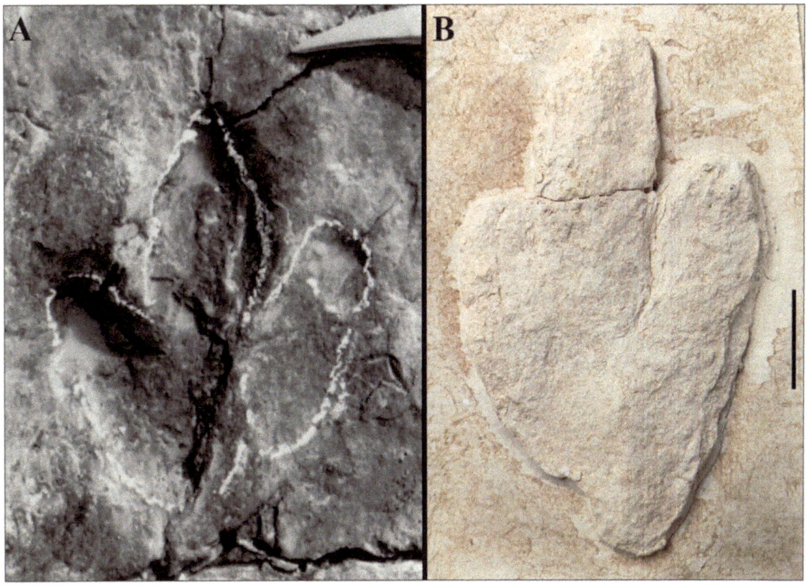

FIGURE 6.101 *Therangospodus.* (A) *T. oncalensis* holotype. No scale provided. (B) *Therangospodus pandemicus*, CU-MWC 186.2, plaster replica from Entrada Formation, Utah. Scale = 5 cm. *(A) From Lockley et al. (1998). (B) Photo by Andrew Milner.*

Type specimen(s). Lockley et al. (1998d) reported the "topotype" specimen of *Therangospodus* as CU-MWC 186.5, a rubber mold and fiberglass replica of five consecutive tracks from a trackway of 13 consecutive tracks at the Moab megatracksite site E.

Diagnosis. Medium-size, elongate, asymmetrical tracks with coalesced (not discrete), elongate, ovoid digital pads. Trackways narrow with little or no rotation of the long axis of digit III from the trackway axis.

Description. See Fig. 6.101. *T. pandemicus* tracks are longer than wide (average length:-width ratio about 1.4) across 33 trackways. Digital pad imprints are always cigar-shaped and generally lack clear boundaries between pads, but enough tracks possess suggestions of such divisions to suggest a 2-3-4 pad formula. Some tracks possess claw imprints. Trackways are narrow (pace angulation averaging 170°).

Discussion. *Therangospodus* tracks are broadly understood to have been made by medium-size theropod dinosaurs (Lockley et al., 1998d). However, Castanera et al. (2013) reassigned *T. oncalensis* from the Lower Cretaceous Huérteles Formation of Spain erected by Lockley et al. (1998d) and discussed further by Barco et al. (2006) to an ornithopod ichnotaxon related to *Iguanodontipus*, and Díaz-Martínez et al. (2015) included them in *Iguanodontipus burreyi*, so presently only *T. pandemicus* is recognized as theropodan.

The earliest occurrence of tracks referred to *Therangospodus* is from the Middle–?Upper Jurassic Isli Formation of Morocco (Gierliński et al., 2009b). Type *Therangospodus* tracks are from the Upper Jurassic Entrada Formation–Summerville Formation transition zone of Utah; they have also been reported from the overlying Summerville Formation–Bluff Sandstone transition zone and Summerville Formation of Utah and Colorado as well as the ?Bell Ranch Formation of Oklahoma (Lockley et al., 1998d, 2014b). Other Late Jurassic specimens attributed to *Therangospodus* isp. were reported from the "Middle Complex" unit of Turkmenistan (Fanti et al., 2013; Lockley et al., 1996b; Meyer and Lockley, 1997) and the Bałtów Coral Limestones of Poland (Gierliński, 2009a). Both cf. *Therangospodus* isp. and *Therangospodus? isp.* tracks were reported from the Reuchenette Formation of Switzerland (Castanera et al., 2018). *Therangospodus*-like tracks have been also reported from the Late Jurassic of Portugal (Lockley et al., 1998d) and Italy (Conti et al., 2005). Slightly younger specimens are known from the Upper Jurassic–Lower Cretaceous Tuchengzi Formation of Hebei, China (Xing et al., 2011b, 2012). Cf. *Therangospodus* tracks were reported from the Upper Jurassic Shangshaximiao Formation of Chongqing, China (Xing et al., 2013d) and Lower Cretaceous Agrio Formation of Neuquén, Argentina (Pazos et al., 2012).

Gierliński et al. (2001b) expressed doubts about whether *Therangospodus* and *Megalosauripus* were distinguishable at the ichnogeneric level based on the presence of an Early Jurassic track from Poland that resembles *Therangospodus* but has discernible digital pads like *Megalosauripus*. Fanti et al. (2013) studied a sample from Turkmenistan and noted differences in the morphometric parameters and qualitative features, concluding that they represent two distinct ichnotaxa. Castanera et al. (2018) also noted the differences (and similarities) in morphometric data and shape between *Megalosauripus* (*M. transjuranicus*) and *Therangospodus*-like tracks from Switzerland and treated them as distinct ichnotaxa. Lockley and Xing (2021) discounted the possibility that *Therangospodus* tracks were made by juveniles of the *Megalosauripus* track maker. As currently understood, identifying *Therangospodus* tracks is complex because of the poor diagnosis of the ichnotaxon and the possibility that it might represent juvenile morphologies of *Megalosauripus*.

Trisauropodiscus
Ellenberger 1970, emended Gierliński et al. 2017a,b
Fig. 6.102

Type ichnospecies. *T. aviforma.*

Valid ichnospecies. *T. aviforma.*

Distribution. Upper Triassic to Lower Jurassic of southern Africa (Elliot Formation: Lesotho). Middle–?Upper Jurassic of North Africa (Imilchil Formation: Morocco).

Type specimen(s). Aviform tracks preserved as shallow natural impressions (concave epi-reliefs) on Upper Triassic and Lower Jurassic sandstone surfaces in Lesotho (from zone A/4 to B/1) as figured in Abrahams and Bordy (2023, Fig. 1). Also recorded as plaster casts LES

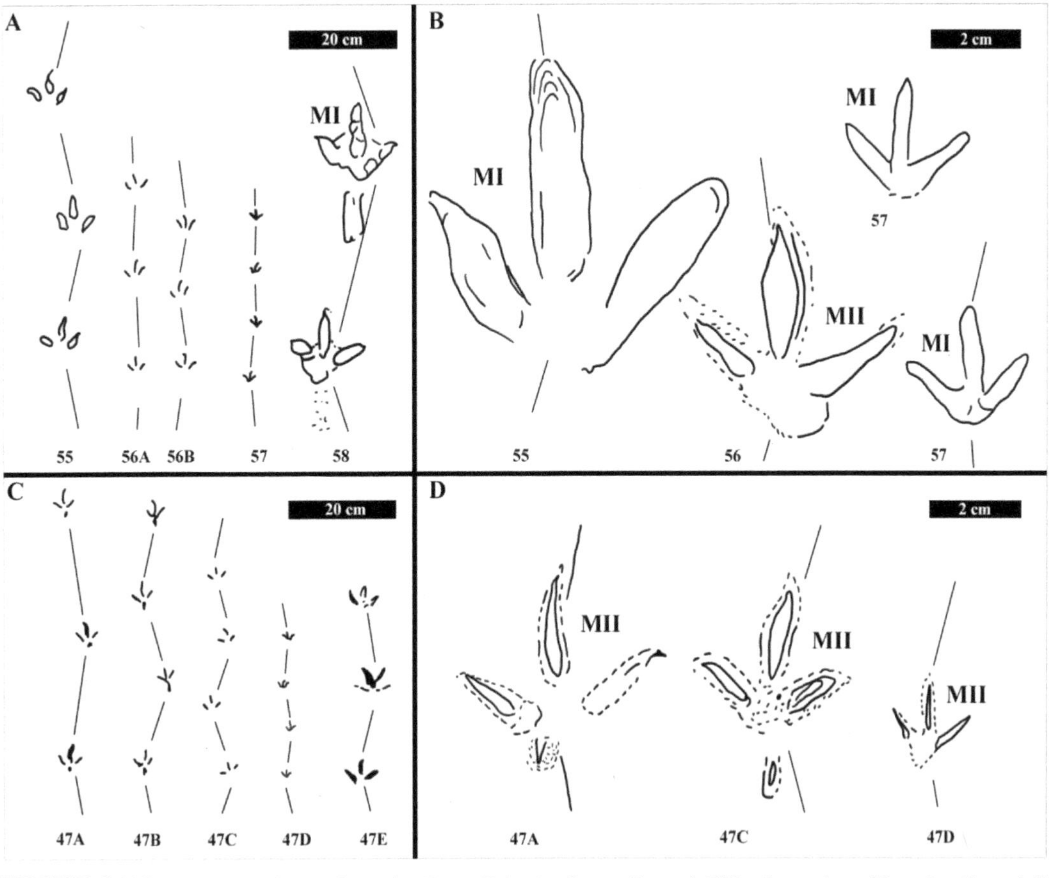

FIGURE 6.102 *Trisauropodiscus*—from the Upper Triassic—Lower Jurassic Elliot formation of Lesotho. (A and C)—trackways. (B and D)—close-ups of tracks. *T. aviforma* (47A–E) is Late Triassic from Maphutseng, whereas *T. galliforma* (55), *T. phasianiforma* (56A, 56B), *T. levis* (57) and *T. popompoi* (58) are Early Jurassic tracks from Thejane. MI, MII—two morphotypes as per Abrahams and Bordy (2023); note that the Jurassic tracks belong to morphotype I, except *T. phasianiforma* (56). For details, including digital models, see Abrahams and Bordy (2023). Redrawn after Ellenberger (1970, 1972) and republished under a CC BY license, with permission from the Geological Society of South Africa and Palaeovertebrata, original copyright 1970 and 1972.

047.1–9 in the PECUoM, and BMG028 in the PECMM&A as well as digital models. Additional specimens are LES 048.2, LES 048.3 in the PECUoM, and BMG041, BM056 in the PECMM&A.

Diagnosis. Modified from Ellenberger (1970, 1972), Gierliński et al. (2017a,b), Abrahams et al. (2017), and Abrahams and Bordy (2023): Slender-toed, aviform tracks and trackways comprising exclusively pes impressions with widely divaricated digits. The symmetrical, digitigrade, wider-than-long tracks resemble a wide V- or U-shape with relatively long, thin digit traces that terminate distally in pointed tips with occasional claw marks. Digital pad impressions are rare to indistinct. Total divarication of Digits II–IV is up to 160°. Strong mesaxony, with Digit III projecting much farther than the subequal Digits II and IV.

Description. See Fig. 6.102. Adapted from Gierliński et al. (2017a,b): The axis of Digit III is offset relative to the proximal end of the digits, where the metatarsophalangeal pad of Digit IV would be expected but is always lacking. Occasionally, a proximal pad is located centrally below the Digit III, and a tiny hallux impression rarely is present; however, neither feature is part of the diagnostic features of these functionally tridactyl tracks.

Seven *Trisauropodiscus* ichnospecies, defined by Ellenberger (1970), belong to two distinct morphotypes, that is, morphotype I (*T. galliforma, T. levis, T. popompoi* and *T. superaviforma)* and morphotype II (*T. aviforma, T. phasianiforma* and *T. superavipes).* Morphotype I is marginally longer than wide, with more robust digits that have a smaller total digit divarication (average ~ 83°), while morphotype II is wider than long, with more slender digits and a larger total digit divarication (average 109°) and thus resembles even better other established avian ichnogenera. The latter morphotype systematically differs from other existing tridactyl ichnogenera, including *Anomoepus*, an asymmetrical track with moderate total divarication (average ~68°) of the short, subequal, cigar-shaped digits.

Trackways have large pace angulations, a stride 6–10 times the track length, and comprise pes impressions that are subparallel to inward rotated relative to the trackway midline, suggesting that the track maker was an obligatory biped.

Discussion. The morphology of *Trisauropodiscus* prompted its classification as a "proto-avian" track. Although the validity of this ichnogenus has faced criticism since its establishment, recent years have witnessed a growing body of support for both the ichnogenus and its associated ichnospecies, *T. aviforma.* (e.g., Klein and Lucas, 2021).

Trisauropodiscus, having been initially described by Ellenberger (1970), is known from both Upper Triassic and Lower Jurassic at several localities in Lesotho (Maphutseng, Thejane, Matsepe, Lephoto, Leribe; also see Abrahams et al., 2017; Abrahams and Bordy, 2023). The Lower Jurassic Matsepe locality yielded *T. superavipes* (Ellenberger, 1970, p. 354, 363, his Fig. 69B) that was reallocated to a new ichnogenus *Trisaurodactylus* by Ellenberger (1974). "*Trisauropodiscus moabensis*" from the Kayenta–Navajo transition near Moab, Utah (Lockley et al., 1992, Figs. 8H, 13, and 14; Lockley and Hunt, 1995, Fig. 4.16) probably represents *Anomoepus* and should be *Anomoepus moabensis* comb. nov., an ichnospecies distinguished by very wide digit divarication between digit trace II–IV up to 120° (Lockley and Gierliński, 2006). *Trisauropodiscus* tracks were reported from the Middle–?Upper Jurassic Isli–Imilchil formation complex of Morocco (Gierliński et al., 2017a,b; Klein et al., 2022).

The overall morphology of *Trisauropodiscus* is reminiscent of modern avian tracks, as first recognized by Ellenberger (1970, 1972, 1974). Given the richness of the coeval body fossil record in southern African, the track maker could be ornithischians or theropods. However, as argued in Abrahams and Bordy (2023), these birdlike tracks, which predate bird fossils by 60 million years, are most likely attributed to a different small-to medium-sized facultatively tridactyl archosaur, which is yet to be found in the body fossil record.

Tuojiangpus
Yang and Yang 1987
Fig. 6.103
Type ichnospecies. *T. shuinanensis.*
Valid ichnospecies. None.
Distribution. Middle Jurassic of Asia (Xintiangou Formation: Sichuan, China).
Type specimen(s). CFZW83, shallow natural mold of a single right pes print.

Diagnosis. From Xing et al. (2021e): tracks of a tridactyl, digitigrade biped. Digit III large, Digit II strong, and Digit IV slender. Digital pad formula 2-3-4. Digits II and III curve inward; Digit IV curves outward. All digits terminate in sharp, curved claw imprints.

Description. See Fig. 6.103. Xing et al. (2021e) noted that the holotype track and others in its trackway are quite shallow, and probably undertracks. The length:width ratio of the anterior triangle of the holotype is 0.48, and the divarication angle between Digits II–IV is 47°. A second track from the same trackway has a ratio of 0.42 and divarication of 56° (Xing et al., 2021a), demonstrating some variation in tracks by the same track maker.

Discussion. *Tuojiangpus* is a theropod dinosaur ichnotaxon. Lockley et al. (2013) declared *Tuojiangpus* a nomen dubium because it was based on poor, undiagnostic material, but Li (2015) and Xing et al. (2021e) assigned it to *Eubrontes* isp., the latter based on similar, moderate degrees of mesaxony and similar morphology to other undertracks of *Eubrontes.*

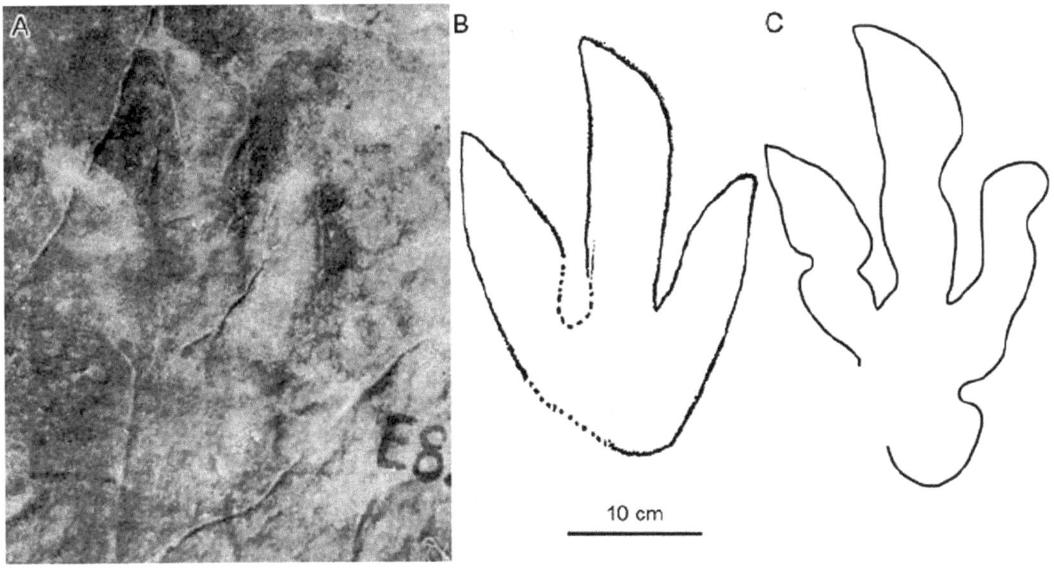

FIGURE 6.103 *Tuojiangpus* holotype CFZW83. (A) Photo of specimen. (B) Original interpretive schematic. Scale = 10 cm. (C) New interpretive schematic by Xing et al. (2021e) drawn at the same scale of B.

Undichna
Anderson 1976
Fig. 6.104

Type ichnospecies. *U. simplicitas.*

Valid ichnospecies. *U. bina, U. britannica, U. unisulca, U. consulca, U. insolentia, U. radnicensis, U. simplicitas, U. gosiutensis,.* See the ichnotaxonomic revision of *Undichna* in Minter and Braddy (2006) and Minter (this volume).

Distribution. *Undichna* ranges from the Devonian through Eocene (e.g., Higgs, 1988; Turek, 1989; Trewin, 2000). Here we discuss only Jurassic occurrences from the Lower Jurassic (Moenave and Kayenta formations: Utah) and Middle Jurassic of North America (Arapien Shale: Utah).

Type specimen(s). Holotype for *U. simplicitas*, BPIPR specimen S.U./K.D.4. Paratypes for *U. simplicitas*, BPIPR specimens S.U./K.D. 10–12, 23, 25, 27, 29–34, 106, 118. All are from the Lower Permian lower Karoo (Ecca Group), Swart Umfolozi locality, KwaZulu-Natal Province, South Africa.

Description. Shallowly incised trails, single sinusoidal incised grooves, paired sinusoidal grooves, and repeated, en-echelon, sigmoidal to arcuate grooves (Fig. 6.104). Trails sometimes form intertwined pairs of horizontal, simple, out of phase, sinusoidal waves with cross-cutting patterns. Simple, exogenic, sinuous furrow (epichnia) may be associated with a pair of slightly raised levees.

FIGURE 6.104 *Undichna.* (A) Uncollected *Undichna* cf. *simplicitas* from the Kayenta Formation, Lake Powell (GLCA), Utah. Scale = 18 cm (ruler). (B) *Undichna* isp. (SGDS 917) from Whitmore Point Member, Moenave Formation, SGDS, Utah. Scale = 10 cm. *(A) Photo by David Slauf. (B) Photo by Andrew Milner.*

Discussion. Getty et al. (2017b) identified two swim trails of *U. unisulca* (de Gibert et al., 1999) on a single slab (SSM,2016/11-3) from the Lower Jurassic East Berlin Formation, Gaulin Tracksite, Holyoke, Massachusetts. de Gibert (2001) described *U. gosiutensis* from the Middle Jurassic Arapien Shale of Utah.

Undichna compares well with modern fish-fin traces (Anderson, 1976). These locomotion traces are made in soft sediment by the ventrally protruding pectoral, pelvic, anal, and/or caudal fins of swimming fishes. The trace morphology is a function of axial undulation of the actively swimming fish and its bauplan. At some sites in the Carboniferous and Triassic, *Undichna* traces are directly associated with limulid trackways (Higgs, 1988; Foster et al., 2001).

Weiyuanpus

See *Eubrontes* (Fig. 6.105).

Wildeichnus

Casamiquela 1964

Fig. 6.106

Type ichnospecies. *W. navesi.*

Valid ichnospecies. *W. navesi.*

Distribution. Middle Jurassic of Asia (Hojedk Formation: Iran) and South America (La Matilde Formation: Santa Cruz, Argentina); Upper Jurassic of Africa (Iouaridène Formation: Morocco) and Europe (Bałtów Coral Limestones and Wierzbica Oolite and Platy Limestones: Poland). Possibly Upper Triassic of Australia (Blackstone Formation). Possibly Lower Jurassic of North America (Navajo Sandstone: Utah). Possibly Middle Jurassic of North America

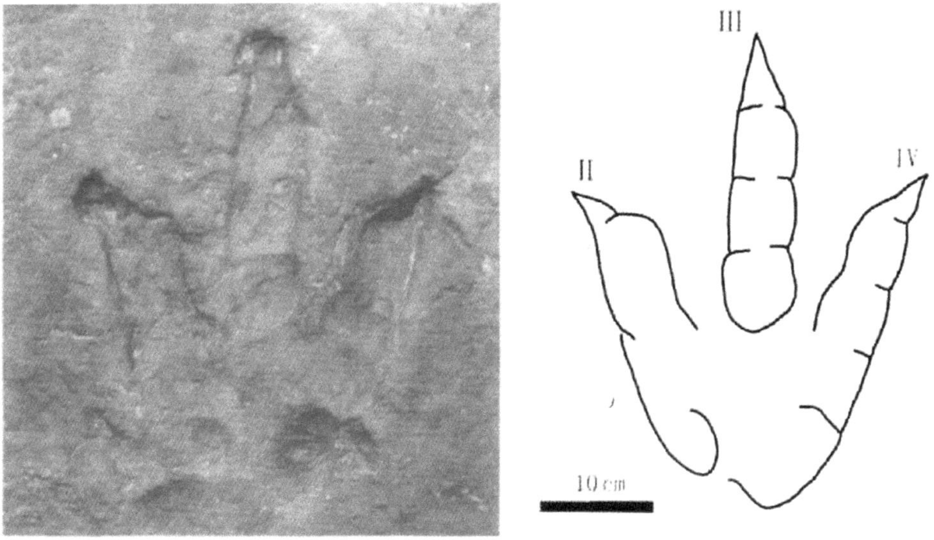

FIGURE 6.105 *Weiyuanpus* holotype ZDM 0032. Scale = 10 cm. *Modified from Gao (2007).*

FIGURE 6.106 *Wildeichnus navesi* UCM 187.30 (replica from the Middle Jurassic LaMatilde Formation, Argentina. Scale = 2 cm. *Photo by Andrew Milner.*

(Curtis Formation: Utah). Possibly Middle–?Upper Jurassic of Africa (Isli–Imilchil formation complex: Morocco).

Type specimen(s). Left pes print MLP 60-X-31–5.

Diagnosis. From de Valais (2011): asymmetrically arranged tracks of a tridactyl biped; rare hallux imprints medially directed when present. Length:width ratios are around 1.5; divarication angles between Digits II–IV> 50°. Trackways have high pace angulation.

Description. See Fig. 6.106. de Valais (2011) described *Wildeichnus navesi* tracks at the type locality as moderately to poorly preserved, which renders comparisons to other theropod ichnotaxa difficult. Only two specimens among the 31 tracks mentioned by de Valais (2011) possess hallux impressions; these have an average Digit I–Digit III divarication angle of about 72°. Digit impressions are narrow and have subparallel margins; they lack clear pad imprints. Digit III is the longest, averaging almost 10 mm longer than Digit II; Digit II is slightly longer than Digit IV. Digit III sometimes bears a claw imprint. Tracks in trackways exhibit virtually no rotation relative to the trackway midline.

Discussion. *Wildeichnus* tracks have been attributed to small theropod dinosaurs (Leonardi, 1989). de Valais (2011) reported that *Wildeichnus* differs from *Grallator, Anchisauripus, Eubrontes,* and *Anomoepus* by having higher average divarication angles between Digits II–IV (62° vs. 40°), a lower length:width ratio (around 1.5 vs. around 2), parameter P of Olsen et al. (1998) > 2, and digit imprints II–IV having similar widths. Rainforth (1997) postulated that

Wildeichnus tracks pertain to juveniles of larger *Grallator* track makers, and de Valais (2011) agreed that this was a possibility.

Unlike many other theropod dinosaur ichnotaxa, *Wildeichnus* tracks are uncommon outside of the type locality. Specimens from the Upper Triassic Blackstone Formation (Ipswich Coal Measures) of Australia initially referred to cf. *Grallator* by Thulborn (1998) were referred to *Wildeichnus* by de Valais (2011), but this was refuted by Romilio et al. (2022), who described the ichnotaxon as poorly diagnosed. Cf. *Wildeichnus* tracks were briefly reported from the Lower Jurassic Navajo Sandstone of Utah (Rainforth, 1997), though no reasoning was given for this attribution; they were attributed instead to *Grallator cursorius* by Hunt and Lucas (2006). *Wildeichnus* isp. tracks were reported from the Middle Jurassic Hojedk Formation (Shemshak Group) of Iran (Abbassi and Madanipour, 2014; Abbassi et al., 2015). Cf. *Wildeichnus* tracks were reported from the Middle Jurassic Moab Member of the Curtis Formation of eastern Utah (Lockley, 2021; Lockley et al., 2007a) and the Middle–?Upper Jurassic Isli–Imilchil formation complex of Morocco (Gierliński et al., 2009b; Klein et al., 2022). Further, Late Jurassic Moroccan specimens were reported from the Iouaridène Formation as well (Belvedere et al., 2010). *Wildeichnus* isp. tracks were also reported from the Upper Jurassic Bałtów Coral Limestones and Wierzbica Oolite and Platy Limestones of Poland (Gierliński et al., 2009a).

Wintonopus
Thulborn and Wade 1984
Fig. 6.107
Type ichnospecies. *W. latomorum.*
Valid ichnospecies. *W. latomorum, W. middletonae.*

FIGURE 6.107 *Wintonopus latomorum*, QMF10319 holotype from the Lower Cretaceous Lark Quarry, Queensland, Australia. *Photo courtesy Anthony Romilio.*

Distribution. Upper Cretaceous of Australia (Winton Formation). Possibly Middle Jurassic of Asia (Zhanghe Formation: Yunnan, China) and Europe (Przysucha Formation: Poland). Possibly Jurassic–Cretaceous boundary of Asia (Anning Formation: Yunnan, China). Possibly Lower Cretaceous of Asia (Hekou Group: Gansu, China) and Australia (Eumeralla Formation).

Type specimen(s). QM F10319, natural mold of a right pes print.

Diagnosis. Thulborn and Wade (1984) did not separate the diagnosis of the ichnogenus from the ichnospecies, and presented the diagnosis as: narrow trackways (pace angulation about 160°) of small–medium, digitigrade, tridactyl bipeds. Tracks slightly wider than long (width:length ratio about 1.15) and exhibit marked positive (inward) rotation. Digit imprints are broad and have rounded or bluntly angular distal tips; distinct digital pad imprints are absent. Digit III longest and has subparallel lateral margins. Digit IV shorter and narrower but closely appressed to Digit III. Digit II is the shortest, and it is widely divergent from Digit III (divarication angle often about 60°); sometimes the imprint of Digit II is entirely disconnected from that of Digit III. Pace length:footprint length (PL:FL) ratio 8.0–13.5; stride length:footprint length (SL:FL) ratio usually 16.0–24.0. Salisbury et al. (2016) emended the diagnosis of *W. latomorum* (not the ichnogenus *Wintonopus* itself) to include many of the aforementioned features, but in addition: trackway pace angulation ranges from 140° to 170°; track width:length ratio ranges from 0.6 to 1.2; the width of each digit imprint is 13%–41% of the total track length; imprint of Digit III usually extends beyond the distal ends of the imprints of Digits II and IV by 0.21–0.52 of the total track length; Digits II and IV extend distally approximately equally relative to the track axis, but the imprint of Digit IV extends farther proximally than either Digit II or III, and the proximal end of the imprint of Digit II lies farther distally than those of Digits III or IV, giving the tracks bilobed, asymmetrical proximal margins; digit divarication angles highly variable (II–III = 26–66°; III–IV = 24–53°); metatarsophalangeal pad imprint rarely present; distinct ungual imprints absent; stride length typically 11–24× maximum track length; and pace length typically 6–12× maximum track length.

Description. See Fig. 6.107. Thulborn and Wade (1984) noted that many *Wintonopus* tracks are not as well defined as the holotype, suffering from disfigurement during withdrawal of the track-making foot. All tracks are deep, but all lack any trace of Digit I. They explicitly noted that the proximal extent of the imprint of Digit IV (a "spur") did not indicate the presence of Digit V on the track-making foot; they inferred that this reflected the presumably ornithopodan metatarsal configuration of the track maker(s). In most specimens, the digit imprints are subequal in width, but in some Digit III is wider than II or IV.

Discussion. *Wintonopus* tracks often have been attributed to ornithopod dinosaurs (Poropat et al., 2021; Romilio et al., 2013; Thulborn, 2017; Thulborn and Wade, 1984). However, several authors have begun to perceive *Wintonopus* tracks as possible swim tracks of various track makers, including theropods and ornithopods (e.g., Milner and Lockley, 2016; Xing et al., 2016a). In addition, it has been considered a senior synonym of *Skartopus*, an ichnotaxon generally considered theropodan (Romilio et al., 2013).

Wintonopus latomorum tracks are definitively known only from the Upper Cretaceous part of the Winton Formation in Queensland, Australia (Poropat et al., 2021; Romilio et al., 2013; Thulborn, 2017; Thulborn and Wade, 1984). *Wintonopus* isp. were also reported from the Lower Cretaceous Broome Sandstone of western Australia by Long

(1998), but this was refuted by McCrea et al. (2011). However, Salisbury et al. (2016) erected the new ichnospecies *W. middletonae* for tracks from the same unit; these differ from *W. latomorum* in having broader digit imprints that are rounded distally and a more symmetrical, bilobed proximal track margin. Tracks referred to cf. *Wintonopus* were reported from the Lower Jurassic Przysucha Formation of Poland (Gierliński and Pieńkowski, 1999). Tracks described as "*Wintonopus*-like" were described from the Middle Jurassic Zhanghe Formation of Yunnan, China (Xing et al., 2023a). A purported cf. *Wintonopus* track was reported from the Middle–Upper Jurassic Walloon Coal Measures of Australia by Turner et al. (2009), but this identification was refuted by Costa da Silva et al. (2012) and Romilio et al. (2021). Tracks tentatively assigned to *Wintonopus* were reported from the Jurassic–Cretaceous Anning Formation of Yunnan, China (Xing et al., 2016a). Tracks from the Lower Cretaceous Hekou Group of Gansu, China that are accompanied by tail-drag impressions were compared, but not assigned, to *Wintonopus* (Li et al., 2006). Additional Early Cretaceous tracks from the Eumeralla Formation of Australia were similarly likened to *Wintonopus* (Romilio and Godfrey, 2021).

Zhengichnus nomen dubium
Zhen, Li, Rao, and Hu 1986
Fig. 6.108

Type ichnospecies. *Z. jinningensis.*

Valid ichnospecies. None.

Distribution. Lower Jurassic of Asia (Fengjiahe Formation: Yunnan, China).

Type specimen(s). BNHM BPV-FP7, a single natural mold? track, presumably of a pes.

Diagnosis. Track of a digitigrade, tridactyl biped shaped like an upside-down T. Digit III much longer than Digits II and IV and thinner proximally than distally. Digit II and IV imprints short and thick. Divarication angles high (II–III 50° and III–IV 53°).

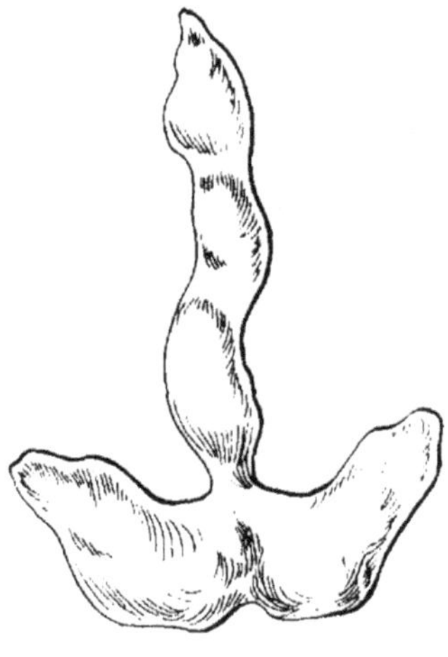

FIGURE 6.108 *Zhengichnus jinningensis*, schematic of holotype BNHM BPV-FP7, Lower Jurassic Fengjiahe Formation, Yunnan, China. *From Zhen et al. (1986a,b).*

Description. See Fig. 6.108. Not much has been described about *Zhengichnus* that is not also covered in the diagnosis. Harris (1998) examined tracks with similar morphologies and hypothesized that they are likely extramorphological variants of more typical theropod dinosaur tracks.

Discussion. *Zhengichnus* was interpreted as a theropod dinosaur ichnotaxon by Zhen et al. (1986a,b), but it was labeled as "Reptilia indet." by Zhen et al. (1989), although it was likened to a gracile "coelurosaur" track. Lockley et al. (2013) and Xing et al. (2023) considered *Zhengichnus* a nomen dubium based on poor, undiagnostic preservation.

***Zizhongpus* (=*Eubrontes*?)**
Yang and Yang 1987
Fig. 6.109
Type ichnospecies. *Z. wumaensis.*
Valid ichnospecies. None.
Distribution. Middle Jurassic of Asia (Xintiangou Formation: Sichuan, China per Chen et al., 2006).
Type specimen(s). CFZW19, natural mold of a single pes print.
Diagnosis. Tracks of a large, tridactyl biped. Digit III longer than II and IV and narrows distally. Digit II curves medially and Digit IV curves laterally. Digit IV lies farther from Digit III than does Digit II. Digital pads slender. Tracks are wider than long.
Description. See Fig. 6.109. Yang and Yang (1987) measured *Zizhongpus* digit divarication angles as: II–III 25–32° and III–IV 32–37°. Trackways are narrow, having high pace angulation angles.
Discussion. *Zizhongpus* tracks were attributed to theropod dinosaurs. Lockley et al. (2003) declared the material too poorly preserved to be diagnostic, but later Lockley et al. (2013) considered the ichnotaxon synonymous with *Kayentapus,* though they retained

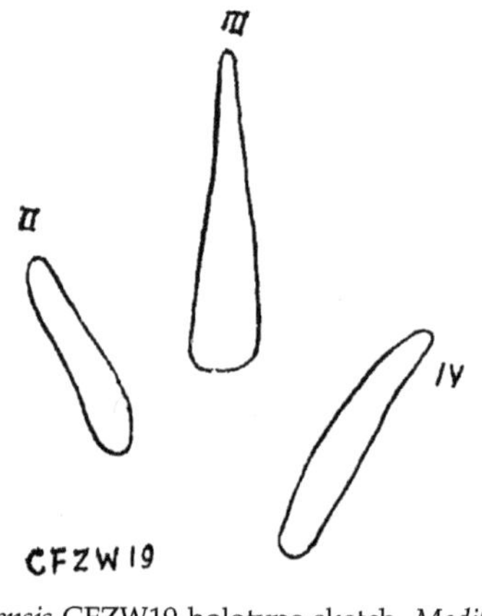

FIGURE 6.109 *Zizhongpus wumaensis* CFZW19 holotype sketch. *Modified from Yang and Yang (1987), Fig. 4.*

K. wumaensis as a distinct ichnotaxon. Xing et al. (2021d) described *Zizhongpus* as lacking a sufficient diagnosis, yet interpreted *Zizhongpus* tracks as undertracks of either *Eubrontes* or *Kayentapus*; they favored the former, assigning the tracks as *Eubrontes* isp. indet.

Zizhoupus
Li and Zhang 2017
Fig. 6.110
Type ichnospecies. *Z. wangi.*
Valid ichnospecies. *Z. wangi.*
Distribution. Middle Jurassic of Asia (Yan'an Formation: Shaanxi, China).
Type specimen(s). NWUV 1404, a natural cast of a right pes print.

Diagnosis. Large, asymmetrical tracks made by a tridactyl biped. Tracks have clear claw and oblong phalangeal pad imprints. Tracks are longer than wide. The average divarication angle between digit imprints III–IV is twice that of II–III. Digit III exhibits inward rotation.

Description. See Fig. 6.110. Tracks are only slightly longer than wide (length:width ratio 1.09). Digit III is the longest. Digit divarication angles in the holotype specimens are 21° between Digits II–III and 46° between Digits III–IV.

Discussion. *Zizhoupus* tracks were attributed to theropod dinosaurs. The only reported occurrence of this ichnogenus is the Middle Jurassic Yan'an Formation of Shaanxi, China (Li and Zhang, 2017). The ichnotaxon is based on a small sample of specimens (five tracks). The ichnotaxon was differentiated from *Changpeipus*, *Eubrontes*, and *Kayentapus* by its larger size, oblong phalangeal pads, and greater divarication angle between Digits II–IV (>60°); it was differentiated from *Chapus* by having smaller divarication angles.

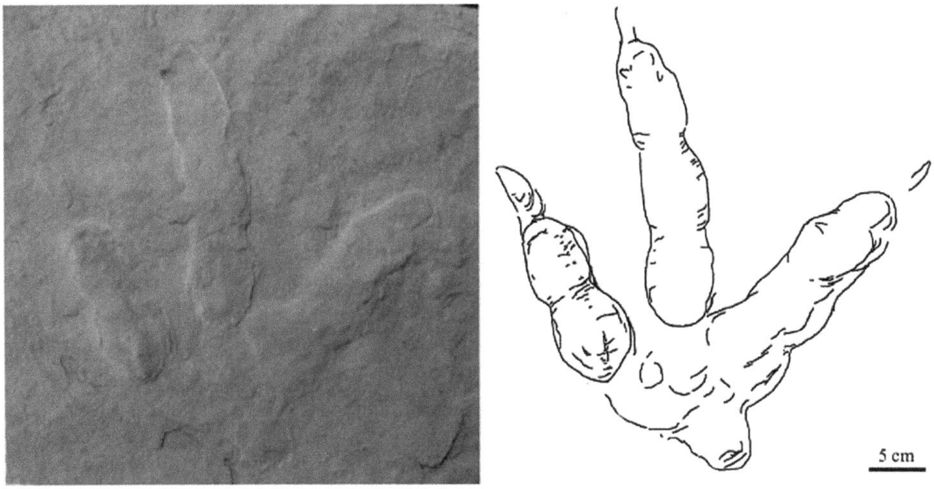

FIGURE 6.110 *Zizhoupus wangi*, NWUV 1404 holotype from the Middle Jurassic, Yan'an Formation, Shaanxi, China. *Modified from Li and Zhang (2017).*

4. Discussion

Interesting patterns can be recognized in the Jurassic ichnological record, particularly among dinosaur ichnotaxa, which dominate the ichnological record from the period. In the Early Jurassic, ichnotaxa exhibit a high degree of cosmopolitanism, especially in the distribution of theropod dinosaur tracks in the *Grallator–Anchisauripus–Eubrontes–Kayentapus* plexus and in a possible *Asianopodus–Changpeipus–Jialingpus* plexus. Because Pangaea was just beginning to break apart in the Early Jurassic, such cosmopolitanism is understandable: closely related track-making taxa could readily spread across continents. Such cosmopolitanism decreases somewhat in the Middle and Late Jurassic, commensurate with the fragmentation of Pangaea, although the track record from these time periods is strongly concentrated in Laurasia, with fewer occurrences for comparison from Gondwana. In both the Middle and Late Jurassic, dinosaurian, crocodylian, and even chelonian ichnotaxa are still shared between North America and Eurasia, as well as with northern Africa, indicating probable terrestrial bridges between all these land masses.

The Jurassic track record largely parallels the body-fossil record in documenting the emergences and spread of mid–late Mesozoic vertebrates, especially dinosaurs. Medium to large theropod tracks and body fossils are rare in the Triassic, but both become more abundant in the Jurassic. Similarly, sauropodomorph dinosaur tracks and body fossils both become more abundant in the Jurassic, although they are known from the Late Triassic as well. Noncerapod ornithischian (i.e., basal ornithischian and thyreophoran) tracks are moderately abundant in the Early Jurassic, matching the body-fossil record of that time; ornithopod tracks questionably appear in the Middle Jurassic and are certainly present by the Late Jurassic, again in step with the body-fossil record.

However, the Jurassic track record contains a few surprises, such as apparently didactyl, larger theropod dinosaur tracks in the Middle Jurassic of Africa and South America, a time when rare skeletal remains of didactyl theropods are of small taxa, known only from East Asia. Wide-gauge sauropod dinosaur ichnotaxa such as *Brontopodus* and *Occitanopodus*, which typically are attributed to macronarian sauropods, appear in the fossil record in the Middle Jurassic in advance of their Late Jurassic–Cretaceous body-fossil diversity. Similarly, narrow-gauge sauropod ichnotaxa such as *Parabrontopodus*, which typically are attributed to diplodocoid sauropods, appear in the Early Jurassic, well in advance of the Middle Jurassic body-fossil record of diplodocoids. Either different gauges evolved independently several times among sauropods, or the track record reflects gaps in our knowledge of the ranges of the attributed track-making taxa.

Ichnotaxonomically, the Jurassic record still needs a great deal of work. For example, many central Asian ichnotaxa do not have adequate holotypes and have never been compared with other ichnotaxa globally, rendering virtually all of them questionable. Also, many Middle and Late Jurassic theropod ichnotaxa, such as *Megalosauripus* and *Therangospodus*, currently lack consensuses as to individual track specimen assignments as well as ichnotaxonomic distinction. Tautologically, the roles of behavior and preservational conditions in distinguishing Jurassic ichnotaxa require further study and ultimately could reduce the currently understood diversity of track and trace types. Simultaneously, however, several track morphologies remain to be studied in detail, and many may pertain to as-yet unnamed ichnotaxa. The Jurassic ichnological record may yet end up as rich and diverse as the preceding Triassic and following Cretaceous records.

Acknowledgments

Thanks, first of all, to our coauthor and mentor, the late Martin Lockley, who directly and indirectly influenced every one of the coauthors significantly over the years. Special thanks to Spencer Lucas, Hendrik Klein, and Adrian Hunt for inviting us to contribute to this project. We express our gratitude to the following persons, whose invaluable assistance and generous sharing of photographs significantly enhanced the quality of this chapter: Michela Contessi (Collezione di Geologia "Museo Giovanni Capellini" – Bologna University Network, Italy); Daniel Marty (Naturhistorisches Museum Basel, Switzerland); Christian A. Meyer (University of Basel, Switzerland); Novella L. Razzolini (Museu de la Conca Dellà, Spain); Vanda F. dos Santos (University of Lisbon, Portugal); Brent Breithaupt (US Bureau of Land Management, Wyoming, USA); Karen Chin (University of Colorado Boulder, Colorado, USA); Loïc Costeur (Naturhistorisches Museum Basel, Switzerland); James Farlow (Purdue University, Indiana, USA); Patrick Getty (deceased; formerly Collin College, Texas, USA); ReBecca Hunt–Foster (Dinosaur National Monument, Utah, USA); Alexander Mudroch (Hannover, Germany); Paul Olsen (Columbia University, New York, USA); Mike Romano (University of Sheffield, UK); Anthony Romilio (The University of Queensland, Brisbane, Australia); Bruno C. Silva (Sociedade de História Natural de Torres Vedras, Portugal); David Slauf (St. George Dinosaur Discovery Site at Johnson Farm, Utah, USA); Jacob Van Veldhuizen (University of Colorado Boulder, Colorado, USA); Martin Whyte (deceased; formerly University of Sheffield, UK); Dan Whalen (St. George Dinosaur Discovery Site at Johnson Farm, Utah, USA); and Joshua Lively (Utah State University Eastern Prehistoric Museum). Thanks also to Jaleesa Buchwitz (St. George Dinosaur Discovery Site at Johnson Farm, Utah, USA) who performed photogrammetry of the *Dilophosauripus* track replica. Thanks for editorial assistance to Ida Dirkes (University of Bonn). Thanks for review comments to Gerard Gierliński and Spencer Lucas. LS acknowledges the financial support of the Swiss National Science Foundation (SNSF) grant (200021_192036). DC is supported by project E04_23R_FOCONTUR (Aragón Regional Government) and Unidad de Paleontología de Teruel (Ministerio de Ciencia e Innovación, Spanish Government).

References

Abbassi, N., Madanipour, S., 2014. Dinosaur tracks from the Jurassic Shemshak group in the central Alborz Mountains (northern Iran). Geol. Carpathica 65 (2), 99–115.

Abbassi, N., D'Orazi Porchetti, S., Wagensommer, A., Dehnavi, M.G., 2015. Dinosaur and crocodylomorph footprints from the Hojedk Formation (Bajocian, middle Jurassic) of North Kerman, central Iran. Ital. J. Geosci. 134 (1), 86–94. https://doi.org/10.3301/IJG.2014.30.

Abel, O., 1926. Amerikafahrt. Jena: Gustav Fischer, 462 pp.

Abrahams, M., 2020. Evaluation of Tridactyl Theropod Tracks in Southern Africa: Quantitative Morphometric Analysis Across the Triassic–Jurassic Boundary. PhD Thesis, p. 208. University of Cape Town, Cape Town.

Abrahams, M., Bordy, E.M., 2023. The oldest fossil bird-like footprints from the upper Triassic of southern Africa. PLoS One 18 (11), e0293021. https://doi.org/10.1371/journal.pone.0293021.

Abrahams, M., Bordy, E.M., Sciscio, L., Knoll, F., 2017. Scampering, trotting, walking tridactyl bipedal dinosaurs in southern Africa: ichnological account of a Lower Jurassic palaeosurface (upper Elliot Formation, Roma Valley) in Lesotho. Hist. Biol. 29 (7), 958–975. https://doi.org/10.1080/08912963.2016.1267164.

Abrahams, M., Sciscio, L., Reid, M., Haupt, T., Bordy, E.M., 2020. Large tridactyl dinosaur tracks from the early Jurassic of southern Gondwana–uppermost Elliot Formation, upper Moyeni, Lesotho. Ann. Soc. Geol. Pol. 90, 1–26. https://doi.org/10.14241/asgp.2020.0.

Abrahams, M., Bordy, E.M., Knoll, F., 2021. Hidden for one hundred years: a diverse theropod ichnoassemblage and tracks preserved in cross-section from the historic Early Jurassic Tsikoane ichnosite (Clarens Formation, northern Lesotho, Southern Africa). Hist. Biol. 33 (10), 2504–2519. https://doi.org/10.1080/08912963.2020.1810681.

Abrahams, M., Rampersadh, A., Mpangala, L., 2023. Riches of the Roma Valley: theropod and ornithischian tracks from the early Jurassic southern Africa. Hist. Biol. https://doi.org/10.1080/08912963.2023.2221306.

Agirrezabala, L.M., Torres, J.A., Viera, L.I., 1985. El Weald de Igea (Cameros—La Rioja). Sedimentología, biostratigrafía y paleoicnología de grandes reptiles (Dinosaurios). Munibe 37, 111—138.

Ahlberg, P.E., Beznosov, P.A., 2022. The significance of the Devonian tetrapods of Timan. Paleontol. J. 56 (9), 1029—1031. https://doi.org/10.1134/S0031030122090039.

Alcalá, L., Cobos, A., Espílez, E., Gascó, F., Mampel, L., Martín Escorza, C., Royo-Torres, R., 2012. Icnitas de dinosaurios en la Formación Villar del Arzobispo de Ababuj (Teruel, España). Geogaceta 51, 35—38.

Alcalá, L., Pérez-Lorente, F., Luque, L., Cobos, A., Royo-Torres, R., Mampel, L., 2014. Preservation of dinosaur footprints in shallow intertidal deposits of the Jurassic—Cretaceous transition in the Iberian Range (Teruel, Spain). Ichnos 21 (1), 19—31. https://doi.org/10.1080/10420940.2013.873721.

Anderson, A.M., 1976. Fish trails from the early permian of South Africa. Palaeontology 19 (2), 397—409.

Antonelli, M., Petti, F.M., Conti, J., Sacco, E., Petruzzelli, M., Spalluto, L., Wagensommer, A., 2023. Lower Cretaceous dinosaur footprints from the Molfetta tracksite (Apulia, southern Italy). Cretac. Res. 142, 105388. https://doi.org/10.1016/j.cretres.2022.105388.

Apaldetti, C., Martinez, R.N., Alcober, O.A., Pol, D., 2011. A new basal sauropodomorph (Dinosauria: Saurischia) from Quebrada del Barro Formation (Marayes-El Carrizal Basin), Northwestern Argentina. PLoS One 6 (11), e26964. https://doi.org/10.1371/journal.pone.0026964.

Arbour, V.M., Larson, D., Vavrek, M., Buckley, L., Evans, D., 2020. An ankylosaurian dinosaur from the Cenomanian Dunvegan Formation of northeastern British Columbia, Canada. Fossil Rec. 23, 179—189. https://doi.org/10.5194/fr-23-179-2020.

Arduini, P., 1996. Early Jurassic tracks from Monte Acuto (Appennino Marchigiano), central Italy. Ichnos 4 (3), 239—240. https://doi.org/10.1080/10420949609380132.

Aurell, M., Bádenas, B., Gasca, J.M., Canudo, J.I., Liesa, C.L., Soria, A.R., Moreno-Azanza, M., Najes, L., 2016. Stratigraphy and evolution of the Galve sub-basin (Spain) in the middle Tithonian—early Barremian: implications for the setting and age of some dinosaur fossil sites. Cretac. Res. 65, 138—162. https://doi.org/10.1016/j.cretres.2016.04.020.

Avanzini, M., Leonardi, G., Mietto, P., 2003. *Lavinipes cheminii* ichnogen., ichnosp. nov., a possible sauropodomorph track from the Lower Jurassic of the Italian Alps. Ichnos 10, 179—193.

Avanzini, M., García-Ramos, J.C., Lires, J., Menegon, M., Piñuela, L., Fernández, L.A., 2005. Turtle tracks from the late Jurassic of Asturias, Spain. Acta Palaeontol. Pol. 50, 743—755. https://doi.org/10.1080/10420940601010943.

Avanzini, M., Garcia-Ramos, J.C., Lires, J., Piñuela, L., Lockley, M.G., 2007. Crocodylomorph tracks from the late Jurassic of Asturias (Spain). Ichnos 14, 143—153. https://doi.org/10.1080/10420940601010943.

Avanzini, M., Piñuela, L., Garcia-Ramos, J.C., 2010a. Preservational morphotypes of *Crocodylopodus* from the late Jurassic of Asturias (Spain). New Mex. Mus. Nat. Hist. Sci. Bullet. 51, 239—244.

Avanzini, M., Piñuela, L., Ruiz-Omeñaca, J.I., Garcia-Ramos, J.C., 2010b. The crocodile track *Hatcherichnus*, from the upper Jurassic of Asturias (Spain). New Mex. Mus. Nat. Hist. Sci. Bullet. 51, 89—92.

Avanzini, M., Piñuela, L., Garcia-Ramos, J.C., 2010c. First report of a late Jurassic lizard-like footprint (Asturias, Spain). J. Iber. Geol. 36, 175—180. https://doi.org/10.5209/rev.

Avanzini, M., Pinuela, L., García-Ramos, J.C., 2012. Late Jurassic footprints reveal walking kinematics of theropod dinosaurs. Lethaia 45, 238—252.

Baird, D., 1957. Triassic reptile footprint faunules from Milford, New Jersey. Harv. College Mus. Compar. Zool. Bullet. 117, 449—520.

Baird, D., 1980. A prosauropod dinosaur trackway from the Navajo Sandstone (Lower Jurassic) of Arizona. In: Jacobs, L.L. (Ed.), Aspects of Vertebrate History. Museum of Northern Arizona Press, Flagstaff, pp. 219—230.

Barbour, E.H., 1892. Notes on a New Order of Gigantic Fossils, vol. 1. University Studies (University of Nebraska, pp. 301—313.

Barco, J.L., Canudo, J.I., Ruiz-Omeñaca, ä J.I., Rubio, J.L., 2005. Evidencia icnológica de un dinosaurio terópodo gigante en el Berriasiense (Cretácico Inferior) de Laurasia (Las Villasecas, Soria, España). Revista Española de Paleontologia, N.E. X, 59—71.

Barco, J.L., Canudo, J.I., Ruiz-Omeñaca, J.I., 2006. New data on *Täerangospodus oncalensis* from the Berriasian Fuentesalvo tracksite (Villar del Río, Soria, Spain): an example of gregarious behaviour in theropod dinosaurs. Ichnos 13 (4), 237—248. https://doi.org/10.1080/10420940600843682.

Barnes, F.A., Lockley, M.G., 1994. Trackway evidence for social sauropods from the Morrison Formation, eastern Utah (USA). Gaia 10, 37—41.

Barrett, P.M., Butler, R.J., Knoll, F., 2005. Small-bodied ornithischian dinosaurs from the middle Jurassic of Sichuan, China. J. Vertebr. Paleontol. 25 (4), 823–834. https://doi.org/10.1671/0272-4634(2005)025[0823:SODFTM]2.0.CO;2.

Barrett, P.M., Butler, R.J., Edwards, N.P., Milner, A.R., 2008. Pterosaur distribution in space and time: an atlas. Zitteliana B28, 61–107.

Belvedere, M., 2008. Ichnological Researches on the Upper Jurassic Dinosaur Tracks in the Louaridène Area (Demnat, Central High-atlas, Morocco). Ph.D. Thesis. Università degli Studi di Padova, Italy, pp. 1–121. https://hdl.handle.net/11577/3426042.

Belvedere, M., Mietto, P., 2010. First evidence of stegosaurian *Deltapodus* footprints in North Africa (Iouaridène Formation, upper Jurassic, Morocco). Palaeontology 53 (1), 233–240. https://doi.org/10.1111/j.1475-4983.2009.00928.x.

Belvedere, M., Mietto, P., Ishigaki, S., 2010. A late Jurassic diverse ichnocoenosis from the siliciclastic Iouaridène Formation (central high atlas, Morocco). Geol. Q. 54 (3), 367–380.

Belvedere, M., Castanera, D., Meyer, C.A., Marty, D., Mateus, O., Silva, B.C., Santos, V.F., Cobos, A., 2019. Late Jurassic globetrotters compared: a closer look at large and giant theropod tracks of North Africa and Europe. J. Afr. Earth Sci. 158, 103547. https://doi.org/10.1016/j.jafrearsci.2019.103547.

Benton, M.J., 1993. Late Triassic terrestrial vertebrate extinctions: stratigraphic aspects and the record of the Germanic Basin. In: Mazin, J.M., Pinna, G. (Eds.), Evolution, Ecology and Biogeography of the Triassic Reptiles, vol. 2. Paleontol. Lombarda Società Italiana di Scienze Naturali e del Museo Civico di Storia Naturale N.S., pp. 19–38

Bernier, P., Barale, G., Bourseau, J.-P., Buffetaut, É., Demathieu, G., Gaillard, C., Gall, J.-C., 1982. Trace nouvelle de locomotion de Chélonienet figures d'émersion associées dans les calcaires lithographiques de Cerin (Kimméridgien supérieur, Ain, France). Geobios 15, 447–467. https://doi.org/10.1016/S0016-6995(82)80068-5.

Bernier, P., Barale, G., Bourseau, J.-P., Buffetaut, E., Demathieu, G., Gaillard, C., Gall, J.-C., Wenz, S., 1984. Decouverte de pistes de dinosaures sauteurs dans les calcaires lithographiques de cerin (Kimmeridgien superieur, ain, France) implications paleoecologiques. Geobios 17, 177–187. https://doi.org/10.1016/S0016-6995(84)80171-0.

Beurlen, K., 1950. Neue Fährtenfunde aus de Fränkischen Trias. Neues Jahrbuch Geol. Palaontol. Monatsh. 1950, 308–320.

Beutler, G., 1998. Keuper. Hallesches Jahrbuch für Geowissenschaften Reihe B 6, 45–58.

Bilbey, S.A., Mickelson, D.L., Hall, J.E., Kirkland, J.I., Madsen, S.K., Blackshear, B., Todd, C., 2005. Vertebrate ichnofossils from the upper Jurassic Stump to Morrison transition –flaming Gorge reservoir, Utah. In: Dehler, C.M., Pederson, J.L., Sprinkel, D.A., Kowallis, B.J. (Eds.), Uinta Mountain Geology, vol. 33. Utah Geological Association Publication, pp. 111–123.

Billon-Bruyat, J.-P., Mazin, J.-M., 2003. The Systematic Problem of Tetrapod Ichnotaxa: The Case Study of *Pteraichnus* Stokes, 1957 (Pterosauria, Pterodactyloidea), vol. 217. Geological Society, London, Special Publications, pp. 315–324. https://doi.org/10.1144/GSL.SP.2003.217.01.19.

Bindellini, G., Dal Sasso, C., 2021. Sauropod teeth from the middle Jurassic of Madagascar, and the oldest record of Titanosauriformes. Papers Palaeontol. 7 (1), 137–161. https://doi.org/10.1002/spp2.1282.

Bock, W., 1952. Triassic reptilian tracks and trends of locomotive evolution. J. Paleontol. 26, 395–433.

Bonaparte, J.F., 1979. Terapsidos, mamiferos Mesozoicos y arcosaurios en la historia biogeografica de America del Sur. Acta Zool. Lilloana 34, 66–82.

Bonaparte, J.F., 1983. Jurassic tetrapods from South America and dispersal routes. In: Jacobs, L.L. (Ed.), Aspects of Vertebrate History: Essays in Honor of Edwin Harris Colbert. Museum of Northern Arizona Press, Flagstaff, pp. 73–98.

Bordy, E.M., 2021. Darting towards Storm Shelter: a minute dinosaur trackway from southern Africa. South Afr. J. Sci. 117 (5/6). https://doi.org/10.17159/sajs/2021/9145.

Bordy, E.M., Sztano, O., Rubidge, B.S., Bumby, A.J., 2011. Early triassic vertebrate burrows from the Katberg Formation of the south-western Karoo Basin, South Africa. Lethaia 44, 33–45.

Bordy, E.M., Sciscio, L., Abdala, F., McPhee, B.W., Choiniere, J.N., 2017. First lower Jurassic vertebrate burrow from southern Africa (upper Elliot Formation, Karoo Basin, South Africa). Palaeogeogr. Palaeoclimatol. Palaeoecol. 468, 362–372. https://doi.org/10.1016/j.palaeo.2016.12.024.

Bordy, E.M., Rampersadh, A., Abrahams, M., Lockley, M.G., Head, H.V., 2020. Tracking the Pliensbachian–Toarcian Karoo firewalkers: trackways of quadruped and biped dinosaurs and mammaliaforms. PLoS One 15 (1), e0226847. https://doi.org/10.1371/journal.pone.0226847.3.

Bordy, E.M., Lockley, M.G., Rampersadh, A., Mukaddam, R., Head, H., 2022. Life and land engulfed in the late Early Jurassic Karoo lavas of southern Gondwana. Geol. Mag. 160, 645–666. https://doi.org/10.1017/S0016756822001169.

Bordy, E.M., Smith, R.M.H., Choiniere, J.N., Rubidge, B.S., 2024. Selected Karoo Geoheritage Sites of Paleontological Significance in South Africa and Lesotho, vol. 543. Geological Society, London, Special Publications. https://doi.org/10.1144/SP543-2022-202.

Boutakiout, M., Masrour, M., Pérez-Lorente, F., 2020. New sauropod morphotype definition in the oriental section 0F Imilchil megatracksite, High Atlas (Morocco). J. Afr. Earth Sci. 161, 103664. https://doi.org/10.1016/j.jafrearsci.2019.103664.

Breithaupt, B.H., Adams, T., Southwell, E.H., Matthews, N.A., 2004. *Carmelopodus* and the Sundance vertebrate ichnofaunal province: insights into the preservational complexities of vertebrate ichnotaxa. Geol. Soc. Am. Abstr. Progr. 36 (5), 93.

Breithaupt, B.H., Chan, M.A., Seiler, W.M., Matthews, N.A., 2021. Weathering pits versus trample marks: a reinterpretation of the "dinosaur dance floor": aJurassic Navajo sandstone surface in the Vermillion Cliffs National Monument, Arizona. Palaios 36, 331–338.

Buck, P.V., Ghilardi, A.M., Fernandes, L.B.R., Fernandes, M.A., 2017b. A new ichnotaxon classification of large mammaliform trackways from the Lower Cretaceous Botucatu Formation, Paraná Basin, Brazil. Palaeogeogr. Palaeoclimatol. Palaeoecol. 485, 377–388. https://doi.org/10.1016/j.palaeo.2017.06.027.

Calvo, J.O., 1991. Huellas de dinosaurios en la Formación Río Limay (Albiano–Cenomaniano?), Picún Leufú, provincia de Neuquén, República Argentina.(Ornithischia-Saurischia: Sauropoda-Theropoda). Ameghiniana 28, 241–258.

Calvo, J.O., 1999. Dinosaurs and other vertebrates of the Lake Ezequiel ramos Mexía area, Neuquén-Patagonia, Argentina. Proc. Sec. Gondw. Dinos. Symp. Nat. Sci. Mus. Monogr. 15, 13–45.

Calvo, J.O., Lockley, M.G., 2001. The first pterosaur tracks from Gondwana. Cretac. Res. 22, 585–590.

Campos-Soto, S., Benito, M.I., Cobos, A., Caus, E., Quijada, I.E., Suarez-Gonzalez, P., Mas, R., Royo-Torres, R., Alcalá, L., 2019. Revisiting the age and palaeoenvironments of the Upper Jurassic–Lower Cretaceous? dinosaur-bearing sedimentary record of eastern Spain: implications for Iberian palaeogeography. J. Iber. Geol. 45, 471–510. https://doi.org/10.1007/s41513-019-00106-y.

Carpenter, K., 1984. Skeletal reconstruction and life restoration of *Sauropelta* (Ankylosauria: Nodosauridae) from the cretaceous of NorthNorth America. Can. J. Earth Sci. 21, 1491–1498. https://doi.org/10.1139/e84-154.

Casamiquela, R.M., 1960. El hallazgo del primer elenco (icnológico) jurásico de vertebrados terrestres de Latinoamérica (noticia). Rev. Asoc. Geol. Argent. 15, 5–14.

Casamiquela, R.M., 1961. Sobre la presencia de un mamífero en el primer elenco (icnológico) de vertebrados del Jurásico de la Patagonia. Physis 22, 225–233.

Casamiquela, R.M., 1964. Estudios icnológicos. Problemas y métodos de la icnología con aplicación al estudio de pisadas Mesozoicas (Reptilia, Mammalia) de la Patagonia. Colegio Ind. Pio IX. Buenos Aires, Argentina, 229 pp.

Casamiquela, R.M., 1966. Algunas consideraciones teóricas sobre los andares de los dinosaurios saurisquios. Implicaciones filogenéticas. Ameghiniana 4, 373–385.

Casamiquela, R.M., Fasola, A., 1968. Ichnolgia. In: Casamiquela, R., Fasola, A. (Eds.), Sobre pisadas de dinosaurios del Cretacico Inferior de Cochagua (Chile), vol. 30. Universidad de Chile Publication, pp. 13–24.

Casanovas, M.L., Ezquerra, R., Fernández, A., Pérez-Lorente, F., Santafé, J.V., Torcida, F., 1995. Dos nuevos yacimientos de icnitas de dinosaurios en La Rioja y en la provincia de Soria (España), vol. 47. Coloquios de Paleonologia, pp. 9–23.

Castanera, D., Pascual, C., Canudo, J., Hernández, N., Barco, J.L., 2012. Gauge variations in sauropod trackways from the Huérteles Formation (Berriasian) of the Cameros basin (Spain): implications for sauropod ichnotaxonomy. In: Royo-Torres, R., Gascó, F., Acalá, L. (Eds.), 10th Annual Meeting of the European Association of Vertebrate Palaeontologists, vol. 20. ¡Fundamental!, pp. 35–37.

Castanera, D., Pascual, C., Razzolini, N.L., Vila, B., Barco, J.L., Canudo, J.I., 2013. Discriminating between medium-sized tridactyl trackmakers: tracking ornithopod tracks in the base of the Cretaceous (Berriasian, Spain). PLoS One 8 (11), e81830. https://doi.org/10.1371/journal.pone.0081830.

Castanera, D., Vila, B., Razzolini, N.L., Santos, V.F., Pascual, C., Canudo, J.I., 2014. Sauropod trackways of the Iberian Penninsula: palaeoetological and palaeoenvironmental implications. J. Iber. Geol. 40 (1), 49–59.

Castanera, D., Colmenar, J., Sauqué, V., Canudo, J.I., 2015. Geometric morphometric analysis applied to theropod tracks from the Lower Cretaceous (Berriasian) of Spain. Palaeontology 58, 183–200.

Castanera, D., Santos, V.F., Piñuela, L., Pascual, C., Vila, B., Canudo, J.I., Moratalla, J.J., 2016a. Iberian sauropod tracks through time: variations in sauropod manus and pes track morphologies. In: Falkingham, P.L., Marty, D., Richter, A. (Eds.), Dinosaur Tracks—The Next Steps. Indiana University Press, Bloomington, pp. 120–137.

Castanera, D., Piñuela, L., García-Ramos, J.C., 2016b. *Grallator* theropod tracks from the Late Jurassic of Asturias (Spain): ichnotaxonomic implications. Spanish J. Paleontol. 31 (2), 283–296.

Castanera, D., Belvedere, M., Marty, D., Paratte, G., Lapaire-Cattin, M., Lovis, C., Meyer, C.A., 2018. A walk in the maze: variation in Late Jurassic tridactyl dinosaur tracks from the Swiss Jura Mountains (NW Switzerland). PeerJ 6, e4579. https://doi.org/10.7717/peerj.4579.

Castanera, D., Malafaia, E., Silva, B.C., Santos, V.F., Belvedere, M., 2020a. New dinosaur, crocodylomorph and swim tracks from the Late Jurassic of the Lusitanian Basin: implications for ichnodiversity. Lethaia 54 (2), 260–270. https://doi.org/10.1111/let.12402.

Castanera, D., Silva, B.C., Santos, V.F., Malafaia, E., Belvedere, M., 2020b. Tracking late Jurassic ornithopods in the Lusitanian Basin of Portugal: ichnotaxonomic implications. Acta Palaeontol. Pol. 65 (2), 399–412. https://doi.org/10.4202/APP.00707.2019.

Castanera, D., Malafaia, E., Silva, B.C., Santos, V.F., Belvedere, M., 2021. New dinosaur, crocodylomorph and swim tracks from the Late Jurassic of the Lusitanian Basin: implications for ichnodiversity. Lethaia 54 (2), 260–270. https://doi.org/10.1111/let.12402.

Castanera, D., Pascual-Arribas, C., Canudo, J.I., Puértolas-Pascual, E., 2022. A new look at *Crocodylopodus meijidei*: implications for crocodylomorph locomotion. J. Vertebr. Paleontol. 41. https://doi.org/10.1080/02724634.2021.2020803.

Castanera, D., Aurell, M., Canudo, J.I., Cuenca-Bescós, G., Gasca, J.M., Bádenas, B., 2023b. Paleoecology and paleoenvironment of the early cretaceous theropod-dominated ichnoassemblage of the Los Corrales del Pelejón tracksite, Teruel Province, Spain. Palaeogeogr. Palaeoclimatol. Palaeoecol. 111761. https://doi.org/10.1016/J.PALAEO.2023.111761.

Castanera, D., Guarido, A., Mampel, L., Espilez, E., Cobos, A., 2023a. New data on the identification of dinosaur trackmakers in the late Jurassic of the Maestrazgo basin (Teruel, Spain). In: Alba, D.M., Marigó, J., Nacarino-Meneses, C., Villa, A. (Eds.), Book of Abstracts of the 20th Annual Conference of the European Association of Vertebrate Palaeontologists, vols. 1–23. Palaeovertebrata Special, p. 54.

Castanera, D., Mampel, L., Cobos, A., 2024. The complexity of tracking stegosaurs and their gregarious behavior. Sci. Rep. 14 (1), 14833.

Charriére, A., Haddoumi, H., Mojon, P.O., 2005. Découverte de Jurassique supérieur et d'un niveau marin du Barrémien dans les «couches rouges» continentales du Haut Atlas central marocain: Implications paléogéographiques et structurales. Comptes Rendus Palevol. 4, 385–394. https://doi.org/10.1016/j.crpv.2005.04.009.

Chen, P.-J., Li, J., Matsukawa, M., Zhang, H., Wang, Q., Lockley, M.G., 2006. Geological ages of dinosaur-track-bearing formations in China. Cretac. Res. 27, 22–32. https://doi.org/10.1016/j.cretres.2005.10.008.

Chure, D.J., Engelmann, G.F., Britt, B.B., Good, T.R., 2014. It's not your parents' erg deposit anymore: fossil management implications of a paleontological study of the Nugget Sandstone in northeastern Utah. In: Proceedings of the 10th Conference on Fossil Resources, Rapid City, South Dakota, 6, pp. 140–154. Dakoterra.

Clark, N.D.L., Barco Rodriguez, J.L., 1998. The first dinosaur trackway from the Valtos Sandstone Formation (Bathonian, Jurassic) of the Isle of Skye, Scotland, UK. Geogaceta 24, 79–82.

Clark, N.D.L., Ross, D.A., Booth, P., 2005. Dinosaur tracks from the Kilmaluag formation (Bathonian, middle Jurassic) of Score Bay, Isle of Skye, Scotland, UK. Ichnos 12, 93–104.

Cobos, A., Royo-Torres, R., Luque, L., Alcalá, L., Mampel, L., 2010. An Iberian stegosaurs paradise: the Villar del Arzobispo Formation (Tithonian–Berriasian) in Teruel (Spain). Palaeogeogr. Palaeoclimatol. Palaeoecol. 293 (1–2), 223–236. https://doi.org/10.1016/j.palaeo.2010.05.024.

Cobos, A., Lockley, M.G., Gascó, F., Royo–Torres, R., Alcalá, L., 2014. Megatheropods as apex predators in the typically Jurassic ecosystems of the Villar del Arzobispo Formation (Iberian Range, Spain). Palaeogeogr. Palaeoclimatol. Palaeoecol. 399, 31–41. https://doi.org/10.1016/j.palaeo.2014.02.008.

Cobos, A., Royo-Torres, R., Torrente, M.A., Alcalá, L., 2015. Nuevos fósiles de dinosaurios en la Formación Villar del Arzobispo de El Castellar y Formiche Alto (Teruel, España). Geogaceta 58, 43–46.

Connely, M.V., 2006. Paleoecology of pterosaur tracks from the upper Sundance and lower Morrison formations in central Wyoming. New Mex. Mus. Nat. Hist. Sci. Bullet. 36, 199–202.

Contessi, M., Fanti, F., 2012. Vertebrate tracksites in the middle Jurassic-upper cretaceous of South Tunisia. Ichnos 19 (4), 211–227.

Contessi, M., 2013. First report of mammal-like tracks from the cretaceous of North Africa (Tunisia). Cretac. Res. 42, 48–54.

Conti, M.A., Morsilli, M., Nicosia, U., Sacchi, E., Savino, V., Wagensommer, A., Gianolla, P., 2005. Jurassic dinosaur footprints from southern Italy: footprints as indicators of constraints in paleogeographic interpretation. Palaios 20 (6), 534–550.

Coria, R.A., Carabajal, P., 2004. Nuevas huellas de Theropoda (Dinosauria: Saurischia) del Jurásico de Patagonia, Argentina. Ameghiniana 41 (3), 393–398.

Costa da Silva, R., Barboni, R., Dutra, T., Godoy, M.M., Binotto, R.B., 2012. Footprints of large theropod dinosaurs and implications on the age of Triassic biotas from southern Brazil. J. S. Am. Earth Sci. 39, 16–23. https://doi.org/10.1016/j.jsames.2012.06.017.

Courel, L., Demathieu, G., 1976. Une ichnofaune reptilienne remarquable dans les Grès triasiques de Largentière Ardèche, France. Palaeontograph. Abteilung 151, 194–216.

Courel, L., Demathieu, G.R., 2000. Une nouvelle ichnoespèce *Coelurosaurichnus grancieri* du Trias supérieur de l'Ardèche, France. Geodiversitas 22 (1), 35–46.

D'Emic, M.D., 2012. The early evolution of titanosauriform sauropod dinosaurs. Zool. J. Linn. Soc. 166 (3), 624–671. https://doi.org/10.1111/j.1096-3642.2012.00853.x.

Dalman, S.G., 2012. New data on small theropod footprints from the early Jurassic (Hettangian) Hartford Basin of Massachusetts, United States. Bull. Peabody Mus. Nat. Hist. 53 (2), 333–353. https://doi.org/10.3374/014.053.0201.

Dalman, S.G., Weems, R.E., 2013. A new look at morphological variation in the ichnogenus *Anomoepus*, with special reference to material from the Lower Jurassic Newark Supergroup: implications for ichnotaxonomy and ichnodiversity. Bull. - Peabody Mus. Nat. Hist. 54 (1), 67–124.

da Silva, R.C., Sedor, F.A., 2017. Mesosaurid swim traces. Front. Ecol. Evol. 5, 22. https://doi.org/10.3389/fevo.2017.00022.

Day, J.J., Norman, D.B., Gale, A.S., Upchurch, P., Powell, H.P., 2004. A Middle Jurassic dinosaur trackway site from Oxfordshire, UK. Palaeontology 47, 319–348. https://doi.org/10.1111/j.0031-0239.2004.00366.x.

Day, J.J., Norman, D.B., Upchurch, P., Powell, H.P., 2002. Dinosaur locomotion from a new trackway. Nature 415, 494–495. https://doi.org/10.1038/415494a.

de Gibert, J.M., 2001. *Undichna gosiutensis*, isp. nov.: a new fish trace fossil from the Jurassic of Utah. Ichnos 8, 15–22.

de Gibert, J.M., Buatois, L.A., Fregenal-Martínez, M.A., Mángano, M.G., Ortega, F., Poyato-Ariza, F.J., Wenz, S., 1999. The fish trace fossil *Undichna* from the Cretaceous of Spain. Palaeontology 42 (1), 409–427.

De Jaime-Soguero, C., Mujal, E., Oms, O., Bolet, A., Dinarès-Turell, J., Ibáñez-Insa, J., Fortuny, J., 2023. Palaeoenvironmental reconstruction of a lower to middle Permian terrestrial composite succession from the Catalan Pyrenees: implications for the evolution of tetrapod ecosystems in equatorial Pangaea. Palaeogeogr. Palaeoclimatol. Palaeoecol. 632, 111837. https://doi.org/10.1016/j.palaeo.2023.111837.

Delair, J.B., 1963. Notes on Purbeck fossil footprints, with descriptions of two hierto unknown forms from Dorset. Proc. Dorset Nat. Hist. Archaeol. Soc. 84, 92–100.

Delair, J.B., Sarjeant, W.A.S., 1985. History and bibliography of the study of fossil vertebrate footprints in the British Isles: supplement 1973–1983. Palaeogeogr. Palaeoclimatol. Palaeoecol. 49 (1–2), 123–160. https://doi.org/10.1016/0031-0182(85)90007-0.

de Lapparent, A.F., Montenat, C., 1967. Les emprintes de pas de reptiles de l'infralias du Veillon (Vendée). Mem. Soc. Geol. Fr. 107, 1–44.

Demathieu, G., 1970. Les empreintes de pas de vertébrés du Trias de la bordure Nord-Est du Massif Central. Cahiers de Paléontologie, Paris, pp. 230–pp.

Demathieu, G., 1990. Problems in discrimination of tridactyl dinosaur footprints, exemplified by the Hettangian trackways, the Causses, France. Ichnos 1 (2), 97–110. https://doi.org/10.1080/10420949009386339.

Demathieu, G., Gand, D., 1972. Les pistes dinosauroides du Trias Moyen du Plateau d'Antully et leur signification paléozoologique. Bulletin Trimestriel de la Société d'Histoire Naturelle Musee d'Autun, N.S. 62, 4–18.

Demathieu, G.R., Sciau, J., 1992. Des pistes de dinosaures et de crocodiliens dans les dolomites de l'Hettangien du Causse du Larzac. Comptes Rendus Acad. Sci. Paris 315, 1561–1566.

Demathieu, G., Gand, G., Sciau, J., Freytet, P., 2002. Les traces de pas de dinosaures et autre archosaures du Lias inferieur des Grands Causses, Sud de la France. Palaeovertebrata 31, 1–143.

dePolo, P.E., Brusatte, S.L., Challands, T.J., Foffa, D., Ross, D.A., Wilkinson, M., Yi, H., 2018. A sauropod-dominated tracksite from Rubha nam Brathairean (Brothers'point), Isle of Skye, Scotland. Scot. J. Geol. 54, 1–12. https://doi.org/10.1144/sjg2017-016.

dePolo, P.E., Brusatte, S.L., Challands, T.J., Foffa, D., Wilkinson, M., Clark, N.D.L., Hoad, J., da Costa Pereira, P.V.L.G., Ross, D.A., Wade, T.J., 2020. Novel track morphotypes from new tracksites indicate increased Middle Jurassic dinosaur diversity on the Isle of Skye, Scotland. PLoS One 15 (3), e0229640. https://doi.org/10.1371/journal.pone.0229640.

de Valais, S., 2009. Ichnotaxonomic revision of *Ameghinichnus*, a mammalian ichnogenus from the middle Jurassic La Matilde Formation, Santa Cruz province, Argentina. Zootaxa 2203, 1–21.

de Valais, S., 2011. Revision of dinosaur ichnotaxa from the La Matilde Formation (middle Jurassic), Santa Cruz province, Argentina. Ameghiniana 48 (1), 28–42.

Díaz-Martínez, I., Pereda-Suberbiola, X., Pérez-Lorente, F., Canudo, J.I., 2015. Ichnotaxonomic review of large ornithopod dinosaur tracks: temporal and geographic implications. PLoS One 10 (2), e0115477. https://doi.org/10.1371/journal.pone.0115477.

Díaz-Martínez, I., González, S.N., De Valais, S., 2017. Dinosaur footprints in the early Jurassic of Patagonia (Marifil Volcanic complex, Argentina): biochronological and palaeobiogeographical inferences. Geol. Mag. 154, 914–922.

Díaz-Martínez, I., Heredia, A.M., González, S.N., Canale, N., de Valais, S., Cónsole-Gonella, C.A., Montes, R.M., Caratelli, M., Urzgasti-Torres, S., Fischer, G., Lecuona, A., Paniceres, P., Salgado, L., Citton, P., 2022. Pterosaur tracks from the upper Cretaceous Anacleto formation (Neuquén basin), northern Patagonia, Argentina: insights into Campanian pterosaur diversity in Gondwana. Diversity 14, 1007. https://doi.org/10.3390/d14111007.

Diedrich, C., 2011. Upper Jurassic tidal flat megatracksites of Germany—coastal dinosaur migration highways between European islands, and a review of the dinosaur footprints. Palaeobiodivers. Palaeoenviron. 91, 129–155. https://doi.org/10.1007/s12549-010-0044-y.

Doody, J.S., James, H., Colyvas, K., McHenry, C.R., Clulow, S., 2015. Deep nesting in a lizard, *déjà vu* devil's corkscrews: first helical reptile burrow and deepest vertebrate nest. Biol. J. Linn. Soc. 116, 13–26.

D'Orazi Porchetti, S., Nicosia, U., 2007. Re-examination of some large Early Mesozoic tetrapod footprints from the African collection of Paul Ellenberger. Ichnos 14 (3–4), 219–245. https://doi.org/10.1080/10420940601049990.

D'Orazi Porchetti, S., Mocke, H.B., Latiano, M., Wagensommer, A., 2014. First record of Otozoum from Namibia. Lethaia 48 (1), 72–82. https://doi.org/10.1111/let.12088.

D' Orazi Porchetti, S., Wagensommer, A., 2015. A vertebrate trackway from the Twefelfontein formation (lower Cretaceous), Damaraland, Namibia. Paläontol. Z. 89, 807. https://doi.org/10.1007/s12542-015-0264-6.

D'Orazi Porchetti, S., Bertini, R.J., Langer, M.C., 2017. Walking, running, hopping: analysis of gait variability and locomotor skills in Brasilichnium elusivum Leonardi, with inferences on trackmaker identification. Palaeogeogr. Palaeoclimatol. Palaeoecol. 465, 14–29.

D'Orazi Porchetti, S., Bertini, R.J., Langer, M.C., 2018. Proposal for ichnotaxonomic allocation of therapsid footprints from the Botucatu Formation (Brazil). Ichnos 25, 192–207.

Dornan, S.S., 1908. Notes on the geology of Basutoland. Geol. Mag. 5 (2), 57–63.

Dutuit, J.-M., Ouazzou, A., 1980. Découverte d'une piste de Dinosaure sauropode sur le site d'empreintes de Demnat (Haut-Atlas marocain). Mémoir Societé Géologique de France, nouvelle série 95–102.

Dzhalilov, M.R., Novikov, V.P., 1992. Fossil traces of dinosaurs in the territory of Tajikistan. In: Vialov, O.S., Fedonkin, M.A. (Eds.), Trace Fossils of Extinct Organisms. Nauka, Moscow, pp. 47–65.

Edgar, K.M., Haller, L., Cashmore, D.D., Dunne, E.M., Butler, R.J., 2023. Stratigraphic and geographic distribution of dinosaur tracks in the UK. J. Geol. Soc. London 180. https://doi.org/10.1144/jgs2023-003.

Elgh, E., Pieńkowski, G., Niedźwiedzki, G., 2019. Pterosaur track assemblages from the Upper Jurassic (lower Kimmeridgian) intertidal deposits of Poland: linking ichnites to potential trackmakers. Palaeogeogr. Palaeoclimatol. Palaeoecol. 530, 32–48.

Ellenberger, P., 1970. Les niveaux paleontologiques de première apparition des mammifères primordiaux en Afrique du Sud et leur ichnologie. Establissement de zones stratigraphiques detaillées dans le Stormberg du Lesotho

(Afrique du Sud) (Trias Supérieur à Jurassique). In: International Union of Geological Sciences, Commission on Stratigraphy (Ed.), Second Gondwana Symposium Proceedings and Papers, pp. 343–370.

Ellenberger, P., 1972. Contribution à la classification des Pistes de Vertébrés du Trias: les types du Stormberg d 'Afrique du Sud (I) [Contribution to the classification of Triassic vertebrate trackways: types found in the Stormberg Series of South Africa. I)]. Paleovertebrata, vol. 152. Memoire Extraordinaire, Montpellier.

Ellenberger, P., 1974. Contribution a la Classification des Pistes de Vértebrés du Trias: les Types du Stormberg d'Afrique du Sud, vol. 1974. Laboratoire de paléontologie des vertébrés, Montpellier, pp. 1–142.

Ellenberger, P., 1975. L'explosion démographique des petits quadrupèdes a allure de mammifères dans le Stormberg supérieur (Trias) d'Afrique du sud aperçu sur leur origine au Permien (France et Karoo). In: Lehman, J.-P. (Ed.), Problèmes Actuels de Paléontologie (Évolution des Vertébrés), Colloq, vol. 218. Internationaux CNRS, pp. 409–439.

Ellenberger, F., Ellenberger, P., 1958. Principaux types de pistes de vertébrés dans les couches du Stormberg au Basutoland (Afrique du Sud) (note préliminaire). In: Comptes-Rendus Sommaire des Séances de la Société Géologique de France 1958, pp. 65–67.

Ellenberger, F., Ellenberger, P., Fabre, J., Mendrez, C., 1963. Deux nouvelles dalles à pistes de Vertébrés fossils découvertes au Basutoland (Afrique du Sud) [Two new slabs of fossil vertebrate trackways discovered in Basutoland (Southern Africa)]. Compte Rendus de la Société Géologique de France 1963, 315–317.

Ellenberger, P., Mossman, D.J., Mossman, A.D., Lockley, M.G., 2005. Bushman cave paintings of ornithopod dinosaurs: paleolithic trackers interpret Early Jurassic footprints. Ichnos 12, 223–226. https://doi.org/10.1080/10420940591008971.

Engelmann, G.F., 2010. An occurrence of remarkably abundant *Brasilichnium* tracks (Nugget sandstone, early Jurassic, dinosaur National Monument) and their environmental context. GSA Denv. Ann. Meeting Abstr. Progr. 42, 642.

Engelmann, G.F., Chure, D.J., 2017. Morphology and sediment deformation of downslope *Brasilichnium* trackways on a dune slipface in the Nugget Sandstone of northeastern Utah, USA. Palaeontol. Electron. 20 (2.22A), 1–21. https://doi.org/10.26879/617. https://palaeo-electronica.org/content/2017/1845-downslope-trackways.

Fanti, F., Contessi, M., Nigarov, A., Esenov, P., 2013. New data on two large dinosaur tracksites from the Upper Jurassic of eastern Turkmenistan (central Asia). Ichnos 20 (2), 54–71. https://doi.org/10.1080/10420940.2013.778845.

Farlow, J.O., 1992. Sauropod tracks and trackmakers: integrating the ichnological and skeletal records. Zubia 10, 89–138.

Farlow, J.O., Pittman, J.G., Hawthorne, J.M., 1989. *Brontopodus birdi*, lower Cretaceous sauropod footprints from the U.S. Gulf coastal plain. In: Gillette, D.D., Lockley, M.G. (Eds.), Dinosaur Tracks and Traces. Cambridge University Press, New York, pp. 165–178.

Farlow, J.O., Coroian, D., Currie, P.J., 2018. Noah's Ravens: Interpreting the Makers of Tridactyl Dinosaur Footprints. Indiana University Press, Bloomington, p. 643.

Fernández-Baldor, F.T., Díaz-Martínez, I., Contreras, R., Huerta, P., Montero, D., Urién, V., 2015. Unusual sauropod tracks in the Jurassic-Cretaceous interval of the cameros basin (Burgos, Spain). J. Iber. Geol. 41 (1), 141–154. https://doi.org/10.5209/rev_JIGE.2015.v41.n1.48660.

Ferrusquía-Villafranca, I., Jiménez-Hidalgo, E., Bravo-Cuevas, V.M., 1996. Footprints of small sauropods from the middle Jurassic of Oaxaca, southeastern Mexico. In: Morales, M. (Ed.), The Continental Jurassic, Museum of Northern Arizona Bulletin, vol. 60, pp. 119–126.

Figueiredo, S.D., Carvalho, I.S., Pereda-Suberbiola, X., Cunha, P.P., Antunes, V., Dıaz-Martínez, I., 2022. New ornithopod footprints from the Areia do Mastro formation (lower Cretaceous), Espichel Cape (Portugal, western Iberia) and their context in the iberian ichnological ornithopod record. Cretac. Res. 131, 105069. https://doi.org/10.1016/j.cretres.2021.105069.

Figueiredo, S.D., de Carvalho, C.N., Cunha, P.P., Duarte, L.V., Fonseca, A., Monteiro, C., Forte, J., 2023. The first dinosaurs in Iberia: a new dinosaur tracksite from the Sinemurian (Lower Jurassic) of Portugal. Hist. Biol. https://doi.org/10.1080/08912963.2023.2256751.

Fillmore, D.L., Lucas, S.G., Simpson, E.L., 2012. Ichnology of the Mississippian Mauch Chunk Formation, eastern Pennsylvania. New Mex. Mus. Nat. Hist. Sci. Bullet. 54, 1–136.

Fiorillo, A.R., Kobayashi, Y., McCarthy, P.J., Tanaka, T., Tykoski, R.S., Lee, Y.-N., Takasaki, R., Yoshida, J., 2019. Dinosaur ichnology and sedimentology of the Chignik Formation (Upper Cretaceous), Aniakchak National

Monument, southwestern Alaska; further insights on habitat preferences of high-latitude hadrosaurs. PLoS One 14 (10), e0223471. https://doi.org/10.1371/journal.pone.0223471.

Foster, J.R., 2015. Theropod dinosaur ichnogenus *Hispanosauropus* identified from the Morrison Formation (upper Jurassic), western North America. Ichnos 22, 183–191. https://doi.org/10.1080/10420940.2015.1059335.

Foster, J.R., Lockley, M.G., 1995. Tridactyl dinosaur footprints from the Morrison Formation of northeast Wyoming. Ichnos 4, 35–41.

Foster, J.R., Lockley, M.G., 1997. Probable crocodilian tracks and traces from the Morrison Formation (upper Jurassic) of Eastern Utah. Ichnos 5 (2), 121–129. https://doi.org/10.1080/10420949709386411.

Foster, J.R., Lockley, M.G., 2006. The vertebrate ichnological record of the Morrison Formation (upper Jurassic, North America). New Mex. Mus. Nat. Hist. Sci. Bullet. 36, 203–216.

Foster, J.R., Lockley, M.G., 2024. The vertebrate ichnological record of the Morrison Formation (upper Jurassic, North America) – Part 2. New Mex. Mus. Nat. Hist. Sci. Bullet.

Foster, J.R., Lockley, M.G., Brockett, J., 1999. Possible Turtle Tracks from the Morrison Formation of Southern Utah: Vertebrate Paleontology in Utah. Utah Geological Survey Miscellaneous Publication 99-1, pp. 185–191.

Foster, J.R., Hamblin, A.H., Lockley, M.G., 2000. The oldest evidence of a sauropod dinosaur in the western United States and other important vertebrate trackways from Grand Staircase-Escalante National Monument, Utah. Ichnos 7, 169–181.

Foster, J.R., Titus, A.L., Winterfeld, G.F., Hayden, M.C., Hamblin, A.H., 2001. Paleontological survey of the Grand Staircase-Escalante National Monument, Garfield and Kane Counties, Utah. Utah Geol. Surv. Spec. Study 99, 1–98.

Francischini, H., Sales, M.A.F., Dentzien-Dias, P., Schultz, C.L., 2017. The presence of ankylosaur tracks in the Guará Formation (Brazil) and remarks on the spatial and temporal distribution of Late Jurassic dinosaurs. Ichnos 25, 177–191.

Friese, H., 1979. Die Dinosaurierfährten von Barkhausen im Wiehengebirge. Veröffentlichungen des Landkreises Osnabrück 1, 1–38.

Fuentes Vidarte, C., 2001. A new species of *Pteraichnus* for the Spanish lower Cretacous: *Pteraichnus cidacoi*. Strata. Série 1. Communications 11, 44–46.

Fuentes Vidarte, C., Meijide Calvo, M., 2001. Primeras huellas de cocodrilo en el Weald de Cameros (Soria, España). Nueva familia: Crocodylopodidae, nuevo icnogénero: *Crocodylopodus*, nueva icnoespecie: *C. meijidei*. In: Colectivo Arqueológico-Paleontológico de Salas, C.S. (Ed.), Actas de Las I Jornadas Internacionales Sobre Paleontología de Dinosaurios y Su Entorno. Burgos, pp. 329–335.

Fuentes Vidarte, C., Meijide Calvo, M., Meijide Fuentes, F., Meijide Fuentes, M., 2003. Rastro de un pequeño tetrápodo de pequeño tamaño en el Weald de Cameros (Sierra de Oncala, Soria, España) nov. icnogen. nov. icnosp: *Emydhipus cameroi*. In: Pérez Lorente, F., Romero Molina, M.M., Ricas Carrera, P. (Eds.), Dinosaurios y Otros Reptiles Mesozoicos En España. Ciencias de La Tierra. Universidad de La Rioja: Instituto de Estudios Riojanos, Logroño, pp. 119–128.

Fuentes Vidarte, C., Meijide Calvo, M., Meijide Fuentes, F., Meijide Fuentes, M., 2004a. Huellas de pterosaurios en la Sierra de Oncala (Soria, España). Nuevas icnoespecies: *Pteraichnus vetustior, Pteraichnus parvus, Pteraichnus manueli*. Celtiberia 54 (98), 471–490.

Fuentes Vidarte, C., Meijide Calvo, M., Meijide Fuentes, F., Meijide Fuentes, M., 2004b. *Pteraichnus longipodus* nov. icnosp. en la Sierra de Oncala (Soria, España). Stud. Geol. Salmant. (40), 103–114.

Fujita, M., Azuma, Y., Lee, Y.-N., Lu, J., Dong, Z., Noda, Y., Urano, K., 2007. New theropod track site from the upper Jurassic Tuchengzi Formation of Liaoning Province, northeastern China. Mem. Fukui Prefect. Dinosaur Mus. 6, 17–25.

Fürsich, F.T., Schneider, S., Werner, W., Lopez-Mir, B., Pierce, C.S., 2022. Life at the continental–marine interface: palaeoenvironments and biota of the Alcobaça Formation (Late Jurassic, Central Portugal), with a formal definition of the unit appended. Palaeobiodivers. Palaeoenviron. 102, 265–329.

Gabuniya, L.K., Kurbatov, V.V., 1982. Jurassic dinosaur tracks of Tashkurgan (Uzbekistan SSR). Abstr. Sci. Sess. Dedic. 60th Anniv. Establ. USSR 20–22.

Gabuniya, L.K., Kurbatov, V.V., 1988. Traces of dinosaurs in Jurassic deposits of southern central Asia. In: Bogdanova, T.N., Khozatsky, L.I., Istchenko, A.A. (Eds.), Fossil Traces of Vital Activity and Dynamics of the Environment in Ancient Biotopes. Nauka Dumka, Kiev, pp. 45–57.

Gaillard, C., Bernier, P., Barale, G., Bourseau, J., Buffetaut, E., Ezquerra, R., Gall, J., De Broin, F.D.L., Renous, S., Wenz, S., 2003. A giant Upper Jurassic turtle revealed by its trackways. Lethaia 36, 315–322. https://doi.org/10.1080/00241160310004611.

Gand, G., 1974a. Les traces de vertébrés Triasiques de l'Autunois et du nord du Charollais. Bulletin de la Société d'Histoire Naturelle d'Autun 69, 7–24.

Gand, G., 1974b. Sur les niveaux a empreintes de pas de vertébrés triasiques. Des carrières de St-Sernin-du.Bois (Saône-et-Loire). Bulletin de la Société d'Histoire Naturelle du Creusot 1974, 12–22.

Gand, G., 1976. *Coelurosaurichnus palissyi*: une nouvelle trace dinosauroide des Grès à Empreintes du Plateau d'Antully (Saône-et-Loire–France). Bulletin Trimestriel de la Société d'Histoire Naturelle Musee d'Autun, N.S. 79, 11–14.

Gand, G., Pellier, J.-F., 1976. *Coelurosaurichnus sabinensis*: trace triasique nouvelle de la bordure N.-E. du Massif Central. Bulletin Trimestriel de la Société d'Histoire Naturelle Musee d'Autun, N.S. 79, 19–22.

Gand, G., Demathieu, G., 2005. The French Middle Triassic dinosauroid trackways: palaeontological result and nomenclatural re-evaluation. Geobios 38 (6), 725–749. https://doi.org/10.1016/j.geobios.2005.04.001.

Gand, G., Vianey-Liaud, M., Demathieu, G., Garric, J., 2000. Deux Nouvelles traces de pas de dinosaurs du Trias Supérieur de la bordure cévenole (La Grand-Combe, sud-est de la France). Geobios 33 (5), 599–624.

Gand, G., Fara, E., Durlet, C., Moreau, J.-D., Caravaca, G., Baret, L., André, D., Lefillatre, R., Passet, A., Wiénin, M., Gély, J.-P., 2018. Les pistes d'archosauriens : *Kayentapus ubacensis* nov. isp. (théropodes) et crocodylomorphes du Bathonien des Grands-Causses (France). Conséquences paléo-biologiques, environnementales et géographiques. Ann. Paleontol. 104, 183–216. https://doi.org/10.1016/j.annpal.2018.06.002.

Gangloff, R.A., May, K.C., Storer, J.E., 2004. An early Late Cretaceous dinosaur tracksite in central Yukon Territory, Canada. Ichnos 11 (3–4), 299–309. https://doi.org/10.1080/10420940490442269.

Gao, Y.-H., 2007. New dinosaur footprints from lower Jurassic of Weiyuan, Sichuan. Vertebr. Palasiat. 45 (4), 342–345.

García-Ramos, J., Piñuela, L., Ruiz-Omeñaca, J.I., 2006. Vertebrate tracks from the late Jurassic (Kimmeridgian) of Asturias (North Spain). J. Vertebr. Paleontol. 26 (Suppl. 3), 65A.

García-Ramos, J.C., Piñuela, L., Lires, J., 2011. Atlas del Jurásico de Asturias, second ed. Ediciones Nóbel, SA, Oviedo, Spain.

Gascón, J.H., Pérez-Lorente, F., 2017. Hoof-like unguals, skin, and foot movements deduced from *Deltapodus* casts of the Galve basin (upper Jurassic–lower Cretaceous, Teruel, Spain). Ichnos 24 (2), 146–161. https://doi.org/10.1080/10420940.2016.1223655.

Gatesy, S.M., Falkingham, P.L., 2020. Hitchcock's Leptodactyli, penetrative tracks, and dinosaur footprint diversity. J. Vertebr. Paleontol. e1781142 (18). https://doi.org/10.1080/02724634.2020.1781142.

Getty, P.R., Sproule, R., Stimson, M.R., Lyons, P.C., 2017a. Invertebrate trace fossils from the Pennsylvanian Rhode Island Formation of Massachusetts, USA. Atl. Geol. 53, 185–206. https://doi.org/10.4138/atlgeol.2017.007.

Getty, P.R., Aucoin, C., Fox, N., Judge, A., Hardy, L., Bush, A.M., 2017b. Perennial lakes as an environmental control on theropod movement in the Jurassic of the Hartford Basin. Geosciences 7 (13), 1–15. https://doi.org/10.3390/geosciences7010013.

Gierliński, G., 1991. New dinosaur ichnotaxa from the early Jurassic of the holy cross Mountains, Poland. Palaeogeogr. Palaeoclimatol. Palaeoecol. 85 (1– 2), 137–148. https://doi.org/10.1016/0031-0182(91)90030-U.

Gierliński, G., 1994. Early Jurassic theropod tracks with the metatarsal impressions. Przeglad Geol. 42 (4), 280–284.

Gierliński, G., 1996a. Avialian theropod tracks from the Early Jurassic strata of Poland. Zubia 14, 79–87.

Gierliński, G., 1996b. Dinosaur ichnotaxa from the lower Jurassic of Hungary. Geol. Q. 40 (1), 119–128.

Gierliński, G., 1996c. Feather-like impressions in a theropod resting trace from the Lower Jurassic of Massachusetts. In: Morales, M. (Ed.), The Continental Jurassic, vol. 60. Museum of Northern Arizona Bulletin, pp. 179–184.

Gierliński, G., 1999. Tracks of a large thyreophoran dinosaur from the Early Jurassic of Poland. Acta Palaeontol. Pol. 44 (2), 231–234.

Gierliński, G.D., 2009. A preliminary report on new dinosaur tracks in the Triassic, Jurassic and Cretaceous of Poland. Actas de las IV Jornadas Internacionales sobre Paleontología de Dinosaurios y su Entorno Salas de los Infantes. Burgos 75–90.

Gierliński, G.D., Potemska, A., 1987. Lower Jurassic dinosaur footprints from Gliniany las, northern slope of the holy cross Mountains, Poland. N. Jb. Geol. Paläont. Abh. 175 (1), 107–120.

Gierliński, G., Ahlberg, A., 1994. Late Triassic and early Jurassic dinosaur footprints in the Höganäs Formation of southern Sweden. Ichnos 3, 99–105.

Gierliński, G., Sabath, K., 1998. Protoavian affinity of the *Plesiornis* trackmaker. J. Vertebr. Paleontol. 18 (Suppl. 3), 46A.

Gierliński, G., Pieńkowski, G., 1999. Dinosaur track assemblages from the Hettangian of Poland. Geol. Q. 43 (3), 329–346.

Gierliński, G., Niedźwiedzki, G., 2002a. Dinosaur footprints from the upper Jurassic of Błaziny. Geol. Q. 46, 463–465.

Gierliński, G., Niedzwiedzki, G., 2002b. Enigmatic dinosaur footprints from the lower Jurassic of Poland. Geol. Q. 46, 467–472.

Gierliński, G., Sabath, K., 2002. A probable stegosaurian track from the Late Jurassic of Poland. Acta Palaeontol. Pol. 47, 561–564.

Gierliński, G., Niedzwiedzki, G., 2005. New saurischian dinosaur footprints from the Lower Jurassic of Poland. Geol. Q. 49, 99–104.

Gierliński, G.D., Nowacki, P., 2008. Middle Jurassic dinosaur track from the polish Jura chain. In: Uchman, A. (Ed.), Second International Congress on Ichnology Abstract Book and the Intra-Congress Field Trip Guidebook. Polish Geological Institute, Warsaw, p. 46.

Gierliński, G.D., Sabath, K., 2008. Stegosaurian footprints from the Morrison Formation of Utah and their implications for interpreting other ornithischian tracks. Oryctos 8, 29–46.

Gierliński, G., Gaździcka, E., Niedźwiedzki, G., Pieńkowski, G., 2001a. New ornithischian dinosaur footprints in the Jurassic of Poland. Geol. Q. 45, 205–210.

Gierliński, G., Niedźwiedzki, G., Pieńkowski, G., 2001b. Gigantic footprint of a theropod dinosaur in the Early Jurassic of Poland. Acta Palaeontol. Pol. 46 (3), 441–446.

Gierliński, G., Pieńkowski, G., Niedźwiedzki, G., 2004. Tetrapod track assemblage in the Hettangian of Sołtyków, Poland, and its paleoenvironmental background. Ichnos 11 (3–4), 195–213. https://doi.org/10.1080/10420940490444861.

Gierliński, G.D., Niedźwiedzki, G., Nowacki, P., 2009a. Small theropod and ornithopod footprints in the Late Jurassic of Poland. Acta Geol. Pol. 59 (2), 221–234.

Gierliński, G.D., Menducki, P., Janiszewska, K., Wicik, I., Boczarowski, A., 2009b. A preliminary report on dinosaur track assemblages from the Middle Jurassic of the Imilchil area, Morocco. Geol. Q. 53 (4), 477–482.

Gierliński, G.D., Lockley, M.G., Niedźwiedzki, G., 2009c. A distinctive crouching theropod trace from the Lower Jurassic of Poland. Geol. Q. 53, 471–476.

Gierliński, G., Szrek, P., Lockley, M.G., 2010. New finds of dinosaur tracks in the Morrison Formation of Moab area, Utah, USA. Earth Sci. Front. 17, 167–168.

Gierliński, G.D., Klein, H., Lagnaoui, A., Saber, H., Oukassou, M., Charrière, A., Adach, L.A., 2017a. Mid-Cretaceous dinosaur track assemblage from the High Moulouya region of Morocco. In: Gierliński, G.D., Lockley, M.G., Milner, A.R.C. (Eds.), Utah Friends of Paleontology Symposium and Annual Meeting Abstracts Book, pp. 15–17.

Gierliński, G.D., Lagnaoui, A., Klein, H., Saber, H., Oukassou, M., Charrière, A., 2017b. Bird-like tracks from the Imilchil formation (middle Jurassic, Bajocian–Bathonian) of the central high atlas, Morocco, in comparison with similar Mesozoic tridactylous ichnotaxa. Boll. Soc. Paleontol. Ital. 56, 207–215. https://doi.org/10.4435/BSPI.2017.19.

Gilmore, C.W., 1927. Fossil footprints from the Grand Canyon: second contribution. Smithsonian Misc. Collect. 80 (3), 1–78.

Ginsburg, L., de Lapparent, A.F., Loiret, B., Taquet, P., 1966. Empreintes de pas de Vertébrés tétrapodes dans les séries continentales à l'Ouest d'Agadès (République du Niger). Comptes Rendus de l'Acadmie des Sciences de Paris Série D 263, 28–31.

Gomes, J.P., 1916. Descoberta de rastros de saurios gigantescos no Jurássico do Cabo Mondego. Comunicacoes dos Serv. Geol. Port. 11, 132–134.

Goodrich, E.S., 1894. On the fossil Mammalia from the Stonesfield slate. Q. J. Microsc. Sci. 35, 407–432. https://doi.org/10.1242/jcs.s2-35.139.407.

Guarido, A., Castanera, D., Cobos, A., 2024. Icnitas de dinosaurios en el Jurásico Superior de La Puebla de Valverde (Teruel, España). Geogaceta 75, 55–58.

Guillaume, A.R.D., Costa, F., Mateus, O., 2022. Stegosaur tracks from the Upper Jurassic of Portugal: new occurrences and perspectives. Cien. Terra 20 (1), 37–60. https://doi.org/10.21695/cterraesj.v20i1.437.

Ha, S., Kim, K.S., Lim, H.S., Lockley, M.G., Yoo, J.S., Lim, J.D., 2022. Diminutive pterosaur tracks and trackways (*Pteraichnus gracilis* ichnosp. nov.) from the lower Cretaceous Jinju Formation, Gyeongsang basin, Korea. Cretac. Res. 131, 105080. https://doi.org/10.1016/j.cretres.2021.105080.

Häntzschel, W., 1975. Trace fossils and Problematica. In: Teichert, C. (Ed.), Treatise on Invertebrate Paleontology, Part W. Miscellanea, Suppl. 1. Geological Society of America, Boulder, Colorado, second ed. University of Kansas Press, Lawrence, Kansas, p. 269.

Harris, J.D., 1998. Dinosaur footprints from Garden Park, Colorado. Mod. Geol. 23, 291–307.

Harris, J.D., Lacovara, K.J., 2004. Enigmatic fossil footprints from the Sundance Formation (upper Jurassic) of Bighorn Canyon National Recreation area, Wyoming. Ichnos 11, 151–166.

Harris, J.D., Milner, A.R.C., 2015. Tracks in Deep Time: The St. George Dinosaur Discovery Site at Johnson Farm. University of Utah Press, Salt Lake City, p. 99.

Harris, J.D., Johnson, K.R., Hicks, J., Tauxe, L., 1996. Four-toed theropod footprints and a paleomagnetic age from the Whetstone falls member of the harebell formation (upper Cretaceous: Maastrichtian), northwestern Wyoming. Cretac. Res. 17, 381–401. https://doi.org/10.1006/cres.1996.0024.

Hasiotis, S.T., 2004. Reconnaissance of Upper Jurassic Morrison Formation ichnofossils, Rocky Mountain region, USA: paleoenvironmental, stratigraphic, and paleoclimatic significance of terrestrial and freshwater ichnocoenoses. Sediment. Geol. 167, 177–268.

Hasiotis, S.T., Wellner, R.W., Martin, A.J., Demko, T.M., 2004. Vertebrate burrows from Triassic and Jurassic continental deposits of North America and Antarctica: their paleoenvironmental and paleoecological significance. Ichnos 11, 103–124.

Hasiotis, S.T., Parrish, J.T., Chan, M.A., 2019. Vertebrate burrows in deposits of an eolian system, Lower Jurassic Navajo Sandstone, Moab, Utah, area, U.S.A. In: 3rd International Conference of Continental Ichnology, Abstract Volume & Field Trip Guide, pp. 32–34.

Hasiotis, S.T., Parrish, J.T., Chan, M.A., 2023. Using the diversity and distribution of vertebrate burrows to reconstruct paleoecological relationships in the Lower Jurassic Navajo Sandstone, southeastern Utah. Anat. Rec. 306 (Suppl. 1), 124–126. https://doi.org/10.1002/ar25219.

Haubold, H., 1968. Die Evolution der Archosaurier in der Trias aus der Sicht ihrer Fährten. Hercynia-Ökologie und Umwelt in Mitteleuropa 6 (1), 90–106.

Haubold, H., 1971a. Ichnia Amphibiorum et Reptiliorum fossilium. Handbuch der Paläoherpetologie Teil 18, 1–124.

Haubold, H., 1971b. Die Tetrapodenfährten des Buntsandsteins. Paläontologische Abhandlungen. Abt. B 4, 395–548.

Haubold, H., 1984. Saurierffihrten, (Neue Brehm-Bficherei, 479) Ziemsen, Wittenberg, 231 pp.

Haubold, H., 1986. Archosaur footprints at the terrestrial Triassic–Jurassic transition. In: Padian, K. (Ed.), The Beginning of the Age of Dinosaurs. Cambridge University Press, London, pp. 189–201.

Hay, O.P., 1902. Bibliography and Catalogue of the Fossil Vertebrata of North America. Government Printing Office, Washington, D.C., p. 868

He, X.L., Wang, C.S., Liu, S.Z., Zhou, F.Y., Liu, T.Q., Cai, K.J., Dai, B., 1998. A new sauropod dinosaur from the Early Jurassic in Gongxian County, south Sichuan. Acta Geol. Sichuan 18 (1), 1–7.

Heller, F., 1952. Reptilfährten-Funde aus dem Ansbacher Sandstein des Mittleren Keupers von Franken. Geol. Blatter Nordost-Bayern Angrenzende Geb. 2, 129–141.

Hendrickx, C., Bell, P.R., Pittman, M., Milner, A.R.C., Cuesta, E., O'Connor, J., Loewen, M., Currie, P.J., Mateus, O., Kaye, T.G., Delcourt, R., 2022. Morphology and distribution of scales, dermal ossifications, and other non-feather integumentary structures in non-avialan theropod dinosaurs. Biol. Rev. 97 (3), 960–1004. https://doi.org/10.1111/brv.12829.

Heredia, A.M., Díaz-Martínez, I., Pazos, P.J., de Valais, S., 2024. Pterosaur tracks from the Upper Cretaceous (Cenomanian) Candeleros Formation of northwestern Patagonia, Argentina: Ichnotaxonomic and palaeoecological perspectives from Gondwana. Palaeogeogr. Palaeoclimatol. Palaeoecol. 112338.

Hernández-Medrano, N., Pascual-Arribas, C., Pérez-Lorente, F., Sesma, R.S., 2014. Icnitas terópodas, ornitópodas y de pterosaurio en la Formación Aguilar del Río Alhama, Grupo de Oncala, Cuenca de Cameros. Zubia (32), 33–71.

Hernández-Medrano, N., Pascual-Arribas, C., Pérez-Lorente, F., 2017. First pterosaur footprints from the Tera Group (Tithonian–Berriasian) Cameros Basin, Spain. J. Iber. Geol. 43, 245–255.

Herrero, N., Pereda-Suberbiola, X., Herrero-Gascón, J., Pérez-Lorente, F., 2017. Pentadactyl ankylosaurian manus tracks from the Lower Cretaceous of Galve (Teruel, Spain): first occurrence of *Tetrapodosaurus* in the Iberian Peninsula. J. Iber. Geol. 43, 319–344. https://doi.org/10.1007/s41513-017-0024-4.
Higgs, R., 1988. Fish trails in the Upper Carboniferous of south-west England. Palaeontology 31 (2), 255–272.
Hips, K., Józsa, S., Nagy, Á., Pataki, Z., 1988. Őshüllők nyomában. Természet Világa 120, 108–111.
Hitchcock, C.H., 1890. Recent progress in ichnology. Proc. Boston Soc. Nat. Hist. 24, 117–127.
Hitchcock, E., 1836. Ornithichnology.—Description of the footmarks of birds (*Ornithichnites*) on New Red Sandstone in Massachusetts. Am. J. Sci. Arts 29, 307–340.
Hitchcock, E., 1837. Ornithichnites in Connecticut. American Joural of Science. Series 1 (31), 174–175.
Hitchcock, E., 1841. Final Report on the Geology of Massachusetts. Adam and J. H. Butler. Amherst and Northampton, 831 pp.
Hitchcock, E., 1843. Description of five new species of fossil footmarks, from the red sandstone of the valley of Connecticut River. Rep. First, Second, and Third Meetings, Assoc. Amer. Geol. Nat. 1843, 254–264.
Hitchcock, E., 1845. An attempt to name, classify, and describe, the animals that made the fossil footmarks of New England. In: Abstracts Proc. 6th Annual Meeting of the Association of American Geol. Naturalists, pp. 23–25.
Hitchcock, E., 1847. Description of two new species of fossil footmarks found in Massachusetts and Connecticut, or, of the animals that made them. Am. J. Sci. Arts (4), 46–57.
Hitchcock, E., 1848. An attempt to discriminate and describe the animals that made the fossil footmarks of the United States, and especially of New England. Memoirs Am. Acad. Arts Sci. N.S. 3, 129–255.
Hitchcock, E., 1856. On a new fossil fish, and new fossil footmarks. Am. J. Sci. 2nd series 20 (1856), 87–100.
Hitchcock, E., 1858. Ichnology of New England: A Report on the Sandstone of the Connecticut Valley, Especially its Fossil Footmarks. Commonwealth of Massachusetts, Boston, 220 pp.
Hitchcock, E., 1865. Supplement to the Ichnology of New England. Wright & Porter, State Printers, Boston, p. 96.
Hornung, J.J., Reich, M., 2014. Metatetrapous valdensis Nopcsa, 1923 and the presence of ankylosaur tracks (Dinosauria: Thyreophora) in the Berriasian (Early Cretaceous) of northwestern Germany. Ichnos 21 (1), 1–18.
Hu, D., Clarke, J.A., Eliason, C.M., Qiu, R., Li, Q., Shawkey, M.D., Zhao, C., D'Alba, L., Jiang, J., Xu, X., 2018. A bony-crested Jurassic dinosaur with evidence of iridescent plumage highlights complexity in early paravian evolution. Nat. Commun. 9, 217. https://doi.org/10.1038/s41467-017-02515-y.
Hudson, J.G., Romano, M., Lomax, D.R., Taylor, R., Woods, M., 2023. A new giant theropod dinosaur track from the Middle Jurassic of the Cleveland Basin, Yorkshire, UK. Proc. Yorks. Geol. Soc. 64. https://doi.org/10.1144/pygs2022-008.
Huene, F. von, 1931. Verschiedenen Wirbeltierreste aus Südamerika. Neues Jahrbuch für Mineralogie. Geol. Paläontol. 66 (B), 181–198.
Huene, F. von, 1932. Die fossile Reptil-Ordnung Saurischia, ihre Entwicklung und Geschichte. Monographien zur Geologie und Paläontologie 4, 1–361.
Huene, F. von, 1941. Die Tetrapoden-Fahrten im toskanischen Verrucano und ihre Bedeutung. Neues Jahrbuch für Mineralogie, Geologie und Paläontologie 1941B, 1–34.
Hunt, A.P., Lucas, S.G., 2006. The taxonomic status of *Navahopus falcipollex* and the ichnofauna and ichnofacies of the Navajo lithesome (Lower Jurassic) of western North America. New Mex. Mus. Nat. Hist. Sci. Bullet. 37, 164–169.
Hunt, A.P., Lucas, S.G., 2007. Tetrapod ichnofacies: a new paradigm. Ichnos 14 (1–2), 59–68. https://doi.org/10.1080/10420940601006826.
Hunt, A.P., Lucas, S.G., Klein, H., 2018. Late Triassic nonmarine vertebrate and invertebrate trace fossils and the patter of the Phanerozoic record of vertebrate trace fossils. In: Tanner, L.H. (Ed.), The Late Triassic World. Springer International Publishing AG, Cham, pp. 447–544.
Hunt-Foster, R.K., Lockley, M.G., Milner, A.R.C., Foster, J.R., Matthews, N.A., Breithaupt, B.H., Smith, J.A., 2016. Tracking dinosaurs in BLM Canyon Country, Utah. Geol. Interm. West 3, 67–100.
International Commission on Zoological Nomenclature (ICZN), 1999. International Code of Zoological Nomenclature, fourth ed. International Trust for Zoological Nomenclature, London. 124 pp.
International Commission on Zoological Nomenclature (ICZN), 2007. Opinion 2182 (Case 3348). *Palamopus* E. Hitchcock, 1845 (Ichnotaxa, Reptilia?): conserved. Bull. Zool. Nomencl. 64 (3), 203–204.
Irby, G.V., 1993. Early Jurassic dinosaur tracksites, northeastern Arizona. In: Boaz, D., Dornan, M. (Eds.), Fossils of Arizona. Proceedings of the 1st Annual Symposium. Mesa Southwest Museum, Mesa, pp. 15–25.

Irby, G.V., 1995. Posterolateral markings on dinosaur tracks, Cameron Dinosaur Tracksite, Lower Jurassic Moenave Formation, northeastern Arizona. J. Paleontol. 69 (4), 779–784.

Ishigaki, S., 1985. Dinosaur footprints from the Atlas Mountains. Nat. Stud. 31.

Ishigaki, S., 1988. Les empreintes de Dinosaures du Jurasique inférieur du Haut Atlas central marocain. Notes Serv. Geol. Maroc 44 (334), 79–86.

Ishigaki, S., 1989. Footprints of swimming Sauropods from Morocco. In: Gillette, D.D., Lockley, M.G. (Eds.), Dinosaur Tracks and Traces. Cambridge University Press., Cambridge, pp. 83–86.

Ishigaki, S., Matsumoto, Y., 2009a. "Off-tracking "-like phenomenon observed in the turning sauropod trackway from the Upper Jurassic of Morocco. Mem. Fukui Prefect. Dinosaur Mus. 8, 1–10.

Ishigaki, S., Matsumoto, Y., 2009b. Re-examination of manus-only and manus-dominated sauropod trackways from Morocco. Geol. Q. 53, 441–448.

Ishigaki, S., Lockley, M.G., 2010. Didactyl, tridactyl and tetradactyl theropod trackways from the Lower Jurassic of Morocco: evidence of limping, labouring and other irregular gait. Hist. Biol. 22, 100–108.

Jenny, J., Jossen, J.-A., 1982. Découverte d'empreintes des pas de dinosauriens dans le Jurassique Inférieur (Pliensbachien) di Haut-Atlas central (Maroc). Comptes Rendu de l'Académie des Science de Paris 294, 223–226.

Jovanović, M., 2013. Amphibia and Reptilia of Permian red sandstones (red-beds) from Glavica hill near Donji Milanovac (eastern Serbia). Permophiles 57, 20–22.

Kaever, M., de Lapparent, A.F., 1974. Les traces de pas de dinosaures du Jurassique de Barkhausen (Basse Saxe, Allemagne). Bull. Soc. Geol. Fr. VII 16, 516–525.

Kim, J.Y., Kim, K.S., Lockley, M.G., Yang, S.Y., Seo, S.J., Choi, H.I., Lim, J.D., 2008. New didactyl dinosaur footprints (*Dromaeosauripus hamanensis* ichnogen. et ichnosp. nov.) from the Early Cretaceous Haman Formation, south coast of Korea. Palaeogeogr. Palaeoclimatol. Palaeoecol. 262 (1–2), 72–78. https://doi.org/10.1016/j.palaeo.2008.02.003.

Kim, K.S., Lim, J.D., Lockley, M.G., Xing, L., Choi, Y., 2017. Korean trackway of a hopping, mammaliform trackmaker is first from the Cretaceous of Asia. Cretac. Res. 74, 188–191. https://doi.org/10.1016/j.cretres.2017.02.005.

King, M.J., Sarjeant, W.A.S., Thompson, D.B., Tresise, G., 2005. A revised systematic ichnotaxonomy and review of the vertebrate footprint ichnofamily Chirotheriidae from the British Triassic. Ichnos 12 (4), 241–299.

King, M.R., Botterill, S.E., Gingras, M.K., Pemberton, S.G., 2020. *Rhizocorallium* and turtle tracks: a late Cretaceous proximal distributary channel trace-fossil assemblage, central Utah. Ichnos 27 (4), 406–427. https://doi.org/10.1080/10420940.2020.1763337.

Kirkland, J.I., Foster, J., Hunt-Foster, R., Liggett, G.A., Trujillo, K., 2014a. Mid-Mesozoic: the Age of Dinosaurs in Transition. Abstr. Vol. Field Trip Guidebook 88.

Kirkland, J.I., Milner, A.R.C., Olsen, P.E., Hargrave, J.E., 2014b. The Whitmore Point Member of the Moenave Formation in its type area in northern Arizona and its correlation with the section in St. George, Utah: Evidence for two major lacustrine sequences. In: MacLean, J.S., Biek, R.F., Huntoon, J.E. (Eds.), Geology of Utah's Far South, vol. 43. Utah Geological Association, pp. 321–356.

Klein, H., Lucas, S.G., 2010. The Triassic footprint record of crocodylomorphs – A critical re-evaluation. New Mex. Mus. Nat. Hist. Sci. Bullet. 51, 55–60.

Klein, H., Niedźwiedzki, G., 2012. Revision of the Lower Triassic tetrapod ichnofauna from Wióry, Holy Cross Mountains, Poland. New Mex. Mus. Nat. Hist. Sci. Bullet. 56, 1–62.

Klein, H., Lucas, S.G., 2021. The Triassic tetrapod footprint record. New Mex. Mus. Nat. Hist. Sci. Bullet. 83, 1–194.

Klein, H., Lagnaoui, A., Gierliński, G.D., Saber, H., Lallensack, J.N., Oukassou, M., Charrière, A., 2018. Crocodylomorph, turtle and mammal tracks in dinosaur-dominated Middle–?Upper Jurassic and mid-Cretaceous ichnoassemblages of Morocco. Palaeogeogr. Palaeoclimatol. Palaeoecol. 498, 39–52. https://doi.org/10.1016/j.palaeo.2018.02.028.

Klein, H., Gierliński, G.D., Oukassou, M., Saber, H., Lallensack, J.N., Lagnaoui, A., Hminna, A., Charrière, A., 2022. Theropod and ornithischian dinosaur track assemblages from Middle to ?Late Jurassic deposits of the Central High Atlas, Morocco. Hist. Biol. 35 (3), 320–346. https://doi.org/10.1080/08912963.2022.2042808.

Kordos, L., 1983. Fontosabb szórványleletek a máfi gerinces-gyűjteményében (8. Közlemény). A Magyar Állami Földtani Intézet Évkönyve Jelentése 1981, 503–511.

Kozu, S., 2017. Dinosaur Footprints from the Khorat Group, Northeastern Thailand. Unpublished Ph.D. dissertation. University of Tsukuba, Tsukuba, 178 pp.

Krapovickas, V., Mancuso, A.C., Marsicano, C.A., Dommanovich, N.S., Schultz, C.L., 2013. Large tetrapod burrows from the Middle Triassic of Argentina: A behavioural adaptation to seasonal semi-arid climate? Lethaia 46, 154–169.

Krapovickas, V., Garrido, A.C., Canale, J.I., 2020. Dinosaur tracks at Cañadón de las Campanas and Villa El Chocón area (Late Cretaceous, Candeleros Formation), Patagonia, Argentina. Rev. Asoc. Geol. Argent. 77, 447–462.

Krummeck, W.D., Bordy, E.M., 2018. *Reniformichnus katikatii* (new ichnogenus and ichnospecies): Continental vertebrate burrows from the Lower Triassic, main Karoo Basin, South Africa. Ichnos 25 (2–3), 138–149. https://doi.org/10.1080/10420940.2017.1292909.

Kuhn, O., 1958a. Die Fährten der vorzeitlichen Amphibien und Reptilien. Bamberg Verlagshaus Meisenbach 64.

Kuhn, O., 1958b. Zwei neue Arten von *Coelurosaurichnus* aus dem Keuper Frankens. Neues Jahrbuch Geol. Palaontol. Monatsh. 1958, 437–440.

Kukihara, R., Lockley, M., Houck, K., 2010. Crocodile footprints from the Dakota Group (Cretaceous), John Martin Reservoir, Bent County, Colorado. New Mex. Mus. Nat. Hist. Sci. Bullet. 51, 121–136.

Lagnaoui, A., Klein, H., Voigt, S., Hminna, A., Saber, H., Schneider, J.W., Werneburg, R., 2012. Late Triassic tetrapod-dominated ichnoassemblages from the Argana Basin (Western High Atlas, Morocco). Ichnos 19 (4), 238–253. https://doi.org/10.1080/10420940.2012.718014.

Lallensack, J.N., Engler, T., Barthel, H.J., 2020. Shape variability in tridactyl dinosaur footprints: the significance of size and function. Palaeontology 63, 203–228. https://doi.org/10.1111/pala.12449.

Lallensack, J.N., Bordy, E.M., Lockley, M.G., Wings, O., 2022. Relaunching the TY tracksite: tridactyl dinosaur footprints from the Lower Jurassic of southern Africa. Hist. Biol. 35 (10), 1782–1793. https://doi.org/10.1080/08912963.2022.2117042.

Lee, Y.-N., Lee, H.-J., Lu, J., Kobayashi, Y., 2008. New pterosaur tracks from the Hasandong Formation (Lower Cretaceous) of Hadong County, South Korea. Cretac. Res. 29, 345–353.

Lee, Y.-N., Azuma, Y., Lee, H.-J., Shibata, M., Lu, J., 2010. The first pterosaur trackways from Japan. Cretac. Res. 31, 263–273.

Le Lœuff, J., Gourrat, C., Landry, P., Hautier, L., Liard, R., Souillat, C., Buffetaut, E., Enay, R., 2006. A Late Jurassic sauropod tracksite from southern Jura (France). Comptes Rendus Palevol 5, 705–709.

Leonardi, G., 1977. On a new occurrence of tetrapod trackways in the Botucatu Formation in the State of São Paulo, Brazil. Dusenia 10, 181–183.

Leonardi, G., 1979. Abundante Icnofauna (Vertebrados e Invertebrados) na Formação Botucatu s.s. em Araraquara, São Paulo. An Acad. Bras Ciências 51, 360–361.

Leonardi, G., 1980. On the discovery of an abundant ichno–fauna (vertebrates and invertebrates) in the Botucatu Formation *s.s.* in Araraquara, São Paulo, Brazil. An Acad. Bras Ciências 52, 559–567.

Leonardi, G., 1981. Novo icnogênero de tetrápode Mesozóico da Formação Botucatu, Araraquara, SP. An Acad. Bras Ciências 53 (4), 793–805.

Leonardi, G., 1987. Glossary and Manual of Tetrapod Footprint Palaeoichnology. Departmento Nacional da Producao Mineral. Brasilia, Brazil, 75 pp.

Leonardi, G., 1989. Inventory and statistics of the South American dinosaurian ichnofauna and its paleobiological interpretation. In: Gillette, D.D., Lockley, M.G. (Eds.), Dinosaur Tracks and Traces. Cambridge University Press, New York, pp. 165–178.

Leonardi, G., 1992. Sulle prime impronte fossili del Paraguay. Paleocronache 1992 (1), 66–67 (Milano).

Leonardi, G., 1994. Annotated Atlas of South America Tetrapod Footprints (Devonian to Holocene). Maio Grafica Editora Ecomercio LTDA, Rio de Janeiro, 246 pp.

Leonardi, G., de Souza Carvalho, I., 2020. Review of the early mammal *Brasilichnium* and *Brasilichnium*-like tracks from the Lower Cretaceous of South America. J. S. Am. Earth Sci. 106, 102940.

Leonardi, G., de Souza Carvalho, I., 2021. Vertebrate trace fossils: the Congo's *Brasilichnium* mammaloid fossil footprints. Ital. J. Geosci. 140 (1), 141–154. https://doi.org/10.3301/LIG.2020.24.

Leonardi, G., de Souza Carvalho, I., Fernandes, M.A., 2007. The desert ichnofauna from Botucatu Formation (Upper Jurassic–Lower Cretaceous), Brazil. In: de Souza Carvalho, I., de Cassia Tardin Cassab, R., de Carvalho, M.S.S., Arai, M., Oliveira, M.E.Q. (Eds.), Paleontologia: Cendrias de Vila, vol. 1. Rio de Janeiro, Interciència, pp. 371–383.

Lerner, A.J., Lucas, S.G., 2015. *Selenichnites* ichnoassociation from Early Permian tidal flats of the Prehistoric Trackways National Monument of south-central New Mexico. New Mex. Mus. Nat. Hist. Sci. Bullet. 65, 141–152.

Lessertisseur, J., 1955. Traces fossiles d'activite animale et leur signification paleobiologique. Mémoires de la Société Géologique de France N.S. 74, 1–150.
Li, D., Azuma, Y., Fujita, M., Lee, Y.-N., Arakawa, Y., 2006. A preliminary report on two new vertebrate track sites including dinosaurs from the Early Cretaceous Hekou Group, Gansu Province, China. J. Paleontol. Soc. Korea 22 (1), 29–49.
Li, J.J., 2015. Footprints of Mesozoic Reptilians and Avians—Palaeovertebr. Sin. Fasc. 8.
Li, R., Lockley, M.G., Makovicky, P.J., Matsukawa, M., Norell, M.A., Harris, J.D., Liu, M., 2007. Behavioral and faunal implications of Early Cretaceous deinonychosaur trackways from China. Naturwissenschaften 95, 185–191. https://doi.org/10.1007/s00114-007-0310-7.
Li, J., Bai, Z., Lockley, M., Zhou, B., Liu, J., Song, Y., 2010. Dinosaur tracks in Wulatezhongqi, Inner Mongolia. Acta Geol. Sin. 84 (5), 723–742.
Li, J., Lockley, M.G., Zhang, Y., Hu, S., Matsukawa, M., Bai, Z., 2012. An important ornithischian tracksite in the Early Jurassic of the Shenmu region, Shaanxi, China. Acta Geol. Sin. 86 (1), 1–10.
Li, Y.-X., Zhang, Y.-X., 2017. Early Middle Jurassic dinosaur footprints from Zizhou County, Shaanxi, China. Vertebr. Palasiat. 55 (3), 276–288.
Li, Y., Yao, H., Foster, W.J., Xing, L., Wang, C., Tahir, A., Khan, J., An, Z., Zhao, H., Wang, J., 2022. First discovery of Middle Jurassic theropod trackways in Changdu Region, Eastern Tibet of China. Earth Sci. 47 (11), 4223–4245.
Lichtig, A.J., Lucas, S.G., Klein, H., Lovelace, D.M., 2018. Triassic turtle tracks and the origin of turtles. Hist. Biol. 30, 1112–1122.
Lingham-Soliar, T., Broderick, T., 2000. An enigmatic Early Mesozoic dinosaur trackway from Zimbabwe. Ichnos 7 (2), 135–148. https://doi.org/10.1080/10420940009380152.
Lockley, M., 1986. A Guide to Dinosaur Tracksites of the Colorado Plateau and American Southwest. University of Colorado at Denver, Denver, p. 56.
Lockley, M.G., 1991. Tracking Dinosaurs: A New Look at an Ancient World. Cambridge University Press, Cambridge.
Lockley, M., 1998. Philosophical perspectives on theropod track morphology" blending qualities and quantities in the science of ichnology. Gaia 15, 279–300.
Lockley, M.G., 2001. Trackways—dinosaur locomotion. In: Briggs, D.E.G., Crowther, P. (Eds.), Paleobiology: ASynthesis. Wiley-Blackwell, Hoboken, New Jersey, pp. 412–416.
Lockley, M.G., 2007. The morphodynamics of dinosaurs, other archosaurs and their trackways: holistic insights into relationships between feet, limbs and the whole body. In: Bromley, R., Melchor, R. (Eds.), Ichnology at the Crossroads: AMultidimensional Approach to the Science of Organism—Substrate Interactions, vol. 88. Society of Economic Paleontologists and Mineralogists, Special Publication, Broken Arrow, Oklahoma, pp. 27–51.
Lockley, M.G., 2009. Some comparisons between dinosaur-dominated footprint assemblages in North America and Europe. In: Actas de las IV Jornadas Internacionales sobre Paleontologıa de Dinosaurios y su Entorno. Salas de los Infantes, Burgos, pp. 121–138.
Lockley, M.G., 2011b. The Ichnotaxonomic status of *Brasilichnium* with special reference to occurrences in the Navajo Sandstone (Lower Jurassic) in the western USA. New Mex. Mus. Nat. Hist. Sci. Bullet. 53, 306–315.
Lockley, M., 2021. Integration of tetrapod ichnofaunas and coastal dynamics for paleocommunity reconstruction in the Curtis and Summerville formations (Jurassic), eastern Utah. Hist. Biol. 34 (5), 784–801. https://doi.org/10.1080/08912963.2021.1946531.
Lockley, M.G., Santos, V.F., 1993. A preliminary report on sauropod trackways from the Avelino site, Sesimbra region, Upper Jurassic Portugal. Gaia 6, 38–42.
Lockley, M.G., Hunt, A.P., 1994. A track of the giant theropod dinosaur *Tyrannosaurus* from close to the Cretaceous/Tertiary Boundary, northern New Mexico. Ichnos 3, 213–218. https://doi.org/10.1080/10420949409386390.
Lockley, M.G., Hunt, A.P., 1995. Dinosaur Tracks and Other Fossil Footprints of the Western United States. Cambridge University Press, New York, p. 338.
Lockley, M.G., Mickelson, D.L., 1997. Dinosaur and pterosaur tracks in the Summerville and Bluff (Jurassic) beds of eastern Utah and northeastern Arizona. In: New Mexico Geological Society Guidebook, 48th Field Conference, Mesozoic Geology and Paleontology of the Four Corners Region, 1997, pp. 133–138.
Lockley, M.G., Hunt, A.P., 1998. A probable stegosaur track from the Morrison Formation of Utah. Mod. Geol. 23, 331–342.

Lockley, M., Meyer, C., 2000. Dinosaur Tracks and Other Fossil Footprints of Europe. Columbia University Press, New York, p. 323.

Lockley, M.G., Wright, J.L., 2003. Pterosaur swim tracks and other ichnological evidence of behavior and ecology. In: Buffetaut, E., Mazin, J.-M. (Eds.), Evolution and Palaeobiology of Pterosaurs, vol. 217. Geological Society, London, Special Publications, pp. 297–313.

Lockley, M.G., Meyer, C., 2004. Crocodylomorph trackways from the Jurassic to Early Cretaceous of North America and Europe: implications for ichnotaxonomy. Ichnos 11, 167–178.

Lockley, M.G., Foster, J.R., 2006. Dinosaur and turtle tracks from the Morrison Formation (Upper Jurassic) of Colorado National Monument, with observations on the taxonomy of vertebrate swim tracks. New Mex. Mus. Nat. Hist. Sci. Bullet. 36, 193–198.

Lockley, M.G., Gierliński, G.D., 2006. Diverse vertebrate ichnofaunas containing *Anomoepus* and other unusual trace fossils from the Lower Jurassic of the Western United States: Implications for paleoecology and palichnostratigraphy. New Mex. Mus. Nat. Hist. Sci. Bullet. 37, 176–191.

Lockley, M.G., Milner, A.R.C., 2006. Tetrapod tracksites from the Shinarump Formation (Chinle Group, Upper Triassic) of southwestern Utah. In: Harris, J.D., Lucas, S.G., Spielmann, J.A., Lockley, M.G., Milner, A.R.C., Kirkland, J.I. (Eds.), The Triassic–Jurassic Transition, vol. 37. New Mexico Museum of Natural History and Science Bulletin, pp. 257–262.

Lockley, M., Matsukawa, M., 2009. A review of vertebrate track distributions in east and southeast Asia. J. Paleontol. Soc. Korea 25 (1), 17–42.

Lockley, M.G., Tedrow, A., 2009. Are *Bipedopus, Semibipedopus, Lacertipus, Navahopus* and *Brasilichnium* distinct ichnogenera? Reevaluating Jurassic tracks from the Western USA. J. Vertebr. Paleontol. 29, 136A.

Lockley, M.G., Foster, J.R., 2010. An assemblage of probable crocodilian traces and associated dinosaur tracks from the lower Morrison Formation (Upper Jurassic) of eastern Utah. New Mex. Mus. Nat. Hist. Sci. Bullet. 51, 93–97.

Lockley, M.G., Gierliński, G.D., 2014. A new *Otozoum*-dominated tracksite in the Glen Canyon Group (Jurassic) of eastern Utah. New Mex. Mus. Nat. Hist. Sci. Bullet. 62, 211–214.

Lockley, M.G., Xing, L., 2015. Flattened fossil footprints: Implications for paleobiology. Palaeogeogr. Palaeoclimatol. Palaeoecol. 426, 85–94.

Lockley, M.G., Xing, L., 2021. A review of the non-avian theropod track record and the implications for the ontogenetic niche shift model. Earth Sci. Rev. 220, 103715. https://doi.org/10.1016/j.earscirev.2021.103715.

Lockley, M.G., Meyer, C.A., 2023. The megatracksite phenomenon: implications for tetrapod palaeobiology across terrestrial–shallow-marine transitional zones. Geol. Soc. 522, 285–324. https://doi.org/10.1144/SP522-2021-164.

Lockley, M.G., Houck, K.J., Prince, N.K., 1986. North America's largest dinosaur trackway site: implications for Morrison Formation paleoecology. Geol. Soc. Am. Bull. 97, 1163–1176.

Lockley, M.G., Yang, S.Y., Matsukawa, M., Fleming, F., Lim, S.K., 1992. The track record of Mesozoic birds: evidence and implications. Phil. Trans. Roy. Soc. Lond. B 336, 113–134.

Lockley, M.G., Farlow, J.O., Meyer, C.A., 1994a. *Brontopodus* and *Parabrontopodus* ichnogen. nov. and the significance of wide- and narrow-gauge sauropod trackways. Gaia 10, 135–145.

Lockley, M.G., Meyer, C.A., Santos, V.F., 1994b. Trackway evidence for a herd of juvenile sauropods from the Late Jurassic of Portugal. Gaia 10, 27–35.

Lockley, M.G., Logue, T.J., Moratalla, J.J., Hunt, A.P., Schultz, R.J., Robinson, J.W., 1995. The fossil trackway *Pteraichnus* is pterosaurian, not crocodilian: implications for the global distribution of pterosaur tracks. Ichnos 4, 7–20.

Lockley, M.G., Meyer, C.A., dos Santos, V.F., 1996a. *Megalosauripus, Megalosauropus* and the concept of megalosaur footprints. In: Morales, M. (Ed.), The Continental Jurassic, pp. 113–118. Museum of Northern Arizona Bulletin 60.

Lockley, M.G., Meyer, C.A., Schultz-Pittman, R., Forney, G., 1996b. Late Jurassic dinosaur tracksites from central Asia: a preliminary report on the world's longest trackways. In: Morales, M. (Ed.), The Continental Jurassic, pp. 137–140. Museum of Northern Arizona Bulletin 60.

Lockley, M., Hunt, A., Paquette, M., Bilbey, S.-A., Hamblin, A., 1998a. Dinosaur tracks from the Carmel Formation, northeastern Utah: implications for Middle Jurassic paleoecology. Ichnos 5 (4), 255–267. https://doi.org/10.1080/10420949809386424.

Lockley, M.G., dos Santos, V.F., Meyer, C., Hunt, A., 1998b. A new dinosaur tracksite in the Morrison Formation, Boundary Butte, southeastern Utah. Mod. Geol. 23, 317–330.

Lockley, M.G., Meyer, C.A., dos Santos, V.F., 1998c. *Megalosauripus* and the problematic concept of megalosaur footprints. Gaia 15, 313–337.

Lockley, M.G., Meyer, C.A., Moratalla, J.J., 1998d. *Therangospodus*: trackway evidence for the widespread distribution of a Late Jurassic theropod with well-padded feet. Gaia 15, 339–353.

Lockley, M.G., Foster, J.R., Hunt, A.P., 1998e. A short summary of dinosaur tracks and other fossil footprints from the Morrison Formation. Mod. Geol. 23, 277–290.

Lockley, M.G., Hunt, A.P., Meyer, C.M., Rainforth, E., Schultz, R., 1998f. A survey of fossil footprints at Glen Canyon National Recreation Area (western USA): a case study in documentation of trace fossil resources. Ichnos 5, 177–211.

Lockley, M.G., Meyer, C.A., Dos Santos, V.F., 2000. *Megalosauripus*, and the problematic concept of megalosaur footprints. Gaia 15, 313–337.

Lockley, M.G., Wright, J., White, D., Matsukawa, M., Jianjun, L., Lu, F., Hong, L., 2002. Cretaceous Research, vol. 23, pp. 363–381.

Lockley, M., Matsukawa, M., Li, J., 2003. Crouching theropods in taxonomic jungles: ichnological and ichnotaxonomic investigations of footprints with metatarsal and ischial impressions. Ichnos 10 (2–4), 169–177. https://doi.org/10.1080/10420940390256249.

Lockley, M.G., Wright, J.L., Thies, D., 2004a. Some observations on the dinosaur tracks at Münchehagen (Lower Cretaceous), Germany. Ichnos 11, 261–274. https://doi.org/10.1080/10420940490428805.

Lockley, M.G., Nadon, G., Currie, P.J., 2004b. A diverse dinosaur-bird footprint assemblage from the Lance Formation, Upper Cretaceous, eastern Wyoming: implications for ichnotaxonomy. Ichnos 11, 229–249.

Lockley, M.G., Lucas, S.G., Hunt, A.P., Gaston, R., 2004c. Ichnofaunas from the Triassic–Jurassic boundary sequences of the Gateway area, western Colorado: implications for faunal composition and correlations with other areas. Ichnos 11, 89–102.

Lockley, M.G., Lucas, S.G., Hunt, A.P., 2006a. *Evazoum* and the renaming of northern hemisphere '*Pseudotetrasauropus*': implications for tetrapod ichnotaxonomy at the Triassic–Jurassic boundary. New Mex. Mus. Nat. Hist. Sci. Bullet. 37, 199–206.

Lockley, M., Matsukawa, M., Ohira, H., Li, J., Wright, J., White, D., Chen, P., 2006b. Bird tracks from Liaoning Province, China: new insights into avian evolution during the Jurassic–Cretaceous transition. Cretac. Res. 27, 33–43.

Lockley, M.G., Mitchell, L., Odier, G.P., 2007a. Small theropod track assemblages from Middle Jurassic eolianites of eastern Utah: paleoecological insights from dune ichnofacies in a transgressive sequence. Ichnos 14 (1–2), 131–142. https://doi.org/10.1080/10420940601010901.

Lockley, M.G., Lires, J., García-Ramos, J.C., Pinuela, L., Avanzini, M., 2007b. Shrinking the world's largest dinosaur tracks: observations on the ichnotaxonomy of *Gigantosauropus asturiensis* and *Hispanosauropus hauboldi* from the Upper Jurassic of Asturias, Spain. Ichnos 14, 247–255. https://doi.org/10.1080/10420940601050048.

Lockley, M.G., Harris, J.D., Mitchell, L., 2008a. A global overview of pterosaur ichnology: tracksite distribution in space and time. Zitteliana B28, 185–198.

Lockley, M.G., Garcia-Ramos, J.C., Piñuela, L., Avanzini, M., 2008b. A review of vertebrate track assemblages from the Late Jurassic of Asturias, Spain with comparative notes on coeval ichnofaunas from the western USA: implications for faunal diversity in siliciclastic facies assemblages. Oryctos 8, 53–70.

Lockley, M.G., McCrea, R.T., Matsukawa, M., 2009. Ichnological evidence for small quadrupedal ornithischians from the basal Cretaceous of SE Asia and North America: implications for a global radiation. In: Buffetaut, E., Cuny, G., Le Lœuff, J., Suteethorn, V. (Eds.), Late Palaeozoic and Mesozoic Ecosystems in SE Asia, vol. 315. Geological Society of London Special Publication, pp. 255–269.

Lockley, M.G., Lucas, S.G., Milàn, J., Harris, J.D., Avanzini, M., Foster, J.R., Spielmann, J.A., 2010a. The fossil record of crocodilian tracks and traces: an overview. New Mex. Mus. Nat. Hist. Sci. Bullet. 51, 1–13.

Lockley, M.G., Li, R., Matsukawa, M., Li, J., 2010b. Tracking Chinese crocodylians: *Kuangyuanpus*, *Laiyangpus*, and implications for naming crocodylians and crocodylian-like tracks and associated ichnofacies. New Mex. Mus. Nat. Hist. Sci. Bullet. 51, 99–108.

Lockley, M.G., Tedrow, A.R., Chamberlain, K.C., Minter, N.J., Lim, J.-D., 2011. Footprints and invertebrate traces from a new site in the Nugget Sandstone (Lower Jurassic) of Idaho: Implications for life in the northern reaches of the great Navajo–Nugget erg system in the western USA. New Mex. Mus. Nat. Hist. Sci. Bullet. 53, 344–356.

Lockley, M.G., Li, J., Li, R., Matsukawa, M., Harris, J.D., Xing, L., 2013. A review of the tetrapod track record in China, with special reference to type ichnospecies: implications for ichnotaxonomy and paleobiology. Acta Geol. Sin. 87 (1), 1–20. https://doi.org/10.1111/1755-6724.12026.

Lockley, M., Cart, K., Martin, J., Prunty, R., Houck, K., Hups, K., Lim, J.-D., Kim, K.S., Gierliński, G., 2014a. A bonanza of new tetrapod tracksites from the Cretaceous Dakota Group, western Colorado: implications for paleoecology. New Mex. Mus. Nat. Hist. Sci. Bullet. 62, 393–409.

Lockley, M., Hups, K., Trujillo, K., Gerwe, S., 2014b. Mesozoic vertebrate tracksites from the Dominguez-Escalante National Conservation Area of western Colorado: implications for paleoecology and management. New Mex. Mus. Nat. Hist. Sci. Bullet. 62, 215–223.

Lockley, M.G., Kukihara, R., Pionek, L., Delgalvis, A., 2014c. A survey of new fossil footprint sites from Glen Canyon National Recreation Area (western USA), with special reference to the Kayenta–Navajo transition zone (Glen Canyon Group, Lower Jurassic). New Mex. Mus. Nat. Hist. Sci. Bullet. 62, 157–179.

Lockley, M.G., Gierlinski, G.D., Houck, K., Lim, J.D., Kim, K.S., Kim, D.-Y., Kim, T.K., Kang, S.H., Hunt Foster, R., Li, R., Chesser, C., Gay, R., Dubicka, Z., Cart, K., Wright, C., 2014d. New excavations at the Mill Canyon Dinosaur Tracksite (Cedar Mountain Formation, Lower Cretaceous) of eastern Utah. New Mex. Mus. Nat. Hist. Sci. Bullet. 62, 287–300.

Lockley, M.G., Harris, J.D., Li, R., Xing, L., van der Lubbe, T., 2016. Two-toed tracks through time: on the trail of "raptors" and their allies. In: Falkingham, P.L., Marty, D., Richter, A. (Eds.), Dinosaur Tracks: The Next Steps. Indiana University Press, Bloomington, pp. 183–200.

Lockley, M.G., Gierlinski, G., Matthews, Xing, L., Foster, J., Cart, K., 2017. New dinosaur track occurrences from the Upper Jurassic Salt Wash Member (Morrison Formation) of southeastern Utah: implications for thyreophoran trackmaker distribution and diversity. Palaeogeogr. Palaeoclimatol. Palaeoecol. 470, 116–121. https://doi.org/10.1016/j.palaeo.2016.12.047.

Lockley, M.G., Foster, J.R., Hunt-Foster, R.K., 2018a. The first North American *Deltapodus* trackway in a diverse *Anomoepus*, theropod, sauropod and turtle track assemblage from the Upper Jurassic Salt Wash Member (Morrison Formation) of eastern Utah. New Mex. Mus. Nat. Hist. Sci. Bullet. 79, 407–416.

Lockley, M., Burton, R., Grondel, L., 2018b. A large assemblage of tetrapod tracks from the Cretaceous Naturita Formation, Cedar Canyon region, southwestern Utah. Cretac. Res. 92, 108–121. https://doi.org/10.1016/j.cretres.2018.08.003.

Lockley, M.G., Cawthra, H.C., De Vynck, J.C., Helm, C.W., McCrea, R.T., Nel, R., 2019. New fossil sea turtle tracway morphotypes from the Pleistocene of South Africa highlight role of ichnology in turtle paleobiology. Quat. Res. 1–15.

Lockley, M.G., Lim, J.D., Park, H.D., Romilio, A., Yoo, J.S., Choi, J.W., Kim, K.S., Choi, Y., Kang, S.-H., Kim, D.H., Kim, T.H., 2020. First reports of *Crocodylopodus* from east Asia: implications for the paleoecology of the Lower Cretaceous. Cretac. Res. 111, 104441. https://doi.org/10.1016/j.cretres.2020.104441.

Lockley, M.G., Kim, K.S., Lim, J.D., 2021. *Tetrapodosaurus* trackways from the Cretaceous of Colorado and Canada: observations on variability in ankylosaurian gaits. New Mex. Mus. Nat. Hist. Sci. Bullet. 82, 227–235.

Lockley, M.G., Lallensack, J.N., Sciscio, L., Bordy, E.M., 2022. The early Mesozoic saurischian trackways *Evazoum* and *Otozoum*: implications for "prosauropod" (basal sauropodomorph) gaits. Hist. Biol. https://doi.org/10.1080/08912963.2022.2163170.

Long, J.A., 1998. Dinosaurs of Australia and New Zealand and Other Animals of the Mesozoic Era. UNSW Press, Sydney, 188 pp.

Lovelace, D.M., Lovelace, S.D., 2012. Paleoenvironments and paleoecology of a Lower Triassic invertebrate and vertebrate ichnoassemblage from the Red Peak Formation (Chugwater Group), central Wyoming. Palaios 27, 636–657.

Lu, J., Azuma, Y., Wang, T., Li, S., Pan, S., 2006. The first discovery of dinosaur footprint from Lufeng of Yunnan Province, China. Mem. Fukui Prefect. Dinosaur Mus. 5, 35–39.

Lu, J., Zhang, X., Jia, S., Hu, W., Wu, Y., Ji, Q., 2007. The discovery of theropod dinosaur footprints from the Middle Jurassic Yima Formation of Yima County, Henan Province. Acta Geol. Sin. 81 (4), 439–444.

Lucas, S.G., 2006. Comment on the proposed conservation of *Palamopus* E. Hitchcock, 1845 (Ichnotaxa, Reptilia?). Bull. Zool. Nomencl. 63 (1), 49–50.

Lucas, S.G., 2007. Tetrapod footprint biostratigraphy and biochronology. Ichnos 14, 5–38.

Lucas, S.G., Dalman, S.G., 2016. The Early Cretaceous Clayton Lake dinosaur tracksite, northeastern New Mexico. New Mex. Mus. Nat. Hist. Sci. Bullet. 74, 127–140.

Lucas, S.G., Lockley, M.G., Hunt, A.P., Tanner, L.H., 2006. Biostratigraphic significance of tetrapod footprints from the Triassic-Jurassic Wingate Sandstone on the Colorado Plateau. New Mex. Mus. Nat. Hist. Sci. Bullet. 37, 109–117.

Lucas, S.G., Spielmann, J.A., Klein, H., Lerner, A.J., 2010. Ichnology of the Upper Triassic (Apachean) Redonda Formation, east-central New Mexico. New Mex. Mus. Nat. Hist. Sci. Bullet. 47, 1–75.

Lull, R.S., 1904. Fossil footprints of the Jura–Trias of North America. Memoirs Boston Soc. Nat. Hist. 5 (11), 461–557.

Lull, R.S., 1915. Triassic life of the Connecticut Valley. State Geol. Nat. Hist. Surv. Connecticut Bullet. 24, 1–285.

Lull, R.S., 1918. Fossil footprints from the Grand Canyon of the Colorado. Am. J. Sci. 4th Ser. 45 (269), 337–346.

Lull, R.S., 1953. Triassic Life of the Connecticut Valley. State Geol. Nat. Hist. Surv. Connecticut Bullet. 81, 1–336.

Maidwell, F.T., 1911. Notes on footprints from the Keuper of Runcorn Hill. Proc. Liverp. Geol. Soc. 11, 140–152.

Malkani, M.S., 2007. Trackways evidence of sauropod dinosaurs confronted by a theropod found from Middle Jurassic Samana Suk Limestone of Pakistan. Sind. Univ. Res. J. (Sci. Ser.) 39 (1), 1–14.

Mampel, L., Cobos, A., Alcalá, L., Espílez, E., Royo-Torres, R., González, A., Gascó, F., 2011. Icnitas de dinosaurios en Aguilar del Alfambra (Teruel, España). Teruel 93, 41–54.

Manni, R., Nicosia, U., Nobili, G., 1999. An unusual tetrapod trackway from Lower Jurassic marine sediments of central Italy: *Accordiichnus natans* n. ichnogen. N. Ichnosp. Geol. Romana 35, 165–187.

Marchetti, L., Collareta, A., Belvedere, M., Leonardi, G., 2021. Ichnotaxonomy, biostratigraphy and palaeoecology of the Monti Pisani tetrapod ichnoassociation (Tuscany, Italy) and new insights on Middle Triassic Dinosauromorpha. Palaeogeogr. Palaeoclimatol. Palaeoecol. 567. https://doi.org/10.1016/j.palaeo.2021.110235.

Marsicano, C.A., Wilson, J.A., Smith, R.M.H., 2014. A temnospondyl trackway from the Early Mesozoic of Western Gondwana and its implications for basal tetrapod locomotion. PLoS One 9, e103255. https://doi.org/10.1371/journal.pone.0103255.

Marty, D., 2008. Sedimentology, taphonomy, and ichnology of Late Jurassic dinosaur tracks from the Jura carbonate platform (Chevenez–Combe Ronde tracksite, NW Switzerland): insights into the tidal-flat palaeoenvironment and dinosaur diversity, locomotion, and palaeoecology. GeoFocus Thèse 1608, 1–275.

Marty, D., Belvedere, M., Meyer, C.A., Mietto, P., Paratte, G., Lovis, C., Thüring, B., 2010. Comparative analysis of Late Jurassic sauropod trackways from the Jura Mountains (NW Switzerland) and the central High Atlas Mountains (Morocco): Implications for sauropod ichnotaxonomy. Hist. Biol. 22. https://doi.org/10.1080/08912960903503345.

Marty, D., Belvedere, M., Razzolini, N.L., Lockley, M.G., Paratte, G., Cattin, M., Lovis, C., Meyer, C.A., 2018. The tracks of giant theropods (*Jurabrontes curtedulensis* ichnogen. & ichnosp. nov.) from the Late Jurassic of NW Switzerland: palaeoecological & palaeogeographical implications. Hist. Biol. 30, 928–956. https://doi.org/10.1080/08912963.2017.1324438.

Masrour, M., Pérez-Lorente, F., 2014. *Otozoum* trackway in Issil-Aït-Arbi (Lower Jurassic, Central High Atlas, Morocco). Geogaceta 56, 107–110.

Masrour, M., Ladel, L., Pérez-Lorente, F., 2015. New theropod and prosauropod ichnites from Issil-Aït-Arbi (Lower Jurassic, Central High Atlas, Morocco). Geogaceta 57, 55–58.

Masrour, M., Boutakiout, M., Herrero Gascón, J., Sáinz Ruiz De Zuazo, J.L., Ochoa Martínez, R., Pérez-Lorente, F., 2020. Footprints of Batrachopus isp. from the Imilchil megatracksite Middle?–Upper Jurassic, central High Atlas (Morocco). J. Afr. Earth Sci. 172, 103980. https://doi.org/10.1016/j.afrearsci.2020.103980.

Matamales-Andreu, R., Mujal, E., Dinarès-Turell, J., Kustatscher, E., Roghi, G., Oms, O., Galobart, À., Fortuny, J., 2022. Early–middle Permian ecosystems of equatorial Pangaea: integrated multi-stratigraphic and palaeontological review of the Permian of Mallorca (Balearic Islands, western Mediterranean). Earth Sci. Rev. 228, 103948. https://doi.org/10.1016/j.earscirev.2022.103948.

Mateus, O., Milàn, J., 2010. First records of crocodile and pterosaur tracks in the Upper Jurassic of Portugal. New Mex. Mus. Nat. Hist. Sci. Bullet. 51, 83–87.

Mateus, O., Milàn, J., Romano, M., Whyte, M.A., 2011. New finds of stegosaur tracks from the Upper Jurassic Lourinhã Formation, Portugal. Acta Palaeontol. Pol. 56 (3), 651–658. https://doi.org/10.4202/app.2009.0055.

Matsukawa, M., Shibata, K., Kukihara, R., Koarai, K., Lockley, M.G., 2005. Review of Japanese dinosaur track localities: implications for ichnotaxonomy, paleogeography and stratigraphic correlation. Ichnos 12 (3), 201–222. https://doi.org/10.1080/10420940591009231.

May, K.C., Druckenmiller, P.S., 2009. A Dinosaurian Ichnofossil Assemblage from the Nanushuk Formation, National Petroleum Reserve–Alaska, p. 32. Alaska Geological Society Technical Conference Abstract vol. 2009.

Mazin, J.M., Hantzpergue, P., Lafaurie, G., Vignaud, P., 1995. Des pistes de ptérosaures dans le Tithonien de Crayssac (Quercy, France). Comptes rendus de l'Académie des sciences. Série 2. Sciences de la Terre et des Planètes 321 (5), 417–424.

Mazin, J.-M., Hantzpergue, P., Bassoullet, J.-P., Lafaurie, G., Vignaud, P., 1997. Le gisement de Crayssac (Tithonien inférieur, Quercy, Lot, France): découverte de pistes de dinosaurs en place et premier bilan ichnologique [The Crayssac site (lower Tithonian, Quercy, Lot, France): discovery of dinosaur trackways in situ and first ichnological results]. C.R. Académie des Sciences Paris. Sciences de la terre et des planets/Earth and Planetary Sciences 325, 733–739.

Mazin, J.M., Billon-Bruyat, J.P., Hantzpergue, P., Lafaurie, G., 2000. Domination reptilienne dans l'écosystème littoral de Crayssac (Tithonien inférieur, Quercy, Lot). Bull. Soc. Herpetol. Fr. 96, 71–81.

Mazin, J.M., Billon-Bruyat, J.P., Hantzpergue, P., Lafaurie, G., 2003. Ichnological evidence for quadrupedal locomotion in pterodactyloid pterosaurs: trackways from the Late Jurassic of Crayssac (southwestern France). Geol. Soc. 217 (1), 283–296.

Mazin, J.M., Billon-Bruyat, J.P., Padian, K., 2009. First record of a pterosaur landing trackway. Proc. Biol. Sci. 276 (1674), 3881–3886.

Mazin, J.-M., Hantzpergue, P., Pouech, J., 2016. The dinosaur tracksite of Loulle (early Kimmeridgian; Jura, France). Geobios 49 (3), 211–228. https://doi.org/10.1016/j.geobios.2016.01.018.

Mazin, J.-M., Hantzpergue, P., Olivier, N., 2017. The dinosaur tracksite of Plagne (early Tithonian, Late Jurassic; Jura Mountains, France): The longest known sauropod trackway. Geobios 50, 279–301. https://doi.org/10.1016/j.geobios.2017.06.004.

Mazin, J.-M., Pouech, J., 2020. The first non-pterodactyloid pterosaurian trackways and the terrestrial ability of non-pterodactyloid pterosaurs. Geobios 58, 39–53. https://doi.org/10.1016/j.geobios.2019.12.002.

McAllister, J.A., 1989. Subaqueous vertebrate footmarks from the upper Dakota Formation (Cretaceous) of Kansas, U.S.A. Occas. Pap. Mus. Nat. Hist. 127, 1–22.

McCrea, R.T., 2000. Vertebrate Palaeoichnology of the Lower Cretaceous (Lower Albian) Gates Formation of Alberta. Unpublished MS thesis. University of Saskatchewan, Saskatoon, 133 pp.

McCrea, R.T., Currie, P.J., 1998. A preliminary report on dinosaur tracksites in the Lower Cretaceous (Albian) Gates Formation near Grande Cache, Alberta. New Mex. Mus. Nat. Hist. Sci. Bullet. 14, 155–162.

McCrea, R.T., Lockley, M.G., Meyer, C.A., 2001. Global distribution of purported ankylosaur track occurrences. In: Carpenter, K. (Ed.), The Armored Dinosaurs. Indiana University Press, Bloomington, pp. 413–454.

McCrea, R.T., Lockley, M.G., Haines, P.W., Draper, N., 2011. Palaeontology Survey of the Broome Sandstone – Browse LNG Precinct Report. Department of State Development, Government of Western Australia, p. 120.

McCrea, R.T., Buckley, L.G., Plint, A.G., Currie, P.J., Haggart, J.W., Helm, C.W., Pemberton, S.G., 2014. A review of vertebrate track-bearing formations from the Mesozoic and earliest Cenozoic of western Canada with a description of a new theropod ichnospecies and reassignment of an avian ichnogenus. New Mex. Mus. Nat. Hist. Sci. Bullet. 62, 5–93.

Meijide Calvo, M., 2001. Pterosaur tracks in Oncala Berriasian (Soria, Spain) new ichnospecies: *Pteraichnus manueli*. Strata. Série 1. Communications 11, 72–74.

Meijide Fuentes, F., 2001. Pterosaur tracks in Oncala Mountain Range (Soria, Spain) a new ichnospecies: *Pteraichnus vetustior*. Strata. Série 1. Communications 11, 70–71.

Melchor, R.N., Rivarola, D.L., Umazano, A.M., Moyano, M.N., Belmontes, F.R.M., 2019. Elusive Cretaceous Gondwanan theropods: the footprint evidence from central Argentina. Cretac. Res. 97, 125–142.

Mensink, H., Mertmann, D., 1984. Dinosaurier-Fährten (*Gigantosauropus asturiensis* n. g. n. sp.; *Hispanosauropus hauboldi* n. g. n. sp.) im Jura Asturiens bei la Griega und Ribadesella (Spanien). Neues Jahrbuch Geol. Palaontol. Monatsh. 405–415.

Meyer, C.A., Pittman, J.G., 1994. A comparison between the *Brontopodus* ichnofacies of Portugal, Switzerland and Texas. Gaia 10, 125–133.

Meyer, C.A., Lockley, M.G., 1997. Jurassic and Cretaceous Dinosaur Tracksites from Central Asia (Uzbekistan and Turkmenistan), vol. 2. Paleontological Society of Korea Special Publication, pp. 77–92.

Meyer, C.A., Monbaron, M., 2002. Middle Jurassic dinosaur tracks from Morocco – Facts and fiction. In: 3rd European Workshop on Vertebrate Palaeontology, p. 27. Abstract Book, Sibiu (Romania).

Meyer, C.A., Thüring, B., 2003. Dinosaurs of Switzerland. Comptes Rendus Palevol. 2 (1), 103–117.

Meyer, C., Lockley, M.G., Robinson, J.W., Santos, V.F., 1994. A comparison of well-preserved sauropod tracks from the Late Jurassic of Portugal and the western United States: evidence and implications. Gaia 10, 57–64.

Meyer, C.A., Thüring, B., Wetzel, A., 2006. The hitch-hikers guide to the Late Jurassic– Basement structures provide clues to dinosaur migration routes. Hantkeniana 5, 38.

Meyer, C.A., Belvedere, M., Englich, B., Lockley, M.G., 2021. A reevaluation of the Late Jurassic dinosaur tracksite Barkhausen (Wiehengebirge, Northern Germany). PalZ 95, 537–558. https://doi.org/10.1007/s12542-021-00562-x.

Meyers, V.L., Breithaupt, B.H., 2014. *Pteraichnus saltwashensis* tracks and trackways in the Upper Jurassic Windy Hill Sandstone of Wyoming. New Mex. Mus. Nat. Hist. Sci. Bullet. 62, 235–248.

Mezga, A., Tesovic, B.C., Bajraktarevic, Z., 2007. First record of dinosaurs in the Late Jurassic of the Adriatic-Dinaridic carbonate platform (Croatia). Palaios 22, 188–199.

Mezga, A., Bucković, D., Šantak, F., 2017. New dinosaur tracksite in the Late Jurassic of kirmenjak quarry (Istria). Riv. Ital. Paleontol. Stratigr. 123, 443–454.

Mickelson, D.L., Lockley, M.G., Bishop, J., Kirkland, J., 2004. A new pterosaur tracksite from the Jurassic Summerville Formation, near Ferron, Utah. Ichnos 11, 125–142.

Mikuláš, R., 2010. *Characichnos* isp., probable crocodile swim traces from the Miocene of the České Budějovice basin, Czech Republic. New Mex. Mus. Nat. Hist. Sci. Bullet. 51, 179–181.

Milàn, J., 2011. New theropod, thyreophoran, and small sauropod tracks from the Middle Jurassic Bagå Formation, Bornholm, Denmark. Bull. Geol. Soc. Den. 59, 51–59.

Milàn, J., Bromley, R.G., 2005. Dinosaur footprints from the Middle Jurassic Bagå Formation. Bornholm Denmark 52, 7–15.

Milàn, J., Bromley, R.G., 2006. True tracks, undertracks and eroded tracks, experimental work with tetrapod tracks in laboratory and field. Palaeogeogr. Palaeoclimatol. Paleoecol. 231, 253–264.

Milàn, J., Loope, D.B., 2007. Preservation and erosion of theropod tracks in eolian deposits: examples from the Middle Jurassic Entrada Sandstone, Utah, U.S.A. J. Geol. 115, 375–386. https://doi.org/10.1086/512758.

Milàn, J., Chiappe, L.M., 2009. First American record of the Jurassic ichnospecies *Deltapodus brodricki* and a review of the fossil record of stegosaurian footprints. J. Geol. 117 (3), 343–348. https://doi.org/10.1086/597363.

Milàn, J., Surlyk, F., 2015. An enigmatic, diminutive theropod footprint in the shallow marine Pliensbachian Hasle Formation, Bornholm, Denmark. Lethaia 48 (4), 429–435. https://doi.org/10.1111/let.12115.

Milàn, J., Loope, D.B., Bromley, R.G., 2008. Crouching theropod and *Navahopus* sauropodomorph tracks from the Early Jurassic Navajo Sandstone of USA. Acta Palaeontol. Pol. 53 (2), 197–205.

Milner, A.R.C., Lockley, M.G., 2016. Dinosaur swim track assemblages: characteristics, contexts, and ichnofacies implications. In: Falkingham, P.L., Marty, D., Richter, A. (Eds.), Dinosaur Tracks: The Next Steps. Indiana University Press, Bloomington, pp. 153–180.

Milner, A.R.C., Lockley, M.G., Johnson, S.B., 2006a. The story of the St. George Dinosaur Discovery Site at Johnson Farm: An important new Lower Jurassic dinosaur tracksite from the Moenave Formation of southwestern Utah. New Mex. Mus. Nat. Hist. Sci. Bullet. 37, 329–345.

Milner, A.R.C., Lockley, M.G., Kirkland, J.I., 2006b. A large collection of theropod dinosaur swim tracks from the Lower Jurassic Moenave Formation, St. George, Utah. New Mex. Mus. Nat. Hist. Sci. Bullet. 37, 315–328.

Milner, A.R.C., Harris, J.D., Lockley, M.G., Kirkland, J.I., Matthews, N.A., 2009. Bird-like anatomy and behavior revealed by an Early Jurassic theropod resting trace. PLoS One 4 (3), e4591. https://doi.org/10.1371/journal.pone.0004591.

Milner, A.R.C., Birthisel, T.A., Kirkland, J.I., Breithaupt, B.H., Matthews, N.A., Lockley, M.G., Santucci, V.L., Gibson, S.Z., DeBlieux, D.D., Hurlbut, M., Harris, J.D., Olsen, P.E., 2011. Tracking Early Jurassic dinosaurs across southwestern Utahand the Triassic–Jurassic Transition. In: Bonde, J.W., Milner, A.R.C. (Eds.), Field Trip Guide Book for the 71st Annual Meeting of the Society of Vertebrate Paleontology, vol. 1. Nevada State Museum Paleontological Papers, pp. 1–107.

Milner, A.R.C., Marsh, A., Sues, H.-D., Buchwitz, J.C., Carter, H.A., Santucci, V., 2023a. Preliminary report of tritylodontid bonebeds in the Lower Jurassic Kayenta–Navajo Transition (Glen Canyon Group) at Glen Canyon National Recreation Area. Soc. Vertebr. Paleontol. Progr. Guide Cincinnati Ohio 303.

Milner, A.R.C., Santucci, V.L., Wood, J.R., Birthisel, T.A., Clites, E., Lockley, M.G., 2023b. The John Wesley Powell fossil track block–theropod tracks with ornithopod-like morphology from the Early Jurassic Navajo Sandstone, Glen Canyon National Recreation Area, Utah. Geol.e Interm. West 10, 185–222.

Milner, A.R.C., Santucci, V.L., Marsh, A.D., King, R., Harris, J.D., Delgalvis, A., Wood, J.R., Buchwitz, J.C., Carter, H.A., Kirkland, J.I., Slauf, D.L., Charobee, A.L., Bennett, C., Rodriguez, M., Tweet, J.S., Clites, E., 2024. Following in the footsteps of Dr. Martin G. Lockley: Another ten years of paleontological investigation within Glen Canyon National Recreation Area, Utah and Arizona. New Mex. Mus. Nat. Hist. Sci. Bullet. 96, 283–326.

Minter, N.J., Braddy, S.J., 2006. The fish and amphibian swimming traces *Undichna* and *Lunichnium*, with examples from the Lower Permian of New Mexico, USA. Palaeontology 49 (5), 1123–1142. https://doi.org/10.1111/j.1475-4983.2006.00588.x.

Mojica, J., Macia, C., 1987. Nota preliminar sobre la ocurrencia de improntas de vertebrados (*Batrachopus* sp.) en sedimentitas de la Formación Saldaña, region de Prado-Dolores, Valle Superior del Magdalena, Colombia. Geol. Colomb. 16, 89–94.

Molina-Pérez, R., Larramendi, A., 2019. Dinosaur Facts and Figures: The Theropods and Other Dinosauriformes. Princeton University Press, Princeton, 288 pp.

Montenat, C., 1968. Empreintes de pas de reptiles dans le Trias moyen du plateau du Daüs près d'Aubenas (Ardèche). Bull. Sci. Bourgogne 25, 369–389.

Montenat, C., Bessonnat, G., 2003. L'ichnofaune reptilienne hettangienne du Veillon (Vendée, France). Le Naturaliste Vendéen 3, 41–45.

Moratalla, J.J., 1993. Restos Indirectos de Dinosaurios del Registro Español: Paleoicnología de la Cuenca de Cameros (Jurásico Superior–Cretácico Inferior) y Paleoología del Cretácico Superior. Unpublished Ph.D. dissertation. Universidad Autonoma de Madrid, Madrid, 729 pp.

Moratalla, J.J., 2009. Sauropod tracks of the Cameros Basin (Spain): Identification, trackway patterns and changes over the Jurassic-Cretaceous. Geobios 42, 797–811.

Moratalla, J.J., Hernan, J., 2008. Evolution of the eastern Cameros vertebrate ichno-faunas through the Oncala and Enciso groups (Upper Jurassic–Lower Cretaceous, Cameros Basin, Spain). Doc. Labor. Géol. 164, 72–73.

Moratalla, J.J., Hernán, J., 2009. Turtle and pterosaur tracks from the Los Cayos dinosaur tracksite, Cameros Basin (Cornago, La Rioja, Spain): tracking the Lower Cretaceous bio-diversity. Rev. Esp. Palaontol. 24 (1), 59–77.

Moreau, J.-D., Gand, G., 2022. New data on the Permian ecosystem of the Rodez Basin: ichnofauna (traces of protostomians, tetrapods and fishes), jellyfishes and plants from Banassac-Canilhac (Lozère, southern France). Geodiversitas 44 (31), 975–987. https://doi.org/10.5252/geodiversitas2022v44a31.

Moreau, J.-D., Fara, E., Néraudeau, D., Gand, G., 2019. New Hettangian tracks from the Causses Basin (Lozère, southern France) complement the poor fossil record of earliest Jurassic crocodylomorphs in Europe. Hist. Biol. 31, 341–352.

Moreau, J.-D., Trincal, V., Fara, E., Baret, L., Jacquet, A., Barbini, C., Flament, R., Wienin, M., Bourel, B., Jean, A., 2020. Middle Jurassic tracks of sauropod dinosaurs in a deep karst cave in France. J. Vertebr. Paleontol. 39 (6), e1728286. https://doi.org/10.1080/02724634.2019.1728286.

Moreau, J.-D., Sciau, J., Gand, G., Fara, E., 2021. Uncommon preservation of dinosaur footprints in a tidal breccia: *Eubrontes giganteus* from the Early Jurassic Mongisty tracksite of Aveyron, southern France. Geol. Mag. https://doi.org/10.1017/S0016756820001454.

Moreno, K., Benton, M.J., 2005. Occurrence of Sauropod Dinosaur Tracks in the Upper Jurassic of Chile (Redescription of *Iguanodonichnus frenki*).

Moreno, K., Pino, M., 2002. Huellas de dinosaurios en la Formación Baños del Flaco (Titoniano-Jurásico Superior), VI Región, Chile: paleoetología y paleoambiente. Rev. Geol. Chile 29 (2), 191–206.

Moreno, K., Pino, M., 2010. Dinosaur footprints in the Baños del Flaco Formation (Tithonian-Upper Jurassic), VI Region, Chile: paleoethology and paleoenvironment. Andean Geol. 29 (2), 191–206.

Moser, M., 2003. *Plateosaurus engelhardti* MEYER, 1837 (Dinosauria: Sauropodomorpha) aus dem Feuerletten (Mittelkeuper; Obertrias) von Bayern. Zitteliana Reihe B 24, 1–186.

Moser, M., 2006. Comment on the proposed conservation of *Palamopus* E. Hitchcock, 1845 (Ichnotaxa, Reptilia?). Bull. Zool. Nomencl. 63 (2), 131–133.

Mudroch, A., Richter, U., Joger, U., Kosma, R., Idé, O., Maga, A., 2011. Didactyl tracks of paravian theropods (Maniraptora) from the ?Middle Jurassic of Africa. PLoS One 6 (2), e14642. https://doi.org/10.1371/journal.pone.0014642.

Mujal, E., Fortuny, J., Oms, O., Bolet, A., Galobart, A., Anadón, P., 2015. Palaeoenvironmental reconstruction and early Permian ichnoassemblage from the NE Iberian Peninsula (Pyrenean Basin). Geol. Mag. 153 (4), 578–600. https://doi.org/10.1017/S0016756815000576.

Mukaddam, R., Bordy, E.M., Lockley, M.G., Chapelle, K.E., 2020. Reviving *Kalosauropus*, and Early Jurassic sauropodomorph track from Southern Africa (Lesotho). Hist. Biol. 33 (11), 2908–2930.

Mustoe, G.E., 2019. Lower Eocene footprints from northwest Washington, USA. Part 1: Reptile tracks. Geosciences 9, 321. https://doi.org/10.3390/geosciences9070321.

Navarro-Lorbés, P., Díaz-Martínez, I., Valle-Melón, J.M., Rodríguez-Miranda, Á., Moratalla, J.J., Ferrer-Ventura, M., San Juan-Palacios, R., Torices, A., 2023. Dinosaur swim tracks from the Lower Cretaceous of La Rioja, Spain: an ichnological approach to non-common behaviours. Cretac. Res. 147, 105516. https://doi.org/10.1016/j.cretres.2023.105516.

Neto de Carvalho, C., 2016. *Psilonichnus* Fürsich, 1981 in its type-locality (Praia do Salgado, western Portugal). Comunicações Geológicas 103 (Especial 1), 13–22.

Niedźwiedzki, G., 2008. Theropod dinosaur tracks from the Rhaetian deposits of Seebergen, Thuringia, Germany. Przeglad Geol. 56, 539–544.

Niedźwiedzki, G., 2011. A Late Triassic dinosaur-dominated ichnofauna from the Tomanová Formation of the Tatra Mountains, Central Europe. Acta Palaeontol. Pol. 56 (2), 291–300.

Niedźwiedzki, G., Remin, Z., 2008. Gigantic theropod dinosaur footprints from the upper Pliensbachian of the Holy Cross Mountains, Poland. Prz. Geology 56, 823–825.

Niedźwiedzki, G., Soussib, M., Boukhalfac, K., Gierliński, G.D., 2017. Middle–Upper Triassic and Middle Jurassic tetrapod track assemblages of southern Tunisia, Sahara Platform. J. Afr. Earth Sci. 129, 31–44. https://doi.org/10.1016/j.jafrearsci.2016.12.006.

Nopcsa, F., 1923. Die Familien der Reptilien. Gebrüder Borntraeger.

Nouri, J., 2007. La paléoichnologie des empreintes de pas dinosauriens, imprimées dans les couches du Jurassique du Haut-Atlas Central du Maroc (PhD dissertation). Université Mohammed V, Faculté des Sciences, AGDAL, Rabat, Morocco.

Nouri, J., Díaz-Martínez, I., Pérez-Lorente, F., 2011. Tetradactyl footprints of an unknown affinity theropod dinosaur from the Upper Jurassic of Morocco. PLoS One 6, e26882. https://doi.org/10.1371/journal.pone.0026882.

Novikov, V.P., Radililovsky, V.V., 1984. New locations of dinosaur tracks in the river basin. Shirkent (Gissar ridge). Rep. Acad. Sci. Republ. Tajikistan 28 (10), 605–608.

Olsen, P.E., 1980. A comparison of the vertebrate assemblages from the Newark and Hartford basins (Early Mesozoic, Newark Supergroup) of eastern North America. In: Jacobs, L.L. (Ed.), Aspects of Vertebrate History: Essays in Honor of Edwin Harris Colbert. Museum of Northern Arizona Press, Flagstaff, pp. 35–53.

Olsen, P.E., 1988. Paleoecology and paleoenvironments of the continental early Mesozoic Newark Supergroup of eastern North America. In: Manspeizer, W. (Ed.), Triassic–Jurassic Rifting and the Opening of the Atlantic Ocean. Elsevier, Amsterdam, pp. 185–230.

Olsen, P.E., 1995. Paleontology and paleoenvironments of Early Jurassic age strata in the Walter Kidde Dinosaur Park (New Jersey, USA). In: Baker, J.E.B. (Ed.), Field Guide and Proceedings of the 12th Annual Meeting of the Geological Association of New Jersey. Geological Association of New Jersey, Wayne, pp. 156–190.

Olsen, P.E., Galton, P.M., 1984. A review of the reptile and amphibian assemblages from the Stormberg of southern Africa, with special emphasis on the footprints and the age of the Stormberg. Palaeontol. Afr. 25, 87–110.

Olsen, P.E., Baird, D., 1986. The ichnogenus *Atreipus* and its significance for Triassic biostratigraphy. In: Padian, K. (Ed.), The Beginning of the Age of Dinosaurs, Faunal Change Across the Triassic–Jurassic Boundary. Cambridge University Press, New York, pp. 61–87.

Olsen, P.E., Padian, K., 1986. Earliest records of *Batrachopus* from the southwestern United States, and a revision of some Early Mesozoic crocodylomorph ichnogenera. In: Padian, K. (Ed.), The Beginning of the Age of Dinosaurs. Cambridge University Press, London, pp. 259–273.

Olsen, P.E., Rainforth, E.C., 2001. The "Age of Dinosaurs" in the Newark Basin, with special reference to the Lower Hudson Valley. New York State Geol. Assoc. Guidebook 59–176.

Olsen, P.E., Rainforth, E.C., 2003. The Early Jurassic ornithischian dinosaurian ichnogenus *Anomoepus*. In: LeTourneau, P.M., Olsen, P.E. (Eds.), The Great Rift Valleys of Pangea in Eastern North America, vol. 2. Columbia University Press, New York, pp. 314–368.

Olsen, P.E., McDonald, N.G., Huber, P., Cornet, B., 1992. Stratigraphy and paleoecology of the Deerfield Rift Basin (Triassic–Jurassic, Newark Supergroup), Massachusetts. In: Robinson, P., Brady, J.B. (Eds.), Guidebook for Field Trips in the Connecticut Valley Region of Massachusetts and Adjacent States, Volume 2. New England Intercollegiate Geological Conference, 84th Annual Meeting, Amherst, Massachusetts, p. 535.

Olsen, P.E., Smith, J.B., McDonald, N.G., 1998. Type material of the type species of the classic theropod footprints genera *Eubrontes, Anchisauripus* and *Grallator* (Early Jurassic, Hartford and Deerfield Basins, Connecticut and Massachusetts, USA). J. Vertebr. Paleontol. 18, 586–601. https://doi.org/10.1080/02724634.1998.10011086.

Olsen, P.E., Kent, D.V., Sues, H.-D., Koeberl, C., Huber, H., Montanari, A., Rainforth, E.C., Fowell, S.J., Szajna, M.J., Hartline, B.W., 2002. Ascent of dinosaurs linked to an iridium anomaly at the Triassic–Jurassic boundary. Science 296, 1305–1307. https://doi.org/10.1126/science.106552.

Ősi, A., Pálfy, J., Makádi, L., Szentesi, Z., Gulyás, P., Rabi, M., Botfalvai, G., Hips, K., 2011. Hettangian (Early Jurassic) dinosaur tracksites from the Mecsek Mountains, Hungary. Ichnos 18, 79–94. https://doi.org/10.1080/10420940.2011.573603.

Ostrom, J.H., 1969. Osteology of *Deinonychus antirrhopus*, an unusual theropod from the Lower Cretaceous of Montana. Peabody Mus. Nat. Hist. Bullet. 30, 1–165.

Otero, A., Reguero, M., 2013. Dinosaurs (Reptilia, Archosauria) at Museo de La Plata, Argentina: annotated catalogue of the type material and Antarctic specimens. Paleontologia Electronica 16 (1), 3T.

Oukassou, M., Klein, H., Lagnaoui, A., Charrière, A., Saber, H., Gierliński, G.D., Lallensack, J.N., Hminna, A., Boumaalif, A., Oussou, A., Ouarhache, D., 2019. *Polyonyx*-like tracks from Middle–?Upper Jurassic red beds of Morocco: implications for sauropod communities on southern margins of Tethys. Palaeogeogr. Palaeoclimatol. Palaeoecol. 536, 109394. https://doi.org/10.1016/j.palaeo.2019.109394.

Oukassou, M., Zafaty, O., Gierliński, G.D., Klein, H., Saber, H., Amzil, M., Charrière, A., 2022. First record of a small stegosaur footprint (cf. *Stegopodus*) from the ?Upper Jurassic–?Lower Cretaceous red beds of the Middle Atlas, Morocco. Ichnos 29, 195–204.

Oussou, A., Falkingham, P.L., Butler, R.J., Boumir, K., Ouarhache, D., Ech-charay, K., Charrière, A., Maidment, S.C.R., 2023. New Middle to ?Late Jurassic dinosaur tracksites in the Central High Atlas Mountains, Morocco. R. Soc. Open Sci. 10. https://doi.org/10.1098/rsos.231091.

Pacheco, J.A.A., 1913. Notas sobre a geologia do Valle do Rio Grande a partir da fóz do Rio Pardo até a sua confluência com o Rio Parahyba. In: Comissão Geográphica e Geológica do Estado de São Paulo. Exploração do Rio Grande e de seus afluentes. São José dos Dourados. Comissão Geográphica e Geológica do Estado de São Paulo, São Paulo, pp. 33–38.

Pacyna, G., Ziaja, J., Barbacka, M., Pieńkowski, G., Jarzynka, A., Niedźwiedzki, G., 2022. Jurassic dinosaur-dominated track assemblages, floristic and environmental changes in the Holy Cross Mountains region, Poland. Geol. Q. 66, 29. https://doi.org/10.7306/gq.1660.

Padian, K., Olsen, P.E., 1984. The fossil trackway *Pteraichnus*: not pterosaurian, but crocodilian. J. Paleontol. 58, 178–184.

Panciroli, E., Romano, M., 2021. Enigmatic vertebrate trackway from the Scalby Formation (Middle Jurassic) Yorkshire, United Kingdom, with discussion of archosaur and 'mammal' trace fossils. Ichnos 28 (2), 97–113. https://doi.org/10.1080/10420940.2021.1930537.

Panciroli, E., Benson, R.B.J., Walsh, S., Butler, R.J., Castro, T.A., Jones, M.E.H., Evans, S.E., 2020. Diverse vertebrate assemblage of the Kilmaluag Formation (Bathonian, Middle Jurassic) of Skye, Scotland. Earth Environ. Trans. Royal Soc. Edinburgh 111, 135–156. https://doi.org/10.1017/S1755691020000055.

Pascual Arribas, C., Latorre Macarrón, P., 2000. Huellas de *Eubrontes* y *Anchisauripus* en Carrascosa de Arriba (Soria, España). Bol. Geol. Min. 111 (1), 21–32.

Pascual Arribas, C., Sanz Pérez, E., 2000. Huellas de Pterosaurios en el grupo Oncala (Soria, España). *Pteraichnus palaciei-saenzi*, nov. icnosp. Estud. Geol. 56 (1–2), 73–100.

Pascual-Arribas, C., Hernández-Medrano, N., 2012. Asociación de huellas de pterosaurio en el yacimiento de Los Tormos (Soria, España): implicaciones icnotaxonómicas. Stud. Geol. Salmant. 48 (1), 7–36.

Pascual-Arribas, C., Hernández-Medrano, N., 2015a. Nuevas huellas de estegosáuridos en el Titoniense–Berriasiense de la Cuenca de Cameros (Formación Magaña). Rev. Soc. Geol. Espana 28, 15–27.

Pascual-Arribas, C., Hernández-Medrano, N., 2015b. Las huellas de tortuga del Grupo Oncala (Berriasiense, Cuenca de Cameros, España). Estud. Geol. 71 (1), e030. https://doi.org/10.3989/egeol.41821.341.

Pascual-Arribas, C., Hernández-Medrano, N., 2016. Huellas de *Pteraichnus* en la Muela (Soria, España): consideraciones sobre el icnogénero y sobre la diversidad de huellas de pterosaurios en la Cuenca de Cameros. Rev. Soc. Geol. Espana 29 (2), 89–105.

Pascual-Arribas, C., Canudo, J.I., Sanz Perez, E., Hernández-Medrano, N., Castanera, D., Barco, J.L., 2015. On the validity of *Pteraichnus palacieisaenzi*, Pascual Arribas and Sanz Pérez 2000: new data from the Huérteles formation. Paläontol. Z. 89, 459–483.

Pascual, C., Canudo, J.I., Hernández, N., Barco, J.L., Castanera, D., 2012. First record of stegosaur dinosaur tracks in the Lower Cretaceous (Berriasian) of Europe (Oncala Group, Soria, Spain). Geodiversitas 34, 297–312. https://doi.org/10.5252/g2012n2a4.

Pazos, P.J., 2009. Sintesis icnológica de unidades Mesozoica marinas de la Cuenca Neuquina, nuevos datos y perspectivas. Rev. Asoc. Geol. Argent. 65 (2), 362–372.

Pazos, P.J., González Estebenet, M.C., Cocca, S.E., Pascua, D., 2019. The oldest record of a thyreophoran track in Gondwana: geological implications of subaerial exposure in the lower part of the Lajas Formation at the Covunco section (Neuquén Basin), Patagonia, Argentina. J. S. Am. Earth Sci. 94, 102198. https://doi.org/10.1016/j.jsames.2019.05.014.

Pazos, P.J., Lazo, D.G., Tunik, M.A., Marsicano, C.A., Fernández, D.E., Aguirre-Urreta, M.B., 2012. Paleoenvironmental framework of dinosaur tracksites and other ichnofossils in Early Cretaceous mixed siliciclastic-carbonate deposits in the Neuquén Basin, northern Patagonia (Argentina). Gondwana Res. 22 (3–4), 1125–1140. https://doi.org/10.1016/j.gr.2012.02.003.

Petti, F.M., Avanzini, M., Antonelli, M., Bernardi, M., Leonardi, G., Manni, R., Mietto, P., Pignatti, J., Piubelli, D., Sacco, E., Wagensommer, A., 2020. Jurassic tetrapod tracks from Italy: a training ground for generations of researchers. In: Romano, M., Citton, P. (Eds.), Tetrapod Ichnology in Italy: The State of the Art, vol. 12, pp. 31–52. Journal of Mediterranean Earth Sciences.

Pieńkowski, G., Niedźwiedzki, G., 2005. Pterosaur tracks from the early Kimmeridgian intertidal deposits of Wierzbica, Poland. Geol. Q. 49 (3), 339–346.

Pieńkowski, G., Niedźwiedzki, G., Brański, P., 2014. Climatic reversals related to the Central Atlantic magmatic province caused the end-Triassic biotic crisis—evidence from continental strata in Poland. In: Keller, G., Kerr, A. (Eds.), Volcanism, Impacts, and Mass Extinctions: Causes and Effects, vol. 505. Geological Society of America Special Paper, p. 13. https://doi.org/10.1130/2014.2505(13).

Piñuela, L., García-Ramos, J.C., Fernández, L.A., Ruiz-Omeñaca, J.I., 2014. Nuevos rastros de estegosaurios en el Jurásico Superior de Asturias. In: Royo-Torres, R., Verdú, F.J., Alcalá, L. (Eds.), XXX Jornadas de Paleontología de La Sociedad Española de Paleontología, vol. 24. ¡Fundamental!, pp. 187–190.

Piñuela, L., García-Ramos, J.C., Romano, M., Ruiz-Omeñaca, J.I., 2016. First record of gregarious behavior in robust medium-sized Jurassic ornithopods: evidence from the Kimmeridgian trackways of Asturias (N. Spain) and some general considerations on other medium–large ornithopod tracks in the Mesozoic record. Ichnos 23, 298–311. https://doi.org/10.1080/10420940.2016.1178640.

Piñuela Suárez, L., 2015. Huellas de dinosaurios y de otros reptiles del Jurásico Superior de Asturias. Unpublished Ph.D. Thesis. Oviedo University, Oviedo, p. 326.

Piubelli, D., 2002. Nuove indagini paleoicnologiche sul sito Giurassico dei Lavini di Marco (Rovereto – Trento). Universit`a di Padova, p. 140 (Ph.D. thesis).

Piubelli, D., Avanzini, M., Mietto, P., 2005. The Early Jurassic ichnogenus *Kayentapus* at Lavini di Marco ichnosite (NE Italy). Global distribution and palaeogeographic implications. Boll. Soc. Geol. It. 124, 259–267.

Pond, S., Lockley, M.G., Lockwood, J.A.F., Breithaupt, B.H., Matthews, N.A., 2014. Tracking dinosaurs on the Isle of Wight: a review of tracks, sites, and current research. Biol. J. Linn. Soc. 113 (3), 737–757. https://doi.org/10.1111/bij.12340.

Popa, M.E., 2000. First find of Mesozoic tetrapod tracks in Romania. Acta Palaeontologica Romaniae 2, 387–390.

Popa, M.E., Kedzior, A., 2006. Preliminary ichnological results on the Steierdorf Formation in Anina, Romania. In: Csiki, Z. (Ed.), Mesozoic and Cenozoic Vertebrates and Paleoenvironments: Tributes to the Career of Prof. Dan Grigorescu. Bucharest, Ars Docendi, pp. 197–201.

Poropat, S.F., White, M.A., Ziegler, T., Pentland, A.H., Rigby, S.L., Duncan, R.J., Sloan, T., Elliott, D.A., 2021. A diverse Late Cretaceous vertebrate tracksite from the Winton Formation of Queensland, Australia. PeerJ 9, e11544. https://doi.org/10.7717/peerj.11544.

Raath, M.A., Kitching, J.W., Shone, R.W., Rossouw, G.J., 1990. Dinosaur tracks in Triassic Molteno sediments: the earliest evidence of dinosaurs in South Africa. Palaeontol. Afr. 27, 89–95.

Rampersadh, A., Bordy, E.M., Sciscio, L., Abrahams, M., 2018. Dinosaur behaviour in an Early Jurassic palaeoecosystem—uppermost Elliot Formation, Ha Nohana, Lesotho. Ann. Soc. Geol. Pol. 88 (2), 163–179.

Rainforth, E.C., 1997. Vertebrate Ichnological Diversity and Census Studies, Lower Jurassic Navajo Sandstone. Unpublished Masters thesis. Department of Geological Sciences, University of Colorado, Boulder, 49 pp.

Rainforth, E.C., 2003. Revision and re-evaluation of the Early Jurassic dinosaurian ichnogenus *Otozoum*. Palaeontology 46, 803–838.

Rainforth, E.C., 2005a. Ichnotaxonomy of the Fossil Footprints of the Connecticut Valley (Early Jurassic, Newark Supergroup, Connecticut and Massachusetts). Unpublished Ph.D. dissertation. Columbia University, New York, 1301 pp.

Rainforth, E.C., 2005b. Case 3348: *Palamopus* E. Hitchcock, 1845 (Ichnotaxa, Reptilia?): proposed conservation. Bull. Zool. Nomencl. 62 (4), 237–239.

Rainforth, E.C., 2007. Ichnotaxonomic updates from the Newark Supergroup. In: Contributions to the Paleontology of New Jersey (II): Field Guide and Proceedings. Geological Association of New Jersey, Trenton, pp. 49–59.

Raisanen, D.C.W., Hasiotis, S.T., 2018. New ichnotaxa of vertebrate burrows from the Salt Wash Member, Upper Jurassic Morrison Formation, south-eastern Utah (Utah). Ann. Soc. Geol. Pol. 88, 181–202.

Rauhut, O.W.M., Piñuela, L., Castanera, D., García-Ramos, J.-C., Cela, I.S., 2018. The largest European theropod dinosaurs: remains of a gigantic megalosaurid and giant theropod tracks from the Kimmeridgian of Asturias, Spain. PeerJ 6, e4963. https://doi.org/10.7717/peerj.4963.

Razzolini, N.L., Oms, O., Castanera, D., Vila, B., Santos, V.F., Galobart, À., 2016. Ichnological evidence of megalosaurid dinosaurs crossing Middle Jurassic tidal flats. Sci. Rep. 6, 31494. https://doi.org/10.1038/srep31494.

Razzolini, N.L., Belvedere, M., Marty, D., Paratte, G., Lovis, C., Cattin, M., Meyer, C.A., 2017. *Megalosauripus transjuranicus* ichnosp. nov. A new Late Jurassic theropod ichnotaxon from NW Switzerland and implications for tridactyl dinosaur ichnology and ichnotaxomy. PLoS One 12 (7), e0180289.

Reolid, J., Reolid, M., 2017. Traces of floating archosaurs: an interpretation of the enigmatic trace fossils from the Triassic of the Tabular Cover of southern Spain. Ichnos 24 (3), 222–233. https://doi.org/10.1080/10420940.2016.1265524.

Reynolds, R.E., 2006. Jurassic tracks in California. In: Reynolds, R.E. (Ed.), Making Tracks across the Southwest, Abstracts from the 2006 Desert Symposium, California State University, Desert Studies Consortium and. LSA Associates Inc., pp. 19–24

Riese, D.J., Hasiotis, S.T., Odier, G.P., 2011. Synapsid burrows and associated trace fossils in the Lower Jurassic Navajo Sandstone, southeastern Utah, U.S.A., indicates a diverse community living in a wet desert environment. J. Sediment. Res. 81, 299–325.

Riguetti, F., Citton, P., Apesteguía, S., Zacarías, G.G., Pereda-Suberbiola, X., 2021. New ankylosaurian trackways (cf. *Tetrapodosaurus*) from an uppermost Cretaceous level of the El Molino Formation of Bolivia. Cretac. Res. 124, 104810. https://doi.org/10.1016/j.cretres.2021.104810.

Rodríguez-de la Rosa, R.A., Velasco-de León, M.P., Arellano-Gil, J., Lozano-Carmona, D.E., 2018. Middle Jurassic ankylosaur tracks from Mexico. Bol. Soc. Geol. Mex. 70, 379–395. https://doi.org/10.18268/BSGM2018v70n2a8.

Romano, M., Whyte, M.A., 2003. Jurassic dinosaur tracks and trackways of the Cleveland Basin, Yorkshire: preservation, diversity and distribution. Proc. Yorks. Geol. Soc. 54, 185–215. https://doi.org/10.1144/pygs.54.3.185.

Romano, M., Whyte, M.A., 2010. Crocodilian and other non-dinosaurian tracks and trackways from the Ravenscar Group (Middle Jurassic) of the Cleveland Basin, Yorkshire, UK. New Mex. Mus. Nat. Hist. Sci. Bullet. 51, 69–81.

Romano, M., Whyte, M.A., 2012. Information on the Foot Morphology, Pedal Skin Texture and Limb Dynamics of Sauropods: Evidence from the Ichnological Record of the Middle Jurassic of the Cleveland Basin, vol. 45. Yorkshire, UK. Zubia.

Romano, M., Whyte, M.A., 2015. Could stegosaurs swim? Suggestive evidence from the Middle Jurassic tracksite of the Cleveland Basin, Yorkshire, UK. Proc. Yorks. Geol. Soc. 60, 227–233. https://doi.org/10.1144/pygs2015-354.

Romano, M., Whyte, M.A., Manning, P.L., 1999. New sauropod dinosaur ichnites from the Saltwick Formation (Middle Jurassic) of the Cleveland Basin, Yorkshire. Proc. Yorks. Geol. Soc. 52, 361–369.

Romano, M., Whyte, M.A., Jackson, S.J., 2007. Trackway ratio: a new look at trackway gauge in the analysis of quadrupedal dinosaur trackways and its implications for ichnotaxonomy. Ichnos 14 (3–4), 257–270. https://doi.org/10.1080/10420940601050014.

Romano, M., Clark, N.D.L., Brusatte, S.L., 2018. A comparison of the dinosaur communities from the Middle Jurassic of the Cleveland (Yorkshire) and Hebrides (Skye) basins, based on their ichnites. Geosciences 8, 327. https://doi.org/10.3390/geosciences8090327.

Romanovsky, G.D., 1882. Geological character of the Sarvadan Lignite Formation in the Zeravshan district. Rec. Imperial St. Petersburg Mineral. Soc. 17, 276–296.

Romilio, A., Godfrey, T., 2021. A new dinosaur tracksite from the Lower Cretaceous (Aptian–Albian) Eumeralla Formation of Wattle Hill, Victoria, Australia: a preliminary investigation. Hist. Biol. 34 (12), 2315–2323. https://doi.org/10.1080/08912963.2021.2014481.

Romilio, A., Tucker, R.T., Salisbury, S.W., 2013. Reevaluation of the Lark Quarry dinosaur tracksite (Late Albian–Cenomanian Winton Formation, central-western Queensland, Australia): no longer a stampede? J. Vertebr. Paleontol. 33 (1), 102–120. https://doi.org/10.1080/02724634.2012.694591.

Romilio, A., Dick, R., Skinner, H., Millar, J., 2020. Archival data provides insights into the ambiguous track-maker gait from the Lower Jurassic (Sinemurian) Razorback beds, Queensland, Australia: evidence of theropod quadrupedalism? Hist. Biol. 33 (9), 1573–1579. https://doi.org/10.1080/08912963.2020.1720014.

Romilio, A., Salisbury, S.W., Jannel, A., 2021. Footprints of large theropod dinosaurs in the Middle–Upper Jurassic (lower Callovian–lower Tithonian) Walloon Coal Measures of southern Queensland, Australia. Hist. Biol. 33 (10), 2135–2146. https://doi.org/10.1080/08912963.2020.1772252.

Romilio, A., Klein, H., Jannel, A., Salisbury, S.W., 2022. Saurischian dinosaur tracks from the Upper Triassic of southern Queensland: possible evidence for Australia's earliest sauropodomorph trackmaker. Hist. Biol. 34 (9), 1834–1843. https://doi.org/10.1080/08912963.2021.1984447.

Romilio, A., Dick, R., Skinner, H., Millar, J., 2024. Uncovering hidden footprints: revision of the Lower Jurassic (Sinemurian) Razorback Beds—home to Australia's earliest reported dinosaur trackway. Hist. Biol. https://doi.org/10.1080/08912963.2024.2320184.

Rowland, S.M., Mercadante, J.M., 2014. Trackways of a gregarious, dunefield-dwelling, Early Jurassic therapsid in the Aztec Sandstone of southern Nevada. Palaios 29, 539–552.

Royo-Torres, R., Cobos, A., Mocho, P., Alcalá, L., 2021. Origin and evolution of turiasaur dinosaurs set by means of a new 'Rosetta' specimen from Spain. Zool. J. Linn. Soc. 191, 201–227. https://doi.org/10.1093/zoolinnean/zlaa091.

Rozhdestvensky, A.K., 1964. New data on the localities of dinosaurs on the territory of Kazakhstan and Middle Asia. In: Nauchnye Trudy Tashkentskogo Gosudarstvennogo Universiteta Imeni V.I. Lenina, Seriya Geologiya, pp. 227–241 (in Russian).

Sabirov, A.A., 2008. Late Jurassic dinosaurs from basin of Shirkent River (Hissar Range). News of the Academy of Sciences of the Republic of Tajikistan. Department of Physical and Mathematical. Chem. Geol. Tech. Sci. 2, 69–79.

Sadlok, G., 2019. *Brasilichnium*-like tracks from the Lower Triassic of the Holy Cross Mountains, Poland. Neues Jahrbuch Geol. Palaontol. Abhand. 292, 103–112.

Salisbury, S.W., Romilio, A., Herne, M.C., Tucker, R.T., Nair, J.P., 2016. The dinosaurian ichnofauna of the Lower Cretaceous (Valanginian–Barremian) Broome Sandstone of the Walmadany area (James Price Point), Dampier Peninsula, western Australia. J. Vertebr. Paleontol. 36 (Suppl. 6), 1–152. https://doi.org/10.1080/02724634.2016.1269539.

Sánchez-Hernández, B., Przewieslik, A.G., Benton, M.J., 2009. A reassessment of the *Pteraichnus* ichnospecies from the Early Cretaceous of Soria Province, Spain. J. Vertebr. Paleontol. 29, 487–497.

Santisteban, C., Suñer, M., Vila, B., 2009. El yacimiento de icnitas de dinosaurios de Cañada Paris, Alpuente, Valencia. In: Actas de las IV Jornadas Internacionales sobre Paleontologia de Dinosaurios y su Entorno. Burgos. Colectivo Arqueológico- Paleontológico de Salas de los Infantes, pp. 301–309.

Santos, V.F., 2008. Pegadas de dinossáurios de Portugal. Museu Nacional de História Natural da Universidade de Lisboa, Lisboa, p. 123.

Santos, V.F., Moratalla, J.J., Royo-Torres, R., 2009. New sauropod trackways from the Middle Jurassic of Portugal. Acta Palaeontol. Pol. 54 (3), 409–422. https://doi.org/10.4202/app.2008.0049.

Santos, V.F., Callapez, P.M., Rodrigues, N.P., 2013. Dinosaur footprints from the Lower Cretaceous of the Algarve Basin (Portugal): new data on the ornithopod palaeoecology and palaeobiogeography of the Iberian Peninsula. Cretac. Res. 40, 158–169. https://doi.org/10.1016/j.cretres.2012.07.001.

Santos, C.M.D., Amorim, D.S., Klassa, B., Fachin, D.A., Nihei, S.S., de Carvalho, C.J.B., Falashi, R.L., Mello-Patiu, C.A., Couri, M.S., Oliveira, S.S., Silva, V.C., Ribeiro, G.C., Capellari, R.S., Lamas, C.J.E., 2016. On typeless species and the perils of fast taxonomy. Syst. Entomol. 41 (3), 511–515. https://doi.org/10.1111/syen.12180.

Sarjeant, W.A.S., 1975. A vertebrate footprint from the Stonesfield Slate (Middle Jurassic) of Oxfordshire. Mercian Geol. 5 (4), 273–277.

Sarjeant, W.A.S., Langston, W., 1994. Vertebrate footprints and invertebrate traces from the Chadronian (Late Eocene) of Trans-Pecos Texas. Texas Memorial Mus. Bullet. 36, 1–86.

Sarjeant, W.A.S., Delair, J.B., Lockley, M.G., 1998. The footprints of *Iguanodon*: a history and taxonomic study. Ichnos 6, 183–202. https://doi.org/10.1080/10420949809386448.

Sciscio, L., Bordy, E.M., Reid, M., Abrahams, M., 2016. Sedimentology and ichnology of the Mafube dinosaur track site (Lower Jurassic, eastern Free State, South Africa): a report on footprint preservation and palaeoenvironment. PeerJ 4, e2285. https://doi.org/10.7717/peerj.2285.

Sciscio, L., Bordy, E.M., Abrahams, M., Knoll, F., McPhee, B.W., 2017. The first megatheropod tracks from the Lower Jurassic upper Elliot Formation, Karoo Basin, Lesotho. PLoS One 12 (10), e0185941. https://doi.org/10.1371/journal.pone.0185941.

Sciscio, L., Belvedere, M., Meyer, C.A., Marty, D., 2022. Sauropod trackway morphometrics: an exploratory study using highway A16 excavation at the Courtedoux-Tchâfouè track site (Late Jurassic, NW Switzerland). Front. Earth Sci. 10, 805442.

Shaffer, A.B., Tweet, J.S., Santucci, V.L., 2024. Colorado National Monument: Paleontological Resource Inventory. Science Report NPS/SR—2024/116. National Park Service, Fort Collins, Colorado. https://doi.org/10.36967/2303613.

Shatalkin, A.I., Galinskaya, T.V., 2017. A commentary on the practice of using the so-called typeless species. ZooKeys 693, 129–139. https://doi.org/10.3897/zookeys.693.10945.

Shibata, K., Matsukawa, M., Lockley, M.G., 2006. Energy flow modeling applied to data from the Lower Jurassic Navajo Sandstone, western North America: implication for ecological replacement between the Late Triassic and Early Jurassic ecosystems. New Mex. Mus. Nat. Hist. Sci. Bullet. 37, 29–34.

Shillito, A.P., Davies, N.S., 2018. Dinosaur-landscape interactions at a diverse Early Cretaceous tracksite (Lee Ness Sandstone, Ashdown Formation, southern England). Palaeogeogr. Palaeoclimatol. Palaeoecol. 514, 593–612. https://doi.org/10.1016/j.palaeo.2018.11.018.

Shuler, E.W., 1935. Dinosaur track mounted in the bad stand at Glen Rose. Texas. Field Lab. 4 (1), 9–13.

Slater, S.M., Wellman, C.H., Romano, M., Vajda, V., 2018. Dinosaur-plant interactions within a Middle Jurassic ecosystem – Palynology of the Burniston Bay dinosaur footprint locality, Yorkshire, UK. Palaeobiodivers. Palaeoenviron. 98, 139–151.

Smith, J.R., 1982. Dinosaurs in Virginia: evidence of two new genera. Lapid. J. 36, 1110–1111.

Smith, R.M., Marsicano, C.A., Wilson, J.A., 2009. Sedimentology and paleoecology of a diverse Early Jurassic tetrapod tracksite in Lesotho, Southern Africa. Palaios 24 (10), 672–684. https://doi.org/10.2110/palo.2008.p08-115r.

Soto-Acuña, S., Otero, R.A., Rubilar-Rogers, D., Vargas, A.O., 2015. Arcosaurios no avianos de Chile. In: Rubilar-Rogers, D., Otero, R., Vargas, A., Sallaberry, M. (Eds.), Vertebrados Fósiles de Chile, vol. 63. Publ. Oc. Museo Nacional de Historia Natural de Chile, pp. 209–263.

Sternberg, C.M., 1932. Dinosaur Tracks from Peace River, British Columbia. National Museum of Canada Annual Report for 1930, Ottawa, pp. 59–85.

Stimpson, W., 1858. Prodromus descriptionis animalium evertebratorum, quae in Expeditione ad Oceanum Pacificum Septentrionalem, a Republica Federata missa, Cadwaladaro Ringgold et Johanne Rodgers Ducibus, observavit et desoripsit. Pars VII. Crustacea Anomoura. Proc. Acad. Nat. Sci. Phila. 10, 225–252.

Stimson, M.R., Miller, R.F., Lucas, S.G., Park, A.F., Hinds, S.J., 2016. Redescription of tetrapod trackways from the Mississippian Mabou Group, Lepreau Falls, New Brunswick, Canada. Atlantic Geosci. 52, 1–19. https://doi.org/10.4138/atlgeol.2016.001.

Stokes, W.L., 1957. Pterodactyl tracks from the Morrison Formation. J. Paleontol. 31, 952–954.

Stokes, W.L., 1978. Animal tracks in the Navajo–Nugget Sandstone. Contrib. Geol. Univ. Wyo. 16 (2), 103–107.

Stokes, W.L., Madsen Jr., J.H., 1979. Environmental significance of pterosaur tracks in the Navajo Sandstone (Jurassic). Grand County Utah Brigham Young Univ. Geol. Stud. 26, 21–26.

Suarez, C.A., Knobbe, T.K., Crowley, J.L., Kirkland, J.I., Milner, A.R.C., 2017. A chronostratigraphic assessment of the Moenave Formation, USA using C-isotope chemostratigraphy and detrital zircon geochronology: Implications for the terrestrial end Triassic extinction. Earth Planet Sci. Lett. 475, 83–93.

Suarez, C., Sharman, G., Park Boush, L., Milner, A.R.C., Stigall, A.L., Antonietto, L.S., Whiteside, J., 2023. Locating the Triassic–Jurassic boundary within the Moenave Formation: implications for climatic, faunal, and environmental change of the lacustrine Whitmore Point Member following the End–Triassic Extinction event. Anat. Rec. 306 (Suppl. 1), 236–238. https://doi.org/10.1002/ar25219.

Talanda, M., Dzieciol, S., Sulej, T., Niedźwiedzki, G., 2011. Vertebrate burrow system from the Upper Triassic of Poland. Palaios 26, 99–105.

Thomson, T.J., Droser, M.L., 2015. Swimming reptiles make their mark in the Early Triassic: delayed ecologic recovery increased the preservation potential of vertebrate swim tracks. Geology 43 (3), 215–218. https://doi.org/10.1130/G36332.1.

Thulborn, R.A., 1993. A tale of three fingers: ichnological evidence revealing the homologies of manual digits in theropod dinosaurs. New Mex. Mus. Nat. Hist. Sci. Bullet. 3, 461–463.

Thulborn, R.A., 1994. Ornithopod dinosaur tracks from the Lower Jurassic of Queensland. Alcheringa 18, 247–258.

Thulborn, R.A., Wade, M., 1984. Dinosaur trackways in the Winton Formation (mid-Cretaceous) of Queensland. Memoir. Queensl. Mus. 21 (2), 413–517.

Thulborn, T., 1989. The gaits of dinosaurs. In: Gillette, D.D., Lockley, M.G. (Eds.), Dinosaur Tracks and Traces. Cambridge University Press, Cambridge, pp. 39–50.

Thulborn, T., 1990. Dinosaur Tracks. Chapman & Hall, London.

Thulborn, T., 1998. Australia's earliest theropods: footprint evidence in the Ipswich Coal Measures (Upper Triassic) of Queensland. Gaia 15, 301–311.

Thulborn, T., 2001. History and nomenclature of the theropod dinosaur tracks *Bueckeburgichnus* and *Megalosauripus*. Ichnos 8, 207–222. https://doi.org/10.1080/10420940109380188.

Thulborn, T., 2017. Behaviour of dinosaurian track-makers in the Winton Formation (Cretaceous, Albian–Cenomanian) at Lark Quarry, western Queensland, Australia: running or swimming? Ichnos 24 (1), 1–18. https://doi.org/10.1080/10420940.2015.1129326.

Torcida Fernández-Baldor, F., Díaz-Martínez, I., Huerta, P., Montero Huerta, Castanera, D., 2021. Enigmatic tracks of solitary sauropods roaming an extensive lacustrine megatracksite in Iberia. Sci. Rep. 11 (1), 16939.

Trewin, N.H., 2000. The ichnogenus *Undichna*, with examples from the Permian of the Falkland Islands. Palaeontology 43, 979–997.

Turek, V., 1989. Fish and amphibian trace fossils from Westphalian sediments of Bohemia. Palaeontology 32, 623–643.

Turner, S., Bean, L.B., Dettmann, M., McKellar, J.L., McLoughlin, S., Thulborn, T., 2009. Australian Jurassic sedimentary and fossil successions: current work and future prospects for marine and non-marine correlation. GFF 131 (1–2), 49–70. https://doi.org/10.1080/11035890902924877.

Valentini, M., Alessandra Conti, M., Mariotti, N., 2007. Lacertoid footprints of the Upper Permian Arenaria di Val Gardena Foramtion (northern Italy). Ichnos 14, 193–218.

Van Dijk, D.E., 1978. Trackways in the Stormberg. Palaeontol. Afr. 21, 113–120.

Viera, L.I., Torres, J.A., 1992. Sobre "Dinosaurios coelúridos gregarious en el yaximiento de Valdevajes (La Rioja, España)". Nota de replica y crítica. Rev. Esp. Palaontol. 7 (1), 93–96.

Vila, B., Castanera, D., Marmi, J., Canudo, J.I., Galobart, À., 2015. Crocodile swim tracks from the latest Cretaceous of Europe. Lethaia 48 (2), 256–266. https://doi.org/10.1111/let.12103.

Wagensommer, A., Latiano, M., Nicosia, U., 2010. First report of dinosaur footprints from Madagascar: Two tracksites from the Middle Jurassic Bemaraha Formation. Ichnos 17, 127–136.

Wagensommer, A., Latiano, M., Leroux, G., Cassano, G., D'Orazi, P., 2012. New dinosaur tracksites from the Middle Jurassic of Madagascar: ichnotaxonomical, behavioural and palaeoenvironmental implications. Palaeontology 55 (1), 109–126.

Wagensommer, A., Latiano, M., Mocke, H.B., D'Orazi Porchetti, S., 2016. Dinosaur diversity in an Early Jurassic African desert: the significance of the Etjo Sandstone ichnofauna at the Otjihaenamaparero locality (Namibia). Neues Jahrbuch Geol. Palaontol. Abhand. 281 (2), 155–182.

Wagensommer, A., Dolch, R., Ratolojanahary, T., Donato, S., D'Orazi Porchetti, S., 2023. Defining the Bemaraha megatracksite: An update on dinosaur ichnology in Madagascar. In: Cónsole-Gonella, C., de Valais, S., Díaz-Martínez, I., Citton, P., Verde, M., McIlroy, D. (Eds.), Ichnology in Shallow Marine and Transitional Environments. Geological Society of, London. https://doi.org/10.1144/SP522-2021-86.

Wang, B., Li, J., Bai, Z., Gao, J., Dong, S., Hu, B., Zhao, S., Chang, J., 2016. Research on dinoasur [sic] footprints in Zizhou, Shaanxi Province, China. Acta Geol. Sin. 90 (1), 1–18. https://doi.org/10.1111/1755-6724.12639.

Weems, R.E., 1987. A Late Triassic Footprint Fauna from the Culpeper Basin Northern Virginia (U.S.A.), vol. 77. Transactions of the American Philosophical Society, pp. 1–79.

Weems, R.E., 1992. A re-evaluation of the taxonomy of Newark Supergroup saurischian dinosaur tracks, using extensive statistical data from a recently exposed tracksite near Culpeper, Virginia. In: Sweet, P.C. (Ed.), Proceedings, 26th Forum on the Geology of Industrial Minerals. Virginia Division of Mining Res. Publ. 119, pp. 113–127.

Weems, R.E., 2003. *Plateosaurus* foot structure suggests a single trackmaker for *Eubrontes* and *Gigandipus* footprints. In: LeTourneau, P.M., Olsen, P.E. (Eds.), The Great Rift Valleys of Pangea in Eastern North America. Volume Two, Sedimentology, Stratigraphy, and Paleontology. Columbia University Press, New York, pp. 293–313.

Weems, R.E., 2019. Evidence for bipedal prosauropods as the likely *Eubrontes* track-makers. Ichnos 26 (3), 187–215. https://doi.org/10.1080/10420940.2018.1532902.

Weems, R.E., Kimmel, P.G., 1993. Upper Triassic reptile footprints and a coelacanth fish scale from the Culpeper Basin, Virginia. Proc. Biol. Soc. Wash. 106 (2), 390–401.

Weems, R.E., Bachman, J.M., 2015. The Lower Cretaceous Patuxent Formation ichnofauna of Virginia. Ichnos 22 (3–4), 208–219. https://doi.org/10.1080/10420940.2015.1063493.

Weems, R.E., Lucas, S.G., 2020. The first record of vertebrate ichnofossils from the Norton Formation (Pennsylvanian, middle Moscovian) in southwestern Virginia, USA. New Mex. Mus. Nat. Hist. Sci. Bullet. 82, 497–504.

Welles, S.P., 1971. Dinosaur footprints from the Kayenta Formation of northern Arizona. Plateau 44 (1), 27–38.

Whyte, M.A., Romano, M., 1993. Footprints of a sauropod dinosaur from the middle Jurassic of Yorkshire. Proc. Geol. Assoc. 104 (3), 195–199. https://doi.org/10.1016/S0016-7878(08)80037-5.

Whyte, M.A., Romano, M., 1994. Probable sauropod footprints from the Middle Jurassic of Yorkshire, England. Gaia 10, 15–26.

Whyte, M.A., Romano, M., 2001a. A dinosaur ichnocoenosis from the Middle Jurassic of Yorkshire, UK. Ichnos 8 (3–4), 223–234. https://doi.org/10.1080/10420940109380189.

Whyte, M.A., Romano, M., 2001b. Probable stegosaurian dinosaur tracks from the Saltwick Formation (Middle Jurassic) of Yorkshire, England. Proc. Geol. Assoc. 112, 45–54.

Whyte, M.A., Romano, M., 2014. First record of the pterosaur footprint *Pteraichnus* from the Saltwick Formation (Aalenian) of the Cleveland Basin, Yorkshire, UK. Proc. Yorks. Geol. Soc. 60, 19–27.

Williams, J.A.J., Milner, A.R.C., Lockley, M.G., 2006. A new Early Jurassic (Hettangian) track-bearing horizon from the Moenave Formation, St. George Dinosaur Discovery Site at Johnson Farm, Washington County, Utah. New Mex. Mus. Nat. Hist. Sci. Bullet. 37, 346–351.

Wilson, J.A., Marsicano, C.A., Smith, R.M.H., 2009. Dynamic locomotor capabilities revealed by early dinosaur track-makers from Southern Africa. PLoS One 4, e7331. https://doi.org/10.1371/journal.pone.0007331.

Wood, J.R., Bozek, M.A., Milner, A.R.C., Mims, A.L., Santucci, V.L., Frost, F., 2021. Structure from motion photogrammetry enhances paleontological resource documentation, research, preservation and education efforts for National Park Service areas. New Mex. Mus. Nat. Hist. Sci. Bullet. 84, 513–523.

Wright, J.L., 2005. Steps in understanding sauropod biology. In: Rogers, K.A., Wilson, J.A. (Eds.), The Sauropods: Evolution and Paleobiology. University of California Press, Berkeley, pp. 252–284.

Wright, J.L., Unwin, D.M., Lockley, M.G., Rainforth, E.C., 1997. Pterosaur tracks from the Purbeck Limestone Formation of Dorset, England. PGA (Proc. Geol. Assoc.) 108, 39–48. https://doi.org/10.1016/S0016-7878(97)80004-1.

Wroblewski, A.F.-J., 2023. The makers of *Rhamphichnus* ispp. reinterpreted as lepidosaurian and crocodilian, not pterosaurian. Geobios 77, 71–79. https://doi.org/10.1016/j.geobios.2022.12.001.

Xing, L.-D., Harris, J.D., Toru, S., Masato, F., Dong, Z.-M., 2009. Discovery of dinosaur footprints from the Lower Jurassic Lufeng Formation of Yunnan Province, China and new observations on *Changpeipus*. Geol. Bull. China 28 (1), 16–29.

Xing, L.D., Harris, J.D., Jia, C.K., Luo, Z.J., Wang, S.N., An, J.F., 2011a. Early Cretaceous bird-dominated and dinosaur footprint assemblages from the northwestern margin of the Junggar Basin, Xinjiang, China. Palaeoworld 20 (4), 308–321. https://doi.org/10.1016/j.palwor.2011.01.001.

Xing, L.D., Harris, J.D., Gierliński, G.D., 2011b. *Therangospodus* and *Megalosauripus* track assemblage from the Upper Jurassic–Lower Cretaceous Tuchengzi Formation of Chicheng County, Hebei Province, China and their paleoecological implications. Vertebr. Palasiat. 49 (4), 423–434.

Xing, L.-D., Gierliński, G.D., Harris, J.D., Divay, J.D., 2012. A probable crouching theropod dinosaur trace from the Tuchengzi Formation in Chicheng area, Hebei Province, China. Geol. Bull. China 31 (1), 20–25.

Xing, L., Lockley, M.G., Zhang, J., Milner, A.R.C., Klein, H., Li, D., Persons, W.S.I.V., Ebi, J., 2013a. A new Early Cretaceous dinosaur track assemblage and the first definite non-avian theropod swim trackway from China. Chin. Sci. Bull. 58, 2370–2378. https://doi.org/10.1007/s11434-013-5802-6.

Xing, L., Lockley, M.G., McCrea, R.T., Gierliński, G.D., Buckley, L.G., Zhang, J., Qi, L., Jia, C., 2013b. First record of *Deltapodus* tracks from the Early Cretaceous of China. Cretac. Res. 42, 55–65. https://doi.org/10.1016/j.cretres.2013.01.006.

Xing, L.D., Lockley, M.G., Klein, H., Zhang, J.P., He, Q., Divay, J.D., Qi, L.Q., Jia, C.K., 2013c. Dinosaur, bird and pterosaur footprints from the Lower Cretaceous of Wuerhe asphaltite area, Xinjiang, China, with notes on overlapping track relationships. Palaeoworld 22 (1–2), 42–51. https://doi.org/10.1016/j.palwor.2013.03.001.

Xing, L.-D., Lockley, M.G., Chen, W., Gierliński, G.D., Li, J.-J., Persons, W.S.I.V., Matsukawa, M., Ye, Y., Gingras, M.K., Wang, C.-W., 2013d. Two theropod track assemblages from the Jurassic of Chongqing, China, and the Jurassic stratigraphy of Sichuan Basin. Vertebr. Palasiat. 51 (2), 107–130.

Xing, L.-D., Klein, H., Lockley, M.G., Wang, S.-L., Chen, W., Ye, Y., Matsukawa, M., Zhang, J.-P., 2013e. Earliest records of theropod and mammal-like tetrapod footprints in the Upper Triassic of Sichuan Basin, China. Vertebr. PalAsiatica 51 (3), 184–198.

Xing, L.-D., Klein, H., Lockley, M.G., Wetzel, A., Li, Z.-D., Li, J.-J., Gierliński, G.D., Zhang, J.-P., Matsukawa, M., Divay, J.D., Zhou, L., 2014a. *Changpeipus* (theropod) tracks from the Middle Jurassic of the Turpan Basin, Xinjiang, northwest China: review, new discoveries, ichnotaxonomy, preservation and paleoecology. Vertebr. Palasiat. 52 (2), 233–259.

Xing, L.D., Lockley, M.G., Klein, H., Gierliński, G.D., Divay, J.D., Hu, S.M., Zhang, J.P., Ye, Y., He, Y.P., 2014b. The non-avian theropod track *Jialingpus* from the Cretaceous of the Ordos Basin, China, with a revision of the type material: implications for ichnotaxonomy and trackmaker morphology. Palaeoworld 23 (2), 187–199. https://doi.org/10.1016/j.palwor.2013.12.001.

Xing, L.-D., Lockley, M.G., Miyashita, T., Klein, H., Wang, T., Persons, W.S.I.V., Pan, S.-G., Zhang, J.-P., Dong, Z.-M., 2014c. Large sauropod and theropod tracks from the Middle Jurassic Chuanjie Formation of Lufeng County, Yunnan Province and palaeobiogeography of the Middle Jurassic sauropod tracks from southwestern China. Palaeoworld 23, 294–303.

Xing, L., Liu, Y.Q., Kuang, H.W., Klein, H., Zhang, J.P., Burns, M.E., Chen, J., Wang, M.W., Hu, J., 2014d. Theropod and possible ornithopod track assemblages from the Jurassic–Cretaceous boundary Houcheng formation, Shangyi, northern Hebei, China. Palaeoworld 23, 200–208. https://doi.org/10.1016/j.palwor.2013.10.002.

Xing, L.D., Peng, G.Z., Ye, Y., Lockley, M.G., McCrea, R.T., Currie, P.J., Zhang, J.P., Burns, M.E., 2014e. Large theropod trackway from the Lower Jurassic Zhenzhuchong Formation of Weiyun County, Suchuan Province, China: Review, new observations and special preservation. Palaeoworld 23, 285–293.

Xing, L., Lockley, M.G., Zhang, J., Klein, H., Marty, D., Peng, G., Ye, Y., McCrea, R.T., Persons, W.S.I.V., Xu, T., 2015a. The longest theropod trackway from East Asia, and a diverse sauropod-, theropod-, and ornithopod-track assemblage from the Lower Cretaceous Jiaguan Formation, southwest China. Cretac. Res. 56, 345–362. https://doi.org/10.1016/j.cretres.2015.05.008.

Xing, L., Lockley, M.G., Tang, Y., Klein, H., Zhang, J., Persons IV, W.S., Dai, H., Ye, Y., 2015b. Theropod and ornithischian footprints from the Middle Jurassic Yanan Formation of Zizhou County, Shaanxi, China. Ichnos 22, 1–11. https://doi.org/10.1080/10420940.2014.985670.

Xing, L.-D., Lockley, M.G., Klein, H., Zhang, J.-P., Wang, T., Persons, W.S., Dong, Z.-M., 2016a. A tetrapod footprint assemblage with possible swim traces from the Jurassic–Cretaceous boundary, Anning Formation, Konglongshan, Yunnan, China. Palaeoworld 25 (3), 444–452. https://doi.org/10.1016/j.palwor.2015.06.002.

Xing, L., McCrea, R.T., Lockley, M.G., Lu, T., Burns, M.E., Peng, G., Zhang, J., 2016b. A possible ankylosaurian (Thyreophora) trackway from the Lower Cretaceous Jiaguan Formation of Emei, southwest China: paleoecological implications. New Mex. Mus. Nat. Hist. Sci. Bullet. 74, 339–343.

Xing, L.D., Lockley, M.G., Klein, H., Falkingham, P.L., Kim, J.Y., McCrea, R.T., Zhang, J., Persons IV, W.S., Wang, T., Wang, Z., 2016c. First Early Jurassic small ornithischian tracks from the Yunnan province, Southwestern China. Palaios 31, 516–524.

Xing, L., Lockley, M.G., Zhang, J., Klein, H., Li, D., Miyashita, T., Li, Z., Kümmell, S.B., 2016d. A new sauropodomorph ichnogenus from the Lower Jurassic of Sichuan, China fills a gap in the track record. Hist. Biol. 28 (7), 881–895. https://doi.org/10.1080/08912963.2015.1052427.

Xing, L., Lockley, M.G., Zhang, J., You, H., Klein, H., Persons IV, W.S., Dai, H., Dong, Z., 2016e. First Early Jurassic ornithischian and theropod footprint assemblage and a new ichnotaxon *Shenmuichnus wangi* ichnosp. nov. from Yunnan Province, southwestern China. Hist. Biol. 28 (6), 721–733. https://doi.org/10.1080/08912963.2015.1020424.

Xing, L.-D., Lockley, M.G., Zhang, J., Klein, H., Kümmell, S.B., Persons, W.S.I.V., Kuang, H.-W., 2017a. Theropod tracks from the Lower Jurassic of Gulin area, Sichuan Province, China. Palaeoworld 26 (1), 115–123. https://doi.org/10.1016/j.palwor.2015.11.003.

Xing, L., Lockley, M.G., Zhang, J., Romilio, A., Klein, H., Wang, Y., Tang, Y., Burns, M.E., Wang, X., 2017b. A diversified vertebrate ichnite fauna from the Dasheng Group (Lower Cretaceous) of southeast Shandong Province, China. Hist. Biol. 31 (3), 353–362. https://doi.org/10.1080/08912963.2017.1370588.

Xing, L., Abbassi, N., Lockley, M.G., 2017c. Enigmatic didactyl tracks from the Jurassic of Iran. Hist. Biol. 30 (8), 1132–1138. https://doi.org/10.1080/08912963.2017.1339700.

Xing, L.-D., Ba, J., Lockley, M.G., Klein, H., Yan, S.-W., Romilio, A., Chou, C.-Y., Persons, W.S.I.V., 2018a. Late Triassic sauropodomorph and Middle Jurassic theropod tracks from the Xichang Basin, Sichuan Province, southwestern China: first report of the ichnogenus *Carmelopodus*. J. Palaeogeogr. 7 (1), 1–13. https://doi.org/10.1016/j.jop.2017.11.004.

Xing, L., Lockley, M.G., Tang, Y., Romilio, A., Xu, T., Li, X., Tang, Y., Li, Y., 2018b. Tetrapod track assemblages from Lower Cretaceous desert facies in the Ordos Basin, Shaanxi Province, China, and their implications for Mesozoic paleoecology. Palaeogeogr. Palaeoclimatol. Palaeoecol. 507, 1–14. https://doi.org/10.1016/j.palaeo.2018.05.016.

Xing, L.-D., Klein, H., Qin, Z.-H., Persons, W.S.I.V., Xu, X., 2018c. Lower Cretaceous theropod tracks with the new ichnogenus and combination *Lockleypus luanpingensis* from the Dabeigou Formation of the Luanping Basin, Hebei Province, China. Palaeoworld 27 (4), 482–489. https://doi.org/10.1016/j.palwor.2018.09.004.

Xing, L., Lockley, M.G., Klein, H., Zeng, R., Cai, S., Luo, X., Li, C., 2018d. Theropod assemblages and a new ichnotaxon *Gigandipus chiappei* ichnosp. nov. from the Jiaguan Formation, Lower Cretaceous of Guizhou Province, China. Geosci. Front. 9 (6), 1745–1754.

Xing, L., Lockley, M.G., Klein, H., Qiu, L., Chou, C., Wang, D., Persons, W.S.I.V., 2019a. A Lower Cretaceous dinosaur track assemblage from the Taoqihe Formation in central Heilongjiang, China. Palaeogeogr. Palaeoclimatol. Palaeoecol. 532, 109275. https://doi.org/10.1016/j.palaeo.2019.109275.

Xing, L., Lockley, M.G., Romilio, A., 2019b. An unusual dinosaur track assemblage from the Jurassic–Cretaceous boundary, Anning Formation, Lufeng Basin, China. Hist. Biol. 33 (4), 514–526. https://doi.org/10.1080/08912963.2019.1642335.

Xing, L., Klein, H., Lockley, M.G., Wu, X.-C., Benton, M.J., Zeng, R., Romilio, A., 2020. Footprints of marine reptiles from the Middle Triassic (Anisian–Ladinian) Guanling Formation of Guizhou Province, southwestern China: the earliest evidence of synchronous style of swimming. Palaeogeogr. Palaeoclimatol. Palaeoecol. 558, 109943. https://doi.org/10.1016/j.palaeo.2020.109943.

Xing, L., Lockley, M.G., Persons, W.S.I.V., Klein, H., Romilio, A., Wang, D., Wang, M., 2021a. Stegosaur track assemblage from Xinjiang, China, featuring the smallest known stegosaur record. Palaios 36 (2), 68–76. https://doi.org/10.2110/palo.2020.036.

Xing, L., Lockley, M.G., Tang, Y., Klein, H., Persons, W.S.I.V., Wang, M., Li, X., Wu, H., 2021b. Unusual dinosaur trackway preservation as clues to paleo-landscape and behavior from the Lower Cretaceous Luohe Formation, Shaanxi Province, China. Geosci. Front. 12 (2), 737–745. https://doi.org/10.1016/j.gsf.2020.09.014.

Xing, L., Lockley, M.G., Tong, B., Klein, H., Persons, W.S.I.V., Peng, G., Ye, Y., Wang, M., 2021c. Theropod footprints from the Lower Cretaceous Cangxi Formation in the northern margin of the Sichuan Basin, China. Biosis Biol. Syst. 2 (1), 174–182. https://doi.org/10.37819/biosis.002.01.0097.

Xing, L., Lockley, M.G., Klein, H., Zhang, L., Romilio, A., Scott Persons IV, W., Peng, G., Ye, Y., Wang, M., 2021d. The new ichnotaxon *Eubrontes nobitai* ichnosp. nov. and other Saurischian tracks from the Lower Cretaceous Jiaguan Formation of Sichuan Province. China J. Paleogeogr. 10, 17.

Xing, L., Lockley, M.G., Peng, G., Ye, Y., Jiang, S., Romilio, A., Persons, W.S.I.V., Wang, M., 2021e. A review of two Middle Jurassic theropod tracksites discovered in the 1980s from Sichuan Basin. Biosis 2 (1), 191–208. https://doi.org/10.37819/biosis.002.01.0094.

Xing, L., Romilio, A., Wang, Y., Yang, L., Wang, D., 2023a. New dinosaur swim tracks from the Middle Jurassic Zhanghe Formation in Yunnan, China. Hist. Biol. https://doi.org/10.1080/08912963.2023.2194906.

Xing, L., Lockley, M.G., Tong, B.-L., Klein, H., Liu, C., Persons, W.S., 2023b. The Middle–Upper Jurassic dinosaur-pterosaur fauna in Shandong Province, China: evidence from ichnology. J. Palaeogeogr. 12, 419–433. https://doi.org/10.1016/j.jop.2023.05.001.

Xing, L., Lallensack, J.N., Romilio, A., Wang, Y., Yang, L., 2023c. Early Jurassic theropod elongate tracks with metatarsal marks from Yunnan, China. Hist. Biol. https://doi.org/10.1080/08912963.2023.2220121.

Yang, X.L., Yang, D.H., 1987. Dinosaur Footprints from Mesozoic of Sichuan Basin. Sichuan Science and Technological Publication, Chengdu.

Young, C.-C., 1943. Note on some fossil footprints in China. Bullet. Geol. Soc. China 13 (3–4), 151–154.

Young, C.-C., 1960. Fossil footprints in China. Vertebr. Palasiat. 4 (2), 53–67.
Young, C.-C., 1966. Two footprints from the Jiaoping Coal Mine of Tungchuan, Shensi. Vertebr. Palasiat. 10 (1), 68–71.
Young, C.-C., 1979. Footprints from Luanping, Hebei. Vertebr. Palasiat. 17 (2), 116–117.
Zhang, Q., Wen, W., Hu, S., Benton, M.J., Zhou, C., Xie, T., Lu, T., Huang, J., Choo, B., Chen, Z.-Q., Liu, J., Zhang, Q., 2014. Nothosaur foraging tracks from the Middle Triassic of southwestern China. Nat. Commun. 5, 3973. https://doi.org/10.1038/ncomms4973.
Zhen, S., Li, J., Zhen, B., 1983. Dinosaur footprints of Yuechi, Sichuan. Mem. Beijing Nat. Hist. Mus. 25, 1–19.
Zhen, S., Li, J., Chenggang, R., 1986a. Dinosaur Footprints of Jinning, Yunnan. Beijing Natural History Museum.
Zhen, S., Li, J., Rao, C., Hu, S., 1986b. Dinosaur footprints of Jinning, Yunnan. Mem. Beijing Nat. Hist. Mus. 33, 1–19.
Zhen, S., Li, J., Chenggang, R., Mateer, N.J., Lockley, M.G., 1989. A review of dinosaur footprints in China. In: Gillette, D.D., Lockley, M.G. (Eds.), Dinosaur Tracks and Traces. Cambridge University Press, Cambridge, pp. 187–197.
Zhen, S., Li, J., Han, Z., 1996. The Study of Dinosaur Footprints in China. Chengdu, Sichuan Scientific and Technological. Publishing House, p. 110.

Further reading

Buck, P.V., Ghilardi, A.M., Peixoto, C.P.M., Fernandes, L.B.R., Fernandes, M.A., 2017a. A new tetrapod ichnotaxon from Botucatu Formation, Lower Cretaceous (Neocomian), Brazil, with comments on fossil track preservation on inclined planes and local paleoecology. Palaeogeogr. Palaeoclimatol. Palaeoecol. 21–37. https://doi.org/10.1016/j.palaeo.2016.11.009.
Leonardi, G., Lockley, M.G., 1995. A proposal to abandon the ichnogenus *Coelurosaurichnus* Huene, 1941–a junior synonym of *Grallator*. J. Vertebr. Paleontol. 15 (Suppl. 3), 40A.
Lockley, M., 2011a. Theropod- and prosauropod-dominated ichnofaunas from the Navajo–Nugget Sandstone (Lower Jurassic) at Dinosaur National Monument: Implications for prosauropod behavior and ecology. New Mex. Mus. Nat. Hist. Sci. Bullet. 53, 316–320.
Mujal, E., Fortuny, J., Bolet, A., Oms, O., López, J.Á., 2017. An archosauromorph dominated ichnoassemblage in fluvial settings from the late Early Triassic of the Catalan Pyrenees (NE Iberian Peninsula). PLoS One 12 (4), e0174693. https://doi.org/10.1371/journal.pone.0174693.
Niedźwiedzki, G., Ptaszyński, T., 2007. Large Chirotheriidae tracks in the Early Triassic of Wióry, Holy Cross Mountains, Poland. Acta Geol. Pol. 57 (3), 325–342.
Rodríguez-de la Rosa, R.A., 2003. Pterosaur tracks from the latest Campanian Cerro del Pueblo Formation of southeastern Coahuila, Mexico. Geol. Soc. 217 (1), 275–282.
Rodríguez-de la Rosa, R.A., Aguillón-Martínez, M.C., López-Espinoza, J., 2004. The fossil record of vertebrate tracks in Mexico. Ichnos 11, 27–37.
Rühle von Lilienstern, H., 1939. Fährten und Spüren im Chirotheriem – Sandstein von Südthüringen. Forschritte der Geologie und Paläontologie 12, 293–387.

CHAPTER

7

Cretaceous vertebrate tracks

Lida Xing[1,2,3], Ignacio Díaz-Martínez[4], Jens N. Lallensack[5], Kyung Soo Kim[6], Lisa G. Buckley[3], Anthony Romilio[7], Laura Piñuela[8], Donghao Wang[1,2], Qiyan Chen[2], Qing Yang[2], Chunyong Chou[2], Hang Yin[2] and Yufei Jin[2]

[1]Frontiers Science Center for Deep-time Digital Earth, China University of Geosciences (Beijing), Beijing, China; [2]School of the Earth Sciences and Resources, China University of Geosciences (Beijing), Beijing, China; [3]Dinosaur Odyssey Science Museum, Guangxi, China; [4]Departamento de Ciencias de la Tierra y Física de la Materia Condensada, Facultad de Ciencias, Universidad de Cantabria, Cantabria, Spain; [5]Departamento de Ciência da Computação, Universidade Federal de Minas Gerais, Belo Horizonte, Brazil; [6]Department of Science Education, Chinju National University of Education, Gyeongnam, South Korea; [7]School of Biological Sciences, The University of Queensland, Brisbane, QLD, Australia; [8]Museo del Jurásico de Asturias (Jurassic Museum of Asturias), Colunga, Spain

1. Introduction

The Cretaceous (about 145 to 66 Ma) is the longest geological period of the Phanerozoic, spanning approximately 79 million years. Fossil evidence indicates that during the Cretaceous witnessed major evolutionary events such as the transformation of nonavian theropods from land to sky and the emergence of terrestrial and flying creatures on Earth (Benton et al., 2022). It also witnessed how biodiversity changes in a greenhouse gas environment and was affected by the Cretaceous–Paleogene extinction event 66 million years ago.

During the late Aptian to early Albian (~110 Ma) of the Early Cretaceous, the Pangean breakup reached its peak, and the following events occurred successively: the breakup of the African plate and the South American plate, opening of the South Atlantic; the separation of the North American plate and the European plate, opening of the North Atlantic; and the separation of the Indian plate from Antarctica and its northward movement, forming the central Indian Ocean (Jackson et al., 2000; Heine et al., 2013; Tugend et al., 2014). Tectonic events had affected cascading changes in geomorphology, sea level, climate, and other factors.

Vertebrate Ichnology
https://doi.org/10.1016/B978-0-443-13837-9.00002-0

The Cretaceous is known for warmth tied to high levels of atmospheric carbon dioxide (Barron and Washington, 1985; Berner, 1990), but had fluctuating temperatures. The earliest Cretaceous (Berriasian–Barremian) transitioned between the Hothouse and Coldhouse (Frakes et al., 1992), with an average global temperature of 17°C (Zhang et al., 2021). Extreme events during the Barremian–Aptian (~128 Ma) shifted to the Hothouse (Scotese et al., 2021), followed by an Aptian–Albian Cold Snap (118 Ma–111 Ma) (O'Brien et al., 2017). The temperature peaked in the Cenomanian to Turonian (94 Ma–93 Ma) at 28°C (Scotese et al., 2021), with potentially ice-free poles. Notably, fossils of tropical plants and dinosaurs have been found in Antarctica (Cantrill and Poole, 2012). Temperatures gradually decreased, stabilizing the hothouse global average at 23°C from the "mid"-Cretaceous to the Paleogene (Scotese et al., 2021). In an overall greenhouse climate, the Cretaceous may have experienced an increase in the frequency and intensity of extreme precipitation events. These occurrences could have had a definite impact on the preservation of tetrapod fossils, including tracks.

The Cretaceous terrestrial ecosystem ushered in a revolution, known as the "Angiosperm Terrestrial Revolution" (Benton et al., 2022), which resulted in a complete reversal of plant patterns on the continents. The angiosperm revolution directly changed the pollinating insect biota (Benton et al., 2022), but it is controversial whether it had a clear impact on the vertebrate biota, especially the dinosaurs that flourished the most in the Cretaceous (Benton et al., 2022). At least in the fossil record, there is no evidence that angiosperms became the exclusive diet of herbivorous dinosaurs (Barrett and Rayfield, 2006; Chin, 2007).

Tectonic, climatic, and vegetational changes affected the Cretaceous fauna to varying degrees. The diversity of terrestrial vertebrates increased significantly compared with the Jurassic (Benton et al., 2022). A number of famous and highly speciose Cretaceous faunas have been discovered, such as the Early Cretaceous Jehol Biota in China (Zhou, 2006; Sullivan et al., 2014) and the Late Cretaceous Hell Creek Formation fauna in North America (Fastovsky and Bercovici, 2016). Tetrapod track records also yielded noteworthy ichnofaunas, such as in the Liujiagou fauna of Gansu, China. An approximately 600 m^2 bedding plane surface in this region has exposed around 30 trackways attributed to pterosaurs, birds, theropods, sauropods, and ornithopods, reflecting a minimum ichnodiversity of eight ichnotaxa (Zhang et al., 2006).

In addition to diversity, tetrapod tracks provide more information than skeletal fossils in terms of restoring biological posture, movement, behavior, etc. Further data analysis can also obtain some paleoecological information, such as that the number of maximum sizes of theropods increased during the Cretaceous (Lockley and Xing, 2021). Here we review the footprint ichnogenera of the Cretaceous, providing an overview of their taphonomy, ichnotaxonomy, ichnofaunal diversity, track-trackmaker correlation, behavior, locomotion, paleoecology, and biostratigraphy.

2. History of study

Fossil tetrapod tracks played a significant role in human culture from the earliest times. Even before the rise of science, humans noticed vertebrate tracks on the surface of rocks and considered them miraculous, and some tribes even used them as totems (Mayor, 2005). Taking China's tracks as examples, many Cretaceous dinosaur tracks have been linked to myths and historical figures. For instance, the Early Cretaceous theropod tracks in Chicheng City, Hebei Province, are still believed by local people to be the traces of a phoenix landing on the rock in recent years (Xing et al., 2011a), and the Early Cretaceous sauropod tracks in Yiwu City, Zhejiang Province, are believed by local people to be the buttocks prints of the Chinese god Pan Gu (Xing et al., 2021a).

The terrestrial strata of the Cretaceous are well developed, and the tetrapod tracks from Cretaceous strata are distributed in all regions except Antarctica. But the research history varies from region to region.

2.1 North America

Fossil tetrapod tracks from North America have been studied in more detail than those of other regions. They were first reported by Mudge from the Republican River tracksite, Kansas, USA, possibly from Dakota Formation (Upper Albian), decided as bird tracks initially but not effectively. The description of Cretaceous tetrapod tracks was not available for study in North America till Snow described the world's first bird track in the same site in 1886 (Mudge, 1866; Snow, 1886; Lockley and Lucas, 2014).Detailed research that has been conducted on tetrapod tracks in North America since 1980, mostly concentrated in the Midwest of the USA (west of the Mississippi River) and Canada (Lockley and Hunt, 1995a; Lockley et al., 2014a); see Lockley and Hunt (1995a) for a comprehensive review. Multiple sites are also known from Mexico, as reviewed by Rodríguez-de la Rosa et al. (2004). Representative examples include:

(1) Cedar Mountain Formation in Utah. This fossil site has strong diversity, with the discovery of nonavian theropod, avian, sauropod, ornithopod, and thyreophoran tracks (Lockley et al., 2014b).
(2) Glen Rose Group in Texas, mainly including large theropod tracks and wide-gauge sauropod tracks (Farlow et al., 2012).
(3) Dakota Group in Colorado and Utah. There are over 120 tracksites in this unit, but no sauropod tracks have been found. Trackmaker taxa include ornithopods, thyreophorans, nonavian theropods, crocodylomorphs, avians, and pterosaurs (Lockley et al., 2014c). In the eastern part of this stratigraphic unit, ornithopod tracks are dominant, while in the western part thyreophoran tracks are dominant (Lockley et al., 2014c).
(4) Mist Mountain Formation in western Canada. The first sauropod track in northern North America was discovered in this unit; in addition to dinosaur, avian, and pterosaur tracks crocodylomorph tracks are also known (McCrea et al., 2001).
(5) Dunvegan Group in Canada. This unit is the transition from the Early Cretaceous to the Late Cretaceous, and a large number of thyreophoran tracks, as well as large ornithopod tracks, various sizes of nonavian theropod, avian, turtle, and crocodylomorph tracks have been discovered (McCrea et al., 2014a).
(6) Cantwell Formation in Alaska, a unique track association that includes hadrosaurs and therizinosaurs. This association was first reported from North America but was subsequently discovered in central Asia (Fiorillo et al., 2018).
(7) San Juan Raya Formation (Aptian) in Mexico. The tracks in this formation are referable to Theropoda, Sauropoda, and Ornithopoda, mainly from the San Martín Atexcal tracksite (Rodríguez-de la Rosa et al., 2004).

2.2 South America

The Cretaceous tetrapod tracks in South America were mainly recorded by Leonardi (1984, 1989, 1994). There are two high-level summaries: Leonardi (1994) reported 115 sites, of which 88 belonged to the Mesozoic Era; 38 belonged to the Cretaceous. Lockley et al. (2012a) added about 16 Cretaceous sites on the basis of the former, making the total number of sites reach about 54. The Rio do Peixe basins in northeastern Brazil (Sousa, Antenor Navarro, and Rio Piranha formations) have a particularly abundant ichnofauna, which was recently summarized in a comprehensive book (Leonardi and Carvalho, 2021c). This area preserves 37 sites on ca. 96 different levels, including 535 dinosaur trackways as well as less common tracks of nondinosaurian tetrapods (Leonardi and Carvalho, 2021c). Important dinosaur tracksites are also found in Late Cretaceous deposits of Bolivia, such as Humaca, which preserves 11 subparallel titanosaur trackways, and Pujacata, which preserves large didactyl tracks referred to dromaeosaurids (Meyer et al., 2021). The most extensive tracksite is Cal Orck'o, with an estimated total of 14,000 tracks, including 40 trackways that are more than 100 m in length, with the longest reaching 620 m. Cal Orck'o is dominated by theropods (75%), followed by sauropods (11%) and ornithopods (9%) (Meyer et al., 2021). The Argentine record primarily focuses on the Neuquén Basin (Calvo, 1991, 1999; González Riga and Tomaselli, 2019; de Valais et al., 2020) and the Yacoraite Formation (Alonso and Maquillas, 1986; Cónsole-Gonella et al., 2017; Díaz-Martínez et al., 2018), which contains numerous dinosaur tracks of sauropods, theropods, and ornithischians. Although more sparse, vertebrate tracks from the Cretaceous Period have also been recorded in Chile (Rubilar-Rogers et al., 2010), Uruguay (Mesa and Perea, 2015), Colombia (Cortés et al., 2023), and Peru (Jaillard et al., 1993).

Of particular importance is also the Early Cretaceous Botucatu Formation in the Paraná Basin of southern Brazil, which represents a paleodesert, with tracks preserved in dune deposits. This ichnofauna is dominated by tracks of early mammals, though theropod tracks, including cursorial forms such as *Farlowichnus*, and the single trackway of a lacertiform reptile are also preserved (Leonardi et al., 2024).

Avian and pterosaur tracks from the Cretaceous period in South America are exceedingly rare. The avian track sites are confined to the Neuquén Group (Coria et al., 2002; de Valais et al., 2020), the Angostura Colorada Formation (Leonardi, 1994; de Valais et al., 2020), and the Yacoraite Formation (de Valais and Cónsole-Gonella, 2019). Pterosaur tracks have only been found in the Neuquen Group (Candeleros and Anacleto formations) of Neuquen and Río Negro provinces, furthermore the avian track *Yacoraitichnus* (Calvo and Lockley, 2001; de Valais et al., 2020; Díaz-Martínez et al., 2022; de Valais and Cónsole-Gonella, 2019). Sauropod, theropod, and ornithischian tracks have also been found in the area (Calvo, 1999; Heredia et al., 2023).

2.3 Europe

The research history in the European region is the longest, especially in Western Europe. *Chirotherium*, the first scientifically named tetrapod ichnotaxon, was described from the Middle Triassic Buntsandstein of Germany in 1835 (Kaup, 1835a, 1835b) and first believed to have been left by mammals or amphibians, later conjectured to come from pseudosuchian archosaurs (Swinton, 1961). Tagert (1846) had first discovered Cretaceous tetrapod tracks from

Lower Cretaceous Wealden group of the southern coast of England and the Isle of Wight, all of which are believed to have been produced by *Iguanodon* (Lockwood et al., 2014). Sarjeant et al. (1998) named *Iguanodontipus* of the Purbeck Group in Dorsetshire, England. In 2014, Lockley et al. (2014e) identified that the ornithopod tracks of the Wealden group belong to *Caririchnium*, and there may also be nonavian theropod, sauropod, and other ornithischian tracks (Lockwood et al., 2014).

In continental Europe, the most productive areas for tetrapod tracks are the Cameros Basin and the region of Burgos, Spain (the Enciso, Urbión, and Oncala groups), Italy and Croatia, and central Germany (Obernkirchen Sandstone). La Rioja in particular preserves the most extensive tetrapod track record in Europe, which is estimated at tens of thousands of tracks (Pérez-Lorente, 2015, p. 3), even though only a fraction has been excavated at more than 100 tracksites. The track record of La Rioja has been comprehensively reviewed by Pérez-Lorente (2015). Other regions of the Iberian Peninsula, such as Portugal, Teruel province, and the Pyrenees, also preserve important sites (e.g., Santos et al., 2008; Vila et al., 2013a; Castanera et al., 2022b). Tracks in Italy and Croatia are preserved in carbonate platforms (the Apulian, Laziale-Abruzzese-Campana, and Periadriatic carbonate platforms) (Dalla-Vecchia 2008; Petti et al., 2020). The first, and most extensive, of these tracksites was discovered near the town of Altamura, which includes about 30,000 dinosaur tracks over a total area of 15,000 m^2 (Nicosia et al., 1999). The Obernkirchen Sandstone yields numerous sites, such as the Münchehagen tracksite, that include theropod, ornithopod, sauropod, and ankylosaurian tracks (Hornung et al., 2012; Lallensack et al., 2016; Richter and Böhme, 2016). The northernmost European dinosaur tracks come from the polar island of Spitzbergen (Lockley and Meyer, 2000). Nondinosaurian tracks from the Cretaceous of Europe include those of pterosaurs, birds, crocodylians, and turtles (see review in Lockley and Meyer, 2000).

2.4 Africa

Research in Africa is still in its infancy, with scattered and scarce records, mainly concentrated in northern Africa (Lockley et al., 2014a). As early as 1880, tridactyl theropod tracks from the mid-Cretaceous were discovered in northern Algeria (Bellair and de Lapparent, 1948; Bassoullet, 1971). Recently, a series of Early Cretaceous dinosaur tracks were discovered in the d'El Bayadh region of northern Algeria (Mahboubi et al., 2004, 2007; Bensalah et al., 2005). The 12 trackways consist of approximately 350 tridactyl and tetradactyl tracks, as well as sauropod tracks. These tridactyl tracks are tentatively assigned to *Grallator* and *Eubrontes*, while the sauropod tracks are assigned to *Brontopodus* (Bessedik et al., 2008). Until now, the Cretaceous tetrapod tracks in Africa have been concentrated mainly in the region of North Africa dominated by Algeria, including Morocco and Tunisia, both of which border Algeria.

Over the past 60 years, a number of Cretaceous tetrapod tracks have been discovered in Morocco, including theropod, sauropod, possible ornithopod, crocodile, possible pterosaur tracks, and turtle swimming tracks (Cavin et al., 2010; Masrour et al., 2017a, 2017b; Belvedere et al., 2013). Contessi and Fanti (2012) reported the discovery of Cretaceous avian tracks in southern Tunisia. There are only three tracks, and they are assigned to *Koreanaornis*, the first record of this genus in Africa, indicating the worldwide distribution of shorebird tracks.

Cameroon, which is in the Midwest of Africa preserved the most southern records. The first Cretaceous dinosaur tracks in Cameroon were discovered in the Koum Basin in 1989. The well-preserved theropod trackway consists of four large tridactyl tracks (Dejax et al., 1989; Jacobs et al., 1989). Martin et al. (2017) reported stratigraphically younger dinosaur trackways from the Mamfe basin in western Cameroon.

2.5 Australia

Despite a limited Cretaceous dinosaur body fossil record in Australia (Long, 1998), the continent's trace fossil record, specifically footprints, offers a richer insight into dinosaur abundance and diversity. Key tracksites are found in four formations: Queensland's Winton Formation, Western Australia's Broome Sandstone, and Victoria's Wonthaggi and Eumeralla Formations. Indigenous peoples have long recognized tridactyl impressions in the Broome Sandstone, traditionally interpreted as the prints of the creator being Marella, or Emu-man (Salisbury et al., 2016). Comprehensive studies reveal a broad diversity of dinosaur tracks in this region, contrasting sharply with the scarce body fossils in Western Australia (Salisbury et al., 2016). The Winton Formation, famous for its track-bearing sites at Lark Quarry, came to attention in the late 1960s when numerous footprint casts were discovered at the nearby Seymour Quarry site by Glen Seymour. Lark Quarry, significant for challenging the then-prevailing view of dinosaurs as large, sluggish creatures, was interpreted by Thulborn and Wade (1984) as a site of a dinosaur stampede, a notion later contested (Romilio and Salisbury, 2011, 2014; Romilio et al., 2013; Romilio, 2023). Snake Creek, also in the Winton Formation, is notable for containing the only known Cretaceous sauropod tracks in eastern Australia (Poropat et al., 2021).

The Wonthaggi and Eumeralla Formations in Victoria, dating to a period of Australia's separation from Antarctica, have yielded nonavian and avian theropod tracks as well as ornithopod prints from multiple track-bearing localities along coastal exposures, with further tracksite discoveries likely in the near future (Romilio and Godfrey, 2022).

2.6 East Asia

The record of tetrapod tracks in East Asia is quite abundant, with China and South Korea being the main sites. A very important international expedition greatly promoted tetrapod track research in Asia. In 1999–2001, The Ministry of Education, Science, Sports and Culture in Japan funded an international study tour, which was entitled: Dinosaur research as a tool for science education, public education, and life-long learning. Asian, North American, and European scientists on the expedition team collected a large amount of track data in China, Korea, Japan, Laos, Mongolia, and Thailand (Matsukawa et al., 2006).

The earliest record of tracks in South Korea came from 1969, when Kim (1969) described the avian track *Koreanaornis*. Since then, Yang (1982) described a series of tracks from the Jindong Formation. In the past 30 years, many scientists have collaborated closely and published a series of important papers in a systematic way (such as Lim et al., 1989, 1994; Lockley et al., 1992a, 2006a, 2008a). The tracks are almost all preserved in the Gyeongsang Supergroup, especially in the Jindong Formation (Cenomanian–Santonian) and Haman

Formation (Albian–Cenomanian, Houck and Lockley, 2006). There are at least six new ichnospecies in the Haman Formation, including avian, dromaeosaurid, sauropod, theropod, and pterosaur (Lockley et al., 2012a). There are hundreds of track sites in the Jindong Formation, mainly ornithopod and avian tracks (Lockley et al., 2014e). These track sites are mostly concentrated along the southern coast of South Korea, and include sauropod, nonavian theropod, avian, and ornithopod tracks, which has been dubbed the Korean Cretaceous Dinosaur Coast (KCDC; Lockley et al., 2012b).

In 1940, Japanese paleobiologist Yabe H. recorded over 4000 theropod tracks in the Liaoning Province and named all 4000 of them *Jeholosauripus s-satoi* (Yabe et al., 1940). This is the first record of Cretaceous dinosaur tracks in China. From 1929 to 1979, Zhongjian Yang (Young, CC) pioneered the study of tetrapod tracks in China, and his research results also included Cretaceous records, such as *Yangtzepus* (Young, 1960). In the 1980s–1990s, the Beijing Natural History Museum conducted systematic research on tetrapod tracks discovered in Sichuan and Yunnan provinces in China and actively engaged in international exchanges. Entering the 21st century, the study of tetrapod tracks in China has emerged with characteristics of internationalization, large-scale, and multidisciplinary. In 2012, Lida Xing convened the Qijiang International Dinosaur Tracks Symposium, Chongqing, China. This was the first dinosaur track symposium held in China.

More than 150 Cretaceous track sites have been reported from China, most of them have been discovered in the last 15 years or so. Among these tracksites, the Jurassic–Cretaceous boundary Tuchengzi Formation (Kimmeridgian–Valanginian) is particularly important because it connects the Jurassic Yanliao fauna and the Cretaceous Jehol fauna. This formation lacks skeletal fossils, but records sauropod, theropod, bird, and ornithopod tracks, with those of saurischians being dominant (Xing and Zhang, 2022; Xing and Lockley, 2021; Xing et al., 2021b).

Early and "Mid"-Cretaceous tracksites include the Dasheng Group in Shandong from East China (Li et al., 2015; Xing et al., 2019a); the Jiaguan, Feitianshan and Xiaoba formations in Sichuan from southwest China (Xing and Lockley, 2016; Xing et al., 2015a, 2015b, 2016a); Hekou Group in Gansu (Zhang et al., 2006; Xing et al., 2013a, 2021c); and the Tugulu Group in Xinjiang (Xing et al., 2013b, 2020, 2021d), from northwestern China. Most of these deposits include dozens of well-described sites. Among them, the Feitianshan Formation records the largest dinosaur tracksite in China: over 9000 m^2 with 1928 tracks (Xing et al., 2021d). The records of the Late Cretaceous are relatively sparse and the sites with relatively higher diversity are from Heilongjiang, from North China (Xing et al., 2019b); and Zhejiang (Xing et al., 2018a), Guangdong (Xing et al., 2017a) and Fujian (Niu and Xing, 2023), from southern China. Most of these sites are dominated by ornithopod tracks.

In addition to the rich fossil record of tetrapod skeletons, Mongolia is also quite rich in tracks. There were at least 20,759 tracks recorded over 20 years ago (Ishigaki et al., 2009). Namnandorzhi (1957) first recorded dinosaur tracks in Mongolia from Lower Cretaceous sandstones of Mt. Sayzhrakh. A series of joint Mongolian and foreign expeditions resulted in numerous discoveries. For example, the Hayashibara Museum of Natural Sciences Mongolian Paleontological Center Joint Paleontological Expedition (HMNS-MPC Expedition) discovered abundant dinosaur tracks in the Upper Cretaceous of Shar Tsav, South Gobi (Suzuki and Watabe, 2000; Watabe and Suzuki, 2000). The Korea–Mongolia International Dinosaur Project recorded a series of theropod tracks from the Nemegt Formation, Bügiin

Tsav, Mongolia (Lee et al., 2018). In comparison, the diversity of tetrapod tracks in Mongolia is relatively low, including at least hadrosaur, theropod, sauropod, and ankylosaur tracks (Currie et al., 2003; Ishigaki et al., 2009). More detailed studies need to be carried out in the future. It is also a fascinating task to compare tracks with skeletal records thanks to the rich skeletal record, such as a protoceratopsid skeleton with an associated track (Niedźwiedzki et al., 2012) and theropod trackways associated with a *Gallimimus* foot skeleton (Lee et al., 2018).

The first record of Cretaceous tetrapod tracks in Thailand was in 1985 in the Phu Luang Wildlife Sanctuary, Loei Province in the northeast (Buffetaut and Ingavat, 1985). Currently, all Cretaceous tracks have been reported in the northeast, including theropod tracks, small ornithischian tracks, sauropod tracks, ornithopod tracks, and crocodilian tracks. Among them, the theropod tracks found in the Khorat Group of Nakhon Ratchasima Province were named *Siamopodus khaoyaiensis* due to their bilobed heel impressions (Lockley et al., 2006b).

Matsukawa and Obata (1985) first reported the discovery of tetrapod tracks in the Cretaceous in Japan, which were found in Sebayashi, Nakazato, and Gunma. Since then, a large number of track sites have been discovered in the famous Tottori Group in central Japan, including theropod, ornithopod, and possibly tetrapod tracks (Matsukawa et al., 2005). Among them, Matsukawa et al. (2005) named the new ichnotaxon *Asianopodus pulvinicalx*, found in the Kuwajima Formation. The assemblage of tracks and skeletal fossils in the Tottori Group is consistent with the fossil assemblage of the same period in East Asia (Matsukawa et al., 2005). Later, theropod tracks were discovered in Toba, Mie (Saka, 2001), and ornithopod tracks were discovered in Mifune, Kumamoto (The board of dinosaur fossil research of Mifune Town, 1998).

3. Ichnotaxonomy

For the purpose of this review, we mainly describe the valid ichnogenera and valid ichnospecies defined in the Cretaceous, with a comparative diagnosis and precise reference to the type material. We also include some valid ichnospecies in the Cretaceous of Triassic and Jurassic ichnogenera. Ichnogenera and ichnospecies that have been clearly revised or reduced are mentioned but not in detail.

There are some differences in our description of the Distribution section. For most ichnogenera with a widespread distribution, we only provide the range of stages and the countries in which they are found, and specific stratigraphic units are only be mentioned selectively due to space limitations.

Institutional abbreviations: AC, Hitchcock Ichnology Collection at the Pratt Museum of Amherst College, Amherst Massachusetts, USA; AMNH, American Museum of Natural History, New York, USA; BCPM, British Columbia Provincial Museum, Canada; BNSS, Bournemouth Natural Science Society, Bournemouth, UK; BYU, Brigham Young University, Utah, USA; CFEC, C (Chinese) F (Footprint) E (Emei) C (Chuanzhu) specimens housed in the Chongqing Museum of Natural History, Chongqing, China; CNUPH, Chonnam National University, Kwangju, South Korea; CU, University of Colorado at Denver, USA; CUE, Chinju National University of Education, Gyeongsang, South Korea; CUGM, Museum of Geology, Chulalongkorn University, Bangkok, Thailand; CU-MWC, University of Colorado at Denver,

Museum of Western Colorado Collection, USA; DGSU, Department of Geology, Seoul University, South Korea; DMNH, Dallas Museum of Natural History, Texas, USA; DMNH, Denver Museum of Natural History, Colorado, USA; DORCM, Dorset County Museum, UK; FGM, Fraser-Fort George Regional Museum, Prince George, Canada; FSA, Université Ibn Zohr, Faculté des Sciences, Agadir, Morocco; GSC, Geological Survey of Canada; GSFC-VP, Goddard Space Flight Center, NASA, USA; GZG, Geoscience Cenetr, University of Göttingen, Germany; IANIGLA, Instituto Argentino de Nivología, Glaciología y Ciencias Ambientales, Argentina; IVPP, Institute of Vertebrate Paleontology and Paleoanthropology, Beijing, China; JQW, Jiuquwan specimens in Hunan Museum of Geology, Changsha, China; KIGAM VP, Korea Institute of Geoscience and Mineral Resources, Vertebrate Paleontology, South Korea; KNUE, Department of Earth Science Education, Korea National University of Education, South Korea; KPE, Kyungpook National University, Earth Science Education Department, Daegu, South Korea; LDRC, Lufeng Dinosaur Research Center, Yunnan, China; LPP, Laboratory of Paleoecology and Paleoichnology, Federal University of São Carlos, São Paulo, Brazil; MCF-PVPH-SB, Museo Carmen Funes, Paleontología de Vertebrados, Plaza Huincul, Sierra Barrosa collection, Argentina; MGUAN-PA, Geological Museum of Universidade Agostinho Neto-PaleoAngola Project (Catoca collection) in Luanda, Angola; MJSN, Natural History Museum of the Canton Jurassica Museum, Porrentruy, Swiss; MNRJ, National Museum of Rio de Janeiro, National Museum of Brazil, Brazil; MNS, Numantine Museum of Soria, Spain; MUCP, Museo de Geología y Paleontología, Universidad Nacional del Comahue, Neuquén, Argentina; NHCG, National Heritage Center Geology, Cultural Heritage Administration, South Korea; NMB, Natural History Museum of Basel, Switzerland; NMC, Canadian Museum of Nature, Ottawa, Canada; PMA, Provincial Museum of Alberta, Canada; PRPRC, Peace Region Paleontology Research Center, Tumbler Ridge, British Columbia, Canada; QJGM, Geological Museum of Qijiang District Bureau of Land Resources, Chongqing, China; QM, Queensland Museum, Brisbane, Australia; RMDRC, Rocky Mountain Dinosaur Resource Center, Colorado, USA; SMU, Shuler Museum of Paleontology, Southern Methodist University, Texas, USA; TGU, Tokyu Gakugei University, Tokyo, Japan; TMM, Texas Memorial Museum, USA; TMP, Royal Tyrrell Museum of Palaeontology, Drumheller, Canada; UCM, University of Colorado Museum of Natural History, Boulder, USA; UQ, The University of Queensland, Brisbane, Australia; WAM, Western Australian Museum, Perth, Australia.

Tracksite abbreviations: AC, Agua del Choique tracksite, Neuquén, Argentina; ACDL, Cava De Lucia, Altamura, Italy; BL, Beilin site, Shandong, China; CHABU, Chabu site, Otog Dinosaur Trace National Nature Reserve, Inner Mongolia, China; CYCD, Yuanjitun, Cangling, Yunnan, China; EVP, El Villar-Poyales site, La Rioja, Spain; GDM, Litan site, Gansu, China; LC, Los Campos, Sierra de Oncala, Soria, Spain; LG, Zhangzhuhewan site, Shandong, China; LRH, Li R collection in Qingdao Institute of Marine Geology, Shandong, China; LS, Jishan site in Linshu County, Shandong, China; LTR, Los Tormos, Santa Cruz de Yanguas, Soria, Spain; NGZ, Nanguzhai site, Jiangsu, China; PGN, Plagne, Gravesiana Subzone, France; QJD, Qianjiadian site, Beijing, China; SOL, Solaris Campground, Istria, Croatia; SOPPb, Passagem das Pedras in the Hazenda Ilha, Sousa, Pariabia State, Brazil; TDGZ, Tangdigezhuang site, Shandong, China; UQL-DP, University of Queensland Dampier Peninsula Field Locality, Australia; XHT, Xiaohutian site, Anhui, China; ZP, Zhongpu site, Gansu, China.

3.1 Sauropods, saurischia

Though both most typical and widespread wide-and narrow-gauge ichnogenera of sauropods (i.e., *Brontopodus* and *Parabrontopodus*, the latter now including the ichnospecies of *Elephantopoides*, an older name for *Parabrontopodus*) occur in the Cretaceous, all ichnospecies considered valid of *Parabrontopodus* (or "*Elephantopoides*") have only named based on Jurassic material. So, the Cretaceous section does not include *Parabrontopodus*, which is treated in the Jurassic section.

3.1.1 **Brontopodus** *Farlow et al., 1989 (Fig. 7.1)*

3.1.1.1 Type ichnospecies

Brontopodus birdi Farlow et al., 1989.

3.1.1.2 Valid ichnospecies in Cretaceous

Brontopodus birdi Farlow et al., 1989.
Brontopodus changlingensis (Chen and Huang, 1993); Lockley et al., 2013.

3.1.1.3 Distribution

Jurassic–Coniacian, worldwide distribution in Algeria, Argentina, Canada, Chile, China, France, Morocco, Poland, Portugal, South Korea, Spain, Switzerland, the United Kingdom, United States, etc.

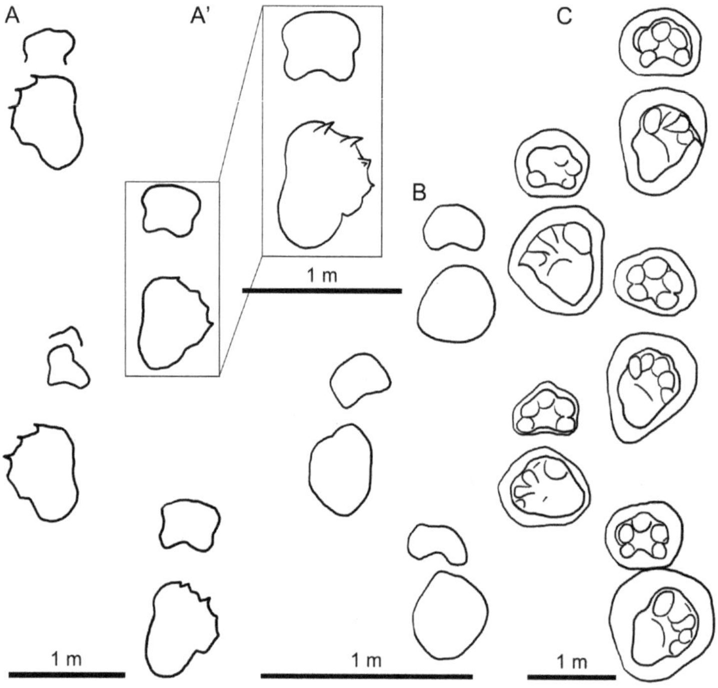

FIGURE 7.1 Tracks of Cretaceous valid ichnospecies of *Brontopodus*. (A) Partial type series of *Brontopodus birdi*; (B) series of *Brontopodus changlingensis*; (C) Partial type series of *Brontopodus plagnensis*. *(A) Partial New York slab; redrawn from Lockley et al. (1994a), Kim and Lockley (2012). (B) Holotype undesignated; redrawn from Lockley et al. (2002). (C) Redrawn from Mazin et al. (2017).*

3.1.1.4 Holotype

AMNH 3065 and TMM 40,638-1. Its type locality is TMM 40,638, Paluxy River tracksite #A3 (Bird Site), which is in an Albian shallow subtidal wackestone/mudstone in the Glen Rose Limestone Formation of Texas.

3.1.1.5 Diagnosis

Sauropod ichnites of small to large size, known pes footprint length ranging from 50 to more than 100 cm. Manus footprint length and width about the same in well-preserved tracks; manus tracks clawless, somewhat U-shaped, with digit impressions I and V slightly separated from the impression of the conjoined Digits II–IV. Pes tracks longer than broad, with large, laterally directed claw marks at Digits I–III (diminishing in size from I to III), a small claw, nail, or callosity mark at Digit IV, and a small callosity or pad mark at Digit V; digit marks IV and V only seen in well-preserved footprints. Manus tracks often (usually?) rotated outward with respect to direction of travel. Manus track medial to a line through pes track long axis, such that manus track centers are somewhat closer to the trackway midline than pes track centers. Trackway is broad, with left and right manus and pes footprints often well away from the trackway midline; trackway width roughly 1–1.5 times pes track length. Outer limits of trackway defined by pes tracks. Manus-pes distance 0.5–1.2 times pes footprint length. Stride length roughly 2–5 times pes track length. Step angle generally 100°–120°. Gleno-acetabular length (a distance for interpreting the size and gait of quadrupedal trackmakers) *c.* 3–4 times pes track length. Tail drag marks rare or absent (diagnosis emended by Marty et al., 2010).

3.1.1.6 General description

Brontopodus birdi was described based on the trackways from the Early Cretaceous Glen Rose Formation of Texas, USA (Farlow et al., 1989). Defined as a wide-gauge trackway with nonclawed pentadactyl manus prints and elongate, outwardly rotated pes prints. Manus outward rotation as low as 10° with heteropody (manus/pes area ratio) area ratio >1:3 (Mazin et al., 2017).

The known species of *Brontopodus* exhibit a relatively wide range of manus track morphology. Both *B. changlingensis* and the type species have a width of manus prints equal to or larger than the width of pes prints, whereas *B. triangularis* is rather different. The feature of elongated pes tracks is therefore stable, at least in the individually named ichnospecies, and the large differences in manus morphology may be related to the nature of the sediment. High heteropody (>1:2) is not a stable feature of the known species of *Brontopodus*. For example, *B. plagnensis* shows relatively weak heteropody (about 1:1.86). The outward rotation of pes and manus tracks is also a relatively constant feature in *Brontopodus*, but it is not significant, and manus and pes tracks vary across ichnospecies. They are similar in *B. changlingensis*, but in *B. plagnensis*, the rotation of the pes print is more pronounced than that for the manus print. Claw mark features are not commonly used to distinguish between different ichnospecies within *Brontopodus*. However, particularly well-preserved pes prints of *B. plagnensis* are characterized by a unique elongate and sharp antero-lateral claw mark that points postero-laterally from Digit IV.

Brontopodus changlingensis was named by Chen and Huang (1993). Its type specimen is CYCD-09 FP8, type locality is Yuanjitun, Cangling tracksite 1, which is in a Turonian/Coniacian terrestrial sandstone in the Jiangdihe Formation of Yunnan, China. *B. changlingensis* is characterized by high heteropody (1: 2.9), width of manus prints equal to or larger than width of pes prints, manus tracks aligned with and partially overprinted by anterior margins of pes tracks, pes and manus tracks rotated slightly outward relative to trackway midline (~17°), and average pace angulation of manus tracks 122°, and of pes tracks 117° (Xing et al., 2019c).

The only Jurassic ichnospecies considered valid is *Brontopodus plagnensis,* which was defined based on the trackway PGN 01 from the Late Jurassic of the Jura Mountains, eastern France (Mazin et al., 2017). Characterized by positive rotation of the pes prints more pronounced than that for the manus prints, subtriangular pentadactyl pes prints with a significant mesial bulge, a unique elongate and sharp antero-lateral claw mark that points posterolaterally from Digit IV on particularly well-preserved pes prints, and high heteropody (index HI [heteropody index] = 1.86).

3.1.1.7 Discussion

3.1.1.7.1 The wide-and narrow-gauge sauropod tracks Sauropod trackways could be distinguished according to trackway width as wide gauge and narrow gauge (Farlow, 1992). The best examples of the wide and narrow-gauge sauropod trackways are *Brontopodus* -type *B. birdi* (Farlow et al., 1989) and *Parabrontopodus* -type *P. mcintoshi* (Lockley et al., 1994a), respectively. Not surprisingly, discussions of these two ichnogenera/morphological types often take place at the same time. The discussion and trackmaker sections on *Brontopodus* here cover the contents related to both, so the corresponding section on *Parabrontopodus* below only provides additional information that is not included in this section, and which is essentially related to *Parabrontopodus*.

Brontopodus was erected and described on the basis of trackways from the Lower Cretaceous Glen Rose Formation of Texas (Farlow et al., 1989). Lockley et al. (1994a) followed the scheme of Farlow (1992) for *Brontopodus* and *Parabrontopodus*, except for the following minor changes. To the list of wide-gauge trackways, they added atleast some of the documented examples from the Upper Jurassic (Kimmeridgian) of Switzerland (Meyer, 1993) and the Upper Jurassic (Portlandian) of Portugal (Lockley and Meyer, 1994; Lockley et al., 1994a), and provisionally assigned both trackway types to *Brontopodus* isp. (Lockley et al., 1994a), and also assigned the Middle Jurassic sauropod trackways from Portugal (Santos et al., 1994) to this ichnotaxa. The latter was attributed to *Polyonyx*, a new ichnogenus erected in 2009 based on their obvious morphological difference with other *Brontopodus* in their largely asymmetric manus and tetradactyl pes prints (Santos et al., 2009).

Globally, *Brontopodus* is more common in the Cretaceous. Lockley et al. (2002a) described sauropod tracks in China for the first time, considering most sauropod trackways in China to be wide- (or medium-) gauge and therefore referred them to the ichnogenus *Brontopodus*, and and indicated most Early Cretaceous sauropod tracks in East Asia could be attributed to the typical *Brontopodus* or *Parabrontopodus*. Lockley et al. (2002a) also considered the type of *Chuxiongpus* to be a junior synonym of *Brontopodus*, referring both *C. changlingensis*, and *C. zheni* to *B. changlingensis*. These *Chuxiongpus* materials are currently quite rare Late Cretaceous sauropod tracks in China and have been re-described (Xing et al., 2019c).

While the use of *Brontopodus* and *Parabrontopodus* as two ichnotaxa that typically distinguish important traits of wide- and narrow-gauge in sauropod tracks remains important, it has been demonstrated that sauropod trackway gauge may sometimes vary from narrow to wide within a single trackway. Examples from the Upper Jurassic of Switzerland (Marty et al., 2010) and the Upper Cretaceous of Bolivia (Meyer et al., 2018) cast doubt on whether the two sauropod ichnotaxa of *Brontopodus* and *Parabrontopodus* can be separated by gauge variability alone. A revision of the known sauropod trackways would be helpful to address these concerns (Meyer et al., 2021).

3.1.1.7.2* The ichnospecies of *Brontopodus Nowadays, the large number of late Mesozoic *Brontopodus* from around the world have been assigned to the ichnogenus level, to *Brontopodus*. Several *Brontopodus* ichnospecies that have been named so far require further discussion, some may be *nomina nuda*, or belong to another ichnogenus; for example, *B. changlingensis* is distinguishable from other *Brontopodus* ichnospecies by having wider manus prints (Xing et al., 2019c). *B. plagnensis* is clearly differentiated by the clawless pedal Digits I–III and the latero-posterior clawlike mark on Digit IV (Mazin et al., 2017).

When *Brontopodus pentadactylus* compared with other *Brontopodus* ichnospecies, the manus tracks of *B. pentadactylus* showed a distinct positive (outward) rotation that varies from 40° to 90° based on the axis of inferred Digit III (Kim and Lockley, 2012). Due to the differences between *B. pentadactylus* and other wide-gauged trackways, such as the interautopodial spacing pattern in *B. pentadactylus*, compared with *B. birdi*, it is very similar to *Oobardjidama* (Salisbury et al., 2016) and *Titanopodus* (González Riga and Calvo, 2009), and the heteropody in *B. pentadactylus* is more similar to *Polyonyx* (Santos et al., 2009) than other wide-gauged ichnotaxa, whereas the pedal trackway ratio as medium gauge is dissimilar enough to *B. birdi* to query it (Salisbury et al., 2016). Salisbury et al. (2016) suggested *B. pentadactylus* must be a new ichnogenus rather than being referable to *Brontopodus*. This ichnospecies is reassigned in this chapter to *Gyeongsangsauropus* ichnogen. nov., which will be described at the end of this section.

Some poorly preserved *Brontopodus* tracks were named and are currently regarded as *nomina dubia* or *nomina nuda*. They at least include *B. oncalensis*. *B. oncalensis* was named by Meijide Fuentes et al. (2004). Its locality is La Revilleja tracksite, Soria, which is in a Berriasian fluvial sandstone in the Huérteles Formation of Spain. The tracks are much smaller (pes tracks mean 48.2 cm in length), oval in shape. There are no digit impressions except for several internal, rounded pits inside the footprint outline. The manus tracks are subrounded in shape, but no typical half-moon outline can be discerned. The manus prints are located very near the pes prints, but no overlapping is observed (Moratalla, 2009). Moratalla (2009) considered *B. oncalensis* as of doubtful validity. The La Revilleja trackway is very poorly preserved and no anatomical features can be observed in the tracks. The trackway pattern is clearly that of a *Brontopodus*-like morphotype (wide-gauge group), but this is not enough to assign it to the ichnogenus *Brontopodus*. The morphology of the manus and pes prints between *B. oncalensis* and the standard trackway described by Farlow et al. (1989) is different in terms of size, shape, arrangement, and relative dimensions. Thus, the La Revilleja trackway has a *Brontopodus*-like sauropod trackway pattern (wide-gauge) but it is morphologically different from the standard described by Farlow (1989). As above, *Brontopodus oncalensis* would perhaps best be considered a nomen dubium (Marchetti et al., 2019; Torcida Fernández-Baldor et al., 2021).

3.1.1.7.3 **The indeterminate ichnospecies of Brontopodus** Given the breadth of the diagnosis of *Brontopodus* in the description of features other than the wide-gauge signature (e.g., morphological features of the manus and pes tracks), a significant proportion of the wide-gauge tracks were initially described as *Brontopodus* isp., but not assigned to specific ichnospecies. Of the three currently named and still valid ichnospecies, *B. plagnensis* was named from the Upper Jurassic of Europe, while the remaining two ichnospecies were named from the Lower Cretaceous or lower Upper Cretaceous: *B. birdi* from southern USA and *B. changlingensis* from southwestern China. These species occur during the time period when *Brontopodus* flourished. However, only *B. birdi* is believed to be found outside the named sites (such as the eastern USA, Stanford et al., 2004 and South Korea, Kim and Lockley, 2012). The other ichnospecies are only found at the named sites. The remaining materials from other regions or periods have not been identified and are largely attributed to cf. *Brontopodus* based on wide-gauge characters, particularly from South America (including Africa, Argentina, and Spain) and the Lower–Middle Jurassic. The track characters for *Brontopodus* are less robust, and they most likely represent trackmakers from a different time or region than the main occurrences of *Brontopodus*.

During the Early—Middle Jurassic, *Brontopodus* and cf. *Brontopodus* had a limited distribution, being found only in small numbers in the western USA and southwestern China. This includes the Lower Jurassic of Sichuan and eastern Tibet, and the Middle Jurassic of Yunnan and Utah (Foster et al., 2000; Xing et al., 2021e, 2023a). According to Xing et al. (2021e), they at least dominate the sauropod tracks in southwest China, accounting for over 70% of the tracks. This statement suggests that trackmakers from different evolutionary positions may exhibit similar wide-gauge features simultaneously, as noted by Xing et al. (2021e). Therefore, basal eusauropods may share the somewhat similar postcranial body plan with much more advanced neosauropods.

3.1.1.8 Trackmaker

The study of Lockley et al. (1994a) provides some clues such as narrow gauge, and strong heteropody to help identify the main track-making groups with more confidence than in previous studies. Moratalla (1994) proposed that the differences between *Brontopodus* and *Parabrontopodus* were not caused by the general body construction as much as by the disposition of the legs. Perhaps, *Brontopodus* trackmakers walked with their legs parallel to the sagittal plane while *Parabrontopodus* trackmakers had their legs angled inward with respect to that plane, like the proboscideans.

Wilson and Carrano (1999) suggested that wide-gauge trackways were the products of wide-gauge sauropods, rather than narrow-gauge sauropods of different sizes or engaging in different behaviors. Based on these morphological differences, Lockley (1999) suggested that narrow-gauge trackways were attributed to more "basal" sauropods, basal macronarians, or Diplodocoidea, while more derived sauropods such as the titanosaurs were the trackmakers of wide-gauge trackways.

Similar-sized individuals indicate that gauge is not a function of body size, specifically, narrow-gauge limb posture, acquired early in sauropod evolution, was retained by all basal sauropods, diplodocoids, and basal macronarians (Day et al., 2002). The Titanosauria, in contrast, developed a number of derived skeletal features, such as an outwardly angled femoral posture and beveled knee condyles, along with a more asymmetrical femoral midshaft (Wilson and Carrano, 1999), indicating that their limbs were held further from the midline of the animal. And the consistent absence of the manus claw impression in the wide-gauge trackways also coincides with titanosaur claws that believed to have been entirely lost. These traits suggests that this advanced clade of largely Cretaceous sauropods was responsible for wide-gauge trackways (Day et al., 2002). By simulating the relationship between relative areas of tracks and the relative fractions of the body weight borne by the forefeet and hindfeet, Henderson (2006) draws further conclusions that *Diplodocus* with more posteriorly placed center of mass, were the makers of narrow-gauge trackways; brachiosaurids and other large sauropods with more anteriorly positioned center of mass, were the makers of wide-gauge trackways.

The trackmaker of the Cretaceous *Brontopodus birdi* has been more specifically speculated to be a brachiosaurid sauropod based on manual impression morphology (Farlow et al., 1989; Farlow, 1992) and has also been regularly linked to the finding of a pes skeleton from the stratigraphically higher Paluxy Formation. This pes, once attributed to "*Pleurocoelus*" (Gallup, 1989), is now referred to *Cedarosaurus weiskopfae*, a brachiosaurid titanosauriform (D'Emic, 2013). Farlow et al. (2012) suggested a plausible candidate for the *Brontopodus* maker is *Paluxysaurus* (previously assigned to Pleurocoelus), known from skeletal (including pedal) material from stratigraphic units of about the same age, from the same region (Langston, 1974; Gallup, 1989; Rose, 2007). Another candidate is *Sauroposeidon* (Wedel et al., 2000), but nothing is known of its pedal skeleton. Given that the wide-gauge sauropod tracks are probably not of monophyletic origin, the trackmaker may be mamenchisaurids belonging to the basal Eusauropoda for *Brontopodus* trackways that appear in the Early–Middle Jurassic before the rise of clade Titanosauriformes in Late Jurassic (Xing et al., 2021a). However, based on the temporal range of body fossils, the current taxonomic problems within *Brontopodus* and *Parabrontopodus*, plus the presence of multiple basal somphospondylians in the same formation as mentioned in D'Emic (2013), this speculation as to the precise trackmaker of these ichnogenera is unreliable.

3.1.2 Chorrokhosauropus *Dzhalilov and Novikov, 1993*

3.1.2.1 Type ichnospecies

Chorrokhosauropus khakimovi Dzhalilov and Novikov, 1993.

3.1.2.2 Valid ichnospecies

Chorrokhosauropus khakimovi Dzhalilov and Novikov, 1993.

3.1.2.3 Distribution

Upper Albian, Tajikistan.

3.1.2.4 Holotype

Shaded tracks of the front and hind legs. Fourth trackway from Babatag-2 tracksite, Albian, Luchak Formation peritidal limestone of Tajikistan.

3.1.2.5 Diagnosis

A quadruped animal with tridactyl manus and pes tracks. The manus and pes tracks are rotated relatively evenly (probably outward, based on the figure) from the trackway midline.

The pes tracks are subtriangular, with length and width of 80 and 70 cm respectively. The heel is sharply curved, the digit traces are wide; the lateral digits are shorter, and the distal ends are rounded. The middle digit is slightly forward, it is the largest (proximal width 42–45 cm, length about 12 cm). The intersection angle between lateral digits is 65°. The manus tracks are rounded, semilunar, with slight short digits. The length of these tracks (35 cm) is shorter than the width of the digits area (52 cm). The middle digit is weaker than the lateral digits, its proximal width 15–17 cm, length 6–8 cm (diagnosis translated from Dzhalilov and Novikov, 1993).

3.1.2.6 General description

Dzhalilov and Novikov (1993) recorded the length of the holotype trackway is about 52.2 m, and the width from 0.8 to 1.5 m. The length of the step (distance from the heel of the pes print to the heel of the manus print) at the beginning of the trackway is 1.1–1.3 m, further reduced to 0.8–1 m. The depth of the manus and pes prints are 6–9 cm respectively, and the pes tracks have an obvious sediment displacement rim. The displacement rim of manus tracks is mainly around digits.

3.1.2.7 Discussion

Dzhalilov and Novikov (1993) considered that the morphology of *Chorrokhosauropus* differs from other sauropod ichnogenera. However, *Chorrokhosauropus* has almost never been officially cited, and there is a lack of photos and detailed outline drawings of specimens. The only clues are the single figure in the original paper. Moreover, the original author did not provide a diagnosis in the paper, and the description lacked some important information, such as the footprint rotation, step, stride, pace angulation. Thus, we can only discuss this ichnotaxon based on the limited information above. The validity of this ichnogenus may only be restrengthened after more morphological information is obtained, thus, we consider *Chorrokhosauropus* as *nomen nudum* here.

3.1.2.8 Trackmaker

The Luchak Formation does not record any sauropod skeleton. *Chorrokhosauropus khakimovi* is attributed to the sauropod, but it remains unclear if they belong to a particular ichnofamily (Nessov, 1995).

3.1.3 Titanosaurimanus *Dalla Vecchia and Tarlao, 2000 (Fig. 7.2A)*

3.1.3.1 Type ichnospecies

Titanosaurimanus nana Dalla Vecchia and Tarlao, 2000.

3.1.3.2 Valid ichnospecies

Titanosaurimanus nana Dalla Vecchia and Tarlao, 2000.

3.1.3.3 Distribution

Upper Albian–Cenomanian, Croatia.

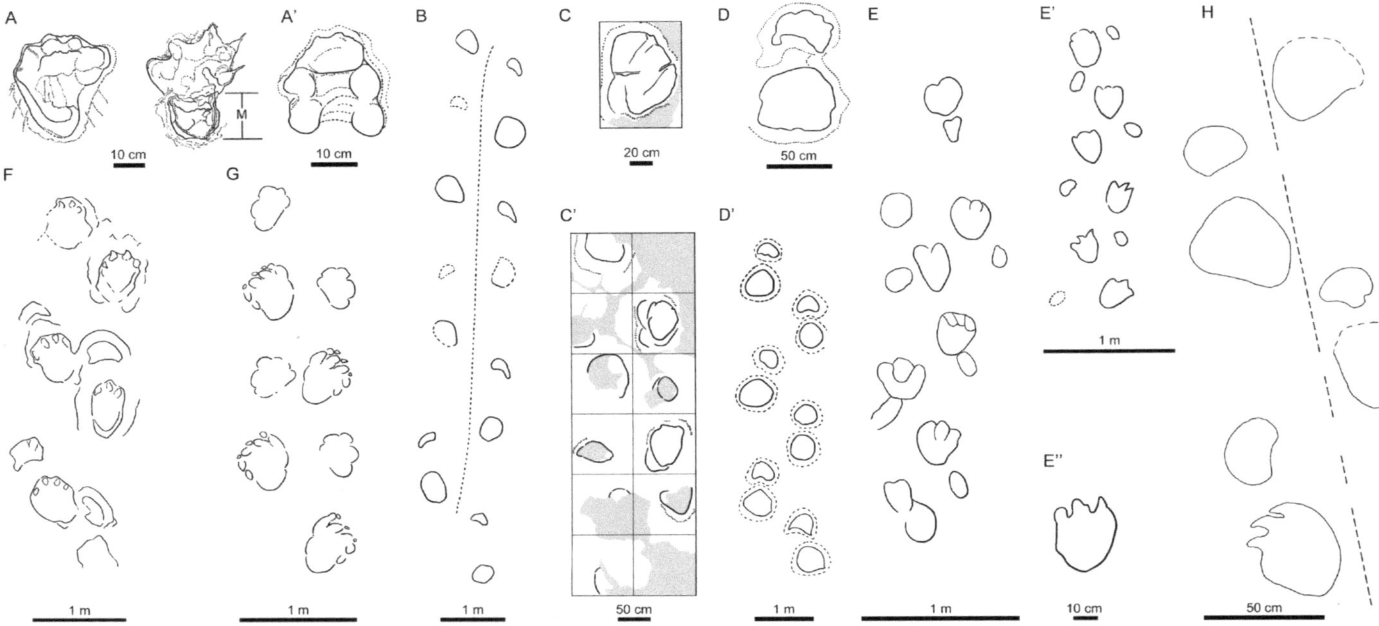

FIGURE 7.2 Tracks of other sauropod trackways. (A) Holotype (manus, A′) and referred pes tracks (A) of *Titanosaurimanus nana*; (B) type series of *Titanopodus mendozensi*; (C) holotype (C) and referred trackway (C′) of *Oobardjidama foulkesi*; (D) referred series (D′) and manus-pes set (D) of *Calorckosauripus lazari*; (E) holotype (E), paratype (E′) and GDM-LT-1 (E″) of *Parvieouspodus refugii* ichnogen. nov. ichnosp. nov. (F) *Iniestapodus burgensis*; (G) *Gyeongsangsauropus pentadactylus* ichnogen. nov. (H) Type series of *Teratopodus malarguensis*. *(A) Redrawn from Dalla Vecchia (2005). (B) Redrawn from González Riga and Calvo (2009). (C) Pes only, redrawn from Salisbury et al. (2016). (D) Trackway S 1150.01; drawn from Meyer et al. (2018). (E) Redrawn from Xing et al. (2015). (F) Redrawn from Torcida Fernández-Baldor et al. (2021) (G) Redrawn from Kim and Lockley (2012). (H) Redrawn from Tomaselli et al. (2022).*

3.1.3.4 Holotype

SOLII-S5m, manual print, limestones with *Neoiraqia insolita*, Solaris Campground, Punta del Dente near Červar village, Istria, Croatia; "plastotype," cast in silicon rubber, stored at the Museo Paleontologico Cittadino of Monfalcone, Gorizia, Italy.

3.1.3.5 Diagnosis

Small sauropod, pes tracks not longer than 50 cm, manus tracks not wider than 25 cm; prints of manual Digits I, II, IV, and V are distinct, rounded and of subequal size; the print of manual Digit III is larger than the others and roughly rectangular; pedal claw prints are present on Digits I and II and possibly on Digit III whereas Digit IV is rounded; Digit I is never larger or longer than the others, never the most projecting digit, and lacks a large claw mark; medium-gauge trackway with complete or nearly complete overprint of the pes track on the manus track (original diagnosis from Dalla Vecchia and Tarlao, 2000).

3.1.3.6 General description

Dalla Vecchia and Tarlao (2000) measured the holotype including 28 tracks, which are identified as pes tracks. The length of these pes tracks ranges from 20.5 cm/24 cm (internal/outer outline) to 42 cm/51.1 cm, and width ranges from 13 /20–32 /41.7 cm. These pes tracks exhibit similar sizes, all being anteroposteriorly elongated with a broader anterior digital part. More or less distinct marks of five digits in each pes track. The Digit I is consistently not larger than the other digits and lacks a prominent claw; Digits II–IV are defined and of approximately equal size, with Digit III often being the longest; as in other sauropod pes tracks, the Digit V is rounded, lobe-like, and faces laterally (Dalla Vecchia and Tarlao, 2000). There is also a poorly preserved trackway ST1 in the same area as holotype, which is composed of 11 pes tracks. The manus tracks are unclear and may be overlapped by pes tracks.

3.1.3.7 Discussion

Dalla Vecchia and Tarlao (2000) emphasized that the most obvious feature of *Titanosaurimanus nana* is the separate state of the manual digits, and it is not evident in manus prints that are less well-preserved than S5m. This is reminiscent of the fact that this feature is easily affected by preservation conditions such as sedimentation. For example, this feature can only be observed in tracks with fairly good preservation conditions. However, in the currently published sauropod morphological data sets, well-preserved manus tracks, like *Brontopodus birdi* from Texas (Lockley et al., 1994a), the digital separation is not so evident (Dalla Vecchia and Tarlao, 2000).

Lužar-Oberiter et al. (2017) used a high-resolution digital model illustrating the holotype specimen of *T. nana*, and emphasized the importance of digitalization.

3.1.3.8 Trackmaker

Due to the small size of *Titanosaurimanus*, Dalla Vecchia and Tarlao (2000) favored that these trackmakers of Solaris site belong to a small-sized taxon. Furthermore, it appears clear that pes traces are more common than manus ones at the site. Based on the predominance of pes traces, Dalla Vecchia and Tarlao (2000) suggested that the trackmaker was more *Diplodocus*-like, with most of its weight carried on the hind legs, than *Brachiosaurus*-like and with little pronounced heteropody. However, Dalla Vecchia (2005, 2008) preferred that the trackmaker of *Titanosaurimanus* is more derived than *Brachiosaurus*, possibly titanosaurians, because it lacked the claw as well as the manual phalanges.

3.1.4 Titanopodus *González Riga and Calvo, 2009 (Fig. 7.2B)*

3.1.4.1 Type ichnospecies

Titanopodus mendozensis González Riga and Calvo, 2009.

3.1.4.2 Valid ichnospecies

Titanopodus mendozensis González Riga and Calvo, 2009.

3.1.4.3 Distribution

Late Campanian–early Maastrichtian, Argentina.

3.1.4.4 Holotype

Original manus and pes tracks in situ, trackway AC-1, Agua del Choique tracksite, Loncoche Formation, Mendoza, Neuquén Basin, Argentina. Cast of manus-pes set from trackway AC-1 is housed under the abbreviation IANIGLA-PV 052.

3.1.4.5 Diagnosis

Sauropod tracks of medium size (track length about 50 cm) characterized by a wide-gauge trackway. Manus tracks without claw impressions and strongly rotated outward (25°–48°), with asymmetrical crescent shape and acuminate and thinner external border; pes tracks with subtriangular to subcircular shape, longer than wide, and rotated slight negative (outward); pronounced heteropody (1:3); outer limits of trackway defined, in some cases, by manus tracks; manus-pes distance about 1.7 times pes footprint length; stride length around 4.9 times pes track length; step angle 92°–11°; tail drag marks absent (original diagnosis from González Riga and Calvo, 2009).

3.1.4.6 General description

The trackway of the holotype of *Titanopodus mendozensis* is 18.5 m long and includes 27 tracks. The total width of the trackway is about 1.25 m, and the inner trackway width is about 40 cm. Both manus and pes prints are well separated from the midline, as in other wide-gauge trackways (Farlow, 1992; Lockley et al., 1994a). Manus tracks rotated outward (25°–48°). This value, measured between the transverse axis of the manus (as compared to a perpendicular line running to the midline), is larger than that of *Brontopodus birdi* (Farlow et al., 1989). Pes tracks are larger than manus tracks. Pedal claw impressions are absent, and some pes tracks show an elongated and distorted contour, mainly in the deeper footprints. Moreover, the pes pace angulation (110°) is larger than that of the manus (104°) (González Riga and Calvo, 2009).

3.1.4.7 Discussion

González Riga and Calvo (2009) named *Titanopodus* based on a fully wide-gauge trackway from the Agua del Choique tracksite. *Titanopodus* differs from other sauropod tracks by typical features both in the manus tracks and trackway. The crescent outline of the manus impressions is present in most sauropod tracks (e.g., Farlow et al., 1989; Lockley, 1991; Santos et al., 1994), but a strongly asymmetrical contour with an acuminate and thinner external border has not been described in other formal ichnotaxa. *Titanopodus* differs from *Calorckosauripus* for the trackway gauge, heteropody, and general shape of the manus and pes (Meyer et al., 2018). *Titanopodus* differs from *Occitanopodus* by the outer limits of the trackway that are defined in some cases by manual prints, smaller tracks, and lower heteropody (1:3), and manual prints strongly rotated outward (i.e., 25°–48°) (Moreau et al., 2019). *Teratopodus* and *Titanopodus* have similar heteropody (1:3) but exhibit substantial differences in the shape of their manus and pedes, *Titanopodus* has asymmetrical manus tracks with an external acuminate border shape and pes tracks that differ from those of *Teratopodus* in that they are oval shaped, with a U-shaped heel impression and a markedly concave anterior border (Tomaselli et al., 2022).

3.1.4.8 Trackmaker

In the Campanian–Maastrichtian strata globally, the only sauropod species known represent derived titanosaurs. The ichnological features of *Titanopodus* indicated that the trackmakers were saltasaurine or aeolosaurine titanosaurs of wide-gauge trackways and middle size (hip height of 2.11–2.29 m, body length 14–16 m) (González Riga and Calvo 2009; González Riga, 2011). The walking speeds of *Titanopodus* trackmakers (4.7 km/hour–4.9 km/hour) are somewhat faster than those previously inferred for most sauropods (González Riga, 2011). González Riga (2011) also suggests that these speeds and some taphonomic features of the tracks indicate the capacity of derived titanosaurs to walk effectively over a very wet and slippery substrate.

3.1.5 Oobardjidama *Salisbury et al., 2016 (Fig. 7.2C)*

3.1.5.1 Type ichnospecies

Oobardjidama foulkesi Salisbury et al., 2016.

3.1.5.2 Valid ichnospecies

Oobardjidama foulkesi Salisbury et al., 2016.

3.1.5.3 Distribution

Valanginian–Barremian, Western Australia.

3.1.5.4 Holotype

A replica of the holotype is WAM 12.1.6, "plastotype," a rigid polyurethane resin replica of the natural mold of a right pes impression, (UQL-DP45-8[rp2]). Broome Sandstone. DP45 in the intertidal zone of the Yanijarri–Lurujarri section, Dampier Peninsula, west Kimberley region, Western Australia.

3.1.5.5 Diagnosis

Pes tracks: large-sized (length: 70–85 cm), near-piriform shape, but broadest across mid-track region, and with a lobed medial margin due to intersection of an asymmetrical heel pad crease; hallucal imprint widened medially by a bulbous cranial pad callosity; external outline of heel region semicircular result of occupation by a medio caudal heel pad callosity;

heel pad callosity medially offset and partitioned from fore-heel region of track by well-defined oblique ridge (~45°–75° to the horizontal); acute fore-heel caudal margin of internal track exclusive of heel pad; pes track angulation between 90° and 100°.

Manus tracks: marginally wider than long (~115%); manus track angulation between 69° and 74°.

Trackway: weakly heteropodous (30%–45%), variably medium-to wide-gauge trackways, with inner trackway width of 30–50 cm; manual tracks positioned midway between fore and hind pedal tracks, and in line with the rear half of contralateral pedal tracks; manual tracks situated relatively medially, within pedal trackway width (original diagnosis from Salisbury et al., 2016).

3.1.5.6 General description

Oobardjidama foulkesi, a single continuous trackway comprising 10 consecutive manual and pedal impressions. The holotypic impression DP45-8 (rp2), with a maximum craniocaudal length of 71 cm and a mediolateral mid-width of 56 cm. DP45-8 (rp2), when compared to other sauropod pedal impressions, stands out due to its unique morphology in internal track floor. It features a distinct triangular subimpression, constituting approximately 75% of the track area. The inner subimpression is widest at the cranial end, displaying traces of the digital arcade, and narrows toward an acute heel end caudally (Salisbury et al., 2016).

3.1.5.7 Discussion

Salisbury et al. (2016) proposed an informal grouping of wide-gauged "ichnoclade" of sauropod tracks, which mainly includes *Brontopodus birdi*, *"B." pentadactylus*, *Polyonyx gomesi*, *Titanopodus mendozensis*, and *Oobardjidama foulkesi*, as well as many dubious ichnotaxa, such as *"Rotundichnus" munchehagensis* that was considered invalid by lacking any diagnostic or morphological details of sauropods, such as claw/digit impressions (Hendricks, 1981; Wright, 2005), but is similar to *Brontopodus* in length/width ratio and the crescent-shape of the manus (Hornung et al., 2012). And the morphological characteristics of *Oobardjidama* are between *Brontopodus*-type and *Titanopodus*, *Oobardjidama* has comparable heteropody and a moderate trackway gauge ratio range, whereas the greater symmetry in the fore and hind interautopodial gap distances renders *Oobardjidama* more similar to the extremely wide-gauged *Titanopodus* (Salisbury et al., 2016). Torcida Fernández-Baldor et al. (2021) compared *Oobardjidama* and *Iniestapodus*, suggesting that a pollex impressionis not preserved in the manus of *Oobardjidama*, which is a key feature in *Iniestapodus*.

3.1.5.8 Trackmaker

Based on the skeleton record of sauropods from various geological periods, Salisbury et al. (2016) believe that the trackmaker of *Oobardjidama* is posited to be either a macronarian, likely a nonlithostrotian titanosauriform, or a diplodocoid. The first option is more likely, but the second cannot be ruled out.

3.1.6 Calorckosauripus *Meyer et al., 2018 (Fig. 7.2D)*

3.1.6.1 Type ichnospecies

Calorckosauripus lazari Meyer et al., 2018.

3.1.6.2 Valid ichnospecies

Calorckosauripus lazari Meyer et al., 2018.

3.1.6.3 Distribution

Middle Maastrichtian, Bolivia.

3.1.6.4 Holotype

Trackway S 1080.01 (trackway segment: cast and holotype NMB SA.M.2). Its type locality is Cal Orcko tracksite, large limestone/packstone in the El Molino Formation, Bolivia.

3.1.6.5 Diagnosis

Pes tracks are entaxonic and always longer than wide (average length: 49 cm, and width: 42 cm; length/width: 1.6), but with a shape that is variable from slightly elliptical to bell-shaped, with a narrow "heel" and a wider anterior area. The phalangeal region is generally the deepest part of the track, although with a certain degree of variability, due to locomotion- or sediment-related preservation issues. Pes impressions are slightly rotated outward, with the left pedes sometimes rotated to a higher angle (S1150.01: 17.5°) than the right (S1150.01: 7.5°). Digit impressions are not present and only one of the tracks analyzed shows a deeper area that might correspond to Digit I impression.

Manus tracks are always wider than long (average length: 34 cm, and width: 29 cm; length/width: 1.17). When not (partially) overprinted and deformed by the pes impression, the manus is horseshoe-shaped, with two clear lobes corresponding to Digits I and V. The lobes are fainter, with only the one corresponding to Digit I discernible. No claw marks are present. Generally, manus tracks are outward rotated (20°−60°), with the deepest part of the track constantly on the external side.

Trackway heteropody is marked, but not extreme (e.g., 1:1.85), with the manus being roughly half of the size of the pes. The gauge is wide/intermediate, with a PTR that varies from 22% to 34%. Despite being highly variable, trackways S1080.01 and S1080.02 show a trend, whereby most of the manus tracks on the left side of the trackway are impressed more internally than on the right side, where the tracks are impressed more externally. The latter could be the results of walking on a palaeoslope along the shoreline.

Gleno-acetabular distance is fairly constant and estimated around 2.8 m. Hip height in S1120.01 (type trackway) reaches 2.24 m and stride averages at 2.29 m (original diagnosis from Meyer et al., 2018).

3.1.6.6 General description

Meyer et al. (2018) described *Calorckosauripus* as having more rounded and axially compressed pes imprints and horseshoe-like manus impressions. The manus shows clear impressions of Digits I and V. Meyer et al. (2021) also attributed 11 trackways from the Humaca tracksite to the ichnotaxon *Calorckosauripus lazari*. These trackways were made by small individuals (field length manus 22 cm; field width manus 25.3 cm; pes length 44.6 cm, field width pes 33.2 cm) and formed a group moving simultaneously in the same direction; this can nicely be seen in trackways 1−3, which turn in the same direction.

3.1.6.7 Discussion

Meyer et al. (2018) analyzed the sauropod tracks from the CalOrck'o tracksite, identifying two distinct morphotypes (E and F). Morphotype F was recognized as belonging to a new ichnotaxon: *Calorckosauripus lazari* (Meyer et al., 2018). This ichnotaxon is similar to *Titanopodus mendozensis* (González Riga and Calvo, 2009) and "*Sauropodichnus giganteus*" (cf. *Brontopodus* isp., Calvo, 1991; Lockley et al., 1994a; Krapovickas et al., 2020) in their loose manus-pes set. However, *T. mendozensis* has a slightly lower trackway gauge (side width/overall width of and pes, *T. mendozensis* reaches from 26 to 31, while *C. lazari* ranges from 27 to 34) and higher heteropody (about 1:3), and "*S. giganteus*" also has stronger heteropody (1:2.9). Besides, both of them differ from *C. lazari* in the general shape of the manus and pes (Meyer et al., 2018).

3.1.6.8 Trackmakers

Titanosaurids, as they are the only sauropods present during the Campanian–Maastrichtian. The presence of two morphotypes, morphotype E and *Calorckosauripus*, indicates the coeval presence of a more basal and a more derived titanosaur in the Late Cretaceous (Meyer et al., 2018).

3.1.7 Iniestapodus *Torcida Fernández-Baldor et al., 2021 (Fig. 7.2F)*

3.1.7.1 Type ichnospecies

Iniestapodus burgensis Torcida Fernández-Baldor et al., 2021.

3.1.7.2 Valid ichnospecies

Iniestapodus burgensis Torcida Fernández-Baldor et al., 2021.

3.1.7.3 Distribution

Tithonian–Berriasian, Spain.

3.1.7.4 Holotype

A complete manus-pes set cataloged as LS7B, 3m, and LS7B, 3p, Rupelo Formation, Quintanilla de las Viñas, Burgos. They are in situ and there is a photogrammetric model of both tracks.

3.1.7.5 Diagnosis

Manus tracks roughly symmetrical, wider than long, and have a semicircular shape. In the best-preserved specimens, three short, broad, and rounded digit impressions (II, III, and IV) are anteriorly oriented, and a short and a subcircular Digit I impression (pollex) is medially located. Pes tracks are longer than wide, with a subrectangular shape and four claw impressions. Claw I impression is smaller and more backwardly located than the others. Claw I and II impressions are forward oriented and impressions III and IV are slightly rotated and curved laterally. The lengths of the pes tracks are more than double those of the manus tracks. The widths of the manus tracks are slightly less than the pes tracks (original diagnosis from Torcida Fernández-Baldor et al., 2021).

3.1.7.6 General description

The morphology of the manus tracks varies from semicircular to crescent-shaped. Several of them are deformed by overprinting of the pes tracks. The holotype manus track is semicircular, and measures 29 cm long and 38 cm wide. The morphological quality (MP), which has been analyzed using the scale proposed by Marchetti et al. (2019), is variable from 0. 5 to 2 (holotype has a value of 2). Pes tracks are subrectangular in shape and the posterior edge is slightly protruding. They reveal four claw impressions occupying the entire width of the anterior zone. There are no impressions of a blunt callosity or a claw impression at the distal-lateral end of Digit IV that might suggest the presence of a Digit V impression. This feature is commonly observed in well-preserved pes tracks of sauropod (Lockley et al., 2002a). The holotype pes track measures 65 cm long and 37 cm wide. Pes tracks are generally deeper than the manus tracks. The heteropody is variable from low to medium. The distances between the pes tracks and the manus tracks (from proximal end of manus to distal end of pes) vary from more than 40 cm to a partial overlap. The pace length, stride length, track rotation, pace angulation, and the distance between manus and pes tracksare variable among the trackways and within the same trackway. The PTR (pes trackway ratio) value is 35.3–47.6, which was calculated as TR = (pesW/eTW)*100 (Romano et al., 2007). The value indicates an intermediate/wide gauge trackway (Torcida Fernández-Baldor et al., 2021).

3.1.7.7 Discussion

Iniestapodus differs from other sauropod ichnotaxa by the tetradactyl pes tracks with evidence of four sharp claw impressions. Torcida Fernández-Baldor et al. (2021) considered the main differences between *Iniestapodus* and all other known sauropod pes tracks are its generally outwardly rotated claw impressions, with a notable decrease in their size from I to IV, with claw IV being very small, and usually blunt, or with no claw trace. *Titanosaurimanus* is known only from a manus track, which has a horseshoe shape (Dalla Vecchia and Tarlao, 2000), different from the semicircular morphology of the manus of *Iniestapodus*. The pes tracks of "*Brontopodus oncalensis*" were described as triangular, with four possible claw impressions and the manus tracks as subcircular to pentagonal (Meijide Fuentes et al., 2004). In *Iniestapodus*, the claw I digit impression is the smallest of the pes track. *Iniestapodus* shares the orientation of the digit pes impressions with *Polyonyx* (Santos et al., 2009), but differs in its general shape, size, and the position of the pes Digit I impressions. Both ichnotaxa are clearly different in manus morphology and especially in the size and orientation of the pollex impression. Moreover, *Iniestapodus* differs from *Polyonyx* in the heteropody and in trackway gauge.

3.1.7.8 Trackmaker

The analysis and comparison carried out between the anatomical characteristics preserved in the tracks (small ungual phalanx in the manual Digit I, four ungual phalanges in the pes with unguals I and II subequal in size), and the bone record in the Iberian Peninsula suggest that titanosauriforms are the most plausible trackmakers (Torcida Fernández-Baldor et al., 2021).

3.1.8 Teratopodus *Tomaselli et al., 2022 (Fig. 7.2H)*

3.1.8.1 Type ichnospecies

Teratopodus malarguensis Tomaselli et al., 2022.

3.1.8.2 Valid ichnospecies

Teratopodus malarguensis Tomaselli et al., 2022.

3.1.8.3 Distribution

Lower-middle Campanian, Anacleto Formation, Argentina.

3.1.8.4 Holotype

Trackway preserved in situ identified as An-1/Tw-1 and composed of Track-1 left pes, Track-2 left manus, Track-3 right pes and Track-4 right manus, Track-5 left pes, Track-6 left manus, and Track-7 right pes.

3.1.8.5 Diagnosis

Sauropod titanosaur manus and pes tracks characterized by the following unique association of characters: (1) manus-pes ratio with a medium degree of heteropody (ca. 1/3); (2) manus tracks with symmetrical kidney-shaped configuration, with the posterior border slightly concave and wider than long (relation MW/ML = 0.76); (3) manus tracks with outward rotation between 43° and 53°; (4) pes tracks with suboval shape and longer than wide; (5) V-shaped heel impression with a PLM° slightly obtuse (95°–100°); (6) pes tracks outward rotation between 22° and 33°; (7) entaxonic pes tracks with three claw impressions corresponding to Digits I, II and III; (8) claw impressions deflected toward the lateral plane; (9) relatively long claw impressions and reduced interdigital spaces; (10) medium-gauge trackway (MTR: 49.2%, PTR: 48.26%, WAP/PL: 1.12) (original diagnosis from Tomaselli et al., 2022).

3.1.8.6 General description

Teratopodus is a recently established titanosaur sauropod ichnogenus, erected based on a unique association of 10 characters. The pes track Tr-1 has well-defined displacement rims, especially in the medial and postero-lateral borders. It is one of the shallowest tracks (1–2.5 cm of depth) and has a pes outward rotation of 22°. Tr-3 has large displacement rims on the medial and postero-lateral borders. The track has a depth of 2–3 cm, and its outward rotation is 30°. Tr-5 presents displacement rims in all borders which are bigger toward the anterior part of the track. This pes track is the deepest one and has a depth of 4–5 cm. Its outward rotation is ~33°. As for Tr-7, however, it shows poor preservation and it is for this reason why neither depth nor outward rotation were measured; the displacement rim of the posterior part is well-preserved. The manus track Tr-2 has a prominent displacement rim along the anterior border, a depth of 1–3 cm and its outward rotation is 53°. Tr-4 has no displacement rims and is the shallowest track of the trackway (1–2 cm); it has an outward rotation of ~43°. Tr-6 is the deepest manus track of the trackway (4–5 cm) and shows an outward rotation of ~45°. Displacement rims are present in all borders and are very prominent (Tomaselli et al., 2022).

3.1.8.7 Discussion

Teratopodus has a slightly V-shaped heel impression, which is also observed in *Parabrontopodus mcintoshi* (Lockley et al., 1994a), *Elephantopoides barkhausensis* (Kaever and de Lapparent, 1974) and *Polyonyx gomesi* (Santos et al., 2009).

The pes track from the late Campanian-early Maastrichtian of Mongolia (Stettner et al., 2018) shows three claw impressions similar to *Teratopodus*, but compared with *Teratopodus*, its claw impressions are shorter, less curved, and shows a larger interdigital space. *Brontopodus birdi* from the USA (Farlow et al., 1989) show claw impressions that are slimmer and shorter than those of *Teratopodus*, and the interdigital space is still narrower. Claw impressions of pes tracks from the Lower Cretaceous Gansu tracksite (Zhang et al., 2006) also differ from those of the *Teratopodus* in their shortness and the mostly straight and anteriorly oriented impression of claw I, while the impressions of claw II and III are slightly laterally deflected.

Manus tracks of *Teratopodus* have a kidney-shaped configuration, which differs from the subquadrangular manus tracks of *Brontopodus birdi* (Farlow et al., 1989), *Polyonyx gomesi* (Santos et al., 2009), and also from the acuminate border manus tracks of *Titanopodus mendozensis* (González Riga and Calvo, 2009).

3.1.8.8 Trackmaker

According to the middle Campanian paleontological record, South American sauropods are limited to Titanosauria. The trackmakers of *Teratopodus* were assigned to small–medium-sized titanosaurs of ca. 11 m in length (holotype specimen) and ca. 13.8 m in length (paratype specimen). The preservation of the tracks indicates that at least two titanosaurs walked from a humid ground to a more flooded one, heading toward 165° southeast (Tomaselli et al., 2022).

3.1.9 Parvieouspodus *ichnogen. nov (Figs. 7.2E and 7.29).*

3.1.9.1 Type ichnospecies

Parvieouspodus refugii ichnogen. nov. ichnosp. nov.

3.1.9.2 Valid ichnospecies

Parvieouspodus refugii ichnogen. nov. ichnosp. nov.

3.1.9.3 Etymology

Parvi and *eous* are Latin, meaning "small" and "eastern," which is used in reference to the relatively small size of these tracks and its occurrences in East Asia. The ichnospecies name is the genitive of the Latin word refugium for refuge, which indicates that in the Early Cretaceous East Asia represented a refuge of Jurassic faunas with a distinct endemic ichnofauna, which are known for the high proportion of sauropod trackways.

3.1.9.4 Holotype

The type specimen LSI-S2 is preserved in situ in the Jishan tracksite in Linshu County, Shandong, China, from the Lower Cretaceous Tianjialou Formation (Albian–Cenomanian). The holotype was preserved in soft sediments, so that good details are preserved. Paratype, LSV-S1 preserved at the same site.

3.1.9.5 Referred materials

The same track types are also known from the Tuchengzi Formation of Beijing (QJD1D-S1 and QJD1U–S3 in Qianjiadian site), the Hekou Group of Gansu (ZPI-S2 in Zhongpu site, GDM-LT-1 in Litan site), Dasheng Group of Shandong (BL-S1, S2 and isolated SIP-6 in Beilin site, LG.1 in Zhangzhuhewan site, TDGZ-S1 and TDGZ-S2 in Tangdigezhuang site) and Jiangsu (NGZ-T2 and T3 in Nanguzhai site). Most of these tracks located in the Dasheng Group were found in the Tianjialou Formation. All of these sites are located in China, nearly at the same latitude around 35° N.

3.1.9.6 Diagnosis

Narrow to medium-gauge sauropod track or trackway of evident relatively small size (pes print length about 25–40 cm, no larger than 45 cm), characterized by pronounced outward rotation of manus (about 50°–80°). Strongly developed heteropody (about 1: 2.2–1: 3.5). Semielliptic manus tracks, narrowing at mid-length at the metacarpophalangeal region. Pes tracks longer than wide with long axis rotated outward. Pes traces with three anteriorly directed digit (I–III) (Summarized from the original description in Xing et al., 2015c).

3.1.9.7 General description

All of these sauropod tracks are from the Lower Cretaceous, and are significantly smaller than other contemporaneous heteropodous ichnotaxa (pes traces usually longer than 50 cm, for example, *Brontopodus* and *Parabrontopodus*), some of which in the Beilin and Qianjiadian sites even fall into the "tiny" (PL < 25 cm) category. Manus traces preserved at the Nanguzhai tracksite show discernable Digits I–V, while pes traces in Zhongpu tracksite exhibit rare possible Digit IV and V impressions that are closer to the lateral side than in other ichnotaxa (e.g., *Brontopodus birdi*). The claw impressions at least show in pes Digit I. The latter also preserves the robust and tapering-ended Digit I and blunt Digit II. In in this ichnotaxon also appears a rare example of turning trackways at the Nanguzhai tracksite, indicating an increase in the footprint rotation angle during the turn. The width of the angulation pattern of the pes (WAP)/Pes length ratio is about 1, while that of the manus is about 1.5–1.95. Outward rotation of manus prints of the atypical material at the Qianjiadian site and the turning trackways at the Tangdigezhuang site can reach about 90°–130° (emended from Xing et al., 2015c).

3.1.9.8 Discussion

Almost all small sauropod tracks were preserved along with medium-sized to large sauropod tracks, many of them are attributed to *Brontopodus* (Xing et al., 2013c). *Parvieouspodus* has often been placed in other ichnotaxa in previous studies: some of them were assigned to *Parabrontopodus* isp. on the basis of the narrow-gauge and the high heteropody (Xing et al., 2010a,b, 2013c). *Parvieouspodus refugii* ichnogen. nov. ichnosp. nov. could be juveniles of the same trackmaker taxon reported in the same site, however, juvenile and adult sauropods may have adopted different feeding and herding strategies that resulted in their segregation (Myers and Fiorillo, 2009; Xing et al., 2015c).

Parvieouspodus refugii ichnogen. nov. ichnosp. nov. is also different from other small sauropod tracks preserved in East Asia: Those recorded in Inner Mongolia (Neimenggu) and China show less heteropody and small outward rotation of the manus (Lockley et al., 2002a), while the trackway in the Lower Cretaceous of South Korea shows distinct smaller outward rotation of the manus prints (~29°) and wide gauge (Lim et al., 1989; Lockley et al., 2006a). Tracks from both sites mentioned above were signed to *Brontopodus*. However, it is notable that the small footprints from Korea have a latitudinal distribution similar to *P. refugii* (see Fig. 5 in Xing et al., 2015c).

3.1.9.9 Trackmaker

The trackmakers are presumed to have been small sauropods. Unfortunately, no body fossils corresponding to a potential quadrupedal dinosaur trackmaker have been discovered from the abovementioned regions, which is unusual for China (Xing et al., 2015c). This situation is most likely due to preservation bias in body fossils. The trackmakers belonging to the same taxon of sauropods have similar body proportions despite large differences in body size (Vila et al., 2013b), which means that it is reasonable to differentiate these trackmakers by gauge if they vary significantly, while their rotation of manus tracks is not as important for categorization due to their nontightly constrained forelimb posture, as their relatively thick cartilage allowed them to move their forelimbs over a wider range without injury (Bonnan et al., 2010, 2013; Holliday et al., 2010). The variable forelimb posture is well indicated in *P. refugii*.

Although gauge may also vary during the traveling of sauropods (Meyer et al., 2018), coupled with the consideration of the strong inheritance of Early Cretaceous ichnofaunas of East Asia at the base of the Jurassic dinosaur fauna (Matsukawa et al., 2006; Lockley et al., 2002a, 2013), it is likely that the narrow to medium-gauge *P. refugii* was also produced by a small basal sauropod trackmaker, including diplodocids and brachiosaurids.

3.1.10 Gyeongsangsauropus *ichnogen. comb. nov (Fig. 7.2G).*

3.1.10.1 Type ichnospecies

Gyeongsangsauropus pentadactylus (="Brontopodus" *pentadactylus*; Kim and Lockley, 2012)

3.1.10.2 Valid ichnospecies

Gyeongsangsauropus pentadactylus (Kim and Lockley, 2012)

3.1.10.3 Etymology

Ichnogenus "*Gyeongsang*" named in reference to Gyeongsang Province in southeastern Korea where the holotype is located, "*saur*" means "lizard" in Greek; "*pus*" means "*foot*" in Greek.

3.1.10.4 Holotype

Holotype dinosaur trackway No.1 is located at the Gajin tracksite, Jinseong-myeon, Jinju City, South Gyeongsang Province, South Korea, from the Haman Formation (Albian–Cenomanian). It is now preserved within the Heritage Pavillion at the Science Education Center, Gyeongsangnam-do Office of Education (formerly Gyeongsangnamdo Institute of Science Education). A plaster replica of the holotype (KNUE-0912001) is housed at the Department of Earth Science Education, Korea National University of Education (KNUE).

3.1.10.5 Referred materials

Trackway No. 2 at the Gajin tracksite as well as other similar tracks with five round shaped digits of manus found at the Jeongchon, tracksite Level 1 from the Jinju Formation (Albian) and the Bidaegol tracksite from the Jeomgok Formation (Albian), now also assigned in *Gyeongsangsauropus*, but not to the type species.

3.1.10.6 Diagnosis

Medium-gauge sauropod trackway of small to medium size (pes print length about 40–50 cm), characterized by wide pentadactyl manus tracks, with very wide blunt manus digits. Manus tracks strongly rotated outwards from the midline of trackway (about 30°–60° in holotype, 85°–125° in another trackway). Pentadactyl pes tracks with outwardly rotated digit claws I–III, also rotated outward (about 30°–50°). Manus is wider than long, and pes is longer than wide. Moderate heteropody for about 1: 2 (diagnosis emended from Kim and Lockley, 2012).

3.1.10.7 General description

Manus tracks are typically composed of five blunt short digit impressions of which only Digits III-V are consistently impressed in the holotype trackway, and the three digits (III–V) are nearly equal in length and width. From the well-preserved holotype, pes Digits I, II, and III have well-defined sharp claw traces, in which Digit I is longest, and Digits IV and V are rounded (subcircular), short, and lack claw traces.

For the referred materials of this ichnogenus, the maximum value of track rotation can reach about 90°. The inner trackway width of the manus (between inside margins of right and left manus tracks) measures up to ~22 cm; for the pes inner trackway, width is up to ~12 cm in the holotype. Pace is about 80–90 cm in pes and 80–100 in manus, approximately 3 times the manus track length and 2 times the pes track length. Stride is ~150 cm on average, about 5 times the manus track length and slightly over 3 times the pes track length. Pace angulation of type ichnospecies is about 110°, but in other materials can reach 135°. Distance between manus and pes is about 50–80 cm. Gleno-acetabular distance is estimated at about 150 cm, approximately 3 times the pes track length. The maximum outer trackway width is 104 cm in the holotype, while in *Gyeongsangsauropus* isp. It is only 79 cm. In addition, different trackways of this ichnogenus do not show the same pace, such as trackway No.2 representing a faster pace compared with trackway No.1 (updated measurements in Table 7.1).

3.1.10.8 Discussion

This ichnospecies was assigned to *Brontopodus* in previous study (Kim and Lockley, 2012), we are now establishing *Gyeongsangsauropus* ichnogen. nov. on the basis of its distinctive manus morphology and its degree of lateral rotation.

TABLE 7.1 Measurements (in cm) of *Gyeongsangsauropus* ichnogen. nov. from Gazin-ri, Jinju, South Korea.

	TL	TW	TL/TW	PL	SL	PA	OTW	ITW	RA	WAP	WAP/TL
Pes 1	50.39	44.33	1.14	89.70	149.13?	111.23?	101.7?	10.31	34.49	50.54?	0.98
Manus 1	32.46	41.36	0.79	97.68?	150.41	101.31	93.93	19.58	50.07	61.16	1.83
Pes 2	39.54	36.43	1.09	84.42	150.00	125.62	75.73	7.78	24.59	38.44	0.97
Manus 2	25.18	34.79?	0.73	84.19?	147.92?	121.86?	65.08	9.23	108.61	41.08?	1.62?

ITW, Inner Trackway Width; *OTW*, Outer Trackway Width; *PA*, Pace Angle; *PL*, Pace Length; *RA*, Rotation Angle; *SL*, Stride Length; *TL*, Track Length; *TW*, Track Width; *WAP*, Width of the Angulation pattern of Pes; Manus 1 and pes 1 are the mean value from the holotype of *Gyeongsangsauropus pentadactylus* (trackway No.1), manus 2 and pes 2 are the mean value from *Gyeongsangsauropus* ichnosp; from the same site (trackway No, 2).

Track rotation is one of the important features for describing dinosaur tracks because it is closely related to locomotory posture and dinosaur behavior, while the rotation angle of manus tracks has received comparatively little attention probably because they are completely different in appearance and usually show no clear indications of separate digits (Kim and Lockley, 2012). Thulborn (1990) mentioned that the manus impressions of sauropod tracks show distinct negative (outward) rotation. Outward rotation also occurs in sauropod pes tracks that show definite negative rotation of an angle between 20° and 30°, which is further emphasized by the outward rotation of pes digit claws I–III (Lockley, 2007a). Rotation angles of manus impressions of sauropod tracks from the Cretaceous Jindong Formation of Korea (Lim et al., 1989), Jurassic of Portugal (Santos et al., 1994), and Jurassic and Cretaceous of Italy and Croatia (Dalla Vecchia, 1994, Dalla Vecchia et al., 2002) are, according to these authors, up to 45°. These are higher values than those reported by Thulborn (1990). However, ichnology still lacks a large systematically collected database on sauropod manus rotation, and Kim and Lockley (2012) presume such data could only be accurately compiled where the position of manus digit impressions I and IV can be observed as a means to verify track orientation.

For the referred materials of *Gyeongsangsauropus*, the maximum value of track rotation can reach about 90°, which is significantly greater than the data in these studies. But it is seemingly quantitative to distinguish *Gyeongsangsauropus* from *Brontopodus* by using the rotation angle of the manus, given that both *Brontopodus* and its type ichnospecies *B. birdi* lack this measurement (Farlow et al., 1989).

3.1.10.9 Trackmaker

The repeated registration of the outer manus digits (III–IV) and the sporadic registration of the inner digits (especially Digit I) suggest that perhaps the animal could have placed more weight on the outer, posterolateral portion of the manus. However, this is difficult to demonstrate, and the distinctive nature of the resultant traces suggests that the trackmaker was likely different from other reported trackmakers, both in morphology (broad manus toes) and gait/behavior (strong outward rotation of the manus) (Kim and Lockley, 2012). Although the pes tracks of *G. pentadactylus* are unmistakably sauropod, the pentadactyl manus has some similarities to certain blunt-toed ankylosaur tracks (cf. McCrea et al., 2001).

The trackmaker of *G. pentadactylus* is consistent with other typical wide-gauge sauropod tracks, which belong to the titanosaurs (see 3.1.1).

3.2 Non-avian theropods, saurischia

Unlike ornithischian, sauropod, or other types of tracks, theropod ichnotaxonomy is complex and troubling. For historical reasons, any revision of theropod ichnotaxonomy will likely be a daunting task, especially in relation to regions where provincial ichnotaxonomy has been prevalent. The most typical example of this is the mid-to-late Mesozoic track records of China, where many ichnogenera have been proposed that are obvious subjective synonyms of *Grallator* and *Eubrontes* (Lockley et al., 2013). These forms have been distributed for a very long time, at least in East Asia, from the Early Jurassic to the "mid" to Late Cretaceous. Likewise, a mature global perspective on the ichnotaxonomy of Cretaceous theropod tracks is also likely to prove difficult until the problem of these classical morphotypes is resolved satisfactorily.

In this section, we limit all the valid ichnospecies and part of the isp. indet. defined in the Cretaceous, especially those classic morphological types (*Grallator*, *Eubrontes* etc.). For the valid ichnospecies of the same ichnogenera defined in the Triassic and Jurassic, please refer to other chapters of this book. For the theropod display trace ichnogenus *Ostendichnus*, we do not count it in the theropod tracks here, please check the chapters in this book that deal with reproduction.

3.2.1 Eubrontes *Hitchcock, 1845 (Figs. 7.3A,B and 7.30A)*

3.2.1.1 Type ichnospecies

Eubrontes giganteus Hitchcock, 1845.

3.2.1.2 Valid ichnospecies in Cretaceous

Eubrontes (?) *glenrosensis* Shuler, 1935 (Fig. 7.3A)
Eubrontes nobitai Xing et al., 2021f (Fig. 7.3B)

3.2.1.3 Distribution in Cretaceous

Berriasian–Cenomanian, Brazil, America, England, China, etc.

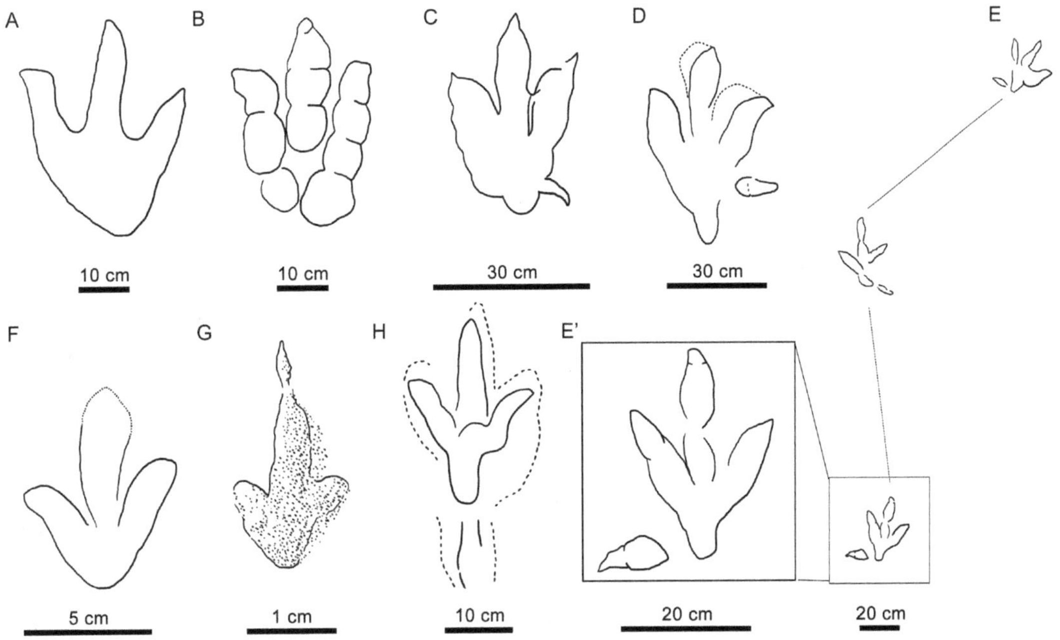

FIGURE 7.3 Cretaceous valid ichnospecies of classical Jurassic ichnogenus *Eubrontes- Gigandipus- Grallator*. (A) Holotype of *Eubrontes* (?) *glenrosensis*; (B) holotype of *Eubrontes nobitai*; (C) holotype of *Gigandipus caudatus*; (D) holotype of *Gigandipus hei*; (E) type series of *Gigandipus chiappei*; (F) holotype of *Grallator ssatoi* (Xing et al., 2024a); (G) holotype of *Grallator emeiensis*; (H) holotype of *Grallator jiuquwanensis*. *(A) Redrawn from Shuler (1935). (B) Redrawn from Xing et al. (2021f). (C) Redrawn from Lull (1904). (D) Redrawn from Xing et al. (2018b). (E) Redrawn from Xing et al. (2018b). (G) Redrawn from Zhen et al. (1996). (H) Redrawn from Xing et al. (2016b).*

3.2.1.4 Holotype

AC 15/3, a natural cast in gray fine sandstone, housed in the Hitchcock Ichnology Collection at the Pratt Museum of Amherst College (abbreviated AC), Amherst Massachusetts.

3.2.1.5 Diagnosis

Large (>25 cm long) bipedal, functionally tridactyl ichnite with a relatively short Digit III, a broad pes, and a hallux which is rarely, if ever, impressed. Divarication of outer digits averaging 25°–40° (emended by Olsen et al., 1998).

3.2.1.6 General description

E. (?) *glenrosensis* was discovered from the Glen Rose Limestone and De Queen formations of North America (early Albian). Large, tridactyl pes tracks, with lengths and widths ranging from 35 to 63.5 cm and 34–43.2 cm, respectively. The track is basically symmetrical around Digit III and the divarication values range 26°–33° (Digits II–III), and 27°–34° (Digits III–IV). The digits are particularly wide (Shuler, 1935).

E. nobitai was discovered from the Jiaguan Formation (Barremian/Aptian–Santonian) of west/southern China. It has only been found in the Cretaceous. Large, tridactyl pes tracks, with length 31.4 cm, and the average aspect ratio about 1.4. The mesaxony is 0.37, and the divarication value is 47° (Digits II–IV). The phalangeal pad of Digit II is strongly developed, which is similar to Digit IV. The phalangeal pads are basically symmetrical along Digit III. The pace length is 2.6 times track length, and the pace angulation is 160° (Xing et al., 2021f).

The related ichnofamily Eubrontidae (Lull, 1904) is also commonly found in track records. However, the ichnogenus-level units included in this ichnofamily require more work. There are some preliminary diagnoses provided (Lull, 1904): large bipedal forms, more or less rounded claws, digits broad with distinct phalangeal pads. Foot functionally, possibly structurally tridactyl as the hallux claw never impresses. No caudal trace.

3.2.1.7 Discussion

Eubrontes is one of the typical large theropod ichnogenera in Jurassic and Cretaceous in Lauraisa (Olsen et al., 1998; Xing et al., 2021f), and also occures in Gondwana (Leonardi and Carvalho, 2021c). It is also one of the earliest named theropod ichnogenera (Hitchcock, 1845). *Eubrontes* is similar to *Anchisauripus* and *Grallator* in the over all morphology except the size range (Olsen et al., 1998). At least 13 valid ichnospecies were established by researchers: *E. minusculus* (Hitchcock, 1837), *E. sillimani* (Hitchcock, 1843), *E. giganteus* Hitchcock, 1845, *E. divaricatus* Hitchcock, 1858, *E. platypus* Lull, 1904, *E.* (?) *glenrosensis* Shuler, 1935, *E. veillonensis* de Lapparent and Montenat, 1967, *E. monax* (Zhen et al., 1986), *E. xiyangensis* (Zhen et al., 1986), *E. nianpanshanensis* (Yang and Yang, 1987), *E. zigongensis* (Gao, 2007), *E. pareschequier* (Xing et al., 2009a), and *E. nobitai* (Xing et al., 2021f). Only two of them, *E.* (?) *glenrosensis* and *E. nobitai*, have been defined in the Cretaceous, and two other ichnospecies, *E. giganteus* and *E. divaricatus*, were identified in Cretaceous stratas (see Adams et al., 2010; Leonardi and Carvalho, 2021c; Xing et al., 2021f).

Eubrontes (?) *glenrosensis* was proposed by Shuler (1935) for a track which currently is embedded in the stone wall of the bandstand in the Somervell County Courthouse square in Glen Rose, Texas (see Adams et al., 2010). The track is deep with a well-defined contour line, but lacks other anatomical details. Its ichnotaxonomy has been studied by several researchers (see Lockley et al., 1998 and references). For these authors, this ichnospecies is valid, but does not belong to *Eubrontes*. *E.* (?) *glenrosensis* differs from other *Eubrontes* ichnospecies originally by their of the lower mesaxony (0.4), and the relatively wide digits (Shuler, 1935), while these features could not be diagnositic after more specimens are included in *Eubrontes* (Xing et al., 2021f). The subsequent works noted that *E.* (?) *glenrosensis* is at least more similar to the Jurassic ichnogenus *Kayentapus* in lower L/W ratio, and the higher divarication angle of digit II to IV (Xing et al., 2021m). *E. nobitai* differs from other *Eubrontes* by the lower mesaxony (0.36), the very developed phalangeal pad of digit II, and the basically symmetrical phalangeal pas along the digit III (Xing et al., 2021f). The aformentioned features cooccur with the low L/W ratio (1.4) and nearly parallaled anterior part of digit II and IV (ungual marks excluded) than in other *Eubrontes*.

3.2.1.8 Trackmaker

Lockley and Xing (2021) suggested that *Eubrontes* is associated with a relatively large theropod compared with the trackmakers of *Grallator*. Shell and Boss (2013) suggested that *E.* (?) *glenrosensis* can be associated with allosaurian *Acrocanthosaurus atokensis*, as the latter is the largest theropod in the same region. Xing et al. (2021f) associated the trackmaker of medium-sized *Eubrontes*, *E. nobitai* (~30 cm), with a kind of 4 m long theropod.

3.2.2 Gigandipus *Hitchcock, 1855; Lull, 1904 (Figs. 7.3C–E and 7.30H)*

3.2.2.1 Type ichnospecies

Gigandipus caudatus Lull, 1904 (Fig. 7.3C)

3.2.2.2 Valid ichnospecies in Cretaceous

Gigandipus chiappei Xing et al., 2018b (Figs. 7.3E and 7.30H)

3.2.2.3 Distribution in Cretaceous

Cretaceous: Barremian/Aptian–Santonian, Jiaguan Formation, China.

3.2.2.4 Holotype

AC 9/9–10, housed in the Hitchcock Ichnology Collection at the Pratt Museum of Amherst College, Amherst Massachusetts.

3.2.2.5 Diagnosis

Large, bipedal, tetradactyl pes track. A medially or caudomedially oriented hallux impression. Tail drag marks are present in the holotype and several referred specimens and have been touted as diagnostic (Lull, 1904; Milner et al., 2009).

3.2.2.6 General description

Gigandipus was only found in America and China. There are three valid ichnospecies of *Gigandipus* (*G. caudatus*, *G. hei*, and *G. chiappei*), but only one ichnospecies was found from the Cretaceous (*G. chiappei*).

The features of *G. chiappei* are provided by Xing et al. (2018b): Large-sized functionally tridactyl, tetradactyl track with a strong and posteromedially directed hallux trace and pes aspect ratio of 1.3. The angle between the axis of the hallux and the track axis is 107°. The position of the proximal hallux is almost equal to the proximal end of the metatarsal pad. The ratio of hallux to track length is 0.3. The metatarsophalangeal pad lies on the midline of the track. Step about 4 times track length, and mean pace angulation is relatively high, which is about 127°.

3.2.2.7 Discussion

Weems (2003) suggested *Gigandipus* is a junior synonym of *Eubrontes* based on the inferred similarities in shape and maximum length of *Eubrontes* and *Gigandipus* and cooccurrence in the same stratigraphic horizons while the major difference, the presence of hallux traces, is possibly due to different walking behaviors or individual differences of the trackmaker. However, some *Eubrontes* tracks lacking hallux impressions are apparently deeper than some *Gigandipus* tracks (Weems, 1992), so foot-substrate interactions cannot universally explain these differences.

As noted above, *Gigandipus* has been reported only once in the Cretaceous, and that is *G. chiappei* from the Jiaguan Formation of Guizhou, China (Xing et al., 2018b). Compared with other *Gigandipus* ichnospecies, *G. chiappei* has a length/width ratio and mesaxony of 1.3 and 0.47, while in *G. hei* it is 1.6 and 0.38 respectively. In both *G. caudatus* and *G. hei*, the ratios of hallux to track length are 0.2, lower than 0.3 in *G. chiappei*. In *G. chiappei*, the position of the proximal hallux is almost equal to the proximal end of the metatarsal pad, and this feature is not only different from *G. caudatus* and *G. hei*, but also different from other footprint tracks, such as *Eubrontes*-like tracks (Xing et al., 2018b).

Another theropod track with a hallux trace from the Cretaceous is *Ordexallopus zhanglifui* (Lockley et al., 2018a). These authors compared this ichnotaxon with *Gigandipus* and showed the differences notably in the much greater divarication angle of *O. zhanglifui*.

3.2.2.8 Trackmaker

The trackmaker of *Gigandipus* was initially thought to be a wading bird due to the birdlike proportions of its foot (Hitchcock, 1856). However, later interpretations suggested that it could have been a gigantic bipedal lizard or amphibian (Hitchcock, 1858). Lull (1904) classified the trackmark-makers as theropods based on an understanding of foot anatomy. Weems (2003) suggested that the trackmaker of *Gigandipus* was associated with *Plateosaurus* based on tail drag marks and the proportions, size, and anatomical specializations of the foot of *Plateosaurus*. Meanwhile, Rainforth (2005) concluded that the trackmaker was a theropod due to strongly rearward rotated hallux impressions.

3.2.3 Grallator *Hitchcock, 1858 (Figs. 7.3F–H and 7.30D)*

3.2.3.1 Type ichnospecies

Grallator parallelus Hitchcock, 1858.

3.2.3.2 Valid ichnospecies in Cretaceous

Grallator ssatoi (Yabe et al., 1940); Zhen et al., 1989; Xing et al., 2024a (Fig. 7.3F)
Grallator emeiensis Zhen et al., 1996 (Figs. 7.3G and 7.30D)
Grallator jiuquwanensis Zeng, 1982; Xing et al., 2016b (Fig. 7.3H)

3.2.3.3 Distribution in Cretaceous

Cretaceous: Berriasian–Maastrichtian, China, South Korea, North America, Argentina etc.

3.2.3.4 Holotype

A trackway of four successive pes impressions on slab AC 4/1a, housed in the Hitchcock Ichnology Collection at the Pratt Museum of Amherst College, Amherst Massachusetts.

3.2.3.5 Diagnosis

Small (<15 cm long) functionally tridactyl ichnite in which the Digit III projection ratio is about 1.3, and the length-to-width ratio is about 2.2. Digit II projects farther along the axis of Digit III and is more robust than Digit IV, which is narrow. Divarication of outer digits averages 28° (emended from Olsen et al., 1998).

3.2.3.6 General description

G. ssatoi was discovered from the J/K boundary, the Third Member of the Tuchengzi Formation in Liaoning, China. Initially, it was believed that this species was only present during the Early Cretaceous. However, Xing et al. (2024a) reviewed the Chinese *Grallator* specimens and provided a new diagnosis: A relatively small tridactyl track with well-developed rounded phalangeal pads, heel pad, and robust triangular claws; 2, 3, and 3 digital pads corresponding to Digits II–IV. The metatarsophalangeal pad of Digit IV emerges at the axis of Digit III. A metatarsophalangeal pad of Digit II is seldom shown. Digit III and IV are the same length, or a III is slightly longer than IV. The pes length/width ratio is 1.6. The mesaxony is 0.8 (medium). The divarication angle is relatively small (about 57°) between Digits II and IV. Digit III is 68% of the total pes length. Step length is about 3.8 x track length, and mean pace angulation is high (about 170°).

G. jiuquwanensis comes from the Albian Shenhuangshan Formation in Hunan Province, China. The holotype is tridactyl and medium in size (26.2 cm in length). The Digit IV impression is the longest, whereas the Digit II is the shortest. Claw impressions are clearly visible and the divarication angle between II and IV is 58° (Zeng, 1982).

G. emeiensis was found from the Jiaguan Formation (Barremian/Aptian–Santonian) in Sichuan, China. It is recorded as small, tridactyl pes tracks without digit I impression, with length 2.7 cm and width 1.6 cm. The claw prints are preserved, the claw print of Digit III was particularly large and sharp with a length 0.5 cm. The pace length is 9.4 cm, the stride length is 18.4 cm, the pace angulation is 154° and the width of trackway is 2.1 cm (Zhen et al., 1996).

3.2.3.7 Discussion

Grallator is one of the earliest named theropod tracks (Hitchcock, 1858). *Grallator*, typically of Upper Triassic–Middle Jurassic rocks (see Triassic and Jurassic Chapters for more information), is similar to *Anchisauripus* and *Eubrontes*. *Grallator* is relatively small, the length is generally less than 15 cm, the Digit III projection ratio (length of the flush part of Digits III and II/length of Digit III protruding from other two) is less than 1.3, the average length/width ratio is higher than 2 (Olsen et al., 1998). At least 16 valid ichnospecies were established by the researchers: *G. tuberous* (Hitchcock, 1836), *G. parallelus* (Hitchcock, 1858), *G. tenuis* (Hitchcock,

1858), *G. rectus* (Hitchcock and Grier, 1927), *G. ssatoi* (Yabe et al., 1940), *G. maximus* (de Lapparent and Montenat, 1967), *G. olonensis* (de Lapparent and Montenat, 1967), *G. variabilis* (de Lapparent and Montenat, 1967), *G. jiuguwanensis* (Zeng, 1982), *G. limnosus* (Zhen et al., 1986), *G. yemiaoxiensis* (Yang and Yang, 1987), *G. sauclierensis* (Demathieu and Sciau, 1992), *G. lescurei* (Demathieu, 1993), *G. emeiensis* (Zhen et al., 1996), *G. adeolensis* Gand et al., 2000, *G. yangi* (Lockley et al., 2015a), and *G. pingchuanensis* (Li et al., 2019).

G. ssatoi was defined originally as *Jeholosauripus s-satoi* by Yabe et al. (1940). Subsequently, Zhen et al. (1989), revised the ichnospecific epithet and classified it within *Grallator* as *G. ssatoi*. It differs from other *Grallator* because of the sub-small size. Compared with *G. parallelus*, the length-to-width ratio is closed to 1 (Yabe et al., 1940). *G. ssatoi* is now a suitable classification that can include the large number of *Grallator* records in China, from the Early Jurassic to the "Mid" Cretaceous (Xing et al., 2024a), while this ichnospecies may cover too much small theropod tracks morphotypes form North China (Xing and Lockley, 2021).

G. emeiensis is the smallest *Grallator* and differs from other *Grallator* because the claw print of Digit III is particularly large and sharp, and the "heel" impression is large (Zhen et al., 1996).

G. yangi used to be named *Paragrallator yangi* (Li and Zhang, 2000). Lockley et al. (2015a) assigned *Paragrallator yangi* to *Grallator yangi* while describing more than 2200 dinosaur tracks from the Huanglonggou site (Yellow Dragon Valley), near Zhucheng, in Shandong, China. Xing et al., 2024a assigned *G. yangi* to *G. ssatoi*, based on the poor preservation of the original specimens and the main measurement data falling within the scope of *G. ssatoi*.

There are some other Cretaceous tracks classified within the ichnogenus *Grallator* isp. in Argentina, Brazil, China, South Korea, and the USA from Berriasian to Maastrichtian (Leonardi, 1981; Leonardi and Carvalho, 2021c; Matsukawa et al., 2006; Lockley et al., 2011, 2012a, 2022b; Xing et al., 2018c, 2019d). On the other hand, Lull (1942) named *Anchisauripus australis* Lull, 1942 to some tracks recorded from the ?Aptian El Jume Formation (Argentina). Melchor et al. (2019) suggested that this ichnotaxon is probably better compared with *Grallator*. Zeng (1982) established *Hunanpus jiuquwanensis*, which is classified within *Grallator* by Xing et al. (2016b).

Besides, the grallatorid-type tracks are classified in ichnofamily Grallatoridae (Lull, 1904). There are some preliminary diagnoses provided: typical tridactyl (occasionally tetradactyl) footprints composing bipedal trackways and showing moderate to marked elongation (footprint length/footprint width ratio >1.50) and moderate to marked mesaxony (anterior triangle length/width ratio >0.6), well-defined digital pads and clear claw marks. Trackway narrow with large stride length (Melchor et al., 2019). However, this loose definition does not currently encompass the overall morphospace of grallatorids (e.g. Xing et al., 2024a). This suggests that systematic comparisons of *Grallator* ichnospecies and other small theropod ichnogenera occurring in different regions are needed to further delineate the morphological boundaries of *Grallator* in future studies.

3.2.3.8 Trackmaker

There is no doubt that these Cretaceous *Grallator* were associated with small theropod trackmkers (see Lockley and Xing, 2021). Spanning so many different regions and stages, *Grallator* trackmakers have quite a few possibilities and span multiple clades. For example, *Grallator* from the Yixian Formation China is suggested to have an affinity with oviraptorosaurs (Xing et al., 2009b) while Triassic–Jurassic remains have been related with ceratosaurs, basal tetanurans, and carnosaurs (Lockley, 2009; Lucas et al., 2006).

3.2.4 Ornithomimipus *Sternberg, 1926 (Fig. 7.4A and B)*

3.2.4.1 Type ichnospecies

Ornithomimipus angustus Sternberg, 1926 (Fig. 7.4A)

3.2.4.2 Valid ichnospecies

Ornithomimipus angustus Sternberg, 1926.
Ornithomimipus jaillardi Ellenberger in Jaillard et al., 1993 (Fig. 7.4B)

3.2.4.3 Distribution

Albian–Cenomanian, Tantalus Formation (Yukon), Canpanian–Maastrichtian, Horseshoe Canyon Formation and Wapiti Formation (Alberta) of Canada; Aptian, Patuxent Formation (Virginia), Campanian, Mesaverde Group (Colorado) of USA; Maastrichtian, Vilquechicho Formation of Peru.

3.2.4.4 Holotype

NMC 8513, a track located at Rumsey tracksite in the Horseshoe Canyon Formation (Alberta) of Canada.

3.2.4.5 Diagnosis

Moderate size; bipedal; tridactyl; digitigrade; mesaxonic; Digits II and IV highly divergent, long and slender; Digit IV typically deeply impressed, Digit II typically detached; phalangeal pads moderately strongly impressed along all three digits; Digit II with fewer pads than other digits; heel short and distinct; metatarsal pads indistinct or absent (Gangloff et al., 2004).

3.2.4.6 General description

Ornithomimipus angustus is about 28 cm in length, originally characterized by its slenderness; no pad for the proximal portion of phalanx one of Digit II and this toe is separated from the rest of the foot; and the phalangeal formula is 2-3-4 for Digits II, III and IV, respectively (Sternberg, 1926; Lockley et al., 2011). *Ornithomimipus jaillardi* is characterized by having a narrow foot with a large median digit, an interdigital tissue or membrane, a light gait, and a heart-like planter sole. It is smaller than the type ichnospecies, and a bit narrower in proportion due to the much closer Digits II–IV (Jaillard et al., 1993). Its length is about 20 cm, the Digit III length beyond the line joining the pits of II and IV for 8 cm; and claws appear to be blunt (Jaillard et al., 1993).

On the whole, *Ornithomimipus* is tridactyl, quite symmetrical, digitigrade, mesaxonic, and longer than wide (30 × 22 cm the holotype of *O. angustus* and about 20 × 15 cm the tracks of *O. jaillardi*). The Digit III impression is the longest one, whereas Digits II and IV are shorter. In *O. angustus*, claw and pad traces are preserved. The "heel" impression is rounded and placed behind the Digit III. The Digit II impression is separated from the rest of the track.

3.2.4.7 Discussion

Lockley et al. (2011) defined the ichnofamily Ornithomimipodidae as Slender-toed, tridactyl, gracile dinosaur tracks as wide or slightly wider than long with well-defined phalangeal pad traces (formula 2-3-4 corresponding to Digits II–III and IV). Divarication between traces

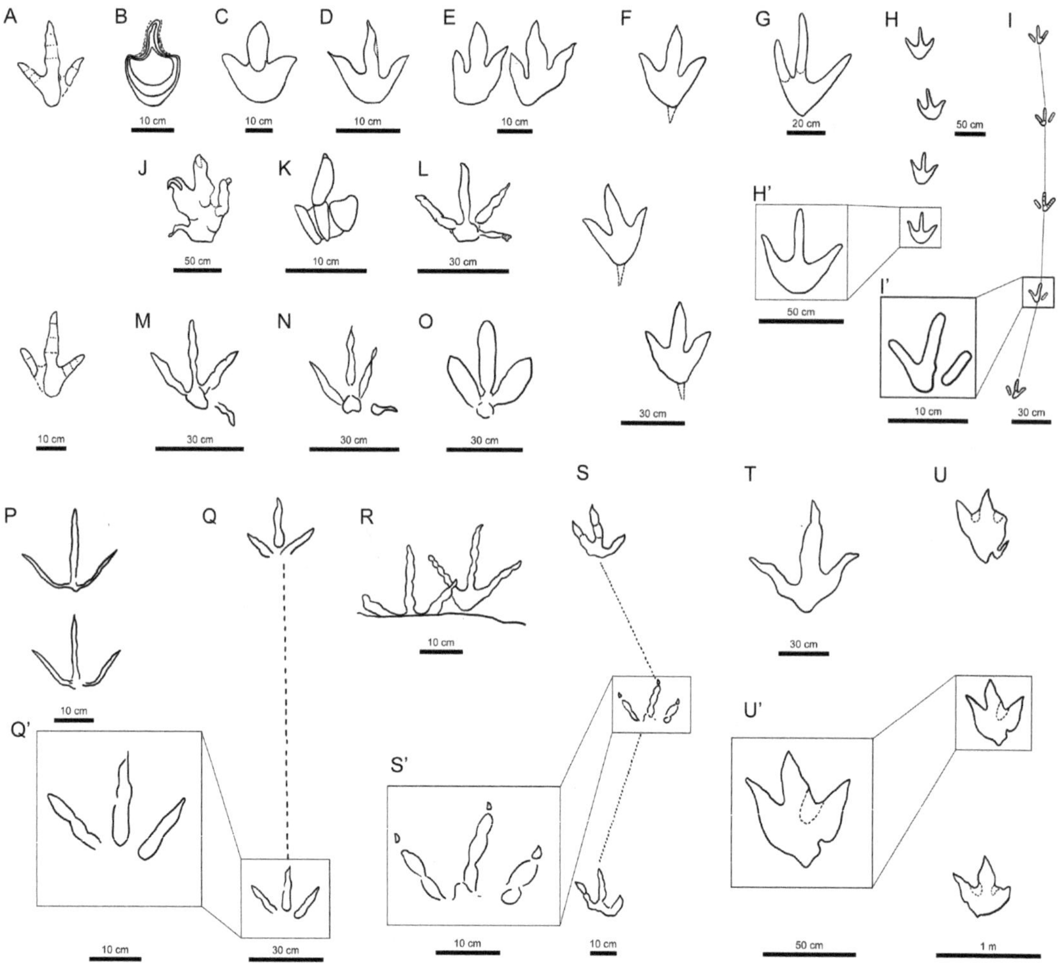

FIGURE 7.4 Theropod tracks in North America from Cretaceous. (A) Holotype of *Ornithomimipus angustus;* (B) holotype of *Ornithomimipus jaillardi;* (C) holotype of *Gypsichnites pacensis;* (D) holotype of *Columbosauripus ungulate;* (E) holotype of *Columbosauripus amouraensis;* (F) type series of *Irenesauripus mclearni;* (G) holotype of *Irenesauripus acutus;* (H) type series of *Irenesauripus occidentalis;* (I) type series of *Irenichnites gracilis;* (J) holotype of *Tyrannosauripus pillmorei;* (K) holotype of *Tyrannosauripus bachmani;* (L) holotype of *Saurexallopus lovei;* (M) holotype of *Saurexallopus zerbsti;* (N) holotype of *Saurexallopus cordata;* (O) holotype of *Fuscinapedis woodbinensis;* (P) holotype and paratype of *Magnoavipes lowei;* (Q) holotype of *Magnoavipes caneeri;* (R) type series of *Magnoavipes denaliensis;* (S) type series of *Magnoavipes asiaticus;* (T) holotype of *Wakinyantanka styxi;* (U) type series of *Bellatoripes fredlundi. (A) Redrawn from Sternberg (1926). (B) Redrawn from Jaillard et al. (1993). (C) Redrawn from Sternberg (1932). (D) Redrawn from Sternberg (1932). (E) Redrawn from Haubold (1971). (F) Redrawn from Sternberg (1932). (G) Redrawn from Sternberg (1932). (H) Redrawn from Sternberg (1932). (I) Redrawn from Sternberg (1932). (J) Redrawn from Lockley and Hunt (1994a). (K) Redrawn from Weems (2021). (L) Redrawn from Lockley et al. (2018a). (M) Redrawn from Lockley et al. (2018a). (N) Redrawn from Lockley et al. (2018a). (O) Redrawn from Lee (1997). (P) Redrawn from Lee (1997). (Q) Redrawn from Lockley et al. (2001). (R) Redrawn from Fiorillo et al. (2011). (S) Redrawn from Matsukawa et al. (2014). (T) Redrawn from Lockley et al. (2014d). (U) Redrawn from McCrea et al. (2014).*

of Digits II and IV variable (~60°–100°), but typically ~75°–80°. Trace of Digit II often separated from the rest of the footprint (traces of Digits III and IV). Heel pad trace usually connected to traces of Digits III and IV, centrally located, and typically separated from the proximal phalangeal pad trace of Digit III by a notable constriction. Trackway narrow: that is, only about as wide or slightly wider than maximum footprint width. They proposed that *Ornithomimipus, Irenichnites, Columbosauripus, and Magnoavipes* belong to the Ornithomimipodidae ichnofamily. Gierliński et al. (2017) and Xing et al. (2018e) added also *Xiangxipus* Zeng, 1982 to this ichnofamily. Moreover, Gierliński et al. (2017) suggested that *Xiangxipus* is comparable to *Saurexallopus*, and being tetradactyl, they would not correspond to ornithomimid tracks.

Weems (2021) commented that *Irenichnites, Columbosauripus, and Magnoavipes* have a Digit III impression only slightly more elongated that Digits II and IV and also present a high divarication angle between Digits II–IV. On the other hand, *Ornithomimipus* has a more elongated Digit III and less divarication.

Sternberg (1926) established *Ornithomimipus* without any formal diagnosis in his work, but only described the trackway he found in the Maastrichtian Horseshoe Canyon Formation in Alberta (Canada). After that, Haubold (1971) proposed the diagnosis of this ichnogenus and also suggested that *Irenichnites gracilis* Sternberg, 1932 is indeed *Ornithomimipus gracilis*.

Ellenberger in Jaillard et al. (1993) established *O. jaillardi* for Campanian-Maastrichtian tracks from Peru. As has happened other times, the authors proposed a diagnosis for the new ichnospecies, but did not review the diagnoses of the ichnogenus and the type ichnospecies. Therefore, *Ornithomimipus* currently should be reviewed.

Weems (2021) pointed out that *Ornithomimipus* encompasses three ichnospecies: *O. angustus, O. yuanmuensis* and *O. jaillardi*. The author differentitates between *O. angustus*, found in Maastrichtian beds of Canada, and *O. jaillardi*, from the Maastrichtian of Peru, by the relative extension of Digit III beyond Digit IV, which is greater in *O. jaillardi*. On the other hand, we have not found any information about *O. yuanmuensis* (China), so it is not evaluated here. Taking into account the descriptions and the graphical data available of *O. angustus*, we consider it valid. On the other hand, *O. jaillardi* is well described, although its graphical information is scarce, so it is advisable to add new data to understand better this ichnospecies.

Other records of *Ornithomimipus* have been found in the Aptian (Weems, 2021), Campanian (Lockley et al., 2011), and Maastrichtian (Lockley et al., 2003) of the USA, and Albian-Cenomanian (Gangloff et al., 2004) of Canada.

3.2.4.8 Trackmaker

Among Cretaceous theropods, the Ornithomimosauria is the only group known to include species with an elongated Digit III as what shown in ichnofamily Ornithomimopodidae (Lockley et al., 2011). Sternberg (1926) originally assigned *Ornithomimipus* (*O. angustus*) to *Ornithomimus* or at least a member of the family Ornithomimidae basing on the similarity of its morphology.

Weems and Bachman (2015) attributed *Ornithomimipus angustus* to the ornithomimosaur *Coelosaurus affinis*, which has been described from skeletal remains found in the overlying Arundel Formation (Makovicky et al., 2004; Weems and Bachman, 2015).

Weems (2021) noted that these two valid ichnospecies can appear in the same tracksite, combined with previous studies of local ornithomimid skeletal remains, it can be concluded that at least two distinctly different types of ornithomimids trackmakers can be attributed to *Ornithomimipus*, which differs in their body size and the lateral flexibility in its outer digits, one of them can be linked with *Arkansaurus fridayi* (Hunt and Quinn, 2018).

3.2.5 Gypsichnites *Sternberg, 1932 (Fig. 7.4C)*

3.2.5.1 Type ichnospecies

Gypsichnites pacensis Sternberg, 1932.

3.2.5.2 Valid ichnospecies

Gypsichnites pacensis Sternberg, 1932.

3.2.5.3 Distribution

Aptian, Gething Formation (British Columbia and Albian), Gates Formation (Alberta) of Canada; Albian, Glen Rose Limestone/Paluxy Formation (Texas) and Aptian, Patuxent Formation (Virginia) of USA; Albian–Cenomanian Roos River Rock, Canada.

3.2.5.4 Holotype

NMC 8553, originally located in Peace River Canyon, lodged at the Canadian Museum of Nature, Ottawa, Ontario. Paratype: Cat. No. 8554, The Geological Survey of Canada (GSC), consists of two tracks.

3.2.5.5 Diagnosis

Small; semidigitigrade; tridactyl; foot short and broad; heel broadly rounded; digits typically broad and terminally rounded, partly enclosed in web; Digit III typically longer than II and IV but appear to carry weight evenly (emended by Gangloff et al., 2004).

3.2.5.6 General description

Gypsichnites pacensis has a very good outline, but no phalangeal pads. All the toes are very broad and end in bluntly pointed hoofs. Digits II and IV are strongly divergent, point outward at their distal ends, but are free for only about one-half of their length. The proximal portions of the toes were doubtless enclosed in a pad or web. In the type, Digit III is more deeply impressed than the others. Digit III is broadest near the distal extremity, whereas the others taper gradually (Sternberg, 1932). Pace is about 90–120 cm, or slightly more than 3–4 times the length of the track. The tracks are almost in a straight line, but point slightly outward from the line of march (Sternberg, 1932; McCrea, 2000). In some specimens, lateral digits separate from the rest of the footprint, and show the terminal claws (McCrea, 2000).

3.2.5.7 Discussion

Originally, Sternberg's (1932) tentatively referred *Gypsichnites* to the theropod ichnofamily Eubrontidae, but also suggested that the footprints had been made by plant-eating ornithopods. McCrea (2000) supported that the length/width ratios of *Gypsichnites* are more typical of ornithopod dinosaurs rather than theropods. Langston (1997) related *Gypsichnites* with *Tenontosaurus*. McCrea (2000) identified claw and pad impressions in *Gypsichnites* tracks. Various authors affirmed that it is a theropod ichnotaxon (Lockley, 2000; McCrea, 2000; Gangloff et al., 2004; Díaz-Martínez et al., 2015b).

Besides, the difference in preservation apparently influences the existence claw and pad traces of *Gypsichnites*. More material is needed to stabilize this ichnotaxon since the limitation in quantity and preservation of *Gypsichnites records*.

Gypsichnites has been also identified in the Aptian–Albian of the USA (Weems and Bachman, 2015), and Albian–Cenomanian of Canada (Gangloff et al., 2004)

3.2.5.8 Trackmaker

Sternberg (1932) suggested that *Gypsichnites pacensis* might well have been made by a *Camptosaurus*-like ornithopod, but in the same research, he also noted that this ichnogenus was similar to Eubrontidae that is considered to have been made by theropods. Some other earlier research also attributed *Gypsichnites* to an ornithopod trackmaker such as *Tenontosaurus* (Langston, 1997) or linked it with other iguanodontids like *Caririchnium* (Weems and Bachman, 2004). McCrea (2000) speculated that at least some of the *G. pacensis* tracks might be the product of theropods with an atypical length/width ratio and claw marks, which was followed by Gangloff et al. (2004). *G. pascensis* was formally attributed to a theropod trackmaker by Díaz-Martínez et al. (2015a) without adding or modifying features. However, McCrea et al. (2017) suggested that *G. pscensis* should be placed in ornithopod trackmakers based on the manual impressions at a specific tracksite. So far, *Gypsichnites* lacks decisive evidence for a trackmaker, while there is a lack of skeletal taxon support for theropod trackmakers (Weems, 2021).

3.2.6 Columbosauripus *Sternberg, 1932 (Fig. 7.4E)*

3.2.6.1 Type ichnospecies

Columbosauripus ungulatus Sternberg, 1932.

3.2.6.2 Valid ichnospecies

Columbosauripus ungulatus Sternberg, 1932.

3.2.6.3 Distribution

Cenomanian (SE Djebel Bou Khahil) of Algeria; Berriasian–Hauterivian, Sousa Formation of Brazil; Aptian, Gething Formation (British Columbia), Albian–Cenomanian, Roos River Rock, and Cenomanian, Dunvegan Formation (British Columbia) of Canada.

3.2.6.4 Holotype

NMC 8551, an original natural mold housed in the Canadian Museum of Nature, Ottawa, Ontario.

3.2.6.5 Diagnosis

Small size; semidigitigrade, Digit I absent (?); tridactyl; pes print mesaxonic; digit impressions distinct and widely divergent; Digits II and IV equal to subequal in length and width, Digit II more highly divergent than IV. Phalangeal pads distinct and confined to Digit III; claws long; heel broad, rounded to indistinct; all digits attached to metatarsal pad (emended by Gangloff et al., 2004).

3.2.6.6 General description

Columbosauripus ungulatus from Canada is 12.5 cm in length and width, and the divarication between Digits II and III is 38°, and between Digits III and IV is 39° (Sternberg, 1932). Yukon *C. ungulatus* also provided more features: a broad shape (low length/width ratio); Digits II and IV highly divergent, symmetrical to asymmetrical; divarication in Digit II may be 10° higher than Digit IV; Digit III with two well-defined phalangeal pad impressions; a third is more proximal and weaker; claws on Digits II and IV acuminate; and Digit III terminus rounded then extended by a distinct narrow slit (Gangloff et al., 2004).

Haubold (1971) provided a new ichnospecies, *Columbosauripus amouraensis*, but no diagnosis or discussion was provided, so it is a nomen nudum. *C. amouraensis* is based on Group A–C trackways from the top of a Cenomanian limestone slab containing mud cracks, located in Ksar Amoura, at the southern edge of the Ouled Nail Mountains in Algeria (Mesle and Peron, 1880; Bellair and de Lapparent, 1948; Chabou et al., 2015). Bellair and de Lapparent (1948) surveyed 140 tridactyl tracks, which were divided into Groups A–C based on size. The tracks were categorized into three size ranges: approximately 23 cm, 9.4 cm, and 8 cm in length (Bellair and de Lapparent, 1948).

Group A tracks are the largest and most numerous, characterized by an elongated, narrow, and quite slender tridactyl track; the length is 24 cm, width is 17 cm, and heel width is 10 cm; the divarication between Digits II and III is 45°, and between Digits III and IV is 30°; on average, the stride length is about 75–80 cm; each digit has two-digit pads, and Digit III has obvious claw marks. The characteristics of Group B tracks are as follows: the middle digit length is 8.5 cm, the footprint width is 6.5 cm, and the divarication between Digits II and III is 43°, while that between Digits II and IV is 30°. Group C tracks have a middle digit length of 8 cm and a track width of 7.5 cm; the divarication between Digits II and III is equal to that between Digits II and IV, both 50°. The digits of Group C tracks are more spread apart than those of the other groups (Bellair and de Lapparent, 1948).

Bellair and de Lapparent (1948) proposed that Group B may represent the juvenile individual of Group A, and Group C may be the manus tracks of A. Group B lacks figures, and the figures of Groups A and C show significant differences in mesaxony values, with A being 0.62 and 0.44, while B is 0.2, 0.19, and 0.4. Group C may exhibit external morphological changes and is more likely to belong to an additional ichnotaxon. We propose that *C. amouraensis* be considered temporarily valid and that Group A be used as the type specimens. The specimens need to be restudied in the future to obtain more diagnostic information before any discussion can take place.

3.2.6.7 Discussion

Lockley et al. (2011) proposed that *Columbosauripus* belongs to Ornithomimipodidae ichnofamily. In the original material, *Columbosauripus* lacks phalangeal and "heel" pads and all the digit impressions are united, features that allow it to be differentiated from *Ornithomimipus* and *Irenichnites*. McCrea (2000) identified three pad traces in the Digit III impression. Gangloff et al. (2004) emended the original diagnosis and provided an ichnospecific diagnosis for *C. ungulatus*.

Haubold (1971) established *C. amouraensis* based on tracks found by Bellair and de Lapparent (1948) in Cenomanian deposits of Algeria. Nevertheless, this ichnotaxon lacks a diagnosis and description. Currie and Koppelhus (2015) used the name *Columbosauripus pacensis* instead of *Gypsichnites pacensis* without explaining whether it is an ichnotaxonomic proposal or is a simple writing error. Just like *Irenichnites*, the *Columbosauripus* record is relatively limited. Moreover, both ichnotaxa present similar features and similar size. New finds are needed to discuss if they are preservational variations of the same type of track, or are different ichnotaxa.

Columbosauripus has a global distribution, such as eastern British Columbia (Sternberg, 1932), Albian-Cenomanian of northeastern British Columbia (Storer, 1975), and Yukon of Canada (Gangloff et al., 2004); Algeria (Bellair and de Lapparent, 1948; Bessedik et al., 2020a,b), and Piau-Caiçara of Brazil (*Columbosauripus*-like tracks Leonardi, 1984).

3.2.6.8 Trackmaker

Bellair and de Lapparent (1948) compared *Columbosauripus* tracks from Algeria with coelurosaurs like the Upper Cretaceous *Struthiomimus*. Haubold (1971) assigned Columbosauripus to the Coelurosauria, family indeterminate. McCrea (2000) proposed that the *Columbosauripus* trackmaker is identified as a coelurosaur, ornithomimid, or some other mid-sized theropod. Gangloff et al. (2004) concluded that a moderate-sized member of the theropods most probably produced the ichnites. *Columbosauripus* also exhibits an ornithomimosaur affinity.

3.2.7 Irenesauripus *Sternberg, 1932 (Fig. 7.4F–H)*

3.2.7.1 Type ichnospecies

Irenesauripus mclearni Sternberg, 1932 (Fig. 7.4F)

3.2.7.2 Valid ichnospecies

Irenesauripus mclearni Sternberg, 1932.
Irenesauripus acutus Sternberg, 1932 (Fig. 7.4G)
Irenesauripus occidentalis Sternberg, 1932 (Fig. 7.4H)

3.2.7.3 Distribution

Tithonian–Berriasian, Mist Mountain Formation (British Columbia), Valanginian, Gorman Creek Formation (British Columbia), Aptian–Albian, Gething Formation and Gates Formation (Alberta and British Columbia), Pasayten Group (British Columbia), and middle Albian–early Cenomanian (Yukon) of Canada; Aptian–Albian, Cedar Mountain Formation (Utah), Paluxy Formation (Texas) of USA; Berriasian–Barremian, Feitianshan Formation (Sichuan) and Barremian/Aptian–Santonian, Jiaguan Formation (Guizhou) of China; Late Campanian of Poland; Berriasian–Hauterivian, Sousa Formation of Brazil.

3.2.7.4 Holotype

NMC 8548, original natural mold, housed at the Canadian Museum of Nature, Ottawa, Ontario, Canada.

3.2.7.5 Diagnosis

Large size (about 28 to over 50 cm); semidigitigrade, Digit I is rarely impressed; tridactyl; pes print is mesaxonic—Digit III is the principal digit; digit impressions typically distinct, moderately well separated—divergent from midline; Digits II and IV equal to slightly subequal in length and maximum width. Phalangeal pads and claw marks may be present on Digits II, III, and IV; claws acuminate; heel typically broad to moderately narrow (emended by Gangloff et al., 2004).

3.2.7.6 General description

Irenesauripus mclearni ranges in length from about 30 to 50 cm (McCrea, 2000). Typically, it exhibits an almost linear arrangement with a stride approximately 3 times the track's length. The heel exhibits moderate narrowness. The toes are well-separated toward their proximal ends without any indication of webbing or extensive padding. They gradually taper from near their proximal ends and terminate in relatively sharp claws; Digit II a bit longer than IV. In addition, the widest breadth of a track spans 31.8 cm, while Digits II–III and III–IV diverge at a similar angle of about 35° (Sternberg, 1932).

Irenesauripus acutus has has a pes length ranging from about 36 cm to over 50 cm with Digit IV a bit longer than II; length longer than width (Sternberg, 1932; Gangloff et al., 2004). The digits are elongate and slender with typically weak to absent phalangeal pad impressions. In particular, there is usually a greater divarication between Digits III and IV than between Digits II and III, resulting in an inward asymmetry toward the midline of the trackway (Gangloff et al., 2004).

Irenesauripus occidentalis exhibits large tracks with a broad heel and narrow but well-spaced toes, resulting in a moderately broad trackway with a short stride. The maximum length of the track is 50 cm; similar length in Digits II and IV with the shortest Digit III; breadth of the track a bit shorter than length (46.5 cm) but wider in proportion than other isp. in this ichnogenus. The angle between Digits II and III is 33°, while that between Digits III and IV is 40°. The stride length ranges from 96 to 100 cm. These measurements suggest that both slow movement during track formation and anatomical features contribute to the observed shortness of stride and width of the trackway (Sternberg, 1932).

3.2.7.7 Discussion

Sternberg (1932) described all three ichnospecies within the ichnogenus *Irenesauripus* from the Peace River Canyon of Gething Formation that are dominated by *I. acutus*, with the type species (medium) smaller than the others (large) (McCrea et al., 2017). Gangloff et al. (2004) suggested that the synonymy between *I. mclearni* and *I. acutus* is possible, however, *I. acutus* also has a narrower heel compared with the type ichnospecies, and Digits II and III of *I. acutus* are more deeply impressed than Digit IV. In addition, *I. acutus* exhibits distinctive characteristics such as narrowness of the digits and asymmetry features (Gangloff et al., 2004). Moreover, the rare forelimb impressions of a theropod were found in *I. mclearni*, consisting of long, narrow scrapes (McCrea et al., 2002).

3.2.7.8 Trackmaker

The size of the tracks and trackways, as well as the presence of large claws, suggest that the trackmaker of *Irenesauripus* was most likely a large (for *I. mclearni* is somewhat middle sized) theropod (Moratalla et al., 1988; McCrea et al., 2017). The length and slenderness of the digits suggest that the prints were made by a gracile "carnosaur," such as an allosaurid or megalosaurid, rather than a robust "carnosaur," which would have shorter and broader digits (Moratalla et al., 1988). It has been suggested that the trackmaker in Texas was *Acrocanthosaurus atokensis*, whose skeletal remains occur in the Trinity Group (Aptian-Albian) (Langston, 1997; Jacobs, 1997), or a "carnosaur" (allosaurid) of similar size and shape (Harris, 1998).

3.2.8 Irenichnites *Sternberg, 1932 (Fig. 7.4H)*

3.2.8.1 Type ichnospecies

Irenichnites gracilis Sternberg, 1932.

3.2.8.2 Valid ichnospecies

Irenichnites gracilis Sternberg, 1932.

3.2.8.3 Distribution

Aptian–Albian, Gates Formation (Alberta), Gething Formation, and Boulder Creek Formation (British Columbia) of Canada; Albian, Mesa Rica Sandstone (New Mexico) of USA.

3.2.8.4 Holotype

NMC 8552, original natural molds, housed in the Canadian Museum of Nature, Ottawa, Ontario, Canada. Paratype: trackway of three tracks on slab, Cat. No. 8558, The Geological Survey of Canada, Gething Formation, British Columbia, Canada.

3.2.8.5 Diagnosis

Functionally bipedal; tridactyl digitigrade; digits well spread, not bound together by pad, and of uniform breadth throughout; claws blunt; Digit II cut away from the rest of foot; phalangeal pads faintly shown; heel impression not complete; tracks small; stride relatively long; trackway narrow (emended by McCrea, 2000).

3.2.8.6 General description

Irenichnites gracilis is characterized by Digit II with two phalangeal pads and a terminal claw; Digit III with three phalangeal pads and a terminal claw; Digit IV with at least three phalangeal pads and a triangular, terminal claw. Digits are of uniform breadth. Tracks are generally longer than wide. The average track length/width ratio is 1.19, but varies considerably (1.1–1.31). Total divarication is low; between 60° and 80°. Pace and stride lengths indicate along-legged animal. The trackway is very narrow, with the tracks positioned in almost a straight line with little or no rotation from the midline (Sternberg, 1932; McCrea, 2000).

3.2.8.7 Discussion

The ichnotaxon *Irenichnites gracilis* was described by Sternberg in 1932 from the Lower Cretaceous Gething Formation near Hudson's Hope, British Columbia, Canada. One specimen of *I. gracilis* described by Sternberg (1932) displays skin impressions associated with the digital pads of Digit III and the metatarsal pad, representing both the first record of skin impressions from tracks attributed to this ichnotaxon and Early Cretaceous skin impressions on small theropod tracks (McCrea et al., 2015a). Lockley et al. (2011) proposed that *Irenichnites* belongs to the Ornithomimipodidae ichnofamily. Haubold (1971) classified *Irenichnites gracilis* as *Ornithomimipus gracilis*.

Irenichnites has the Digit II impression separated from the III–IV, similar to *Ornithomimipus*, but *Combosauripus* has all the digits united. *Ornithomimipus* presents a more developed the Digit III impression than *Irenichnites*. Currie (1981) identified five trackways of *Irenichnites* in the Aptian Gething Formation, and McCrea (2000) found three at the same formation. Moreover, McCrea et al. (2015a) also identified in the Albian of Canada a pair of tracks that belong to this ichnotaxon (McCrea et al., 2015a). We consider that *Irenichnites* is a well-preserved ichnotaxaon, but currently is poorly known with few studied specimens.

3.2.8.8 Trackmaker

Sternberg (1932) observed that the tracks of *Irenichnites gracilis* exhibit a long stride, indicating the presence of a long-legged animal. Thulborn (1990) identified ornithomimid dinosaurs as the trackmakers of *Irenichnites*, based on the consistent separation of Digit II impression from the rest of the track, suggesting its higher attachment to the foot compared with other weight-bearing digits, an arrangement observed in ornithomimid skeletons. The phalangeal pattern of these tracks also corresponds well with the phalangeal formula for ornithomimid pes skeletons, which is 2:3:4:5:0 (Barsbold and Osmólska, 1990). The ornithomimid that produced the *Irenichnites* tracks would have been approximately the size of *Dromiceiomimus*, whose pes skeleton length (Barsbold and Osmólska, 1990) falls within the range of the *Irenichnites* tracks observed at Grande Cache and Peace River (McCrea, 2000).

3.2.9 Satapliasaurus *Gabuniya, 1951 (Fig. 7.5A–C)*

3.2.9.1 Type ichnospecies

Satapliasaurus tschabukianii Gabuniya, 1951 (Fig. 7.5A)

3.2.9.2 Valid ichnospecies

Satapliasaurus tschabukianii Gabuniya, 1951.
Satapliasaurus dsocenidzei Gabuniya, 1951 (Fig. 7.5B)
Satapliasaurus kandelakii Gabuniya, 1951 (Fig. 7.5C)

3.2.9.3 Distribution

Jurassic: Upper Aalenian, Peak Alum Quarries (Yorkshire) of England; Cretaceous: lower Neocomian deposits (Sataplia) of Georgia.

3.2.9.4 Holotype

A replica of the holotype is PC 4237 housed in the Department of Geology, University of Nottingham.

3.2.9.5 Diagnosis

Small, slender, functionally bipedal, digitigrade dinosaurs. The pes has three functioning digits, equipped with sharp claws. The digits are rather close-set, giving the pes a narrow shape. The weakly developed Digit I left little trace in the footprints. The tail balanced, or lightly touched the ground when walking (Gabuniya, 1951).

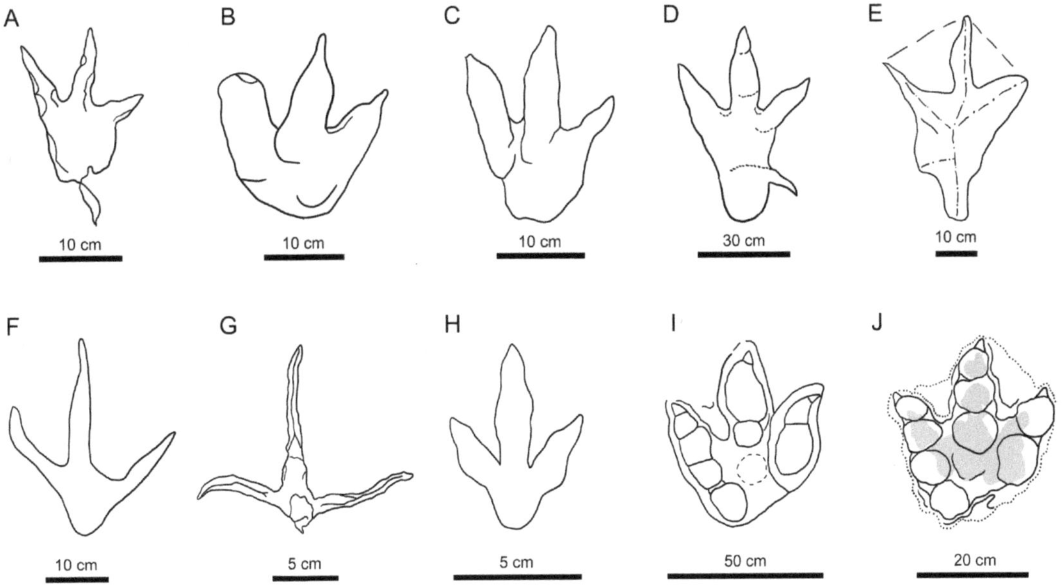

FIGURE 7.5 Theropod tracks in Europe from Cretaceous. (A) Holotype of *Satapliasaurus tschabukianii;* (B) holotype of *Satapliasaurus kandelakii;* (C) holotype of *Satapliasaurus dsocenidzei;* (D) holotype of *Bueckeburgichnus maximus;* (E) holotype of *Theroplantigrada encisensis;* (F) holotype of *Filichnites oncalensis;* (G) holotype of *Archaeornithipus meijidei;* (H) holotype of *Kalohipus bretunensis;* (I) holotype of *Jurabrontes curtedulensis;* (J) holotype of *Jurabrontes melphicticus. (A) Redrawn from Gabuniya (1951). (B) Redrawn from Gabuniya (1951). (C) Redrawn from Gabuniya (1951). (D) Redrawn from Kuhn (1958). (E) Redrawn from Casanovas et al. (1993a). (F) Redrawn from Moratalla (1993). (G) Redrawn from Fuentes Vidarte (1996). (H) Redrawn from Fuentes Vidarte and Meijide Calvo (1998). (I) Redrawn from Marty et al. (2018). (J) Redrawn from Antonelli et al. (2023).*

3.2.9.6 General description

Of the three ichnospecies from Neocomian marine limestone, Georgia, originally placed in *Satapliasaurus* by Gabuniya (1951), only *S. tschabukianii* shows an indication of the presence of the Digit I (hallux) and has a narrow-shaped pes; both of *S. kandelakii* and *S. dsocenidzei* are typical tridactyl tracks (Sarjeant, 1970).

The features of *S. tschabukianii* are provided by Gabuniya (1951): small, bipedal, digitigrade dinosaurs, tetradactyl; Digit I poorly developed, and directed backward and to the side (posteromedial); digits equipped with pointed claws; no tail traces; track length 25 cm, track width 16 cm, step length 70 cm. The Digit I in *S. tschabukianii* is a distinguishing characteristic. According to Gabuniya (1951), it appears only when the trackmaker comes into contact with soft, wet soil.

The features of *S. dsocenidzei* included: small, biped, digitigrade dinosaurs. The pes has a narrow shape and comparatively short digits, with strongly developed claws. On the imprint of the pes is a well-marked footprint (of the postphalangeal section of the "heel"). Prints of the first digit and tail are lacking. The length is 22 cm, width 15.5 cm, and step length 60 cm (Gabuniya, 1951; and translated by Sarjeant, 1970).

The features of *S. kandelakii* include small, bipedal, digitigrade dinosaurs; tridactyl, relatively wide in shape; narrow and pointed claws in each digit; without traces of Digit I, manus or tail traces; imprint of the subphalangeal section is poorly outlined; track length: 23 cm, track width 21 cm, step length: ~65 cm (Gabuniya, 1951).

3.2.9.7 Discussion

Sarjeant (1970) pointed out that the ichnogeneric diagnosis presents several points of difficulty. A revision of the diagnosis, possibly involving removal of the two latter ichnospecies to a separate ichnogenus, appears urgently necessary: a clarification of the features distinguishing *Satapliasaurus* from the older ichnogenus *Anomoepus*. The diagnosis of *Satapliasaurus* lacks crucial information, such as trackways and additional measurement data. This article treats it as a valid ichnotaxon, but further data and discussions are necessary to confirm its validity.

Sarjeant (1970) also described *Satapliasaurus* cf. *dsocenidzei* from the Upper Aalenian in Yorkshire, England, which was termed as cf. *Grallator* by Lockley and Meyer (2000). The specimens measured a maximum of 24 cm in length and 23 cm in width. They have three functional digits that spread quite widely, with blunt claws. The front part of the heel is deeply impressed. The Yorkshire specimens are similar to *S. dsocenidzei*, with some limited differences, such as the sharpness of the claws and the clarity and depth of the heel (Sarjeant, 1970).

3.2.9.8 Trackmaker

Gabuniya (1951) classified *Satapliasaurus* as a carnivorous dinosaur. Sarjeant (1970) considered *Satapliasaurus* to be that of a bipedal, herbivorous ornithopod dinosaur, and compared it with *Camptosaurus* (Marsh, 1885). The track maker of *Satapliasaurus* is undoubtedly a theropod dinosaur, as evidenced by the sharp claw marks and typical theropod track pattern.

3.2.10 Bueckeburgichnus *Kuhn, 1958 (Fig. 7.5D)*

3.2.10.1 Type ichnospecies

Bueckeburgichnus maximus Kuhn, 1958.

3.2.10.2 Valid ichnospecies

Bueckeburgichnus maximus Kuhn, 1958.

3.2.10.3 Distribution

Bückeburg Formation, Wealden Group (Bückeburg) of Germany.

3.2.10.4 Holotype

Specimen B4B.9, housed in the Münster Museum of Geology and Paleontology, Münster, Germany.

3.2.10.5 Diagnosis

Large tetradactyl theropod track with small hallux (Digit I). Digit II wide and well-padded. Digit III parallel sided proximally but strongly tapering distally. Digit IV narrow with traces of discrete digital pads (emended by Lockley, 2000)

3.2.10.6 General description

Lockley (2000) described the holotype as a left natural cast of a large theropod tetradactyl footprint, which is longer (56 cm) than wide (42 cm) including the heel. The hypices between

Digits II and III and between Digits III and IV are located exactly half the distance (28 cm) between the anterior point of Digit III and the posterior edge of the "heel." The Digit I impression measures 12 cm in length and is located medially with a slight posterior curvature at the distal end. It is located 9 cm behind the medial groove that marks the posterior margin of Digit II. The hallux is not as deep as Digits II and IV. Digit II is wide (14 cm) and is closed with a drop-shaped pad and separated from the postero-medial "heel" by a strong wrinkle. Digit III is narrow (8 cm) proximally and narrows abruptly (4 cm) toward the distal end. The Digit IV is also narrow with visible wrinkles marking the boundaries of two distal pads.

3.2.10.7 Discussion

These tracks were published first by Ballerstedt (1905), then by Abel (1935), and later by Lessertisseur (1955). The diagnosis proposed by Kuhn (1958) for this ichnogenus is the same as the description made by Abel (1935) for that same track. Kuhn (1958) used the term *Bückeburgichnus*, which should be *Buckeburgichnus* (without the umlaut). Lockley (2000) corrected the name *Buckeburgichnus* to *Bueckeburgichnus*. Haubold (1971) suggested that *Bückeburgichnus* could be *Megalosauripus* Lessertisseur, 1955 or *Tyrannosauripus* Lessertisseur, 1955. Thulborn (2001) stated that *Megalosauripus* is a senior synonym of *Bueckeburgichnus*. Thulborn (2001) commented that the holotype used by Lockley (2000) is a natural cast of a left footprint, and that the holotype designated by Kuhn (1958) and drawn by Ballerstedt (1905) and Abel (1935) is a natural cast of a right footprint.

Hornung et al. (2012) identified two morphotypes of theropod tracks in the Bückeberg Formation and admitted the usefulness of using *Bueckeburgichnus* to differentiate *Megalosauripus*. Haderer and Neef (2004) reported new material of *Bueckeburgichnus* maximus in this formation.

3.2.10.8 Trackmaker

Lockley (2000) suggested that the trackmaker was a theropod and that it is not possible to label the trackmaker of *Bueckeburgichnus* as a megalosaur with any certainty either on morphological grounds or on the basis of its age.

3.2.11 Yangtzepus *Young, 1960 (Fig. 7.6A and B)*

3.2.11.1 Type ichnospecies

Yangtzepus yipingensis Young, 1960 (Fig. 7.6A)

3.2.11.2 Valid ichnospecies

Yangtzepus yipingensis Young, 1960.
Yangtzepus clarkei Salisbury et al., 2016 (Fig. 7.6B)

3.2.11.3 Distribution

Barremian/Aptian–Santonian, Jiaguan Formation (Sichuan) of China; Valanginian–Barremian, Broome Sandstone (Dampier Peninsula) of western Australia.

3.2.11.4 Holotype

IVPP V2473, three natural molds on three slabs, housed in the Institute of Vertebrate Paleontology and Paleoanthropology, China.

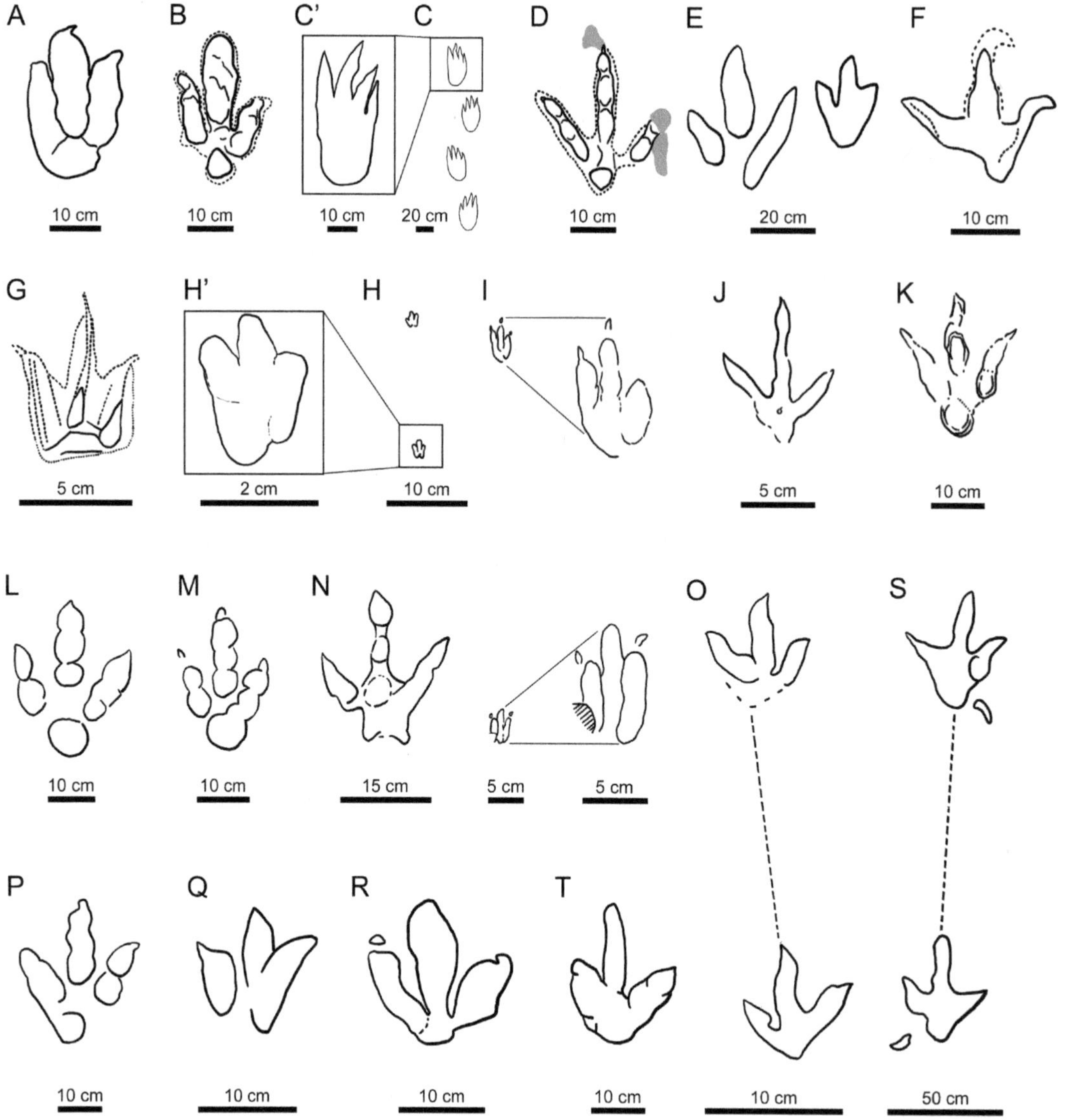

FIGURE 7.6 Theropod tracks in Asia and Australia from Cretaceous. (A) Holotype of *Yangtzepus yipingensis*; (B) holotype of *Yangtzepus clarkei*; (C) type series (C) and (C′) of *Macropodosauropus gravis*; (D) holotype of *Megalosauropus broomensis*; (E) holotype of *Megalosauropus brionensis*; (F) holotype of *Xiangxipus chenxiensis*; (G) holotype of *Skartopus australis*; (H) holotype and type series of *Minisauripus chuanzhuensis*; (I) type series of *Minisauripus zhenshuonani*; (J) holotype of *Toyamasauripus masuiae*; (K) holotype of *Asianopodus pulvinicalx*; (L) holotype of *Asianopodus robustus*; (M) type specimen of *Asianopodus wangi*; (N) holotype of *Siamopodus khaoyaiensis*; (O) holotype of *Siamopodus xui*; (P) holotype of *Chapus lockleyi*; (Q) holotype of *Corpulentapus lilasia*; (R) holotype of *Paracorpulentapus zhangsanfengi*; (S) holotype of *Ordexallopus zhanglifui*; (T) holotype of *Lockleypus luanpingensis*. *(A) Redrawn from Xing et al. (2009c). (B) Redrawn from Salisbury et al. (2016). (C) C: redrawn from Zakharov (1964); C′: Sennikov (2021). (D) Redrawn from Salisbury et al. (2016). (E) Redrawn from Lockley et al. (2000). (F) Redrawn from Xing et al. (2016b). (G) Redrawn from Romilio et al. (2013). (H) Redrawn from Zhen et al. (1996). (I) Lockley et al. (2008a). (J) Redrawn from Matsukawa et al. (1997). (K) Redrawn from Matsukawa et al. (2005). (L) Redrawn from Li et al. (2011). (M) Redrawn from Xing et al. (2021b). (N) Redrawn from Lockley et al. (2006). (O) Redrawn from Xing et al. (2014b). (P) Redrawn from Li et al. (2006). (Q) Redrawn from Li et al. (2011). (R) Redrawn from Xing et al. (2014c). (S) Redrawn from Lockley et al. (2018a). (T) Redrawn from Xing et al. (2018e).*

3.2.11.5 Diagnosis

Medium-sized, tridactyl theropod tracks (originally attributed to an ornithopod), lacking manus and tail traces. The divarication angle between Digits II and III is wider than that between Digits III and IV. Digit III is directed forward and Digit IV is similar in length to Digit II. The long axis of the digit is parallel to the track axis. A metatarsophalangeal pad is oval in shape (emended by Xing et al., 2009c).

3.2.11.6 General description

Yangtzepus yipingensis has the elongated shape with a length/width ratio of about 1.5–1,9; the divarication angles between Digits II–IV both about 20°; relatively narrow impressions of digits along with claw marks; oval metatarsophalangeal pad impression sits at the caudal end of Digit III; the pad impression is separated from Digit III by a distinct line, but for the other two digits, such demarcation lines are not well-defined (Xing et al., 2009c; Díaz-Martínez et al., 2015a).

Yangtzepus clarkei has pedal tracks of medium to large size (proximodistal length 24–32.9 cm, mediolateral width 19.7–20.5 cm); tridactyl, mesaxonic, longer than wide, with an average maximum-length-to-maximum width ratio of approximately 1.2–1.6. The individual digital impressions are proportionately elongated and broad, with Digit III being the widest and longest. Digit II extends farther distally than Digit IV in relation to the main track axis. Digits II and IV have a total divarication angle between their axes of 30°–34°, while the divarication of the axes between Digits III and IV is greater (20°–23°) than that between Digits II and III (10°–11°). The metatarsodigital pad impression is immediately proximal to the proximal end of the Digit IV impression, in line with the principal track axis. It has a width approximately equivalent to the Digit III impression, while the hallux impression is absent (Salisbury et al., 2016).

3.2.11.7 Discussion

Yangtzepus clarkei can be distinguished from *Yangtzepus yipingensis* on the basis of the clear separation and greater angle of the digital impressions, and a well-defined metatarsophalangeal pad impression that is separate from the proximomedial and proximolateral portions of the digital pad impressions. Furthermore, in *Y. clarkei,* most of the medial margins of Digits II and IV impressions contact Digit III impressions to a lesser extent than in *Y. yipingensis,* and the divarication angle between the axes of Digit II and III impressions is smaller than that of the Digit III and IV impressions (Salisbury et al., 2016).

3.2.11.8 Trackmaker

Yangtzepus yipingensis was named based on tracks from the Early Cretaceous of China, that were originally assigned to ornithopods due to the close proximity of a small, tridactyl track with the larger pedal tracks (Young, 1960; Zhen et al., 1989). However, Xing et al. (2009c) interpreted the tracks as theropod tracks similar to those found in the USA (Pittman, 1989; Farlow, 2001). Based on the morphology of the tracks (longer than wide, with elongated, narrow digit impressions and claw marks), Díaz-Martínez et al. (2015a) also classified them as theropod tracks.

Salisbury et al. (2016) agreed with Xing et al. (2009c) that the *Y. yipingensis* trackmaker was most likely a theropod, and they also considered *Y. clarkei* to be of probable theropod affinity using Wright (2005) criteria of high track length to track width ratio, overall track asymmetry, and the presence of narrow ungual impressions.

3.2.12 Macropodosauropus *Zakharov, 1964 (Fig. 7.6C)*

3.2.12.1 Type ichnospecies

Macropodosauropus gravis Zakharov, 1964.

3.2.12.2 Valid ichnospecies

Macropodosauropus gravis Zakharov, 1964.

3.2.12.3 Distribution

Albian, Shirabad Formation of Tajikistan, Coniacian of Morocco, Campanian of Poland, and Turonian of USA.

3.2.12.4 Holotype

Nos. 1–3 (*in situ*), from Shirkent I tracksite, which is in an Albian peritidal grainstone/evaporite in the Shirabad Formation of Tajikistan.

3.2.12.5 Diagnosis

Large massive plantigrade dinosaurs. When moving, lizards used their hind limbs, which had four digits (I–IV), of which three (II–IV) were mainly functional: these three Digits were connected by strong membranes and had claws. The first digit served only as additional support when walking on sticky ground. The maximum length of the footprint, accepted as a type, is 43 cm, together with the claw on the Digit III, 50 cm. The width of the footprint (maximum) is 29 cm. The claws are narrow, moderately curved, and 6–7 cm in length. Pace length 72–75 cm. The tail is short, does not touch the ground even when walking calmly (translated from Russian by Zakharov, 1964).

3.2.12.6 General description

Macropodosauropus presents tetradactyl, mesaxonic, and digitigrade well-preserved tracks. They are longer, about 50 cm, than wide; the maximum width at the base of the first digit is 28–30 cm. The tracks are deeply imprinted (to a depth of 3–5 cm), and are surrounded by wide expulsion rims, mostly in their anterior part (Sennikov, 2021). The trackway is narrow gauge, with a pace angulation of about 150°–160°. Pace length is about 75 cm, stride length about 146 cm (Sennikov, 2021).

3.2.12.7 Discussion

Macropodosauropus tracks are known from the Cretaceous of Central Asia (Tajikistan), Central Europe (Poland) (Gierliński, 2009, 2015), North Africa (Morocco) (Masrour et al., 2017a) and North America (USA) (Gierliński and Lockley, 2013b). The type *M. gravis* tracks from Tajikistan are 50 cm long and 29–30 cm wide. All other known tracks are considerably smaller: 35 by 20 cm in Poland, 30 by 15 cm in the USA, and 26 by 16 cm in Morocco. Polish tracks are wider among *Macropodosauropus*; USA tracks shows somewhat narrower metatarsal impression, and better differentiated in the outline; Moroccan tracks have better pronounced and better differentiated (by depth) metatarsal and digital impressions, while the tips of Digit II–IV impressions are blunt without visible claw impressions (Sennikov, 2021).

Gierliński and Lockley (2013b) suggested that *Macropodosauropus* and *Otozoum* are different in having well-developed digital pads. The general similarity in foot morphology between the two ichnogenera suggests a degree of convergence between the footprints attributed to two completely different saurischian trackmakers. Another similar ichnogenus is "*Haenamichnus*" (Kim et al., 2012a; very likely to be considered invalid due to the poor preservation) but they are relatively (length/width) longer, have a characteristic subtriangular shape and a thinner back termination that distinguishes them from those of the *Macropodosauropus*. Moreover, in *Macropodosauropus* the "heel" shape is rounded or subrectangular while in *Haenamichnus gainensis* it is subtriangular (Masrour et al., 2017a).

3.2.12.8 Trackmaker

Zakharov (1964) believed the trackmakers of *Macropodosaurus* to have been bipedal, massive, plantigrade dinosaurs with relatively short limbs and a short stride, that is, slow-moving. These dinosaurs were therefore likely not to have been true carnivores like most theropods, but had become omnivores and inhabited coastlines and other shores (Zakharov, 1964).

Sennikov (2006) reevaluated *Macropodosaurus*, making a detailed case for therizinosaur affinity by illustrating the tracks with the foot skeletons of *Therizinosaurus* and the related genus *Erlikosaurus* superimposed, and proposed that the trackmaker was probably a therizinosauroid.

Through a comparison of *Macropodosaurus* tracks with those of other dinosaurs and by a juxtaposition of track morphology and pes skeletal morphology, Sennikov (2021) argued that they belong to an aberrant group of theropods, the segnosaurs.

3.2.13 Megalosauropus *Colbert and Merrilees, 1967 (Figs. 7.6D,E and 7.33D,E)*

3.2.13.1 Type ichnospecies

Megalosauropus broomensis Colbert and Merrilees, 1967 (Figs. 7.6D and 7.33E)

3.2.13.2 Valid ichnospecies

Megalosauropus broomensis Colbert and Merrilees, 1967.
Megalosauropus brionensis Haubold, 1971 (Fig. 7.6E)

3.2.13.3 Distribution

Valanginian–Barremian, Broome Sandstone (Dampier Peninsula) of western Australia.

3.2.13.4 Holotype

WAM 66.2.51, housed in Western Australian Museum.

3.2.13.5 Diagnosis

Medium- to large-sized (proximodistal length 28.6–44.3 cm, mediolateral width 22.5–35.5 cm), tridactyl, mesaxonic (digital impression extension to track length ratio 0.30–0.47), longer than wide, with an average maximum-length-to-maximum-width ratio approximately 1.0–1.3; individual digital impressions proportionately elongated and narrow, and of similar width (the approximate width of each digital impression is approximately one-seventh the total track length),with the central digital impression (Digit III) being the longest (basal digital impression length 42%–72% the total length), and the impressions of Digits II and IV often extending distally to approximately the same level relative to the principal track axis; on shallow tracks, the impression of Digit II shortens proximally relative to the length of the impression of Digit IV; axes of the impressions of Digits III and IV typically intersect distal to the intersection of the axes of the impressions of Digits II and III; total divarication angle between the axes of impressions of Digits II and IV 66°–81°; divarication of axes of impressions of Digits III and IV (31°–45°) typically greater than divarication of axes of impressions of Digits II and III (27°–39°); digital pad impressions may be present, with formula 3-4-3 for Digits II, III and IV, respectively; a possible fourth pad impression may also be present proximally on the impression of Digit III (visible only on the holotype); ungual impressions may be present distally on all digital impressions; a separate, circular metatarso-digital pad impression immediately proximal to the proximal end of the impression of Digit IV and in line with the principal track axis on the majority of shallow tracks, with a maximum width that is slightly greater than that of each digital impression; on deeper tracks, the diameter of the single metatarso-digital pad impression increases, becoming contiguous with the proximal end of each digital impression; separation between the impression of Digit II and that of the metatarso-digital pad results in proximomedial indentation on shallow tracks; proximolateral indentations between the impression of Digit IV and the metatarso-digital pad shorter than the corresponding proximomedial indentations; hallucal impression absent on all tracks, irrespective of depth. Trackway: pace angulation 140°–160°; typical stride length approximately 7 times the maximum pedal track length; typical pace approximately 3 times the maximum pedal track length (amended by Salisbury et al., 2016).

3.2.13.6 General description

Haubold (1971) erected the *Megalosauropus brionensis* for Lower Cretaceous tracks from the former Yugoslavia. Such usage implies a close morphological match between ichnites and the type *Megalosauropus* material from Australia. But such a match has not been demonstrated (Haubold, 1971). Salisbury et al. (2016) reconsidered *Megalosauropus* by reanalyzing the holotype and the referred topotype tracks from Minyirr, and including the new specimens from intertidal zone of the Yanijarri–Lurujarri section of the Dampier Peninsula. The latter shows consistent with the proportions of the topotype tracks (see the emended diagnosis), while the basal digital length (42%–49%) values less than the holotype (69%) and paratype (60%, 71%), perhaps due to the lack of internal details to the former.

3.2.13.7 Discussion

This ichnogenus exists two spellings: "Megalosauropus" and "Megalosauripus". It has been originally named as "Megalosauripus" by Lessertisseur (1955) informally, basing on the Late Jurassic theropod tracks from Portugal, then Kuhn (1958) redefined this ichnotaxon with the addition of similar Cretaceous specimens with large size and hallux impressions. Till 1967, Colbert and Merrilees (1967) firstly named "Megalosauripus" formally, while with the name "Megalosauropus". Colbert and Merrilees (1967) defined "Megalosauropus" basing on the Lessertisseur (1955) specimens as holotype, and linked the some other Laurasian speciemens from the similar chronostratigraphic levels (close to the Oxfordian–Kimmeridgian boundary) to this ichnogenus. In accordance with the principles of biological nomenclature, the earliest formally named "Megalosauropus" should be designated as the ichnogenus name.

Ever since Colbert and Merrilees (1967) established *Megalosauropus*, the ichnotaxon has been riddled with taxonomic issues (Salisbury et al., 2016). A large number of track records have been assigned to this ichnogenus, such as tracks from the Upper Jurassic of Uzbekistan and Turkmenistan (*Megalosauropus uzbekistanicus*; Lockley et al., 1996a) and Germany (*M. teutonicus*; Kaever and de Lapparent, 1974) and the Lower Cretaceous of Portugal (*M* (? *Eutynichnium*) *gomesi*; Antunes, 1976). These tracks show little if any resemblance to the type ichnospecies from the Broome Sandstone (Salisbury et al., 2016). The consensus of a suite of studies (Lockley et al., 1996b, 2000; Lockley, 2000; Thulborn, 2001; Salisbury et al., 2016) is that *Megalosauropus* should be restricted to the type species from the Broome Sandstone.

3.2.13.8 Trackmaker

Colbert and Merrilees (1967) suggested that trackways may be regarded as representative of a large megalosaur, perhaps even of *Megalosaurus* itself.

3.2.14 Moraesichnium *Leonardi, 1979a (Fig. 7.7A)*

3.2.14.1 Type ichnospecies

Moraesichnium barberenae Leonardi, 1979a.

3.2.14.2 Valid ichnospecies

Moraesichnium barberenae Leonardi, 1979a.

3.2.14.3 Distribution

Berriasian–Hauterivian, the Sousa Formation (Paraíba) of Brazil.

3.2.14.4 Holotype

SOPPb trackway, Passagem das Pedras in the Hazenda Ilha, 4 km NW of Sousa, Paraíba State, Brazil.

3.2.14.5 Diagnosis

Trackway of a medium-sized bipedal dinosaur, with a high to very high-pace angle, extremely variable oblique pace, and stride values, the index stride: footprint length can reach a value of 12.5 cm; inner trackway width always with a negative value and external trackway width small; axis of the footprints is differently inclined slightly outwards or inwards in relation to the axis of the track. Tridactyl, symmetrical, wedge-shaped footprints, slender, provided with a "plant" that covers more than half the length of the foot; functional prevalence in the phalangeal-metatarsal pads row, and in Digits III and IV: relatively long and slender digits, being Digit III the longest, and IV slightly prevailing over II, total divergence (II–IV) very high; partial divergences subequal. Pads generally visible on Digit III, more rarely on the others; small claws rarely visible. No tail traces (translated from Portuguese of Leonardi, 1979a).

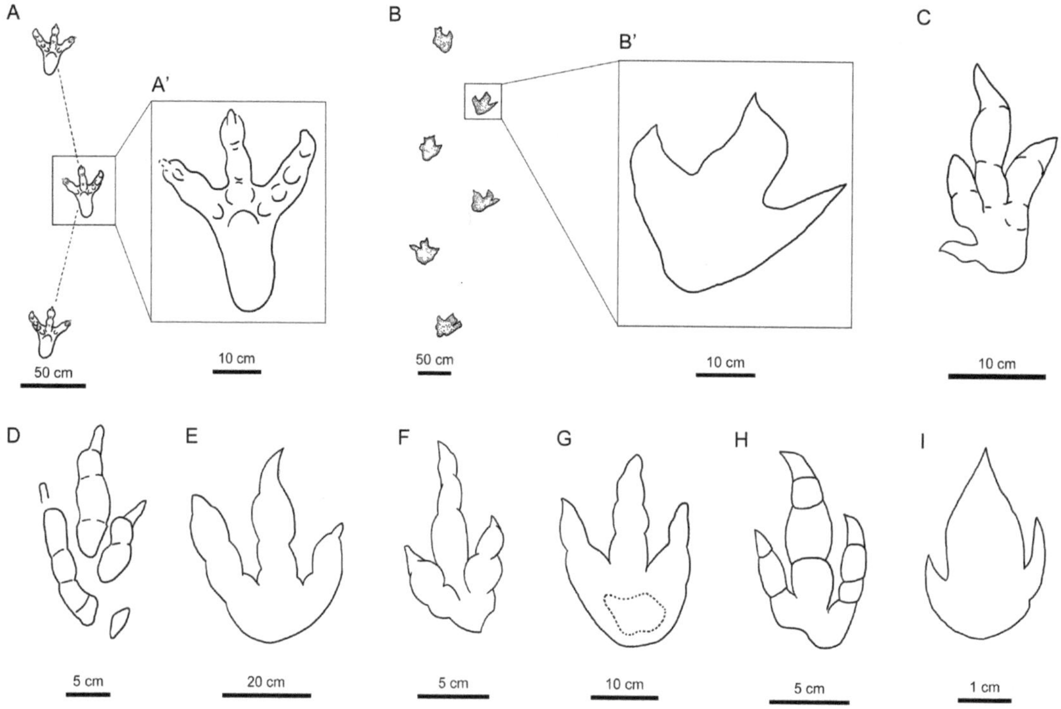

FIGURE 7.7 Theropod tracks in South America from Cretaceous. (A) Type series of *Moraesichnium barberenae*; (B) type series of *Salfitichnus mentoor*; (C) holotype of *Picunichnus benedettoi*; (D) holotype of *Picunichnus quijadaensis*; (E) holotype of *Abelichnus astigarrae*; (F) holotype of *Deferrariischnium mapuchensis*; (G) holotype of *Bressanichnus patagonicus*; (H) holotype of *Candeleroichnus canalei*; (I) holotype of *Farlowichnus rapidus*. *(A) Redrawn from Leonardi (1979a). (B) Redrawn from Alonso and Marquillas (1986). (C) Redrawn from Calvo (1991). (D) Redrawn from Melchor et al. (2019). (E) Redrawn from Calvo (1991). (F) Redrawn from Calvo and Rivera (2018). (G) Redrawn from Calvo and Rivera (2018). (H) Redrawn from Calvo and Rivera (2018). (I) Redrawn from Leonardi et al. (2024).*

3.2.14.6 General description

The paratype of *Moraesichnium barberenae* consists of 34 excavated footprints on the right side of the river and a further 13 footprints on the left side. The total excavated length of the trackway is now 56 m. The footprints are tridactyl, symmetrical, with a long sole or "plant" that constitute more than half the footprint length and area, and an arrowhead shape. The Digit III impression is more deeply impressed than those of Digits IV and II. The digit impressions are long and narrow. They preserve digital pads in Digit III but sometimes in Digit IV. The Digit III impression is the longest, and Digit IV is slightly longer than Digit II. The total divarication is variable from 55° to 88°. Claw impressions only preserved in Digit III.

3.2.14.7 Discussion

Leonardi (1979) suggested that *Moraesichnium* is similar to *Camptosaurichnus* (Casamiquela and Fasola, 1968) from Chile. However, he identified some differences between *Camptosaurichnus* and *Moraesichnium*: (1) none of the *Moraesichnium* trackways are quadrupedal or semi–

quadrupedal, unlike the Chilean record; (2) the pace and stride lengths are always larger than in *Camptosaurichnus* and vary greatly; (3) the total and partial divergence angles are much greater than in the tracks from Chile.

Recently, Lallensack et al. (2022b) studied the formation and preservation of dinosaur tracks with an elongated "heel." They concluded that elongate tracks were usually produced by deep penetrations of the foot in soft sediment, so "metatarsal marks should only be used for ichnotaxonomic assignments and trackmaker identifications if they can be demonstrated to be resting traces or to record plantigrade locomotion." In this case, the holotype of *Moraesichnium* (see Leonardi and Carvalho, 2021c, Fig. 4.71), which is the better-preserved track of this ichnotaxon, has a clear outline, and pad and claw impressions. Nevertheless, we are not sure if it can be given a value of 2 on the numerical preservation scale proposed by Marchetti et al. (2019), and that its shape is not conditioned by extramorphological features.

3.2.14.8 Trackmaker

Due to the peculiar posture of the trackmaker, *Moraesichnium* has been speculated to be of multiple origins. Originally it was assigned to "Hypsilophodontidae" by Leonardi (1979); to Ornithopoda by Bonaparte (1980); to Carnosauria by Carvalho (1995); and to Theropoda by Romero Molina et al. (2003) and Leonardi and Carvalho (2021c).

3.2.15 *Xiangxipus* Zeng, 1982 (Fig. 7.6F)

3.2.15.1 Type ichnospecies

Xiangxipus chenxiensis Zeng, 1982.

3.2.15.2 Valid ichnospecies

Xiangxipus chenxiensis Zeng, 1982.

3.2.15.3 Distribution

Albian, Shenhuangshan Formation (Hunan) of China.

3.2.15.4 Holotype

JQWI-HV003-4, housed in the Hunan Museum of Geology, Changsha, China.

3.2.15.5 Diagnosis

A medium-sized, slender-toed tridactyl track, pes length/width ratio of 0.8, with weak mesaxony (0.33). A metatarsophalangeal pad trace from Digit IV is located lateral to the long axis of Digit III. The track is nearly symmetrical, with a wide (97°) divarication angle. Step length is roughly 3 times track length. Mean pace angulation is relatively high, about 159° (emended from Xing et al., 2016b).

3.2.15.6 General description

Xiangxipus chenxiensis has a length/width ratio of 1.0. Digit IV is the longest, and Digits II and III are nearly equal in length. The metatarsophalangeal pad of Digit IV is lateral to the long axis of Digit III. Distal claw traces are also clearly visible. The phalangeal pads of all three digits are undifferentiated and indistinct. The divarication angle between Digits II–IV (90°) is high. Pace angulations of the trackway are between 149° and 164°, and the ratio of track length to pace length is 1:3. Some of the specimens show double traces, possibly due to the wet and soft sediments in which they were made (Xing et al., 2016b).

3.2.15.7 Discussion

Zeng (1982) initially described two ichnospecies of *Xiangxipus*, *X. chenxiensis*, and *X. youngi*. However, Lockley et al. (2013) suggested that the differences between the larger *X. chenxiensis* and the smaller *X. youngi* may be sufficient to place the latter form in a different ichnogenus, as *X. chenxiensis* has wider digits and wider or more robust digit traces; and pointed out that *X. youngi* is similar in size and digit divarication to *Wupus agilis* (Xing et al., 2007, 2015d). After the redescription in Xing et al. (2016b), the specimens of *X. youngi* were classified as *Wupus* primarily due to the relatively thick toes. Matsukawa et al. (2006) and Lockley et al. (2011) suggested that the trackmaker of *Xiangxipus chenxiensis* should be assigned to the Ornithomimipodidae.

3.2.15.8 Trackmaker

Xiangxipus, which is part of a larger trackway sequence, is characterized by very widely spread toes, firstly assigned to a coelurosaurian dinosaur such as an ornithomimid (Matsukawa et al., 2006).

3.2.16 Skartopus *Thulborn and Wade, 1984 (Fig. 7.6G)*

3.2.16.1 Type ichnospecies

Skartopus australis Thulborn and Wade, 1984.

3.2.16.2 Valid ichnospecies

Skartopus australis Thulborn and Wade, 1984.

3.2.16.3 Distribution

Cenomanian, Winton Formation (Central W. Queensland) of Australia.

3.2.16.4 Holotype

Single left track, preserved as natural mold; QM F10330.

3.2.16.5 Diagnosis

Trackway of small digitigrade biped, with pace angulation about 150°. Track size index (SI) between 2.9 and 5.7 cm. No imprints of hand or tail. Tracks tridactyl (Digits II, III, and IV), slightly longer than broad (ratio width/length about 0.95) and showing distinct positive rotation. Digit imprints narrow, straight, and sharply pointed, without indications of phalangeal pads. Digit III longest; Digits II and IV about equal in length, and almost equally divergent from Digit III (both interdigital angles between 25° and 30°). Imprint of Digit IV extends slightly farther back than imprint of Digit II, but does not form a posterior salient or "spur." Traces of small interdigital webs sometimes present. Posterior margin of track is an oblique line (posterolateral to anteromedial), either straight or arched forward. In some examples there is an imprint of the metapodium: this is subrectangular in outline and roughly equivalent in length to Digit 3. Ratio PLIFL (pace length to footprint length) usually between 5.5 and 8.5, rarely as low as 5.2 or as high as 9.1; ratio SLIFL usually between 11.0 and 16.0, rarely as low as 10.6 or as high as 17.3 (Thulborn and Wade, 1984).

3.2.16.6 General description

According to Thulborn and Wade (1984), all tracks referred to *Skartopus* are tridactyl and almost bilaterally symmetrical. The digit impressions are straight, relatively long and narrow, and about equally broad. Digit III impression is the longer, while the Digits II and IV are roughly equal in length. When the claw impressions are preserved, they are quite sharply pointed. Even the deepest footprints lack a Digit I impression. Generally, the digit impressions are parallel, but when the three digits are deeply impressed, as in the holotype, it may be seen that their posterior ends do not fall on a straight line. Digits II and IV extend slightly farther back than Digit III, to form a curve (convex forward) that presumably reflects the natural arch formed by the distal ends of the metatarsals. The interdigital angles are small, and in some tracks Digits III and IV are subparallel. There are no definite indications of phalangeal pads in any of the tracks.

3.2.16.7 Discussion

Skartopus and *Wintonopus* were defined based on hundreds of small tracks found on the same tracking surface. They were interpreted as the effect of a stampede (Thulborn and Wade, 1984). These authors differentiated between both ichnotaxa mainly emphasizing the shape of the digit impressions (thin vs. robust), relating *Skartopus* to theropod trackmakers and *Wintonopus* to ornithopods. Romilio et al. (2013) suggested that these tracks were impressed by swimming dinosaurs. Moreover, they analyzed the variability of these 100 tracks and concluded that all are a taphonomical continuum and the differences are preservational bias. Finally, they proposed that *Skartopus* was a junior synonym of *Wintonopus*, and we agree with this proposal.

3.2.16.8 Trackmaker

Thulborn and Wade (1984) suggested that *Skartopus* tracks were very probably made by coelurosaurs—small representatives of the dinosaur suborder Theropoda although theropod body fossils have not yet been reported from Queensland. Haubold (1971) stated that the *Skartopus* footprints are comparable with those attributed elsewhere to coelurosaurs in their size and general appearance. Finally, Romilio et al. (2013) suggested that *Wintonopus*, and therefore the tracks originally classified as *Skartopus*, were impressed by ornithopod trackmakers.

3.2.17 Salfitichnus *Alonso and Marquillas, 1986 (Fig. 7.7B)*

3.2.17.1 Type ichnospecies

Salfitichnus mentoor Alonso and Marquillas, 1986.

3.2.17.2 Valid ichnospecies

Salfitichnus mentoor Alonso and Marquillas, 1986.

3.2.17.3 Distribution

Maastrichtian—Danian; Yacoraite Formation (Salta) of Argentina.

3.2.17.4 Holotype

Material in situ. Subvertical stratum containing a trace composed of six hollow ichnites, oriented from W to E in a length of 4.60 m, medium to good conservation.

3.2.17.5 Diagnosis

Traces of a bipedal animal, tridactyl, of medium size, with ichnites barely longer than wide, marked digits, and noticeable presence of claws in all three digits (translated from Spanish of Alonso and Marquillas, 1986).

3.2.17.6 General description

Alonso and Marquillas (1986) commented that the ichnites are loosely printed and are scarcely longer than wide (40 cm by 38 cm) with a clear impression of the Digits II, III, and IV and the absence of the Digit I. The digit impressions are relatively short and wide, with sharp claw traces. The pace length is 105 cm and the stride varies between 145 and 190 cm with an average of 164 cm. The pace angle is about 100° on average. The interdigital angles are highly variable. In the track 5, the divarications are 28° and 38° between the Digits II–III and III–IV respectively.

3.2.17.7 Discussion

Salfitichnus is known only by a trackway found in the type locality. Alonso and Marquillas (1986) pointed out that the tracks are very shallow, and Masrour et al. (2017) suggested that the diagnosis of *Salfitichnus* does not provide sufficiently discriminative characters as they are deduced from the drawing with which these authors represent the ichnogenus (wide digits, without pads or lateral constrictions, filiform protrusion on the backward-facing heel, very wide footprint of the main area). In our opinion, it is necessary new field information of this ichnotaxon is necessary to discuss its ichnotaxonomic status, but we will provisionally consider it as *nomen dubium*.

3.2.17.8 Trackmaker

Alonso and Marquillas (1986) proposed that *Salfitichnus mentoor* corresponds to a carnosaur, and they leaned toward megalosauriforms as the trackmakers of *Salfitichnus mentoor*, which would have left smaller and more graceful traces by comparing the characteristics of both groups.

3.2.18 **Picunichnus** *Calvo, 1991 (Figs. 7.7C,D and 7.30B)*

3.2.18.1 Type ichnospecies

Picunichnus benedettoi Calvo, 1991 (Fig. 7.7C)

3.2.18.2 Valid ichnospecies

Picunichnus benedettoi Calvo, 1991.
Picunichnus quijadaensis Melchor et al., 2019 (Figs. 7.7D and 7.30B)

3.2.18.3 Distribution

Albian–Cenomanian, Candeleros Formation (Neuquén), Albian, Lagarcito Formation (San Luis) of Argentina.

3.2.18.4 Holotype

Relief footprint deposited in the Museum of Natural Sciences of the Universidad Nacional del Comahue with the acronym MUCPv 72.

3.2.18.5 Diagnosis

Medium-sized, mesaxonic, tridactyl, or tetradactyl footprint composing bipedal trackways with high-pace angulation. Digit imprints commonly show distinct pads. The footprint length/width ratio ranges between 1.5 and 2.15, the anterior triangle length/width ratio ranges between 0.6 and 0.9, and divarication of Digits II–IV is usually in the range 23°–42°. Occasional hallux impression lacking claw mark that is distinctive of the ichnogenus. Distinguished from *Grallator* by a less pronounced mesaxony, lower footprint elongation, and higher divarication; and from *Kayentapus* by a more pronounced mesaxony, higher footprint elongation, and lower digit divarication (Melchor et al., 2019).

3.2.18.6 General description

Picunichnus is tridactyl to tetradactyl, mesaxonic and longer than wide track. The digit III impression is more forwardly projected than Digits II and IV. The tracks preserve pad and claw traces. The tracks of *P. benedettoi* have the hallux impression directed backward, instead in *P. quijadaensis* is medially oriented. In *P. benedettoi* the Digit IV impression is the longer one, but in *P. quijadaensis* Digit III and IV impressions are subequal and longer than Digit II. Finally, the footprint length/footprint width ratio is higher in *P. benedettoi* than in *P. quijadaensis*.

3.2.18.7 Discussion

Picunichnus benedettoi was defined based on a unique natural cast (Calvo, 1991). Gierliński (1994) proposed that *Picunichnus* is a junior synonym of *Grallator*. Calvo and Rivera (2018) emended the original diagnosis with information of new material found in the same area. Melchor et al. (2019) pointed out that *Picunichnus* was poorly defined by Calvo (1991) with scarce material, but the new material is important to understand better this ichnogenus. They defined a new ichnospecies of *Picunichnus, P. quijadensis* Melchor et al., 2019, with well-preserved material of the Lagarcito Formation (Albian, Argentina). The authors gave a diagnosis to the ichnogenus, a diagnosis to the new ichnospecies and emended the diagnosis of *P. benedettoi*. Moreover, they suggested to change the type ichnospecies from *P. benedettoi* to *P. quijadensis* following the recommendation 69.A of the ICZN (ICZN, 1999) because the latter is an ichnospecies that is adequately described, illustrated and of which material is easily obtained. We maintain the ichnospecies *P. benedettoi* as the type ichnospecies. Melchor et al. (2019) differentiated *Picunichnus* from *Grallator* by a lower mesaxony and footprint elongation, and higher digit divarication. In our opinion, *Picunichnus* is similar in shape with other coeval ichnogenera such as *Deferrariischnium* and *Candeleroichnus* and their ichnotaxonomical relationships should be discussed in the future.

Outside of Argentina, footprints classified as cf. *Picunichnus* have been found in the Upper Cretaceous of Jordan (Klein et al., 2020).

3.2.18.8 Trackmaker

Melchor et al. (2019) noted that the closest trackmakers are the smaller basal coelurosaur *Aniksosaurus darwini* (Cenomanian) and the similar–sized abelisaurid *Majungasaurus crenatissimus* (Maastrichtian). *Picunichnus quijadaensis* is the one indication of the former presence of a group of Early Cretaceous basal coelurosaurs or abelisaurs in the center of Argentina, in a region lacking any other evidence of Cretaceous theropods.

3.2.19 Abelichnus *Calvo, 1991 (Fig. 7.7E)*

3.2.19.1 Type ichnospecies

Abelichnus astigarrae Calvo, 1991.

3.2.19.2 Valid ichnospecies

Abelichnus astigarrae Calvo, 1991.

3.2.19.3 Distribution

Albian–Cenomanian, Candeleros Formation of western Argentina.

3.2.19.4 Holotype

A trackway with four tracks, preserved in situ at the Peninsula Nueva tracksite of Argentina; a replica, MUCPv 74, is housed in the Museum of Natural Sciences of the National University of Comahue.

3.2.19.5 Diagnosis

Large, tridactyl, bipedal dinosaur. Digits long and thick with pads, slightly longer than wide. Digit III more developed than the lateral ones, presence of claws on the digits, deeper in Digit III and displaced inward. Digit IV longer than Digit II with claws displaced outwards. Pace angle less than 150°, palm with subrounded edge and reduced surface (translated from Spanish of Calvo, 1991).

3.2.19.6 General description

Calvo (1991) pointed out that the footprints have medium to good preservation. They are tridactyl with the presence of claws at their ends, the one on Digit III is larger and curved inwards, and the shorter lateral digits have claws displaced outwards. The phalangeal pads are longer than they are wide. The holotype measures 43 cm in length and 35 cm in width. The divergence of the digits is close to 25°. The "heel" is subrounded and has a reduced surface. The pace angle is 150°.

3.2.19.7 Discussion

Abelichnus differs from *Megalosauropus* because of the divarication between the digits, and Digit III more developed than Digits II and IV (Calvo, 1991). Compared with *Abelichnus* and *Bressanichnus,* the Digits II and IV are subequal, the Digit IV is more robust than Digit II, the divarication between Digits II and IV are lower and the heel of *Bressanichnus* is smaller and round (Calvo and Rivera, 2018).

This ichnotaxon was established with one trackway, but Calvo and Mazzetta (2004) and Calvo and Rivera (2018) referred several remains to it. Melchor et al. (2019) related *Abelichnus* with the ichnofamily Eubrontidae.

3.2.19.8 Trackmaker

Calvo and Mazzetta (2004) stated that this ichnotaxon had been left by *Giganotosaurus carolinii,* because it is the largest dinosaur at that location. Calvo and Rivera (2018) inferred that the trackmakers are large theropod dinosaurs because of the presence of claws, which indicates that the tracks were made by carnivores.

3.2.20 Deferrariischnium *Calvo, 1991 (Fig. 7.7F)*

3.2.20.1 Type ichnospecies

Deferrariischnium mapuchensis Calvo, 1991.

3.2.20.2 Valid ichnospecies

Deferrariischnium mapuchensis Calvo, 1991.

3.2.20.3 Distribution

Cenomanian, Candeleros Formation, Neuquen Group of Argentina; Midelt Formation of Morocco.

3.2.20.4 Holotype

Trackway with nine footprints in situ, Picún Leufú, Neuquén Province, Argentina. Three artificial casts of the holotype deposited in the Museum of Natural Sciences of the Universidad Nacional del Comahue, MUCPv 62 (Calvo, 1991). After that, Calvo and Rivera (2018) proposed to change the holotype to the specimen MUCPv 67, which is a natural cast.

3.2.20.5 Diagnosis

Almost symmetrical tridactyl footprints with the impression of a long Digit III, almost half the length of the footprint, displaced internally and with a very marked and sharp claw impression. Digit II is slightly shorter than IV, showing evidence of small claws. Digits II and IV diverging by no more than 25° from Digit III. Basal phalanges are thick, so they can amalgamate, giving a massive appearance. The heel is small and acute and connects laterally with the digits (emended from Calvo and Rivera, 2018; translated from the Spanish of Calvo and Rivera, 2018).

3.2.20.6 General description

Calvo and Rivera (2018) pointed out that MUCPv 97 is 15 cm long and 7.5 cm wide with a Digit III that is the longest (125 mm), and similar Digit II (60 mm) and Digit IV (75 mm). The second digital pad of Digit III is thick and as long as is the first pad of Digit II. The "heel" is acute, displaced outwards, and very small.

3.2.20.7 Discussion

Calvo (1991) suggested that it is similar to *Grallator*. Krapovickas (2010) proposed that *Deferrariischnium mapuchensis* should be synonymous with *Irenesauripus* or *Anchisauripus*. Gierliński et al. (2017) classified *Deferrariischnium* as a member of the ichnofamily Grallatoridae. We consider that *Deferrariischnium*, *Picunichnus* and *Candeleroichnus* are very similar in shape and their relationships must be solved. *Deferrariischnium* isp. has been also identified from the Midelt Formation (mid-Cretaceous) of Morocco (Gierliński et al., 2017).

3.2.20.8 Trackmaker

Gierliński et al. (2017) suggested an abelisauroid as trackmaker of *Deferrariischnium*.

3.2.21 Bressanichnus *Calvo, 1991 (Fig. 7.7G)*

3.2.21.1 Type ichnospecies

Bressanichnus patagonicus Calvo, 1991.

3.2.21.2 Valid ichnospecies

Bressanichnus patagonicus Calvo, 1991.

3.2.21.3 Distribution

Albian–Cenomanian, Candeleros Formation (Neuquén) of Argentina.

3.2.21.4 Holotype

A trackway with 21 footprints in situ (Picún Leufú, Neuquén Province, Argentina), and with 11 molds (MUCPv 60), housed in the Museum of Natural Sciences of the Universidad Nacional del Comahue.

3.2.21.5 Diagnosis

Medium-sized footprints, with the Digit III more developed than the lateral ones and all have claw impressions. Each footprint has a length that ranges between 20 and 25 cm and the ratio between length and width is 1.5. The impression of the Digit III is internally curved and the heel is subrounded and small. The footprints are slightly asymmetrical. The hypex between Digits II and III is posteriorly located with respect to the hypex between Digits III and IV. The pace angle is approximately 175° (translated from Spanish of Calvo, 1991).

3.2.21.6 General description

Calvo (1991) suggested that the holotype is represented by a trackway of 21 very well-preserved footprints. They are tridactyl, mesaxonic, digitigrade, and almost symmetrical, with a subrounded posterior edge and longer than wide. The Digit III impression is the longest one, while the Digit II is longer than IV. The digit impressions present acuminate claw traces. They have a small "heel." The trackway length is 176 cm. The pace angulation is 175°–180°. The trackway width between the inner edges gives negative values. The midline is practically straight.

3.2.21.7 Discussion

To the material described by Calvo (1991), others have been added recently (Calvo and Rivera, 2018; Krapovickas et al., 2020), and all are from the Candeleros Formation. Originally, Krapovickas (2010) proposed that *Bressanichnus* should be synonymous with *Irenesauripus* or *Anchisauripus*. Subsequently, Krapovickas et al. (2020) differentiated *Bressanichnus* and *Irenesauripus* by the size and shape of the Digit III impression. Moreover, they found that the "heel" in *Bressanichnus* is very variable, and therefore has not been given ichnotaxonomic significance.

3.2.21.8 Trackmaker

Bressanichnus shows strong affinities with large theropods. There were many large theropods discovered from the Candeleros Formation, for instance Carcharodontosauridae, *Giganotosaurus* (Calvo and Salgado, 1995), and *Ekrixinatosaurus* (Calvo et al., 2004). Krapovickas et al. (2020) related this ichnotaxon to medium- to large-sized, nonavian theropods.

3.2.22 Theroplantigrada *Casanovas et al., 1993a (Fig. 7.5E)*

3.2.22.1 Type ichnospecies

Theroplantigrada encisensis Casanovas et al., 1993a.

3.2.22.2 Valid ichnospecies

Theroplantigrada encisensis Casanovas et al., 1993a.

3.2.22.3 Distribution

Aptian, Enciso Group, La Rioja, Spain.

3.2.22.4 Holotype

Track 13 of trackway A, a trace fossil, located at El Villar-Poyales tracksite (EVP) in northern Spain.

3.2.22.5 Diagnosis

Elongated, fan-shaped, plantigrade ichnite, consisting of a profound parasagittal part, from which three grooves emerge. The parasagittal part is formed by Digit III and the metatarsal; the grooves are formed by Digits II and IV in front and Digit I behind. The direction of Digit I is subperpendicular to the axis of the track, with a tendency toward backward. In the center of the digits and on the metatarsus, there is an irregular band, possibly due to the collapse of the mud produced when the foot was lifted; this band has the same characteristics as the grooves mentioned for the digits. The digits have no toe pads. Digit length is 26 cm, 30 and 26 cm for Digits II–IV, respectively if the lines of the supposed collapse (hallux, 12 cm) are measured, and 17 cm, 18 cm, and 13 cm, if the hypex is considered. The angles between II–III and III–IV are both 51°, and the angle between the axis of the foot and the hallux is 105°. The foot is elongated, length 53 cm; the width ranges from 40 cm at the front of the foot to 17 cm at the hallux. The maximum depth of the footprint is reached at the point where the front digits (II–IV) meet (7 cm). The height of the mud that rises around the footprint is on the order of a millimeter. Between the four digits, there is an interdigital membrane print that is manifested by a depressed area that not only extends between Digits II, III, and IV, but also from Digits I to II (translated from the Spanish of Casanova et al., 1993a).

3.2.22.6 General description

Given that it only occurs at just one tracksite, general description is all based on its holotype. Casanovas et al. (1993a) considered the presence of an interdigital web as the main diagnostic feature of *Theroplantigrada*, which is a rare case among dinosaurs (Conti et al., 2005). The main diagnostic features would also be the presence of hallux and metatarsal impressions corresponding to the skeletal structure on the premise that supported webbing in the track is extramorphological (Conti et al., 2005).

3.2.22.7 Discussion

Theroplantigrada was named by Casanovas et al. (1993a). It is considered a form taxon. Kuban (1989) reviewed the literature on elongate tracks and also collected unpublished information. From the list of ichnites, Casanova et al. (1993a) only drew comparisons with the

Barranco de la Sierra del Palo site of Brancas et al. (1979) and that of Farlow (1989). The morphology of *Theroplantigrada* differs from that of the others: it is a bipedal dinosaur; there is no sign of pads; the distal end is webbed; the hallux is imprinted. *Theroplantigrada* also has notable differences from *Macropodosaurus* (Zakharov, 1964), a webbed theropod. The digits of *Theroplantigrada* are all directed forward; digits are parallel; and the heel is wide and rounded (Casanovas et al., 1993a).

Discussions about *Theroplantigrada* have focused more on its unique interdigital webs since then. Conti et al. (2005) stated that it a character at the ichnospecies level. If the above rationale is accepted, the Mattinata material described by Conti et al. (2005), trackways C and G described by Aguirrezabala and Viera (1980), trackway D of Casanovas Cladellas et al. (1993), and the trackway ascribed to *Theroplantigrada encisensis* all could be included in the same ichnogenus. Pérez-Lorente (2015) also suggested that the interdigital webs are more likely to be morphological, instead of deformations induced by the digits, even bearing in mind the conclusions on the deformation of the interdigital spaces of the footprints (Manning, 2004). However, it is necessary to have more information about this ichnotaxon to discuss its ichnotaxonomic validity.

3.2.22.8 Trackmaker

Theroplantigrada was assigned to Tyrannosauroidea by Casanovas et al. (1993a). Although the *Theroplantigrada* tracks in Spain differ slightly in having a higher gait angulation (narrower interpedal distance), the Mattinata material described by Conti et al. (2005) from the Upper Jurassic Sannicandro Formation of Italy is classified as *Theroplantigrada*. As a point of reference, all tracks in the Mattinata material are attributed to medium-sized theropod trackmakers (Conti et al., 2005), with somewhat developed webs on their pes (Conti et al., 2005; Pérez-Lorente, 2015).

3.2.23 Filichnites *Moratalla, 1993 (Fig. 7.5F)*

3.2.23.1 Type ichnospecies

Filichnites oncalensis Moratalla, 1993.

3.2.23.2 Valid ichnospecies

Filichnites oncalensis Moratalla, 1993.

3.2.23.3 Distribution

Berriasian, Huérteles Formation (Soria) of Spain.

3.2.23.4 Holotype

LTR-F1 footprint, concave epirelief of the site from Los Tornos, Santa Cruz de Yanguas, Soria, Spain.

3.2.23.5 Diagnosis

Tridactyl ichnite, measuring between 12 and 30 cm in length. Very thin and elongated digits, with Digits II and III of similar development. Digits are occasionally curved. Hypex symmetrical and very open. Very small plantar surface, with a very slight notch in the medial area. BL3/z index very high (between 2.7 and 3.4). Relatively low length/width, between 1 and 1.2. Bipedal and narrow trackway (translated from the Spanish of Moratalla, 1993).

3.2.23.6 General description

The tracks are tridactyl, mesaxonic, longer than wide. The most notable character is the relative digital slenderness together with a reduced plantar surface.

3.2.23.7 Discussion

Filichnites was defined by Moratalla (1993) in his PhD thesis. Subsequently, it has not been formally proposed. For this reason, it is considered *nomen nudum*. Furthermore, the morphology of these footprints (slender digits and reduced plantar area) is closely related to extramorphological characteristics (mud collapse).

3.2.23.8 Trackmaker

Filichnites was related to a theropod dinosaur that was digitigrade and probably had little support from the distal metatarsal area.

3.2.24 Tyrannosauripus *Lockley and Hunt, 1994a (Figs. 7.4J,K and 7.31A).*

3.2.24.1 Type ichnospecies

Tyrannosauripus pillmorei Lockley and Hunt, 1994a (Fig. 7.4J).

3.2.24.2 Valid ichnospecies

Tyrannosauripus pillmorei Lockley and Hunt, 1994a.
Tyrannosauripus bachmani Weems, 2021 (Figs. 7.4K and 7.31A)

3.2.24.3 Distribution

Maastrichtian interval of the Raton Formation (New Mexico, USA) and Albian, Patuxent Formation of USA. Barremian–Aptian, Hekou Formation (Jiangxi) of China.

3.2.24.4 Holotype

A cast, CU-MWC 225.1, is housed in a joint collection of the University of Colorado Denver and the Museum of Western Colorado, USA. The original holotype is in the field.

3.2.24.5 Diagnosis

Large, slightly asymmetrical, mesaxonic, tetradactyl theropod track, about two thirds as wide as long, with slender hallux impression and variable claw impressions (Lockley and Hunt, 1994a).

3.2.24.6 General description

Tyrannosauripus pillmorei is tetradactyl, slightly asymmetrical, digitigrade, mesaxonic, and longer (86 cm) than wide (64 cm) natural cast. The Digit I impression is short (17 cm), narrow, and recurved. The Digit III impression is the longest one (57 cm), whereas the Digit II impression measures 36 cm and Digit IV is 27 cm. The digit impressions terminate in a claw trace that is sharp or rounded. Pad impressions are preserved in Digit IV. The posterior portion of the "heel" impression is fairly straight with rounded edges (Lockley and Hunt, 1994a). McCrea et al. (2014b) pointed out that *T. pillmorei* also has indistinct digital pad impressions and a hallux that is directed medially.

3.2.24.7 Discussion

Peterson (1924) described some footprints found in a Campanian coal mine (USA) as footprints of *Tyrannosaurus*. Lessertisseur (1955) proposed that if these footprints were produced by a *Tyrannosaurus* they should be called "Tyrannosauripus".

Haubold (1971) defined the ichnogenus *Tyrannosauropus* for the tracks studied by Peterson (1924). Lockley (1991) and Lockley and Hunt (1994) suggested that this material presents ichnotaxonomic problems (inadequately described, undiagnostic figure, damaged holotype, etc.) and could have been produced by a hadrosaur, so we prefer to not include it in this chapter. Moreover, Lockley and Hunt (1994) and Lockley et al. (2000) affirmed that *Tyrannosauripus* is a nomen nudum because Lessertisseur (1955), when he proposed the ichnogenus, did not assign a type ichnospecies, so it was available to be adopted. The footprint with which the ichnogenus *Tyrannosauripus* has been described is not the same one used by Lessertisseur (1955) or Haubold (1971).

Weems (2021) established the ichnospecies *Tyrannosauripus bachmani* based on two tracks found in the Albian Patuxent Formation, USA. He proposed an ichnospecific diagnosis, nevertheless, he did not add an ichnogeneric diagnosis. Weems (2021) added those tracks to *Tyrannosauripus* due to the fact that they present a robust Digit II impression and small Digit IV, but they are different from *T. pillmorei* because they are smaller, lack an elongated "heel" impression, and the Digit III impression is much longer.

The holotype of *T. bachmani* is poorly preserved and undiagnostic, so we propose that this ichnospecies is a nomen dubium. The holotype of *Tyrannosauripus pillmorei* is not well-preserved following the proposal of Marchetti et al. (2019), so new material is needed in the future to evaluate the ichnotaxonomic significance of this ichnogenus.

McCrea et al. (2014b) proposed that *Tyrannosauripus* belongs to the Tyrannosauripodidae McCrea et al., 2014b ichnofamily. Xing et al. (2019e) referred *Tyrannosauripus* isp. to large theropod tracks from the Maastrichtian Hekou Formation, China.

3.2.24.8 Trackmaker

Taking into account the age (Maastrichtian), the shape (theropodian), and the size (giant), Lockley and Hunt (1994) and Xing et al. (2019e) proposed a tyrannosaurid as the possible trackmaker.

3.2.25 *Minisauripus Zhen et al., 1996 (Fig. 7.6H, I and 7.30F)*

3.2.25.1 Type ichnospecies

Minisauripus chuanzhuensis Zhen et al., 1996 (Figs. 7.6H,I and 7.30F)

3.2.25.2 Valid ichnospecies

Minisauripus chuanzhuensis Zhen et al., 1996 (Figs. 7.6H and 7.30F)
Minisauripus zhenshuonani Lockley et al., 2008a (Fig. 7.6I)

3.2.25.3 Distribution

Barremian/Aptian–Santonian, Jiaguan Formation (Sichuan), late Berriasian–Albian, Feitianshan Formation (Sichuan), Albian–Cenomanian, Tianjialou Formation (Shandong) of China; Albian, Jinju Formation, Albian–Cenomanian, Haman Formation (Changseon, Sinsu, Gae Je, Sangcheon, Jinju) of Korea.

3.2.25.4 Holotype

CFEC-A-1, housed in the Chongqing Museum of Natural History, Chongqing, China.

3.2.25.5 Diagnosis

Small tridactyl track with subparallel, elongate, well-padded digits with blunt distal terminations connected to narrow distal claw traces. Digit III only slightly longer than IV, which is slightly longer than Digit II. Phalangeal formula of 2–3–?4 for Digits II–IV, respectively discernible in well-preserved examples. Pace long, 3–10 times track length. Trackway narrow (emended by Xing et al., 2016c).

3.2.25.6 General description

The type of *Minisauripus chuanzhuensis* (Zhen et al., 1996) originates from the Emei track site near Chuanzhu town, Emei County, Sichuan. Associated tracks include *Grallator emeiensis*, *Velociraptorichnus sichuanensis*, and "*Iguanodonopus xinduensis*" *M. chuanzhuensis* was based on 21 specimens, the length was given as between 2.1 and 3.0 cm and the width as between 1.6 and 1.8 cm. By 2019, there were currently at least ~92 *Minisauripus* tracks in total, representing at least 55 trackways from more than eight localities in Korea and China. All unambiguously identified *Minisauripus* tracks fall within the size range of ~1.0–6.1 cm (Lockley et al., 2008a; Kim et al., 2019a).

Lockley et al. (2008a) described a second ichnospecies of *Minisauripus* from Shandong, *M. zhenshuonani*, and gave a diagnosis: small, elongate, tridactyl track with parallel digits with conspicuous claw traces. Track narrower than in *M. chuanzhuensis* with digits less divergent, and Digit II relatively shorter. Pace and stride about 10 times footprint length and typically longer than in *M. chuanzhuensis*. Trackway very narrow (Lockley et al., 2008a).

Kim et al. (2019a) described *Minisauripus* skin impressions. The only known example for such a small dinosaur, these exquisitely preserved skin traces with densely packed, reticulated arrays of small (<0.5 mm) polygonal scales covering the entire plantar surface of the foot.

3.2.25.7 Discussion

Many details of *M. zhenshuonani* morphology reveal significant differences between the type *M. chuanzhuensis* materials. These differences include evidence of pad morphology, claw traces, and track length/width ratios. For example, the larger tracks tend to be more elongate than the small tracks, with more prominent evidence of long slender claws (Lockley et al., 2008a).

Minisauripus is abundant in Korea, originating from over five localities in the Jinju and Haman formations. The Korean specimens are well preserved and have been tentatively assigned to *M.* cf. *zhenshuonani* or *M.* cf. *chuanzhuensis* (Lockley et al., 2008a; Kim et al., 2019a).

Lockley et al. (2008a) suggested that although the distribution of *Minisauripus* now suggests that it is relatively widespread in East Asia, it may still represent a distinctive small vertebrate ichnofauna containing relatively rare elements of small size. The relatively consistent association of *Minisauripus* and dromaeopodid tracks may suggest a close ecological association between the makers of these two track types (Xing et al., 2016c).

3.2.25.8 Trackmaker

Kim et al. (2016) documented the traces left by *Minisauripus* scraping the substrate, and suggested that *Minisauripus* trackmakers were considered adults rather than juveniles, as the scrapes indicated that they were engaged in courtship behavior. Although Xing et al. (2016c) emphasized that the proposed affinity of *Minisauripus* to small adult theropods cannot currently be proven with certainty, and that a juvenile origin cannot be ruled out. However, due to the emergence of courtship records, this conclusion should be more inclined toward the *Minisauripus* trackmaker being a small adult theropod. The minimum body length of the trackmaker of *Minisauripus* is probably only about 12 cm (Xing et al., 2016c).

Xing et al. (2016c) pointed out that compsognathids and oviraptorosaurs are known from abundant specimens of the same age and location, which would make these two groups the most likely candidates for trackmakers. They pointed out that the high running speed of the *Minisauripus* trackmaker from the Yangmozu sample, estimated at up to 22.5 km/hour (=6.2 m/second), is consistent with the rapid locomotion of *Compsognathus*. Therefore, they tentatively suggested that *Minisauripus* tracks could indicate compsognathids with good cursorial ability.

3.2.26 *Velociraptorichnus* Zhen et al., 1996 (Figs. 7.8A,B and 7.32A)

3.2.26.1 Type ichnospecies

Velociraptorichnus sichuanensis Zhen et al., 1996 (Figs. 7.8A and 7.32A)

3.2.26.2 Valid ichnospecies

Velociraptorichnus sichuanensis Zhen et al., 1996
Velociraptorichnus zhangi Xing et al., 2015e (Fig. 7.8B)

3.2.26.3 Distribution

Aptian–Santonian, Xiaoba Formation (Sichuan), Kimmeridgian–Valanginian, Tuchengzi Formation (Hebei, Beijing), Albian–Cenomanian, Tianjialou Formation (Shandong), Cenomanian–Coniacian, Shaxian Formation (Fujian), Barremian/Aptian–Santonian, Jiaguan Formation (Sichuan), Turonian–Coniacian, Jinhua Formation (Zhejiang) of China; Late Campanian of Poland, etc.

3.2.26.4 Holotype

CFEC-B-1, positive, housed in the Chongqing Natural History Museum, China.

3.2.26.5 Diagnois

Bipedal, digitigrade form, didactyl structure in walking, with claws at the end of the digits, the ratio of pace to foot length is 4.4:1, without manus or caudal impressions (Zhen et al., 1996).

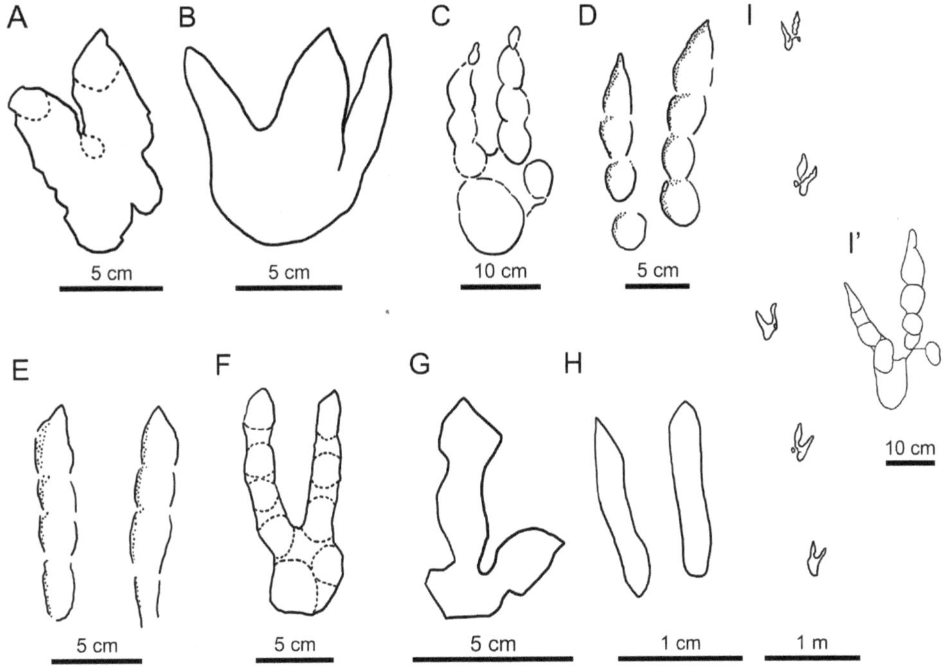

FIGURE 7.8 Deinonychosaur ichnotaxa from Cretaceous. (A) Holotype of *Velociraptorichnus sichuanensis*; (B) holotype of *Velociraptorichnus zhangi* from Sichuan; (C) holotype of *Dromaeopodus shandongensis* from Shandong; (D) holotype of *Dromaeosauripus hamanensis* from Korea; (E) holotype of *Dromaeosauripus jinjuensis* from Korea; (F) holotype of *Dromaeosauripus yongjingensis* from Gansu; (G) holotype of *Menglongipus sinensis* from Hebei; (H) holotype of *Dromaeosauriformipes rarus* from Korea; (I) type series (I) and enlarged track (I′) of *Fujianipus yingliangi* from Fujian. *(A) Redrawn from Xing et al. (2009d). (B) Redrawn from Xing et al. (2015e). (C) Redrawn from Li et al. (2008). (D) Redrawn from Kim et al. (2008). (E) Redrawn from Kim et al. (2012b). (F) Redrawn from Xing et al. (2013a). (G) Redrawn from Xing et al. (2009d). (H) Redrawn from Kim et al. (2018b). (I) Redrawn from Xing et al. (2024b).*

3.2.26.6 General description

Velociraptorichnus sichuanensis has the divarication angles between Digits III and IV in the range of 21°–28°. They are indeed narrow, didactyl tracks with subequal Digit III and IV lengths. The ratio of the length of Digit III to that of Digit IV is 1.3 (6.2 cm/4.7 cm). Digit III is slightly wider than Digit IV, and both bear discernible claw marks. Digital pads are indistinct to nonexistent. Digit II is represented only by a short, round impression that is at least partially embedded in the impression of Digit III at its proximomedial edge. The large metatarsophalangeal region is hemispherical and not separated from the digits by a distinct border. The average ratio of the length to the width of the entire sample is 1.77 (Zhen et al., 1996; Xing et al., 2009d). The length of the type specimen of *V. sichuanensis* is 11 cm, and that of *Velociraptorichnus* from Shandong is ~10 cm (Li et al., 2008). High-pace angulation (more than 170°) with track axis subparallel, or slightly inward, to trackway axis (Tsukiji et al., 2021).

Velociraptorichnus zhangi is small (~10 cm in length and width) tridactyl theropod track with extremely low mesaxony. It is tridactyl with Digit III slightly longer than other digits or subequal with Digit II. The digits have sharp distal ends, indicating the presence of acuminate claws, but the digit pad traces are faint or indistinct and still show residues of infilling sediment. Digit II trace is very narrow compared with other digits, about half the width of Digit III; large semicircular metatarsophalangeal region. The divarication angle of Digits II and IV is 52°, with the divarication angle between Digits II and III being much smaller (17°) than that between Digits III and IV, also smaller (35°). Some of the heel trace is prominent and probably formed by the dynamics of the foot registering on the sediment (Xing et al., 2015e). *V. sichuanensis* exhibits subequal lengths of Digits III and IV with indistinct or absent digital pad traces, and the length-to-width ratio for *V. sichuanensis* is 1.8, whereas for *V. zhangi* it ranges from 1.5 to 1.6 (Xing et al., 2015e). However, the length/width ratio of *Velociraptorichnus* can vary significantly depending on preservation quality (Xing et al., 2009d).

3.2.26.7 Discussion

Velociraptorichnus was the first documented instance of deinonychosaurian tracks (Zhen et al., 1996; Xing et al., 2015e). In Zhen et al. (1996), *Velociraptorichnus* was also included in the ichnofamily Dromaeopodidae. This ichnogenus is distributed in the Early Cretaceous of China, with small amounts occurring in the Late Cretaceous (Zhen et al., 1996; Xing et al., 2015e; Niu and Xing, 2023). Two didactyl tracks in a partial trackway attributed to *Velociraptorichnus* also have been reported from Poland (Gierliński et al., 2008), but later claimed to be more similar to *Dromaeopodus* (Gierliński, 2009).

It is widely recognized that the dromaeosaurs possessed a prominent blade-like claw on its second digit of the pes, rendering this digit nonfunctional for walking and leaving only impressions from Digits III and IV on the ground (Zhen et al., 1996). All previously identified deinonychosaurian tracks are didactyl; however, two specimens of *V. zhangi* are considered tridactyl versions of *Velociraptorichnus* (Xing et al., 2015e). Xing et al. (2015e) also proposed that the narrow impression made by Digit II likely resulted from the large raptorial claw since this highly mobile digit could occasionally make contact with the ground to leave a distinct trace within a tridactyl trackway context. The difference in preservation of the Digit II claw impression may also be attributed to different substrate conditions (Xing et al., 2015e).

3.2.26.8 Trackmaker

Zhen et al. (1996) initially attributed the of *V. sichuanensis* to a dromaeosaur-type trackmaker based on the morphology of the didactyl foot (Xing et al., 2015e), so it could be at least a relative of *Deinonychus*, or may potentially originate from a *Velociraptor*. Given the potential morphological diversity among deinonychosaurians of similar size to the *Velociraptorichnus* trackmakers reported from East Asia, it is plausible that different species exhibited significant variation in the degree of retraction of pedal Digit II, potentially contributing to a wide range of track morphotypes (Xing et al., 2015e). To some of the specific localities without body fossils assignable to Deinonychosauria, it is difficult to judge whether it could be linked to a dromaeosaurid or troodontid trackmaker (Tsukiji et al., 2021).

Besides, Tsukiji et al. (2021) provided the estimated speed of the trackmaker as 6.1 or 4.6 km/hour based on both dromaeosaurid and troodontid h/FL values, which shows a walking gait, with the best *Velociraptoichnus* specimen for estimating the speed and gait found in Zhejiang.

3.2.27 Saurexallopus *Harris et al., 1996 (Figs. 7.4L–N and 7.30G)*

3.2.27.1 Type ichnospecies

Saurexallopus lovei Harris et al., 1996 (Fig. 7.4L)

3.2.27.2 Valid ichnospecies

Saurexallopus lovei Harris et al., 1996.
Saurexallopus zerbsti Lockley et al., 2003 (Figs. 7.4M and 7.30G)
Saurexallopus cordata McCrea et al., 2014 (Fig. 7.4N)

3.2.27.3 Distribution

Campanian, Hunter Canyon Formation (Colorado), Cantwell Formation (Alaska) and Blackhawk Formation (Utah); Maastrichtian, Laramie Formation (Colorado), Harebell Formation and Lance Formation (Wyoming) of USA; Aptian, Gething Formation and Campanian, Wapiti Formation (British Columbia) of Canada; Campanian of Poland.

3.2.27.4 Holotype

DMNH 5989, housed in the Denver Museum of Natural History, Denver, Colorado.

3.2.27.5 Diagnosis

Tetradactyl track with slender toe impressions, and prominent hallux trace extended medially to postero–medially. The phalangeal formula is 2-2-3-?4 for Digits I–IV, respectively. Metapodial impressions subtriangular and slightly asymmetrical with anterior apex converging with proximal portion of digit three. Digit IV connected to metapodial proximally, whereas Digit II connects distally. Trackway narrow with relatively short step (emended by Lockley et al., 2003).

3.2.27.6 General description

It should be noted that all the previously named ichnospecies of *Saurexallopus* are about the same size (about 30 cm excluding hallux), while most of the other *Saurexallopus* isp. are smaller (≤22 cm).

S. lovei shares the similar diagnosis as the ichnogenus, however, three of the tracks suggest an additional uncertain webbing between the digits: areas between Digits II and III and between III and IV have greater relief than the surrounding substrate, but less relief than the digital impressions, which might indicate webbing between the toes extended horizontally between the digits. The webbing impressions begin at the hypex and terminate roughly one-third the way up the toe (Harris et al., 1996).

S. zerbsti is characterized by a tetradactyl track with very slender toe impressions, and prominent hallux trace extended medially, and its different shapes of the metatarsal traces and the configuration of the hypices are also notable (Lockley et al., 2003, 2018a). Metapodial impressions subtriangular and slightly asymmetrical with anterior apex converging with distal portion of digit three, and posteromedial corner oriented toward hallux impression: posterolateral corner of metapodial impression connects proximally with Digit IV, whereas Digit II connects distally, resulting in a much deeper posterior recess (hypex) for the interdigital angle between III and IV than between II and III. Trackway narrow with relatively short step (78 cm). Foot length/pace length ratio about 1:2.5 (Lockley et al., 2003).

S. cordata is characterized by possessing distinctly heart-shaped metatarsal pads, forming a bilobed "heel" impression in the holotype. The holotype print also shows evidence of distinct phalangeal pad impressions for Digits II–III (? -2-3-3-x) and long, slender, acuminate claw impressions; and a distinct difference in digit divarication between Digits II–III (23°) and III–IV (34°) (McCrea et al., 2014a).

3.2.27.7 Discussion

Saurexallopus was first described by Harris et al. (1996) as *Exallopus*. But, to avoid homonymy, Harris (1997) recombined *E. lovel* as *S. lovei.* The only non-American site was described by Gierliński (2009), and the only site from the Early Cretaceous was described by Lockley et al. (2022a). However, all are reported from western North America from somewhat similar palaeoenvironments associated with the well-vegetated coastal plains bordering the western shores of the Cretaceous Interior Seaway (Lockley et al., 2022a).

Saurexallopus is mainly characterized by its tetradactyl tracks, but exhibits considerable variation in other traits, such as size (with foot length about 12–33 cm, small to medium), divarication angles (57° in *S. cordata*, about 83°–115° in other materials), Digit I orientation (mostly middle to postero-medial) and morphology (curved/uncurved) (Lockley et al., 2022a). According to Lockley et al. (2018), in *S. lovei, S. cordata,* and *Saurexallopus* isp, the hallux is directed medially almost exactly along the transverse axis that bisects the Digit III axis perpendicularly at the metapodial phalangeal pad, reaching as far medially as the distal trace of Digit II. Only in *S. zerbsti* is the hallux trace rotated postero-medially at an angle of ~30° to the transverse line.

3.2.27.8 Trackmaker

As an midsized ichnogenus, the trackmaker of *Saurexallopus* is still under discussion. Harris et al. (1996) did not directly identify a trackmaker, but they suggested that it may have been made by a medium-sized theropod having a reduced and elevated hallux from its size and the functionality of Digit I, and that it's more parsimonious that it was made by a theropod rather than a bird. Gierliński (2009) added that the only morphologically similar trackmakers that fit in the size of *Saurexallopus* (>20 cm) are oviraptorosaurs, while tetradactyl avialans are much smaller in size, and the presence of a hallux imprint excludes ornithomimosaurs. However, the posterior position of the hypex of Digits III–IV is more similar to that of *Ornithomimipus* (Lockley et al., 2011, 2018a), and with the addition of smaller tracks in the Utah and Canada, the wide range of track sizes attributed to *Saurexallopus* (<20 cm) is not inconsistent with the size range of oviraptosaurids in the body fossil record (Lockley et al., 2022a).

Lockley et al. (2021) included *Saurexallopus* in the discussion of large bird and or birdlike tracks from the Cretaceous, along with *Archaeornithipus* and *Wupus,* which are considered flightless bird tracks, and *Magnoavipes* with probably of avian origin but lacks skeletal equivalents. In comparison with *Magnoavipes*, *Saurexallopus* with hallux prints is larger in size, which gives them more uncertainty on trackmaker (Lockley et al., 2021). Hong et al. (2024) further used the data from the ichnogenera of Mesozoic bird and other theropod tracks. With the morphological comparisons, they concluded that *Saurexallopus* share high morphological similarities with Mesozoic bird ichnospecies excluding their size, which suggests their large avian origin along with the concluded that *Saurexallopus* share high morphological similarities with Mesozoic bird ichnospecies. This researchers increased the likelihood of large bird trackmakers of *Saurexallopus* (especially the smaller *Saurexallopus* isp.). However, none of the aforementioned studies were able to assign the various sizes of *Saurexallopus* to the same type of avian trackmaker, and size remains an important criterion for determining avian trackmakers (Lockley et al., 2021; Hong et al., 2024). This may mean requiring follow-up studies to disentangle this ichnogenus.

3.2.28 Archaeornithipus *Fuentes Vidarte, 1996 (Fig. 7.5G)*

3.2.28.1 Type ichnospecies

Archaeornithipus meijidei Fuentes Vidarte, 1996.

3.2.28.2 Valid ichnospecies

Archaeornithipus meijidei Fuentes Vidarte, 1996.

3.2.28.3 Distribution

Berriasian, Huérteles Formation (Villar del Río) of Spain.

3.2.28.4 Holotype

No. 97 on slab SRT−3c in sandstone in situ.

3.2.28.5 Diagnosis

Anisodactyl tracks, with long and narrow toes, provided with claws and deeply keeled; Digit I is distinctively marked, forming an angle of 120°−180° with the axis of the Digit III; the Digit II and the Digit IV have a similar development, almost symmetrical with respect to the Digit III, which is the largest, with which they form respectively maximum angles of 60° and 65°, and between all a maximum angle of 150°; central plantar impression (translated from the Spanish of Fuentes Vidarte, 1996).

3.2.28.6 General description

A. meijidei was discovered from the Huérteles Formation (Berriasian) at Villar del Río, Spain. Only found at this site. Small, pes tracks, with length of 12.1 cm and width of 12 cm. The values of divarication angles are 118° (Digits II−IV) and 165° (Digits I−III). The claws of the digits are well-preserved, Digits II and IV have the same length, and Digit III is the longest. Digit I is preserved but very small (Fuentes Vidarte, 1996).

3.2.28.7 Discussion

Archaeornithipus was only found in Spain and had only one valid ichnospecies. *Archaeornithipus* is similar to *Wupus*, *Magnoavipes* and *Saurexallopus*. The size of *Wupus* and *Archaeornithipus* is the same, but *Wupus* lacks the Digit I. The average divarication between Digits II and IV of *Wupus* (97.5°) is less than *Archaeornithipus* (118°). *Magnoavipes* and *Saurexallopus* are almostly 2 times larger than *Archaeornithipus*, the average divarication between Digits II and IV of *Saurexallopus* (150°) is the largest, and that of *Magnoavipes* (93°) is the smallest (Lockley et al., 2021). Castanera et al. (2016) commented that some of the footprints at the tracksite show evidence of substrate deformation such as mud collapse. Therefore, it is necessary to have more data on these tracks to ensure that their avian morphological characteristics are not extramorphological features.

3.2.28.8 Trackmaker

Fuentes Vidarte (1996) suggested that *A. meijidei* is an avian ichnotaxon. But, the author doubted this conclusion because of the size of the tracks. Lockley and Harris (2010) suggested that *A. meijidei* is associated with gracile, bird-like, but nonavian theropod dinosaurs such as the appropriately named ornithomimosaurs.

3.2.29 Fuscinapedis *Lee, 1997 (Figs. 7.40 and 7.32B)*

3.2.29.1 Type ichnospecies

Fuscinapedis woodbinensis Lee, 1997.

3.2.29.2 Valid ichnospecies

Fuscinapedis woodbinensis Lee, 1997.

3.2.29.3 Distribution

Cenomanian, Woodbine Formation (Texas) of USA.

3.2.29.4 Holotype

SMU74651, housed in the Shuler Museum of Paleontology, Southern Methodist University, Texas, USA.

3.2.29.5 Diagnosis

Bipedal tracks with three long, slender functional pedal digit impressions, a small circular metatarsal impression, and no trace of the hallux. Track length greater than width. Digit impressions of uniform width; Digit II width greater than Digits III and IV. Phalangeal pads not well defined, but at least two pad impressions on Digit II. Claw impression on each digit. Average divarication of Digits II and IV is 70° (Lee, 1997).

3.2.29.6 General description

The footprints are tridactyl, mesaxonic, symmetrical, digitigrade, and longer than wide (the holotype is 38 cm in length and 35 cm in width). The digit impressions are well separated from each other, present in pointed small claw impressions, and the phalangeal pads are not well defined. Total divarication is 69°. The metatarsophalangeal pad is circular and is located in alignment with the track axis.

3.2.29.7 Discussion

Lee (1997) pointed out that mid-Cretaceous theropod tracks in North America are mostly from the Dakota Group (Aptian–Cenomanian), Colorado and New Mexico, and the Gething Formation (Aptian) and Dunvegan Formation (Cenomanian), Canada (Lee, 1997). *Fuscinapedis* differs from *Irenesauripus* and *Columbosauripus* based on the large metatarsal pads (Lee, 1997) and from *Irenichnites* because it has the Digit II impression united with Digits III and IV. *Fuscinapedis* presents an overall shape similar to Cretaceous Ornithomimipodidae ichnotaxa (*Ornithomimipus, Columbosauripus, Irenichnites,* and *Magnoavipes* sensu Lockley et al., 2011) and large tridactyl ratites.

3.2.29.8 Trackmaker

Farlow (2001) suggested that *Fuscinapedis* is comparable in size to large moa footprints and Lee (1997) attributed *Fuscinapedis* to a "theropod," but it has a marked metatarsal impression.

3.2.30 Magnoavipes *Lee, 1997 (Fig. 7.4P–S)*

3.2.30.1 Type ichnospecies

Magnoavipes lowei Lee, 1997 (Fig. 7.4P)

3.2.30.2 Valid ichnospecies

Magnoavipes lowei Lee, 1997.
Magnoavipes caneeri Lockley et al., 2001a (Fig. 7.4Q)
Magnoavipes denaliensis Fiorillo et al., 2011 (Fig. 7.4R)
Magnoavipes asiaticus Matsukawa et al., 2014 (Fig. 7.4S)

3.2.30.3 Distribution

Cretaceous of Canada, China, Mexico, and USA.

3.2.30.4 Holotype

DMNH 918, housed in the Dallas Museum of Natural History, from Woodbine Formation, Cenomanian, Denton County, Texas.

3.2.30.5 Diagnosis

Narrow-toed tridactyl footprints of a biped. Footprint length-to-width ratio 1.0–0.8. Digit divarication angles wide (average 85°) but highly variable (range 65°–105°). Track slightly asymmetrical (emended by Lockley et al., 2001a).

3.2.30.6 General description

All the *Magnoavipes* tracks are small to medium (about 20 cm in length) tridactyl and mesaxonic tracks, longer than wide or as long as wide, and have high divarication angles.

M. lowei is based on one trackway of poorly preserved tracks with very slender digit impressions (Lee, 1997). *M. caneeri* is based on a series of 15 trackways (Lockley et al., 2001a) from the Platte Formation (Dinosaur Ridge, Colorado) and has shallow, but well-defined digital pad impressions.

M. denaliensis, based on two specimens from Tattler Creek (Denali National Park, AlaskaDMNH) has slender digit impressions with sharp claw impressions on all digits (Fiorillo et al., 2011). *M. asiaticus,* found in the Lower Cretaceous Fenjiashan Formation of China, is an asymmetrical tridactyl track with slightly tapering of digit traces, and divarication angles between Digits II and III larger than angles between Digits III and IV (Matsukawa et al., 2014).

3.2.30.7 Discussion

Magnoavipes was named by Lee (1997), and type ichnospecies *M. lowei* was described as a bird trackway. Lockley et al. (2001a) suggested that *Magnoavipes* may not be a valid avian ichnotaxon because of the divarication angle <110° and related it to an ornithomimosaur-like theropod. They emended the ichnogeneric diagnosis of *Magnoavipes* and established a new ichnospecies *M. caneeri,* which can be distinguished from *M. lowei* by shallow, but well-defined digital pad impressions. On the other hand, Lockley et al. (2001a) suggested that "extreme slenderness of *M. lowei* is, in our opinion, an extramorphological artifact of poor preservation, due to collapse and inflow of sediment into the digit impressions." Maybe Lockley et al. (2001a) should propose that *Magnoavipes* was a nomen dubium and establish a new ichnogenus for the new ichnospecies. Fiorillo et al. (2011) named *M. denaliensis* based on two isolated tracks and used the divarication angle range and length width measurements as important ichnotaxobases. *M. denaliensis* has more pronounced and more robust digit impressions than *M. lowei* (Lee, 1997), while *M. caneeri* has far more robust toes and much narrower divarication angles (Lockley et al., 2001a) than *M. denaliensis.* Matsukawa et al. (2014) defined *M. asiaticus* and indicated it is different from *M. lowei* by possessing a very small "heel" impression posterior to the trace of Digit II and more tapering Digit III. *M. asiaticus* is also distinguished from *M.denaliensis,* because of asymmetrical outline and wider divarication angle between Digits II and IV, and digit impressions of the former ichnospecies are wider than in the latter ichnospecies. The differences between *M. asiaticus* with *M. caneeri* are: (1) divarication angle between Digits II and III is larger than between III and IV in the former ichnospecies, but vice versa in the latter ichnospecies, and (2) digit impressions of the former ichnospecies are wider than in the latter ichnospecies.

Lockley et al. (2011a) assigned ichnogenus *Magnoavipes* to the ichnofamily Ornithomimipodidae. McCrea et al. (2014a) reported a probable Cenomanian *Magnoavipes* track from British Columbia, Canada.

3.2.30.8 Trackmaker

Lee (1997) suggested that the trackmaker of *M. lowei* was a long-legged wading bird adapted to shallow waters, like modern cranes (Gruiformes). Lockley et al. (2001a) proposed the trackmaker of *M. lowei* was probably a nonavian dinosaur, most likely an ornithomimid theropod. Fiorillo et al. (2011) returned to the idea of Lee (1997) and proposed an avian trackmaker for *M. denaliensis*. Matsukawa et al. (2014) referred *M. asiaticus* to a probable ornithomimid theropod track.

3.2.31 Toyamasauripus *Matsukawa et al., 1997 (Fig. 7.6J)*

3.2.31.1 Type ichnospecies

Toyamasauripus masuiae Matsukawa et al., 1997.

3.2.31.2 Valid ichnospecies

Toyamasauripus masuiae Matsukawa et al., 1997.

3.2.31.3 Distribution

Valanginian–Aptian, Atotsugawa, and Nagatogawa formations of western Japan.

3.2.31.4 Holotype

TGU SE-DT 1001, housed in the Tokyo Gakugei University, Japan.

3.2.31.5 Diagnosis

Narrow trackway of a small biped, typically 3–10 cm in length, with high-pace angulation. Tridactyl track consists of slender digits with variably preserved pad impressions and projecting small heel, and is slightly asymmetrical, mesaxonic, and about as long as wide. Divarication between Digits II and IV averaging about 90° (range 60°–136°). End of Digit III curves inward (Matsukawa et al., 1997).

3.2.31.6 General description

T. masuiae has only been found in Cretaceous strata. Small, tridactyl pes tracks, with three slender digits and a small heel and the lengths and widths ranging from 3 to 9.4 cm and 3.2–7.8 cm, respectively. Small specimens, less than 6.5 cm in length, have a tendency to be wider than long, but large specimens are longer than wide. The Digit III is the longest, and its end curves inward. Digit IV has a sharp claw mark and two robust phalangeal pads. The metatarsophalangeal pad of Digit IV is located at the axis of digit III. The divarication between Digits II–IV is 60°–127°. The track is slightly asymmetrical and has a projecting small heel. There is a slight posterior indentation along the posterior margin of Digit II (Matsukawa et al., 1997).

3.2.31.7 Discussion

Matsukawa et al. (1997) claimed that *Toyamasauripus* has characteristics of both theropod and/or ornithopod tracks (Thulborn, 1990; Lockley, 1991): symmetrical, as wide, or wider than long in small specimens, without well-defined claw impressions, with wide digit divarication angle, slight inward rotation of Digit III and lack of well-defined pad impressions. But it was different from ornithopod tracks because of the former has a smaller pace angulation and positive rotation of Digit III. It was also similar to bird tracks (Lockley et al., 1992), but *Toyamasauripus* is larger and has a wider divarication between Digits II–IV.

3.2.31.8 Trackmaker

Matsukawa et al. (1997) suggested that *Toyamasauripus* was named by the small theropods from East Asia, and the trackways showed estimated speeds ranging from 1.21 to 1.50 km/hour.

3.2.32 Kalohipus *Fuentes Vidarte and Meijide Calvo, 1999 (Fig. 7.5H)*

3.2.32.1 Type ichnospecies

Kalohipus bretunensis Fuentes Vidarte and Meijide Calvo, 1999.

3.2.32.2 Valid ichnospecies

Kalohipus bretunensis Fuentes Vidarte and Meijide Calvo, 1999.

3.2.32.3 Distribution

Berriasian, Huérteles Formation, Oncala Group (Cameros Basin, Soria) of Spain.

3.2.32.4 Holotype

Ichnite no. 21, trackway V. In a Berriasian terrestrial siltstone/shale in the Huérteles Formation of Spain.

3.2.32.5 Diagnosis

Tridactyl tracks, 5.5–7.5 cm long. They belong to a digitigradal biped dinosaur. Digits relatively robust, acuminate and with claw and plantar pads. Digits II and IV of similar development and the Digit III much longer; hypex practically symmetrical. Plantar surface relatively large. Heel oval, prolonged and with strong notches. Divarication angle II–IV low (69° average). Length/width variable between 1.2 and 1.7. Bipedal and narrow trackway (gauge 5–7). Pace angle between 170° and 175° (translated from the Spanish of Fuentes Vidarte and Meijide Calvo, 1999).

3.2.32.6 General description

Tracks are tridactyl, mesaxonic, digitigrade, longer than wide (length/width ratio = 1.37–1.66) and small (about 7 cm in length). The digits are slender. They have an acuminated ends that represent claw traces preserved in some of the tracks. The Digit III impression is clearly the longest, while Digits II and IV impressions are subequal in length. The divarication angles are low (about 60° in angle II–IV), and the hypices quite symmetrical. The metatarsophalangeal area is oval to subtriangular in shape, and there is a small medial notch in Digit II (Castanera et al., 2015).

3.2.32.7 Discussion

Gierliński et al. (2009) suggested that *Kalohipus bretunensis* is very similar to *Wildeichnus* and hence *Kalohipus* might be a junior synonym of *Wildeichnus*. Castanera et al. (2015) assigned to *Kalohipus* several very well-preserved tracks published by Pascual-Arribas and Hernández-Medrano (2011) and considered them grallatorid tracks. Melchor et al. (2019) do not consider *Kalohipus* within this ichnofamily. Castanera et al. (2018) classified as cf. *Kalohipus* isp. some tracks found in the Upper Jurassic of Switzerland.

3.2.32.8 Trackmaker

For the possible trackmakers of the *K. bretunensis* tracks, some authors proposed that grallatorid tracks from the Late Triassic—Early Cretaceous have traditionally been associated with ceratosaurs (Lockley 2009; Lockley and Gierliński, 2006; Petti et al., 2011). According to the osteological record (Carrano and Sampson, 2008), the trackmaker of *K. bretunensis* might be a theropod belonging to this clade, although osteological remains have not yet been found in the Berriasian strata of the Iberian Peninsula. The presence of this group of theropods has indeed been described in older (late Kimmeridgian) deposits in Portugal (Malafaia et al., 2015).

3.2.33 Asianopodus *Matsukawa et al., 2005 (Figs. 7.6K–M and 7.30E)*

3.2.33.1 Type ichnospecies

Asianopodus pulvinicalx Matsukawa et al., 2005 (Fig. 7.6K)

3.2.33.2 Valid ichnospecies

Asianopodus pulvinicalx Matsukawa et al., 2005.
Asianopodus robustus Li et al., 2011 (Fig. 7.6L)
Asianopodus wangi Xing et al., 2021b (Figs. 7.6M and 7.30E)

3.2.33.3 Distribution

Valanginian–Albian, Tugulu, and Hekou Groups, and Jingchuan (often miswritten as Jianchuan) and Tuchengzi formations of China; Khok Kruat Formation of northeastern Thailand; Jinju Formation of southern South Korea; Kuwajima Formation of western Japan.

3.2.33.4 Holotype

A replica of the holotype is TGU SE–DT1004, housed in the Tokyo Gakugei University as rubber molds of original specimen, Japan.

3.2.33.5 Diagnosis

Small- to medium-sized tridactyl, mesaxonic, and subsymmetrical track with distinct bulbous heel impression. Track is longer than wide. The digital divarication angles are rather narrow (Matsukawa et al., 2005).

3.2.33.6 General description

A. robustus was discovered from the Jingchuan Formation (Aptian–Albian) in Otog Qi, China. Only found in the Cretaceous. Large, tridactyl pes tracks, with a length of 34.1 cm, and a width of 26.4 cm; the average aspect ratio is about 1.3. The values of digit divarication are 18° (Digits II–III) and 30° (Digits III–IV). The heel pad is distinctly developed, and the digits are robust. The track is asymmetrical (Li et al., 2011).

A. wangi was discovered from the Tuchengzi Formatian (Kimmeridgian–Valanginian) in Hebei, China. Only found in the Cretaceous. Large, tridactyl pes tracks, with the average aspect ratio about 1.6, and the medium mesaxony is 0.65. The value of divarication is 50° (Digits II–IV). Digit III is 62% of total pes length. The pace length is 4 times track length, and the pace angulation is 160° (Xing et al., 2021b).

3.2.33.7 Discussion

Asianopodus was only found in east Asia, and first found in Japan (Matsukawa et al., 2005). At least four ichnospecies were established by the researchers: *A. pulvinicalx* (Matsukawa et al., 2005), *A. robustus* (Li et al., 2011), *A. niui* (Li et al., 2020), and *A. wangi* (Xing et al., 2021b). Xing et al. (2021b) proposed that *A. niui* is a nomen dubium. Therefore, three valid ichnospecies and at least six *Asianopodus* isp. were retained after re-examination and confirmation. Its most notable character is the large and clear heel pad.

A. robustus differs from *A. pulvinicalx* because of the larger width and the smaller average aspect ratio ($1.3 < 1.5$). *A. wangi* differs from other *Asianopodus* because of the higher mesaxony and a longer Digit III.

In addition, there are other Cretaceous tracks classified within the ichnogenus *Asianopodus* in China (Hekou Group, Xing et al., 2014a), Thailand (Khok Kruat Formation, Matsukawa et al., 2005), and South Korea (Jinju Formation, Kim et al., 2018a). Heredia et al. (2020) classified as aff. *Asianopodus* some trackways from the Candeleros Formation (Argentina).

3.2.33.8 Trackmaker

Xing et al. (2021b) associated *A. wangi* with medium-sized to large coelurosaurs. Lockley and Xing (2021) suggested that *Asianopodus* should be associated with a medium-sized theropod.

3.2.34 Siamopodus *Lockley et al., 2006b (Fig. 7.6N and O)*

3.2.34.1 Type ichnospecies

Siamopodus khaoyaiensis Lockley et al., 2006b (Fig. 7.6N)

3.2.34.2 Valid ichnospecies

Siamopodus khaoyaiensis Lockley et al., 2006b.
Siamopodus xui Xing et al., 2014b (Fig. 7.6O)

3.2.34.3 Distribution

Berriasian, Phra Wihan Formation (Nakhon Ratchasima) of Thailand; late Berriasian–Albian, Feitianshan Formation (Sichuan) of China.

3.2.34.4 Holotype

CUGM F0020 (the largest specimen, also designated as CU 214.46), housed in the University of Colorado at Denver collections.

3.2.34.5 Diagnosis

Small- to medium-sized tridactyl theropod track with slender toes and subsymmetric bilobed heel. Inner hypex between Digits II and III situated posterior to outer hypex between Digits III and IV (Lockley et al., 2006b).

3.2.34.6 General description

The footprints are tridactyl, mesaxonic, and digitigrade. The inner hypex between Digits II and III is situated posterior to the outer hypex between Digits III and IV. They are more or less as wide as long. In *S. khaoyaiensis* the length ranges 14–30 cm and the width 11–25 cm, whereas in *S. xui* 7.1 and 8.0 cm respectively. The metatarsophalangeal pad impression is located in alignment with the track axis, but in *S. khaoyaiensis* is bilobed and in *S. xui* rounded.

3.2.34.7 Discussion

This ichnogenus presents two ichnospecies: *S. khaoyaiensis* and *S. xui*. Lockley et al. (2006b) defined the first ichnospecies and pointed out that the bilobed heel impression is unusual in nonlarge ornithopod tracks, so that justifies its erection. Subsequently, Xing et al. (2014b) proposed *S. xui* based on an ichnospecific diagnosis. The problem arises that when the second ichnospecies of an ichnogenus is proposed, the authors should propose an ichnogeneric diagnosis, emend the diagnosis of the type ichnospecies and propose the ichnospecific diagnosis of the new ichnotaxon. So, an ichnotaxonomical revision of *Siamopodus* is needed.

3.2.34.8 Trackmaker

Xing et al. (2014b) argue that *S. xui* is different from either the grallatorid and the anomoepid morphotypes, but nevertheless shows diagnostic theropodan attributes. Its most diagnostic feature, the configuration of the hypices, allies it most closely with *Siamopodus*. Differences are noteworthy, but it is premature to erect a new ichnogenus, especially in the light of recent efforts to reduce the number of indistinguishable theropodan ichnogenera reported from the ichnological literature on the Mesozoic of China (Lockley et al., 2013). In this regard, however, it has been stated that Cretaceous theropod ichnotaxa are more differentiated, distinct, and diverse than those in the Jurassic. This inference requires further testing by careful comparative study.

3.2.35 Chapus *Li et al., 2006 (Fig. 7.6P)*

3.2.35.1 Type ichnospecies

Chapus lockleyi Li et al., 2006.

3.2.35.2 Valid ichnospecies

Chapus lockleyi Li et al., 2006.

3.2.35.3 Distribution

Aptian–Albian, Jingchuan Formation of Northern China.

3.2.35.4 Holotype

CHABU-8-42, CHABU-8-39, CHABU-8-41, preserved in situ in the Otog Dinosaur Trace National Nature Reserve.

3.2.35.5 Diagnosis

Bipedal, tridactyl tracks, claw tracks at the distal end of digits, relatively low anterior triangles (0.48), low length of digit III/track length (0.60), divarication between Digits II, III and IV are II 13°, III and 35°, IV. Pace length is 135 cm and stride length 233 cm, pace angulation is 133° (emended by Xing et al., 2021b).

3.2.35.6 General description

The holotype are consecutive tracks. The length of the CHABU-42 is 58 cm and the width is 42.6 cm; the pace length of the left tracks (115.7 cm) is shorter than that of the right tracks (137.5 cm) (Li et al., 2006).

3.2.35.7 Discussion

There are at least three ichnogenera of medium and large theropod tracks in the Otog area, including *Chapus* (Li et al., 2006), *Asianopodus* (Li et al., 2011), and *Ordexallopus* (Lockley et al., 2018a). *Chapus* was originally assigned to the ichnofamily Eubrontidae (Li et al., 2006). Lockley et al. (2018a) considered *Chapus lockleyi* to be tentatively valid, and suggested that a careful study of all known samples is necessary to test our suggestion that *Chapus* may be a subjective junior synonym of *Asianopodus*.

Xing et al. (2021b) considered the large theropod track *Chapus* characterized by relatively low anterior triangles (0.48) and low length of digit III/track length ratio (0.60). These values are different from the 0.58 and 0.67 ratios of *Eubrontes giganteus* (type) (Xing et al., 2021b). Lockley et al. (2018a) considered *Chapus* to be different from *Ordexallopus*. The consistently tridactyl *Chapus* tracks are more elongate (mean length/width ratio 1.33, compared with 1.12 for *Ordexallopus*), and with a mean divarication angle of 52.7° (compared with 83° for *Ordexallopus*) between Digits II and IV.

3.2.35.8 Trackmaker

Li et al. (2006) documented an underlying pathological situation. In the left tracks (CHABU-42, CHABU-41), the digit II inclines to Digit III, whereas the right track (CHABU-39) does not have such a feature. Therefore, it is presumed that the trackmaker had its Digit II of the left foot wounded by some unknown cause (Li et al., 2006). However, unlike typical pathological evidence (such as *Megalosauripus* isp. Xing et al., 2021g), the difference between the left and right footprints of CHABU specimens is very limited and cannot be ruled out as due to preservation or other factors.

3.2.36 Dromaeopodus *Li et al., 2008 (Figs. 7.8C and 7.32C)*

3.2.36.1 Type ichnospecies

Dromaeopodus shandongensis Li et al., 2008.

3.2.36.2 Valid ichnospecies

Dromaeopodus shandongensis Li et al., 2008.

3.2.36.3 Distribution

Albian–Cenomanian, Tianjialou Formation (Shandong), Barremian/Aptian–Santonian, Jiaguan Formation (Sichuan), late Berriasian–Albian, Feitianshan Formation (Sichuan) of China; Maastrichtian, El Molino Formation of Bolivia.

3.2.36.4 Holotype

Trackway 7, preserved in situ at Houzuoshan Dinosaur Park in Shandong, China, and the replica (CU 214.111), housed in the University of Colorado at Denver.

3.2.36.5 Diagnosis

Tracks made by a biped that are functionally didactyl, elongate, and exhibit the following characteristics: Digit III and IV traces subparallel, subequal in length, displaying laterally convex curvature, and possessing three-digit pads each as well as sharp claw impressions; suboval heel pad forming about one-third of the posterior length; proximal portion of Digit II impressed as a large, hemispherical depression medial to the anterior margin of the heel pad and posteromedial to the margin of Digit III; step lengths from 92 to 103 cm; pace angulation high (170°); Digit III axis shows slight inward rotation (~10°) relative to sagittal plane (Li et al., 2008).

3.2.36.6 General description

The holotype trackway seven contained four tracks, including two left tracks and two right tracks. Pes length is 28.5 cm, pes width is 12.5 cm, step length is 93 cm, stride length is 186 cm, and pace angulation is 170°. The features of *Dromaeopodus* tracks are consistent with the anatomy of articulated deinonychosaur foot fossils: the abbreviated impression of Digit II, indicating retraction of its claw; the subequal lengths of Digits III and IV; and their nearly parallel orientation and slight inward curvature (Ostrom, 1969; Norell and Makovicky, 1997; Li et al., 2008).

Xing et al. (2016d) reported the largest dromaeosaurid track that was from the upper member of the Jiaguan Formation, Leibei site, Gulin County of Sichuan, China. The length is

29.5 cm, and the length/width ratio is 2.8. Its Digits III and IV are strong, elongated, parallel, and almost equal length. Lockley and Xing (2015) interpreted this to mean that the tracks were extruded into two layers of thick, noncompactable sand layers. During burial, the sand-filled tracks (casts) are flattened or widened as the ductile layers were compressed.

Another dromaeosaurid track from the Lower Cretaceous Feitianshan Formation of Zhaojue County of Sichuan, China was described by Xing et al. (2016d). The track length is 24.8 cm and the pes track has two slender, elongated prints of Digits III and IV, a short, round print, apparently representing the proximal part of Digit II, and a round, developed heel area. Xing et al. (2016d) diagnosed the tracks as *Dromaeopodus* isp.

Meyer et al. (2021) discovered tracks and trackways from the Puca Group (Coniacian–Late Maastrichtian) of the eastern Cordillera of Bolivia, which has a close resemblance to the *Dromaeopodus* due to the subequal length of Digit III and Digit IV and nearly parallel with no visible imprint of Digit II. But Meyer et al. (2021) suggested the tracks they reported point to a different ichnotaxon. They refrained from erecting a new ichnotaxon until they have more thorough documentation at hand.

3.2.36.7 Discussion

Dromaeopodus is clearly morphologically distinct from other deinonychosaurian ichnogenera, such as *Velociraptorichnus* (Zhen et al., 1996), *Menglongipus* (Xing et al., 2009d) and *Dromaeosauripus* (Kim et al., 2008). The curved digits of *Dromaeopodus* distinguish the ichnotaxon from the smaller *Velociraptorichnus*, which has straight digits (Li et al., 2008). The details of the *Dromaeopodus* pad imprint and traces of the short, proximal part of Digit II are not visible in *Dromaeosauripus* (Lockley et al., 2016). Digit IV of *Menglongipus* is short compared with Digit III, and it differs from *Velociraptorichnus*, *Dromaeopodus*, and *Dromaeosauripus*, in which Digits III and IV are subequal in length (Xing et al., 2016a,d).

3.2.36.8 Trackmaker

Li et al. (2008) inferred that trackmakers of *Dromaeopodus* are most parsimoniously a large dromaeosaurid, based on the phylogenetic understanding of body size evolution (Turner et al., 2007) and considerations of geological age and paleogeography. Li et al. (2008) suggested that the trackways of *Dromaeopodus* showed estimated speeds lower than 4.1 km/hour.

3.2.37 Dromaeosauripus *Kim et al., 2008 (Figs. 7.8D–F and 7.32D,E)*

3.2.37.1 Type ichnospecies

Dromaeosauripus hamanensis Kim et al., 2008 (Figs. 7.8D and 7.32E)

3.2.37.2 Valid ichnospecies

Dromaeosauripus hamanensis Kim et al., 2008.
Dromaeosauripus jinjuensis Kim et al., 2012b (Fig. 7.8E)
Dromaeosauripus yongjingensis Xing et al., 2013a (Figs. 7.8F and 7.32D)

3.2.37.3 Distribution

Albian–Cenomanian, Haman Formation, late Aptian–early Albian, Jinju Formation of South Korea; Hauterivian–Aptian, Yanguoxia Formation (or fifth informal formation-level unit) (Gansu) and Aptian–Albian, Jingchuan Formation (Inner Mongolia) of China; Albian, South Platte Formation (Colorado) and late Aptian–Albian, Cedar Mountain Formation (Utah) of USA.

3.2.37.4 Holotype

KNUE (Korea National University of Education) 060,801, a trackway composed of five consecutive tracks, Chu Island, Changseonmyeon, Namhae area, Korea.

3.2.37.5 Diagnosis

Medium-sized (~15 cm long and ~9.0 cm wide) didactyl tracks with slender digit impressions, very narrow (5–10°) divarication angles between Digits III and IV. Digit III slightly longer than IV. Digital pad impressions well-developed, sharp claw impressions distinct. Individual pes impressions show little rotation with respect to the trackway axis, that is, Digit III is directed cranially and nearly parallel to the trackway axis (Kim et al., 2008).

3.2.37.6 General description

D. hamanensis is characterized by slightly curved digit impressions with very slight divarication angles (ranging from 5° to 10° between Digits III and IV); digital pad impressions are distinct; sharp claw impressions are observed at the ends of the digit traces. The trackway is nearly straight and the pace angulation is thus nearly 180° (Kim et al., 2008).

D. jinjuensis is characterized by relatively small size (pes tracks average 9.3 cm in length and 6.8 cm in width), slender digit impressions, and narrow divarication angle (8°–14°) between Digits III and IV. Digit III slightly longer than IV. Sharp claw impressions are distinct and slightly inward. The trackway is nearly straight, and the pace angulation is thus nearly 180°, with an average pace of 40.1 cm and average stride of 80.4 cm (Kim et al., 2012b).

D. yongjingensis is described as medium–sized (~14.8 cm in length and ~6.4 cm in width) didactyl theropod tracks. Digital pad impressions well-developed and with a formula of 1–3–4 for Digits II, III, and IV, respectively; sharp claw impressions absent; proximal portion of Digit II contacts the anterior margin of a large, rounded metatarsophalangeal pad; mean divarication angle between Digits III and IV is 19°; step lengths range from 35.7 to 37.5 cm and pace angulations from 143° to 180° (Xing et al., 2013a).

3.2.37.7 Discussion

Dromaeosauripus was named on the basis of medium-sized tracks that differs from other didactyl tracks in having two slender digital impressions that are nearly parallel and separated (Kim et al., 2008), without heel impressions as seen in *Dromaeopodus* (Kim et al., 2012b), which is similar to but much larger than the later-named *Dromaeosauriformipes* (Kim et al., 2018b). *D. jinjuensis* measurements differ significantly from *D. hamanensis* in length (only 60% of the latter), width (81%), length/width ratio (0.72 in *D. hamanensis*), and digit thickness (55%) (Kim et al., 2012b). In *D. yongjingensis*, the divarication angles between Digits III–IV are greater than in other ichnospecies, and the digital pad formula also differentiates it from others (formula x-1-4-4-x; Kim et al., 2008, 2012; Xing et al., 2013a).

Most studies of didactyl tracks have focused on the deinonychosaurian (dromaeosaurid and troodontid) clade of theropod dinosaurs, and most dromaeosaurids tracks have been reported from East Asia (Lockely et al., 2016; also see the discussion in *Dromaeosauriformipes* section). Furthermore, it is interesting to stress that *Dromaeosauriformipes* and *Dromaeosauripus*, which are morphologically very similar except for size, occur in similar areas and in the same formation (Kim et al., 2018b). Apart from the implications of preservation bias, these traces may help revealing the difference in morphology and modes of action in the dissimilar ontogenetic stages of small-bodied dinosaurs.

3.2.37.8 Trackmaker

The most probable trackmaker of *Dromaeosauripus* is a member of clade Deinonychosauria (probably subclade Dromaeosauridae), which includes theropods such as *Deinonychus*, *Dromaeosaurus*, *Utahraptor*, and *Velociraptor* (Kim et al., 2008). Furthermore, van der Lubbe et al. (2009, 2011) argued on osteological grounds that dromaeosaurids would produce tracks with subequal lengths between Digits III and IV, whereas troodontids possess a relatively shorter Digit IV, based on which Xing et al. (2013a) considered that *D. yongjingensis* (and probably all *Dromaeosauripus* tracks) were likely made by a dromaeosaurid.

Besides, the estimated height of *D. jinjuensis* trackmaker at the hip (h) is 0.4 m. The relative stride length ($SL/h = 1.98$) represents a walking gait, and estimated speed is ~0.76 m/second (Kim et al., 2012b), while the holotype of the type ichnospecies shows running and trotting in both dromaeosaurid and troodontid trackways (Tsukiji et al., 2021).

3.2.38 Menglongipus *Xing et al., 2009d (Fig. 7.8G)*

3.2.38.1 Type ichnospecies

Menglongipus sinensis Xing et al., 2009d.

3.2.38.2 Valid ichnospecies

Menglongipus sinensis Xing et al., 2009d.

3.2.38.3 Distribution

Jurassic: Aalenian–Bajocian, Dansirit Formation, Iran; Cretaceous: Berriasian–Valanginian, Tuchengzi Formation (Hebei) and Albian, Dasheng Group (Shandong) of China; late Campanian–early Maastrichtian, lower Cantwell Formation (Alaska) of USA.

3.2.38.4 Holotype

A complete natural mold cataloged as T.A.1 from the Nijiagou tracksite. An artificial mold of the track is stored at the Lufeng Dinosaur Research Center where it is cataloged as LDRC–v.x.8.The original track remains in the field.

3.2.38.5 Diagnosis

Small, didactyl theropod track that lacks manus and tail traces. The average length ratio of Digit III and IV is 1.8 (holotype 2.36), differentiating the ichnotaxon from *Velociraptorichnus*, *Dromaeopodus*, and *Dromaeosauripus*, in which Digit IV is almost as long as Digit III. Digit II, when present, is represented only by a short round impression located at the proximomedial margin of Digit III. The divarication angle between Digits III and IV ranges from 40° to 44° and is wider than in *Velociraptorichnus*, *Dromaeopodus*, or *Dromaeosauripus* (Xing et al., 2009d).

3.2.38.6 General description

Menglongipus sinensis has a trackway that exhibits a very narrow stance with a pace angulation of 180°, also supporting that the tracks represent a single trackway registered by a single track maker. The length/width ratio of the holotype is 1.67. Discernible claw marks were observed on Digits III and IV. Digital pads are indistinct but based on the photogrammetric maps it appears that Digit III has at least two pads and Digit IV at least one. Digit impressions do not substantially taper distally. The metatarsophalangeal region is small; no distinct border demarcating a pad divides this region from either Digit III or IV (Xing et al., 2009d).

3.2.38.7 Discussion

Menglongipus was originally defined based on a poorly preserved trackway that provides new evidence for the existence and distribution of deinonychosaurians around the J-K boundary (Xing et al., 2009d), which has been supplemented by later-discovered tracksites (Xing et al., 2018d). Among the didactyl tracks associated with deinonychosaurs, *Menglongipus* shares the similar distinctly unparallel impressions of Digits III and IV and relatively small size (expected in *Dromaeosauriformipes*) with *Velociraptorichnus*, but differs in Digit IV that is much shorter than Digit III (Xing et al., 2018d; Tsukiji et al., 2021).

While deinonychosaur tracks are widely distributed in the Cretaceous and J-K boundary of the Northern Hemisphere (Lockley et al., 2016), currently only *Dromaeosauripus* and *Menglongipus* have been found outside of East Asia (Xing et al., 2024b, Table 2). Moreover, the possible *Menglongipus* type is the only Cretaceous didactyl track type seen before the Late Jurassic, that may indicate the trackmaker is of early deinonychosaurian affinity or convergence (Xing et al., 2018d).

3.2.38.8 Trackmaker

Xing et al. (2009d) suggested that *Menglongipus sinensis* tracks demonstrate that deinonychosaurians lived in the same region roughly 15 million years earlier, which includes a variety of dromaeosaurids and troodontids, but it is hard to determine whether *Menglongipus* tracks could belong to one of these taxa. The estimated body length of the *M. sinensis* trackmaker is 66.6 cm (Henderson, 2003; Xing et al., 2009d), which is very similar to that (about 65 cm long) estimated for a basal paravian by Turner et al. (2007). The gaits indicated for the holotype of *Menglongipus* are running and trotting of both dromaeosaurid and troodontid trackmakers (Xing et al., 2009d), while cf. *Menglongipus* isp., with the estimated speed of 7.16 km/second–8.64 km/second (Xing et al., 2018d), could indicate a trotting gait of a dromaeosaurid trackmaker and walking gait of a troodontid trackway (Tsukiji et al., 2021).

3.2.39 Corpulentapus *Li et al., 2011 (Fig. 7.6Q)*

3.2.39.1 Type ichnospecies

Corpulentapus lilasia Li et al., 2011.

3.2.39.2 Valid ichnospecies

Corpulentapus lilasia Li et al., 2011.

3.2.39.3 Distribution

Barremian, Longwangzhuang Formation of eastern China; Aptian, Jinju Formation of Korea.

3.2.39.4 Holotype

CU 214.174, CU 214.172, right and left tracks are preserved in situ at the Huanglonggou tracksite, China, and replicas are housed in the University of Colorado Museum of Natural History.

3.2.39.5 Diagnosis

Robust tridactyl, "lily" or "fleur de lis" -shaped track of a biped, almost as wide as long, with three wide digits (II–IV), low digit divarication, and weak mesaxony. Digit traces not separated by a hypex for most of their proximal length. Trackway narrow with a long step (Li et al., 2011)

3.2.39.6 General description

Traces of three, wide, robust digits (II–IV) are typically not separated by a hypex proximally. Track length only slightly greater than width (mean 11.8 and 10.8 cm respectively) giving a mean l/w ratio of 1.09. Anterior triangle, with parasagittal apex height only one third of baseline width indicates very weak mesaxony. Distal ends of digit traces tend to rotate away from track long axis (parasagittal plane) giving tracks a flared appearance. Similarly blunt claw traces tend to project laterally and medially rather than anteriorly. Step long (mean 66.4 cm) averaging more than 5.6 times track length (Li et al., 2011).

3.2.39.7 Discussion

Li et al. (2011) named *Corpulentapus*, which is readily differentiated from any other known theropod track. Its characteristics includes typical weak mesaxony, distinctive "fleur de lis" shape, robust, fleshy digit traces, lacking well-defined digital pads separated by recognizable creases, and a relatively short Digit III showing little anterior projection beyond the tips of Digits II and IV (Li et al., 2011). Li et al. (2011) compared *Corpulentapus* with common theropod tracks in the same area. *Paragrallator* is characterized by an elongate shape, narrow digits, and distinctive pad impressions (Xing et al., 2010a). *Eubrontes* tracks are consistently more elongate than *Corpulentapus* with more widely separated digits (Hitchcock, 1845; Lull, 1953; Olsen et al., 1998). *Grallator* consistently has very strong mesaxony, well-defined pad impressions, and sharp distal claw traces (Olsen et al., 1998). *Minisauripus* is more elongate and the differential length of Digits II and IV is more pronounced leading to a characteristic asymmetry (Lockley et al., 2008a).

Lockley et al. (2015) measured 25 *Corpulentapus* trackways at the Huanglonggou site in Shandong of China, and the speed of *Corpulentapus* was estimated by the formula of Alexander (1976). Speed estimates range from 1.7 to 3.2 m/second ($N = 25$), and these values translate into between 6.2 and 11.4 km/hour.

Kim et al. (2018a) reported *Corpulentapus* from the Lower Cretaceous Jinju Formation of the Jinju, including a trackway with five tracks and an isolated track. *Grallator* and *Asianopodus* were also reported in this area. The new assemblage of theropod tracks (*Grallator*, *Corpulentapus* and *Asianopodus*) is remarkably similar to the Huanglonggou site in China.

3.2.39.8 Trackmaker

There are no corresponding skeleton fossils in this area that would provide direct clues to the trackmaker of *Corpulentapus*. Li et al. (2011) suggested that decrease in foot length and degree of mesaxony may correspond to increase in limb length, at least among bipedal dinosaurs. Therefore, the *Corpulentapus* trackmaker is likely to be significantly different from the *Grallator* type, it may have smaller body size and has longer hind legs.

3.2.40 Wakinyantanka *Lockley et al., 2014d (Fig. 7.4T)*

3.2.40.1 Type ichnospecies

Wakinyantanka styxi Lockley et al., 2014d.

3.2.40.2 Valid ichnospecies

Wakinyantanka styxi Lockley et al., 2014d.

3.2.40.3 Distribution

Maastrichtian; Hell Creek Formation of South Dakota, USA.

3.2.40.4 Holotype

RMDRC97-001-01-08 and RMDRC97-001-10-16, housed in the Rocky Mountain Dinosaur Resource Center in Colorado, USA.

3.2.40.5 Diagnosis

Large, slender-toed, functionally tridactyl theropod tracks about as wide as long. Digits are parallel sided with abrupt distally tapering claw impressions lacking well-defined digital pads. Hypex between Digits II and III is much deeper (more posterior) than between Digits III and IV. Digit divarication is wide, averaging 90°. Small, posteromedial hallux is present in some of the more deeply impressed tracks. Metatarsal impression is long and slender. Trackway is narrow.

3.2.40.6 General description

Wakinyantanka is a large tetradactyl-tridactyl ichnotaxon, which has slender digit impressions, sharp claw impressions and lacks digital pads. The digit I impression is present in deep impressions and is narrow and short. It is as long (55–60 cm, excluding metatarsal and hallux) as wide ~60 cm. The interdigital angle between Digits II and IV is about 90°. The metatarsal impression is elongated, narrow, and variable in shape (Lockley et al., 2014d).

3.2.40.7 Discussion

Lockley et al. (2014d) pointed out that the track shape of *Wakinyantanka* presents a significant variation within the type trackway. They suggested that this variability is caused by locomotion over a soft substrate and by the effects of substrate collapse of the substrate. The absence of digital pad impressions can also be a preservational bias. Taking into account that the morphology of *Wakinyantanka* is conditioned by extramorphological features and there is only one trackway assigned to this ichnotaxon, we consider that it is necessary to revise it in the future.

3.2.40.8 Trackmaker

Lockley et al. (2014d) introduced a new ichnotaxon *Wakinyantanka styxi*, and, in contrast to other known tyrannosaurid tracks, they show less robust digits, and therefore may belong to a more gracile, smaller form such as *Nanotyrannus* or some albertosaurines, or not be tyrannosaur tracks (Xing et al., 2019e).

3.2.41 Paracorpulentapus *Xing et al., 2014c (Fig. 7.6R)*

3.2.41.1 Type ichnospecies

Paracorpulentapus zhangsanfengi Xing et al., 2014c.

3.2.41.2 Valid ichnospecies

Paracorpulentapus zhangsanfengi Xing et al., 2014c.

3.2.41.3 Distribution

Barremian/Aptian–Santonian, Jiaguan Formation (Sichuan), Albian–Cenomanian, Mawu Formation (Guangdong) and Campanian–Maastrichtian, Xiaoyan Formation (Anhui) of China.

3.2.41.4 Holotype

A complete pes of a natural cast track, cataloged as XHT-30, from the Xiaohutian tracksite, Anhui, China. The original specimens remain in the field.

3.2.41.5 Diagnosis

Medium-sized (~16.0 cm in length and ~15.4 cm in width), robust tridactyl theropod tracks, almost as wide as long; mean divarication between Digits II and IV 74°; weak mesaxony; digit traces relatively short and "fleshy" with indistinct creases only between pads and with blunt claws; digit traces separated by a hypex for most of their proximal length; Digit IV always the narrowest; digit traces II and III proximally with thin interspace area; metatarsophalangeal pad of Digit IV located close to the axis of Digit III forming a short rounded "heel"; trackway narrow with short steps; step lengths 2.7 times as much as the track length (Xing et al., 2014c)

3.2.41.6 General description

P. zhangsanfengi is characterized by length/width ratio of 1.1; longest Digit III; subequal Digits II and IV with blunt claws; Digit II with two robust phalangeal pads; no distinct border between the phalangeal pads of Digits III and IV. The metatarsophalangeal pad of Digit IV is located close to the axis of Digit III. Digits II and III are connected posteriorly. Proximally, Digits II and III show a thinner inter-pad space. The divarication between Digits II–III is larger than that between Digits III–IV. The trackway of *P. zhangsanfengi* shows a slight outward rotation of the footprints. The trackway pattern with the position of the fourth (right) imprint crossing the trackway midline possibly indicates a slight turn to the left. In the footprints, the distal end of Digit III tends to rotate outward away from the long axis of the footprint. The step is more than 2.7 times larger than footprint length in the type trackway while *Paracorpulentapus* isp. could reach over 5 times the length (Xing et al., 2014c, 2023b).

3.2.41.7 Discussion

Paracorpulentapus has been compared with *Corpulentapus* in shape and size and shares most of the other defining characteristics of *Corpulentapus*, such as robust, "fleshy" digit traces, an absence of well-defined digital pads separated by recognizable creases, weak mesaxony, and similar divarication between Digits II and IV (58–74° vs. 65°). But, *Paracorpulentapus* is larger than *Corpulentapus* in absolute size (15–16 cm vs. 11.8 cm) and slightly stronger in mesaxony (0.37–0.47 vs. 0.32), as well as does not show the "fleur de lis" shape, which indicates digits are more strongly separated at the distal end of Digits II and IV (Xing et al., 2014c, 2021, 2023b). Also, the trackway is proportionately wider with shorter step lengths (Xing et al., 2014c).

For the distribution of this ichnotaxon, Xing et al. (2021h, 2023b) mentioned that the latitudes of tracksites with *Paracorpulentapus* vary from 29°45′N to 23°54′N during the "mid" Cretaceous to 29°48′N in the Late Cretaceous. This evidence suggests that *Paracorpulentapus*-grallatorid-eubrontid assemblages may have been relatively widespread in Cretaceous South China and lasted until the Late Cretaceous, which could help in reconstructing the "mid" Cretaceous dinosaur faunas (Xing et al., 2023b).

3.2.41.8 Trackmaker

Xing et al. (2014c) pointed out that *Paracorpulentapus* has long and narrow claw marks, and a strong indentation behind Digit II, which are major characteristics of theropod tracks (Lockley, 1991), while earlier researchers attributed these tracks to ornithopods (Yu, 1999). Lockley (2009) considered that there is convergence in the pes development of some short-toed theropods and ornithopods especially in the Cretaceous, that may indicate robust tridactyl *Paracorpulentapus* may have been produced by a short-toed theropod that had long legs (Xing et al., 2014c).

3.2.42 Bellatoripes *McCrea et al., 2014b (Figs. 7.4U and 7.31B)*

3.2.42.1 Type ichnospecies

Bellatoripes fredlundi McCrea et al., 2014b.

3.2.42.2 Valid ichnospecies

Bellatoripes fredlundi McCrea et al., 2014b.

3.2.42.3 Distribution

Campanian–Maastrichtian Wapiti Formation (British Columbia) of Canada.

3.2.42.4 Holotype

In situ specimen PRPRC2011.11.001 (Trackway A) natural mold with three pes prints (two left, one right); also replicas as silicone mold (PRPRC, 2011.11.001M) and fiberglass reinforced (FGR95) plaster cast (PRPRC2011.11.001 MC) stored at the Peace Region Palaeontology Research Center (PRPRC) collection, Tumbler Ridge, British Columbia. Print#2 (right) is designated as the holotype footprint.

3.2.42.5 Diagnosis

Bipedal trackway of a large, functionally tridactyl, digitigrade track-maker. Footprints longer than wide with footprint length (FL) over 50 cm. Footprint area to length ratio greater than 2:1. Prints are mesaxonic and Digits II and IV are of similar length. Digit impressions are wide, lacking obvious digital pad impressions. Digits are thick proximally and taper strongly distally. The free digit length (FDL, Digit III protrudes from the other two) of Digit III is short compared with the length of the footprint. Total divarication is variable, but is generally no greater than 80°. Divarication between Digits III and IV is normally greater than that between Digits II and III. Metatarsal pad impression presents with a wide caudal margin. Pace length nearing 175 cm; stride length nearing 350 cm. Footprints in trackway are in line, with a pace angulation nearing 180°. Trackway width is narrow, with footprints overlapping the midline of the trackway. Footprint rotation is slight, but generally toward the trackway midline (McCrea et al., 2014b).

3.2.42.6 General description

The trackway is 4 m long, featuring three distinct footprints—two of which are left and one right. Notably, both left footprints exhibit a conspicuously abbreviated second digit, suggestive of a pathological condition. The right footprint, containing three intact digit impressions (II–IV), has been identified as the holotype. Each footprint reveals entry striations and impressions of skin tubercles. Alongside Digit III, striations and occasional skin tubercle impressions are observable on either or both margins across all footprints, with additional skin tubercle impressions present in the metatarsal pad region of the third print. The discernible skin impressions substantiate the authenticity of these tracks, distinguishing them from mere undertracks (McCrea et al., 2014b).

3.2.42.7 Discussion

McCrea et al. (2014b) defined the ichnofamily Tyrannosauripodidae as large, functionally tridactyl, mesaxonic tracks with distal metatarsal pad impression; may have craniomedially directed hallux impression; footprint length greater than width; robust footprint with thick digits; generally lacking distinct digital pad impressions. Digit impressions thickened proximally, strongly tapering distally and terminating in acuminate claw impressions. Trackway narrow with slight inward rotation of pes toward midline; pace length close to 175 cm or greater; stride length close to 350 cm or greater. They proposed that *Bellatoripes* and *Tyrannosauripus* belong to the Tyrannosauripodidae ichnofamily.

Bellatoripes is based on three parallel trackways, one of them, the largest one with three tracks, presents the evidence of a paleopathology. McCrea et al. (2014b) suggested that *Bellatoripes* is composed of true tracks based on their entry and/or exit striations and skin tubercle impressions on many of the prints. They lack digital pad impressions, but presents a metatarsophalangeal pad trace associated with a posteriorly elongated posteriorly depression. As happened with *Tyrannosauripus*, we think that these tracks are poorly preserved (less than 2 value in the Marchetti et al., 2019 proposal) and their ichnotaxonomic significance should be discussed with new findings.

3.2.42.8 Trackmaker

Taking into account the age (Campanian-Maastrichtian), the shape (theropodian), and the size (large), McCrea et al. (2014b) and Enriquez et al. (2020) proposed a tyrannosaurid dinosaur as possible trackmaker.

3.2.43 Ordexallopus *Lockley et al., 2018a (Fig. 7.6S)*

3.2.43.1 Type ichnospecies

Ordexallopus zhanglifui Lockley et al., 2018a.

3.2.43.2 Valid ichnospecies

Ordexallopus zhanglifui Lockley et al., 2018a.

3.2.43.3 Distribution

Aptian–Albian; Jingchuan Formation (Inner Mongolia) of China.

3.2.43.4 Holotype

Trackway T1 at Chabu site 8D: UCM tracing T1440, housed in the University of Colorado Museum of Natural History, which is the Jingchuan Formation of Inner Mongolia, China.

3.2.43.5 Diagnosis

Tetradactyl theropod track with unusually long hallux directed posteromedially for between half and a third of track width from center of metatarsophalangeal Digit IV heel pad. Tridactyl portion of track almost as wide as long with Digit II–IV divarication ~80°–90° (Lockley et al., 2018a).

3.2.43.6 General description

They are tetradactyl, mesaxonic, and longer than wide (averaging 42.7 and 38.0 cm wide giving a mean length-width ratio of 1.12). The divarication angle between Digits II and IV is 83°, between II and III is 38.5°, and between III and IV is 44.5°. The Digit I impression is long and posteromedially directed with the distal portion curved toward the posterior and generally not connected proximally to the tridactyl portion of the track comprising traces of Digits II, III, and IV. The length of the Digit I impression (hallux) is between 50% and 66% of track width, with mean angle of posterior rotation from transverse track axis ~20° (Lockley et al., 2018a).

3.2.43.7 Discussion

Lockley et al. (2018a) suggest that the presence of a hallux impression in *Ordexallopus zhanglifui* is an important ichnotaxobase that differentiates it from all the tridactyl ichnotaxa. For other tetradactyl ichnogenera, Lockley et al. (2018a) pointed out differences in size and divarication of the hallux impression, but only compared *Ordexallopus* with the Cretaceous ichnotaxa *Bueckeburgichnus* and *Saurexallopus*.

In our opinion, *Ordexallopus* presents a particular hallux trace that can justify its erection, nevertheless, the preservation of these tracks does not reach 2 on the Marchetti et al. (2019) scale, so new finds are needed to stabilize this ichnogenus.

3.2.43.8 Trackmaker

Lockley et al. (2018a) suggest that the presence of a large tetradactyl theropod trackmaker of probable arctometatarsalian affinity registered tracks named *Ordexallopus zhanglifui*. Moreover, they pointed out that they could have been impressed by oviraptosaurids (Lockley et al., 2018a).

3.2.44 Dromaeosauriformipes *Kim et al., 2018b (Fig. 7.8H)*

3.2.44.1 Type ichnospecies

Dromaeosauriformipes rarus Kim et al., 2018b.

3.2.44.2 Valid ichnospecies

Dromaeosauriformipes rarus Kim et al., 2018b.

3.2.44.3 Distribution

Albian, Jinju Formation of South Korea.

3.2.44.4 Holotype

CUE JI−2E Dr001, housed in the Chinju National University of Education, also represented by University of Colorado Museum of Natural History replica UCM 214.309, South Korea.

3.2.44.5 Diagnosis

Didactyl tracks of a diminutive biped more than twice as long as wide, with two equidimensional, elongate parallel traces. Trackway about as narrow as individual tracks, with step length highly variable. Differs from *Dromaeosauripus* in size and degree of footprint elongation, and in the absence of any trace of Digit II characteristic of other dromaeosaurid ichnotaxa (Kim et al., 2018b)

3.2.44.6 General description

Dromaeosauriformipes rarus is characterized by a very narrow trackway, with correspondingly high-pace angulation values (168°−188°). The total tracks are about 6−17.3 mm in length and 2.3−5.2 mm in width, with length/width of about 2.5−2.8 in the holotype and paratype (Kim et al., 2018b).

3.2.44.7 Discussion

Dromaeosauriformipes shares the similar subparallel Digits III and IV with *Dromaeopodus* and *Dromaeosauripus* in a didactyl track inferred to represent the dromaeosaurid clade (Tsukiji et al., 2021), especially similar with the latter that has separate toe traces that are not joined at the heel, which differs from the united Digits III and IV seen in some *Velociraptorichnus* and *Dromaeopodus* (Kim et al., 2018b). In this clade, *Dromaeosauriformipes* is an order of magnitude smaller than the others, but very similar to *Dromaeosauripus jinjuensis* in morphology (Kim et al., 2018b).

3.2.44.8 Trackmaker

Dromaeosauriformipes rarus tracks are attributed to small hatchlings of a species of unknown adult size or very small species representative of that clade, such as several of the smaller microraptorine dromaeosaurid dinosaurs known from the Asian fossil record. The step and stride length of its holotype and paratype could respectively indicate a step about 4.5 times footprint length and slow progression (0.6 m/second, or 2.16 km/hour) or steps about 26.49 times footprint length and much higher estimated speed of 10.5 m/second, or

37.8 km/hour (Kim et al., 2018b). The former could be interpreted as a trot gait in a dromaeosaurid condition (Tsukiji et al., 2021). Besides, the lack of heel traces could be interpreted as either a more digitigrade gait, in which case the explanation is behavioral, or substrate influences (Kim et al., 2018b).

3.2.45 Candeleroichnus *Calvo and Rivera, 2018 (Fig. 7.7H)*

3.2.45.1 Type ichnospecies

Candeleroichnus canalei Calvo and Rivera, 2018.

3.2.45.2 Valid ichnospecies

Candeleroichnus canalei Calvo and Rivera, 2018.

3.2.45.3 Distribution

Albian–Cenomanian, Candeleros Formation (Neuquén) of Argentina.

3.2.45.4 Holotype

MUCPv 76, housed in the Museum of Natural Sciences of the National University of Comahue.

3.2.45.5 Diagnosis

Tridactyl footprints of small size. Digit III well developed and displaced inwards. Digits II and IV thinner than the central one and elongated. Digit II with claw impression displaced slightly inward. The interdigital divergence generally does not exceed 10°. The hypex between Digits II and III is displaced posteriorly. The pace angle ranges between 175° and 180°. The heel is small and moved outward. Footprints are digitigrade (translated from Spanish of Calvo and Rivera, 2018).

3.2.45.6 General description

Calvo and Rivera (2018) pointed out that *Candeleroichnus* is tridactyl, asymmetrical, mesaxonic, digitigrade, and longer (average 11 cm) than wide (average 6 cm). The Digit III impression is the longest and robust of the three. The hypex of III and IV is posteriorly displaced with respect to II and III. The footprints preserve clear claw and pad impressions. The "heel" impression is short, almost nonexistent and displaced outward.

3.2.45.7 Discussion

Calvo and Rivera (2018) suggested that *Candeleroichnus* is similar to *Grallator* in size, shape, ratio between stride length and pace length and position of the digit. However, there are differences in the heel pad impression between *Candeleroichnus* and *Grallator*. *Candeleroichnus* is similar to *Deferrariichnium* and *Picunichnus* but it presents less developed Digit III impression.

3.2.45.8 Trackmaker

Calvo and Rivera (2018) suggested the *Candeleroichnus* can be associated with the small theropod from the Candeleros Formation, for instance: *Alnashetri* (Makovicky et al., 2004). More evidence is needed to determine the specific trackmakers.

3.2.46 Lockleypus *Xing et al., 2018e (Fig. 7.6T)*

3.2.46.1 Type ichnospecies

Lockleypus luanpingensis Xing et al., 2018e.

3.2.46.2 Valid ichnospecies

Lockleypus luanpingensis Xing et al., 2018e.

3.2.46.3 Distribution

Hauterivian–Barremian, Dabeigou Formation of northern China.

3.2.46.4 Holotype

IVPP TCL, housed in the Paleozoological Museum of China, Institute of Vertebrate Paleontology and Paleoanthropology, Beijing, China.

3.2.46.5 Diagnosis

Medium-sized (~25 cm in length and ~19 cm in width), robust tridactyl theropod tracks, length slightly larger than width; mean divarication between Digits II and IV = 66°; medium mesaxony; the free length of Digit IV is twice as long as the free length of Digit II; metatarsophalangeal pad of Digit IV located at the axis of Digit III forming a rounded metatarsophalangeal pad impression (Xing et al., 2018e).

3.2.46.6 General description

L. luanpingesis was discovered from Dabeigou Formation in Luanping City, Hebei, China. Only found in Cretaceous. Tridactyl pes tracks, without Digit I print, and the average length/width ratio was about 1.3. The Digit III was the longest, and the Digits II and IV were the subequal. Digit IV had a sharp claw mark and two robust phalangeal pads. The metatarsophalangeal pad of Digit IV is located at the axis of Digit III. The value divarications were 26° (Digits II–III) and 40° (Digits III–IV). The inner hypex between Digits II and III was situated anterior to the outer hypex between Digits III and IV (Xing et al., 2018e).

3.2.46.7 Discussion

The most different character of *Lockleypus* was which the free length of Digit IV was twice as long as Digit II compared with *Grallator*, *Kayentapus* and *Eubrontes*, and *Minisauripus*. Bivariate analysis of the aspect ratio versus anterior triangle length/width ratio showed *Lockleypus* was similar to the ornithomimipodid ichnotaxon *Ornithomimipus* (Xing et al., 2018e).

3.2.46.8 Trackmaker

Xing et al. (2018e) suggested *L. luanpingesis* can be associated with 3.2 m long theropod, most likely ornithomimosaur, but the morphological affinity between the features of *L. luanpingesis* and ornithomimosaur is not sufficient.

3.2.47 Jurabrontes *Marty et al., 2018 (Fig. 7.5I and J)*

3.2.47.1 Type ichnospecies

Jurabrontes curtedulensis Marty et al., 2018 (Fig. 7.5I)

3.2.47.2 Valid ichnospecies

Jurabrontes curtedulensis Marty et al., 2018.
Jurabrontes melphicticus Antonelli et al., 2023 (Fig. 7.5J)

3.2.47.3 Distribution

The type ichnospecies was only found from Upper Jurassic, Reuchenette Formation Formation of Northwestern Switzerland. Albian, Calcare di Bari Formation (Apulia) of Italy.

3.2.47.4 Holotype

SCR1500-T1-L8, Courtedoux tracksites, Canton Jura, NW Switzerland. The original specimen (track) was recovered on a limestone slab from the field and is now stored under the collection number MJSN-SCR011-553.

3.2.47.5 Diagnosis

Giant tridactyl tracks; longer than wide; Digits II–IV with a clear phalangeal pad 2-3-4 configuration; peculiar and isolated position of the proximal pad PIII1 of Digit III; broad subtriangular, pointed claw marks present on the tips of all three Digits II–IV; narrow and slightly asymmetrical interdigital divarication angles (II–III < III–IV); small anterior triangle and weak mesaxony; asymmetrical heel region; broad and massive digits with a blunt aspect; lack of a hallux impression.

The three trackways SCR1500-T1 and BSY1050-TR1 and TR2 have a regular configuration with similar left and right pace lengths. Pace angulation ranges between 151° and 172° and is correlated with a variation in estimated locomotion speed from 5.4 km/hour to 8.1 km/hour and in the [WAP (width of the angulation pattern)/PL (pes length)] ratio from 0.19 to 0.58. The [WAP/PL]-ratio characterizes all trackways as (very) narrow. In all three trackways, both left and right pes tracks are outwardly and inwardly rotated, whereas outwardly rotated tracks are overall about twice as frequent as inwardly rotated tracks. Mean track rotations are on the order of several degrees only (Marty et al., 2018).

3.2.47.6 General description

There are two valid ichnospecies of *Jurabrontes* (*J. curtedulensis* and *J. melphicticus*), but only one ichnospecies has been found in the Cretaceous (*J. melphicticus*).

J. melphicticus was discovered from Calcare di Bari Formation (Albian) in Apulia, Southern Italy. The features of *J. melphicticus* include medium to large-sized theropod tracks, with the length range from 28 to 34 cm, slightly longer than wide (aspect ratio: 1.03–1.16) and weakly mesaxonic. Typical 2-3-4 phalangeal pad configuration on robust Digits II, III, and IV, respectively, with the frequent occurrence of a further proximal (metatarsal-phalangeal) impression on Digit III. Short, subtriangular and blunt claw traces recognizable on each digit. Digit length increasing from Digit II to Digit IV; Digit II generally broader and more deeply impressed than Digits III and IV. Slightly asymmetrical interdigital angles (Digits III–IV > II–III), with divarication (II–IV) up to 74°. Proximal portion of Digit IV positioned backward with respect to Digit II, resulting in an asymmetrical and concave heel region. The pace angle is between 150° and 170°, and stride length ranges from 174 to 196 cm (Antonelli et al., 2023).

3.2.47.7 Discussion

Marty et al. (2018) erected and described a new ichnogenus and ichnospecies of a giant, robust theropod dinosaur, *J. curtedulensis* which is the type ichnospecies of *Jurabrontes*. It is characterized by massive and blunt digits, weak mesaxony, a clear 2-3-4 phalangeal pad formula, and a distinctive detached proximal pad of Digit III. The track BSY1050-TR2-R3, with a pace length of 78.5 cm, is one of the largest theropod tracks known.

J. melphicticus tracks show features comparable to the older tracks from other sites on the Apulia Carbonate Platform (Mattinata and Borgo Celano), but differs from them mainly in its larger size, the higher robustness, and lack of hallux impression. The large-sized, weakly mesaxonic, and robust theropod footprints only partially match *Jurabrontes* from the Upper Jurassic to Lower Cretaceous.

3.2.47.8 Trackmaker

Antonelli et al. (2023) suggested that the *J. melphicticus* track morphology and robustness indicate a megatheropod trackmaker and correlated it to a variable estimated speed ranging from 2.95 km/hour to 6.48 km/hour (using Alexander, 1976) or from 1.83 km/hour to 4.76 km/hour (using Weems, 2006)

3.2.48 Farlowichnus *Leonardi et al., 2024 (Figs. 7.7I and 7.30C)*

3.2.48.1 Type ichnospecies

Farlowichnus rapidus Leonardi et al., 2024.

3.2.48.2 Valid ichnospecies

Farlowichnus rapidus Leonardi et al., 2024.

3.2.48.3 Distribution

Neocomian, Botucatu Formation of Brazil.

3.2.48.4 Holotype

LPP–IC–0200, housed in the Laboratory of Paleoecology and Paleoichnology of the Federal University of São Carlos (UFSCar), São Paulo, Brazil.

3.2.48.5 Diagnosis

The trackway has a very high step angle, long paces and strides and a very narrow internal and external trackway width. The internal width values are always negative. The stride length/track length ratio is high. The tracks are almost parallel or only slightly divergent from the midline; in this last case, they show low positive and negative track rotations. No tail marks or drags are visible (Leonardi et al., 2024).

The footprints are tridactyl and strongly mesaxonic, with low digital divergence (inherently narrow divergence), and acute hypices. No hallux is visible. They are longer than they are wide. The posterior margin of the pes is U-shaped.The footprint contour presents a waterdrop-like appearance, especially when the outer digits are very close to the central digit. An evident heterodactyly is observed, with the Digit III being wide, long, and relevant, usually with a triangular outline, and outer digits (II and IV) proportionally very short.

The Digit II is longer (almost the double) than Digit IV. So that the digit-free lengths are III > II > IV. The claw marks of Digit III are often slightly turned inward. The third digit is always very wide at its base, and then goes progressively sharpens toward the tip. Often, however, it swells slightly soon after the hypices. The outer digits are more often triangular more rarely, rather elongated, with parallel margins, and always with an acuminate termination. The Digit II is narrow and deep, with a blade-like aspect in the holotype. The proximal notch of Digit II is rarely evident. The most striking feature of the outer digits is that the medial (or preaxial) digits are considerably more developed, that is, longer and more clearly imprinted, than the lateral (or postaxial) digits, both in the left and right foot.

The area of the anterior triangle (AT) is rather greater than the area of the backfoot triangle (BFT). These tracks are theropod footprints, sensu Masrour et al. (2023).

3.2.48.6 General description

Farlowichnus tracks are tridactyl, mesaxonic, and digitigrade. They are very small (<5 cm) and longer than wide. The Digit III impression is longer, wider, and larger than II and IV. Claw impressions are present in Digit III and lack pad traces. The "heel" area is U-shaped and presents an evident displacement rim.

3.2.48.7 Discussion

Farlowichnus is preserved in the foresets of dry dunal deposits of the Botucatu Formation. Leonardi et al. (2024) suggested that the particular shape of the Digit III, as well as the two external digits, depends on the anatomy of the trackmaker's autopod, and not on of the trackmaker's behavior or substrate properties. Moreover, the authors pointed out that the shape and structure of the posterior displacement rim represent the sand displaced by the animal's foot during its progress across the slopes of a dune. Nevertheless, the tracks do not preserve anatomical characters beyond impressions of digits and claws. We consider that these footprints have a low value within the scale for morphological preservation of Marchetti et al. (2019). So, future finds of better-preserved tracks are needed to consolidate the ichnotaxonomicstatus of this ichnotaxon.

3.2.48.8 Trackmaker

Leonardi et al. (2024) suggested that the tracks of *Farlowichnus rapidus* are assigned to small theropods adapted to desert life with a mainly cursorial gait, probably precursors of clades such as noasaurs and velocisaurs.

3.2.49 Fujianipus *Xing et al., 2024b (Fig. 7.8I)*

3.2.49.1 Type ichnospecies

Fujianipus yingliangi Xing et al., 2024b.

3.2.49.2 Valid ichnospecies

Fujianipus yingliangi Xing et al., 2024b.

3.2.49.3 Distribution

Cenomanian–Coniacian, Shaxian Formation of China.

3.2.49.4 Holotype

LXIU-T3, a trackway composed of five tracks, four of which are consecutive.

3.2.49.5 Diagnosis

Large functionally didactyl track, up to ~36.0 cm long with Digit III and IV traces straight and divergent. Proximal trace of Digit II is subrounded to elliptical with the proximal pad not clearly connected to the heel print. Diagnostically distinct from *Dromaeopodus*, the largest previously known deinonychosaurian track, which has curved, subparallel Digit III and IV traces and larger, more robust heel pad traces (Xing et al., 2024b).

3.2.49.6 General description

Fujianipus yingliangi was discovered from Shaxian Formation in Longyan City, Fujian, China. Only found in Cretaceous deposits. *F. yingliangi* belongs to the ichnofamily Deinonycosauripodidae (Xing et al., 2024b). There is only one valid ichnospecies of *Fujianipus*.

Didactyl pes tracks with slender digit print, the length is less than 36.4 cm, and the width is less than 16.9 cm. The formula of digital pad prints is x-1-3-4-x which are well preserved. Digit III is the longest which is 1.2 times longer than Digit IV. Digit II is an elliptic print. The distal end of Digit II is near the proximal end of Digit III and the proximal end of Digit II is not connected to the heel print. The heel of the pes is one-fourth of the whole print. The divarication angle between Digits III and IV is 27°.

3.2.49.7 Discussion

Li et al. (2008) proposed the ichnofamily Dromaeopodidae. This ichnofamily attributes to dromaeosaurs. But other deinonychosaurids tracks are needed be classified. Xing et al. (2024b) proposed the new superichnofamily Deinonychosauripodidae and ichnofamily Deinonycosauripodidae. They also amended Dromaeopodidae to the subichnofamily Dromaeopodinae which belong to Deinonychosauripodidae (Xing et al., 2024b).

Fujianipus yingliangi is the largest deinonychosaurid track. In addition, the larger ratio between the anatomically curved Digits III and IV, the more robust heel pad prints, and the subparallel Digit III and IV traces are among the differences of *F. yingliangi*.

3.2.49.8 Trackmaker

Xing et al., 2024b suggested that *Fujianipus yingliangi* can be associated with a giant troodontid whose hip height is 156–197 cm. The speed of *F. yingliangi* from the Longxiang site is 7.34 km/hour Xing et al. (2024b).

3.3 Avian theropods, saurischia

According to our survey, there are 17 Cretaceous avian ichnogenera, including 26 ichnospecies. There is only one avian ichnogenus in the non-Cretaceous, including one ichnospecies from the Cretaceous. Several ichnogenera such as *Archaeornithipus* (Fuentes Vidarte, 1996) and *Magnoavipes* (Lee, 1997), named to imply true avian affinity, are now considered to be traces of nonavian theropods (ornithomimiosaurid or struthiomimid; Gierliński and Lockley, 2013a; Lockley et al., 2021).

Ichnofamily Ignotornidae Lockley et al., 1992a (Fig. 7.9)

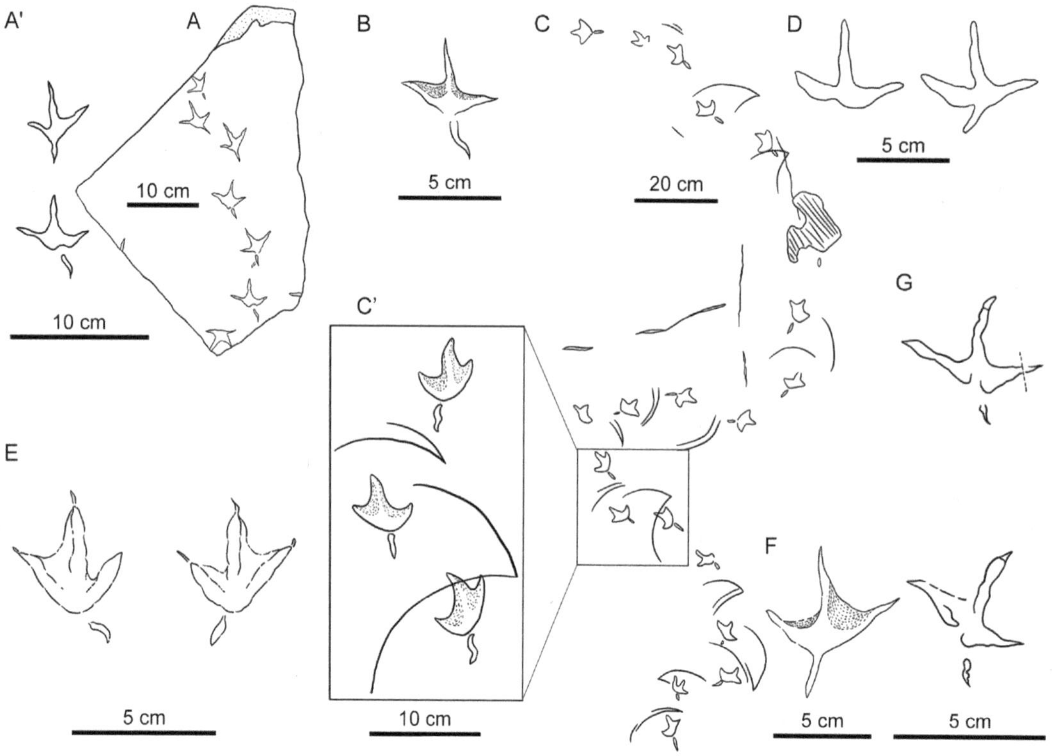

FIGURE 7.9 Tracks of Ignotornidae. (A and A′) Type series and enlarged tracks of *Ignotornis mcconnelli*; (B) holotype of *Ignotornis yangi*; (C and C′) type series and enlarged tracks of *Ignotornis gajinensis*; (D) holotype of *Ignotornis canadensis*; (E) holotype of *Ignotornis seoungjoseoi*; (F) holotype of *Hwangsanipes choughi*; (G) holotype of *Goseongornipes markjonesi*. *(A and A′) Redrawn from Lockley et al. (1992a). (B) Redrawn from Kim et al. (2006). (C and C′) Redrawn from Kim et al. (2012c). (D) Redrawn from Buckley et al. (2018). (E) Redrawn from Kang et al. (2021). (F) Redrawn from Kim et al. (2013). (G) Redrawn from Lockley et al. (2006a).*

Type Ichnogenus *Ignotornis* Mehl, 1931 emend. Lockley et al., 1992a.

Diagnosis.

Tetradactyl bird tracks showing prominent, posteromedially directed hallux impressions comprising about one-third of the total track length. Hypex between Digits III and IV is more anteriorly situated than hypex between Digits II and III, with a tendency for the development of an asymmetrical, semipalmated web that is more strongly developed between Digits III and IV. Digit divarication between II and IV, wide usually averaging at least 110°–120°. Digit pad impressions are variable, but sometimes show a 2-3-4 phalangeal formula corresponding to Digits II, III, and IV. Asymmetrical semipalmate webbing traces attached more mesially on digit impressions III–IV than on digit impressions II–IV. Step typically short with a strong tendency toward inward rotation (diagnosis emended by Kim et al., 2006).

Remarks.

Lockley et al. (1992a) established a new ichnofamily, Ignotornidae, but tis membership was subsequently adjusted (Kim et al., 2006; Xing et al., 2018g). Ignotornidae at least included

Ignotornis mcconnelli, *I. yangi*, *I. gajinensis*, *I. canadensis*, *I. seoungjoseoi*, *Hwangsanipes choughi*, *Goseongornipes markjonesi* and *Goseongornipes* isp. (Buckley et al., 2016; Kang et al., 2021). This ichnofamily has a board distribution. It was reported from Colorado, USA (Mehl, 1931; Lockley et al., 2009a), South Korea (Lockley and Harris, 2010; Kim et al., 2012c; Lockley et al., 2012c), Canada (Buckley et al., 2018) and China (Xing et al., 2018g).

3.3.1 Ignotornis *Mehl, 1931 (Figs. 7.9A–E and 7.34A)*

3.3.1.1 Type ichnospecies

Ignotornis mcconnelli Mehl, 1931 (Figs. 7.9A and 7.34A)

3.3.1.2 Valid ichnospecies

Ignotornis mcconnelli Mehl, 1931.
Ignotornis yangi Kim et al., 2006 (Fig. 7.9B)
Ignotornis gajinensis Kim et al., 2012c (Fig. 7.9C)
Ignotornis canadensis Buckley et al., 2018 (Fig. 7.9D)
Ignotornis seoungjoseoi Kang et al., 2021 (Fig. 7.9E)

3.3.1.3 Distribution

Aptian–Cenomanian; the upper part of the Dakota Group, Colorado; Albian–Cenomanian, Haman Formation, Albian, Jinju Formation of southern coast of Korea; lower Albian, Gates Formation, Canada; middle Berriasian–Albian Tugulu Group, Xinjiang, China.

3.3.1.4 Holotype

UCM-17,614, housed in Hunter Museum at the University of Colorado at Boulder, USA.

3.3.1.5 Diagnosis

Tetradactyl bird tracks with prominent, posteromedially directed hallux impressions comprising about one-third of total track length. Length with hallux is ~5.5 cm; width, ~4 cm. Hypex between Digits III and IV is more anteriorly situated than hypex between Digits II and III. Semipalmate web is slightly more developed between Digits III and IV than between II and III. For digit divarication between II and IV, width varies between 105° and 125°. Step is short with average inward rotation of ~15° (diagnosis emended by Kim et al., 2012c).

3.3.1.6 General description

Ignotornis mcconnelli is the first bird track ichnogenus and ichnospecies described from the Mesozoic. *I. mcconnelli* tracks are quite deep, suggesting progression over a soft substrate, and all four-digit impressions are clear (Lockley et al., 1992a). Lockley et al. (1992a) considered *Ignotornis* is similar to *Koreanaornis* (Kim, 1969) and *Jindongornipes* (Lockley et al., 1992a) in general morphology, differing mainly in size and the variability of preservation of details such as hallux and digital pad impressions.

Ignotornis yangi is based on 80 tracks from Korea National University of Education (KNUE) 040417 (Kim et al., 2006). Tetradactyl and anisodactyl bird tracks with posteromedially directed hallux impression and fine acuminate claw traces on all four digits, displaying semipalmated webbing between Digits II and IV. Web traces are asymmetrical. The average width of the tracks is approximately 4.5 cm, and the average length of the tracks, excluding the hallux, is 3.3 cm, and with hallux is 5.1 cm. The average angle of divarication between Digits II and IV is 123°. The pace length ranges from 7.5 to 11.5 cm, stride from 15.8 to 22.9 cm, pace width from 2.1 to 3.6 cm, and pace angulation from 139° to 161° in each trackway (Kim et al., 2006).

Ignotornis gajinensis is based on a trackway composed of eight consecutive tracks from the Haman Formation, Fossil Heritage Hall I, Gyeongsangnamdo Institute of Science Education (Kim et al., 2012c). *I. gajinensis* is a tetradactyl, semipalmate webbed track with length including prominent postero-directed hallux impressions about 6.8 cm and width about 5.5 cm. Divarication angle between Digits II and IV is about 130°; angles between Digits II and III and III and IV are approximately 70° and 60°, respectively, indicating slight asymmetry. Pace and stride length are highly variable, 9.6–11.2 cm and 17.4–34.2 cm, respectively. Pace angulations range from 135° to 160°, and external trackway width varies from 7.7 to 11.1 cm (Kim et al., 2012c).

Ignotornis canadensis is on an *ex situ* block from the Grande Cache Member of the Gates Formation, lower Albian in Alberta. Buckley et al. (2018) proposed a diagnosis: anisodactyl tetradactyl avian tracks of large size (tracks length 3.9–5.7 cm; length with hallux 6.4–7.2 mm), hallux impression posteomedially directed and larger than 25% of Digit III impression length. Digit II and IV impressions weakly curve toward Digit III impression. Claws short and acuminate. Asymmetrical semipalmate webbing, impressions with larger webbing impressions between Digits III–IV than between II–III. Regarding the relationship between tracks and sediment, Buckley et al. (2018) proposed that the apparent sinusoidal Digit III impressions, while common in Ignotornidae, are due to preservational variation rather than the morphology of Digit III.

Ignotornis seoungjoseoi is based on a set of standing tracks T0L and T0R in the Specimen CUE Bito_2013_WB001 from the Lower Cretaceous Jinju Formation, Sindong Group; Bito Island, southern coast of South Korea. Specimen locality on file at the Natural Heritage Center, National Research Institute of Cultural Heritage. Tetradactyl tracks average 3.4 cm in length without hallux impression (range 2.9–4 cm), track lengths including the hallux impressions average 4.7 cm (range 4.4–5 cm). Digit III impressions are the longest, with Digit II (average 2.7 cm) and IV (average 2.6 cm) approximately equal in length, track width average 4.2 cm (ranges from 3.5 to 5.4 cm). Digit I impressions vary in length average 1.5 cm (1.2–1.6 cm). Where preserved, Digit I impressions are greater than 42.2% of the total track length (without the hallux impression). Digit I impressions are posteromedially oriented, with Digit I–II divarication between 55.3° and 105°. Total divarication between Digits II–IV is average 99.3° (86.3°–112°), with the proximal edge of the track (where the digit impressions II and IV converge) appearing pointed rather than rounded. The webbing impressions are extensive but not to the extent of appearing as impressions of palmate webbing. Webbing impressions between Digits III–IV impressions extend more distally on the digit impressions than do the webbing impressions between Digits II–III. No visible webbing between Digits I and II. No trackways present (Kang et al., 2021).

3.3.1.7 Discussion

Ignotornis yangi is very similar to *I. mcconnelli* (Mehl, 1931), and *Hwangsanipes choughi* (Yang et al., 1995), with the main distinction being that the latter two are larger, with widths of 5 and 6.26 cm, respectively (Kim et al., 2006). When comparing *I. mcconnelli* with *I. yangi,* the webbing in *I. mcconneli* is only slightly developed in the proximal (posterior) part of the hypex between Digits II and III and slightly more pronounced in the hypex between Digits III and IV (Kim et al., 2006). In addition, *H. choughi* exhibits a well-developed web between Digits III and IV.

In general, *Ignotornis gajinensis* is very similar to *Hwangsanipes choughi* (Yang et al., 1995) and *Ignotornis mcconnelli* (Mehl, 1931), and differs from the former in that *I. gajinensis* has a smaller size and *H. choughi* has a larger divarication angle between Digits II and IV, and more developed web impressions. *I. gajinensis* is also smaller than *I. mcconnelli* in size, and slightly smaller divarication angle between Digits II and IV (Kim et al., 2012c). As the morphological features of the tracks are not prominent, Kim et al. (2012c) emphasized that the most striking feature of *I. gajinensisis* is the repeated association of arcuate, double-grooved impressions, which are thought to be attributed to sweeping the opened bill for catching invertebrates in the substrate under shallow water. This is the first unequivocal example of extensive bill traces reported from the Cretaceous (Kim et al., 2012c).

Ignotornis canadensis has been compared with *I. mcconnelli, I. yangi, I. gajinensis, Goseongornipes markjonesi, Hwangsanipes choughi, Jindongornipes kimi,* and *Gyeongsangornipes lockleyi* in terms of size, divarication between Digits I–II and III–IV, the degree of symmetry in webbing impressions, the posterior direction of the hallux impression, multivariate analyses, and other aspects. Overall, *I. canadensis* is larger than other reported Ignotornidae, yet it possesses digit divarication similar to that of *Goseongornipes* and *Hwangsanipes* (Buckley et al., 2018).

Ignotornis seoungjoseoi is most similar to *I. gajinensis,* with the main differences focusing on size and webbing impressions, such as the track width of *I. seoungjoseoi* being 59.2% of *I. gajinensis.* Kang et al. (2021) also discussed the relationship between webbing impressions and sediment, the expression of their morphology in track records, and the paleogeographic factors influencing bird tracks in East Asia and North America. During the Early Cretaceous, Laurasia remained sufficiently connected, providing opportunities for volant shorebirds and wading birds (trackmakers of Avipedidae, Koreanaornipodidae, Ignotornidae, Limiavipedidae, and Paxavipedidae) to access most parts of the continent. Due to the absence of significant barriers hindering the spread of flying birds across Laurasia during the Early Cretaceous, numerous bird tracks were found in both East Asia and North America dating to that time period (Kang et al., 2021).

Huangyangquan morphotype A specimens described by Xing et al. (2021j) clearly show all the characteristics of *Ignotornis* (Mehl, 1931; Lockley et al., 2009a). Huangyangquan *Ignotornis* isp, together with *Aquatilavipes dodsoni* (Xing et al., 2011b, 2021j), constitute evidence of shorebirds from the easternmost edge of Asia.

3.3.1.8 Trackmaker

Trackmakers of *Ignotornis mcconnelli* are interpreted as plover-like birds (Charadriiformes) because of their small size and high density of preserved tracks (Lockley et al., 1992a). The webbing of *Ignotornis, Hwangsanipes,* and *Sarjeantopodus* is more pronounced than in most sandpipers and plovers, though it is convergent with patterns found in semipalmated plovers (Kim et al., 2006).

Arcuate, parallel, double grooves frequently occur with tracks of *Ignotornis gajinensis*. They are attributed to sweeping movements from side to side of the opened bill (Kim et al., 2012c). Such distinctive feeding traces are highly convergent with those produced by modern spoonbills (Swennen and Yu, 2005; Lockley and Harris, 2010; Kim et al., 2012c).

Overall, the trackmakers of *Ignotornis* may have a close relationship with Aequorlitornithes. Aequorlitornithes is a clade of waterbirds and comprises the clades Charadriiformes (waders and shorebirds), Mirandornithes (flamingos and grebes), and Phaethoquornithes (Eurypygimorphae and Aequornithes) (Prum et al., 2015).

3.3.2 Hwangsanipes *Yang et al., 1995 (Fig. 7.9F)*

3.3.2.1 Type ichnospecies

Hwangsanipes choughi Yang et al., 1995.

3.3.2.2 Valid ichnospecies

Hwangsanipes choughi Yang et al., 1995.

3.3.2.3 Distribution

Campanian, Uhangri Formation, Hwangsan Basin, southwestern South Korea.

3.3.2.4 Holotype

KPE 50,101, Kyungpook National University, Earth Science Education Department, Daegu, South Korea.

3.3.2.5 Diagnosis

Tetradactyl bird tracks with prominent, posteromedially directed hallux impressions comprising about one-third of the total track length. Length with hallux is ~7.5–8.0 cm, without hallux average is 4.9 cm; width is ~6–6.5 cm. Hypex between Digits III and IV is more anteriorly situated than hypex between Digits II and III. Semipalmate web is more strongly developed between Digits III and IV than between II and III. Digit divarication between Digits II and IV averages at least 110°–120°. Step is short, with a strong tendency toward inward rotation (diagnosis emended by Lockley et al., 2012c)

3.3.2.6 General description

Hwangsanipes choughi is based on a single trackway consisting of four consecutive, tetradactyl bird tracks showing full webbing between Digits II and IV. Digits II and IV are curved slightly inward. The trackway displays a high-pace angulation value. The average step and stride lengths of this trackway are 5.1 and 10.8 cm, respectively. As with *Uhangrichnus chuni*, the web impressions display a concave anterior margin (Yang et al., 1995).

3.3.2.7 Discussion

Yang et al. (1995) considered that *H. choughi* is clearly distinguishable from *U. chuni* in its larger size, the wider average divarication between the Digits II–IV, and the presence of distinct hallux impressions.

Lockley et al. (2012c) suggested that *Hwangsanipes* has partial web traces indicating a semi-palmate trackmaker somewhat similar to *Ignotornis*. Their common feature is the web trace between Digits III and IV, which is more developed than between digit traces II and III. However, *Hwangsanipes* is larger with more developed web traces than *Ignotornis* (Kim et al., 2006).

3.3.2.8 Trackmaker

The slab specimen on which the *H. choughi* trackway is found, also contains abundant *U. chuni* tracks and a partial track of a dinosaur. Yang et al. (1995) considered the trackmaker of *H. choughi* may be a larger, less common shorebird or wading bird that used larger foraging areas.

3.3.3 Goseongornipes *Lockley et al., 2006a (Fig. 7.9G)*

3.3.3.1 Type ichnospecies

Goseongornipes markjonesi Lockley et al., 2006a.

3.3.3.2 Valid ichnospecies

Goseongornipes markjonesi Lockley et al., 2006a.

3.3.3.3 Distribution

Cenomanian–Santonian, Jindong Formation, Dukmeongri section 1 (bed 27), South Korea; Albian–Cenomanian, Tianjialou/Mengtuan formations, Dasheng Group, China.

3.3.3.4 Holotype

KPE 50,010, Kyungpook National University, Earth Science Education Department, Daegu, South Korea.

3.3.3.5 Diagnosis

Slightly semipalmate tridactyl bird track with three widely splayed digits (II–IV) and a small posteriorly directed hallux. The length with the hallux between 4.1 and 4.5 cm, without hallux 3.0–3.5 cm, and 4.2–4.5 cm in width. The length of Digit II is shorter than IV. Hallux length between 22% and 25% of foot length. The wide digit divarication angle between Digits II and IV is 140°–150°, and divarication between Digits II and III and between III and IV sub-equal (about 70°). Angle between hallux and Digit II about 75°. Web traces subtle, confined to proximal hypex between Digits III and IV (diagnosis emended from Kim et al., 2012c).

3.3.3.6 General description

Goseongornipes also occurs in China, which is the first occurrence of Ignotornidae from the Lower Cretaceous China (Xing et al., 2018g). Small to medium-sized, subsymmetrical, functionally tetradactyl bird tracks with three widely splayed digits (II–IV) and a posteriorly directed hallux. The average length/width ratio of these tracks is 0.8 (range 0.5–1.0 cm). Average maximum length including hallux impressions is 3.8 cm, and the value not including hallux impressions is 2.8 cm; average maximum width is 3.5 cm. Average divarication angle between Digits II and IV is 124°. The average pace angulation is 157°. Pace length is about half of the stride length. Tracks in the trackway exhibit slight positive (inward) rotation. Morphological and multivariate statistical analyses demonstrate that these functionally tetradactyl, semipalmate bird tracks are most similar to *Goseongornipes* isp. (Xing et al., 2018g).

3.3.3.7 Discussion

One of the reasons Lockley et al. (2006a) established *Goseongornipes* was the realization that there is a significant size gap between the small size of *Koreanaornis hamanensis* and the large size of *Jindongornipes kimi.*

As a typical avian ichnogenus in Korea, *Goseongornipes* appears in a variety of assemblages. Lockley et al. (2006a) believe that, at least in Jindong Formation, *Koreanaornis hamanensis, Goseongornipes markjonesi,* and *Jindongornipes kimi* are clearly different and distinctive ichnospecies. The widths of *Koreanaornis, Goseongornipes,* and *Jindongornipes* are in the range of about 2.5–3.0 cm, 4.0–4.5 cm. and 6.5–7.5 cm respectively. They also show progressively larger (longer) hallux impressions corresponding to this size increase. Compared with *Jindongornipes, Goseongornipes* is always much larger with a longer hallux that is directed even more strongly toward the posterior parasaggital plane (trackway axis) (Lockley et al., 2006a) Two new first-order units of ichnofamilies were also established based on these three ichnogenera, becoming Koreanaornipodidae, Ignotornidae, and Jindongornipodidae (also see Kim et al., 2006).

Kim et al. (2012a) established a morphological series based on the order of increasing size and degree of development of the semipalmated web between Digits III and IV (Kim et al., 2006; Lockley, 2007b): *Goseongornipes- Ignotornis- Hwangsanipes* in the ichnofamily Ignotornidae. The length with hallux of this series is in the range of about 4.2 4.5 cm, ~5.5 cm, and ~7.5–8.0 cm (Kim et al., 2012a).

Goseongornipes is only about half the size of *Hwangsanipes,* and *Ignotornis* is intermediate in size. All these ichnogenera have a well-developed hallux, directed posteromedially in line with the anterolateral orientation of Digit IV (Kim et al., 2012c). The hallux trace of *Goseongornipes* is less developed and more frequently separated from the traces of Digits II, III, and IV in the smaller ichnogenus. Also, in all three ichnogenera the divarication angle between Digits II and III are more (on average, 5°–15°) than the angles between Digits III and IV (Kim et al., 2012c).

Kim et al. (2006) compared the Changseon *Ignotornis yangi* with *Goseongornipes markjonesi.* The two ichnospecies are similar in size, but *G. markjonesi* has a much shorter hallux, consisting of less than one quarter of track length, and little or no evidence of webbing.

3.3.3.8 Trackmaker

The posteromedially hallux, semipalmate, web trace is found between Digits III and IV only, as is the most common feature among present-day semipalmate shorebirds (Lockley et al., 2006a). *Goseongornipes* is typical of shorebird-like trackways.

Ichnofamily Avipedidae Sarjeant and Langston, 1994 (Fig. 7.10)

Type Ichnogenus *Avipeda* Vialov, 1965.

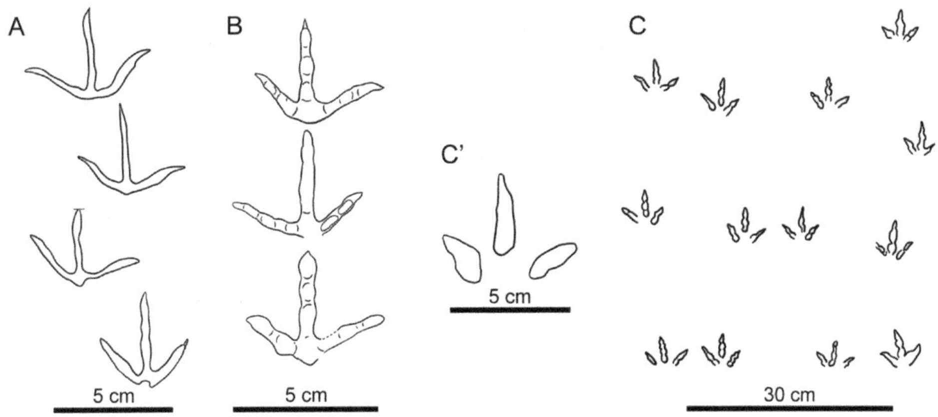

FIGURE 7.10 Tracks of Avipedidae. (A) Holotype (the second track from the front) within the referred trackway of *Aquatilavipes swiboldae*; (B) holotype (the second track from the top) and referred tracks of *Aquatilavipes izumiensis*; (C) holotype (C) and referred series (C′) of *Aquatilavipes dodsoni. (A) Redrawn from Currie (1981). (B) Redrawn from Azuma et al. (2002). (C) Redrawn from Xing et al. (2011b, 2021j).*

Diagnosis. Avian tracks showing three digits, all directed forward. Digits united or separate proximally. Webbing lacking or limited to the most proximal part of the interdigital angles (original diagnosis by Sarjeant and Lagnston 1994).

3.3.4 Aquatilavipes *Currie, 1981 (Fig. 7.10)*

3.3.4.1 Type ichnospecies

Aquatilavipes swiboldae Currie, 1981 (Fig. 7.10A)

3.3.4.2 Valid ichnospecies in Cretaceous

Aquatilavipes swiboldae Currie, 1981.
Aquatilavipes izumiensis Azuma et al., 2002 (Fig. 7.10B)
Aquatilavipes dodsoni (Xing et al., 2011b) Xing et al., 2021j (Fig. 7.10C)

3.3.4.3 Distribution

Aptian, Gething Formation, northeastern British Columbia; Berriasian–Valanginian, Tetori Group, Central Japan; middle Berriasian–Albian Tugulu Group, Xinjiang, China; ?Aptian to ?Barremian, Lakota Group of South Dakota; Aptian, Poison Strip Sandstone Member Cedar Mountain Formation, Utah; unknown stage Upper Cretaceous, South Korea.

3.3.4.4 Holotype

Track No.76 on counterpart blocks. Positive impression BCPM 744, British Columbia Provincial Museum. Negative PMA P79.23.37, Provincial Museum of Alberta.

3.3.4.5 Diagnosis

Made by a bipedal animal with three slender functional pedal digits. Width greater than length; average divarication of Digits II and IV greater than 100°. Digit IV longer than Digit II and shorter than Digit III. Sharp claw impression on each digit. No hallux impression. Faint digital pad impressions numbering at least 3–4 in Digit III under suitable substrate conditions (diagnosis emended by Lockley et al., 1992a)

3.3.4.6 General description

The holotype is 3.8 cm long and 4.7 cm wide. Digit II is 2.2. cm long and 0.3 cm at its widest point. The length of Digit III is 3.3 cm and the greatest width is 0.4 cm. The Digit IV is 2.4 cm long and 0.4 cm wide. Divarication was measured from the point where the longitudinal axis of Digit III intersects the posterior outline of the track and the digit tips. The total divarication is 118°, which is the sum of 48° between Digits II and III, and 70° between III and IV. The outer digits of the animal that made the track curve so that the distal ends point away from the central axis of the foot. The deepest impression of each toe is at the base, and all digits end in sharp claws. The claw of Digit III is almost 0.4 cm long (Currie, 1981).

Aquatilavipes izumiensis from the Tetori Group, Central Japan is tridactyl and small in size, and each digit is slender with digital pad impressions. The holotype was made by a left foot and is 3.9 cm long by 4.5 cm wide. Digit II is 1.8 cm long and 0.4 cm at its widest point. The Digit III is 3.1 cm in length and 0.5 cm at its widest point. Digit IV is 2.5 cm long and 0.5 mm wide. The divarication between the Digits II and IV is 111°, between Digits II and III is 45°, and between III and IV is 65°. Digit II curves inward anterolaterally, but Digit IV curves outward laterally. Small triangular claw impressions are preserved on the ends of all digits (Azuma et al., 2002).

Aquatilavipes dodsoni is from Tugulu Group, Xinjiang, China. We provide a new diagnosis here: tridactyl tracks lack hallux impressions. average length is 5.0 cm, length/width ratio is 0.9, and mesaxony is 0.42. The three digits are separated in most cases. Digit III is the longest digit, Digits II and IV are subequal in length. Digit II is broader than Digits III and IV. Digits II and III have two and three pads, respectively, and the pads of Digit IV are unidentifiable. The divarication angles between Digits II and III are smaller than those between Digits III and IV. The average divarication angle between Digits II and IV is 92° (range 89°–99°). The pace angulation is 160°. Tracks in the trackway exhibit slight positive (inward) rotation.

3.3.4.7 Discussion

Azuma et al. (2002) took length-width ratios as one of the main characteristics and compared *Aquatilavipes izumiensis* with *A. swiboldae*, *Koreanaornis*, *Hwangsanipes* and other ichnotaxa. *A. izumiensis* is also clearly distinguishable from *A. curriei* by smaller size, narrower average divarication between Digits II and IV, and the absence of a groove on the center of each digit (Azuma et al., 2002).

Zonneveld et al. (2011) elaborated on the history of *Aquatilavipes* and compared and discussed it with *Koreanaornis*. Because *A. wallacei* is from the Oligocene, this article will not discuss it in detail.

Xing et al. (2011b) named avian tracks as *Koreanaornis dodsoni* at the Huangyangquan site Xinjiang, China. Lockley et al. (2013) suggested that *K. dodsoni* is more likely to belong to *Aquatilavipes*. In multivariate analyses, *K. dodsonii* shows no overlap in morphospace with *K. hamanensis* (Buckley et al., 2016). However, Xing et al. (2021j) supported the view after detailed discussion based on more specimens from the same location. The *Aquatilavipes dodsoni* holotype (length 4.4 cm, width 5.7 cm) is more typical of *Aquatilavipes*. Features of

Aquatilavipes dodsoni that distinguish it from *Koreanaornis hamanensis* include smaller divarication angles between Digits II and IV, greater overall track length and width, and persistent absence of Digit I (Xing et al., 2021j).

Xing et al. (2011b) also classified avian theropod tracks as *Moguiornipes robusta* at the Huangyangquan site and suggested that *Moguiornipes* tracks may represent bird with lobed feet. However, such wide digit traces could also be the product of the widening of digit traces by dissolution of traces that were originally narrower (Xing et al., 2021j). Also, there are no statistical difference between *Moguiornipes robusta* and *Aquatilavipes dodsoni* base on the multivariate analysis (Buckley et al., 2016). Therefore, we regard it as a nomen dubium and an extramorphological state of *A. dodsoni.*

Aquatilavipes is also known from the basal Cretaceous Tuchengzi Formation (Xing et al., 2020a); purportedly Aptian to possibly Barremian aged Lakota Group of South Dakota (Lockley et al., 2001a); Poison Strip Sandstone Member (Aptian) of the Cedar Mountain Formation, Utah (Lockley et al., 2015b); the Upper Cretaceous (?77 Ma to 65 Ma) Yeosu Islands, South Korea (Huh et al., 2012).

3.3.4.8 Trackmaker

Aquatilavipes, a track type morphologically similar to those of modern shorebirds, occurs in lacustrine facies and coastal plain deposits. Lockley et al. (2015b) pointed out that, at least in North America, *Aquatilavipes* may be relatively common in the middle of the Early Cretaceous (Barremian to Aptian), based on British Columbia, South Dakota, and Utah sites.

Ichnofamily Jindongornipodidae Lockley et al., 2006a (Fig. 7.11)

Type ichnogenus *Jindongornipes* Lockley et al., 1992a.

Diagnosis. Medium-sized (6.5–7.5 cm wide) tetradactyl tracks with slender digit impressions, and very wide (125°–160°) divarication between Digits II and IV. Digit II shorter than IV. Moderately long hallux usually impressed and directed posteriorly about 180° away from Digit III. Digital pad impressions (excluding sharp claw traces) mostly indistinct but with the probable 2-3-4 phalangeal formula corresponding to Digits II, III, and IV. Trackway shows little rotation, that is, Digit III directed anteriorly parallel to trackway axis (original diagnosis from Lockley et al., 2006a).

3.3.5 Jindongornipes *Lockley et al., 1992a (Fig. 7.11A)*

3.3.5.1 Type ichnospecies

Jindongornipes kimi Lockley et al., 1992a.

3.3.5.2 Valid ichnospecies

Jindongornipes kimi Lockley et al., 1992a.

3.3.5.3 Distribution

Cenomanian–Santonian, Jindong Formation, Kyeongsangnam Province, South Korea.
Albian, Jinju Formation, coast of South Korea.

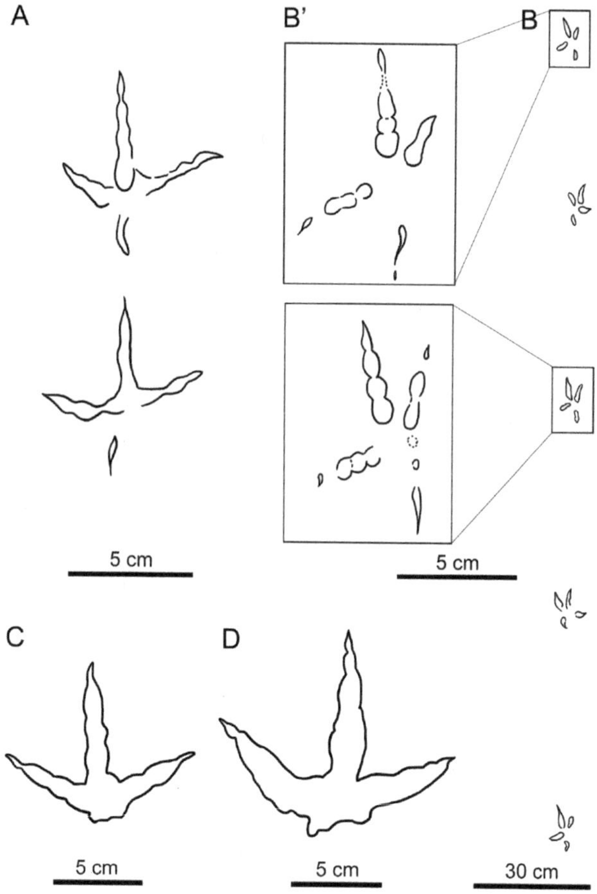

FIGURE 7.11 Tracks of Jindongonipodidae, Shandongornipedidae and Limiavipedidae. (A) Partial holotype tracks of *Jindongornipes kimi*; (B and B′) type series and enlarged tracks of *Shandongornipes muxiai*; (C) holotype of *Limiavipes curriei*; (D) individual track from holotype slab of *Wupus agilis*. *(A) Redrawn from Lockley et al. (1992a). (B and B′) Redrawn from Li et al. (2005). (C) Redrawn from Herrero et al. (2023). (D) Redrawn from Herrero et al. (2023).*

3.3.5.4 Holotype

KPE 50,006, Kyungpook National University, Earth Science Education Department, Daegu, South Korea.

3.3.5.5 Diagnosis

Moderately large tetradactyl bird tracks. Track length, with hallux, up to 8 cm. Hallux impression is well-developed, often bent suggesting two distal phalanges in contact with the substrate. Up to four phalangeal pad impressions in Digit III trace. Number of pad impressions in Digits II and IV unknown, but at least three. Divarication angle between Digits II and IV very pronounced, ranging 125°–150°. Digit IV impression often wide proximally with a posteriorly directed heel that gives track a pronounced asymmetry. Track width ranging between 6.5–7.5 cm. Trackway variable but with only slight positive rotation and pace angulation about 140° (original diagnosis from Lockley et al., 1992a).

3.3.5.6 General description

The characteristics of *Jindongornipes kimi* are special enough but there are not many specimens and descriptions. Some *Jindongornipes* isp. also occur in other locations, such as specimens from the Lower Cretaceous Jinju Formation, Bito Island, Sacheon City, southern coast of South Korea (Kang et al., 2021).

The Bito specimens are described by Kang et al. (2021) and provide the following morphological characteristics: natural mold, functionally tetradactyl tracks; tracks are an average 4.6 cm in length without the hallux impression; seven tracks have hallux impressions, with an average track length including hallux impression of 6.3 cm; hallux impression is on average 59.3% the length of the Digit III; the length of the Digit I impression is highly variable (0.44%–0.97% length of Digit III); hallux impression oriented posteriorly to posteromedially from the long axis of the Digit III impression; small webbing impressions are inconsistently preserved between digit impressions III–IV; digital pads present but inconsistently preserved.

3.3.5.7 Discussion

Jindongornipes kimi and *Koreanornis hamanensis* were originally included in the ichnofamily Ignotornidae (Lockley et al., 1992a). Kim et al. (2006) amended the Ignotornidae, providing a definition that restricts the concept to bird tracks that have a slight proximal web and a well-developed hallux that is directed posteromedially. Lockley et al. (2006a) removed *J. kimi* and *K. hamanensis* from the Ignotornidae and assigned them to their new ichnofamilies, Koreanaornipodidae and Jindongornipodidae.

3.3.5.8 Trackmaker

Jindongornipes kimi is the largest bird track reported from the Lower Cretaceous. Examples are typically twice the size of *Koreanaornis hamanensis* from the Jindong Formation (Lockley et al., 1992a, 2006a). *Jindongornipes* represent birds with a well-developed hallux, while *Koreanaornis* has a small hallux (Lockley et al., 1992a). They usually occur separately, which may be related to the combined effects of size, water depth, and feeding habits. Lockley et al. (1992a) suggested that the trackmaker of *Koreanaornis* frequented the shoreline whereas the *Jindongornipes* trackmaker waded in the shallows.

Ichnofamily Koreanaornipodidae (=Koreanornipodidae, misspelled) Lockley et al., 2006a (Fig. 7.12)

Type Ichnogenus *Koreanaornis* Kim, 1969.

Diagnosis. Small, wide (typically 2.5–3.0 cm) subsymmetric functionally tridactyl tracks with slender digit impressions, and wide (90°–115°) divarication between Digits II and IV. Small halluxs are occasionally present, and posteromedially directed about 180° away from Digit IV. Digital pad impressions (excluding sharp claw traces) show a 2-3-4 phalangeal formula corresponding to Digits II, III, and IV. Trackway shows positive (inward rotation) (original diagnosis from Lockley et al., 2006a).

3.3.6 Koreanaornis *Kim, 1969 (Fig. 7.12A–C)*

3.3.6.1 Type ichnospecies

Koreanaornis hamanensis Kim, 1969 (Fig. 7.12A)

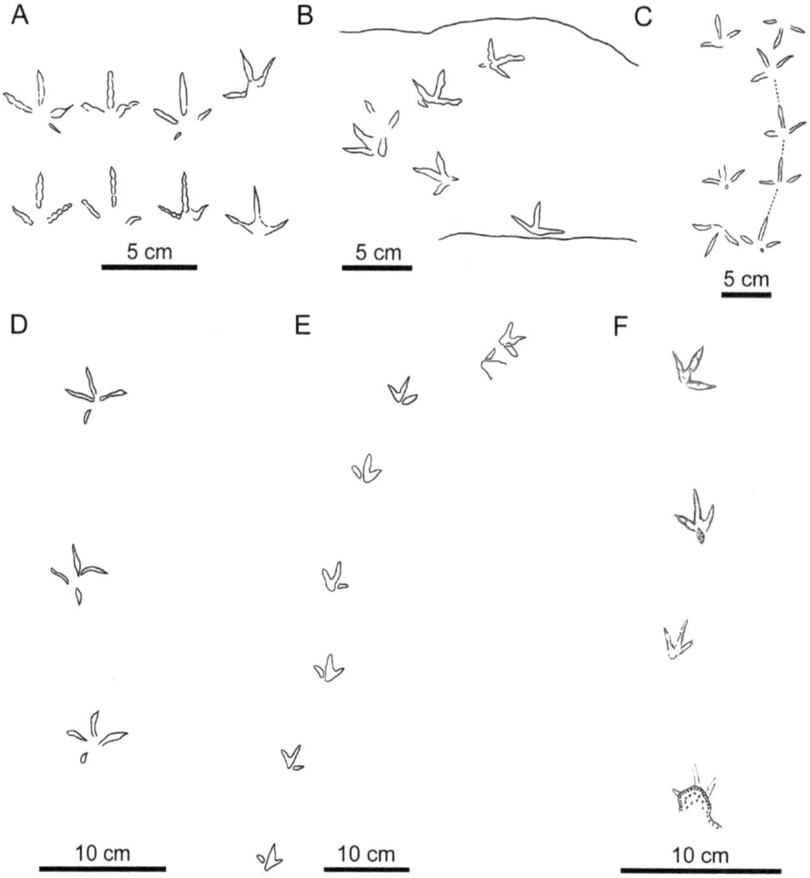

FIGURE 7.12 Tracks of Koreanaornipodidae and Paxavipedidae. (A) Tracks from the holotype slab of *Koreanaornis hamanensis*; (B) holotype (the first track from the top) and referred series of *Koreanaornis anhuiensis*; (C) type series of *Koreanaornis lii*; (D) partial type series of *Pullornipes aureus*; (E) referred series of *Barrosopus slobodai*; (F) holotype (the first track from the top) and type series of *Paxavipes babcockensis. (A) Redrawn from Lockley et al. (1992a). (B) Redrawn from Xing et al. (2018h). (C) Redrawn from Xing et al. (2016e). (D) Redrawn from Lockley et al. (2006c). (E) In MCF PHPV-SB 337, redrawn from Coria et al. (2002). (F) Redrawn from McCrea et al. (2015a).*

3.3.6.2 Valid ichnospecies

Koreanaornis hamanensis Kim, 1969.
Koreanaornis anhuiensis (Jin and Yan, 1994); Lockley et al., 2013 (Fig. 7.12B)
Koreanaornis lii Xing et al., 2016e (Fig. 7.12C)

3.3.6.3 Distribution

Cenozoic: Early Miocene, Gendulain Formation, Navarre, Spain.

Cretaceous: Albian–Cenomanian, Haman Formation, southern coast of Korea; Hauterivian–Aptian, Yanguoxia Formation of Hekou Group, Gansu, China; Albian, Qiuzhuang Formation, Anhui, China; Barremian/Aptian–Santonian, Jiaguan Formation,

Chongqing, China; Albian–Cenomanian Dakota Formation in northeastern Utah, USA; Cenomanian, Zebbag Formation in Tunisia.

3.3.6.4 Holotype

KPE 50,001 and DGSU (Department of Geology, Seoul University) unnumbered specimen, South Korea.

3.3.6.5 Diagnosis

Small, functionally tridactyl bird tracks, with slender and mostly separate digit traces, sometimes ending in sharp claw traces. Tracks wider than long, with widths ranging from 2.5 to 4.4 cm. Faint hallux traces are rarely preserved. Digital pad impressions 2-3-4 correspond to Digits II, III, and IVoccur in well-preserved tracks. Digit divarication averages 120°. Trackways show inward rotation of Digit III (diagnosis emended by Kim et al., 2012c).

3.3.6.6 General description

Xing et al. (2018h) described *K. anhuiensis* in detail. *K. anhuiensis* from the Lower Cretaceous Qiuzhuang Formation, Mingguang tracksite, Anhui, China. Small (track length ~3.3 cm, width ~3.6 cm), subsymmetrical, functionally tridactyl tracks distinguished by the following combination of characters: a consistently wide average Digit II–IV divarication; consistently preserved, the midline of small hallux impression and Digit III trace are aligned, and more than one third of the length of Digit III trace; Digit divarication I–II 101°–105°; divarication II–IV 112°–138° (Xing et al., 2018h).

Koreanaornis lii from the Hekou Group, Gansu, China (Xing et al., 2016e) has the following characteristics: small-sized (~3.2 cm in length, ~4.6 cm in width), subsymmetrical, functionally tridactyl tracks with a consistently wide average Digit II–IV divarication; lateral digit traces form a posterior margin (when preserved) that ranges from an obtuse angle to nearly perpendicular to the long axis of the Digit III impression; strong curvature of medial and lateral digit impressions away from Digit III impression; inconsistently preserved, small posteromedially directed hallux impression less than one third of the length of Digit III trace; digit divarication I–II 134°–160°, divarication II–IV is 78°–177°; tracks in trackway exhibit slight inward rotation.

Koreanaornis cf. *hamanensis* reported from Utah (Anfinson et al., 2009) displays medial and lateral digit traces (Digits II and IV) that curve strongly away from the Digit III trace, as they do in *Koreanaornis lii*. However, none of the tracks from Utah possess an impressed Digit I, so it is difficult to determine whether the Utah tracks are more similar to *K. hamanensis* or *K. lii* (Xing et al., 2016e).

3.3.6.7 Discussion

Koreanaornis (Kim, 1969) was the second named Mesozoic bird track, which opened the prelude to the discovery of Cretaceous avian ichnology from the southern Korean peninsula.

Koreanaornis is similar to *Aquatilavipes* (Currie, 1981). It is important to note the size distinction between small type *Koreanaornis* (width ranges 2.5–4.4 cm) and larger type *Aquatilavipes* (4–5 cm wide and long) because among modern shorebirds such size differences clearly represent different trackmaking species, even though size is not a criterion for differentiation of otherwise identical ichnites/ichnotaxa (Xing et al., 2021j) Moreover, multivariate analysis clearly separates Koreanaornipodidae (mostly *Koreanaornis*) and Avipedidae (including *Aquatilavipes*) (Buckley et al., 2016).

Tracks of *Koreanaornis lii* are morphologically most similar to those of Koreanaornipodidae given their slender digit traces, small, posteromedially directed Digit I impression, and wide divarication. Given their close morphologic similarity to *K. hamanensis*, such as slender digit impressions, wide divarication, small but inconsistently preserved hallux impression, Xing et al. (2016e) concurred that these tracks are an ichnospecies of *Koreanaornis*. The strong outward curvature of digit traces II and IV away from the Digit III trace is unique to this track morphology, warranting the uniqueness of *Koreananornis lii* (Xing et al., 2016e).

Koreanaornis is too often used as a catch-all ichnotaxon (Anfinson et al., 2009). Jin and Yan (1994) named "Aquatilavipes anhuiensis" based on a single trackway. Lockley et al. (2013) reclassified "*A. anhuiensis*" as *Koreanaornis anhuiensis. K. anhuiensis* shares many features with *K. hamanensis*, and *K. lii*. However, significant differences exist. *K. hamanensis*, and *K. lii* have digit traces that do not join proximally on most tracks, but digit traces of *K. anhuiensis* connect proximally to a metatarsophalangeal pad (Xing et al., 2018h).

Zhen et al. (1996) reported a Cretaceous ichnofauna from the Emei region of Sichuan, China that included one new avian ichnospecies: *Aquatilavipes sinensis. A. sinensis* is more similar to type *Koreanaornis* than to type *Aquatilavipes*, and therefore has been transferred to *Koreanaornis sinensis* (Lockley et al., 2008a, 2012c, 2013). However, *Koreanaornis sinensis* is based on only two not well-preserved tracks, and lacks trackway information and information on intraspecific morphological changes. The morphological characters of *K. sinensis* are likely to be only within the general range of variation in known *Koreanaornis* samples and are likely to have no practical ichnotaxonomic utility. A better approach would be to assign *K. sinensis* to the *Koreanaornis* isp. track rather than increasing ichnospecies level complexity.

Koreanaornis is well known from tracksites in South Korea. It was first discovered and named by Kim (1969) on the basis of abundant tracks from the Haman Formation (Albian–Cenomanian). Subsequently, the ichnotaxon was reported from the Jindong Formation (Cenomanian–Santonian) of Goseong County (Lockley et al., 1992a) and other Korean localities (Kim, 2008). In addition to the records from Sichuan, Anhui and Gansu described above, *Koreanaornis* was also described from the Early Cretaceous Tianjialou Formation (Albian–Cenomanian) Jishan tracksite (Xing et al., 2019a) and Houzuoshan Dinosaur Park (Li et al., 2015) in Shandong, Tibet (Xing et al., 2017c), and Hebei (Xing et al., 2020a). Among them, the Hebei specimens are from the basal Cretaceous Tuchengzi Formation and probably represents the oldest record of *Koreanaornis* (Xing et al., 2020a). In general, *Koreanaornis* was mainly widely distributed in northern, southwestern and central China in the late Early Cretaceous in addition to being abundant in Korea. Records outside Asia mainly included Utah *Koreanaornis* cf. *hamanensis* from the Dakota Formation (Albian-Cenomanian) (Anfinson et al., 2009). Based on the present findings, *Koreanaornis* is found in the late Early Cretaceous (especially Albian) of East Asia, mainly Korea (Lockley et al., 2012c) and China, as well as in deposits of broadly similar age in North America (Anfinson et al., al., 2009).

A series of Cenozoic small shorebird-like tracks have also been assigned to *Koreanaornis*, such as lower Miocene specimens from Cenicero, La Rioja, Spain (Díaz-Martínez et al., 2015b), and the late Miocene–Pliocene Agha Jari Formation specimens from the Persian Gulf region, Iran (Abbassi and Dashtban, 2021). These are beyond the scope of this article.

3.3.6.8 Trackmaker

Kim (1969) suggested that the tracks resemble those of modern plovers such as the Long-billed Plover (*Charadrius placidus*), the Little-ringed Plover (*C. bulius*) and the Kentish Plover (*C. alexandrinus*). This is a convergence in foot morphology and has nothing to do with the taxonomic relationships of the trackmaker. Huh et al. (2012) discovered that the cooccurrence of *Koreanaornis* with *Cochlichnus* (multiple nematode trails) suggests a feeding association.

3.3.7 Pullornipes *Lockley et al., 2006c (Fig. 7.12D)*

3.3.7.1 Type ichnospecies

Pullornipes aureus Lockley et al., 2006c.

3.3.7.2 Valid ichnospecies

Pullornipes aureus Lockley et al., 2006c.

3.3.7.3 Distribution

Kimmeridgian–Valanginian, Tuchengzi Formation, Kangjiatun, Liaoning, China.

3.3.7.4 Holotype

A replica of the holotype is CU 212.21 and 212.22, "plastotype" of four consecutive tracks (1–4) and a further seven consecutive tracks (30–36) from the same trackway.

3.3.7.5 Diagnosis

Small, subsymmetrical functionally tridactyl track with wide digit divarication and small posteromedially directed hallux. Trackway narrow with only slight positive (inward) rotation (original diagnosis from Lockley et al., 2006c).

3.3.7.6 General description

Tetradactyl tracks that are slightly wider than long (average 4.1 cm in length and 4.4 cm in width). Average digit divarication between II and III is 53° (range 28°–90°); between III and IV is 61° (range 42°–90°); between II and IV is 115 degrees (range 88°–141°). The hallux is short and directed medially (i.e., toward the trackway mid-line) more or less at 180° to Digit IV. The hallux impression is separate from the impressions of Digits II, III, and IV. These impressions in turn are mostly separate from one another but are more frequently connected between Digits II and III than between III and IV. Trackway narrow, average step 15.6 cm and average stride 31.2 cm, with slight inward rotation of Digit III (Lockley et al., 2006c).

3.3.7.7 Discussion

Lockley et al. (2006c) compared *Pullornipes aureus* with Cretaceous avian ichnotaxa from North America, South Korea, Argentina and China. Based on size, hallux impression and web traces, *Pullornipes* was most similar to *Koreanaornis*. But the hallux of *Koreanaornis* is significantly smaller than that of *Pullornipes*. Lockley et al. (2006c) also note that type, topotype and paratype *Koreanaornis* show digital pad impressions in some cases (Lockley et al., 1992a, 2006c), and this feature is not seen clearly in *Pullornipes*. Therefore, *Pullornipes* is included in the ichnofamily Koreanaornipodidae.

3.3.7.8 Trackmaker

The track-maker of *Pullornipes* aureus may be related to modern Scolopacidae. Scolopacidae is a large family of shorebirds (Charadriformes) that includes the sandpipers (Lockley et al., 2006c)

Ichnofamily Limiavipedidae McCrea et al., 2014a (Fig. 7.11)

Type Ichnogenus *Limiavipes* McCrea et al., 2014a

Diagnosis.

Trackway of a large, long-legged avian track-maker. Functionally tridactyl pes track with no obvious webbing. No hallux impression. Digits with sharp, tapering claws. Digital pad impressions show phalangeal formula of ?-2-3-4-0, with Digit I impression likely not preserved. Mean of the total digit divarication over 100°. Pes width greater than length. Pace and stride short compared with similar-sized theropod ichnotaxa, but long compared with other Mesozoic avian ichnotaxa. Average pace greater than 20 cm, average stride greater than 38 cm. Strong rotation of the tracks toward the midline of the trackway (original diagnosis from McCrea et al., 2014a)

3.3.8 **Limiavipes** ***McCrea et al., 2014a (Fig. 7.11C)***

3.3.8.1 Type ichnospecies

Limiavipes curriei McCrea et al., 2014a.

3.3.8.2 Valid ichnospecies

Limiavipes curriei McCrea et al., 2014a.

3.3.8.3 Distribution

Lower Albian, Lower Cretaceous Grande Cache Member of the Gates Formation, Grande Cache, Alberta; Aptian, Gething Formation, Elisabeth Creek, northern British Columbia, Canada.

3.3.8.4 Holotype

Specimen no. TMP 98.89.11; cast of isolated left pes taken from Grid H/G16. Lodged in the Royal Tyrrell Museum of Palaeontology, Drumheller, Alberta.

3.3.8.5 Diagnosis

Large avian track with three, moderately thick, functional pedal digits; no hallux impression. Digits II and IV approximately equal in length and shorter than Digit III. Sharp, triangular claw on each digit. Divarication between Digits II and IV greater than 100°. Track width greater than length. Pace and stride normally short. Individual tracks show a strong rotation toward the midline of the trackway (original diagnosis from McCrea et al., 2014a).

3.3.8.6 General description

The holotype specimen is 7.9 cm in length, 9.5 cm in width; length of Digits II, 4.5 cm: III, 6.7 cm: IV, 5.0 cm. Robust digits, the thickness of which are around 10% of their length. Digits terminate in narrow, sharp claws, with those of Digits II and IV inclined slightly axially toward Digit III. Total interdigital span varies between 120° and 135° according to original substrate conditions; the angle between Digits II and III on average greater than between Digits

III and IV. Digital pads often visible: three on Digit III, two on Digits II and IV. The center of each digit impression may show a groove, continuous or discontinuous; this is not evident in shallower imprints. Based on the orientation of Digit III, the tracks are rotated slightly inward toward the center of the trackway; the trackway is quite broad, and the pace and stride, although variable, are consistently short (average 23.5 and 48.4 cm, respectively) (from McCrea et al., 2014a).

3.3.8.7 Discussion

McCrea and Sarjeant (2001) based on an isolated left pes taken from Grid H/G16 named a new ichnospecies *Aquatilavipes curriei*. McCrea et al. (2014a) reassigned *A. curriei* to the ichnogenus: *Limiavipes*. There are distinct differences between *Limiavipes curriei* and all the *Aquatilavipes* ichnospecies: (1), the most obvious is the track length of *L. curriei* (8 cm) and *A. swiboldae* and *A. izumiensis* (combined mean = 3.5 cm); (2), the digits of *Aquatilavipes* are slender, while those assigned to *Limiavipes* are much more robust; (3), Digit IV of *Aquatilavipes* is longer than Digit II, creating an asymmetry with outer digit lengths. The outer digits lengths of *Limiavipes* specimens are nearly equivalent creating a symmetrical print. The multivariate analyzes by Buckley et al. (2016) also support this new classification (McCrea et al., 2014a).

3.3.8.8 Trackmaker

Xing et al. (2015d) drew an analogy between tracks from Limiavipedidae with those of extant herons (Ardeidae). Lockley et al. (2022a) contend that an analogy with extant cranes (Gruidae) is more appropriate: although both families are comprised of long-legged birds, cranes are functionally tridactyl, whereas herons produce a tetradactyl track with a distinctively oriented and very long hallux impression. Unquestionably, there is currently no body fossil evidence that modern cranes and herons existed in the Cretaceous, so these analogies are based on morphological convergence rather than taxonomic affinity (Lockley et al., 2022a); the same situation is true for shorebird tracks in South Korea.

3.3.9 Wupus *Xing et al., 2007 (Fig. 7.11D)*

3.3.9.1 Type ichnospecies

Wupus agilis Xing et al., 2007.

3.3.9.2 Valid ichnospecies

Wupus agilis Xing et al., 2007.

3.3.9.3 Distribution

Barremian/Aptian–Santonian, Jiaguan Formation, Chongqing, China.
Albian–Cenomanian, Daegu Formation, Ulsan, South Korea.

3.3.9.4 Holotype

At least 187 tracks represented by QJGM-T4-4 come from the Lotus site, Chongqing, China.

3.3.9.5 Diagnosis

A large tridactyl avian track with no discernible hallux; digital pads distinct with phalangeal formula of ?-2-3-4-0; well-preserved prints display a subrounded metatarsal impression that is closer to Digit II than Digit IV; webbing indistinct; track length to track width ratio 0.89; divarication angle between Digits II and III is approximately 50°, with a mean total divarication of under 100°, although total divarication can range 67°–132°; pace angulation is 180°; track length:pace length ratio 1:3.62 (diagnosis emended by Xing et al., 2015d).

3.3.9.6 General description

In well-preserved prints of *Wupus agilis* the proximal margin of the pes is asymmetrically bi-lobed, displaying a short, posteromedially protuberance that may correlate with the metatarsal pad for an unimpressed hallux, however, no halluces were identified in 187 documented prints (Xing et al., 2015d).

Wupus-like theropod track (8.3 cm in length, with a length/width ratio of 1.0) also recorded from the Late Cretaceous Zhutian Formation (Nanxiong Group) near Nanxiong, Guangdong, China (Xing et al., 2017a)

Wupus was not just recorded in China. First *Wupus* in North America and outside of China is the Ninesting Creek tridactyl tracks (track length ~10 cm) (Lockley et al., 2022a) Ninesting *Wupus* is probably the largest Cretaceous avian track from North America. *Wupus*-like tracks from the Daegu Formation (Albian–Cenomanian), Ulsan Metropolitan City, South Korea are the largest bird tracks in South Korea (Hong et al., 2024). As a large avian track, Korean *Wupus* has a length of 9.5 cm and a width of 11 cm with no hallux or webbing impression, the mean total divarication angle is 94° (Hong et al., 2024).

3.3.9.7 Discussion

There are obvious similarities between *Wupus agilis* and *Limiavipes curriei*, and in addition to their overall morphology, the trackway data are also similar. In four trackways, the average pace and stride lengths of *W. agilis* are 38.7 cm (23–63 cm) and 75.9 cm (48.5–113.5 cm), respectively, and a track length to pace length ratio of 0.3 (0.2–0.6). This is similar to *L. curriei*, where average pace and stride lengths are 23.9 cm (18–31.5 cm) and 46.5 cm (36.5–60.0 cm), and a track length/pace length of 0.3 (0.2–0.4). *L. curriei* and *W. agilis* also have enough differences, such as (1), the average track length of *L. curriei* is 7.9 cm (6.3–10.1 cm), while *W. agilis* is larger, with an average length of 10.3 cm (7.0–17.0 cm); (2), the *W. agilis* possess an average length/width of 0.9 (0.6–1.3), while those of *L. curriei* display a larger splay and possesses a length/width of 0.75 (0.6–0.9); (3), the average total divarication (II–IV) of *W. agilis* is 96.9° (67°–132°), which is smaller than *L. curriei* (123°) (Xing et al., 2015d).

3.3.9.8 Trackmaker

Wupus was originally identified as the track of a small theropod (Xing et al., 2007). It was assigned to the ichnofamily Limiavipedidae (McCrea et al., 2014a) due to similarity to *Limiavipes* from the Early Cretaceous of western Canada. *Wupus* represents a larger track-maker in both track length and leg length, with more narrowly splayed pedes than those of *Limiavipes* (Xing et al., 2015d). Reanalysis of *Wupus* indicates that it represents the traces of a relatively large avian track-maker, analogous to extant herons (Xing et al., 2015d) or Gruidae (Lockley et al., 2022a).

Ichnofamily Paxavipedidae McCrea et al., 2015a (Fig. 7.12)

Type Ichnogenus *Paxavipes* McCrea et al., 2015a.

Diagnosis Small tridactyl avian tracks with digit traces united proximally with the metatarsal pad and lacking webbing and hallux. Divarication between Digit II and Digit III traces greater than that between Digit III and Digit IV. Trackway characters consistent with a small, short-legged track-maker (original diagnosis from McCrea et al., 2015a).

3.3.10 Barrosopus *Coria et al., 2002 (Fig. 7.12E)*

3.3.10.1 Type ichnospecies

Barrosopus slobodai Coria et al., 2002.

3.3.10.2 Valid ichnospecies

Barrosopus slobodai Coria et al., 2002.

3.3.10.3 Distribution

Campanian, Anacleto Formation, Neuquén Province, Argentina.

3.3.10.4 Holotype

MCF-PVPH-SB 415-17c, footprint #87, Argentina.

3.3.10.5 Diagnosis

Tridactylous track of a small-sized avian theropod characterized by having conspicuous separation of Digit II impression from the other two digits. Divarication angle between Digits II and IV ranges between 100° and 120°. Trackways show a long stride, suggesting the track-maker was long-legged (original diagnosis from Coria et al., 2002).

3.3.10.6 General description

Barrosopus slobodai is small with an average length of 3.1 cm, and an average width of 3.6 cm. None of the tracks has either a hallux or metatarsal pad impressions, and the divarication angle between the Digit II and IV is 118°. The Digit II impression is consistently separated from the others. The narrow toes all end in claw impressions. A single trackway includes nine tracks, and has an average pace of 9.9 cm and an average stride of 19.8 cm (Coria et al., 2002).

3.3.10.7 Discussion

Barrosopus slobodai has also been recorded in the same locality as cf. *Aquatilavipes* (average length is 4.7 cm, the average width is 5.5 cm) and cf. *Ignotornis* (average length is 5 cm excluding the hallux, the average width is 5.8 cm) tracks. *B. slobodai* tracks are smaller than the other two ichnogenera, the divarication angle between Digits II and IV is also slightly smaller, and show a consistent separation of Digit II from the rest of the tracks. The average stride of the *B. slobodai* trackway is 6.4 times longer than average track length, compared with 5.0 times in cf. *Aquatilavipes* and 5.4 times in cf. *Ignotornis* (Coria et al., 2002).

3.3.10.8 Trackmaker

Coria et al. (2002) considered *Barrosopus slobodai* as a trace from an avian theropod due to the widely spread Digit II and IV, but otherwise it seems similar to a nonavian theropod. Similar specimens from Canada, *Paxavipes* indicates their affinity with Aves (McCrea et al., 2015a).

3.3.11 Paxavipes *McCrea et al., 2015a (Fig. 7.12F)*

3.3.11.1 Type ichnospecies

Paxavipes babcockensis McCrea et al., 2015a.

3.3.11.2 Valid ichnospecies

Paxavipes babcockensis McCrea et al., 2015a.

3.3.11.3 Distribution

Middle to ?late Albian, Boulder Creek Formation, south of Tumbler Ridge, northeastern British Columbia, Canada.

3.3.11.4 Holotype

PRPRC 2005.15.001.014 and PRPRC 2005.15.001.014 MC, single natural cast print and trackway C (PRPRC.15.001.011–014 inclusive)

3.3.11.5 Diagnosis

Trackway of a biped consisting of small tridactyl tracks, without webbing and hallux. Individual tracks are generally wider than long (foot length ~2.7 cm, width ~3 cm); length/width ratio approximately 0.91. Divarication between digit traces II–III consistently greater than that between Digits III–IV. Inward rotation of pes pronounced (original diagnosis from McCrea et al., 2015a).

3.3.11.6 General description

Trackway of a small biped lacking hallux or web traces. Tracks are ranging 2.0–3.4 cm in length and 2.1–3.9 cm in width. The divarication angles between Digits II and III are averaging 57° and greater than the divarication angles between Digits III and IV which average 43°. All digit traces are narrow (around 4 mm thick), approximately 22% of the length of Digit II, 15% of the length of Digit III, and 19% of the Digit IV length. Digital pads are evident on some tracks, but are not consistently preserved. There is pronounced inward rotation of prints toward the midline of the trackway. The trackway is narrow with short pace and stride (McCrea et al., 2015a).

3.3.11.7 Discussion

Paxavipes babcockensis and *Barrosopus slobodai* share a number of common characteristics, such as consistently higher divarication between Digits II and III than is found between Digits III and IV while lacking visible webbing and a hallux impression (McCrea et al., 2015a). However, *P. babcockensis* digits are joined proximally via a metatarsal pad that is absent in *B. slobodai*. *B. slobodai* also shows a significantly higher total divarication, and trackways have slightly greater pace and stride lengths than those of *P. babcockensis*. Digit widths of *P. babcockensis* are more uniform than those of *B. slobodai* prints. The outer digits of *P. babcockensis* have a greater equality in length than those of *B. slobodai* where Digit IV is longer than Digit II and almost as long as Digit III (McCrea et al., 2015a).

3.3.11.8 Trackmaker

Paxavipes babcockensis trackways were evidently produced by small short-legged birds (McCrea et al., 2015a). McCrea et al. (2015a) also noted that *Paxavipes* are remarkably similar to the tracks produced by extant members of Charadriidae, particularly those of the modern Killdeer (*Charadrius vociferous*) (Elbroch and Marks, 2001).

Ichnofamily Shandongornipedidae Lockley et al., 2007 (Fig. 7.11).

Type Ichnogenus *Shandongornipes* Li et al., 2005 emend. Lockley et al., 2007.

3.3.11.9 Diagnosis

Tetradactyl, paraxonic, zygodactyl tracks (i.e., with cranially directed Digits II and III in opposition to caudally directed Digits I and IV). Trackway narrow (about 7.0 cm), only marginally wider than track width (about 6.0 cm), with long step (about 44 cm) averaging about 5 times track length (about 8.5 cm) (diagnosis emended by Lockley et al., 2007).

3.3.12 Shandongornipes *Li et al., 2005 (Figs. 7.11B and 7.34B)*

3.3.12.1 Type ichnospecies

Shandongornipes muxiai Li et al., 2005.

3.3.12.2 Valid ichnospecies

Shandongornipes muxiai Li et al., 2005.

3.3.12.3 Distribution

Albian–Cenomanian, Tianjialou Formation, Lower Cretaceous Dasheng Group, Shandong, China.

3.3.12.4 Holotype

LRH-DH01. Five tracks were preserved as negative epireliefs on a bedding plane. An artificial plaster cast of tracks LRH-dz67 is in the Qingdao Institute of Marine Geology.

3.3.12.5 Diagnosis

Tetradactyl, paraxonic, zygodactyl tracks in narrow (6–7 cm in width) trackway with long (41–46 cm) steps. Elongate hallux oriented caudally, subparallel to trackway axis, and in opposition to the hypex between cranially directed Digits II and III, which subtend a narrow mean angle of about 32°. Digit IV rotated caudolaterally to subtend a mildly acute mean angle of about 77° with the hallux. Phalangeal pad formula 2-3-3 for Digits II, III, and IV, respectively (diagnosis emended by Lockley et al., 2007).

3.3.12.6 General description

The only known specimen of *Shandongornipes muxiai* is a single trackway of five consecutive well-preserved tracks. *S. muxiai* has an average track length (with hallux) of 8.5 cm (range 7.8–8.8 cm) and width (measured perpendicular to trackway axis) of 5.3 cm (range

4.4–6.0 cm). All digit plantar surface traces are separate from one another, indicating that their proximal connection to the metatarsus was typically elevated above the substrate. All digit traces end in narrow, acuminate claw marks. The hallux trace is short, consisting of a single elongate claw mark averaging 1.4 cm long and separated proximally from all other digit traces by at least 2.0 cm. Digit II consists of two proximal pads and a claw trace (mean 3.5 cm) directed craniomedially toward the trackway midline at an acute angle of 10°–15°. Digit III is the longest (mean 4.2 cm) and composed of three proximal pads and one distal claw impression; the digit is directed craniolaterally away from the trackway midline at an acute angle of 10–15°. Digit IV is caudolaterally directed at a 10°–25° angle caudal to a line perpendicular to the trackway axis; the digit impression comprises three short pads plus a claw trace (mean 3 cm). The average divarication angles are 153° for Digits I and II, 32 degrees for II and III, 101° for III and IV, and 77° for I–IV. The average step and stride lengths of the trackway are 44 and 87 cm, respectively. Pace angulation values are high (175°) (Lockley et al., 2007).

3.3.12.7 Discussion

The initial description interpreted the *Shandongornipes muxiai* trackway as if it were an anisodactyl form tending toward heterodactyl and, thus, interpreted Digit II as IV (Li et al., 2005). The detailed analysis by Lockley et al. (2007) indicated a zygodactyl, paraxonic form in which the cranially oriented Digits II and III are bisected by the parasagittal plane of the trackway axis, and lateral Digit IV is easily distinguished by its short pads compared with those of Digit II. The relative lengths of the digit impressions, Digit III longer than Digit IV, both longer than Digit II, are remarkably similar to those of tracks made by the greater roadrunner (Cuculidae: Neomorphinae: *Geococcyx californianus*), which are also very similar in track size and step length. This is likely the result of convergent evolution, given the huge age difference between them (Lockley et al., 2007).

3.3.12.8 Trackmaker

Initial studies regarded *Shandongornipes muxiai* as a shorebird-like taxon (Li et al., 2005). Lockley et al. (2007) believe the trackmaker of *S. muxiai* was a relatively fast-moving cursorial biped, with cursorial ability similar to present-day *Geococcyx*.

Based on the classic speed formula of Alexander (1976), Lockley et al. (2007) derived a speed estimate of 2.23 m/second for the *Shandongornipes* trackway. The large stride length values are typical of higher speed gaits, the alternative explanation is a particularly long hip height for such a short foot length. These details are used to support a *Geococcyx*-analog trackmaker of *S. muxiai*.

3.3.12.9 Unassigned to ichnofamilies

3.3.13 Gruipeda *Panin and Avram, 1962 (Fig. 7.13F)*

3.3.13.1 Type ichnospecies

Gruipeda maxima Panin and Avram, 1962.

3.3.13.2 Valid ichnospecies in Cretaceous

Gruipeda vegrandiunus Fiorillo et al., 2011.

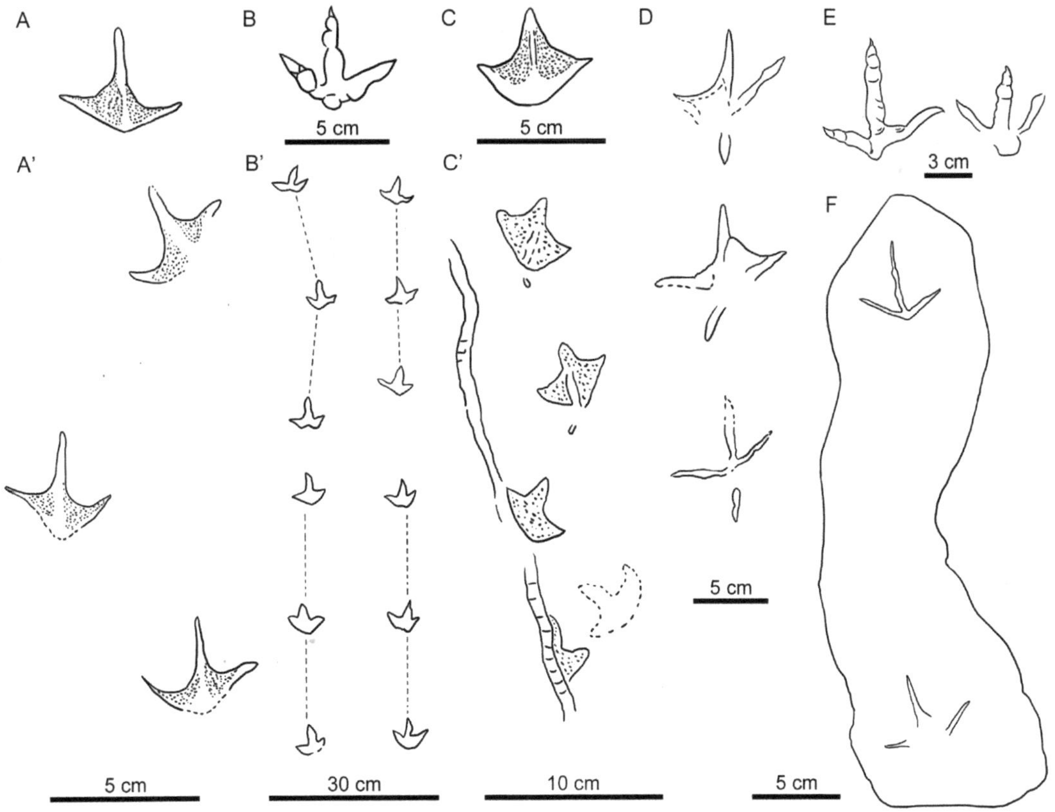

FIGURE 7.13 Avian tracks unassigned to ichnofamilies. (A) holotype and A′ paratypes of *Gyeongsangornipes lockleyi*; (B) holotype and B′ related trackway of *Tatarornipes chabuensis*; (C and C′) track and trackway from the holotype slabof *Uhangrichnus chuni*; (D) holotype (in the middle) and referred series of *Sarjeantopodus semipalmatus*; (E) separated tracks on holotype slab of *Yacoraitichnus avis*; (F) holotype of *Gruipeda vegrandiunus*. *(A) Redrawn from Kim et al. (2013). (B) Redrawn from Lockley et al. (2012). (C and C′) Redrawn from Kim et al. (2013), Lockley et al. (2012c). (D) Redrawn from Lockley et al. (2014a). (E) Redrawn from Alonso and Marquillas (1986). (F) Redrawn from Fiorillo et al. (2011).*

3.3.13.3 Total distribution

Cenozoic: Late Quaternary, Korea; Pleistocene, Argentina and New Mexico (USA); Pliocene, California (USA); Miocene, Iran, Romania, Spain, Argentina and California (USA); Neogene, Argentina; Oligocene, Spain, Argentina and Switzerland; Eocene, Argentina, Iran, Utah (USA), King George Island (Antarctica) and Ellesmere Island (Canada).

Cenozoic-Meszoic boundary: Maastrichtian–Danian, Argentina.

Cretaceous: late Campanian or early Maastrichtian, Alaska(USA).

3.3.13.4 Holotype of the ichnogenus

The specimen number is unknown, based on linear drawing of the type material in Fig. 25 (pl. 7) of Panin and Avram (1962).

3.3.13.5 Ichnogeneric diagnosis

Tracks showing four digits, three of which (II to IV) are directed forward and larger, the fourth (I), directed backward, spur-like, and short. The interdigital angles between Digits II and III and between Digits III and IV are commonly less than 70°. The axis of Digit I does not correspond with that of Digit III, the interdigital angle between Digits I and II being smaller than that between Digits I and IV. When present, phalangeal pad marks display the relationship I: 2, II: 2, III: 3, IV: 4. Webbing absent (diagnosis emended by De Valais and Melchor, 2008).

Another diagnosis is revised based on the premise that *Yacoraitichnus* belongs to *Gruipeda filiportatis*: Footprints showing four-digit imprints, three of which (II to IV) are directed forward and larger, the fourth (I), directed backward, spur-like and short. The interdigital angles between Digits II and III and between Digits III and IV are commonly less than 70°. The hallux imprint is posteromedially directed; the interdigital angle between Digits I and II being smaller than that between Digits I and IV. When present, digital pad traces display the relation I: 2, II: 2, III: 3, IV: 4. Webbing trace absent (emended diagnosis from De Valais and Cónsole-Gonella, 2019).

3.3.13.6 Ichnospecific diagnosis of *Gruipeda vegrandiunus*

Small tracks that are ~3.5 cm long and are ~4 cm wide. The length: width ratio for tracks is ~0.88. Divarication angles consistently 108°–109°. Pad impressions are indistinct. Full stride measurements are unavailable. Half stride distance is 22 cm. The orientation of Digit III effectively parallels median line (original diagnosis from Fiorillo et al., 2011).

3.3.13.7 General description

The ichnogenus *Gruipeda* was originally described by Panin and Avram (1962), based on bird tracks from Miocene Romania. Subsequently, *Gruipeda* was recorded in California, Iran, and other places (Remeika, 1999; Ataabadi and Khazaee, 2004; Lucas et al., 2023). De Valais and Melchor (2008) described a series of Late Triassic–Early Jurassic bird-like tracks from northwest Argentina, the most common of which is *Gruipeda dominguensis*. The age of *G. dominguensis* was subsequently corrected to the Late Eocene (Melchor et al., 2013).

It is worth mentioning that it is rare to assign Mesozoic specimens to a Cenozoic ichnotaxon (Anfinson et al., 2009; Lockley and Harris, 2010). *G. maxima* is revised and an emendation of its diagnosis is suggested (de Valais and Melchor, 2008). de Valais and Melchor (2008) listed 12 proposed *Gruipeda* ichnospecies: *G. maxima* Panin and Avram, 1962, *G. intermedia* Panin, 1965, *G. abeli* Lambrecht, 1938; Sarjeant and Langston, 1994, *G. minor* Panin, 1965; Sarjeant and Langston, 1994, *G. minima* Panin and Avram, 1962; Sarjeant and Langston, 1994, *G. calcarifera* Sarjeant and Langston, 1994, *G. disjuncta* Panin and Avram, 1962; Sarjeant and Langston, 1994, *G. filiportatis* Vialov, 1965; Sarjeant and Langston, 1994, *G. becassi* Panin and Avram, 1962; Sarjeant and Langston, 1994, *G. diabloensis* Remeika, 1999, *G. lambrechti* Ataabadi and Khazaee, 2004, *G. dominguensis* de Valais and Melchor, 2008. Lockley and Harris (2010) suggested that some of these species are affected by extramorphological preservational effects, but these discussions are beyond the scope of this article.

There is only one Cretaceous case of *Gruipeda*, *Gruipeda vegrandiunus* from the Upper Cretaceous Cantwell Formation (Campanian–Maastrichtian) of Denali National Park, Alaska

(Fiorillo et al., 2011). Alaska *Gruipeda* DMNH 18,352, are small tracks that are 3.5 cm in length and are 4 cm in width. The length/width ratio is 0.88. Divarication angles consistently 108°–109°. Pad impressions are indistinct. Full stride measurements are unavailable. Half stride distance is 22 cm. The orientation of Digit III effectively parallels the median line (Fiorillo et al., 2011).

3.3.13.8 Discussion

The size range of *Gruipeda vegrandiunus* falls within the overall size range of tracks described for *Gruipeda* (Ellenberger, 1972; Lockley et al., 1992a; Fiorillo et al., 2011). The orientation of the tracks to the median line of a trackway in *G. vegrandiunus* is virtually parallel, and this feature is basically consistent with most ichnospecies of *Gruipeda* (Lockley et al., 1992a; Fiorillo et al., 2011). Fiorillo et al. (2011) emphasized that the divarication angles of *G. vegrandiunus* are somewhat different from other ichnospecies of this ichnotaxon.

3.3.13.9 Trackmaker

Gruipeda are virtually indistinguishable from tracks made by post-Jurassic inferred shorebirds (Lockley et al., 1992a; Fiorillo et al., 2011).

3.3.14 Yacoraitichnus *Alonso and Marquillas, 1986* (=Yacoriteichnus, *misspelled*) *(Fig. 7.13E)*

3.3.14.1 Type ichnospecies

Yacoraitichnus avis Alonso and Marquillas, 1986.

3.3.14.2 Valid ichnospecies

Yacoraitichnus avis Alonso and Marquillas, 1986.

3.3.14.3 Distribution

Maastrichtian, Yacoraite Formation (Salta Group), northwestern Argentina.

3.3.14.4 Holotype

There is no specific specimen number, originally from from Quebrada del Tapon, Valle del Tonco locality and the single slab is deposited in the Departamento de Paleontología of the Universidad Nacional de Salta, Salta province, Argentina.

3.3.14.5 Diagnosis

Tridactyl tracks with well-marked claws at the distal end of the digits, Digit III much larger than Digit II and Digit IV; Digit II and Digit IV form variable angles with Digit III (diagnosis emended from Alonso and Marquillas, 1986).

3.3.14.6 General description

The original paper by Alonso and Marquillas (1986) is in Spanish, and the specimen does not appear to have been described again. The specimens are less well-preserved, and the morphological characteristics are basically consistent. The tracks are very overlapping which makes measurements difficult. In the original paper, the authors provided the following details: (1), the length of *Yacoraitichnus avis* is a little longer than the width (8 cm in length, and

7 cm in width), which suggests relatively large divarication angles; (2) very variable interdigital angles (54° and 78° in the holotype); (3) the digits join posteriorly forming a taloniforme pad/heel; (4), Digit III longer than Digit II and IV; (5) hallux absent; (65), presence of well-marked nails and plantar calluses.

However, based on the figure provided by Alonso and Marquillas (1986; Fig. 4 from plate II), the divarication angle between Digits II and IV is 113°. The original diagnosis also includes "specimen with pronounced lateral bulges" which may refer to the digit pad impressions of the digits. Identifiable digit pads are obvious on the digits II and III.

3.3.14.7 Discussion

Alonso and Marquillas (1986) did not compare *Yacoraitichnus avis* with other bird ichnotaxa. This may be because there were rather limited avian ichnotaxa in the 1980s, now there is no doubt that *Y. avis* is an avian track. De Valais and Cónsole-Gonella (2019) suggest that *Yacoraitichnus* should be reclassified into the Miocene ichnospecies *Gruipeda filiportatis* (Vialov, 1965; see Figs. 6, 14 and 15 in Lucas, 2007). We recommend that this ichnogenus needs more specimens or further description in the future.

3.3.14.8 Trackmaker

Alonso and Marquillas (1986) contended that the trackmaker of *Yacoraitichnus avis* belonged to Neornithes, but the possibility of Enantiornithes could not be ruled out.

3.3.15 Uhangrichnus *Yang et al., 1995 (Fig. 7.13C)*

3.3.15.1 Type ichnospecies

Uhangrichnus chuni Yang et al., 1995.

3.3.15.2 Valid ichnospecies

Uhangrichnus chuni Yang et al., 1995.

3.3.15.3 Distribution

Campanian, Uhangri Formation, Hwangsan Basin, southwestern South Korea.
Turonian–Coniacian, Jinhua Formation, Dongyang City, Zhejiang, China.

3.3.15.4 Holotype

A replica of the holotype is KPE 50,101, silicon rubber mold and plaster replicas of original in situ bedding plane.

3.3.15.5 Diagnosis

Small, functionally tridactyl track of a web-footed bird with small, posteromedially directed hallux trace sporadically preserved. Web configuration palmate (i.e., well-developed) and equally developed in the hypex between Digits II and III and III and IV. Track, excluding hallux, wider than long, averaging 3.7 and 4.6 cm, respectively, length/width = 0.81, but track length with hallux slightly longer than wide (length/width = 1.1). Trackway narrow with short step and stride (7.8 and 15.7 cm, respectively) and strong inward rotation (mean 20°) of Digit III relative to trackway midline (emended diagnosis from Lockley et al., 2012c).

3.3.15.6 General description

The holotype of *U. chuni* is KPE 50,101 based on a large sample of isolated tracks, measured to give an average length (without hallux) and width of 3.7 cm (N = 85) and 4.6 cm (N = 70), respectively. The average angle of divarication between Digits II and III is 53° (N = 38, right tracks) and 58° (N = 34, left track); between Digits III and IV the average angle is 59° and 51°. The total average divarication angles between Digits II–IV is 110°–111°.

3.3.15.7 Discussion

Uhangrichnus chuni found in the Uhangri Formation was first reported by Chun (1990) and formally named by Yang et al. (1995). Lockley and Harris (2010) added features of the ichnogenus, including the clear trackway, characteristic inward rotation, and the evidence of a hallux trace (Lockley et al., 2012b). *U. chuni* tracks reveal palmate web traces, which are equally well developed between Digits II and III and between III and IV. Kim et al. (2006) compared *U. chuni* tracks and semipalmate forms such as *Ignotornis* and *Hwangsanipes*.

Uhangrichnus has also been recorded in China. Azuma et al. (2013) named "Dongyangornipes sinensis" based on about 12 tracks from Late Cretaceous Jinhua Formation, Dongyang City, Zhejiang, China. This was the first record of a webbed bird track in China (Azuma et al., 2013). The features of "Dongyangornipes" mainly include relatively small tridactyl tracks with well-developed web structures; the length and width of the tracks are approximately 3.5 and 4 cm respectively; the length/width ratio is 0.87; the total divarication is nearly 100°; the divarication angle between Digits II and III is slightly smaller than that between Digits III and IV; the web impression between Digits II and III joins the apex of Digit II and the distal third of Digit III, and the impression between Digits III and IV is linked from the apex of Digit IV to the middle of Digit III (Azuma et al., 2013). *Dongyangornipes* is similar to *U. chuni* in size and general morphology, the only difference is the position of the web impression connected three digits (Buckley et al., 2016). Buckley et al. (2016) considered these differences could be explained by preservational variation: it is not uncommon for webbed tracks of one trackmaker to present a variety of webbing conditions. Therefore, *Dongyangornipes sinensis* a subjective junior synonym of *Uhangrichnus chuni* (Buckley et al., 2016).

3.3.15.8 Trackmaker

Uhangrichnus chuni cooccurs with the distinctive ichnospecies *Hwangsanipes choughi*. Both ichnotaxa were the first tracks with web traces reported from the Mesozoic (Yang et al., 1995; Lockley et al., 2012b). *Uhangrichnus* was made by a web-footed water bird or shorebird of relatively small size. Yang et al. (1995) suggested that from the perspective of size and morphology, the *Uhangrichnus* tracks resemble those of ducks such as the modern Mallard.

3.3.16 Sarjeantopodus *Lockley et al., 2004a (Fig. 7.13D)*

3.3.16.1 Type ichnospecies

Sarjeantopodus semipalmatus Lockley et al., 2004a.

3.3.16.2 Valid ichnospecies

Sarjeantopodus semipalmatus Lockley et al., 2004a.

3.3.16.3 Distribution

Maastrichtian, Lance Formation, Zerbst Ranch, Niobrara County, Wyoming.

3.3.16.4 Holotype

CU-MWC 224.4, rubber molds, plaster, and fiberglass replicas in the University of Colorado at Denver collection.

3.3.16.5 Diagnosis

Medium-size, semipalmated, charadriiform-like bird track with well-developed hallux. Web extends for more than half the lengths of Digits III and IV (original diagnosis from Lockley et al., 2004a).

3.3.16.6 General description

Natural cast of medium-size right track (9.5 cm in length, including hallux, by 9.2 cm in width). Digits narrow without distinct phalangeal pad impressions, terminating in blunt points, without distinct claw impressions. Web between Digits III and IV connects to digits about 1.5 cm from their distal ends. Web between Digits II and III connects proximally at 3 and 4 cm, respectively, from the distal ends. Divarication between Digits II and IV is 150° (i.e., 95° between Digits II and III, and 50° between Digits III and IV). Hallux, medium length, rotated posterior-medially: that is, subtending a 165° angle to Digit III. Trackway unknown.

3.3.16.7 Discussion

Sarjeantopodus semipalmatus is unique in that it represents the first North American Cretaceous report of a well-preserved semipalmate form (Lockley et al., 2004a). *S. semipalmatus* is morphologically similar to *Hwangsanipes*, but *Hwangsanipes* is much smaller than *Sarjeantopodus* (about 32%) and appears to be asymmetrical and semipalmate, though this characteristic is not brought out clearly in the original descriptions (Yang et al., 1995; Lockley et al., 2004a).

3.3.16.8 Trackmaker

The semipalmate feet are found in avocets, dotterels, willets, semipalmated plovers, certain other plover and sandpiper species, and certain herons (Kochan, 1994). Lockley et al. (2004a) discussed the similarity of *Sarjeantopodus semipalmatus* to American avocet, Curlew, and black-necked stilt feet.

3.3.17 Gyeongsangornipes *Kim et al., 2013 (Fig. 7.13A)*

3.3.17.1 Type ichnospecies

Gyeongsangornipes lockleyi Kim et al., 2013.

3.3.17.2 Valid ichnospecies

Gyeongsangornipes lockleyi Kim et al., 2013.

3.3.17.3 Distribution

Cenomanian–Santonian, Jindong Formation, northeast of Donghae-myeon, Goseong County, South Korea.

3.3.17.4 Holotype

A replica of the holotype is KNUE 090,204, housed at the Department of Earth Science Education, Korea National University of Education, South Korea.

3.3.17.5 Diagnosis

Asymmetrical semipalmate, strong mesaxonic bird track with three widely splayed thin digits impressions (II–IV) with sharp end. Asymmetrical web traces proximal to mesial between Digits II–III–IV (original diagnosis from Kim et al., 2013).

3.3.17.6 General description

The track length of *Gyeongsangornipes lockleyi* is about 3.1–3.2 cm and track width is about 4–4.1 cm on average. Track length/width ratio is about 0.77–0.78. Divarication between Digits II and III is about 69°–71°, which is much larger than that (about 56°–58°) between Digits III and IV, resulting in remarkable asymmetry. Divarication between Digits II and IV is about 130° and 125°. Digit III impression is straight and distal parts of the impressions of Digits II and IV are slightly curved toward the Digit III impression. Phalangeal pad impressions are rarely seen. Thickness of digit impressions is up to 3 mm. Mean values of length of Digits II, III, and IV are 2.3 cm, 3.1–3.2 cm, and 2.5 cm, respectively, emphasizing the prominent mesaxonic characteristics of this track (Kim et al., 2013).

Semipalmate webbing traces of *G. lockleyi* clearly developed between Digits II and III, and III and IV are proximal to mesial. The webbing impression is evidently asymmetrical; web between Digits II and III is proximal and web between Digits III and IV is, however, mesial. Hallux impressions are not observed (Kim et al., 2013).

3.3.17.7 Discussion

Kim et al. (2013) compared *Gyeongsangornipes lockleyi* with webbed bird tracks: (1), *G. lockleyi* is slightly smaller in size than *Ignotornis mcconnelli* (Mehl, 1931) excluding hallux impressions, and is extremely different from *I. mcconnelli* and *I. yangi* (Kim et al., 2006), which are characterized by semipalmate webbing, a prominent hallux and acuminate claw impressions. (2), The development and asymmetry of webbing of *G. lockleyi* are similar to those of *Hwangsanipes choughi* (Yang et al., 1995). However, *G. lockleyi* is distinguished from *H. choughi* because the latter is 1.5 times larger in size than the former and shows prominent hallux impressions. (3), *G. lockleyi* clearly differs in symmetry, divarication between Digits II and IV, web development, tapering thin digit impressions, shape of heel margin, and size from *Uhangrichnus chuni* (Yang et al., 1995). Though *U. chuni* lacks hallux impressions in most cases, *U. chuni* does show a small hallux impression in some cases. (4), *Sarjeantichnus semipalmatus* (Lockley et al., 2004a) strikingly differs from *G. lockleyi*. The former shows clear hallux impressions and the very asymmetrical outline, and track size of the former is about 2 times that of the latter. (5), *G. lockleyi* is different from *Goseongornipes markjonesi* (Lockley et al., 2006a). The latter shows clear hallux impressions, web impressions developed only between Digits III and IV, and a much larger divarication between Digits II and IV (140°–150°), at least in the type specimen (Kim et al., 2013).

3.3.17.8 Trackmaker

Gyeongsangornipes lockleyi was left by web-footed birds. Measurement data indicate that at least two age classes of web-footed birds walked on the sediments of a lake margin environment (Kim et al., 2013).

3.3.18 Tatarornipes *Lockley et al., 2012d (Fig. 7.13B)*

3.3.18.1 Type ichnospecies

Tatarornipes chabuensis Lockley et al., 2012d.

3.3.18.2 Valid ichnospecies

Tatarornipes chabuensis Lockley et al., 2012d.

3.3.18.3 Distribution

Aptian–Albian, Jingchuan Formation, Chabu site 4, Inner Mongolia, China.

3.3.18.4 Holotype

EG BT 001. A plaster replica of the holotype is available for comparative study as specimen CU 214.184.

3.3.18.5 Diagnosis

Medium-sized subsymmetrical functionally tridactyl track with wide digit divarication and robust digit traces that are wider proximally than distally. Pad traces may be well defined. Trackway narrow with a relatively long step and little inward rotation. Both the robust width of the proximal digit traces and the long step with little inward rotation distinguish *Tatarornipes* from other avian ichnotaxa (original diagnosis from Lockley et al., 2012d).

3.3.18.6 General description

Lockley et al. (2012d) provide the following additional details: tracks tridactyl subsymmetric, but with III–IV hypex slightly more anterior than II–III hypex. Phalangeal formula of 2–3–4 corresponding to Digits II, III, and IV, partially visible in some specimens, and apparent from composite tracings. Digit traces wide, robust, widest proximally, and tapering slowly, sometimes ending in a sharp claw trace. Overall track shape slightly wider than long (average 5.2 cm in length and 6 cm in width). Average digit divarication between II and IV is 110°, between II and III is 51° and is less than between III and IV which is 60°. Digit traces between II and III and between III and IV are always connected. Hallux traces are apparently absent. Trackway narrow, average step (20.5 cm) and stride (41 cm) long for four trackways with little rotation of Digit III.

3.3.18.7 Discussion

Lockley et al. (2012d) believed that the most striking feature of *Tatarornipes chabuensis* is the tracks with robust (wide) digit impressions, which in many cases show clear pad impressions. These features are different from avian track records in South Korea.

Tatarornipes is larger in size than *Aquatilavipes* and *Koreananornis*, the most common avian track records from the Early Cretaceous in China, and, both *Aquatilavipes* and *Koreananornis* have very slender digit traces. *Koreananornis* also reveals small hallux traces in rare cases. The only other avian tracks from the Cretaceous of China are *Pullornipes* (Lockley et al., 2006c) which has slender digits and consistently clear and well-developed hallux traces (Lockley et al., 2012d).

3.3.18.8 Trackmaker

Lockley et al. (2012d) made two suggestions, (1) the trackmaker of the bird track *Tatarornipes chabuensis* is different from common Cretaceous shorebird tracks, such as *Aquatilavipes* and *Koreanaornis*, because it ischaracterized by robust digits that are unusually wide proximally; (2) *T. chabuensis* is typically found in fluviolacustrine settings in association with saurischian (theropod and sauropod) dominated track assemblages.

3.4 Ornithischia

The phylogeny of ornithischians has undergone numerous changes in recent years (Butler et al., 2008; Maidment et al., 2008; Dieudonné et al., 2021; Norman et al., 2022). Nevertheless, the main clades are quite stable. To organize this section with an ichnological sense, we have divided it into five groups: (1) Basal ornithischians/basal ornithopods: in this group the tracks of Cretaceous small ornithischian trackmakers (i.e., heterodontosaurids, basal ornithischians, hypsilodontids) are summarized. One of the great challenges of ichnology, and not resolved yet, is to differentiate between the traces of these phylogenetically different groups. (2) Ankylopollexia: in this group, we have included the called "large ornithopod footprints," that is basal ankylopollexians, basal iguanodontoids, and hadrosaurids. (3) Ceratopsia: we have not proposed the most inclusive group Marginocephalia because to date no traces of pachycephalosaurids have been identified yet (see Díaz-Martínez et al., 2023). Finally, the clade Thyreophora is divided in two groups: (4) Stegosauria; and (5) Ankylosauria.

BASAL ORNITHISCHIA/BASAL ORNITHOPODA (FIG. 7.14)

3.4.1 Hypsiloichnus *Stanford et al., 2004 (Fig. 7.14A)*

3.4.1.1 Type ichnospecies

Hypsiloichnus marylandicus Stanford et al., 2004.

3.4.1.2 Valid ichnospecies

Hypsiloichnus marylandicus Stanford et al., 2004.

3.4.1.3 Distribution

Aptian, USA.

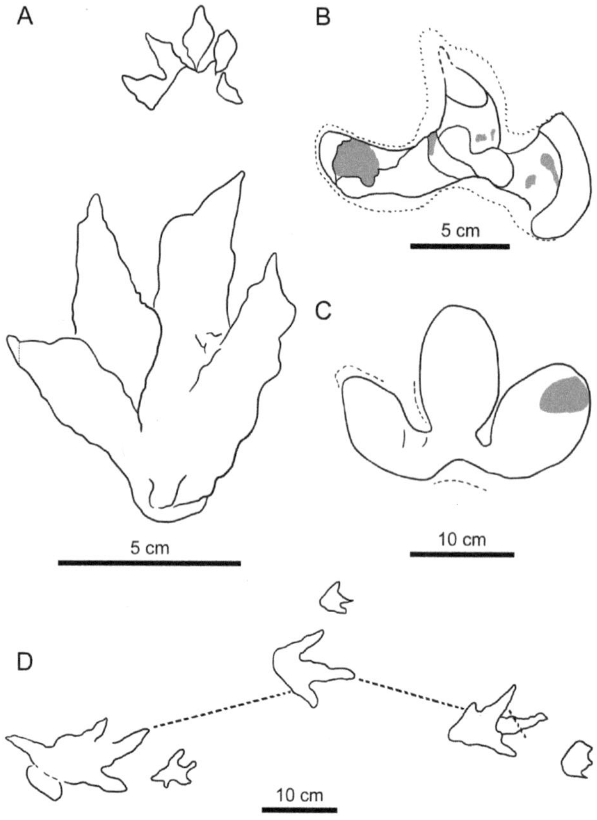

FIGURE 7.14 Tracks of basal ornithischians/basal ornithopods. (A) Holotype of *Hypsiloichnus marylandicus;* (B) holotype of *Wintonopus latomorum;* (C) holotype of *Wintonopus middletonae.* (D) holotype of *Neoanomoepus. (A) Redrawn from Stanford et al. (2004). (B) Redrawn from Romilio et al. (2013). (C) Redrawn from Salisbury et al. (2016). (D) redrawn from Lockley et al. (2009b).*

3.4.1.4 Holotype

CU 199.46, cast of a manus/pes pair preserved on a small block of siderite-cemented sandstone. Its type locality is Prince Georges beltway tracksite, which is in an Aptian delta plain sandstone in the Patuxent Formation of Maryland.

3.4.1.5 Diagnosis

Manus relatively small, length only about 10% of pes length. Manus print wider than long, pentadactyl, with digit tips defining a bluntly arcuate anterior border. Digit III slightly longer than other Digits; Digit II slightly longer than IV; Digits I and V slightly shorter than digits II and IV and of nearly equal length; Digit V markedly more divergent than other digits. Manus digit divarication values range from 30° to 31° (I–II), 22°–35° (II–III), 22°–21° (III–IV), and 74°–105° (IV–V). Pes print longer than wide, tetradactyl, with digit tips II to IV defining a much more acute anterior border arc than the digits of the manus. Axis of foot lies along Digit III, which is the longest. Digits II and IV symmetrically placed on either side of Digit III, of nearly equal length, and moderately shorter than Digit III. Digit I much shorter than other digits and placed well behind the impression of Digit II. Distally, digit tips blunt with stubby claws. Pes digit divarication values range from 26° to 31° (I–II), 27°–28° (II–III), and 33°–22°

(III–IV). Maximum known length of manus 1.5 cm, maximum known length of pes 15 cm (original diagnosis from Stanford et al., 2004).

3.4.1.6 General description

Manus tracks are pentadactyl, wider than long (mean length 3.6 cm; mean width 6.0 cm) with digit impressions quite defined. The Digit III impression is the longest, followed by Digit II, Digit IV, and Digits I and V, which are subequal. While Digit I–IV impressions are oriented forwardly, the Digit V impression is in the lateral direction. Manus length is about 10% of pes length. Pes track is tetradactyl, longer than wide (mean length 13.2 cm excluding short heel or metatarsal trace; mean width 11.1 cm) and mesaxonic. Digit III impression is the longest. Digit II and IV impressions are subequal in length and Digit I impression is the shortest one. Pes tracks present blunt claw impressions. No trackways have yet been found.

3.4.1.7 Discussion

First, Stanford et al. (2004) assigned the tracks to a taxonomic group, in this case hypsilophodontid dinosaurs. Later, they compared the tracks only with other tracks related to hypsilophodontids, that is, with a record in Spain. They claimed that these tracks were a new ichnotaxon because of they differed from the Spanish tracks. Nevertheless, they did not compare *Hypsiloichnus* with other similar ichnotaxa. For instance, *Hypsiloichnus*, *Anomoepus* and *Neoanomoepus* present a pentadactyl manus tracks instead in *Delatorrichnus* it is tridactyl, in *Atreipus* is tridactyl-tetradactyl, and *Dinehichnus* lacks manus impressions. On the other hand, the *Hypsiloichnus* heteropody is much greater than in *Anomoepus* and *Neoanomoepus*. The pes track of *Atreipus*, *Delatorrichnus*, and *Dinehichnus* is tridactyl, whereas *Hypsiloichnus*, *Anomoepus*, and *Neoanomoepus* are tetradactyl. Of the latter ichnogenera, *Hyspiloichnus* has the largest Digit I. According to this comparison, the combination of features present in *Hypsiloichnus* is unique and justified its erection.

3.4.1.8 Trackmaker

Stanford et al. (2004) suggested that only ornithischian dinosaurs in the Cretaceous had in combination a five-toed manus and a four-toed pes that touched the ground in the Cretaceous. The long middle digit trace on the pes track and the great difference in size between the manus and pes tracks were not seen in thyreophorans (ankylosaurs or stegosaurs) or ceratopsians. Moreover, among ornithischian tracks, ornithopods have great heteropody, thus *Hypsiloichnus* was made by an ornithopod dinosaur. These prints are too small for any adult ornithopod from the families Iguanodontidae or Hadrosauridae, and none of them had a functional digit I in the pes. The family Dryosauridae (e.g.,: *Dryosaurus*, *Valdosaurus*) has two genera that are similar in size to the trackmaker. The well-known one, *Dryosaurus altus* from the Late Jurassic, only had a tiny metatarsal I, and *Valdosarus* does not preserve the pes. Moreover, Stanford et al. (2004) pointed out that *Tenontosaurus tilletti*, *Hypsilophodon foxii*, and *Zephyrosaurus schaffi* were identified in rocks of the same age and close to the type locality. These tracks are too small for *Tenontosaurus*, which also had a different manus-to-pes-size ratio than what is seen in both of the manus/pes tracks shown in *Hypsoloichnus*. Finally, these authors concluded that *Hypsilophodon* or *Zephyrosaurus* could be good candidates to trackmaker based on its size. Gierliński and Sabath (2008) proposed basal ornithischians and basal ornithopods as possible trackmakers of *Hypsiloichnus*.

3.4.2 Wintonopus *Thulborn and Wade, 1984 (Figs. 7.14B,C and 7.33A–C)*

3.4.2.1 Type ichnospecies

Wintonopus latomorum Thulborn and Wade, 1984 (Figs. 7.14B and 7.33A–C)

3.4.2.2 Valid ichnospecies

Wintonopus latomorum Thulborn and Wade, 1984.
Wintonopus middletonae Salisbury et al., 2016 (Fig. 7.14C)

3.4.2.3 Distribution

Valanginian–Cenomanian, Australia.

3.4.2.4 Holotype

QM F10319, a footprint. Its type locality is Lark Quarry dinosaur tracksite (QM), which is in a Cenomanian lacustrine sandstone/claystone in the Winton Formation of Australia.

3.4.2.5 Diagnosis

Small- to medium-sized (<0.3 m long), tridactyl, mesaxonic pes impressions that are wider than long. Digit impressions are cranially directed and are short in length, with Digit III being the longest, and Digit IV being equivalent to or longer than Digit II. The proximal track margin is more concave proximomedially than proximolaterally, with the proximal margin of the Digit IV impression more proximally positioned relative to the impression of Digit II. Both Digit II and IV impressions extend farther proximally than the Digit III impression. A metatarsophalangeal pad impression is absent, and claw impressions, when present, are pointed to rounded (emended from Romilio et al., 2013).

3.4.2.6 General description

Wintonopus tracks are tridactyl, mesaxonic, and wider than long. The "heel" pad impression is absent, but a bilobed proximal track margin is visible, which represents the proximal margins of the digital impressions of II and IV. *W. latomorum* ranges from very small to large in size (proximodistal length 5–26 cm, mediolateral width 3.5–32 cm). Digit impressions are elongated and narrow, and lack pad impressions. *W. middletonae* is small to medium size (proximodistal length 19–25 cm, mediolateral width 28–36.6 cm). Digit impressions are moderately elongated, broad, and oval in shape, with usually single-digit pad impressions of each digit impression (see Salisbury et al., 2016: Fig. 41).

3.4.2.7 Discussion

Thulborn and Wade (1984) defined the ichnogenus *Skartopus australis* and *Wintonopus latomorum* at Lark Quarry tracksite (Winton Formation, late Albian-Cenomanian) for theropod and ornithopod trackmakers respectively. Romilio et al. (2013) suggested that *Skartopus* is a preservational variation of *Wintonopus* and should be a junior synonym of *Wintonopus*, and emended the *Wintonopus* diagnosis. Moreover, Thulborn and Wade (1984) considered that the morphology of the tracks found at Lark Quarry, and the arrangement of its trackways, are the evidence of a terrestrial stampede. Nevertheless, Romilio et al. (2013) pointed out that many of these tracks, including the holotype of *Skartopus*, do not reflect the morphology of the trackmaker's plantar anatomy and instead record trackmakers that were partially buoyed at the time of track registration. This information is relevant for an ichnotaxonomical point of view, since most tetrapod ichnotaxa are left by animals walking through mud, and only a few have been defined for swimming animals (*Characichnos*, or *Paravipes*). Interestingly, Romilio et al. (2013;Figs. 7 and 12) pointed out tracks and trackways in association with *Wintonopus* that could (and perhaps should) be ichnologically attributable to *Characichnos*.

In addition to tracks of this ichnogenus being found from Queensland, Australia, *Wintonopus*-like tracks have been described more recently within the Lower Cretaceous (Aptian-Albian) Eumeralla Formation in Victoria (Romilio and Godfrey, 2022). Significantly, tracks of the same *W. latomorum* ichnospecies have been described from the Lower Cretaceous (Valanginian–Barremian) Broome Sandstone in Western Australia, in addition to the occurrence of a new ichnospecies within *Wintonopus*: *W. middletonae* (Salisbury et al., 2016). The Broome *Wintonopus* ichnospecies are quite different from *W. latomorum*, with *W. middletonae*, due to their shared absence of a metatarsophalangeal pad impression. *Wintonopus middletonae* has broad digital impressions that are rounded at the tips, and a symmetrical bilobed proximal track margin. There exists a similarity of its well-rounded (albeit more oval-shaped) digit impressions with those of the Broome Sandstone ornithopodan ichnotaxon *Walmadanyichnus hunteri*, with the latter being considerably larger. Salisbury et al. (2016) discarded that both ichnotaxa were ontogenetic variants associated with a single type of trackmaker. In our review, we consider *Walmadanyichnus hunteri* a nomen dubium and likely attributable to either *Caririchninium* or *Hadrosauropodus*. However, *Caririchninium* or *Hadrosauropodus* of a size comparable to *Wintonopus middletonae* have all been observed to possess a metatarsophalangeal pad impression.

3.4.2.8 Trackmaker

A potential trackmaker of the Queensland tracks could be an as-yet undescribed ornithopod species from the Upper Albian deposits of the Winton Formation in the vicinity of Isisford (Salisbury et al., 2016), which matches the estimated size of the presumed trackmakers at Lark Quarry (see "*Skartopus*" entry in this manuscript). Salisbury et al. (2016) also considered an ornithopod as trackmaker of *W. middletonae*, albeit it one with broad pedal digits.

3.4.3 Neoanomoepus *Lockley et al., 2009b (Figs. 7.14D and 7.35)*

3.4.3.1 Type ichnospecies

Neoanomoepus perigrinatus Lockley et al., 2009b.

3.4.3.2 Valid ichnospecies

Neoanomoepus perigrinatus Lockley et al., 2009b.

3.4.3.3 Distribution

Berriasian–Valanginian, Canada, and Thailand.

3.4.3.4 Holotype

FGM 002.01.20a, a footprint. Its type locality is Fording River Operations coal mine, which is in a Berriasian terrestrial horizon in the Mist Mountain Formation of Canada.

3.4.3.5 Diagnosis

Small trackway of a quadruped with tetradactyl pes larger than pentadactyl manus. Pes axis inwardly rotated and pes Digit I short and anteromedially directed. Manus outwardly rotated and situated lateral to pes Digit III or IV. Step short and trackway irregular (original diagnosis from Lockley et al., 2009b).

3.4.3.6 General description

Pes tracks are tetradactyl, longer than wide (mean length 13.2 cm excluding short heel or metatarsal trace; mean width 11.1 cm) and mesaxonic. Digit III impression is the longest and Digit I the shortest. Digit II–IV impressions are elongated, narrow, and forwardly oriented, while Digit I is anteromedially oriented. The digit traces present claw traces but lack pad impressions. The "heel" region is elongated and narrow in its final part. Pes tracks are longer than manus tracks. The manus tracks are pentadactyl, wider than long (mean length 3.6 cm; mean width 6.0 cm), outwardly rotated and situated lateral to pes Digit III or II. Outer digit impressions (I and V) are oriented lateral and medially, and the inner (II–IV) are forwardly directed. Digit impressions are short, narrow and present sharp end (see Lockley et al., 2009b).

3.4.3.7 Discussion

Neoanomoepus differs from the other basal ornithischian/ornithopod ichnotaxa.

For instance, *Atreipus* and *Delatorrichnus* have tridactyl pes and tridactyl-tetradactyl manus tracks, while *Neoanomoepus* has tetradactyl pes and pentadactyl manus tracks. This same configuration is also in *Anomoepus* and *Hypsiloichnus*, but the latter has a large pes Digit I impression. *Neoanomoepus* and *Anomoepus* are similar in shape and size. Lockley et al. (2009b) pointed out the tracks of *Neoanomoepus* lack distinct pad traces on the manus and pes tracks in most cases, which are present in *Anomoepus*. This could be explained by poor preservation of the tracks, so the absence of pad traces could be an extramorphological feature, but the tracks preserve very fine details, such as the digits of the manus tracks, so it is assumed that it is not a preservation issue (Lockley et al., 2009b).

Neoanomoepus has been identified in the Early Cretaceous of Canada and Thailand. Lockley et al. (2009b) determined also as *Neoanomoepus*-like other material of Late Jurassic–Early Cretaceous age from Spain, Zimbabwe, and Canada. McCrea et al. (2014a) referred a trackway found in the Minnes Group (Berriasian–Valanginian) of Canada to *Neoanomoepus*.

3.4.3.8 Trackmaker

The trackmaker identity is often proposed by temporal and/or geographical correlations. In this case, ornithischian tracks with marked heteropody have been related to basal ornithischians or ornithopods (Lockley et al., 2009b).

ANKYLOPOLLEXIA (FIGS. 7.14–7.16)

The ichnofamily Iguanodontipodidae was proposed by Vialov (1988) without diagnosis or ichnogenus type. This ichnofamily was practically forgotten until Lockley et al. (2014a), who after reviewing the Cretaceous large ornithopod tracks, recovered its use, and provided a diagnosis. Díaz-Martínez et al. (2015a) revised all the published large ornithopod ichnotaxa, chose *Iguanodontipus* Sarjeant et al., 1998 as the type for this ichnofamily and emended the diagnosis as "*Mesaxonic, tridactyl, subsymmetrical pes tracks that are as wide as or wider than long; one pad impression in each digit and one in the heel; digit pads longer than wide; well-developed notches in the proximal part of the Digit II and IV impressions; manus tracks occasionally present and much smaller than the pes tracks.*" These authors also analyzed 34 ichnogenera and 44 ichnospecies and provided an ichnotaxonomic opinion on all of them. Moreover, they added the scheme of all the holotypes and the diagnosis in the original language. They considered valid only three ichnogenera: *Iguanodontipus*, *Caririchnium* Leonardi, 1984, and *Hadrosauropodus* Lockley et al., 2004a. Other authors also used in their works *Amblydactylus* Sternberg, 1932 (e.g., Lockley et al., 2014a; McCrea et al., 2014a; Herrero et al., 2016; Tsukiji et al., 2018), *Ornithopodichnus* Kim et al., 2009 (e.g., Lockley et al., 2012e; Abbassi et al., 2018; Xing et al., 2021c,d), *Jiayinosauropus* Dong et al., 2003; Lockley et al., 2014a, and *Sousaichnium* Leonardi, 1979a; Leonardi and Carvalho, 2021c. In this section, we will follow mainly the synthetic proposals of Lockley et al. (2014a) and Díaz-Martínez et al. (2015a) discussing new opinions and finds.

Iguanodontipodidae ichnotaxa.

A total of 35 ichnogenera referred to large ornithopod tracks have been recorded in Upper Jurassic-Cretaceous strata. These are listed below in alphabetical order. The symbol ⊠ indicates an invalid ichnotaxon, the reasons are summarized between parentheses and discussed in Lockley et al. (2014a) and Díaz-Martínez et al. (2015a). If there are more modern discussions or other relevant opinions, the reference is also indicated.

⊠ *Akmechetosauropus* Dzhalilov and Novikov, 1993 (nomen dubium)
  ⊠ *Akmechetosauropus makhkamovi* Dzhalilov and Novikov, 1993 (nomen dubium)
⊠ *Amblydactylus* Sternberg, 1932 (nomen dubium)
  ⊠ *Amblydactylus gethingi* Sternberg, 1932) (nomen dubium)
  ⊠ *Amblydactylus kortmeyeri* Currie and Sarjeant, 1979 (= *Caririchnium kortmeyeri*)
Apulosauripus Nicosia et al., 1999 (non Iguanodontipodidae)
  Apulosauripus federicianus Nicosia et al., 1999 (non Iguanodontipodidae)
⊠ *Babatagosauropus* Dzhalilov and Novikov, 1993 (nomen dubium)
  ⊠ *Babatagosauropus bulini* Dzhalilov and Novikov, 1993 (nomen dubium)
⊠ *Bonaparteichnium* Calvo, 1991 (nomen dubium)
  ⊠ *Bonaparteichnium tali* Calvo, 1991(nomen dubium)
⊠ "*Brachyguanodonipus*" Moratalla, 1993 (nomen nudum)
  ⊠ "*Brachyguanodonipus prejanensis*" Moratalla, 1993 (nomen nudum)
⊠ *Camptosaurichnus* Casamiquela and Fasola, 1968 (nomen dubium)
  ⊠ *Camptosaurichnus* fasolae Casamiquela and Fasola, 1968 (nomen dubium)
⊠ *Camptosauropus* Gabunia and Kurbatov, 1988 (nomen dubium)
  ⊠ *Camptosauropus vialovii* Gabunia and Kurbatov, 1988 (nomen dubium)
Caririchnium Leonardi, 1984
  Caririchnium magnificum Leonardi, 1984

⊠ *Caririchnium leonardii* Lockley, 1987 (= *Hadrosauropodus leonardii*)
Caririchnium kortmeyeri (Currie and Sarjeant, 1979)
⊠ *Caririchnium protohadrosaurichnos* Lee, 1997 (nomen dubium)
Caririchnium billsarjeanti (Meyer and Thüring, 2003)
Caririchnium lotus Xing et al., 2007
⊠ *Caririchnium kyoungsookimi* Lim et al., 2012 (= *Hadrosauropodus kyoungsookimi*)
⊠ *Caririchnium liucixini* Xing et al., 2016f (= *Caririchnium* isp.)
Caririchnium yeondongensis Kim et al., 2016
⊠ *Gigantoshiraminesauropus* Azuma and Takeyama, 1991 (nomen dubium)
⊠ *Gigantoshiraminesauropus matsuoi* Azuma and Takeyama, 1991 (nomen dubium)
⊠ *Goseongosauripus* Kim, 1986 (nomen nudum)
⊠ *Goseongosauripus kimi* Kim, 1986 (nomen nudum)
⊠ *Gypsichnites* Sternberg, 1932 (non Iguanodontipodidae)
⊠ *Gypsichnites pacensis* Sternberg, 1932 (non Iguanodontipodidae)
⊠ *Hadrosaurichnoides* Casanovas et al., 1993b (nomen dubium)
⊠ *Hadrosaurichnoides igeensis* Casanovas et al., 1993b (nomen dubium)
⊠ *Hadrosaurichnus* Alonso, 1980 (nomen dubium)
⊠ *Hadrosaurichnus australis* Alonso, 1980 (nomen dubium)
⊠ *Hadrosaurichnus titicaensis* Jaillard et al., 1993 (nomen dubium)
⊠ *Hadrosauripeda* Vialov, 1988 (nomen nudum)
⊠ *Hadrosauripeda hauboldi* Vialov, 1988 (nomen nudum)
Hadrosauropodus Lockley et al., 2004a
Hadrosauropodus langstoni Lockley et al., 2004a
⊠ *Hadrosauropodus nanxiongensis* Xing et al., 2009c (nomen dubium)
Hadrosauropodus leonardii (Lockley, 1987)
Hadrosauropodus kyoungsookimi (Lim et al., 2012)
⊠ *Iguanodonichnus* Casamiquela and Fasola, 1968 (nomen dubium)
⊠ *Iguanodonichnus frenkii* Casamiquela and Fasola, 1968 (nomen dubium)
⊠ "*Iguanodonipus*" Moratalla, 1993 (nomen nudum)
⊠ "*Iguanodonipus cuadrupedae*" Moratalla, 1993 (nomen nudum)
⊠ *Iguanodonopus* Zhen et al., 1996
⊠ *Iguanodonopus xingfuensis* Zhen et al., 1996 (non Iguanodontipodidae Wright, 1999; nomen dubium Xing et al., 2009c)
Iguanodontipus Sarjeant et al., 1998
Iguanodontipus burreyi Sarjeant et al., 1998
⊠ *Iguanodontipus billsarjeanti* Meyer and Thüring, 2003 (= *Caririchnium billsarjeanti*)
⊠ *Jianyinosauropus* Dong et al., 2003 (nomen dubium)
⊠ *Jianyinosauropus johnsoni* Dong et al., 2003 (nomen dubium)
⊠ *Kharkushosauropus* Dzhalilov and Novikov, 1993 (nomen dubium)
⊠ *Kharkushosauropus kharkushensis* Dzhalilov and Novikov, 1993 (nomen dubium)
⊠ *Limayichnus* Calvo, 1991 (nomen dubium; non Iguanodontipodidae Apesteguía and Gallina, 2011)
⊠ *Limayichnus major* Calvo, 1991 (nomen dubium; non Iguanodontipodidae)
⊠ *Orcauichnites* Llompart et al., 1984 (nomen dubium)
⊠ *Orcauichnites garumniensis* Llompart et al., 1984 (nomen dubium)

☒ *Ornithopodichnites* Llompart et al., 1984 (nomen dubium)
 ☒ *Ornithopodichnites magna* Llompart et al., 1984 (nomen dubium)
☒ *Ornithopodichnus* Kim et al., 2009 (nomen dubium)
 ☒ *Ornithopodichnus masanensis* Kim et al., 2009 (nomen dubium)
☒ *Shiraminesauropus* Azuma and Takeyama, 1991 (nomen dubium)
 ☒ *Shiraminesauropus reini* Azuma and Takeyama, 1991 (nomen dubium)
 ☒ *Shiraminesauropus hayashidaniensis* Azuma and Takeyama, 1991 (nomen dubium)
☒ *Sinoichnites* Kuhn, 1958 (nomen dubium)
 ☒ *Sinoichnites youngi* Kuhn, 1958 (nomen dubium)
☒ *Sousaichnium* Leonardi, 1979a (nomen dubium)
 ☒ *Sousaichnium pricei* Leonardi, 1979a (nomen dubium; see Leonardi and Carvalho, 2021c for other opinion)
 ☒ *Sousaichnium monettae* Calvo, 1991 (nomen dubium)
☒ *Staurichnium* Leonardi, 1979a (nomen dubium; see also Leonardi and Carvalho, 2021c)
 ☒ *Staurichnium diogenis* Leonardi, 1979a (nomen dubium; see also Leonardi and Carvalho, 2021c)
☒ *Taponichnus* Alonso and Maquillas, 1986 (nomen dubium)
 ☒ *Taponichnus donottoi* Alonso and Maquillas, 1986 (nomen dubium)
☒ *Telosichnus* Alonso and Maquillas, 1986 (nomen dubium)
 ☒ *Telosichnus saltensis* Alonso and Maquillas, 1986 (nomen dubium)
☒ *Walmadanyichnus* Salinsbury et al., 2016 (nomen dubium, herein)
 ☒ *Walmadanyichnus hunter* Salisbury et al., 2016 (nomen dubium, herein)
☒ *Wealdenichnites* Kuhn, 1958 (nomen dubium)
 ☒ *Wealdenichnites iguanodontoides* Kuhn, 1958 (nomen dubium)
☒ *Yangtzepus* Young, 1960 (non Iguanodontipodidae)
 ☒ *Yangtzepus yipingensis* Young, 1960 (non Iguanodontipodidae)

Regardless of their ichnotaxonomic valuation, the tracks assigned to "*Brachyguanodonipus*," *Gigantoshiraminesauropus*, *Hadrosaurichnoides*, *Hadrosauripeda*, "*Iguanodonipus*," *Jiayinosauropus*, *Limayichnus*, *Shiraminesauropus*, *Ornithopodichnus*, *Sousaichnium*, *Staurichnium*, *Walmadanyichnus* and *Wealdenichnites* can be classified within Iguanodontipodidae.

As commented before, after the revisions of Lockley et al. (2014a) and Díaz-Martínez et al. (2015a) the number of ichnogenera belonging to Iguanodontipodidae is quite reduced. For instance, Lockley et al. (2013, 2014a) considered that the Upper Cretaceous *Jiayinosauropus* as an available ichnogenus, but they also pointed out that the holotype is incomplete, and the paratype is very partial. In our opinion, the morphology of *Jiayinosauropus* is insufficiently known but consistent with Iguanodontipodidae and it is considered as a nomen dubium.

On the other hand, the ichnogenus *Ornithopodichnus* has been used recurrently in South Korea (Kim et al., 2009; Lockley et al., 2012e), China (Xing and Lockley, 2014a; Li et al., 2015; Xing et al., 2021c, 2021d; Xu et al., 2021) and Iran (Abbassi et al., 2018) for mainly Lower Cretaceous tracks. Niu and Xing (2023) also identified this ichnogenus in the lower Upper Cretaceous of China. The abovementioned authors suggested that *Ornithopodichnus* is different from other large ornithopod tracks because it has a distinctively weak mesaxonic and broad, transverse pes imprints that are wider than long. Lockley et al. (2014a) pointed out that the preservation of the type series is suboptimal, with some tracks still partially filled. The tracks classified as *Ornithopodichnus* are very similar to the *Caririchnium* morphology (sensu Díaz-Martínez et al., 2015a). It would be necessary to undertake a complete comparison between *Ornithopodichnus* tracks (type series and referred material) and all the available *Caririchnium* ichnospecies. Therefore, for the time being, we considered that *Ornithopodichnus* is a nomen dubium and we classified the type series as *Caririchnium* isp.

Leonardi and Carvalho (2021c) presented an extensive work about dinosaur tracks of Brazil, where new new photos and information related to previous works were included. They pointed out that *Sousaichnus pricei* is an available ichnotaxon different from *Caririchnium magnificum*, and justify why it is not a nomen dubium (sensu Díaz-Martínez et al., 2015a). This ichnotaxon is defined based on a trackway with a variable track morphology. The better-preserved tracks present a quadripartite distribution and some of them have manus impressions. The shape of *Sousaichnus pricei* is typical of *Caririchnium* (i.e., large and rounded "heel" impression, short and wide digit traces), but in the preservation scale of Marchetti et al. (2019), these tracks do not reach 2. So, we considered that *S. pricei* is a nomen dubium and we classified them as *Caririchnium* isp.

The ichnogenus *Amblydactylus* is used by ichnologist to identify upper Lower Cretaceous-Upper Cretaceous large ornithopod tracks (e.g., Gangloff et al., 2004; Romilio and Salinsbury, 2013; McCrea et al., 2014a; Tsukiji et al., 2018, 2019). There are two defined ichnospecies: *A. gethingi* Sternberg, 1932 and *A. kortmeyeri* Currie and Sarjeant, 1979. Some authors (see Díaz-Martínez et al., 2015a for a complete discussion; Tsukiji et al., 2018) accept that the type ichnospecies of *Amblydactylus*, *A. gethingi*, is a nomen dubium (see Lockley et al., 2014a for another opinion). Indeed, Tsukiji et al. (2018) proposed to assign *A. kortmeyeri* as the type ichnospecies of *Amblydactylus*. The ichnotaxonomical problem is that if we consider that the type ichnospecies of an ichnogenus is a nomen dubium, the ichnogenus is no longer available. In this situation, "*Amblydactylus*" *kormeyeri* would be a nomen nudum and it would be necessary to erect a new ichnogenus in which to include this ichnospecies. In this context, Díaz-Martínez et al. (2015a,b) proposed including this ichnospecies in the ichnogenus *Caririchnium* as *C. kortmeyeri* to make it available.

In the last years, two ichnospecies of *Caririchnium*, *Caririchnium liucixini* Xing et al., 2016f and *Caririchnium yeondongensis* Kim et al., 2016 (see discussion in the *Caririchnium* section), and a new ichnogenus and ichnospecies *Walmadanyichnus hunter* Salisbury et al., 2016 have been described. Except for the holotype of *Walmadanyichnus*, the rest of the referred tracks are not very well-preserved. Even so, none of the *Walmadanyichnus* tracks has a value of two or higher in the preservation scale of Marchetti et al. (2019). In addition, the authors suggested that the "heel" may be rounded or bilobed. In this sense, either it is conditioned by extramorphological features, or in this area there are two different ichnotaxa, one *Caririchnium*-like and another *Hadrosauropodus*-like. In this sense, we consider that the traces could be classified within one of the aforementioned ichnogenera, and *Walmadanyichnus* is a nomen dubium.

We will delve into the three ichnogenera within Iguanodontipodidae used by Díaz-Martínez et al. (2015a,b) taking into account the shape of the heel and digit impressions: *Iguanodontipus* (small, rounded "heel," with elongate, narrow digits), *Caririchnium* (large, rounded "heel," with short, wide digits) and *Hadrosauropodus* (large, bilobed "heel," and short, wide digits). Maybe this proposal is an oversimplification of the record, but it can be a good base to work on in the future.

3.4.4 Iguanodontipus *Sarjeant et al., 1998 (Fig. 7.15)*

3.4.4.1 Type ichnospecies

Iguanodontipus burreyi Sarjeant et al., 1998.

3.4.4.2 Valid ichnospecies

Iguanodontipus burreyi Sarjeant et al., 1998.

3.4.4.3 Distribution

Berriasian–Valanginian, England, Germany, Spain.

3.4.4.4 Holotype

BNSS 33,793b, seven natural casts preserved on three slabs, the slab bearing a right and a left pes. Paratypes: BNSS 33,793a and c.

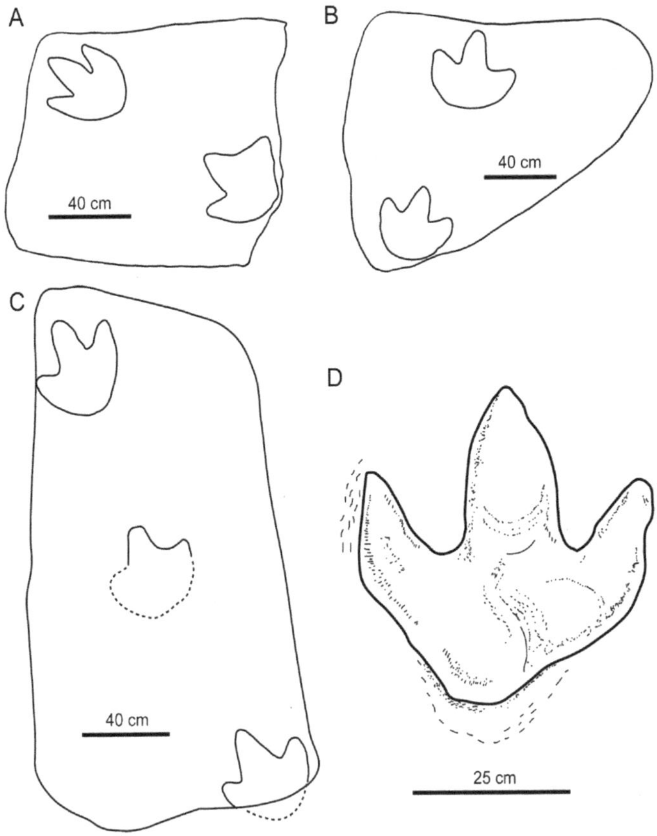

FIGURE 7.15 Tracks of *Iguanodontipus*. (A–C) Type series of *Iguanodontipus burreyi*; (D) referred track of *Iguanodontipus burreyi*. *(A–C) Redrawn from Sarjeant et al. (1998). (D) Redrawn from Beckles (1852).*

3.4.4.5 Diagnosis

Tracks belonging to Iguanodontipodidae with a small heel impression that is rounded, centered, and narrow (as wide as the width of the proximal part of the Digit III impression); long, narrow digit impressions with sharp distal ends (Díaz-Martínez et al., 2015a).

3.4.4.6 General description

The type series consists of seven poorly preserved natural casts from the Berriasian of England. These tracks present a large, rounded heel and three short, wide digits, lacking pad and claw impressions (Sarjeant et al., 1998). On the other hand, some referred tracks to *Iguanodontipus*, found also in the Berriasian of England, appear undistorted and differ from those in the type series, featuring a narrow, rounded heel and long, separate digit impressions with pointed distal ends. The wide, rounded heel impression seen in the *I. burreyi* holotype is attributed to the absence of the notch in the proximal part of Digits II and IV. The triangular appearance of the digits is likely due to the distal shift of the hypex in poorly preserved footprints (Díaz-Martínez et al., 2015a).

3.4.4.7 Discussion

The description of *Iguanodontipus* by Sarjeant et al. (1998) aimed to formally group tracks that were previously attributed to *Iguanodon*. Some authors pointed out that the type series trackway in the Bournemouth Museum of England, is formed by a special type of deep undertracks that look like ornithopod tracks but are produced by a theropod (e.g., Meyer and Thüring, 2003; Piñuela et al., 2016). The diagnosis proposed by Sarjeant et al. (1998) reflects the shape of the tracks in the type series, but the authors also included other footprints with a more complex morphology under *Iguanodontipus*, which did not align with these diagnostic features. Indeed, Gierliński et al. (2008) and Lucas et al. (2011) have illustrated *Iguanodontipus* using a footprint of "*Ornithoidichnites*" that was studied originally by Beckles and was subsequently referred to *Iguanodontipus* by Sarjeant et al. (1998). In this context, some authors questioned that this ichnogenus is available due to the poorly state of preservation of the material and even questioned whether the tracks were made by an ornithopod (DallaVecchia et al., 2002; Diedrich, 2004; Piñuela et al., 2016). If *Iguanodontipus* was no longer used, a new ichnogenus would have to be erected to include the aforementioned and well-preserved material. Appealing to the stability of the system, as the ICZN recommends, Díaz-Martínez et al. (2015a,b) decided to modify the diagnosis using the features of additional material.

Iguanodontipus has only one ichnospecies, *I. burreyi,* and has been identified only in the Berriasian–Valanginian of Europe (England, Germany, and Spain; see Díaz-Martínez et al., 2015a; Castanera et al., 2022b). Nevertheless, the specimens of Spain and Germany are slightlydifferent from each other and with respect to the type (see Hornung et al., 2012; Castanera et al., 2013b), therefore, a single ichnospecies may be an underestimate, and all this material should be carefully reviewed.

Based on Swiss Aptian tracks, Meyer and Thüring (2003) erected *Iguanodontipus billsarjeanti* as a new ichnospecies. But we think these tracks match better with *Caririchnium* with respect to the heel and digit features than *Iguanodontipus* (see below). Nevertheless, the trackway is poorly preserved, each footprint is different from others and two tracks are almost missing, so a revision is needed to discuss the validity of this ichnospecies. On the other hand,

Figueiredo et al. (2022a) classified as *Iguanodontipus* isp. several tracks from the Barremian of Portugal. These tracks are poorly preserved and probably can be assigned to Iguanodontipodidae indet. Noè et al. (2020) identified *Iguanodontipus* in the upper Valanginian–lower Hauterivian of Colombia, but probably these tracks are theropod in origin. Finally, the track classified as *Iguanodontipus* by Penn et al. (2020) in the Barremian–Aptian of England is poorly preserved and could be related to Iguanodontipodidae indet. or even an undetermined theropod track.

There is also material found in Upper Jurassic rocks similar to *Iguanodontipus*. For instance, Piñuela et al. (2016) reported medium- to large-sized ornithopod tracks from the Late Jurassic of Asturias, Spain. They noted that these tracks differ from known ichnogenera but are not well enough preserved to propose a new one. Castanera et al. (2020) described several large ornithopod tracks from the Late Jurassic of Portugal, and although they point out their similarity to *Iguanodontipus*, they also see clear differences. Following the observations of these authors, the Upper Jurassic footprints could belong to a new ichnogenus, but new material has to be found to demonstrate that.

3.4.4.8 Trackmaker

Iguanodontipus has been historically related to *Iguanodon* (Sarjeant et al., 1998). Nevertheless, this assignment is not possible because the chronology of both species of *Iguanodon*, *I. bernissartensis* and *I. galvensis*, is Barremian-early Aptian (see Verdú et al., 2015), and *Iguanodontipus* is Berriasian-Valanginian.

Probably, this kind of track is associated with basal members of Ankylopollexia or Styracosterna. Possible candidates are *Owenodon hoggii* and have been found in the Berriasian Purbeck Beds of England (Galton, 2009), and *Barilium dawsoni* or *Hypselospinus fittoni*, both from the Valanginian Lower Wealden Group of England (Norman, 2010).

3.4.5 Caririchnium *Leonardi, 1984 (Figs. 7.16 and 7.36A)*

3.4.5.1 Type ichnospecies

Caririchnium magnificum Leonardi, 1984 (Fig. 7.16A and B)

3.4.5.2 Valid ichnospecies

Caririchnium magnificum Leonardi, 1984.
Caririchnium kortmeyeri (Currie and Sarjeant, 1979) (Fig. 7.16C and D)
Caririchnium billsarjeanti (Meyer and Thüring, 2003) (Fig. 7.16E and F)
Caririchnium lotus Xing et al., 2007 (Figs. 7.16G,H and 7.36A)
Caririchnium yeongdongensis Kim et al., 2016 (Fig. 7.16I and J)

3.4.5.3 Distribution

Berriasian–Albian, Brazil, Portugal, Spain, England, Switzerland, South Korea, USA, Canada, Japan, China.

3.4.5.4 Holotype

Trackway about 25 m long. Cast of the first manus and pes impressions in the Museu Câmara Cascudo Federal University of Rio Grande do Norte, Natal, Brazil.

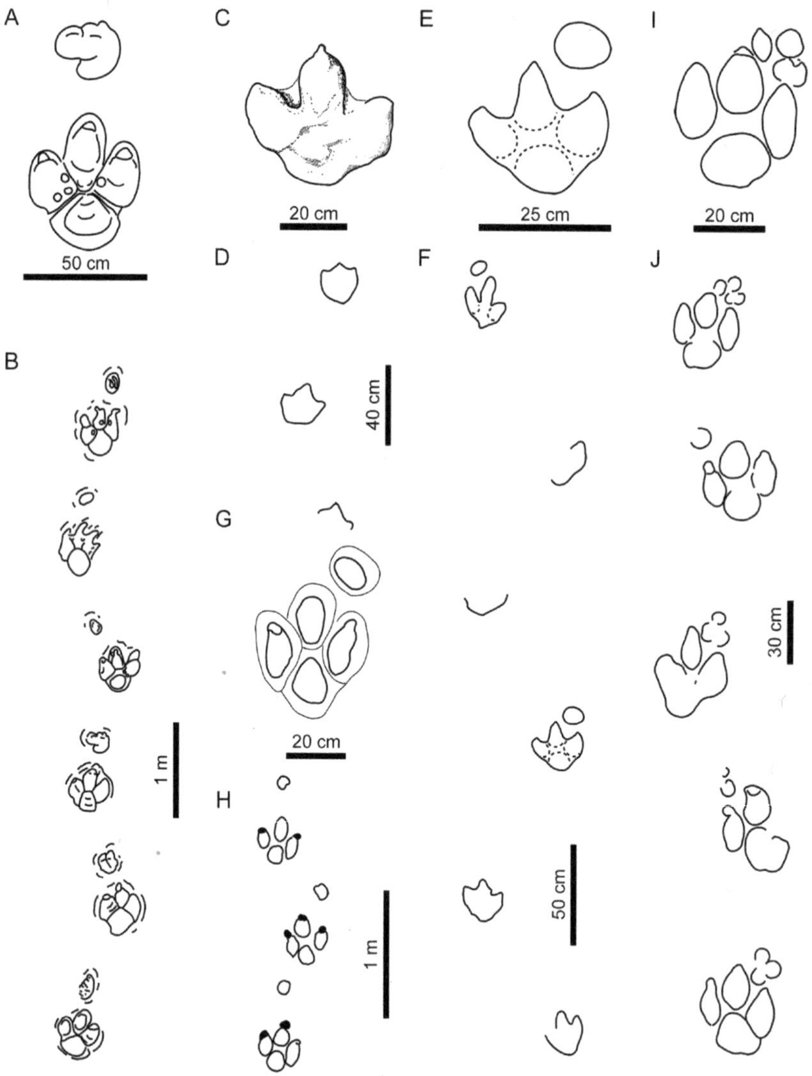

FIGURE 7.16 Tracks of *Caririchnium*. (A and B) Holotype and type series of *Caririchnium magnificum*; (C and D) type series of *Caririchnium kortmeyeri*; (E and F) type series of *Caririchnium billsarjeanti*; (G and H) holotype and type series of *Caririchnium lotus*; (I and J) type series of *Caririchnium yeongdongensis*. *(A and B) Redrawn from Leonardi (1987). (C and D) Redrawn from Currie and Sarjeant (1979). (E and F) Redrawn from Meyer and Thüring (2003). (G and H) Redrawn from Xing et al. (2015f). (I and J) Redrawn from Kim et al. (2016).*

3.4.5.5 Diagnosis

Pes tracks belonging to Iguanodontipodidae, with a large heel impression that is rounded, centered and wide (wider than the width of the proximal part of the Digit III impression); short, wide digit impressions (Díaz-Martínez et al., 2015a,b).

3.4.5.6 General description

Caririchinium is characterized by bipedal and quadrupedal trackways. Quadrupedal trackways present strong heteropody. Pes tracks are tridactyl and mesaxonic. They have three short and wide digit prints with blunt distal ends preserved as large elongated pads, and one broad, centered, rounded to triangular "heel" pad. *Caririchnium* has a large "heel" that is wider than the maximum width of the proximal part of Digit III. The ratio between the length and width of the "heel" pad is different among the ichnospecies. In general, the pes tracks are as wide as or wider than long. The size is very variable, the usual pes tracks are between 30 and 60 cm long (Leonardi, 1987; Meyer and Thüring, 2003; Xing et al., 2007; Razzolini et al., 2016), but there may be smaller or larger examples (e.g., Lockwood et al., 2014; Xing et al., 2016f, 2018c; Lee et al., 2023).

Manus tracks, which are much smaller than pes tracks, are wider than long. Their morphology is very distinctive and useful to differentiate among ichnospecies: elliptical, rectangular, and clover-leaflike subtriangular (Leonardi, 1987; Meyer and Thüring, 2003; Xing et al., 2007; Kim et al., 2016). Manus tracks are located in front of the pes track, between the Digits III and IV (see Fig. 7 in Díaz-Martínez et al., 2015a,b).

3.4.5.7 Discussion

Leonardi (1984) presented two distinct diagnoses for the ichnogenus and ichnospecies when he defined *Caririchnium*. Later, Lockley (1987) revised the ichnogeneric diagnosis, essentially translating Leonardi's original Italian diagnosis into English. In addition, Lee (1997) revised the *Caririchnium* diagnosis based on the distinct features of trackways made by quadrupedal and bipedal trackmakers. Finally, Díaz-Martínez et al. (2015a,b) emended the ichnogeneric diagnosis and discussed the validity of the assigned ichnospecies.

According to Currie (1995), junior synonyms of "*Amblydactylus*" include *Caririchnium*, "*Hadrosaurichnus*," and "*Ornithopodichnites*." However, the latter ichnogenera were considered nomina dubia by Díaz-Martínez et al. (2015a,b). Lucas et al. (2011) pointed out that there are no characteristics that set *Caririchnium* and *Hadrosauropodus* apart, and that their diagnoses are comparable. As a result, they suggested that *Hadrosauropodus* and *Caririchnium* are synonyms. This proposal is probably based on the "heel" shape due to some authors (Hunt and Lucas, 1998; Lim et al., 2012) considering that *Caririchnium* exhibits a bilobed heel imprint. However, as was previously mentioned, the *Hadrosauropodus* "heel" impression is bilobed, whereas the *Caririchnium* "heel" is rounded. Therefore, we follow the proposal of Lockley et al. (2014a) and Díaz-Martínez et al. (2015a,b) and consider both ichnogenera valid.

Originally, *Caririchnium* had seven ichnospecies: *C. magnificum*, *C. leonardii*, *C. protohadrosaurichnos*, *C. lotus*, *C. kyoungsookimi*, *C. liucixini* and *C. yeongdongensis*. The ichnospecies *C. magnificum* is featured by subtriangular "heel" pad impressions, which are wider than long, have blunt distal ends of digit impressions and wider than long elliptical manus tracks (see Leonardi and Carvalho, 2021c). On the other hand, *C. leonardii* presents bilobed "heel" impressions, thus Díaz-Martínez et al. (2015a,b) included it within *Hadrosauropodus* as *H. leonardii* (see below). *C. protohadrosaurichnos* is considered as a nomen dubium (see Díaz-Martínez et al., 2015a,b) because it was defined based on the position of the manus impression (Lee, 1997), which is not a valid ichnotaxonomic feature because it is conditioned by the dinosaur's behavior (Hunt and Lucas, 1998; Lim et al., 2012). The ichnospecies *C. lotus* has a subtriangular distal part of the "heel" pad, which is longer than wide, digit impressions with blunt distal end and manus tracks rectangular, smoothly concave at the proximal part and wider than long (Xing et al., 2015f). As *C. kyoungsookimi* shows a bilobed "heel" impression, it is also included in *Hadrosauropodus* as *H. kyoungsookimi* (Díaz-Martínez et al., 2015a,b, see below). *C. yeongdongensis* presents a subrounded to subtriangular "heel" pad that is wider than long, digit impressions with blunt distal end and manus tracks with three suboval to subrounded digit traces forming a clover-leaflike subtriangular arrangement (Kim et al., 2016). The pes tracks are similar to those of *C. magnificum* and *C. billsarjeanti*, but the manus impressions are different. As suggested by Lim et al. (2012) the shape of the manus impressions is a very important feature in the ichnotaxonomy of large ornithopod tracks. Although the specimens assigned to *C. yeongdongensis* are still scarce, we consider that it is an available ichnospecies that should be reaffirmed with new finds. The ichnospecies *Caririchnium liucixini* shows pes tracks with subrounded to triangular "heel" pad impression, which is approximately as wide as or wider than long, digit impressions with blunt distal end and lacks manus traces. The pes tracks are similar in shape to *C. magnificum*, *C. billsarjeanti*, and *C. yeongdongensis*. The main ichnotaxobase to differentiate among these three ichnospecies is the manus impression. Therefore, until manus impressions are found associated with this ichnotaxon, we consider these tracks as *Caririchnium* isp. *C. billsarjeanti*, previously classified within *Iguanodontipus*, has a rounded distal part of the "heel" pad, which is approximately as wide as or wider than long, digit impressions with blunt distal end and manus tracks elliptical and wider than long. Whereas, *C. kortmeyeri*, previously classified within "*Amblydactylus*," presents a subtriangular distal part of the "heel" pad that is approximately as wide as or wider than long and a sharp distal end of the digit impressions.

Caririchnium is a quite a frequent ichnogenus in Early Cretaceous faunas. It is very abundant in the Hauterivian–Albian (see Lucas and Dalman, 2016; Tsukiji et al., 2018; Xing et al., 2021c,d; Castanera et al., 2022a; García-Cobeña et al., 2023; Figueiredo et al., 2022b), although it is also identified in the earliest Cretaceous (Shillito and Davies, 2019; Xu et al., 2021).

Many large ornithopod ichnotaxa– "*Brachyguanodonipus*," "*Iguanodonipus*," "*Sousaichnium*," "*Staurichnium*," "*Limayichnus*," and "*Ornithopodichnus*"—that were deemed non available in the current study had morphologies comparable to *Caririchnium*.

3.4.5.8 Trackmaker

Leonardi (1987) did not initially attribute it to an ornithopod trackmaker, which raises the interesting question of whether he made sufficient comparisons with other ichnotaxa, such as "*Amblydactylus*." He suggested, nevertheless, that it was definitely of ornithischian affinity. Lockley (1987) deemed his endeavors to juxtapose *Caririchnium* with other ichnotaxa more than sufficient, considering the literature accessible during that period, which contained minimal mention of the tracks of massive quadrupedal ornithopods.

Probably the trackmaker of *Caririchinium* was a basal ankylopollexian or styracosternan (see Díaz-Martínez et al., 2015a,b). In Spain, García-Cobeña et al. (2023) proposed medium- to large-sized styracosternans, such as *Iguanodon*, as a trackmaker of El Castellar Formation *Caririchnium* tracks. It is curious that in South America, where *C. magnificum* was defined, there is no skeletal record of large representatives of these clades in the Early Cretaceous (but see Bandeira et al., 2024).

3.4.6 Hadrosauropodus *Lockley et al., 2004a (Figs. 7.17 and 7.36D)*

3.4.6.1 Type ichnospecies

Hadrosauropodus langstoni Lockley et al., 2004a (Fig. 7.17 A and B)

3.4.6.2 Valid ichnospecies

Hadrosauropodus leonardii Lockley, 1987 (Fig. 7.17C)
Hadrosauropodus langstoni Lockley et al., 2004a.
Hadrosauropodus kyoungsookimi Lim et al., 2012 (Figs. 7.17D and 7.36D)

3.4.6.3 Distribution

Aptian–Maastrichtian, South Korea, Canada, USA, Mexico, China, Spain and Argentina.

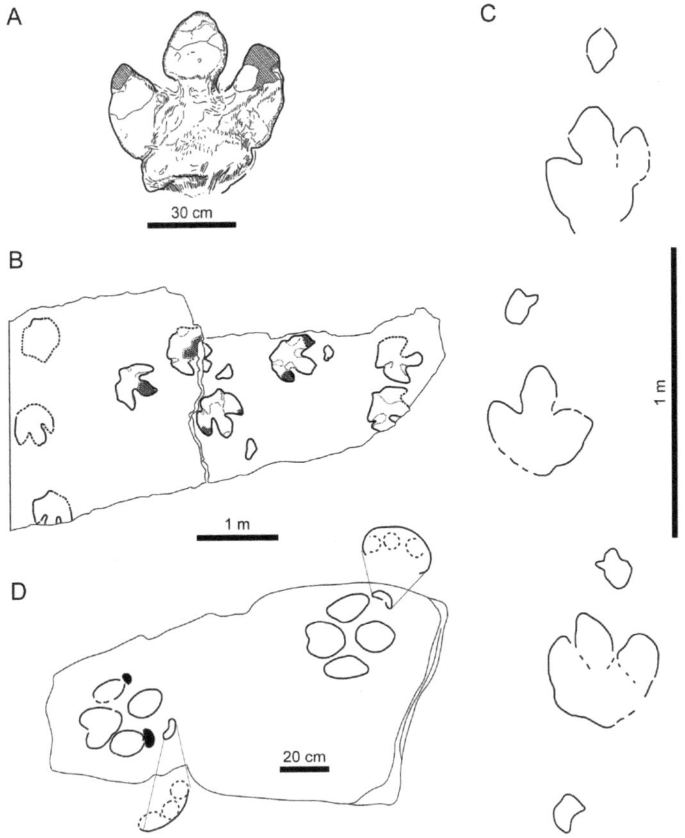

FIGURE 7.17 Tracks of *Hadrosauropodus*. (A) Holotype and (B) type series of *Hadrosauropodus langstoni*; (C) type series of *Hadrosauropodus leonardii*; (D) type series of *Hadrosauropodus kyoungsookimi*. *(B) Redrawn from Lockley et al. (2004a). (C) Redrawn from Lockley (1987). (D) Redrawn from Lim et al. (2012).*

3.4.6.4 Holotype

TMP 87.76.7, a pes track found with associated manus and topotype/paratype trackway segments on the same surface in field.

3.4.6.5 Diagnosis

Tracks belonging to Iguanodontipodidae with a large heel impression that is bilobed, centered, and wide (wider than the width of the proximal part of the Digit III impression); pad of Digit III shorter than those of Digits II and IV; short, wide digit impressions with blunt distal ends (Díaz-Martínez et al., 2015a,b).

3.4.6.6 General description

In general, *Hadrosauropodus* pes tracks have a quadripartite configuration, with three short and wide digit prints with blunt distal ends preserved as large elongated pads, and one broad, centered, bilobed "heel" pad. The width of the bilobed "heel" impression and the place of Digits II and IV notches with respect to the proximal part of the Digit III pad impression are distinctive among the ichnospecies. The pes tracks are as wide as or wider than long. Very small specimens are known (Fiorillo et al., 2014), but usually they measure between 30 and 60 cm long (e.g., Lockley, 1987; Lockley et al., 2004a; Lim et al., 2012; Enriquez et al., 2022).

Manus tracks are much smaller than pes ones. They are wider than long and have a different shape in each ichnospecies: rectangular, crescent-shaped, and triangular (Lockley, 1987; Lockley et al., 2004a; Lim et al., 2012). Occasionally, manus tracks are less deeply impressed than the pes (Lim et al., 2012). They are located in front of the pes track, between the Digits III and IV (see Fig. 8 in Díaz-Martínez et al., 2015a,b).

3.4.6.7 Discussion

Hadrosauropodus is an ichnogenus of ornithopod dinosaur tracks with a bilobed "heel impression," defined by Lockley et al. (2004a) based on tracks from the Maastrichtian of Canada, previously regarded as "hadrosaur footprints" and named as *Hadrosauropodus langstoni*. They also assigned tracks from the same age in the USA to *Hadrosauropodus* isp. Xing et al. (2009c) erected *Hadrosauropodus nanxiongensis* for some Maastrichtian tracks in the Chinese Zhutian Formation, and referred more tracks from the same formation to *Hadrosauropodus* isp. *H. nanxiongensis* is a nomen dubium because it is based on poorly preserved tracks, but its bilobed heel indicates that it belongs to *Hadrosauropodus*. Later studies reported *Hadrosauropodus* tracks from the Campanian–Maastrichtian of Spain, Canada, China and Argentina (Suñer et al., 2008; Xing et al., 2009c, 2017a; Fanti et al., 2013; Vila et al., 2013a; Fiorillo et al., 2014; McCrea et al., 2014a; Díaz-Martínez et al., 2016; Herrero et al., 2016), indicating a wide geographical but very limited temporal distribution.

Nevertheless, *Hadrosauropodus* is not the first ichnotaxon to which footprints with bilobed "heel" impressions have been assigned. Other tracks from earlier stages, such as the Albian–Cenomanian and the Campanian, were assigned to different ichnotaxa, such as *Caririchnium* and *Amblydactylus* (Lockley, 1987, 1988; Currie et al., 1991; Lim et al., 2012). However, some authors consider that *Caririchnium* has a rounded heel impression and is distinct from *Hadrosauropodus* (see Vila et al., 2013a; Díaz-Martínez et al., 2015a,b) and thus reassign some *Caririchnium* species to *Hadrosauropodus*. Examples include the ichnospecies *C. leonardii* and

C. kyoungsookimi from the mid-Cretaceous of the USA and South Korea (Lockley, 1987; Lim et al., 2012), respectively, which were referred to *Hadrosauropodus* by Díaz-Martínez et al. (2015a,b). The three ichnospecies of *Hadrosauropodus—H. langstoni, H. leonardii,* and *H. kyoungsookimi*—are distinguished by the "heel" morphology, proximal digit impressions, and the manus imprint. For instance, *H. langstoni* has a wider, more bilobed "heel" impression than the other ichnospecies, and the notches of Digits II and IV are placed more proximally. *H. leonardii* and *H. kyoungsookimi* pes tracks are very similar and they differ in manus morphology: rectangular in *H. leonardii*; crescent-shaped in *H. kyoungsookimi*; and triangular in *H. langstoni*. More studies are needed to examine the variation in the manus tracks within the *Hadrosauropodus* ichnospecies to know its ichnotaxonomic significance. Recently, Li et al. (2023) published several *Hadrosauropodus* tracks from the Late Cretaceous of China. They pointed out that they have unique features and thus might represent a new ichnospecies of *Hadrosauropodus*, but more material is needed to confirm it.

H. leonardii and *H. kyoungsookimi* represent the earliest records of *Hadrosauropodus* (Aptian-Cenomanian) and *H. langstoni* is identified in Campanian-Maastrichtian beds.

3.4.6.8 Trackmaker

The ichnogenus *Hadrosauropodus* (Aptian–Maastrichtian) shows a temporal distribution that matches the skeletal fossil record of Hadrosauroidea. Hadrosauroidea includes all the taxa that are more closely related to *Edmontosaurus* than to *Iguanodon*, and encompasses the hadrosaurids, which were the most abundant and diverse large vertebrates of Laurasia in the latter half of the Late Cretaceous. The first hadrosauroids appeared in the late Early Cretaceous, and they were especially diverse in Asia and to a lesser degree in North America. Based on stratigraphy, the footprints of *Hadrosauropodus kyoungsookimi* from the Barremian-Albian of Asia and *H. leonardii* from the Albian-Cenomanian of North America might have been produced by nonhadrosaurid hadrosauroids (such as *Altirhinus, Bolong, Equijubus, Jinzhousaurus, Probactrosaurus* and *Xuwulong* from the Barremian-Albian of Asia, and *Eolambia* and *Protohadros* from the Cenomanian of North America), while those of *H. langstoni* and *Hadrosauropodus* isp. from the Maastrichtian of North America, Europe South America and Asia might have been produced by advanced hadrosaurids (Díaz-Martínez et al., 2015a,b).

STEGOSAURIA (FIG. 7.18)

While during the Jurassic, the tracks of thyreophorans were dominated by stegosaurs, during the Cretaceous, these tracks are quite residual and the tracks of ankylosaurs are more abundant (see Pereda-Suberbiola et al., 2015; Maidment et al., 2020). This pattern is also visible in the ichnological record (see McCrea et al., 2001; Xing et al., 2013d; and references).

There are some ichnogenera related with stegosaurian trackmakers: *Deltapodus, Stegopodus, Shenmuichnus, Garbina,* and *Luluichnus*. Of them, *Deltapodus, Stegopodus, Shenmuichnus* have been described from Jurassic material, so they are more extensively discussed in the Jurassic chapter of this volume. Herein, *Deltapodus curriei, Garbina,* and *Luluichnus* are examined.

3.4.7 Deltapodus *Whyte and Romano, 1994 (Fig. 7.18A)*

3.4.7.1 Type ichnospecies

Deltapodus brodricki Whyte and Romano, 1994.

3.4.7.2 Valid ichnospecies

Deltapodus brodricki Whyte and Romano, 1994.

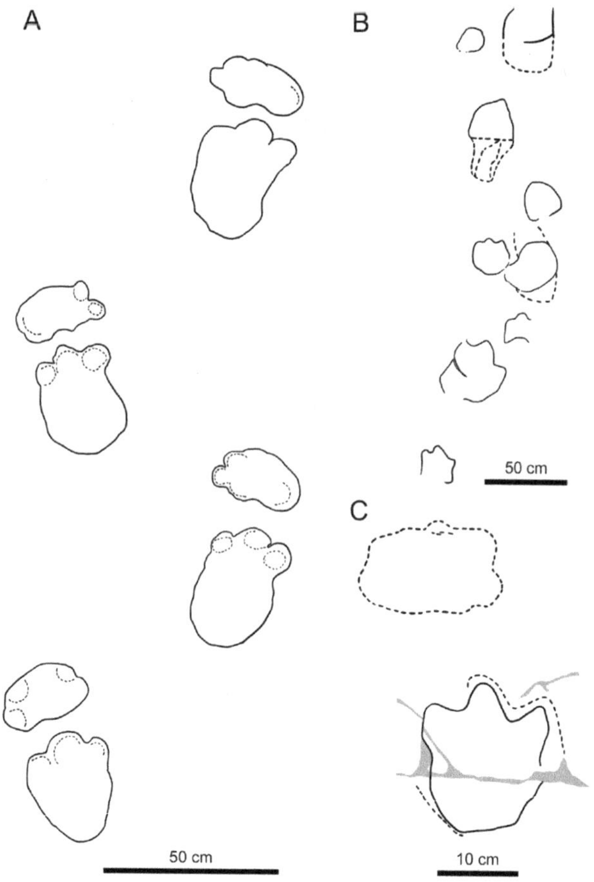

FIGURE 7.18 Tracks of Stegosauria. (A) Type series of *Deltapodus curriei*; (B) holotype of *Garbina roeorum*; (C) holotype of *Luluichnus*. *(A) Redrawn from Xing et al. (2013d). (B) Redrawn from Salisbury et al. (2016). (C) Redrawn from Salisbury et al. (2016).*

Deltapodus curriei Xing et al., 2013d (Fig. 7.18A)
Deltapodus ibericus Cobos et al., 2010.

3.4.7.3 Distribution

Cretaceous: Berriasian to the top of the Maastrichtian, China, France, India, Spain, USA.

3.4.7.4 Holotype

A replica of the holotype is F00768, latex mold and corresponding plaster cast of a left manus-pes couple. Deposited in the Paleoecological Collection, Department of Earth Sciences, University of Sheffield.

3.4.7.5 Diagnosis

Manus: (1) entaxonic footprints, (2) broadly crescent-shaped in outline, and anteriorly convex, and (3) occasional development of an inwardly (or medially) directed digit (pollex) impression. Pes: (4) generally oval, subtriangular in outline, (5) internal side slightly concave,

(6) pes prints mesaxonic, with three-digit impressions as short projections or bluntly rounded points. Trackway: (7) medium gauge (internal trackway width moderately wide: approx. 10–30 cm) (emended from Cobos et al., 2010).

3.4.7.6 General description

The pes track shape of all *Deltapodus* material is very similar. They are tridactyl, mesaxonic, and generally present a triangular outline. The digit impressions are short and blunt with a maximum width across the base of the digit impressions. The Digit II is the most robust. The "heel" impression is elongated, making the tracks about 35%–50% as long as wide (see Whyte and Romano, 1994; Milan and Chiappe, 2009; Xing et al., 2013d). Pes tracks are larger than manus tracks.

Manus tracks are more variable among the *Deltapodus* ichnospecies. For instance, *D. brodricki* (Whyte and Romano, 1994) has a crescentic manus impression that lacks clear digit traces, *D. ibericus* (Cobos et al., 2010) is kidney-shaped and has a regular, broadly crescent-shaped outline, with no clear digit traces, and *D. curriei* (Xing et al., 2013d) presents an oval shape with well-defined traces of manual Digits I and II. The manual tracks are wider than long.

3.4.7.7 Discussion

Jurassic *Deltapodus* ichnospecies are *Deltapodus brodricki*, originally described from the Middle Jurassic of England, and *Deltapodus ibericus* from the Tithonian (age sensu Campos-Soto et al., 2017) of Spain. Both ichnospecies are described and discussed in the Jurassic chapter. The Lower Cretaceous *Deltapodus curriei* was erected by Xing et al. (2013d) in China with the diagnosis: "The manus track is entaxonic and oval-shaped, although somewhat irregular. The manus is approximately twice as wide as long. The impression of manual Digit I is distinct and robust. The medial metacarpophalangeal pad region is slightly concave, while the lateral region is slightly convex. The length of the pes track slightly exceeds its width by 35%. The pes tracks have three functional digits (II–IV) that are distinct but short, bluntly rounded impressions. These digit impressions are subequal in length and display a higher degree of entaxony than other species of *Deltapodus*. The heel region is elongate and has smoothly curved margins; the divarication between Digits II and III appears slightly greater than that between Digits III and IV, but this is difficult to measure accurately in tracks with such short, blunt toe traces. The trackway (both manus and pes) exhibits slight outward rotation."

D. curriei differs from the other ichnospecies of *Deltapodus* in the manus track. The former has well-defined traces of manual Digits I and II, while *D. brodricki* has a crescentic manus impression that lacks clear digit traces, and *D. ibericus* has a kidney-shaped manus impression that lacks also digit traces (Cobos et al., 2010; Xing et al., 2013d).

The record of *Deltapodus* in Cretaceous beds is scarce. It has been identified in the Lower Cretaceous of Spain (Pascual et al., 2012; Herrero-Gascón and Pérez-Lorente, 2017), China (Xing et al., 2013d, 2021k), USA (Milan et al., 2015; Lucas and Dalman, 2016) and Iran (Abbassi et al., 2018). Mateus et al. (2011) called a single track from the Upper Cretaceous of India (Mohabey, 1986) *Deltapodus*-like, and said that it might be very young, and younger than any known stegosaurian trackmakers.

3.4.7.8 Trackmaker

Deltapodus was considered initially as a sauropod ichnotaxon (Whyte and Romano, 1994). Subsequently, the same authors saw that many characteristics that do not conform to sauropod tracks, such as (1) the pes track is triangular, while sauropod pes tracks are mostly round or oval; (2) the shape of the pes track is axially symmetrical, and in sauropods it is asymmetrical; (3) the pes tracks have three claw traces, while sauropod tracks, if they have preserve digit impressions, generally show four to five short digit traces; and (4) the digit impressions of pes tracks appear on the front edge, whereas in sauropod tracks, they are often curved (Whyte and Romano, 2001). Finally, these authors concluded that the *Deltapodus* tracks are stegosaur tracks.

3.4.8 Garbina *Salisbury et al., 2016 (Fig. 7.18B)*

3.4.8.1 Type ichnospecies

Garbina roeorum Salisbury et al., 2016.

3.4.8.2 Valid ichnospecies

Garbina roeorum Salisbury et al., 2016.

3.4.8.3 Distribution

Valanginian–Barremian, Australia.

3.4.8.4 Holotype

Replicas of the holotype are WAM12.1.19 and WAM12.1.20, polyurethane resin replicas from associated natural molds of a left manus and a left pes, respectively (track locality references UQL-DP141, UQL-DP14-1), Broome Sandstone (Dampier Peninsula) of Western Australia.

3.4.8.5 Diagnosis

Manual tracks: medium- to large-sized (proximodistal length 11.9–21.4 cm, mediolateral width 17.2–25.4 cm), tetradactyl, paraxonic, typically wider than long, with a maximum proximodistal length to maximum mediolateral width ratio of 0.6–1.1; individual manual digital impressions proportionately short and broad (the maximum width of each digital impression 16%–50% of the total track length); apex of each digital impression blunt to slightly tapered; impressions of Digits II and III extend subequally (i.e., paraxonic) distally beyond the impression of Digits I and IV by about one-third the maximum track length; impression of Digit I often positioned more distally relative to the impression of Digit IV; impressions of Digits II, III, and IV subequal in size, with the impression of Digit I typically narrower; total divarication angle between the axes of the impressions of Digits I and IV 137°–170°; divarication of axes between digital impressions variable (12°–60°, 39°–78°, and 26°–54° for I–II, II–III, and III–IV, respectively); proximal track margin variable in shape; metacarpodigital pad impression may or may not be present; impressions of digital pads and ungual impressions absent; manual impressions occur craniolateral to pedal tracks, but placement can vary; manual impressions may be absent, only faintly impressed, or as deeply impressed as associated pedal impressions. Pedal tracks: medium-to very large-sized (length of unified digital-metatarsodigital portion of track 25.8–44 cm; total length of tracks that include a metatarsal pad impression 42–80 cm; width 32.5–70 cm), tridactyl, mesaxonic (toe extension to track length ratio 0.14–0.21), typically wider than long for the unified digital-metatarsodigital portion of the track, with a maximum proximodistal length to maximum mediolateral width ratio of 0.54–0.9, or longer than wide when the metatarsal pad is impressed with a maximum-length-to-maximum-width ratio of 1.1–1.7; individual digital impressions proportionately short (the maximum length of each digital impression 17%–51% of the total track length excluding the metatarsal pad impression) and broad (the maximum width of each digital impression 17%–40% of the total track length excluding the metatarsal pad impression), and rounded distally; impression of Digit III extends distally beyond the impressions of Digits II and IV by 14%–21% of the total track length excluding the metatarsal pad impression; hypex of the impressions of Digits III and IV usually broader than that of Digits II and III, and more pronounced in bipedal pedal track; total divarication angle between the axes of impressions of Digits II and IV 57°–93°; divarication of axes between impressions of Digits II and III (27°–46°) and III and IV (30°–50°) similar; when present, the metatarsal pad impression is wide and irregularly shaped; when absent, the metatarsal pad impression is gently convex distally, and continuous with the proximal margin of the impression of the unified digital metatarsodigital pad; impressions of digital pads and unguals absent. Trackway (quadrupedal): manual track pace length approximately 2–6 times maximum manual track length; manual track stride length approximately 3–5 times maximum manual track length; manual tracks inwardly rotated, or directed cranially relative to the trackway orientation; metatarsal pad impression present in "quadrupedal" tracks; pace angulation of pedal track is approximately 133° (129°–142°); pedal pace length (excluding metatarsal pad) approximately 3 times pedal track length; pedal track stride length approximately 3–5 times the maximum pedal track length (excluding metatarsal pad). Trackway (bipedal): the metatarsal pad impression absent; pedal track pace length approximately 2 times the maximum pedal track length; pedal tracks inwardly rotated relative to trackway orientation; trackway is "narrow"-gauge, with the medial margin of the pedal tracks crossing the trackway midline (Salisbury et al., 2016).

3.4.8.6 General description

The best-preserved manus tracks are tetradactyl, wider (17.2–25.4 cm) than long (11.9–21.4 cm), and paraxonic. The outer digit impressions (I and IV) are oriented laterally and medially, while the inner digits are forwardly oriented. The digit impressions are short, wide, and blunt. They lack digital pads. Pes tracks are tridactyl, longer (42–80 cm) than wide (32.5–70 cm), and mesaxonic. Digit impressions are short, broad, and blunt. The metatarsal area is large and elongated. Metatarsal pad is preserved in some specimens, but lacks digital pad impressions. Pes tracks are much larger than manus tracks (Salisbury et al., 2016).

3.4.8.7 Discussion

Salisbury et al. (2016) compared *Garbina* with the main thyreophoran ichnotaxa. Tridactyl pes tracks are present in *Deltapodus* and *Stegopodus*. The authors pointed out that *Deltapodus* pes tracks lack pad impressions, but *Garbina* has a clear gap between the metatarsophalangeal and digital pad impressions. Salisbury et al. (2016) observed several similarities between *Garbina* and *Stegopodus*, but the former pes track is more symmetrical and presents a greater divarication angle. Moreover, the digital manus tracks of both ichnogenera have a different arrangement.

Romilio et al. (2021) reported a manus track classified as cf. *Garbina* from the middle Oxfordian of Queensland, which is the oldest stegosaurian fossil in Australia.

3.4.8.8 Trackmaker

The trackway and pes/manus track parameters (quadrupedal animal with tridactyl pes and great heteropody) are typical of ornithischians (see Salisbury et al., 2016). These authors affirmed that the shape of the tracks shows that they were made by a stegosaurian armoued dinosaur that moved mainly with its hindlimb and was capable of both quadrupedal and bipedal stances.

3.4.9 Luluichnus *Salisbury et al., 2016 (Fig. 7.18C)*

3.4.9.1 Type ichnospecies

Luluichnus mueckei Salisbury et al., 2016.

3.4.9.2 Valid ichnospecies

Luluichnus mueckei Salisbury et al., 2016.

3.4.9.3 Distribution

Valanginian–Barremian, Australia.

3.4.9.4 Holotype

A replica of the holotype is WAM15.12.701, polyurethane resin replicas from associated natural molds of a left manus and a left pes within a trackway (track locality reference UQL-DP45-6); NMVP230370-B, a plaster replica of the left pedal impression (UQL-DP45-6), Broome Sandstone (Dampier Peninsula) of Western Australia.

3.4.9.5 Diagnosis

Manus tracks: medium- to large-sized (proximodistal length 11–14 cm, mediolateral width 19.5–22 cm), typically wider than long, with a proximodistal length to maximum mediolateral width ratio of approximately 0.6; overall track morphology oval with indistinct/absent digital impressions; digital pad impressions absent. Pedal tracks: small- to medium-sized (proximodistal length 15–22.5 cm, mediolateral width 16–20.7 cm), tridactyl, mesaxonic, longer than wide (maximum proximodistal length to maximum mediolateral width ratio of 1.0–1.3); individual digital impressions proportionately short (maximum length of each digit impression to maximum proximodistal track length ratio of 0.13–0.35); individual digit impressions fairly broad (maximum width of each digit impression to maximum proximodistal track length ratio of 0.2–0.37); Digit II and IV impressions extend subequally relative to the principal track axis; total divarication angle between the axes of the impressions of Digits II and IV 30°–55°; divarication of axes between impressions of Digits II and III (21°–35°) and Digits III and IV (15°–39°) approximately the same. Trackway (quadrupedal): pace angulation of pedal track stride length approximately 5.8 times the maximum pes length; pedal track pace length approximately 3.3 times the pedal track length. Pedal tracks can be laterally or medially rotated relative to the trackway midline; manual tracks laterally rotated, and directed craniolateral relative to the trackway orientation.

3.4.9.6 General description

Manus tracks are oval in shape without digit or pad impressions. They are wider (19.5–22 cm) than long (11–14 cm). Pes tracks are tridactyl, mesaxonic, and longer (15–22.5 cm) than wide (16–20.7 cm). Digit impressions are short, sharp, and broad, and the "heel" region is elongated and wide. Pes and manus tracks are similar in size (Salisbury et al., 2016).

3.4.9.7 Discussion

Salisbury et al. (2016) suggested that *Luluichnus* is different from *Garbina* because it does not have clear digit traces on the manus tracks, and it has sharp digit impressions on the pes tracks. The pes and manus tracks are also similar in size, unlike *Garbina* where the pes tracks are much larger than the manus tracks. *Stegopodus* and *Deltapodus* also present tridactyl pes tracks. The pes digit impressions of *Stegopodus* are long and elongated, those of *Deltapodus* short and blunt and those of *Luluichnus* short and subtriangular. The manus tracks of *Luluichnus* are similar to some *Deltapodus* tracks, but present greater heteropody (Salisbury et al., 2016).

3.4.9.8 Trackmaker

Salisbury et al. (2016) pointed out that *Luluichnus* most likely pertains to a thyreophoran trackmaker. Moreover, they proposed that the tridactyl pedal impressions with acuminate digital impressions, with large, oval-shaped manus tracks, lacking clear digital impressions, are typically linked to stegosaurians.

ANKYLOSAURIA (FIG. 7.19)

The Ankylosauria is a diverse group of quadrupedal ornithischian armored dinosaurs that are known from Middle Jurassic to latest Cretaceous deposits worldwide and are divided into at least two main clades, the Ankylosauridae and the Nodosauridae (Thompson et al., 2012; Arbour and Currie, 2016). Its ichnological record is known since the first half of the 20th century, although in these pioneering works, the trackmaker was not exactly known. Ballerstedt (1922) and Nopcsa (1923) studied a trackway found in the Berriasian beds of Germany and erected the ichnogenus *Metatetrapous* Nopcsa, 1923 (see below). Nopcsa (1923) proposed a quadrupedal thyreophoran as track producer. Subsequently, Sternberg (1932) described *Tetrapodosaurus* Sternberg, 1932 and referred it to a ceratopsian trackmaker. Later, these types of tracks were largely forgotten until the conscientious work of McCrea et al. (2001), which laid the foundations for the study of ankylosaur tracks. In addition to the two ichnogenera previously referred to, in this section we have included the ichnogenera *Apulosauripus* Nicosia et al., 1999 and *Ligabueichnium* Leonardi, 1984 because it has been proposed by several authors that they were left by ankylosaurs.

On the other hand, Sternberg (1932) described the ichnofamily Tetrapodosauridae in which he only included *Tetrapodosaurus*. The original diagnosis is: *Habitually quadrupedal; medium sized; bluntly pointed hoofs; trackway wide; stride short; no caudal impression. Tracks of this family were doubtlessly made by quadrupedal, predentate dinosaurs.* This ichnofamily is very little used (see Haubold, 1971; McCrea, 2000, unpublished MSc; Rodríguez-de la Rosa et al., 2018).

In addition to the aforementioned ichnotaxa, the ichnogenus *Qijiangpus* Xing et al., 2007 has also been related to ankylosaur footprints. Xing et al. (2015f) recently reexamined the *Qijiangpus* type locality and declared this ichnogenus as a nomen dubium, considered a poorly preserved version of *Caririchnium*, an ichnogenus related to ornithopods. Finally, Kappus et al. (2011) documented Cretaceous ankylosaur tracks from New Mexico, USA, and noted similarity to "*Ankylosauripus*" and "*Tetrapodosauripus*," but these names are not recognized (see Rodríguez-de la Rosa et al., 2018).

3.4.10 Metatetrapous *Nopcsa, 1923 (Fig. 7.19A and B)*

3.4.10.1 Type ichnospecies

Metatetrapous valdensis Nopcsa, 1923 (Fig. 7.19A)

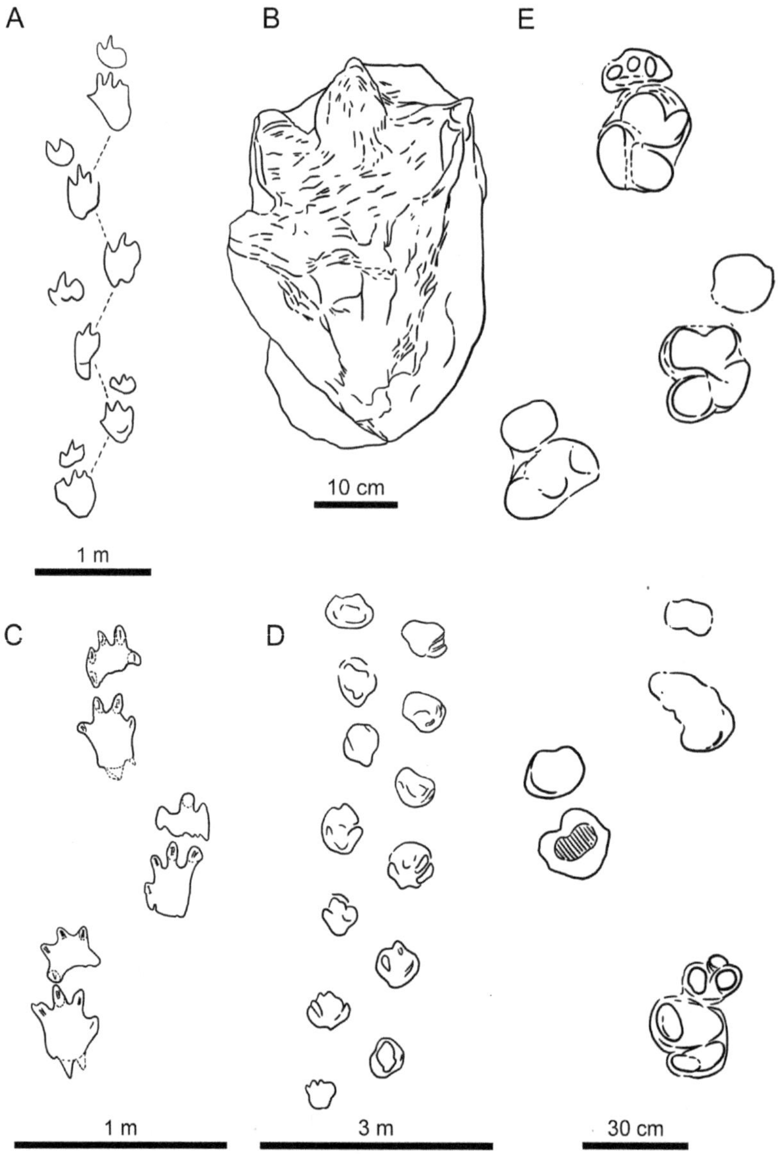

FIGURE 7.19 Tracks of Ankylosauria. (A) Type series of *Metatetrapous valdensis*; (B) details of *Metatetrapous* track; (C) holotype of *Tetrapodosaurus borealis*; (D) *Ligabueichnium bolivianum*; (E) *Apulosauripus federicianus*. *(A) Redrawn from Haubold (1971). (B) GZG.BA.0050; redrawn from Hornung and Reich (2014). (C) Redrawn from Sternberg (1932). (D) Redrawn from Leonardi (1984). (E) Redrawn from Nicosia et al. (1999).*

3.4.10.2 Valid ichnospecies

Metatetrapous valdensis Nopcsa, 1923.

3.4.10.3 Distribution

Late Jurassic Brazil; late Berriasian, Germany.

3.4.10.4 Holotype

Preserved in situ, the holotype is an epichnial trackway of 10.70 m length, with 15 manus and 16 pes prints (by monotypy; Ballerstedt, 1921, 1922, Fig. 1, no. 3), Late Jurassic Guará Formation, Paraná Basin, southern Brazil. Other referred material: GZG.BA.0049, GZG.BA.0050 resides today at the Geoscience Cenetr, University of Göttingen; specimens from late Berriasian, middle Obernkirchen Member, Bückeburg Formation, Germany.

3.4.10.5 Diagnosis

Moderately sized, quadrupedal dinosaur (probably ankylosaurian) trackway, manus tetradactyl, pes tetradactyl. Manus imprints short, arcuate, digitigrade, mesaxonic or slightly entaxonic, transversely wide (~80% of pes width). Manus digits short and triangular; Digit I directed posteromedially. Manus imprints separated from pes imprints and situated slightly anterolateral to the latter. Pes imprints semiplantigrade, mesaxonic, with well-developed and separated Digits I to IV. While Digits II to IV point into the cranial direction and converge proximally nearly symmetrical relative to the base of Digit III, Digit I shows a distinct posterior offset, is smaller, and divaricates medially from Digit II in a wide angle. No distinct phalangeal pads present. Digits with short, slender, pointed claw marks. Longitudinal axis of manus and pes imprints point slightly outward from the trackway midline (Hornung and Reich, 2014).

3.4.10.6 General description

Manus tracks are tetradactyl, digitigrade, and wider than long, with short, and triangular digit impressions. The Digit I is directed posteromedially, while the Digits II–IV are directed anteriorly located. Manus track is slightly anterolateral to the pes track. Pes tracks are tetradactyl, semiplantigrade, and longer than wide, with well-developed and independent digit impressions, and elongated "heel." The Digits II–IV are forwardly oriented, while the Digit I point in the posterior direction. There are no clear digit pads, and they have short, thin, sharp claw traces (Hornung and Reich, 2014).

According to the sketches of Ballerstedt (1922), Haubold (1971) and the new material of Hornung and Reich (2014), manus tracks measure about 20–25 cm in length and 30–35 cm in width, and pes tracks about 40–50 cm and 30–35 cm, respectively.

3.4.10.7 Discussion

Metatetrapous was defined by Nopcsa (1923) based on the original work of Ballerstedt (1922). Gierliński and Sabath (2008) argued that this ichnospecies is a nomen nudum, because Nopcsa (1923) had relied solely on the sketches of Ballerstedt (1922) to define the type species, according to Haubold (1971). Hornung and Reich (2014) re-studied *Metatetrapous*, added a differential ichnogeneric diagnosis and an ichnospecific emended diagnosis, and pointed out that the trackway based on which it was defined is lost, but there are two pes tracks preserved as natural casts that probably come from the same locality and have the same shape as the original. Hornung and Reich (2014) affirmed that this is an available ichnotaxon because Nopcsa (1923) gave "a short but sufficiently precise description referring to a published figure, which unambiguously shows the specimen in question, Nopcsa satisfied the requirements for establishing an (ichno) taxon before 1931 (ICZN, 1999, Articles 11, 12)." Francischini et al. (2018) agreed with the Hornung and Reich (2014) proposal and identified *Metatetrapous* in the Upper Jurassic of Brazil.

The most diagnostic feature is a tetradactyl instead of a pentadactyl manus, which is unique among formally described ankylosaurian ichnotaxa (e.g., *Tetrapodosaurus*). This character was depicted by Ballerstedt (1922) and explicitly noted by Nopcsa (1923). However, the newly assigned material of Hornung and Reich (2014) has no preserved manus tracks, and it is possible that a fifth digit, which is commonly faintly impressed in ankylosaur tracks, was overlooked on the original track and cannot be ruled out without the original material.

Hornung and Reich (2014) also considered at least three ichnogenera are very similar to *Metatetrapous*: (1) *Tetrapodosaurus* (Sternberg, 1932), (2) *Apulosauripus* (Nicosia et al., 1999), and (3) *Qijiangpus* (Xing et al., 2007). In addition to the tetradactyl manus mentioned above, the trackway gauge is narrower in *Metatetrapous* than in *Tetrapodosaurus* (Sternberg, 1932; McCrea, 2000; McCrea et al., 2001; Harington et al., 2005). However, the meaning and potential variation in trackway gauge in this type of tracks is not well understood (Hornung and Reich, 2014). *Metatetrapous* differs from *Apulosauripus* in having four digits on the pes instead of three (Gierliński and Sabath, 2008). *Qijiangpus* is invalid, based on a specimen of *Caririchnium* (see above), and there is no need to discuss it (Xing et al., 2015f). Despite what was mentioned, and following Hornung and Reich (2014) and Francischini et al. (2018), we tentatively consider *Metatetrapous* an available ichnotaxonomic name.

3.4.10.8 Trackmaker

In contrast to the definition given by Nopcsa (1923), Thyreophora does not include the marginocephalians according to modern views. Therefore, to identify the maker of the tracks more precisely, it is necessary to compare not only thyreophorans and sauropods, but also different groups of four-legged ornithischians. The fossil evidence shows that having four toes on each foot is an ancestral trait for Thyreophora, with five bones in the ankle and four in each toe (Pereda-Suberbiola et al., 2005). McCrea et al. (2001) showed the possible features to tell apart the trackways of ankylosaurs from those of other four-legged dinosaurs. The analysis is the only one that uses osteological comparisons of the ankylosaur hand and foot and how they match the supposed ankylosaur tracks. To sum up, McCrea et al. (2001) suggested that the following features can identify ankylosaur tracks: (1) moderate to large size, (2) quadrupedal gait, (3) tetradactyl pes, (4) metapodial and metacarpal regions relatively short compared with free digit length, (5) digits on manus and pes well separated and not webbed, (6) pes Digit I distinctly shorter than Digit II, (7) pes asymmetrical, with

Digit IV located distinctly anterior to Digit I, (8) pentadactyl manus, (9) manus imprint width reaching 60%–80% of pes imprint width, (10) trail of manus imprints slightly medial to or in front of pes imprints, and (11) manus strongly arcuate with an angle of 180°–250° between Digits I and IV or V, respectively. Of the above-listed characters, all except (8) are in concordance with *Metatetrapous*, so it probably was impressed by an ankylosaur (Hornung and Reich, 2014). Nevertheless, Gierliński and Sabath (2008) questioned the diagnostic significance of some of the previously mentioned features, noting that the footprints (especially of the hind foot) of ankylosaurs and basal ceratopsians are expected to be very similar.

3.4.11 Tetrapodosaurus *Sternberg, 1932 (Fig. 7.19C)*

3.4.11.1 Type ichnospecies

Tetrapodosaurus borealis Sternberg, 1932

3.4.11.2 Valid ichnospecies

Tetrapodosaurus borealis Sternberg, 1932

3.4.11.3 Distribution

Middle Jurassic (Bajocian)–Cretaceous (Maastrichtian), the Cretaceous records described herein are in Barremian/Aptian–Late/Upper Cenomanian, Canada, USA, England, and Spain.

3.4.11.4 Holotype

NMC 8556 (6 footprints: three manus, three pes; replica cast: cast and mold). Its type locality is 1.5 miles upstream from Gething's mine (B), Peace River Canyon, which is in an Aptian terrestrial sandstone/claystone in the Gething Formation of Canada.

3.4.11.5 Diagnosis

Diagnosis from Sternberg (1932): Quadrupedal; toes enclosed in pad or web; manus impressions in front of and completely separated from those of the pes; manus short and broad, five toes, digitigrade; pes of medium length, semiplantigrade, with four toes.

Emended Diagnosis from McCrea (2000): Footprints of a medium-large quadrupedal animal; manual and pedal digits are not enclosed in a pad or web; footprints that appear to have this morphology are likely the result of deformation of the substrate caused by the weight of the animal. The degree to which this occurs largely depends on the consistency and composition of the substrate. The manus is wider than long with a concave posterior margin.

The manus is approximately two-thirds the size of the pes. There are five manual digits. The pes is slightly longer than wide. There are four pedal digits.

3.4.11.6 General description

Manus tracks are pentadactyl, although some less perfectly preserved prints may show fewer digits. They are wider than long, digitigrade, and the digit traces join proximally in a small manual pad. Outer digits (I and V) are directed laterally or even posteriorally, while the inner digits (II–IV) face forward. Digit I impression is long, slender, and oriented perpendicular to the midline of the trackway. Pes tracks are tetradactyl, semiplantigrade, and longer than wide. Digit impressions are elongate and slender. Digit I is short and positioned medially, while Digits II–IV are longer and anteriorly positioned. Pes tracks are larger than manus tracks (McCrea, 2000).

The ichnogenus *Tetrapodosaurus* has specimens from diminutive to larger than 55 cm (McCrea et al., 2014a)

3.4.11.7 Discussion

The type locality has been flooded by the waters behind the Peace River dam (Currie, 1989), but fortunately, a replica of part of the trackway is kept at the National Museum of Canada, Ottawa (Carpenter and Kirkland, 1998).

McCrea (2000) revised the initial definition of this ichnogenus to exclude the digits of the manus and pes from being surrounded by a pad or web. He reached this conclusion by examining hundreds of manus and pes prints in trackways at the Smoky River Coal Mine near Grande Cache, Alberta. He inferred that Sternberg probably confused substrate deformation with a primary morphological feature due to the animal's weight.

As mentioned earlier, the main distinction between *Metatetrapous* and *Tetrapodosaurus* is the presence of four digits in the hand impressions of the former. Also, the trackway gauge is narrower in *Metatetrapous* than in *Tetrapodosaurus* (Sternberg, 1932; McCrea, 2000; McCrea et al., 2001; Harington et al., 2005). However, the significance and possible variation of trackway gauge in this kind of tracks is not well known, making its application for ichnotaxonomy doubtful at the current state of knowledge (Hornung and Reich, 2014).

In general, the temporal distribution of *Tetrapodosaurus* is Barremian–Cenomanian (see McCrea et al., 2001) and it was usually found in Canada and the USA. In the last years, both ranges have been increased with new finds. For instance, Pond et al. (2014) referred some tracks found in the Barremian-Aptian rocks of England to *Tetrapodosaurus*, as well as Herrero et al. (2017), who classified within this ichnogenus several manus tracks from the Barremian of Spain. Out of this temporal range, there are also tracks classified as *Tetrapodosaurus*: Rodríguez-de la Rosa et al. (2018) published a manus-pes set preserved as natural casts from the Middle Jurassic of Mexico. They assigned these tracks to *Tetrapodosaurus*. Shillito and Davies (2019) assigned to *Tetrapodosaurus* some Berriasian–Valanginian natural casts from England. Tracks related to *Tetrapodosaurus* have been also identified in the Coniacian-Santonian of the USA (Panascí et al., 2023). Fanti et al. (2013) and Fiorillo et al. (2019) included in *Tetrapodosaurus* Campanian-Maastrichtian tracks from Canada and the and Riguetti et al. (2021) assigned to cf. *Tetrapodosaurus* tracks found in the Maastrichtian of Bolivia. If these records are correctly determined, the temporal and geographical ranges of *Tetrapodosaurus* could be broader than expected.

3.4.11.8 Trackmaker

Tetrapodosaurus was related to ceratopsians by Sternberg (1932). Carpenter (1984) argued persuasively that this type of tracks belonged to nodosaurids, particularly *Sauropelta*.

Gierliński and Sabath (2008) pointed out that *Tetrapodosaurus* is restricted to the mid-Cretaceous (Aptian through Cenomanian) of Cordilleran North America and would support the Aptian migrations of early neoceretopsians from Asia to North America postulated by Chinnery et al. (1998). The relationship between *Tetrapodosaurus* and ankylosaur trackmakers is broadly accepted (e.g., McCrea et al., 2001, 2014a; Francischini et al., 2018; Fiorillo et al., 2019).

3.4.12 Ligabueichnium *Leonardi, 1984 (Fig. 7.19D)*

3.4.12.1 Type ichnospecies

Ligabueichnium bolivianum Leonardi, 1984.

3.4.12.2 Valid ichnospecies

Ligabueichnium bolivianum Leonardi, 1984.

3.4.12.3 Distribution

Maastrichtian, Bolivia.

3.4.12.4 Holotype

Trackway of 34 tracks in situ. Its type locality is the Toro Toro tracksite, Pista de Danzas, El Molino Formation, Bolivia. They is also another trackway as paratype.

3.4.12.5 Diagnosis

Massive, quadruped trackway with totally overlapping, apparently bipedal appearance. Manus tracks probably smaller than the pes tracks. Low-pace angle (~44°–95°; average 66°); low stride/footprint length index (~2); relatively wide trackway width (120–170 cm); internal width with always positive values (25–55 cm); posterior footprints slightly wider than long (length/width index = average ~0.99), with a general subround contour, with the anterior margin slightly concave and the posterior one provided with a "heel"; functional prevalence in the posterior portion of the foot. Probably pentadactyl or tetradactyl, and digits with flat, rounded claws in the front and sharp on the edges (translated from the Italian).

3.4.12.6 General description

As is noted by Leonardi (1984), the trackways were made by a quadrupedal dinosaur,although it is difficult to distinguish between pes and manus tracks because of the overlapping. Pes tracks are tetradactyl-pentadactyl. Claw traces on the anterior margin are present. The manus-pes set is subrounded in shape and is about 40–60 cm in diameter.

3.4.12.7 Discussion

Originally, Campbell (1983) related these tracks to a sauropod. Leonardi (1984) suggested that *Ligabueichnium* could belong to either an ankylosaur or a ceratopsian. This ichnotaxon is composed of two trackways, the holotype and the paratype. Meyer et al. (2021) pointed out it is possible to identify manus and pes tracks, and the trackway arrangement resembles those of other ankylosaur trackways, but the type trackway is poorly preserved and lacks clearly visible details. They agree with Leonardi (1984, 1994) and McCrea et al. (2001) that an ankylosaur made these tracks, but it is necessary to find better-preserved ones to confirm if it is valid or a synonym of a previous ichnotaxon.

3.4.12.8 Trackmaker

As commented before, Leonardi (1984) suggested that this ichnogenus could be made by an ankylosaur or a ceratopsian. Subsequently, Leonardi (1994) indicated that *Ligabueichnium* could be attributed by to an unusually large ankylosaur, but if the pes tracks are tetradactyl instead of pentadactyl, a ceratopsian might be the trackmaker as well. According to McCrea et al. (2001) and Meyer et al. (2021), the most likely explanation is that ankylosaurs made the *Ligabueichnium* tracks, since no ceratopsian fossils have been discovered in South America.

3.4.13 Apulosauripus *Nicosia et al., 1999 (Fig. 7.19E)*

3.4.13.1 Type ichnospecies

Apulosauripus federicianus Nicosia et al., 1999.

3.4.13.2 Valid ichnospecies

Apulosauripus federicianus Nicosia et al., 1999.

3.4.13.3 Distribution

Early/Lower Santonian, Italy.

3.4.13.4 Holotype

ACDL 99/3, a trackway of 13 manus-pes sets. Its type locality is Cava De Lucia, Altamura, which is in a Santonian offshore shelf sandstone in the Calcare de Altamura Formation of Italy.

3.4.13.5 Diagnosis

Trackway of a quadruped dinosaur, with very low-pace angulation. Pes impressions slightly divergent with respect to the midline (positive rotation), while manus impressions are forward directed and parallel to the midline. Pes and manus both (functionally) tridactyl. Manus less than half the size of the pes. The manus impressions are placed forward and medial to the pedes. No tail mark. Trackway relatively broad with a space between the interior margins of the footprints nearly 1/3 of the maximum width (wide gauge). Pace angulation very low both for pedes and manus. Stride/foot length 4:1.

3.4.13.6 General description

Apulosauripus presents three-digit impressions in both manus and pes tracks, the manus is much smaller than pes. The digit impressions are about the same length, usually short and curved. The pes tracks have a transversal ridge that divides the heel from the toe marks. In the type series, the mean length of the pes track is about 25 cm and the mean width is 26 cm, while in the manus, tracks are 14 and 18 cm, respectively. Manus tracks located in front and slightly inside of the pes tracks, and in some cases part of the pes track overlaps the manus track. The pace angle is low in both manus and pes tracks, around 91 and 100° on average (see Nicosia et al., 1999).

3.4.13.7 Discussion

Apulosauripus is defined based on a long quadrupedal trackway found in the Santonian of Italy. The tracks are quite well-preserved, and the diagnosis is accurate, therefore it is an available ichnotaxon (Díaz-Martínez et al., 2015a,b). The main problem with this ichnotaxon is the paleobiological attribution (see below), because it has been related to hadrosaurs, basal iguanodonts, ankylosaurs, and thyreophorans (Nicosia et al., 1999; Gierliński and Sabath, 2008; Dalla Vecchia, 2008; Petti et al., 2010; Contessi, 2013). In addition to the type tracksite, Contessi (2013) classified as cf. *Apulosauripus federicianus* tracks from the late Albian of Tunisia.

3.4.13.8 Trackmaker

Originally, *Apulosauripus* was assigned to a hadrosaurian dinosaur (Nicosia et al., 1999). However, Gierliński et al. (2005) suggested that the *Apulosauripus* ichnogenus belonged to an ankylosaurian dinosaur, because of the large manus tracks (larger than other four-legged ornithopods), the great trackway width and absence of inward rotation of the pes tracks. Petti et al. (2010) disagreed with this idea, because some ankylosaurs may have tridactyl hands (possibly *Panoplosaurus*) while others may have tridactyl feet (*Euoplocephalus*) and the combination of both characters has yet to be reported. Dalla Vecchia (2008) hypothesized a possible basal iguanodontian origin. Contessi (2013) suggested that due to the osteological record found in coeval strata, the trackmaker could be an hadrosaurian. Finally, Xing et al. (2021k) considered that *Apulosauripus* is a thyreophoran ichnotaxon.

CERATOPSIA (FIG. 7.20)

Ceratopsia dinosaurs were herbivorous dinosaurs with parrot-like beaks, bony frills, and horns on the skull, that lived in what is now North America, Europe, and Asia, during the Cretaceous period, but their ancestors lived earlier, in the Jurassic (see Sampson and Loewen, 2010; Brown and Henderson, 2015). Although more basal members such as *Psittacosaurus* or *Protoceratops*, or more derived members such as *Triceratops* or *Chasmosaurus*, are relatively frequent in the fossil record of Asia and North America, finds of ceratopsian tracks are really scarce. Apart from the tracks associated with the ichnogenus *Ceratopsipes* (see below), the record is restricted to few Cretaceous remains from China, the USA, and Mongolia (Fiorillo et al., 2010; Niedźwiedzki et al., 2012; Xing et al., 2013c; Lockley et al., 2018d).

3.4.14 Ceratopsipes *Lockley and Hunt, 1995b (Fig. 7.20)*

3.4.14.1 Type ichnospecies

Ceratopsipes goldenensis Lockley and Hunt, 1995b.

3.4.14.2 Valid ichnospecies

Ceratopsipes goldenensis Lockley and Hunt, 1995b.

3.4.14.3 Distribution

Santonian–Maastrichtian, USA.

FIGURE 7.20 Tracks of Ceratopsia. (A) Idealized ceratopsid trackway based on the holotype of *Ceratopsipes* (CU-MWC 220.1); (B) pedal (CU-MWC 220.6) and manual tracks (CU-MWC 220.5) of larger ceratopsid individuals. *(B) Redrawn from Lockley and Hunt (1995b).*

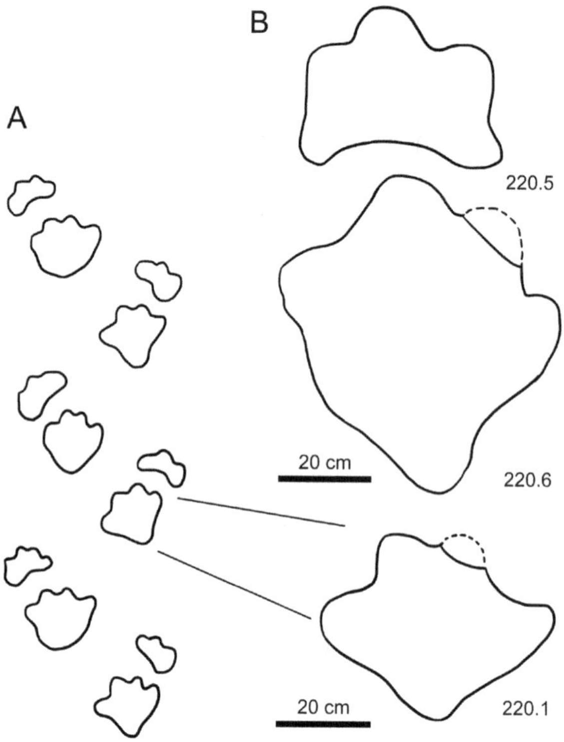

3.4.14.4 Holotype

CU-MWC 220.1, a footprint. Its type locality is 12th Street tracksite (lower), which is in a Maastrichtian terrestrial horizon in the Arapahoe Formation of Colorado.

3.4.14.5 Diagnosis

Large quadrupedal dinosaur track that differs from all others in having four blunt pedal digit impressions and four manual digit impressions, pedal imprint about twice the size of that of the manus, and manual impressions with more concave posterior margins than pedal ones.

3.4.14.6 General description

Pes tracks are tetradactyl, wider than long, blunt, wide, and have short digit impressions and elongated, rounded, and symmetrical "heel" traces. Inner digit impressions (II and III) are directed forward and outer digits (I and IV) are directed laterally or even posterolaterally. Manus tracks are also tetradactyl, wider than long, with flattened to concave heel. Digit II–IV impressions are oriented forwardly and Digit I medially. Pes tracks are larger than manus tracks.

3.4.14.7 Discussion

Ceratopsipes is similar to the ankylosaurid ichnotaxa *Metatetrapous* and *Tetrapodosaurus*. All have tetradactyl pes tracks, but ankylosaurid tracks present a smaller and more backward Digit I impression than in *Ceratopsipes*. The manus impression of *Ceratopsipes* is tetradactyl, the same as in *Metatetrapous*, but in *Tetrapodosaurus* is pentadactyl. Lockley and Hunt (1995b) compared ceratopsid and ankylosaur tracks and suggested that *Ceratopsipes* has a relatively smaller manus track, blunter pedal digit impressions, a more concave posterior margin to the manus track and a narrower trackway.

In addition to the material referred by Lockley and Hunt (1995b), *Ceratopsipes* tracks have been cited in the Santonian–Maastrichtian of the USA (Crowell and Shimer, 2021; Stewart et al., 2023).

3.4.14.8 Trackmaker

McCrea et al. (2001) pointed out that ceratopsians and ankylosaurs have a convergent anatomy of the autopodia, as it is observed in the osteology of *Sauropelta* (Ankylosauria) and *Centrosaurus* (Neoceratopsia) autopodia (see Lockley and Gierliński, 2014). According to Lockley and Hunt (1995b), both morphologies are quite different, and adding the geographical and temporal information of *Ceratopsipes*, these tracks pertain to ceratopsian dinosaurs similar to *Triceratops*.

3.5 Pterosauria

3.5.1 **Pteraichnus** *Stokes, 1957 (Fig. 7.21)*

3.5.1.1 Type ichnospecies

Pteraichnus saltwashensis Stokes, 1957 (Fig. 7.21A)

3.5.1.2 Valid ichnospecies

Pteraichnus saltwashensis Stokes, 1957.
Pteraichnus stokesi Lockley et al., 1995 (Fig. 7.21B)
Pteraichnus palacieisaenzi Pascual-Arribas and Sanz-Pérez, 2000 (Fig. 7.21C)
Pteraichnus parvus Fuentes Vidarte et al., 2004a (Fig. 7.21D)
Pteraichnus longipodus Fuentes Vidarte et al., 2004b (Fig. 7.21E)
Pteraichnus koreanensis Lee et al., 2008 (Fig. 7.21F)
Pteraichnus nipponensis Lee et al., 2010 (Fig. 7.21G)
Pteraichnus gracilis Ha et al., 2022 (Fig. 7.21H)

3.5.1.3 Distribution

Upper Jurassic in America, Spain, France, Portugal, and Poland; Cretaceous record: Berriasian–Maastrichtian, Dunvegan Formation of Western Canada; Blackhawk and North Horn formations of Utah; Muerto Formation of northern Mexico; Candeleros Formation of del Neuquén, Argentina; Huérteles Formation and Enciso Group of Spain; Tugulu and Hekou Groups, and Jiaguan, Feitianshan, Jinhua, Zhutian and Qugezhuang formations of China; Daegu, Haman and Hasandong Formations of South Korea; Kitadani Formation of Japan.

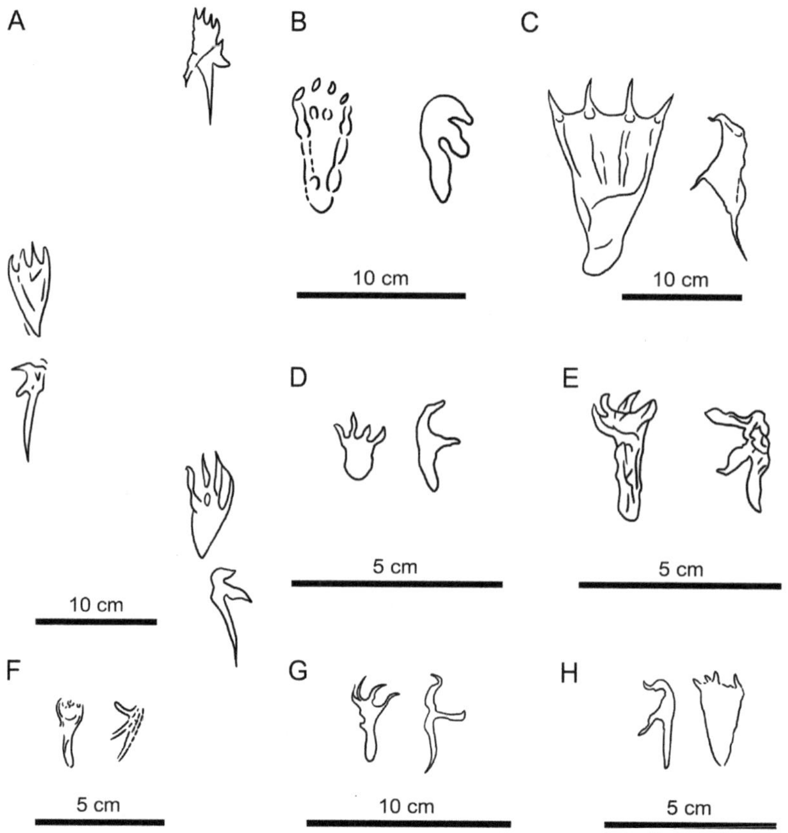

FIGURE 7.21 Tracks of pterosaurian *Pteraichnus*. (A) Type series of *Pterdichnus saltwashensis*; (B) series of *Pteraichnus stokesi* from Sundance Formation, Wyoming; (C) holotype of *Pteraichnus palacieisaenzi*; (D) holotype of *Pteraichnus parvus*; (E) holotype of *Pteraichnus longipodus*; (F) holotype of *Pteraichnus koreanensis*; (G) holotype of *Pteraichnus nipponensis*; (H) holotype of *Pteraichnus gracilis*. *(B) Redrawn from Stokes (1957). (C) Redrawn from Pascual-Arribas et al. (2015). (D) Redrawn from Meijide Calvo (2001). (E) Redrawn from Fuentes et al. (2004). (F) Redrawn from Lee et al. (2008). (G) Redrawn from Lee et al. (2010). (H) Redrawn from Ha et al. (2022).*

3.5.1.4 Holotype

A trackway consisting of nine pairs of consecutive prints on four slabs of sandstone, which are housed in the Geology Museum, University of Utah.

3.5.1.5 Diagnosis

Quadruped track; manus prints on the same axis or more laterally to the pes prints; pes print anterior to the ipsilateral manus print; manus prints as or more deeply impressed than pes prints; elongate, subtriangular, plantigrade and tetradactyl pes print; Digit II and III prints are slightly longer than I and IV, Digit I–IV prints are clawed; elongate, asymmetrical, digitigrade and tridactyl manus print; Digit I print anterior or anterolateral (generally with claw mark), Digit II print anterolateral to posterolateral (rarely with claw mark), Digit

III print posterior (exceptionally with claw mark), digit prints increasing respectively in length; Digit IV is rarely marked and limited to the print of the proximal part of the fourth manus digit (oriented posteromedially); rounded print in the medial margin of the manus print (print of the fourth metacarpo-phalangeal joint); elongate penultimate phalanges can be observed in some exceptionally well-preserved manus and pes prints (inferred from digital pads when available) (emmeded by Billon-Bruyat and Mazin, 2003).

3.5.1.6 General description

P. saltwashensis was the first described *Pteraichnus*. Stokes (1957) reported nine pairs of consecutive prints from the Morrison Formation (Kimmeridgian–Tithonian) in Arizona, USA. The manus prints are deeper than the pes prints and may indicate an anterior shift of weight on the head and front part of the body. It consists of slender digits, elongated triangular pes prints, and a gradually curved direction of the manus digit prints (Lockley et al., 1995).

P. stokesi was discovered from the Sundance Formation (Oxfordian) in Wyoming, USA. The holotype is wide, with symmetrical metatarsal prints rotated outward, and asymmetrical, tridactyl, outwardly rotated manus prints (Lockley et al., 1995). The width of the outer track is about 27 cm, or about 6 times pes width, and the inner manus trackway is about 16 cm wide, or about 4 times pes width. Manus-to-manus and pes-to-pes stride (pace) about 30 cm. Manus-to-manus and pes-to-pes span is approximately 42 cm. Manus-to-manus and pes-to-pes pace angles are approximately 90°, indicating a regular alternation of right and left prints (Lockley et al., 1995).

P. palacieisaenzi was discovered from the Huérteles Formation (Berriasian) in Soria, Spain. Only found in the Cretaceous. The average manus length is 130 mm and slightly greater than the width of the pes. The values of divarication are 90° (I–II) and 35°–60° (II–III). The average length of the pes prints is 150 mm, and the average length/width ratio is 1.29. The digit prints of pes are similar in size (Digits II and III are slightly longer) and slightly divergent. The divarication value is 36°–40° (Digits I–IV) (Pascual-Arribas et al., 2015).

P. parvus was also discovered from the Huérteles Formation in Soria, Spain. Only found in the Cretaceous. The digit ratio of the holotype manus prints is 1: 1.73: 3.79, with an average length of 24.63 mm and an average width of 12 mm (average manus length/width = 2.05). The values of divarication are 64° (I–II), 58° (II–III) and 122° (I–III). The average length of the pes prints is 15.3 mm, the average width is 11.43 mm, and the average length/width ratio is 1.34. The imprints on the soles are distinctive, occupying between 2/3 and 3/4 of the length of the pes (compared with about 1/2 for *P. saltwashensis*). The digits are slender and straight, with the Digit II longer than the other three. All digits are pointed (Sánchez-Hernández et al., 2009).

P. longipodus was discovered from the Huérteles Formation in Soria, Spain. Only found in Cretaceous. The digit ratio of the holotype manus prints is 1: 1.2: 4.8, with an average length of 24.5 mm and an average width of 17 mm (average manus length/width = 1.44–1.46). The divarication of Digits I–III is 101°. The average length of the pes prints is 34 mm, the average width is 17.4 mm (average manus length/width = 1.95). The values of divarication is 47° (I–II), 49° (II–III), and 96° (I–III). The pes prints show four short digits joined at the base to form two diverging groups. The four-digit prints vary in size, with the Digit IV being longer than the others. A Digit V was preserved in one of the pes prints. The pes prints are notable for having a narrow metatarsal print with a constriction between the metatarsal prints that is more than 4.5 times the length of the longest digit print. There is no constriction between the metatarsal and heel prints (Sánchez-Hernández et al., 2009).

P. koreanensis was discovered in the Hasandong Formation (Aptian–Albian) in Gyeongsang, South Korea. Only found in the Cretaceous. The average digit ratio of the manus prints is 1:1.2:1.73, with an average length of 25.6 mm and an average width of 12.3 mm (average manus length/width = 2.08). The values of divarication are 68.5° (I–II), 58° (II–III), and 116.2° (I–III). The metacarpo-phalangeal joint depression is deeper than the digital prints and no webbing is present between the digits. The average length and width of the pes are 25.7 and 12.8 mm (average pes length/width = 2.01). Although Digit I (average length 8.9 mm) is the shortest, and Digit III is the longest (average length 10.8 mm), the lengths of the four digits are approximately equal. The metatarsals are long and slender, moderately bifurcated like the handle of a fork, and always longer than the digits (average length 17.5 mm) (Lee et al., 2008).

P. nipponensis was discovered in the Kitadani Formation (Aptian) in Fukui, Japan. Only found in the Cretaceous. The average digit ratio of the manus prints is 1:1:1.5, with an average length of 22.6 mm. The values of divarication are 80.9° (I–II), 71.9° (II–III), and 155.4° (I–III). The interphalangeal webbing extends from the metacarpophalangeal joints to the bases of the claws. The metatarsals are long and slender, moderately forked, and always longer than the digits. The values of divarication of Digits I and II prints are smaller than those between II and III, III and IV; the digit I is the shortest, and other three digits are roughly equal (Lee et al., 2010).

P. gracilis was discovered in the Jinju Formation (Albian) in Gyeongsang, South Korea. Only found in the Cretaceous. The average digit ratio of the manus prints is 1: 2: 2.8, with an average length of 25.6 mm and an average width of 12.3 mm (average manus length/width = 3.51). The values of divarication are 60.16° (I–II), 51.02° (II–III), and 111.4° (I–III). The average length and width of the pes are 25.48 and 12.21 mm (average pes length/width ~ 3.04) (Ha et al., 2022).

3.5.1.7 Discussion

After Stokes (1957) erected *Pteraichnus*, at least 14 ichnospecies were established by the researchers: *P. saltwashensis* Stokes, 1957, *P. stokesi* Lockley et al., 1995, *P. palacieisaenzi* Pascual-Arribas and Sanz-Pérez, 2000, *P. cidacoi* Fuentes Vidarte, 2001, *P. vetustior* Meijide Fuentes, 2001, *P. parvus* Meijide Calvo et al., 2001, *P. longipodus* Fuentes Vidarte et al., 2004b, *P. manueli* Meijide Calvo, 2001, *P. yanguoxiaensis* Peng et al., 2004, *P. koreanensis* Lee et al., 2008, *P. nipponensis* Lee et al., 2010, *P. dongyangensis* Chen et al., 2013, *P. wuerhoensis* Li et al., 2021, *P. gracilis* Ha et al., 2022. However, eight valid ichnospecies were retained after re-examination and confirmation by several researchers: *P. saltwashensis*, *P. stokesi*, *P. palacieisaenzi*, *P. parvus*, *P. longipodus*, *P. koreanensis*, *P. nipponensis*, *P. gracilis*. The history of these revisions includes Sánchez-Hernández et al. (2009), who considered *P. palacieisaenzi*, *P. cidacoi*, *P. vetustior*, and *P. manueli* as *nomina dubia*. Li et al. (2015) considered *P. yanguoxiaensis* as a nomen dubium. Xing et al. (2018a) considered *P. dongyangensis* as a nomen dubium. Li et al. (2021) identified five differences between *P. saltwashensis* and the tracks from the Wuerhe region: (1) size; (2) more elongate metatarsal print; (3) much wider; (4) the ratio of pes digits; and (5) the divarication of the pes. A study by Xing et al. (2021j) described more specimens from the Wuerhe region and compared the trackways with *P. saltwashensis*, attributing tracks from the Wuerhe region to *P. saltwashensis*. Therefore, *P. wuerhoensis* is also a nomen dubium.

P. stokesi differs from *P. saltwashensis* pes in that pes has a greater degree of externally rotation relative to the midline of the trackway, about 45°; Digit III of the pes is relatively shorter and wider than that of manus, and a few tracks have prints of Digit V. In addition, the manus of *P. stokesi* has a smaller pace angulation of 90°. The pes of *P. stokesi* is longer than the manus, in contrast to *P. saltwashensis* (Sánchez-Hernández et al., 2009). *P. stokesi* has been widely accepted as valid for multiple comparisons, while considered invalid by Bennett (1997); Billon-Bruyat and Mazin (2003) (Masrour et al., 2018), as lacking the diagnostic features as the undistorted manus track morphologies(Billon-Bruyat and Mazin, 2003).

P. palacieisaenzi has a medium-sized pes, and the value of divarication between Digits I and IV is 36–40°. The average pes length/width is 1.29, compared to 2.5 for *P. saltwashensis*. And the average length of pes/length of the manus is 1.18. The value of divarication is greater than 90°, and lower than 180° (Digits I–III) (Pascual-Arribas et al., 2015). This ichnotaxon has been consider invalid by (Billon-Bruyat and Mazin (2003), Lee et al. (2008), Masrour et al. (2018), Sánchez-Hernández et al. (2009)), as the distinct measurements among the referred materials, and the unavailablle holotype.

P. parvus is the smallest pterosaur track (one-fifth the size of the holotype of *P. saltwashensis*). The sole is relatively long, and digits of both pes and manus are short. The average manus length/width is 2.0 and that of *P. saltwashensis* is 2.5 (Sánchez-Hernández et al., 2009). This ichnotaxon has been consider valid by (Masrour et al., 2018; Sánchez-Hernández et al., 2009), mainly due to its small size, low L/W ratio and relatively longer manus than pes (Sánchez-Hernández et al., 2009). However, due to the holotype of *P. parvus* is not in sufficient quantity and various in morphology of both manus and pes, the L/W ratio may indicate the pes track of *P. parvus* has been shortened and widen by the substrate condition and behaviour of trackmaker.

P. longipodus is distinguished from other ichnospecies of *Pteraichnus* by the very short manus (length/width ratio is 1.46, compared to 2.0 in *P. parvus* and 2.5 in *P. saltwashensis*), and the intermediate pes (length/width ratio is 1.95, compared to 1.30 in *P. parvus* and 2.5 in *P. saltwashensis*). Further, the pes imprint includes a substantial metatarsal imprint, narrower than the digit imprints, and extending back for some four-fifths of the pes imprint length. The manus imprint is similar to *P. saltwashensis*. The manus of *P. longipodus* is very short, the average manus length/width is 1.46, compared to 2.0 for *P. parvus* and 2.5 for *P. saltwashensis*. In the intermediate pes, the average pes length/width is 1.95, compared to 1.3 for *P. parvus* and 2.5 for *P. saltwashensis*. In addition, the pes includes a substantial metatarsal print, narrower than the digit print that extends backward for about four-fifths of the length of the pes. The manus print is similar to that of *P. saltwashensis* (Sánchez-Hernández et al., 2009). This ichnotaxon is considered invalid by Kim and Lockley (2012).

P. koreanensis is about 3 times smaller than *P. saltwashensis*. The metatarsal prints of *P. koreanensis* are long and elongate, resembling the handle of a fork. The divarication of the pes is 10° lower than that of *P. saltwashensis*. The average digit length/pes length is 2.6, compared to 1.9 for *P. saltwashensis*. In *P. koreanensis*, pes angulation is 11° greater than that of *P. saltwashensis* (Lee et al., 2008). *P. koreanensis* is comparable to small *Pteraichnus* ichnospecies in size, specifically *P. parvus* (see Fig. 6 of Ha et al., 2022). However, the diacritical diagnostic features between at least *P. parvus* and *P. koreanensis* (Lee et al., 2008), is very likely extramorphological.

P. nipponensis is distinguished from other ichnospecies of *Pteraichnus* by high divarication of Digit I and III of the manus (Lee et al., 2010), while the divarication angel of manus digits may not be an effective diagnostic feature for pterosaur tracks, at least for tracks in poor preservation (i.e., suffer from much extramorphological distortion; Masrour et al., 2018; Mazin et al., 2003). The divarication angle of digit I–III ranges from 121.7° to 193.1° for *P. nipponensis* holotype, and the excessive range may not fit the role as a diagnostic feature.

The holotype of *P. gracilis* is the best preserved among the small *Pteraichnus* (<5 cm in length). In *P. gracilis*, the average manus length/width is much greater than in the other small *Pteraichnus* (*P. parvus*, *P. longipodus*, *P. koreanensis*, *P. nipponensis*). *P. gracilis* has wider metatarsal parts with a V-shaped heel. The trackway is much narrower (Ha et al., 2022).

It needs to be emphasized that *Pteraichnus* is an overly large ichnogenus, and the morphological differences of some valid ichnospecies may have exceeded ichnospecies-level differences. On the other hand, some scholars hope to control the scale of the ichnogenus and avoid oversplitting (such as Lockley et al., 2008b; Masrour et al., 2018). Overall, this ichnogenus needs to be recompared and discussed in the future. In particular, there is a need to discuss whether they need to be differentiated by size.

3.5.1.8 Trackmaker

Peters (2011) tried to map the tracks and the skeleton of the pes, and he suggested that the *Pteraichnus* sp. discovered in Crayssac, France (Mazin et al., 1995) can be associated with *Pterodaustro* because of splayed metatarsals, short digits with Digits II–IV the longest and tiny unguals; *P. saltwashensis* is associated with *Ctenochasma gracile* because of metatarsal II the longest, unguals I–III aligned and digits half the length of the pes and bent medially; *P. stokesi* is associated with *Pterodactylus elegans* because of Digits II and III are subequal and Digit V is vestigial. Xing et al. (2021j) suggested that the *P. saltwashensis* discovered in Wuerhe, China is associated with the dsungaripteroid from the Xinjiang region. Masrour et al. (2018) supposed the Cretaceous large Pteraichnus may be attribute to azhdarchoids, while the complete pes anatomy of this clade remained uncertain.

Pascual-Arribas et al. (2015) suggested that *P. palacieisaenzi* is associated with a ctenochasmatid, which could have been *Ardeadactylus* because of the large size of the pes, the similar length of the digits, the average length/width ratio of pes, and the location of the ungual marks. The trackmakers of the species from East Asia (*P. koreanensis*, *P. nipponensis*, *P. gracilis*) are thought to be a small dsungaripteroids from the Jehol Biota (Lee et al., 2008, 2010; Ha et al., 2022).

3.5.2 Purbeckopus *Delair 1963 (Fig. 7.22A)*

3.5.2.1 Type ichnospecies

Purbeckopus pentadactylus Delair 1963.

3.5.2.2 Valid ichnospecies

Purbeckopus pentadactylus Delair 1963.

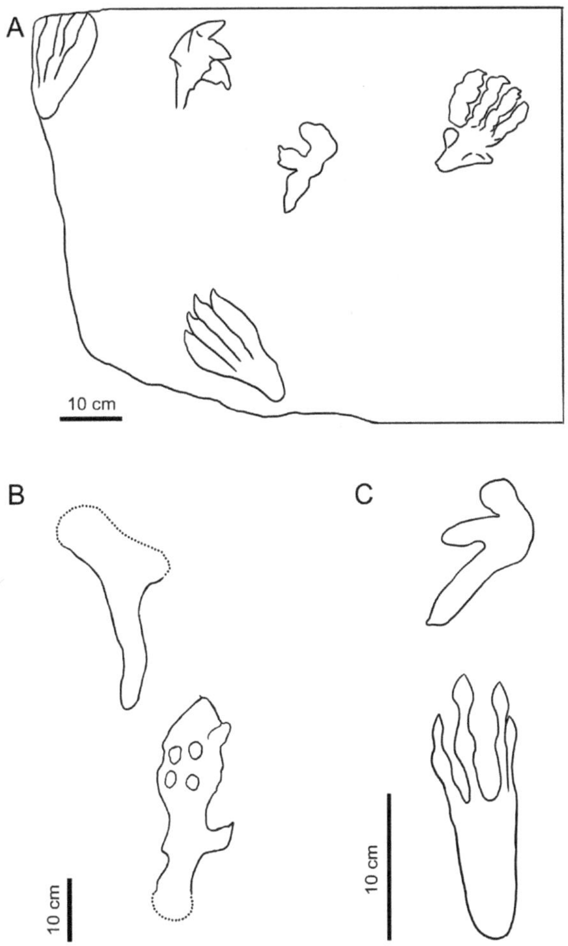

FIGURE 7.22 Other ichnogenus of pterosaurian. (A) Holotype of *Purbeckopus pentadactylus;* (B) holotype of *Agadirichnus elegans;* (C) holotype of *Haenamichnus uhangriensis. (A) Redrawn from Delair (1963). (B) Redrawn from Ambroggi and de Lapparent (1954). (C) Redrawn from Hwang et al. (2002).*

3.5.2.3 Distribution

Berriasian, England, Germany.

3.5.2.4 Holotype

DORCM G. 6664, housed in the Dorset County Museum with two manus prints and three pes prints, Durlston Formation of southern England.

3.5.2.5 Diagnosis

Tracks of a quadrupedal animal with elongate (approximately twice as long as wide), subtriangular, symmetrical, functionally tetradactyl, plantigrade pes prints; elongate, asymmetrical tridactyl manus prints may also be present. The digits of the pes are subequal in length and curved slightly inwards, the curvature being most pronounced in the outermost digit

(Digit IV). Digits II and III of the pes are slightly longer than the outer Digits I and IV. Manus prints, if present, lie outside the pes prints (Wright et al., 1997).

3.5.2.6 General description

Purbeckopus pentadactylus was discovered from Durlston Formation (Berriasian) in Chinchen's Quarry, southern England, and Bückeberg Formation (Berriasian) in Lower Saxony, northwestern Germany. The length of the manus prints is slightly shorter than the pes, and the width is half of the pes. The length of the pes prints varies from 187 to 225 mm, the width varies from 98 to 123 mm, and the average length/width ratio is about 2. The digit ratio of the manus prints is 1: 0.95: 1.23. The digit prints of the pes are similar in size (the two middle digits are slightly longer). The values of divarication are 50° (I–II), 100° (II–III), 150 (I–III). Digits I and II have the phalangeal pad, but it is not so distinct in Digit III. The average length is 130 mm and slightly greater than the width of the pes. The digit prints, especially Digit IV, and the claw prints curved inwards to the midline of trackway. The digits of the pes are wide, and the spacing between the digits is narrow. The phalangeal pads are obvious. Most *Purbeckopus pentadactylus* have three digits, but one specimen has a faint print extending posteriorly that may be the print of the Digit IV. The manus prints are located outside or 10–14 cm posterior to the pes. The claws are blunt and short (Wright et al., 1997; Hornung and Reich, 2013).

3.5.2.7 Discussion

Purbeckopus is remarkable similarity to *Pteraichnus* (Wright et al., 1997). It is much larger than all *Pteraichnus*. The digits of *Purbeckopus* are much thicker than *Pteraichnus* and there is proximal offset of the metapodial-phalangeal joint in Digit IV of the pes. The manus of *Purbeckopus* has a more diversified morphology than the pes, as in *Pteraichnus*. Digits I and II of the manu,s and all digits of pes have phalangeal pads (Wright et al., 1997).

The German specimen only preserves the manus, and the most important difference is the depth of substrate intrusion of the manus compared with the holotype, because of softer sediment (Hornung and Reich, 2013).

3.5.2.8 Trackmaker

Wright et al. (1997) and Hornung and Reich (2013) suggested that the trackmaker of *Purbeckopus pentadactylus* holotype had a wingspan of about 6 m and it may be associated with the pterosaurs from the Hastings Beds of England (Valanginian). The holotype shows that very large pterosaurs had already appeared by the end of the Jurassic. Peters (2011) suggested that the *Purbeckopus pentadactylus* holotype is associated with *Sinopterus* because of the similar metatarsal/digit ratio.

3.5.3 Haenamichnus *Hwang et al., 2002 (Fig. 7.22B)*

3.5.3.1 Type ichnospecies

Haenamichnus uhangriensis Hwang et al., 2002.

3.5.3.2 Valid ichnospecies

Haenamichnus uhangriensis Hwang et al., 2002.

3.5.3.3 Distribution

Campanian, South Korea.

3.5.3.4 Holotype

CNUPH.P2, in natural cast of right manus and pes prints, site P2, Uhangri Formation of Jeollanam, South Korea.

3.5.3.5 Diagnosis

Pes impressions elongate with narrow heel and postero-lateral impression of Digit V rounded. Digital pad impressions sometimes present. Manus tridactyl, occasionally tetradactyl, and usually behind and slightly outside pes (Hwang et al., 2002).

3.5.3.6 General description

Haenamichnus uhangriensis was discovered from the Uhangri Formation in Jeollanam, South Korea. The length of the manus prints is 330 mm and the width of the manus prints is 110 mm. The length/width ratio is 3. The manus prints have a broad knuckle print. The manus is asymmetrical, rotated outward at almost a right angle to the midline of the trackway. The pes has no separation of the digits except Digit V. The Digits II and III are the longest. Digit V of the pes is thick and short. Digit III of the manus has a clear outline and is much longer and deeper than Digits I and II. There are also large interphalangeal pads on Digits II and III of the pes (Hwang et al., 2002; Jung et al., 2022).

3.5.3.7 Discussion

Haenamichnus uhangriensis is a typical pterosaur track. However, it is very different from other pterosaur tracks reported previously because of the thick and short prints of the digits, large interphalangeal pad prints, no visible prints of digits and claws, the narrow and elongate shape with protruding and rounded heel print and no separation of the digits of the pes except Digit V. Besides, *Haenamichnus uhangriensis* is the largest pterosaur track; it is 2 times larger than those of *Purbeckopus pentadactylus* (Hwang et al., 2002; Jung et al., 2022; Ha et al., 2022).

Due to its poor preservation that lacks stable shape in the type series, we suggest to consider it unvalid or redescribe this ichnogenus with more specimens.

3.5.3.8 Trackmaker

Hwang et al. (2002) suggested that *P. pentadactylus* is associated with azhdarchids with wingspans of 5–6 m because of the size, proportions, and features of the tracks. Azhdarchids were widespread during the Late Cretaceous, almost worldwide. In eastern Asia, azhdarchids are found in China, Mongolia, and Japan, but only *Zhejiangopterus linhaiensis* was identified at the generic level (Witton and Naish, 2015).

3.5.4 Agadirichnus *Ambroggi and de Lapparent 1954 (Fig. 7.22C)*

3.5.4.1 Type ichnospecies

Agadirichnus elegans Ambroggi and de Lapparent 1954

3.5.4.2 Valid ichnospecies

Agadirichnus elegans Ambroggi and de Lapparent 1954

3.5.4.3 Distribution

Maastrichtian, Morocco.

3.5.4.4 Holotype

A slab bearing two pes prints described by Ambroggi and Lapparent (1954) but lost. Masrour et al. (2018) described a neotype specimen, FSA.G.54, which is an isolated pes print from the Tagragra Formation of Agadir, Morocco, and housed in the Université Ibn Zohr, Faculté des Sciences, Agadir, Morocco.

3.5.4.5 Diagnosis

Large (77–170 mm in length), elongate, tetradactyl, and plantigrade pes prints with rounded heel and subparallel edges. Pes digits are elongate, well separated, and often incompletely marked in the median portion. Central pes digits (II and III) longer than external digits (I and IV). Robust tridactyl manus prints, with short and rounded Digit I, without claw mark (Masrour et al., 2018).

3.5.4.6 General description

Agadirichnus elegans was discovered from the Tagragra Formation in Agadi, Morocco. It was the first reported pterosaur track. The length of the manus prints is 50–100 mm. The Digit I of the manus is the shortest without claw mark or curvature. The Digit II of the manus is thick, and the Digit III is the longest and massive. The length of the pes prints is 77–170 mm. The pes is elongate, and the average pes length/width ratio is 3 with a rounded heel. The Digits II and III of pes are longer than Digits I and IV. The marks of the claws are distinct. The heel and metacarpals are also distinct, but the digits of the pes are not sharp (Masrour et al., 2018).

3.5.4.7 Discussion

Agadirichnus elegans is different from *Pteraichnus* because of the elongate pes print with a rounded heel, the long digits of thepes, and the short and rounded Digit I of the manus (Masrour et al., 2018).

3.5.4.8 Trackmaker

Masrour et al. (2018) suggested that *Agadirichnus elegans* is associated with unknown azhdarchids because of the size and proportions of tracks. No pterosaur specimens have been found that match the tracks (Masrour et al., 2018).

3.6 Choristodera and other aquatic reptiles

3.6.1 Champsosaurichnus *Lockley and Hunt 1995b (Fig. 7.23A)*

3.6.1.1 Type ichnospecies

Champsosaurichnus parfeti Lockley and Hunt 1995b.

3.6.1.2 Valid ichnospecies

Champsosaurichnus parfeti Lockley and Hunt 1995b.

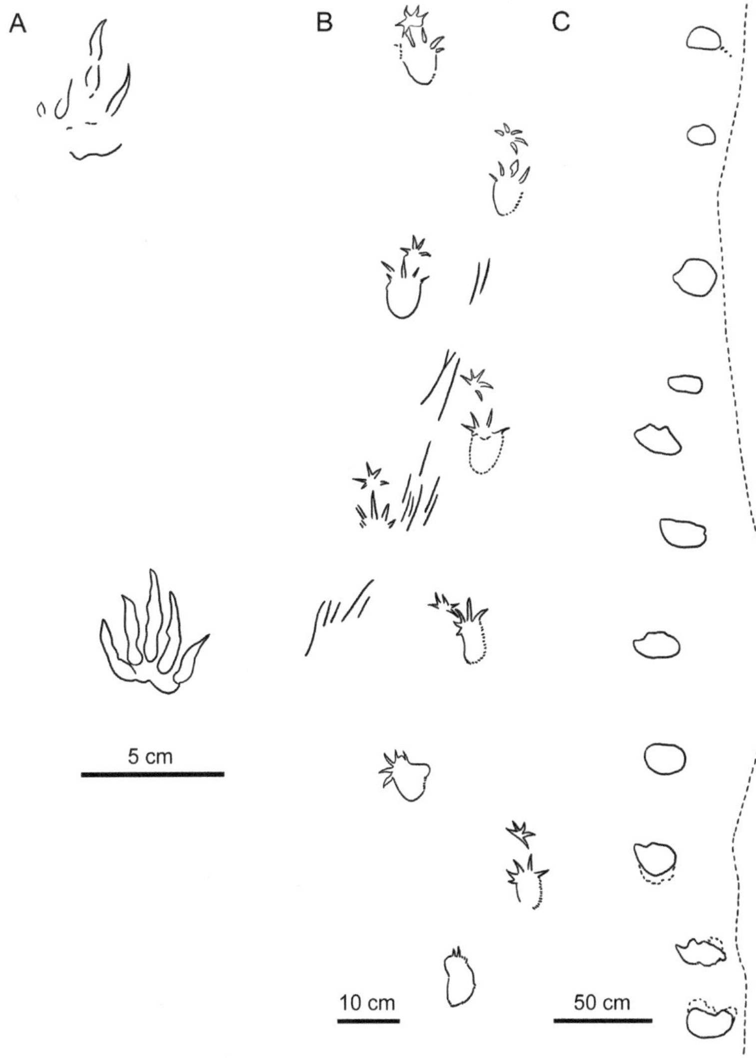

FIGURE 7.23 Tracks of Cretaceous aquatic reptile trackways. (A) Holotype (in the bottom) and a referred track of *Champsosaurichnus parfeti*; (B) type series of *Novapes ulsanensis*; (C) type series *Coneroichnus marinus*. *(A) Redrawn from Lockley and Hunt (1995b). (B) Redrawn from Lee et al. (2020). (C) Redrawn from Natali and Leonardi (2023); only perserves left front limb.*

3.6.1.3 Distribution

Maastrichtian, USA.

3.6.1.4 Holotype

A replica of the holotype is CU-MWC 223.2, plaster cast, yielded from the quarry of the Parfet Clay Pits, Laramie Formation of Colorado in the USA.

3.6.1.5 Diagnosis

Champsosaurichnus differs from other quadrupedal tracks in having small, elongate, symmetrical pentadactyl manual and pedal impressions with transverse heel and digit impressions which taper to fine points; in possessing the combination of Digit III impression longest, with Digits II and IV impressions subequal in length and shorter than III impression, and with Digits I and V impressions being subequal in length and shorter and II and IV impressions (Lockley and Hunt, 1995b).

3.6.1.6 General description

The digit impressions directed forward and have indistinct phalangeal pad impressions. The Digits II and IV are approximately 75% of the length of Digit III, while the lengths of Digits I and V are about 50% of Digit III. The divarication between digits is low, with 15°−20° divarication for Digits I and II, II and III, III and IV, and IV and V, for a total divarication of approximately 60°. The maximum length of the holotype is 4 cm, and the width is 3.5 cm. The second track is located approximately 18.5 cm in front of the holotype (original diagnosis from Lockley and Hunt, 1995b).

3.6.1.7 Discussion

Lockley and Hunt (1995b) indicated that the two isolated tracks of *Champsosaurichnus parfeti* are part of a trackway, but details such as step, stride, or pace angulation cannot be determined from them. It is also uncertain the holotype is a manus or pes track, and whether the two "consecutive" tracks actually represent a step, or a stride with the possibility of tracks missing on the other side. However, based on skeletal and ichnological similarities, including overall symmetry, with Digit III being the longest, Digits II and IV being shorter but equal in length, and Digits I and V being the shortest and equal in length, it is suggested that *C. parfeti* matches the symmetrical foot of champsosaurs (Erickson, 1972). Compared with the asymmetrical foot of crocodilians, the foot of *Champsosaurus* is remarkably symmetrical. Morphologically, the closest approximation is *Gwyneddichnium* from the Late Triassic (Lockley et al., 1992b) also attributed to aquatic reptiles. Therefore, *Champsosaurichnus* is highly distinctive and different from any previously reported tracks from the late Mesozoic, and should be named (Lockley and Hunt, 1995b).

3.6.1.7 Trackmaker

Choristoderes are generally thought to be derived neodiapsids that are close relatives or members of Sauria (Sues, 2019). They are usually semiaquatic reptiles that ranged from the Middle Jurassic, or possibly Triassic, to the Miocene. The best-known members being the crocodile-like neochoristoderes such as *Champsosaurus*. *Champsosaurichnus* was interpreted as the first recognized champsosaur ichnotaxon on the basis of skeletal and ichnological similarities (Lockley and Hunt, 1995b). However, Lee et al. (2020) pointed out that the fore- and hind-foot outlines of *Champsosaurus laramiensis* are not symmetrical, with Digit IV being equal to Digit III rather than Digit II, which is different from the digit character of *C. parfeti*.

3.6.2 Novapes *Lee et al., 2020 (Fig. 7.23B)*

3.6.2.1 Type ichnospecies

Novapes ulsanensis Lee et al., 2020.

3.6.2.2 Valid ichnospecies

Novapes ulsanensis Lee et al., 2020.

3.6.2.3 Distribution

Albian, South Korea.

3.6.2.4 Holotype

NHCG-a10952, Daegu Formation (Albian–Cenomanian) in the vicinity of Ulsan Metropolitan City, and housed in National Heritage Center Geology, Cultural Heritage Administration in South Korea.

3.6.2.5 Diagnosis

Quadrupedal tracks with a pronounced heteropody; pentadactyl manus impression with claw marks and semisymmetrical outline, wider than longer; divergence between Digit I and V imprints ranges 180°–210°; the Digit IV imprint slightly longer than Digit II; entaxonic; pentadactyl pes impression with claw marks and asymmetrical outline (i.e., lateral digits are more developed), longer than wide; webbing between the proximal portion of slender digits; the subequal Digits III and IV imprints longer than others (the Digit I imprint only 30% in the length of the Digit IV imprint); the sole pad impression is elongate with a U-shaped "heel"; entaxonic (original diagnosis from Lee et al., 2020).

3.6.2.6 General description

Novapes ulsanensis consists of nine manus-pes tracks, and these tracks are preserved as positive depressions with clear impressions of manus and pes digits, sharp claw marks, and tail traces. The acropodial portions of the manus and pes are deeper than the metapodial and basipodial portions. The manus digit impression is separated from the small, centered sole pad. The average length and width are 2.9 and 5.3 cm, respectively. The Digit I print is the shortest of the manus digit prints. Because of poor preservation of the Digit III print on the right manus track, it makes the track tetradactyl. There is no webbing between digits. Mean stride and pace lengths are 35 and 20.8 cm, respectively, with pace angulation of 120° on average.

The pes tracks are fully plantigrade, and the divergence between Digits I–V did not exceed 180°. The average length and width of the pes tracks are 9.9 and 6.9 cm, respectively. Mean stride and pace lengths are 33.1 and 22.8 cm, respectively, with pace angulation of 93° on average.

The length ratio of manus to pes tracks is 1: 3.36, and the width ratio is 1: 1.3. The trackway width is approximately three times the pes width. All manus tracks are closely anterior to the pes tracks, with their outer trackway width (average 18.8 cm) narrower than that of the pes (average 21.5 cm). The tail traces are finely striated and neither symmetrical nor close to the midline of the trackway (Lee et al., 2020).

3.6.2.7 Discussion

In the Mesozoic, webbed tracks have occasionally been reported in various quadrupedal animals, such as freshwater turtles, amphibians, crocodilians, and pterosaurs (Lee et al., 2020). Lee et al. (2020) stated that *N. ulsanensis* does not match any previously reported webbed tracks, and made the following comparison: (1), the pentadactyl pes tracks of *N. ulsanensis* can be easily distinguished from these of crocodyliform ichnotaxa like *Crocodylopodus* (Simpson et al., 2010; Meijide Calvo and Fuentes Vidarte, 2001; Castanera et al., 2022a) and *Batrachopus* (Hitchcock, 1845) by the divergence of Digits I—IV: crocodyliform 20°—55° (Kubo, 2008), while *N. ulsanensis* almost 180°; (2), the nonpterodactyloid pterosaurian *Rhamphichnus* (Mazin and Pouech, 2020) also exhibits similar pentadactyl pes tracks, but its manus tracks are tridactyl; (3), turtles typically have short and symmetrical pentadactyl manus and pes tracks but lack distinct heteropody (Avanzini et al., 2005), and the tracks are always arranged in parallel (Lichtig et al., 2018); (4), the pentadactyl pes of amphibians usually are without claw marks (Foster, 2001); (5), the "typical" lizard trackways show obvious heteropody, with pentadactyl manus and pes, such as *Sauripes* (Lee et al., 2018) and *Neosauroides* (Kim et al., 2017a, 2019a), but lack webbing between the digits (Pianka and Vitt, 2003), and the imprint of Digit V is clearly separated from other digits, connecting to the rear of the heel trace (Lee et al., 2018); (6), *N. ulsanensis* is distinct from *Champsosaurichnus parfeti* (Lockley and Hunt, 1995b) as well, as the latter has a small transverse heel on the pes tracks and lacks interdigital webbing.

3.6.2.8 Trackmaker

Lee et al. (2020) considered *Novapes ulsanensis* was left by a webbed trackmaker with five functional digits, not only matching the foot skeletons and body size of *Monjurosuchus* (Gao et al., 2000) but also aligning with the fossil record of East Asian choristoderes (Matsumoto et al., 2007, 2019). It can be inferred that *N. ulsanensis* may have been produced by a *Monjurosuchus*-like choristoderan, representing the first potential choristoderan trackway in Asia (Lee et al., 2020). *N. ulsanensis* also suggests that semiaquatic choristoderans adopted a semi-erect, high walk posture when moving on land and utilized tail dragging for locomotion, resembling the movement pattern of crocodilians (Reilly and Elias, 1998; Farlow et al., 2018; Lee et al., 2020).

3.6.3 Coneroichnus *Natali and Leonardi, 2023 (Fig. 7.23C)*

3.6.3.1 Type ichnospecies

Coneroichnus marinus Natali and Leonardi, 2023.

3.6.3.2 Valid ichnospecies

Coneroichnus marinus Natali and Leonardi, 2023.

3.6.3.3 Distribution

Berriasian—Hauterivian, Italy.

3.6.3.4 Holotype

No specimen number, in situ, protected by the Conero Natural Park, in whitish micritic limestone of the Maiolica Formation of Italy. A replica of the holotype is housed in the Museum System, Museum of Sciences of the University of Camerino.

3.6.3.5 Diagnosis

The whole general structure of the trackway is almost straight; it has a very high step angle and a very narrow internal and external trackway width. The breadth between tracks has negative values; as long as the trackway is to be considered complete, as it was found, and as it currently appears. In the first part of the trackway, the tracks are arranged somewhat zigzag, then toward the end they are arranged in a straighter way. This implies a somewhat variable swimming and punting progression. The imprints are rarely roundish, and almost always ellipsoid, with the longest axis transverse to the midline. No indications of digits were seen, but the imprints are generally characterized by a smooth and slightly concave surface at the deepest point of the tracks. So, the animal could not possibly perform a quadrupedal gait. That is, it did not realize a "punting gait" or "punting swim," which is a contact propulsion with all four appendages (original diagnosis from Natali and Leonardi, 2023).

3.6.3.6 General description

Coneroichnus marinus is a marine tetrapod trackway consisting of 11 imprints, with an alternating pace averaging about 50 cm (range 30–70 cm) and an average stride of approximately 100 cm (range 74–123 cm). Almost all imprints have a "V" shape with respect to the layer surface in most of the longitudinal sections, with the anterior side of the prints steeper than the posterior side. Some imprints have a pointed tip on their left or right margins, giving them a slightly pyriform appearance. The imprints are relatively small, with an average length of 12.73 cm (range 8–16 cm) and width of 21 cm (range 15–26 cm). The long axes of almost all prints are roughly parallel and inclined with an acute angle facing forward relative to the midline of the trackway, resembling a fish-bone pattern. No tail traces (Natali and Leonardi, 2023).

3.6.3.7 Discussion

Natali and Leonardi (2023) compared the *Coneroichnus marinus* tracks with those of turtles, marine reptiles tracks, showing the uniqueness of *C. marinus*. Manni et al. (1999) described the Lower Jurassic marine reptile tracks *Accordiichnus natans* somewhat resembling the *C. marinus*. Although their size is comparable, the *A. natans* has a low step angle (averaging 49°), while the step angle of *C. marinus* is almost 180° (Natali and Leonardi, 2023).

3.6.3.8 Trackmaker

The ichnotaxa of tetrapods preserved on the sea bottom are rare and are interpreted as traces of an aquatic reptile softly interfering with the sediment-water interface, such as marine turtles (Gaillard et al., 2003), protorosaurs (Lucas et al., 2014), nothosaurs (Zhang et al., 2014), placodonts and/or saurosphargids *Anshunpes* (Xing et al., 2020b). However, the shortcomings in the preservation of *Coneroichnus* make it uncertain who created the traces. Natali and Leonardi (2023) believe that if the traces represent only the left side of the trackway, with the right side not preserved, the complete trackway might have been produced by a more fully built, flatter-bodied animal. Its gait can be defined as a half swimming manner or a punting and foraging movement (Natali and Leonardi, 2023).

3.7 Terrestrial lizards

3.7.1 Neosauroides *Kim et al., 2017a (Figs. 7.24A,C,D and 7.37)*

3.7.1.1 Type ichnospecies

Neosauroides koreaensis Kim et al., 2017a (Figs. 7.24C and 7.37)

3.7.1.2 Valid ichnospecies

Neosauroides koreaensis Kim et al., 2017a.
Neosauroides innovatus Kim et al., 2019b (Fig. 7.24A)

3.7.1.3 Distribution

Albian–Cenomanian, South Korea.

3.7.1.4 Holotype

CUE 130,304L, Haman Formation, housed in the Chinju National University of Education, South Korea.

3.7.1.5 Diagnosis

Small lizard-like trackway with ectaxonic lacertiform pes and pes trackway width being wider than the manus trackway width. Manus pentadactyl, distinctly mesaxonic and symmetric to subsymmetrical, smaller than the pes and less outwardly rotated (original diagnosis from Kim et al., 2017a).

3.7.1.6 General description

Neosauroides koreaensis from the Haman Formation is the first lizard-like (lacertiform) trackway reported from the Mesozoic of Asia, and the first known example in the global Cretaceous track record, which consists of four consecutive manus-pes sets of natural casts. Pes traces exhibit an incomplete tridactyl pattern, comprising only the distal traces of Digits II, III, and IV, with a progressive increase in length. The recorded segment of the pes track length (excluding the proximal heel) measures approximately 1.90 cm, and the width is around 1.0 cm when measured parallel to the long axis of Digit IV. The pes traces show an outward rotation of about 60 ° along the axis of Digit IV with regard to the trackway midline, or approximately 20° along the axis of Digit III. Notably, the distal claw traces of Digits III and IV curve inward toward the trackway axis. Divergence between pes Digits II–III and III–IV average 8.3° and 41.3°, respectively ($N = 3$). In contrast, Digit III of the manus traces is the longest and Digits I and V are the shortest (III > IV > II > V=I). Manus are positioned anteromedial to pes within the trackway. Digit traces have fine distal claw traces, while heel traces are faint. The axis of Digit III is subparallel to the trackway axis. The overall length of the manus averages about 1.9 cm, with a range of 1.5–2.4 cm depending on the visibility of

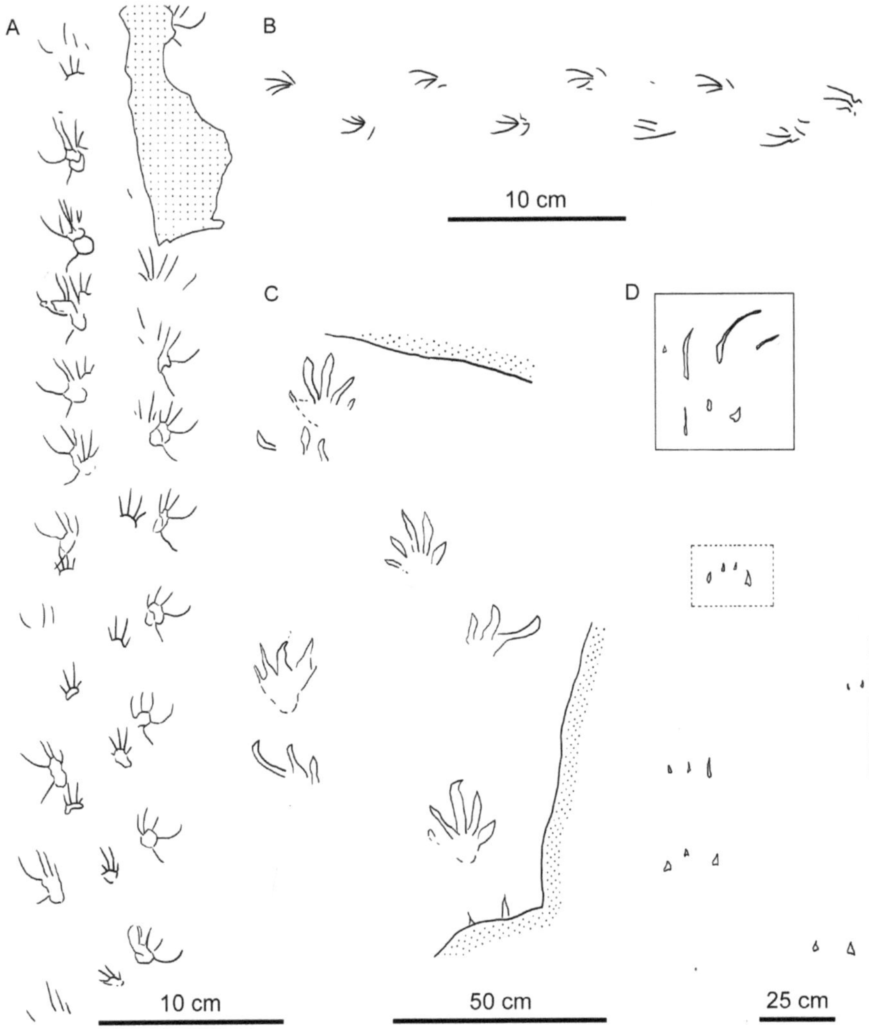

FIGURE 7.24 All occurrence of terrestrial lizard trackways Cretaceous. (A) Partial type series of *Neosauroides innovatus*; (B) trackway A from the holotype slab of *Sauripes hadongensis*; (C) type series of *Neosauroides koreaensis*; (D) the only occurrence of lizard-like track in Cretaceous outside Korea: ?*Neosauroides* isp. indet. From Early Cretaceous of Gansu (China). *(A) Redrawn from central part; Kim et al. (2019). (B) Redrawn from Lee et al. (2018). (C) Redrawn from Kim et al. (2017a). (D) Redrawn from Xing et al. (2021l).*

the heel trace. Manus width averages approximately 1.59 cm and shows less variability (1.5–1.7 cm). Digit divarication (I–V) averages about 83°, ranging from 76° to 86° ($N = 4$ for all preceding measurements). Manus digit traces I–V are essentially equally spaced, with the following average divarication angles: I–II 19.5° ($N = 2$), II–III 26° ($N = 4$), III–IV 19.5° ($N = 4$), and IV–V 20.3° ($N = 4$) (original diagnosis from Kim et al., 2017a).

Neosauroides innovatus from the Jinju Formation (Albian) is characterized by prominent heteropody, with strong internal pes trackway widths and less pronounced internal manus trackway widths. The large, splayed pes is pentadactyl ectaxonic, where Digits I–III are typically straighter than the Digit IV trace, with an increasing length (IV=III > II > I) and a postero-laterally oriented inclination. Manus is smaller but more elongate, slightly ectaxonic, and pentadactyl, oriented predominantly forward and situated closer to the trackway midline. The manus digit traces are slightly divergent, with Digit IV being the longest (IV > III > II > V > I) (Kim et al., 2019b).

3.7.1.7 Discussion

Peabody (1955) proposed that "a trackway of a tetrapod is as a rule generically distinctive relative to zoological classifications." The pes tracks of *N. koreaensis* show similarities with most ichnospecies of the Triassic ichnogenus *Rhynchosauroides*, but the manus tracks are notably different. The manus tracks of *Rhynchosauroides* are distinctly ectaxonic (Avanzini et al., 2010), unlike *Neosauroides*, which has a mesaxonic manus. In contrast, the manus tracks of *N. koreaensis* are more similar to those of extant lizards, such as *Sceloporus graciosus* (Halfpenny, 1998). This difference strongly justifies the establishment of the new ichnogenus *Neosauroides*. Therefore, according to ichnological classification criteria, this track should be considered distinct at the genus level (Kim et al., 2017a).

Neosauroides innovatus exhibits significant differences compared with *Sauripes hadongensis* (Lee et al., 2018): (1) *N. innovatus* exhibits distinct pes and manus tracks, indicating pronounced heteropody; (2) The pes tracks of the Jinju trackmaker T1–T3 are larger and more widely splayed; (3) the manus is smaller and less splayed compared with the pes; and (4) The trackways is wider, suggesting a more sprawling gait. *N. nnovates* also differs from *N. koreaensis*: (1) The manus of *N. innovatus* exhibits slight ectaxony rather than mesaxony; and (2) The position of manus is closer to the midline in *N. innovatus* than in *N. koreaensis*, resulting in larger pace angulation values. These morphological features distinguish *N. innovatus* from the previously reported *S. hadongensis* and *N. koreaensis* (Kim et al., 2019b).

Xing et al. (2021l) based on considerations of track size, slenderness of digits, pronounced heteropody, and both inner and outer widths of the trackway, suggested that the newly discovered tetrapod tracks, referred to as *Neosauroides* isp., bear similarities to *Neosauroides innovatus* (Kim et al., 2019b) from the Cretaceous of Korea. However, due to the incomplete nature of the trackway, further ichnotaxonomic description could not be undertaken (Xing et al., 2021l).

3.7.1.8 Trackmaker

Kim et al. (2017a) propose that *Neosauroides* is the trackway of a lacertiform, lizard-like trackmaker. The manus of *Neosauroides* bear a striking resemblance to the tracks of some modern lizards, as illustrated in various field guides, for example, the sagebrush lizard *Sceloporus graciosus* (Halfpenny, 1998). However, the differences in the foot morphology of body fossils compared with the track morphology hinder the inference of a perfect or near-perfect "Cinderella" match (cf. "Cinderella Syndrome" of Lockley, 1998).

Xing et al. (2021l) propose that the trackmaker was a small quadrupedal animal. The exclusion of turtles is based on considerations of size and heteropody, and multiple track and trackway features support the interpretation that the trackmaker is a lizard.

3.7.2 Sauripes *Lee et al., 2018 (Fig. 7.24B)*

3.7.2.1 Type ichnospecies

Sauripes hadongensis Lee et al., 2018.

3.7.2.2 Valid ichnospecies

Sauripes hadongensis Lee et al., 2018.

3.7.2.3 Distribution

Aptian, South Korea.

3.7.2.4 Holotype

KIGAM VP 201,501, Hasandong Formation, housed in Korea Institute of Geoscience and Mineral Resources, Vertebrate Paleontology, South Korea.

3.7.2.5 Diagnosis

Quadrupedal tracks; plantigrade and pentadactyl pes prints are longer than wide; the digit length progressively increasing from Digits I to IV (ectaxonic); Digit V is oriented more laterally and offset from other digits; digit imprint IV is more than twice the length of the metatarsal impression. Relatively narrow-gauge with about the same width as the pes (diagnosis emended from Kim et al., 2019b)

3.7.2.6 General description

Sauripes hadongensis represents the oldest crown-group lizard trackways, which is the first direct evidence of bipedal locomotion in fossil lizards. The slab is preserved with 29 lizard tracks that are preserved as depressions. The best-preserved trackway A has an average pes stride length of 7.9 cm, pace length of 4.8 cm, and pace angulation of 113°.

The majority of tracks are pes tracks (N = 25), averaging 2.2 cm in length and 1.2 cm in width, and showing ectaxonic and asymmetrical impressions. The divarication of Digits I and V is less than 90°. The metatarsal impression is longitudinal and located posterior to Digits I–IV, and slightly raised compared with the digit impressions, with the deepest impressions in the area of the metatarsophalangeal joint. Lacks evident tail marks (Lee et al., 2018) or complete manus tracks by comparing with *Neosauroides* (Kim et al., 2019b).

3.7.2.7 Discussion

The movement of lizard is significantly affected by body shape, length, and the specific characteristics of habitat (Rocha-Barbosa et al., 2008). Despite this, Snyder (1952) recognized four forms of gait for locomotion in lizards: a quadrupedal walk at low speeds, a quadrupedal fast gait, a diagonal run at high speeds, and the bipedal run (e.g., *Basiliscus basiliscus*). When lizards move at slower speeds, they adopt a sprawling limb posture with laterally oriented plantigrade feet (Brinkman, 1981), as observed in *Neosauroides koreanesis* (Kim et al., 2017a). In contrast, *Sauripes hadongensis* displays pes-dominant trackways with distinctive features such as long strides, large pace angulation, and digitigrade prints throughout the locomotion sequences. During high-speed running, lizards tend to leave digitigrade tracks instead of plantigrade ones (Irschick and Jayne, 1999; Lee et al., 2018). Kim et al. (2019b) expressed doubts about the validity of the previous considered manus-pes, and they also note that the relative length of the stride and manus tracks registered inside the pes supposedly indicates relatively slow speeds and a sprawling limb posture.

3.7.2.8 Trackmaker

Lee et al. (2018) concluded that the tracks of *Sauripes hadongensis* were likely produced by a lizard running bipedally, based on the predominant pes tracks, long pes stride length, narrow trackway width, digitigrade manus and pes tracks, and the anteriorly oriented long axis of the pes Digit IV. This particular lizard is thought to belong to an extinct family or represent a stem member of Iguania, determined by examining the paleobiogeographic distribution of existing lizard families (Evans and Manabe, 2008; Clemente, 2014). The estimated snout-vent length (SVL) of the trackmaker is approximately 6.8 cm, derived from the allometric plot between foot length and SVL observed in the iguanian *Tropidurus torquatus* (Kohlsdorf et al., 2001; Lee et al., 2018).

3.8 Crocodiles (Crocodylomorpha)

Crocodylian tracks have been reported from the Mesozoic and Cenozoic and assigned to at least 10 ichnogenera, five of which occur in the Cretaceous (*Hatcherichnus, Crocodylopodus, Mehliella, Batrachopus, Angolaichnus*). There are only four valid ichnospecies of crocodylomorph tracks named in the Cretaceous, *Mehliella jeffersonensis, Crocodylopodus meijidei, Batrachopus grandis* and *Angolaichnus adamanticus*. There are also records of *Hatcherichnus* isp. (or ? *Hatcherichnus*) in the Cretaceous strata.

Lockley et al. (2010) reported crocodylian tracks from Dakota Group of Colorado; McCrea et al. (2014a) reported crocodylian tracks from the Dunvegan Formation, the Gates Formation, the and Mist Mountain Formation of Canada; Lockley et al. (2018b) reported crocodylian tracks from the Blackhawk Formation of the USA; Lee et al. (2019) reported crocodylian tracks from the Bayanshiree Formation of Mongolia, and these authors assigned the crocodylian tracks to *Hatcherichnus* isp. or ? *Hatcherichnus*.

Both *Hatcherichnus* and *Characichnos* are thought to be animal swimming tracks, however, the trackmakers of the former is crocodiles, and the latter is theropods or other tetrapods (Foster and Lockley, 1997; Whyte and Romano, 2001). Vila et al. (2015) pointed out that the major difference between them is that *Characichnos* has longer scratch or scrape marks. For example, the trackmakers of different sizes of crocodylian tracks appear to live in different environments. For example, the trackmakers of the large crocodylian tracks *Hatcherichnus* live in fluvial settings, while the trackmakers of small crocodylomorphs tracks *Crocodylopodus* live in lacustrine settings (Lockley et al., 2020).

3.8.1 **Batrachopus** *Hitchcock, 1845 (Fig. 7.25D)*

3.8.1.1 Type ichnospecies

Batrachopus deweyi Hitchcock, 1843.

3.8.1.2 Valid ichnospecies in Cretaceous

Batrachopus grandis Kim et al., 2020.

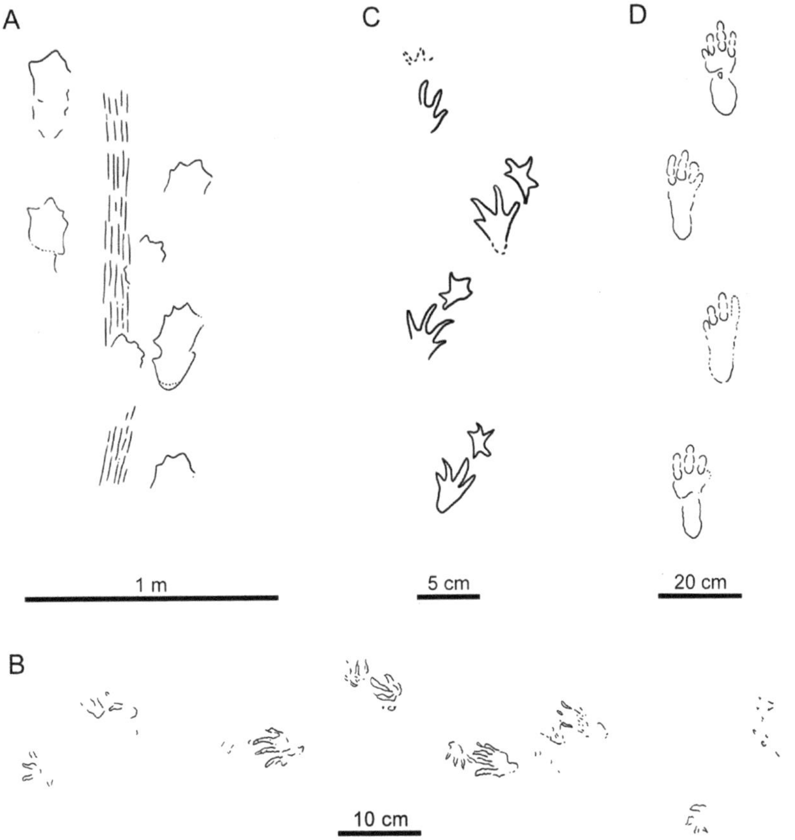

FIGURE 7.25 Tracks of crocodiles (Crocodylomorpha). (A) Holotype (the complete track from the most bottom on the left side of the tail trace) within a series of *Mehliella jeffersonensis*; (B) type series of *Angolaichnus adamanticus*; (C) type series of *Crocodylopodus meijidei*; (D) type series of *Batrachopus grandis*. *(A) Redrawn from Lockley (2010). (B) Redrawn from Mateus et al. (2017). (C) MNS2002/96/2bis; redrawn from Castanera et al. (2022a,b). (D) Redrawn from Kim et al. (2020).*

3.8.1.3 Total distribution

Triassic: Norian, Gettysburg Formation, Pennsylvania (USA); Norian-Rhaetian, Passaic Formation, New Jersey (USA).

Lower Jurassic, eastern and western North America and Europe;

Cretaceous: Aptian–Albian, South Korea and Thailand.

3.8.1.4 Holotype of the ichnogenus

AC 26/5 (type trackway) and 26/6, housed in the Hitchcock Ichnology Collection at the Pratt Museum of Amherst College, Amherst Massachusetts.

3.8.1.5 Ichnogeneric diagnosis

Batrachopus is a small quadrupedal archosaurian ichnite with a pronounced heteropody. The manus print is much smaller than the pes print, manus pentadactyl, digitigrade, and is usually rotated so that Digit II points forward, Digit IV points laterally, and Digit V points posteriorly. The pes is functionally tetradactyl and digitigrade, and outwardly rotated (25–33°). Digit V of the pes, when impressed, is reduced to an oval pad posterior to and nearly on the same line with Digit III. Digits I–IV are nearly parallel or slightly divaricated (III and IV). Digit III of the pes is the longest, and Digit I is the shortest. The pes length, from the base of Digit I to the tip of Digit IV, ranges between approximately 2 and 8 cm. The trackway is narrow, with the inner margin of the pedal print falling on or near the trackway midline. The placement of the manus is also close to the trackway midline and typically immediately in front of the pes (diagnosis emended by Masrour et al., 2020).

3.8.1.6 Ichnospecific diagnosis of *Batrachopus grandis*

Batrachopus grandis is much larger than all previously described *Batrachopus* ichnospecies and morphotypes, mostly "lumped" under the Lower Jurassic ichnogenus *B. deweyi*, and differs from them in consistently displaying a well-defined heel making up on average ~40% of total pes length. B. grandis is also less outwardly rotated than type *Batrachopus* (*B. deweyi*): that is, in the former morphotype Digit III parallels the trackway midline, not Digit II as in the latter ichnospecies with a more outwardly rotated pes. No other *Batrachopus* ichnospecies has been described in such detail, from such a large and well-preserved sample, even with skin traces, as Korean *B. grandis*. This allows for the definition of anatomy-consistent and diagnostic features such as the relative heel length and relative digit divarication angles with III–IV consistently greater than I–II and II–III: that is, ∠III–IV>∠I–II and ∠II–III (original diagnosis from Kim et al., 2020).

3.8.1.7 General description

Hitchcock (1843) first described *Batrachopus* and included six ichnospecies (Rainforth, 2007). Olsen and Padian (1986) synonymized *B. bellus*, *B. gracilior* and *B. gracilis* with the type species, *B. deweyi*, but those authors had not examined the holotype of *B. deweyi*; their synonymization was rejected by Rainforth (2005). The specimen of *B. deweyi* (AC 26/5 and 26/6) is an excellent trackway with seven manus-pes sets that show all the characters typical of the genus and species. The two most posterior manus-pes impressions show a well-defined pad for Digit V (Olsen and Padian, 1986).

Kim et al. (2020) reported crocodylian tracks from the Jinju Formation (Albian) of South Korea and named *Batrachopus grandis*. The trackway of *B. grandi* has a tetradactyl pes. Pes twice as long as wide with a narrow heel, separated from the wider anterior portion, with digit traces, by a transverse, bilobed crease with anteriorly convex sulcus. Total track length averaging 20.7 cm (range 18.0–24.0 cm), along the axis of longest Digit III, with mean lengths of Digits I, II, and IV respectively as 14.7 cm, 18.6 cm, and 19.4 cm (from posterior margin of heel). Thus, III > IV > II > I. Digit traces up to ~2.0 cm wide, with well-defined pad impressions, generally in the better-preserved specimens indicating a phalangeal pad formula of 1-2-3-4 corresponding to Digits I, II, III, and IV. Length of that portion of the track anterior to the transverse crease averaging 12.2 cm (range 11.0–15.1 cm): thus L/W ratio of the anterior digital portion of the pes is 1.27 (12.2 cm/9.6 cm), with the maximum track width situated anteriorly near distal margins of Digits I and IV. Total I–IV digit divarication averaging 31.2° (range 28.0°–38.0°) with mean divarication values for I–II, II–III, and III–IV 9.3°, 9.4°, and 12.3° respectively. Skin traces consist of angular polygonal scale traces about 1.0 cm in diameter and are presently known in the heel region. Trackway narrow without manus traces on surfaces where pes traces registered. Step averaging 37.3 cm (range 29.8–47.0 cm); stride averaging 69.3 cm (range 50.8–81.0 cm); pace angulation averaging 148.5° (range 135°–165°); outer trackway width averaging 18.7 cm and inner trackway width averaging 0.6 cm and ranging from 5.5 to 4.0 cm.

Le Loeuff et al. (2010) reported two trackways of *Batrachopus* isp. from the Khok Kruat Formation (Aptian–Albian), Huai Dan Chum, northeastern Thailand. The trackways are wider than usual batrachopodid trackways. In the two trackways, the trackway width is 2.61 times and 3.27 times the track width.

3.8.1.8 Discussion

There are significant distinctions between *Batrachopus* and *Hatcherichnus,* the two widespread crocodylian ichnogenera in terms of age, trackmaker morphology, facies connection, and behavior. With the exception of occurrence in the Lower Cretaceous of Thailand (Le Loeuff et al., 2010) and South Korea (Kim et al., 2020), *Batrachopus* is mainly known from the Lower Jurassic, where it is found primarily in eastern and western North America and Europe (Olsen and Padian, 1986; Lockley and Meyer, 2004; Lockley et al., 2018c, 2020), and in association with lacustrine deposits in rift valley and other terrestrial basin settings. *Batrachopus,* like *Crocodylopodus,* is recognized almost entirely for its rather narrow trackways with no tail imprints left by small creatures walking on emerging surfaces. It is assumed to represent a tiny terrestrial species.

The high density of *Batrachopus* tracks suggests either a high density of crocodylomorph trackmakers, possibly allied to *Protosuchus* (Colbert and Mook, 1951), or high levels of activity by a small number of individuals in a given area and short time duration (Lockley et al., 2018c). Heel traces occur sporadically in type *Batrachopus,* that is, *B. deweyi* from the Lower Jurassic of the Connecticut Valley region (Olsen and Padian, 1986), and are regarded as possibly incorporating a reduced Digit V, never illustrated as a separate digit trace (Olsen and Padian, 1986; Hunt et al., 2018). However, they are inconsistently and incompletely registered, thus appearing in only two of seven consecutive tracks in the type trackway, and never with skin traces. There is general agreement that there are similarities between *Batrachopus* and *Antipus,* a possible junior synonym (Coombs, 1996). But *Batrachopus,* while similar to *Crocodylopodus* in lacking tail traces, is different in trackway gauge, pace angulation and footprint rotation etc., some of which indicate a more erect limb posture in semi-erect posture than *Crocodylopodus* (Castanera et al., 2022a). Both *Batrachopus* and *Antipus* are known exclusively from the Lower Jurassic with the exception of the aforementioned Asian occurrences (Le Loeuff et al., 2010).

Kim et al. (2020) named *Batrachopus grandis* from the Jinju Formation of South Korea. These tracks were first assigned to bipedal pterosaurs (Kim et al., 2012a) and then to large bipedal crocodylomorphs (Kim et al., 2020). They are much larger than previously described *Batrachopus* ichnospecies and morphotypes, mostly "lumped" under the Lower Jurassic ichnospecies *B. deweyi,* and differ from them in consistently displaying a well-defined heel making up on average ~40% of total pes length. In the case of the type of the *B. "gracilis"* trackway (Olsen and Padian, 1986) there are no heel traces (Kim et al., 2020). The manus traces were not found but register as pentadactyl traces in other *Batrachopus* ichnospecies. The creases that differentiate the pad traces are not always clearly registered, especially in Digit IV.

3.8.1.9 Trackmaker of Cretaceous *Batrachopus*

Kim et al. (2020) inferred that the trackmaker of *Batrachopus grandis* is different from those of extant crocodilians. They concluded that the trackmaker was a terrestrial or semiterrestrial form, with the weight of evidence supporting bipedality. *B. grandis* trackways have a narrow gauge. Sennikov (2021) explained that is not characteristic of crocodylomorphs but typical for bipedal dinosaurs. There is a similarity between *B. grandis* and segnosaur tracks and trackways, especially *Macropodosaurus* from the USA (Gierliński and Lockley, 2013b). Therefore, Sennikov (2021) suggested that the possibility of attributing *B. grandis* to some segnosaur group should be entertained, especially since these tracks differ from the type ichnospecies *B. deweyi*.

Le Loeuff et al. (2010) described crocodile tracks from Huai Dan Chum of Thailand. These tracks were assigned to *Batrachopus* isp., and the trackmaker was considered to be a dwarf crocodyliform (adult skull length: 63 mm). Teeth of similar dwarf crocodiles are known from the Khok Kruat Formation (Lauprasert et al., 2011) and such a dwarf atoposaurid might be the maker of the very small tracks of Huai Dan Chum.

3.8.2 Mehliella *Strand, 1932 (Fig. 7.25A)*

3.8.2.1 Type ichnospecies

Mehliella jeffersonensis Strand, 1932.

3.8.2.2 Valid ichnospecies

Mehliella jeffersonensis Strand, 1932.

3.8.2.3 Distribution

Cenomanian, USA.

3.8.2.4 Holotype

Replicas of the holotype belong to UCM 18,056, two plaster "molds" with the same number, Dakota Group of Colorado; specimen preserved in the University of Colorado Museum of Natural History.

3.8.2.5 Diagnosis

Large, quadrupedal, plantigrade; manus pentadactylate; pes tetradactyl, Digits II and III are longest and have distal claw traces, Digit IV is slightly shorter, and Digit I the shortest, shows a slight double trace of Digit III; interdigital webbing traces can be ascertained and more fully developed in pes; Digit I of manus and pes widely separated from Digit II; heel and base of palm broad; trackway wide, about equal length of stride; tail trace can be observed (revised diagnosis based on the comments from Lockley, 2010).

3.8.2.6 General description

Mehliella is not only the first crocodylian trackway ever found or named from the Cretaceous, it also represents the largest individual (estimated length of the trackmaker about 4 m) to have left a diagnostic trackway (Lockley, 2010).

Lockley (2010) pointed that the conclusions of Mehl (1931) are problematic for several reasons: (1) Mehl (1931) incorrectly reversed the manus and pes, the larger track is the pes and the pentadactyl track is the manus, it was the pes overstepping the manus secondarily; (2) Mehl (1931) also incorrectly gave the pes an extra Digit V, in fact, the best-preserved pes track is tetradactyl; (3) The relative digit length formula "$1 < 5 < 4 = 3 = 2$" Mehl described for the pes and "$1 < 2 < 3 = 5 < 4$" for the manus are impossible, at least for extant crocodylians, which have a tetradactyl pes with formula $1 < 2 = 4 < 3$ and manus of $5 < 1 < 4 = 2 < 3$ (Farlow and Elsey, 2010; Kumagai and Farlow, 2010); (4) the actual measurement of track UCM 18,056 is 23.5 cm, which suggests an error in the scale of the diagram (one-10th instead of one-eighth); (5) Mehl (1931) inferred that the tracks of both the manus and pes were fully webbed, while actual illustrations showed that webbing was more developed in larger tracks; (6) the "faint" original trace described by Mehl (1931), due to the narrowness of the marks (6.0 cm), are most likely a tail trace rather than a belly trace.

3.8.2.7 Discussion

Mehl (1931) discovered the trackway and named this ichnotaxon *Walteria jeffersonensis*. In 1932, the trackway was renamed *Mehliella jeffersonensis* Strand, 1932 because the genus *Walteria* was preoccupied by a sponge. The holotype specimen was not found in the University of Missouri collections and evidently lost. Two plaster molds made by Maurice Mehl, which were located in the UCM collections, revealed that it is a crocodylian track that corresponds very well with the left pes shown in the original illustration of Mehl (1931) (Lockley, 2010).

Lockley (2010) pointed that the conclusions of Mehl (1931) are problematic for several reasons. (1) UCM 18056 is clearly a tetradactyl left pes mold, not a manus as Mehl suggested, (2) the actual measurement of track UCM 18056 is 23.5 cm, which suggests an error in the scale of the diagram.

Houck et al. (2010) report a Dakota Group crocodylian trackway designated as the North Golden Trackway. It is not far from the type locality of *Mehliella jeffersonensis*. However, the new discovery is clearly not the same trackway as the type of *M. jeffersonensis*. The North Golden trackway represents a smaller animal than the *Mehliella* trackmaker, and one that was not dragging its tail or belly in such a way as to leave a trace. The differences between the two trackways are also reflected in track morphology, step length, stride length, formation environment, trackway width, body length, and so on.

3.8.2.8 Trackmaker

Mehl (1931) proposed that from the standpoint of geological occurrence and limb development *Mehliella* (his *Walteria*) may well have stood in the direct ancestral line of the mosasaurs. *Mehliella* was also attributed to a turtle (Kuhn, 1963) and an ankylosaur (Haubold, 1971; Thulborn, 1990).

Lockley (2010) proposed that *Mehliella* is an example of an ichnotaxon somewhat representative of "high walk" of gait, except for manus-pes spacing, which represents the only example from the Dakota Group of a crocodylian walking on land. It provides a fascinating case study of the problems involved in identifying trackmakers from incomplete tracks.

3.8.3 *Crocodylopodus Meijide Calvo and Fuentes Vidarte, 2001 (Fig. 7.25C)*

3.8.3.1 Type ichnospecies

Crocodylopodus meijidei Meijide Calvo and Fuentes Vidarte, 2001

3.8.3.2 Valid ichnospecies

Crocodylopodus meijidei Meijide Calvo and Fuentes Vidarte, 2001

3.8.3.3 Distribution

Jurassic: Kimmeridgian–Tithonian, Spain. Cretaceous: Berriasian–Albian, Spain and South Korea.

3.8.3.4 Holotype

Lectotype: MNS2002/96/2bis, Paralectotypes: MNS2003/92/8a; MNS2002/96/4; from Huérteles Formation (Berriasian) of Soria, housed in Numantine Museum of Soria, Spain.

3.8.3.5 Diagnosis

Star-shaped pentadactyl manual prints with slender digital impressions whose lengths vary as follows: III ≥ IV =II > V = I. Pedal track with digital impressions with the following length variations: DIII > DII ≥ DIV > DI. Digit III is clearly the longest and Digit I the shortest. Interdigital divarication varies from medium to high. Pronounced to medium heteropody (heteropody index = 30%–40%). Manual tracks are laterally rotated whereas the pedal tracks are slightly medially rotated. Intermediate-gauge trackway. Absence of tail, belly, or any other drag marks (diagnosis emended by Castanera et al., 2022a).

3.8.3.6 General description

Crocodylopodus meijidei includes four sets of coupled manus and pes tracks. The length of the manus is 1.5–2.2 cm, with 2.5–3 cm of width and length/width ratio of 0.6–0.75. I–II is the lowest angle (36°–41°), the other angles (II–III, III–IV, IV–V) being higher and variable (43°–66°). The interdigital divarication angle in the manus is 209°–218°. Generally, Digits I–II and IV–V are oriented medially/anteromedially and laterally/posterolaterally respectively, and Digit III has an anterior orientation. Digits I and V are to a large extent pointing in opposite directions. These orientations are variable because of the variability of the footprint rotation, which is lateral (15°–36° outwards) in all the manual prints. No clear claw marks are identified in the manual tracks. The length/width ratio of pes tracks is 1.3–1.42. The digit impressions are variable in width from 0.3 to 0.6 cm. Four of them have acuminate end, showing clear evidence of claw marks associated with the first three digits, Digit IV having a less acuminate end. The orientation of the digit impressions is medial (Digits I and II), anteromedial (Digit III), and anterolateral (Digit IV), with an average total divarication of 56°–57°. I–II (10°–15°) is the lowest angle, II–III (21°–24°), and III–IV (18°–24°) being quite similar. The heel impression is oval to subtriangular and quite symmetrical, and is shallower than the anterior part of the footprint. No clear evidence for the presence of webbing in both manual and pedal tracks. The manual–pedal impression distance is short at 3.8–4.5 cm. Pace length shows few variations between the manus and pes tracks of 9.5–11.6 cm. Stride length is 17–20 cm for both the manus and pes tracks. Pace angulation is high but variable (from the center of the track = 118°–137° and from tip of Digit III = 116°–137° for the pes tracks; same location 105°–140° and 102°–129° for the manus tracks). The manus-pes track ratio is about 1:3. The estimated gleno-acetabular distances range from 11.5 to 14.2 cm (Castanera et al., 2022a).

3.8.3.7 Discussion

Three holotypes (Rastro A, B, and C = MNS2002/96/2bis, MNS2003/92/8a, and MNS2002/96/4, respectively) were designated in the original description by Fuentes Vidarte and Meijide Calvo (1999) and such a procedure is not permitted by the ICZN (1999). Castanera et al. (2022) designate specimen MNS2002/96/2bis as a lectotype and specimens MNS2003/92/8a and MNS2002/96/4 as paralectotypes.

Fuentes Vidarte and Meijide Calvo (1999) suggested that *C. meijidei* material was different enough to define the new ichnofamily Crocodylopodidae. Lockley and Meyer (2004) judged that there were not enough differences to define a new ichnofamily so they included *C.s. meijidei* in Batrachopodidae Lull, 1904. Castanera et al. (2022) suggested that although the procedure might be correct, *C. meijidei* does not fit in the revised diagnosis of Batrachopodidae proposed by Lockley and Meyer (2004). The difference between *Crocodylopodus* and other crocodylomorph ichnotaxa is mainly reflected in anatomical factors, especially locomotor and behavior factors (e.g., manus/pes morphology, lengths of digital impressions in the pes, heteropody, track rotation or trackway gauge), and some substrate-related factors (Castanera et al., 2022a).

Lockley et al. (2020) described another large sample of *Crocodylopodus* tracks from the Lower Cretaceous (Albian, Jinju Formation) of South Korea. These trackways show several features that are different from *C. meijidei* especially the varied length/width ratio, the relative digital impression lengths, and the more anterior orientation, much lower heteropody, the lower pace angulation and wider-gauge trackway than in *C. meijidei*. It also shows no evidence of tail drag marks. There are also some cf. *Crocodylopodus* and ?*Crocodylopodus* found from the same formation as *C. meijidei* (Pascual-Arribas et al., 2005; Lockley et al., 2010, 2020), the Upper Cretaceous Wahweap Formation of Utah (Simpson et al., 2010; Lockley et al., 2020) and the Upper Cretaceous of Spain (Vila et al., 2015).

3.8.3.8 Trackmaker

Pascual-Arribas et al. (2005) suggest that Goniopholidae (Dollo, 1883) is the most suitable trackmaker of tracks classified as cf. *Crocodylopodus* because the size of its species, as well as the environment in which they lived (river and lake environments), are consistent with the trackmaker of the Valdelavilla tracks. Simpson et al. (2010) consider that the likely trackmaker of the Wawap tracks were either an alligatoroid or a juvenile goniopholid or deinosuchid. Lockley et al. (2020) thought that there is no evidence that the Korean trackmakers of *Crocodylopodus* were swimming. In contrast, they appear to be active along lake shoreline in a niche similar to that reported for Lower Jurassic *Batrachopus*.

Castanera et al. (2022) stated that considering the paleogeographic and geochronological context of *C. meijidei*, only some noneusuchian neosuchian taxa are considered candidates for trackmakers, as the oldest eusuchian Hylaeochampsa is known from the Barremian of England; there is no record of protosuchians, notosuchians, gobiosuchids, or sphenosuchians in the European Berriasian. Therefore, the producer of *C. meijidei* likely belongs to noneusuchian neosuchian crocodylomorphs such as a goniopholidids, atoposaurid, or bernissartids. Its center of mass may have been positioned more anteriorly during locomotion, with limbs exhibiting a more erect posture than extant crocodylomorphs. Therefore, the trackmaker may have been more adapted to terrestrial movement than modern crocodylians (Castanera et al., 2022a).

3.8.4 Angolaichnus *Mateus et al., 2017 (Fig. 7.25B)*

3.8.4.1 Type ichnospecies

Angolaichnus adamanticus Mateus et al., 2017.

3.8.4.2 Valid ichnospecies

Angolaichnus adamanticus Mateus et al., 2017.

3.8.4.3 Distribution

Late Aptian, Angola.

3.8.4.4 Holotype

Trackway one in MGUAN-PA 600, Calonda Formation, housed in the Geological Museum of Universidade Agostinho Neto-PaleoAngola Project (Catoca collection) in Luanda, Angola.

3.8.4.5 Diagnosis

Narrow trackway of a small–medium-sized crocodylomorph. The manus is mesaxonic, functionally tetradactyl, and plantigrade, wider than longer (length is 3.0 cm, width is 3.4 cm). Manus digits are slender, with claw marks, and increase in length from Digit I (1.1 cm) to Digit V (1.8 cm), with a total divarication I–IV of 79.5°. The pes is mesaxonic, functionally tetradactyl and plantigrade, longer than wider (5.3 cm in length, 3.7 cm in width). Pes digits are slender, with claw marks, tend to increase in length from Digit I (2.6 cm) to Digit III (3.0 cm), with Digit IV being the shortest (2.4 cm) and a total divarication I–IV of 38°; Digits II and III are slightly bent outward. Manus pace angulation of 148°, and manus pace and stride lengths of respectively 13.9 and 23.8 cm. Pes pace angulation is of 145°, and pes pace and stride lengths of 14 and 25.1 cm, respectively. Manus tracks present an average extreme outward rotation of 118° with respect to the pes (original diagnosis from Mateus et al., 2017).

3.8.4.6 General description

Angolaichnus adamanticus trackway one comprises track No.26–32. Manus average digit angles are I–II 29°, II–III 17.5°, III–IV 33.5°. Digit III in the left manus print 29 slightly bends medially, but in the right manus print 28, it does not. Pes average divarication angles are I–II 10°, II–III 11°, III–IV 17°. Trackway 2, as well as isolated tracks 33, 35, 51, and 62, exhibit morphological features similar to the crocodyliform type described by the tracks in trackway 1. They show mesaxonic, functionally tetradactyl and plantigrade pes with slender, slightly curved digits and narrow digital divarication. Trackway two is composed of left track 1A, right track 02, and track 1B. The average length and width of tracks 1A and two are 6.8 and 4.1 cm, respectively, with a length/width ratio of 1.65. The average length for isolated tracks is 4.6 cm, average width is 3.4 cm, and the length/width ratio is 1.35. The average length of Digit I is 1.5 cm, Digit II is 2.3 cm, Digit III is 2.9 cm, and Digit IV is 2.2 cm. The average divarication angles are I–II 20°, II–III 16°, III–IV 15°, with an average total divarication of 51° (Mateus et al., 2017).

3.8.4.7 Discussion

Mateus et al. (2017) suggest that due to its functionally tetradactyl and plantigrade pes, the length value of the foot (ranging from 3.7 to 7.6 cm), and the presence of claw marks, *Angolaichnus adamanticus* can be attributed to the ichnofamily Batrachopodidae. The total digital divarication in *A. adamanticus* resembles the *Batrachopus-Antipus* morphotype (<40°), while the specific tracks No.33 and No.62 in MGUAN-PA601 (with total digital divarication of 62° and 72°, respectively) are similar to the Early Cretaceous morphotype *Crocodylopodus*. *Angolaichnus adamanticus* trackway is comparable to other known batrachopodid trackways, such as its narrowness and pes pace angulation of 145° (Lockley and Meyer, 2004; Mateus et al., 2017).

However, *A. adamanticus* exhibits a distinctive manus functional morphology, being tetradactyl and plantigrade, whereas the manus tracks of batrachopodids are usually pentadactyl and functionality digitigrade (Lockley et al., 2004b; Avanzini et al., 2007). In addition, the unusual outward rotation features in the manus tracks of *A. adamanticus* distinguish it from other trackways attributed to crocodylomorph ichnospecies, classifying it as a new ichnogenus (Mateus et al., 2017).

3.8.4.8 Trackmaker

Mateus et al. (2017) state that the crocodylomorph trackways and tracks are attributed to new batrachopodid ichnospecies *Angolaichnus adamanticus*, were made by small- to medium-sized crocodylomorphs. These tracks were produced by crocodylomorph trackmakers, exhibiting unique characteristics of tetradactyl and plantigrade manus print.

3.9 Turtles

3.9.1 Emydhipus *Fuentes Vidarte et al., 2003 (Fig. 7.26A and B)*

3.9.1.1 Type ichnospecies

Emydhipus cameroi Fuentes Vidarte et al., 2003 (Fig. 7.26A)

3.9.1.2 Valid ichnospecies

Emydhipus cameroi Fuentes Vidarte et al., 2003.
Emydhipus cerinensis Bernier et al., 1982; Avanzini et al., 2005 (Fig. 7.26B)

3.9.1.3 Distribution

Jurassic: Kimmeridgian–Tithonian, Spain, France; Cretaceous: Berriasian–Albian, Spain, Germany, China, Morocco.

3.9.1.4 Holotype

MNS-2002-96-38, housed in the Numantino Museum of Soria, Spain.

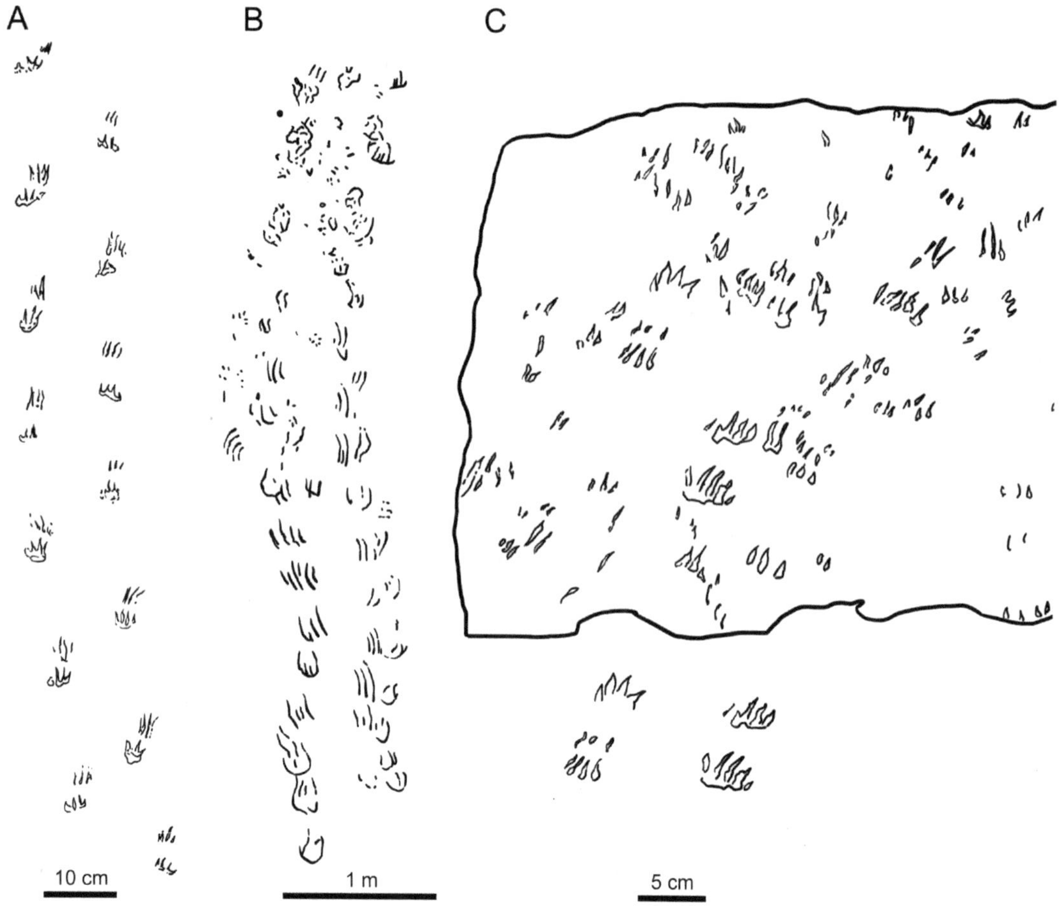

FIGURE 7.26 Tracks of turtles. (A) Type series of *Emydhipus cameroi;* (B) type series of *Emydhipus cerinense;* (C) holotype slab of *Chelonipus liui. (A) Redrawn from Avanzini et al. (2005). (B) Redrawn from Avanzini et al. (2005). (C) Redrawn from Lockley et al. (2019).*

3.9.1.5 Diagnosis

The manual print with four elongated claw marks parallel to each other and to the midline of the trackway. The pedal print is plantigrade with four clawed digits and a short, rounded sole. Both manual and pedal prints lack rotation with respect to the trackway midline (emended diagnosis by Klein et al., 2018).

3.9.1.6 General description

Fuentes Vidarte et al. (2003) based on a well-preserved trackway from the Middle–Late Berriasian, Lower section of the Oncala Group of Soria, Spain named *Emydhipus cameroi.* They are interpreted as having been made by turtles partially buoyed by water and by turtles walking in wet or firm mud in a subaerial environment (Avanzini et al., 2005). Manual track with four elongated and parallel claw marks in the axis with the trackway midline (1.7 cm in length, 2 cm in width). Pedal track plantigrade (1 cm in length, 1.5 cm in width) with four clawed digits. Pedal Digits II and III are nearly equal in length, and I and IV are shorter and nearly divergent with respect to the pes long axis. The sole is short and rounded. The trackway width is about 4.5 cm, and the pace angle is 63°–85°. Manual and pedal prints without rotation with respect to the trackway midline clearly show the track of four digits in decreasing length (Fuentes Vidarte et al., 2003; Avanzini et al., 2005).

Avanzini et al. (2005) suggested *Chelonichnium* is a nomen dubium, and referred *C. cerinense* to *Emydhipus cerinensis*. Bernier et al. (1982) originally named *Chelonichnium cerinense* from the late Kimmeridgian lithograph limestones of Cerin, France. Bernier et al. (1982) described the diagnosis of *C. cerinense* as pentadactyl (sometimes tri- or tetradactyl), slightly longer than wide, with hallux shifted inward in relation to the other digits. Short sole, fairly large claws. Manual tracks are represented only by the traces of the three main claws, curved inwards and parallel, the middle trace slightly longer. Pes tracks farther apart than those of the manus. Ratio of the stride to the length of the pes of the order of four, that of the stride to the width of the track is 0.6 in normal walking.

3.9.1.7 Discussion

Pterosaurs and crocodylians have different manus and pes morphologies, but turtle swimming tracks have similar manus and pes tracks. For example, the turtle manus and pes are typically short and pentadactyl with well-developed webbing. In comparison with crocodilian swim tracks, which are usually tridactyl and tetradactyl and much larger, turtle tracks are typically short, wide, and often pentadactyl. Likewise, the distinction between turtle and pterosaur tracks is also easily made in most places. Pterosaur manus tracks are highly distinctive in shape and rarely occur in swim track assemblages, while pterosaur pes tracks are elongated and often show elongated scratch marks produced by a different swimming action (Lockley et al., 2014c).

According to Avanzini et al. (2005), among the ichnotaxa attributed to turtles, there are only two valid ichnogenera: *Chelonipus* and *Emydhipus*. Based on the trackways, *Chelonipus* has pes tracks closer to the trackway midline and closer to the front of the pes tracks than *Emydhipus*. *Emydhipus* differs from *Chelonipus* by showing manual tracks with evident and parallel ungual traces, slightly internal and apparently always separated with respect to the pedal tracks, whereas in the latter the manus tracks are positioned external and posterior to the pedal tracks (Xing et al., 2019f). Lockley et al. (2018e) conclude that *Chelonipus* represents incomplete "toe-tip" tracks, evidently created while swimming, whereas *Emydhipus* represents both manus and pes tracks in alternating (left and right) sequences and evidently represents walking. Avanzini et al. (2005) suggested that the different positions of the manual tracks in the trackways (*Emydhipus* and *Chelonipus*) could be related to a different trackmaker's anatomy and possibly to different vertebrate taxa.

Emydhipus cameroi was named by Fuentes Vidarte et al. (2003). According to Avanzini et al. (2005), is valid, though similar to *Chelonipus*. It is often not possible to determine which ichnogenus is represented by a given set of tracks (Lockley et al., 2019). Fuentes Vidarte et al. (2003) did not compare their type of *Emydhypus* (*E. cameroi*) with the type material of *Chelonipus* (*C. torquatus*), their ichnotaxonomic comparisons are incomplete and open to question. In China, due to the lack of a trackway, most *Emydhipus* records are difficult to distinguish from *Chelonipus* (Xing et al., 2023c).

Chelonichnium cerinensis, with the clawed manual prints slightly internal to those of the pedes and without apparent overstep, shows a closer comparison with *Emydhipus* rather than with *Chelonipus*, and such traces are compatible also with movement underwater (bottom walking) in a shallow water environment (Avanzini et al., 2005). So Avanzini et al. (2005) suggested *Chelonichnium* is a nomen dubium, and referred *C. cerinense* to *Emydhipus cerinensis*. The trackway width of *E. cameroi* is approximately 4.5 cm (Avanzini et al., 2005), while Mossman and Grantham (2008) revisited *E. cerinensis* and found that the trackway width is 7.3 cm. It is suggested that the trackmaker of *E. cameroi* had a wider gait than that of *E. cerinensis*.

Klein et al. (2018) reported *Emydhipus* isp. from the lower part of the Midelt Formation (upper Barremian−?Aptian) of the High Moulouya region, Morocco. These tracks are similar to *E. cameroi* reported by Avanzini et al. (2005), but comparing the trackway of both, they have larger pace angulation, 88°−95° and 63°−85°, respectively. Moratalla and Hernán (2009) described *Emydhipus* isp. from the Enciso Group (late Barremian−Early Aptian) of Cornago township, La Rioja Province, Spain. These tracks are consistent with *E. cameroi* and have certain similarities, which are mainly reflected the general tridactyl track morphology, the presence of parallel digit marks, and the relative curved shape of the external digital impressions. But the size of the Asturian tracks is larger, 5−8 cm in length.

3.9.1.8 Trackmaker

Avanzini et al. (2005) interpreted the trackmaker of *Emydhipus cameroi* as having been made by turtles partially buoyed by water or by turtles walking in a slightly wet subaerial environment. The dimensions of the trackmakers inferred by their trackways probably varied from 15 to 40 cm of plastron width and from 20 to 60 cm of maximum length.

Pascual-Arribas and Hernández-Medrano (2015) reported *Emydhipus cameroi* from Spain whose morphology is similar to those caused by present-day terrestrial turtles on soft sediment (Fiorillo, 2005) and aquatic or semiaquatic turtles on wet sediment (Páez et al., 2012). Klein et al. (2018) reported *Emydhipus* isp. from Morocco is nearly identical in morphology to the trackways of recent semiaquatic turtles, such as the European pond turtle *Emys* (Avanzini et al., 2005).

Xing et al. (2014e) described turtle tracks from the Huangyangquan area of Gansu, China. These specimens were assigned to *Emydhipus* isp., and the trackmaker was considered to be *Xinjiangchelys* sp. or *Wuguia* sp. Xing et al. (2019f) based on the specimen from Gansu, pointed out the length of the carapace is proportional to the width of the track and trackway. By comparing size and foot shape, Gansu *Emydhipus* isp. could have been made by any of the Early Cretaceous Gansu turtles, such as *Sinemys*. cf. *brevispinus*, *Kirgizemys kansuensis*, *Sinochelys* sp. and *Changmachelys* (Xing et al., 2019f).

3.9.2 Chelonipus *Rühle von Lilienstern, 1939 (Fig. 7.26C)*

3.9.2.1 Type ichnospecies

Chelonipus torquatus Rühle von Lilienstern, 1939 emended. Sarjeant and Langston, 1994.

3.9.2.2 Valid ichnospecies in Cretaceous

Chelonipus liui Young, 1960; Lockley et al., 2019.

3.9.2.3 Total distribution

Triassic: middle Buntsandstein, Germany; Jurassic: Bathonian–Tithonian, China, USA; Cretaceous: Berriasian–Maastrichtian, China, South Korea, Canada, USA; Eocene: USA.

3.9.2.4 Holotype of the ichnogenus

Holotype trackway (B. Rühle XII) housed in the Museum für Naturkunde, Berlin.

3.9.2.5 Ichnogeneric diagnosis

Quadrupedal tracks with trackways that are broad and stride short; tetradactyl to pentadactyl manus prints, normally with four well recognizable digits, Digits III and IV being the longest, and connected by a fleshy, arched structure; pes prints are tetradactyl with relatively longer digits than in the manus. The manus prints are not significantly angled toward the midline, instead having their transverse (width) axis perpendicular to the trackway midline; the pes prints slightly closer to the midline than the manus prints (emended diagnosis by Lichtig et al., 2018).

3.9.2.6 Ichnospecific diagnosis of *Chelonipus liui*

Tridactyl to pentadactyl tracks, and they are wider (2.7–4.0 cm, mean 3.5 cm) than long (2.0–3.2 cm, mean 2.65 cm), the ratio of length to width is 0.76, with lateral digit traces shorter than central digit traces, with sharply pointed finger traces (revised diagnosis based on the comments from Lockley et al., 2019).

3.9.2.7 General description

Rühle von Liliestern (1939) named *Chelonipus torquatus* from a unit called the "Thüringischer Chirotheriensandstein" at the top of the Lower Triassic Solling Formation (Middle Buntsandstein). *Chelonipus* is diagnosed by trackway width (up to 25 cm), a pace angle of 50°–70°, digitigrade or semiplantigrade pes (2 cm in length, 5.5 cm in width) with three or four digits with thin claws, and manus (2 cm in length, 4 cm in width) with four well recognizable digits (Haubold, 1971; Avanzini et al., 2005). The sole of the pes can be posteriorly concave or convex and sometimes narrows distinctly to a rounded proximal margin (Lichtig et al., 2018).

The tetrapod track *Laiyangpus liui*, from the Lower Cretaceous of Shandong in China, has been the subject of misunderstanding and misinterpretation, and attributed to nonavian theropod trackmaker, since it was discovered and named in 1960. Lockley et al. (2019) revealed that *L. liui* is sufficiently similar, morphologically, to *Chelonipus* to be considered a subjective junior synonym and suggested that the trackmaker of these tracks is a turtle. But, Lockley et al. (2019) did not provide a specific diagnosis for *Chelonipus liui*.

Individual *Chelonipus liui* tracks cannot represent left or right tracks, or manus or pes tracks (Lockley et al., 2019). Several *C. liui* track casts have transverse posterior margins or overhangs on the lower side of the track-bearing slab. These posterior margins represent where the digits dug in, downward, and with posterior motion into the substrate before the impressions were cast by infilling sediment.

3.9.2.8 Discussion

Rühle von Lilienstern (1939) named the ichnogenus *Chelonipus* to accommodate two new ichnospecies *C. torquatus* and *C. cuneiformis* from the early Middle Triassic (Anisian) of Germany. Kuhn (1958) considered that *C. torquatus* and *C. cuneiformis* are synonyms (Avanzini et al., 2005). Haubold (1971) synonymized the latter ichnospecies under *C. torquatus* and named another ichnospecies, *C. plieningeri*, based on Upper Triassic (Carnian) tracks, also from Germany. Turtles walk with an understep ranging from an extreme, full understep in which the manus is overstepped by the pes of the next cycle of steps, to a more standard understep walk in which the manus is placed just behind the pes of the same series of steps (Lichtig et al., 2018). This extreme understep is likely the basis for the variation in turtle tracks given the name *C. plieningeri* (Haubold, 1971). So, Lichtig et al. (2018) consider *C. plieningeri* to be a junior subjective synonym of *C. torquatus*.

Chelonipus has been described from strata dating from the Early Triassic, the Late Triassic, the Middle Jurassic, and the Late Jurassic (Foster et al., 1999; Lockley and Foster, 2006; Lovelace and Lovelace, 2012; Lichtig et al., 2018; Xing et al., 2022), Cretaceous (Lockley et al., 2018e,f, 2019), Middle Eocene (Sarjeant and Langston, 1994; Mustoe, 2019) and even the Middle Miocene (Krapovickas et al., 2023). Lockley et al. (2018f) discussed Cretaceous turtle tracks from 15 tracksites of North America and assigned them to cf. *Chelonipus*. Jurassic records of *Chelonipus* are not only found in the USA but have also been found in China. Xing et al. (2022) reported turtle tracks from the Zhiluo Formation, Shaanxi, China.

Chelonipus liui is similar to *C. torquatus* (in Haubold, 1971, named *C. plieningeri*) and *Chelonipus* isp. (Lockley et al., 2019). These trackways were labeled *Chelonipus* on the basis of the similarity of individual tracks to those reported for *Chelonipus*, but also on the regularity of the trackway. However, when comparing these trackways, the difference is the space between tracks, what would be called steps and strides in the trackway of a walking animal. The stride of *C. torquatus* from Germany and *Chelonipus* isp. from Utah is between 20.0 and 43.0 cm, respectively, for tracks 5.0 and 7.0 cm wide (Lockley et al., 2019).

Zhang et al. (2023) reported turtle tracks preserved in situ at Huanglonggou, Zhucheng City, Shandong, China. By comparison with *C. liui*, Zhang et al. (2023) found that the tracks from Huanglonggou have a larger size, the distance between the digits is larger, the length of the claw marks is larger, and the ratio of the distance between two adjacent claw marks is larger. So, they established a new ichnogenus (*Zhuchengichnites*) and ichnospecies (*Zhuchengichnites perfectus*). However, these features are similar to those of *Chelonipus* described by Lichtig et al. (2018), and we suggest that *Z. perfectus* should be assigned to *Chelonipus* isp., and *Z. perfectus* is a nomen dubium.

3.9.2.9 Trackmaker

The value of pace angulation of the trackway is similar to that of the extant turtle *Rhinoclemmys pulcherima* (Avanzini et al., 2005). Lockley et al. (2018e) concluded that *Chelonipus* isp. of the Cretaceous in North America and the trackmakers are interpreted as turtles partially buoyed up by water but essentially walking along the bottom. Kim and Lockley (2016) reported a partial trackway of *Chelonipus* from the Jinju Formation (Sindong Group), and they inferred a trackmaker that was partially buoyant or bottom walking while progressing in a down-current direction.

3.10 Mammal

3.10.1 Brasilichnium *Leonardi, 1981 (Fig. 7.27F–H)*

3.10.1.1 Type ichnospecies

Brasilichnium elusivum Leonardi, 1981 (Fig. 7.27F)

3.10.1.2 Valid ichnospecies in Cretaceous

Brasilichnium elusivum Leonardi, 1981.
Brasilichnium saltatorium Buck et al., 2017a (Fig. 7.27G)
Brasilichnium anaiti D'Orazi Porchetti et al., 2017 (Fig. 7.27H)

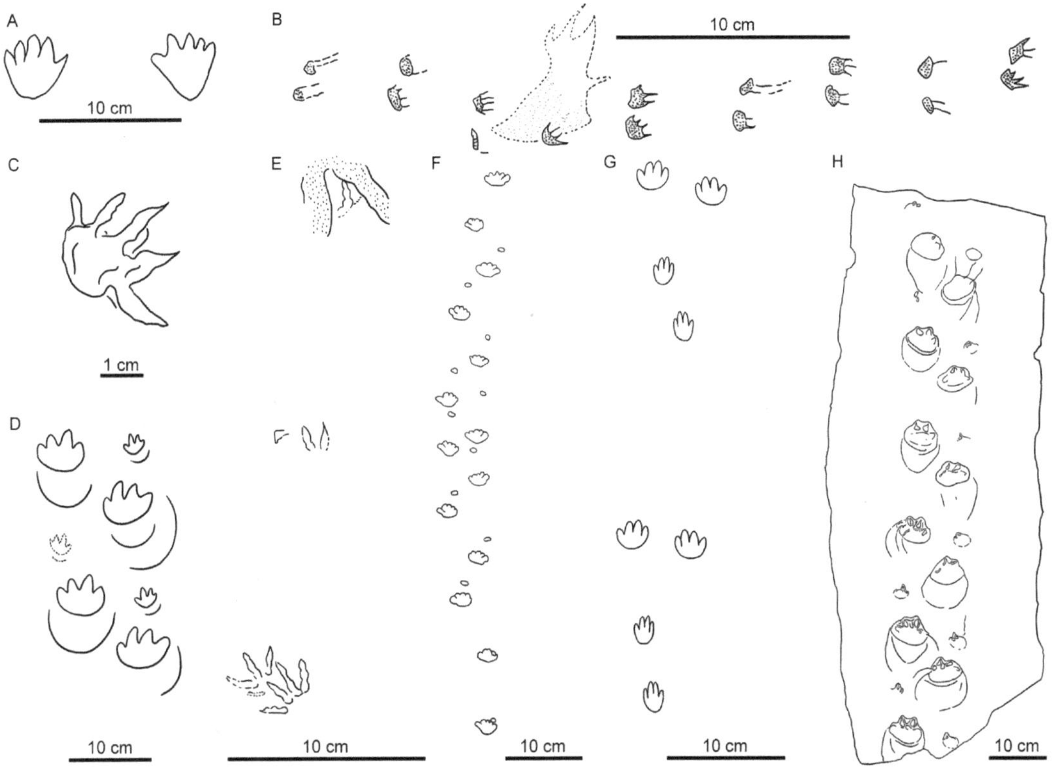

FIGURE 7.27 Tracks of Cretaceous mammal. (A) Holotype of *Sederipes goddardensis*; (B) type series of *Koreasaltipes jinjuensis*; (C) holotype of *Catocapes angolanus*; (D) partial series on holotype slab of *Aracoaraichnium leonardii*; (E) *Schadipes crypticus*; (F) type series of *Brasilichnium elusivum*; (G) partial type series of *Brasilichnium saltatorium*; (H) type series of *Brasilichnium anaiti*. *(A) Redrawn from Stanford et al. (2018). (B) Redrawn from Kim et al. (2017b). (C) Redrawn from Mateus et al. (2017). (D) Redrawn from Buck et al. (2017b). (E) Redrawn from Lockley and Foster (2003). (F) Redrawn from Leonardi (1981). (G) Redrawn from Buck et al. (2017a). (H) Redrawn from D'Orazi Porchetti et al. (2017).*

3.10.1.3 Distribution

Cretaceous: Berriasian–Campanian, Brazil, Paraguay, Mexico, Tunisia, Namibia, R.D. Congo;

Jurassic: lower Jurassic, Lesotho, USA, Italy.

3.10.1.4 Holotype

MNRJ 3902-V (ARSB 78), housed in the paleontological collections of the National Museum of Rio de Janeiro, Vertebrate Section.

3.10.1.5 Diagnosis

Quadrupedal trackway, with small dimensions; mean gleno-acetabular distance of 7.5 cm. Manus significantly smaller than the pes; manus-pes distance increases when the walking speed decreases. Pace angulation is relatively high for a quadruped. Heteropody is directly proportional to the pes longitudinal axis positive rotation (values $> 75°$). Pes with the longitudinal axis parallel to the trackway axis and with pace angulation N125°. Do not show heteropody. Pes impressions are elliptical with transverse axis almost larger and anteroposterior axis slightly directed inwards; short toes, generally rounded, with possible 2-3-3-3-3 phalangeal formula; the posterior autopodium is ectaxonic and tetradactyl (Digits II, III, IV and V) in semiplantigrade condition. There is a greater hypex digit in slight abduction. Anterior autopodium has at least four digits with nails; nails may be apparent or not, by virtue of preservation. Tail impressions/drag are always absent (diagnosis emended by Buck et al., 2017a).

3.10.1.6 General description

Brasilichnium elusivum was originally found in the aeolian deposits of Botucatu Formation in the Paraná Basin, Brazil.It exhibits only four digits (Digits II, III, IV, and V) in its hind footprints, not five, which is considered a characteristic of *B. elusivum*. If the digit I were to exist, it should be positioned higher relative to the other digits. The negative rotation of the hind footprints, observed in cases with smaller step angles, may indicate that the animal exerted more effort in moving across the dunes of the ancient desert (Fernandes and Carvalho, 2008).

Brasilichnium saltatorium from the Botucatu Formation, Brazil, is characterized by a hopping locomotion in phases. During the locomotion cycle, the pes surpassed the manus. Pes are arranged in a nearly parallel manner with horizontal spacing; manus' are vertically aligned with horizontal spacing, and the distance to the trackway midline is minimal. Taking the trackway midline as a reference, the right pes and manus are positioned behind the left side (Buck et al., 2017a).

Brasilichnium anaiti from ?Middle Jurassic–Early Cretaceous of Brazil, exhibits discontinuity in the proximal portion of the phalangeal portion of the foot, typically separated from the metapodials by a stenosis or groove, especially on the central couple of digit marks. First and fourth pes digit marks are blunt at the distal end, while those of the second and third appear distally sharp. Digit marks I–III are more deeply impressed compared with the lateral one, with the deepest area corresponding to the metatarsal-phalangeal joints. Manus print has at least three-digit marks, always preceding the footprints (D'Orazi Porchetti et al., 2017).

3.10.1.7 Discussion

Brasilichnium saltatorium differs from *B. elusivum* mainly in the locomotion pattern. It is also different from other ichnotaxa in that it indicates hopping progression regarding limb morphology and track configuration. *B. saltatorium* exhibits genuine biological variations arising from differences in locomotive behavior, presenting an asymmetrical hopping gait that distinctly differs from the typical coarse locomotion described in *B. elusivum*, reflecting a different limb configuration (Buck et al., 2017a).

The general morphology of *Brasilichnium anaiti* exhibits ichnotaxonomic similarity to *B. elusivum*, but there are significant differences at the ichnospecies level in the following characters: in comparison with *B. elusivum*, the digit marks II and III of *B. anaiti* functionally dominate in load-bearing; the pes shape of *B. anaiti* is more symmetrical than *B. elusivum*, and pes digit marks in *B. anaiti* are more distinctly separated from the sole marks; in *B. anaiti*, the central digit marks commonly show divergent distal ends, whereas in *B. elusivum*, digit marks I–III are usually iso-oriented; the average size of *B. anaiti* is 4 times larger than that of *B. elusivum*, with no overlap in the footprint sizes between the two (D'Orazi Porchetti et al., 2017).

3.10.1.8 Trackmaker

Pacheco (1913) attributed the trackway to "a reptile"; Huene (1931) attributed it to "more probably a Cynodont"; Mezzalira (1966) to a "?cynodont or archosaur"; Casamiquela (1975) to "a mammaloid"; Leonardi (1979b) wrote that the trackmaker could be "perhaps a therapsid, more probably a small primitive mammal" (Leonardi and Carvalho, 2021a).

Leonardi (1981) considered that attributing these new tracks to primitive mammals is more reasonable, with a glenoid-acetabular distance of ~8 cm. It is a quadrupedal animal but with a noticeable trend of reduced anterior body support, perhaps leading to (or has already led to) a bipedal jumping gait. It has very small front forelimbs and relatively larger hind legs, possibly with a phalangeal formula of 2-3-3-3-3. In a desert or semiarid environment, it is likely that this type of small animal could be found in the ecological niche of a small predator.

Buck et al. (2017a) suggested that *B. elusivum* and *B. saltatorium* might be produced by the same mammalian organism, representing alternating locomotive behaviors. Compared with modern organisms living in similar environments and conditions, the ichnogenus *Brasilichnium* producer might be a small, omnivorous animal with fossorial habits. It could adopt hopping locomotion to gain speed during hunting, escape from predators, or facilitate progression on inclined and hot sand.

D'Orazi Porchetti et al. (2017) suggest that a comparison of the time ranges of *Brasilichnium* and Mesozoic therapsids indicates that potential trackmakers may belong to nonmammalian mammaliamorphs or early mammals. However, the definitive features observed in *B. anaiti* do not provide clear distinctions between these two groups. Furthermore, because *B. elusivum* exhibits a series of gaits unknown in *B. anaiti*, the two should not be considered expressions of the same trackmaker population; the trackmaker of *B. elusivum* was a Mammaliamorpha with a semierect leg posture (D'Orazi Porchetti et al., 2017).

Leonardi and Carvalho (2021b) thought that the trackmakers of the ichnogenus *Brasilichnium* would better be attributed to Mammaliaformes Rowe, 1988, a clade that includes mammals and their extinct relatives.

3.10.2 Schadipes *Lockley and Foster, 2003 (Fig. 7.27E)*

3.10.2.1 Type ichnospecies

Schadipes crypticus Lockley and Foster, 2003.

3.10.2.2 Valid ichnospecies

Schadipes crypticus Lockley and Foster, 2003.

3.10.2.3 Distribution

Maastrichtian, USA.

3.10.2.4 Holotype

The holotype preserved in situ, in Laramie Formation of Colorado. A replica of the holotype is CU-MWC 220.33, rubber mold and replica of original specimen, housed in University of Colorado at Denver, Museum of Western Colorado.

3.10.2.5 Diagnosis

Small, slender-toed tetradactyl and pentadactyl tracks with central digits (II–IV) subequal in length and only slightly divergent. Lateral digits (I and V) shorter and more divergent when preserved at all. Trackway pattern variable. Differs from lacertilian tracks in being more symmetrical with more equidimensional digits (Lockley and Foster, 2003).

3.10.2.6 General description

Pes of *Schadipes crypticus* is likely pentadactyl. Length of Digits II–IV ranges from 1.5–2 cm. Digit I is the shortest. Digit V may be longer (approximately 1 cm) than Digit I and diverges from Digit IV at a high angle (50°–60°). Digits II–V sometimes exhibit up to three clear pad impressions. Manus is smaller and shorter, likely tetradactyl. Trackway has inter-footprint distances (?steps) of up to 12 cm (original diagnosis by Lockley and Foster, 2003).

3.10.2.7 Discussion

According to the initial discovery specimen (CU-MWC 220.28), these tracks were thought to be made by a lizard trackmaker (Lockley et al., 2001b). The holotype CU-MWC 220.33 reveals digit pad impressions consistent with a 2-3-3-3-3 mammalian phalangeal formula, rather than the lizard/lacertilian formula of 2-3-4-5-3(4) (*sensu* Leonardi, 1987). This confirms the mammalian origin of these tracks, and contributes to a better understanding of the morphologies of individual tracks (Lockley et al., 2002c; Lockley, 2003), allowing for a more precise definition of the ichnotaxon (*Schadipes crypticus*). CU-MWC 220.28 is referred to as *Schadipes* isp (Lockley and Foster, 2003).

3.10.2.8 Trackmaker

Lockley and Foster (2003) proposed that based on the shape and size of the foot, as well as the gait pattern, the trackmaker of *Schadipes crypticus* belongs to the class Mammalia. It is capable of altering its gait between walking, running, and jumping. The footprint morphology is similar to those of modern rodents, but based on body fossil records, it is likely to be related to marsupials or multituberculates.

3.10.3 Koreasaltipes *Kim et al., 2017b (Fig. 7.27B)*

3.10.3.1 Type ichnospecies

Koreasaltipes jinjuensis Kim et al., 2017b.

3.10.3.2 Valid ichnospecies

Koreasaltipes jinjuensis Kim et al., 2017b.

3.10.3.3 Distribution

Albian, South Korea.

3.10.3.4 Holotype

CUE E4M001, Jinju Formation of the Jinju City area, housed in Chinju University of Education; UCM 214.305, a replica of CUE E4M001, housed in Colorado Museum, South Korea.

3.10.3.5 Diagnosis

Symmetrical trackway of a small hopping, bipedal trackmaker with slightly transverse (wider than long) functionally tetradactyl pes traces. Digit traces subequal in length. Gait characterized by narrow trackway and hop distance about 4 times pes length (original diagnosis by Kim et al., 2017b).

3.10.3.6 General description

Koreasaltipes jinjuensis includes nine pairs of pes prints. Trackway averaging 21.3 mm outer width (range 19.0–24.0 mm) and 5.0 mm inner width. The average width and length of *K. jinjuensis* are 4.76 and 8.06 mm respectively. R4 is the best-preserved track with total length (with digit traces) of 9.0 mm and width of 11.0 mm (length/width is 0.82). R2 and L3 are also well-preserved, with length/width of 0.78 for R2 (length 7.0 mm, width 9.0 mm) and 1.28 for L3 (length 9.0 mm, width 7.0 mm). The average total length/width values for all tracks are considered meaningless due to the elongation of some digit traces caused by dragging during the foot's exit from the soft substrate. Digit traces taper gradually at the anterior end, constituting approximately 50% of the track length. No tail traces were observed (Kim et al., 2017b).

3.10.3.7 Discussion

Koreasaltipes jinjuensis is the first and only instance of footprints attributed to a small hopping mammal in Asia. It is commonly compared with *Ameghinichnus* from the Jurassic of Argentina and *Musaltipes* from the Cenozoic of North America. *Ameghinichnus patagonicus*, as redefined by de Valais (2009), and *Musaltipes occidens*, as described by Lockley and Milner (2014), both represent a quadrupedal hopping mode of progression, contrasting with the bipedal hopping gait of *K. jinjuensis*. In addition, both *A. patagonicus* and *M. longidigitus* exhibit pentadactyl characteristics, excluding the influence of preservation factors, even though the digit traces in well-preserved *K. jinjuensis* tracks do not exceed four. *Ranipes laci* (Lockley and Milner, 2014) is a set of frog tracks that exhibit a hopping gait, but it is morphologically very different from *K. jinjuensis*. Buck et al. (2017a) described footprints attributed to a small mammalian trackmaker with a hopping gait, but these footprints are much larger than *K. jinjuensis*. The footprint morphology of *K. jinjuensis* differs from these ichnogenera in digit number, digit proportions, and trackway configuration. Moreover, it lacks tail traces. Therefore, *Koreasaltipes* is considered an independent ichnogenus (Kim et al., 2017b).

3.10.3.8 Trackmaker

De Valais (2009) proposed seven characteristics of mammalian/mammaliform track affinity. These criteria include (1) similarity to conservative mammal form; (2) generally pentadactyl; (3) digit traces subequal in size and shape; (4) or mesaxonic; (5) no sprawling; (6) advanced gait, for example, hopping; and (7) small size. *Koreasaltipes jinjuensis* essentially conforms to these criteria, representing a small, mouse-sized hopping mammal (Kim et al., 2017b).

3.10.4 Catocapes *Mateus et al., 2017 (Fig. 7.27C)*

3.10.4.1 Type ichnospecies

Catocapes angolanus Mateus et al., 2017.

3.10.4.2 Valid ichnospecies

Catocapes angolanus Mateus et al., 2017.

3.10.4.3 Distribution

Aptian, Angola.

3.10.4.4 Holotype

MGUAN-PA601 Track 13, Calonda Formation, housed in the Geological Museum of Universidade Agostinho Neto-PaleoAngola Project (Catoca collection).

3.10.4.5 Diagnosis

Relatively large tracks of a mammaliamorph. Mesaxonic, functionally pentadactyl, and plantigrade track, wider than longer (length is 2.7 cm, width is 3.2 cm). Digits are short, straight, distally rounded, with no claw marks. The orientation of digits is autapomorphic, with medial Digits II–IV projecting anteriorly and lateral Digits I and V more divergent and anterolaterally projected. Divarication angle I–II is 43°, II–IV is 43.5°, and IV–V is 33°. Digits II and III are the longest (1.5 cm), while Digit V is the shortest (1.0 cm) (original diagnosis by Mateus et al., 2017).

3.10.4.6 General description

Catocapes angolanus is the largest mammaliamorph track known from the Early Cretaceous. It comprises at least 45 mammaliamorph tracks, preserved in the form of concave epireliefs, with an overall poor preservation condition. The average Digits I–V lengths are 1.3, 1.5, 1.5, 1.4, and 1.0 cm, respectively. The average total divarication is 118°, with an average digital divarication of 25° between II–III and 19° between III–IV (Mateus et al., 2017).

3.10.4.7 Discussion

Mateus et al. (2017) indicate that despite similarities in size, digits length, and overall digital divarication, *Catocapes* is comparable to *Ameghinichnus* (reported from the Early Jurassic of USA to the Middle–Late Jurassic of Argentina; Casamiquela, 1961, 1964a; Leonardi and Oliveira, 1990; Olsen, 2002). However, the morphotype of *Catocapes angolanus* is not associated with any known Cretaceous mammal-like track morphotype. Therefore, it can be classified as a new ichnogenus: *Catocapes* (Mateus et al., 2017). However, Stanford et al. (2018) pointed out that the utility of the ichnotaxon *Catocapes* is debatable, because it is based on type material that cannot be identified as pes or manus and no mammaliamorph trackway configurations were described from it.

3.10.4.8 Trackmaker

Mateus et al. (2017) propose that the trackmaker of *Catocapes angolanus* resembles a particularly large-bodied Early Cretaceous Gondwanan mammaliamorph with distinctive tracks (see also Marzola et al., 2015). The size of the trackmaker is unmatched to coeval skeletal fossil records and is comparable to modern raccoons (Mateus et al., 2017).

3.10.5 Aracoaraichnium *Buck et al., 2017b (Fig. 7.27D)*

3.10.5.1 Type ichnospecies

Aracoaraichnium leonardii Buck et al., 2017b.

3.10.5.2 Valid ichnospecies

Aracoaraichnium leonardii Buck et al., 2017b.

3.10.5.3 Distribution

Neocomian, Brazil.

3.10.5.4 Holotype

LPP–IC–0015, Botucatu Formation, housed in the Laboratory of Paleoecology and Paleoichnology, Federal University of São Carlos, São Paulo, Brazil.

3.10.5.5 Diagnosis

Trackway of a quadruped characterized by walking locomotion. Narrow trackway with the interior margin of the pes very close to the midline. Footprints are semiplantigrade, in a mesaxonic condition, with Digits III and IV more anteriorly registered. Manus are significantly smaller than the pes. When the manus is imprinted, it always registers anterior to the pes in the locomotion cycle. Both manus and pes are tetradactyl, with short digits. The proximal traces of the digits are wide, but taper distally, ending in nail or claw traces. Digit III is the longest, followed by Digit IV, which is subequal in length. Digits II and V are smaller, but also subequal to each other in length. Footprints are wider than they are long, with an oval heel pad, sometimes with the anteroposterior axis directed slightly inward. When the anteroposterior axis is directed inwards, it occurs for both the manus and pes in a similar degree. Tail drags are always absent (original diagnosis by Buck et al., 2017b).

3.10.5.6 General description

The average widths of the pes and manus are 6.1 ± 0.4 cm and 2.2 ± 0.7 cm, respectively, and the average lengths are 5 ± 0.3 cm and 2 ± 0.5 cm, respectively. Measured from the internal autopodium midpoint, the average total distance between manus and pes impressions

(DMP) is 5.9 cm; the average distance between anterior extremity of the pes and the posterior extremity of the manus (IDMP) is 2.7 cm. The average width of the trackway is 12.7 cm for the pes and 13 cm for the manus. The average values for pes stride and oblique pace are 19.1 and 11.7 cm, respectively. Measurements for the manus are 18.7 and 13.8 cm, respectively. The average pace angle is 112.4° ± 12.7° for the pes and 88.8° ± 10° for the manus, which is in good agreement with the measurements given by Leonardi et al. (2007) (Buck et al., 2017b).

3.10.5.7 Discussion

Buck et al. (2017b) identified the ichnogenus *Aracoaraichnium* as belonging to the ichnofamily Chelichnopodidae Lockley, 2011. Despite some similarities in features, distinctive characteristics such as the number and morphology of digits, track morphology, heteropody degree, pace angle, stride length, locomotion parameters, and distribution set *Aracoaraichnium* apart from the other two track ichnotaxa within the Chelichnopodidae: *Chelichnus* (Jardine, 1850; McKeever and Haubold, 1996) and *Brasilichnium* (Leonardi, 1981; Fernandes and Carvalho, 2008), as well as other well-known tetrapod ichnotaxa. This suggests that *Aracoaraichnium* is formed by different taxa (Buck et al., 2017b).

3.10.5.8 Trackmaker

Buck et al. (2017b) inferred that the trackmaker of *Aracoaraichnium* belongs to the Mammaliamorpha, based on its autopodial morphology, gait, and locomotion characteristics. The more specific affinities are yet to be determined, but it might be associated with the Tritylodontidae or Mammaliaformes. The trackmaker likely inhabited areas near deserts, occasionally venturing into desert regions, engaging in both diurnal and nocturnal activities, and consuming invertebrates and/or plant material.

3.10.6 Sederipes *Stanford et al., 2018 (Fig. 7.27A)*

3.10.6.1 Type ichnospecies

Sederipes goddardensis Stanford et al., 2018.

3.10.6.2 Valid ichnospecies

Sederipes goddardensis Stanford et al., 2018.

3.10.6.3 Distribution

Aptian, USA.

3.10.6.4 Holotype

Pair of tracks designated as m1 and m2 on specimen GSFC-VP1, in situ, Patuxent Formation of Maryland, housed in the Goddard Space Flight Center, NASA.

3.10.6.5 Diagnosis

Small pentadactyl pes tracks with Digit I trace shorter than equidimensional traces Digits II–V arranged parasagittally, in symmetrical pairs indicative of sitting posture (Stanford et al., 2018).

3.10.6.6 General description

Sederipes goddardensis has consistent length and width, with Digit I approximately half the length of Digits II–V. The hypex between Digit I and Digit II is wider than the hypices between Digits II–III, III–IV, and IV–V. Digits II–IV are approximately half the length of the track, and the posterior half of the track is undifferentiated padding with a broad, subcircular, posteriorly convex margin to the heel (original diagnosis by Stanford et al., 2018).

3.10.6.7 Discussion

Stanford et al. (2018) state that due to the rarity of mammal tracks representing eutherians and/or metatherians in the Mesozoic, it is challenging to identify them or assign taxonomic labels. Mesozoic mammal tracks are seldom preserved in trackways, and available ichnotaxa indicate hopping gaits (Lockley and Foster, 2003; Kim et al., 2017b). The Patuxent sample presents symmetrical pairs but not long trackway sequences, revealing the position of the right and left pes in a sitting posture. The sitting-on-haunches posture appears to be a characteristic feature of the Patuxent sample, and based on previously unreported small mammal posture (behavior) diagnostic ichnological records, the naming of the new ichnotaxon (*Sederipes goddardensis*) is considered justified.

3.10.6.8 Trackmaker

The well-preserved pair of *Sederipes* exhibits an almost perfect mirror image morphology, with Digit I shorter than Digits II–V, representing a diagnostic crown mammalian pattern. In addition, it is suggested that the trackmaker assumed a temporarily "sedentary" sitting posture, a characteristic sitting position observed in extant small mammals (such as mice and squirrels) as well as some small insectivores and carnivores (Stanford et al., 2018).

3.11 Amphibia

3.11.1 Ranipes *Lockley and Milner, 2014 (Fig. 7.28)*

3.11.1.1 Type ichnospecies

Ranipes laci Lockley and Milner, 2014.

3.11.1.2 Valid ichnospecies

Ranipes laci Lockley and Milner, 2014.
Ranipes saokensis Park et al., 2018.

3.11.1.3 Distribution

Albian–Campanian, South Korea, USA; late Paleocene–early Eocene, USA.

3.11.1.4 Holotype

BYU 3874, housed in Brigham Young University, with replica UCM 230.41, housed in University of Colorado Museum of Natural History.

3.11.1.5 Diagnosis

Highly symmetrical, narrow, straight trackway of a small hopping vertebrate, with four tracks (two pes and two manus) in quadripartite "hop set" configuration. Distances between

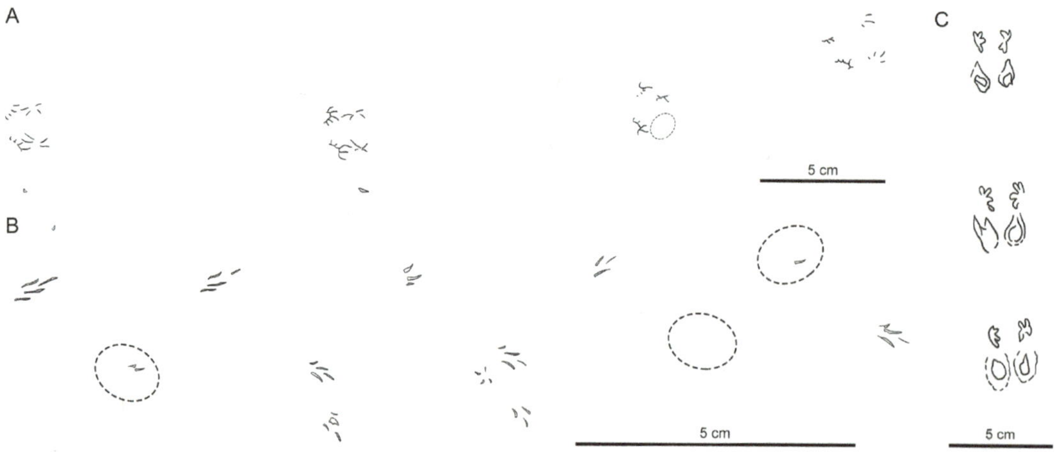

FIGURE 7.28 Typical trackways belong to Cretaceous amphibian. A and B belongs to *Ranipes saokensis*, dotted circles represent missing or incomplete tracks whose locations are reliabl inferred from position sin clearly-defined trackways. (A) The type series that shows the sequence hop sets of *Ranipes saokensis*; (B) the possible walking trace of *Ranipes saokensis*; (C) series on the holotype slab of *Ranipes laci*. *(A) Redrawn from Park et al. (2018). (B) Redrawn from Park et al. (2018). (C) Redrawn from Lockley and Milner (2014).*

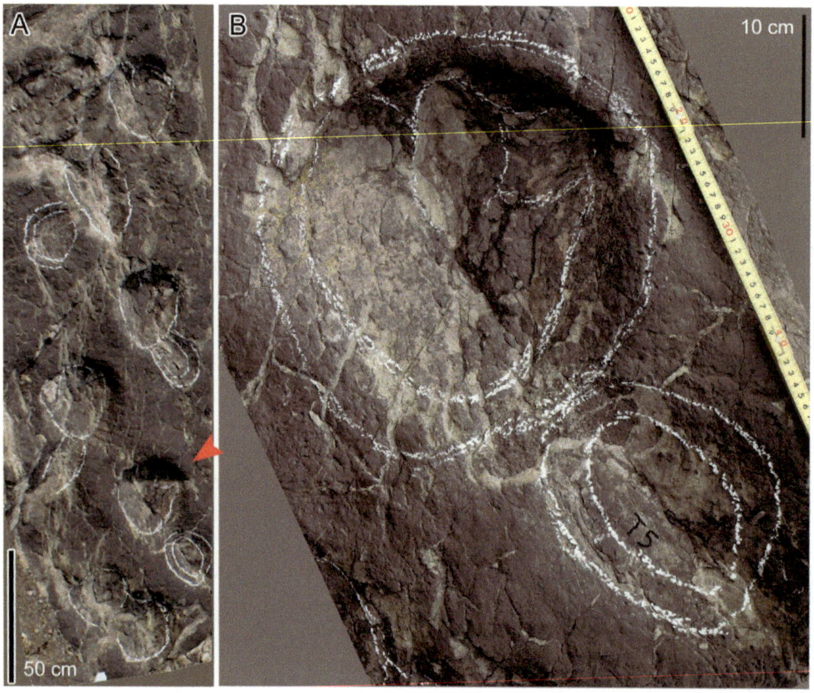

FIGURE 7.29 Photograph of holotype (LSI-S2) of *Parvieouspodus refugii* ichnogen. nov. ichnosp. nov. this holotype is preserved in situ in the Jishan tracksite in Linshu County, Shandong, China. *Photo by Lida Xing.*

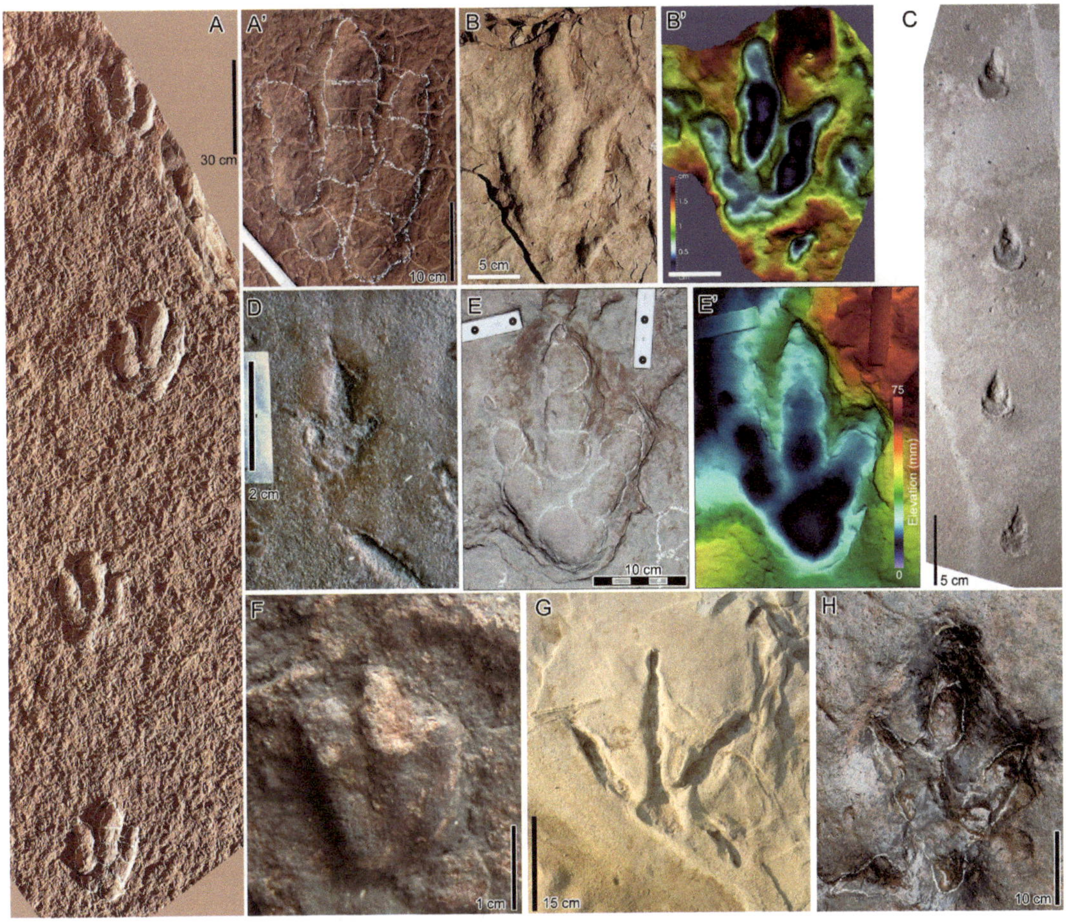

FIGURE 7.30 Photographs and depth-color images of medium–small sized tridactyle theropod ichnotaxa from Cretaceous. (A) Type series (A) and enlarged tracks (A′) of the holotype of *Eubrontes nobitai*; (B) Photograph (B) and depth-color image (B′) of the holotype of *Picunichnus quijadaensis;* (C) photograph of *Farlowichnus rapidus*; (D) photograph of the holotype of *Grallator emeiensis*; (E) photograph (E) and depth-color image (E′) of *Asianopodus wangi* (Xing et al., 2021b); (F) holotype of *Minisauripus chuanzhuensis*; (G) holotype of *Saurexallopus zerbsti*; (H) one of the holotype of *Gigandipus chiappei. (A) From Figs. 2 and 3 in Xing et al. (2021f). (B) From Fig. 7 in Melchor et al. (2019). (C) From Fig. 2B in Leonardi et al. (2024). (D) Photo provided from Martin G. Lockley. (F) Photo provided by Wei Chen. (G) Photo provided from Martin G. Lockley. (H) Photo by Lida Xing.*

quadripartite groups along trackway axis variable up to 6–7 times length of an individual hop set. Outer trackway width about 3 times manus and pes width. Inner trackway width spacing between pairs of manus and pes traces is similar and about 1/3 outer trackway width. Pes footprints pentadactyl, ectaxonic with heel trace length variable. Manus tracks tetradactyl with strong inward rotation, and respective digit traces II and V, tending to posterior (II) and anterior (V) parasagittal alignment. Respective orientations of Digit III and IV traces are postero-medial and antero-medial (emended from Park et al., 2018).

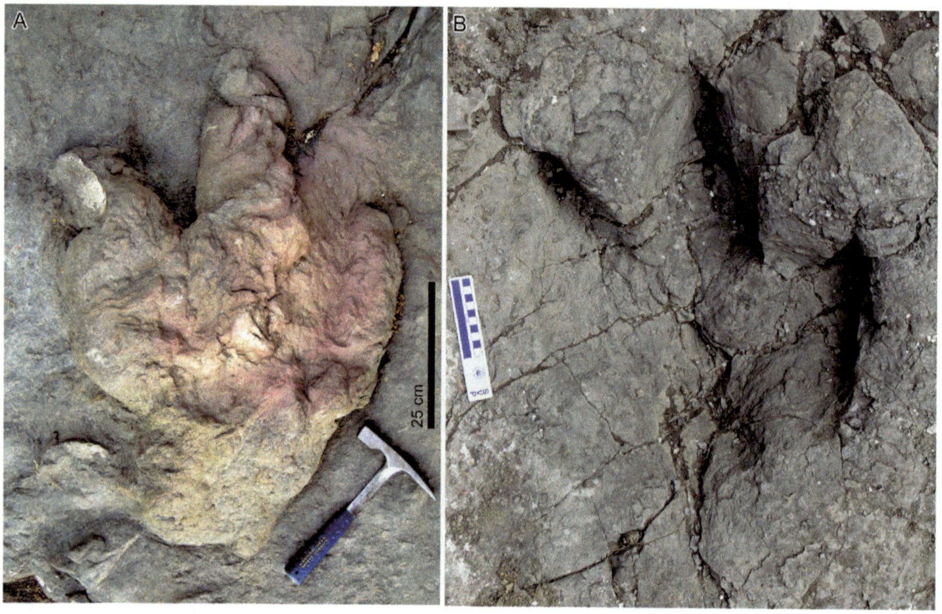

FIGURE 7.31 Photographs of large tridactyle theropods ichnotaxa from Cretaceous. (A) Holotype of *Tyrannosauripus bachmani*; (B) one of the holotypes of *Bellatoripes fredlundi*. *(A) Photo provided by Spencer Lucas. (B) Photo by Lida Xing.*

3.11.1.6 General description

Ranipes laci exhibits a clear trackway configuration. Manus tracks are clearer than pes tracks. The average length of the quadripartite groups is 2.5 cm, with spacing between groups ranging from 2.8 to 3.6 cm, and the hopping length varies between 5.4 and 6.0 cm. Left pes two displays an elongated groove inferred to represent a single-digit trace. The length of the tracks (~1.0 cm) is greater than the width (~0.5 cm), with measurements taken parallel to the trackway axis (Lockley and Milner, 2014).

Ranipes saokensis from Santonian–Campanian, unnamed strata, Saok Island, represents the first report of Mesozoic frog tracks from Asia, exhibiting typical features of modern anurans (frogs) such as pentadactyl pes and tetradactyl manus, as well as the characteristic hopping locomotion. Pes has an average length of ~0.9 cm and a width of ~0.7 cm, usually with only narrow distal digit traces preserved. The manus traces, situated in front of the pes traces, are ~0.9 cm long and ~0.4 cm wide, with respective digit traces II and V tending to posterior (II) and anterior (V) parasagittal alignment. The hop set of the type trackway measures 2.1 cm in width and 1.8 cm in length. The outer trackway width, which is comparable to the hop set width, is around 2 cm, while the inner trackway width ranges between 0.5 and 0.7 cm (Park et al., 2018).

3.11.1.7 Discussion

The trace fossil of Amphibian only occurs in the USA and South Korea (Barcelos and dos Santos, 2023). The first report of frog tracks in the fossil record comes from the Upper

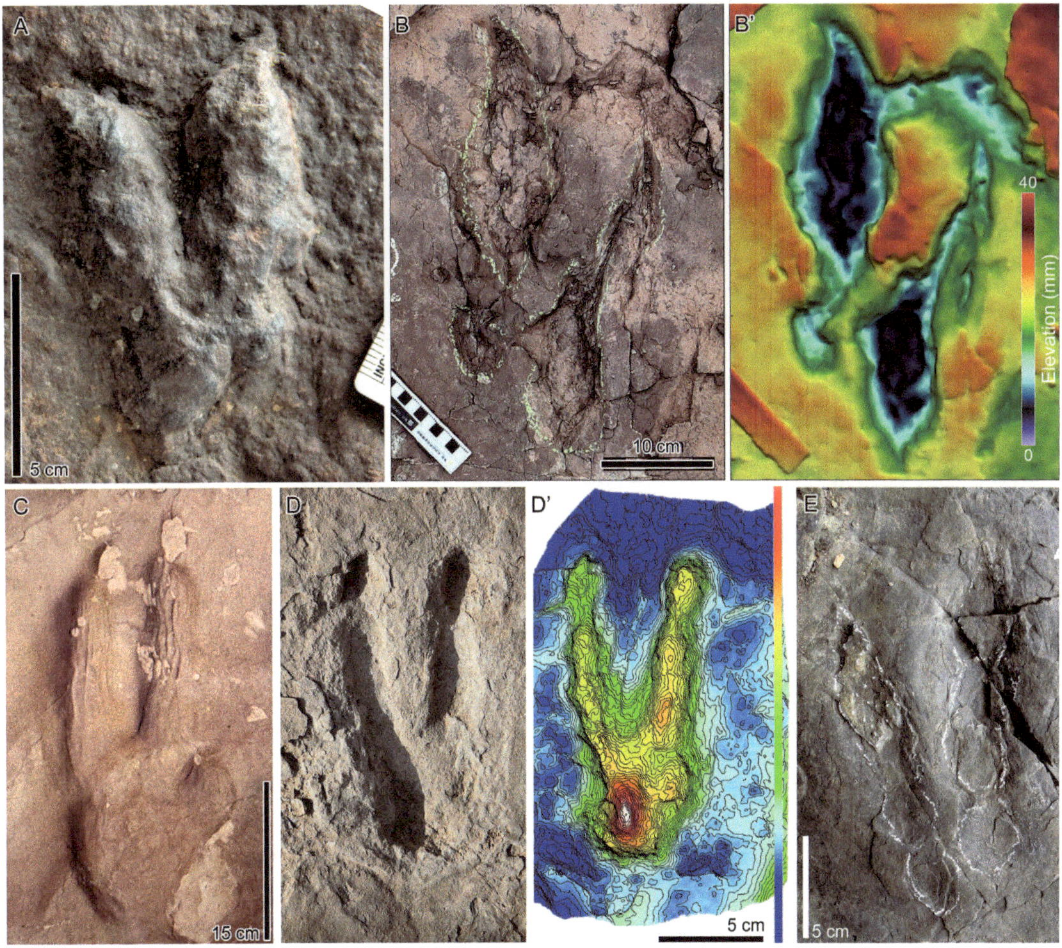

FIGURE 7.32 Photographs and depth-color images of deinonychosaurian ichnotaxa from Cretaceous. (A) Holotype of *Velociraptorichnus sichuanensis*; (B) Photograph (B) and depth-color image (B′) of *Fujianipus yingliangi* (LXIU-T3-R5, Xing et al., 2024b); (C) holotype of *Dromaeopodus shandongensis* from Shandong; (D) photograph (D) and depth-color image (D′) of holotype of *Dromaeosauripus yongjingensis* from Gansu (Xing et al., 2013a); (E) Track L2 of the holotype of *Dromaeosauripus hamanensis* from Korea (Kim et al., 2008). *(A) Photo provided by Wei Chen. (C) Photo provided from Rihui Li.*

Cretaceous Blackhawk Formation (Campanian) in Utah, USA. Initially illustrated by Robison (1991) and classified into an unnamed category, these tracks were later reillustrated by Lockley and Hunt (1995a). After a reexamination by Lockley et al. (2018b), it was confirmed that these tracks could be assigned to the ichnogenus *Ranipes* isp. The tracks consist of a pair of manus tracks and partial pes tracks, symmetrically arranged about the trackway midline, suggesting a hopping gait or possibly a crouching or sitting position (Robison, 1991; Park, et al., 2018)

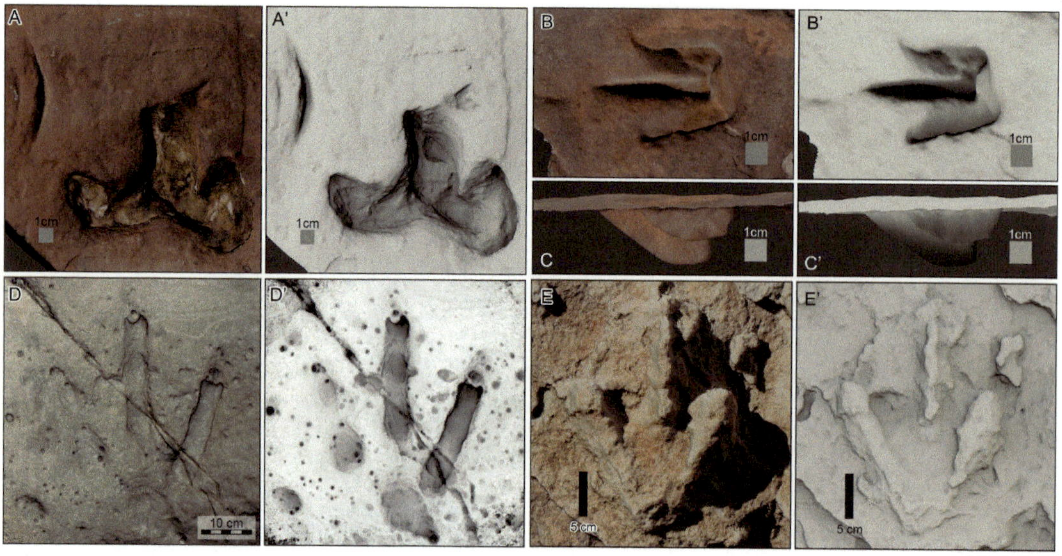

FIGURE 7.33 Photographs and occlusion images of Australian dinosaur tracks in Cretaceous. (A) Photograph (A) and occlusion image (A′) of *Wintonopus latomorum* from the mid-Cretaceous (late Albian-Cenomaniin) Winton Formation, Queensland. The large tridactyl footprint is the *W. latomorum* holotype QMF10319, while are *Characichnos* or *W. latomorum* category one swim trace (sensu Romilio et al., 2013); (B) Photograph (B) and occlusion image (B′) of '*Skartopus australis*' (junior synonym of *Wintonopus latomorum*) holotype QMF10330 from the mid-Cretaceous (late Albian-Cenomaniin) Winton Formation, Queensland; (C) Photograph (C) and occlusion image (C′) of inside view of B; (D) Photograph (D) and occlusion image (D′) of *Megalosauropus* isp. From the lower Cretaceous (Aptian-Albian) eumeralla Formation, Victoria; (E) Photograph € and occlusion image (E′) of *Megalosauropus broomensis* from the lower Cretaceous (Valanginian-Barremian) Broome Formation, western Australia (all photos provided by Anthony Romilio).

Lockley and Milner (2014) described the ichnospecies *Ranipes laci* based on the diagnostic features of the inwardly rotated tetradactyl manus. However, *R. laci* did not reveal complete pes digit traces. In contrast, the tracks of *R. saokensis* demonstrate that the pes is pentadactyl and ectaxonic, consistent with the characteristics of all described modern anurans (frogs). Furthermore, the original diagnosis indicated a short hop length, while the tracks of *R. saokensis* imply that the hop lengths can reach 6–7 times the length of the hop sets. Therefore, Park et al. (2018) proposed a revised diagnosis for *Ranipes* and assigned the material from Saok Island, South Korea, to the new ichnospecies *R. saokensis*.

Kim et al. (2019c) discovered frog tracks in the Jinju Formation in South Korea, which are the oldest frog tracks reported in the fossil record. The Jinju tracks are similar to *R. saokensis* in terms of size and morphology but are poorly preserved. Therefore, Kim et al. (2019c) referred them only as *Ranipes* isp. Jinju specimens have relatively short hop lengths, ranging from ~5 to ~10 cm. In addition, the configurations of quadripartite hop sets in both the Jinju and other trackways exhibit significant similarities.

3.11.1.8 Trackmaker

Sperry (1980), the collector of the type specimen of *R. laci*, described these traces as "similar to trackways produced by modern, small hopping frogs" Lockley and Hunt (1995a)

FIGURE 7.34 Photographs of bird ichnotaxa from Cretaceous. (A) Type series (A) and enlarged tracks (A′) of *Ignotornis mcconnelli*; (B) holotype of *Shandongornipes muxiai*. *(A) Photo provided by Martin G. Lockley. (B) LRH-dz67, photo provided by Martin G. Lockley.*

independently interpreted it as being made by a frog. Robison (1991) considered that the trackmaker of Utah *Ranipes* isp. is not a climbing or tree frog because there are no pads or discs on the "fingers."

Park et al. (2018) pointed out that the pes digit trace is dominant in the tracks of *Ranipes saokensis*, and the heel trace is rare, which is consistent with the known locomotion pattern of frogs. In addition, the Cretaceous environment of Saok Island, which belonged to a wet aquatic habitat rather than a dry terrestrial one, and the relatively small size and strong jumping ability of the trackmakers, suggest that the trackmakers of *R. saokensis* are likely frogs

FIGURE 7.35 Photograph of basal ornithischians or ornithopods ichnotaxon from Cretaceous. Holotype of *Neoanomoepus. FGM 002.01.20 and CU 199.22, Fig. 6 of Lockley et al. (2009b), photo provided from Martin G. Lockley.*

rather than toads. However, due to the difficulty in accurately determining the length of the hops, attempting to classify the trackmakers as hoppers or jumpers is speculative.

The Jinju *Ranipes* isp. with relatively short hop lengths suggests that the frog trackmakers were likely a hopping species rather than a jumping species (Kim et al., 2019c).

4. Ichnotaphonomy and ichnotaxonomy

In a highly desirable state, tetrapod tracks could correspond to the osteological record that is approximately isochronous and colocated (time accurate to age), thus refining its taxonomic framework. The Cretaceous Period is known for its abundance of terrestrial fossil skeletons and tracks. This allows for more accurate identification of trackmakers at different taxonomic scales. For example, Cretaceous *Brontopodus* have been attributed to titanosauriform neosauropods (Lockley et al., 1994a,b; Lockley, 1999; Xing, et al., 2021), while *Tyrannosauripus* has been linked to *Tyrannosaurus* (Lockley and Hunt, 1994a; McCrea et al., 2014b). However, due to preservation issues and the diachronism of skeletons, achieving an ideal

FIGURE 7.36 Photographs and depth-color image of ankylopollexian ichnotaxa from Cretaceous. (A) Photograph (A) and depth-color image (B) of holotype of *Caririchnium lotus* (Fig, 6 of Xing et al., 2015f); (B) type series of *Hadrosauropodus kyoungsookimi. (B) NHCG 10194, photo provided from Martin G. Lockley.*

FIGURE 7.37 Photograph of terrestrial lizard ichnotaxa from Cretaceous. Cast of holotype of *Neosauroides koreaensis* (Fig. 3 of Kim et al., 2017a).

level of correspondence between most of these tracks and their trackmakers remains difficult. In addition, the classification of tetrapod tracks is also affected by the following problems.

4.1 Extramorphological distortion

The nature of the substrate could affect track morphology as the limbs of tetrapods are inserted into and withdrawn from the sedimentary basement, resulting in the convergence in morphology of such tracks due to similar substrates rather than the osteological features (Marchetti et al., 2019). For example, slipping tracks that were left on a soft, wet, and slippery surface including biostabilised tidal flats and muddy river banks. The morphological deformations of this type of tracks include sinuous elongate tracks, drag marks, the uncommon direction inward and outward relative to the midline (Xing et al., 2014e; Heredia et al., 2023), on the better preserved biostabilised substrate with additional messy trampling marks (Heredia et al., 2023). Similarly, possible sliding tracks in desert environments are not unique to the Mesozoic, such as *Sarmientichnus*, which was originally found in the Jurassic of Argentina, and also in the Lower Cretaceous of Shaanxi, which represent elongate, straight and narrow digits drag marks (Casamiquela, 1964b; Xing et al., 2018f). In addition to the obvious elongation, the more prone to deformity of silt-to-very fine sandstone can also hide some features including claw marks, as well as increases the instability of the track morphology, so producing the high variability and lack of discrete margins and landmarks (Lallensack et al., 2022a).

Moreover, Lockley and Xing (2015) found apparent flattening on natural casts of some theropods, such as *Paracorpulentapus zhangsanfeng* and *Yangtzepus yipingensis* from China, as well as in some cases from the Jurassic of North America. Lockley and Xing (2015) interpret this to mean that the tracks were extruded into two layers of thick, noncompactable sand layers. During burial, the sand-filled tracks (casts) are flattened or widened as the ductile layers are compressed, which also clearly affects their morphology.

4.2 The similar morphology of the trackmaker

The stable similarity between some tracks of different taxa should not be interpreted as an extramorphological distortion. This has left some ichnotaxa with a longstanding dilemma in identifying the trackmakers. In this aspect, the Cretaceous mainly involves intermediate tridactyl tracks with morphologies between typical theropod and ornithischian tracks (especially ornithopod) (Lockley et al., 2009a; Lallensack et al., 2020; Castanera et al., 2013a,b), as well as relatively large intermediate ichnotaxa between avian and nonavian theropod tracks (Lockley et al., 2021; Hong et al., 2024). Regarding the former, there are several ichnotaxa that have been the subject of controversy, such as *Skartopus* and *Wintonopus* (Romilio et al., 2013), *Gypsichnites* (Sternberg, 1932; McCrea, 2000; McCrea et al., 2017), and *Yangtzepus* (Young, 1960; Xing et al., 2009c), all of which were introduced in Section 3. Although some tracks are incomplete (e.g., the digit only/dominated tracks, such as *Skartopus* and *Wintonopus*) or distorted (e.g., the flattened tracks such as *Yangtzepus*), the trend toward widening of the foot and shortening of the anterior triangle is evident across different theropod clades over time (Lockley et al., 2009). This makes Cretaceous theropod tracks more similar to typical ornithischian tracks. For the latter, the discussion that has focused on the lack of skeleton support for relatively large pentadactyl birdlike tracks (10–30 cm), which mainly includes *Wupus*, *Magnoavipes*, and *Saurexallopus* (Xing et al., 2015d; Lockley et al., 2021; Hong et al., 2024).

Unfortunately, even with abundant skeletal material, it is still not feasible to restore the pes morphology of the trackmaker due to the lack of complete pes skeletons (Lallensack et al., 2020; Hong et al., 2024). Furthermore, the convergence and conservativeness arise from a strong functional signal rather than phylogenetic information, as demonstrated by the foot shape of tridactyl dinosaurs and birds. These similarities across phylogenetic relationships make it challenging to use morphology-based taxonomy to identify the numerous tracks of the Cretaceous Period, and could result in an overestimation of the number of distinct tridactyl dinosaurian ichnotaxa (Farlow, 2018; Lallensack et al., 2020).

In addition to taxonomic controversy and overestimation, morphological ambiguity also contributes to differences in the accuracy of classifying ichnotaxa across skeletal groups, which is particularly evident in the contrast between nonavian theropods and avian tracks. After excluding the effect of size, the other Mesozoic bird tracks occupy a morphological space that is only slightly larger than that of four intermediate ichnospecies and significantly smaller than that of theropods (see Fig. 8 in Hong et al., 2024), and there is no such difference in the number of ichnogenera (see Section 3).

4.3 Solutions and perspectives

Part of the extramorphological influences can be rectified by photogrammetric 3D modeling, as shown in Figueiredo et al. (2017), which can be helpful to get the final position of the trackmaker's foot resting inside the track. For some of the morphological influences, another type of commonly used methods is statistical analyses, including the classic multivariate analysis (MVA, multiple bivariate analysis is included; Moratalla et al., 1988; Romilio and Salisbury, 2013; Castanera et al., 2013a), as well as the fresher usage of principal component analysis (PCA) and neural networks (Lallensack et al., 2022a; Hong et al., 2024).

MVA and neural networks have been effectively used to classify theropod and ornithischian dinosaur tracks. The reattribution of the trackmaker to *Iguanodontipus*? *oncalensis* (formerly *Therangospodus oncalensis*; Castanera et al., 2013a) and *Amblydactylus* of Lark Quarry (formerly cf. *Tyrannosauropus*; Romilio and Salisbury, 2011) is reasonable. However, the effectiveness of this method depends on the researcher's selection of features and the stability of the outline drawings within the chosen dataset. Neural networks, on the other hand, are more capable of dealing with ichnotaxa, which form a continuum of shapes rather than discrete classes. And future research on neural networks could also involve databases of three-dimensional models of trackways and/or trackway parameters, which can resolve morphologically ambiguous specimens to a greater extent (Lallensack et al., 2022a). But, herein lies the problem with the latter compared with artificial classification: at least the Cretaceous theropods and ornithischians tracks need to be taxonomically dichotomized, while neural networks provide actually continuous results. The ambiguous part of them does not have much practical significance as transition taxa in phylogeny. In addition, as noted by Lallensack et al. (2022a), the model is trained using previous identifications of tracks as either "theropod" or "ornithischian." Future studies may attempt to address these issues.

In contrast, PCA is not suitable for addressing the above issues (Lallensack et al., 2020). Currently, it is mainly effective for analyzing intermediate ichnospecies within theropods (Xing et al., 2015, 2015d; Hong et al., 2024); at least *Wupus* is reliably assigned to bird tracks in the lack of such a skeleton. Yet, in both studies, the predominant screening criteria in PC1 and PC2 are relatively homogeneous (stride length in Xing et al., 2015d; foot length, foot length/foot width, and total divarication in Hong et al., 2024) and strongly related to the feature selected (e.g., Hong et al., 2024 did not select stride length as one of the criteria). The successful categorization of *Wupus* is closely related to its specific gait that differs from theropod, whereas for *Saurexallopus* and *Magnoavipes* it is more ambiguous. As mentioned in both articles, the ambiguity in the track taxonomy is due to the fact that the boundaries between their trackmakers are not clear. It may be useful to test the machine learning method in future studies.

5. Posture, locomotion and behavior

Unlike skeletons, tetrapod tracks provide direct evidence for posture, locomotion, and behavior. It should be noted here that the paleontological framework for classifying behavior was developed independently of the behavior of extant animals (Seilacher, 1953; Vallon et al., 2016), and has little overlap with concepts underlying behavioral biology, thus its existing framework for classifying behaviors is rarely known outside the paleontological community, and has little overlap with concepts in behavioral biology of the extant animals in earlier studies (Plotnick and Smith, 2012). However, in recent years, there has been a gradual shift toward studies that classify paleontological behaviors under the framework of behavioral biology (Hsieh and Plotnick, 2020).

As the tetrapod track is quite limited in the types of behavior it displays, and more commonly shows different dynamics, the behaviors presented below continue to use the classical framework for classifying paleontological behaviors.

5.1 Posture and locomotion

5.1.1 Ceratopsids

Lockley and Hunt (1995b) described the *Ceratopsipes* from the Colorado Laramie Formation in the USA, and concluded that the ceratopsid posture was closer to that of an erect quadrupedal mammal rather than to that of crocodilian walking with a sprawling gait based on the insufficient distances between the manus and pes prints and the midline of trackway. This is consistent with the conclusion reached by Fujiwara (2009) based on the analysis of skeletal fossils.

5.1.2 Sauropods

Cretaceous sauropod trackways tend to be wider gauged than those of the Jurassic. This trend has been correlated with anatomical innovations in titanosauriforms and, in particular, titanosaurs, where the femur is angled outwards so that the feet stand further apart (Wilson and Carrano, 1999). However, while gauge width is partly controlled by anatomy, it has also been shown to vary considerably within single trackways (Meyer et al., 2018; Mannion and Upchurch, 2010; Castanera et al., 2012). Long and complete trackways can help us understand how tetrapods walked. Lallensack and Falkingham (2022) developed a method to calculate the gait (i.e., the relative timings of the footfalls, known as limb phase) based on the variation within long trackways. This method compares different gleno-acetabular distances that each assume different gaits; the gleno-acetabular distance that shows the least variation within the trackway can be considered to reflect the correct gait. The study found that three wide-gauged sauropods from the Lower Cretaceous De Queen Formation, Texas, employed a lateral sequence diagonal couplet walk, that is, the sauropods used the diagonal limb pairs (the front and back feet on opposite sides of the body moving simultaneously) in a gait similar to hippos. This may be related to their body width and stability.

Nonstraight trackways also provide important information. Xing et al. (2015g) described a 180-degree turn of *Parabrontopodus*-like trackways in the Early Cretaceous in Shandong, China. This trackway has a clear inward movement, with the manus tracks clearly deviating from the trackway centerline, which is similar to other turning tracks (Ishigaki and Matsumoto, 2009), indicating that sauropods have the same locomotion strategy when turning.

Abundant "manus-only trackways," in which only the manus tracks are visible but not the much larger pes tracks, led to much debate. Bird (1944) famously interpreted such a trackway from Mayan Ranch, Texas, as evidence of a swimming sauropod in which only the forelimbs touched the ground while the hindquarters floated. Although this "punting hypothesis" cannot be entirely excluded in all cases (Farlow et al., 2020), manus-only trackways are now thought to form as a consequence of differential foot loading: If peak pressure exerted on the manus is only slightly larger than that of the pes, the yield strength of the sediment may be exceeded, causing the manus to leave a track while the pes does not sink in (Falkingham et al., 2011; Vila et al., 2005).

5.1.3 *Ornithopods*

Ornithopods, the most common ornithischians of the Cretaceous, have been the subject of ongoing debate regarding their locomotion, whether they were quadrupedal or bipedal. At least in fossil tracks, there are both trackways with manus impressions and trackways lacking manus impressions. Brett-Surman (1997) believed that at least large ornithopods were quadrupedal or facultatively quadrupedal. However, due to problems such as sedimentary substrate, center of gravity during locomotion, and surface area of the manus, the manus impressions are often shallow or very small, easily eroded by weathering, and may not always be found (Diedrich, 2004; Castanera et al., 2013). Such "pes-only trackways" are also common in sauropods (Falkingham et al., 2011a,b).

5.1.4 *Theropods*

The bed of the Paluxy River (Glen Rose Formation, Texas) preserves abundant theropod tracks with elongated metatarsal marks, so-called "elongate tracks." Because these metatarsal marks are often on a level with the footprint proper, they are traditionally interpreted as evidence for a facultative plantigrade walking posture, where the metatarsal is pressed against the ground to increase traction or to lower the body (Kuban, 1989). Similar elongate tracks have been reported from many other localities of the world. A recent reevaluation demonstrated that such tracks represent deep sinking of the foot into soft sediment while the metatarsus remained at an angle to the foot (Lallensack et al., 2022b).

5.1.5 *Pterosaurs*

The posture and locomotion of pterosaurs are also controversial. Clark et al. (1998) believed that this was related to the lack of three-dimensionally preserved fossils. The discovery of pterosaur tracks provides a new view. According to the earliest discovered pterosaur tracks, Lockley et al. (1995) found that pterosaur tracks showed: the forelimbs supported more weight; quadrupedal walking; semierect; the palms and toes are all flatly, which is also supported by the three-dimensional preserved fossil bones described by Clark et al. (1998); and lateral rotation of the forelimbs. Regarding posture, there are still controversies, Bennett (1997) believed that pterosaurs were erect. Padian (1983) and Chatterjee and Templin (2004) demonstrated that pterosaurs walked quadrupedally but used bipedal running to take off and land. There are also a few examples of this in pterosaur tracks (Mazin et al., 2009).

5.2 Speed

Estimating the speed of trackmakers using measurements from vertebrate trackways is a commonly used method, especially in dinosaur trackways. The most classic formulas are Alexander (1976), Thulborn (1990) and González Riga (2011, for sauropod) which are often applied. The animals whose speeds are most often calculated include theropods, followed by sauropods and ornithopods.

Most of the Cretaceous sauropods were walking, while theropods were both walking and trotting, just a few running trackways have been reported. Navarro-Lorbés et al. (2021) calculated speeds of Early Cretaceous medium-sized theropod trackways from Spain ranging from 10.5 to 11.6 m/s. Li (2015) mentioned that an Early Cretaceous theropod trackway from Etuoke (Otok) area of Inner Mongolia. Each track is 0.28 m in length and the speed was 12.2 m/second. However, if follow the formulas of Navarro-Lorbés et al. (2021), this value needs to be adjusted to 10.7 m/second. Both of these trackmakers were the fastest-running theropods of the Cretaceous.

5.3 Behavior

In this section, we are mainly interested in swimming, foraging, gregariousness, and crouching as special forms of locomotion, locomotor states, and behaviors in the narrower sense. As for trace of excreting and possible courtship behavior, these will be discussed in other articles in this book.

5.3.1 *Swimming*

Long and parallel, slender scratch marks and the preservation of only the distal portions of digits impressions in trackways typically indicate swimming behavior (Ezquerra et al., 2007). Among these forms, the relatively common ichnogenus is *Characichnos* (Whyte and Romano, 2001). It has also been recorded from the Cretaceous of China (e.g., Xing et al., 2013e). Hunt and Lucas (2007) proposed the *Characichnos* ichnofacies represents shallow lacustrine (and tidal) environments. Some other ichnogenera are also considered to be related to swimming behavior because of such characteristics, such as *Wintonopus*, *Skartopus*, and *Characichnos* from Australia that shows difference in partial weight-bearing to full buoyancy (Romilio et al., 2013), but there is still some controversy. Other Cretaceous trackmakers of swimming tracks include pterosaur (Xing et al., 2021j), turtles (Pascual Arribas et al., 2005; Xing et al., 2021j); crocodylians tracks (Vila et al., 2015), and marine reptiles (Natali and Leonardi, 2023). Navarro-Lorbés et al. (2023) has catagorized the subparalelled swimming tracks into two main types by the different behaviors: the "Buoyancy track" for with partial or complete buoyancy. as only the clows contact the substrate; and "punting track" for bottom walking, like modern turtles and crocodiles

Except for the aformentioned paralelled tracks, the digit only tracks are less commom and harder for interpretation. As previously mentioned, in 1944, Bird, R.T. described almost exclusively of sauropod manus prints in the Cretaceous of Texas, USA, and considered them as evidence of swimming in sauropods. Lockley et al. (1994c) identified these tracks as pes undertracks. Farlow et al. (2020) believed that the possibility of unusual behavior cannot be ruled out. Xing et al. (2015h) described only digit traces in the Cretaceous of Gansu, China, and explained that these tracks were made by the maker walking on a silt and mud substrate above the sandstone layer into which the pes claws penetrated from above, rather than swimming.

There are similar characteristics (digit-only tracks) to those ornithopod tracks (Gierliński and Potemska, 1987; McAllister, 1989; Thulborn and Wade, 1989; Xing et al., 2015f; Fujita et al., 2012). Fujita et al. (2012) thought these tracks represent swimming behavior. Xing et al. (2015f; 2017a) refuted that with the same reasons as the above (Lockley, 1991; Xing et al., 2015f).

5.3.2 *Foraging*

Falk et al. (2010) described avian tracks from the Early Cretaceous in South Korea and associated small, elongate, and double-oval traces. By comparing their behaviors with living birds, they identified them as clustered pecking and probing.

Li et al. (2015) documented a large number of pterosaur forefoot tracks preserved together with insect traces in shallow lacustrine deposits from the Early Cretaceous of China, revealing scenes of pterosaurs feeding on invertebrates in the lake.

5.3.3 *Gregarious behavior*

Paralleled trackways confirmed that sauropods displayed gregarious behavior. The most typical case comes from North America and Europe, such as the Upper Jurassic Dinosaur Lake site, USA (Lockley et al., 1986; Schumacher and Lockley, 2014) and Upper Jurassic Lagosteiros Bay, Portugal (Lockley et al., 1994d; Castanera et al., 2014; García-Ortiz and Pérez-Lorente, 2014). Similar records also appear in the Cretaceous, for example, Late Cretaceous Humaca and Cal Orcko sites from Sucre Bolivia (Lockley et al., 2002b). Parallel trackways from two Lower Cretaceous sites in the Glen Rose Formation of Texas indicate the passage of a herd of 12 large individuals and 23 individuals of variable size from the Paluxy and Davenport Ranch sites respectively (Bird, 1939, 1941, 1944, 1954, 1985). Lockley et al. (2002a) documented parallel trackways of sauropods in the Otog of Inner Mongolia, China, where the central trackways were smaller and the lateral trackways were larger, suggesting a migrating group of sauropods with smaller individuals protected by larger adults.

Similar parallel trackways have also been observed in ornithopod, nonavian theropod, and avian trackways (Currie, 1983; Xing et al., 2014b, 2018d), including some of the parallel scratch marks of swim trace found both in isolation and within complex trackways (Romilio et al., 2013). It is noteworthy that some clusters of parallel, small ornithopod trackways may represent juvenile ornithopod groups that formed independent activities (Lockley and Hunt, 1995a; Yoon et al., 2021).

5.3.4 *Crouching*

Dinosaur crouching traces represent the behavior of sitting or resting, generally preserving metatarsal impressions, and associated ischial, and possibly pubic, callosity traces. Most crouching traces come from the Jurassic, the most famous of which are *Anomoepus major* and *Sauropus barratti* (Hitchcock, 1837; Lull, 1915) from the Early Cretaceous in the USA. Xing et al. (2012) recorded an Early Cretaceous crouching trace in Hebei, China.

5.4 Pathologies

Some pathological theropod tracks have also been preserved in Cretaceous deposits, representing foot swelling, deformities, fractures, dislocations, osteoproliferation, tumors, injuries, and limping (McCrea et al., 2015b). The most intuitive examples are the pedal claw impression in lateral profile of the large theropod tracks in the Late Cretaceous of Alberta, Canada, indicating an extreme dislocation of the distal end of Digit IV (Sweet et al., 1989); and the abnormal swelling of the Digit II of an Early Cretaceous theropod in Colorado, USA (McCrea et al., 2015b).

Notably, trackways showing signs of limping have been identified for both sauropod, theropod, and ornithopod trackmakers (Lockley et al., 1994e; Xing et al., 2017b). These trackways are characterized by a consistent asymmetry in the step length—the distance between a pair of left and right impressions measured parallel to the stride direction (Romilio and Shao, 2023; Razzolini et al., 2016). One side of the trackway is consistently shorter than the other, suggesting that the trackmaker was favoring one side of its body over the other, likely due to an injury, thereby supporting the notion of an injured limb at the time of trackway registration.

6. Track paleoecology

Skeletal fossils often undergo transportation prior to burial, which can pose challenges in reconstructing the actual paleoecology of ancient organisms. In contrast, trackway fossils, being preserved in situ, provide valuable information for paleoecology reconstructions, as the trackmakers are integral components of the fauna. Cretaceous tetrapod track records occur in a considerable number of depositional environments. The correlation of ichnotaxa with depositional environments is a topic of interest. Based on the characteristics of specific ichno-assemblages in different depositional environments, Hunt and Lucas (2005, 2007) suggested a series of tetrapod ichnofacies, and named five archetypal vertebrate ichnofacies for nonmarine environments, and all of these ichnofacies are related to the Cretaceous: *Chelichnus, Batrachichnus, Brontopodus, Grallator,* and *Characichnos* ichnofacies, and are associated with aeolian, tidal flat-alluvial plain, coastal plain, lacustrine margin, and shallow lacustrine paleoenvironments. Considering the complexity of the terrestrial environment and relatively fast ichnofaunal replacement through time, these ichnofacies have been considered controversial (Santi and Nicosia, 2008). Lockley (2007c) emphasized the importance of ichnofacies and also pointed out that tetrapod ichnofacies are inherently more diverse and differentiated than invertebrate ichnofacies. In the last 10 years or so, with the discovery of hundreds of sites in East Asia and other places, the provincial character of tetrapod tracks is more obvious, and the ichnofacies that warrant further review.

Of the five major tetrapod ichnofacies (Hunt and Lucas, 2007), aeolian is relatively rare (Loope and Milàn, 2016). Xing et al. (2018f) described the track assemblages from Shaanxi in Early Cretaceous deposits, which is the first reported from an "aeolian dominant" desert facies in Asia. The occurrence of the mammaliamorph track *Brasilichnium* isp. makes it possible to classify it into *Brasilichnium* subichnofacies or the *Chelichnus* ichnofacies (Lockley et al., 1994f; Hunt and Lucas, 2007). *Brasilichnium* tracks are abundant in the aeolian Early

Cretaceous Botucatu Formation of Brazil, but absent, rare, or little known from other facies (Fernandes and Carvalho, 2008).

After the five-fold archetypal tetrapod ichnofacies were proposed, some new types were also described, such as the Patuxent ichnofauna from wetland deposits in the Lower Cretaceous of Maryland, representing a high diversity ironstone-wetland ichnocoenosis, might loosely be subsumed in the coastal plain ichnofacies (Stanford et al., 2018).

The relationship between the similar Cretaceous tetrapod ichnotaxa and the environment may be quite different in different regions. The Early Cretaceous track assemblages of North America and East Asia are a typical example. The track assemblages in North America are mainly associated with extensive, coastal-plain systems (Lockley et al., 2012a). This includes the lower-latitude, carbonate facies *Brontopodus* ichnofacies, and the higher-latitude, locally coal-bearing, siliciclastic facies *Caririchnium* ichnofacies. That is to say, sauropod and ornithopod tracks are rarely recorded together. However, in East Asian track assemblages, especially in China, sauropods and ornithopods frequently appear together in isolated, inland, fluvio-lacustrine basins (Lockley et al., 2012a).

The high abundance and endemic nature of the Cretaceous avian ichnofauna in East Asia suggest the presence of a distinct and thriving avian fauna (Lockley et al., 2012a). The most diverse records are from the Korean Cretaceous Dinosaur Coast (KCDC) of the southern part of the Korean Peninsula (Lockley et al., 2012b). The bird tracks of the KCDC usually occur in the lacustrine deposits in which thinly interlaminated fine-grained sandstones and siltstones–mudstones prevail (Huh et al., 2003; Lim and Kong, 2008). If one considers the rich bird skeletal record (ornithurine and enantiornithine) of the Jehol biota in Liaoning, China, and the bird tracks records of lacustrine deposits in Inner Mongolia and Xinjiang, the latter is currently the most abundant in China (Xing et al., 2021j). It can be roughly inferred that there was a Cretaceous bird record belt (including skeletal and tracks) at ~34°N–46°N. This region may have been an endemic center for this evolutionary radiation (Lockley et al., 2012a).

Overall, both qualitative and quantitative research on Cretaceous tetrapods needs more work in the future, such as considering the structural differences of different ichnofaunas, emphasizing the comparative study of skeleton fossils from the same formation and the same region, or using ecological statistical methods of skeleton fossils, such as the Ontogenetic Niche Shift (ONS) model.

7. Track biostratigraphy and biochronology

Tracks not only provide information about the diversity of organisms but also serve as evidence for long-distance stratigraphic correlations (Sarjeant, 1975). Scholars have attempted to use Cretaceous tetrapod tracks for stratigraphic correlations, as biostratigraphy and biochronology (palichnostratigraphy) (Lockley et al., 1994f, 2008b, 2012a; Lockley, 1998; Carrano and Wilson, 2001; Lucas, 2007). Palichnostratigraphy is based on recurrent trace fossil assemblages (=associations, ichnofaunas or ichnocoenoses) (Lockley and Gierliński, 2006), and the more places these assemblages are found, the more valuable they are for comparison.

Therefore, some well-preserved but isolated occurrences of ichnotaxa are not accessible for discussion in this section.

Lucas (2007) provided a comprehensive review of this field and proposed three factors: (1), the validity or invalidity of ichnotaxa, (2), relatively slow apparent evolutionary turnover rates, and (3), facies restrictions. When considering the appearance or disappearance of ichnogenera during the Cretaceous Period, one should carefully evaluate whether it is attributed to factors 2 or 3. In other words, certain track morphologies may be relatively conservative and persist for a considerable duration under different depositional conditions.

Cretaceous tetrapod tracks are widespread on all continents except Antarctica. Overall, the Early Cretaceous biochron is characterized by sauropod and ornithopod tracks; the Late Cretaceous biochron is characterized by fewer sauropod tracks but adds more numerous tracks of large ornithopods, plus tracks of tyrannosaurids and ceratopsians (Lockley et al., 2012a).

Lucas (2007) considers that the Cretaceous global vertebrate fauna began to exhibit increasingly pronounced provincialization, and therefore adopted provincial rather than global vertebrate biochronologies during the Cretaceous. After 2007, more and more unique ichnofaunas, especially the ichnofaunas from the Early–Mid Cretaceous in Asia, support this statement, that is, the Cretaceous ichnofaunas indeed show some degree of provincialism (Lockley et al., 2012a). Lockley et al. (2012a) analyzed the biostratigraphy of North America (Lockley and Hunt, 1995a), East Asia (Matsukawa et al., 2006; Lockley and Matsukawa, 2009), Europe (Lockley and Meyer, 2000), South America (Leonardi, 1994), as well as Africa and Australia. Among them, the most provincial characteristics of palichnostratigraphy are North America and East Asia, especially the latter. Most other records are too scattered to suggest any meaningful patterns in the distribution of ichnofaunas.

Most of the significant ichnofaunas from the Early Cretaceous in North America are found in the western part of the continent, such as in Western Canada, Colorado, and Texas (Lockley et al., 2012a). In this region, post-Neocomian, there is a noticeable contrast between the sauropod *Brontopodus* ichnofacies are associated with lower-latitude, carbonate facies, and the ornithopod- or *Caririchnium* ichnofacies are associated with higher-latitude, locally coal-bearing, siliciclastic facies (Lockley and Hunt, 1994b; Lockley, 2007c; Lockley et al., 1994f, 2012a). The siliciclastic facies feature abundant bird tracks, including the *Ignotornis*, *Aquatilavipes*, and *Koreanaornis* (Lockley et al., 2012a). The patterns of these post-Neocomian siliciclastic and carbonate facies seem to be most typical in the Early Cretaceous deposits of the western USA.

The Late Cretaceous track records in North America are primarily found in the Campanian–Maastrichtian deposits from western Canada and the USA (Lucas, 2007). Lockley et al. (1994f) suggested an ornithopod ichnofacies, or rather a hadrosaurid ichnofacies, with larged sized hadrosaurid (*Hadrosauropodus*) dominated ichnofacies being quite common in this region. This study appends a Campanian–Maastrichtian *Tyrannosauripus*-Ornithomimipodidae (Lockley et al., 2011) biochron, identified by tyrannosaurid and coelurosaurians tracks.

Many areas in East Asia lack standard fossils of marine strata, or tuffs as materials for radioactive dating. It is quite difficult to divide the various stages of the Cretaceous, but with the advancement of technology, this dilemma is easing. In China, there is a certain degree of difference between the stratigraphic age information provided by Chen et al. (2006) and the recent information (Cao, 2013; Xi et al., 2019). We try to differentiate between Early and Late Cretaceous Asian ichnofaunas, but it is likely that some of these will change in the future as geological ages are updated.

Many East Asian tetrapod ichnofaunas are dominated by small and medium-sized theropods, consisting of both avian and nonavian theropod ichnotaxa during the Early Cretaceous (Matsukawa et al., 2006; Xing and Lockley, 2016). The latter is predominantly characterized by small-sized grallatorid and *Eubrontes* type/*Asianopodus* tracks. This situation can be traced back at least to the Jurassic–Cretaceous boundary, especially in the Tuchengzi Formation (156 Ma–139 Ma, Wan et al., 2020) in China (Xing and Lockley, 2021).

The Early–"Mid" Cretaceous of Korea recorded extremely rich avian tracks assemblages such as *Koreanaornis, Jindongornipes, Ignotornis*, and *Goseongornipes* (Kim et al., 2006, 2012c; Lockley et al., 1992a). Among them, *Koreanaornis* holds significant global comparative significance. Korea has also recorded the ornithopod tracks *Caririchnium* and *Ornithopodichnus*, sauropod *Brontopodus*, and pterosaur *Pteraichnus*. These track types are reflected in the same or similar strata in China, especially in the Jiaguan Formation (Barremian/Aptian–Santonian), Sichuan and Tianjialou Formations (Albian–Cenomanian), Shandong (Xing and Lockley, 2016). However, there are regional exceptions, such as only a single record of a grallatorid in the post-Neocomian of Korea (Lockley et al., 2022b), as well as some distinct ichnotaxa with palichnostratigraphic attributes, such as *Minisauripus* and *Velociraptorichnus*, both of which are found exclusively in Asia (Lockley et al., 2012a; Xing and Lockley, 2016).

In East Asia, three or four footprint biochrons may be recognizable (Lockley et al., 2012a). Based on new records and geological ages, these biochrons have severe overlap and also have longer time distributions, whose accuracy is weakened, but this may be limited by geochronological issues in continental stratigraphy: (1) a Berriasian–Albian grallatorid-*Asianopodus* biochron; (2) a Berriasian–?Albian *Minisauripus*-deinonychosaur-*Koreanaornis* biochron. The question mark mainly comes from the age of the Jiaguan Formation, which may extend to the Santonian; and (3) a less distinctive Coniacian-Maastrichtian hadrosaurid (large ornithopod) tracks biochron (Lockley et al., 2012a).

Large ornithopod tracks are also distributed in Asia during the late Late Cretaceous. Large ornithopod tracks were recorded in the Nemegt Formation in Mongolia (Currie et al., 2003), and in southern Guangdong, China, where they were classified as *Hadrosauropodus* isp. (Xing et al., 2017a), which is similar to hadrosaurid ichnofacies in the Late Cretaceous of the North America. Similar to North America, some large theropod tracks are also recorded in these areas, such as tyrannosaurid track from Hekou Formation (?Coniacian–Santonian), southern China (Xing et al., 2019e).

8. Track paleobiogeography

The Cretaceous saw a continued separation of landmasses, with the Atlantic separating the Americas from Europe and Africa, the split of the northern from the southern landmasses, and the break-up of the southern landmasses (South America, Australia–Antarctica, India, and Madagascar). This created more coastal habitats, possibly an increased habitat diversity, and stronger disparity of faunas (the separated faunas evolving largely independently from each other). The Cretaceous, and particularly the Late Cretaceous, was also a time of high sea levels and extensive epicontinental seas (Carrano, 2012). Tracks have been used to interpret distribution, migration, and local endemism of Cretaceous dinosaur faunas.

8.1 Carbonate platforms

The extent of carbonate platforms continued to increase during the Cretaceous with rising sea levels, expanding into areas at higher latitudes as far as Europe–Central Asia in the Northern Hemisphere, where it left a massive Cenomanian–Campanian carbonate record so that the number of tracksites in this phase reaches the highest level (Kiessling et al., 2003; Torsvik and Cocks, 2016). The Laurasia–Gondwana transition region, centered on Europe, is an important location for this phase in both paleogeography and paleobiogeography during the Cretaceous, including Europe, North and South America, Africa (Madagascar), and the Eastern Levant (Kiessling et al., 2003; Porchetti et al., 2016; Lallensack et al., 2023), forming the distinctive *Brontopodus* ichnofacies (Lockley et al., 2012a). These tracks, especially those of large sauropods and theropods, occur in areas quite far from the coast. These findings suggest that, in areas that were previously thought to be shallow marine, there were emergent lands capable of supporting dinosaur populations with vegetation cover (Dalla Vecchia, 2008). Such carbonate platforms show that pre-Barremian Europe was connected to the Gondwana region, leading to faunal exchange between the two realms (Ezcurra and Agnolín, 2012).

8.2 The polar fauna and their migration

Due to the significant warming and flattening of the latitudinal zonation of climate in major parts of the Cretaceous (Hay and Floegel, 2012; O'Brien et al., 2017), the temperate climate in polar regions led to an abundance of tracks.

Tracks in the far north have been studied to a much greater extent than the Antarctic region. In 1960, ornithopod tracks discovered in the Lower Cretaceous of the Arctic Island of Spitsbergen provided the first evidence that dinosaurs were able to live in polar habitats (de Lapparent, 1960, 1962; Hurum et al., 2006, 2016). Diverse polar dinosaur faunas have since been discovered in Alaska, Canada, Australia, and Antarctica. Such discoveries either imply that dinosaurs were able to overwinter under harsh climatic conditions, or undertook substantial seasonal migrations; both possibilities strongly suggest some degree of endothermy (Bell and Snively, 2008). Simultaneously, similarities between the North American and Asian dinosaur faunas during the Late Cretaceous suggest the presence of a connection between Alaska and Asia, the Bering land bridge (Brouwers et al., 1987; Cifelli et al., 1997; Wen et al., 2016); these faunal similarities are also reflected by the similar ichnofaunas, especially in dinosaurs and avians (Fiorillo et al., 2011; Kang et al., 2021). The ecological information contained in the traces of some taxa (e.g., hadrosaurus) from far north can also be used to compare the growth patterns, and the environmental preferences that summarized of their relatives from middle and low latitudes (Fiorillo et al., 2012; Flaig et al., 2018).

Migration in dinosaurs is indicated by extensive bone beds, but has also been suggested by extensive megatracksites such as the "Dinosaur Freeway" of the Dakota Sandstone. However, it cannot be confirmed that such sites actually record long-distance migrations, which may indicate the migration route to or from the polar region (Lockley, 1991; Bell and Snively, 2008).

Antarctica contains no tracks to date, and Western Australia has abundant but restricted tracksites. The latter is almost the only Early Cretaceous dinosaur fossil site in Western Australia, showing similarities to Western Gondwana taxa (Salisbury et al., 2016; Thulborn, 2017).

8.3 The comparison of ichnofaunas of different regions

Ichnofaunas of different regions may be compared with draw inferences on distribution and disparity of the actual faunas. For example, the Early Cretaceous of East and Southeast Asia has been argued to show a distinct theropod and bird ichnofauna, with many ichnotaxa that are found nowhere else (Lockley and Matsukawa, 2009). Problematically, however, such conclusions rely on ichnotaxonomy, which is highly unstable, and probably oversplit, especially in the case of tridactyl tracks. This may result in circular arguments: tracks tend to be ascribed to ichnotaxa typical for the region even if ichnotaxa from other continents may qualify as well. If such biases cannot be excluded, ichnotaxonomy cannot serve as the basis for paleobiogeographical interpretations.

9. Evolutionary events

9.1 A path to gigantism

The Cretaceous Period witnessed a remarkable evolution of gigantism, which was prevalent among the major Cretaceous tetrapods, especially in nonmarine archosaurs that shows more exaggerated size than mammals. Their gigantism was fueled by increasing evolutionary pressures during this era, coupled with the abundance of food resources and productive habitats due to climatic and geographic patterns (Sander et al., 2011).

The most striking degree of gigantization occured in sauropods (Sander et al., 2011). However, this group had already shown considerable body mass in basal taxa of mamenchisaurids in the Jurassic. The derived macronarian in the Cretaceous underwent more evolutionary innovations in morphology and function than continued elongation (Sander, 2013). The morphological changes are mainly reflected in the increase in the proportion of wide-gauge sauropod tracks, a transition described in detail above in "3.1.1. *Brontopodus*" Theropods and sauropods are relatively similar to some extent: the maximum size of these taxa in the fossil record (including body fossil and trace fossil) increased insignificantly after the J–K boundary (Rauhut et al., 2018). But at least in trace fossils, the amount of very large tracks increases with time (Rauhut et al., 2018; Lockley and Xing, 2021; Castanera et al., 2022b).

As for the ornithischians, although their size is not remarkable among dinosaurs, the general increase in track size of the dominant type during the Late Jurassic–Early Cretaceous ornithopod faunal changes shown in Europe (*Dinehichnus*-like/*Iguanodontipus*-like/*Caririchnium*-like, Castanera et al., 2022b) indicates that at least the ornithopod record was influenced by increased resources in the Cretaceous and that there was a general increase in body size. Pterosaurs follow a similar trend, which shows clearly a general trend toward increase in maximum track size from the Jurassic to Cretaceous, that continued to rise through the Cretaceous, showing a concentrated occurrence of their very large manus tracks in the Late Cretaceous (Lockley et al., 2008b).

9.2 Jurassic–Cretaceous insights: A case study of the Tuchengzi Formation, East Asia

The early filamentous integumentary coverings (i.e., protofeathers) of dinosaurs and even Avemetatarsalia are a special feature within reptiles that served as one of the factors that allowed them to survive the Late Triassic mass extinction and subsequently to emerge (Barrett et al., 2015; Yang et al., 2019; Olsen et al., 2022). Early records of feather differentiation in theropods are preserved in the Jurassic Yanliao biota of northeastern China (Xu et al., 2016), but the Cretaceous Jehol biota of the same region yielded much more diverse taxa, including basal ornithischians and large coelurosaurs (Ksepka, 2020). The Tuchengzi Formation, which lies between the two, is relatively devoid of body fossils, but the large number of tracks provides the potential to bridge the evolutionary history of northern China's most well-studied and understood Mesozoic terrestrial faunas, especially the feathered dinosaurs (Xing and Lockley, 2021).

The relatively diverse Tuchengzi dinosaur ichnofauna, which is saurischian-dominated and includes tridactyl and didactyl nonavian theropods, birds, sauropods, and possible small ornithopods. In terms of track morphology, Tuchengzi theropod tracks tend to be more homogeneous and conservative compared with the Jehol Biota, and their trackmakers can be matched with fossilized skeletons from both the preceding and following biotas. From the body fossil record, the genera of the seven dinosaurs found in the Yanliao Biota all belong to the basal Paraves (Li et al., 2023). However, fossil tracks from the Tuchengzi Formation show brontozoid ichnites that may have belonged to oviraptorosaurids, and other tridactyl tracks from various potential coelurosaurian trackmakers (Xing et al., 2009b, 2015i; Xing and Lockley, 2021). A small number of large theropod tracks (*Megalosauripus*) may record the occurrence of feathered theropods similar to *Yutyrannus*, which is now thought to have first appeared in the Early Cretaceous (Xing and Zhang, 2022). In addition, a small number of poorly diversified small didactyl tracks from this formation (Zhen et al., 1996; Xing et al., 2009b) may record the presence of small pre-Cretaceous dromaeosaurids, a hypothesis that may be supported by possible fossil dental evidence from Europe (Wills et al., 2023) and follows logically from the rich microraptorian and other fossil record in the Jehol Biota (Xing and Zhang, 2022).

9.3 The Cretaceous–Tertiary mass extinction event

The Cretaceous–Paleogene (K–Pg) extinction event, also known as the Cretaceous–Tertiary (K–T) extinction, was a sudden mass extinction of three-quarters of the living plant and animal species living on the Earth, approximately 66 million years ago (Harris, 2023). On terrestrial stratigraphic evidence of the K–T boundary, the contrast between dinosaur-(including birds) dominated Cretaceous tetrapod ichnofaunas and mammal-dominated Paleogene tetrapod ichnofaunas is clear, paralleling the underlying extinction of dinosaurs at the end of the Cretaceous and subsequent Paleogene diversification of mammals (Lockley et al., 2012a). The Cretaceous track records correspond with the body fossil records for the terminal Cretaceous extinction of dinosaurs across the K–T boundary. For example, the stratigraphically youngest dinosaur tracks from the Upper Maastrichtian Raton Basin in New Mexico, USA, provide biostratigraphic data points as valid as those from the stratigraphically highest dinosaur bones (Lockley et al., 2012a). A comprehensive review and study of the rich dinosaur track record of the Tremp Formation in the southern Pyrenees of southwestern Europe has demonstrated that the vertebrate-rich Upper Maastrichtian continental sites, containing hadrosauroid ornithopods, titanosaurid sauropods, and theropods, as well as egg sites and tracks, have been replaced by similar sedimentological facies characterized by the virtual absence not only of dinosaurs but also of any vertebrate remains throughout the Lower Paleocene. And isolated tracks and dinosaur bones were also found very close to where the K–Pg boundary extinction event is thought to have occurred (Canudo et al., 2016; Pérez-Pueyo et al., 2021). In terms of survival behaviors, with the exception of a few avian, crocodylian, and turtle lineages, there is no evidence that Late Maastrichtian pterosaurs or nonavian dinosaurs could burrow, swim, or dive (Robertson et al., 2004), suggesting that they were unable to protect themselves from the worst of the environmental stresses that occurred at the K–Pg boundary.

10. Conclusions and perspectives

At least 112 Cretaceous tetrapod ichnogenera are considered valid (= only contains part of the Cretaceous ichnospecies), including 10 of sauropods, at least 48 of nonavian theropods, 18 of avian theropods, 14 of ornithischians, four of pterosaurs, three of aquatic reptiles, two of terrestrial lizards, four of crocodiles, two of turtles, six of mammals and one of amphibians.

Tracks attributed to sauropods include *Brontopodus, Chorrokhosauropus, Parabrontopodus, Titanosaurimanus, Titanopodus, Oobardjidama, Calorckosauripus, Iniestapodus, Parvieouspodus* and *Gyeongsangsauropus,* of which *Parvieouspodus* and *Gyeongsangsauropus* are ichnogen. nov. here. The ichnospecies are in need of revision.

Tracks attributed to nonavian theropods include *Eubrontes, Gigandipus, Grallator, Ornithomimipus, Gypsichnites, Columbosauripus, Irenesauripus, Irenichnites, Satapliasaurus, Bueckeburgichnus, Yangtzepus, Macropodosaurus, Megalosauropus, Moraesichnium, Xiangxipus, Skartopus, Salfitichnus, Picunichnus, Abelichnus, Deferrariischnium, Bressanichnus, Theroplantigrada, Filichnites, Tyrannosauripus, Minisauripus, Velociraptorichnus, Saurexallopus, Archaeornithipus, Fuscinapedis, Magnoavipes, Toyamasauripus, Kalohipus, Asianopodus, Siamopodus, Chapus, Dromaeopodus, Dromaeosauripus, Menglongipus, Corpulentapus, Wakinyantanka, Paracorpulentapus, Bellatoripes, Ordexallopus, Dromaeosauriformipes, Candeleroichnus, Lockleypus, Jurabrontes* and *Farlowichnus*: both ichnogenera and ichnospecies are in need of revision;

Tracks attributed to avian theropods include *Ignotornis, Hwangsanipes, Goseongornipes, Aquatilavipes, Jindongornipes, Koreanaornis, Pullornipes, Limiavipes, Wupus, Barrosopus, Paxavipes, Shandongornipes, Gyeongsangornipes, Tatarornipes, Uhangrichnus, Sarjeantopodus, Yacoriteichnus* and *Gruipeda,* and the ichnospecies are in need of revision.

Tracks attributed to ornithischian include *Hypsiloichnus, Wintonopus, Neoanomoepus, Iguanodontipus, Caririchnium, Hadrosauropodus, Deltapodus, Garbina, Luluichnus, Metatetrapous, Tetrapodosaurus, Ligabueichnium, Apulosauripus* and *Ceratopsipes,* the ichnospecies are in need of revision.

Tracks attributed to pterosaurs include *Pteraichnus, Purbeckopus, Haenamichnus* and *Agadirichnus,* and the ichnospecies are in need of revision.

Tracks attributed to aquatic reptiles include *Champsosaurichnus*, *Novapes*, and *Coneroichnus*.

Tracks attributed to terrestrial lizard include *Neosauroides* and *Sauripes*.

Tracks attributed to crocodiles include *Batrachopus*, *Mehliella*, *Crocodylopodus*, and *Angolaichnus*, the ichnospecies are in need of revision.

Tracks attributed to turtles include *Emydhipus* and *Chelonipus*, the ichnospecies are in need of revision.

Tracks attributed to mammals include *Brasilichnium*, *Schadipes*, *Koreasaltipes*, *Catocapes*, *Aracoaraichnium*, and *Sederipes*: the ichnospecies are in need of revision.

Tracks attributed to amphibian include *Ranipes*.

Extramorphological distortion and similar pes morphology of trackmaker both have an influence on Cretaceous ichnotaxonomy, which could be partially resolved by statistical methods. The accuracy of classifying ichnotaxa across skeletal groups is in need of further revision.

Cretaceous tetrapod tracks provide direct evidence of posture, locomotion, and behavior for all clades of dinosaurs, as well as for pterosaurs, crocodylomorph, turtles, terrestrial lizards, mammals, amphibians, and marine reptiles. The kinematic features include gait on different substrates and the speed of different gaits, while behavioral characteristics include swimming, foraging, gregariousness, and crouching. Special pathological dinosaur tracks show a unique limping gait and nonextramorphological distortion.

Cretaceous tetrapod track records are found in various depositional environments, resulting in different ichno-assemblages across regions. The correspondence mainly consists of fivefold archetypal tetrapod ichnofacies.

Cretaceous ichnofaunas are almost found globally but exhibit pronounced provincialism among Mesozoic, especially in North America and East Asia. Based on the assemblage of endemic tetrapod ichnogenera, it is possible to establish biostratigraphic standards within a certain range that gains significance as the absolute age of the stratigraphic units is ascertained.

An increase in habitat and habitat diversity has led to the breadth of tetrapod track distributions in the Cretaceous, such as occurrences on off-shore carbonate platforms and in polar regions, which have been used to interpret the distribution, migration, and local endemism of terrestrial faunas. The influence of probably oversplit ichnotaxonomy should be considered.

The tetrapod tracks offer additional evidence for the evolution of Cretaceous terrestrial faunas, including the gigantism of dinosaurs and the K–Pg extinction event. They also fill gaps in the skeletal record at key points, such as the Asian feathered dinosaur record at the J-K boundary.

Acknowledgments

These authors (Lida Xing, Ignacio Díaz Martínez, Jens N. Lallensack, Kyung Soo Kim, Donghao Wang, Qiyan Chen) contributed equally to this work. The authors thank Spencer G. Lucas and Diego Castanera Andres for their helpful considerations and constructive feedback. Dan Liang, Ruisha Xue, Junyan Fu, Keyi Fu, Hao Li, Qi Qi, Lin Liu from Lida Xing Lab contributed to the literature review of systematic taxonomy, and Chang Liu formatted the reference list. This research was funded by the "Deep-time Digital Earth" Science and Technology Leading Talents Team Funds for the Central Universities for the Frontiers Science Center for Deep-time Digital Earth, China University of Geosciences (Beijing) (Fundamental Research Funds for the Central Universities; grant number: 2652023001), the 111 project (B20011), Fundamental Research Funds for Central Universities (265QZ201903) and Project PID2022-140205NA-

I00 funded by Ministerio de Ciencia e Innovación (Spanish Government). And the material is a contribution to the Deep-time Digital Earth (DDE) Big Science Program. Ignacio Díaz-Martínez is supported by a Ramón y Cajal fellowship (RYC-2022) and PID2022-140205NA-100 project granted by the Ministerio de Ciencia e Innovación (Spanish Government).

References

Abbassi, N., Alimohammadian, H., Shakeri, S., Boroumand, S., Boroumand, A., 2018. Aptian small dinosaur footprints from the Tirgan formation, Kopet-Dagh region, northeastern Iran. New Mex. Mus. Nat. Hist. Sci. Bullet. 80, 5–14.

Abbassi, N., Dashtban, H., 2021. Vertebrate footprints from the Agha Jari Formation (late Miocene–Pliocene), Zagros Mountains, a review of the cenozoic vertebrate ichnites in the Persian Gulf region. Bull. Geosci. 96 (2), 159–179.

Abel, O., 1935. Vorzeitliche Lebensspuren, 664 pp.. Gustav Fischer Verlag, Jena.

Adams, T.L., Strganac, C., Polcyn, M.J., Jacobs, L.L., 2010. High resolution three-dimensional LaserScanning of the type specimen of *Eubrontes* (?) *glenrosensis* Shuler, 1935, from the Comanchean (lower cretaceous) of Texas: implications for digital archiving and preservation. Palaeontol. Electron. 13 (3).

Aguirrezabala, L.M., Viera, L.I., 1980. Icnitas de dinosaurios en Bretún, Soria. Munibe n° 32 (3/4), 257–279.

Alexander, R.M., 1976. Estimates of speeds of dinosaurs. Nature 261 (5556), 129–130.

Alonso, R.N., 1980. Icnitas de dinosaurios (Ornithopoda, Hadrosauridae) en el Cretácico Superior de norte de Argentina. Acta Geol. Lilloana 15, 55–63.

Alonso, R.N., Marquillas, R.A., 1986. Nueva localidad con huellas de dinosaurios y primer hallazgo de huellas de aves en la Formación Yacoraite (Maastrichtiense) del Norte Argentino. Actas 4 Congreso Argentino de Paleontología y Bioestratigrafía 2, 33–41.

Ambroggi, R., de Lapparent, A.F., 1954. Les empreintes de pas fossiles du Maestrichtien d'Agadir. Notes Serv. Geol. Maroc 10, 43–46.

Anfinson, O.A., Lockley, M.G., Kim, S.H., Kim, K.S., Kim, J.Y., 2009. First report of the small bird track *Koreanaornis* from the Cretaceous of North America: implications for avian ichnotaxonomy and paleoecology. Cretac. Res. 30, 885–894.

Antonelli, M., Petti, F.M., Conti, J., Sacco, E., Petruzzelli, M., Spalluto, L., Wagensommer, A., 2023. Lower Cretaceous dinosaur footprints from the Molfetta tracksite (Apulia, southern Italy). Cretac. Res. 142, 105388.

Antunes, M.T., 1976. Dinossáurios Eocretácicos de Lagosteiros; *Ciências da Terra*, vol. 1. Universidade Nova de Lisboa, Lisboa, pp. 1–36.

Apesteguía, S., Gallina, P.A., 2011. Tunasniyoj, a dinosaur tracksite from the Jurassic-Cretaceous boundary of Bolivia. An. Acad. Bras. Ciências 83 (1), 267–277.

Arbour, V.M., Currie, P.J., 2016. Systematics, phylogeny and palaeobiogeography of the ankylosaurid dinosaurs. J. Syst. Palaeontol. 14 (5), 385–444.

Ataabadi, M.M., Khazaee, A.R., 2004. New Eocene mammal and bird footprints from Birjand area, eastern Iran. Ichnos 11 (3–4), 363–370.

Avanzini, M., García-Ramos, J.C., Lires, J., Menegon, M., Pinuela, L., Fernández, L.A., 2005. Turtle tracks from the late Jurassic of Asturias, Spain. Acta Palaeontol. Pol. 50 (4), 743–755.

Avanzini, M., García-Ramos, J.C., Lires, J., Piñuela, L., Lockley, M.G., 2007. Crocodylomorph tracks from the late Jurassic of Asturias (Spain). Ichnos 14 (1–2), 143–153.

Avanzini, M., Pinuela, L., García-Ramos, J.C., 2010. First report of a lizard-like footprint from the late Jurassic (Asturias, Spain). J. Iber. Geol. 36, 175–180.

Azuma, Y., Arakawa, Y., Tomida, Y., Currie, P.J., 2002. Early Cretaceous bird tracks from the Tetori group, Fukui Prefecture, Japan. Mem. Fukui Prefect. Dinosaur Mus. 1, 1–6.

Azuma, Y., Lv, J., Jin, X., Noda, Y., Shibata, M., Chen, R., Zheng, W., 2013. A bird footprint assemblage of early late cretaceous age Dongyang city Zhejiang province, China. Cretac. Res. 40, 3–9.

Azuma, Y., Takeyama, K., 1991. Dinosaur footprints from the Tetori group, central Japan. Research of dinosaurs from the Tetori group (4). Bullet. Fukui Pref. Mus. 4, 33–51.

Ballerstedt, M., 1905. Über Saurierfährten der Wealdenformation Bückeburgs. Naturwissenschaftliche Wochenschrift, pp. 481–485.

Ballerstedt, M., 1921. Dinosaurierfährten im Wealdensandstein des Harrl bei Bückeburg und eine zurzeit freiliegende Spur eines, vierfüßigen. Z. Dtsch. Geol. Ges. 72, 231–233.

Ballerstedt, M., 1922. Über Schreckensaurier und ihre Fußspuren. Kosmos 19 (3), 77–80.

Bandeira, K.L., Navarro, B.A., Pêgas, R.V., Brilhante, N.S., Brum, A.S., de Souza, L.G., da Silva, R.C., Gallo, V., 2024. A reassessment of the historical fossil findings from Bahia State (Northeast Brazil) reveals a diversified dinosaur fauna in the Lower Cretaceous of South America. Hist. Biol. 1–42.

Barcelos, L.A., dos Santos, R.O., 2023. The Lissamphibian fossil record of South America. Palaeobiodivers. Palaeoenviron. 103 (2), 341–405.

Barrett, P.M., Rayfield, E.J., 2006. Ecological and evolutionary implications of dinosaur feeding behaviour. Trends Ecol. Evol. 21 (4), 217–224.

Barrett, P.M., Evans, D.C., Campione, N.E., 2015. Evolution of dinosaur epidermal structures. Biol. Lett. 11 (6), 20150229.

Barron, E.J., Washington, W.M., 1985. Warm Cretaceous Climates: High Atmospheric CO2 as A Plausible Mechanism. AGU, Washington, DC.

Barsbold, R., Osmólska, H., 1990. Ornithomimosauria. In: Weishampel, D.B., Dodson, P., Osmólska, H. (Eds.), The Dinosauria. University of California Press, Berkeley, pp. 169–209.

Bassoullet, J.P., 1971. Découverte d'empreintes de pas de Reptiles dans l'Infralias de la région d'Ain-Sefra (Atlas saharien-Algérie). Compte rendu sommaire des séances de la Société géologique de Frances 7, 358–359.

Beckles, S.H., 1852. On the Ornithoidichnites of the Wealden. Quart. J. Geol. Soc. 8 (1–2), 396–397.

Bell, P.R., Snively, E., 2008. Polar dinosaurs on parade: a review of dinosaur migration. Alcheringa 32, 271–284.

Bellair, P., de Lapparent, A.F., 1948. Le Crétacé et les empreintes de pas de Dinosauriensd'Amoura (Algérie). Bull. Soc. Hist. Nat. Afr. 32, 168–175.

Bellair, P., de Lapparent, A.F., 1948. Le Crétacé et les empreintes de pas de Dinosauriens d'Amoura (Algérie). Bull. Soc. Hist. Nat. Afr. Nord 39, 168–175.

Belvedere, M., Jalil, N.E., Breda, A., Gattolin, G., Bourget, H., Khaldoune, F., Dyke, G.J., 2013. Vertebrate footprints from the Kem Kem beds (Morocco): a novel ichnological approach to faunal reconstruction. Palaeogeogr. Palaeoclimatol. Palaeoecol. 383, 52–58.

Bennett, S.C., 1997. Terrestrial locomotion of pterosaurs: a reconstruction based on Pteraichnus trackways. J. Vertebr. Paleontol. 17 (1), 104–113.

Bensalah, M., Adaci, M., Hebib, H., Bessedik, M., Belkebir, L., Mahboubi, M., Mansouri, M.E.H., Mammeri, C., Mansour, B., 2005. Presence d'empreintes de pas de dinosauriens dans le Cretace au nord d'el Bayadh (Djebel Amour, Algerie). Sciences Technologie. B, Sciences de l'ingénieur 107–109.

Benton, M.J., Wilf, P., Sauquet, H.S., 2022. The Angiosperm Terrestrial Revolution and origins of modern biodiversity. New Phytol. 233, 2017–2035.

Berner, R.A., 1990. Atmospheric carbon dioxide levels over Phanerozoic time. Science 249, 1382–1386.

Bernier, P., Barale, G., Bourseau, J.P., Buffetaut, É., Demathieu, G., Gaillard, C., Gall, J.C., 1982. Trace nouvelle de locomotion de Chélonienet figures d'émersion associées dans les calcaires lithographiques de Cerin (Kimméridgien supérieur, Ain, France). Geobios 15 (4), 447–467.

Bessedik, M., Mammeri, C., Belkebir, L., Mahboubi, M., Adaci, M., Hebib, H., Bensalah, M., Mansour, B., Mansouri, M.E.H., 2008. Nouvelles données sur les ichnites de dinosaures de la région d'El Bayadh (Crétacé inférieur, Algérie). Palaeovertebrata 36 (1–4), 7–35.

Bessedik, M., Mahboubi, M., Mammeri, C., Belkebir, L., 2020a. Les restes de dinosaures et leurs traces en Algérie. Mémoire du Service Géologique de l'Algérie 21, 19–28.

Bessedik, M., Mammeri, C., Belkebir, L., Mahboubi, M., 2020b. Les empreintes de pieds tri et tétradactyles de dinosaures (Rhétien-Hettangien à Cénomanien) de l'Atlas saharien (Algérie). Mémoire du Service Géologique de l'Algérie 21, 29–50.

Billon-Bruyat, J.P., Mazin, J.M., 2003. The systematic problem of tetrapod ichnotaxa: the case study of *Pteraichnus* Stokes, 1957 (Pterosauria, Pterodactyloidea). Geol. Soc. 217 (1), 315–324.

Bird, R.T., 1939. Thunder in his footsteps. Nat. Hist. 43, 254–261.

Bird, R.T., 1941. A dinosaur walks into the museum. Nat. Hist. 47, 74–81.

Bird, R.T., 1944. Did Brontosaurus ever walk on land? Nat. Hist. 53, 61–67.

Bird, R.T., 1954. We captured a live brontosaur. Natl. Geogr. Mag. 105, 707–722.

Bird, R.T., 1985. Bones for Barnum Brown. Adventures of a Dinosaur Hunter. Texas Christian University Press, Forth Worth, Texas, p. 225.

Bonaparte, J.F., 1980. Jurassic tetrapods from South America and Dispersa- routes. In: Jacobs, L.L. (Ed.), Aspects of Vertebrate History, Essays in Honor of Edwin Harris Colbert. Museum of Northern Arizona Press, pp. 73–98.

Bonnan, M.F., Sandrik, J.L., Nishiwaki, T., Wilhite, D.R., Elsey, R.M., Vittore, C., 2010. Calcified cartilage shape in archosaur long bones reflects overlying joint shape in stress-bearing elements: implications for nonavian dinosaur locomotion. Anat. Rec. Adv. Integr. Anat. Evol. Biol. 293 (12), 2044–2055.

Bonnan, M.F., Wilhite, D.R., Masters, S.L., Yates, A.M., Gardner, C.K., Aguiar, A., 2013. What lies beneath: sub-articular long bone shape scaling in eutherian mammals and saurischian dinosaurs suggests different locomotor adaptations for gigantism. PLoS One 8 (10), e75216.

Brancas, R., Martínez, J., Blaschke, J., 1979. Huellas de dinosaurio en Enciso. In: de Berceo, G. (Ed.), Diputación Logroño, vol. 2, p. 96.

Brett-Surman, M.K., 1997. Ornithopods. The Complete Dinosaur, 1st. Indiana University Press, pp. 330–346.

Brinkman, D., 1981. The hind limb step cycle of Iguana and primitive reptiles. J. Zool. 193 (1), 91–103.

Brouwers, E.M., Clemens, W.A., Spicer, R.A., Ager, T.A., Carter, L.D., Sliter, W.V., 1987. Dinosaurs on the North Slope, Alaska: high latitude, latest Cretaceous environments. Science 237 (4822), 1608–1610.

Brown, C.M., Henderson, D.M., 2015. A new horned dinosaur reveals convergent evolution in cranial ornamentation in Ceratopsidae. Curr. Biol. 25 (12), 1641–1648.

Buck, P.V., Ghilardi, A.M., Peixoto, B. de C.P., Peixoto, M., Fernandes, L.B. dos R., Fernandes, M.A., 2017a. A new tetrapod ichnotaxon from Botucatu Formation, Lower Cretaceous (Neocomian), Brazil, with comments on fossil track preservation on inclined planes and local paleoecology. Palaeogeogr. Palaeoclimatol. Palaeoecol. 466, 21–37.

Buck, P.V., Ghilardi, A.M., dos Reis Fernandes, L.B., Fernandes, M.A., 2017b. A new ichnotaxon classification of large mammaliform trackways from the Lower Cretaceous Botucatu Formation, Paraná Basin, Brazil. Palaeogeogr. Palaeoclimatol. Palaeoecol. 485, 377–388.

Buckley, L.G., McCrea, R.T., Lockley, M.G., 2016. Analyzing and resolving Cretaceous avian ichnotaxonomy using multivariate statistical analyses: approaches and results. In: Dinosaur Tracks: The Next Steps, pp. 258–308.

Buckley, L.G., McCrea, R.T., Xing, L.D., 2018. First report of Ignotornidae (aves) from the lower cretaceous Gates Formation (Albian) of western Canada, with description of a new ichnospecies of *Ignotornis, Ignotornis canadensis* ichnosp. nov. Cretac. Res. 84, 209–222.

Buffetaut, E., Ingavat, R., 1985. The mesozoic vertebrates of Thailand. Sci. Am. 253 (2), 80–87.

Butler, R.J., Upchurch, P., Norman, D.B., 2008. The phylogeny of the ornithischian dinosaurs. J. Syst. Palaeontol. 6 (1), 1–40.

Calvo, J.O., 1991. Huellas de dinosaurios en la Formación Río Limay (Albiano-Cenomaniano?), Picún Leufú, provincia de Neuquén, República Argentina. Ameghiniana 28, 241–258.

Calvo, J.O., Salgado, L., 1995. *Rebbachisaurus tessonei* sp. nov. a new Sauropoda from the Albian-Cenomanian of Argentina; new evidence on the origin of the Diplodocidae. Gaia 11, 13–33.

Calvo, J.O., 1999. Dinosaurs and other vertebrates of the lake Ezequiel ramos Mexía area, Neuquén-Patagonia, Argentina. In: Proceedings of the Second Gondwanan Dinosaur Symposium, vol. 15. National Science Museum Monographs, pp. 13–45.

Calvo, J.O., Lockley, M.G., 2001. The first pterosaur tracks from Gondwana. Cretac. Res. 22 (5), 585–590.

Calvo, J.O., Mazzetta, G.V., 2004. Nuevos hallazgos de huellas de Dinosaurios en la Formación Candeleros (Albiano–Cenomaniano), Picún Leufú, Neuquén, Argentina. Ameghiniana 41 (4), 545–554.

Calvo, J.O., Rubilar-Rogers, D., Moreno, K., 2004. A new abelisauridae (Dinosauria: Theropoda) from northwest Patagonia. Ameghiniana 41 (4), 555–563.

Calvo, J.O., Rivera, C., 2018. Huellas de dinosaurios en la costa oeste del embalse Ezequiel Ramos Mexía y alrededores (Cretácico Superior, Provincia de Neuquén, República Argentina). Bol. Soc. Geol. Mex. 70 (2), 449–497.

Campbell, K., 1983. Trackways, clues to passing dinosaurs. Terra Mus. Nat. Hist. Los Angeles County 6, 12–13.

Campos-Soto, S., Cobos, A., Caus, E., Benito, M.I., Fernández-Labrador, L., Suarez-Gonzalez, P., Quijada, I.E., Mas, R., Royo-Torres, R., Alcalá, L., 2017. Jurassic Coastal Park: a great diversity of palaeoenvironments for the dinosaurs of the Villar del Arzobispo Formation (Teruel, eastern Spain). Palaeogeogr. Palaeoclimatol. Palaeoecol. 485, 154–177.

Cantrill, D.J., Poole, I., 2012. The Vegetation of Antarctica through Geological Time. Cambridge University Press, Cambridge, 480 pp.

Canudo, J.I., Oms, O., Vila, B., Galobart, À., Fondevilla, V., Puértolas-Pascual, E., Sel-lés, A.G., Cruzado-Caballero, P., Dinarès-Turell, J., Vicens, E., Castanera, D., Company, J., Burrel, L., Estrada, R., Marmi, J., Blanco, A., 2016. The upper maastrichtian dinosaur fossil record from the southern pyrenees and its contribu-tion to the topic of the cretaceous-palaeogene mass extinction event. Cretac. Res. 57, 540–551.

Cao, K., 2013. Cretaceous terrestrial stratigraphic correlation in China. Geol. Rev. 59 (1), 24–40.

Carpenter, K., 1984. Skeletal reconstruction and life restoration of Sauropelta (Ankylosauria: Nodosauridae) from the cretaceous of North America. Can. J. Earth Sci. 21 (12), 1491–1498.

Carpenter, K., Kirkland, J.I., 1998. Review of lower and middle cretaceous ankylosaurs from North America. New Mex. Mus. Nat. Hist. Sci. Bull. 14, 249–270.

Carrano, M.T., Wilson, J.A., 2001. Taxon distributions and the tetrapod track record. Paleobiology 27, 564–582.

Carrano, M.T., Sampson, S.D., 2008. The phylogeny of ceratosauria (Dinosauria: Theropoda). J. Syst. Palaeontol. 6 (2), 183–236.

Carrano, M.T., 2012. Dinosaurian faunas of the later mesozoic. In: The Complete Dinosaur. Indiana University Press, Bloomington, pp. 1003–1026.

Carvalho, I.S., 1995. As pegadas de Dinosauros da Bacia de uirauna-Brejo das Freiras (Cretaceo Inferior, Estado da Paraiba): Boletim do 4° Simposio sobre o Cretaco do Brasil, pp. 115–121.

Casamiquela, R.M., 1961. Sobre la presencia de un mamífero en el primer elenco (icnológico) de vertebrados del Jurásico de la Patagonia. Physis 22, 225–233.

Casamiquela, R.M., 1964a. Estudios Icnologicos. Colegio Industrial Pio IX, p. 229. Buenos Aires.

Casamiquela, R.M., 1964b. Ichnological Studies: Problems and Methods of Ichnology with Application to the Study of Mesozoic Footprints (Reptilia, Mammalia) of Patagonia. Ministerio de Asuntos Sociales de la Province del Rio Negro, Burenos Aires, 229 pp.

Casamiquela, R.M., Fasola, A., 1968. Sobre pisadas de dinosaurios del Cretácico Inferior de Colchagua (Chile), vol. 30. Universidad de Chile, Departamento de Geología, pp. 1–24.

Casamiquela, R.M., 1975. Nuevo material y reinterpretación de las icnitas mesozoicas (Neotriásicas) de Los Menucos, Provincia de Río Negro (Patagonia). Actas del 1 Congreso Argentino de Paleontología y Bioestratigrafía 1, 555–580.

Casanovas, M.L., Ezquerra Miguel, R., Fernández Ortega, A., Pérez-Lorente, F., Santafé Llopis, J.V., Torcida Fernández, F., 1993a. Icnitas digitígradas y plantígradas de dinosaurios en el afloramiento de El Villar-Poyales (La Rioja, España) [Digitigrade and plantigrade dinosaur tracks in the El Villar-Poyales outrop (La Rioja, Spain)]. Zubia 5, 135–163.

Casanovas, M.L., Ezquerra, R., Fernández, A., Pérez-Lorente, F., Santafé, J.V., Torcida, F., 1993b. Tracks of a herd of webbed ornithopods and other footprints found in the same site (Igea, La Rioja, Spain). Rev. Paléobiol. 7, 29–36.

Castanera, D., Pascual, C., Canudo, J.I., Hernandez, N., Barco, J.L., 2012. Ethological variations in gauge in sauropod trackways from the Berriasian of Spain. Lethaia 45 (4), 476–489.

Castanera, D., Pascual, C., Razzolini, N.L., Vila, B., Barco, J.L., Canudo, J.I., 2013a. Discriminating between medium-sized tridactyl trackmakers: tracking ornithopod tracks in the base of the Cretaceous (Berriasian, Spain). PLoS One 8 (11), e81830.

Castanera, D., Vila, B., Razzolini, N.L., Falkingham, P.L., Canudo, J.I., Manning, P.L., Galobart, A., 2013b. Manus track preservation bias as a key factor for assessing trackmaker identity and quadrupedalism in basal ornithopods. PLoS One 8 (1), e54177.

Castanera, D., Vila, B., Razzolini, N.L., Santos, V.F., Pascual, C., Canudo, J.I., 2014. Sauropod trackways of the iberian peninsula: Palaeoetological and palaeoenvironmental implications. J. Iber. Geol. 40 (1), 49–59.

Castanera, D., Colmenar, J., Sauque, V., Canudo, J.I., 2015. Geometric morphometric analysis applied to theropod tracks from the Lower Cretaceous (Berriasian) of Spain. Palaeontology 58 (1), 183–200.

Castanera, D., Canudo, J., Barco, J., Rauhut, O., 2016. Avian or dinosaur tracks? The serrantes tracksite (Huérteles Fm., Berriasian, Soria, Spain) revisited. In: *VII Jornadas Internaciones sobre Paleontología de Dinosaurios y su Entorno.* Salas de los Infantes Burgos, pp. 47–48.

Castanera, D., Belvedere, M., Marty, D., Paratte, G., Lapaire-Cattin, M., Lovis, C., Meyer, C.A., 2018. A walk in the maze: variation in Late Jurassic tridactyl dinosaur tracks-A case study from the Late Jurassic of the Swiss Jura Mountains (NW Switzerland). PeerJ Preprints, e3506v1.

Castanera, D., Silva, B.C., Santos, V.F., Malafaia, E., Belvedere, M., 2020. Tracking late Jurassic ornithopods in the Lusitanian Basin of Portugal: ichnotaxonomic implications. Acta Palaeontol. Pol. 65 (2), 399–412.

Castanera, D., Pascual-Arribas, C., Canudo, J.I., Puértolas-Pascual, E., 2022a. A new look at Crocodylopodus meijidei: implications for crocodylomorph locomotion. J. Vertebr. Paleontol. 41 (5), e2020803.

Castanera, D., Bádenas, B., Aurell, M., Canudo, J.I., Gasca, J.M., 2022b. New ornithopod tracks from the Lower Cretaceous El Castellar Formation (Spain): implications for track preservation and evolution of ornithopod footprints. Palaeogeogr. Palaeoclimatol. Palaeoecol. 591, 110866.

Cavin, L., Tong, H., Boudad, L., Meister, C., Piuz, A., Tabouelle, J., Aarab, M., Amiot, R., Buffetaut, E., Dyke, G., Hua, S., 2010. Vertebrate assemblages from the early Late Cretaceous of southeastern Morocco: an overview. J. Afr. Earth Sci. 57 (5), 391–412.

Chabou, M.C., Laghouag, M.Y., Bendaoud, A., 2015. Dinosaur Track Sites in Algeria: A Significant National Geological Heritage in Danger. (Les sites des empreintes de pas de dinosaures en Algérie: un patrimoine national en danger). In: From Geoheritage to Geoparks. Case Studies from Africa and beyond, pp. 157–166.

Chatterjee, S., Templin, R.J., 2004. The Geological Society of America. Special Paper 376, Boulder, Colorado, 64 + iv p.

Chen, P.J., Li, J., Matsukawa, M., Zhang, H., Wang, Q., Lockley, M.G., 2006. Geological ages of dinosaurtrackbearing formations in China. Cretac. Res. 27 (1), 22–32.

Chen, R.J., Lv, J.C., Zhu, Y.X., Azuma, Y., 2013. Pterosaur tracks from the early late cretaceous of Dongyang city, Zhejiang province, China. Geol. Bull. China 32 (5), 693–698.

Chen, S., Huang, X., 1993. Preliminary study of dinosaur tracks in changling, Chuxiong Prefecture. J. Yunn. Geol. 12 (3), 266–276.

Chin, K., 2007. The paleobiological implications of herbivorous dinosaur coprolites from the Upper Cretaceous Two Medicine Formation of Montana: why eat wood? Palaios 22, 554–566.

Chinnery, B.J., Lipka, T.R., Kirkland, J.I., Parrish, J.M., Brett-Surman, M.K., 1998. Neoceratopsian teeth from the Lower to Middle Cretaceous of North America. New Mex. Mus. Nat. Hist. Sci. Bull. 14, 297–302.

Chun, S.S., 1990. Sedimentary processes, depositional environments and tectonic settings of the Cretaceous Uhangri Formation. In: Ph.D. Dissertation, Department of Oceanography. Seoul National University, Korea.

Cifelli, R.L., Kirkland, J.I., Weil, A., Deino, A.L., Kowallis, B.J., 1997. High-precision 40Ar/39Ar geochronology and the advent of North America's Late Cretaceous terrestrial fauna. Proc. Natl. Acad. Sci. USA 94 (21), 11163–11167.

Cladellas, M.L.C., Miguel, R.E., Ortega, A.F., Lorente, F.P., Llopis, J.V.S., Torcida Fernández-Baldor, F.T., 1993. Icnitas digígradas y plantígradas de dinosaurios en el afloramiento de El Villar-Poyales (La Rioja, España). Zubia (5), 135–163.

Clark, J.M., Hopson, J.A., Hernandez, R., Fastovsky, D.E., Montellano, M., 1998. Foot posture in a primitive pterosaur. Nature 391 (6670), 886–889.

Clemente, C.J., 2014. The evolution of bipedal running in lizards suggests a consequential origin may be exploited in later lineages. Evolution 68, 2171–2183.

Cobos, A., Royo-Torres, R., Luque, L., Alcalá, L., Mampel, L., 2010. An Iberian stegosaurs paradise: The Villar del Arzobispo Formation (Tithonian–Berriasian) in Teruel (Spain). Palaeogeogr. Palaeoclimatol. Palaeoecol. 293 (1–2), 223–236.

Colbert, E.H., Mook, C.C., 1951. The ancestral crocodilian *Protosuchus*. Am. Mus. Nat. Hist. Bull. 97 (3), 149–182.

Colbert, E.H., Merrilees, D., 1967. Cretaceous dinosaur footprints from Western Australia. J. Roy. Soc. West Aust. 50 (1), 21–25.

Cónsole-Gonella, C., de Valais, S., Marquillas, R.A., Sánchez, M.C., 2017. The Maastrichtian–Danian Maimará tracksite (Yacoraite Formation, Salta Group), Quebrada de Humahuaca, Argentina: environments and ichnofacies implications. Palaeogeogr. Palaeoclimatol. Palaeoecol. 468, 327–350.

Contessi, M., Fanti, F., 2012. First record of bird tracks in the late Cretaceous (Cenomanian) of Tunisia. Palaios 27, 455–464.

Contessi, M., 2013. A new dinosaur ichnofauna from Tunisia: implications for the palaeobiogeography of peri-Adriatic carbonate platforms in the mid-Cretaceous. Palaeogeogr. Palaeoclimatol. Palaeoecol. 392, 302–311.

Conti, M.A., Morsilli, M., Nicosia, U., Sacchi, E., Savino, V., Wagensommer, A., Maggio, L.D., Gianolla, P., 2005. Jurassic dinosaur tracks from southern Italy: tracks as indicators of constraints in paleogeographic interpretation. Palaios 20 (6), 534–550.

Coombs, W.P., 1996. Redescription of the ichnospecies Antipus flexiloquus Hitchcock, from the early Jurassic of the Connecticut valley. J. Paleontol. 70 (2), 327–331.

Coria, R.A., Currie, P.J., Eberth, D., Garrido, A., 2002. Bird footprints from the Anacleto Formation (late cretaceous) in Neuquén province, Argentina. Ameghiniana 39 (4), 453–463.

Cortés, D., Parra-Ruge, M.L., Parra-Ruge, J.D.D., Deers-Potvin, A., Smith, A., Hoai-Nam, B., Larsson, H., 2023. Theropod dinosaur footprints from the Early Cretaceous of Colombia: walking dynamics and paleoenvironmental implications. Revista Facies 9, 1–17.

Crowell, J.K., Shimer, G.T., 2021. Late cretaceous dinosaur tracks from the iron springs formation, iron county, Utah. Compass Earth Sci. J. Sigma Gamma Epsilon 91 (1), 1.

Currie, P.J., Sarjeant, W.A., 1979. Lower cretaceous dinosaur footprints from the Peace River Canyon, British Columbia, Canada. Palaeogeogr. Palaeoclimatol. Palaeoecol. 28, 103–115.

Currie, P.J., 1981. Bird footprints from the *Gething Formation* (Aptian, lower cretaceous) of northeastern British Columbia, Canada. J. Vertebr. Paleontol. 1 (3–4), 257–264.

Currie, P.J., 1983. Hadrosaur trackways from the lower cretaceous of Canada. Acta Palaeontol. Pol. 28, 63–73.

Currie, P.J., 1989. Dinosaur tracksites of western Canada. In: Gillette, D.D., Lockley, M.G. (Eds.), Dinosaur Tracks and Traces. Cambridge University Press, New York, pp. 293–300.

Currie, P.J., Nadon, G.C., Lockley, M.G., 1991. Dinosaur footprints with skin impressions from the Cretaceous of Alberta and Colorado. Can. J. Earth Sci. 28 (1), 102–115.

Currie, P.J., 1995. Ornithopod trackways from the lower cretaceous of Canada. In: Vertebrate Fossils and the Evolution of Scientific Concepts, p. 431.

Currie, P.J., Badamgarav, D., Koppelhus, E.B., 2003. The first late cretaceous footprints from the Nemegt locality in the Gobi of Mongolia. Ichnos 10 (1), 1–13.

Currie, P.J., Koppelhus, E.B., 2015. The significance of the theropod collections of the Royal Tyrrell Museum of Palaeontology to our understanding of Late Cretaceous theropod diversity. Can. J. Earth Sci. 52 (8), 620–629.

D'Emic, M.D., 2013. Revision of the sauropod dinosaurs of the Lower Cretaceous Trinity Group, southern USA, with the description of a new genus. J. Syst. Palaeontol. 11 (6), 707–726.

D'Orazi Porchetti, S., Bertini, R.J., Langer, M.C., 2017. Proposal for ichnotaxonomic allocation of therapsid footprints from the Botucatu Formation, Brazil. Ichnos 1–15.

Dalla Vecchia, F.M., 1994. Jurassic and Cretaceous Sauropod Evidence in the Mesozoic Carbonate Platforms of the Southern Alps and Dinarids, pp. 65–73.

Dalla Vecchia, F.M., Tarlao, A., 2000. New dinosaur tracks sites in the Albian (early cretaceous) of the Istrian peninsula (Croatia). Parte II—Paleontology. Mem. Sci. Geol. 52, 227–292.

Dalla Vecchia, F.M., 2005. Between Gondwana and Laurasia: cretaceous sauropods in an intraoceanic carbonate platform. In: Tidwell, V., Carpenter, K. (Eds.), Thunder-lizards. The Sauropodomorph Dinosaurs. Indiana University Press, Bloomington, pp. 395–429.

Dalla Vecchia, F.M., 2008. The impact of dinosaur palaeoichnology in palaeoenvironmental and palaeogeographic reconstructions: the case of the Periadriatic carbonate platforms. Oryctos 8, 89–106.

Dalla Vecchia, F.M., Vlahović, I., Posocco, L., Tarlao, A., Tentor, M., 2002. Late Barremian and late Albian (early cretaceous) dinosaur track sites in the main Brioni/Brijun island (SW Istria, Croatia). Nat. Nascosta 25, 1–36.

Day, J.J., Upchurch, P., Norman, D.B., Gale, A.S., Powell, H.P., 2002. Sauropod trackways, evolution, behavior. Science 296, 1659.

de Lapparent, A.F., 1960. Decouverte de traces de pas de dinosauriens dans le Crétacé de Spitsberg. C. R. Acad. Sci. 251, 1399–1400.

de Lapparent, A.F., 1962. Footprints of Dinosaurs in the Lower Cretaceous of Vestspitsbergen-Svalbard. Norsk Polarinstitutt, Årbok, pp. 14–21.

de Lapparent, A.F., Montenat, C., 1967. Les empreintes de pas de reptiles de l'Infralias de Veillon (Vendée) [Reptile footprints from the lower Lias of Veillon (Vendée)]. Mémoires de la Société géologique de France, nouvelle série 107, 1–44.

de Valais, S., Melchor, R.N., 2008. Ichnotaxonomy of bird-like footprints: an example from the Late Triassic-Early Jurassic of northwest Argentina. J. Vertebr. Paleontol. 28 (1), 145–159.

de Valais, S., 2009. Ichnotaxonomic revision of Ameghinichnus, a mammalian ichnogenus from the middle Jurassic La Matilde formation, Santa Cruz province, Argentina. Zootaxa 2203 (1), 1–21.

de Valais, S., Cónsole-Gonella, C., 2019. An updated review of the avian footprint record from the Yacoraite Formation (Maastrichtian-Danian), Northwestern Argentina. Ichnos 26 (3), 224–241.

de Valais, S., Díaz Martínez, I., Citton, P., Cónsole Gonella, C.A., 2020. Vertebrate tracks of the Río Negro province, Patagonia, Argentina: stratigraphy, palaeobiology and environmental contexts. Rev. Asoc. Geol. Argent. 77, 1–63.

Dejax, J., Michard, J.G., Brunet, M., Hell, J., 1989. Empreintes de pas de dinosauriens datées du Crétacé inférieur dans le bassin de Babouri-Figuil (Fossé de la Bénoué, Cameroun). Neues Jahrbuch für Geologie und Paläontologie. Abhandlungen 178 (1), 85–108.

Delair, J.B., 1963. Notes on Purbeck fossil footprints, with descriptions of two hitherto unknown forms from Dorset. In: Proceedings of the Dorset Natural History and Archaeological Society, vol. 84, pp. 92–100.

Demathieu, G.R.S., Sciau, J., 1992. Des pistes de dinosaures et de crocodiliens dans les dolomies de l'Hettangien du Causse du Larzac. ComptesRendus de l'Académie des Sciences à Paris, Série II 315, 1561–1566.

Demathieu, G.R.S., 1993. Empreintes de pas de dinosaures dans les Causses (France). Zubia 5, 229–252.

Díaz-Martínez, I., Pereda-Suberbiola, X., Perez-Lorente, F., Canudo, J.I., 2015a. Ichnotaxonomic review of large ornithopod dinosaur tracks: temporal and geographic implications. PLoS One 10 (2), e0115477.

Díaz-Martínez, I., Suarez-Hernando, O., Martinez-Garcia, B., Hernandez, J.M., Fernandez, S.G., Perez-Lorente, F., Murelaga, X., 2015b. Early Miocene shorebird-like footprints from the Ebro Basin, La Rioja, Spain: paleoecological and paleoenvironmental significance. Palaios 30 (5), 424–431.

Díaz-Martínez, I., de Valais, S., Cónsole-Gonella, C., 2016. First evidence of Hadrosauropodus in Gondwana (Yacoraite Formation, maastrichtian-Danian), northwestern Argentina. J. Afr. Earth Sci. 122, 79–87.

Díaz-Martínez, I., de Valais, S., Cónsole-Gonella, C., 2018. New sauropod tracks from the Yacoraite Formation (Maastrichtian–Danian), Valle del Tonco tracksite, Salta, northwestern Argentina. J. Iber. Geol. 44, 113–127.

Díaz-Martínez, I., Heredia, A.M., González, S.N., Canale, N., de Valais, S., Cónsole-Gonella, C.A., Montes, R.M., Caratelli, M., Urzagasti-Torres, S., Fischer, G., Lecuona, A., 2022. Pterosaur tracks from the upper cretaceous Anacleto Formation (Neuquén Basin), northern Patagonia, Argentina: insights into campanian pterosaur diversity in Gondwana. Diversity 14 (11), 1007.

Díaz-Martínez, I., Citton, P., Castanera, D., 2023. What do their footprints tell us? Many questions and some answers about the life of non-avian dinosaurs. J. Iber. Geol. 1–22.

Diedrich, C., 2004. New important iguanodontid and theropod trackways of the tracksite obernkirchen in the berriasian of NW Germany and megatracksite concept of central Europe. Ichnos 11, 215–228.

Dieudonné, P.E., Cruzado-Caballero, P., Godefroit, P., Tortosa, T., 2021. A new phylogeny of cerapodan dinosaurs. Hist. Biol. 33 (10), 2335–2355.

Dollo, L., 1883. Première note sur les crocodiliens de Bernissart. Bull. Musee R. Hist. Nat. Belg. 2, 309–338.

Dong, Z.M., Zhou, Z.L., Wu, S.Y., 2003. Note on hadrosaur footprint from Heilongjiang River area of China. Vertebr. Palasiat. 41, 324–326.

Dzhalilov, M.R., Novikov, V.P., 1993. Fossil traces of the dinosaurs in the territory of Tadjikistan. In: Vialov, O.S., Fedonkin, M.A. (Eds.), Trace Fossils of Extinct Organisms. Nauka, Moscow, pp. 47–65.

Elbroch, M., Marks, E.M., 2001. Bird Tracks Sign: A Guide to North American Species. Stackpole Books, p. 456.

Ellenberger, P., 1972. Contribution a la classification des pistes de vertebras du Trias: les types du Stormberg d'Afrique du Sud (I). Palaeovertebr. Mem. Extraordinaire 1–152.

Enriquez, N.J., Campione, N.E., Brougham, T., Fanti, F., White, M.A., Sissons, R.L., Sullivan, C., Vavrek, M.J., Bell, P.R., 2020. Exploring possible ontogenetic trajectories in tyrannosaurids using tracks from the Wapiti Formation (upper Campanian) of Alberta, Canada. J. Vertebr. Paleontol. 40 (6), e1878201.

Enriquez, N.J., Campione, N.E., White, M.A., Fanti, F., Sissons, R.L., Sullivan, C., Vavrek, M.J., Bell, P.R., 2022. The dinosaur tracks of tyrants aisle: an Upper Cretaceous ichnofauna from unit 4 of the Wapiti Formation (upper Campanian), Alberta, Canada. PLoS One 17 (2), e0262824.

Erickson, B.R., 1972. The lepidosaurian reptile *Champsosaurus* in North America. Monogr. Sci. Mus. Minnesota 1, 1.

Evans, S.E., Manabe, M., 2008. An early herbivorous lizard from the Lower Cretaceous of Japan. Palaeontology 51, 487–498.

Ezcurra, M.D., Agnolín, F.L., 2012. A new global palaeobiogeographical model for the late Mesozoic and early Tertiary. Syst. Biol. 61 (4), 553–566.

Ezquerra, R., Doublet, S., Costeur, L., Galton, P., Pérez-Lorente, F., 2007. Were non-avian theropod dinosaurs able to swim? Supportive evidence from an Early Cretaceous trackway, Cameros Basin (La Rioja, Spain). Geology 35 (6), 507.

Falk, A.A., Hasiotis, S.T., Martin, L.D., 2010. Feeding traces associated with bird tracks from the lower cretaceous Haman Formation, Republic of Korea. Palaios 25, 730–741.

Falkingham, P.L., Bates, K.T., Margetts, L., Manning, P.L., 2011a. The 'Goldilocks' effect: preservation bias in vertebrate track assemblages. J. R. Soc. Interface 8 (61), 1142–1154.

Falkingham, P.L., Bates, K.T., Margetts, L., Manning, P.L., 2011b. Simulating sauropod manus-only trackway formation using finite-element analysis. Biol. Lett. 7 (1), 142–145.

Fanti, F., Bell, P.R., Sissons, R.L., 2013. A diverse, high-latitude ichnofauna from the late cretaceous Wapiti Formation, Alberta, Canada. Cretac. Res. 41, 256–269.

Farlow, J.O., 1989. Brotopodus birdi Lower Cretaceous sauropod footprints from the US Gulf coastal plain. In: Dinosaur Tracks and Traces.

Farlow, J.O., Pitiman, J.G., Hawthorne, J.M., 1989. *Brontopodus birdi,* Lower Cretaceous sauropod footprints from the U.S. Gulf coastal plain. In: Gillette, D.D., Lockley, M.G. (Eds.), Dinosaur Tracks and Traces. Cambridge University Press, Cambridge, pp. 371–394.

Farlow, J.O., 1992. Sauropod tracks and trackmakers: integrating the ichnological and skeletal records. Zubia 10, 89–138.

Farlow, J.O., 2001. Acrocanthosaurus and the maker of Comanchean large-theropod footprints. In: Mesozoic Vertebrate Life, vol. 408. Indiana University Press, Bloomington.

Farlow, J.O., Elsey, R., 2010. Footprints and trackways of the American alligator, Rockefeller Wildlife refuge, Louisiana. Crocodyle Tracks Traces Bull. 51, 31.

Farlow, J.O., O'Brien, M., Kuban, G.J., Dattilo, B.F., Bates, K.T., Falkingham, P.L., Piñuela, L., 2012. Dinosaur tracksites of the Paluxy River valley (Glen Rose Formation, lower cretaceous), Dinosaur Valley State Park, Somervell county, Texas. In: *Proceedings of the V International Symposium about Dinosaur Palaeontology and their Environment*, Burgos: Salas de los Infantes, pp. 41–69.

Farlow, J.O., 2018. Noah's Ravens: Interpreting the Makers of Tridactyl Dinosaur Footprints. Indiana University Press, Bloomington, 651pp.

Farlow, J.O., Robinson, N.J., Kumagai, C.J., Paladino, F.V., Falkingham, P.L., Elsey, R.M., Martin, A.J., 2018. Trackways of the American crocodile (*Crocodylus acutus*) in northwestern Costa Rica: implications for crocodylian ichnology. Ichnos 25 (1), 30–65.

Farlow, J.O., Bakker, R.T., Dattilo, B.F., Everett Deschner, E., Falkingham, P.L., Harter, C., Solis, R., Temple, D., Ward, W., 2020. Thunder lizard handstands: manus-only sauropod trackways from the glen Rose Formation (lower cretaceous, Kendall county, Texas). Ichnos 27 (2), 167–199.

Fastovsky, D.E., Bercovici, A., 2016. The Hell Creek Formation and its contribution to the Cretaceous-Paleogene extinction: a short primer. Cretac. Res. 57, 368–390.

Fernandes, M.A., Carvalho, I.S., 2008. Revisão diagnóstica para a icnoespécie de tetrápode Mesozóico *Brasilichnium elusivum* (Leonardi, 1981) (Mammalia) da Formação Botucatu, Bacia do Paraná, Brasil. Ameghiniana 45 (1), 167–173.

Figueiredo, S.D., Dinis, P., Belo, J., Rosina, P., Bachtsevanidou Strantzali, I., 2017. A new record of a possible ornithopod footprint from the Lower Cretaceous of Cabo Espichel (Sesimbra, Portugal). Boll. Soc. Paleontol. Ital. 56 (2), 217–231.

Figueiredo, S.D., Cunha, P.P., Suberbiola, X.P., de Carvalho, C.N., Carvalho, I.S., Buffetaut, E., Tong, H., Sousa, M.F., Antunes, V., Anastácio, R., 2022a. The dinosaur tracksite from the lower Barremian of Areia do Mastro Formation (Cabo Espichel, Portugal): implications for dinosaur behavior. Cretac. Res. 137, 105219.

Figueiredo, S.D., Carvalho, I.S., Pereda-Suberbiola, X., Cunha, P.P., Antunes, V., Díaz-Martínez, I., 2022b. New ornithopod footprints from the Areia do mastro formation (lower cretaceous), Espichel Cape (Portugal, western Iberia) and their context in the iberian ichnological ornithopod record. Cretac. Res. 131, 105069.

Fiorillo, A.R., 2005. Turtle tracks in the Judith river formation (upper cretaceous) of South-central Montana. Palaeontol. Electron. 8 (1), 1.

Fiorillo, A.R., Decker, P.L., LePain, D.L., Wartes, M., McCarthy, P.J., 2010. A probable neoceratopsian manus track from the Nanushuk Formation (Albian, northern Alaska). J. Iber. Geol. 36 (2), 165–174.

Fiorillo, A.R., Hasiotis, S.T., Kobayashi, Y., Breithaupt, B.H., Mccarthy, P.J., 2011. Bird tracks from the Upper Cretaceous Cantwell Formation of Denali National Park, Alaska, USA: a new perspective on ancient northern polar vertebrate biodiversity. J. Syst. Palaeontol. 9 (1), 33–49.

Fiorillo, A.R., Adams, T.L., Kobayashi, Y., 2012. New sedimentological, palaeobotanical, and dinosaur ichnological data on the palaeoecology of an unnamed Late Cretaceous rock unit in Wrangell-St. Elias National Park and Preserve, Alaska, USA. Cretac. Res. 37, 291–299.

Fiorillo, A.R., Hasiotis, S.T., Kobayashi, Y., 2014. Herd structure in Late Cretaceous polar dinosaurs: a remarkable new dinosaur tracksite, Denali National Park, Alaska, USA. Geology 42 (8), 719–722.

Fiorillo, A.R., McCarthy, P.J., Kobayashi, Y., Tomsich, C.S., Tykoski, R.S., Lee, Y.N., Tanaka, T., Noto, C.R., 2018. An unusual association of hadrosaur and therizinosaur tracks within Late Cretaceous rocks of Denali National Park, Alaska. Sci. Rep. 8 (1), 11706.

Fiorillo, A.R., Kobayashi, Y., McCarthy, P.J., Tanaka, T., Tykoski, R.S., Lee, Y.N., Takasaki, R., Yoshida, J., 2019. Dinosaur ichnology and sedimentology of the Chignik formation (upper cretaceous), Aniakchak national monument, southwestern Alaska; further insights on habitat preferences of high-latitude hadrosaurs. PLoS One 14 (10), e0223471.

Flaig, P.P., Hasiotis, S.T., Fiorillo, A.R., 2018. A paleopolar dinosaur track site in the Cretaceous (Maastrichtian) Prince Creek Formation of Arctic Alaska: track characteristics and probable trackmakers. Ichnos 25 (2–3), 208–220.

Foster, J.R., Lockley, M.G., 1997. Probable crocodilian tracks and traces from the Morrison Formation (upper Jurassic) of eastern Utah. Ichnos 5 (2), 121–129.

Foster, J.R., Lockley, M.G., Brockett, J., 1999. Possible turtle tracks from the Morrison Formation of southern Utah. Vertebr. Paleontol. Utah 99, 185–191.

Foster, J.R., Hamblin, A.H., Lockley, M.G., 2000. The oldest evidence of a sauropod dinosaur in the western USA and other important vertebrate trackways from grand staircase-escalante national monument, Utah. Ichnos 7 (3), 169–181.

Foster, J.R., 2001. Salamander tracks (ambystomichnus?) from the cathedral bluffs tongue of the Wasatch Formation (Eocene), northeastern Green River Basin, Wyoming. J. Paleontol. 75 (4), 901–904.

Frakes, L.A., Francis, J.E., Syktus, J.I., 1992. Climate Modes of the Phanerozoic. Cambridge University Press, Cambridge, 274pp.

Francischini, H., Sales, M.A., Dentzien-Dias, P., Schultz, C.L., 2018. The presence of Ankylosaur tracks in the Guará Formation (Brazil) and remarks on the spatial and temporal distribution of Late Jurassic Dinosaurs. Ichnos 25 (2–3), 177–191.

Fuentes Vidarte, C., 1996. Primeras huellas de aves in en Weald de Soria (España). Nuevo icnogenero, *Archaeornithipus* y nueva icnoespecie *A. meijidei*. Estud. Geol. 52, 63–75.

Fuentes Vidarte, C., Meijide Calvo, M., 1998. Icnitas de dinosaurios terópodos en el Weald de Soria (Espana). Nuevo icnogénero *Kalohipus*. Estud. Geol. 54 (3–4), 147–152.

Fuentes Vidarte, C., Meijide Calvo, M., 1999. Primeras huellas de cocodrilo en el Weald de Cameros (Soria, España). Nueva familia: Crocodylopodidae, nuevo icnogénero: Crocodylopodus, nueva icnoespecie: C. meijidei. In: Colectivo Arqueológico-Paleontológico, de S. (Ed.), Actas de las I Jornadas Internacionales sobre Paleontología de Dinosaurios y su Entorno, pp. 329–338.

Fuentes Vidarte, C., 2001. A new species of Pteraichnus of the Spanish Lower Cretaceous. Pteraichnus cidacoi. Strata 11, 44–46.

Fuentes Vidarte, C., Meijede Calvo, M., Meijede Fuentes Calvo, F., Meijede Fuentes Calvo, M., 2003. Rastro de un tetrápodo de pequeño tamaño en el Weald de Cameros (Sierra de Oncala, Soria, España) nov. icnogen. nov. icnosp. Emydhipus cameroi. In: Pérez-Lorente, F. (Ed.), Dinosaurios y otros reptiles mesozoicos en España. Ciencias de la Tierra, vol. 26. Instituto de Estudios Riojanos, Logroño, pp. 119–128.

Fuentes Vidarte, C., Meijide Calvo, M., Meijide Fuentes, F., Meijide Fuentes, M., 2004b. *Pteraichnus longipodus* nov. icnosp. en la Sierra de Oncala (Soria, España). Studia Geologica Salmanticensia, pp. 103–114.

Fuentes Vidarte, C., Meijide Calvo, M., Meijide Fuentes, M., Meijide Fuentes, F., 2004a. Huellas de pterosaurios en la Sierra de Oncala (Soria, España). Nuevas icnoespecies: Pteraichnus vetustior, Pteraichnus parvus, Pteraichnus manueli. Celtiberia 54 (98), 471–490.

Fujita, M., Lee, Y.N., Azuma, Y., Li, D., 2012. Palaios 27 (8), 560–570.

Fujiwara, S., 2009. A reevaluation of the manus structure in Triceratops (Ceratopsia: Ceratopsidae). J. Vertebr. Paleontol. 29 (4), 1136–1147.

Gabunia, L.K., Kurbatov, V., 1988. Jurassic Dinosaur Tracks in the South of Central Asia. in: Fossils Traces of Vital Activity and Dynamics of the Environment in Ancient Biotopes, Trans. XXX Session Union Paleontol. Soc. and VII session Ukrainian Paleontological Society, p. 202.

Gabuniya, L.K., 1951. O sledakh dinozavrov iz nizhnemelovykh otlozheniy zapadnoy Gruziy. Dokl. Akad. Nauk SSSR 81 (5), 917–919.

Gaillard, C., Bernier, P., Barale, G., Bourseau, J.P., Buffetaut, E., Ezquerra, R., Gall, J.C., de Lapparent, F., Renous, S., Wenz, S., 2003. A giant Upper Jurassic turtle revealed by its trackways. Lethaia 36, 315–322.

Gallup, M.R., 1989. Functional morphology of the hindfoot of the Texas sauropod *Pleurocoelus* sp. indet. In: Farlow, J.O. (Ed.), *Paleobiology of the Dinosaurs.* Special Paper 238. Geological Society of America, pp. 71–74.

Galton, P.M., 2009. Notes on neocomian (lower cretaceous) ornithopod dinosaurs from England—Hypsilophodon, Valdosaurus, "Camptosaurus", Iguanodon—and referred specimens from Romania and elsewhere. Rev. Paléobiol. 28, 211–273.

Gand, G., Vianey–Liaud, M., Demathieu, G., Garric, J., 2000. Deux nouvelles traces de pas de Dinosaures du Trias-supérieur de la bordure cévenole (La Grand–Combe, Sud–Est de la France) [Two new dinosaur footprints from the Upper Triassic of the Cevennes border (La Grande–Combe, southeastern France)]. Geobios 33 (5), 599–624.

Gangloff, R.A., May, K.C., Storer, J.E., 2004. An early late cretaceous dinosaur tracksite in Central Yukon Territory, Canada. Ichnos 11 (3–4), 299–309.

Gao, K., Evans, S., Ji, Q., Norell, M.A., Ji, S., 2000. Exceptional fossil material of a semi-aquatic reptile from China: the resolution of an enigma. J. Vertebr. Paleontol. 20, 417–421.

Gao, Y.H., 2007. New dinosaur footprints from Lower Jurassic of Weiyuan, Sichuan. Vertebr. Palasiat. 45 (4), 342–345.

García-Cobeña, J., Cobos, A., Verdú, F.J., 2023. Ornithopod tracks and bones: paleoecology and an unusual evidence of quadrupedal locomotion in the Lower Cretaceous of eastern Iberia (Teruel, Spain). Cretac. Res. 144, 105473.

García-Ortiz, E., Pérez-Lorente, F., 2014. Palaeoecological inferences about dinosaur gregarious behaviour based on the study of tracksites from La Rioja area in the Cameros Basin (Lower Cretaceous, Spain). J. Iber. Geol. 40 (1), 113.

Gierliński, G., Potemska, A., 1987. Lower Jurassic dinosaur footprints from Gliniani las, northern slope of the holy Cross Mountains, Poland. Neues Jahrbuch Geol. Palaontol. Abhand. 175, 107–120.

Gierliński, G., 1994. Early Jurassic theropod tracks with the metatarsal impressions. Przeglad Geol. 42 (04), 280–284.

Gierliński, G., Mossbrucker, T., Sabath, K., 2005. Stegosaurian footprints from the Morrison Formation of Western USA and their implications for other finds. In: Le Loeuff, J., Galobart, A., Oms, O., Vila, B., Marmi, J. (Eds.), International Symposium on Dinosaurs and other Vertebrates Palaeoichnology. Fumanya/Sant Corneli, pp. 32–33.

Gierliński, G., Ploch, I., Gawor-Biedawa, E., Niedzwiedzki, G., 2008. The first evidence of dinosaur tracks in the Upper Cretaceous of Poland. Oryctos 8, 107–113.

Gierliński, G.D., Sabath, K., 2008. Stegosaurian footprints from the Morrison Formation of Utah and their implication for interpreting other ornitischian tracks. Oryctos 8, 29–46.

Gierliński, G.D., 2009. A preliminary report on new dinosaur tracks in the Triassic, Jurassic and Cretaceous of Polans. In: Actas de las IV Jornadas Internacionales sobre Paleontología de Dinosaurios y su Entorno: celebrado en Salas de los Infantes (Burgos, España) en septiembre de 2007. Colectivo Arqueológico y Paleontológico de Salas, pp. 75–90.

Gierliński, G.D., Niedzwiedzki, G., Nowacki, P., 2009. Small theropod and ornithopod footprints in the Late Jurassic of Poland. Acta Geol. Pol. 59 (2), 221–234.

Gierliński, G., Lockley, M.G., 2013a. A trackmaker for Saurexallopus: ichnological evidence for oviraptorosaurian tracks from the Upper Cretaceous of western North America. In: At the Top of the Grand Staircase: The Late Cretaceous of Southern Utah. Indiana University Press, Bloomington, pp. 526–529.

Gierliński, G., Lockley, M.G., 2013b. First report of probable therizinosaur (cf. *Macropodosaurus*) tracks from North America, with notes on the neglected vertebrate ichnofauna of the Ferron Sandstone. In: At the Top of the Grand Staircase. Indiana University Press, Bloomington, pp. 530–535.

Gierliński, G.D., 2015. New dinosaur footprints from the Upper Cretaceous of Poland in the light of paleogeographic context. Ichnos 22 (3–4), 220–226.

Gierliński, G.D., Lagnaoui, A., Klein, H., Saber, H., Oukassou, M., Charriere, A., 2017. Bird-like tracks from the Imilchil Formation (Middle Jurassic, Bajocian-Bathonian) of the Central High Atlas, Morocco, in comparison with similar mesozoic tridactylous ichnotaxa. Boll. Della Soc. Paleontol. Italiana 56, 207–215.

González Riga, B.J., Calvo, J.O., 2009. A new wide-gauge sauropod track site from the Late Cretaceous of Mendoza, Neuquén Basin, Argentina. Palaeontology 52 (3), 631–640.

González Riga, B.J., 2011. Speeds and stance of titanosaur sauropods: analysis of *Titanopodus* tracks from the Late Cretaceous of Mendoza, Argentina. An Acad. Bras Ciências 83 (1), 279–290.

González-Riga, B.J., Tomaselli, M.B., 2019. Different trackway patterns in titanosaur sauropods: analysis of new Titanopodus tracks from the Upper Cretaceous of Mendoza, Neuquén Basin, Argentina. Cretac. Res. 93, 49–59.

Ha, S., Kim, K.S., Lim, H.S., Lockley, M.G., Yoo, J.S., Lim, J.D., 2022. Diminutive pterosaur tracks and trackways (*Pteraichnus gracilis* ichnosp. nov.) from the lower cretaceous Jinju Formation, Gyeongsang basin, Korea. Cretac. Res. 131, 105080.

Haderer, F., Neef, C., 2004. Ein weiterer Nachweis der Theropodenfahrte *Bueckeburgichnus* aus dem Obernkirchener Sandstein von Nordwestdeutschland. Ber. Naturhist. Ges. Hann. 146, 1–6.

Halfpenny, J.C., 1998. Scats and Tracks of the Rocky Mountains. Falcon, Publishing Inc, Helena, Montana, p. 145.

Harington, C.R., Foster, H., Holmes, R., Currie, P.J., 2005. Photographic catalogue of trackways in the Canadian museum of nature. Syllogeus 74, 1–151.

Harris, J.D., Johnson, K.R., Hicks, J., Tauxe, L., 1996. Four-toed theropod footprints and a paleomagnetic age from the whetstone falls member of the Harebell Formation (upper cretaceous: maastrichtian), northwestern Wyoming. Cretac. Res. 17 (4), 381–401.

Harris, J.D., 1997. Four-toed theropod footprints and a paleomagnetic age from the whetstone falls member of the Harebell Formation (Upper Cretaceous: Maastrichtian), northwestern Wyoming: a correction. Cretac. Res. 18 (1), 139.

Harris, J.D., 1998. Large, early cretaceous theropods in North America. New Mex. Mus. Nat. Hist. Sci. Bull. 14, 225–228.

Harris, S., 2023. Evolution of the climate of the earth. J. Civil Eng. Res. Technol. 5, 1–15.

Haubold, H., 1971. Ichnia amphibiorum et reptiliorum fossilium: *Encyclopedia of Paleoherpetology*, vol. 18, p. 124.

Hay, W.W., Floegel, S., 2012. New thoughts about the Cretaceous climate and oceans. Earth Sci. Rev. 115 (4), 262–272.

Heine, C., Zoethout, J., Müller, R.D., 2013. Kinematics of the South Atlantic rift. Solid Earth 4 (2), 215–253.

Henderson, D.M., 2003. Footprints, trackways, and hip heights of bipedal dinosaurs—testing hip height predictions with computer models. Ichnos 10 (2–4), 99–114.

Henderson, D.M., 2006. Burly gaits: centers of mass, stability, and the trackways of sauropod dinosaurs. J. Vertebr. Paleontol. 26 (4), 907–921.

Hendricks, A., 1981. Die Saurierfährte von Münchehagen bei Rehburg-Loccum (NW-Deutschland). In: Westfälisches Landesmuseum für Naturkunde.

Heredia, A.M., Díaz-Martínez, I., Pazos, P.J., Comerio, M., Fernández, D.E., 2020. Gregarious behaviour among non-avian theropods inferred from trackways: a case study from the Cretaceous (Cenomanian) Candeleros Formation of Patagonia, Argentina. Palaeogeogr. Palaeoclimatol. Palaeoecol. 538, 109480.

Heredia, A.M., Pazos, P.J., Fernández, D.E., 2023. Unusual sauropod slipping tracks preserved on a biostabilized tidal flat from the Lower Cretaceous of northern Patagonia, Argentina. Geol. Soc. 522 (1), 215–235.

Herrero, C., Herrero, E., Martín-Chivelet, J., Pérez-Lorente, F., 2016. Contribution to knowledge of the last dinosaur footprints in Europe. Persistence of ornithopods in the upper Maastrichtian of SE Spain. Cretac. Res. 57, 490–507.

Herrero, N., Pereda-Suberbiola, X., Herrero-Gascón, J., Pérez-Lorente, F., 2017. Pentadactyl ankylosaurian manus tracks from the lower cretaceous of Galve (Teruel, Spain): first occurrence of *Tetrapodosaurus* in the iberian peninsula. J. Iber. Geol. 43, 319–344.

Herrero, C., Herrero, E., Martín-Chivelet, J., Pérez-Lorente, F., 2023. Avian ichnofauna from Sierra de las Cabras tracksite (late Miocene, Jumilla, SE Spain). J. Iber. Geol. 49 (1), 31–46.

Herrero-Gascón, J., Pérez-Lorente, F., 2017. Hoof-like unguals, skin, and foot movements deduced from *Deltapodus* casts of the Galve basin (Upper Jurassic-Lower Cretaceous, Teruel, Spain). Ichnos 24 (2), 146–161.

Hitchcock, C.H., Grier, N.M., 1927. The Hitchcock lecture upon ichnology, and the Dartmouth College ichnological collection. Am. Midl. Nat. 10 (7), 161–197.

Hitchcock, E., 1836. Ornithchnology. Description of the footmarks of birds (ornithoidichnites) on new red sandstone m Massachusetts. Am. J. Sci. 29, 307–340.

Hitchcock, E., 1837. Fossil footsteps in sandstone and graywacke. Am. J. Sci. 32, 174–176.

Hitchcock, E., 1843. Description of five new species of fossil footmarks, from the red sandstone of the valley of Connecticut River. Trans. Assoc. Am. Geol. Nat. 254–264.

Hitchcock, E., April 1845. An attempt to name, classify, and describe the animals that made the fossil footmarks of New England. In: Proceedings of the 6th Annual Meeting of the Association of American Geologists and Naturalists, New Haven, Connecticut, vol. 6, pp. 23–25.

Hitchcock, E., 1855. Shark remains from the coal formation of Illinois, and bones and tracks from the Connecticut river sandstone. Am. J. Sci. 2, 416–417.

Hitchcock, E., 1856. On a new fossil fish, and new fossil footmarks. Am. J. Sci. 20, 87–100.

Hitchcock, E., 1858. Ichnology of New England. A Report on the Sandstone of the Connecticut Valley, Especially its Fossil Footmarks, Made to the Government of the Commonwealth of Massachusetts, Xii–214.

Holliday, C.M., Ridgely, R.C., Sedlmayr, J.C., Witmer, L.M., 2010. Cartilaginous epiphyses in extant archosaurs and their implications for reconstructing limb function in dinosaurs. PLoS One 5 (9), e13120.

Hong, S.Y., Lee, Y.N., Kong, D.Y., Jung, S.H., 2024. The discovery of *Wupus agilis* in South Korea and a new quantitative analysis of intermediate ichnospecies between non-avian theropods and birds. Cretac. Res. 155, 105785.

Hornung, J.J., Böhme, A., van der Lubbe, T., Reich, M., Richter, A., 2012. Vertebrate tracksites in the Obernkirchen Sandstone (late Berriasian, Early Cretaceous) of northwest Germany—their stratigraphical, palaeogeographical, palaeoecological, and historical context. Paläontol. Z. 86, 231–267.

Hornung, J.J., Reich, M., 2013. The first record of the pterosaur ichnogenus *Purbeckopus* in the late Berriasian (Early Cretaceous) of Northwest Germany. Ichnos 20, 164–172.

Hornung, J.J., Reich, M., 2014. *Metatetrapous valdensis* Nopcsa, 1923 and the presence of ankylosaur tracks (Dinosauria: Thyreophora) in the Berriasian (Early Cretaceous) of Northwestern Germany. Ichnos 21 (1), 1–18.

Houck, K.J., Lockley, M.G., 2006. Life in an active volcanic arc: petrology and sedimentology of dinosaur track beds in the Jindong Formation (Cretaceous), Gyeongsang Basin, South Korea. Cretac. Res. 27 (1), 102–122.

Houck, K., Lockley, M.G., Caldwell, M., Clark, B., 2010. A crocodile trackway from the Dakota group (cretaceous) of the golden area, Colorado. New Mex. Mus. Nat. Hist. Sci. Bull. 51, 115–120.

Hsieh, S., Plotnick, R.E., 2020. The representation of animal behaviour in the fossil record. Anim. Behav. 169, 65–80.

Huene, F.von, 1931. Verschiedenen Wirbeltierreste aus Südamerika. Neues Jahrbuch für Mineralogie, Geologie und Palaontologie 66 (B), 181–198.

Huh, M., Hwang, K.G., Paik, I.S., Chung, C.H., Kim, B.S., 2003. Dinosaur tracks from the Cretaceous of South Korea: distribution, occurrences and paleobiological significance. Isl. Arc 12 (2), 132–144.

Huh, M., Lockley, M.G., Kim, K.S., Kim, J.Y., Gwak, S.G., 2012. First report of Aquatilavipes from Korea: new finds from Cretaceous strata in the Yeosu Islands Archipelago. Ichnos 19 (1–2), 43–49.

Hunt, A.P., Lucas, S.G., 1998. Tetrapod ichnofaunas from the Lower Cretaceous of Northeastern New Mexico. New Mex. Mus. Nat. Hist. Sci. Bull. 14, 163–167.

Hunt, A.P., Lucas, S.G., 2005. Tetrapod ichnofacies and their utility in the Paleozoic. In: Buta, R.J., Rindsberg, A.K., Kopaska-Merkel, D.C. (Eds.), Pennsylvanian Footprints in the Black Warrior Basin of Alabama, vol. 1. Alabama Paleontological Society Monograph, pp. 113–119.

Hunt, A.P., Lucas, S.G., 2007. Tetrapod ichnofacies: a new paradigm. Ichnos 14 (1–2), 59–68.

Hunt, A.P., Lucas, S.G., Klein, H., 2018. Late Triassic nonmarine vertebrate and invertebrate trace fossils and the pattern of the Phanerozoic record of vertebrate trace fossils. Late Triass. World Earth Time Trans. 447–544.

Hunt, R.K., Quinn, J.H., 2018. A new ornithomimosaur from the Lower Cretaceous Trinity Group of Arkansas. J. Vertebr. Paleontol. 38 (1), e1421209.

Hurum, J.H., Milàn, J., Hammer, O., Midtkandal, I., Amundsen, H., Sæther, B., 2006. Tracking polar dinosaurs-new finds from the Lower Cretaceous of Svalbard. Nor. Geol. Tidsskr. 86 (4), 397.

Hurum, J.H., Druckenmiller, P.S., Hammer, Ø., Nakrem, H.A., Olaussen, S., 2016. The theropod that wasn't: an ornithopod tracksite from the helvetiafjellet formation (lower cretaceous) of Boltodden, Svalbard. Geol. Soc. 434 (1), 189–206.

Hwang, K.G., Huh, M.I.N., Lockley, M.G., Unwin, D.M., Wright, J.L., 2002. New pterosaur tracks (*Pteraichnidae*) from the late cretaceous Uhangri Formation, southwestern Korea. Geol. Mag. 139 (4), 421–435.

ICZN, 1999. International Code of Zoological Nomenclature, fourth ed. The International Trust for Zoological Nomenclature, London, UK, p. 306.

Irschick, D.J., Jayne, B.C., 1999. Comparative three-dimensional kinematics of the hindlimb for high-speed bipedal and quadrupedal locomotion of lizards. J. Exp. Biol. 202, 1047–1065.

Ishigaki, S., Matsumoto, Y., 2009. "Off-tracking"-like phenomenon observed in the turning sauropod trackway from the Upper Jurassic of Morocco. Mem. Fukui Prefect. Dinosaur Mus. 8, 1–10.

Ishigaki, S., Watabe, M., Tsogtbaatar, K., Saneyoshi, M., 2009. Dinosaur footprints from the upper cretaceous of Mongolia. Geol. Q. 53 (4), 449–460.

Jackson, M.P.A., Cramez, C., Fonck, J.M., 2000. Role of subaerial volcanic rocks and mantle plumes in creation of South Atlantic margins: implications for salt tectonics and source rocks. Mar. Petrol. Geol. 17, 477–488.

Jacobs, L.J., 1997. Dinosaur trackmakers in the cretaceous of central Texas. In: Hastings, R.J., Jacobs, L.L., Clarke, R.T. (Eds.), Dinosaur Tracks in the Cretaceous Glen Rose Formation of Central Texas, vol. 3. Dallas Geological Society Guidebook, Field Trip, pp. 1–4.

Jacobs, L.L., Flanagan, K.M., Brunet, M., Flynn, L.J., Dejax, J.E.A.N., Hell, J.V., 1989. Dinosaur footprints from the Lower Cretaceous of Cameroon, West Africa. In: Gillette, D.D., Lockley, M.G. (Eds.), Dinosaur Tracks and Traces. Cambridge University Press, Cambridge, pp. 349–351.

Jaillard, E., Cappetta, H., Ellenberger, P., Feist, M., Grambast-Fessard, N., Lefranc, J.P., Sigé, B., 1993. Sedimentology, palaeontology, biostratigraphy and correlation of the late cretaceous vilquechico group of southern Peru. Cretac. Res. 14 (6), 623–661.

Jardine, W., 1850. Note to Mr. Harkness's paper "on the position of the impressions of footsteps in the bunter sandstones of dumfriesshire.". Ann. Mag. Nat. Hist. 6, 208–209.

Jin, Y., Yan, H., 1994. Bird footprints discovered in the cretaceous red bed of the Gunpey basin, Anhui. Geol. Anhui 4, 57–61.

Jung, J.Y., Min, H., Unwin, D.M., Smyth, R.S.H., Hwang, K.G., Kim, H.J., Choi, B.D., Xing, L.D., 2022. Evidence for a mixed-age group in a pterosaur footprint assemblage from the early Upper Cretaceous of Korea. Sci. Rep. 12 (1), 1–13.

Kaever, M., de Lapparent, A.F., 1974. Les traces de pas de dinosaures du Jurassique de Barkhausen (Basse Saxe, Allemagne). Bull. Soc. Geol. Fr. 7 (5), 516–525.

Kang, S.H., Buckley, L.G., McCrea, R.T., Kim, K.S., Lockley, M.G., Lim, J.D., Lim, H.S., Kim, C.B., 2021. First report of bird tracks (*Ignotornis seoungjoseoi* ichnosp. nov.) from the Jinju Formation (Lower Cretaceous), Sacheon city, Korea. Cretac. Res. 127, 104899.

Kappus, E.J., Lucas, S.G., Langford, R., 2011. The Cerro de Cristo Rey Cretaceous dinosaur tracksites, Sunland Park, New Mexico, USA, and Chihuahua, Mexico. New Mex. Mus. Nat. Hist. Sci. Bull. 53, 272–288.

Kaup, J.J., 1835a. Über Thierfährten bei Hildburghausen: Neues Jahrbuch für Mineralogie. Geologie und Paläontologie 1835, 327–328.

Kaup, J.J., 1835b. Fährten von Beuteltieren. In: Das Tierreich, pp. 246–248.

Kiessling, W., Flügel, E., Golonka, J.A.N., 2003. Patterns of Phanerozoic carbonate platform sedimentation. Lethaia 36 (3), 195–225.

Kim, H.M., 1969. A study of several sole marks in the Haman Formation. J. Geol. Soc. Korea 4, 243–258.

Kim, H.M., 1986. New early cretaceous dinosaur tracks from the Republic of Korea. In: Gillette, D.D. (Ed.), First International Symposium Dinosaur Tracks Traces, Albuquerque, New Mexico, p. 17.

Kim, J.Y., Kim, S.H., Kim, K.S., Lockley, M.G., 2006. The oldest record of webbed bird and pterosaur tracks from South Korea (Cretaceous *Haman Formation*, Changseon and Sinsu Islands): more evidence of high avian diversity in East Asia. Cretac. Res. 56–69.

Kim, J.Y., Kim, K.S., Lockley, M.G., Yang, S.Y., Seo, S.J., Choi, H.I., Lim, J.D., 2008. New didactyl dinosaur footprints (*Dromaeosauripus hamanensis* ichnogen. et ichnosp. nov.) from the Early Cretaceous Haman Formation, south coast of Korea. Palaeogeogr. Palaeoclimatol. Palaeoecol. 262 (1–2), 72–78.

Kim, J.Y., Lockley, M.G., Kim, H.M., Lim, J.D., Kim, K.S., 2009. New dinosaur tracks from Korea, *Ornithopodichnus masanensis* ichnogen. et ichnosp. nov. (Jindong Formation, Lower Cretaceous): implications for polarities in ornithopod foot morphology. Cretac. Res. 30 (6), 1387–1397.

Kim, J.Y., Lockley, M.G., 2012. New sauropod tracks (*Brontopodus pentadactylus* ichnosp. nov.) from the Early Cretaceous Haman Formation of Jinju Area, Korea: implications for sauropods manus morphology. Ichnos 19 (1–2), 84–92.

Kim, J.Y., Lockley, M.G., Kim, K.S., Seo, S.J., Lim, J.D., 2012a. Enigmatic giant pterosaur tracks and associated ichnofauna from the Cretaceous of Korea: implication for the bipedal locomotion of pterosaurs. Ichnos 19 (1–2), 50–65.

Kim, J.Y., Lockley, M.G., Woo, J.O., Kim, S.H., 2012b. Unusual didactyl traces from the Xiju formation (Early Cretaceous, South Korea) indicate a new ichnospecies of *Dromaeosauripus*. Ichnos 19 (1–2), 75–83.

Kim, J.Y., Lockley, M.G., Seo, S.J., Kim, K.S., Kim, S.H., Baek, K.S., 2012c. A paradise of Mesozoic birds: the world's richest and most diverse Cretaceous bird track assemblage from the Early Cretaceous Haman Formation of the Gajin Tracksite, Jinju, Korea. Ichnos 19, 28–42.

Kim, J.Y., Kim, M.K., Oh, M.S., Lee, C.Z., 2013. A new semi-palmate bird track, *Gyeongsangornipes lockleyi* ichnogen. et ichnosp. nov., and *Koreanaornis* from the Early Cretaceous Jindong Formation of Goseong County, southern coast of Korea. Ichnos 20, 72–80.

Kim, J.Y., Lockley, M., 2016. First report of turtle tracks from the Lower Cretaceous of Korea. Cretac. Res. 64, 1–6.

Kim, J.Y., Lockley, M.G., Chun, H.Y., 2016. New dinosaur tracks from the Lower Cretaceous (Valanginian-Hauterivian) Saniri Formation of Yeongdong area, central Korea: implications for quadrupedal ornithopod locomotion. Cretac. Res. 61, 5–16.

Kim, K.S., Lockley, M.G., Lim, J.D., Pinuela, L., Xing, L.D., Moon, H.W., 2017a. First report of lacertiform (lizard) tracks from the Cretaceous of Asia. Cretac. Res. 69, 62–70.

Kim, K.S., Lim, J.D., Lockley, M.G., Xing, L., Choi, Y., 2017b. Korean trackway of a hopping, mammaliform trackmaker is the first from the Cretaceous of Asia. Cretac. Res. 74, 188–191.

Kim, K.S., Lim, J.D., Lockley, M.G., Xing, L.D., Ha, S.J., Kim, C.B., Paik, I.S., Ahn, J.H., 2018a. First reports of a distinctive theropod track assemblage from the Jinju Formation (Lower Cretaceous) of Korea provides strong correlations with China. Cretac. Res. 81, 26–35.

Kim, K.S., Lim, J.D., Lockley, M.G., Xing, L., Kim, D.H., Piñuela, L., Romilio, A., Yoo, J.S., Kim, J.H., Ahn, J., 2018b. Smallest known raptor tracks suggest *microraptorine* activity in lakeshore setting. Sci. Rep. 8 (1), 16908.

Kim, K.S., Lockley, M.G., Lim, J.D., Xing, L.D., 2019a. Exquisitely-preserved, high-definition skin traces in diminutive theropod tracks from the Cretaceous of Korea. Sci. Rep. 9, 2039.

Kim, K.S., Lockley, M.G., Lim, J.D., Hee Kim, D., Piñuela, L., Yoo, J.S., 2019b. Largest Cretaceous lizard track assemblage, new morphotypes and longest trackways comprise diverse components of an exceptional Korean Konservat-Lagerstätten ichnofauna. Sci. Rep. 9 (1), 13278.

Kim, K.S., Lockley, M.G., Lim, J.D., Kim, D.H., 2019c. The oldest known anuran (frog) trackways from the Jinju Formation, Lower Cretaceous, Korea. Cretac. Res. 96, 142–148.

Kim, K.S., Lockley, M.G., Lim, J.D., Bae, S.M., Romilio, A., 2020. Trackway evidence for large bipedal crocodylomorphs from the Cretaceous of Korea. Sci. Rep. 10 (1), 8680.

Kim, S.H., 2008. Stratigraphic Diversity of Bird and Diminutive Dinosaur Tracks from the Cretaceous Haman Formation, South Gyeongsang Province of Korea: Implication for Ichnostratigraphy. Unpublished Ph. D. dissertation. Korea National University of Education, Chungbuk, Korea, 141 pp.

Klein, H., Lagnaoui, A., Gierliński, G.D., Saber, H., Lallensack, J.N., Oukassou, M., Charrière, A., 2018. Crocodylomorph, turtle and mammal tracks in dinosaur-dominated Middle–? Upper Jurassic and mid-Cretaceous ichnoassemblages of Morocco. Palaeogeogr. Palaeoclimatol. Palaeoecol. 498, 39–52.

Klein, H., Gierliński, G., Lallensack, J.N., Abu Hamad, A., Al-Mashakbeh, H., Alhejoj, I., Konopka, M., Błoński, M., 2020. First Upper Cretaceous dinosaur track assemblage from Jordan (Middle East): preliminary results. Ann. Soc. Geol. Pol. 90 (3), 331–342.

Kochan, J.B., 1994. Birds feet and legs. Stackpole Books, Mechanicsburg, Pennsylvania, pp. 1–81.

Kohlsdorf, T., Garland Jr., T., Navas, C.A., 2001. Limb and tail lengths in relation to substrate usage in *Tropidurus* lizards. J. Morphol. 248, 151–164.

Krapovickas, V., 2010. El rol de las trazas fósiles de tetrápodos en los modelos de icnofacies continentales en ambientes de climas áridos-semiáridos (Doctoral dissertation, Universidad de Buenos Aires. Facultad de Ciencias Exactas y Naturales).

Krapovickas, V., Garrido, A.C., Canale, J.I., 2020. Dinosaur tracks at Cañadón de las. Campanas and Villa El Chocón area (Late Cretaceous, Candeleros Formation), Patagonia, Argentina. Rev. Asoc. Geol. Argent. 77 (3), 447–462.

Krapovickas, V., Vera, R.B., Farina, M.E., Piana, L.R.F., Koelewijn, A., 2023. A new vertebrate ichnological association sheds light on the small metatherian record of the Middle Miocene in South America. J. S. Am. Earth Sci. 129, 104529.

Ksepka, D.T., 2020. Feathered dinosaurs. Curr. Biol. 30 (22), R1347–R1353.

Kuban, G.J., 1989. Elongate dinosaur tracks. In: Gillette, D.D., Lockley, M.G. (Eds.), Dinosaur Tracks and Traces. Cambridge University Press, Cambridge, pp. 57–72.

Kubo, T., 2008. In quest of the *Pteraichnus* trackmaker: comparisons to modern crocodilians. Acta Palaeontol. Pol. 53 (3), 405–412.

Kuhn, O., 1958. Die fährten der vorzeitlichen Amphibien und reptilien. Meisenbach. 64 pp.

Kuhn, O., 1963. Ichnia Tetrapodorum. Fossilum Catalogus I (Animalia): Pars 101. Junk, s'Gravenhagc, 176 pp.

Kumagai, C.J., Farlow, J.O., 2010. Observations on traces of the American crocodile (Crocodylus acutus) from northwest Costa Rica. New Mex. Mus. Nat. Hist. Sci. Bull. 51, 41–49.

Lallensack, J.N., van Heteren, A.H., Wings, O., 2016. Geometric Morphometric Analysis of Intratrackway Variability: A Case Study on Theropod and Ornithopod Dinosaur.

Lallensack, J.N., Engler, T., Barthel, H.J., 2020. Shape variability in tridactyl dinosaur footprints: the significance of size and function. Palaeontology 63 (2), 203–228.

Lallensack, J.N., Falkingham, P.L., 2022. A new method to calculate limb phase from trackways reveals gaits of sauropod dinosaurs. Curr. Biol. 32, 1–6.

Lallensack, J.N., Romilio, A., Falkingham, P.L., 2022a. A machine learning approach for the discrimination of theropod and ornithischian dinosaur tracks. J. R. Soc. Interface 19, 20220588.

Lallensack, J.N., Farlow, J.O., Falkingham, P.L., 2022b. A new solution to an old riddle: elongate dinosaur tracks explained as deep penetration of the foot, not plantigrade locomotion. Palaeontology 65 (1), e12584.

Lallensack, J.N., Owais, A., Falkingham, P.L., Breithaupt, B.H., Sander, P.M., 2023. How to verify fossil tracks: the first record of dinosaurs from Palestine. Hist. Biol. 35 (6), 924–934.

Lambrecht, K., 1938. *Urmiornis abeli* n. sp., eine Pliozäne Vogelfährte aus Persien. Palaeobiologica 6 (2), 242–245.

Langston Jr., W., 1974. Nonmammalian comanchean tetrapods. Geosci. Man 8, 77–102.

Langston Jr., W., 1997. Lower cretaceous dinosaur tracks near glen Rose, Texas. In: Hastings, R.J., Jacobs, L.L., Clarke, R.T. (Eds.), Dinosaur Tracks in the Cretaceous Glen Rose Formation of Central Texas, vol. 3. Dallas Geological Society Guidebook, Field Trip, pp. 5–33.

Lauprasert, K., Laojumpon, C., Saenphala, W., Cuny, G., Thirakhupt, K., Suteethorn, V., 2011. Atoposaurid crocodyliforms from the Khorat group of Thailand: first record of *Theriosuchus* from Southeast Asia. Paläontol. Z. 85, 37–47.

Le Loeuff, J., Lauprasert, K., Suteethorn, S., Souillat, C., Suteethorn, V., Buffetaut, E., 2010. Late Early Cretaceous crocodyliform trackways from northeastern Thailand. New Mex. Mus. Nat. Hist. Sci. Bull. 51, 175–178.

Lee, H.J., Lee, Y.N., Fiorillo, A.R., Lu, J., 2018. Lizards ran bipedally 110 million years ago. Sci. Rep. 8 (1), 2617.

Lee, J.H., Lockley, M.G., Jung, S., Ha, Y., Cho, K.R., Lim, J.D., Kim, K.S., 2023. Dinosaur tracksites from the cretaceous Sagok Formation in the Uiseong regional Geopark, South Korea: historical and palaeobiological perspectives. Hist. Biol. 35 (9), 1656–1664.

Lee, Y.N., 1997. Bird and dinosaur footprints in the Woodbine Formation (Cenomanian), Texas. Cretac. Res. 18 (6), 849–864.

Lee, Y.N., Lee, H.J., Lu, J., Kobayashi, Y., 2008. New pterosaur tracks from the Hasandong Formation (lower cretaceous) of Hadong county, South Korea. Cretac. Res. 29 (2), 345–353.

Lee, Y.N., Azuma, Y., Lee, H.J., Shibata, M., Lv, J.C., 2010. The first pterosaur trackways from Japan. Cretac. Res. 31 (2), 263–273.

Lee, Y.N., Lee, H.J., Kobayashi, Y., Paulina-Carabajal, A., Barsbold, R., Fiorillo, A.R., Tsogtbaatar, K., 2019. Unusual locomotion behaviour preserved within a crocodyliform trackway from the Upper Cretaceous Bayanshiree Formation of Mongolia and its palaeobiological implications. Palaeogeogr. Palaeoclimatol. Palaeoecol. 533, 109239.

Lee, Y.N., Kong, D.Y., Jung, S.H., 2020. The first possible choristoderan trackway from the Lower Cretaceous Daegu Formation of South Korea and its implications for choristoderan locomotion. Sci. Rep. 10 (1), 14442.

Leonardi, G., 1979a. Nota preliminar sobre seis pistas de dinossauros Ornithischia da Bacia do Rio de Peixe, em Sousa, Paraíba, Brasil. An Acad. Bras Ciências 71 (3), 501–516.

Leonardi, G., 1979b. Abundante icnofauna (vertebrados e invertebrados) na Formação Botucatu s.s. em. Araraquara, São Paulo. An Acad. Bras Ciências 51, 360–361.

Leonardi, G., 1981. Novo icnogênero de tetrápode mesozóico da Formação Botucatu, Araraquara, SP. Anais da Academia brasileira de Ciências 53 (4), 793–805.

Leonardi, G., 1984. Le impronte fossili di Dinosauri. In: Bonaparte, J.F., Colbert, E.H., Currie, P.J., de Ricqlès, A.J., Keilan-Jaworowska, Z., Leonardi, G., Morello, F. (Eds.), Sulle Orme dei Dinosauri. Erizzo Editrice, Venice, pp. 164–186.

Leonardi, G., 1987. Glossary and Manual of Tetrapod Footprint Ichnology. Ministry of Mines and Energy, Brasilia, Brazil, p. 75.

Leonardi, G., 1989. Inventory and statistics of the South American dinosaurian ichnofauna and its paleobiological interpretation. In: Gillette, D.D., Lockley, M.G. (Eds.), Dinosaur Tracks and Traces. Cambridge University Press, Cambridge, pp. 165–178.

Leonardi, G., Oliveira, F.H., 1990. A revision of the Triassic and Jurassic tetrapod footprints of Argentina and a new approach on the age and meaning of the *Botucatu Formation* footprints (Brazil). Rev. Bras. Geociencias 20, 216–229.

Leonardi, G., 1994. Annotated Atlas of South America tetrapod footprints (Devonian to Holocene). In: Rio de Janeiro: República Federativa do Brasil, Ministério de Minas e Energia, *Companhia de Pesquisa de Recursos Minerais*, pp. 1–247.

Leonardi, G., Carvalho, I.S., Fernandes, M.A., 2007. The desert ichnofauna from Botucatu Formation (upper- Jurassic-lower cretaceous). In: Carvalho, I. (Ed.), Paleontologia: Cenários de Vida. Editora Interciência, Rio de Janeiro, pp. 379–392.

Leonardi, G., Carvalho, I.S., 2021a. Review of the early mammal *Brasilichnium* and *Brasilichnium*-like tracks from the lower cretaceous of SouthSouth America. J. S. Am. Earth Sci. 106, 102940.

Leonardi, G., Carvalho, I.S., 2021b. Vertebrate Trace Fossils: the Congo's *Brasilichnium* mammaloid fossil footprints. Ital. J. Geosci. 140 (1), 141–154.

Leonardi, G., Carvalho, I.S., 2021c. Dinosaur Tracks from Brazil: A Lost World of Gondwana. Indiana University Press, Bloomington.

Leonardi, G., Fernandes, M.A., Carvalho, I.S., Schutzer, J.B., da Silva, R.C., 2024. *Farlowichnus rapidus* new ichnogen., new ichnosp.: a speedy and small theropod in the Early Cretaceous Botucatu paleodesert (Paraná Basin), Brazil. Cretac. Res. 153, 105720.

Lessertisseur, J., 1955. Traces fossiles d'activité animale et leur signification paléobiologique. In: Mémoires de la Société Géologique de France, Nouvelle Edition. Tome XXXIV. Fasc. 4 Feuilles 15 à 24 Mémoire n 74. 150 pp.

Li, D.Q., Xing, L.D., Lockley, M.G., Piñuela, L., Zhang, J.P., Dai, H., Kim, J.Y., Persons, W.S.I.V., Kong, D.L., 2015. A manus dominated pterosaur track assemblage from Gansu, China: implications for behavior. Sci. Bull. 60 (2), 264–272.

Li, D.Q., Xing, L.D., Lockley, M.G., Romilio, A., Yang, J.T., Li, L.F., 2019. The first theropod tracks from the Middle Jurassic of Gansu, Northwest China: new and rare evidence of quadrupedal progression in theropod dinosaurs. J. Palaeogeogr. 8 (1), 1–11.

Li, J., Zhang, W., Hu, B., Gao, L., 2006. A new type of dinosaur tracks from lower cretaceous of Chabu, Otog Qi, inner Mongolia. Acta Palaeontol. Sin. 45 (2), 221–234.

Li, J.J., 2015. Footprints of mesozoic reptilians and avians. In: Li, J.L., Zhou, Z.H. (Eds.), Palaeovertebrata Sinca. Science Press, Beijing, pp. 1–270.

Li, R., Zhang, G., 2000. New dinosaur ichnotaxon from the early cretaceous laiyang Group,in the laiyang basin, Shandong province. Geol. Rev. 46, 605–610.

Li, R., Lockley, M.G., Liu, M., 2005. A new ichnotaxon of fossil bird track from the early cretaceous Tianjialou Formation (Barremian-Albian), Shandong province, China. Chin. Sci. Bull. 50, 1149–1154.

Li, R., Lockley, M.G., Makovicky, P.J., Matsukawa, M., Norell, M.A., Harris, J.D., Liu, M., 2008. Behavioral and faunal implications of Early Cretaceous deinonychosaur trackways from China. Naturwissenschaften 95, 185–191.

Li, R., Lockley, M.G., Matsukawa, M., Wang, K., Liu, M., 2011. An unusual theropod track assemblage from the Cretaceous of the Zhucheng area, Shandong Province, China. Cretac. Res. 32 (4), 422–432.

Li, R., Lockley, M.G., Matsukawa, M., Liu, M., 2015. Important Dinosaur-dominated footprint assemblages from the Lower Cretaceous Tianjialou Formation at the Houzuoshan Dinosaur Park, Junan County, Shandong Province, China. Cretac. Res. 52, 83–100.

Li, Y., Wang, X., Jiang, S., 2015. A new pterosaur tracksite from the lower cretaceous of Wuerho, Junggar Basin, China: inferring the first putative pterosaur trackmaker. PeerJ 9, e11361.

Li, Y., Jiang, S., Wang, X., 2020. The largest species of *Asianopodus* footprints from Junggar basin, Xinjiang, China. Chin. Sci. Bull. 65 (18), 1875–1887.

Li, Y., Wang, X., Jiang, S., 2021. A new pterosaur tracksite from the Lower Cretaceous of Wuerho, Junggar Basin, China: inferring the first putative pterosaur trackmaker. PeerJ 9, e11361.

Li, Y., Chang, S.C., Zhang, H., Wang, J., Pei, R., Zheng, D., Fang, Y., Hemming, S.R., 2023. A chronostratigraphic and biostratigraphic framework for the Yanliao Biota of northeastern China: implications for Jurassic terrestrial ecosystems and evolution. Palaeogeogr. Palaeoclimatol. Palaeoecol. 630, 111818.

Lichtig, A.J., Lucas, S.G., Klein, H., Lovelace, D.M., 2018. Triassic turtle tracks and the origin of turtles. Hist. Biol. 30 (8), 1112–1122.

Lim, J.D., Kong, D.Y., 2008. Korean cretaceous dinosaur coast: preservation and monitoring plans. J. Korean Nat. 1 (1), 127–131.

Lim, J.D., Lockley, M.G., Kong, D.Y., 2012. The trackway of a quadrupedal ornithopod from the Jindong Formation (cretaceous) of Korea. Ichnos 19 (1–2), 101–104.

Lim, S.K., Yang, S.Y., Lockley, M.G., 1989. Large dinosaur footprint assemblages from the cretaceous Jindong Formation of southern Korea. In: Gillette, D.D., Lockley, M.G. (Eds.), Dinosaur Tracks and Traces. Cambridge University Press, Cambridge, pp. 33–336.

Lim, S.K., Lockley, M.G., Yang, S.Y., Fleming, R.F., Houck, K.J., 1994. Preliminary report on sauropod tracksites from the cretaceous of Korea. Gaia Revista de Geociencias, Museu Nacional de Historia Natural, Lisbon, Portugal 10, 109–117.

Llompart, C., Casanovas, M.L., Santafé, J.V., 1984. Un nuevo yacimiento de icnitas de dinosaurios en las facies garumnienses de la Conca de Tremp (Lleida, España). Acta Geol. Hisp. 19, 143–147.

Lockley, M.G., Houck, K.J., Price, N.K., 1986. North America's largest dinosaur trackway site: implications for Morrison Formation paleoecology. Geol. Soc. Am. Bull. 97 (10), 1163–1176.

Lockley, M.G., 1987. Dinosaur footprints from the Dakota Group of eastern Colorado. Mt. Geol. 24, 107–122.

Lockley, M.G., 1988. Dinosaurs near denver. In: Holden, G.S., Tafoya, R.E. (Eds.), Geological Society of America Field Trip Guidebook, vol. 12. Professional Contributions-Colorado School of Mines, Centennial Meeting-Denver, Colorado, pp. 288–299.

Lockley, M.G., 1991. Tracking Dinosaurs: A New Look at an Ancient World. Cambridge University Press, Cambridge, 252pp.

Lockley, M.G., Yang, S.Y., Matsukawa, M., Fleming, F., Lim, S.K., 1992a. The track record of Mesozoic birds: evidence and implications. Phil. Trans. Biol. Sci. 336 (1277), 113–134.

Lockley, M.G., Conrad, K., Paquette, M., Hamblin, A., 1992b. Late Triassic Vertebrate Tracks in the Dinosaur National Monument Area, vol. 92. Utah Geological Survey Miscellaneous Publication, p. 3.

Lockley, M.G., Meyer, C., 1994. Trackway evidence for a herd of juvenile sauropods from the Late Jurassic of Portugal. Gaia Revista de Geociências 10, 27.

Lockley, M.G., Farlow, J.O., Meyer, C.A., 1994a. *Brontopodus* and *Parabrontopodus* ichnogen. nov. and the significance of wide-and narrow-gauge sauropod trackways. Gaia 10, 135–145.

Lockley, M.G., Hunt, A.P., 1994a. A track of the giant theropod dinosaur *Tyrannosaurus* from close to the Cretaceous/Tertiary boundary, northern New Mexico. Ichnos 3 (3), 213–218.

Lockley, M.G., Hunt, A.P., 1994b. A review of Mesozoic vertebrate ichnofaunas of the Western Interior USA: evidence and implications of a superior track record. Rocky Mountain Section (SEPM). In: Caputo, M.V., Peterson, J.A., Franczyk, K.J. (Eds.), Mesozoic Systems of the Rocky Mountain Region, USA, pp. 95–108.

Lockley, M.G., Meyer, C.A., Hunt, A.P., Lucas, S.G., 1994b. The distribution of sauropod tracks and trackmakers. Gaia 10, 233–248.

Lockley, M.G., Pittman, J.G., Meyer, C.A., Santos, V.F., 1994c. On the common occurrence of manus-dominated sauropod trackways in Mesozoic carbonates. Gaia 10, 119–124.

Lockley, M.G., Meyer, C.A., Santos, V.F., 1994d. Trackway evidence for a herd of juvenile sauropods from the Late Jurassic of Portugal. Gaia 10, 27–35.

Lockley, M.G., Hunt, A.P., Moratalla, J., Matsukawa, M., 1994e. Limping dinosaurs? Trackway evidence for abnormal gaits. Ichnos 3, 193–202.

Lockley, M.G., Hunt, A.P., Meyer, C., 1994f. Vertebrate tracks and the ichnofacies concept: implications for paleoecology and palichnostratigraphy. In: Donovan, S.K. (Ed.), The Paleobiology of Trace Fossils. Wiley Sons, New York, pp. 241–268.

Lockley, M.G., Logue, T.J., Moratalla, J.J., Hunt, A.P., Schultz, R.J., Robinson, J.W., 1995. The fossil trackway *Pteraichnus* is pterosaurian, not crocodilian: implications for the global distribution of pterosaur tracks. Ichnos 4, 7–20.

Lockley, M.G., Hunt, A.P., 1995a. Dinosaur Tracks and Other Fossil Footprints of the Western USA. Columbia University Press, New York, 338 pp.

Lockley, M.G., Hunt, A.P., 1995b. Ceratopsid tracks and associated ichnofauna from the Laramie Formation (upper cretaceous: maastrichtian) of Colorado. J. Vertebr. Paleontol. 15 (3), 592–614.

Lockley, M.G., Meyer, C.A., Schultz-Pittman, R., Forney, G., 1996a. Late Jurassic dinosaur tracksites from Central Asia: a preliminary report on the world's longest trackways. In: Morales, M. (Ed.), The Continental Jurassic. Museum of Northern Arizona Bulletin, vol. 60, pp. 137–140.

Lockley, M.G., Meyer, C.A., Santos, V.F., 1996b. *Megalosauripus, Megalosauropus* and the concept of megalosaur footprints. In: Morales, M. (Ed.), The Continental Jurassic. Museum of Northern Arizona Bulletin, vol. 60, pp. 113–118.

Lockley, M.G., 1998. The vertebrate track record. Nature 396 (6710), 429–432.

Lockley, M.G., Meyer, C., Santos, V.F., 1998. Megalosauripus and the problematic concept of megalosaur footprints. Gaia: Revista de Geociências 15, 313.

Lockley, M.G., 1999. The Eternal Trail: A Tracker Looks at Evolution. Basic Books, 352 pp.

Lockley, M.G., 2000. An amended description of the theropod footprint *Bueckeburgichnus maximus* Kuhn 1958, and its bearing on the megalosaur tracks debate. Ichnos 7, 217–225.

Lockley, M.G., Meyer, C., 2000. Dinosaur Tracks and Other Fossil Footprint of Europe. Columbia University Press, New York, 323pp.

Lockley, M.G., Meyer, C.A., Santos, V.F., 2000. *Megalosauripus* and the problematic concept of megalosaur footprints. Gaia 15, 313–337.

Lockley, M.G., Wright, J.L., Matsukawa, M., 2001a. A new look at *Magnoavipes* and so-called big bird tracks from Dinosaur Ridge (Cretaceous, Colorado). Mt. Geol. 38 (3), 137–146.

Lockley, M.G., Wright, J., Giblin, R., Manning, M., 2001b. "Fossil trace" trace fossils. New finds of cretaceous tracks from the Laramie Formation Clay pits and golf course in golden. Friends Dinosaur Ridge Ann. Rep. 2001 Morrison Colorado 18–26.

Lockley, M.G., Wright, J., White, D., Matsukawa, M., Jianjun, L., Lu, F., Hong, L., 2002a. The first sauropod trackways from China. Cretac. Res. 23 (3), 363–381.

Lockley, M.G., Schulp, A.S., Meyer, C.A., Leonardi, G., Kerumba, D., 2002b. Titanosaurid trackways from the upper cretaceous of Bolivia: evidence for large manus, wide-gauge locomotion and gregarious behaviour. Cretac. Res. 23, 383–400.

Lockley, M.G., Foster, J., Manning, M., 2002c. Late cretaceous mammal tracks from Colorado. In: Friends of Dinosaur Ridge Annual Report for 2002, Morrison, Colorado, pp. 18–24.

Lockley, M.G., 2003. A guide to the fossil footprints of the dinosaur ridge and fossil trace areas. Friends Dinosaur Ridge Morrison Colorado 66.

Lockley, M.G., Foster, J.R., 2003. Late cretaceous mammal tracks from North America. Ichnos 10, 269–276.

Lockley, M.G., Matsukawa, M., Li, J., 2003. Crouching theropods in taxonomic jungles: ichnological and ichnotaxonomic investigations of footprints with metatarsal and ischial impressions. Ichnos 10 (2–4), 169–177.

Lockley, M.G., Meyer, C., 2004. *Crocodylomorph* trackways from the Jurassic to early cretaceous of North America and Europe: implications for ichnotaxonomy. Ichnos 11 (1–2), 167–178.

Lockley, M.G., Nadon, G., Currie, P.J., 2004a. A diverse dinosaur-bird footprint assemblage from the Lance Formation, Upper Cretaceous, eastern Wyoming: implications for ichnotaxonomy. Ichnos 11 (3–4), 229–249.

Lockley, M.G., Kirkland, J., Milner, A.R., 2004b. Probable relationships between the Lower Jurassic crocodilomorph trackways *Batrachopus* and *Selenichnus*: evidence and implications based on new finds from the St. George Area southwestern Utah. Ichnos 11 (1–2), 143–149.

Lockley, M.G., Foster, J.R., 2006. Dinosaur and turtle tracks from the Morrison Formation (Upper Jurassic) of Colorado National Monument, with observations on the taxonomy of vertebrate swim tracks. New Mex. Mus. Nat. Hist. Sci. Bull. 36, 193–198.

Lockley, M.G., Gierliński, G.D., 2006. Diverse Vertebrate Ichnofaunas containing Anomoepus and other unusual trace fossils from the Lower Jurassic of the western USA: implications for paleoecology and palichnostratigraphy. New Mex. Mus. Nat. Hist. Sci. Bull. 37, 176–191.

Lockley, M.G., Houck, K., Yang, S.Y., Matsukawa, M., Lim, S.K., 2006a. Dinosaur-dominated footprint assemblages from the cretaceous Jindong Formation, Hallyo Haesang national park area, Goseong county, South Korea: evidence and implications. Cretac. Res. 27 (1), 70–101.

Lockley, M.G., Matsukawa, M., Sato, Y., Polahan, M., Daorerk, V., 2006b. A distinctive new theropod dinosaur track from the Cretaceous of Thailand: implications for theropod track diversity. Cretac. Res. 27, 139–145.

Lockley, M.G., Matsukawa, M., Ohira, H., Li, J., Wright, J., White, D., Chen, P., 2006c. Bird tracks from Liaoning Province, China: new insights into avian evolution during the Jurassic-Cretaceous transition. Cretac. Res. 27 (1), 33–43.

Lockley, M.G., Li, R., Harris, J.D., Matsukawa, M., Liu, M., 2007. Earliest zygodactyl bird feet: evidence from Early Cretaceous roadrunner-like tracks. Naturwissenschaften 94, 657–665.

Lockley, M.G., 2007a. The Morphodynamics of Dinosaurs, Other Archosaurs, and Thier Trackways: Holistic Insights into Relationships between, Feet, Limbs, and the Whole Body, vol. 88. Society for Sedimentary Geology, special publication, pp. 27–51.

Lockley, M.G., 2007b. A 25-year anniversary celebration of the discovery of fossil footprints of South Korea. In: Proceedings of the Haenam Uhangri International Dinosaur Symposium, pp. 41–62.

Lockley, M.G., 2007c. A tale of two ichnologies: the different goals and potentials of invertebrate and vertebrate (tetrapod) ichnotaxonomy and how they relate to ichnofacies analysis. Ichnos 14, 39–57.

Lockley, M.G., Kim, J.Y., Kim, K.S., Kim, S.H., Matsukawa, M., Rihui, L., Jianjun, L., Yang, S.Y., 2008a. *Minisauripus*—the track of a diminutive dinosaur from the Cretaceous of China and South Korea: implications for stratigraphic correlation and theropod foot morphodynamics. Cretac. Res. 29, 115–130.

Lockley, M.G., Harris, J.D., Mitchell, L., 2008b. A global overview of pterosaur ichnology: tracksite distribution in space and time. Zitteliana 185–198.

Lockley, M.G., 2009. New perspectives on morphological variation in tridactyl footprints: clues to widespread convergence in developmental dynamics. Geol. Q. 53 (4), 415–432.

Lockley, M.G., Matsukawa, M., 2009. A review of vertebrate track distributions in East and Southeast Asia. J. Paleontol. Soc. Korea 25, 17–42.

Lockley, M.G., Chin, K., Houck, K., Matsukawa, M., Kukihara, R., 2009a. New interpretations of *Ignotornis*, the first-reported Mesozoic avian footprints: implications for the paleoecology and behavior of an enigmatic Cretaceous bird. Cretac. Res. 30 (4), 1041–1061.

Lockley, M.G., McCrea, R.T., Matsukawa, M., 2009b. Ichnological evidence for small quadrupedal ornithischians from the basal Cretaceous of SE Asia and North America: implications for a global radiation. Geol. Soc. 315 (1), 255–269.

Lockley, M.G., 2010. A solution to the *Mehliella* mystery: tracking, naming, identifying and measuring the first crocodylian trackway reported from the Cretaceous (Dakota Group, Colorado). New Mex. Mus. Nat. Hist. Sci. Bull. 51, 157–164.

Lockley, M.G., Fanelli, D., Honda, K., Houck, K., Mathews, N.A., 2010. Crocodile waterways and dinosaur freeways: implications of multiple swim track assemblages from the Cretaceous Dakota Group, Golden area, Colorado. New Mex. Mus. Nat. Hist. Sci. Bull. 51, 137–156.

Lockley, M.G., Harris, J.D., 2010. On the trail of early birds: a review of the fossil footprint record of avian morphological and behavioral evolution. In: Ulrich, P.K., Willett, J.H. (Eds.), Trends in Ornithology Research. Nova Science Publishers Inc., Hauppauge, New York, p. 63.

Lockley, M.G., Cart, K., Martin, J., Milner, A.R.C., 2011. New theropod tracksites from the Upper Cretaceous 'Mesaverde' Group, western Colorado: implications for ornithomimosaur track morphology. New Mex. Mus. Nat. Hist. Sci. Bull. 53, 321–329.

Lockley, M.G., 2011. The ichnotaxonomic status of *Brasilicnhinum* with special reference to occurrences in the Navajo Sandstone (Lower Jurassic). New Mex. Mus. Nat. Hist. Sci. Bull. 53, 306–315.

Lockley, M.G., Lucas, S.G., Matsukawa, M., Harris, J.D., 2012a. Cretaceous tetrapod footprint biostratigraphy, biochronology, and ichnofacies. J. Stratigr. 36, 503–550.

Lockley, M.G., Huh, M., Kim, J.Y., Lim, J.D., Kim, K.S., 2012b. Recent advances in Korean vertebrate ichnology: the KCDC comes of age. Ichnos 19, 1–5.

Lockley, M.G., Lim, J.D., Kim, J.Y., Kim, K.S., Huh, M., Hwang, K.G., 2012c. Tracking Korea's early birds: a review of Cretaceous avian ichnology and its implications for evolution and behavior. Ichnos 19, 17–27.

Lockley, M.G., Li, J.J., Matsukawa, M., Li, R., 2012d. A new avian ichnotaxon from the Cretaceous of Nei Mongol, China. Cretac. Res. 34, 84–93.

Lockley, M.G., Huh, M., Kim, B.S., 2012e. *Ornithopodichnus* and pes-only sauropod trackways from the Hwasun tracksite, Cretaceous of Korea. Ichnos 19 (1–2), 93–100.

Lockley, M.G., Li, J.J., Li, R.H., Matsukawa, M., Harris, J.D., Xing, L.D., 2013. A review of the tetrapod track record in China, with special reference to type ichnospecies: implications for ichnotaxonomy and paleobiology. Acta Geol. Sin. 87 (1), 1–20.

Lockley, M.G., Gierliński, G.D., 2014. Notes on a new ankylosaur track from the Dakota group (cretaceous) of northern Colorado. New Mex. Mus. Nat. Hist. Sci. Bull. 62, 301–306.

Lockley, M.G., Lucas, S.G., 2014. Tracking dinosaurs and other tetrapods in the wild west of North America. New Mex. Mus. Nat. Hist. Sci. Bull. 62, 1–4.

Lockley, M.G., Milner, A.R.C., 2014. The ichnotaxonomy of hopping vertebrate trackways from the Cenozoic of the western USA. New Mex. Mus. Nat. Hist. Sci. Bull. 62, 489–500.

Lockley, M.G., Xing, L., Lockwood, J.A., Pond, S., 2014a. A review of large Cretaceous ornithopod tracks, with special reference to their ichnotaxonomy. Biol. J. Linn. Soc. 113 (3), 721–736.

Lockley, M.G., Gierliński, G.D., Dubicka, Z., Breithaupt, B.H., Matthews, N.A., 2014b. A preliminary report on a new dinosaur tracksite in the Cedar Mountain Formation (Cretaceous) of eastern Utah. New Mex. Mus. Nat. Hist. Sci. Bull. 62, 279–285.

Lockley, M.G., Cart, K., Martin, J., Prunty, R., Houck, K., Hups, K., Lim, J.D., Kim, K.S., Houck, K., Gierliński, G., 2014c. A bonanza of new tetrapod tracksites from the Cretaceous Dakota Group, western Colorado: implications for paleoecology. Fossil Footpr. Western North Am. Bull. 62, 393–409.

Lockley, M.G., Triebold, M., Janke, P.R., 2014d. Dinosaur tracks from the Hell Creek Formation (upper cretaceous, maastrichtian), South Dakota. Fossil footprints of western North America. New Mex. Mus. Nat. Hist. Sci. Bull. 62, 459–468.

Lockley, M.G., Xing, L.D., Kim, J.Y., Matsukawa, M., 2014e. Tracking early cretaceous dinosaurs in China: a new database for comparison with ichnofaunal data from Korea, the Americas, Europe, Africa and Australia. Biol. J. Linn. Soc. 113, 770–789.

Lockley, M.G., Xing, L.D., 2015. Flattened fossil footprints: implications for paleobiology. Palaeogeogr. Palaeoclimatol. Palaeoecol. 426, 85–94.

Lockley, M.G., Li, R.H., Matsukawa, M., Xing, L.D., Li, J.J., Liu, M.W., Xu, X., 2015a. Tracking the yellow dragons: implications of China's largest dinosaur tracksite (Cretaceous of the Zhucheng area, Shandong Province, China). Palaeogeogr. Palaeoclimatol. Palaeoecol. 423, 62–79.

Lockley, M.G., Buckley, L.G., Foster, J.R., Kirkland, J.I., DeBlieux, D.D., 2015b. First report of bird tracks (*Aquatilavipes*) from the Cedar Mountain Formation (lower cretaceous), eastern Utah. Palaeogeogr. Palaeoclimatol. Palaeoecol. 420, 150–162.

Lockley, M.G., Xing, L.D., Matthews, N.A., Breithaupt, B.H., 2016. Didactyl raptor tracks from the cretaceous, Plainview sandstone at dinosaur ridge. Cretac. Res. 61, 161–168.

Lockley, M.G., Li, J., Xing, L.D., Guo, B., Matsukawa, M., 2018a. Large theropod and small sauropod trackmakers from the lower cretaceous Jingchuan Formation, inner Mongolia, China. Cretac. Res. 92, 150–167.

Lockley, M.G., Gierliński, G., Adach, L., Schumacher, B., Cart, K., 2018b. New tetrapod ichnotaxa from the Blackhawk Formation (upper cretaceous), Utah. New Mex. Mus. Nat. Hist. Sci. Bull. 79, 469–480.

Lockley, M.G., Burton, R., Grondel, L., 2018c. A large assemblage of tetrapod tracks from the Cretaceous Naturita Formation, Cedar Canyon region, southwestern Utah. Cretac. Res. 92, 108–121.

Lockley, M.G., Hirschfeld, S.E., Simmons, B., 2018d. A new dinosaur track locality in the late cretaceous (maastrichtian) Laramie Formation of Colorado. Fossil Rec. 6, 395–406.

Lockley, M.G., Smith, J.A., King, M.R., 2018e. First reports of turtle tracks from the Williams Fork Formation ('Mesaverde'Group), Upper Cretaceous (Campanian) of western Colorado. Cretac. Res. 84, 474–482.

Lockley, M.G., McCrea, R.T., Buckley, L.I.S.A., King, M.R., 2018f. Some observations on the variation and spatiotemporal distribution of turtle tracks in the Cretaceous of North America. New Mex. Mus. Nat. Hist. Sci. Bull. 79, 451–461.

Lockley, M.G., Xing, L., Xu, X., 2019. The "lost" holotype of *Laiyangpus liui* (Lower Cretaceous, Shandong Province, China) is found: implications for trackmaker identification, ichnotaxonomy and interpretation of turtle tracks. Cretac. Res. 95, 260–267.

Lockley, M.G., Lim, J.D., Park, H.D., Romilio, A., Yoo, J.S., Choi, J.W., Kim, K.S., Choi, Y., Kang, S.H., Kim, D.H., Kim, T.H., 2020. First reports of Crocodylopodus from East Asia: implications for the paleoecology of the lower cretaceous. Cretac. Res. 111, 104441.

Lockley, M.G., Abbassi, N., Helm, C.W., 2021. Large, unwebbed bird and bird-like footprints from the Mesozoic and Cenozoic: a review of ichnotaxonomy and trackmaker affinity. Lethaia 54, 969–987.

Lockley, M.G., Xing, L.D., 2021. A review of the non-avian theropod track record and the implications for the Ontogenetic Niche Shift model. Earth Sci. Rev. 220, 103715.

Lockley, M.G., Helm, C.W., Lawfield, A.M., Sharman, K.J., 2022a. New evidence for avian and small non-avian theropod ichnotaxa from the Lower Cretaceous of Canada: implications for theropod ichnodiversity. Cretac. Res. 138, 105292.

Lockley, M.G., Kim, S.H., Kim, K.S., Bae, S.M., Kim, J.Y., Xing, L., 2022b. A high-density *grallator* assemblage from the Haman Formation (cretaceous), Korea: implications for cretaceous distribution of grallatorids in east Asia. Hist. Biol. 34 (12), 2430–2438.

Lockwood, J.A.F., Lockley, M.G., Pond, S., 2014. A review of footprints from the Wessex formation (Wealden group, lower cretaceous) at Hanover Point, the Isle of Wight, southern England. Biol. J. Linn. Soc. 113 (3), 707–720.

Long, J., 1998. Dinosaurs of Australia and New Zealand. Sydney: University of New South Wales Press, Sydney, 198pp.

Loope, D.B., Milàn, J., 2016. Dinosaur tracks in eolian strata: new insights into track formation, walking kinetics and trackmaker behavior. In: Falkingham, P.L., Marty, D., Richter, A. (Eds.), Dinosaur Tracks: The Next Steps. Indiana University Press, Bloomington, pp. 358–364.

Lovelace, D.M., Lovelace, S.D., 2012. Paleoenvironments and paleoecology of a lower Triassic invertebrate and vertebrate ichnoassemblage from the Red Peak Formation (Chugwater group), central Wyoming. Palaios 27 (9), 636–657.

Lucas, S.G., Klein, H., Lockley, M.G., Spielmann, J.A., Gierliński, G., Hunt, A.P., Tanner, L.H., 2006. Triassic–Jurassic stratigraphic distribution of the theropod footprint ichnogenus Eubrontes. New Mex. Mus. Nat. Hist. Sci. Bull. 37, 86–93.

Lucas, S.G., 2007. Tetrapod footprint biostratigraphy and biochronology. Ichnos 14, 5–38.

Lucas, S.G., Sullivan, R.M., Jasinski, S.E., Ford, T.L., 2011. Hadrosaur footprints from the Upper Cretaceous Fruitland Formation, San Juan Basin, New Mexico, and the ichnotaxonomy of large ornithopod footprints. New Mex. Mus. Nat. Hist. Sci. Bull. 53, 357–362.

Lucas, S.G., Szajna, M.J., Lockley, M.G., Fillmore, D.L., Simpson, E.L., Klein, H., Boyland, J., Hartline, B.W., 2014. The Middle-Late Triassic tetrapod footprint ichnogenus *Gwyneddichnium*. New Mex. Mus. Nat. Hist. Sci. Bull. 62, 135–155.

Lucas, S.G., Dalman, S.G., 2016. The Early Cretaceous Clayton Lake dinosaur tracksite, northeastern New Mexico. Fossil Rec. 74, 127–140.

Lucas, S.G., Abbassi, N., Francischini, H., Denzien-dias, P., Knight, P., 2023. Bird-dominated Pliocene tracksite from the Santa Fe group, Rio Rancho. New Mex. Mus. Nat. Hist. Sci. Bull. 94, 399–410.

Lull, R.S., 1904. Fossil footprints of the Jura–Trias of North America. Mem. Boston Soc. Nat. Hist. 5 (2), 461–557.

Lull, R.S., 1915. Triassic life of the Connecticut valley. State Connecticut, State Geol. Nat. Hist. Surv. Bull. 81, 1–336.

Lull, R.S., 1942. Triassic footprints from Argentina. Am. J. Sci. 240 (6), 421–425.

Lull, R.S., 1953. Triassic life of the Connecticut valley. Bull. Connecticut Geol. Nat. Hist. Surv. 181, 1e331.

Lužar-Oberiter, B., Kordić, B., Mezga, A., 2017. Digital modelling of the late Albian *Solaris* dinosaur tracksite (Istria, Croatia). Palaios 32 (12), 739–749.

Mahboubi, M., Bessedik, M., Belkebir, L., Bensalah, M., Adaci, M., 2004. Découverte des premières empreintes digitales de dinosauriens crétacés dans la région d'El Bayadh. Colloque international Terre et eau, Annaba, pp. 92–95.

Mahboubi, M., Bessedik, M., Belkebir, L., Adaci, M., Hebib, H., Bensalah, M., Mammeri, C., Mansour, B., Mansouri, M.E.H., 2007. Première découverte d'empreintes de pas de dinosaures dans le Crétacé inférieur de la région d'El Bayadh (Algérie). Bull. Serv. Geol. Algerie 18 (2), 127–139.

Maidment, S.C., Norman, D.B., Barrett, P.M., Upchurch, P., 2008. Systematics and phylogeny of Stegosauria (dinosauria: Ornithischia). J. Syst. Palaeontol. 6 (4), 367–407.

Maidment, S.C., Raven, T.J., Ouarhache, D., Barrett, P.M., 2020. North Africa's first stegosaur: implications for Gondwanan thyreophoran dinosaur diversity. Gondwana Res. 77, 82–97.

Makovicky, P.J., Kobayashi, Y., Currie, P.J., 2004. Ornithomimosauria. In: Weishampel, D.B., Dodson, P., Osmólska, H. (Eds.), The Dinosauria, second ed. University of California Press, Berkeley, pp. 137–150.

Malafaia, E., Ortega, F., Escaso, F., Silva, B., 2015. New evidence of ceratosaurus (dinosauria: Theropoda) from the late Jurassic of the Lusitanian basin, Portugal. Hist. Biol. 27 (7), 938–946.

Manni, R., Nicosia, U., Nobili, G., 1999. An unusual tetrapod trackway from Lower Jurassic marine sediments of Central Italy: *Accordiichnus natans* n. ichnogen., n. ichsp. Geol. Rom. 35, 167–187.

Manning, P.L., 2004. A new approach to the analysis and interpretation of tracks: examples from the dinosauria. Geol. Soc. 228 (1), 93–123.

Mannion, P.D., Upchurch, P., 2010. A quantitative analysis of environmental associations in sauropod dinosaurs. Paleobiology 36 (2), 253–282.

Marchetti, L., Belvedere, M., Voigt, S., Klein, H., Castanera, D., Díaz-Martínez, I., Marty, D., Xing, L., Feola, S., Melchor, R.N., Farlow, J.O., 2019. Defining the morphological quality of fossil footprints. Problems and principles of preservation in tetrapod ichnology with examples from the Palaeozoic to the present. Earth Sci. Rev. 193, 109–145.

Marsh, O.C., 1885. Names of extinct reptiles. Am. J. Sci. 29, 169.

Martin, J., Menkem, E.F., Djomeni, A., Fowe, P.G., Ntamak-Nida, M., 2017. Dinosaur trackways from the early Late Cretaceous of western Cameroon. J. Afr. Earth Sci. 134, 213–221.

Marty, D., Belvedere, M., Meyer, C.A., Mietto, P., Paratte, G., Lovis, C., Thüring, B., 2010. Comparative analysis of late Jurassic sauropod trackways from the Jura Mountains (NW Switzerland) and the central high Atlas Mountains (Morocco): implications for sauropod ichnotaxonomy. Hist. Biol. 22, 109–133.

Marty, D., Belvedere, M., Razzolini, N.L., Lockley, M.G., Paratte, G., Cattin, M., Lovis, C., Meyer, C.A., 2018. The tracks of giant theropods (*Jurabrontes curtedulensis* ichnogen. ichnosp. nov.) from the Late Jurassic of NW Switzerland: palaeoecological palaeogeographical implications. Hist. Biol. 30 (7), 928–956.

Marzola, M., Mateus, O., Schulp, A.S., Jacobs, L.L., Polcyn, M.J., Pervov, V., Gonçalves, A.O., Morais, M.L., 2015. Comparative anatomy and systematics of cretaceous mammal tracks of Angola. In: Jago, J.W.M., Hebda, G., Mitrus, S., Jagt-Yazykova, E., Bodzioch, A., Konietzko-Meier, D., Kardynal, K., Gruntmejer, K. (Eds.), 13th Annual Meeting of the European Association of Vertebrate Paleontologists, Opole, Poland, p. 35, 169 pp.

Masrour, M., Lkebir, N., Pérez-Lorente, F., 2017a. Anza palaeoichnological site. Late Cretaceous. Morocco. Part II. Problems of large dinosaur trackways and the first African *Macropodosaurus* trackway. J. Afr. Earth Sci. 134, 776–793.

Masrour, M., Pascual-Arribas, C., de Ducla, M., Hernández-Medrano, N., Pérez-Lorente, F., 2017b. Anza palaeoichnological site. Late Cretaceous. Morocco. Part I. The first African pterosaur trackway (manus only). J. Afr. Earth Sci. 134, 766–775.

Masrour, M., Ducla, M., Billon-Bruyat, J.P., Mazin, J.M., 2018. Rediscovery of the Tagragra tracksite (Maastrichtian, Agadir, Morocco): *Agadirichnus elegans* Ambroggi and de Lapparent 1954 is pterosaurian ichnotaxon. Ichnos 25 (4), 285–294.

Masrour, M., Boutakiout, M., Herrero Gascón, J., Sáinz Ruiz De Zuazo, J.L., Ochoa Martínez, R., Pérez-Lorente, F., 2020. Footprints of *Batrachopus* isp. from the imilchil megatracksite. Middle? - Upper Jurassic, central High Atlas (Morocco). J. Afr. Earth Sci. 172, 103980.

Masrour, M., Boutakiout, M., Ceniceros, J.M., Rolland, T., Pérez-Lorente, F., 2023. Morphometric and biomorphic variability of theropod footprint characters from Imilchil tracksites (Mid-? Late Jurassic, Central High Atlas, Morocco). J. Afr. Earth Sci. 207, 105048.

Mateus, O., Milàn, J., Romano, M., Whyte, M.A., 2011. New finds of stegosaur tracks from the upper Jurassic Lourinhã Formation, Portugal. Acta Palaeontol. Pol. 56 (3), 651–658.

Mateus, O., Marzola, M., Schulp, A.S., Jacobs, L.L., Polcyn, M.J., Pervov, V., Gonçalves, A.O., Morais, M.L., 2017. Angolan ichnosite in a diamond mine shows the presence of a large terrestrial mammaliamorph, a crocodylomorph, and sauropod dinosaurs in the Early Cretaceous of Africa. Palaeogeogr. Palaeoclimatol. Palaeoecol. 471, 220–232.

Matsukawa, M., Obata, I., 1985. Dinosaur footprints and other indentations in the Cretaceous Sebayashi Formation, Sebayashi Japan. Bull. Natl. Sci. Mus. 11 (1), 9–36.

Matsukawa, M., Hamuro, T., Mizukami, T., Fujii, S., 1997. First trackway evidence of gregarious dinosaurs from the lower cretaceous Tetori group of eastern Toyama Prefecture, central Japan. Cretac. Res. 18 (4), 603–619.

Matsukawa, M., Shibata, K., Kukihara, R., Koarai, K., Lockley, M.G., 2005. Review of Japanese dinosaur track localities: implications for ichnotaxonomy, paleogeography and stratigraphic correlation. Ichnos 12 (3), 201–222.

Matsukawa, M., Lockley, M., Li, J., 2006. Cretaceous terrestrial biotas of East Asia, with special reference to dinosaur-dominated ichnofaunas: towards a synthesis. Cretac. Res. 27, 3–21.

Matsukawa, M., Lockley, M.G., Hayashi, K., Korai, K., Peiji, C., Haichun, Z., 2014. First report of the ichnogenus *Magnoavipes* from China: new discovery from the lower cretaceous inter-mountain Basin of Shangzhou, Shaanxi province, central China. Cretac. Res. 47, 131–139.

Matsumoto, R., Evan, S.E., Manabe, M., 2007. The choristoderan reptile *Monjurosuchus* from the early cretaceous of Japan. Acta Palaeontol. Pol. 52, 329–350.

Matsumoto, R., Tsogtbaatar, K., Ishigaki, S., Tsogtbaatar, C., Enkhtaivan, Z., Evans, S.E., 2019. Revealing body proportions of the enigmatic choristodere reptile *Khurendukhosaurus* from Mongolia. Acta Palaeontol. Pol. 64 (2), 363–377.

Mayor, A., 2005. Fossil Legends of the First Americans. Princeton University Press, Princeton, 488pp.

Mazin, J.M., Hantzpergue, P., Lafaurie, G., Vignaud, P., 1995. Des pistes de ptérosaures dans le Tithonien de Crayssac (Quercy, France). Comptes rendus de l'Académie des sciences. Série 2. Sciences de la terre et des planètes 321 (5), 417–424.

Mazin, J.M., Billon-Bruyat, J.P., Hantzpergue, P., Lafaurie, G., 2003. Ichnological evidence for quadrupedal locomotion in pterodactyloid pterosaurs: Trackways from the Late Jurassic of Crayssac (southwestern France). In: Buffetaut, E., Mazin, J.M. (Eds.), Evolution and Palaeobiology of Pterosaurs, vol. 217. Geological Society, London.
Mazin, J.M., Billon-Bruyat, J.P., Padian, K., 2009. First record of a pterosaur landing trackway. Proc. Biol. Sci. 276 (1674), 3881–3886.
Mazin, J.M., Hantzpergue, P., Olivier, N., 2017. The dinosaur tracksite of Plagne (early Tithonian, Late Jurassic; Jura Mountains, France): the longest known sauropod trackway. Geobios 50 (4), 279–301.
Mazin, J.M., Pouech, J., 2020. The first non-pterodactyloid pterosaurian trackways and the terrestrial ability of non-pterodactyloid pterosaurs. Geobios 58, 39–53.
Mcallister, J.A., 1989. Dakota Formation tracks from Kansas: implications for the recognition of Tetrapod subaqueous traces. In: Gillette, D.D., Lockley, M.G. (Eds.), Dinosaur Tracks and Traces. Cambridge University Press, Cambridge, pp. 343–348.
McCrea, R.T., 2000. Vertebrate Palaeoichnology of the Early Cretaceous (Lower Albian) Gates Formation of Alberta. University of Saskatchewan. Unpublished MSc. Thesis. 120 pp.
McCrea, R.T., Lockley, M.G., Meyer, C.A., 2001. Global distribution of purported ankylosaur track occurrences. In: Carpenter, K. (Ed.), The Armored Dinosaurs. Indiana University Press, Bloomington, pp. 413–454.
McCrea, R.T., Sarjeant, W.A.S., 2001. New ichnotaxa of bird and mammal footprints from the lower cretaceous (Albian) Gates Formation of Alberta. In: Tanke, D.H., Carpenter, K. (Eds.), Mesozoic Vertebrate Life: New Research Inspired by the Paleontology of Philip J. Currie. Indiana University Press, Bloomington, pp. 453–478.
McCrea, R.T., Currie, P.J., Pemberton, S.G., 2002. Forelimb impressions associated with a large theropod trackway from the Gates Formation (Lower Cretaceous, Albian) of western Canada. J. Vertebr. Paleontol. 22 (Suppl. 3), 86A.
McCrea, R.T., Buckley, L.G., Pint, A.G., Currie, P.J., Haggart, J.W., Helm, C.W., Pemberton, S.G., 2014a. A review of vertebrate track-bearing formations from the Mesozoic and earliest Cenozoic of western Canada with a description of a new theropod ichnospecies and reassignment of an avian ichnogenus: fossil footprints of western North America. New Mex. Mus. Nat. Hist. Sci. Bull. 62, 5–93.
McCrea, R.T., Buckley, L.G., Farlow, J.O., Lockley, M.G., Currie, P.J., Matthews, N.A., Pemberton, S.G., 2014b. A 'terror of tyrannosaurs': the first trackways of tyrannosaurids and evidence of gregariousness and pathology in Tyrannosauridae. PLoS One 9 (7), e103613.
McCrea, R.T., Buckley, L.G., Plint, A.G., Lockley, M.G., Matthews, N.A., Noble, T.A., Xing, L.D., Krawetz, J.R., 2015a. Vertebrate ichnites from the Boulder Creek Formation (Lower Cretaceous: middle to ?upper Albian) of northeastern British Columbia, with a description of a new avian ichnotaxon, *Paxavipes babcockensis* ichnogen. et isp. nov. Cretac. Res. 55, 1–18.
Mccrea, R.T., Tanke, D.H., Buckley, L.G., Lockley, M.G., Farlow, J.O., Xing, L.D., Matthews, N.A., Helm, C.W., Pemberton, S.G., Breithaupt, B.H., 2015b. Vertebrate ichnopathology: pathologies inferred from dinosaur tracks and trackways from the mesozoic. Ichnos 22, 235–260.
McCrea, R.T., Buckley, L.G., Lockley, M.G., Xing, L.D., Matthews, N., Pigeon, T., Amos, L., 2017. The "Six Peaks Dinosaur Track Site": A Diverse Fossil Vertebrate Track Site from the Gaylard Member of the Gething Formation (Early Cretaceous: Aptian) from the Carbon Creek Basin, Northeastern British Columbia, vol. 31. Hosted by University of Alberta.
McKeever, P.J., Haubold, H., 1996. Reclassification of vertebrate trackways from the Permian of Scotland and related from Arizona and Germany. J. Paleontol. 70 (6), 1011–1022.
Mehl, M.G., 1931. Additions to the vertebrate record of the Dakota Sandstone. Am. J. Sci. 5 (21), 441–452.
Meijide Calvo, M., 2001. Pterosaur tracks in Oncala berriasian (Soria, Spain). New ichnospecies: pteraichnus manueli. Strata 11, 72–74.
Meijide Calvo, M., Fuentes Vidarte, C., 2001. Huellas de pterosaurios en el Weald de Soria (España). In: Ca, S. (Ed.), Actas de las I Jornadas Internacionales sobre Paleontología de Dinosaurios y su Entorno. Salas de los Infantes, Burgos, pp. 397–406.
Meijide Calvo, M., Meijide Fuentes, F., Fuentes Vidarte, C., Meijide Fuentes, M., 2001. Huellas de pterosaurio en la Sierra de Oncala (Soria, España). Nueva icnospecies Pteraichnus parvus. In: Ca, S. (Ed.), Actas de las II Jornadas Internacionales sobre Paleontologíca de Dinosaurios y su Entorno. Salas de los Infantes, Burgos, pp. 397–406.
Meijide Fuentes, F., 2001. Pterosaur tracks in Oncala mountain range (Soria, Spain) a new ichnospecies: pteraichnus vetustior. Strata - Ser. 1 Commun. 11, 70–71.

Meijide Fuentes, F., Fuentes Vidarte, C., Meijide Calvo, M., Meijide Fuentes, M., 2004. Rastro de un dinosaurio Saurópodo en el Weald de Soria (España) *Brontopodus oncalensis* nov. Icnsp. Celtiberia 98, 501–515.

Melchor, R.N., Buchwaldt, R., Bowring, S., 2013. A late Eocene date for late triassic bird tracks. Nature 495, E1–E2.

Melchor, R.N., Rivarola, D.L., Umazano, A.M., Moyano, M.N., Belmontes, F.R.M., 2019. Elusive Cretaceous Gondwanan theropods: the footprint evidence from central Argentina. Cretac. Res. 97, 125–142.

Mesa, V., Perea, D., 2015. First record of theropod and ornithopod tracks and detailed description of sauropod trackways from the Tacuarembó Formation (Late Jurassic-? Early Cretaceous) of Uruguay. Ichnos 22 (2), 109–121.

Mesle, G., Peron, P.A., 1880. Sur des empreintes de pas d'oiseaux observées par M. le Mesle dans le Sud de l'Algérie. Association Française pour l'Avancement des Sciences. Congrès de Reims 1–6.

Meyer, C., Thüring, B., 2003. The first iguanodontid dinosaur tracks from the Swiss Alps (Schrattenkalk Formation, Aptian). Ichnos 10 (2–4), 221–228.

Meyer, C.A., 1993. A sauropod dinosaur mepatracksite from theLate Jurassic of Northern Switzerland. Ichnos 3 (1), 29–38.

Meyer, C.A., Marty, D., Belvedere, M., 2018. Titanosaur trackways from the late cretaceous el molino Formation of Bolivia (Cal orck'o, sucre). Ann. Soc. Geol. Pol. 88 (2), 223–241.

Meyer, C.A., Marty, D., Thüring, B., Thüring, S., Belvedere, M., 2021. The Late Cretaceous dinosaur track record of Bolivia-Review and perspective. J. S. Am. Earth Sci. 106, 102992.

Mezzalira, S., 1966. In: Os fósseis do Estado de. São Paulo, vol. 45. Boletim do Instituto Geográ co. e Geológico, São Paulo, pp. 559–567.

Milan, J., Chiappe, L.M., 2009. First American record of the Jurassic ichnospecies *Deltapodus brodricki* and a review of the fossil record of stegosaurian footprints. J. Geol. 117 (3), 343–348.

Milan, J., Chiappe, L.M., Loope, D.B., Kirkland, J.I., Lockley, M.G., 2015. First report on dinosaur tracks from the Burro Canyon Formation, San Juan County, Utah, USA: evidence of a diverse, hitherto unknown Lower Cretaceous dinosaur fauna. Ann. Soc. Geol. Pol. 85, 515–525.

Milner, A.R., Harris, J.D., Lockley, M.G., Kirkland, J.I., Matthews, N.A., 2009. Bird-like anatomy, posture, and behavior revealed by an Early Jurassic theropod dinosaur resting trace. PLoS One 4 (3), e4591.

Mohabey, D.M., 1986. Note on a dinosaur footprint from Kheda District, Gujarat. J. Geol. Soc. India 27, 456–459.

Moratalla, J.J., Sanz, J.L., Jimenez, S., 1988. Multivariate analysis on Lower Cretaceous dinosaur footprints: discrimination between ornithopods and theropods. Geobios 21 (4), 395–408.

Moratalla, J.J., 1993. Restos indirectos de dinosaurios del registro español: Paleoicnología de la Cuenca de Cameros (Jurásico superior-Cretácico inferior) y Paleoicnología del Cretácico superior. Tesis Doctoral. PhD dissertation. Universidad Complutense de Madrid, Spain, 727 pp.

Moratalla, J.J., 1994. Sauropod Trackways from the Lower Cretaceous of Spain.

Moratalla, J.J., 2009. Sauropod tracks of the Cameros Basin (Spain): identification, trackway patterns and changes over the Jurassic-Cretaceous. Geobios 42, 797–811.

Moratalla, J.J., Hernán, J., 2009. Turtle and pterosaur tracks from the Los Cayos dinosaur tracksite, Cameros Basin (Cornago, La Rioja, Spain): tracking the lower cretaceous biodiversity. Rev. Esp. Palaontol. 24 (1), 59–77.

Moreau, J.D., Trincal, V., Fara, E., Baret, L., Jacquet, A., Barbini, C., Flament, R., Wienin, M., Bourel, B., Jean, A., 2019. Middle Jurassic tracks of sauropod dinosaurs in a deep karst cave in France. J. Vertebr. Paleontol. 39 (6), e1728286.

Mossman, D.J., Grantham, R.G., 2008. Eochelysipus horni, a new vertebrate trace fossil from the Tournaisian Horton Bluff Formation, Nova Scotia. Atl. Geol. 44, 69–77.

Mudge, B.F., 1866. Discovery of fossil foot-marks in the Liassic (?) Formation in Kansas. Am. J. Sci. 41 (122), 174–176.

Mustoe, G.E., 2019. Lower Eocene footprints from northwest Washington, USA. Part 1: reptile tracks. Geosciences 9 (7), 321.

Myers, T.S., Fiorillo, A.R., 2009. Evidence for gregarious behavior and age segregation in sauropod dinosaurs. Palaeogeogr. Palaeoclimatol. Palaeoecol. 274 (1–2), 96–104.

Namnandorzhi, O., 1957. Traces of huge lizards in Mongolia (in Russian). Priroda 44, 110–111.

Natali, L., Leonardi, G., 2023. Coneroichnus marinus ichnogenus et ichnospecies nov., a fossil trackway of marine reptile in the Maiolica Formation (Upper Jurassic-Lower Cretaceous) from Monte Conero, Marche, Italy. Rev. Bras. Palaontol. 26 (3), 156–171.

Navarro-Lorbés, P., Ruiz, J., Díaz-Martínez, I., Isasmendi, E., Sáez-Benito, P., Viera, L., Pereda-Suberbiola, X., Torices, A., 2021. Fast-running theropods tracks from the early cretaceous of La Rioja, Spain. Sci. Rep. 11, 23095.

Navarro-Lorbés, P., Díaz-Martínez, I., Valle-Melón, J.M., Rodriguez-Miranda, A., Moratalla, J.J., Ferrer-Ventura, M., San Juan-Palacios, R., Torices, A., 2023. Dinosaur swim tracks from the Lower Cretaceous of La Rioja, Spain: An ichnological approach to non-common behaviours. Cretac. Res. 147, 105516.

Nessov, L.A., 1995. Dinosaurs of the Northern Eurasia: New Data about Assemblages, Ecology, and Palaeobiogeography. University of Sankt-Petersburg, Institute of Earth Crus, 156 pp., 14 plates.

Nicosia, U., Marino, M., Mariotti, N., Muraro, C., Panigutti, S., Petti, F.M., Sacchi, E., 1999. The Late Cretaceous dinosaur tracksite near Altamura (Bari, southern Italy). Geol. Rom. 35, 231–236.

Niedźwiedzki, G., Singer, T., Gierliński, G.D., Lockley, M.G., 2012. A protoceratopsid skeleton with an associated track from the Upper Cretaceous of Mongolia. Cretac. Res. 33 (1), 7–10.

Niu, K., Xing, L.D., 2023. The first dinosaur track assemblages from the upper cretaceous Shaxian Formation, Fujian province, southeastern China. Cretac. Res. 146, 105486.

Noè, L.F., Gómez–Pérez, M., Rodríguez, J.V., Corrales–García, A., Caranton–Mateus, W.G., 2020. Dinosaur footprints from the lower cretaceous, Batá formation, Colombia (South America), and the possible interchange of large ornithopods between southern Laurasia and northern Gondwana. Geol. Colomb. 2, 375–401.

Nopcsa, F., 1923. Die Familien der Reptilien. Fortschritte der Geologie und Palaeontologie 2, 1–210.

Norell, M., Makovicky, P.J., 1997. Important features of the dromaeosaur skeleton: information from a new specimen. Am. Mus. Novit. 3215, 1–28.

Norman, D.B., 2010. A taxonomy of iguanodontians (dinosauria: Ornithopoda) from the lower Wealden group (cretaceous: valanginian) of southern England. Zootaxa 2489, 47–66.

Norman, D.B., Baron, M.G., Garcia, M.S., Müller, R.T., 2022. Taxonomic, palaeobiological and evolutionary implications of a phylogenetic hypothesis for Ornithischia (Archosauria: dinosauria). Zool. J. Linn. Soc. 196 (4), 1273–1309.

O'Brien, C.L., Robinson, S.A., Pancost, R.D., Damsté, J.S.S., Schouten, S., Lunt, D.J., Alsenz, H., Bornemann, A., Bottini, C., Brassell, S.C., Farnsworth, A., 2017. Cretaceous sea-surface temperature evolution: constraints from TEX86 and planktonic foraminiferal oxygen isotopes. Earth Sci. Rev. 172, 224–247.

Olsen, P., Sha, J., Fang, Y., Chang, C., Whiteside, J.H., Kinney, S., Sues, H.D., Kent, D., Schaller, M., Vajda, V., 2022. Arctic ice and the ecological rise of the dinosaurs. Sci. Adv. 8 (26), eabo6342.

Olsen, P.E., Padian, K., 1986. Earliest records of *Batrachopus* from the southwestern USA, and a revision of some Early Mesozoic crocodylomorph ichnogenera. In: The Beginning of the Age of Dinosaurs, Faunal Change across the Triassic-Jurassic Boundary, vol. 44. Symposium. Society of Vertebrate Paleontology. Annual Meeting, pp. 259–273.

Olsen, P.E., Smith, J.B., McDonald, N.G., 1998. Type material of the type species of the classic theropod footprint genera Eubrontes, Anchisauripus, and Grallator (Early Jurassic, Hartford and Deerfield basins, Connecticut and Massachusetts, USA). J. Vertebr. Paleontol. 18 (3), 586–601.

Olsen, P.E., 2002. Field guide for non-marine boundary events in the Newark Basin (New Jersey, Pennsylvania, and Connecticut), Eastern USA and their litho-, chrono-and biostratigraphic context. Guidebooks Field Workshops Int. Geol. Corr. Progr. 458, 181.

Ostrom, J.H., 1969. Osteology of Deinonychus Antirrhopus, an Unusual Theropod from the Lower Cretaceous of Montana, vol. 30. Peabody Museum of Natural History, Yale University Bulletin, pp. 1–165.

Pacheco, J.A.A., 1913. Notas sobre a geologia do vale do rio grande a partir da foz do rio Pardoaté a sua confluência com o rio Parahyba. In: Relatório Comissão Geographica e Geológica do Estado de São Paulo. Exploração do Rio Grande e de seus afluentes, São José dos Dourados, São Paulo, pp. 33–38.

Padian, K., 1983. A functional analysis of flying and walking in pterosaurs. Paleobiology 9 (3), 218–239.

Páez, V.P., Morales-Betancourt, M.A., Lasso, C.A., Castaño-Mora, O.V., Bock, B.C., 2012. Biology and conservation of the continental turtles of Colombia. Serie Editorial Recursos Hidrobiológicos y Pesqueros Continentales de Colombia. Alexander von Humboldt Biological Resources Research Institute (IAvH), Bogotá, D.C., Colombia, p. 528.

Panascí, G., Varricchio, D.J., Martin, A.J., Dyman, T., 2023. Dinosaur tracks from the Frontier Formation, Montana: preservation, distribution and palaeoecological significance for the middle Cretaceous terrestrial ecosystems of North America. Hist. Biol. 1–24.

Panin, N., Avram, E., 1962. Noi urme de vertebrate in Miocenul Subcarpatilor Rominesti. Stud. Cercet. Geol. 7, 455–484.

Panin, N., 1965. Coexistence de traces de pas de vertébrés et de mécanoglyphes dans le Molasse Miocène des Carpates orientales. Revue Roumaine de Géologie, Géophysique, et Géographie 7, 141–163.
Park, W.M., Lockley, M.G., Kim, J.Y., Kim, K.S., 2018. Anuran (frog) trackways from the Cretaceous of Korea. Cretac. Res. 86, 135–148.
Pascual, C., Canudo, J.I., Hernández, N., Barco, J.L., Castanera, D., 2012. First record of stegosaur dinosaur tracks in the lower cretaceous (berriasian) of Europe (Oncala group, Soria, Spain). Geodiversitas 34 (2), 297–312.
Pascual-Arribas, C., Sanz Pérez, E., 2000. Huellas de Pterosaurios en el grupo Oncala (Soria, España). *Pteraichnus palacieisaenzi*, nov. icnosp. Estud. Geol. 56, 73–100.
Pascual-Arribas, C., Hernández Medrano, N., Latorre Macarrón, P., Sanz Pérez, E., 2005. Nuevo rastro de icnitas de cocodrilo en la Aloformación Huérteles de la Cuenca de Cameros. Stud. Geol. Salmant. 41, 75–91.
Pascual-Arribas, C., Hernández-Medrano, N., 2011. Posibles huellas de crías de terópodo en el yacimiento de Valdehijuelos (Soria, España). Stud. Geol. Salmant. 47 (1), 77–110.
Pascual-Arribas, C., Canudo, J.I., Sanz Perez, E., Hernández-Medrano, N., Castanera, D., Barco, J.L., 2015. On the validity of *Pteraichnus palacieisaenzi*, Pascual Arribas Sanz Pérez 2000: new data from the Huérteles formation. Paläontol. Z. 89, 459–483.
Pascual-Arribas, C., Hernández-Medrano, N., 2015. Las huellas de tortuga del Grupo Oncala (Berriasiense, Cuenca de Cameros, España). Estud. Geol. 71 (1) e030–e030.
Peabody, F.E., 1955. Taxonomy and the footprints of tetrapods. J. Paleontol. 29, 915–918.
Peng, B.X., Du, Y.S., Li, D.Q., Bai, Z.C., 2004. The first discovery of the early cretaceous pterosaur track and its significance in Yanguoxia, Yongjing County, Gansu province. Earth Sci. J. China Univ. Geosci. 29, 21–24.
Penn, S.J., Sweetman, S.C., Martill, D.M., Coram, R.A., 2020. The Wessex Formation (Wealden group, lower cretaceous) of Swanage Bay, southern England. Proc. Geologists' Assoc. 131 (6), 679–698.
Pereda-Suberbiola, X., Dantas, P., Galton, P.M., Sanz, J.L., 2005. Autopodium of the holotype of Dracopelta zbyszewskii (dinosauria, Ankylosauria) and its type horizon and locality (Upper Jurassic: Tithonian, western Portugal). Neues Jahrbuch Geol. Palaontol. Abhand. 175–196.
Pereda-Suberbiola, X., Díaz-Martínez, I., Salgado, L., de Valais, S., 2015. Síntesis del registro fósil de dinosaurios tireóforos en Gondwana. Publicación Electrónica de la Asociación Paleontológica Argentina 15 (1), 90–107.
Pérez-Lorente, F., 2015. Dinosaur Footprints and Trackways of La Rioja. Indiana University Press, Bloomington, 363 pp.
Pérez-Pueyo, M., Cruzado-Caballero, P., Moreno-Azanza, M., Vila, B., Castanera, D., Gasca, J.M., Puértolas-Pascual, E., Bádenas, B., Canudo, J.I., 2021. The tetrapod fossil record from the uppermost Maastrichtian of the Ibero-Armorican Island: an integrative review based on the outcrops of the western Tremp Syncline (Aragón, Huesca Province, NE Spain). Geosciences 11 (4), 162.
Peters, D., 2011. A catalog of pterosaur pedes for trackmaker identification. Ichnos 18, 114–141.
Peterson, W., 1924. Dinosaur tracks in the roofs of coal mines. Nat. Hist. 24, 388–397.
Petti, F.M., Porchetti, S.D.O., Sacchi, E., Nicosia, U., 2010. A new purported ankylosaur trackway in the Lower Cretaceous (lower Aptian) shallow-marine carbonate deposits of Puglia, southern Italy. Cretac. Res. 31 (6), 546–552.
Petti, F.M., Bernardi, M., Ferretti, P., Tomasoni, R., Avanzini, M., 2011. Dinosaur tracks in a marginal marine environment: the Coste dell'Anglone ichnosite (early Jurassic, Trento platform, NE Italy). Ital. J. Geosci. 130 (1), 27–41.
Petti, F.M., Antonelli, M., Citton, P., Mariotti, N., Petruzzelli, M., Pignatti, J., D'Orazi Porchetti, S., Romano, M., Sacchi, E., Sacco, E., Wagensommer, A., 2020. Cretaceous tetrapod tracks from Italy: a treasure trove of exceptional biodiversity. J. Mediterranean Earth Sci. 12, 167–191.
Pianka, E.R., Vitt, L.J., 2003. Lizards: Windows to the Evolution of Diversity, vol. 5. Univ of California Press.
Piñuela, L., Garcia-Ramos, J.C., Romano, M., Ruiz-Omenaca, J.I., 2016. First record of gregarious behavior in robust medium-sized Jurassic Ornithopods: evidence from the Kimmeridgian trackways of Asturias (N. Spain) and some general considerations on other medium-large ornithopod tracks in the Mesozoic record. Ichnos 23 (3–4), 298–311.
Pittman, J.G., 1989. Stratigraphy, lithology, depositional environment, and track type of dinosaur track-bearing beds of the Gulf Coastal Plain. In: International Symposium on Dinosaur Tracks and Traces, vol. 1, pp. 135–153.
Plotnick, R.E., Smith, D.M., 2012. Exceptionally preserved fossil insect ears from the Eocene Green River Formation of Colorado. J. Paleontol. 86 (1), 19–24.
Pond, S., Lockley, M.G., Lockwood, J.A., Breithaupt, B.H., Matthews, N.A., 2014. Tracking dinosaurs on the Isle of Wight: a review of tracks, sites, and current research. Biol. J. Linn. Soc. 113 (3), 737–757.

Porchetti, S.D.O., Bernardi, M., Cinquegranelli, A., Santos, V.F., Marty, D., Petti, F.M., Caetano, P.S., Wagensommer, A., 2016. A review of the dinosaur track record from Jurassic and Cretaceous shallow marine carbonate depositional environments. In: Falkingham, P.L., Marty, D., Richter, A. (Eds.), Dinosaur Tracks: The Next Steps. Indiana University Press, Bloomington, pp. 380–392.

Porchetti, S.D.O., Bertini, R.J., Langer, M.C., 2017. Walking, running, hopping: analysis of gait variability and locomotor skills in Brasilichnium elusivum Leonardi, with inferences on trackmaker identification. Palaeogeogr. Palaeoclimatol. Palaeoecol. 465, 14–29.

Poropat, S., White, M.A., Ziegler, T., Pentland, A., Beeston, S.L., Duncan, R., Sloan, T., Elliott, D.A., 2021. A diverse late cretaceous vertebrate tracksite from the Winton Formation of Queensland, Australia. PeerJ 9 (e11544), 1–87.

Prum, R.O., Berv, J.S., Dornburg, A., Field, D.J., Townsend, J.P., Lemmon, E.M., Lemmon, A.R., 2015. A comprehensive phylogeny of birds (Aves) using targeted next-generation DNA sequencing. Nature 526 (7574), 569–573.

Rainforth, E.C., 2005. Ichnotaxonomy of the Fossil Footprints of the Connecticut Valley (Early Jurassic, Newark Supergroup, Connecticut and Massachusetts). Ph.D. thesis. Columbia University, 1302 pp.

Rainforth, E.C., 2007. Ichnotaxonomic updates from the newark Supergroup. In: Contributions to the Paleontology of New Jersey (II): Field Guide and Proceedings. Trenton: Geological Association of New Jersey, pp. 49–59.

Rauhut, O.W., Piñuela, L., Castanera, D., García-Ramos, J.C., Cela, I.S., 2018. The largest European theropod dinosaurs: remains of a gigantic megalosaurid and giant theropod tracks from the Kimmeridgian of Asturias, Spain. PeerJ 6, e4963.

Razzolini, N.L., Vila, B., Díaz-Martínez, I., Manning, P.L., Galobart, À., 2016. Pes shape variation in an ornithopod dinosaur trackway (Lower Cretaceous, NW Spain): new evidence of an antalgic gait in the fossil track record. Cretac. Res. 58, 125–134.

Reilly, S.M., Elias, J.A., 1998. Locomotion in *Alligator mississippiensis*: kinematic effects of speed and posture and their relevance to the sprawling-to-erect paradigm. J. Exp. Biol. 201, 2559–2574.

Remeika, P., 1999. Identification, stratigraphy, and age of neogene vertebrate footprints from the vallecito-fish creek basin, anza-borrego desert state park, California. San Bernardino County Mus. Assoc. Quarter. 46 (2), 37–46.

Richter, A., Böhme, A., 2016. Too many tracks: preliminary description and interpretation of the diverse and heavily dinoturbated Early Cretaceous 'chicken yard' ichnoassemblage (Obernkirchen Tracksite, Northern Germany). In: Falkingham, P.L., Marty, D., Richter, A. (Eds.), Dinosaur Tracks: The Next Steps. Indiana University Press, Bloomington, pp. 334–357.

Riguetti, F., Citton, P., Apesteguía, S., Zacarías, G.G., Pereda-Suberbiola, X., 2021. New ankylosaurian trackways (cf. *Tetrapodosaurus*) from an uppermost Cretaceous level of the El Molino Formation of Bolivia. Cretac. Res. 124, 104810.

Robertson, D.S., McKenna, M.C., Toon, O.B., Hope, S., Lillegraven, J.A., 2004. Survival in the first hours of the Cenozoic. Geol. Soc. Am. Bull. 116 (5–6), 760–768.

Robison, S.F., 1991. Bird and frog tracks from the late cretaceous Blackhawk Formation in east central Utah. In: Chidsey, T.C. (Ed.), 1991 Field Symposium. Utah Geological Association, Salt Lake City, pp. 325–334.

Rocha-Barbosa, O., Loguercio, M.F.C., Velloso, A.L.R., Bonates, A.C.C., 2008. Bipedal locomotion in *Tropidurus torquatus* (Wied, 1820) and *Liolaemus lutzae* Mertens, 1938. Braz. J. Biol. 68, 649–655.

Rodríguez-de la Rosa, R.A., Aguillón-Martínez, M.C., López-Espinoza, J., Eberth, D.A., 2004. The fossil record of vertebrate tracks in Mexico. Ichnos 11 (1–2), 27–37.

Rodríguez-de la Rosa, R.A., Velasco-de León, M.P., Arellano-Gil, J., Lozano-Carmona, D.E., 2018. Middle Jurassic ankylosaur tracks from Mexico. Bol. Soc. Geol. Mex. 70 (2), 379–395.

Romano, M., Whyte, M.A., Jackson, S.J., 2007. Trackway Ratio: a new look at trackway gauge in the analysis of quadrupedal dinosaur trackways and its implications for icnotaxonomy. Ichnos 14, 257–270.

Romero Molina, M.M., Pérez-Lorente, F., Rivas Carrera, P., 2003. Análisis de la parataxonomía utilizada con las huellas de dinosaurio [Analysis of the parataxonomy used with dinosaur footprints]. In: Pérez Lorente, F. (Ed.), Dinosaurios y Otros Reptiles Mesozóicos de España, pp. 13–32.

Romilio, A., Salisbury, S.W., 2011. A reassessment of large theropod dinosaur tracks from the mid-Cretaceous (late Albian-Cenomanian) Winton Formation of Lark Quarry, central-western Queensland, Australia: a case for mistaken identity. Cretac. Res. 32 (2), 135–142.

Romilio, A., Salisbury, S.W., 2013. Three–dimensional analysis of the largest tracks from the Lark Quarry dinosaur tracksite, central–western Queensland. In: Prideaux, G. (Ed.), *CAVEPS 2013: 14th Conference on Australasian*

Vertebrate Evolution, Palaeontology and Systematics, Flinders University, Adelaide, 30th September–4th October 2013. Flinders University, Adelaide, Australia, pp. 74–75.

Romilio, A., Tucker, R.T., Salisbury, S.W., 2013. Re-evaluation of the Lark quarry dinosaur tracksite (late Albian–Cenomanian Winton Formation, central-western Queensland, Australia): no longer a stampede? J. Vertebr. Paleontol. 33 (1), 102–120.

Romilio, A., Salisbury, S.W., 2014. Large dinosaurian tracks from the Upper Cretaceous (Cenomanian-Turonian) portion of the Winton Formation, Lark Quarry, central-western Queensland, Australia: 3D photogrammetric analysis renders the 'stampede trigger' scenario unlikely. Cretac. Res. 51, 186–207.

Romilio, A., Salisbury, S.W., Jannel, A., 2021. Footprints of large theropod dinosaurs in the Middle-Upper Jurassic (lower Callovian-lower Tithonian) Walloon coal measures of southern Queensland, Australia. Hist. Biol. 33 (10), 2135–2146.

Romilio, A., Godfrey, T., 2022. A new dinosaur tracksite from the lower cretaceous (Aptian-Albian) eumeralla formation of wattle hill, Victoria, Australia: a preliminary investigation. Hist. Biol. 34 (12), 2315–2323.

Romilio, A., 2023. Bringing dinosaur trackmaker behaviour to life: 3D modelling and animation techniques using Lark Quarry as a case study. In: 4th Palaeontological Virtual Congress, Virtual, *9–24 May 2023*.

Romilio, A., Shao, C., 2023. Analysing trackway-based speed calculations to infer dinosaur locomotive capabilities and behaviours. Hist. Biol. 1–10.

Rose, P.J., 2007. A new titanosauriform sauropod (Dinosauria: Saurischia) from the Early Cretaceous of central Texas and its phylogenetic relationships. Palaeontol. Electron. 10 (2), 1.

Rowe, T.S., 1988. Definition, diagnosis, and origin of Mammalia. J. Vertebr. Paleontol. 8, 241–264.

Rubilar-Rogers, D., Moreno, K., Blanco, N., Calvo, J.O., 2010. Theropod dinosaur trackways from the lower cretaceous of the Chacarilla formation, Chile. Andean Geol. 35 (1).

Rühle von Lilienstern, H., 1939. Fährten und Spuren im *Chirotherium*-sandstein von Südthüringen. Fortschritte der Geologie und Paläontologie 12, 293–387.

Saka, Y., 2001. Chichibu terrane of the Shima peninsula, southwest Japan. In: Dinosaur Research Group of Mie Prefecture, *A Sauropod Dinosaur from Toba*, pp. 33–47.

Salisbury, S.W., Romilio, A., Herne, M.C., Tucker, R.T., Nair, J.P., 2016. The dinosaurian ichnofauna of the lower cretaceous (Valanginian-Barremian) Broome sandstone of the Walmadany area (James Price Point), Dampier Peninsula, Western Australia. J. Vertebr. Paleontol. 36 (Suppl. 1), 1–152.

Sampson, S.D., Loewen, M.A., 2010. Unraveling a radiation: a review of the diversity, stratigraphic distribution, biogeography, and evolution of horned dinosaurs (Ornithischia: Ceratopsidae). In: New Perspectives on Horned Dinosaurs: The Royal Tyrrell Museum Ceratopsian Symposium, vol. 405. Indiana University Press, Bloomington, p. 427.

Sánchez-Hernández, B., Przewieslik, A.G., Benton, M.J., 2009. A reassessment of the pteraichnus ichnospecies from the early cretaceous of Soria Province, Spain. J. Vertebr. Paleontol. 29 (2), 487–497.

Sander, P.M., Christian, A., Clauss, M., Fechner, R., Gee, C.T., Griebeler, E.M., Gunga, H.C., Hummel, J., Mallison, H., Perry, S.F., Preuschoft, H., Rauhut, O.W.M., Remes, K., Tütken, T., Wings, O., Witzel, U., 2011. Biology of the sauropod dinosaurs: the evolution of gigantism. Biol. Rev. 86 (1), 117–155.

Sander, P.M., 2013. An evolutionary cascade model for sauropod dinosaur gigantism-overview, update and tests. PLoS One 8 (10), e78573.

Santi, G., Nicosia, U., 2008. The ichnofacies concept in vertebrate ichnology. Studi Trentini di Scienze Naturali, Acta Geologica 83, 223–229.

Santos, V.F., Lockley, M.G., Meyer, C.A., Carvalho, J., Galopim de Carvalho, A.M., Moratalla, J.J., 1994. A new sauropod tracksite from the Middle Jurassic of Portugal. Gaia 10, 5–13.

Santos, V.F., da Silva, R.C., Rodrigues, L.A., 2008. Dinosaur track sites from Portugal: scientific and cultural significance. Oryctos 8, 77.

Santos, V.F., Moratalla, J.J., Royo-Torres, R., 2009. New sauropod trackways from the middle Jurassic of Portugal. Acta Palaeontol. Pol. 54 (3), 409–422.

Sarjeant, W.A.S., 1970. Fossil footprints from the middle Triassic of Nottinghamshire and the middle Jurassic of Yorkshire. Mercian Geol. 3 (3), 269–282.

Sarjeant, W.A.S., 1975. Fossil tracks and impressions of vertebrates. In: Frey, R.W. (Ed.), The Study of Trace Fossils. Springer-Verlag, Berlin and New York, pp. 283–324.

Sarjeant, W.A.S., Langston, W., 1994. Vertebrate footprints and invertebrate traces from the Chadronian (late Eocene) of west Texas. Texas Mem. Mus. Bull. 36, 1–86.

Sarjeant, W.A.S., Delair, J.B., Lockley, M.G., 1998. The footprints of *Iguanodon*: a history and taxonomic study. Ichnos 6 (3), 183–202.

Schumacher, B.A., Lockley, M.G., 2014. Newly documented trackways at "dinosaur. Lake," the Purgatoire valley dinosaur tracksite. New Mex. Mus. Nat. Hist. Bull. 62, 261–267.

Scotese, C.R., Song, H.J., Mills, B.J.W., Van der Meer, D.G., 2021. Phanerozoic paleotemperatures: the earth's changing climate during the last 540 million years. Earth Sci. Rev. 215, 103503.

Seilacher, A., 1953. Studien zur palichnologie. I. Uber die methoden der palichnologie. Neues Jahrbuch fur Geologie und Palaontologie, Abhandlungen 96, 421–452.

Sennikov, A.G., 2006. Reading segnosaur tracks. Priroda 5, 58–67.

Sennikov, A.G., 2021. The plantigrade *Segnosaurians*: sloth dinosaurs or bear dinosaurs? Paleontol. J. 55, 1158–1185.

Shell, R., Boss, S.K., 2013. Morphometric analysis of dinosaur tracks from southwest Arkansas. J. Arkansas Acad. Sci. 67 (1), 121–130.

Shillito, A.P., Davies, N.S., 2019. Dinosaur-landscape interactions at a diverse early cretaceous tracksite (lee Ness sandstone, Ashdown Formation, southern England). Palaeogeogr. Palaeoclimatol. Palaeoecol. 514, 593–612.

Shuler, E.W., 1935. Dinosaur track mounted in the bandstand at Glen Rose, Texas. Field Lab. 4, 9–13.

Simpson, E.L., Hilbert-Wolf, H.L., Wizevich, M.C., Lucas, S.G., Tester, E., Tindall, S.E., Bernard, J.J., 2010. A crocodylomorph track in the upper cretaceous capping sandstone member of the Wahweap Formation, Grand Staircase-Escalante national monument, Utah, USA. New Mex. Mus. Nat. Hist. Sci. Bull. 51, 165–170.

Snow, F.H., 1886. On the discovery of a fossil bird-track in the Dakota Sandstone. In: Transactions of the Annual Meetings of the Kansas Academy of Science, vol.10. Kansas Academy of Science, pp. 3–6.

Snyder, R.C., 1952. Quadrupedal and bipedal locomotion of lizards. Copeia (2), 64–70.

Sperry, S.W., 1980. The Flagstaff Formation: Depositional Environment and Paleoecology of Clastic Deposits Near Salina Utah, vol. 27. Brigham Young University Geology Studies, pp. 153–173.

Stanford, R., Weems, R.E., Lockley, M.G., 2004. A new dinosaur ichnotaxon from the lower cretaceous Patuxent Formation of Maryl Virginia. Ichnos 11, 251–259.

Stanford, R., Lockley, M.G., Tucker, C., Godfrey, S., Stanford, S.M., 2018. A diverse mammal-dominated, footprint assemblage from wetland deposits in the Lower Cretaceous of Maryland. Sci. Rep. 8 (1), 741.

Sternberg, C.M., 1926. Dinosaur tracks from the Edmonton Formation of Alberta. Geol. Surv. Can. Bull. 44, 85–87.

Sternberg, C.M., 1932. Dinosaur tracks from the Peace River, British Columbia. In: Annual Report of the National Museum of Canada for 1930, pp. 59–85.

Stettner, B., Persons IV, W.S., Currie, P.J., 2018. A giant sauropod footprint from the Nemegt Formation (upper cretaceous) of Mongolia. Palaeogeogr. Palaeoclimatol. Palaeoecol. 494, 168–172.

Stewart, D.G., Druckenmiller, P.S., Erickson, G.M., Benowitz, J.A., Capps, D.M., Knight, C.L., May, K.C., McCarthy, P.J., 2023. Vertebrate ichnology and palaeoenvironmental associations of Alaska's largest known dinosaur tracksite in the cretaceous Cantwell Formation (maastrichtian) of Denali national park and preserve. Hist. Biol. 1–17.

Stokes, W.L., 1957. Pterodactyl tracks from the Morrison Formation. J. Palaeontol. 31, 952–954.

Storer, J.E., 1975. Dinosaur tracks, *Columbosauripus ungulatus* (Saurischia: Coelurosauria), from the Dunvegan Formation (Cenomanian) of northeastern British Columbia. Can. J. Earth Sci. 12, 1805–1807.

Strand, E., 1932. Miscellanea nomenklatorica zoologica et palaeontologica, 111.IV. Folia Zoologica et Hydrobiologica 4, 133–147.

Sues, H.D., 2019. Archosauromorpha: Choristodera. In: The Rise of Reptiles: 320 Million Years of Evolution. Johns Hopkins University Press, Baltimore, MD, pp. 167–168.

Sullivan, C., Wang, Y., Hone, D., Wang, Y.Q., Xu, X., Zhang, F.C., 2014. The vertebrates of the Jurassic Daohugou biota of northeastern China. J. Vertebr. Paleontol. 34 (2), 243–280.

Suñer, M., Poza, B., Vila, B., Santos-Cubedo, A., 2008. Síntesis del registro fósil de dinosaurios en el Este de la Península Ibérica. Palaeontologica Nova 8, 397–420.

Suzuki, S., Watabe, M., 2000. Report on the Japan-Mongolia joint paleontological expedition to the Gobi desert, 1995. Hayashibara Mus. Nat. Sci. Res. Bull. 1, 45–57.

Sweet, A., Ricketts, B.D., Cameron, A.R., Norris, D.K., 1989. An Integrated Analysis of the Brackett Coal Basin, Northwest Territories, vol. 89. Geological Survey of Canada, pp. 85–99.

Swennen, C.K., Yu, Y.T., 2005. Food and feeding behavior of the black-faced spoonbill. Waterbirds 28 (1), 19–27.
Swinton, W.E., 1961. The history of Chirotherium. Geol. J. 2 (3), 443–473.
Tagert, E., 1846. On markings in the Hastings sands near Hastings, supposed to be the footprints of birds. Q. J. Geol. Soc. Lond. 2, 267.
The board of dinosaur fossil research of Mifune Town, 1998. Report of the research on the distribution of important fossils in Kumamoto Prefecture. In: Dinosaurs from the Mifune Group, Kumamoto Prefecture, Japan. The Board of Education of Mifune Town, 99 pp.
Thompson, R.S., Parish, J.C., Maidment, S.C., Barrett, P.M., 2012. Phylogeny of the ankylosaurian dinosaurs (Ornithischia: Thyreophora). J. Syst. Palaeontol. 10 (2), 301–312.
Thulborn, R.A., Wade, M.J., 1984. Dinosaur trackways in the Winton Formation (mid-Cretaceous) of Queensland. Memoir. Queensl. Mus. 21 (2), 413–517.
Thulborn, R.A., Wade, M., 1989. A footprint as a history of movement. In: Gillette, D.D., Lockley, M.G. (Eds.), Dinosaur Tracks and Traces, vol. 3. Cambridge University Press, Cambridge, pp. 9–50.
Thulborn, T., 1990. Dinosaur Tracks. Chapman and Hall, London, 410pp.
Thulborn, T., 2001. History and nomenclature of the theropod dinosaur tracks *Bueckeburgichnus* and *Megalosauripus*. Ichnos 8, 207–222.
Thulborn, T., 2017. Behaviour of dinosaurian track-makers in the Winton Formation (cretaceous, Albian–Cenomanian) at Lark quarry, western Queensland, Australia: running or swimming? Ichnos 24 (1), 1–18.
Tomaselli, M.B., Ortiz David, L.D., González Riga, B.J., Coria, J.P., Mercado, C.R., Guerra, M., Sánchez Tiviroli, G., 2022. New titanosaurian sauropod tracks with exceptionally well-preserved claw impressions from the Upper Cretaceous of Argentina. Cretac. Res. 129, 104990.
Torcida Fernández-Baldor, T.F., Díaz-Martínez, I., Huerta, P., Montero Huerta, D., Castanera, D., 2021. Enigmatic tracks of solitary sauropods roaming an extensive lacustrine megatracksite in Iberia. Sci. Rep. 11 (1), 16939.
Torsvik, T.H., Cocks, L.R.M., 2016. Earth History and Palaeogeography. Cambridge University Press.
Tsukiji, Y., Azuma, Y., Shiraishi, F., Shibata, M., Noda, Y., 2018. New ornithopod footprints from the lower cretaceous Kitadani Formation, Fukui, Japan: ichnotaxonomical implications. Cretac. Res. 84, 501–514.
Tsukiji, Y., Sakai, Y., Azuma, Y., 2019. Ichnotaxonomic revision of dinosaur tracks from the lower cretaceous Tetori group, Japan. Mem. Fukui Prefect. Dinosaur Mus. 18, 1–20.
Tsukiji, Y., Jin, X., Du, T., Azuma, Y., Hattori, S., Nakada, K., Nakayama, K., Noda, Y., Zheng, W., 2021. First discovery of a deinonychosaur trackway from the lower Upper Cretaceous of southeastern China. Cretac. Res. 125, 104874.
Tugend, J., Manatschal, G., Kusznir, N.J., Masini, E., Mohn, G., Thinon, I., 2014. Formation and deformation of hyperextended rift systems: insights from rift domain mapping in the Bay of Biscay-Pyrenees. Tectonics 33 (7), 1239–1276.
Turner, A.H., Pol, D., Clarke, J.A., Erickson, G.M., Norell, M.A., 2007. A basal dromaeosaurid and size evolution preceding avian flight. Science 317 (5843), 1378–1381.
Vallon, L.H., Rindsberg, A.K., Bromley, R.G., 2016. An updated classification of animal behaviour preserved in substrates. Geodin. Acta 28 (1–2), 5–20.
Van der Lubbe, T., Richter, A., Böhme, A., 2009. Velociraptor's sisters: first report of troodontid tracks from the Lower Cretaceous of northern Germany. J. Vertebr. Paleontol. 29 (3), 43A.
Van der Lubbe, T., Richter, A., Böhme, A., Sullivan, C., Hübner, T.R., Reich, M., 2011. Sorting out the sickle claws: how to distinguish between dromaeosaurid and troodontid tracks. Dinosaur Tracks 14–17.
Verdú, F.J., Royo-Torres, R., Cobos, A., Alcala, L., 2015. Perinates of a new species of *Iguanodon* (Ornithischia: Ornithopoda) from the lower Barremian of Galve (Teruel, Spain). Cretac. Res. 56, 250–264.
Vialov, O.S., 1965. Stratigrafiya Neogenovix Molass Predcarpatskogo Progiba. Naukova Dumka, Kiev, Ukraine, p. 191 part K.
Vialov, O.S., 1988. On the classification of dinosaurian traces. Ezhegodnik Vsesoyuznogo Paleontol. Obshchestva 31, 322–325.
Vila, B., Oms, O., Galobart, À., 2005. Manus-only titanosaurid trackway from Fumanya (Maastrichtian, Pyrenees): further evidence for an underprint origin. Lethaia 38 (3), 211–218.
Vila, B., Oms, O., Fondevilla, V., Gaete, R., Galobart, A., Riera, V., Canudo, J.I., 2013a. The latest succession of dinosaur tracksites in Europe: hadrosaur ichnology, track production and palaeoenvironments. PLoS One 8 (9), e72579.

Vila, B., Oms, O., Galobart, À., Bates, K.T., Egerton, V.M., Manning, P.L., 2013b. Dynamic similarity in titanosaur sauropods: ichnological evidence from the Fumanya dinosaur tracksite (southern Pyrenees). PLoS One 8 (2), e57408.

Vila, B., Castanera, D., Marmi, J., Canudo, J.I., Galobart, À., 2015. Crocodile swim tracks from the latest Cretaceous of Europe. Lethaia 48 (2), 256–266.

Wan, X.Q., Sun, L.X., Li, W., 2020. Terrestrial fossil assemblages of Tuchengzi Formation and Jurassic-Cretaceous boundary chronostratigraphy in Yanliao area. Acta Palaeontol. Sin. 59 (1), 1–12.

Watabe, M., Suzuki, S., 2000. Report on the JapaneMongolia joint paleontological expedition to the Gobi desert, 1997. In: Hayahsibara Museum of Natural Sciences, Research Bulletin, vol. 1, pp. 69–82.

Wedel, M.J., Cifelli, R.L., Sanders, R.K., 2000. *Sauroposeidon proteles*, a new sauropod from the Early Cretaceous of Oklahoma. J. Vertebr. Paleontol. 20, 109–114.

Weems, R.E., 1992. A re-evaluation of the taxonomy of Newark Supergroup saurischian dinosaur tracks, using extensive statistical data from a recently exposed tracksite near Culpeper. Virginia Richmond Virginia Div. Min. Res. 119, 113.

Weems, R.E., 2003. *Plateosaurus* foot structure suggests a single trackmaker for *Eubrontes* and *Gigandipus* footprints. Great Rift Valleys Pangea Eastern North Am. 2, 293–313.

Weems, R.E., Bachman, J.M., 2004. A dinosaur-dominated ichnofauna from the lower cretaceous (Aptian) Patuxent Formation of Virginia. In: Geological Society of America Abstracts with Programs, Denver, vol. 116.

Weems, R.E., 2006. Locomotor speeds and patterns of running behavior in non-maniraptoriform theropod dinosaurs. New Mex. Mus. Nat. Hist. Sci. Bull. 77, 379e389.

Weems, R.E., Bachman, J.M., 2015. The lower cretaceous Patuxent Formation ichnofauna of Virginia. Ichnos 22 (3–4), 208–219.

Weems, R.E., 2021. Additions and a taxonomic update to the dinosaur ichnofauna from the Patuxent Formation in Virginia, USA. Fossil Rec. 7 (82), 475.

Wen, J., Nie, Z.L., Ickert-Bond, S.M., 2016. Intercontinental disjunctions between eastern Asia and western North America in vascular plants highlight the biogeographic importance of the Bering land bridge from late Cretaceous to Neogene. J. Systemat. Evol. 54 (5), 469–490.

Whyte, M.A., Romano, M., 1994. Probable sauropod footprints from the middle Jurassic of Yorkshire, England. Gaia 10, 15–26.

Whyte, M.A., Romano, M., 2001. A dinosaur ichnocoenosis from the middle Jurassic of Yorkshire, UK. Ichnos 8 (3–4), 223–234.

Wills, S., Underwood, C.J., Barrett, P.M., 2023. Machine learning confirms new records of maniraptoran theropods in Middle Jurassic UK microvertebrate faunas. Papers Palaeontol. 9 (2), e1487.

Wilson, J.A., Carrano, M.T., 1999. Titanosaurs and the origin of "wide-gauge" trackways: a biomechanical and systematic perspective on sauropod locomotion. Paleobiology 25 (2), 252–267.

Witton, M., Naish, D., 2015. Azhdarchid pterosaurs: water-trawling pelican mimics or 'terrestrial stalkers'? Palaeontol. Pol. 60 (3), 651–660.

Wright, J., 2005. Steps in Understanding Sauropod Biology.

Wright, J.L., Unwin, D.M., Lockleyl, M.G., Rainforth, E.C., 1997. Pterosaur tracks from the Purbeck Limestone Formation of Dorset, England. Proc. Geologists' Assoc. 108 (1), 39–48.

Wright, J.L., 1999. Ichnological evidence for the use of the forelimb in iguanodontians. In: Unwin, D.M. (Ed.), Cretaceous Fossil Vertebrates, vol. 60. Special Papers in Palaeontology, pp. 209–219. Palaeontological Association.

Xi, D., Wan, X., Li, G., Li, G., 2019. Cretaceous integrative stratigraphy and timescale of China. Sci. China Earth Sci. 62, 256–286.

Xing, L.D., Wang, F.P., Pan, S.G., Chen, W., 2007. The discovery of dinosaur footprints from the middle cretaceous Jiaguan Formation of Qijiang county, Chongqing city. Acta Geol. Sin. 81 (11), 1591–1602.

Xing, L.D., Harris, J.D., Toru, S., Masato, F., Dong, Z.M., 2009a. Discovery of dinosaur footprints from the lower Jurassic Lufeng Formation of Yunnan province, China and new observations on *Changpeipus*. Geol. Bull. China 28 (1), 16–29.

Xing, L.D., Harris, J.D., Feng, X., Zhang, Z., 2009b. Theropod (dinosauria: Saurischia) tracks from lower cretaceous Yixian Formation at Sihetun village, Liaoning province, China and possible track makers. Geol. Bull. China 28 (6), 705–712.

Xing, L.D., Harris, J.D., Dong, Z.M., Lin, Y.L., Chen, W., Guo, S.B., Ji, Q., 2009c. Ornithopod (dinosauria: Ornithischia) tracks from the upper cretaceous Zhutian Formation in the Nanxiong basin, Guangdong, China and general observation on large Chinese ornithopod footprints. Geol. Bull. China 28 (7), 829–843.

Xing, L.D., Harris, J.D., Sun, D.H., Zhao, H.Q., 2009d. The earliest known deinonychosaur tracks from the Jurassic–Cretaceous boundary in Hebei Province, China. Acta Palaeontol. Sin. 48 (4), 662–671.

Xing, L.D., Harris, J.D., Wang, K.B., Li, R.H., 2010a. An early cretaceous non–avian dinosaur and bird footprint assemblage from the Laiyang group in the Zhucheng basin, Shandong province, China. Geol. Bull. China 29 (8), 1105–1112.

Xing, L.D., Harris, J.D., Jia, C.K., 2010b. Dinosaur tracks from the Lower Cretaceous Mengtuan Formation in Jiangsu, China and morphological diversity of local sauropod tracks. Acta Palaeontol. Sin. 49 (4), 448–460.

Xing, L.D., Mayor, A., Chen, Y., Harris, J.D., Burns, M.E., 2011a. The Folklore of dinosaur trackways in China: impact on Paleontology. Ichnos 18 (4), 213–220.

Xing, L.D., Harris, J.D., Jia, C.K., Luo, Z.J., Wang, S.N., An, J.F., 2011b. Early cretaceous bird–dominated and dinosaur footprint assemblages from the northwestern margin of the Junggar basin, Xinjiang, China. Palaeoworld 20, 308–321.

Xing, L.D., Gierliński, G.D., Harris, J.D., Divay, J.D., 2012. A probable crouching theropod dinosaur trace from the Tuchengzi Formation in Chicheng area, Hebei Province, China. Geol. Bull. China 31 (1), 20–25.

Xing, L.D., Li, D., Harris, J.D., Bell, P.R., Azuma, Y., Fujita, M., Lee, Y.N., Currie, P.J., 2013a. A new deinonychosaurian track from the lower cretaceous Hekou group, Gansu province, China. Acta Palaeontol. Pol. 58 (4), 723–730.

Xing, L.D., Lockley, M.G., Klein, H., Zhang, J.P., He, Q., Divay, J.D., Qi, L.Q., Jia, C.K., 2013b. Dinosaur, bird and pterosaur footprints from the Lower Cretaceous of Wuerhe asphaltite area, Xinjiang, China, with notes on overlapping track relationships. Palaeoworld 22, 42–51.

Xing, L.D., Lockley, M.G., Marty, D., Klein, H., Buckley, L.G., McCrea, R.T., Zhang, J.P., Gierliński, G.D., Divay, J.D., Wu, Q.Z., 2013c. Diverse dinosaur ichnoassemblages from the lower cretaceous Dasheng group in the Yishu fault zone, Shandong province, China. Cretac. Res. 45, 114–134.

Xing, L.D., Lockley, M.G., McCrea, R.T., Gierliński, G.D., Buckley, L.G., Zhang, J.P., Qi, L.Q., Jia, C.K., 2013d. First record of *Deltapodus* tracks from the early cretaceous of China. Cretac. Res. 42, 55–65.

Xing, L.D., Lockley, M.G., Zhang, J.P., Milner, A.R.C., Klein, H., Li, D.Q., Persons, W.S.I.V., Ebi, J.F., 2013e. A new Early Cretaceous dinosaur track assemblage and the first definite non-avian theropod swim trackway from China. Chin. Sci. Bull. 58, 2370–2378.

Xing, L.D., Lockley, M.G., 2014. First report of small *Ornithopodichnus* trackways from the lower cretaceous of sichuan, China. Ichnos 21 (4), 213–222.

Xing, L.D., Niedźwiedzki, G., Lockley, M.G., Zhang, J.P., Cai, X.F., Persons IV, W.S., Ye, Y., 2014a. *Asianopodus*-type footprints from the Hekou group of Honggu District, Lanzhou city, Gansu, China and the "heel" of large theropod tracks. Palaeoworld 23 (3–4), 304–313.

Xing, L.D., Lockley, M.G., Zhang, J.P., Klein, H., Persons, W.S.I.V., Dai, H., 2014b. Diverse sauropod–, theropod–, and ornithopod–track assemblages and a new ichnotaxon *Siamopodus xui* ichnosp. nov. from the Feitianshan Formation, Lower Cretaceous of Sichuan Province, southwest China. Palaeogeogr. Palaeoclimatol. Palaeoecol. 414, 79–97.

Xing, L.D., Lockley, M.G., Zhang, J., Klein, H., Kim, J.Y., Persons IV, W.S., Matsukawa, M., Yu, X., Li, J., Chen, G., Hu, Y., 2014c. Upper Cretaceous dinosaur track assemblages and a new theropod ichnotaxon from Anhui Province, eastern China. Cretac. Res. 49, 190–204.

Xing, L.D., Avanzini, M., Lockley, M.G., Miyashita, T., Klein, H., Zhang, J., He, Q., Qi, L., Divay, J.D., Jia, C., 2014e. Early cretaceous turtle tracks skeletons from the Junggar basin, Xinjiang, China. Palaios 29 (4), 137–144.

Xing, L.D., Lockley, M.G., Zhang, J.P., Klein, H., Marty, D., Peng, G.Z., Ye, Y., Mccrea, R.T., Persons, W.S.I.V., Xu, T., 2015a. The longest theropod trackway from East Asia, and a diverse sauropod-, theropod-, and ornithopod-track assemblage from the Lower Cretaceous Jiaguan Formation, southwest China. Cretac. Res. 56, 345–362.

Xing, L.D., Lockley, M.G., Marty, D., Piñuela, L., Klein, H., Zhang, J.P., Persons, W.S.I.V., 2015b. Re-description of the partially collapsed Early Cretaceous Zhaojue dinosaur tracksite (Sichuan Province, China) by using previously registered video coverage. Cretac. Res. 52, 138–152.

Xing, L.D., Lockley, M.G., Bonnan, M.F., Marty, D., Klein, H., Liu, Y., Zhang, J., Kuang, H., Burns, M.E., Li, N., 2015c. Late Jurassic–Early Cretaceous trackways of small-sized sauropods from China: new discoveries, ichnotaxonomy and sauropod manus morphology. Cretac. Res. 56, 470–481.

Xing, L.D., Buckley, L.G., McCrea, R.T., Lockley, M.G., Zhang, J.P., Piñuela, L., Klein, H., Wang, F.P., 2015d. Reanalysis of *Wupus agilis* (Early Cretaceous) of Chongqing, China as a large avian trace: differentiating between large bird and small non-avian theropod tracks. PLoS One 10 (5), e0124039.

Xing, L.D., Lockley, M.G., Yang, G., Xu, X., Cao, J., Klein, H., Persons, W.S.I.V., Shen, H.J., Zheng, X.M., 2015e. Unusual deinonychosaurian track morphology (*Velociraptorichnus zhangi* n. ichnosp.) from the lower cretaceous Xiaoba Formation, sichuan province, China. Palaeoworld 24 (3), 283–292.

Xing, L.D., Lockley, M.G., Marty, D., Zhang, J.P., Wang, Y., Klein, H., McCrea, R.T., Buckley, L.G., Belvedere, M., Mateus, O., Gierliński, G.D., Piñuela, L., Persons, W.S.I.V., Wang, F.P., Ran, H., Dai, H., Xie, X.M., 2015f. An ornithopod-dominated tracksite from the lower cretaceous Jiaguan Formation (Barremian–Albian) of Qijiang, south-Central China: new discoveries, ichnotaxonomy, preservation and palaeoecology. PLoS One 10 (10), e0141059.

Xing, L.D., Marty, D., Wang, K.B., Lockley, M.G., Chen, S.Q., Xu, X., Liu, Y.Q., Kuang, H.W., Zhang, J.P., Ran, H., Persons, W.S.I.V., 2015g. An unusual sauropod turning trackway from the Early Cretaceous of Shandong Province, China. Palaeogeogr. Palaeoclimatol. Palaeoecol. 437, 74–84.

Xing, L.D., Li, D.Q., Falkingham, P.L., Lockley, M.G., Benton, M.J., Klein, H., Zhang, J.P., Ran, H., Persons, W.S., Dai, H., 2015h. Digit-only sauropod pes trackways from China-evidence of swimming or a preservational phenomenon? Sci. Rep. 6, 21138.

Xing, L.D., Zhang, J., Lockley, M.G., McCrea, R.T., Klein, H., Alcala, L., Buckley, L.G., Burns, M.E., Kümmell, S.B., He, Q., 2015i. Hints of the early Jehol biota: important dinosaur footprint assemblages from the Jurassic-cretaceous boundary Tuchengzi Formation in Beijing, China. PLoS One 10 (4), e0122715.

Xing, L.D., Lockley, M.G., 2016. Early Cretaceous dinosaur and other tetrapod tracks of southwestern China. Sci. Bull. 61 (13), 1044–1051.

Xing, L.D., Lockley, M.G., Zhang, J.P., 2016a. Early Cretaceous Dinosaur and Other Tetrapod Tracks of Southwestern China. Ningbo Publishing House, Ningbo, pp. 1–410.

Xing, L.D., Lockley, M.G., Hu, N., Li, G., Tong, G., Matsukawa, M., Klein, H., Ye, Y., Zhang, J., Persons, W.S., 2016b. Saurischian track assemblages from the lower cretaceous Shenhuangshan Formation in the Yuanma basin, southern China. Cretac. Res. 65, 1–9.

Xing, L.D., Lockley, M.G., Yang, G., Cao, J., Benton, M., Xu, X., Zhang, J.P., Klein, H., Persons, W.S.I.V., Kim, J.Y., Peng, G.Z., Ye, Y., Ran, H., 2016c. A new *Minisauripus* site from the Lower Cretaceous of China: tracks of small adults or juveniles? Palaeogeogr. Palaeoclimatol. Palaeoecol. 452, 28–39.

Xing, L.D., Lockley, M.G., Klein, H., Peng, G.Z., Ye, Y., Jiang, S., Zhang, J.P., Persons, W.S.I.V., Xu, T., 2016d. A theropod track assemblage including large deinonychosaur tracks from the Lower Cretaceous of Asia. Cretac. Res. 65, 213–222.

Xing, L.D., Buckley, L.G., Lockley, M.G., Zhang, J., Marty, D., Wang, Y., Li, J., McCrea, R.T., Peng, C., 2016e. A new bird track, *Koreanaornis lii* ichnosp. nov., from the lower cretaceous Hekou group in the Lanzhou-Minhe basin, Gansu, northwest China, and implications for early cretaceous avian diversity. Cretac. Res. 66, 141–154.

Xing, L.D., Lockley, M.G., Marty, D., Klein, H., Yang, G., Zhang, J.P., Peng, G.Z., Ye, Y., Persons IV, W.S., Yin, X.Y., Xu, T., 2016f. A diverse saurischian (theropod-sauropod) dominated footprint assemblage from the Lower Cretaceous Jiaguan Formation in the Sichuan Basin, southwestern China: a new ornithischian ichnotaxon, pterosaur tracks and an unusual sauropod walking pattern. Cretac. Res. 60, 176–193.

Xing, L.D., Lockley, M.G., Li, D.L., Klein, H., Persons, W.S.I.V., Ye, Y., Zhang, J.P., Ran, H., 2017a. Late Cretaceous Ornithopod-dominated, theropod, and pterosaur track assemblages from the Nanxiong Basin, China: new discoveries, ichnotaxonomy, and palaeoecology. Palaeogeogr. Palaeoclimatol. Palaeoecol. 466, 303–313.

Xing, L.D., Liu, Y., Marty, D., Kuang, H., Klein, H., Persons IV, W.S., Lyu, Y., 2017b. Sauropod trackway reflecting an unusual walking pattern from the Early Cretaceous of Shandong Province, China. Ichnos 24 (1), 27–36.

Xing, L.D., Chou, C., Lockley, M.G., Yangchen, D., Chen, X., Ma, X., Zhou, W., Xu, H., 2017c. First report of avian tracks from the cretaceous of Tibet, China. Acta Geol. Sin.

Xing, L.D., Chou, C.Y., Lockley, M.G., Romilio, A., Klein, H., Li, N., Persons, W.S.I.V., 2018a. Lower Cretaceous sauropod trackways from Lishui City and an overview of dinosaur-dominated track assemblage from Zhejiang Province, China. J. Palaeogeogr. 7, 9.

Xing, L.D., Lockley, M.G., Klein, H., Zeng, R., Cai, S., Luo, X., Li, C., 2018b. Theropod assemblages and a new ichnotaxon *Gigandipus chiappei* ichnosp. nov. From the Jiaguan Formation, lower cretaceous of Guizhou province, China. Geosci. Front. 9 (6), 1745–1754.

Xing, L.D., Lockley, M.G., Romilio, A., Klein, H., Zhang, J.Q., Chen, H.Y., Zhang, J.J., Burns, M.E., Wang, X.L., 2018c. Diverse sauropod-theropod-dominated track assemblage from the Lower Cretaceous Dasheng Group of Eastern China: testing the use of drones in footprint documentation. Cretac. Res. 84, 588–599.

Xing, L.D., Lockley, M.G., Guo, Y., Klein, H., Zhang, J., Zhang, L., Persons, W.S., Romilio, A., Tang, Y., Wang, X., 2018d. Multiple parallel deinonychosaurian trackways from a diverse dinosaur track assemblage of the Lower Cretaceous Dasheng Group of Shandong Province, China. Cretac. Res. 90, 40–55.

Xing, L.D., Klein, H., Qin, Z.H., Persons IV, W.S., Xu, X., 2018e. Lower cretaceous theropod tracks with the new ichnogenus and combination *Lockleypus luanpingensis* from the Dabeigou Formation of the luanping basin, Hebei province, China. Palaeoworld 27, 482–489.

Xing, L.D., Lockley, M.G., Tang, Y.Z., Romilio, A., Xu, T., Li, X.W., Tang, Y., Li, Y.Z., 2018f. Tetrapod track assemblages from lower cretaceous desert facies in the Ordos basin, Shaanxi province, China, and their implications for mesozoic paleoecology. Palaeogeogr. Palaeoclimatol. Palaeoecol. 507, 1–14.

Xing, L.D., Buckley, L.G., Lockley, M.G., McCrea, R.T., Tang, Y., 2018g. Lower cretaceous avian tracks from Jiangsu province, China: a first Chinese report for ichnogenus *Goseongornipes* (Ignotornidae). Cretac. Res. 84, 571–577.

Xing, L.D., Hu, Y.C., Huang, J.D., He, Q., Lockley, M.G., Burns, M.E., Fang, J., 2018h. A redescription of the ichnospecies *Koreanaornis anhuiensis* (aves) from the lower cretaceous Qiuzhuang Formation at Mingguang city, Anhui province, China. J. Palaeogeogr. 7 (1), 58–65.

Xing, L.D., Lockley, M.G., Zhang, J.Q., Romilio, A., Klein, H., Wang, Y., Tang, Y.G., Burns, M.E., Wang, X.L., 2019a. A diversified vertebrate ichnite fauna from the Dasheng group (lower cretaceous) of southeast Shandong province, China. Hist. Biol. 31 (3), 353–362.

Xing, L.D., Lockley, M.G., Klein, H., Qiu, L., Chou, C.Y., Wang, D.H., Persons, W.S.I.V., 2019b. A lower cretaceous dinosaur track assemblage from the Taoqihe Formation in the central Heilongjiang, China. Palaeogeogr. Palaeoclimatol. Palaeoecol. 532, 109275.

Xing, L.D., Lockley, M.G., Matsukawa, M., Klein, H., Zhang, J.P., Wang, T., Persons IV, W.S., Hu, J., Dai, H., Hu, H.J., 2019c. Review and detailed description of sauropod-dominated trackways from the upper cretaceous Jiangdihe Formation of Yunnan, China. Ichnos 26 (2), 108–118.

Xing, L.D., Lockley, M.G., Tong, K., Peng, G., Romilio, A., Klein, H., Persons, W.S., Ye, Y., Jiang, S., 2019d. A diversified dinosaur track assemblage from the Lower Cretaceous Xiaoba Formation of Sichuan Province, China: implications for ichnological database and census studies. Cretac. Res. 96, 120–134.

Xing, L.D., Niu, K.C., Lockley, M.G., Klein, H., Romilio, A., Persons, W.S.I.V., Brusatte, S.L., 2019e. A probable tyrannosaurid track from the Upper Cretaceous of Southern China. Sci. Bull. 64 (16), 1136–1139.

Xing, L.D., Li, D., Klein, H., Lockley, M.G., Liang, Q., Yang, J., Li, L., You, W., 2019f. Lower cretaceous turtle tracks from Hekou group of northwest China. Cretac. Res. 99, 269–274.

Xing, L.D., Lockley, M.G., Du, T., Zhang, L., Klein, H., Romilio, A., Persons IV, W.S., Wang, K., Li, Z., Wan, X., 2020a. Dinosaur tracks from the Jurassic-Cretaceous boundary Tuchengzi Formation (Hebei Province, China) used as building stones in the Chengde imperial summer resort: age, ichnology, and history. Cretac. Res. 107, 104310.

Xing, L.D., Klein, H., Lockley, M.G., Wu, X.C., Benton, M.J., Zeng, R., Romilio, A., 2020b. Footprints of marine reptiles from the middle Triassic (Anisian–Ladinian) Guanling Formation of Guizhou province, southwestern China: the earliest evidence of synchronous swimming. Palaeogeogr. Palaeoclimatol. Palaeoecol. 558, 109943.

Xing, L.D., Lockley, M.G., 2021. Theropod tracks from the Jurassic-cretaceous boundary, Tuchengzi Formation, Chengde, China: review and new observations. Biosis Biol. Syst. 2 (2), 271–282.

Xing, L.D., Lockley, M.G., Klein, H., Chou, C.Y., Wang, D.H., Persons, W.S., 2021a. A mid-Cretaceous dinosaur track assemblage from the Hengshan Formation in central Zhejiang, China. Hist. Biol. 33 (7), 973–980.

Xing, L.D., Lockley, M.G., Mao, Z., Klein, H., Gu, Z., Bai, C., Qiu, L., Liu, Y., Romilio, A., Persons IV, W.S., Wan, X., 2021b. A new dinosaur track site from the earliest Cretaceous (Berriasian) part of the Tuchengzi Formation, Hebei Province, China: implications for morphology, ontogeny and paleocommunity structure. Palaeogeogr. Palaeoclimatol. Palaeoecol. 580, 110619.

Xing, L.D., Peng, C., Lockley, M.G., Wang, Y.D., Li, D.Q., Klein, H., Yang, J.T., Li, L.F., Persons, W.S.I.V., Wang, M.Y., 2021c. A diversified tetrapod ichnite fauna from the lower cretaceous Hekou group of Gansu province, China. Hist. Biol. 33 (11), 3018–3030.

Xing, L.D., Lockley, M.G., Klein, H., Persons IV, W.S., Peng, G., Ye, Y., Wang, M., 2021d. Largest dinosaur tracksite in China (Cretaceous, Zhaojue area, Sichuan Province): on the verge of destruction. Geosci. Front. 12 (5), 101181.

Xing, L.D., Lockley, M.G., Klein, H., Wang, T., 2021e. Sauropod tracks from the middle Jurassic Chuanjie Formation of Yunnan province and the pre-cretaceous sauropodomorph trackways from China. Palaeoworld 30 (3), 495–502.

Xing, L.D., Lockley, M.G., Klein, H., Zhang, L.J., Romilio, A., Persons, W.S.I.V., Peng, G.Z., Ye, Y., Wang, M.Y., 2021f. The new ichnotaxon *Eubrontes nobitai* ichnosp. nov. and other Saurischian tracks from the Lower Cretaceous Sichuan Province and a review of Chinese *Eubrontes* type tracks. J. Palaeogeogr. 10, 17.

Xing, L.D., Lockley, M.G., Klein, H., Peng, G.Z., Ye, Y., Jiang, S., Wang, M.Y., 2021g. An historic theropod-dominated track assemblage from the Upper Jurassic of Sichuan, China. Hist. Biol. 33 (11), 2822–2828.

Xing, L.D., Lockley, M.G., Ren, Z., Liu, C., Persons IV, W.S., Peng, G., Ye, Y., Jiang, S., 2021h. Dinosaur tracks from Tang Dynasty grottoes area in Sichuan province, China. Biosis Biol. Syst. 2 (2), 283–292.

Xing, L.D., Lockley, M.G., Tang, Y.Z., Klein, H., Persons, W.S.I.V., Wang, M.Y., Li, X.W., Wu, H., 2021i. Unusual dinosaur trackway preservation as clues to paleo-landscape and behavior from the lower cretaceous Luohe formation, Shaanxi province, China. Geosci. Front. 12 (2), 737–745.

Xing, L.D., Lockley, M.G., Jia, C.K., Klein, H., Niu, K.C., Zhang, L.J., Oi, L.Q., Chou, C.Y., Romilio, A., Wang, D.H., Zhang, Y., Persons, W.S.I.V., 2021j. Lower cretaceous avian-dominated, theropod, thyreophoran, pterosaur and turtle track assemblages from the Tugulu Group, Xinjiang, China: ichnotaxonomy and palaeoecology. PeerJ 9, e11476.

Xing, L.D., Lockley, M.G., Persons IV, W.S., Klein, H., Wang, M., 2021k. Stegosaur track assemblage from Xinjiang, China, featuring the smallest known stegosaur record. Palaios 36 (2), 68–76.

Xing, L.D., Li, D., Lockley, M.G., Persons, W.S., Klein, H., Yang, J., Li, L., Peng, C., 2021l. Saurischian-dominated tracksites first cretaceous lizard trackway from China from the Hekou group (lower cretaceous) of Lanzhou-Minhe basin, Gansu, northwest China. Hist. Biol. 33 (12), 3703–3713.

Xing, L., Lockley, M.G., Zhang, L., et al., 2021. Tetrapod track assemblages from midwestern Inner Mongolia, China: review and new observations. New Mex. Mus. Nat. Hist. Sci. Bull. 82, 525–541.

Xing, L., Zhang, Y., 2022. A preliminary review of dinosaur track assemblages from the Tuchengzi Formation in North China. Biosis Biol. Syst. 3 (1) e002–e002.

Xing, L.D., Lockley, M.G., Liu, X., Zhang, Z., Zhang, L., Ma, L., Hui, R., Klein, H., Persons IV, W.S., Li, X., 2022. Theropod and turtle tracks from the middle Jurassic Zhiluo Formation in Shaanxi, China: implications for a newly defined *Chelonipus* ichnocoenosis. Hist. Biol. 34 (4), 596–603.

Xing, L.D., Klein, H., Lockley, M.G., Li, H., Tong, B., Ye, Y., Dai, H., Chou, C.Y., Wang, D., Romilio, A., Persons IV, W.S., 2023a. New records of dinosaur tracks in eastern Tibet a review of Middle Jurassic dinosaur faunas from the eastern Tethys, southwest China. Hist. Biol. 1–15.

Xing, L.D., Lockley, M.G., Klein, H., Zhang, X., Liu, C., Persons, W.S., 2023b. The first record of dinosaur tracks from the Cretaceous Zhagang Group of the northwestern Guangdong Province, China. Hist. Biol. 35 (1), 102–107.

Xing, L.D., Romilio, A., Chunyong, C., Lockley, M.G., 2023c. Turtle tracks from the middle Jurassic Yaopo formation in Beijing, China. Hist. Biol. 1–6.

Xing, L.D., Lockley, M.G., Klein, H., Romilio, A., Persons, W.S., Wan, X.Q., 2024. First grallator ichnospecies from the Tuchengzi Formation of North China: review and new observations on grallatorid footprints in China. New Mex. Mus. Nat. Hist. Sci. Bull. 95, 403–433.

Xing, L.D., Niu, K.C., Lockley, M.G., Romilio, A., Deng, K., Persons IV, W.S., 2024b. Deinonychosaur trackways in southeastern China record a possible giant troodontid. iScience.

Xu, H., Liu, Y., Kuang, H., Peng, N., Xu, K., Chen, J., 2021. Diverse preserved dinosaur footprint assemblage from Jurassic–Cretaceous transition eolian dune deposits of western Shandong Province, China. Cretac. Res. 121, 104733.

Xu, X., Zhou, Z., Sullivan, C., Wang, Y., Dong, R., 2016. An updated review of the middle-late Jurassic Yanliao biota: chronology, taphonomy, Paleontology and paleoecology. Acta Geol. Sin. English Edition 90, 2229–2243.

Yabe, H., Inai, Y., Shikama, T., 1940. Discovery of dinosaurian footprints from the cretaceous (?) of Yangshan, Chinchou. Preliminary note. Proc. Imperial Acad. Tokyo 15, 560–563.

Yang, S.Y., 1982. On the dinosaur's footprints from the upper Cretaceous Gyeongsang Group, Korea. J. Geol. Soc. Korea 18, 138–142.

Yang, S.Y., Lockley, M.G., Greben, R., Erickson, B.R., Lim, S.K., 1995. Flamingo and duck-like bird tracks from the late cretaceous and early Tertiary: evidence and implications. Ichnos 4, 21–34.

Yang, X.L., Yang, D.H., 1987. Dinosaur Footprints from Mesozoic of Sichuan Basin. Sichuan Scientific and Technological Publishing House, Chengdu, 30pp.

Yang, Z., Jiang, B., McNamara, M.E., Kearns, S.L., Pittman, M., Kaye, T.G., Orr, P.J., Xu, X., Benton, M.J., 2019. Pterosaur integumentary structures with complex feather—like branching. Nat. Ecol. Evol. 3, 24—30.

Yoon, H.S., Lee, Y.N., Jung, S.H., Kong, D.Y., Kim, S.H., Son, M., 2021. A juvenile ornithopod tracksite from the lower cretaceous Haman Formation, South Korea. Cretac. Res. 125, 104877.

Young, C.C., 1960. Fossil footprints in China. Vertebr. Palasiat. 4 (2), 53—66.

Yu, X.Q., 1999. The occurrence features of dinosaur's fossils in the Xiuning area, south Anhui. Geol. Anhui 9 (2), 94—99.

Zakharov, S.A., 1964. On the Cenomanian dinosaur trackmaker of the tracks found in the Shirkent valley, paleontologiya tadzhikistana (Paleontology of Tajikistan). In: Reiman, V.M. (Ed.), Dushanbe, Akad. Nauk Taj. SSR, pp. 31—35.

Zeng, X.Y., 1982. Dinosaur tracks from the red beds of the Yuanma Basin in northwestern Hunan. Hunan Geol. 1, 57—58.

Zhang, J.P., Li, D.Q., Li, M.L., Lockley, M.G., Bai, Z., 2006. Diverse dinosaur-, pterosaur-, and bird-track assemblages from the Hakou formation, lower cretaceous of Gansu province, northwest China. Cretac. Res. 27 (1), 44—55.

Zhang, L.M., Yin, Y., Wang, C.S., 2021. High-altitude and Cold habitat for the early cretaceous feathered dinosaurs at Sihetun, western Liaoning, China. Geophys. Res. Lett. 48 (14), e2021GL094370.

Zhang, Q., Wen, W., Hu, S., Benton, M.J., Zhou, C., Xie, T., Lu, T., Huang, J., Choo, B., Chen, Z.Q., Liu, J., 2014. Nothosaur foraging tracks from the Middle Triassic of southwestern China. Nat. Commun. 5 (1), 3973.

Zhang, Y.X., Sun, Y.J., Chen, J., Cui, W.S., Chen, S.Q., 2023. A new turtle footprint from the early cretaceous strata in Zhucheng, Shandong province. Geol. Rev. 69 (2), 719—726.

Zhen, S.N., Li, J., Rao, C., 1986. Dinosaur footprints of Jinning, Yunnan. Mem. Beijing Mus. Nat. Hist. 33, 1—18.

Zhen, S.N., Li, J.J., Mateer, N.J., Lockley, M.G., 1989. A review of dinosaur footprints in China. Int. Symp. Dinos. Tracks Traces 1, 187—197.

Zhen, S.N., Li, J., Han, Z., Yang, X., 1996. The Study of Dinosaur Footprints in China. *Sichuan Scientific and Technological* Publishing House, 110pp.

Zhou, Z.H., 2006. Evolutionary radiation of the Jehol Biota: chronological and ecological perspectives. Geol. J. 41 (3—4), 377—393.

Zonneveld, J.P., Zaim, Y., Rizal, Y., Ciochon, R.L., Bettis, E.A., Aswan, A., Gunnell, G.F., 2011. Oligocene shorebird footprint s, Kandi, Ombilin basin, Sumatra. Ichnos 18 (4), 221—227.

CHAPTER

8

Ichnotaxonomy of Paleogene and Neogene tetrapod tracks

Nasrollah Abbassi[1], *Ricardo Néstor Melchor*[2,3] *and Ignacio Díaz-Martínez*[4]

[1]Department of Geology, Faculty of Sciences, University of Zanjan, Zanjan, Iran; [2]Instituto de Ciencias de la Tierra y Ambientales de La Pampa (CONICET and Universidad Nacional de La Pampa), Santa Rosa, La Pampa, Argentina; [3]Departamento de Geología, Universidad Nacional de La Pampa, Santa Rosa, La Pampa, Argentina; [4]Departamento de Ciencias de la Tierra y Física de la Materia Condensada, Facultad de Ciencias, Universidad de Cantabria, Cantabria, Spain

1. Introduction

The Cenozoic (66 Ma—Present) was a time when a series of major mountain-building episodes occurred, such as Himalayas, Zagros, Alps, and Pyrenees, and is also when the Earth's climate changed in several steps from ice-free poles and warm conditions to ice-covered poles and cold conditions (e.g., François and Goddéris, 1998; Mudelsee et al., 2014).

Late Campanian to early Danian trends in climate changes show a global weathering and indicate a long-term trend toward a more humid climate which also culminated during the K-T transition. For the same interval, middle and high southern latitude oxygen isotope records indicate long-term climate cooling. This suggests that during the Maastrichtian humidity increased in low latitudes whereas high latitudes cooled (Adatte et al., 2002). The short-term warming just below the K-T boundary may be linked to Deccan volcanism (Caldeira and Rampino 1990; Barnet et al., 2018) or by the Chicxulub asteroid impact at the Cretaceous-Paleogene (K-Pg) boundary (Vellekoop et al., 2018). Cenozoic climatic oscillations are considerable, for instance, during the Paleocene—Eocene there was a warming trend (Paleocene—Eocene Thermal Maximum, PETM), in the Eocene—Oligocene the tendency was to be cooling (Eocene—Oligocene Transition), and there was another change to warming in the Miocene—Pliocene (Middle Miocene Climate Optimum) (see Mudelsee et al., 2014 and references therein).

Vertebrate Ichnology
https://doi.org/10.1016/B978-0-443-13837-9.00005-6

The Cretaceous–Paleogene (K–Pg) mass extinction, 66 million years ago, was characterized by a worldwide ecological devastation and rapid species turnover (see Hughes et al., 2021). At the level of biodiversity, this event is relevant since it marks the limit between the so-called "Age of Reptiles" of the Mesozoic (Witton, 2017; Pérez-García, 2022) and the "Age of Mammals" of the Cenozoic (Hughes et al., 2021). Although several lineages of reptiles, amphibians, and birds continued after the extinction (see Tambussi and Degrange, 2012; Gardner and Nydam, 2013; Puértolas-Pascual et al., 2020), the surviving mammalian groups would have undergone rapid morphological diversification, mainly in the early Cenozoic after the disappearance of nonavian dinosaurs from global ecosystems, occupying a wide range of empty ecological niches (Hughes et al., 2021; Goswami et al., 2022). Indeed, many of the mammal groups present today developed during the Paleogene, such as perissodactyls, artiodactyls, and carnivorans, among others (see O'Leary et al., 2013).

The record of Cenozoic tetrapod tracks, though, is not as abundant as the osteological record and is well represented by more than 500 bibliographic references (McDonald et al., 2007). Almost all the identified ichnogenera were originally defined and identified only in the Cenozoic (e.g., *Pecoripeda* Vyalov, 1965; *Albertasuchipes* McCrea et al., 2004; *Phoenicopterichnum* (Aramayo and Manera de Bianco, 1987)). Nevertheless, there are some ichnogenera defined in Mesozoic rocks that were also identified in the Cenozoic (e.g., *Aquatilavipes* Currie, 1981; *Jindongornipes* Lockley et al., 1992) and ichnogenera erected with Cenozoic specimens later identified in the Mesozoic as well (e.g., *Leptoptilostipus* Payros et al., 2000; *Ranipes* Lockley and Milner, 2014).

In the present chapter, the ichnotaxa defined for Paleogene/Neogene tetrapod tracks have been revised (Quaternary tracks are analyzed in the chapter by Neto de Carvalho et al. in this volume). Identified Paleogene/Neogene tracks are those of amphibians, reptiles, birds, and mammas (including artiodactyls, perissodactyls, paenungulates, carnivorans, creodonts, anagalids, xenarthrans, and insectivorans) (hominids are described in the Chapter 10). The diagnosis of all ichnogenera is provided in English. The schematic drawings of all holotypes are also included. Furthermore, the ichnotaxonomic validity of the ichnogenera and their ichnospecies is analyzed based on the indicated criteria and the opinions of other authors.

2. Tetrapods

The Devonian Period, some 400 million years ago, is known as the Age of Fishes (Coates, 2009; Burrow and Turner, 2012), which comprises development of the bony fishes and fish-like animals. These initial vertebrates evolved limbs and were ancestors of all tetrapods, which include amphibians, salamanders, caecilians, turtles, crocodiles, lizards, birds, and mammals (Kupriyanova, 2009). In other words, tetrapods include any animal with four legs or whose ancestor had four legs their relatives the snakes (Clack, 2012). Nevertheless, there may not be consensus on the tetrapod definition (e.g., Laurin, 2002). Essentially, total-group tetrapods include all those taxa more closely related to living tetrapods than to their nearest living sister group (lungfishes), whereas crown-group tetrapods consist of the last common ancestor of all living tetrapods plus all the fossil and living descendants of that ancestor. Stem-group tetrapods form a paraphyletic assemblage equivalent to the total group minus the crown group (Coates et al., 2008).

Today, the descendants of these early pioneers are divided into two major groups: amphibians (frogs, salamanders, and caecilians), and amniotes (mammals, turtles, crocodiles, birds, and lizards). Tetrapods colonized land during the Carboniferous, which is considered to be one of the key episodes in the history of life. Sarcopterygians (lobe-finned bony fishes) have segmental inner limb skeleton, forming in the embryo by successive splitting (bifurcation) of primordia (Clack, 2009). In these fishes, fore- and hind limb are generally of similar structure resulted by common developmental process, so that the first element (humerus in the arm, femur in the leg) is long and single, followed by two elements, includes radius and ulna in the arm, and tibia plus fibula in the leg (Coates and Ruta, 2007; Clack, 2009; Schoch, 2014). However, in tetrapods, radials are absent, but there are digits as segmented and flexible outgrowths.

The autopod, including the mesopodium and the acropodium, is the most distal part of the tetrapod limb, and developmental mechanisms of autopod formation serve as a model system of pattern formation during development (Tamura et al., 2008). Primitive tetrapods had pentadactyl autopod, but reduction of digits in tetrapods was common and occurred repeatedly, up to the complete loss of limbs (e.g., caecilians, snakes, and amphisbaenians) (Clack, 2009). The tetrapod limb size and shape and digit number variations reflect the remarkable diversity resulting from the 400 million years experiment of natural selection. Loss of limbs and elements of pes and manus is extreme morphological changes in the tetrapods, which are observed in all major groups of tetrapods (Lande, 1978). It seems that pentadactyly to be acquired relatively early in the history of tetrapods, and changes in acropodium elements are considered as digit number reduction. There are variations in the patterns of digit numbers in the ancient and modern tetrapod groups that will be discussed in the following sections.

3. Paleogene/Neogene vertebrate track preservation

Footprints are biogenic sedimentary structures made by the autopods of vertebrates modifying a substrate. The final morphology of a footprint is affected by three main factors: trackmaker behavior, autopodia anatomy, and substrate features (e.g., McKee, 1947; Laporte and Behrensmeyer, 1980; Scrivner and Bottjer, 1986; Falkingham, 2014; Melchor, 2015). Therefore, differences in any of these factors can cause the same individual to produce quite different tracks, and different biological taxa to leave similar tracks (Basan, 1978; Howard, 1978; Vallon et al., 2016).

The current review of Paleogene / Neogene ichnotaxa is based on a morphological comparison, the analysis of the holotype material and its preservational quality, and sample size. Regulations proposed by the ICZN for ichnotaxa have been considered. The preservation scale of Marchetti et al. (2019a) was applied, considering that a grade 2 or 3 is recommended as type material for erecting new ichnotaxa. It is suggested that ichnotaxa be based on a large sample size that allows us to observe the morphological variability. Qualitative criteria are recommended to distinguish ichnotaxa (Bertling et al., 2006). Application of morphometric techniques to support distinction of ichnotaxa or to compare already defined ichnotaxa can be a useful tool (e.g., Buckley et al., 2016). However, distinction of ichnotaxa should be based on clear morphological differences and not on subtle morphological contrasts that require statistical studies to define the assignment of new material (Melchor et al., 2010; Bertling et al., 2022). A common issue is that diagnoses are commonly not differential, as required by the ICZN. New ichnotaxa should be compared with morphologically similar ichnotaxa to highlight their differences, although this is not a common practice.

Substrate properties regulate the footprint morphology during footprint registration. The main properties consist of the texture of sediments, moisture content, thickness of sediments, topography of substrates, and the degree to which the feet sink in the sediments. Deformation of the sediment surface and underlayers during footprint formation has been reproduced experimentally by several authors (e.g., Gatesy, 2003; Milàn and Bromley, 2007; Bennett et al., 2019; Abbassi, 2022). Allen (1997), in laboratory experiments, used slotted punches and plasticine to model the sediment deformation during the different gaits of the hooves of mammals. This study showed a variety of strain markers and underlayer deformation as folds and microfaults, which are associated with the dead region and shear zone beneath the footprint. However, deformation of sediments during the sinking of the feet of artiodactyls may not conform to a simple or unique model. Miocene cameloid tracks deepened in silt- and sand-sized volcanic ash present a bilobed form in cross section (Lockley and Delgado, 2007; Houck et al., 2009).

Scrivner and Bottjer (1986) classified four different major preservation morphologies of artiodactyl footprints in the Copper Canyon Formation (Eocene), California, USA. These categories range from wet sediments with an ovate, deep depression described as a "squelch mark" to the distorted morphology of partially upward pulled footprints; and the best-preserved footprints that were impressed in a moist or "slightly damp" substrate and, finally, shallow impressions (usually less than 1 cm deep) with poor resolution of morphological detail. Abbassi et al. (2021) identified five morphotypes of modern artiodactyl footprints that were controlled by the moisture of the fine-grained clayey sediment in a small surface: with increasing water content of the sediment, footprints change from obscure and shallow depressions to deeply depressed with strongly damaged rims. In the mostly wet sediments, footprints are impressed as very deep, oval-shaped traces plus sliding imprints.

The anatomical characteristics of the "heel," which includes bones, muscles, and hooves or skin features, form the sole, wall, and clefts of the hooves, thus controlling the morphology of artiodactyl footprints. Plantar pressure distributions of artiodactyl feet are traceable in the morphology of the footprints in both static and dynamic stages. The body size of hoofed mammals influences limb morphology and function, foot design, and locomotor behavior (Warner et al., 2013; Hutchinson, 2021). Large and heavy artiodactyls such as camel, lama, giraffe, and cattle have large and relatively flattened feet, with broad heels and lacking a sole or with a small one. Small artiodactyls, however, have small, narrow hooves, usually with sharp tips and with well-developed hoof walls. These changes accompany behavioral variation of artiodactyls such as speed and mobility. In the phases of the artiodactyl walking cycle, usually the hoof starts to touch on sediment proximally, and after hoof landing completely, hooves lift up distally. Thus, the heel of the hooves first touches on the substrate and endures the main weight force of the animal. The tip of the hooves applies forces sheerly at the end of the gait cycle. The mechanism of gait cycles of artiodactyls varies in the speed and kind of locomotion, and it is related to the body size and configuration of the animal (Alexander, 2003; Hildebrand, 1985; Clifford, 2010), which are registered in the footprint completeness, presence of secondary features (e.g., tail/body impression), and anomalous multidirectional aspects of the locomotion of footprints and trackways (Marchetti et al., 2019a).

In avian footprints, both in fossil and modern ones, the recognition of an interdigital web imprint is sometimes uncertain and may be not recorded in all footprints of the same morphology due to substrate properties (see Falkingham et al., 2009). Deeply set footprints can display a deformation between the front digits that should not be confused with an interdigital web imprint. The registration of the Digit I (hallux) in some anisodactyl footprints may not be uniform due to substrate properties. The occurrence of tri- and tetradactyl footprints with an overall similar morphology and size is taken as indicative of a single tetradactyl morphotype.

On the contrary, it is possible to consider as different ichnotaxa tracks made by the same type of producer but representing different behavior. That is the case with crocodile tracks, where swimming imprints are classified as *Albertasuchipes*, *Indosuchipes* Rajkumar et al., 2015, and *Anticusuchipes* Mustoe, 2019 (Lockley et al. 2020; Castanera et al., 2021), whereas *Borealosuchipu*s Erickson, 2005 would be the evidence of crawling or a basking behavior (Erickson, 2005).

4. Ichnotaxonomy

The following revision of Paleogene and Neogene ichnotaxa has been conducted by IDM for amphibian and reptilian footprints, RNM for avian footprints, and NA for mammalian footprints.

4.1 Amphibia

4.1.1 Amphibian autopodium structure and tracks

Amphibia is a paraphyletic group that represents all nonamniotic tetrapods (Benton, 1990; Steyer, 2011). They are ectothermic, with gill respiration during the larval phase and lung respiration when reaching the adult state; and undergo a transformation during their development, the metamorphosis (Duellman and Trueb, 1994). It is composed of three main subdivisions, Lepospondyli, Temnospondyli, and Lissamphibia, of which only the lissamphibians lived during the Cenozoic (Hime et al., 2021). Lissamphibia arose in the Early Triassic according to the paleontological record (Rage and Roček, 1989; but see San Mauro, 2010 for other interpretations based on molecular data), and to this day they have evolved diverse body plans that differ in modes of locomotion and habitats (Duellman and Trueb, 1994; Pough et al., 1998). For instance, Caudata (living salamanders) presents a developed tail and proportionally paired limbs (Zardoya and Meyer, 2001). Although most of its basal and extant taxa had four digits on their forelimbs and five on their hind limbs (Fig. 8.1A and B), several groups of extant salamanders present a reduction to three digits on their hands and four on their feet (Shubin and Wake, 2003), and even several of them have interdigital webbing (Jaekel and Wake, 2007). In contrast, gymnophionans (modern caecilians) are adapted to a fossorial lifestyle and completely limbless (Zardoya and Meyer, 2001). Finally, anurans (extant frogs and toads) are characterized by having short and powerful hind limbs,

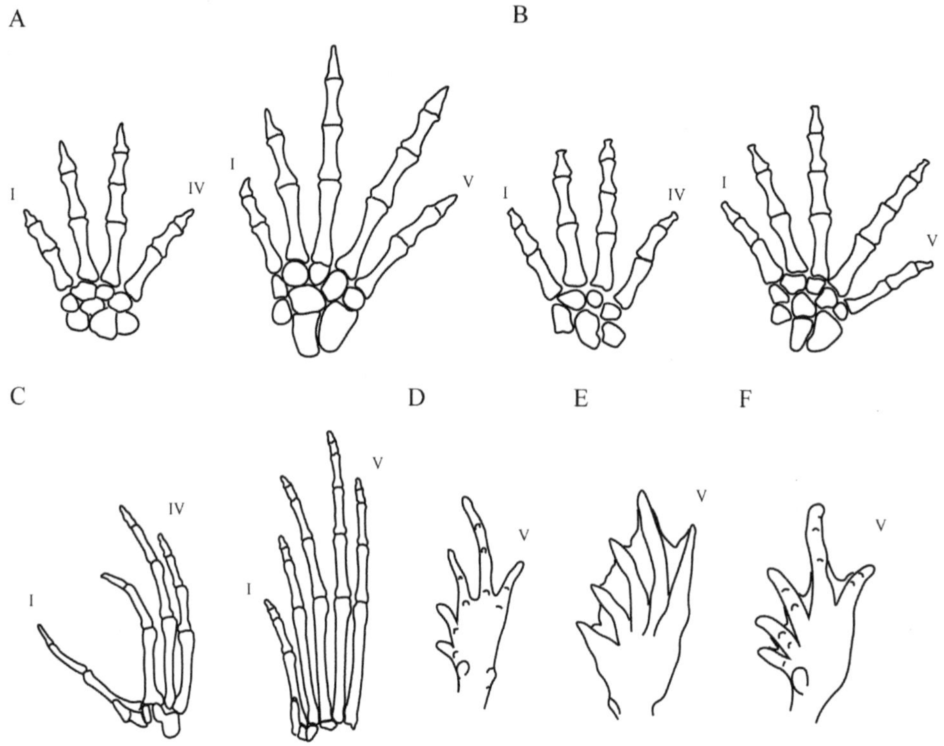

FIGURE 8.1 Skeletal pattern of manus and pes of lissamphibian taxa, (A) *Dicamptodon*, (B) *Rhyacotriton*, and (C) *Xenohyla*. Morphology of anuran feet, (D) *Pleurodema*, (E) *Lepidobatrachus*, and (F) *Melanophryniscus*. Figures not to scale. *I*, digit I; *IV*, digit IV; *V*, digit V. *Modified from (A and B) Shubin and Wake (2003); (C) Turazzini and Gómez (2023); and (D–F) Goldberg and Fabrezi (2008).*

a stiffened vertebral column and lack a tail, which are adaptations for jumping (Zardoya and Meyer, 2001). They have a tetradactyl manus and pentadactyl pes (Fig. 8.1C) adapted to different lifestyles such as aquatic, terrestrial, or arboreal, and consequently, they present particular characteristics in their autopodia (interdigital webbing, claw morphology, etc.) (see Goldberg and Fabrezi, 2008 and references therein) (Fig. 8.1D–F).

Amphibian tracks are very scarce in the fossil record (Braddy et al., 2003). During the Carboniferous and Permian, tracks attributed to lepospondyls and temnospondyls have been identified at least in North America, South America, Africa, and Europe (e.g., Melchor and Sarjeant, 2004; Petti et al., 2014; Marchetti et al., 2019b; Cisneros et al., 2020). In the Mesozoic, they are more abundant in the Triassic (see Mujal and Schoch, 2020; Klein and Lucas, 2021) than in the Jurassic and Cretaceous (Kim et al., 2019). The Cenozoic record is distributed in few salamander and frog tracks (see below).

In general, pes tracks are pentadactyl, ectaxonic, present a developed "heel" and have individualized digit impressions. Manus tracks are subdigitigrade and tetradactyl, although some basal amphibians were pentadactyl (see Marchetti et al., 2018), and the digit impressions are similar in length. The trackways represent diverse locomotion modes: swimming, crawling, hopping, etc. (Marsicano et al., 2014; Park et al., 2018; Mujal and Schoch, 2020).

Due to the environment in which the amphibian lives, the track shape is very conditioned by changes in the substrate grain size, moisture, and slope (Brand, 1996). Indeed, Marchetti et al. (2019b) pointed out that several Paleozoic amphibian ichnotaxa are *nomina dubia* because their morphologies are conditioned by the paleoenvironmental features instead of the trackmaker autopod morphology.

4.1.2 Amphibian ichnotaxa

Two ichnogenera have been related to lissamphibian trackmakers:

Ambystomichnus Peabody, 1954
Ambystomichnus montanensis (Gilmore, 1928a)
Ranipes Lockley and Milner, 2014
Ranipes laci Lockley and Milner, 2014

4.1.2.1 Description and discussion of ichnogenera

Ichnogenus *Ambystomichnus* Peabody, 1954.
Fig. 8.2A.
Type ichnospecies: *Ambystomichnus montanensis* (Gilmore, 1928a).

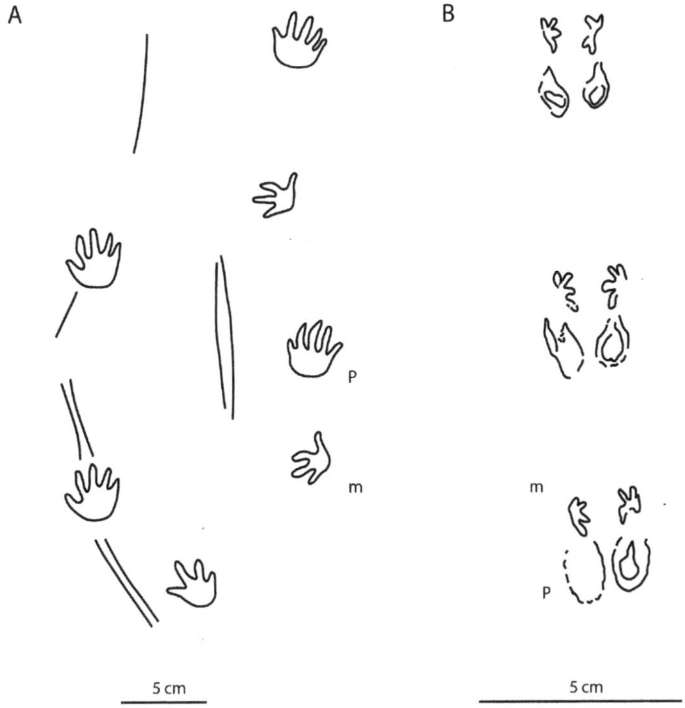

FIGURE 8.2 Outline sketch of the amphibian Paleogene/Neogene ichnotaxa. (A) *Ambystomichnus montanensis*, scale estimated. (B) *Ranipes laci*. *m*, manus; *p*, pes. *(A) Modified from Gilmore (1928a). (B) Modified from Lockley and Milner (2014).*

Diagnosis: Trackways of a large urodele amphibian indicating an individual comparable in body form with the living ambystomid genus *Dicamptodon*, the marbled salamander, but twice as large and slightly longer in the trunk region. Details of the footprints agree closely with footprints of living and Pliocene ambystomids, and the bilobate palm of the manus is exactly as in the living *Dicamptodon ensatus*. Stride 18 cm maximum; pace angulation (defined below) 77° maximum; manus 3.1 cm wide, pes 3.7 cm wide overall; both relatively distant from midline (Peabody, 1954).

Discussion: The holotype of *Ambystomichnus montanensis* was originally referred to as *Ammobatrachus montanensis* by Gilmore (1928a) and was found in the Paleocene of Montana (USA). The ichnogenus *Ammobatrachus* was erected by Gilmore (1928b) for Pennsylvanian (Carboniferous) tracks. Peabody (1954) renamed the specimen as *Ambystomichnus montanensis* due to similarities with modern ambystomatid salamander tracks. Marchetti et al. (2019b) suggested that the Paleozoic *Ammobatrachus* is a *nomen dubium*. Other records related to *Ambystomichnus* have been cited in the Eocene (Foster, 2001) and Pleistocene (Weems, 2021).

Ichnogenus *Ranipes* (Lockley and Milner, 2014).

Fig. 8.2B.

Type ichnospecies: *Ranipes laci* Lockley and Milner, 2014.

Diagnosis: Highly symmetrical, narrow, straight trackway of a small hopping vertebrate, with four tracks (two pes and two manus) in quadripartite "hop set" configuration. Distances between quadripartite groups along trackway axis variable up to 6–7 times length of an individual hop set. Outer trackway width about three times manus and pes width. Inner trackway width spacing between pairs of manus and pes traces is similar and about 1/3 outer trackway width. Pes footprints pentadactyl, ectaxonic with heel trace length variable. Manus tracks tetradactyl with strong inward rotation, and respective digit traces II and V, tending to posterior (II) and anterior (V) parasagittal alignment. Respective orientations of Digit III and IV traces are postero-medial and antero-medial (emended by Park et al., 2018).

Discussion: Lockley and Milner (2014) defined *Ranipes laci* for a trackway found in outcrops of the Eocene Flagstaff Formation (USA). Subsequently, Park et al. (2018) proposed a new ichnospecies of *Ranipes*, *R. saokensis* Park et al., 2018, for Late Cretaceous anuran tracks of Korea. Moreover, they revised the original diagnosis of *Ranipes*, and emended it with information on pes tracks. The record of frog tracks is very scarce. The oldest, *Ranipes* isp., come from the Lower Cretaceous rocks of Korea (Kim et al., 2019). Moreover, frog tracks have been also identified in the Late Cretaceous of the USA (Robinson, 1991) classified as *Ranipes* isp. by Lockley and Milner (2014). Finally, Weems (2021) cited the presence of *Ranipes laci* in the Pleistocene of the USA. Almost all the above-cited anuran tracks indicate hopping progression (see Park et al., 2018; Kim et al., 2019).

4.2 Reptilia

4.2.1 Reptilian autopodium structure and tracks

Reptilia is a group of amniotic tetrapods (see Gauthier, 1994; Modesto and Anderson, 2004). Its taxonomic definition has been the subject of numerous discussions. The emerging consensus divided amniotes into two main groups: synapsids (e.g., mammals) on one hand, and all remaining amniotes on the other (Gauthier et al., 1988; Laurin and Reisz,

1995). The latter group was historically called Reptilia, and some authors pointed out that it is paraphyletic because the Aves should be excluded (de Queiroz and Gauthier, 1990). Other researchers proposed the term Sauropsida for nonsynapsid amniotes including Aves (see Benton, 2005). Modesto and Anderson (2004) synonymized Reptilia and Sauropsida, maintaining the former because is far better known and should have priority. Reptilia is defined as "The most inclusive clade containing *Lacerta agilis* and *Crocodylus niloticus*, but not *Homo sapiens*" (Modesto and Anderson, 2004). Among the different and abundant groups within the Reptilia (e.g., Dinosauria, Rhynchosauria, Pterosauria), those that survived the K-Pg extinction are Lepidosauria, Choristodera, Crocodyliformes, Aves (see next section), and Testudinata.

Lepidosauria is a clade that includes two main subdivisions: Squamata with lizards, snakes, and amphisbaenians (Pyron et al., 2013); and Rhynchocephalia with tuataras (Herrera-Flores et al., 2017). The oldest known Lepidosaurian fossils are from the Middle Triassic (Jones et al., 2013). Squamates and rhynchocephalians are typically pentadactyl (both manus and pes, Fig. 8.3A and B), but some clades present particular limb reduction (Greer, 1991; Russell and Bauer, 2008; Roscito et al., 2014; Wagner et al., 2018). For instance, snakes, amphisbaenians, and some lizards (scincids) have undergone extreme reduction or loss of the limbs (Wiens et al., 2006; Feldman et al., 2016).

Choristodera is a group of diapsid reptiles with semiaquatic lifestyles (Sues, 2019). They are crocodile-mimics and ranged from the Middle Jurassic to the Miocene (Evans and Klembara, 2005; Matsumoto and Evans, 2010). Choristoderans exhibited a pentadactyl manus and pes (Liu, 2004; Matsumoto et al., 2013), and their ichnological record is up to now limited to the Cretaceous (Lockley and Hunt, 1995; Lee et al., 2020).

Testudinata is an abundant and diverse group of tetrapods that has a wide temporal distribution (Early Triassic to present, see Lichtig et al., 2018). It is characterized by a special shell developed mainly from their ribs composed of two halves, the plastron and the carapace (Nagashima et al., 2012). Only three clades apparently survived the K–Pg extinction: Sichuanchelyidae (Middle Jurassic–Paleocene); Meiolaniidae (Early Cretaceous–Holocene); and Testudines (Middle Jurassic–present) (see Pérez-García, 2020). Sichuanchelyidae, which is a Laurasian clade, only has one Cenozoic taxon and is terrestrial (Pérez-García, 2020). Meiolaniids have a Gondwanan distribution, show strong terrestrial adaptations, heavily armored heads, and coexisted with human populations before their extinction (Sterli, 2015). Testudines are the only living turtles and present a global distribution. They are divided into two groups: Pleurodira (side-necked turtles) and Cryptodira (hidden-necked turtles). Currently, Pleurodires are restricted to freshwater environments of South America, Africa–Madagascar, and Australia, but, in the past, they had a broad distribution, being found also on continents of the Northern Hemisphere, and in marine environments (see Ferreira et al., 2018). Cryptodira presents a global distribution and diverse habitats. This clade includes among their species freshwater turtles, terrestrial tortoises, and marine turtles (see Pérez-García, 2017). Testudinates present a primitive pentadactyl pes and manus condition adapted to different lifestyles, but the number of phalanges and their proportions change within the main groups (see Crumly and Sánchez-Villagra, 2004; Sánchez-Villagra et al., 2007; Sheil and Portik, 2008). For instance, terrapins present long and elongated digits (Fig. 8.3C and D), marine turtles have hyperelongated phalanges (Fig. 8.3E and F) and terrestrial tortoises have short phalanges with long and robust claws (Fig. 8.3G and H). Avanzini et al.

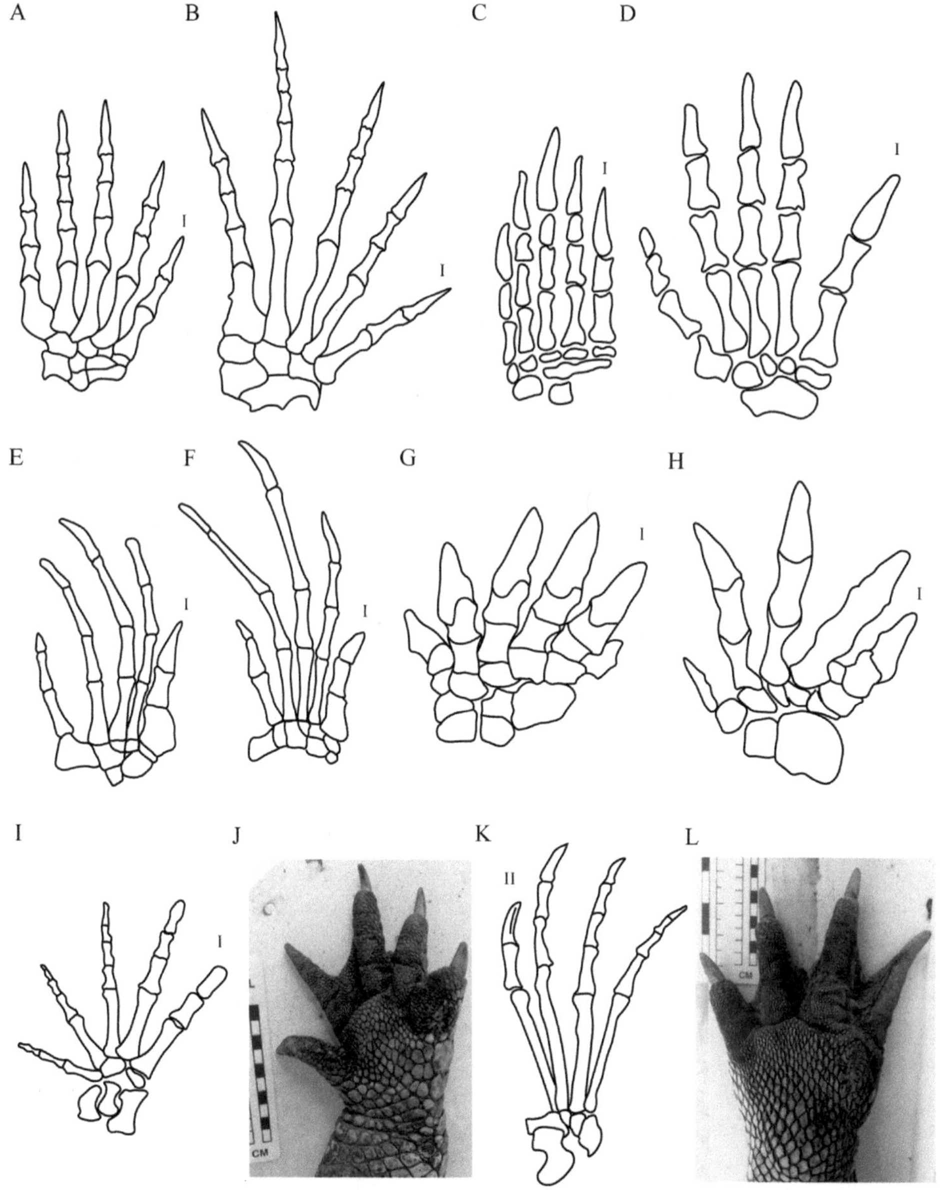

FIGURE 8.3 Skeletal pattern of manus and pes of reptilian taxa, (A and B) *Sphenodon*, (C and D) *Pseudemys*, (E and F) *Chelonia*, (G and H) *Malacochersus*, (I and J) *Caiman* and (K and L) *Alligator*. Figures not to scale. *I*, digit I; *II*, digit II. *Modified from (A and B) Bayer (1885). (C and D) Sánchez-Villagra et al. (2008). (E and F) Sánchez-Villagra et al. (2007). (G and H) Mautner et al. (2017); and (I and K) Reynolds (1897). Photographs of J and L have been provided by James Farlow.*

(2005) suggested that in terrestrial turtles the limbs are stout, rounded structures; in the freshwater forms this has happened to a lesser extent; and in marine forms, the manus and pes evolved into broad, flattened paddles.

Crocodyliformes is a diverse group of archosaurs that arose in the Triassic and has members that have survived to the present (see Nesbitt, 2011). Three clades lived during the Cenozoic: Sebecosuchia, Dyrosauridae, and Eusuchia. Sebecosuchians (Middle Jurassic?–Miocene, Dal Sasso et al., 2017) were notosuchians that presented a terrestrial and semi-aquatic lifestyle (Kellner et al., 2014), and a carnivorous diet (see Martin, 2014). On the contrary, dyrosaurids (Maastrichtian–Eocene, Hill et al., 2008) were marine or estuarine neosuchians with mainly piscivorous habits (Buffetaut, 1979; de Andrade and Sayão, 2014). Crocodylia is, along with Aves, the only known survivors of living archosaurs (see Brochu, 2001). The oldest paleontological remains of this group of eosuchians date from the Cenomanian (Late Cretaceous, Mateus et al., 2019), and tip-dated analyses also suggest this age for the origin of the clade (Darlim et al., 2022). A simplified classification of Crocodylia pointed out the presence of two main groups (see Rio and Mannion, 2021, fig. 3): Longirostres with Crocodyloidea and Gavialoidea; and Alligatoridae with Alligatorinae and Caimaninae. They are predatory and have a semiaquatic lifestyle. In general, Crocodyliformes have a pentadactyl manus and tetradactyl pes (Riff, 2007) (Fig. 8.3I–L).

Reptiles were the dominant terrestrial animals during the Late Paleozoic and Mesozoic, and that is seen in the abundance of their tracks. Nevertheless, during the Cenozoic, the ichnological record is scarce compared with mammalian or avian tracks.

Lepidosaurian tracks are generally represented by the ichnogenus *Rhynchosauroides* Maidwell, 1911. This ichnotaxon has been historically related to lacertoid trackmakers, and it has been identified from the Permian to the Jurassic (e.g., Valentini et al., 2007; Avanzini et al., 2010a) and there is also a questionable record in the Cretaceous (see Buck et al., 2022). The main chronological issue is that Lacertidae is a group of squamates that originated in the Paleogene (Čerňanský et al., 2020). The ichnogenus *Neosauroides* Kim et al., 2017, defined with Cretaceous specimens, has been cited in the Pleistocene of the USA (Weems, 2021). The typical lepidosaurian tracks present: pentadactyl and semiplantigrade, manus very asymmetric and mesaxonic; tridactyl to pentadactyl (Digit I and V are usually absent) pes, digitigrade, asymmetric and ectaxonic (Avanzini and Renesto, 2002; Kim et al., 2017).

Changes in the substrate consistency and type and/or speed can modify the track morphology. An interesting study in this topic was performed by Padian and Olsen (1984) with an extant Komodo monitor. On the contrary, recently the trackway of an apodal lepidosaurian (snake) has been identified in the Pleistocene of South Africa (Helm et al., 2023a).

As it is commented before, choristoderan tracks have only been identified in Cretaceous beds (see Lockley and Hunt, 1995; Lee et al., 2020). They have pentadactyl and entaxonic manus and pes tracks (Lee et al., 2020). Although similar in shape, they differ from crocodylomorph tracks because the latter have tetradactyl pes impressions.

Almost all the tracks of testudinates are similar in shape since the Early Triassic (Lichtig et al., 2018; Lockley et al., 2019a). There are two ichnofamilies within the testudinate tracks: Chelonipedidae (Sarjeant and Langston, 1994); and Marineropodidae (Lockley et al., 2019a). The first one represents terrestrial or freshwater trackmakers (see Avanzini et al., 2005; Lichtig et al., 2018). According to Sarjeant and Langston (1994), *Chelonipus* Rühle von Liliestern

(1939) is the only ichnogenus within the Chelonipedidae. Karl et al. (2012) and King et al. (2020) included also *Emydhipus* Fuentes Vidarte et al., 2003 in this ichnofamily, but it lacks justification in their text. The tracks of Chelonipedidae present quadrupedal trackways, a wide gait, sharp claw impressions, and five-toed hind limbs (Avanzini et al., 2005). Marineropodidae tracks were made by marine turtles when they moved out of the water on the beach. Anatomically, pes and manus present a paddle shape, without individual digits. Mainly, they are paddle traces deeper than the trace of the ventral plastron (Lockley et al., 2019a).

The ichnological record of Cenozoic turtles is very scarce. Indeed, in the Paleogene–Neogene only the holotypes of *Chelonipus chadronicus* Sarjeant and Langston, 1994 and *Chelonipus parvus* Sarjeant and Langston, 1994 were found. In Quaternary deposits (see Chapter 9), *Australochelichnus agulhasii* Lockley et al., 2019a, *Marinerichnus latus* Lockley et al., 2019a, *Chelonipus* isp. (Mustoe, 2019), *Emydhipus* isp. (Weems, 2018), and undetermined tortoise tracks (Helm et al., 2023b) were cited.

Although neoichnological studies with turtles are not common, the works of Avanzini et al. (2005) with freshwater and terrestrial individuals and Lockley et al. (2019a) with marine turtles are remarkable. In the former study, three different species of turtles walked on a substrate with different moisture, evidencing differences in the shape and arrangement of the footprints.

Crocodyliform tracks present a very conservative morphology since the Triassic. They were classified within two ichnofamilies: Batrachopodidae Lull, 1904 and Crocodylopodidae Fuentes Vidarte and Meijide Calvo, 2001. Lockley and Meyer (2004) included the ichnogenus *Crocodylopodus* Fuentes Vidarte and Meijide Calvo, 2001 (ichnogenus type of Crocodylopodidae) in Batrachopodidae, so Crocodylopodidae would be a junior synonym of Batrachopodidae.

Although crocodyliform tracks of Cenozoic age are not common (Lockley et al., 2010a), some authors pointed out the needed revision of their ichnotaxonomy. Several ichnotaxa, such as *Albertasuchipes*, *Indosuchipes*, and *Anticusuchipes*, represent swim tracks (Lockley et al., 2020; Castanera et al., 2021); instead, *Borealosuchipus* was impressed during crawling or a basking behavior (Erickson, 2005). Lockley et al. (2020) suggested that the Cenozoic *Albertasuchipes* and *Anticusuchipes* resemble the Mesozoic *Hatcherichnus* Foster and Lockley, 1997 and *Crocodylopodus*, respectively. Conversely, Mustoe (2019) argued that *Anticusuchipes* is a different ichnogenus because its tracks are larger than the Mesozoic ones. In that case, the discussion would be centered on whether size is a valid ichnotaxobase (see Bertling et al., 2022). Furthermore, these ichnotaxa have been defined practically by a single trackway and have not been cited outside their type locality.

On the contrary, some Cenozoic crocodyliform tracks have been related within typical Mesozoic ichnotaxa. For instance, Mikuláš (2010) classified as *Characichnos* isp., an ichnogenus defined for swimming theropod dinosaurs, several swimming crocodiles track from the Miocene of Czech Republic. Abbassi (2022) cited the presence of *Crocodylopodus* and *Hatcherichnus* in the Miocene of Iran. Moreover, Weems (2018, 2021) and Weems et al. (2021) assigned some tracks found in the Pleistocene of the USA to *Crocodylopodus*.

Several neoichnological studies have been performed with crocodiles (see Farlow et al., 2018a,b). Kubo (2010) examined the effect of different locomotion postures and gaits on crocodylian trackways, and Kumagai and Farlow (2010) documented examples of swim tracks in estuarine environments, similar to others found in the Cretaceous. Moreover, there are several examples of current crocodile tracks impressed in different substrates, under different moisture and or by different behavior. Farlow et al. (2018a) identified that in the extant crocodylian pes tracks the claw traces of Digits I, II, and III are sometimes dug into the substrate, thus not reflecting real anatomical lengths, and lack a Digit IV claw trace. This last feature was also identified in Mesozoic tracks (Castanera et al., 2021). The neoichnological information will be very useful in future ichnotaxonomical studies.

4.2.2 *Reptilian ichnotaxa*

Seven ichnogenera with eight ichnospecies have been related to reptilian trackmakers.

Albertasuchipes McCrea et al., 2004
  Albertasuchipes russellia McCrea et al., 2004
Anticusuchipes Mustoe, 2019
  Anticusuchipes amnis Mustoe, 2019
Borealosuchipus Erickson, 2005
  Borealosuchipus hanksi Erickson, 2005
Chelonipus Rühle von Liliestern, 1939
  Chelonipus chadronicus Sarjeant and Langston, 1994
  Chelonipus parvus Sarjeant and Langston, 1994
Gandopodichnus Abbassi, 2022
  Gandopodichnus caesellum Abbassi, 2022
Indosuchipes Rajkumar et al., 2015
  Indosuchipes manipurensis Rajkumar et al., 2015
Merthykhuwaripus Abbassi, 2022
  Merthykhuwaripus conicus Abbassi, 2022

4.2.2.1 Description and discussion of reptilian ichnogenera

Ichnogenus *Albertasuchipes* McCrea et al., 2004.

Fig. 8.4A.

Type ichnospecies: *Albertasuchipes russellia* McCrea et al., 2004.

Diagnosis: Trackway of quadruped with pronounced heteropody. Manus: tridactyl, digitigrade, approximately one-third length of pes. Impressions wider than long (Footprint Length, FL = 54.0 mm; Footprint Weight, FW = 82.0 mm); FL/FW ratio approximately 0.66. Manual digits range in thickness between 16.0 and 26.0 mm. Manus digits are nearly parallel to axis of print and spaced approximately 20 mm apart. No digital pads. Digits II and III longest, Digit IV less than half length of Digits II and III; claw impressions robust (6.0–12.0 mm thick) and terminate in obtuse to acuminate points; pace of manual prints 540 mm. Placement of manus is typically immediately anterior to pes. Pes: tridactyl and semidigitigrade. Impressions longer than wide (FL = 152.5 mm; FW = 69.0 mm); FL/FW ratio 2.21. Digits nearly parallel to main axis of print with slight curve of digits toward midline of trackway, starting at position half to three-quarters of length of the footprint and extending anteriorly to tips. Digits spaced approximately 20 mm apart and range in thickness between 19.0 and 30.0 mm. No digital pads. Digit III longest in pes; Digit II longer than Digit IV. Digits terminate in robust (6.0–9.0 mm thick), obtuse claws. No true heel impression, posterior portion of print roughly truncate. Pace length (pes) approximately 500 mm (3 times pes

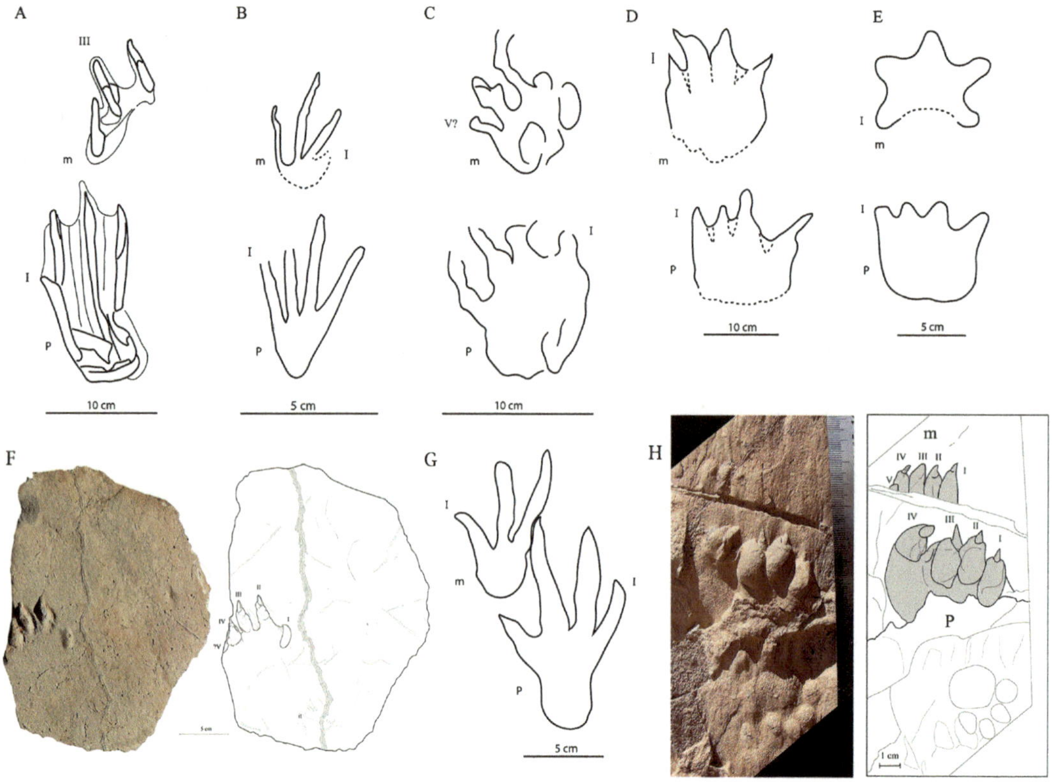

FIGURE 8.4 Outline sketch of the reptilian Paleogene/Neogene ichnotaxa, (A) *Albertasuchipes russellia*. (B) *Anticusuchipes amnis*. (C) *Borealosuchipus hanksi*. (D) *Chelonipus chadronicus*. (E) *Chelonipus parvus*. (F) *Gandopodichnus caesellum*, (G) *Indosuchipes manipurensis*, (H) *Merthykhuwaripus conicus*. *I*, digit I; *III*, digit III; *m*, manus; *p*, pes; *V?*, probably digit V. *(A) Modified from McCrea et al. (2004). (B) Modified from Mustoe (2019). (C) Modified from Erickson (2005). (D) Modified from Sarjeant and Langston (1994). (E) Modified from Sarjeant and Langston (1994). (F) From Abbassi (2022). (G) Modified from Rajkumar et al. (2015). (H) From Abbassi (2022). F and H with permission.*

length); stride length just over 900 mm (6 times pes length); pes pace angulation approximately 130 degrees. Trackway narrow (pes: inner trackway Weight, TWi = 122.0 mm; outer trackway Weight, TWo = 245.0 mm), with inner margin of pes tracks near trackway midline, manus tracks overlapping trackway midline or just lateral to it. Manus rotation variable. Pes rotation variable, but usually toward midline.

Discussion: *Albertasuchipes* tracks were impressed by a crocodile that propels itself through the water without making contact with the substrate (bottom walking) (McCrea et al., 2004). The tracks were found in the Paskapoo Formation (late Paleocene, Canada). Lockley et al. (2010) pointed out that these tracks are similar to examples of *Kuangyuanpus*, *Laiyangpus*, *Saltosauropus*, and *Hatcherichnus*; so *Albertasuchipes* could conceivably have been assigned to one of these four ichnogenera. Nevertheless, taking into account that *Laiyangpus* lacks accessible type, and there are also differences in digit count (whether the result of preservation, behavior, or other factors), the *Albertasuchipes* track type can be sustained. Lockley et al.

(2020) pointed out that *Albertasuchipes* resembles *Hatcherichnus* in shape, robustness, and large size of digit traces, and represents a swimming crocodile. On the contrary, Farlow et al. (2018a) differentiated *Albertasuchipes* from other crocodyliform swim tracks, taking into account that they were made by bottom walking and possibly scoring based on an unusually long stride for their size.

Ichnogenus *Anticusuchipes* Mustoe, 2019.

Fig. 8.4B.

Type ichnospecies: *Anticusuchipes amnis* Mustoe, 2019.

Diagnosis: Tetradactyl tracks with slender elongate digits with apices that range from rounded to sharply pointed. Claws are not distinctly evident at the digital apices. Heel pads are rounded. Tracks are plantigrade to digitigrade. Manus impressions are much smaller than pes impressions. The Chuckanut Formation footprints are distinctly different from previously described crocodylian tracks. Footprints of *Crocodylopodus* Avanzini et al., 2010b and *Batrachopus* (Hitchcock, 1845; Olsen and Padian, 1986) are much smaller than *Anticusuchipes; Mehliella* Lockley, 2010, *Albertasuchipes* (McCrea et al., 2004), and an unnamed track from the USA (Falkingham et al., 2010) are much larger. The shapes of all of these ichnogenera are unlike that of *Anticusuchipes*. Mesozoic crocodylian tracks that are comparable in size are *Hatcherichnus* (Lockley and Foster, 2010) and *Kuangyuanpus* Young, 1943, but these footprints are much broader and have shorter, wider digit impressions than *Indosuchipes* (Rajkumar et al., 2015) or *Anticusuchipes* (Lockley et al., 2010b). The most notable difference between *Indosuchipes* and *Anticusuchipes* is the shape of the heel pad, which is broadly rounded and prominent in the former ichnogenus, and smaller and narrow for *Anticusuchipes*. An unnamed track from Portugal is more similar in shape, but digit impressions of the latter fossil are short and triangular, compared with the slender, elongated digits characteristic of *Indosuchipes* and *Anticusuchipes* (emended from Mustoe 2019).

Discussion: *Anticusuchipes* tracks were identified in the lower Eocene fluvial strata of the Chuckanut Formation (USA). Lockley et al. (2020) suggested that *Anticusuchipes* resembles *Crocodylopodus*, nevertheless, Mustoe (2019) argued that they are different ichnogenera because they are larger than the Mesozoic ones. In this case and tacking into account that size is not a valid ichnotaxobase, *Anticusuchipes* could be a junior synonym of *Crocodylopodus*. The diagnosis by Mustoe (2019) contained a *lapsus calami* in the ichnogenus name ("Anticusichipes") that is corrected herein.

Ichnogenus *Borealosuchipus* Erickson, 2005.

Fig. 8.4C.

Type ichnospecies: *Borealosuchipus hanksi* Erickson, 2005.

Diagnosis: No formal diagnosis was identified but a description provided amounts to a diagnosis. The description is as follows: Natural casts of a trackway showing fore- and hind feet and body traces are preserved in hyporelief at the base of a sandstone layer. The trackway is present on four separate pieces of calcareous sandstone. Only prints of the right fore- and hind feet are preserved. Imprints of the right manus and pes clearly show the presence of Digits I–IV; however, the tracks lack impressions of the complete digits and are too obscured to provide detailed measurements of the feet. A maximum length of the manus imprint is 9.0 cm measured along Digit III, and the maximum breadth of the imprint is 8.5 cm. The maximum length of the pes imprint is 11.5 cm, and the maximum breadth is 11.0 cm. Angles of convergence and divergence of the feet from midline cannot be measured because there is no midline reference due to the incompleteness of the specimen. The trackway shows belly drag marks. Spacing of the individual foot tracks is irregular, and the trackway lacks a "stride". Footprint size, however, suggests that the trackmaker was close in size to a mounted skeleton of *Borealosuchus formidabilis* Erickson, which has a manus maximum length of 18.3 cm and breadth (measured at the heel) of 7.6 cm, and a pes maximum length of 34.0 cm and a breadth (measured at the heel) of 11.0 cm. The gleno-acetabular distance of this skeleton is 89 cm, and the total skeletal length is 3.71 m, which is considered average adult size within the Wannagan Creek crocodile population.

Discussion: *Borealosuchipus* is based on basking crocodilian tracks with singular manus, pes, and body traces found in late Paleocene Bullion Creek Formation, USA (Erickson, 2005). The tracks of *Borealosuchipus*, probably impressed by a eusuchian crocodile, are similar to footprints made by extant *Crocodylus acutus* (see Cuvier, 1807; Kumagai and Farlow, 2010).

Ichnogenus *Chelonipus* Rühle von Lilienstern, 1939.

Fig. 8.4D and E.

Type ichnospecies: *Chelonipus torquatus* Rühle von Lilienstern, 1939.

Diagnosis: Quadruped trackways with wide gauge, pace angle between 50 degree and 70 degree; hands and feet digitigrade, very low longitudinal expansion of the impressions (emended by Haubold, 1971b).

Discussion: *Chelonipus* is an ichnogenus erected by Rühle von Liliestern (1939) to discriminate Lower Triassic turtle tracks. He defined two ichnospecies: *C. torquatus* Rühle von Lilienstern, 1939 and *C. cuneiformis* Rühle von Lilienstern, 1939. Kuhn (1958) synonymized the latter ichnospecies under *C. torquatus* and Haubold (1971b) named another Triassic ichnospecies, *C. plieningeri* Haubold, 1971b and give a diagnosis for the ichnogenus (see Avanzini et al., 2005). *C. plieningeri* shows variations in morphology and in gait and it was considered as synonym of *C. torquatus* Lichtig et al., 2018.

Sarjeant and Langston (1994) defined two new ichnospecies within *Chelonipus* for late Eocene tracks from the USA: *C. chadronicus* and *C. parvus*. Finally, Lockley et al. (2019b) proposed transferring the Early Cretaceous *Laiyangpus liui* Young, 1960 to *Chelonipus liui*, comb. nov. As summary, *Chelonipus* would present six valid ichnospecies, four Mesozoic and two Cenozoic. *Chelonipus* is similar to *Emydhipus*, but the latter has more parallel ungual traces in the manus, turned lightly inward, and apparently always away with respect to the pes (Avanzini et al., 2005).

The ichnospecies *C. chadronicus* (Fig. 8.4D) is defined as "*Quadrupedal footprints exhibiting a broad trackway and short stride. Footprints vary with the animal's behavior from digitigrade to semi-plantigrade. In both manus and pes, four digits are normally impressed, the fifth (V) rarely being impressed. All digits are clawed. Pollex short; Digit II almost twice as long as pollex and curved inwardly; Digits III and IV of intermediate length, Digit III directed forward and Digit IV outward. Digits I to III of pes directed forward, the hallux and Digit II shorter than Digit III; Digit IV outwardly directed and with a longer claw. Webbing present between all these digits. Tail-drag impressions lacking*" (original diagnosis of Sarjeant and Langston, 1994). The authors pointed out that the rather sprawling and uncertain gait of *C. chadronicus* suggests an amphibious turtle rather than a habitually terrestrial one. On the contrary, the diagnosis of *C. parvus* (Fig. 8.4E) is "*Quadrupedal footprints of small size, exhibiting a broad trackway and short stride. Impressions of manus directed outward, those of pes forward. Manus digitigrade, pes typically semiplantigrade or plantigrade. Manus digits short and clustered in a stellate pattern, with the pollex and Digit*

II directed forward, Digit III outward, and Digits IV and V backward; pes digits very short and blunt, with I–III directed forward and IV directed outward. Claws not indicated; no tail-drag impression" (original diagnosis of Sarjeant and Langston, 1994). The authors suggest that the trackmaker was a rather inefficient pedestrian because of its broad trackway and the outwardly directed manus impressions. Moreover, they observed that the tracks are small, so the trackmaker may have been either a young turtle or one only attaining small adult size.

Ichnogenus *Gandopodichnus* Abbassi, 2022.

Fig. 8.4F.

Type ichnospecies: *Gandopodichnus caesellum* Abbassi, 2022.

Diagnosis: Pentadactyl, digitigrade in a moderate size crocodile footprint with five-digit impressions, Digit I impression is the smallest, four forward Digits II–V are elongated ovoidal in shape, and nearly uniform in size. Digit I imprint is further back than Digits III–IV.

Heel imprint is absent. Digit imprints may show at least three digital pads in Digits II–IV; large conical claw impressions are present, contiguous to digits (Abbassi, 2022).

Discussion: *Gandopodichnus* have been identified in Miocene red beds of Central Iran. It is known by large claw imprints in Digits II–IV, which are connected to the digits. *Gandopodichnus* differs from *Hatcherichnus* by lack of metatarsal/metapodium imprints and presence of clear large claw imprints. *Gandopodichnus* differs from *Kuangyuanpus* by size and shape and arrangement of digit imprints. *Gandopodichnus,* however, is based on a single, isolated footprint and seems that it belongs to a quadruped. Based on its distinctive morphology, we consider it a distinctive ichnogenus.

Ichnogenus *Indosuchipes* Rajkumar et al., 2015.

Fig. 8.4G.

Type ichnospecies: *Indosuchipes manipurensis* Rajkumar et al., 2015.

Diagnosis: Tetradactyl tracks with slender elongate digits with apices that range from sharply pointed to rounded. Heel pad broadly rounded, with width approximately one-third the width of the span of Digits I and IV. Tracks are plantigrade, and claws are not distinctly evident at the digital apices. Track lengths are approximately 10 cm, with I–IV span width of approximately 6 cm. The Laisong Formation footprints are distinctively different from previously described crocodilian tracks. Footprints of *Crocodylopodus* (Avanzini et al., 2010b) and *Batrachopus* (Olsen and Padian, 1986) are much smaller than *Indosuchipes; Mehliella* (Lockley, 2010), *Albertasuchipes* (McCrea et al., 2004), and an unnamed track from the USA (Falkingham et al., 2010) are much larger. The shapes of all of these ichnogenera are unlike that of *Indosuchipes*. Mesozoic crocodilian tracks that are comparable in size are *Hatcherichnus* (Lockley and Foster, 2010) and *Kuangyuanpus* (Lockley et al., 2010b), but these footprints are much broader and have shorter, wider digit impressions than *Indosuchipes*. An unnamed track from Portugal (Mateus and Milan, 2010) is more similar in shape, but digit impressions of the latter fossil are short and triangular, compared with the slender, elongate digits characteristic of *Indosuchipes* (Rajkumar et al., 2015).

Discussion: *Indosuchipes* was found in Oligocene rocks of the Laisong Formation (India). Mustoe (2019) pointed out that the main difference between *Indosuchipes* and *Anticusuchipes* is the shape of the "heel" pad, which is broadly rounded and prominent in the former ichnogenus, and smaller and narrow for *Anticusuchipes*. Rajkumar et al. (2015) suggested that *Crocodylopodus* and *Batrachopus* tracks are smaller than *Indosuchipes*, and the Mesozoic crocodilian tracks similar to *Hatcherichnus* and *Kuangyuanpus* are different because footprints are much broader and have shorter, wider digit impressions than *Indosuchipes*.

Ichnogenus *Merthykhuwaripus* Abbassi, 2022.

Fig. 8.4H.

Type ichnospecies: *Merthykhuwaripus conicus* Abbassi, 2022.

Diagnosis: Tetradactyl pes imprint with a large monopod digit and large conical claw imprints. Pes imprint is digitigrade, with fat digits. Digits I–III are relatively equal in size and stick together, and Digit IV is slightly larger and curved toward Digits I–III. The claw imprint has a large conical form, connected to digits with a distinctive epidermal digit pad rim imprint. Manus imprint includes small tridactyl digit imprints with or without claw imprints. Digits of manus do not make a row and overlap each other (Abbassi, 2022).

Discussion: *Merthykhuwaripus* has been introduced from Miocene red beds of Central Iran. *Merthykhuwaripus* consists of fat digits, large claw imprints, and is completely digitigrade, thus it is distinct from *Batrachopus dispar* and *Batrachopus gracilis*.

4.3 AVES

4.3.1 Avian autopodium structure and tracks

Birds commonly have four digits (Digits I to IV) that are homologous to the pentadactyl appendage. They have retained the phalangeal formula of their bipedal reptilian ancestors, namely, 2, 3, 4, 5, for the first four digits, the fifth digit having been lost. Some birds have three digits (II–IV), and only the extant ostrich has two digits (Lucas and Stettenheim, 1972; King and McLelland, 1984). The bird foot is typically digitigrade (only the digits touch the ground), and they commonly join in a metatarso-phalangeal pad (Fig. 8.5) that may or

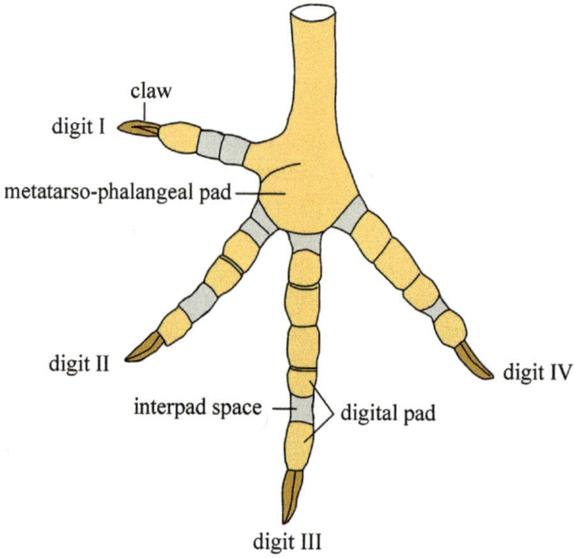

FIGURE 8.5 Morphological features of the avian foot. *Modified from Lucas and Stettenheim (1972).*

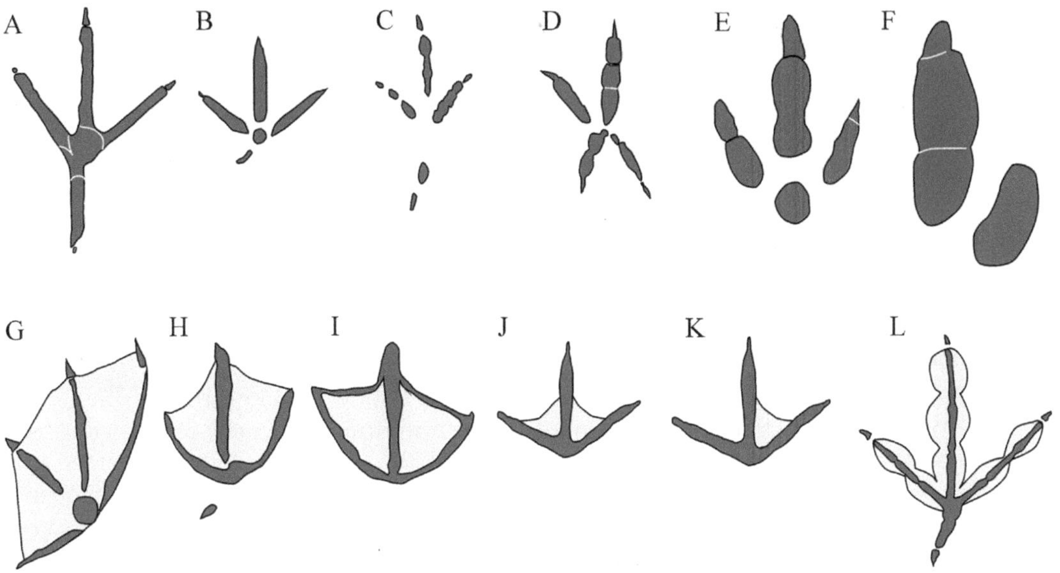

FIGURE 8.6 Morphology of avian footprints. Always represented by right footprints, (A–C) Anisodactyl. (D) Zygodactyl. (E) Tridactyl. (F) Didactyl. (G) Totipalmate. (H and I) Palmate. (J) Semipalmate. (K) Semipalmate asymmetrical. (L) Lobate. *(A–D, G, H) Modified from Elbroch and Marks (2001). (F) Drawn from Schaller et al. (2011). (J–L) Modified from Brown et al. (1987).*

may not touch the ground. The plantar surface of the digits exhibits digital pads and interpad spaces, and digits end in claws of different sizes and shape. Bird feet and the corresponding footprints are commonly classified according to digit arrangement and the presence/absence of interdigital web, and the extent of the interdigital web. We will focus on the foot morphologies that can be recorded in fossil footprints (according to substrate preferences), and especially those of walking and wading birds (Fig. 8.6).

Digit arrangement is related to the main function of the foot (Howell, 1944; Lucas and Stettenheim, 1972; King and McLelland, 1984; Lovette and Fitzpatrick, 2016). The anisodactyl foot is characterized by three digits pointing forward and one (Digit I) backward (Fig. 8.6A–C), as is the case for passerines and *Archaeopteryx*. This is used for perching or holding prey. Another foot type with four digits is the zygodactyl configuration that is typified by two (Digits II and III) directed forward and the other two (I and IV) backward (Fig. 8.6D) as in woodpeckers, toucans, cuckoos, and parrots. The zygodactyl foot reflects an adaptation to climbing and also a more effective holding and manipulation of prey or food than the anisodactyl foot. The tridactyl foot lacks the hallux (or it is kept raised) and is commonly an adaptation to walking and running in many ground-dwelling and aquatic birds (Fig. 8.6E). The didactyl foot reflects an extreme adaptation to running and is only recorded in extant ostriches (Fig. 8.6F). Seriemas display a marked reduction of Digit II as an adaptation for running. The degree of webbing between digits also varies, and the presence of a well-developed interdigital web is an adaptation to foot-propelled swimming (King and McLelland, 1984; Lovette and Fitzpatrick, 2016). The interdigital web between Digits I and II is medial, between Digits II and III is intermediate, and between Digits III and IV is lateral. In a palmate foot (Fig. 8.6H and I) the front digits are webbed together (II–IV), whereas the hallux lacks the web (or it is absent); this is typical of ducks, gulls, and flamingoes. A variation of the last configuration is an incised palmate foot, in cases where the interdigital web reaches the tip of the digits, but describes a marked curvature. In the totipalmate foot (Fig. 8.6G), the web also includes Digit I, as occurs in pelicans. The semipalmated foot is typical of many shorebirds and is typified by a partial interdigital web at the base of the three front digits (Fig. 8.6J). A particular type of semipalmated web (Fig. 8.6K) is that present only between the lateral digits (III–IV). The lobate foot is a particular configuration of coots and grebes in which each digit is bordered by its own webbing (Fig. 8.6L).

4.3.1.1 Ichnotaxonomy of avian footprints

The first reports of avian tracks can be traced to Desnoyers (1859a,b) who described, although he did not illustrate, the footprint of a large bird from the Eocene of the Paris basin, France (see also Buffetaut, 2004; Lockley and Harris, 2010). The first formally named bird track was from the Early Cretaceous of Colorado (USA), and corresponds to the ichnogenus *Ignotornis* Mehl, 1931. A few years later the first Cenozoic bird track was named, *Urmiornis abeli* Lambrecht, 1938, from the Pliocene of Iran. This ichnospecies was later renamed under *Iranipeda abeli* by Vyalov (1989), because *Urmiornis* was preoccupied by a biological taxon. A detailed account of early studies on Cenozoic bird footprints is provided by Sarjeant and Reynolds (2001) and Lockley and Harris (2010).

Oleg Stepanovich Vyalov was influential in the definition of several aspects of the application of binomial nomenclature to fossil vertebrate footprints (see also Lucas, 2007). However, the first to apply systematically these principles to the naming and description of fossil footprints were Panin and Avram (1962), who described material from the Miocene of the Carpathian fold belt of Romania. Panin and Avram (1962) explicitly mentioned a written communication with Vyalov (in 1961) and indicated their adherence to the suggestions of this author. In particular, Panin and Avram (1962) adopted the application of a binomial nomenclature in Latin, and the premise that this nomenclature does not represent the units of a biological classification system but a "conventional" one (i.e., parataxonomy) applied to the "vital activity" of organisms, and also advanced suggestions for the formation of (ichno)generic names. These authors adopted the proposal of Vyalov to include all avian footprints under Avipedia (with a rank of "order"). Panin and Avram (1962) further considered the utility of comparing fossil footprints with those of extant groups and explained the limitations and potential contribution of paleoichnology to the evolution of different animal groups. For the Miocene, Panin and Avram (1962) advocated the naming of the different categories for fossil footprints using a "root" related to the taxonomic category of the extant producers of similar tracks, and the suffix -ipedia (order), -ipedae (family) and -ipeda (genus). Also stated that the designation of (ichno)species must follow the internationally accepted rules for the formation of such names. No definition for (ichno)families was provided. The nomenclatural scheme proposed by Panin and Avram (1962) for avian footprints was as follows:

Order Avipedia Vyalov, 1961 (no longer in use)
 (Ichno)family Ardeipedae

(Ichno)genus *Ardeipeda*
Ardeipeda egretta
Ardeipeda gigantea
Ardeipeda incerta
(Ichno)family Anatipedae
(Ichno)genus *Anatipeda*
Anatipeda anas
(Ichno)family Charadriipedae
(Ichno)genus *Charadriipeda*
Charadriipeda recurvirostrioidea
Charadriipeda minima
Charadriipeda disjuncta
Charadriipeda becassi
(Ichno)family Gruipedae
(Ichno)genus *Gruipeda*
Gruipeda maxima

Probably not being aware of the publication of Panin and Avram (1962), Vyalov proposed in 1965 and then described in detail in 1966, the ichnogenus *Avipeda* to embrace all avian traces, except those that already received names like *Ignotornis* and *Urmiornis* (Mehl, 1931; Lambrecht, 1938). Vyalov (1966) argued that in his collection from the Miocene of the Carpathians there was very little morphological variation to allow for further division. However, he envisaged that this can be done in the future with additional findings. Vyalov (1966) proposed a detailed account of *Avipeda*, including type species, synonymy list, holotype, diagnosis, and a detailed discussion, thus paralleling the currently accepted format for ichnotaxonomic descriptions (Lucas, 2007). The proposal of Vyalov (1965) included the following names.

Avipedia (avian tracks)
(Ichno)genus *Avipeda*
Avipeda phoenix
Avipeda sirin
Avipeda filiportatis

The schemes by Panin and Avram (1962) and Vyalov (1965) were revised by Sarjeant and Langston (1994) and Sarjeant and Reynolds (2001).

The currently accepted categories for ichnotaxa are ichnofamily, ichnogenus, and ichnospecies (ICZN). It is not recommended to use the higher rank categories suggested by Vyalov (1966) including ichnorder, ichnoclass, and ichnodivision, which are still used in some recent publications (Oliva and Arregui, 2018; Campos-Medina et al., 2022). As clearly stated by Panin and Avram (1962), it is also not appropriate to fit ichnotaxa under a higher biological category (but see recent usage of this procedure in Buckley et al., 2016 as an example).

4.3.1.2 Measurements on avian footprints

It is advisable to follow the general conventions used by Leonardi (1987) and Buckley et al. (2016) for measurements on individual footprints and trackways. One of the main sources of

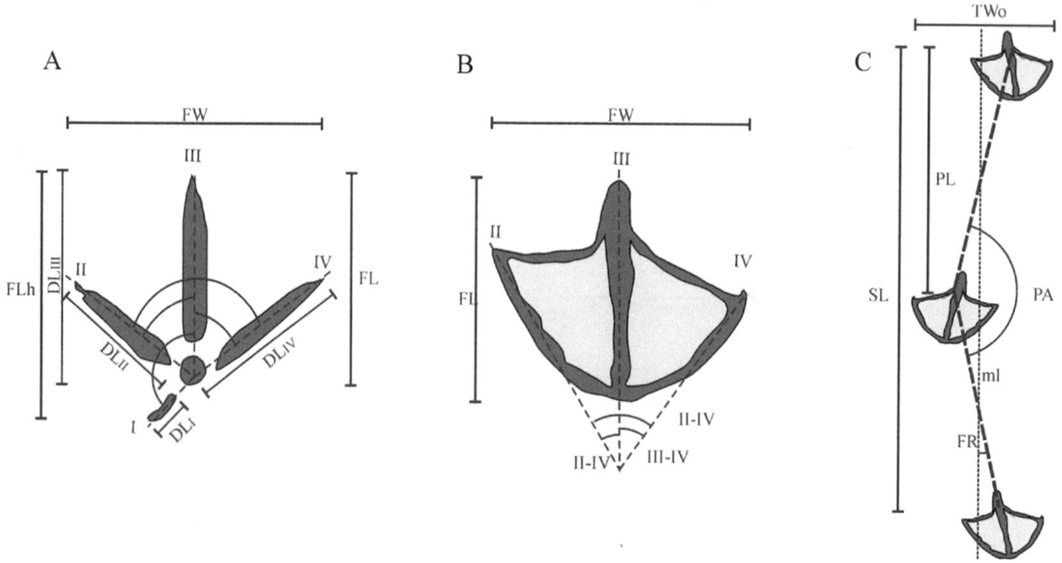

FIGURE 8.7 Measurements on avian footprints, (A) Example of a tetradactyl footprint with straight digits. (B) Example of a tridactyl palmate footprint with curved lateral digits. (C) Trackway measurements. DL_I *to* DL_{IV}, length of digit I to IV; *FL*, footprint length; FL_h, footprint length including the hallux; *FR*, footprint rotation from the midline; *FW*, footprint width; *I to IV*, digit I to IV; *I–III, II–III, III–IV, II–IV*, Angle between the digits I–III, II–III, III–IV and II–IV; *ml*, midline; *PA*, pace angulation; *PL*, pace length; *SL*, stride length; *TWo*, trackway external width.

differences in the measurements taken by different researchers is related to the curvature of the lateral digits and if the imaginary axis of digits converges in a single point or footprint center (Buckley et al., 2016) or not (Sarjeant and Reynolds, 2001). Footprints with curved lateral digits can be measured adjusting the digit axis to the distal (recommended) or proximal part of the digit, thus resulting in differences in the angles between digits (Fig. 8.7B). Common measurements include footprint length (FL), footprint width (FW), length of digit impressions (DL_I to DL_{IV}) measured with reference to the footprint center (a reference point where the digit axes join), width of digits (DW), and divarication of Digits II and IV (II–IV), II and III (II–III), and III and IV (III–IV) (Fig. 8.7A and B). The position of the hallux can be estimated by measuring the angle with respect to the axis of Digit III (I–III in Fig. 8.7A). For tetradactyl footprints, also the footprint length including the hallux (FL_h) and the ratio of the footprint length to hallux length (FL_h/DL_I) can be estimated (Fig. 8.7A).

For trackways (i.e., at least three consecutive footprints for bipedal animals) the pace length (PL), pace angulation (PA), stride length (SD), rotation with respect to the midline (FR) and the trackway external width (TWo) can be measured (Leonardi, 1987) (Fig. 8.7C).

4.3.1.3 Ichnotaxobases for avian footprints

As is common in vertebrate ichnology, there is no a general agreement on what are the most adequate ichnotaxobases to distinguish ichnogenera and ichnospecies of avian footprints (de Valais and Melchor, 2008; Lockley and Harris, 2010; Díaz-Martínez et al., 2011). This situation is the result of a set of factors, including those that are related to the morphological variation of avian footprints and some others related to the methodological approach. The latter includes the discipline of education of the researcher, age of the hosting rocks and inferences about the presumed producer and, in the past century, a poor communication of research results.

A compilation of ichnotaxobases used in the erection of Cenozoic avian ichnotaxa suggests that a total of 29 ichnotaxobases have been used in 33 papers, since 1938 (Fig. 8.8). Among them, the more frequently employed (accounting for 45% of the entries) are the number of digits, footprint size, presence/absence of webbing, total digit divarication, and relative length of digits. Most authors did not distinguish between ichnogeneric and ichnospecific taxobases, probably due to the fact that when describing monospecific ichnotaxa the diagnoses of ichnogenus and ichnospecies match.

A comparison of the works that distinguished the criteria used for ichnogenera and ichnospecies (Currie, 1981; Demathieu et al., 1984; Lockley et al., 1992, 2021; Sarjeant and Langston, 1994; Doyle et al., 2000; McCrea and Sarjeant, 2001; de Valais and Melchor, 2008; Kim et al., 2013) suggest that: (1) ichnogenera are mainly diagnosed considering number of digits, total digit divarication, relative digit length, and presence/absence of webbing. Other important features are overall footprint shape, position of the hallux, digit shape, presence/absence of metatarsal pad, and presence/absence of phalangeal pads. (2) ichnospecies are mainly distinguished by digit divarication, footprint size, and relative digit length, whereas other relevant features are digit shape, presence/absence of claw traces, and shape of metatarsal pads. Trackway parameters (i.e., pace length, trackway width, rotation from the midline, and stride length) have been used both to distinguish ichnogenera and ichnospecies. Recommended ichnogeneric and ichnospecific taxobases for avian footprints are listed in Table 8.1.

The criteria for definition of ichnofamilies should be very broad and follow, in principle, the categories of bird foot outlined above (e.g., Sarjeant and Langston, 1994; Lockley et al., 1992, 2006). Criteria for definition of ichnofamilies include number of digits, arrangement of digits and presence/absence and extent of webbing (Melchor et al., 2010; Bertling et al., 2022).

4.3.2 *Avian ichnotaxa*

A total of 36 ichnogenera and 77 ichnospecies have been recorded in Paleogene/Neogene strata. These are listed below in alphabetical order. The symbol ☒ indicates an invalid ichnotaxon, and the reasons are summarized between parentheses and discussed below. Ichnotaxa with an asterisk below are defined on Mesozoic material but were also identified in the Cenozoic.

- ☒ *Alaripeda* Sarjeant and Reynolds, 2001
  - ☒ *Alaripeda lofgreni* Sarjeant and Reynolds, 2001
  - ☒ *Alaripeda bristolia* Reynolds, 2012 (junior synonym of *Gruipeda dominguensis*)
- *Anatipeda* (Panin and Avram, 1962) Sarjeant and Reynolds, 2001.
  - *Anatipeda anas* Panin and Avram, 1962
  - ☒ *Anatipeda californica* Sarjeant and Reynolds, 2001 (junior synonym of *Roepichnus*)
  - ☒ *Anatipeda alfi* Sarjeant and Reynolds, 2001 (junior synonym of *Roepichnus*)
- ☒ *Antarctichnus* Covacevich and Lamperein, 1970 (synonymized under *Gruipeda* by de Valais and Melchor, 2008).

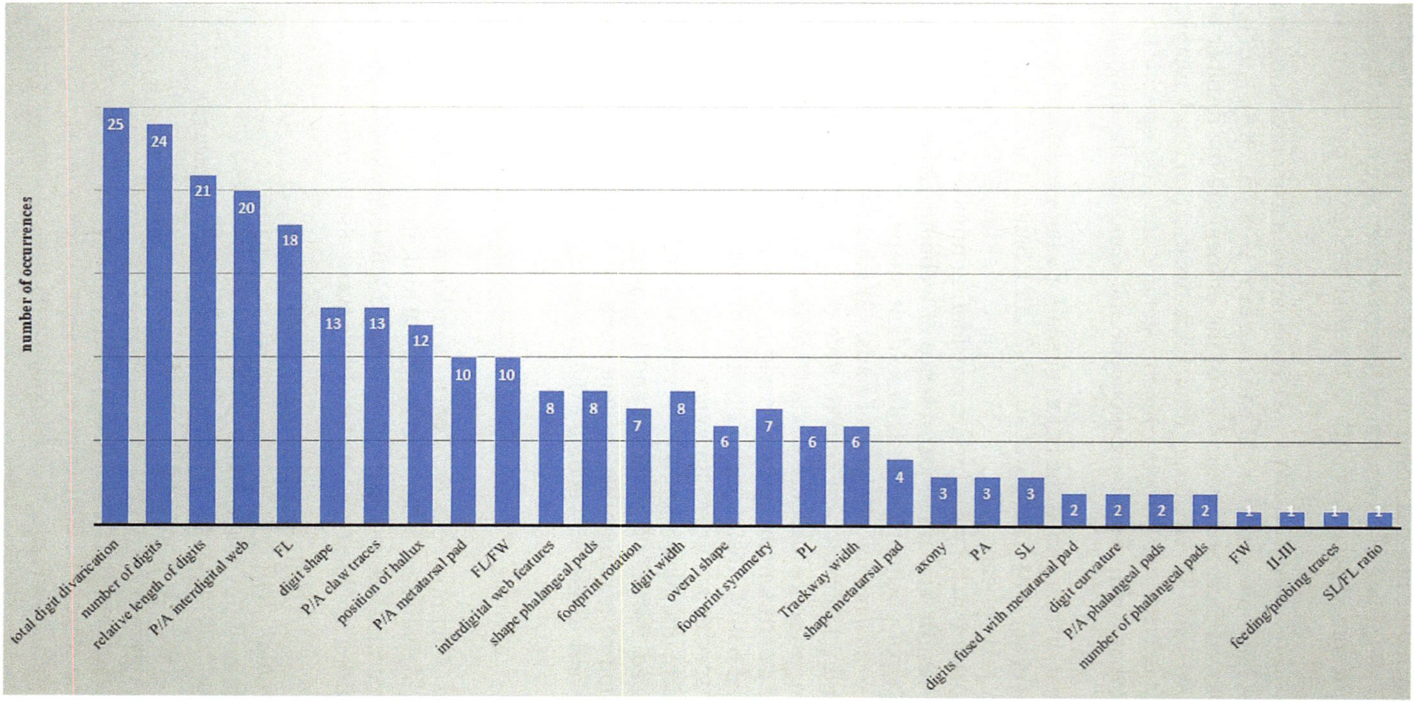

FIGURE 8.8 Histogram showing the frequency of individual features proposed as distinctive criteria for bird ichnotaxa. *FL*, footprint length; *FW*, footprint width; *PA*, pace angulation; *PL*, pace length; *P/A*, presence/absence of a feature; *SL*, stride length. *Data compiled from Lambrecht (1938), Panin and Avram (1962), Panin (1965), Panin et al. (1966), Vyalov (1966), Covacevich and Lamperein (1970), Ellenberger (1980), Currie (1981), Demathieu et al. (1984), Kordos (1985), Alonso and Marquillas (1986), Aramayo and Manera de Bianco (1987), Lockley et al. (1992), Rădan and Brustur (1993), Li and Zhen (1994), Sarjeant and Langston (1994), Yang et al. (1995), Fuentes Vidarte, 1996, Doyle et al. (2000), Payros et al. (2000), Sarjeant and Reynolds (2001), McCrea et al. (2004), Remeika (2001), Mirzaie Ataabadi and Khazaee (2004), Lockley and Delgado (2007), de Valais and Melchor (2008), Mustoe et al. (2012), Díaz-Martínez et al. (2011, 2012), Kim et al. (2013), Aramayo et al. (2015), Farina et al. (2021), Lockley et al. (2021), and Abbassi and Dashtban (2021).*

TABLE 8.1 Suggested ichnogeneric and ichnospecific taxobases to be used with avian fossil footprints.

Ichnogeneric taxobases	Ichnospecific taxobases
Overall shape	Footprint size
Number of digits	Shape metatarsal pad
Presence/absence interdigital web	Range of angle between digits
Footprint symmetry	Digit shape
Footprint length/footprint width ratio	Digit curvature
Relative length of digits and axony	Presence/absence claw traces
Total digit divarication	Presence/absence of phalangeal pads
Position of hallux	Number of phalangeal pads
Extent and shape of interdigital web	Shape phalangeal pads
Presence/absence metatarsal pad	Pace/stride length
Connection (or lack of) between metatarsal pad and digits	Pace angulation
Overall digit width	Trackway width
	Rotation from the midline
	Presence of feeding/probing traces

Antarctichnus fuenzalidae Covacevich and Lamperein, 1970

Aquatilavipes Currie, 1981

*_Aquatilavipes swiboldae_ Currie, 1981

☒ *_Aquatilavipes izumiensis_ Azuma et al., 2002 (junior synonym of _A. swiboldae_)

☒ _Aquatilavipes wallacei_ Zonneveld et al., 2011 (nomen dubium)

☒ *_Aquatilavipes curriei_ McCrea and Sarjeant, 200) (transferred to _Limiavipes_ by McCrea et al., 2014

Aramayoichnus Aramayo et al., 2015

Aramayoichnus rheae Aramayo et al., 2015

Ardeipeda (Panin and Avram, 1962) Sarjeant and Langston 1994

Ardeipeda egretta (Panin and Avram, 1962) Sarjeant and Langston, 1994

Adeipeda gigantea Panin and Avram, 1962

☒ _Ardeipeda incerta_ Panin and Avram, 1962 (junior synonym of _A. egretta_)

☒ _Aviadactyla_ Kordos, 1985

☒ _Aviadactyla media_ Kordos, 1985 (junior synonym of _Ornithotarnocia_)

☒ _Aviadactyla panini_ (Kordos in Kordos and Prakfalvi, 1990) Sarjeant and Reynolds, 2001; comb. nov. (Combination rejected)

☒ _Aviadactyla Vyalovi_ (Kordos in Kordos and Prakfalvi, 1990) Sarjeant and Reynolds, 2001; comb. nov. (Combination rejected)

Avipeda (Vyalov, 1965) Sarjeant and Langston, 1994

Avipeda phoenix (Vyalov, 1965)

☒ *Avipeda sirin* Vyalov, 1965 (junior synonym of *A. phoenix*)
☒ *Avipeda filiportatis* Vyalov, 1965 (= *Gruipeda filiportatis* Sarjeant and Langston, 1994)
Avipeda adunca Sarjeant and Langston Jr, 1994
☒ *Avipeda ipolyensis* (Kordos, 1985) Sarjeant and Langston, 1994 (=*Passeripedia ipolyensis* Kordos, 1985).
Avipeda thrinax Sarjeant and Reynolds, 2001
Avipeda gryponyx Sarjeant and Reynolds, 2001
Avipeda circumontis Lockley et al., 2022
☒ *Carpathipeda* Kordos in Kordos and Prakfalvi, 1990 (= junior synonym of *Avipeda*)
☒ *Carpathipeda panini* Kordos in Kordos and Prakfalvi, 1990 (junior synonym of *Avipeda phoenix*)
☒ *Carpathipeda vyalovi* Kordos in Kordos and Prakfalvi, 1990 (nomen dubium)
Charadriipeda (Panin and Avram, 1962) Sarjeant and Reynolds, 2001
Charadriipeda recurvirostrioidea Panin and Avram, 1962 (= *Charadriipeda recurvirostra* Panin, 1965; *lapsus calami*)
☒ *Charadriipeda minima* Panin and Avram, 1962 (= *Gruipeda minima* Sarjeant and Langston, 1994)
☒ *Charadriipeda disjuncta* Panin and Avram, 1962 (= *Gruipeda disjuncta* Sarjeant and Langston, 1994)
☒ *Charadriipeda becassi* Panin and Avram, 1962 (= *Gruipeda becassi* Sarjeant and Langston, 1994)
☒ *Charadriipeda minor* Panin, 1965 (=*Gruipeda minor*?)
☒ *Charadriipeda limosa* Rădan and Brustur, 1993 (= *Gruipeda limosa* herein)
☒ *Culcitapeda* Sarjeant and Reynolds, 2001 (junior synonym of *Phoenicopterichnum*)
☒ *Culcitapeda ascia* Sarjeant and Reynolds, 2001 (nomen nudum)
☒ *Culcitapeda tridens* Sarjeant and Reynolds, 2001 (nomen nudum)
☒ *Culcitapeda eccentrica* Sarjeant and Reynolds, 2001 (nomen dubium)
☒ *Fildesichnus* Li and Zhen, 1994 (junior synonym of *Gruipeda*)
☒ *Fildesichnus sheni* Li and Zhen, 1994 (junior synonym of *Gruipeda fuenzalidae*)
☒ *Fuscinapeda* Sarjeant and Langston, 1994 (nomen nudum)
☒ *Fuscinapeda sirin* (Vyalov, 1965) Sarjeant and Langston, 1994 (= *Avipeda sirin* Vyalov, 1965:112)
☒ *Fuscinapeda meunieri* Sarjeant and Langston, 1994 (nomen nudum)
☒ *Fuscinapeda texana* Sarjeant and Langston, 1994 (nomen nudum)
Gruipeda (Panin and Avram, 1962)
Gruipeda maxima Panin and Avram, 1962
Gruipeda becassi (Panin and Avram, 1962)
☒ *Gruipeda disjuncta* (Panin and Avram, 1962) (=*Avipeda*?)
☒ *Gruipeda minor* (Panin, 1965) (=*Avipeda*?)
Gruipeda intermedia Panin, 1965
Gruipeda filiportatis (Vyalov, 1965) Sarjeant and Langston, 1994
☒ *Gruipeda grus* Panin et al., 1966 (nomen nudum)
☒ *Gruipeda abeli* (Lambrecht, 1938) (=*Iranipeda abeli*)
Gruipeda calcarifera Sarjeant and Langston, 1994
☒ *Gruipeda diabloensis* Remeika, 1999 (nomen dubium)

Gruipeda lambrechti Mirzaie Ataabadi and Khazaee, 2004
Gruipeda dominguensis de Valais and Melchor, 2008
Gruipeda fuenzalidae (Covacevich and Lamperein, 1970) Mansilla et al., 2012
☒ **Gruipeda vegrandiunus* Fiorillo et al., 2011 (=*Avipeda*)
☒ *Gruipeda limosa* Rădan and Brustur, 1993 (=*Charadriipeda limosa* Rădan and Brustur, 1993) (nomen dubium)
Gyeongsangornipes Kim et al., 2013
**Gyeongsangornipes lockleyi* Kim et al., 2013
Iranipeda (Vyalov, 1989)
Iranipeda abeli (Lambrecht, 1938) Abbassi et al., 2023
☒ *Iranipeda millumi* Doyle et al., 2000 (nomen dubium)
☒ *Iranipeda intermedia* (Panin, 1965) Doyle et al., 2000 (= *Gruipeda intermedia*)
**Jindongornipes* Lockley et al., 1992
**Jindongornipes kimi* Lockley et al., 1992
Jindongornipes falkbuckleyi Lockley et al., 2021
**Koreanaornis* (Kim, 1969)
**Koreanaornis hamanensis* (Kim, 1969)
Leptoptilostipus Payros et al., 2000
Leptoptilostipus pyrenaicus Payros et al., 2000
☒ *Ludicharadripodiscus* Ellenberger, 1980 (nomen dubium)
☒ *Ludicharadripodiscus edax* Ellenberger, 1980 (nomen dubium)
Ornithoformipes Patterson and Lockley, 2004
Ornithoformipes controversus Patterson and Lockley, 2004
Ornithotarnocia Kordos, 1985
Ornithotarnocia lambrechti Kordos, 1985
☒ Passeripedia Kordos, 1985 (nomen dubium)
☒ *Passeripedia ipolyensis* Kordos, 1985 (nomen dubium)
Pavoformipes Lockley and Delgado, 2007
Pavoformipes pintasotoi Lockley and Delgado, 2007
Persiavipes Abbassi and Dashtban, 2021
Persiavipes gulfi Abbassi and Dashtban, 2021
Phoenicopterichnum Aramayo and Manera de Bianco, 1987
Phoenicopterichnum rector (Aramayo and Manera de Bianco, 1987) Melchor et al., 2012
☒ *Phoenicopterichnum lucioi* Farina et al., 2021 (nomen dubium)
☒ *Phoenicopterichnum vinchinaensis* Farina et al., 2021 (nomen dubium)
Presbyorniformipes Yang et al., 1995
Presbyorniformipes feduccii Yang et al., 1995
☒ *Pulchravipes* Demathieu et al., 1984
☒ *Pulchravipes magnificus* Demathieu et al., 1984
Quadridigitus Lockley et al., 2021
Quadridigitus semimembranus Lockley et al., 2021
Reyesichnus Alonso et al., 1980
Reyesichnus punensis Alonso et al., 1980
Rionegrina Melchor et al., 2023
Rionegrina pozosaladensis Melchor et al., 2023

Rivavipes Mustoe et al., 2012
Rivavipes giganteus Mustoe et al., 2012
Roepichnus Doyle et al., 2000
Roepichnus grahami Doyle et al., 2000
☒ *Tetraornithopedia* Kordos, 1985 (nomen nudum)
☒ *Tetraornithopedia tasnadii* Kordos, 1985 (nomen nudum)
**Uhangrichnus* (Yang et al., 1995) Buckley et al., 2016
**Uhangrichnus chuni* (Yang et al., 1995)
Uvaichnites Díaz-Martínez et al., 2012
Uvaichnites riojana Díaz-Martínez et al., 2012
☒ *Yacoraitichnus* Alonso and Marquillas, 1986
☒ *Yacoraitichnus avis* Alonso and Marquillas, 1986 (=*Gruipeda filiportatis* de Valais and Cónsole-Gonella, 2019)

4.3.2.1 Description and discussion of avian ichnogenera

Ichnogenus *Alaripeda* Sarjeant and Reynolds, 2001.
Fig. 8.9A.
Type ichnospecies: *Alaripeda lofgreni* Sarjeant and Reynolds, 2001.

Diagnosis: Avian footprints showing three or, often, four digits. The central digit (III) is directed forward, but may curve quite sharply; Digit I is short, less than half the length of Digit III, often oriented reverse of the axis of Digit III but sometimes deviating up to 20°. The other digits (II and IV) are directed laterally and may also curve. Digits united or separate proximally. Length of Digit III comparable to (or less than 25% greater than) that of Digits II and IV. Webbing lacking; no indication of a metatarsal pad.

Discussion: This ichnogenus was erected from Miocene rocks of California, although no precise location was provided. The type material of the ichnogenus consists of five nearly complete footprints and three incomplete ones preserved in a single slab (preservational grade = 1–2). Thin digits (leptodactylous footprints), lack of details of claw traces and phalangeal pads, and a very large divarication between Digits II–IV (almost 180 degrees) suggest that these are penetrative tracks (Gatesy and Falkingham, 2020) produced in a soft substrate (see also Lockley and Harris, 2010). Relatively short Digit III imprints are also unusual and related to substrate consistency. As a consequence, this is considered a taphotaxon (as suggested by Lockley et al., 2021). A second proposed ichnospecies of the ichnogenus, *A. bristolia*, is considered comparable with *Gruipeda dominguensis*.

Ichnogenus *Anatipeda* (Panin and Avram, 1962).
Fig. 8.10A.
Type ichnospecies: *Anatipeda anas* Panin and Avram, 1962.

Diagnosis: Avian footprints showing four digits, three of which (II to IV) are large and directed forward, the fourth (I) directed backwards, short and spurlike; it is not always imprinted. The three principal digits (II–IV) are of fairly uniform size; total interdigital angle averaging 90 degrees in the proximal part and about 40 degrees if measured at the distal half of the digits. Digits II–IV are united proximally and linked by webbing, almost to their tips; Digit I may be united to the others or separate. Digits acuminate or bearing claws (emended from Sarjeant and Reynolds, 2001).

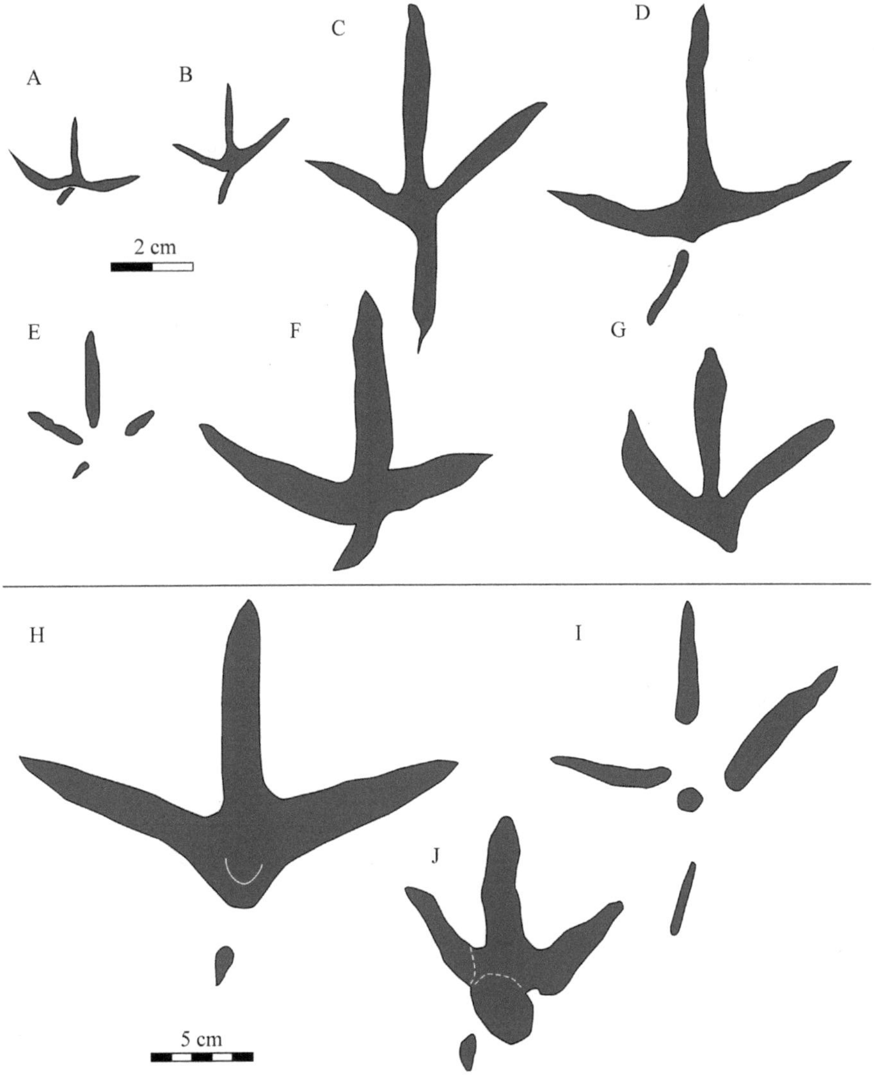

FIGURE 8.9 Outline drawings of holotype material (right footprints) of tetradactyl bird ichnotaxa, referred to the type ichnospecies, (A) *Alaripeda lofgreni.* (B) *Antarctichnus fuenzalidae.* (C) *Ardeipeda egretta.* (D) *Jindongornipes kimi.* (E) *Koreanaornis hamanensis.* F. *Pulchravipes magnificus.* (G) *Tetraornithopedia tasnadii.* (H) *Gruipeda maxima.* (I) *Iranipeda abeli.* (J) *Pavoformipes pintasotoi. Modified from Sarjeant and Reynolds (2001), Covacevich and Lamperein (1970), Panin and Avram (1962), Lockley et al. (1992), Kim (1969), Demathieu et al. (1984), Kordos (1985), Abbassi et al. (2023), and Lockley and Delgado (2007). Note that parts H-J have a different scale.*

Discussion: This is a very distinctive ichnogenus that was originally proposed for tracks from the Miocene of Romania, was revised by Sarjeant and Langston (1994) and Sarjeant and Reynolds (2001), and was also recognized in recent works (Camens and Worthy, 2019; Abbassi, 2022). Abbassi et al. (2015) proposed to synonymize *Charadriipeda* under *Anatipeda,*

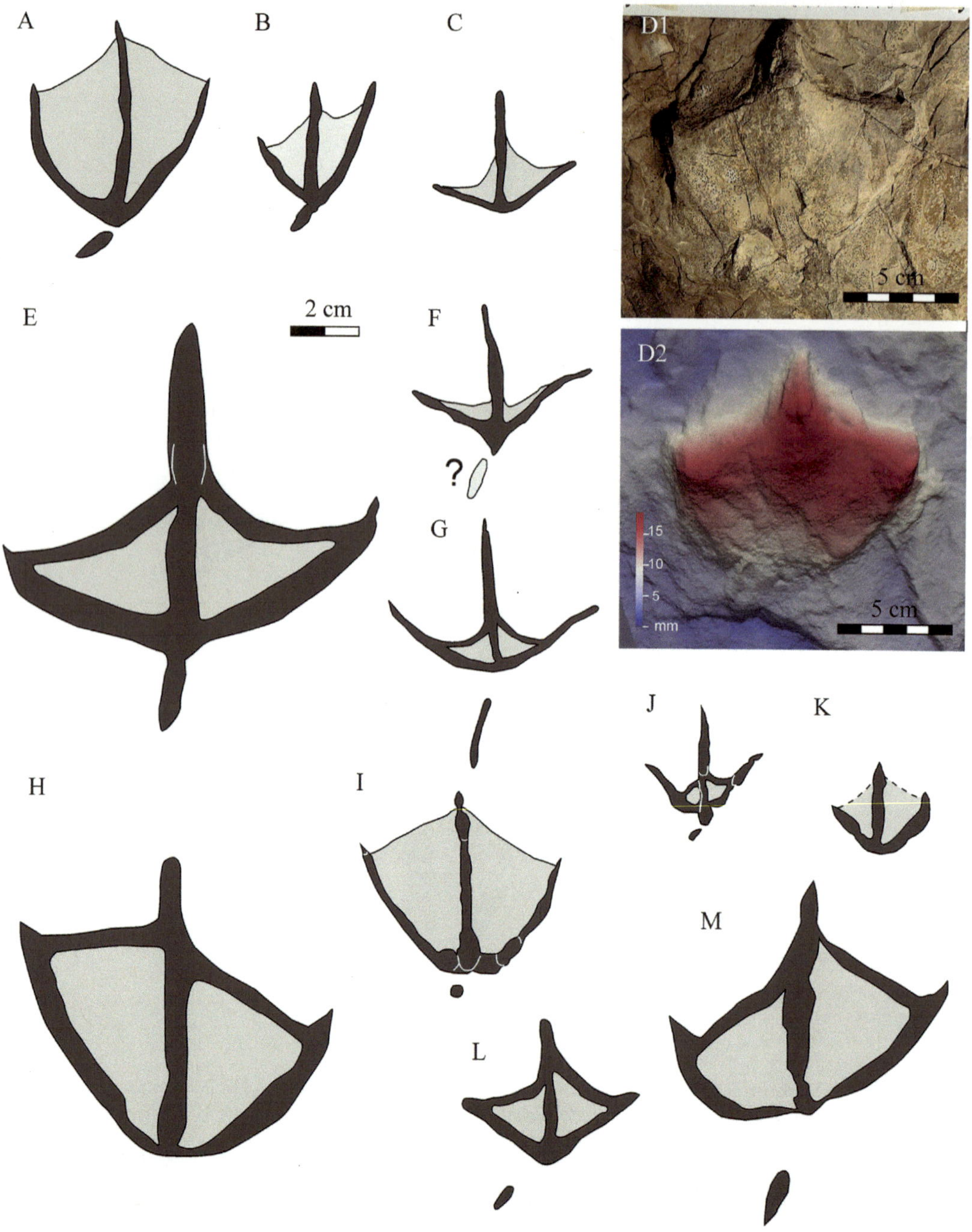

FIGURE 8.10 Outline drawings and photograph of holotype material (right footprints) of webbed bird ichnotaxa, referred to the type ichnospecies. (A) *Anatipeda anas*. (B) *Charadriipeda recurvirostrioidea*. (C) *Gyeongsangornipes lockleyi*. (D1-D2) Photograph and digital elevation model of *Culcitapeda ascia*. (E) *Leptoptilostipus pyrenaicus*. (F) *Ludicharadripodiscus edax*. (G) *Persiavipes gulfi*. (H) *Phoenicopterichnum rector*. (I) *Roepichnus grahami*. (J) *Quadridigitus semimembranus*. (K) *Reyesichnus punensis*. (L) *Uhangrichnus chuni*. (M) *Presbyorniformipes feduccii*. D1-D2 is specimen A500-8177 of San Bernardino County Museum, California, USA; reproduced with permission. Note that D1-D2 have a different scale. *Modified from Panin and Avram (1962), Kim et al. (2013), Sarjeant and Reynolds (2001), Payros et al. (2000), Ellenberger (1980), Abbassi and Dashtban (2021), Aramayo and Manera de Bianco (1987), Doyle et al. (2000), Lockley et al. (2021), Alonso et al. (1980), and Yang et al. (1995).*

a proposal that is not followed herein. One of the aspects of *Anatipeda* is a nearly orthogonal angle between Digits II–IV in the proximal part, whereas Digits II and IV curve to become almost parallel to Digit III in the distal part. This is clarified in the emended diagnosis proposed herein based on the holotype from Panin and Avram (1962). *Anatipeda* is distinguished by a similar length of Digits II–IV and an interdigital web distal margin forming an almost straight angle with Digit III. In *Chraradriipeda* Digit III is longer, and the interdigital web is shaped after a catenary between Digits II–III and III–IV. *Anatipeda alfi* and *Anatipeda californica* are considered junior synonyms of *Roepichnus grahami*. Sarjeant and Reynolds (2001) argued that *Presbyorniformipes* is similar to *A. alfi*. However, the former is clearly distinguished by a marked curvature of the lateral digits over all of its length.

Ichnogenus *Antarctichnus* Covacevich and Lamperein, 1970.

Fig. 8.9B.

Type ichnospecies: *Antarctichnus fuenzalidae* Covacevich and Lamperein, 1970.

Diagnosis: Footprints belonging to a small bird, with a bipedal upright gait and tetradactyl arrangement. Leptodactyl digits, movable and of different size; extended in a plane and lacking interdigital web; short and straight claws. Three digits directed forward and one backward. Length of footprints: 2.8–2.9 cm, width: 2.5 cm, pace angulation: 150 degrees–155 degrees. Pace length variable (translated from the Spanish herein from Covacevich and Lamperein, 1970).

Discussion: The ichnogenus was proposed for tracks from the Eocene beds of King George Island, Antarctica. Li and Zhen (1994) proposed *Fildesichnus sheni* based on a single isolated footprint from the topotype area of *Antarctichnus*, which was considered a junior synonym of *Gruipeda (Antarctichnus) fuenzalidae* by Mansilla et al. (2012). *Antarctichnus* was recognized in the Miocene of Spain, and also an emended diagnosis was proposed (Doyle et al., 2000). de Valais and Melchor (2008) considered *Antarctichnus* a junior synonym of *Gruipeda*, and Mansilla et al. (2012) proposed the new combination *Gruipeda fuenzalidae*. *Gruipeda dominguensis* and *Antarctichnus fuenzalidae* are very similar in size (the latter is a bit smaller), FL/FW ratio, relative divarication angle between lateral and medial digits (II–III and III–IV), position of the hallux with respect to the axis of Digit III, and relative lengths of digits. Some of the differences are the presence of digital pads, larger divarication (II–IV) and smaller length of hallux and its inconsistent presence in *G. dominguensis*. *Gruipeda dominguensis* exhibits phalangeal pad imprints with a relation I: 2, II: 2, III: 3, IV: 4, whereas *A. fuenzalidae* lack phalangeal pads. Further differences between *A. fuenzalidae* and *G. dominguensis* are digit divarication (II–IV) and proportion of the length of hallux in the total length of footprint (FLh/DL_I) (Table 8.2).

Ichnogenus *Aquatilavipes* Currie, 1981.

Fig. 8.11A.

Type ichnospecies: *Aquatilavipes swiboldae* Currie, 1981.

Diagnosis: Footprints less than 4.5 cm in length; average width 26% greater than length; average divarication of Digits II and IV is 113 degrees. Digit III about 50% longer than Digit II and 40% longer than Digit IV.

Discussion: This ichnogenus was proposed for material from the Gething Formation (Aptian, Lower Cretaceous), British Columbia, Canada. Both McCrea and Sarjeant (2001) and Zonneveld et al. (2011) proposed an emendation of the original diagnosis of this ichnogenus. McCrea et al. (2014) suggested abandoning the modifications introduced by McCrea

TABLE 8.2 Comparison of measurements of the type material of *Antarctichnus fuenzalidae*, *Gruipeda fuenzalidae* and *Gruipeda dominguensis*.

	Antarctichnus fuenzalidae	*Gruipeda fuenzalidae*	*Gruipeda dominguensis*
FL_h	27–31 (28.6)		27–46
FL	22.5	22–33 (24.6)	17–33.5 (26.4)
FW	24–26 (25.4)	21.3–31.3 (26.7)	20.5–42.7 (33.3)
FL/DL_I	2.86	3.8	4.47
FL/FW	(0.89)	(0.9)	0.7–0.9 (0.8)
II–III	45–70 (55) n = 4	36–65 (51)	36–78 (60)
III–IV	42–49 (45) n = 4	45–65 (54)	28–77 (56)
II–IV	87–118 (103) n = 15	90–117 (103)	90–135 (115)
I–III	141–172 (158) n = 7	130–162 (151)	107–188 (159)
Relative digit length	I < II < IV < III 10 < 15.2 < 17.2 < 20.7	I < II < IV < III 6.4 < 12.3 < 13.5 < 17.2	I < II < IV < III 5.9 < 13.7 < 16.3 < 19.4
FR	No data	No data	11.4
PA	150–155 n = 2	(155)	150–182 (170)
PL	28–55 (39.4)	(73.8)	53–120 (84.4)
SL	63–95 (77)	(181)	90–230 (170)
SL/FL	–	–	2.5–5
Digital pad impressions	No to poorly defined	–	yes

Measurements in millimeters and degrees. When available, the range and average (between parentheses) measurements are indicated.

DL_I, length of digit I; *FL*, footprint length; *FL_h*, footprint length including the hallux; *FR*, footprint rotation from the midline; *FW*, footprint width; *I to IV*, digit I to IV; *I–III, II–III, III–IV, II–IV*, angle between the digits I–III, II–III, III–IV and II–IV; *PA*, pace angulation; *PL*, pace length; *SL*, stride length.

Data compiled from Covacevich and Lamperein (1970), Mansilla et al. (2012) and de Valais and Melchor (2008).

and Sarjeant (2001). When proposing their emendation, Zonneveld et al. (2011) also followed partially McCrea and Sarjeant (2001). For these reasons, the original diagnosis of the type ichnospecies is followed herein. *Aquatilavipes* is distinguished from *Koreanaornis* because the latter is tetradactyl (Lockley et al., 1992). *Aquatilavipes* is very similar to *Avipeda phoenix*, although this ichnospecies is distinguished by a smaller size, given that Vyalov (1966) explicitly indicated that these are very small footprints (Digit III length 16 mm). Although it is morphologically similar to *Avipeda*, *Aquatilavipes* was erected on a large sample size (n ~ 200), and there is no overlap in footprint length between *A. phoenix* and *Aquatilavipes swiboldae* (see Fig. 8.12A of Currie, 1981). *Aquatilavipes izumiensis* is indistinguishable from *A. swiboldae* in the range of FL, FL/FW, total divarication, relative angles between lateral digits and relative digit length. As a consequence, it is considered to be a junior synonym of *A. swiboldae*. Similarly, *A. wallacei* overlaps in footprint length, FL/FW ratio, total

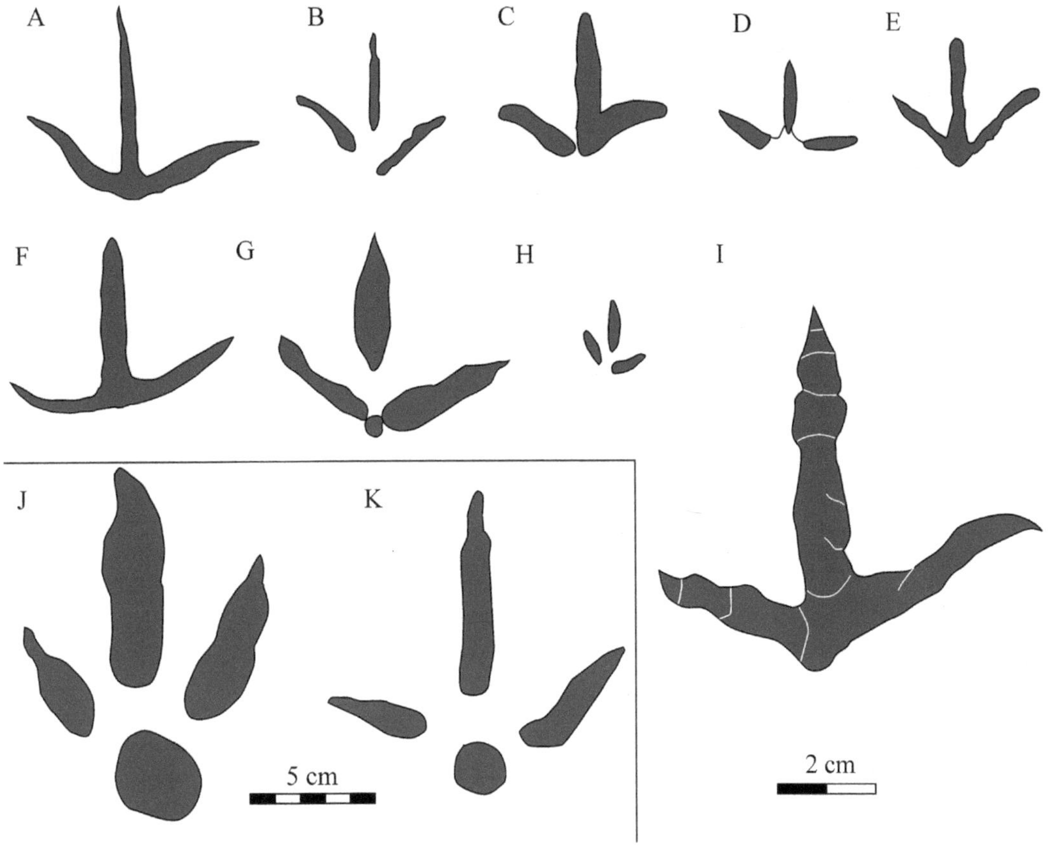

FIGURE 8.11 Outline drawings of holotype material (right footprints) of tridactyl bird ichnotaxa, referred to the type ichnospecies, (A) *Aquatilavipes swiboldae.* (B) *Aviadactyla media.* (C) *Avipeda phoenix.* (D) *Carpathipeda panini.* (E) *Fildesichnus sheni.* (F) *Fuscinapeda sirin.* (G) *Ornithotarnocia lambrechti.* (H) *Passeripedia ipolyensis.* (I) *Yacoraitichnus avis.* (J) *Aramayoichnus rheae.* (K) *Uvaichnites riojana.* Note that parts J-K have a different scale. *Modified from Currie (1981), Kordos (1985), Vyalov (1965), Kordos and Prakfalvi (1990), Li and Zhen (1994), Alonso and Marquillas (1986), Aramayo et al. (2015), and Díaz-Martínez et al. (2016).*

divarication, and relative divarication of lateral digits with *A. swiboldae. Aquatilavipes wallacei* was erected on the basis of a moderate number of footprints (n = 18) with preservation quality in the range grade 1-2 and common leptodactylous digit impressions, with curved lateral digits, and a common poorly defined footprint boundary. As a consequence, *A. wallacei* is considered a nomen dubium.

Ichnogenus *Aramayoichnus* Aramayo et al., 2015.

Fig. 8.11J.

Type ichnospecies: *Aramayoichnus rheae* Aramayo et al., 2015.

Diagnosis: Bipedal trackway of high-pace angulation with mesaxonic tridactyl footprints of moderate size (footprint length usually greater than 10 cm), footprint length/width ratio less than 2, and footprints with small positive rotation with the midline. Footprints with three thick digits directed forward showing clear phalangeal pad impressions and claw marks. At least half of the length of Digit III impression projecting forward of the imaginary line formed by the tip of lateral digits, Digit IV impression longer than that of Digit II. Divarication of Digits II–IV is commonly smaller than 80 degrees. Digits proximally separated from or joined to a deep, circular metatarsal–phalangeal pad impression. Interdigital web trace absent. Differs from *Anchisauripus* by the presence of a metatarsal–phalangeal pad impression, greater digit divarication, and a lower footprint length/width ratio (Aramayo et al., 2015).

Discussion: This ichnogenus was originally proposed for Late Pleistocene deposits from Pehuen Co site (Buenos Aires Province, Argentina), although it was also recorded in Neogene units from Argentina (Tineo et al., 2018; Rojas-Manriquez et al., 2024). Footprint morphology and size are highly distinctive and very similar to modern rheid footprints (Aramayo et al., 2015). The revision of the ichnofauna of the Late Pleistocene Pehuen Co site was initiated by Silvia Aramayo and Teresa Manera de Bianco, shortly before the death of the former. In consequence, this is a posthumous work of S. Aramayo and the co-authors proposed the ichnogenus *Aramayoichnus*, although no differential authorship was declared originally (Aramayo et al., 2015).

Ichnogenus *Ardeipeda* (Panin and Avram, 1962).

Fig. 8.9C.

Type ichnospecies: *Ardeipeda egretta* (Panin and Avram, 1962).

Diagnosis: Avian footprints showing four digits, three (II to IV) directed forward and large, the fourth (Digit I) backward and somewhat smaller. The interdigital angles between Digits II and III and between Digits III and IV are less than 70 degrees. The axis of Digit I corresponds, or almost corresponds, with that of Digit III, the interdigital angle between Digits I and II being almost equal to that between Digits I and IV. Webbing absent (emended from Sarjeant and Langston, 1994).

Discussion: The material for proposal of this ichnogenus came from the Miocene of Romania. Distinctive features of this ichnogenus are a large ($FL/DL_I = 2.8$), backward-directed hallux imprint positioned approximately in the same axis as Digit III, as well as large divarication between Digits II and IV (at least 110 degrees). The original material referred to this ichnogenus are also large footprints: footprint length for the type ichnospecies was 80–85 mm, and 230 mm for *Ardeipeda gigantea*. Lockley and Harris (2010) suggested the placement of *Avipeda filiportatis* under *Ardeipeda*, which is not followed here. *Ardeipeda incerta* was of similar size ($FL = 88$ mm) and very similar Digit I length ($FL/DL_I = 2.9$) as *A. egretta*, and was separated by the likely presence of a reduced interdigital membrane impression. *Ardeipeda incerta* can be considered as a junior synonym of *A. egretta*. It is also a name that has not apparently been used since its erection.

Ichnogenus *Aviadactyla* Kordos, 1985.

Fig. 8.11B.

Type ichnospecies: *Aviadactyla media* Kordos, 1985.

Diagnosis: Bird footprint of small to medium size consisting of three toes. The prints of all three toes are thin, stick-like, shallowly imprinted. Longest is the middle toe, to be followed by the gradually shorter inner and outer toes. The distal end of the inner toe print is, in the normal case, farther away from the basic line (the line normal to the middle toe) than the end of the middle toe print. Consequently, it is slightly asymmetric (Kordos, 1985).

Discussion: The ichnogenus was proposed for material from the Lower Miocene Sandstone of Ipolytarnóc, Hungary. Sarjeant and Reynolds (2001) proposed an emended diagnosis for *Aviadactyla* to embrace also the ichnogenus *Carpathipeda* (Kordos in Kordos and Prakfalvi, 1990), which is not followed here (see also under *Carpathipeda*). After the analysis of the original diagnosis and illustrated material, *Aviadactyla* is herein considered a junior synonym of *Ornithotarnocia* because it is similar and was based on a larger sample size (n = 137; whereas *Aviadactyla* was n = 22). Measurements of FL/FW ratio, symmetry of footprints (similar divarication angles II–III and III–IV) and total digit divarication are overlapping between *Aviadactyla* and *Ornithotarnocia* (Table 8.3). Footprint size overlaps (as inferred from the length of Digit III, because footprint length was not measured by Kordos, 1985), but the author named the smaller footprints under *Aviadactyla* and the larger under *Ornithotarnocia*.

Ichnogenus *Avipeda* (Vyalov, 1965).

Fig. 8.11C.

Type ichnospecies: *Avipeda phoenix* (Vyalov, 1965).

Diagnosis: Avian footprints of small size (footprint length <30 mm), showing three short, thick digits, with distinct claws. Length of central digit (III) less than 25% greater than that of the lateral digits. Total interdigital span 95 degrees or more. Digits closely convergent or united proximally; webbing lacking or limited to the most proximal part of the interdigital angles (modified from Sarjeant and Langston, 1994).

Discussion: Vyalov (1965, 1966) did not offer a diagnosis for *Avipeda*, which was intended to embrace all avian tracks. However, in 1966 he explicitly indicated that the type ichnospecies was *A. phoenix* and diagnosed it as "small, tridactyl tracks up to 1.6 cm long, with short and relatively broad digits and an obtuse angle between the digits" (see also Lucas, 2007). Sarjeant and Langston (1994) proposed a diagnosis for the ichnogenus that established a maximum total interdigital span (II–IV) not larger than 95 degrees, which does not fit the type ichnospecies, which has an obtuse angle (average 122 degrees, Table 8.3). Sarjeant and Langston (1994) also included large footprints under the ichnogenus, whereas the known range for footprint length from type material is 20–30 mm (Table 8.3). For this reason, an emended diagnosis of the ichnogenus is proposed herein.

A total of nine species have been proposed for *Avipeda*: *A. phoenix*, *A. sirin*, *A. filiportatis*, *A. ipolyensis*, *A. adunca*, *A. thrinax*, *A. rastini* and *A. circumontis*. *Avipeda sirin* is herein considered a junior synonym of *A. phoenix*. Vyalov (1966) erected *A. sirin* on the basis of three tracks that were slightly larger than the type ichnospecies. *Avipeda filiportatis* was correctly reassigned to *Gruipeda* by Sarjeant and Langston (1994) because it is tetradactyl and *A. ipolyensis* is herein considered a nomen nudum (see under *Passeripedia*). Lockley and Harris (2010) suggested to include *A. filiportatis* under *Ardeipeda*, but the latter ichnogenus is typified by longer hallux impressions (see also de Valais and Cónsole-Gonella, 2019). For this reason, the suggestion by Lockley and Harris (2010) is not accepted. Among the remaining ichnospecies, *A. gryponyx* is very similar to *A. phoenix*, and both are distinguished from other ichnospecies by a high II–IV angle (>115 degrees) and relatively small FL/FW (<0.80). On the contrary, *A. adunca*, *A. thrinax*, and *A. circumontis* have a smaller divarication angle (<90 degrees) and larger FL/FW (>0.80), and should be probably located under a different ichnogenus (Table 8.3, Fig. 8.12). *A. rastini* was proposed for tridactyl footprints with slender digit imprints, and has a morphology similar to *Avipeda* (Abbassi, 2022). The wide mesaxonic morphology of *A. rastini* is a typical feature of shorebird tracks (Lockley, 2009).

TABLE 8.3 Comparison of measurements of the type material of *Avipeda* and other similar tridactyl ichnotaxa.

	Avipeda phoenix	*Avipeda sirin*	*Aviadactyla media*	*Ornithotarnocia lambrechti*	*Carpathipeda panini*	*Paxavipes babcockensis*	*Barrosopus slobodai*
n	10?	3	22	137	13	42	9
FL	22	28.2	24.8–28.1 (26.8)	40–38	16.5–22.7 (19.7)	19.7–33.6 (27.1)	28–40 (32)
FW	27.8	33.1	24.7–31 (27.6)	48–38	22.4–28.3 (25.7)	20.6–38.6 (30.2)	31–44 (37.2)
FL/FW	0.63–0.87 (0.74)	0.85	0.86–1.13 (0.98)	0.75–0.98 (0.88)	0.66–0.89 (0.76)	0.78–1.20 (0.91)	0.75–1 (0.87)
II–III	63	69	36–64 (46.6)	49.7	76	33–75 (57)	48
III–IV	49	63	45–61 (49.5)	48.9	64	31–56 (43)	63
II–IV	109–139 (122)	128	77–95 (83)	89–113 (98.5)	110–142 (129)	71–125 (99)	100–140 (118)
Relative digit length			II < IV < III 13.6 < 15.9 < 20.1	IV < II < III 21.8 < 25.1 < 30.4	IV ≅ II < III 17.6 ≅ 17.8 < 20.01	II < IV < III 19.2 < 21.8 < 27.8	No data
FR						(17)	
PA						153–173 (162)	
PL						82.2–100 (88)	30–135 (97)
SL				185		161–197 (173)	135–230 (198)

Measurements in millimeters and degrees. When available, the range and average (between parentheses) measurements are indicated.
FL, footprint length; *FR*, footprint rotation from the midline; *FW*, footprint width; *I to IV*, digit I to IV; *I–III, II–III, III–IV, II–IV*, angle between the digits I–III, II–III, III–IV and II–IV; *n*, number of specimens; *PA*, pace angulation; *PL*, pace length; *SL*, stride length.
Data compiled from Vyalov (1966) and Lucas (2007) for A. phoenix *and* A. sirin, *Kordos (1985) for* A. media *and* O. lambrechti, *Kordos and Prakfalvi (1990) for* C. panini, *McCrea et al. (2015) for* Paxavipes babcockensis, *and Coria et al. (2002) for* Barrosopus slobodai.

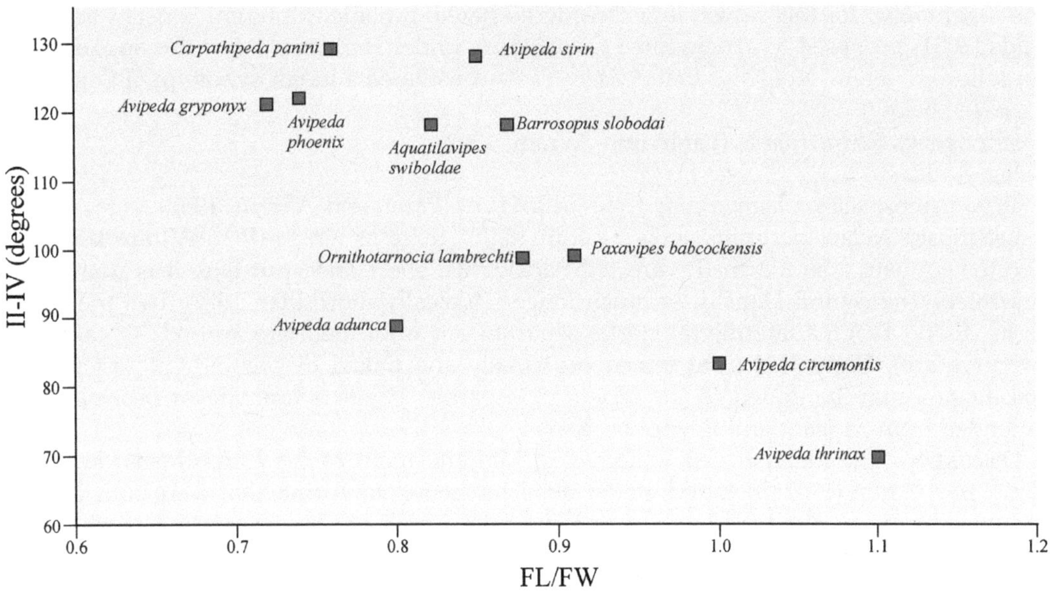

FIGURE 8.12 Scatter plot of footprint length/footprint width ratio (FL/FW) versus total divarication (II–IV) of *Avipeda* ichnospecies and other similar tridactyl ichnotaxa. See also Table 8.3.

The holotype of *A. thrinax* is a trackway of three footprints with very poor preservation and is best considered a *nomen nudum* (taphotaxon). The latter ichnospecies is similar to *A. circumontis*. *Avipeda phoenix* overlaps *Aquatilavipes swiboldae* and *Barrosopus slobodai* (Coria et al., 2002) in FL/FW ratio and II–IV angle, whereas the angles of lateral digits differ (II–III larger in *Avipeda* and the opposite in the remaining ichnotaxa) (Table 8.3, Fig. 8.12).

Ichnogenus *Carpathipeda* Kordos in Kordos and Prakfalvi, 1990.

Fig. 8.11D.

Type ichnospecies: *Carpathipeda panini* Kordos in Kordos and Prakfalvi, 1990.

Diagnosis: No formal diagnosis was provided. The following description is from the type ichnospecies. "Small-to-medium size tridactylous bird's footprint. The middle digit is the longest one, whereas one of the two extreme digits is, in most cases, shorter and thinner. The imprint of digits has the shape of a grain of oat, with their front end slightly inclining. The imprint of sole rim between the digits is found slightly more forward than the imprint of rear end of the digits. The imprints of the three digits show an approximately symmetrical arrangement, including a wide angle with respect to one another" (Kordos and Prakfalvi, 1990).

Discussion: Originally proposed for material collected in Miocene rocks of the Carpathians, Romania. *Carpathipeda* is considered a junior synonym of *Avipeda phoenix* because it shares the overall shape and similar footprint length, footprint length-to-width ratio, total divarication (I–IV), and the divarication between Digits II–III and III–IV (Table 8.3 and Fig. 8.12). Kordos (in Kordos and Prakfalvi, 1990) recognized that *Carpathipeda panini* is similar to *A. phoenix*. *Carpathipeda vyalovi* is like *C. panini,* although about 50% larger, and was based on a single complete footprint with no distinctive features and illustrated by a line diagram, so for this reason it is considered herein a nomen dubium. Sarjeant and Reynolds (2001) proposed to synonymize *Carpathipeda* under *Aviadactyla*, but this suggestion is not followed herein. Note also that *Aviadactyla* is considered a junior synonym of *Ornithotarnocia* (see below).

Ichnogenus *Charadriipeda* (Panin and Avram, 1962).

Fig. 8.10B.

Type ichnospecies: *Charadriipeda recurvirostrioidea* (Panin and Avram, 1962).

Diagnosis: Avian footprints showing four digits, three of which (II to IV) are large and directed forward, the fourth (I) directed backwards, short and spur-like; it is not always imprinted. The central Digit III is much longer (typically more than 25%) than the lateral digits (II and IV); the interdigital angles separating it from them are around 70° (allowing for curvature). Digits II–IV are united proximally and linked by webbing almost to their tips: Digit I may be united to the others or separate. Digits acuminate or bearing claws (emended from Sarjeant and Reynolds, 2001).

Discussion: The material was collected in Miocene rocks of the Carpathians, Romania. Panin and Avram (1962) included, under this ichnogenus, footprints that were both webbed and unwebbed and also bearing three or four digits. These authors originally described four ichnospecies (*C. recurvirostrioidea*, *C. minima*, *C. disjuncta*, and *C. becassi*), and a fifth ichnospecies (*C. minor*) was proposed by Panin (1965), along with an extended description of *C. recurvirostrioidea*. More recently, a sixth ichnospecies, *Charadriipeda limosa*, was proposed by Rădan and Brustur (1993). Sarjeant and Langston (1994) considered *Charadriipeda recurvirostrioidea* as the type ichnospecies, and proposed a diagnosis for the ichnogenus (not originally proposed), which was then emended by Sarjeant and Reynolds (2001). Considering that the type ichnospecies is tetradactyl and webbed, Sarjeant and Langston (1994) combined most of the ichnospecies (except *C. limosa*) under *Gruipeda*. *Charadriipeda limosa* was poorly described and illustrated, and includes tetradactyl, unwebbed footprints that probably fit under *Gruipeda*.

Ichnogenus *Culcitapeda* Sarjeant and Reynolds, 2001.

Fig. 8.10D.

Type ichnospecies: *Culcitapeda ascia* Sarjeant and Reynolds, 2001.

Diagnosis: Avian footprints whose tridactyl character is made evident only by forward prolongations (sometimes inconspicuous) from a foot-pad or web. The print showed slight deepening beneath the positions of the bones of the digits, but these are not otherwise indicated. Posterior outline of footprint semicircular or nearly so; prolongation of central digit (III) more pronounced anteriorly and posteriorly than those of the lateral digits (II and IV) (Sarjeant and Reynolds, 2001).

Discussion: *Culcitapeda* was based on scarce material (seven isolated footprints, ascribed to three separate ichnospecies) of tridactyl webbed footprints from the Early Miocene of the Whipple Mountains, San Bernardino County, California, USA. The preservation of the material is moderate to poor, and the tracks commonly lack the details of digit impressions, which is also reflected in the diagnosis (the preservation grade of the holotype is one in the scale of Marchetti et al., 2019a), as reflected in Fig. 8.10D1-D2 that includes an orthomosaic and digital elevation model of the holotype. Sarjeant and Reynolds (2001) argued that *Culcitapeda* was of lesser length and lesser degree of digital definition than *Phoenicopterichnum*, in addition to lacking a metatarso-phalangeal pad impression. A comparison between *Phoenicopterichnum* and *Culcitapeda* suggests that they exhibit minor morphological differences that are not of ichnotaxonomic value and are linked to the poor preservation and scarcity of the available material of the latter ichnogenus. As a consequence, *Culcitapeda* is considered a junior synonym of *Phoenicopterichnum* (see also Melchor et al., 2012). Sarjeant and Reynolds (2001) proposed two additional ichnospecies: *Culcitapeda tridens* and *Culcitapeda eccentrica.* The type material of the former is an incomplete footprint (lacking one digit and the rear end), which does not fit the recommended requirements for erection of an ichnotaxon, and the latter is based on two poorly preserved and isolated footprints.

Ichnogenus *Fildesichnus* Li and Zhen, 1994.

Fig. 8.11E.

Type ichnospecies: *Fildesichnus sheni* Li and Zhen, 1994.

Diagnosis: Footprints with three pedal digits, enlarged on the distal part of Digits II and III; the width greater than the length; the divarication of the digits being II-45-III-52-IV; no claw or hallux impression.

Discussion: This monotypic ichnogenus was proposed for a single, isolated, and poorly preserved footprint associated with *Gruipeda fuenzalidae,* from Eocene lacustrine deposits on King George Island of Antarctica. According to Mansilla et al. (2012), it is undistinguishable from the mentioned ichnospecies and considered a junior synonym of *Gruipeda.*

Ichnogenus *Fuscinapeda* Sarjeant and Langston, 1994.

Fig. 8.11F.

Type ichnospecies: *Fuscinapeda sirin* (Vyalov, 1965).

Diagnosis: Tridactyl footprints of small to large size, showing three digits united proximally, most often in a metatarsal pad ("heel"); webbing and hallux lacking. Digits moderately thick to thick, their maximum width exceeding 15% of their length: Digit III is more than 25% longer than the lateral digits. Total interdigital span greater than 95 degrees and often exceeds 120 degrees. Length of Digits II and III may be similar, but Digit IV is frequently somewhat larger. All digits clawed, the claw frequently showing inward flexure in relation to the digit axis. Digital pad impressions are visible on better-preserved molds or casts, three or four on Digit III, two on Digits II and IV (emended from McCrea and Sarjeant, 2001).

Discussion: Sarjeant and Langston (1994) erected the ichnogenus *Fuscinapeda* to include *Avipeda sirin* (Vyalov, 1965) as type ichnospecies, arguing that the long central digit and high interdigital span were distinctive of *Fuscinapeda.* However, no discussion about the distinction from *Avipeda* was provided. The original diagnosis (Sarjeant and Langston, 1994) considered a reduced interdigital web, which was not mentioned in the emendation by McCrea and Sarjeant (2001). *Fuscinapeda* includes three ichnospecies that also differ in footprint size and footprint length/width ratio. *Fuscinapeda sirin* and *Fuscinapeda meunieri* are of similar size (FL = 28–32 mm) and FL/FW ratio (0.7–0.85); however, *Fuscinapeda texana* is much larger (FL = 96–100 mm) and displays a contrasting FL/FW ratio (1.17). It is herein recommended to abandon the usage of *Fuscinapeda,* because its type species fits under *Avipeda.* The proposed type specimen of *Fuscinapeda meunieri* was a slab figured by Meunier (1906), a procedure that contradicts Recommendation 73B of the ICZN (1999), so this ichnospecies should not be used. The material illustrated by Meunier (1906) probably is assignable to *Avipeda.* The type material of *Fuscinapeda texana* does not fit *Avipeda* and displays moderate to poor preservation and a marginal rim that can be confused with an interdigital web.

Ichnogenus *Gruipeda* (Panin and Avram, 1962).

Fig. 8.9H.

Type ichnospecies: *Gruipeda maxima* Panin and Avram, 1962.

Diagnosis: Footprints showing four-digit imprints, three of which (II to IV) are directed forward and larger, the fourth (I), directed backward, spur-like and short. The interdigital angles between Digits II and III and between Digits III and IV are commonly less than 70 degrees. The hallux imprint is posteromedially directed; the interdigital angle between Digits I and II being smaller than that between Digits I and IV. When present, digital pad traces display the relationship I: 2, II: 2, III: 3, IV: 4. Webbing trace absent (emended from de Valais and Cónsole-Gonella, 2019).

Discussion: Panin and Avram (1962) did not offer a diagnosis for the ichnogenus, Sarjeant and Langston (1994) proposed a diagnosis, and de Valais and Melchor (2008) and de Valais and Cónsole-Gonella (2019) introduced minor modifications in the diagnosis. The type material was collected in Miocene rocks of the Carpathians, Romania. This is the avian ichnogenus with the greater number of ichnospecies: a total of 17 among those originally proposed or transferred to *Gruipeda*, including *Gruipeda limosa* (originally under *Charadriipeda*), which is proposed tentatively herein. Among them, there are three tridactyl ichnospecies (i.e., they lack a hallux impression) that are thus reassigned to *Avipeda*: *Gruipeda* (formerly *Charadriipeda*) *disjuncta* (the type material is probably represented by undertracks), *Gruipeda minor* (with a probable reduced interdigital membrane that needs re-evaluation) and *Gruipeda vegrandiunus*. Among the remaining ichnospecies, *Gruipeda grus* is considered herein a nomen nudum, and *G. limosa* and *Gruipeda diabloensis* were incompletely described and considered of dubious affinity. *Gruipeda* is in need of an in-depth revision, however, it is possible to distinguish three groups of ichnospecies considering the footprint length (including the hallux) (Table 8.4, Fig. 8.13): (1) *G. lambrechti, G. intermedia, G. filiportatis* and *G. maxima* that are large ($FL_h = 120-172$ mm); (2) *G. becassi* and *G. limosa* have an intermediate size ($FL_h = 55-65$ mm); and (3) *G. fuenzalidae, G. calcarifera, G. dominguensis,* and *G. diabloensis* that are small ($FL_h = 27-35$ mm).

Ichnogenus *Gyeongsangornipes* Kim et al., 2013.

Fig. 8.10C.

Type ichnospecies: *Gyeongsangornipes lockleyi* Kim et al., 2013.

Diagnosis: Asymmetrical semipalmate, strong mesaxonic bird track with three widely splayed, thin digit impressions (II–IV) with sharp end. Asymmetrical web traces proximal to mesial between Digits II–III–IV (Kim et al., 2013).

Discussion: *Gyeongsangornipes* is the unique ichnotaxon available for tridactyl, semipalmate avian footprints, which has been proposed for the Early Cretaceous of Korea (Kim et al., 2013), although this is a common footprint morphology among modern shorebirds. *Uhangrichnus* Yang et al., 1995 is also a semipalmate bird ichnogenus originally defined as tridactyl but recently considered tetradactyl (Buckley et al., 2016; Kim et al., 2013). Melchor et al. (2020) recognized *Gyeongsangornipes* isp. in Eocene deposits from Svalbard, which is about 2.5 times larger than the type material of *Gyeongsangornipes lockleyi*, the type and only recognized ichnospecies.

Ichnogenus *Iranipeda* (Lambrecht, 1938).

Fig. 8.9I.

Type ichnospecies: *Iranipeda abeli* (Lambrecht, 1938).

TABLE 8.4 Comparison of measurements of the type material of *Gruipeda* ichnospecies grouped by size.

	large				intermediate		small			
	Gruipeda maxima	*Gruipeda intermedia*	*Gruipeda filiportatis*	*Gruipeda lambrechti*	*Gruipeda becassi*	*Gruipeda limosa*	*Gruipeda calcarifera*	*Gruipeda dominguensis*	*Gruipeda fuenzalidae*	*Gruipeda diabloensis*
n =	1	1	3	1	4	3	5	~140	17	7
FL_h	180	158	169	119	59	65	28.8	27–46 (34.1)	27–31 (28.6)	35
FL	172	111	123	92	38.5	58	21.8	17–33.5 (26.4)	22.5	30.5
FW		157	153	105	58	67	26.6	20.5–42.7 (33.3)	24–26 (25.4)	49
FL/DL_I	11.47	7.5	4.2	4.6	3.75	7.4	4.50	4.47	2.86	3.66
FL/FW	0.95	0.7	0.8	0.88	0.72	0.84	0.82	0.7–0.9 (0.8)	(0.89)	0.62
FL_h/FW	0.95	1.00	1.1	1.13	1.03	0.96	1.08	1.02	1.12	0.73
II–III	60	80	59	74	66	56	64	36–78 (60)	(55) 45–70 n = 4	70
III–IV	62	65	66	79	71	50	51	28–77 (56)	(45) 42–49 n = 4	64
II–IV	120	145	123	158	134	98	112	90–135 (115)	87–118 (103) n = 15	134
I–III	167	167	169	150	166	166	163	107–188 (159)	141–172 (158) n = 7	139
Relative digit length	I < IV < II < III 15 < 105 < 130 < 140	I < II < IV < III 21 < 74 < 88 < 105	I < II < IV < III 41 < 78 < 91 < 113	I < II < IV < III 26 < 43 < 60 < 60	I < II < IV < III 16 < 27 < 30 < 37	I < IV < II < III 9 < 42 < 44 < 55	I < IV ≅ II < III 6.6 < 15.8≅ 15.9 < 20.6	I < II < IV < III 5.9 < 13.7 < 16.3 < 19.4	I < II < IV < III 10 < 15.2 < 17.2 < 20.7	I < II ≅ IV < III 9 < 22≅ 22 < 23
FR	–	–	–	–	–	–	–	11.4	No data	–
PA	–	–	–	–	–	–	–	150–182 (170)	150–155 (n = 2)	–
PL	–	–	–	–	–	–	–	53–120 (84.4)	28–55 (39.4)	–
SL	–	–	–	–	–	–	–	90–230 (170)	181	97

Measurements in millimeters and degrees. When available, the range and average (between parentheses) measurements are indicated.
DL_I, length of digit I; *FL*, footprint length; FL_h, footprint length including the hallux; *FR*, footprint rotation from the midline; *FW*, footprint width; *I to IV*, digit I to IV; *I–III, II–III, III–IV, II–IV*, angle between the digits I–III, II–III, III–IV and II–IV; *n*, number of specimens; *PA*, pace angulation; *PL*, pace length; *SL*, stride length.
Data compiled from Panin and Avram (1962), Panin (1965), Vyalov (1966), Mirzaie Ataabadi and Khazaee (2004), Rădan and Brustur (1993), Sarjeant and Langston. (1994), de Valais and Melchor (2008), Mansilla et al. (2012), and Remeika (2001).

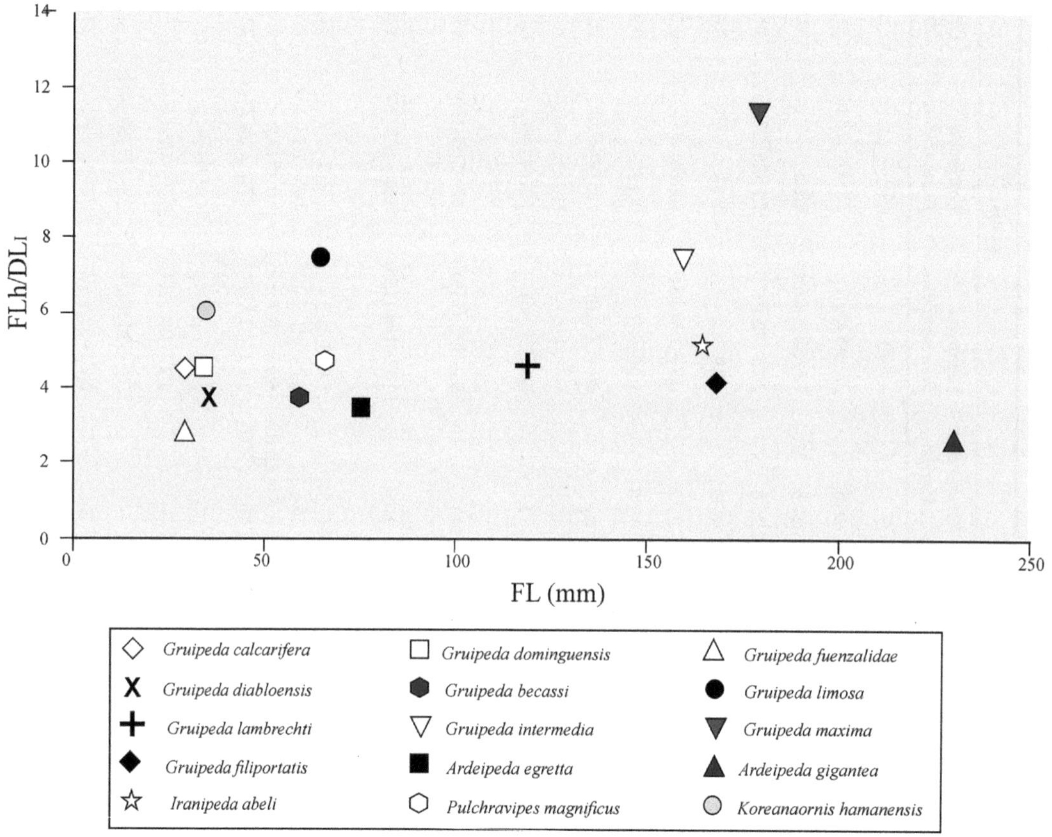

FIGURE 8.13 Scatter plot of footprint length (FL) versus footprint length with hallux/digit I length (FL_h/DL_I) for *Gruipeda, Ardeipeda, Pulchravipes, Koreanaornis* and *Iranipeda*. See also Table 8.4.

Diagnosis: Large tetradactyl avian track with prominent broad, consistently diverging and acutely tapering digit impressions. Prominent heel pad and hallux impressions are present. Digit imprints usually not connected each other or to heel pad. Interdigital angle between Digits II and IV is greater than 120 degrees. Hallux imprint is long and oblique to the axis of footprint. There are no traces of webbing (emended by Abbassi et al., 2023).

Discussion: This ichnogenus has a complex history and has been recently reviewed by Abbassi et al. (2023). The type material came from the Agha Jari Formation (Mio–Pliocene) in the Zagros Mountains, southwest Iran. Doyle et al. (2000) considered *Gruipeda intermedia* under *Iranipeda*, and also erected a new ichnospecies: *Iranipeda millumi*. The first combination (*I. intermedia*) was not supported by any argument, and we consider that *Gruipeda intermedia* is valid. *Iranipeda millumi* was defined on the basis of two isolated (one incomplete) footprints that have poor preservation quality and with an indistinct hallux impression, and for this reason is considered a nomen dubium. Large ichnospecies of *Gruipeda* are very similar to *Iranipeda* and assignment to one or the other ichnogenus is sometimes doubtful (Fig. 8.13).

Ardeipeda is also akin to the former, although distinguished by having a proportionally larger impression of Digit I, in connection with the metatarso-phalangeal pad.

Ichnogenus *Jindongornipes* Lockley et al., 1992.

Fig. 8.9D.

Type ichnospecies: *Jindongornipes kimi* Lockley et al., 1992.

Diagnosis: No formal diagnosis was provided. The description is as follows: Moderately large tetradactyl bird tracks with well-developed hallux impression. Up to four phalangeal pad impressions in Digit III trace. Number of pad impressions in Digits II and IV unknown, but at least three. Hallux impression often bent suggesting two distal phalanges in contact with substrate. Divarication between Digits II and IV very pronounced, ranging from 125 to 150 degrees. Digit IV impression often wide proximally with a posteriorly directed heel that gives track a pronounced asymmetry. Track width ranging between 65–75 mm (measured between distal ends of Digits II and IV). Track length, with hallux, up to 80 mm. Trackway variable but with only slight positive rotation and pace angulation about 140 degrees.

Discussion: No ichnogeneric diagnosis is currently available. The ichnogenus has only two ichnospecies: *J. kimi*, the type ichnospecies from the Cretaceous of Korea, and *J. falkbuckleyi*, which was identified in the Eocene of USA. *J. falkbuckleyi* displays a proximal interdigital web imprint, which is not present in the type ichnospecies. This ichnospecies needs revision and appears more akin to the ichnogenus *Ignotornis*.

Ichnogenus *Koreanaornis* (Kim, 1969).

Fig. 8.9E.

Type ichnospecies: *Koreanaornis hamanensis* (Kim, 1969).

Diagnosis: Small, functionally tridactyl bird tracks with slender and mostly separate digit traces, sometimes ending in sharp claw traces. Faint hallux traces are rarely preserved. Digital pad impressions correspond to Digits II, III, and IV and number 2-3-4 in well-preserved tracks. Digit divarication averages 120 degrees. Trackways show inward rotation of Digit III (emended from Kim et al., 2012).

Discussion: Kim (1969) described the ichnogenus, Lockley et al. (1992) emended the description, and Kim et al. (2012) proposed a diagnosis. The original description distinguished tridactyl footprints, however, Lockley et al. (1992) noted that the topotypic material has the occasional presence of a small hallux impression. As a consequence, this ichnogenus is tetradactyl. A total of four ichnospecies have been proposed, all from the Lower Cretaceous of Korea and China: *K. hamanensis*, *K. dodsoni*, *K. lii*, and *K. anhuiensis*. *K. dodsoni* lacks a hallux impression, so its inclusion in this ichnogenus is doubtful. *Koreanaornis* has been also recorded in Neogene strata (Díaz-Martínez et al., 2015; Camens and Worthy, 2019; Abbassi and Dashtban, 2021). *Koreanaornis hamanensis* is morphologically very close to *Gruipeda dominguensis* and *G. fuenzalidae*, including the size, divarication between Digits II–III, III–IV, and II–IV, relative digit length and the footprint length (without considering the hallux) to width ratio (Table 8.5). Potential differences are the lateral placement of Digit I in *K. hamanensis* (I–III = 133 degrees in *K. hamanensis* vs. 158–159 degrees in *G. dominguensis* and *G. fuenzalidae*) and the fact that the digit impressions are not united in a metatarso-phalangeal pad (Table 8.5).

Ichnogenus *Leptoptilostipus* Payros et al., 2000.

Fig. 8.10E.

TABLE 8.5 Comparison of measurements of the type material of selected small tetradactyl ichnospecies from the ichnogenera *Gruipeda* (three ichnospecies), *Koreanaornis* and *Pulchravipes*.

	Gruipeda becassi	*Gruipeda dominguensis*	*Gruipeda fuenzalidae*	*Koreanaornis hamanensis*	*Pulchravipes magnificus*
n	4	143	41	5	8
FL_h	59	27–46 (34.1)	27–31 (28.6)	34.5	66.7
FL	38.5	17–33.5 (26.4)	22.5	27.6	54.7
FW	58	20.5–42.7 (33.3)	24–26 (25.4)	32.6	69.7
FL/DL_I	3.75	4.47	2.86	5.96	4.7
FL/FW	0.72	0.7–0.9 (0.8)	(0.89)	0.85	0.79
FL_h/FW	1.03	1.02	1.12	1.04	1.03
II–III	66	36–78 (60)	(55) 45–70 n = 4	58	68
III–IV	71	28–77 (56)	(45) 42–49 n = 4	53	69
II–IV	134	90–135 (115)	87–118 (103) n = 15	113	133
I–II	–	–	(103)	–	–
I–III	166	107–188 (159)	141–172 (158) n = 7	133	148
Relative digit length	I < II < IV < III 16 < 27 < 30 < 37	I < II < IV < III 5.9 < 13.7 < 16.3 < 19.4	I < II < IV < III 10 < 15.2 < 17.2 < 20.7	I < II ≅ IV < III 6 < 18.1≅18.5 < 24.5	I < IV < II < III 15.4 < 33.1 < 36.1 < 46.8
FR	–	11.4	No data	–	–
PA	–	150–182 (170)	150–155 (n = 2)	–	–
PL	–	53–120 (84.4)	28–55 (39.4)	–	–
SL	–	90–230 (170)	181	–	–
SL/FL	–	2.5–5		–	–

Measurements in millimeters and degrees. When available, the range and average (between parentheses) measurements are indicated.
DL_I, length of digit I; *FL*, footprint length; *FL_h*, footprint length including the hallux; *FR*, footprint rotation from the midline; *FW*, footprint width; *I to IV*, digit I to IV; *I–III, II–III, III–IV, II–IV*, angle between the digits I–III, II–III, III–IV and II–IV; *n*, number of specimens; *PA*, pace angulation; *PL*, pace length; *SL*, stride length.
Data compiled from Panin and Avram (1962), de Valais and Melchor (2008), Mansilla et al. (2012), Kim (1969), and Demathieu et al. (1984).

Type ichnospecies: *Leptoptilostipus pyrenaicus* Payros et al., 2000.

Diagnosis: Large and slender footprints (overall length from 80 to 113 mm); slightly asymmetrical; tetradactyl, with three digits (II, III, and IV) pointing forward and a short, medially rotated hind digit (I); central digit (III) longer than the lateral digits (II and IV), the latter being slightly curved forwards; angle between Digits II and IV about 120"; mesial interdigital web commonly visible extending from the tips of the lateral digits to the middle part of the central digit (Payros et al., 2000).

Discussion: This monospecific ichnogenus has been recorded in the Eocene of Spain and Maastrichtian of Mexico (Payros et al., 2000; Astibia et al., 2007; Serrano-Brañas et al., 2022). The morphology is akin to *Charadriipeda recurvirostrioidea*. Both are tetradactyl with a mesial interdigital web imprint, and a marked projection of Digit III well beyond the interdigital web. Potential differences are a smaller footprint length (40 mm vs. 80–113 mm), a smaller II–IV angle (80 degrees vs. 130 degrees) and marked curvature of lateral digits in *C. recurvirostrioidea*.

Ichnogenus *Ludicharadripodiscus* Ellenberger, 1980.

Fig. 8.10F.

Type ichnospecies: *Ludicharadripodiscus edax* Ellenberger, 1980.

Diagnosis: Footprints with the three anterior digits particularly divergent (115 degrees). The nail of the median digit tends to orient slightly to the external side. Interdigital membrane reduced to only the proximal part of the digits. Marked depression of the phalangeal-metatarsal joint. Hallux very short oriented toward the anterior to approximately 10 degrees of the internal side (translated of the French from Ellenberger, 1980).

Discussion: The type material includes four nearly complete footprints and two incomplete ones, that came from the Late Eocene of Gard, France. The distinctive features like hallux impression and interdigital web are apparently only present in one of the footprints. The type material was represented by a drawing. Payros et al. (2000) considered that *Ludicharadripodiscus* was a junior synonym of *Charadriipeda*. Herein, *Ludicharadripodiscus* is considered a nomen dubium, and the type material is comparable with Ignotornidae (see Kim et al., 2006).

Ichnogenus *Ornithoformipes* Patterson and Lockley, 2004.

Fig. 8.14A.

Type ichnospecies: *Ornithoformipes controversus* Patterson and Lockley, 2004.

Diagnosis: Large tridactyl track with digit impression subequal in length. Digit impressions parallel-sided with distal, crescentic-sub circular ungual impressions. Heel robust and subsymmetrical (Patterson and Lockley, 2004).

Discussion: The ichnogenus was erected based on a single footprint showing moderate preservation, from the Franklin Series of the Eocene Puget Group, Washington, USA. A distinctive feature is the presence of impressions of hoof-like ungual phalanges. A perissodactyl origin is considered but a relationship with a large ground-dwelling bird (*Diatryma*-like bird) is preferred by Patterson and Lockley (2004). The overall morphology does not match any other known avian footprint (compare Figs. 8.9–8.11).

Ichnogenus *Ornithotarnocia* (Kordos, 1985).

Fig. 8.11G.

Type ichnospecies: *Ornithotarnocia lambrechti* (Kordos, 1985).

Diagnosis: Avian footprint of moderate to large size, composed of three digital impressions. Digital impressions broad, tapering, without distinct digital pads or interpad spaces and without distinct claws. Central digit (III) longest, characteristically more than 25% longer than the lateral digits (II and IV). Total digital span exceeds 95 degrees. Digits united proximally, but without a distinct metatarsal pad (emended by Sarjeant and Reynolds, 2001).

Discussion: This monospecific ichnogenus was erected by Kordos (1985), and an emended diagnosis was proposed by Sarjeant and Reynolds (2001). The type material is from the Lower Miocene Sandstone of Ipolytarnóc, Hungary. This ichnogenus is represented by tridactyl footprints that commonly lack a metatarso-phalangeal pad and with lateral digits that are symmetrical with respect to Digit III. This ichnogenus is represented by a large sample size and needs a revision, including elucidating the possible affinity with *Avipeda* and *Koreanaornis*. One of the figured specimens (Kordos, 1985: plate I, fig. 3) displays a hallux impression, which is also a feature occasionally seen in *Koreanaornis* (Lockley et al., 1992).

Ichnogenus *Passeripedia* Kordos, 1985.

Fig. 8.11H.

Type ichnospecies: *Passeripedia ipolyensis* Kordos, 1985.

Diagnosis: Impression of small size consisting of three independent digit prints. The outline of the middle digit resembles an oat grain, that of the extreme digits rather like a stick. A little asymmetric, the outer and the middle digit prints stand closer to each other than the middle digit to the inner one (Kordos, 1985).

Discussion: This ichnogenus was based on three isolated diminutive footprints, represented by outline drawings. The type material is from the Lower Miocene Sandstone of Ipolytarnóc, Hungary. The diagnosis lacks any distinctive feature, and there is no description. For this reason, it is considered a nomen dubium. Sarjeant and Langston (1994) proposed to combine this ichnospecies under *Aviadactyla*, a proposal that is not followed here.

Ichnogenus *Pavoformipes* Lockley and Delgado, 2007.

Fig. 8.9J.

Type ichnospecies: *Pavoformipes pintasotoi* Lockley and Delgado, 2007.

Diagnosis: Functionally tridactyl, tetradactyl, subsymmetric track, about as wide as long, with small, postero-medially directed hallux and rounded heel pad.

Discussion: This monospecific ichnogenus was erected using a single isolated footprint of moderate preservation from Miocene lacustrine volcaniclastic deposits near Durango, Mexico. The ichnogenus is distinguished from large ichnospecies of *Gruipeda* by thicker digit imprints (15 mm at the hypex) and smaller total divarication (II–IV = 80 degrees in *Pavoformipes* and >120 degrees for large species of *Gruipeda*, see Table 8.4).

Ichnogenus *Persiavipes* Abbassi and Dashtban, 2021.

Fig. 8.10G.

Type ichnospecies: *Persiavipes gulfi* Abbassi and Dashtban, 2021.

Diagnosis: Tetradactyl bird footprints, medium in size, composed of three slender, acuminate digit imprints, with claw marks. Digits connect to each other proximally with a semipalmate webbing. Digit III is longer than the lateral digits (II, IV), and the interdigital angle between II–III is smaller than the III–IV angle. A thin and delicate Digit I imprint is long and oblique to the midline of the footprint. Digit I imprint does not touch the other digit imprints. Footprints slightly inward rotated (Abbassi and Dashtban, 2021).

Discussion: Abbassi and Dashtban (2021), in the evaluation of vertebrate footprints in the Persian Gulf region, documented mammal and bird footprints and established *Persiavipes gulfi*. The ichnogenus was proposed based on a moderate sample size (42 footprints that belong to seven trackways and isolated footprints) with a holotype trackway of very good morphological preservation (Grade 3 on the scale of Marchetti et al., 2019a). Differences from other tetradactyl semipalmated ichnogenera like *Ignotornis* were not discussed. The material was collected from the Upper Miocene–Pliocene Agha Jari Formation, Iran.

Ichnogenus *Phoenicopterichnum* (Aramayo and Manera de Bianco, 1987).

Fig. 8.10H.

Type ichnospecies: *Phoenicopterichnum rector* (Aramayo and Manera de Bianco, 1987).

Diagnosis: Large tridactyl palmate bird footprints with a fan-shaped outline and a length/width ratio less than 1. Imprint of Digit III projects anteriorly from the digital web impression, in an amount of approximately a quarter of the footprint length. In the anterior margin, two clear curved lines join the lateral digit imprints with that of Digit III. Posterior margin of footprint slightly convex (emended from Melchor et al., 2012).

Discussion: The ichnogenus was proposed by Aramayo and Manera de Bianco (1987) from the Late Pleistocene deposits of Buenos Aires Province, Argentina. The original diagnosis of the ichnogenus was revised by Melchor et al. (2012) considering the type series, which is followed by Feola (2022). Melchor et al. (2012) recognized the ichnogenus in the Neogene of northwestern Argentina. Farina et al. (2021) proposed a new diagnosis of the type ichnospecies, although the authors did not analyze the type series, but a reduced set of footprints from another geological unit (the Miocene Vinchina Formation from northwest Argentina). This procedure does not follow the recommendations of the ICZN. In addition, Farina et al. (2021) proposed two further ichnospecies, *P. lucioi* and *P. vinchinaensis*, which are here considered nomina dubia. The reasons for this proposal are as follows: (1) the FL/FW of all ichnospecies recognized by Farina et al. (2021) are similar (average values range from 0.81–0.84); (2) the main differences between *P. lucioi* and *P. vinchinaensis* are related to angles of lateral digits (II–III vs. III–IV) but all of the data set described by Farina et al. (2021) are isolated footprints, so assignment to the right or left footprint is uncertain; and (3) there is a considerable overlap among key morphological features like FW and total divarication in the dataset analyzed by Farina et al. (2021).

Ichnogenus *Presbyorniformipes* Yang et al., 1995.

Fig. 8.10M.

Type ichnospecies: *Presbyorniformipes feduccii* Yang et al., 1995.

Diagnosis: No formal diagnosis was provided. The description of the holotype is as follows: natural mold of trackway of six consecutive tetradactyl tracks of a single bird displaying palmate webbing between Digits II–IV, with prominent but postero-laterally directed hallux impressions, separated from the other digit impressions and showing no indication of webbing. Faint claw impressions on all digits. Footprints average 7.9 cm wide for the left foot, and 7.6 cm for the right foot. Footprint lengths are 9.5 cm for the left foot and 9.2 cm for the right, including the hallux. Excluding the hallux, track lengths average 6.2 and 6.4 cm for the left and right footprints, respectively. Average step and stride lengths are 9.75 and 18.4 cm, respectively, and the average pace angulation is approximately 154 degrees. Maximum trackway width, including the dabble pattern, is 15.0 cm. Without the dabble pattern the width is 10.5 cm. The average divarication angle between Digits II and

IV is 100.7 degrees with the angle between Digits II and III, and III and IV, respectively, averaging 50.8 degrees and 49.3 degrees. Digits II and IV are curved slightly inward. Phalangeal pad impressions are faintly visible on Digit III in several of the more well-preserved tracks. The anterior margin of the web impression is slightly concave. The web impressions also show some wrinkling, probably due to displacement of the muddy substrate from the pressure of the bird's body weight. The distinctive dabble marks reveal a series of closely spaced prod marks resulting from short, shallow stabs of a beak that was oval at the end, not sharply pointed. The width of each dabbles mark ranges from 2 to 5 mm. The feeding trail meanders across the trackway and extends up to 80 mm from the midline (modified from Yang et al., 1995).

Discussion: The material for erection of this ichnogenus came from the Early Eocene Green River Formation in the Uinta Basin, Utah, USA. *Presbyorniformipes* (sometimes misspelled as *Presbyornithiformipes*, for example, Lockley and Harris, 2010) was erected for tetradactyl palmate footprints, which display an overall similar morphology to *Phoenicopterichnum*, *Anatipeda*, and *Roepichnus* (Fig. 8.10A, H and I). *Presbyorniformipes* is distinguished from *Phoenicopterichnum* by the presence of the hallux impression and smaller size. *Presbyorniformipes* has an FL/FW ratio smaller than 1, whereas in *Anatipeda* is larger than 1. *Roepichnus* is of similar size and FL/FW ratio as *Presbyorniformipes*, although it is potentially distinguished by thick phalangeal pad impressions. In addition, the holotype of *Presbyorniformipes* is a trackway associated with prod marks aligned in a curved to meandering path that were interpreted as dabble or probing traces. In consequence, these traces are considered part of the ichnogenus and are a further difference from *Phoenicopterichnum*, *Anatipeda* and *Roepichnus*.

Ichnogenus *Pulchravipes* Demathieu et al., 1984.

Fig. 8.9F.

Type ichnospecies: *Pulchravipes magnificus* Demathieu et al., 1984.

Diagnosis: Tetradactyl print whose three median digits present the overall shape of an inverted T; the hallux, very short, is directed obliquely toward the back and the interior of the track. This one is almost straight; the feet are directed inwards (translated from the French from Demathieu et al., 1984).

Discussion: The type material was recorded in the Oligocene of Saignon, France. *Pulchravipes* represents tetradactyl footprints with a reduced hallux impression that may be missing in some specimens. In this sense, it is close to some ichnospecies of *Gruipeda* and *Koreanaornis*. In particular, *Koreanaornis* is about half of the size of *Pulchravipes*. *Pulchravipes magnificus* is herein considered a junior synonym of *Gruipeda becassi* (Table 8.5). A potential difference is that *Pulchravipes* seems to have a Digit II impression that is longer (on average) than Digit IV; the opposite is true in most *Gruipeda* ichnospecies. In *G. calcarifera* and *G. diabloensis*, the lateral digits are of similar size.

Ichnogenus *Quadridigitus* Lockley et al., 2021.

Fig. 8.10J.

Type ichnospecies: *Quadridigitus semimembranus* Lockley et al., 2021.

Diagnosis: Small tetradactyl tracks of a slender-toed semipalmate bird with well-developed posterior hallux trace and small proximal web trace between Digits III and IV. Semispherical heel pad located axially, posterior to Digit III trace. Trackways show slight inward rotation of pes. Digit divarication averages 90 degrees (range 77–130 degrees). Belongs in Ignotornidae with *Ignotornis* and *Goseongornipes*, as semipalmate tetradactyl form. Slightly smaller than *Goseongornipes,* which has a relatively shorter hallux and much more pronounced digit divarication. Similar to *Ignotornis* in relative length of hallux, digit divarication, and semipalmate web, but *Ignotornis* is about twice as long and wide with a much more developed web. This differs from other small non ignotornid avian footprints such as *Avipeda* in exhibiting traces of an elongate hallux, semipalmate web, and relatively low digit divarication (Lockley et al., 2021).

Discussion: This ichnogeneric name was previously used to designate Triassic footprints from Morocco by Biron and Dutuit (1981). This was considered a junior synonym of *Chirotherium* Kaup, 1835 by Klein and Lucas (2021), although it is not clear if it is an available name. Lockley et al. (2021) proposed *Quadridigitus* on the basis of material from the Eocene Green River Formation, Soldier Summit, Utah, USA. The distinction with *Ignotornis* is based on a smaller size and less developed interdigital web in *Quadridigitus*. A comparison with average values of the type material indicates that *Ignotornis* is about 50% longer and wider than *Quadridigitus* (Lockley et al., 2009, 2021).

Ichnogenus *Reyesichnus* Alonso et al., 1980.

Figs. 8.10K and 8.15.

Type ichnospecies: *Reyesichnus punensis* Alonso et al., 1980.

Diagnosis: Small tridactyl bird footprints with digits arranged similar to the modern charadriiform Andean Lapwing (*Vanellus resplendens*) but having shorter digits and larger interdigital angle (translated and adapted from the Spanish diagnosis by Alonso et al., 1980).

Discussion: The material for erection of this ichnogenus was collected from the Miocene Pozuelos Formation, Salar del Hombre Muerto, Catamarca Province, Argentina. The footprints were described as tridactyl without webbing based on a single slab containing six footprints. The morphology of the footprints (tridactyl and symmetrical) allows comparison with *Avipeda* and related ichnogenera. However, the illustration of the type material displays a possible interdigital web imprint (Fig. 8.15). A revision of the holotype is necessary to better define its morphology.

Ichnogenus *Rionegrina pozosaladensis* Melchor et al., 2023.

Fig. 8.14B.

Type ichnospecies: *Rionegrina pozosaladensis* Melchor et al., 2023.

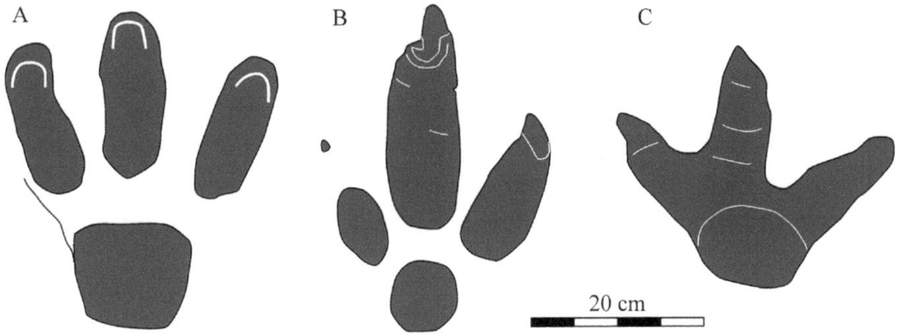

FIGURE 8.14 Outline drawings of holotype material (right footprints) of very large bird ichnotaxa, referred to the type ichnospecies. (A) *Ornithoformipes controversus*. (B) *Rionegrina pozosaladensis*. (C) *Rivavipes giganteus*. *Modified from Patterson and Lockley (2004), Melchor et al. (2023), and Mustoe et al. (2012).*

FIGURE 8.15 Holotype of *Reyesichnus punensis*. Specimen CNS—V 10,001 from Cátedra de Paleontología general from the Universidad Nacional de Salta, Salta Province, Argentina. *Photograph courtesy of Valeria Aquino.*

Diagnosis: Functionally didactyl footprints of moderate to large size (footprint length >300 mm) composing a bipedal trackway. Footprints markedly mesaxonic with footprint length/width ratio ~1.5. Thick, straight, deep, and fusiform Digit III impression, smaller slightly curved Digit IV, and Digit II composing a short and shallow elliptical impression. Digit impressions commonly lacking digital pads and not united to a shallow subcircular metatarso-phalangeal pad impression. Well-developed and thick subtriangular claw trace in digit impressions III and IV. Tip of claw imprint II disconnected from digit impression occasionally preserved. Average divarication II—III smaller than III—IV. Narrow trackway (external trackway breadth/footprint width ~1.6) with high-pace angulation (>160 degrees), and negative (inward) rotation of footprints. Differs from *Velociraptorichnus* by a much larger size, marked mesaxony, and separation between digital and metatarso-phalangeal pads. Distinguished from *Dromaeopodus* by larger III—IV divarication, longer and straight Digit III impression, lack of well-defined phalangeal pads, and proportionally smaller metatarso-phalangeal impression. Differences with *Dromaeosauripus* are a much larger size, presence of a metatarso-phalangeal pad impression, straight Digit III impression that is much larger than that of Digit IV, and absence of well-marked phalangeal pad imprints (Melchor et al., 2023).

Discussion: The type material was recorded in the Upper Miocene—Lower Pliocene Río Negro Formation, Río Negro Province, Argentina. *Rionegrina* exhibits a set of morphological features that differs from any named ichnotaxa and previously recorded fossil or extant footprint. For comparison with other functionally didactyl ichnogenera see Melchor et al. (2023).

Ichnogenus *Rivavipes* Mustoe et al., 2012.

Fig. 8.14C.

Type ichnospecies: *Rivavipes giganteus* Mustoe et al., 2012.

Diagnosis: Large plantigrade tridactyl bird tracks showing robust oval heel pad and three elongate digits. Hallux impression is not present. Phalanges in the shape of elongate triangles, rather than parallel sided. Digit III is elongate relative to Digits II and IV, which are approximately equal in length. Heel pad deeper proximally than distally (Mustoe et al., 2012).

Discussion: The ichnogenus was proposed from the Early Eocene Chuckanut Formation, Washington, USA. The ichnogenus was erected for moderately preserved, isolated (i.e., not composing trackways), large (FL = 245–285 mm), symmetrical, and mesaxonic tridactyl footprints. An avian origin was assumed, and its producer compared with Gastornithidae (Mustoe et al., 2012). However, no alternative interpretations were discussed by Mustoe et al. (2012), including a possible mammalian origin.

Ichnogenus *Roepichnus* Doyle et al., 2000.

Fig. 8.10I.

Type ichnospecies: *Roepichnus grahami* Doyle et al., 2000.

Diagnosis: Tetradactyl avian tracks comprising pes with impressions of prominent phalange joint nodes and curving outer digits; digits terminate in claws; hallux rarely preserved; full interdigital web impressions variably preserved (Doyle et al., 2000).

Discussion: The material for erection of this ichnogenus came from the Upper Miocene (Messinian) Sorbas Member, Caños Formation, southern Spain. Although diagnosed as tetradactyl, the holotype of the ichnogenus lacks a hallux impression, and only a weakly developed impression is assigned to the hallux in the paratype (fig. 12 in Doyle et al., 2000). *Roepichnus* is similar to *Anatipeda*, although it displays a lower FL/FW ratio (0.88 vs. 1.09), and has thick digital pads; the latter feature is considered distinctive by Doyle et al. (2000).

Ichnogenus *Tetraornithopedia* Kordos, 1985.

Fig. 8.9G.

Type ichnospecies: *Tetraornithopedia tasnadii* Kordos, 1985.

Diagnosis: Bird footprint in which the three fore-digits are well developed, the hind digit print, though rudimentary, being quite distinct. The print of the anterior middle digit (III) is straight, widening gradually toward the proximal end and having a pointed end. The inner digit (II) is also straight, parallel-edged, of a character similar to that of the middle digit print. The print of the outer digit (IV) is heavily arched, widening toward the center and then ending in a pointed apex. Its distal sole junction is the widest compared with the others. The hind digit (I) is a little bit displaced toward the inner side, and, after a little widening, it shows a droplike narrowing and a rounded end (Kordos, 1985).

Discussion: Material for erection of this ichnogenus was recorded from the Lower Miocene Sandstone of Ipolytarnóc, Hungary. The ichnogenus was based on five isolated footprints that lack any diagnostic feature, are poorly preserved, and were documented by outline drawings. The main difference with co-occurring *Ornithotarnocia* is the indistinct presence of a hallux imprint, as recognized by Kordos (1985). *Tetraornithopedia* is herein considered a nomen nudum.

Ichnogenus *Uhangrichnus* Yang et al., 1995.

Fig. 8.10L.

Type ichnospecies: *Uhangrichnus chuni* Yang et al., 1995.

Diagnosis: Small, functionally tridactyl tracks of a web-footed bird with small, posteromedially directed hallux trace sporadically preserved. Web configuration palmate, with the posterior margins of the webbing ranging from equally developed between Digits II and III and Digits III and IV, to connecting at the middle to the posterior third of Digit III. Footprint, excluding hallux, wider (W) than long (L), averaging 3.70 and 4.58 cm, respectively (L/W = 0.81), but footprint length with hallux slightly longer than wide (L/W = 1.1). Trackway narrow with short step and stride (7.8 and 15.7 cm, respectively) and strong inward rotation (mean 20 degrees) of Digit III relative to trackway midline (emended by Buckley et al., 2016).

Discussion: This ichnogenus has been commonly reported from Mesozoic rocks, although Mansilla et al. (2012) tentatively recognized its occurrence in the Eocene of Antarctica. *Uhangrichnus* was originally described as palmate and tridactyl (Yang et al., 1995), but Lockley et al. (2012) emended the description as a tetradactyl palmate footprint, and later Buckley et al. (2016) emended the diagnosis to include these features.

Ichnogenus *Uvaichnites* Díaz-Martínez et al., 2012.

Fig. 8.11K.

Type ichnospecies: *Uvaichnites riojana* Díaz-Martínez et al., 2012.

Diagnosis: Very large to enormous (more than 100 mm long) mesaxonic tridactyl bird tracks with a prominent central pad and digits not proximally joined. Total interdigital span is large (more than 100°). Angular asymmetry (angle between Digits II and III smaller than between III and IV). Web trace absent. Slender digit impressions with well-defined claw marks when they are preserved. Central digit impression (III) is the longest. Digit II is smaller than Digit IV. Two digital pad impressions on Digits II and IV and three on Digit III (emended by Díaz-Martínez et al. (2016)).

Discussion: The type material was recorded in the Lower Miocene beds of the Ebro Basin, La Rioja, Spain. This is one of the largest ichnotaxa of tridactyl bird footprints, with a size that is close to those of footprints of large ground birds, like *Aramayoichnus*. The original diagnosis was emended by Díaz-Martinez et al. (2016) in the light of new findings.

Ichnogenus *Yacoraitichnus* Alonso and Marquillas, 1986.

Fig. 8.11I.

Type ichnospecies: *Yacoraitichnus avis* Alonso and Marquillas, 1986.

Diagnosis: Tridactyl footprint, with well-marked digits displaying claw traces, Digit III much larger than II and IV; II and IV display variable angle with III; Digit III with marked digital pads (translated and adapted from the original by Alonso and Marquillas, 1986, in Spanish).

Discussion: This ichnogenus was recorded in the Maastrichtian-Danian Yacoraite Formation (Salta Province, Argentina). The type material is apparently lost and would have a hallux impression (de Valais and Cónsole-Gonella, 2019). As a consequence, de Valais and Cónsole-Gonella (2019) proposed that it is a junior synonym of *Gruipeda filiportatis*.

4.4 Mammalia

The appearance of the mammal-like reptiles and mammals occurred during the Pennsylvanian to Late Triassic, and this event added a new group of tetrapod trackmakers to the continental substrates. Although mammals were not the main trackers among the Mesozoic fauna, some mammal ichnotaxa were named and reported from the Mesozoic sediments, such as *Ameghinichnus* Casamiquela, 1961 from the Middle Jurassic of Argentina (de Valais, 2009), or *Brasilichnium* Leonardi, 1981 from the Cretaceous of Brazil. Here we focus on the mammal ichnotaxa that were erected from Paleogene/Neogene sediments and do not include the pre-Paleogene or post-Neogene ichnotaxa. A total of 70 ichnogenera and 138 ichnospecies of mammal footprints have been recorded and evaluated based on the main groups of mammals that were their presumed producers.

4.4.1 *Artiodactyla*

4.4.1.1 Artiodactyl autopodium structure and tracks

Unguligrade mammals include orders Artiodactyla, Perissodactyla, Notoungulata, and Litopterna (Benton, 2005). Cetartiodactyla was proposed name for Artiodactyla based on molecular phylogenetic studies that showed that cetaceans were closely related to terrestrial artiodactyls, and to the Hippopotamidae in particular (Montgelard et al., 1997). However, the implications of the name are confusing, because Cetacea and Artiodactyla are not sister-taxa. Instead, the evidence clearly shows that cetaceans are a group embedded within Artiodactyla, not a sister-taxon of equal rank (Prothero et al., 2022). Artiodactyls are even-toed ungulates that appeared in the Eocene, and they include diverse and widespread terrestrial species during the Neogene and Quaternary (Marsh, 1984; Theodor et al., 2007). Artiodactyls are known to have hoofed, distal-most phalanges of Digits III and IV, which bear weight and the gait of the animal. The order of Artiodactyla can be classified into the suborders of Tylopoda (camels, llamas, guanacos, alpacas, and vicuñas), Suina (pigs and peccaries), Ruminantia (cattle, goats, sheep, deer, and antelope), and Hippopotamidae (hippopotamus) (Montgelard et al., 1998; Groves and Grubb, 2011) as well as Cetancodontamorpha (Montgelard et al., 1997; Tsagkogeorga et al., 2015; Geisler et al., 2007; Vislobokova, 2013). Living species, in addition to several other fossil families of Artiodactyla, have different digit numbers in the feet, so that families Cervidae, Dromomerycidae, Antilocapridae, Bovidae, Giraffidae, Entelodontidae, and Camelidae have two digits in the feet and lateral toes that are reduced or lost (Clifford, 2010) (Fig. 8.16). On the contrary, the weight of these animals is supported by the third and fourth digits of the feet; and Digits I, II, and V, have been reduced or lost. Clifford (2010) described ways of changes in the numbers of digits in the manus of artiodactyls. Thev pentadactyl artiodactyl manus shows reduced Digit I and nearly equally sized Digits II–V with enlargements of III and IV. The tetradactyl artiodactyl manus, however, has an important change in the anatomy as the loss of the pollex, and f Digits II–V and III–IV are equal in size (Fig. 8.16A). The manus of didactyl artiodactyls and the manus of their ancestral tetradactyla, show that further reduction of the metacarpal elements or all phalangeal bones of Digits II and IV occurred, and led to the loss of Digits II and IV (Fig. 8.16B–E). The external structure of the feet is unique between the members of didactyl artiodactyls, but differences resulted in the shape and size of digit pads (cushion), interdigital space and angle, difference between manus and pes hooves, size and position of the nails, presence or absence of the lateral wall of the toes, size and position of callus to the heel or sole, and size and position of dewclaws (Fig. 8.17). Camel feet include large, heart-shaped symmetrical hooves (Fig. 8.17A). In living camels, the forefoot is larger than the hindfoot. Digit pad III and IV

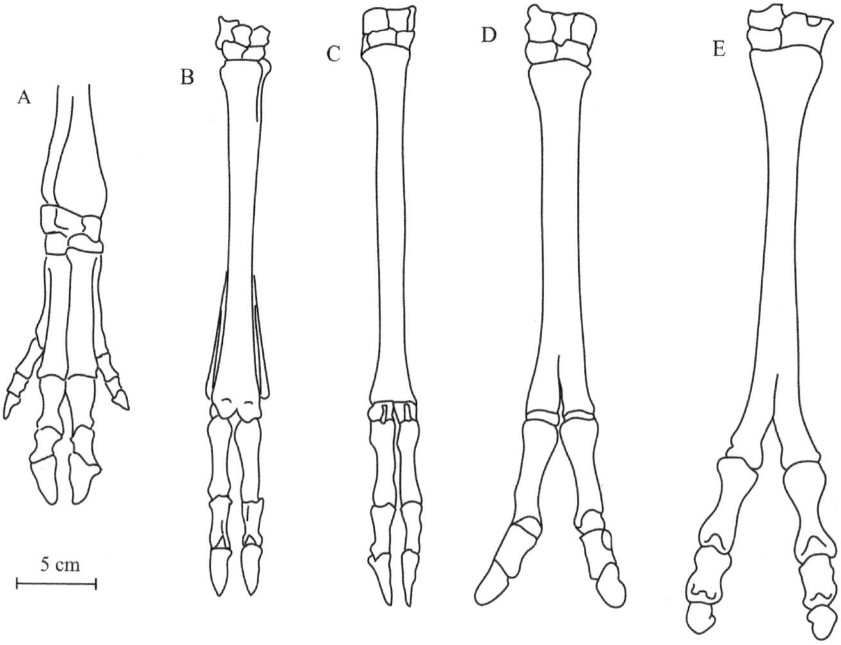

FIGURE 8.16 Skeletal pattern of manus the didactyl artiodactyl taxa, (A) pig, (B) deer, (C) pronghorn, (D) llama, (E) camel. *(A) Modified from* The Encyclopedia Britannica, New Warner Edition *(New York, NY: The Werner Company, 1893. (B–E) Modified from Clifford (2010).*

connect in the mid part of the hooves with an interdigital sulcus (sole concavity). Digits separate in the front and back as anterior and posterior clefts (Lucas and Hunt, 2007). Llamas, however, have a long hoof with sharp and distinct nails (Anderson et al., 2013) (Fig. 8.17B). Elongate digit pads develop in the heel. Digits are slender with an extended anterior cleft, so that the digits have a short contact. Dewclaws of camels and llamas are not developed and confined to the upper part of the manus and pes. Pigs and peccaries have well-developed dewclaws, as Digits II and V (Fig. 8.17C). Hooves include Digits III and IV, and digits are equal in size and symmetrical, with distinct walls in the front of hooves. The heel part is extended and sole limited to the front of the hooves. The outer outline on the hooves is convergent convex, and the inner outline is straight as a long V-shape. Usually, bovids have large, separated hooves equal in size in the manus and pes (Fig. 8.17D). Hooves of Digits III and IV are kidney-shaped and are equal in size. The outlines of digits are convex in outer and slightly concave in the inner part of the hooves. The tips of the digits are sharp and convergent. The walls of the hooves are distinct, and the heel is more protrusive then the sole. Dewclaws in bovid feet are distinct, horny, and conical in shape. The hooves of giraffe's pes and manus are large and approximately equal in size, digits are symmetrical and connected along the length of the hooves (Fig. 8.17E). Walls of hooves present but they are not raised. The heels of the digits are large and slightly convex, and the sole is relatively convex. Dewclaws are not distinct. Outer outline of the digits is convex, and the inner outline is straight. Deer feet include two separate small to large hooves (Fig. 8.17F). The wall of the

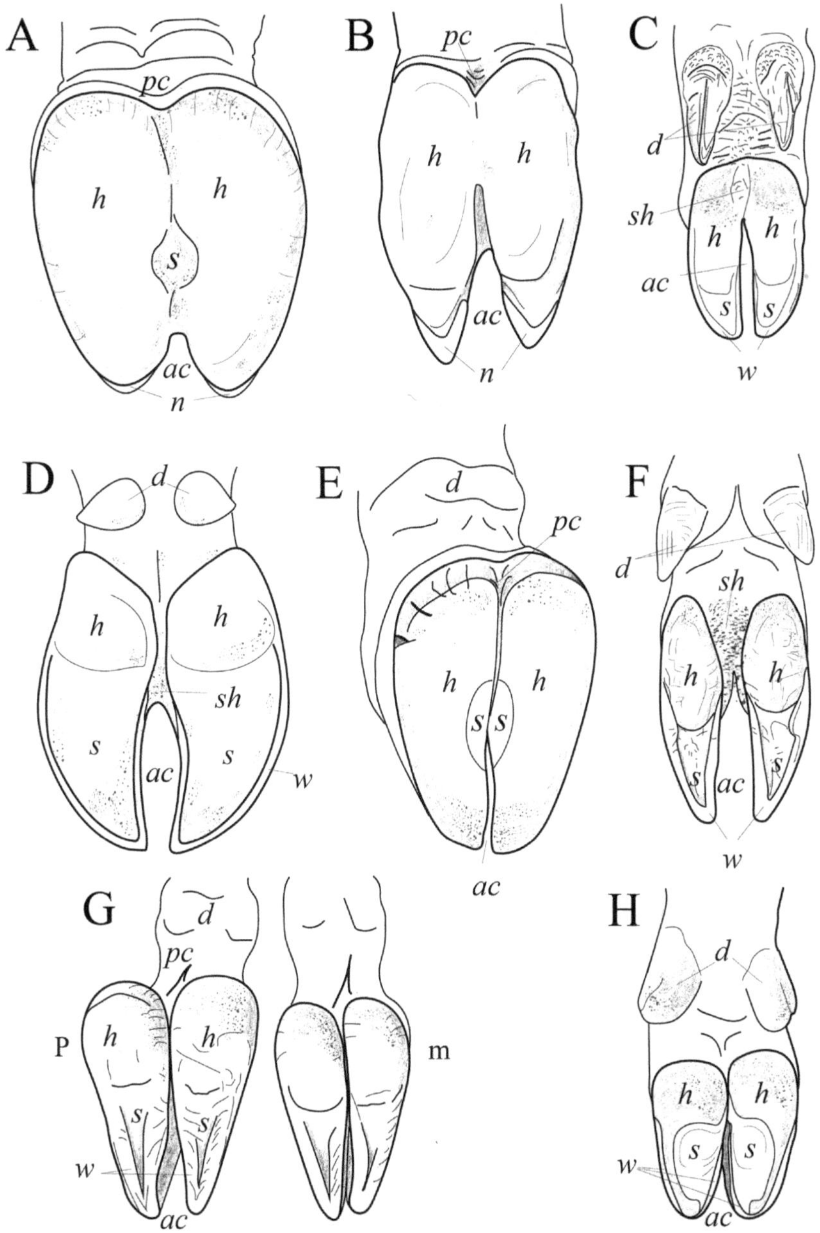

FIGURE 8.17 Morphology of the sole of the foot of living artiodactyls, (A) hind limb of camel, (B) llama, (C) wild boar, *Sus scrofa*, (D) bovine, (E) giraffe, (F) deer, (G) pronghorn, (H) Mountain goat, *Oreamnos americanus*. *ac*, anterior cleft; *d*, dewclaw; *h*, heel; *m*; manus; *n*, nail; *p*, pes; *pc*, posterior cleft; *s*, sole and sulcus; *sh*, space between the heels; *w*, wall. *(A) Modified from https://stichtingdalel.org/en/the-camel. (B) Modified from http://www.shagbarkridge.com/info/feet.html. (C) Modified from https://www.mindenpictures.com/stock-photo-close-up-of-foot-of-wild-boar-sus-scrofa-naturephotography-image90677794.html. (D) Modified from www.producer.com. (E) Modified from www.africahunting.com. (F) Modified from www.northernwoodlands.org. (G) Modified from www.dirttime.ws. (H) Modified from www.mindenpicture.com.*

hooves is well developed and distinct in the front. Digits III and IV are convergent with sharp tips, heels extend two-thirds and the sole is confined to the front of the hooves. Dewclaws are large, conical, and slightly differ in size. Hooves of mountain goats include two symmetrical digits, III and IV, with two round dewclaws (Fig. 8.17G and H). The heel extent half of foot length, and the lateral wall of the hooves is narrow and developed in the front half of feet. Sole limited in the front and inner part of hooves. Tip of the hooves is round. Estimating and measuring of interdigital angle in the feet of living artiodactyls show parallel or posteriorly convergent angles in tips of toes, and only the interdigital angle of mountain goat hooves is proximally divergent.

External morphology and details of the lower part of feet, that contact to the substrate, show differences between the artiodactyl members. Relative size of the heel to sole, thickness and position of the wall of the hooves, extent of the anterior and posterior clefts, size of dewclaws and their position in the metatarsus of Digits III and IV are the main features that control the morphology of artiodactyl footprints.

Artiodactyls leave distinct even-toed tracks that share morphological similarities among the members of the group. In neoichnology, the geographical distribution of artiodactyl animals helps researchers in the identification of trackmakers (Murie, 1982; Elbroch and McFarland, 2019), but its application is restricted, and in some cases is difficult, in the fossil record. As with other animal track registration processes, there are numerous factors that control the morphology of artiodactyl footprints. Substrate properties such as wetness and plasticity at the time of impression, foot anatomy, individual or social behavior of track-makers such as crowding in a tracksite, and postregistration processes are the main factors that control the morphology of artiodactyl footprints.

Quality of preservation of artiodactyl footprints is important in the parataxonomy, and morphology of the footprints is not exactly equal to the morphology of the autopodium. Marchetti et al. (2019a) discussed comprehensively the principles of preservation in tetrapod footprints and their ichnotaxonomy. Morphology of the footprints is the main ichnotaxobase in general, but there are arguments for the determination of details of morphology during the footprint registration, which lead to creation of an ideal footprint for establishing new ichnotaxa (Marchetti et al., 2019a; Belvedere et al., 2018; Lockley, 2009; Milàn and Bromley, 2007).

Anatomy of the feet of artiodactyls discussed in the previous section and includes features of the skeleton, musculature, and hooves. As mentioned earlier, anatomical characteristics of the heel, including the sole, wall, and clefts of the hooves control the morphology of artiodactyl footprints. Most ichnologists think that vertebrate footprint ichnotaxonomy is essentially based on the morphological features (Bromley, 1981; Lucas, 2005; Lockley, 2007; Marchetti et al., 2019a) and behavioral inferences play a secondary role (Magwood, 1992; Melchor and Genise, 2004).

Ichnotaphonomy of the footprints is another process that affects the morphology of footprints, which involves all processes of preservation and how they affect information in the ichnologic record, including the process of trace registration (Marchetti et al., 2019a). Thus, morphology of the footprints resulted from anatomy, preservation, and ethology of the trackmakers.

Evaluation of the diagnosis of named artiodactyl ichnogenera reveals that there are no unique criteria for erection of ichnotaxa. Numerous ichnogenera have been introduced for artiodactyl footprints (the next section provides the details of ichnotaxonomy and discussion). Evaluation of artiodactyl ichnogenera based on the ethology-preservation-anatomical criteria shows that different ichotaxobases have been used for identifications of footprints.

4.4.1.2 Artiodactyl ichnotaxa

After the proposal of *Pecoripeda* from the Burdigalian layers of the Dobrotov Formation (Добротовская свита) in the vicinity of Delyatina (Делятин), Ivano-Frankivsk Region, west Ukraine (Vyalov, 1965), as a first name for artiodactyl footprints, numerous hooved footprints were studied and reported with a scientific name. Vyalov (1965, 1966), considered three (ichno)subgenera for *Pecoripeda* based on the attributed track-makers, and with six (ichno)species as following as:

(Ichno)genus *Pecoripeda* Vyalov, 1965
(Ichno)subgenus *Gazellipeda* Vyalov, 1965
  Pecoripeda (*Gazellipeda*) *gazella* Vyalov, 1965
  Pecoripeda (*Gazellipeda*) *amalphaea* Vyalov, 1965
(Ichno)subgenus *Ovipeda* Vyalov, 1965
  Pecoripeda (*Ovipeda*) *satyri* Vyalov, 1965
  Pecoripeda (*Ovipeda*) *diaboli* Vyalov, 1965
  Pecoripeda (*Ovipeda*) *djali* Vyalov, 1965
(Ichno)subgenus *Cervipeda* Vyalov, 1965
  Pecoripeda (*Cervipeda*) *dicroceroides* Vyalov, 1965

Vyalov (1965) preferred to use only one name, *Pecoripeda,* for all of the artiodactyl footprints (except giraffes), because of difficulties in distinguishing footprints, and lack of information on other groups of artiodactyls such as Tragulidae, Hippotraginae or Alcelaphinae. Although Vyalov (1965) named all footprints of the order Artiodactipedida under the ichnogenus *Pecoripeda,* subsequent researchers gradually introduced or used other ichnotaxa, which currently involve 21 ichnogenera and 46 ichnospecies. The validity of nomenclature is discussed here, and invalid names are indicated by symbol ☒.

Anoplotheriipus Ellenberger, 1980
  A. compactus Ellenberger, 1980
  A. lavocati Ellenberger 1980
  A. similicommunis Ellenberger, 1980
  A. zeuctus Sarjeant and Langston, 1994
Bifidipes Demathieu et al., 1984
  ☒ B. *aeolis* Fornós et al., 2002 *nomen invalidum*
  B. velox Demathieu et al., 1984
Bijugopeda Sarjeant and Reynolds, 1999
  B. anterofossa McNeil et al., 2007
  B. bisymmetrica McNeil et al., 2007
  ☒ *B. simplex* Sarjeant and Reynolds, 1999 nomen dubium (*nomen vanum*)
Bothriodontipus Casanovas-Cladellas and Santafé-Llopis, 1982
  ☒ *B. agramunti* Casanovas-Cladellas and Santafé-Llopis, 1982 *ichnospecies inquirenda*
  B. rovirai Santamaria et al., 1989

Bovinichnus Neto de Carvalho et al., 2022
Bovinichnus uripeda Neto de Carvalho et al., 2022
Camelipeda Vyalov, 1983
Camelipeda turkomanica Vyalov, 1983
Cervipus Matsukawa et al., 2007
Cervipus rex Matsukawa et al., 2007
Corymbipes Sarjeant and Langston, 1994
Corymbipes superstes Sarjeant and Langston, 1994
Diplartiopus Ellenberger, 1980
Diplartiopus longipes Ellenberger, 1980
Dizygopodium Sarjeant and Reynolds, 1999
Dizygopodium dorydium Sarjeant and Reynolds, 1999
☒ *Dizygopodium elachistum* Sarjeant and Reynolds, 1999 synonym with *L. etoromorphum*
Dizygopodium quadracordatum Sarjeant and Reynolds, 1999
☒ *Elaphuruformipes* Matsukawa et al., 2007 is synonym *Entelodontipus*
☒ *Elaphuruformipes tamaensis* Matsukawa et al., 2007 nomen nudum
Entelodontipus Casanovas-Cladellas and Santafe-Llopis, 1982
Entelodontipus viai Casanovas-Cladellas and Santafe-Llopis, 1982
Fustinianapodus Díaz-Martínez et al., 2020
Fustinianapodus arriazui Díaz-Martínez et al., 2020
☒ *Gambapes* Sarjeant and Langston, 1994, synonym of *P.* (*Ovipeda*) *abeli* Vyalov, 1986
☒ *Gambapes hastatus* Sarjeant and Langston, 1994, nomen nudum
Lamaichnum Aramayo and Manera de Bianco, 1987
Lamaichnum alfi Sarjeant and Reynolds, 1999
Lamaichnum borregoensis Remeika, 1999
Lamaichnum etoromorphum Sarjeant and Reynolds, 1999
Lamaichnum guanicoe Aramayo and Manera de Bianco, 1987
Lamaichnum macropodum Sarjeant and Reynolds, 1999
Lamaichnum obliquiclavum Sarjeant and Reynolds, 1999
Lamaichnum sarjeanti McNeil et al., 2007
Lamaichnum tulipensis Aramayo and Manera de Bianco, 1987
Megapecoripeda Kordos, 1985
Megapecoripeda miocaenica Kordos, 1985
☒ *Megapecoripeda velox* Costeur et al., 2009 *invalid combination.*
Pazhanipeda Abbassi and Kazemi, 2023
Pazhanipeda kiani Abbassi and Kazemi, 2023
Pecoripeda Vyalov, 1965
Pecoripeda (Cervipeda) dicroceroides Vyalov, 1965
Pecoripeda (Gazellipeda) gazella Vyalov, 1965
☒ *Pecoripeda (Gazellipeda) tasnadii* Vyalov, 1986, synonym with *Entelodontipus viai* Casanovas-Cladellas and Santafé-Llopis, 1982
Pecoripeda (Ovipeda) abeli Vyalov, 1986 *Gambapes hastatus* Sarjeant and Langston, 1994 is synonym of *P. (Ovipeda) abeli*
Pecoripeda (Ovipeda) diaboli Vyalov, 1965

Pecoripeda (Ovipeda) djali Vyalov, 1965
Pecoripeda (Ovipeda) hamori Vyalov, 1986
Pecoripeda (Ovipeda) satyri Vyalov, 1965
Pecoripeda amalphaea Vyalov, 1965
Pecoripeda commune Aramayo and Manera de Bianco, 1987
Suidichnus Neto de Carvalho et al., 2020
Suidichnus galani Neto de Carvalho et al., 2020
☒ *Suipeda* Vyalov, 1965

4.4.1.2.1 Description and discussion of ichnogenera

Ichnogenus *Anoplotheriipus* Ellenberger, 1980.

Fig. 8.18A–D.

Type ichnospecies: *Anoplotheriipus lavocati* Ellenberger, 1980.

Diagnosis: Very large artiodactyl footprint, Autopod exceeding 4/3 that of *Anoplotherium commune*. Hooves are subparallel. Digit III bulges more distally than Digit IV. Relatively, width of footprint is larger proximally (translated from the French from Ellenberger, 1980).

Discussion: In the short description, Ellenberger (1980) introduced *Anoplotheriipus* as a large artiodactyl footprint with round both in proximal digit tips and distal outlines, and was established from the lower to middle Ludian (=Priabonian, Ludian is the latest Eocene stage in the Paris Basin regional stratigraphy) of France. Two hooves are not always conspicuous, and the division between the hooves is not clear. *Anoplotheriipus* differs from *Lamaichnum* by rounded distal tips of digits, and indistinct interdigital space. Aramayo and Manera de Bianco (1987) named *Odocoileinichnum* for small-to medium-sized (50–120 mm long), hoofed didactyl footprints, digit imprints with flat or slightly concave surface; separation between digits are commonly absent or poorly developed (Fig. 8.19J). The anterior border is acutely marked by both hooves; lateral sides are straight while the posterior border is slightly convex (Aramayo et al., 2015). Thus, *Anoplotheriipus* differs from *Odocoileinichnum* by the pointed anterior end of the hooves in the latter. Sarjeant and Langston (1994) considered *Entelodontipus* as a synonym of *Anoplotheriipus*, based on close similarity in size and shape. *Entelodontipus*, however, has a long medial separation furrow between the hooves, a feature that distinguishes it from *Anoplotheriipus*. The ichnotaxonomic position of *Entelodontipus* will be discussed later in this section. Ellenberger (1980) compared *Anoplotheriipus* to the *Anoplotherium* foot and established three ichnospecies for *Anoplotheriipus*, include *A. compactus*, *A. lavocati*, and *A. similicommunis*. These ichnospecies are distinguished by footprint size, digital imprint protrusility and posterior outlines. Sarjeant and Langston (1994) added the small-sized *A. zeuctus* as a fourth ichnospecies for two hooves imprints from the late Eocene of Texas, which they attributed to a young entelodont individual.

Ichnogenus *Bifidipes* Demathieu et al., 1984.

Fig. 8.18E–F.

Type ichnospecies: *Bifidipes velox* Demathieu et al., 1984.

Diagnosis: Didactyl track made by two hooves that are seamlessly conjoined posteriorly. Anteriorly, the divergent and separate digit impressions are generally pointed. Posteriorly, digit separations are indistinct. When the impression is complete, the length always exceeds the width (translated from the French from Demathieu et al., 1984).

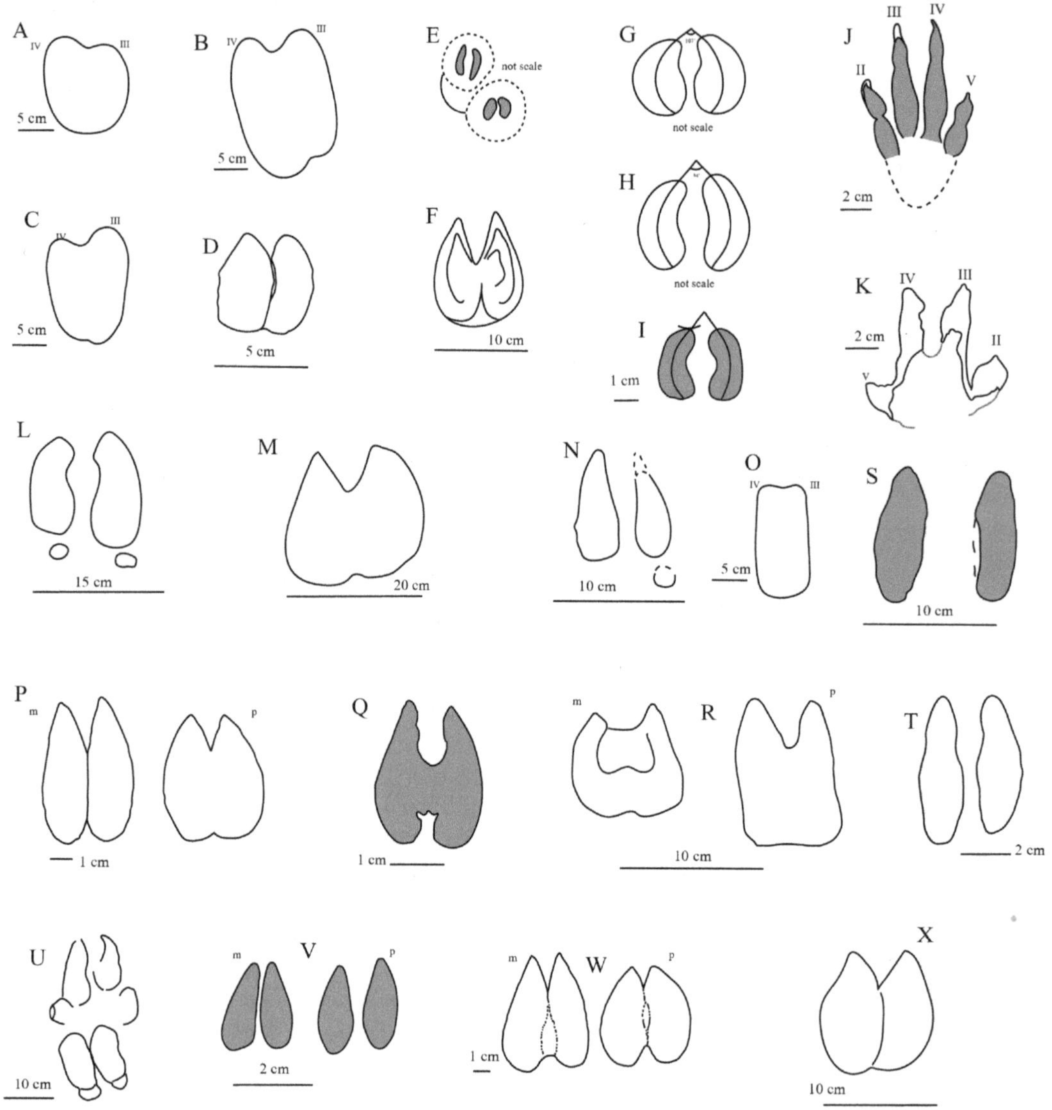

FIGURE 8.18 Modified outline sketch and photograph of the artiodactyl ichnotaxa, invalid names indicated by the gray full color, (A–D) *Anoplotheriipus* Ellenberger (1980), A. *A. compactus* Ellenberger (1980), B. *A. lavocati* Ellenberger (1980), C. A. *similicommunis* Ellenberger (1980), D. *A. zeuctus* Sarjeant and Langston (1994); (E and F) *Bifidipes* Demathieu et al. (1984), E. *B. aeolis* Fornós et al., 2002 *nomen invalidum*, F. *B. velox* Demathieu et al., 1984; (G–I) *Bijugopeda* Sarjeant and Reynolds 1999, G. *B. anteriofossa* McNeil et al., 2007, H. *B. bisymmetrica* McNeil et al., 2007, I. *B. simplex* Sarjeant and Reynolds, 1999 nomen dubium (*nomen vanum*); (J and K) *Bothriodontipus* Casanovas-Cladellas and Santafé-Llopis, 1982, J. *B. agramunti* Casanovas-Cladellas and Santafé-Llopis, 1982, *ichnospecies inquirenda*, K. *B. rovirai* Santamaria et al., 1989; (L) *Bovinichnus uripeda* Neto de Carvalho et al., 2022; (M) *Camelipeda turkomanica* Vyalov, 1983; (N) *Cervipus rex* Matsukawa et al., 2007; (O) *Diplartiopus longipes* Ellenberger (1980); (P–R) *Dizygopodium* Sarjeant and Reynolds 1999, P. *D. dorydium* Sarjeant and Reynolds, 1999, Q. *D. elachistum* Sarjeant and Reynolds, 1999 (synonym with *L. etoromorphum*), R. *D. quadracordatum* Sarjeant and Reynolds, 1999; (S) *Elaphuruformipes* Matsukawa et al., 2007 (synonym *Entelodontipus* and *Elaphuruformipes tamaensis* Matsukawa et al., 2007 nomen nudum); (T) *Entelodontipus viai* Casanovas-Cladellas and Santafe-Llopis, 1982; (U) *Fustinianapodus arriazui* Díaz-Martínez et al., 2020; (V) *Gambapes hastatus* Sarjeant and Langston (1994) (synonym of *P* (*Ovipeda*) *abeli* Vyalov, 1986); W-AD. *Lamaichnum* (1987), W. *L. alfi*, Sarjeant and Reynolds, 1999, X. *L. borregoensis* Remeika, 1999, Y. *L. etoromorphum* Sarjeant and Reynolds (1999), Z. *L. guanicoe* Aramayo and Manera de Bianco, 1987, photo by Ricardo N. Melchor, AA. *L. macropodum* Sarjeant and Reynolds, 1999, AB. *L. obliquiclavum* Sarjeant and Reynolds, 1999, AC. *L. sarjeanti* McNeil et al., 2007, AD. *L. tulipensis* Aramayo and Manera de Bianco, 1987 new combination; AE. *Megalamaichnum* Aramayo and Manera de Bianco, 1987 (synonymous of *Lamaichnum*) and *M. albus* Remeika, 1999 (synonymous of *L. macropodum* Sarjeant and Reynolds 1999); AF-AG. *Megapecoripeda* Kordos, 1985, AF. *M. miocaenica* Kordos, 1985; AG. *M. velox* Costeur et al., 2009 (invalid combination). *m*, manus; *p*, pes; *I–V*, digit coding.

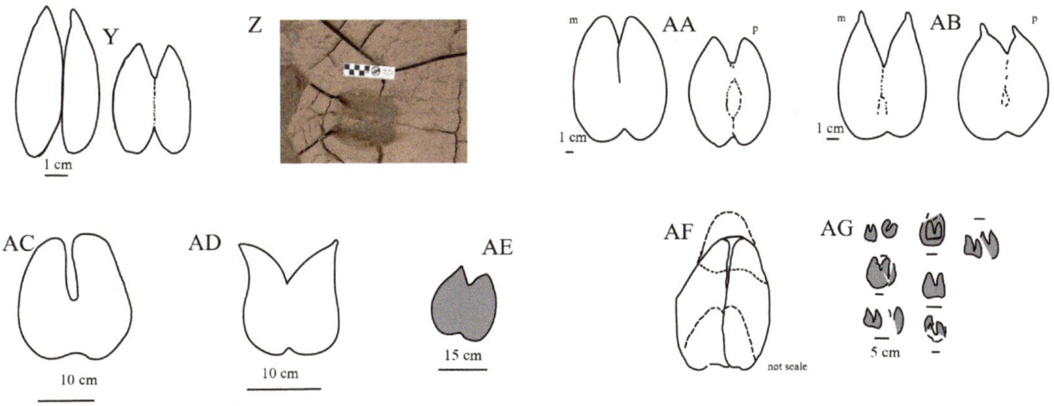

FIGURE 8.18 cont'd

Discussion: *Bifidipes* has been described from the Oligocene of Saignon, southeast Apt, southeastern France and has been attributed to Cervidae (Demathieu et al., 1984), with a greater load of the track-maker body on the foreleg interpreted by the larger size (area) of the manus imprints (Demathieu et al., 1984). *Bifidipes* differs from *Lamaichnum* by pointed anterior ends of the imprint and typically rounded posterior ends of proximally connected hooves, having the digits diverge anteriorly, without claw or sulcus imprints. Two ichnospecies were proposed for *Bifidipes*: *B. velox* and *B. aeolis*.

B. velox is known as homopodous footprints of medium size with length/width ratio that varies from 1.21 ± 0.1 and very variable divergence in the anterior digit impressions, and maximum width of the impression lies in the vicinity of the angle that is formed by the anterior digit impressions (Demathieu et al., 1984). It is difficult to consider *B. aeolis* as a valid name because, as mentioned in its diagnosis, it was created during substrate deformation and by legs immerging in the sediments so that there was a complete disturbance of the surrounding substrate, including microfolding and microfaulting in aeolian sediments (Fornós et al., 2002). Thus, not only is the poorly preserved morphology of the feet, but also the sedimentary environmental characters are involved in its diagnosis. Thus *B. aeolis* is considered a nomen dubium.

Ichnogenus *Bijugopeda* Sarjeant and Reynolds, 1999.

Fig. 8.18G–I.

Type ichnospecies: *Bijugopeda anterofossa* McNeil et al., 2007 (proposed by Abbassi et al., herein).

Emended diagnosis: Large-sized bidigital footprints, digitigrade, showing individual digit impressions. Interdigital trough variably indicated, characterized by a large medial pocket, open both anteriorly and posteriorly. Claws poorly marked or absent. Imprints round overall, are roughly symmetrical across anterior-posterior axes and across lateral axes, which have anterior ends reduced and posterior margins broad. The axes of impressions are convergent anteriorly and posteriorly; anterior interdigital angle is acute to obtuse (emended by Abbassi et al., herein).

Discussion: Sarjeant and Reynolds (1999) named *Bijugopeda* as monospecific with *B. simplex* for incomplete bidigital footprints with arcuate impression and concave sides

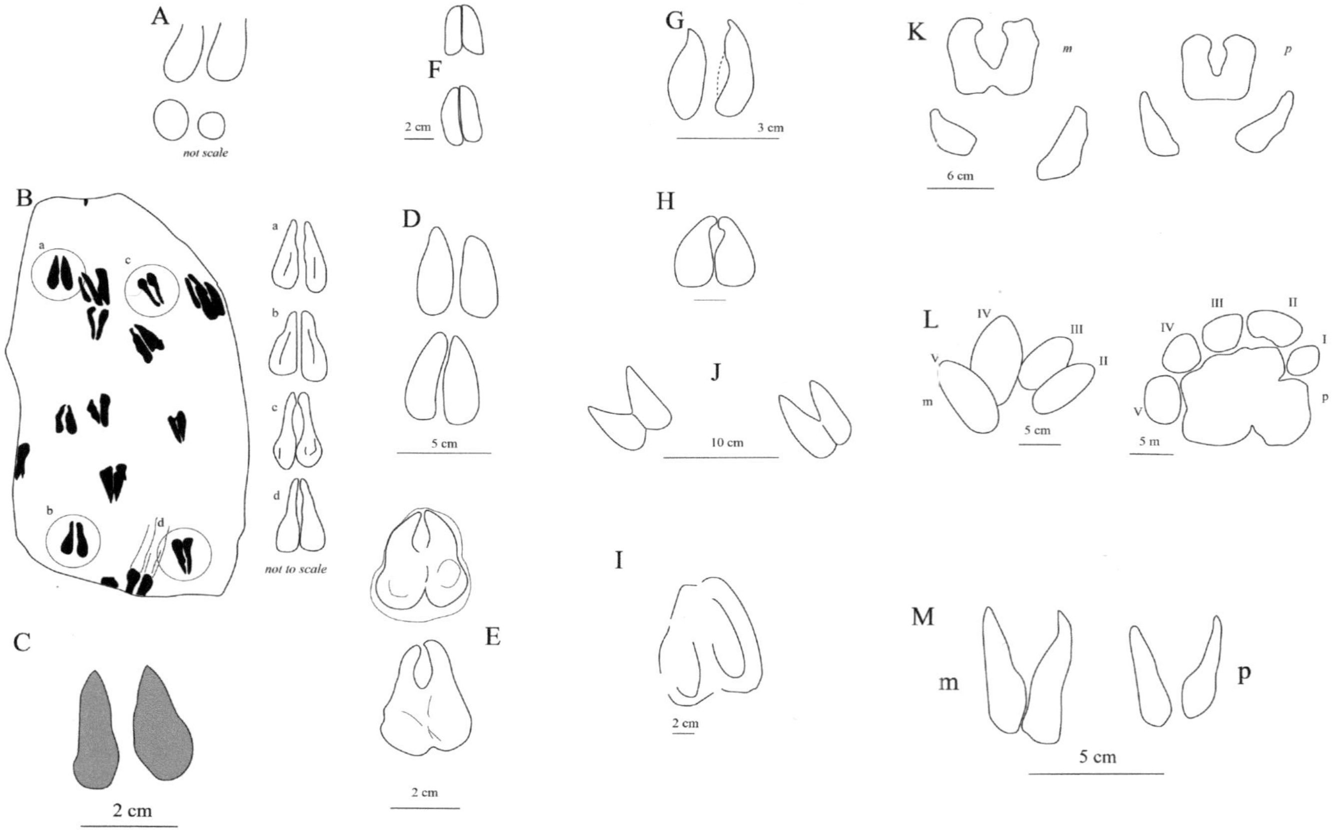

FIGURE 8.19 Modified outline sketch and photograph of the artiodactyl ichnotaxa, invalid name indicated by the gray full color, (A–J) *Pecoripeda* Vyalov, 1965, A. *P. (Cervipeda) dicroceroides* Vyalov, 1965, B. *P. (Gazellipeda) gazella* Vyalov, 1965, C. *P. (Gazellipeda) tasnadii* Vyalov, 1986 (synonym with *Entelodontipus viai*, Casanovas-Cladellas and Santafe-Llopis, 1982), D. *P. (Ovipeda) abeli* Vyalov, 1986 (*Gambapes hastatus* Sarjeant and Langston, 1994 is synonym of *P. (Ovipeda) abeli*), E. *P. (Ovipeda) diaboli* Vyalov, 1965, F. *P. (Ovipeda) djali* Vyalov, 1965, G *P. (Ovipeda) hamori* Vyalov, 1986, H. *P. (Ovipeda) satyri* Vyalov, 1965, I. *P. amalphaea* Vyalov, 1965, J. *P. commune* (Aramayo and Manera de Bianco, 1987) (= *Odocoileinichnum commune* Aramayo and Manera de Bianco, 1987) (synonym of *Pecoripeda* and new combination as *Pecoripeda commune* by Aramayo et al., 2015).; (K) *Suidichnus galani* (Neto de Carvalho et al., 2020). (L) *Corymbipes superstes* (Sarjeant and Langston, 1994), (M) *Pazhanipeda kiani* Abbassi and Kazemi (2023). *m*, manus; *p*, pes. *(A–C, E–F and H–I) Modified from Vyalov (1965), (D) Modified from* Abel (1935), *(G) Modified from Vyalov (1986), (J) Modified from Aramayo and Manera de Bianco (1987), (K) Modified from Neto de Carvalho et al. (2020), (L) Modified from Sarjeant and Langston (1994).*

facing each other. Lucas and Hunt (2007) rejected this identification as an incomplete undertrack and thus an extramorphological variant of questionable ichnotaxonomic significance and regarded it as a nomen dubium (*nomen vanum*) or that it should simply be regarded as a junior synonym of *Lamaichnum*. We agree that *B. simplex* is not a valid name, but the ichnogenus *Bijugopeda* is valid, because McNeil et al. (2007) used *Bijugopeda* for a very similar morphology of artiodactyl footprints at the same issue of the published paper of Lucas and Hunt (2007), and established *B. anterofossa* and *B. bisymmetrica*. Based on the new ichnospecies diagnosis, we would rather revise the diagnosis of the *Bijugopeda*. The presence of the large middle separation between the thick crescent-shaped hooves imprint of *Bijugopeda*, which is called a large middle pocket, is the main feature that distinguishes it from *Lamaichnum*.

Ichnogenus *Bothriodontipus* Casanovas-Cladellas and Santafé-Llopis, 1982.

Fig. 8.18J and K.

Type ichnospecies: *Bothriodontipus agramunti* Casanovas-Cladellas and Santafé-Llopis, 1982.

Diagnosis: Tetradactyl footprints, including slender digits with acuminate ending. The central digits are symmetrical and equidimensional, forming a variable angle between 5 degrees and 30 degrees with the longitudinal axis of the footprint. Lateral digits are symmetrical or slightly asymmetrical with slightly different lengths from each other and generally shorter than the central ones. In them, the angle with the longitudinal axis varies between 20 and 98 degrees. Phalanges are visible or not (emended by Santamaria et al., 1989).

Discussion: Casanovas-Cladellas and Santafé-Llopis (1982) reported tetradactyl footprints from the Oligocene of Agramunt, Ebro basin, north Spain. They recognized *Bothriodontipus* by monotypy (*B. agramunti*) with two central digit imprints of equal length, and two unequal lateral digit imprints, all as digit imprints with tapered tips. Thereafter, Santamaria et al., 1989 added *B. rovirai* as a tetradactyl ichnite, with slender digits with acuminate distal ends. Central digits are symmetrical and equidimensional, and lateral digits are equally symmetrical, with relatively obtuse angles. Provided outlines of these ichnospecies are not the very similar, as the digits of *B. rovirai* were drawn as thick digits. Anatomically, tetradactyl feet of artiodactyls include the difference in the ratio of the lengths of metacarpals III and IV to metacarpals II and V (Clifford, 2010), and the tips of digits are rounded. Santamaria et al. (1989) reported *B. agramunti* in different outlines from Casanovas-Cladellas and Santafé-Llopis (1982: fig. 1) as digitigrade tetradactyl imprints with sharp tips without digital phalanges. It seems that the identification of the track-makers of the *B. agramunti* was wrong, and we believe *B. agramunti* of Casanovas-Cladellas and Santafé-Llopis (1982) is a crocodile footprint and it is not artiodactyl footprints:

1. Artiodactyl footprints usually show thick digit imprints without phalanges or sharp and distinct claw imprints.
2. Based on the outlines of *B. agramunti* (Casanovas-Cladellas and Santafé-Llopis, 1982, fig. 1) it has a proximally rounded hoof. However, most of the artiodactyl footprints are heart-shaped in the back outline of the hooves (except, *Bifidipes* and *Diplartiopus*).
3. There are reports about the alligator body fossils from the Oligocene of Agramunt region (Kälin and Peyer, 1936; Piras and Buscalioni, 2006) and it is possible that these crocodiles were track-makers of the tetradactyl footprints of Agramunt.

Farlow et al. (2018a) identified in the extant crocodylian pes tracks the claw traces of Digits I, II, and III are sometimes dug into the substrate, thus not reflecting real anatomical lengths, and lack a Digit IV claw trace. According to the above, this ichnospecies needs to be revised. Thus, the ichnospecies *B. agramunti* is an *ichnospecies inquirenda* and the ichnogenus *Bothriodontipus* comprises one ichnospecies, *B. rovirai*. The ichnospecies *B. rovirai* differs from *Suidichnus galani* by Digits III and IV symmetrical, sharp toe tip traces, and lateral (dewclaws) II and V impressions conical to comma-shaped, wider than long, smaller than the main toes III and IV (see Neto de Carvalho et al., 2020).

Ichnogenus *Bovinichnus* Neto de Carvalho et al., 2022.

Fig. 8.18L.

Type ichnospecies: *Bovinichnus uripeda* Neto de Carvalho et al., 2022.

Diagnosis: Tetradactyl, paraxonic and subsymmetrical tracks that are as wide as long; central (III–IV) digit impressions are subparallel, with broad distal end pointing forward, and a narrow, linear interdigital space; internal surfaces of hoofprints flat to slightly concave, external surfaces slightly to markedly convex; hoofprints widest near the heel, but tapering only to minor degree between the heel and the apex; rounded-to-rectangular posterior dewclaw (Digits II and V) prints that are much smaller than central digits, printed next to or close to the heel; manus and pes of closely similar form and size (Neto de Carvalho et al., 2022).

Discussion: Valid tetradactyl footprints of artiodactyls include *Cervipeda*, *Bothriodontipus* (partially), *Cervipus*, *Fustinianapodus*, and *Suidichnus*. *Bovinichnus* has two large hoofprints with two posteriorly-located circular dewclaws and it differs from *Bothriodontipus*, which shows four, forward-direct digit prints. The width of the hooves of *Bovinichnus* is considerably larger than *Cervipeda* and *Cervipus* and its size is larger than *Suidichnus* and *Fustinianapodus*. Lucas and Williamson (1993) described tetradactyl artiodactyl footprints from the Eocene Baca Formation of New Mexico, that are the same in shape as *Bovinichnus* with slender hooves imprint and two posteriorly-located circular dewclaws. *Bovinichnus* was considered an artiodactyl ichnotaxon of bovids.

Ichnogenus *Camelipeda* Vyalov, 1965.

Fig. 8.18M.

Type ichnospecies: *Camelipeda turkomanica* Vyalov, 1983.

Diagnosis: Symmetrical footprints, with suboval outline and with a somewhat wider heel imprint. Maximum width is commonly placed at about mid length. Toe imprints are narrow with a subtriangular excavation [sulcus] between them, slightly exceeding one-third of footprint length. A very shallow depression is clearly recognizable on the heel imprint. There is a shallow cleavage confluent between toes. Several similar footprints were observed on the other slabs. While sizes of observed footprints are variable, all of them definitely belong to the same species. A footprint of the anterior foot is always somewhat larger than that of the posterior one. Furthermore, their sizes are variable due to differences in the age of the animals (Vyalov, 1983).

Discussion: Vyalov (1965) introduced the name *Camelipeda* as a general ichnogeneric name for all camel tracks in the study of Miocene footprints from Ukraine but did not designate a type ichnospecies or refer to any specific specimens (Lucas and Hunt, 2007). In the report of footprints of Paratethys sediments of west Turkmenistan, Vyalov (1983) described camel footprints as a new ichnospecies *Camelipeda turkomanica*. However, Lucas and Hunt (2007), without referring to ichnospecies *C. turkomanica*, considered *Camelipeda* as name a *nomen oblitum* because it was made without reference to any specimens, so it lacks any indication of a type or referred specimen, and they said that the name has not been used in last decades since 1965. They did not purposely use the name *Camelipeda* as an ichnogeneric name for camel footprints (Lucas and Hunt, 2007). Although the name *Camelipeda* was erected in 1965 without specifying type material or type ichnospecies (Vyalov, 1965), this name later was attended by Vyalov (1966, 1983) and specified its type ichnospecies *C. turkomanica*, and was mentioned in some lectures (e.g., Alekseev 1987; Herrero et al., 2022). Based on the photographs of the holotype of *Camelipeda* (Vyalov, 1983, plate, fig. 5), it has large, slightly asymmetrical digits imprints, with relatively long digits. It resembles the syntype of *Lamaichnum obliquiclavum* (Lucas and Hunt, 2007, fig. 5E), but differs by digit divergence in the anterior cleft. On the contrary, Lucas and Hunt (2007) considered *Megalamaichnum* and *Dizygopodium* as synonyms of *Lamaichnum* and *Bijugopeda* as a nomen dubium (see *Bijugopeda*, *Dizygopodium* and *Lamaichnum* herein, for more discussion).

Vyalov (1983) attributed most likely C. *turkomanica* to *Paracamelus* because of stratigraphical position (the middle Akchagyl Formation in the Paratethys basin, Caspian Sea region, Pliocene) and occurrence of *Paracamelus* by migration from North America to Asia in the mid-Pliocene and disappeared in the early Pleistocene. Camel footprints are distinguishable from other artiodactyl footprints by (Lucas and Hunt, 2007):

a. having posteriorly broad digit imprints.
b. parallel to divergent anterior digit tips
c. (in most specimens) incomplete separation of the digit imprints.

Ichnogenus *Corymbipes* Sarjeant and Langston, 1994.
Fig. 8.19L.
Type ichnospecies: *Corymbipes superstes* Sarjeant and Langston, 1994.

Diagnosis: Footprints showing a digitigrade to semi digitigrade forefoot and a plantigrade hind foot; all digits have hooves. Four digits (II to V) of the manus are impressed; manual hooves moderately large, Digits II to III directed inward, IV forward and V outward. Five pedal digits, together with the whole heel, are impressed; tarsals and metatarsals are sometimes distinguishable in good impressions. Pedal hooves smaller than those of the manus; I is smallest and directed inward, II to IV are directed forward and V outward (Sarjeant and Langston, 1994).

Discussion: The ichnogenus *Corymbipes* described with one ichnospecies, *Corymbipes superstes*, and general form of the footprints, size of digits, the relationship of digit imprints, and digital characteristics (interdigital angle, digital weight borne), were considered for the diagnosis of *C. superstes*. *Corymbipes* was attributed to the hoofed Mesonychia, an extinct ungulate group (Sarjeant and Langston, 1994). Mesonychia were large, small-to large-sized carnivorous ungulates related to artiodactyls having osteological representation in the late Eocene or Oligocene. They had paraxonic feet, in which the weight of the body is transmitted along an imaginary line between Digits III and IV.

Ichnogenus *Cervipus* Matsukawa et al., 2007.
Fig. 8.18N.
Type ichnospecies: *Cervipus rex* Matsukawa et al., 2007.

Diagnosis: Large, cloven-hoofed track, with two tear-drop-shaped impressions, narrowing gently at one third of length from the posterior of hoof to a sharply pointed tip. Hooves convergent posteriorly, divergent anteriorly. Dew claw impression circular if impressed. Trackway is narrow, subequally-long and outwardly-convex (Matsukawa et al., 2007).

Discussion: Matsukawa et al. (2007) established the ichnogenus *Cervipus* for large artiodactyl footprints from the Plio–Pleistocene of Tokyo, Japan. Wider angles between hooves is a distinctive feature of *Cervipus*, so that digit imprints are anteriorly divergent. Usually, most artiodactyl footprints have parallel axes of hoof imprints or forward directed convergence. *Diplartiopus*, *Lamaichnum*, *Entelodontipus*, and *Bifidipes* are either parallel or divergent in digit imprints. *Pecoripeda* and *Gambapes*, however, show convergent tips. Digital angle in the artiodactyl footprints may differ from manus and pes imprints, exactly. Fore tracks of *Cervipus* show a wider angle, and the hind track has parallel hoof imprints (Matsukawa et al., 2007, fig. 13G and H). *Pecoripeda commune* (= *Odocoileinichnum commune* Aramayo and Manera de Bianco, 1987) also has wider digit imprints in the fore track than the hind track (Aramayo et al., 2015, fig. 13). It thus seems that the interdigital angle is one of the valuable features for the distinction of ichnogenera of artiodactyl footprints.

Ichnogenus *Diplartiopus* Ellenberger, 1980.

Fig. 8.18O.

Type ichnospecies: *Diplartiopus longipes* Ellenberger, 1980.

Diagnosis: Medium sized, slender artiodactyl. The two hooves are completely parallel and end with much more elongated toes. Two hooves are subequal. The width of the sole imprint of the pes is narrow and same in the length of the imprint; depth of footprint increases forward in a regular gradient (approximately 20 degrees) (translated from the French from Ellenberger, 1980).

Discussion: *Diplartiopus* is an antero-posteriorly rectangular track with a very shallow anterior cleft and bears only a limited resemblance to a camel track thought to be made by nonruminants (Ellenberger, 1980; Lucas and Hunt, 2007). *Diplartiopus* is monospecific from the upper Eocene of the Gars, southern France, and has not been reported from outside of France or from other geological time intervals. Based on the extant artiodactyl hooves and their footprint morphologies (e.g., Murie, 1982), the length/width ratio ranges from 1.08 to 1.60, and, in the fossil, footprints range from 0.7 to 1.86 (mean 1.27); in the *Diplartiopus* this ratio reaches 2.16, the highest ratio of the artiodactyl footprints. Ellenberger (1980), however, did not show the photographs of footprints but outline drawing of the footprints. There are some reports about fossil footprints resulting from the sliding of hooves on a slippery substrate (Astibia et al., 1994, fig. 6; Lucas and Hunt, 2007, fig. 6A), and it is possible that *Diplartiopus* is the result of this type of preservation.

Ichnogenus *Dizygopodium* Sarjeant and Reynolds, 1999.

Fig. 8.18P–R.

Type ichnospecies: *Dizygopodium elachistum* Sarjeant and Reynolds, 1999.

Diagnosis: Bidigital digitigrade, the impressions in lateral contact; digital cushions are indicated below the middle phalanges. Claws present, though sometimes poorly marked. Imprints ovoid overall, with reduced anterior end and broad posterior margins. Interdigital trough variably indicated, characterized by a medial pocket positioned anteriorly and close to the interclavular gap (in contrast to *Lamaichnum* where the medial pocket is located centrally posteriorly and ope.ns to the posterior). The posterior gap is smaller and does not reach the medial pocket (Sarjeant and Reynolds, 1999).

Discussion: In the comprehensive study on the Miocene vertebrate tracks from California, Sarjeant and Reynolds (1999) introduced new ichnotaxa of camel and horse footprints. These ichnotaxa include four ichnospecies of *Lamaichnum* (*L. macropodum, L. alfi, L. etoromorphum, L. obliquiclavum*), one ichnospecies for *Bijugopeda* (*B. simplex*), three ichnospecies of *Dizygopodium* (*D. elachistum, D. quadracordatum, D. dorydium*), and three ichnospecies of *Hippipeda* (*H. araiochelata, H. absidata, H. gyripeza*). The ichnotaxonomic position of the *B. simplex* was discussed above as a nomen dubium (*nomen vanum*). Lucas and Hunt (2007) proposed to synonymize all ichnospecies of *Dizygopodium* and *Lamaichnum* of the Miocene of California under one ichnospecies *L. guanicoe*. Reasons for this synonymy considered by Lucas and Hunt (2007) were deficiency and insufficiency of materials and illustrations of the type material of *Dizygopodium* ichnospecies and nonconcurrence between description of footprints with the illustrated specimens. Among the ichnospecies of *Dizygopodium*, the pes imprints of *D. dorydium* show a similar shape as elongate and ellipsoidal outline with distinct claw imprint and interphalangeal furrow (Fig. 8.18P), which is similar to the pes imprint of *L. etoromorphum*. These ichnospecies differ in the mauns imprints, so that the interdigital furrow is not clear in the manus of *D. dorydium*. *D. dorydium* is a synonym of *L. etoromorphum*. But, sketches of the two other ichnospecies of *Dizygopodium* (Sarjeant and Reynolds, 1999, figs. 17 and 18) show distinct features from *Lamaichnum* and they are considered valid names.

Ichnogenus *Elaphuruformipes* Matsukawa et al., 2007.

Fig. 8.18S.

Type ichnospecies: *Elaphuruformipes tamaensis* Matsukawa et al., 2007.

Diagnosis: Cloven-hoofed tracks, consisting of pair of elongated kidney shape impressions. Blunt tips of both hooves. Wide space between two hooves. Functionally didactyl, quadruped trackway with subparallel, subequally-long and outwardly-convex impressions (Matsukawa et al., 2007).

Discussion: Matsukawa et al. (2007) described *Elaphuruformipes* as even-toed ungulate footprints with a blunt tip and wide space between the hooves as a morphological characteristic distinctive from those of other artiodactyls. This ichnogenus is monospecific and known as *Elaphuruformipes tamaensis*. Matsukawa et al. (2007) compared *Elaphuruformipes* with *Lamaichnum* and *Pecoripeda* and *Cervipus*, and with possible elaphurid tracks from the Plio–Pleistocene Kobiwako Group, and Onma Formation, Japan. They did not characterize that the studied footprints have dew imprints, although specimens from the Onma Formation have dew claw impressions. *Entelodontipus* is didactyl, long and parallel imprints (Astibia et al., 1994, 2007). However, based on the illustrated material, *Elaphuruformipes* is larger, but it is morphologically the same and is considered a junior synonym of *Entelodontipus*.

Ichnogenus *Entelodontipus* Casanovas-Cladellas and Santafé-Llopis, 1982.

Fig. 8.18T.

Type ichnospecies: *Entelodontipus viai* Casanovas-Cladellas and Santafé-Llopis, 1982.

Diagnosis: Didactyl footprint includes two impressions of equal length, subparallel, with a subelliptical outline, wider in the distal than the proximal area, which gives rise to a slight divergence. Long axis of the footprint oriented antero-posteriorly (translated from the Spanish from Casanovas-Cladellas and Santafé-Llopis, 1982).

Discussion: *Entelodontipus* is known as two separate hoof imprints of equal size and mostly with parallel axes of hooves imprints. Although Casanovas-Cladellas and Santafé-Llopis (1982) evaluated the hoof imprints of *Entelodontipus* as slightly divergent, in the other reports (Astibia et al., 1994, fig. 3) the proximal or distal divergent/convergent of the axis of the hooves are not present. *Tayassuichnum felixarangutii* and *Elaphuruformipes tamaensis* are synonyms of *Entelodontipus*. *Entelodontipus* differs from *Gambapes* by the heart shape of the footprint of *Gambapes*, parallel axis of hooves imprints and little changes in the width of the hooves. Thus, *Entelodontipus viai* is the only ichnospecies of *Entelodontipus*.

Ichnogenus *Fustinianapodus* Díaz-Martínez et al., 2020.

Fig. 8.18U.

Type ichnospecies: *Fustinianapodus arriazui* Díaz-Martínez et al., 2020.

Diagnosis: Tetradactyl, paraxonic and subsymmetrical manus tracks that are as wide as or wider than long; central digit impressions are subparallel, longer than wide with acuminate distal end rotated inwardly and rounded posterior surfaces; lateral (crew) digit impression is smaller than central digits, longer than wide, elongated, and projected laterally from the posterior surface of central digit impressions. Tetradactyl, paraxonic and subsymmetrical pes tracks that are longer than wide; central digit impressions are subparallel, longer than wide with acuminate distal end and rounded posterior surfaces; lateral (crew) digit impressions are smaller than central digits, as long as wide, subtriangular-shaped and located behind the central digits (Díaz-Martínez et al., 2020).

Discussion: Tetradactyl footprints of artiodactyls are limited to a few ichnotaxa, and belong to a group of artiodactyls with impressions of large Digits II and V on the substrates. These footprints include: *Bothriodontipus rovirai, Pecoripeda (Cervipeda) dicroceroides,*

Cervipus rex, Suidichnus galani, and *Bothriodontipus rovirai*. The latter is tetradactyl with a thick lateral digit in the same shape with forward digits (III and IV). *Suidichnus galani* is a tetradactyl footprint with a trapezoidal shape and symmetrical Digits III and IV. *Fustinianapodus arriazui* differs from the *Suidichnus galani* by smaller lateral digit imprints.

Ichnogenus *Gambapes* Sarjeant and Langston, 1994.

Fig. 8.18V.

Type ichnospecies: *Gambapes hastatus* Sarjeant and Langston, 1994.

Diagnosis: Artiodactyl footprints of wedge shape, indicating the presence of two hooves in both manus and pes. Manus and pes of similar form, though sometimes of different size. The medial (III) and lateral (IV) hoofprints are always distinct, with axial surfaces sometimes in median or posterior contact, more often separated by a continuous interdigital space. The medial and lateral hooves are mirror images in outline; each is broadest near the heel, tapering to an apex of angular or sharply parabolic outline. Apices of hoofprints directed forward; axes convergent, parallel or divergent, according to pace. Maximum breadth of each hoofprint greater than 35% of its length (Sarjeant and Langston, 1994).

Discussion: The ichnogenus *Gambapes* is known as small artiodactyl hoofprints with short hooves and lacking accompanying impressions of claws (Sarjeant and Langston, 1994) and differs from *Pecoripeda* by shorter hooves. Lucas and Hunt (2007), however, considered *Gambapes* a junior subjective synonym of *Pecoripeda*, because the illustrations of *Gambapes* were not enough and significant. Sarjeant and Reynolds (2001) considered *Pecoripeda satyri* as *Gambapes satyri* in the new combination. The differences between *Pecoripeda satyri* and *Gambapes* include presence of a distinctive relief tip of hooves (in cast) and completely convergent tips of hooves, different pes and manus imprints and smaller size. On the contrary, Vyalov (1986) had introduced *Pecoripeda* (*Ovipeda*) *abeli* based on the two small didactyl imprints from the Lower Miocene of Ipolytarnóc, Hungary, which were illustrated by Abel (1935, fig. 142) (Fig. 8.19D). Not only the morphology of the pes and manus of *Gambapes* is same as *Pecoripeda* (*Ovipeda*) *abeli* but also, they are more and less equal in size. Thus, *Gambapes hastatus* is junior subjective synonym of *Pecoripeda* (*Ovipeda*) *abeli*, although the name *Gambapes* has been used by some authors (e.g., Morgan and Williamson, 2000; Abbassi and Mashhadizadeh Maleki, 2020).

Ichnogenus *Lamaichnum* Aramayo and Manera de Bianco, 1987.

Fig. 8.18W-AD.

Type ichnospecies: *Lamaichnum guanicoe* Aramayo and Manera de Bianco, 1987.

Diagnosis: Subdigitigrade, bidigital tracks with pointed anterior ends and rounded posterior ends that range in length from 40 to 260 mm and that are distinguished from the tracks of other ruminant artiodactyls (ichnogenus *Pecoripeda* sensu lato, see above) by typically not having the digits converge anteriorly (they are either parallel or divergent), being relatively broad posteriorly ("heart-shaped") and frequently having the two digit imprints connected or confluent medially (at most separated by a shallow sulcus) for part or most of their length (Lucas and Hunt, 2007).

Discussion: *Lamaichnum* is well-known artiodactyl footprints that were attributed to the Camelidae and introduced from the Pleistocene Pehuen Co area, Atlantic coast, Argentina (Aramayo and Manera de Bianco 1987). As mentioned above, Lucas and Hunt (2007) considered *Dizygopodium* and *Megalamaichnum* (Fig. 8.18AE) and their ichnospecies as synonyms of *Lamaichnum*. They divided *Lamaichnum* into two ichnospecies based on the size: the smaller *L. guanicoe*, which includes *L. alfi*, *L. obliquiclavum*, *L. etoromorphum*, *L. borregoensis*, *D. elachistum*, *D. quadracordatum*, *D. dorydium*; and larger *L. macropodum* that includes *Megalamaichnum albus*. This new combination was not satisfactory, because the size alone is not significant and enough for ichnotaxonomy, and ichnospecies of *Lamaichnum* and *Dizygopodium* show distinct features from the original description of *Lamaichnum*. Among the ichnospecies of the *Dizygopodium*, *D. elachistum* (Sarjeant and Reynolds, 1999) is considered a synonym of *L. etoromorphum*. In the amended diagnosis, the size and morphology of the footprints of *Lamaichnum* have been considered by Lucas and Hunt (2007), and a comparative description of the morphology of the manus and pes imprints has been emphasized by Aramayo et al. (2015).

Ichnogenus *Megapecoripeda* Kordos, 1985.

Fig. 8.18AF-AG.

Type ichnospecies: *Megapecoripeda miocaenica* Kordos, 1985.

Diagnosis: Footprints of a large artiodactyl of composite stature. The print of the inner hoof is usually smaller than is the case with the outer hoof, being shifted distally. In case of [calmly] gait and a horizontal, even surface this asymmetry will disappear. The hoof prints grow proximally wider, deviating by 10–20° from the axial line (Kordos, 1985).

Discussion: The Lower Miocene vertebrate tracksite of Ipolytarnóc, Hungary includes numerous artiodactyl footprints, described by Kordos (1985), and one these footprints was named as *Megapecoripeda*, which differs from *Pecoripeda* by a larger size, and it is not camel footprint (Lucas and Hunt, 2007). Kordos (1985) mentioned *Megapecoripeda* as footprints with composite stature, so that print of the inner hoof is usually smaller than is the case with the outer hoof. The outlines of footprints show parallel to convergent axis of hooves, and photographs, however, show divergent hooves (Kordos 1985, plate XIV). Based on this feature, Costeur et al. (2009) revised *B. velox* Demathieu et al. (1984) as synonym of *Megapecoripeda* and introduced a new combination as *Megapecoripeda velox*. But, the morphology of *Bifidipes* differs from the outline of *Megapecoripeda,* so this combination is not valid, and *Megapecoripeda* has remained monospecific.

Ichnogenus *Pazhanipeda* Abbassi and Kazemi, 2023.

Fig. 8.19M.

Type ichnospecies: *Pazhanipeda kiani* Abbassi and Kazemi, 2023.

Diagnosis: Small, asymmetrical cloven-hoofed footprints, with two gracile digit impressions, with sharply pointed tips. Hooves of either pes or manus footprint are convergent posteriorly, completely divergent anteriorly. Manus footprint is larger than pes, and its hooves are relatively thicker posteriorly. Pes imprint is smaller than manus with shorter hoof imprints. Interdigital angle is greater than 15 degrees and reach to 50 degrees, 30 degrees in mean. Dewclaw imprint is not present. Footprint sequence and footprints point outwards, away from the midline of the footprint sequence (trackway) (negative rotation) (Abbassi and Kazemi, 2023).

Discussion: Recently, Abbassi and Kazemi (2023) introduced *Pazhanipeda* from the Miocene sediments of the Zagros Mountains, Iran. *Pazhanipeda* is distinguished by a wide interdigital angle and differs from *Cervipus rex* by smaller size, more divergent digits, absence of dewclaw imprint and distinction of pes and manus imprints. *Pazhanipeda* differs from *Pecoripeda* by its divergent interdigital angle and tip of imprint of digits not convergent with each other.

Ichnogenus *Pecoripeda* Vyalov, 1965.

Fig. 8.19A–J.

Type ichnopecies: *Pecoripeda gazella* Vyalov, 1965.

Diagnosis: Artiodactyl footprints of elongated wedge shape, indicating the presence of two hooves in both manus and pes. Manus and pes are of closely similar form, though sometimes of different size. The medial (III) and lateral (IV) hoofprints are always distinct, with axial surfaces sometimes in median or posterior contact, more often separated by a continuous interdigital space. The medial and lateral hooves are exact or mirror images in outline; each is broadest near the heel, tapering to an apex of angular or sharply parabolic outline. Apices of hoofprints are directed forward; axes are convergent, parallel or divergent according to pace. Maximum breadth of each hoofprint is less than 35% of its length (emended by Sarjeant and Langston, 1994).

Discussion: *Pecoripeda* is a well-known ichnogenus for a group of artiodactyl footprints that comprise a variety of hoofed didactyl artiodactyls, including traguloids, cervoids and bovoids, and it was divided by Vyalov (1965) into ichnosubgenera (*Gazellipeda, Ovipeda, Cervipeda*) on the basis of presumed affinity rather than morphology. Sarjeant and Langston (1994), however, pointed out that these ichnosubgenera are not valid nomenclature, because the typical ichnosubgenus *Gazellipeda* must be changed to *Pecoripeda,* the same as the ichnogenus name based on Article 44 of the International Code of Zoological Nomenclature. The ichnosubgenera names commonly have not been used by other authors, and here we advocate the common use of ichnogenus and ichnospecies names.

Sarjeant and Langston (1994) have revised *Pecoripeda satyri* and *Pecoripeda dicroceroides* and combined them as *Gambapes satyri* Vyalov, 1965 and *Cervipeda dicroceroides* Vyalov, 1965. *Tayassuichnum* and *Camelopichnum* are not valid names because they were established without a proper description. Bravo-Cuevas et al. (2019), however, introduced *Tayassuichnum felixarangutii* as a new ichnospecies of *Tayassuichnum* and attributed it to extant peccaries from the Plio–Pleistocene Pie de Vaca Formation of southcentral Puebla in central Mexico.

Ichnogenus *Suidichnus* Neto de Carvalho et al., 2020.

Fig. 8.19K.

Type ichnospecies: *Suidichnus galani* Neto de Carvalho et al., 2020.

Diagnosis: Tetradactyl, paraxonic, and subsymmetrical tracks that are longer than wide, with a general trapezoidal shape; Digits III and IV symmetrical, with rounded-to-squarish outline and sharp or blunt toe tip traces, wider in the back of the hoofprint, with narrow interdigital ridge but splayed in the anterior part of the fore print; lateral (dewclaws II and V) impressions conical to comma-shaped, smaller than the main toes III and IV, longer than wide, and projected laterally from behind the main toes, with a wider angle in the fore print than in the hind print (Neto de Carvalho et al., 2020).

Discussion: *Suidichnus* has two large and distinct dewclaws II and V impressions than the other tetradactyl artiodactyl footprints; *Bothriodontipus* shows four slender digits ending with sharp claw impressions. *Suidichnus* differs from *Bovinichnus*, by smaller size, elongate and dewclaws projected laterally from behind the main digit impressions.

Ichnogenus *Suipeda* Vyalov, 1965.

Type ichnospecies: Not identified.

Diagnosis: No formal diagnosis was provided.

Discussion: Vyalov (1965) named mammal footprints based on the nearest track-maker(s) so the footprints Suipedidae footprints introduced as *Suipeda*, but it is a nomen nudum (Neto de Carvalho et al., 2020), which was used later by some authors in the neoichnological (Buynevich, 2015) or paleoichnological (Herrero et al., 2022) studies. *Suipeda* lacks diagnosis, figure or holotype, and thus is a nomen nudum (Neto de Carvalho et al., 2020).

4.4.2 *Perissodactyla and Paenungulata*

4.4.2.1 Perissodactyla and Paenungulata feet structure and tracks

Based on anatomical evidence, the "modern" ungulates appear to be divisible into two major clades, the artiodactyls and cetaceans; and the perissodactyls, hyracoids, and tethytheres (sirenians, proboscideans, and their extinct kin). Orders Perissodactyla (horses, tapirs, rhinos, chalicotheres, titanotheres) and Paenungulata with suborders Hyracoidea, and Tethytheria [include Infraorders Embrithopoda, Sirenia (sea cows, dugongs) and Proboscidea (elephants)] is known as odd toes feet. Among these chalicotheres, titanotheres and Embrithopoda are extinct (Rose, 2006).

Perissodactyls appeared in the early Eocene, probably originated in the Paleocene, and had a much wider Cenozoic radiation and were common Holarctic fauna that included several of the largest land mammals ever known. It seems that the Cenozoic evolution of perissodactyls is considered to have been influenced by global climate change (Bai et al., 2020). Horses, tapirs, and rhinoceroses, as medium- to large-sized terrestrial herbivores, are recent odd-toed ungulates, and among them, tapirs show the least changes in skeleton anatomy from the primitive condition (Rose, 2006).

Perissodactyls have a wide body size of ungulate mammals and include specialized cursorial forms; for example, it is confirmed by often elongated limbs (Figs. 8.20 and 8.21). Perissodactyls are sometimes called mesaxonians, referring to the mesaxonic symmetry of their metapodials. In other words, the axis of symmetry of the mesaxonic foot passes through the third digit. In the tapir and rhinoceros, the main weight passes directly through the center of the median phalanx (Digit III), but in the tetradactyl titanotheres the main weight passes between Digit III and Digit IV (Osborn, 1929).

The foot macroevolution changes in the Perissodactyls are considerable especially in the Equidae. Despite increasing body size and deepening of the cheek teeth from small, leaf-crushing molars to the high-crowned (hypsodont) grass-grinders of modern horses, the number of toes from four (front) and three (back) in the *Hyracotherium* (first horse) changed to three in *Mesohippus* and one in *Pliohippus* and, independently, in modern *Eqüus* (Figs. 8.20A and 8.21A).

The superfamily Tapiroidea first appeared in the early Eocene of North America, with the genus *Homogalax* Radinsky, 1969. Today tapirs are primitive ungulates with a tetradactyl manus and tridactyl pes (Figs. 8.20B and 8.21D), which is like the condition seen in the first member of the Tapiroidea.

The Rhinocerotoidea are the most diverse group of perissodactyls and include amynodonts, hyracodonts and rhinoceroses. Among these groups, rhinoceroses began to diversify by the early Oligocene and dominated in the northern hemisphere and Africa during the Miocene (Prothero et al., 1989; Prothero, 2013). Rhinocerotoidea was limited to Eurasia and Africa during the Plio–Pleistocene, and today, with five genera living in Africa and Asia. The size of Rhinocerotoidea differs from sheep-sized to the largest terrestrial mammal that ever lived. Rhinocerotoidea has tetradactyl (*Acerorhinus* of Aceratheriini, *Brachydiceratherium* and *Diceratherium* of Teleoceratini) to tridactyl manus and almost tridactyl pes (Heissig, 1989) (Fig. 8.21B).

Chalicotheres are Pleistocene, extinct horse-like creatures with huge, fissured claws and had greatly elongated limbs and reduced hindlimbs, representing an extreme adaptation for bipedal high browsing, and giving a gorilla-like stance (Zapfe, 1979; Hooker and Dashzeveg, 2004). Both the pes and manus of Chalicotheres were tridactyl (Fig. 8.20C).

The feet of the titanotheres pass through a lower Eocene tapir-like phase, retained four distinct digits in the forefoot, except that in Digits II, III, IV, V were all of nearly equal size, which is adapted to carrying a moderate amount of weight (Fig. 8.20D and E). Titanothere feet are followed by a middle and upper Eocene rhinoceros-like phase, and finally, they attain a structure similar to that of the rhinoceroses (Osborn, 1929).

Hyracoidea was much more diversified in the early Cenozoic, and some forms dispersed to Europe and Asia in the Miocene that includes five families (four of them extinct) (Pickford et al., 1997), and its phylogenetic position was discussed by authors as Afrotheria (Asher et al., 2017). The forefeet have four digits with flattened claws resembling hooves, the hind foot has three digits, the Digit II has long curved claws and another digit has short, flattened, hooflike claws (Fig. 8.21C). The soles have naked and chubby pads with deep incisions.

Embrithopoda, an extinct group of Afrotheria (Prothero, 2016), is from the late Paleocene to late Eocene of Asia, eastern Europe, and northern Africa. Despite their bizarre huge skull with two massive conical curved bony big pair horns on their forehead, they are known for pillarlike limbs analogous to those of elephants (Rose, 2006). Embrithopoda have pentadactyl hoofed feet.

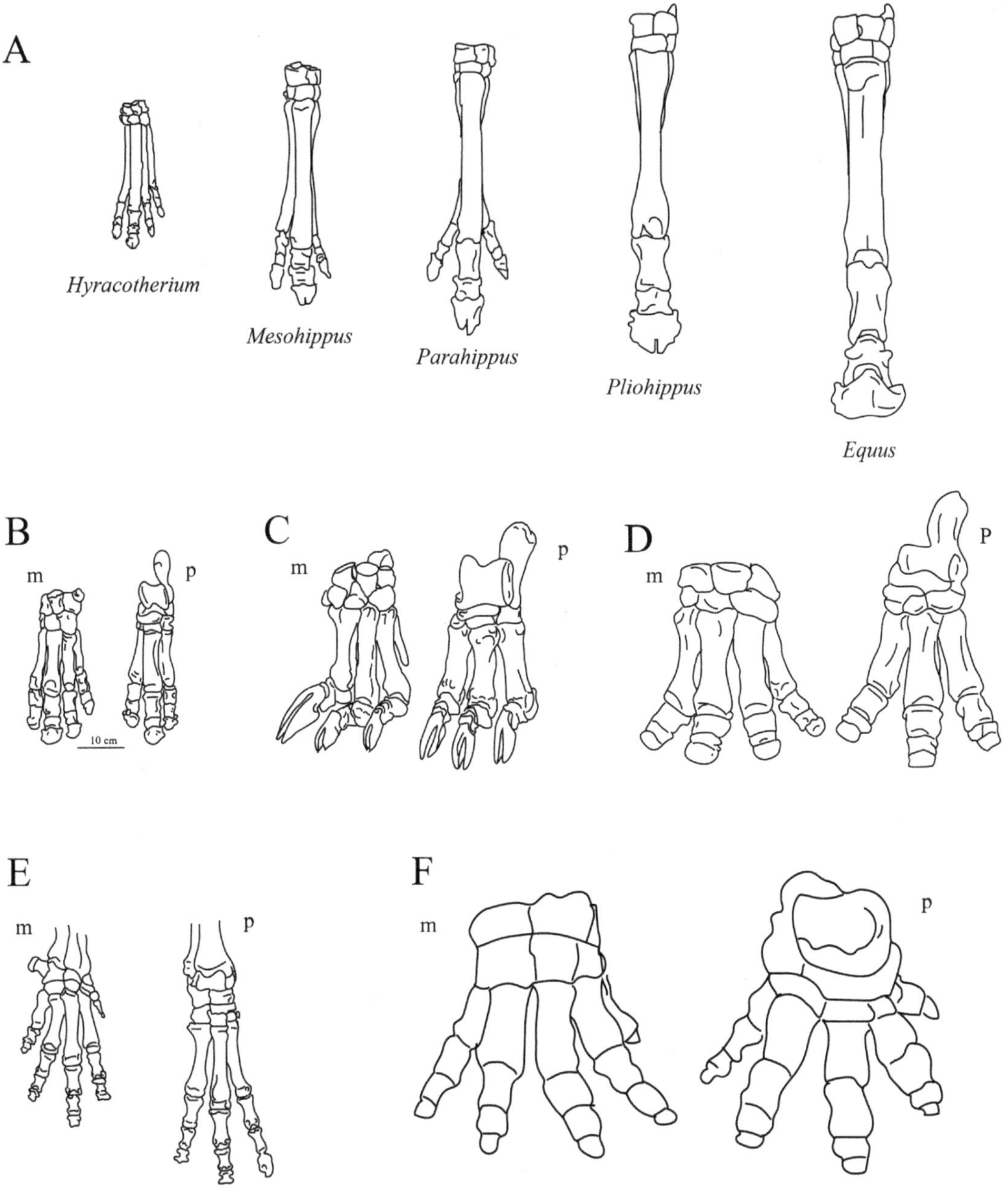

FIGURE 8.20 Skeletal structure of manus (m) and pes (p) of the main groups of perissodactyls, (A) changed number of toes of Equidae, four toes in front limb of *Hyracotherium*, three toes in *Mesohippus* and one in *Pliohippus* and *Equus*, (B) four toes in manus and three toes in pes of Malayan tapir (*Tapirus indicus*), (C) pes and manus of Chalicothere (*Moropus elatus*), (D) pes and manus of titanotheres (*Brontotherium*), (E) Titanothere (*Hyrax*), (F) African elephant (*Loxodonta africana*). *(B) Modified from www.wikipedia.com. (C) Modified from Peterson (1907). (D) Modified from Osborn (1929). (E) Modified from Cope (1882). (F) Modified from Shoshani (1996).*

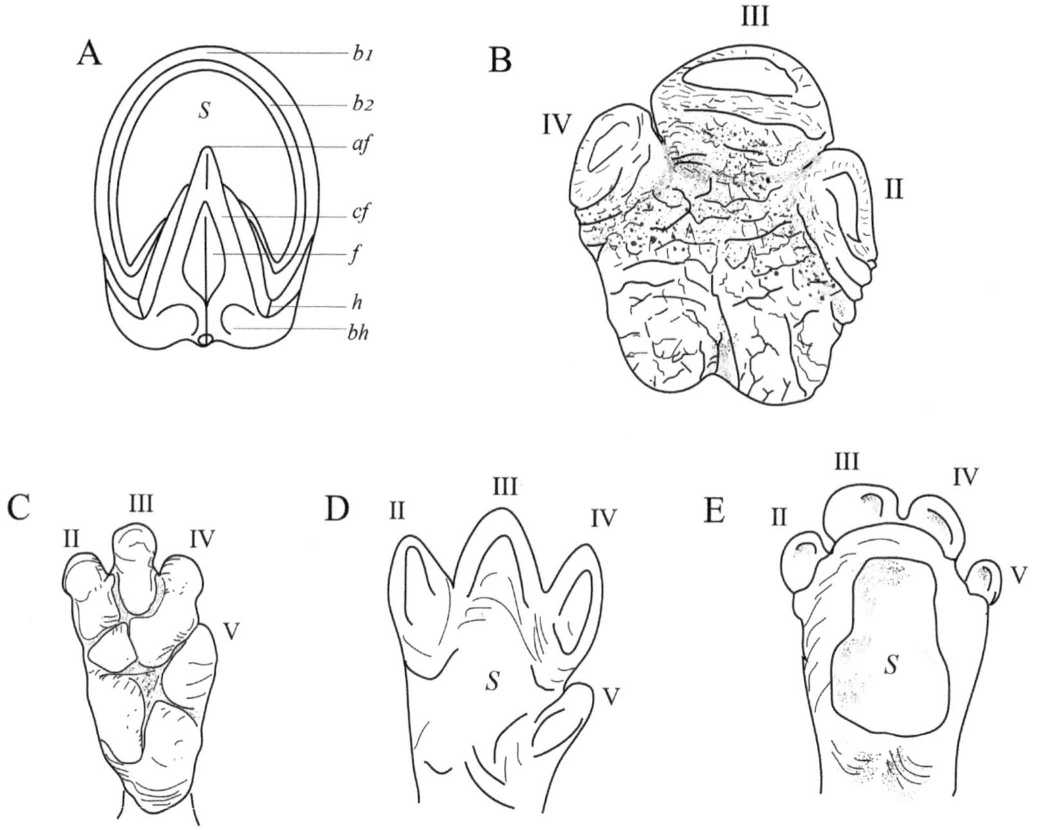

FIGURE 8.21 Morphology of the sole of the foot of the extant perissodactyls, (A) horse sole structure, (B) White rhino (*Ceratotherium simum*) left hind foot, showing the plantar cushion and three toes; (C) Close up of the foot pads on a Rock hyrax *Procavia capensis*, (D) sole of the left fore foot of a tapir (*Tapirus terrestris*), (E) sole of the left fore foot of an Eocene titanothere (*Mesatirhinus petersoni*). *II–V*, digit numbers; *af*, apex of frog; b_1, bar (Parietal horn); b_2 bar (Coronary horn); *bh*, bulbs of the heel; *cf*, crura of frog; *f*, frog; *h*, heel; *S*, sole. *(A) Modified from Budras et al. (2009). (B) Modified from Alibhai et al. (2008). (C) Modified from https://commons.wikimedia.org/wiki/File:Hyraxfootcloseup.jpg. (D) D and E modified from Osborn (1929).*

Sirenians, or sea cows, today comprising the manatees and dugongs, are large, aquatic herbivorous mammals that appeared in the lower to middle Eocene and dispersed worldwide in tropical marine waters, such as the Tethyan region (Domning et al., 2010). Extant sirenians have paddlelike front limbs, no hind limbs, and a horizontal tail fin. The earliest Sirenians (Eocene, families Prorastomidae and Protosirenidae), however, were pig-sized, four-legged amphibious creatures, digitigrade, and most of them probably had tridactyl autopods. It, however, could not confirm that *Pezosiren* had tridactyl feet because there are not any foot bones known; and it may have four or five digits on each foot (Domning, 2001; Daryl P. Domning, pers. comm.). After the appearance of the Dugongidae, at the end of Eocene, sirenians became completely aquatic, with a streamlined body form, featuring flipper-like front legs, no hind legs, and a powerful tail with a horizontal caudal fin (Domning, 2009). Based on the aquatic lifestyle of the sirenians all five toes of the forelimb are encased together by soft tissue and have a flapper shape. Three claws are visible in the dorsal rim of the flappers. The manatee manus is slightly modified as Digit I lacks one phalanx, Digit IV is the longest and most robust, and phalanges on the ends of the digits are irregularly shaped and flattened. These animals used walking forelimbs for movement in contact with the seafloor and registered a feeding trail or flipper imprints in the substrate surface (Keith-Diagne et al., 2022).

Extant elephants were named Proboscidea by Illiger (1811). The order Proboscidea was initially divided into two families: Elephantidae and Mastodontidae, but in recent revisions (Shoshani and Tassy, 2005; Cucina et al., 2022), the order was divided into early Proboscidea (Paleocene–Oligocene), Deinotheriidae (Miocene–Pleistocene), Mammutidae (Miocene–Holocene), Gomphotheriidae (Miocene–Pleistocene), Stegodontidae (Late Pliocene–Pleistocene) and Elephantidae (Miocene–Present) families. Many groups of Proboscidea have become extinct, and only members of the family Elephantidae are still alive, including the Asian elephant (*Elephas*) and the African elephant (*Loxodonta*).

Proboscidea is mainly known to have pentadactyl feet, both in manus and pes. The manus is slightly (Fig. 8.20F) larger than the pes and anatomically with differences in the number and shape of carpus and tarsus (Shoshani, 1996). There is a difference in the number of toenails or nail-like structures between the savanna (African) and forest (Asian) species. Generally, forest elephants have five nails on the front feet and four on the rear, whereas in the savanna species, the front feet have four and the rear three (Turkalo and Barnes, 2013). Usually, Proboscidea has a large, circular, flattened sole, full of deep cracks or fissures and wrinkles.

4.4.2.2 Perissodactyla and Paenungulata ichnotaxa

They involve 16 ichnogenus names and 28 ichnospecies names; the validity of nomenclature is discussed here, and invalid names are indicated by symbol ☒:

Apoxypus Sarjeant and Langston, 1994
  Apoxypus tessellatus Sarjeant and Langston, 1994
☒ *Caballichnus* Casamiquela in Angulo and Casamiquela, 1982
  ☒ *Caballichnus impersonalis* Casamiquela in Angulo and Casamiquela, 1982
Dicerotinichnus Guérin and Demathieu, 1993
  Dicerotinichnus laetoliensis Guérin and Demathieu, 1993
  Dicerotinichnus serengetiensis Guérin and Demathieu, 1993
Hippipeda (Vyalov, 1966)
  Hippipeda absidata Sarjeant and Reynolds, 1999
  Hippipeda araiochelata Sarjeant and Reynolds, 1999
  Hippipeda aurelianis Vyalov, 1966
  Hippipeda cardstoni McNeil et al., 2007
  Hippipeda downsi Remeika, 1999
  Hippipeda gyripeza Sarjeant and Reynolds, 1999
  Hippipeda parva Kulchytzky and Kulchytzky, 1980
Llastaya Vera and Krapovickas, 2022

Llastaya yesera Vera and Krapovickas, 2022
Lophiopus Ellenberger, 1980
Lophiopus rapidus Ellenberger, 1980
Lophiopus lalus Ellenberger, 1980
Macrauchenichnus Casamiquela in Angulo and Casamiquela, 1982
Macrauchenichnus rector Casamiquela in Angulo and Casamiquela, 1982
Moropopus Abbassi et al., 2017
Moropopus elongatus Abbassi et al., 2017
Oplidactylapes Mustoe and Hopkins, 2013
Oplidactylapes eocenica Mustoe and Hopkins, 2013
Palaeotheriipus Ellenberger, 1980
Palaeotheriipus similimedius Ellenberger, 1980
Palaeotheriipus sarjeanti Mirzaie Ataabadi and Kahzaee, 2004
Palimmecopus Sarjeant and Langston, 1994
Palimmecopus praecursor Sarjeant and Langston, 1994
Plagiolophustipus Santamaria et al., 1989
Plagiolophustipus montfalcoensis Santamaria et al., 1989
Proboscipeda Panin and Avram, 1962
Proboscipeda enigmatica (Panin and Avram, 1962)
Proboscipeda panfamilia McNeil et al., 2007
Rhinoceripeda Vyalov, 1965
Rhinoceripeda tasnadyi Vyalov, 1965
Rhinoceripeda voconcense Demathieu et al., 1984
☒*Ronzotherichnus* Demathieu et al., 1984
R. voconcense (Demathieu et al., 1984), new comb. *Rhinoceripeda voconcense*
Thrinaxopus Sarjeant and Langston, 1994
Thrinaxopus hoplephoreus Sarjeant and Langston, 1994

4.4.2.2.1 Description and discussion of ichnogenera

Ichnogenus *Apoxypus* Sarjeant and Langston, 1994.
Fig. 8.22A.
Type ichnospecies: *Apoxypus tessellatus* Sarjeant and Langston, 1994.

Diagnosis: Plantigrade to semiplantigrade footprints having three principal pointed hooves (II to IV) of relatively large size, with a much reduced fifth hoof (V) on the manus. The pedal axis runs through hoof III, which is longest; hooves II and IV of the manus and II of the pes point inward toward it, IV of the pes being almost parallel to HI. Palm and sole are impressed. Manus spade-shaped; pes ovate (Sarjeant and Langston, 1994).

Discussion: Sarjeant and Langston (1994) considered *Apoxypus* from the Eocene of Texas as tapiroidea perissodactyl footprints based on overlapping of the longitudinal axes of the footprint on the enlarged Digit III, plantigrade nature and size. The sharp-tipped hoof impressions and extensive sole pads on *Apoxypus* are features distinct from the hooves of Tapiroidea. Sarjeant and Langston (1994), however, tried to justify sharp-tipped hoof impressions by comparing *Apoxypus* with the ungual phalanges of the extinct genus of tapir-type herbivore (*Heptodon*). *Oplidcatylapes* are very different from *Apoxypus tessellatus*, differences include the larger size of *Apoxypus* footprints, the relatively short length of the digits, and

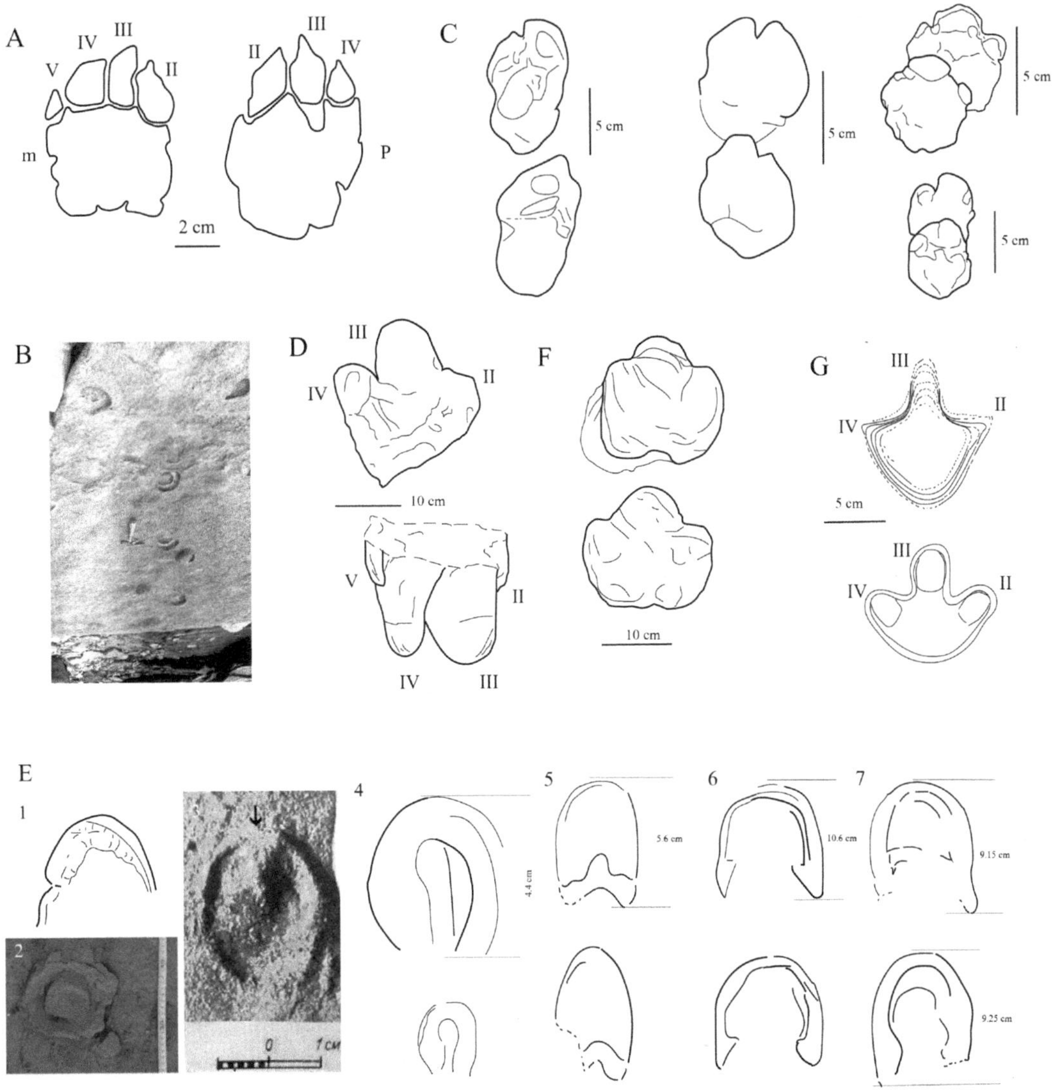

FIGURE 8.22 Modified outline sketch and photograph of the perissodactyl ichnotaxa, (A) *Apoxypus tessellatus* (Sarjeant and Langston, 1994), (B) *Caballichnus impersonalis* Casamiquela in Angulo and Casamiquela, 1982, (C) *Dicerotinichnus laetoliensis* (Guièrin and Demathieu, 1993) (left), *Dicerotinichnus serengetiensis* (Guièrin and Demathieu, 1993) (middle and right), (D) *Llastaya yesera* (Vera and Krapovickas, 2022), (E) *Hippipeda* (Vyalov, 1966), includes 1. *Hippipeda downsi* (Remeika, 1999), 2. *Hippipeda cardstoni* (McNeil et al., 2007), 3. *Hippipeda parva* Kulchytzky and Kulchytzky, 1980, 4. *Hippipeda aurelianis* (Vyalov, 1966) sketch (above) and photograph (below), 5. *Hippipeda absidata* (Sarjeant and Reynolds, 1999), 6. *Hippipeda araiochelata* (Sarjeant and Reynolds, 1999), 7. *Hippipeda gyripeza* (Sarjeant and Reynolds, 1999), (F) *Eumacrauchenichnus patachonicus* (Aramayo and Manera de Bianco, 1987). (G) *Lophiopus rapidus* (Ellenberger, 1980) (above), *Lophiopus lalus* (Ellenberger, 1980) (below). *(B) and (E2-3) with permission.*

the spade-shaped to ovate shapes of the sole pads (Mustoe and Hopkins, 2013). Lockley et al. (1999) compared small tridactyl vertebrate footprints from the Cretaceous and Paleogene of China, Peru, North America and Europe. They concluded that in all regions the track assemblages are mammalian (i.e., of ungulate affinity). The northern hemisphere tracks can be attributed to perissodactyls, but the South American tracks probably represent native ungulate groups such as the Notoungulata and Litopterna.

Ichnogenus *Caballichnus* Casamiquela in Angulo and Casamiquela, 1982.

Fig. 8.22B.

Type ichnospecies: *Caballichnus impersonalis* Casamiquela in Angulo and Casamiquela, 1982.

Diagnosis: Monodactyl footprints of the medium-sized quadrupedal mammal, similar to *Ichnhippus cotaposi* Casamiquela and Chong (1975) and *Macrauchenichnus rector* Casamiquela in Angulo and Casamiquela (1982). Differs from the former by the subcircular outline (in well-preserved footprints) and wing-shaped anterior part, and from the latter by the absence of three lobes (digit imprints) (translated and adapted from the Spanish from Angulo and Casamiquela, 1982)

Discussion: Based on the description of Angulo and Casamiquela (1982), most samples of the Miocene–Pliocene Río Negro Formation of Argentina are poorly preserved and may be tridactyl footprints. This ichnospecies is likely attributed to Equidae or *Macrauchenichnus*. Based on the reexamination on the formation by one of us (RM), these footprints are deep undertracks of quadrupedal mammals. This is considered a nomen dubium based on poorly preserved and deep undertracks of quadrupedal mammals.

Ichnogenus *Dicerotinichnus* Guérin and Demathieu, 1993.

Fig. 8.22C.

Ichnospecies: *Dicerotinichnus laetoliensis* Guérin and Demathieu, 1993.

Diagnosis: Mesaxonic tridactyl impression with three-lobed anterior and rounded trapezoidal posterior outline. Broad and long median digit perpendicular to sagittal plane; small lateral digits are very oblique on the sagittal plane; Digit II is more prominent than that of the Digit IV. Imprint of the sole has less depth than the digit. Autopodia are normally longer than wide, hand wider than foot. In normal walking, the foot lands partly on the front of the imprint of the hand (translated from the French from Guérin and Demathieu, 1993).

Discussion: Guérin and Demathieu (1993) established *Dicerotinichnus* from the Pliocene of Laetoli, Tanzania. Two ichnospecies have been identified for *Dicerotinichnus*, and include *D. laetoliensis,* which was attributed to extant *Ceratotherium praecox,* and *D. serengetiensis* was attributed to a Pliocene extinct species of rhinoceros, *Diceros bicornis* Guérin and Demathieu 1993. Guérin and Demathieu (1993) displayed only outline sketches of footprints which are not significant, and it is not possible to explore the details of the footprints. Based on the other data and figures of tridactyl footprints from Laetoli (Guérin and Demathieu, 1993, fig. 6), *Dicerotinichnus* is a synonym of *Rhinoceripeda*.

Ichnogenus *Hippipeda* (Vyalov, 1966).

Fig. 8.22E.

Ichnospecies: *Hippipeda aurelianis* Vyalov, 1966.

Diagnosis: Footprints functionally of single-digit hooved herbivores, having a hemiellipsoidal to hemispherical or hemiovidal hoof wall, with or without traces of a frog. One or two lateral digits (II and IV) may also be impressed in trackways during a fast gait, but these are much smaller than the central impression of Digit III. Trackway narrow; stride and pace long (emended by Sarjeant and Reynolds, 1999).

Discussion: The horse family experienced an interesting evolution in size and number of foot digits, so that it involves a reduction of the digit number from the five digits autopodium to one digit (III). Naturally, attribution of the penta or tetradactyl footprints from Paleogene records to Equidae is limited and is not citable. The new group of Equidae, the *Hipparion* Group, appeared in the Miocene, with enlarged Digit III as single-toed equids (MacFadden, 2005; Janis and Bernor, 2019; Bernor et al., 2021) and contemporaneously monodactyl footprints are found in the lower Miocene of Ukraine (Vyalov, 1966). Attribution of tri- or tetradactyl perissodactyl footprints to the Palaeothere Group of Equidae is impossible or may be involved in incorrect trackmaker identification. There are however reports of large U- or V- shape imprints of Digit III, with two posterior holes of those two much smaller digits (II and IV) (Herrero et al., 2022; Pérez-Lorente et al., 1999). Among the perissodactyl footprints, ichnogenus *Hippipeda* comprises more ichnospecies, including *H. absidata, H. araiochelata,*

H. aurelianis, H. cardstoni, H. downsi, H. gyripeza and *H. parva.* Variations of the morphology of ichnospecies of *Hippipeda* is dependent on the morphology of the bottom surface of hooves of equids, gaits of the animals and substrate characteristics. The triangular outline of the frog is most often subtle at the back of the hoof print and usually the frog impression is not left because it is filled up or covered over with sediment (Vincelette et al., 2023). The distal phalanx may not comprise the median cleft and show an uninterrupted arc anteriorly (Solounias et al., 2018). Ichnospecies of *Hippipeda* from the Miocene of California (Sarjeant and Reynolds, 1999), including *H. absidata, H. araiochelata* and *H. gyripeza,* show well developed U-shaped anterior outline, and their frog imprints are not clear. On the contrary, *H. downsi* shows only the anterior rim imprint of the hoof. *H. parva* is relatively well preserved and shows well developed frog imprints and the anterior rim of the footprint. *H. cardstoni* from the Pleistocene of Wally's Beach, however, is a poorly preserved imprint and hardly can be identified as a new and distinct ichnospecies.

Ichnogenus *Llastaya* Vera and Krapovickas, 2022.

Fig. 8.22D.

Type ichnospecies: *Llastaya yesera* Vera and Krapovickas, 2022.

Diagnosis: Ungulate and functional tridactyl manus and pes impressions, except when impressed deeply, manus impressions are tetradactyl, denoting a dewclaw impression. Tridactyl to tetradactyl footprints with subtriangular outline, straight lateral margins and a notch on the posterior margin of the footprint. The lateral digit impressions are short and broad with sharp tips pointing anteriorly. Lateral digits are up to 3/4 the length and 1/2 the width of Digit III impression. Digit III is the longest and the widest with blunt tip. Trackways are very narrow, the manus impressions are slightly larger than pedes and are placed in front of the pedes. Footprints have a low divergence angle relative to the midline and are anteriorly directed (Vera and Krapovickas, 2022).

Discussion: Vera and Krapovickas (2022) considered that morphological characters include more triangular shape of *Llastaya yesera* as a feature distinct from the other tridactyl footprints such as *Macrauchenichnus* and *Eumacrauchenichnus* (Fig. 8.22F), and, moreover, it has a tetradactyl manus that includes a dewclaw mark when impressed deeply. In *Macrauchenichnus rector* and *Eumacrauchenichnus patachonicus,* the lateral digit impressions are short and stout, resembling undulations in a subcircular outline. Early toxodontids were probable trackmakers of the *L. yesera* (Vera and Krapovickas, 2022).

Ichnogenus *Lophiopus rapidus* Ellenberger, 1980.

Fig. 8.22G.

Type ichnospecies: *Lophiopus rapidus* Ellenberger, 1980.

Diagnosis: Footprint of small Perissodactyls, [imprints belong to] three-toed foot, that shows a longer [middle digit, which] supports the entire pedal sole on the ground. The foot[print] is smaller. The lateral Digit II and IV imprints are symmetrical to each other, and are set farther back from the middle Digit III. These digit imprints end distally in sharp tips. The trackway is narrow, which denotes a much faster and lighter mode of locomotion (translated and adapted from the French from Ellenberger, 1980).

Discussion: *Lophiopus* includes two ichnospecies, *L. lalus* and *L. rapidus,* both named from the Eocene of France (Ellenberger, 1980). *L. lalus* has a marked "heel mark" and elongate Digit III, whereas *L. rapidus* is wider, and its lateral or central digits are more rounded in the tips. *Macrauchenichnus* is known as a plantar pad tridactyl mammal footprint, its digit imprints are short, and FL/FW tends to 1. *Lophiopus* differs from *Rhinoceripeda,* which is circular in outline, and comprises three distinct oval digit imprints that separate by a margin from the "heel." *Lophiopus* has been reported only from the Eocene of France; the report of *Lophiopus* lacking adequate figure and description from the Eocene of Alborz Mountain of north Iran (Davoodi et al., 2016) needs further investigation.

Ichnogenus *Macrauchenichnus* Casamiquela in Angulo and Casamiquela (1982).

Fig. 8.23A.

Type ichnospecies: *Macrauchenichnus rector* Casamiquela in Angulo and Casamiquela, 1982.

Diagnosis: Imprints corresponding to a medium sized, quadrupedal and functionally tridactyl mammal, apparently provided with a plantar pad (elephant-like), which forms a trilobed outline, on the basis of a subcircular-oblong or piriform model (emended by Vera and Krapovickas, 2022).

Discussion: *Macrauchenichnus* is a footprint of a homopod tridactyl ungulate, with subquadrangular outline. It presents a marked notch on the posterior margin and a small notch on each lateral margin. All digit imprints have rounded tips. *Macrauchenichnus* comprises two ichnospecies, *M. rector* and *M. troyana* (Vera and Krapovickas, 2022), both from the Miocene of Argentina. Ichnogenera *Macrauchenichnus, Llastaya, Palaeotheriipus* (Ellenberger, 1980), and *Rhinoceripeda* (Vyalov, 1965) have roughly similar morphology and are distinguished on different features of the digit/heel, thus making problematic their identification. On the contrary, the morphology of these footprints may pertain to the quality of preservation. For correctly distinguishing them the geometric criterion may be useful. If there were the same substrate properties, the pedal elements played different roles in body weight support. They show different axes of maximum load during locomotion by analyzing the differential depth of footprint and gait mechanics. *Rhinoceripeda* shows maximum depth on the distal portion of the digits, while the Digits II and III are deeper than the Digit IV or the proximal part of the heel (Vyalov, 1965; Abel, 1935, fig. 140). *Llastaya yesera* has the same depth in the sole with slightly more distal inclination in Digit III and has a wider locomotion gait (Vera and Krapovickas, 2022, fig. 5A–D). *Macrauchenichnus rector* exhibits the same depth of penetration pattern of digits and general outline (Angulo and Casamiquela, 1982, plate VIII;

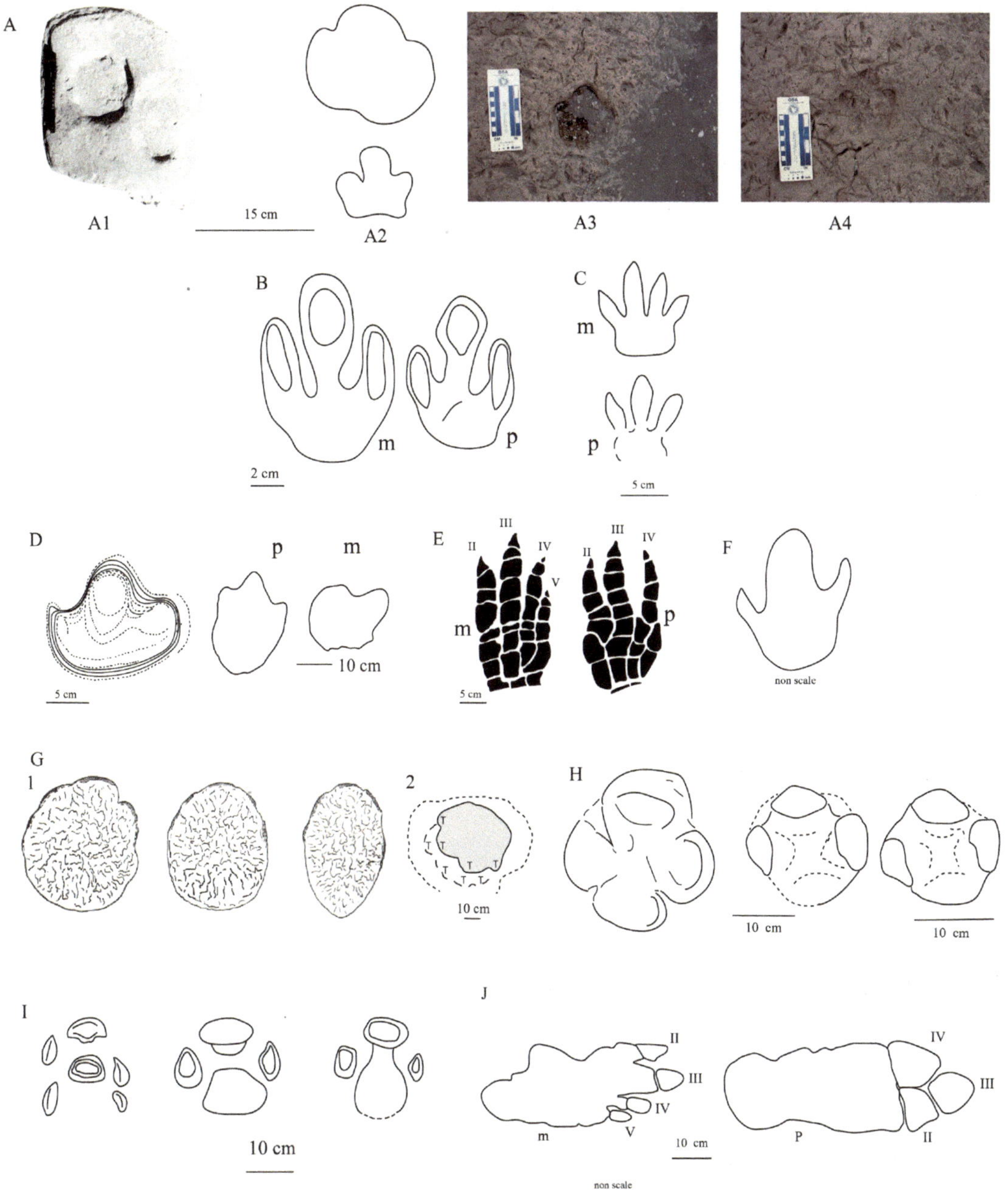

FIGURE 8.23 Modified outline sketch and photograph of the perissodactyl ichnotaxa (continue), (A) A1-*Macrauchenichnus rector* Casamiquela in Angulo and Casamiquela, 1982, A2-sketches, A3-A4 from the type locality and unit photographs by Ricardo N. Melchor, (B) *Moropopus elongatus* (Abbassi et al., 2017), (C) *Oplidactylapes eocenica* (Mustoe and Hopkins, 2013), (D) *Palaeotheriipus similimedius* (Ellenberger, 1980) (left) and *Palaeotheriipus sarjeanti* (Mirzaie Ataabadi and Kahzaee, 2004) (right), (E) *Palimmecopus praecursor* (Sarjeant and Langston, 1994), (F) *Plagiolophustipus montfalcoensis* (Santamaria et al., 1989), (G) *Proboscipeda enigmatica* (Panin and Avram 1962) (1, three footprints), *Proboscipeda panfamilia* (McNeil et al., 2007) (2), (H) *Rhinoceripeda tasnadyi* (Vyalov, 1965) (left, Abel, 1935, fig. 140; right, Kordos 1985; text-fig. 14), (I) *Rhinoceripeda voconcense* (Demathieu et al., 1984), (J) *Thrinaxopus hoplephoreus* (Sarjeant and Langston, 1994). (C) And hind-limb right (D). *(A1), (A3), (A4) With permission, (A2) Modified from Vera and Krapovickas (2022).*

Krapovickas et al., 2009, fig. 8A). *Macrauchenichnus troyana* includes a very narrow trackway with more depth in the lateral Digit [? Digit II, Vera and Krapovickas (2022, fig. 8C and D)]. *Palaeotheriipus* is a tridactyl footprint with a large and deep Digit III (Ellenberger, 1980, fig. 8). A notch on the posterior margin of the footprints is a common feature in *Macrauchenichnus* and *Llastaya,* and it is not in *Palaeotheriipus* and *Rhinoceripeda. Macrauchenichnus* has been reported from Argentina and Peru (Angulo and Casamiquela, 1982; Vera and Krapovickas, 2022; Rojas-Manriquez et al., 2024).

Ichnogenus *Moropopus* Abbassi et al., 2017.

Fig. 8.23B.

Type ichnospecies: *Moropopus elongatus* Abbassi et al., 2017.

Emended diagnosis: Medium-sized quadrupedal tridactyl footprints with elongated oval to fusiform digit imprints. Manus imprints is larger than the pes imprint. Lateral digits (Digits II and IV) are smaller than the middle digit (Digit III) and are curved toward the front. Digit III imprint is wider in the front. Digit tips are completely curved without claw imprints. Metatarsus imprints include two metatarsal pads with a complete rounded back, or with two lobes (emended herein).

Discussion: The Late Eocene footprint assemblage form the Alborz Mountains, North Iran, includes numerous tridactyl footprints. *Moropopus elongatus* is an ichnogenus/ichnospecies introduced by its long digit imprints, which widen in the distal portion. *Moropopus elongatus* is differentiated from *Tricorynopus elaphrus* by its much larger size and the absence of claw imprints. *Moropopus elongatus* is different from *Oplidcatylapes eocenica,* both in its tridactyl pes and manus and the rounded ends of the digits (Abbassi et al., 2017). *Moropopus* is attributed to the infraorder Moropomorpha, which includes tapiromorph animals. Abbassi et al. (2017) did not describe *Moropopus* as a quadrupedal footprint and considered the larger footprint as the pes imprint. Usually, the pes imprint is smaller than the manus imprint in mammals, and its diagnosis is emended here.

Ichnogenus *Oplidactylapes* Mustoe and Hopkins, 2013.

Fig. 8.23C.

Type ichnospecies: *Oplidactylapes eocenica* Mustoe and Hopkins, 2013.

Diagnosis: Mammal trackways containing plantigrade, four-toed manus and three-toed pes impressions, having maximum widths of 5–10 cm. Digits have apices that are deeply impressed in the substrate, with clearly defined margins that indicate hooflike morphology. Tracks have FL/FW ratios that are approximately equal. Heel pad impressions are broad and deep, with indistinct margins that suggest soft tissue, in contrast to the clearly incised distal margins of digits. Manus: Four-toed tracks are plantigrade, with broad heel pad. Outer Digits I and IV are slightly curved toward the centerline of the track. Digits II and III are straight. Manual digits are hoof-like, have acute triangular shapes with clearly defined margins. Pes: Three-toed plantigrade tracks, with broad heel pad. Digits have broadly rounded shape, but with distinct margins that indicate hoof-like morphology (Mustoe and Hopkins, 2013).

Discussion: Mustoe and Hopkins (2013) introduced *Oplidcatylapes eocenica* from the Eocene of northwest Washington and compared it with *Apoxypus tessellatus* from the Eocene of Texas. Key differences include the larger size of *Apoxypus* footprints, the relatively short length of the digits, and the spade-shaped to ovate shapes of the sole pads. Both *Oplidactylapes eocenica* and *Apoxypus tessellatus* were described as tapiroid tracks. There are misspelling of *Oplidactylapes* in Mustoe and Hopkins (2013), in some places typed as *Oplidcatylapes.*

Ichnogenus *Palaeotheriipus* Ellenberger, 1980.

Fig. 8.23D.

Type ichnospecies: *Palaeotheriipus similimedius* Ellenberger, 1980.

Diagnosis: Footprint of large perissodactyls. Tridactyl pes print shows very short sole and deep Digit III imprint. Lateral Digits II and IV are flat, like Digit III, but display lesser sinking in substrate. All three-digit imprints end in round nails. The size of footprint very closely corresponds to *Palaeotherium medium* or *P*. cf. *crissum*, so that their foot skeleton completely corresponds to the footprints.

Discussion: The mentioned diagnosis was translated from French (Ellenberger, 1980) and considered shape and size of the digit and pad imprints. At first *Palaeotheriipus* was erected as monospecific from the Eocene of France, and later Mirzaie Ataabadi and Khazaee (2004) added *P. sarjeanti* as large tridactyl footprint with digital impression III subrounded and much broader and longer than the lateral Digits II and IV. Lateral digital impressions are almost identical in size and shape. Pad impression is almost hemispherical in shape and well developed, and manus less elongate than pes, with median Digit III also less developed. *P. similimedius* has a much less complete posterior portion of the track (metacarpal impression) than *P. sarjeanti*, while the interdigital angles in the latter ichnospecies are much smaller than in *P. similimedius*. Lateral digits are less distinguishable in *P. sarjeanti*, but the median digit is impressive. Mirzaie Ataabadi and Khazaee (2004) and Abbassi et al. (2015), however, considered a different diagnosis for *Palaeotheriipus* without mentioning the emended diagnosis. Here we would rather consider the original diagnosis of Ellenberger (1980). *Palaeotheriipus* differs from *Plagiolophustipus* by relatively small and broad digit imprints. *Lophiopus* is smaller in size and has divergent lateral digit imprints.

Ichnogenus *Palimmecopus* Sarjeant and Langston, 1994.

Fig. 8.23E.

Type ichnospecies: *Palimmecopus praecursor* Sarjeant and Langston, 1994.

Diagnosis. Plantigrade footprints of medium to large size, with long digits terminating in claws. Manus longer and larger than pes, but having a similar, shovel-like shape; the entire palm and sole are impressed. The manus has four digits (II to V). Digits II to IV are long and parallel for much of their length; the foot axis passes through Digit III, which is longest. Digit V is greatly reduced, parallel to the proximal end of IV and much shorter. The pes has three digits (II to IV), more widely spread than the manual digits. The pedal axis passes through Digit III, which is longer and thicker than the others (Sarjeant and Langston, 1994).

Discussion: *Palimmecopus* has peculiar morphology include large size, long and clawed digits, and pes mesaxony, which make difficult to identify potential trackmaker(s). Sarjeant and Langston (1994) placed *Palimmecopus* uncertainly as Creodonta or Perissodactyla. Krapovickas and Nasif (2011) established *Tacheria troyana* for tetradactyl manus and pes impressions of subequal size with long and robust digits with strong elongated claws. It has an elongated sole imprint with digitigrade footprints. *Tacheria troyana* was attributed to a large rodent or rodent-like mammal. *Palimmecopus praecursor* differs from *Tacheria troyana* by lesser number of digit imprints of pes and by relatively thicker digit imprints.

Ichnogenus *Plagiolophustipus* Santamaria et al., 1989.

Fig. 8.23F.

Type ichnospecies *Plagiolophustipus montfalcoensis* Santamaria et al., 1989.

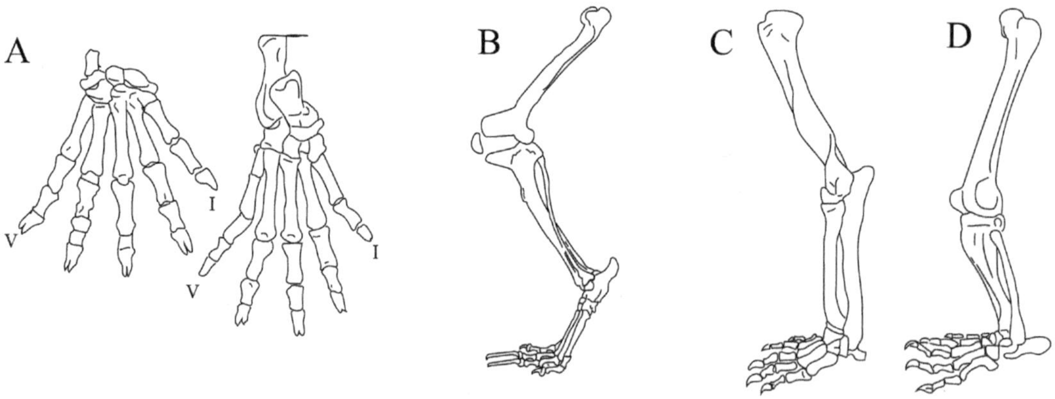

FIGURE 8.24 Skeletal anatomy of Creodonta (A), and digitigrade felid (B) and plantigrade bear (C–D). *(A) Modified from Denison (1938). (B) Modified from Le Conte (1891), and (C-D) Modified from https://www.donsmaps.com/bear.html, forelimb left (C) and hind-limb right (D).*

Diagnosis: Tridactyl ichnites with the central digit much wider and longer than the lateral ones, which are slightly asymmetrical. Total length of the variable footprint between 5 and 6 cm (translation from the original in Spanish by Linares Montes et al., 2022).

Discussion: This ichnogenus was erected based on a single isolated footprint. An elongate, small sized tridactyl footprint with long Digit III and small lateral digits is known as *Plagiolophustipus,* which was reported from the Oligocene of Spain, and not elsewhere (Santamaria et al., 1989; Murelaga et al., 2000; Astibia et al., 2007; Linares Montes et al., 2022). Owing to its morphology, it was assigned to Equidae (Murelaga et al., 2000). Linares Montes et al. (2022) compared *Plagiolophustipus* with other tridactyl mammal footprints and discussed the morphologic differences between them. They appraised perissodactyls as mesaxonic so that the middle digit bears most of the animal's weight. It is possible that the lateral digits of the three-toed perissodactyls partly bear the animal's weight during the movement. Certainly, *Plagiolophustipus* with a large middle digit, shows mesaxonic function of the autopod. *Plagiolophustipus* includes only one ichnospecies, *P. montfalcoensis,* and it was reported from the other areas of the Ebro Basin with a great variety of size and morphology (Santamaría et al., 1989; Murelaga et al., 2000; Astibia et al., 2007; de Gibert and Sáez, 2009; Díaz-Martínez et al., 2018; Linares Montes et al., 2022).

Ichnogenus *Proboscipeda* (Panin and Avram, 1962).

Fig. 8.23G.

Type ichnospecies: *Proboscipeda enigmatica* (Panin and Avram, 1962) (designation by Abbassi et al., 2017).

Diagnosis: Large oval to subcircular footprints, large and flat sole/palm surface either ornamented or smooth, three to five short and blunt digit impressions pointing anteriorly, deep footprints with a noticeable marginal ridge (emended by Aramayo et al., 2015).

Discussion: *Proboscipeda* shows simple morphology as large, circular to oval, shallow to deep depressions with or without digit imprints. These characters make easy identification of the proboscidean footprints. There are, however, other proposed ichnogeneric names for proboscideans with little differences from *Proboscipeda*, so that they usually have been considered as synonyms of *Proboscipeda*. Lucas et al. (2007) regarded *Stegomastodonichnum* as a junior subjective synonym of *Proboscipeda. Mammuthichnum* was coined without diagnosis of the ichnogenus or designation of a type ichnospecies; therefore, by ICZN rules *Mammuthichnum* is a nomen nudum (Lucas et al., 2007). Most records of proboscidean footprints (>80%) have been reported from the Pleistocene (McDonald et al., 2007). *Proboscipeda* includes three ichnospecies, *P. enigmatica, P. panfamilia* and *P. australis*. The ichnospecies *P. panfamilia* was established only by pes imprints because manus impressions are overprinted by the pes in typical locomotion. *P. panfamilia* is wider, and larger with wide short impressions than *P. enigmatica* (McNeil et al., 2007). *Proboscipeda australis* was erected for the Late Pleistocene of the Buenos Aires Province, Argentina and is distinguished by a sole and palm that is rough and cracked, a smaller size, and manus and pes can be distinguished considering the relative size and number of digit imprints (Aramayo et al., 2015).

Ichnogenus *Rhinoceripeda* Vyalov, 1965.

Fig. 8.23H.

Type ichnospecies: *Rhinoceripeda tasnadyi* Vyalov, 1965.

Diagnosis: Footprints of a large rhinoceros deriving from a male, a female and a juvenile individual. On the proximal side of the footprints there are three oval hoofprints, of which the two lateral ones are approximately equal, the middle one being larger. The hoof impressions are close to one another, their margins being often in contact (after Kordos 1985).

Discussion: Based on reference to Abel (1935, fig. 140), Vyalov (1965) established *Rhinoceripeda* from the lower Miocene vertebrate tracksite of Ipolytarnóci, Hungary, without giving a holotype, description and diagnosis. Thereafter, Kordos (1985) proposed a neotype and described *Rhinoceripeda tasnadyi. Rhinoceripeda* is recognized by its large oval shape and short digit imprints and round back outline with small heel imprint. These features distinguish it from the other large to intermediate-sized tridactyl footprints like *Palaeotheriipus, Macrauchenichnus* and *Llastaya*. Other tridactyl footprints, *Moropopus* and *Plagiolufostipus,* are smaller with different morphology than *Rhinoceripeda. Rhinoceripeda* comprises two ichnospecies, includes *R. tasnadyi* as type ichnospecies and *R. voconcense* as a new combination of *Ronzotherichnus voconcense* by Costeur et al. (2009). *Rhonzotherichnus* has no validity and differs from *Rhinoceripeda* only by the type of preservation.

Ichnogenus *Ronzotherichnus* Demathieu et al., 1984.

Fig. 8.23I.

Type ichnospecies: *Ronzotherichnus voconcense* Demathieu et al., 1984.

Diagnosis: Mesaxonic imprint characterized by four impressions of different sizes: the hoof III with an elliptical shape whose long axis is perpendicular to that of the autopodium. Two others, the impressions of hoofs II and IV, approximately symmetrical in relation to the axis of the foot, elliptical in shape, are located behind the impression of the hoof III. For the hand, the plantar sole combined with the hoof V is represented by a trapezoidal trace of very variable length whose axis of symmetry merges with that of the autopodium. For the foot, the sole is subcircular. Hand and foot are of comparable sizes, In the trackways, the pace angulation is large (≥130 degrees) (translated from the French from Demathieu et al., 1984).

Discussion: Costeur et al. (2009) considered *Ronzotherichnus* as a different preservation of *Rhinoceripeda* and combined it as *Rhinoceripeda voconcense. Ronzotherichnus voconcense,* however, is poorly preserved and it seems that its materials are underprints.

Ichnogenus *Thrinaxopus* Sarjeant and Langston, 1994.
Fig. 8.23J.
Type ichnospecies: *Thrinaxopus hoplephoreus* Sarjeant and Langston, 1994.
Diagnosis: Plantigrade footprints showing four digits in the forefoot, with Digit V much reduced, and three digits in the hind foot. The pedal axes pass through Digit III on both feet, and all digits bear hooves. The whole palm and sole are impressed, as may be the wrist and ankle. The hooves of the manus are all relatively small; those of the pes are larger and compressed together, so that hoof III is in front of hooves II and IV (Sarjeant and Langston, 1994).
Discussion: Mesaxonic foot imprints with different sizes and arrangement of digit imprints have caused *Thrinaxopus* to be attributed to a rhinocerotoid (Sarjeant and Langston, 1994). *Thrinaxopus* has been reported only from the late Eocene of Texas.

4.4.3 *Carnivora and Creodonta*

4.4.3.1 Carnivora and Creodonta autopodium structure and tracks

Carnivora is divided into two major suborders (Flynn et al., 2010): Feliformia (including cats, linsangs, civets, mongooses, fossas, falanoucs, and hyaenas) and Caniformia (encompassing dogs, bears, seals, sea lions, walruses, the red panda, raccoons, skunks, weasels, badgers, otters, and wolverines). There are seven families of Feliformia including Felidae (cat-like species), Eupleridae (Malagasy carnivorans), Herpestidae (mongooses), Hyaenidae (hyenas), Nandiniidae (African palm civet), Prionodontidae (Asiatic linsangs) and Viverridae (civets and genets); and nine families of Caniformia as follows: Canidae (dog-like species), Ailuridae (red panda), Mephitidae (skunks and stink badgers), Mustelidae (weasels, badgers, martens, otters, etc.), Odobenidae (walrus), Otariidae (eared seals), Phocidae (earless seal), Procyonidae (raccoons, coatis, kinkajous, etc.) and Ursidae (bears).

Carnivorans are widespread as native terrestrial animals in all continents (except Australia and Antarctica) and show a wide range locomotion style in aquatic, semi-aquatic and nonaquatic environments, with cursorial, arboreal, fossorial, and swimming behaviors (Goswami, 2010).

Creodonta lived from the early Paleocene to the late Miocene epochs in North America, Europe, Asia and Africa, and, during the Paleogene, "creodont" species were the most abundant terrestrial carnivores in the Old World (Gunnell and Gingerich, 1991). Indeed, *Sarkastodon* is known as one of the largest mammalian land predators of all time (Sorkin, 2008). Creodonta comprises two families, Hyaenodontidae and Oxyanenidae, and many of these families are related to Carnivora but are not their direct ancestors (Muizon and Lange-Badré, 2007).

Limbs of Creodonta are mesaxonic (with the axis of the foot provided by the middle of their five digits), and their method of locomotion ranged from plantigrade to digitigrade (Fig. 8.24A). The tips of phalanges comprise claws (Gunnell, 1998).

Usually, carnivoran fossils have been reported from the whole of the Cenozoic, and the oldest known fossils of Carnivora have been found in the late Paleocene; they belong to the extinct families Miacidae and Viverravidae (Goswami, 2010; Spaulding and Flynn, 2012). However, the earliest fossil representatives of the living families of Carnivora appeared in the late Eocene. Size of terrestrial carnivores differs from less than a decimeter in *Mustela nivalis* (weasels) as the smallest member of the order Carnivora (Zub et al., 2009), to gigantic species, *Arctotherium angustidens*, from the early Pleistocene of Argentina (Soibelzon and Schubert, 2011).

The forefeet of carnivores always have five digits, and often Digit I is short and vestigial (the dewclaw), which is not used in walking and usually is not in contact with the ground, however, digits are equipped with a hooked claw that is used for predation. There are three small carpal bones (scaphoid, semilunar, and central) that merge into the single bone ("scapholunar") of the wrist. The hind feet of carnivores can have four or five digits depending on the gait adopted by the species. Plantigrade species have five digits and digitigrade species four. Felids and canids, as cursorial animals, are digitigrade; meanwhile Ursidae and Procyonidae have legs suitable to their plantigrade gait (Bellani, 2019) (Fig. 8.24C and D).

Eupleridae is an endemic Madagascar carnivore family with seven genera comprising 10 living species, commonly known as Malagasy mongooses or Malagasy carnivorans. Feet of Eupleridae include five digits in both manus and pes with distinct claws and different size pads on the sole. For example, *Salanoia durrelli* has broad feet, and large naked palmar and plantar surfaces, with large foot pads. On the fore feet, palmar pads three and four below Digits III to V and the hypothenar pad are large. On the hind feet, plantar pads 1, three and 4, below Digits I and III to V are large, and the thenar and hypothenar pads are elongated. Five digits are present on fore and hind feet with long claws (Durbin et al., 2010) (Fig. 8.25A).

Herpestidae is a clade of feliforms, with relatively small size and mostly carnivorous feeding members that are known as snake killers. Herpestidae are terrestrial animals with different habits ranging from solitary living in some members to complex social systems in mongooses, semifossorial in meerkats, or semi-aquatic, in marsh mongoose. The fore and hindlimbs of the Herpestidae comprise five slender digits with relatively exposed claws. In the mongooses the size of the feet differs in manus and pedes. Their claws are moderately long or very long, curved to a comparatively small extent. The digital pads are small. The plantar pad is well developed, cushion-like and trilobed, and the area between it and the digital pads, whether webbed or not, is naked. The hallux and pollex, when present, are short or very short and set above the plantar pad; and the hallucal and pollical lobes of the plantar pad are small or obsolete, and detached from the plantar pad. A single or double carpal pad is always present and separated by a naked tract from the plantar pad (Pocock, 1916) (Fig. 8.25B).

The forefeet of the African palm civet, *Nandinia*, show developed webs at least to the proximal ends of the digital pads, in the extent to which the toes are capable of separation and in the confluence of the plantar and carpal pads to form a single mass, of which the component elements are defined merely by grooves. A very narrow area of skin, sometimes giving off short streaks toward the digital pads of the II, III, IV and V digits, is naked. The claws are retractile but are not guarded by small skin lobes. The hind feet, so far as the webbing and the structure of the plantar pad are concerned, differ from those mentioned in connection with the fore foot. The digital pads, however, of the third and fourth digits are widely separated (Pocock, 1915) (Fig. 8.25C).

Prionodontidae include two species as spotted linsang (*Prionodon pardicolor*) and banded linsang (*Prionodon linsang*), living in the southeast Asia, were previously classified within the Viverridae, but diverged from *Panthera* and *Felis* ca. 33.3 million years (Myr) ago, and are considered "living fossils" (Gaubert and Vèron, 2003; Jennings and Veron, 2015).

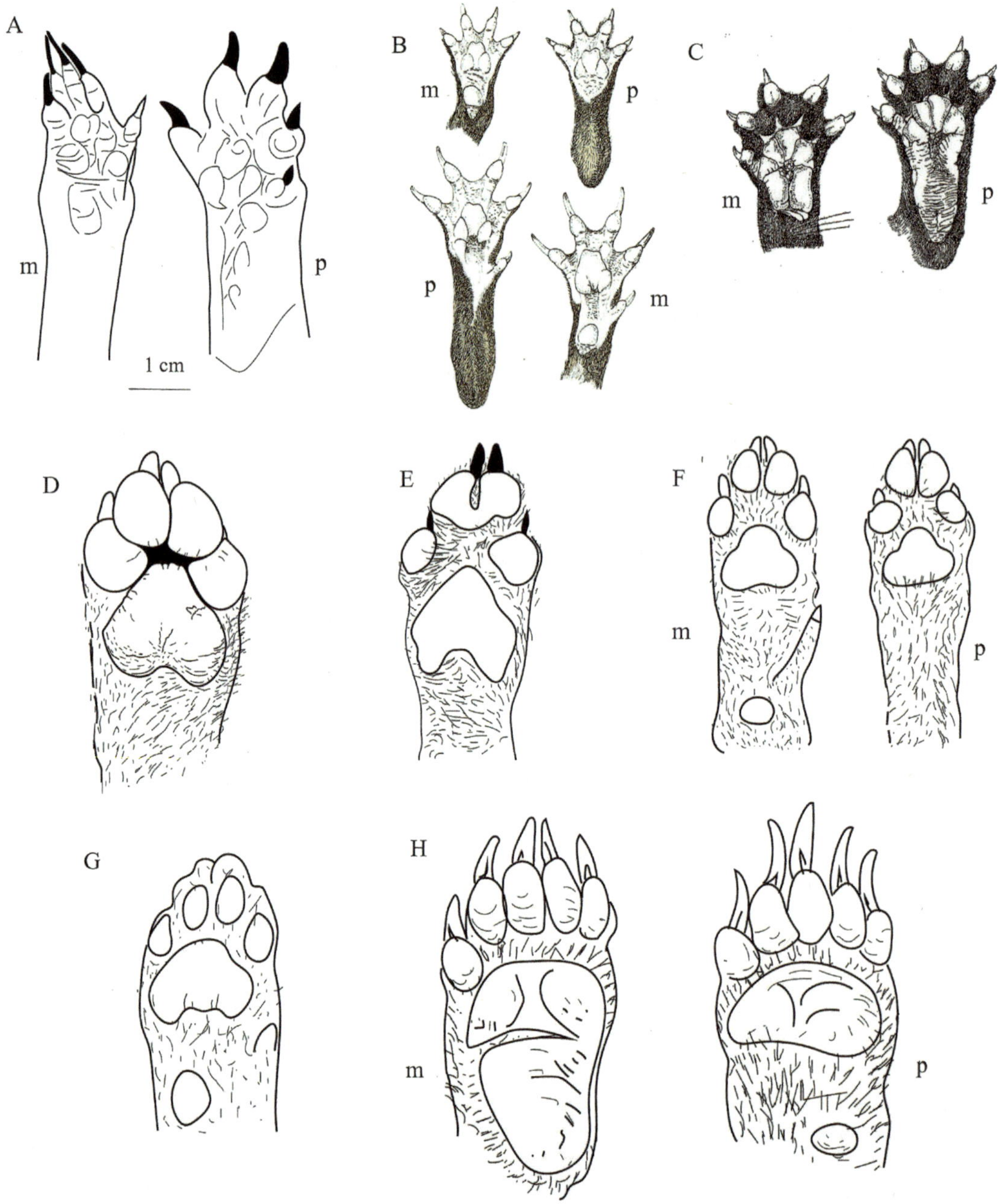

FIGURE 8.25 Morphology of the sole of the foot of the some live carnivores, (A) fore (m) and hind (p) feet of *Salanoia durrelli* (Eupleridae), (B) feet of *Bdeogale puisa* (top) and *Ichneumia albicauda* (lower), both mongooses, (C) feet of *Nandinia binotata* (Nandiniidae), Pocock (1915); (D) hind foot of Hyaena; (E) fore foot of wild dog; (F) feet of dog; (G) foot of tiger, (H) feet of black bear; *m*, manus and *p*, pes. All not to scale. *(A) Modified from Durbin et al. (2010). (B) Modified from Pocock (1916). (D) Modified from https://www.animalspot.net/hyena. (E) Modified from https://africageographic.com. (H) Modified from https://drawingref.com/bear-paws/.*

Prionodontidae members are arboreal and have a plantigrade forefoot with a relatively short metacarpal (Taylor, 1988).

Hyaenidae includes diverse and abundant fossil forms during the last 25 million years, and were most diverse in the late Miocene (Turolian) (Werdelin and Solounias, 1991). Hyaenas have four digits in both front and hind feet, with large, slightly elongated and closely connected paws and with a deep back depression and two metacarpal pads (Fig. 8.25D). Hyaena feet comprise nonretractable, blunt claws, so that the tips of the claws touch the ground.

Families of Caniformia have diverse feet anatomy in toe number, size and shape of sole pads or digit claws, and digitigrade or plantigrade manner of walking. The soles of the feet of canids are thickly furred, and the toe pads are completely covered by hair with four digits in both fore and hind feet (Fig. 8.25E and F). Canids are digitigrade, and the tarsal and carpal regions never touch the ground during walking. There are four functional digits on each limb, and the two middle digits are considerably longer than the rest. On the forelimb the first digit is vestigial and is represented by a claw and small pad higher up on the foot than the four functional pads (in the African Wild Dog this digit is totally absent). Unlike in the Felidae, the claws are not protractile and are relatively blunt and not sharply compressed laterally. The last phalanges are not capable of flexing far back and upward and have normal articulating surfaces (Castelló, 2018). Thus, the ungual phalanges commonly touch the ground.

Felidae is abundant and diverse and includes eight major lineages with 41 extant species belonging to 14 genera and divided into 92 subspecies, and originated in the late Miocene, inhabiting all the continents except Antarctica (Johnson et al., 2006). Felidae have four digits in the feet, and the paw of the fore foot is slightly larger than the hind foot (Fig. 8.25G). They have retractable claws with ellipsoidal digit pads and three lobate metacarpal pads, with two furrows between metacarpal pads.

Red pandas (Ailuridae) are the sister clade to other musteloids, and appeared in the Middle Miocene of Europe and Early Pliocene of North America, but were extinct in both regions by the Late Pliocene and are currently only found in Asia. They are plantigrade with five digits in both fore and hind feet with nonretractable claws. It is interesting that the name panda means claw, and in Nepali the root is "paja" and in Persian it is "panjeh," which means claw.

Mephitidae and Mustelidae are small carnivores with skeletons that tend to be long bodied and short legged. Foot structure ranges from semi plantigrade to digitigrade, and five toes are present on each foot (Baskin, 1998).

Odobenidae (walrus), Otariidae (eared seals), Phocidae (earless seal) are members of Pinnipeds, belonging to the clade Caniformia, which are characterized as fin-footed, semiaquatic, and mostly marine mammals. Pinnipeds are closely related to musteloids and earliest fossil pinniped, reported from the early Miocene of western North America (Hammond et al., 2012; Rybczynski et al., 2009).

Procyonidae, distributed in South and North America, appeared from the Late Miocene to Recent. Morphofunctionally, forelimbs of procyonids participate not only in locomotion (including swimming and/or climbing), but also in other activities such as digging and foraging, including prey capture and manipulation (Tarquini et al., 2017).

Ursidae is a large mammal carnivore family ranging back to the late Eocene and includes three groups: Amphicynodontinae (late Eocene to the early Oligocene), Hemicyoninae (early Oligocene to the late Pliocene) and Ursinae (from the Early Miocene).

The limbs of the Ursidae are strong, short, and plantigrade. Bears usually have slow locomotion; they can stand and walk on their hind limbs, resting on their entire foot. The forelimbs are used for climbing and manipulating prey. The hind limbs are shortened relative to the front ones or are of the same length. Fore and hind feet are wide, with five digits almost equal in size. Ursidae are plantigrade with long claws. The digit pads are arranged as slightly curved lines, with the pad of Digit I slightly shifted back relative to the pad of Digit II (Fig. 8.25H). The plantar pad is large and wide, its width is almost two times its length.

4.4.3.2 Carnivora and Creodonta ichnotaxa

Numerous ichnotaxa have been named and introduced for carnivorans since the 1960s. Different morphologic features were considered as ichnotaxobases for these footprints. Melchor et al. (2019) suggested that features that are envisaged as generic ichnotaxobases for canid and felid footprints, include: (1) number of digital pads, (2) position of the foot (grady), (3) footprint length/width ratio (distinguishing between values about 1, > 1 or < 1), (4) presence/absence of claw marks, (5) outline and degree of asymmetry of the metapodial pad impression, (6) shape and relative size of digital pad impressions and, (7) degree of separation between digital pad and metapodial pad impressions. Most canid ichnotaxa have been erected based on a single or a couple of footprints.

Here, we introduce and discuss on the ichnogenera assigned to Carnivoran, the ichnotaxa marked by symbol ☒ have change(s) in nomenclature or ichnotaxonomic position:

Axiciapes Sarjeant and Langston, 1994
- *Axiciapes curvidigitatus* Sarjeant and Langston, 1994
- *Axiciapes ferox* Sarjeant and Langston, 1994

Bestiopeda (Vyalov, 1965)
- ☒ *Bestiopeda amphicyonides* Thenius, 1967
- ☒ *Bestiopeda böckhi* Vyalov, 1985
- *Bestiopeda gracilis* Vyalov, 1965
- *Bestiopeda guloides* Thenius, 1967
- ☒ *Bestiopeda hungarica* Vyalov, 1985
- ☒ *Bestiopeda maxima* Kordos, 1985
- ☒ *Bestiopeda sanguinolenta* Vyalov, 1966
- ☒ *Bestiopeda turkomanica* Vyalov, 1983

Canipeda (Panin and Avram, 1962)
- ☒ *Canipeda amphicyonides* (Thenius, 1967) by Sarjeant et al. (2002), not valid
- *Canipeda gracilis* Vyalov, 1966
- *Canipeda longigriffa* Panin and Avram, 1962
- *Canipeda sanguinolenta* Vyalov, 1966
- ☒ *Canipeda turkomanica* (Vyalov, 1983), *numen nudum* by Abbassi (2022).

Carnotipus López et al., 1992
- *Carnotipus solitarius* López et al., 1992

Carnivoripeda Kordos, 1985

Carnivoripeda nogradensis Kordos, 1985
Carnivoripeda sudamericana Krapovickas et al., 2023
Creodontipus Santamaria et al., 1989
☒ *Creodontipus mongayensis* (Santamaria et al., 1989) is synonym of *Canipeda gracilis* (Vyalov, 1965).
Creodontipus almenarensis Santamaria et al., 1989
☒ *Chelipus* Sarjeant and Langston, 1994 is synonym of *Canipeda*
☒ *Chelipus gracilis* Vyalov, 1965is new combination as *Canipeda gracilis*
☒ *Chelipus therates* Remeika, 1999 in new combination *Canipeda gracilis*
Dischidodactylus Sarjeant and Wilson, 1988
Dischidodactylus stevensi Sarjeant and Wilson, 1988
Falcatipes Sarjeant and Langston, 1994
Falcatipes floriformis Sarjeant and Langston, 1994
Felipeda (Vyalov, 1966)
Felipeda bestia Vyalov, 1966
Felipeda biancoi Aramayo and Manera de Bianco, 1987
Felipeda bottjeri Sarjeant et al., 2002
Felipeda dugesii Rodríguez-de la Rosa and Guzmán-Gutiérrez, 2012, new comb. from *Mitsupes dugesii*
Felipeda felis Panin, 1965
Felipeda lynxi Panin and Avram, 1962
☒ *Felipeda maxima* Kordos, 1985 was combined as *Platykopus maxima* (Kordos 1985) by Botfalvai et al. (2023).
Felipeda milleri Remeika, 1999
Felipeda parvula Anton et al., 2004
☒ *Felipeda sanguinolenta* Vyalov, 1966 belong to *Canipeda sanguinolenta* by Melchor et al. (2019).
Felipeda scrivneri Sarjeant et al., 2002
Hyaenodontipus Ellenberger, 1980
Hyaenodontipus praedator Ellenberger, 1980
Hirpexipes Sarjeant et al., 2002
Hirpexipes alfi Sarjeant et al., 2002
Khafipus Abbassi et al., 2015
Khafipus khadari Abbassi et al., 2015
☒ *Mitsupes* Rodríguez-de la Rosa and Guzmán-Gutiérrez, 2012, synonym *Felipeda*
☒ *Mitsupes dugesii* Rodríguez-de la Rosa and Guzmán-Gutiérrez, 2012
Mustelipeda Kordos, 1985
Mustelipeda punctata Kordos, 1985
☒ *Pantheraichnus* Diedrich, 2011b
☒ *Pantheraichnus bottropensis* Diedrich, 2011b synonym of the *Felipeda biancoi* Aramayo and Manera de Bianco, 1987.
Phacelopus Sarjeant and Langston, 1994
Phacelopus therates Sarjeant and Langston, 1994
Platykopus Sarjeant et al., 2002
Platykopus ilycalcator Sarjeant et al., 2002

☒ *Platykopus maxima* Kordos 1985 by Botfalvai et al. (2023).
Platykopus stuartjohnstoni Lucas and Schultz, 2007
Pycnodactylopus Sarjeant et al., 2002
Pycnodactylopus achras Sarjeant et al., 2002
Quiritipes Sarjeant et al., 2002
Quiritipes impendens Sarjeant et al., 2002
Sarjeantipes McCrea et al., 2004
Sarjeantipes whitea McCrea et al., 2004
Sarcotherichnus Demathieu et al., 1984
Sarcotherichnus enigmaticus Demathieu et al., 1984
Tetrastoibopus Sarjeant and Langston, 1994
Tetrastoibopus phoros Sarjeant and Langston, 1994
Ursichnus (Diedrich, 2011a)
☒ *Ursichnus amphicyonides* (Thenius, 1967), new combination here
Ursichnus europaeus Diedrich, 2011a
Ursichnus sudamericanus Aramayo et al., 2015
Zanclonychopus Sarjeant and Langston, 1994
Zanclonychopus cinicalcator Sarjeant and Langston, 1994

4.4.3.2.1 Description and discussion of ichnogenera

Ichnogenus *Axiciapes* Sarjeant and Langston, 1994.

Fig. 8.26A and B.

Type ichnospecies: *Axiciapes ferox* Sarjeant and Langston, 1994).

Diagnosis: ?Digitigrade or semiplantigrade to plantigrade footprints, with five digits represented in the forefoot and four (II to V) in the hind foot. Digits of manus widely spread; claws incompletely impressed and probably fully or partly retractile. Digits of pes with claws fully extended; II and III pointing forward, IV and V outward (Sarjeant and Langston, 1994).

Discussion: Sarjeant and Langston (1994) described *Axiciapes* from the late Eocene of Presidio County (west Texas, USA) and discussed the trackmaker of *Axiciapes*. The ichnogenus *Axiciapes* was tentatively attributed to the extinct family of terrestrial carnivorans, Amphicyonidae (genus *Daphoenus*), based on the size and morphological features. Amphicyonids are a common element of Cenozoic carnivoran faunas, appeared in North America in the middle Eocene, and spread to Europe, Asia and Africa during the late Eocene to Early Miocene and became extinct in the latest Miocene (Viranta, 1996; Hunt, 1998). *Axiciapes* comprises two ichnospecies. The first is *Axiciapes ferox,* which has five digits on the manus and only four on the pes, and the second is *Axiciapes curvidigitatus* that is characterized by a strong inward curvature of Digits I to IV.

Ichnogenus *Bestiopeda* (=*Felipeda*) (Vyalov, 1965).

Fig. 8.27A.

Type ichnospecies: *Bestiopeda bestia* (Vyalov, 1965).

Diagnosis: Plantigrade to semiplantigrade footprints exhibiting four digits (II to V), each with a spheroidal to ovoidal or elongate digital pad forming a semicircle in front of, or about the front portion of, sole or heel pads. Digital pads are of equal or similar size. Impressions of claw tips may be present but are usually absent (emended by Sarjeant and Langston, 1994).

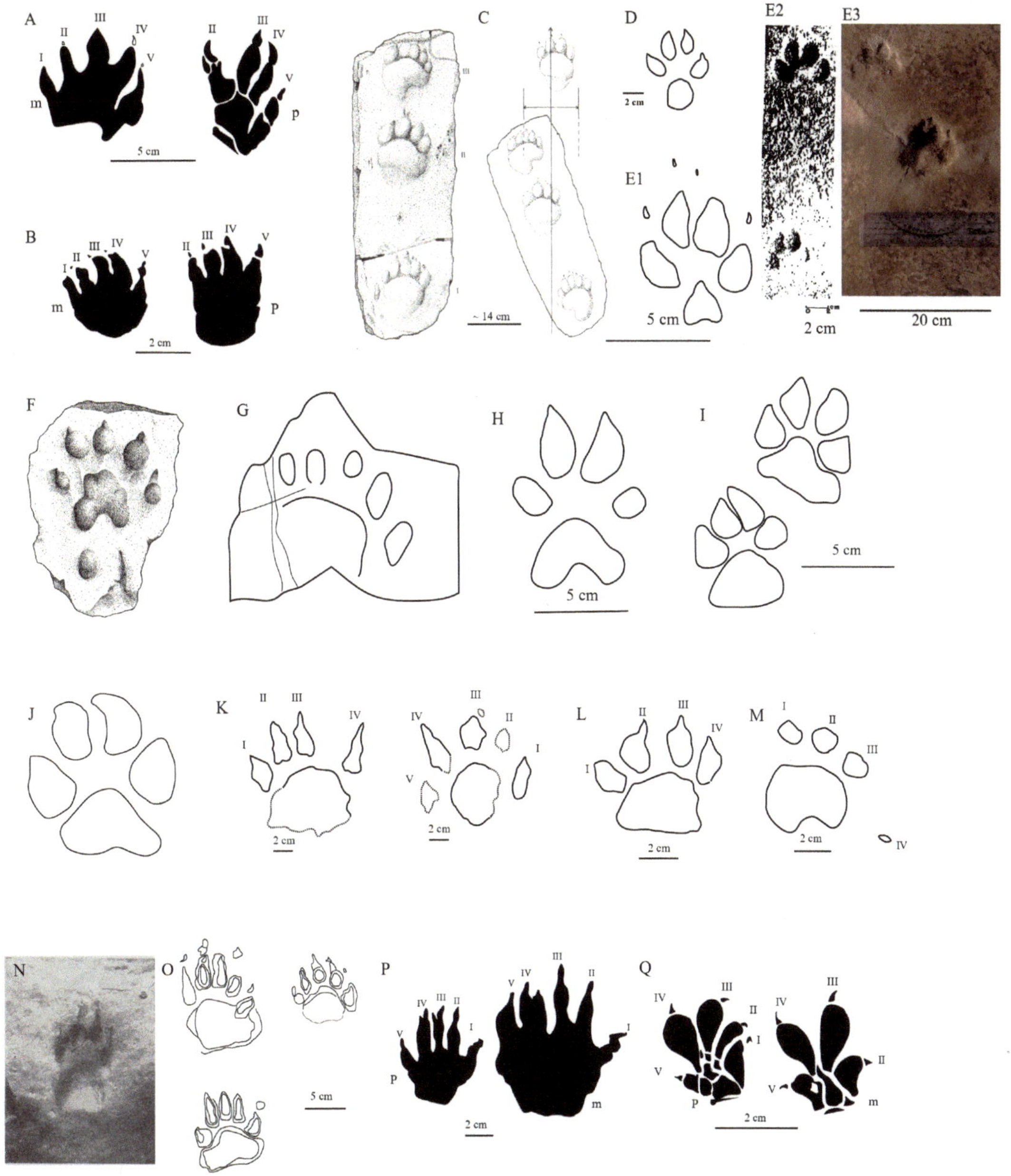

FIGURE 8.26 Modified outline sketches and photographs of the Carnivora ichnotaxa, (A) *Axiciapes ferox* Sarjeant and Langston, 1994; (B) *Axiciapes curvidigitatus* Sarjeant and Langston, 1994; (C) *Ursichnus amphicyonides* Thenius, 1967 (= *Bestiopeda amphicyonides* Thenius, 1967) revised here, as new combination; (D) *Felipeda gracilis* (Vyalov, 1966) (= *Bestiopeda gracilis*); (E) *Pehuencoichnum gracilis* (Aramayo and Manera de Bianco, 1987) synonym of *Canipeda gracilis* (Vyalov, 1965) by Melchor et al. (2019); E1, E2 from Aramayo and Manera de Bianco, 1987 and E3 photograph by Ricardo N. Melchor; (F) *Bestiopeda guloides* Thenius, 1967; (G) *Platykopus maxima* (Kordos 1985) (= *Bestiopeda maxima* Kordos, 1985, new combination by Botfalvai et al., 2023), (with permission); (H) *Canipeda sanguinolenta* Vyalov, 1966 (= Bestiopeda *sanguinolenta*) new combination by Melchor et al., 2019; (I) *Canipeda turkomanica* (Vyalov, 1983) (= *Bestiopeda turkomanica*), new combination by Abbassi, 2022; (J) *Chelipus therates* Remeika, 1999, nomen nudum, (= *Canipeda gracilis*); (K) *Carnotipus solitarius* López et al., 1992; (L) *Creodontipus mongayensis* (Santamaria et al., 1989); (M) *Creodontipus almenarensis* (Santamaria et al., 1989); (N) *Canipeda longigriffa* (Panin and Avram, 1962), (with permission); (O) *Carnivoripeda nogradensis* Kordos, 1985; (P) *Dischidodactylus stevensi* Sarjeant and Wilson, 1988; (Q) *Falcatipes floriformis* Sarjeant and Langston, 1994. *(C), (F) and (N) With permission, (E) Modified from Melchor et al., 2019.*

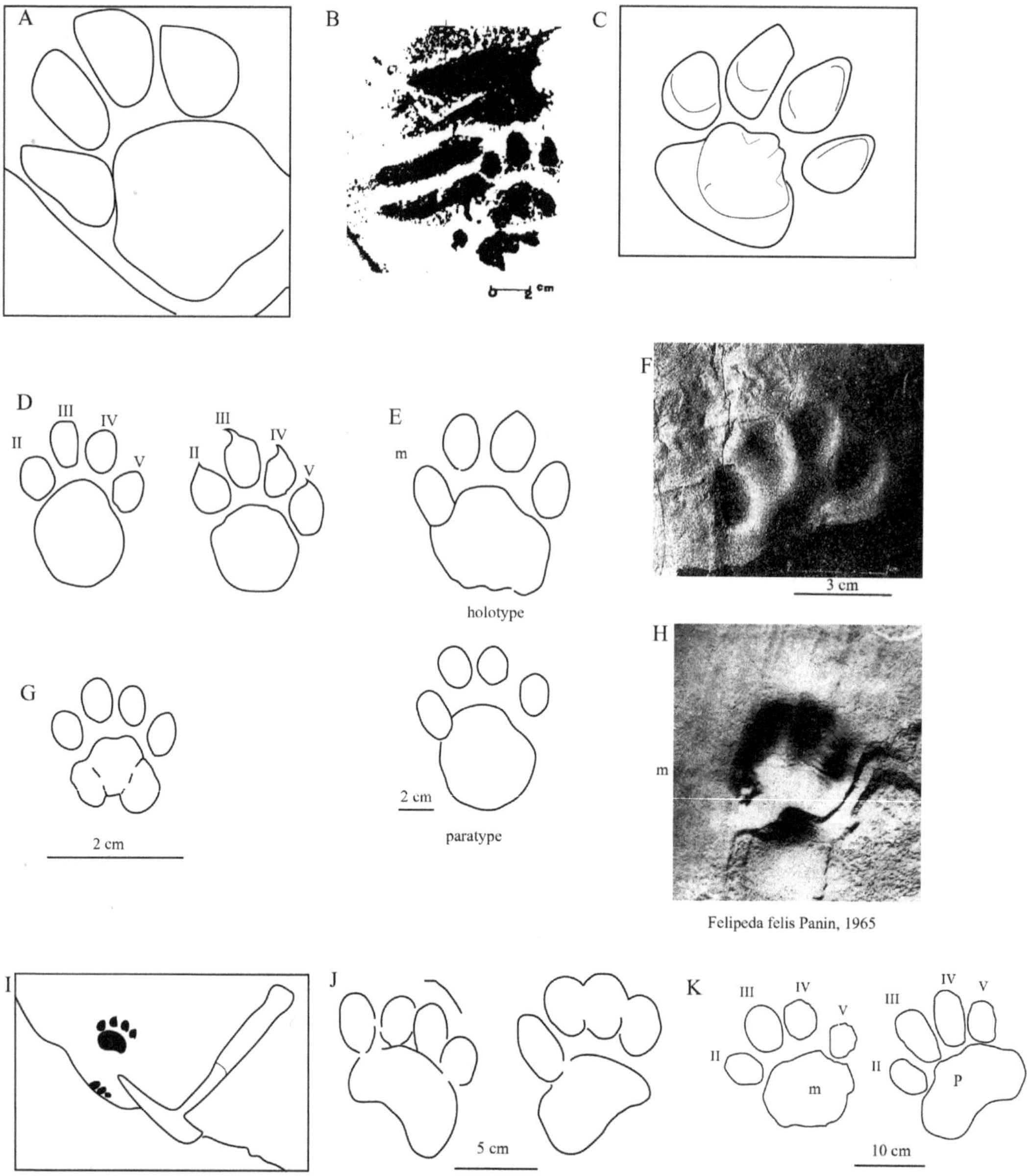

FIGURE 8.27 Modified outline sketches and photographs of the Carnivora ichnotaxa, (A) *Bestiopeda* (=*Felipeda*) *bestia* (Vyalov, 1966); (B) *Felipeda biancoi* (Aramayo and Manera de Bianco, 1987) (=*Pumaeichnum biancoi* Aramayo and Manera de Bianco, 1987 by Sarjeant et al., 2002); (C) *Felipeda milleri* (=*Pumaeichnum milleri* Remeika, 2001 by Sarjeant et al., 2002); (D) *Felipeda bottjeri* Sarjeant et al., 2002 (with permission); (E) *Felipeda scrivneri* Sarjeant et al., 2002; (F) *Felipeda lynxi* Panin and Avram, 1962 (with permission); (G) *Felipeda parvula* Anton et al., 2004; (H) *Felipeda felis* Panin, 1965 (with permission); (I) *Pumaeichnum stouti* synonym of *Felipeda biancoi*, (J) *Mitsupes dugesii* Rodríguez-de la Rosa and Guzmán-Gutiérrez, 2012, (K) *Pantheraichnus bottropensis* (Diedrich 2011) is a synonym of *Felipeda biancoi* (Aramayo and Manera de Bianco, 1987). *Modified from Remeika, 2006.*

Discussion: Vyalov (1965, 1966) proposed the single name *Bestiopeda* for the footprints of predators, and his description comprises canids or felids. Thus, this name was used by other researchers for all carnivorans footprints with different morphologies. Vyalov (1965) described *Bestiopeda* to include both tetradactyl and pentadactyl footprints in pes and manus imprints, so *Bestiopeda* involves diverse morphotypes of carnivoran footprints. Differentiation between these footprints started by Kordos (1985) by proposing *Carnivoripeda* to refer to homopodial pentadacyl, carnivore footprints, that differ from *Bestiopeda* as a homopodial tetradactyl carnivore footprint (Melchor et al., 2019). Sarjeant and Langston (1994) restricted the name *Bestiopeda* to felid footprints because of the absence of claw imprints and considered *Pumaeichnum* as a junior synonym of *Bestiopeda*. Melchor et al. (2019) summarized key features for distinct ichnogenera, *Canipeda, Creodontipus, Felipeda, Quiritipes, Tetrastoibopus*, and *Bestiopeda*: (1) *Canipeda* presents footprints with a large metapodial pad and smaller digital pad impressions with claw marks. Footprint longer than wide, digital pads of similar size and metapodial pad symmetrical and well separated from digital pads, and manus tends to be larger than pes, small to intermediate claw marks. (2) *Creodontipus* has footprints with a large metapodial pad and smaller digital pad impressions, lacking claw marks, with wider than long, acuminate digital pads. (3) *Felipeda* is characterized by footprints with a large metapodial pad and smaller digital pad impressions, lacking claw marks wider than long, subcircular digital pads. (4) *Quiritipes* are footprints with a large metapodial pad and smaller digital pad impressions without claw marks, footprint longer than wide, digital pads of similar size and metapodial pad symmetrical and well separated from digital pads and manus tends to be smaller than pes, large claw marks. (5) *Tetrastoibopus* are footprints with a large metapodial pad and smaller digital pad impressions with claw marks and about as long as wide, lateral digits (II, V) larger than the central ones (III–IV). (6) *Bestiopeda* presents footprints with a large metapodial pad and smaller digital pad impressions with claw marks, and footprint longer than wide, digital pads of similar size and metapodial pad asymmetrical, massive footprint.

Bestiopeda involve numerous ichnospecies, the validity and synonymy of which were discussed by several authors, so that:

- *Bestiopeda amphicyonides* (Fig. 8.26C) combined as *Canipeda amphicyonides* by Sarjeant et al. (2002) but *B. amphicyonides* is pentadactyl and differs from *Canipeda*, and the combination as *Canipeda amphicyonides* is not valid. *Bestiopeda amphicyonides* is a heteropodous trackway with pentadactyl manus and tetradactyl pes, so we consider *Ursichnus amphicyonides* a new combination here (see ichnogenus *Ursichnus* Diedrich, 2011a).
- *Bestiopeda bestia* was combined under *Felipeda* by Sarjeant et al. (2002).
- *Bestiopeda böckhi* is a junior synonym of *B. maxima* (Kordos, 1985).
- *Bestiopeda gracilis* proposed by Vyalov (1965). *Bestiopeda gracilis* was revised as *Chelipus gracilis* by Sarjeant and Langston (1994) and then *Felipeda gracilis* by Sarjeant et al. (2002). *Pehuencoichnum* was originally described as lacking claw marks and metapodial pad impressions, but reexamination of original photographs by Melchor et al. (2019) suggests that most of the footprints do display claw marks, making it indistinguishable from *Canipeda*. Metapodial impressions are shallow but can be distinguished. Thus, *Pehuencoichnum gracilis* (Fig. 8.26E) was synonymized under *Canipeda gracilis* (Vyalov, 1965) by Melchor et al. (2019).

- *Bestiopeda guloides* (Fig. 8.26F) was attributed to a mustelid-like impression. Costeur et al. (2009) mentioned an isolated pentadactyl footprint (named *Bestiopeda* sp.) from the Oligocene sediments of the Regional Natural Park of Luberon, France, showing resemblances in morphology and size with *Bestiopeda guloides*
- *Bestiopeda maxima* (Fig. 8.26G) was first considered as *Felipeda maxima* but recently it was combined as *Platykopus maxima* by Botfalvai et al. (2023).
- *Bestiopeda sanguinolenta* (Fig. 8.26H) is synonym of *Canipeda sanguinolenta* after Melchor et al. (2019) and is placed in *Felipeda* after Sarjeant et al. (2002).
- *Bestiopeda turkomanica* (Fig. 8.268I) is a synonym *Canipeda turkomanica* after Abbassi (2022).
- *Bestiopeda hungarica* is a junior synonym of *Carnivoripeda nogradensis.*
- *Bestiopeda tarnocensis* is a valid name for the tetradactyl carnivore footprints of Ipolytarnóc, which were characterized as *Bestiopeda* isp. in Kordos (1985).

Ichnogenus *Canipeda* Panin and Avram, 1962.
Fig. 8.26N.
Type ichnospecies: *Canipeda longigriffa* Panin and Avram, 1962.

Diagnosis: Digitigrade to semidigitigrade, tetradactyl, paraxonic, longer than wide footprints; arranged in quadrupedal and homopodial trackways. Elliptical, similar-sized and clawed digital pads form an arc in front of, and are well separated from, a large metapodial pad. Metapodial pad rounded to triangular or trapezoidal. *Felipeda* is distinguished by having footprints wider than long, lacking claw marks and with digital pad impressions that tend to be subcircular (emended by Melchor et al., 2019).

Discussion: Both *Canipeda* and *Felipeda* comprise tetradactyl footprints in quadrupedal and homopodial trackways. Thus, other ichnogenera and ichnospecies referred to *Canipeda* or *Felipeda* should be revised as belonging to other ichnogenera. Presence or absence of claw imprints, footprint length/width ratio, shape of digital pad impressions, and shape of metapodial pad, symmetry or asymmetry of digits, arrangements of toes as X, H or C forms in the front of metacarpal imprint, homopodial character and quadrupedality, as well as basic trackway features are the ichnotaxobase for distinction of *Canipeda* (Melchor et al., 2019). Melchor et al. (2019) discussed the ichnotaxonomic position of *Canipeda* with other carnivoran ichnotaxa. Based on revising the diagnosis of *Bestiopeda*, these ichnospecies separated from *Bestiopeda*. The footprints were revised for ichnogenus *Felipeda*, include *Bestiopeda amphicyonides, Bestiopeda gracilis, Bestiopeda sanguinolenta,* and *Bestiopeda turkomanica. B. amphicyonides* is pentadactyl and differs from *Canipeda,* thus this combination is not valid.

Vyalov (1983) introduced *Bestiopeda turkomanica* for tetradactyl canid footprints from the Pliocene of Turkmenistan. Based on the morphologies of the Turkmenistan footprints, which show large digit and metapodial pad imprints (Vyalov, 1983, fig. 4), Abbassi (2022) combined *Bestiopeda turkomanica* under *Canipeda* as the new combination *C. turkomanica.*

Ichnogenus *Carnotipus* López et al., 1992.
Fig. 8.26K.
Ichnospecies *Carnotipus solitarius* López et al., 1992.

Diagnosis: Large, paraxonic tetradactyl or mesaxonic pentadactyl footprints. The impression of third phalanges is elongated, with the distal end pointed. Digit imprint arrangement is more or less radial around the metacarpal/metatarsal pad, whose morphology is variable (translated here from the Spanish from López et al., 1992).

Discussion: *Carnotipus* comprises diverse morphology as tetradactyl or pentadactyl imprints. López et al. (1992) did not distinguish if the pes or manus is pentadactyl, however, larger imprints can be attributable to the manus imprint and are pentadactyl (López et al., 1992, Figs. 8.11 and 8.12). Botfalvai et al. (2023), however, considered *Carnotipus* as a junior synonym of *Carnivoripeda* (Kordos, 1985), based on many similarities in their diagnostic characters (e.g., size, elongated and radially orientated digital imprints, and claw impressions).

Ichnogenus *Carnivoripeda* (Kordos, 1985).

Fig. 8.26O.

Ichnospecies: *Carnivoripeda nogradensis* (Kordos, 1985).

Diagnosis: Plantigrade footprints with five elongated, sharply clawed digits on the manus and pes. Imprints of the manus and pes are longer than they are wide. Digits III–V point forwards, while Digits I and II outwards in the case of the manus. In the case of the pes, the central digits (II, III, IV) are parallel, and the outer digits (I and V) are offset, at an angle pointing slightly outwards. Digits II and IV of the pes curve toward Digit III. Imprints of claws are distinct. Digit III represents the largest digital imprint. The palm and sole are of greater width than length and form a relatively deep, kidney-shaped impression. The impressions of the hind foot imprints never cover the impressions of the forefoot (emended by Botfalvai et al., 2023).

Discussion: Kordos (1985) named *Carnivoripeda* for large pentadactyl footprints from the Miocene of the Ipolytarnóc track site of Hungary. He compared the size of *Carnivoripeda* with other tetra- and pentadactyl footprints such as *Felipeda lynx*, *Bestiopeda amphicyonides* and *Canipeda sanguinolenta*. *Bestiopeda amphicyonides* is revised here and considered as new combination of *Ursichnus amphicyonides* (see ichnogenus *Ursichnus* Diedrich, 2011a) and *Bestiopeda sanguinolenta* was combined as *Canipeda sanguinolenta*. Botfalvai et al. (2023) reexamined the holotype and other materials of *Carnivoripeda* of Ipolytarnóc and revised its diagnosis. Botfalvai et al. (2023) compared *Carnivoripeda nogradensis* with similar footprints including *Zanclonychopus cinicalcator*, *Dischidodactylus stevensi*, *Sarcotherichnus enigmaticus*, *Carnotipus solitarius* (which were considered junior subjective synonyms of *Carnivoripeda*), *Axiciapes ferox*, *Hirpexipes alfi*, *Ursichnus europaeus*, *Platykopus*, *Phacelopus therates*, *Bestiopeda guloides*, and *Mustelidichnum* (*M. enigmaticum*, and *M. vallecitoensis*; see Fig. 8.28D). *Carnivoripeda* remained monospecific for many years, but recently Krapovickas et al. (2023) introduced a second ichnospecies of *Carnivoripeda*, *C. sudamericana* from the middle Miocene of Argentina, as tetradactyl to pentadactyl homopod footprints with oval digital pad impressions separated from the metapodial pad. *C. sudamericana* is consistent with the overall features of *Carnivoripeda*, but they differ in terms of the shape of the digital impressions, which are oval, and the absence of claw marks (Krapovickas et al., 2023). In the description of *Carnivoripeda*, Botfalvai et al. (2023) described *Carnivoripeda* as pentadactyl homopod footprints, and it is considered that the assignment of *C. sudamericana* to *Carnivoripeda* is questionable. Morphological characters suggest that the *C. nogradensis* footprints belong to a mustelid-like carnivore.

Ichnogenus *Creodontipus* Santamaria et al., 1989.

Fig. 8.26L.

Type ichnospecies: *Creodontipus mongayensis* Santamaria et al., 1989.

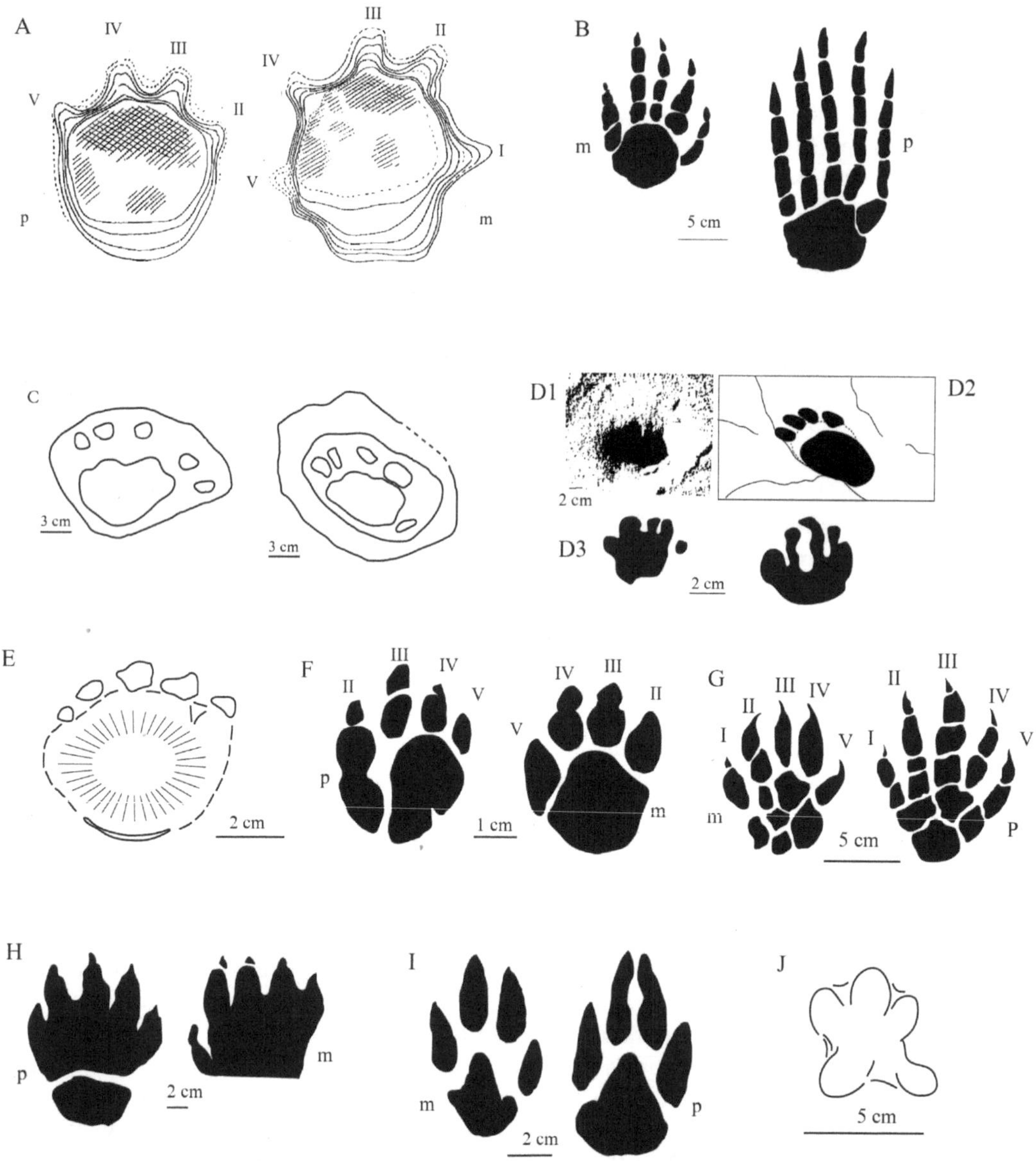

FIGURE 8.28 Modified outline sketches and photographs of the Carnivora ichnotaxa, (A) *Hyaenodontipus praedator* (Ellenberger, 1980); (B) *Hirpexipes alfi* (Sarjeant et al., 2002); (C) *Khafipus khadari* (Abbassi et al., 2015); (D1) *Mustelidichnum enigmaticum* (Aramayo and Manera de Bianco, 1987) (with permission); (D2) *Mustelidichnum vallecitoensis* (Remeika, 2001); (D3) *Mustelidichnum vallecitoensis* in Weems, 2018; (E) *Mustelipeda punctata* (Kordos, 1985); (F) *Pycnodactylopus achras* (Sarjeant et al., 2002); (G). *Phacelopus therates* (Sarjeant and Langston, 1994); (H) *Platykopus ilycalcator* (Sarjeant et al., 2002); (I) *Quiritipes impendens* (Sarjeant et al., 2002); (J) *Sarcotherichnus enigmaticus* (Demathieu et al., 1984); (K) *Tetrastoibopus phoros* (Sarjeant and Langston, 1994); (L) *Ursichnus europaeus* (Diedrich, 2011); (M) *Ursichnus sudamericanus* (Aramayo et al., 2015); (N) *Zanclonychopus cinicalcator* (Sarjeant and Langston, 1994). (O) *Sarjeantipes whitea* (McCrea et al., 2004).

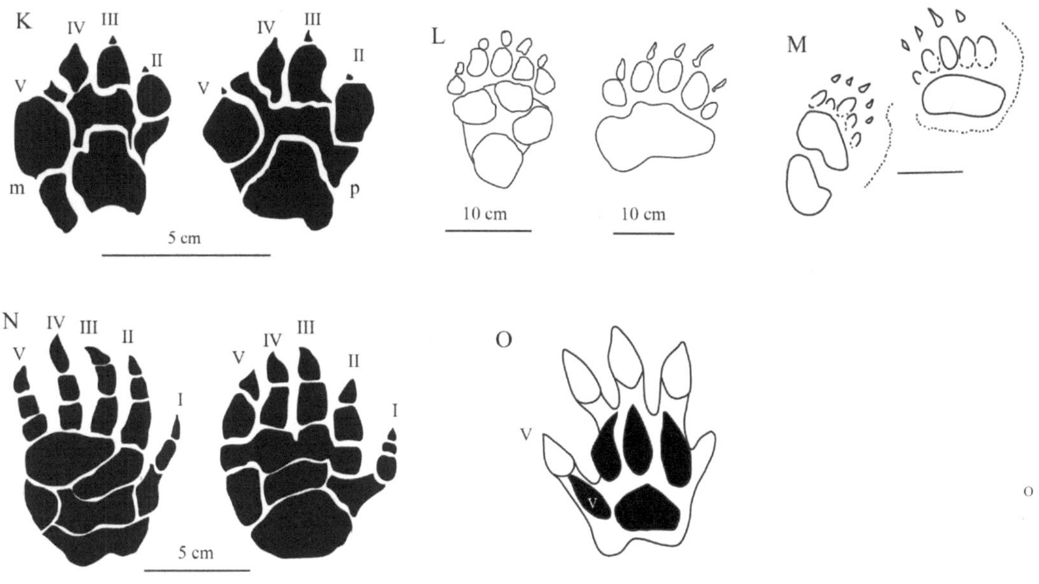

FIGURE 8.28 cont'd

Diagnosis: Medium sized, digitigrade tetradactyl footprints, three of the digits are round and the fourth digit imprint is small, completely separate from other digit imprints. Digit tips are pointed or rounded. Digit imprints are arranged radially in front of the metapodial pad. Metapodial pad is crescentic in shape and has anteriorly convex and posteriorly concave or rectilinear outlines (Santamaria et al., 1989, Translated from Spanish).

Discussion: Santamaria et al. (1989) established *Creodontipus* from the Oligocene of Spain for tetradactyl homopodial manus and pes prints with two ichnospecies, *Creodontipus mongayensis* and *Creodontipus almenarensis*. The type and only known material are a couple or a single footprint, respectively. *Creodontipus mongayensis* are imprints with relatively elongate digit imprints and sharp tips. *Creodontipus mongayensis* has a slightly broad metapodial imprint but it is synonym of *Canipeda gracilis* (Fig. 8.26D). *Creodontipus almenarensis* has a completely separated Digit IV imprint (Fig. 8.26M). *Creodontipus* is distinguished from *Carnivoripeda nogradensis*, *Platykopus maxima* (= *Bestiopeda maxima* and *Felipeda maxima*) and *Carnotipus solitarius* by the absence of the imprint of Digit V. *Creodontipus almenarensis* is distinct from *Pantheraichnus bottropensis* by size and the separate position of the Digit IV imprint. Santamaria et al. (1989) attributed *Creodontipus* to Creodonta and likely to Hyaenodontidae.

Ichnogenus *Chelipus* Sarjeant and Langston, 1994.

Type ichnospecies: *Chelipus gracilis* (Vyalov 1965).

Diagnosis: Plantigrade to semiplantigrade footprints exhibiting four digits (II to V), each with a spheroidal to ovoidal or elongate digital pad and each giving rise to a prominent claw (permanently extended). The digits form a semicircle in front of, or about the front portion of, sole or heel pads. Digital pads are of equal or similar size. The trackway narrow; stride long (Sarjeant and Langston, 1994).

Discussion: Sarjeant and Langston (1994) considered *Bestiopeda* as felid footprints based on Vyalov's (1965) inference that claws are normally retracted in *Bestiopeda*, and separated *Bestiopeda gracilis* in a new ichnogenus *Chelipus* for canid-like footprints, which later was called *Felipeda gracilis* by Sarjeant et al. (2002). Based on the diagnosis provided for *Chelipus* and the selection of the type ichnospecies by Sarjeant and Langston (1994), *Chelipus* is considered a junior subjective synonym of *Canipeda* (Sarjeant et al., 2002; Melchor et al., 2019). Abbassi and Amini (2008) reported poorly preserved felid footprints from the Miocene Upper Red Formation of the Eyvan-e-Key section in east Tehran, Iran, as *Chelipus* isp. Based on morphology and the lack of claw imprints, these reports are revised as *Felipeda* isp.

Remeika (1999) considered a tetradactyl footprint with large pad digit imprints and attributed it to extinct coyote and named it as *Chelipus therates* (Fig. 8.26J), but it was considered as *Canipeda* and as a synonym of *Canipeda gracilis* (Vyalov 1965) by Melchor et al. (2019).

Ichnogenus *Dischidodactylus* Sarjeant and Wilson, 1988.

Fig. 8.26P.

Type ichnospecies: *Dischidodactylus stevensi* Sarjeant and Wilson, 1988.

Diagnosis: Plantigrade mammalian footprints, with pes larger than manus. Pollex and hallux shortened and offset; fifth digit of manus reduced. The fourth digit of pes divided distally and bearing two claws; third and fourth digits of manus sometimes also distally divided, with or without doubling of claws (Sarjeant and Wilson, 1988).

Discussion: *Dischidodactylus* has been named for pentadactyl footprints from the late Eocene (Duchesnean) of Trans-Pecos Texas (Sarjeant and Wilson, 1988; Sarjeant et al., 2002). *Dischidodactylus* is monospecific, but Sarjeant and Wilson (1988) proposed a separate diagnosis for the only ichnospecies *Dischidodactylus stevensi.* On the contrary, the diagnosis of *D. stevensi* must be summarized and combined with the ichnogenus diagnosis. Sarjeant and Wilson (1988) believed that *Simidectes* is most probably the trackmaker of *Dischidodactylus*, therefore *Dischidodactylus* is not a carnivoran footprint. Sarjeant and Wilson (1988) mention an extramorphological feature of the pes imprint as fourth digit of pes divided distally, and this may be related to repetition of Digit IV imprints during walking, anatomical polydactyly of the trackmaker or overprint of more than one footprint.

Ichnogenus *Falcatipes* Sarjeant and Langston, 1994.

Fig. 8.26Q.

Type ichnospecies: *Falcatipes floriformis* Sarjeant and Langston, 1994.

Diagnosis: Small, semidigitigrade footprints with prominent phalangeal pads, broadest medially or distally. The presumed pes is larger than the presumed manus and has only four digits (II to V). Digits II to IV curve inward. The presumed manus has five digits, with Digit I reduced and inconspicuous; Digits I to IV curve inward. All digits bear sharp claws, those of Digits I to IV of the manus and II to IV of the pes curving inward (Sarjeant and Langston, 1994).

Discussion: The ichnogenus *Falcatipes*, displaying an exclusive morphology, is a small footprint with distinct digit pads and claws. It is only known from four moderately preserved and deep tracks and it is not clear if the particular morphology is due to substrate properties (volcanic ash) at the time of impression. It was tentatively attributed to Late Eocene rodents or miacid carnivores.

Ichnogenus *Felipeda* (Panin and Avram, 1962).

Fig. 8.27F.

Type ichnospecies: *Felipeda lynxi* (Panin and Avram, 1962).

Diagnosis: Plantigrade to semiplantigrade footprints exhibiting four digits (II to V), each with a spheroidal to ovoid or elongate digital pad. The pads form a semicircle in front of, or in about the front portion of, the metatarsal pads. Digital pad III is often the most anterior, and there never appears to be lateral symmetry on an anterior/posterior (AP) line between digital pads III and IV. Digital pads may be of equal or similar size or may show limited dimensional variation. Impressions of claw tips may be present but are usually absent (emended by Sarjeant et al., 2002).

Discussion: *Felipeda* is a common carnivore footprint in the Neogene record, which is known as tetradactyl footprints with round digit pads without claw imprints. Based on these features, *Felipeda* is distinguished from *Bestiopeda* or *Canipeda*. There are, however, other tetradactyl footprints that resemble *Felipeda*. Sarjeant et al. (2002) considered *Pumaeichnum* (Fig. 8.278I) as a junior synonym of *Felipeda*. There are no morphological differences between *Pantheraichnus* and *Felipeda*, so it is another junior synonym of *Felipeda* (Abbassi et al., 2021). *Mitsupes* Rodríguez-de la Rosa and Guzmán-Gutiérrez (2012) has been introduced for medium-sized felid footprints with asymmetrical metatarsal imprints. This asymmetry distinguishes this ichnogenus from the bilaterally symmetrical *Felipeda* (Abbassi et al., 2021). *Quiritipes* was considered as a possible synonym of *Canipeda* by Melchor et al. (2019), and its elongate digit pads are distinct from *Felipeda*. *Felipeda* differs from *Pycnodactylopus* in lacking broadly ovoid and broad digital impressions with two phalangeal pads. Among the carnivoran fossil footprints, *Felipeda* comprises diverse ichnospecies: *Felipeda bestia* (Fig. 8.27A), *Felipeda biancoi* (Fig. 8.27B), *Felipeda maxima*, *Felipeda milleri* (Fig. 8.27C), and *Felipeda sanguinolenta* are attributed to *Felipeda* as a new combination; *Felipeda bottjeri* (Fig. 8.27D), *Felipeda scrivneri* (Fig. 8.27E), *Felipeda lynxi* (Fig. 8.27F), *Felipeda parvula* (Fig. 8.27G), and *Felipeda felis* (Fig. 8.27H) are accepted ichnospecies of *Felipeda*. Another Late Pleistocene ichnospecies is *Felipeda miramarensis* (Agnolín et al., 2019) (see Chapter 9).

Felipeda maxima was recently combined as *Platykopus maxima* by Botfalvai et al. (2023). *Bestiopeda sanguinolenta* was considered an ichnospecies of *Felipeda* by Sarjeant et al. (2002), but *Bestiopeda sanguinolenta* shows sharp digit tips and is distinct from *Felipeda* and belongs to *Canipeda* (Melchor et al., 2019).

Diedrich (2011b) considered the new ichnogenus and ichnospecies name *Pantheraichnus bottropensis* for the Late Pleistocene lion *Panthera leo spelaea* from Bottrop and other open-air sites in northern Germany. Based on the description and figures of Diedrich (2011b), *Pantheraichnus bottropensis* is a synonym of *Felipeda biancoi* (Aramayo and Manera de Bianco, 1987).

Remeika (1999) used a new ichnospecies name for tetradactyl footprints and considered it as ichnospecies of the *Pumaeichnum stuoti* (Fig. 8.27I). He discussed t the track maker and attributed it to the extinct puma-size cheetah *Acinonyx* (*Miracinonyx*) *studeri* as the likely candidate trackmaker, without discussion about the distinct features between *Pumaeichnum stuoti* and *Felipeda milleri* (Remeika, 1999) (Fig. 8.27C and I). The photographs of the holotype of *Pumaeichnum stuoti* are not significant, so we prefer to consider it as *Felipeda biancoi*.

Ichnogenus *Hyaenodontipus* Ellenberger, 1980.

Fig. 8.28A.

Type ichnospecies: *Hyaenodontipus praedator* Ellenberger, 1980.

Diagnosis: Footprint belongs to a very large carnivore. Both proximal and distal sides of footprints are discoidal in shape. The imprint is circular to the bottom of the trace, but claw imprints are visible in the front rim of the footprint, generally five claw imprints in the front limb and four in the hind limb. The trackway is very wide when the animal stops and narrow when in motion (translated from the French from Ellenberger, 1980).

Discussion: Ellenberger (1980) linked *Hyaenodontipus* to Hyaenodontidae and discussed the members of this family as the main predators during the Eocene–Miocene. Contrasting with other carnivoran footprints, *Hyaenodontipus* shows short digit imprints, which were ascribed to the claw imprints. This type of preservation can also be linked to substrate properties, so limbs deeply set into soft sediment resulted in poor preservation of digit imprints (Abbassi, 2022) and it may be related to undertracks (Sarjeant and Wilson, 1988). Like the other ichnotaxa in the same paper, Ellenberger (1980) did not assigned holotype, and the type of material(s) of *Hyaenodontipus* needs reexamination, if the holotype was deposited in a museum or somewhere else. Otherwise, this name is nomen nudum.

Ichnogenus *Hirpexipes* Sarjeant et al., 2002.

Fig. 8.28B.

Type ichnospecies: *Hirpexipes alfi* Sarjeant et al., 2002.

Diagnosis: Semidigitigrade to semiplantigrade mammalian footprints. Pes impressions markedly larger and more elongate than those of the manus. All digits exhibit long, sharp claws. The manual digits each have three pads, though fusion of adjacent pads may occur.

The tips of all digits are directed forward; however, Digit IV shows slight, and Digit V strong, initial outward curvature. The metacarpal pads are variably impressed, showing indistinct indications of the positions of the carpals. Four of the pedal digits show four pads, the other (I) only three; all digits are directed forward, though again Digit IV shows slight, and Digit V strong, initial outer curvature. The metatarsal pads are variably impressed and may show indications of the positions of the tarsals. The trackway relatively narrow; stride and pace long (Sarjeant et al., 2002).

Discussion: *Hirpexipes* is a pentadactyl imprint and exhibits the strong, sharp claws characteristic of predators with elongate digit imprints. *Tacheria* Krapovickas and Nasif (2011) from the late Oligocene of Argentina includes tetradactyl manus and pes impressions of subequal size with long and robust digits with strong, elongated claws. Analyzing the morphology of *Tacheria*, it was suggested that *Tacheria* was produced by a rodent or rodent-like mammal of medium to large size (Krapovickas and Nasif, 2011; Krapovickas et al., 2018). The morphology of *Tacheria*, however, is reminiscent of *Hirpexipes*, and it differs from *Hirpexipes* by number of digit imprints and lesser digit number and slightly slender digit imprints.

Ichnogenus *Khafipus* Abbassi et al., 2015.

Fig. 8.28C.

Type ichnospecies: *Khafipus khadari* Abbassi et al., 2015.

Diagnosis: Large plantigrade footprints that have five dispersed digit imprints in the pes or manus. Both heel and digit imprints are disconnected from each other and enclosed by a circular ridge. All digit imprints are round. Digits II–V are oriented forward, and Digit I is on the lateral side. The heel imprint consists of a circular hole with a planar floor and may show some forward lobes. Middle Digits II–IV are larger than lateral Digits I and V (Abbassi et al., 2015).

Discussion: Abbassi et al. (2015) established *Khafipus* from the Oligocene of eastern Iran and for large pentadactyl footprints so that digits and sole imprints are surrounded by a distinctive rim, and compared it with the pentadactyl mammalian ichnogenera. They considered nonclawed, round, and mono-pad digits that are disconnected from the metapodium imprint as distinctive features of *Khafipus*. Footprints of similar morphology were reported from the Miocene of Jumilla, Spain by Vilas et al. (2009), which were attributed to *Agriotherium*, ursine-like animals. Based on local information, the holotype of *Khafipus* has been lost, and new type material should be proposed from the Oligocene of eastern Iran.

Ichnogenus *Mitsupes* Rodríguez-de la Rosa and Guzmán-Gutiérrez, 2012.

Fig. 8.27J.

Type ichnospecies: *Mitsupes dugesii* Rodríguez-de la Rosa and Guzmán-Gutiérrez, 2012.

Diagnosis: Felid tracks of medium size (9–11 cm long, 9–10 cm wide), with an interdigital pad of asymmetric partition, slightly subtriangular to strongly reniform. Manus imprint is strongly impressed, despite a shallower pes imprint, and indicates heavy weight on the front of the body (Rodríguez-de la Rosa and Guzmán-Gutiérrez, 2012).

Discussion: *Mitsupes* is distinguished by the asymmetrical kidney-shape of the metapodial imprint with its four rounded to ellipsoid digit imprints, while Digit V is touching the metacarpal. These features are closely similar to *Felipeda scrivneri*. Indeed, *Mitsupes dugesii* is a form of *Felipeda* and differs from *Felipeda scrivneri* by its kidney-shaped metatarsal/metacarpal imprints. It is proposed that *Mitsupes* is synonym of *Felipeda*, and *Mitsupes dugesii* be combined under *Felipeda* as *Felipeda dugesii*.

Ichnogenus *Mustelipeda* (Kordos, 1985).

Fig. 8.28E.

Ichnospecies *Mustelipeda punctata* (Kordos, 1985).

Diagnosis: Small circular sole print with the imprints of the five toes closely attached to it.

Discussion: *Mustelipeda* is another ichnogenus from the Miocene tracksite of Ipolytarnóc of Hungary. The form of *Mustelipeda* is simple circular imprints as five holes in a curved row. This ichnogenus seems to be undertracks and not true tracks, and there are no data about quality of preservation of the footprints (the material is illustrated by line drawings). *Mustelipeda* differs from *Khafipus* by smaller size, poorly preserved metapodium imprints and closely spaced digit imprints.

Ichnogenus *Pantheraichnus* Diedrich, 2011b.

Fig. 8.27K.

Type ichnospecies: *Pantheraichnus bottropensis* Diedrich, 2011b.

Diagnosis: No formal diagnosis was provided. The description from Diedrich (2011b) is as follows: Large plantigrade felid footprints, each with four round-oval digit pad II–V imprints in front of the large pad. This large pad is oval shaped in the manus and more kidney-shaped or rectangular in the pes, in the best-preserved imprints. The distances between the digit pad impressions are larger in the manus than in the pes imprints, in ideal cases. The footprint shapes are highly variable depending on the substrate moisture and their state of preservation, as can be seen from examining the complete trackway. The manus is much longer than its width (142/136 mm), whereas the pes imprints are similar in length and width (143/143 mm). The stride of the trackway is, on average, 175.8 mm, and the trackway width averages 317 mm, but both depend on the size of the individual age and sex, which in this case was a fully grown adult but unclear if a lion or lioness.

Discussion: Diedrich (2011b) did not discuss the distinctive characters of *Pantheraichnus bottropensis* among the remaining Felidae footprints. Here we considered *Pantheraichnus bottropensis* to be a synonym of *Felipeda biancoi* (Aramayo and Manera de Bianco, 1987); see also under *Felipeda*.

Ichnogenus *Phacelopus* Sarjeant and Langston, 1994.

Fig. 8.28G.

Type ichnospecies *Phacelopus therates* Sarjeant and Langston, 1994.

Diagnosis: Plantigrade to semiplantigrade footprints, with five clawed digits on each foot. Manus somewhat smaller than pes. The digits and claws curve inward toward an axis between Digits II and III. Digits of manus with one or two phalangeal pads; digits of pes with two or three phalangeal pads. Impressions of metacarpals and some carpals, and of metatarsals and some tarsals, may be distinguishable. Total interdigital span of manus (up to 70 degrees) less than that of pes (up to 80 degrees) (Sarjeant and Langston, 1994).

Discussion: *Phacelopus* is distinguished from other pentadactyl footprints by its large and curved claw imprints, and it was likely attributed to the musteloids or to *Hesperocyon*. *Phacelopus* was only reported from late Eocene of Trans-Pecos, Texas (Sarjeant and Langston, 1994).

Ichnogenus *Platykopus* Sarjeant et al., 2002.

Fig. 8.28H.

Type ichnospecies: *Platykopus ilycalcator* Sarjeant et al., 2002.

Diagnosis: Large plantigrade footprints with digits close to manual/pedal pad. Manus width is similar to that of the pes, with the pes being slightly narrower and elongated due to a metatarsal pad. Five digits on manus and pes are clawed, and Digits II–V are of equal length on a quadratic manus with Digit I offset. Digits I–V form a symmetrical arc around the pes, with Digits I and V being shorter than II and IV (Sarjeant et al., 2002).

Discussion: The ichnogenus *Platykopus* was erected for large pentadactyl footprints, which were attributed to creodonts, amphicyonids, and to ursids that were reported from the Oligocene (Abbassi et al., 2015), Miocene (Sarjeant et al., 2002; Lucas and Schultz, 2007; Abbassi, 2010) and Pliocene (Martinez and Farke, 2018). Sarjeant et al. (2002) established *Platykopus* as monospecific, but later *Platykopus stuartjohnstoni* was added (Lucas and Schultz, 2007). *P. stuartjohnstoni* differs from *P. ilycalcator* in having: manus narrower than pes, clear separation of metatarsal/metacarpal pads from digital pads, Digits II–V of subequal size on the pes, metatarsal pad on pes wide and short and only four pes digit imprints (Lucas and Schultz, 2007). Thus, attributing material of *Platykopus stuartjohnstoni* to *Platykopus* is doubtful. Recently, Botfalvai et al. (2023) revised *Bestiopeda maxima* as *Platykopus maxima* by comparison with other relevant Cenozoic pentadactyl footprints. Thus, *Platykopus* comprises three valid ichnospecies.

Ichnogenus *Pycnodactylopus* Sarjeant et al., 2002.

Fig. 8.28F.

Type ichnospecies: *Pycnodactylopus achras* Sarjeant et al., 2002.

Diagnosis: Semidigitigrade mammalian footprints of moderate size and broadly ovoidal outline. Four short digits on all feet, lacking claws. Phalangeal impressions rather stout (especially so in the presumed pes), their width being two-thirds of their length. Digits II to IV are of pyriform to scutiform outline exhibiting two digital pads (though the division between the pads may not be clear) Digit V is reduced (especially in the presumed manus) and with pad undivided. Proximally the phalangeal impressions are set closely about metacarpal pads and metatarsal pads, with which they may be in actual contact. The trackway is relatively narrow (Sarjeant et al., 2002).

Discussion: *Pycnodactylopus* differs from *Felipeda* by variable form of the phalangeal impressions and the close proximity of their bases to metacarpal pads and metatarsal pads, and attributed likely to felids by lack of both bilateral symmetry and impressions of claws (Sarjeant et al., 2002). *Pycnodactylopus* was named from the Miocene of California and reported from the Eocene of Iran (Abbassi and Mashhadizadeh Maleki, 2020). *P. achras* is distinguished from *Platykopus ilycalcator* by overall shape and number of digits. *P. ilycalcator* shows short digit imprints with claws. *Hirpexipes alfi* Sarjeant et al. (2002) shows very long digit imprints with sharp claws and is thus distinct from *Platykopus ilycalcator* and *Pycnodactylopus achras* (Abbassi and Mashhadizadeh Maleki, 2020).

Ichnogenus *Quiritipes* Sarjeant et al., 2002.

Fig. 8.28I.

Type ichnospecies: *Quiritipes impendens* Sarjeant et al., 2002.

Diagnosis. Semidigitigrade to semiplantigrade mammalian footprints of ovoidal to lanceolate outline and small to moderate size. Pes somewhat larger than manus. Four digits are imprinted; all are acuminate (though not acute), without claws. Digits III and IV are of similar length and have almost parallel axes, though in the pes their distal extremities converge somewhat. Digits II and V are shorter (pedal Digit II markedly so) and directed forward, but slightly pronounced outward. Digital pads are not normally distinguishable. A space separates the digital impressions from metacarpal pads and metatarsal pads; these latter are most deeply impressed, and triangular, in their anterior portion. The trackway narrow, and stride and pace vary according to gait but quite long; the pedal imprints are commonly superimposed upon the manual imprints (Sarjeant et al., 2002).

Discussion: The monospecific ichnogenus *Quiritipes* was described as nonclawed tetradactyl, and was attributed to the Creodonta or Felidae (Sarjeant et al., 2002). The lack of claw impressions suggests felid-like footprints, but the relatively elongated digit imprints markedly differentiate *Quiritipes* from *Felipeda,* and are typical of canid footprints. Melchor et al. (2019), however, re-evaluated the photographs of *Quiritipes* (Sarjeant et al., 2002, plate 3c; Hunt and Lucas, 2007, fig. 4) and distinguished claw imprints in several footprints. Although *Quiritipes* differs from footprints of extant canids, by larger pes imprints than the manus imprints, overlap of the pes on the manus in some specimens, and the disputable original interpretation that digital impressions do not include claw marks, Melchor et al. (2019) believed that *Quiritipes* is possible synonym of *Canipeda.*

Ichnogenus *Sarjeantipes* McCrea et al., 2004.

Fig. 8.28O.

Type ichnospecies: *Sarjeantipes whitea* McCrea et al., 2004.

Diagnosis: Trackway of quadruped with pronounced heteropody; manus tetradactyl, semidigitigrade and about 2/3 length of pes. Manus impressions with distinct triangular to pentangular main pad. Manus Digits II–V and pes Digits II–V with single distinct phalangeal pad with rounded posterior margins. Phalangeal pads of manus separated by space from metacarpal pad. Manus digit impressions terminating in acuminate claws. Manus slightly wider than long. Pes pentadactyl, plantigrade with distinctly concave, bi-lobed heel. Pes digits terminating in rounded claw impressions. Pes slightly wider than long with distinct phalangeal pads. Metatarsal pad not seen due to overprinting. Pes Digits II and IV are approximately 5% shorter than Digit III. Pes Digit V is short, being only 80% of the mean length of Digit III. Pes Digit I is the shortest, only 65% of the length of Digit III. Digits III and IV are the longest in the manus, with Digit II only slightly shorter than Digit IV. Digit V is shortest in the manus, 70%–80% of lengths of Digits III and IV, respectively. Total divarications of digits for both manus and pes are approximately 108°. Good deal of variability of divarication between digits of manus and pes, although more so in manus. The pace length of the pes is approximately one-third meter and stride length just over one-half meter. The trackway narrow, about one pes width; pes pace angulation approximately 130°. Inner margin of pes tracks near trackway axis; pes rotation low ($X = +1.4°$; outward) based on alignment of Digit III (longitudinal axis), in relation to trackway midline. Placement of the manus is to outside of trackway midline; manus rotation is variable (McCrea et al., 2004).

Discussion: The *Sarjeantipes whitea* trackway was produced by a short-bodied animal walking in a straight line in the first part of the trackway (McCrea et al., 2004). Heteropody, position of the manus imprint, pentadactyly of pes and semiplantigrade to plantigrade character are distinct features of *Sarjeantipes,* which distinguish it from *Quiritipes impendens.*

Ichnogenus *Sarcotherichnus* Demathieu et al., 1984.

Fig. 8.28J.

Type ichnospecies: *Sarcotherichnus enigmaticus* Demathieu et al., 1984.

Diagnosis: Pentadactyl footprints. Digit II, III, and IV imprints transversely aligned and with slightly protrusion of central digit imprint. Imprints of the lateral digits (I and V) far back, and one digit imprint (?I) located just behind lateral middle digit imprint (?II), and the other (V) smaller and more offset toward the outside (translated from the French from Demathieu et al., 1984).

Discussion: Demathieu et al. (1984) described one isolated small footprint with ambiguous digital coding as a new ichnogenus and ichnospecies, *Sarcotherichnus enigmaticus,* from Oligocene of south France, that was likely attributed to the canids or felids. Compared with small animal footprints, the pes imprint of extant opossum, has lateral digits (I and V) with a wide interdigital angle, so that the Digit I imprint is behind the metatarsal imprint (Elbroch, 2003). Opossum tracks, however, include partly overlapped pes and manus imprints.

Ichnogenus *Tetrastoibopus* Sarjeant and Langston, 1994.

Fig. 8.28K.

Type ichnospecies: *Tetrastoibopus phoros* Sarjeant and Langston, 1994.

Diagnosis: Small plantigrade to semiplantigrade footprints exhibiting four digits (II to V), each with a digital pad. Digits II to IV bear small pads that occupy a quadrant in front, and on the inner side, of the imprints of sole or heel; in the manus, they are almost spheroidal and of almost uniform size, in the pes they are more elongate and less regular in form. In both manus and pes, Digit V bears a much larger pad of ovoidal to subhexagonal shape, set on the front outer lateral side of sole or heel. Central digits (III and IV) almost parallel; outer digits (II and V) proximally divergent at angles of less than 25 degrees; Digit II curving inward toward Digit III. Impressions of claw tips may be discernible on some or all digits. Trackway moderate; stride long (Sarjeant and Langston, 1994).

Discussion: *Tetrastoibopus* comprises small tetradactyl footprints that were reported from the late Eocene of Texas and differs from *Canipeda* by having lateral digit impressions (II and V) significantly larger than the central ones (III–IV) and showing only occasional claw marks (Melchor et al., 2019). This ichnogenus was tentatively assigned to miacid carnivores (Sarjeant and Langston, 1994).

Ichnogenus *Ursichnus* (Diedrich, 2011a).

Fig. 8.28L and M.

Type ichnospecies: *Ursichnus europaeus* Diedrich, 2011a.

Diagnosis: Quadrupedal trackway composed of plantigrade, pentadactyl manus, and pes imprints of different morphologies, manus nearly circular and pes of ellipsoidal shape, and manus/pes length ratio of about 1:1.25. In normal gait, the manus is placed close and in front of the pes footprint. The heel of the manus footprint is absent, and the palm is ovoid to kidney-shaped and small, while the large sole of the pes footprint is roughly triangular and shows a well-developed heel. Manus and pes footprints with short, oval to rounded digit imprints with clear claw marks. Digit imprints not connected with heel impression (emended by Aramayo et al., 2015).

Discussion: The large pentadactyl mammal footprints of *Ursichnus* are attributed to the bear or bear-like trackmakers. Presence of straight to curved claw imprints, large round or elongate digit pad imprints, which connected together, usually with metapodial imprint, and presence or absence of heel imprint are the main features of ursid footprints. *Ursichnus* differs from the bear-like footprint *Platykopus* by its smaller size, its digital pads are all within a distance of the metapodial pads and are larger and singular, and the pes is larger than the manus (Diedrich, 2011a). Also, in *Platykopus*, the hallux in the manus is offset posteriorly, indicating a more primitive mammalian trackmaker (Sarjeant et al., 2002).

Diedrich (2011a) proposed *Ursichnus* as a monospecific ichnogenus (*U. europaeus*). Later, Aramayo et al. (2015) added a new ichnospecies from the Late Pleistocene, *Ursichnus sudamericanus* (Fig. 8.28M), and emended the ichnogenus diagnosis. *U. sudamericanus* is distinguished from *U. europaeus* by the absence of a heel impression in the manus, and by having the Digit III imprint longer than the remaining digits (in *U. europaeus* the three central digits are longer than the lateral digits) (Aramayo et al., 2015).

Thenius (1967) introduced *Bestiopeda amphicyonides* for large pentadactyl footprints from the Pliocene of northern Austria and attributed it to the bear dogs, amphicyonids. Sarjeant et al. (2002) combined *Bestiopeda amphicyonides* as *Canipeda amphicyonides*, but *Canipeda* is tetradactyl and differs from the heteropodous pentadactyl character of the former. Thenius (1967) argued that ursids can be discarded as producers, by comparing it with live bear footprints and the lack of complete heel imprints. We, however, would rather combine *Bestiopeda amphicyonides* as *Ursichnus amphicyonides*, because of the circular pad (the heel) at the posterior edge of the tracks that may or may not show in live bear tracks (Elbroch, 2003).

Ichnogenus *Zanclonychopus* Sarjeant and Langston, 1994.

Fig. 8.28N.

Type ichnospecies: *Zanclonychopus cinicalcator* Sarjeant and Langston, 1994.

Diagnosis: Plantigrade footprints, with manus and pes of similar size and with all digits strongly clawed. Digits II to IV most deeply impressed and forming, with palm or sole, an approximately oval shape (more marked in the pes); Digit I set off to one side and impressed lightly or not at all. Tips and claws of inner digits curving outward, of outer digits curving inward. Inner phalangeal pads of pes fused; those of manus free (Sarjeant and Langston, 1994).

Discussion: *Zanclonychopus* is the footprint of a large plantigrade mammal with mesaxonic feet, and Sarjeant and Langston (1994) considered the Creodonta as trackmaker. The ichnotaxon *Zanclonychopus* differs from *Platykopus* by long and slender digit imprints, and both ichnogenera have Digit I placed behind the other digits, which is interpreted as indicating a more primitive mammalian trackmaker (Sarjeant et al., 2002; Diedrich, 2011a).

4.4.4 *Anagalida*

4.4.4.1 Anagalida autopodium structure and tracks

Anagalida is the grandorder name for a diverse and abundant group of mammals with different sizes, as small as Pygmy Jerboa to hippo-sized *Phoberomys* from the Late Miocene of Venezuela (Sánchez-Villagra et al., 2003). Anagalida involves Cretaceous Zalambdalestidae, early Paleogene Anagalidae and Pseudictopidae, and Glires (rabbits, beavers, and rodents) with chisel-like gnawing front teeth. The phylogenetic relationships between the members of Anagalida are open, under continuous discussion and also disputed, so there are different classifications for superorder Anagalida (Rose, 2006). The difference in classification is related to the great variety and abundance of Anagalida, so that, only rodents make up about 44% of the living mammal species, and there are, at least, five times as many extinct rodent genera and species as there are living genera and species (Wolff and Sherman, 2008; Prothero, 2017).

Cretaceous Zalambdalestidae is a possible ancestor of Glires; they have longer hind limbs than forelimbs, the tibia and fibula are fused, and they display many specializations for running and hopping, including a deeply grooved astragalus (McKenna, 1975; Korth, 1994; Rose, 2006). The tibia and fibula are fused high up on the shaft, the astragalus is deeply grooved, and the metatarsals are very elongate, with a reduced hallux locomotion probably similar to rabbits or elephant shrews (Rose, 2006).

Anagalids were an endemic family in Asia, during the late Paleocene to Oligocene and are closely related to Glires (McKenna, 1975). Anagalids were terrestrial quadrupeds, capable of digging with the forelimbs, and with short and stout manual phalanges; and long, curved, deeply fissured ungual phalanges (Rose, 2006) (Fig. 8.29A).

Pseudictopidae have a pentadactyl hindlimb, probably semiplantigrade or semidigitigrade. The claw phalanges on the hindlimb are fissured and with proximal enlargement. The calcaneum is antero-posteriorly elongated with a well-developed tuber calcis. The metatarsal bones II–IV are long (the longest being metatarsal Ill), while metatarsal bones I and V are reduced in length. The cuboideum and navicular are similar to the astragalus-primitive insectivore-carnivorous type (Sulimski, 1969).

The front foot of rabbits has I–V digits, of which the inward facing is the vestigial dewclaw that grows higher on the leg and does not reach the ground. Each hind foot has Digits II–V (Craigie, 1969) (Fig. 8.29B).

Beavers are large, semiaquatic rodents of Eurasia and North America, along with Heteromyidae (kangaroo rat and kangaroo mice), and the gophers, belong to the rodent suborder Castorimorpha. Beavers are agile swimmers with their large webbed hind feet having powerful muscles. The front feet are more suitable for digging and holding objects. Both hind and front feet are pentadactyl with clawed digits. The metacarpal is large and elongated and the metatarsal is smaller and includes round pads.

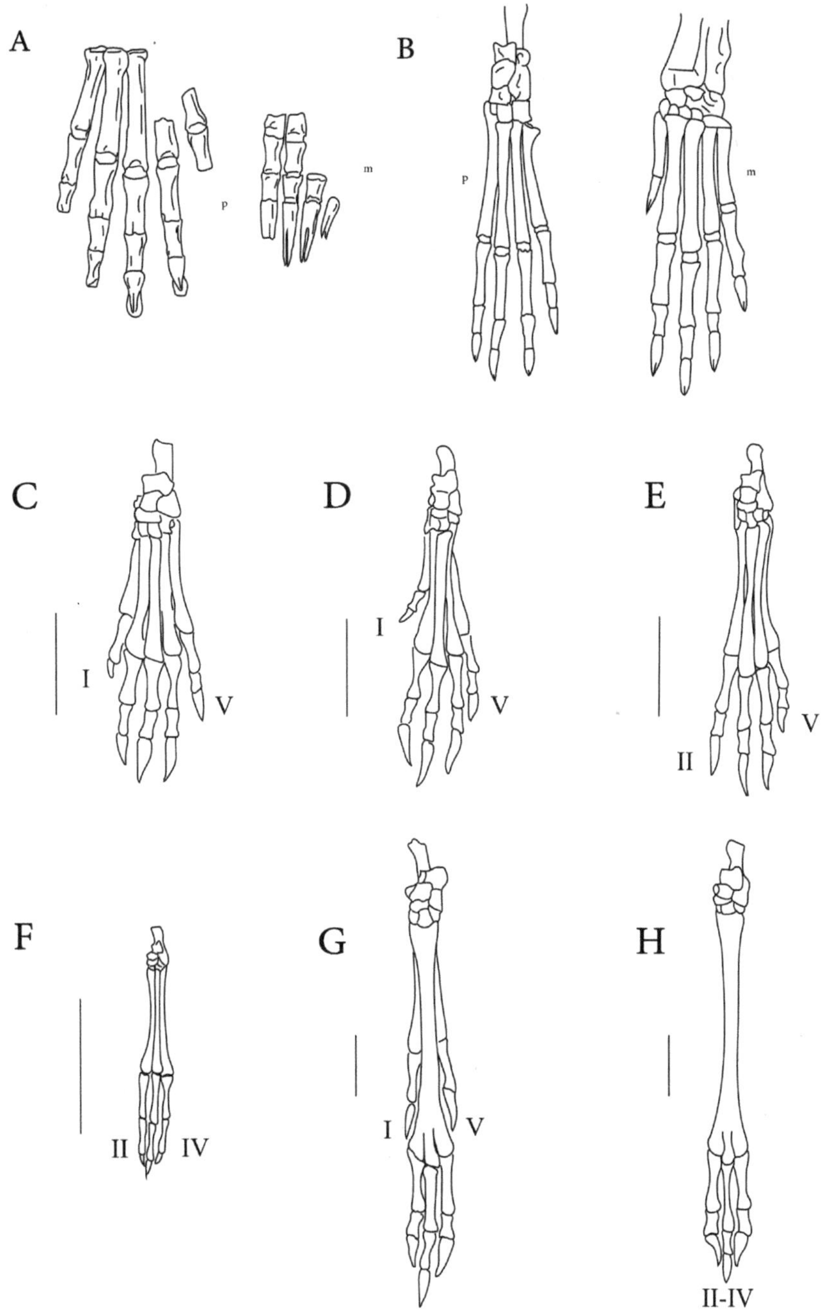

FIGURE 8.29 Osteology images of the limbs of Anagalida (A, *Anagale*) and Lagomorpha (B), (C–H). Reductions and fusions of elements of the foot exhibited by bipedal rodents. (C) No reductions or fusions. Condition exhibited by *Meriones* sp., (D) Reduction of the hallux. *Dipodomys ordii.* (E) Loss of one digit, but no fusions among remaining elements, *D. merriami.* (F) Loss of two digits, no fusions among remaining elements, *Salpingotus michaelis.* (G) Reduction of two digits, extensive fusions, *Alkactuga* sp. and (H) Loss of two digits, extensive fusions, *Jaculus orientalis.* A–B, no scale, C–H, scales equal 1 cm. *(A) Modified from Simpson (1931). (B) Modified from Craigie (1969). (H) Modified from Craigie (1969).*

Rodents are as old as late Paleocene in North America and Asia. The early success of rodents is indicated by their Eocene abundance in Eurasia and North America and by their rapid radiation. Rodents have diverse habitats, and their foot structures differ. Usually, rodents have different numbers of digits in the pes and manus. For example, South American *Dasyprocta* sp. (Dasyproctidae) has atridactyl hind foot, and Family Gliridae has four toes in the manus, and five toes in the pes. In members of the subfamily Cardiocraniinae, the toes vary in number from three to five, and the metatarsals are not fused (Vaughan et al., 2013). Reductions and fusions of elements of the foot are exhibited by bipedal rodents (Berman, 1985) (Fig. 8.29C–H). In the rodent families adapted for bipedal locomotion (Heteromyidae, Pedetidae, Muridae, Zapodidae, and Dipodidae) forelimb function is for grooming, burrowing, and manipulation of food or other objects. On the contrary, the longer hindlimb is specialized for saltation and has undergone foot elongation by an increase in metatarsal length.

4.4.4.2 Anagalida ichnotaxa

The following list includes the ichnotaxa of Anagalida, in particular, ichnogenera attributed to Glires.

Musaltipes Lockley and Milner, 2014
- *Musaltipes occidens* Lockley and Milner, 2014
- *Musaltipes longidigitus* Lockley and Milner, 2014
- *Musaltipes taromi* Abbassi et al., 2017

Porcellusignum Casamiquela in Angulo and Casamiquela, 1982.
- *Porcellusignum conculcator* Angulo and Casamiquela, 1982

Ptyariopus Sarjeant and Langston, 1994
- *Ptyariopus aichmanticheirus* Sarjeant and Langston, 1994

Tricorynopus Sarjeant and Langston, 1994
- *Tricorynopus elaphrus* Sarjeant and Langston, 1994

Tacheria (Krapovickas and Nasif, 2011)
- *Tacheria troyana* (Krapovickas and Nasif, 2011)

4.4.4.2.1 Description and discussion of ichnogenera

Ichnogenus *Musaltipes* Lockley and Milner, 2014.
Fig. 8.30A–C.
Type ichnospecies: *Musaltipes occidens* Lockley and Milner, 2014.
Diagnosis: Symmetrical trackways of a small, heteropodous, hopping rodent or rodent-like quadruped with elongate, triangular pes and smaller, shorter manus. The quadripartite configuration of two pedals and two manual traces typically indicates quadrupedal progression. Pes traces typically triangular with anterior portion wider, sometimes showing splayed pentadactyl digit trace configuration and posterior portion with narrow, elongate heel. Manus traces subcircular, and often indistinct. Pes and manus traces may overlap, and, in some cases, only the pes traces are registered, suggesting bipedal hopping. A mid-line tail trace may be present. Toe-to-toe or heel-to-heel hopping distance measured as distance between paired pes, or paired pes and manus sets and typically between two and four pes track lengths (Lockley and Milner, 2014).

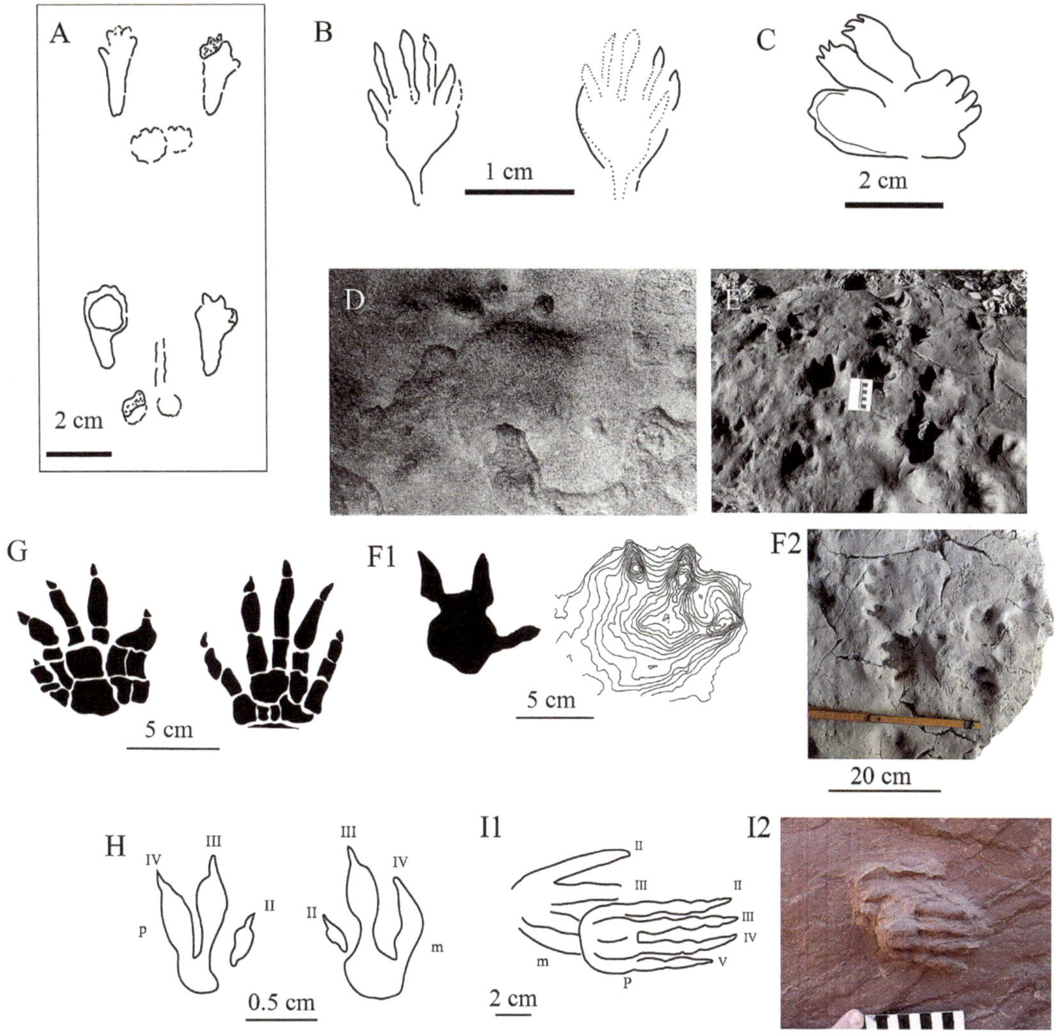

FIGURE 8.30 Modified outline sketch and photograph of the Anagalida ichnotaxa, (A) *Musaltipes occidens* Lockley and Milner, 2014, (B) *Musaltipes longidigitus* Lockley and Milner, 2014, (C) *Musaltipes taromi* Abbassi et al., 2017. (D–F) *Porcellusignum conculcator* Casamiquela in Angulo and Casamiquela, 1982, (G) *Ptyariopus aichmanticheirus* Sarjeant and Langston, 1994, (H) *Tricorynopus elaphrus* Sarjeant and Langston, 1994, (I) *Tacheria troyana* Krapovickas and Nasif (2011). *(D–F) with permission and photograph of F2 by Ricardo N. Melchor. (I) Photograph of I2 by Ricardo N. Melchor*

Discussion: Hopping locomotion of small vertebrates is known in fossil footprints from the Permian to Miocene. *Schadipes crypticus* Lockley and Foster (2003) is a mammal track from the Upper Cretaceous of Colorado that morphologically resembles those of some modern rodents. Numerous specimens from the Miocene of Colorado and Utah have been collected and described as hopping, small mammal footprints without ichnotaxonomic naming (Lockley et al., 2007; Reynolds and Milner, 2007). Lockley and Milner (2014) named these footprints and established *Musaltipes* with two ichnospecies, *M. longidigitus* and *M. occidens* from the Miocene of Utah, and Abbassi et al. (2017) added a third ichnospecies *M. taromi* from the Eocene of the Alborz Mountains, Iran (Fig. 8.30A–C). *M. longidigitus* and *M. taromi* are quadrupedal, and *M. occidens* is bipedal. *Ranipes laci* is another hopping track from the Eocene Flagstaff Formation, Utah, which was attributed to a frog hopping trackway (Lockley and Foster, 2003). *Ranipes* is a highly symmetrical trackway of a small hopping vertebrate, with slight outward rotated and with strong inward rotated tetradactyl front footprints (compare with Fig. 8.2B).

Ichnogenus *Porcellusignum* Casamiquela in Angulo and Casamiquela, 1982.

Fig. 8.30D–F.

Type ichnospecies: *Porcellusignum conculcator* Casamiquela in Angulo and Casamiquela, 1982.

Diagnosis: Digitigrade impressions of a medium-sized mammal, functionally tridactyl but occasionally tetradactyl. Manus and pes subequal in dimensions; digital impressions acuminate and large, sometimes transversely or diagonally elongated, grouped in sets of three, with the central digit in a more advanced position and asymmetrically located. Footprints are randomly distributed, suggesting high mobility of the producer. Tail trace absent (translated from Casamiquela in Angulo and Casamiquela, 1982, by Citton et al., 2022).

Discussion: The ichnotaxon *Porcellusignum conculcator* has been established for digitigrade footprints with tridactyl and occasionally tetradactyl impressions recorded in the Upper Miocene–Lower Pliocene Río Negro Formation, at Balneario El Cóndor, Río Negro Province, Argentina (Angulo and Casamiquela, 1982). The original diagnosis and illustrations by Angulo and Casamiquela (1982) depicted dominantly tridactyl tracks, in some cases deep undertracks, but recognized eventually as a tetradactyl posture. Working with specimens from the type locality, Aramayo (2007) proposed that the manus was tetradactyl and the pes tridactyl and illustrated better-preserved material. Recently, abundant and well-preserved footprints, including a trackway, have been documented (Rojas-Manriquez et al., 2024). It seems that *Porcellusignum conculcator* is restricted to Argentina and was primarily described from the Río Negro Formation (Upper Miocene–Lower Pliocene), Río Negro Province, Argentina (Aramayo, 2007; Rojas-Manriquez et al., 2024). However, it was also tentatively recorded from the Upper Miocene Las Flores Formation of San Juan Province (Melchor, 2022 personal observations) and in the Late Pleistocene of the Buenos Aires Province (Agnolín et al., 2019), both from Argentina. Hydrochoeridae has been purported to be the trackmaker (Angulo and Casamiquela, 1982; Aramayo, 2007).

Ichnogenus *Ptyariopus* Sarjeant and Langston, 1994.

Fig. 8.30G.

Type ichnospecies: *Ptyariopus aichmanticheirus* Sarjeant and Langston, 1994.

Diagnosis: Plantigrade footprints of moderate size, in which all five digits are represented. Pes somewhat longer than manus. In the manus, Digit I is reduced and opposed; Digits II to IV are slender and longer; and Digit V is very much reduced. All five digits of the pes are slender, radiating in fan-like fashion with the hallux somewhat opposed; Digits I and V are somewhat reduced. All digits bear claws. The trackway narrow; stride moderately long (Sarjeant and Langston, 1994).

Discussion: This track was attributed to a large rodent with unwebbed feet from the late Eocene of Texas, USA. The rodent footprint *Ptyariopus* may resemble footprints of carnivores, and Sarjeant and Langston (1994) enumerated differences between *Ptyariopus* and Carnivora footprints. These differences include great reduction of Digits I and V in the manus, though not in the pes; the wide spread of the digits; and the irregular pattern of the footprints, with the hind feet impressions placed besides, on top of, or ahead of those of the forefeet.

Ichnogenus *Tricorynopus* Sarjeant and Langston, 1994.

Fig. 8.30H.

Type ichnospecies: *Tricorynopus elaphrus* Sarjeant and Langston, 1994.

Diagnosis. Very small digitigrade footprints in which only three digits, presumably II to IV, are imprinted. The digits diverge and broaden distally and bear distinct claws. Manus and pes are closely similar in size (Sarjeant and Langston, 1994).

Discussion: This small tridactyl digitigrade footprint shows sharp digit tips with wide digits, which are considered as a distinctive feature of mammal origin as opposed to a reptilian origin (Sarjeant and Langston, 1994). Sarjeant and Langston (1994) interpreted the tridactyl nature of *Tricorynopus* as result of tracks produced by running at speed, although the trackway pattern was not determined.

Ichnogenus *Tacheria* (Krapovickas and Nasif, 2011).

Fig. 8.30I.

Type ichnospecies: *Tacheria troyana* Krapovickas and Nasif, 2011.

Diagnosis: Tetradactyl manus and pes impressions of subequal size, with long and robust digits with strong elongated claw marks. The manus footprints are asymmetrical. The Digit III impression is the longest, while Digits II and IV impressions are subequal and following in size, and finally, Digit V is the shortest and at times does not imprint. They present well developed phalangeal pads, but generally lack metacarpal pads, resulting in almost digitigrade impressions. In the pes imprints, the lateral digits (II–V) are shorter than central digits (III–IV), resulting in a symmetrical arrangement. Elongated phalangeal and metapodial pad impressions are present, occasionally imprinting two heel pads. Trackways are narrow with footprint impressions near the midline, and pes impressions are placed in front of, and more lateral to the midline, than the manus impressions (emended by Krapovickas et al., 2018).

Discussion: *Tacheria* was introduced from the Miocene Vinchina Formation, La Rioja Province, Argentina. Both living and extinct South American rodents and rodent-like mammals of medium to large size with tetradactyl feet were considered as track makers of *Tacheria troyana,* and notoungulates are possible producers, and among South American rodents, the Hystricognathi (caviomorphs) are the most likely producer candidates of *T. troyana*. Krapovickas and Nasif (2011) used the spelling of ichnogenus *Tacheria* in the establishing the new ichnogenus, but in some part of the text used "*Tracheria*" in spelling, which was repeated in Krapovickas et al. (2018) and by others (e.g., Valencia Fermin, 2020).

4.4.5 Xenarthra

4.4.5.1 Xenarthra autopodium structure and tracks

The Xenarthra, which appeared in South America during the Paleocene, are unusual and the most primitive placental mammals. Xenarthra are also known in Central and North America during the Neogene and Pleistocene. Anteating animals, such as sloths, pangolins, and aardvarks possess a robust, fossorial skeleton. Although various taxa have been considered as Edentata, most of these were later removed from the concept of a natural Edentata, and Xenarthra is the formal name of the remaining group of "edentates" (Rose, 2006; Prothero, 2017).

There are 31 living species of xenarthrans: 6 sloths, 4 anteaters, and 21 species of armadillos, and these species are divisible into two main groups: Cingulata (such as extant armadillos), and Pilosa, which comprise two groups, Vermilingua, or anteaters, characterized by an elongate, tubular skull, the absence of any teeth, and a prehensile tongue used in acquiring prey, and the Phyllophaga, or sloths. Fossil forms of the Xenarthra comprise more than 150 identified genera, which are remarkable by the tremendous body size of some forms. These giant forms went extinct between the end of the Pleistocene and the beginnings of the Holocene, possibly because of their large size (e.g., Fariña et al., 2012).

The xenarthran hind foot has four toes, and the forefoot has two or three prominent claws; these are used to dig up insect nests and are specialized for climbing. In addition to different digit numbers, each of these types also has distinct fusions of the medial carpal and metacarpal elements. The morphology of the manus is important both phylogenetically and for behaviors such as feeding and locomotory posture (Tito and de Iuliis, 2003).

Ground sloths (Folivora) as large, clawed herbivores have generally large ungual phalanges, especially those of Digit III of the manus and pes. The lateral digits of both manus and pes are usually larger than the medial digits. Anterior body weight was probably carried on the ulnar side of the manus in the manner of *Myrmecophaga* (Coombs, 1983) (Fig. 8.31A).

The manus of Megatheriinae includes three morphological types (Fig. 8.31B–D): the tridactyl manus, characterized by *Eremotherium laurillardi*; the tetradactyl manus, characterized by most other Megatheriinae taxa, such as *Megatherium americanum*; and the pentadactyl manus, characterized by *Mylodon robustus* Tito and de Iuliis, 2003. *Panochthus* has tetradactyl feet with thick Digits II–IV (Fig. 8.31E). The extant armadillo *Zaedyus* (Cingulata) has a pentadactyl manus with long distal phalanges (Fig. 8.31F).

4.4.5.2 Xenarthra ichnotaxa

Numerous footprints from the Cenozoic of South America have been attributed to the Xenarthra. The Quaternary ichnotaxa are treated in Chapter 9, whereas the Neogene ichnotaxa include:

Falsatorichnum Casamiquela in Angulo and Casamiquela (1982).
 Falsatorichnum calceocannabius Casamiquela in Angulo and Casamiquela (1982).
Megatherichnum (Casamiquela, 1974)
 Megatherichnum oportoi Casamiquela, 1974
Venatoripes Frenguelli, 1950
 Venatoripes riojanus Frenguelli, 1950

4.4.5.2.1 Description and discussion of ichnogenera Ichnogenus *Falsatorichnum* Casamiquela in Angulo and Casamiquela, 1982.

Fig. 8.32A.

Type ichnospecies: *Falsatorichnum calceocannabius* Casamiquela in Angulo and Casamiquela, 1982.

Diagnosis: Impressions of the bipedal plantigrade mammal, they are morphologically similar to *Megatherichnum oportoi*, but different in size (*Falsatorichnum* half than *Megatherichnum*). Step length ±80 cm (translated from the Spanish from Angulo and Casamiquela, 1982).

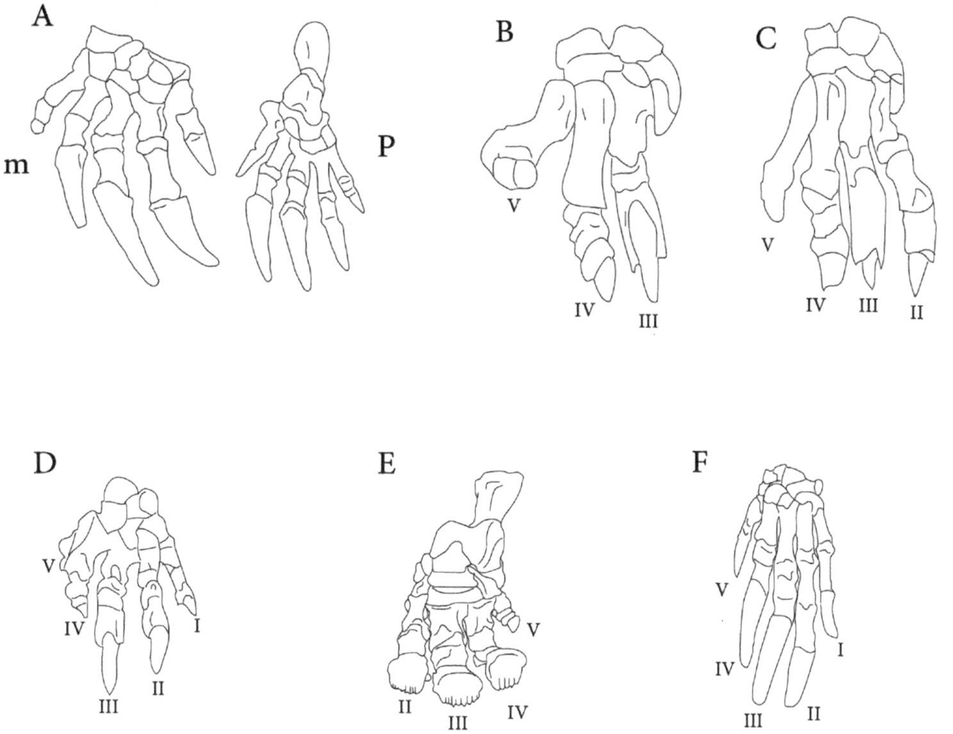

FIGURE 8.31 Skeletal structure of manus (m) or pes (p) of the main groups of Xenarthra, (A) Dorsal view of manus and pes of *Megalocnus*, B–C. Dorsal view of right manus of Megatheriinae, *Eremotherium laurillardi* (B), *Megatherium americanum* (C), and *Mylodon robustus* (D), E–F. Dorsal view of manus of Cingulata, *Panochthus* (E), and *Zaedyus* (F). Not to scale. *(A) Modified from Coombs (1983). (D) Modified from Tito and de Iuliis (2003). (E) Modified form Barbosa et al. (2017). (F) Modified from Marshall et al. (2021).*

Discussion: Angulo and Casamiquela (1982) proposed this new ichnotaxon from the Miocene–Pliocene Río Negro Formation of Argentina. They described it as similar to *Megatherichnum oportoi* but smaller in size and discussed its trackmakers as tentatively related to Megatheriidae. Based on morphological information *Falsatorichnum calceocannabius* is ellipsoid and proximally wider (Angulo and Casamiquela, 1982; plate I, fig. 2 and plate II, fig. 1), but *Megatherichnum* is distinctly distally curved roughly like the paisley pattern. Similar footprints from the type area of *Falsatorichnum* were compared with the Quaternary ichnogenus *Mylodontidichnum* (Melchor et al., 2013). Ground sloth ichnotaxonomy needs a thorough revision.

Ichnogenus *Megatherichnum* (Casamiquela, 1974).

Fig. 8.32B.

Type ichnospecies: *Megatherichnum oportoi* Casamiquela, 1974.

Diagnosis: Quadrupedal trackway composed of large subelliptical footprints, footprint long axis parallel to trackway midline, moderate pace angulation, breadth between footprints about one third of trackway width. Manus and pes footprints of similar size, at least 0.5 m

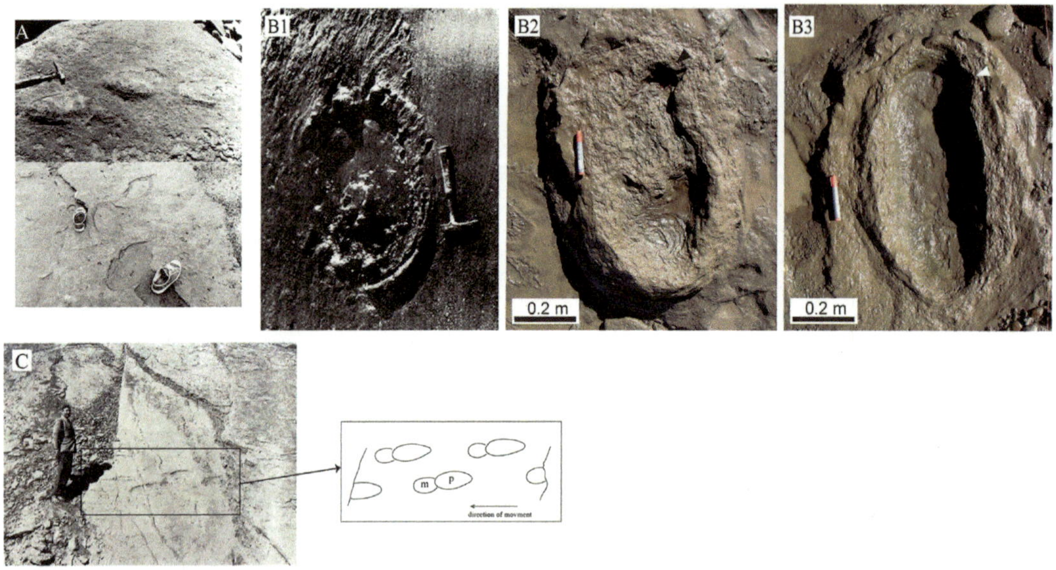

FIGURE 8.32 Outline sketch and photograph of the Xenarthra ichnotaxa, (A) *Falsatorichnum calceocannabius* Casamiquela in Angulo and Casamiquela, 1982, (B1-3) *Megatherichnum oportoi* Casamiquela, 1974, (C) *Venatoripes riojanus*. *(B1) From Casamiquela (1974) with permission, (B2-3) Photograph by Ricardo N. Melchor. (C) Modified from Frenguelli (1950).*

long, lacking digit impressions, footprint width about half its length, flanked by a clear marginal ridge. Manus footprint of broadly falciform outline, large inwardly directed claw mark, rear of footprint slightly wider than anterior part. Pes footprint kidney-shaped, distinct but smaller claw mark. *Megatherichnum* pes footprint outline similar to *Neomegatherichnum* but of smaller size. *Neomegatherichnum* manus print about one third the length of the associated pes print with a marked claw mark (emended by Melchor et al., 2015).

Discussion: *Megatherichnum* is the first proposed ground sloth ichnogenus, and *Pyramiodontherium* (Megatheriinae) is a candidate as trackmaker (Casamiquela, 1974; Melchor et al., 2015). *Megatherichnum* has been only reported from the Upper Miocene–Lower Pliocene of Río Negro Formation of Río Negro and Buenos Aires provinces of Argentina (Casamiquela, 1974; Aramayo, 2007; Melchor et al., 2015; Krapovickas and Vizcaíno, 2016). The pes imprint of *M. oportoi* is kidney-shaped, and the manus is falciform, both showing a distinct and inwardly directed claw mark. Melchor et al. (2015) restudied *Megatherichnum* and concluded that *M. oportoi* is a quadrupedal trackway without overlap of the pes on the manus. Pleistocene ground sloth trackways (ichnogenus *Neomegatherichnum* Aramayo and Manera de Bianco, 1987) are distinctive because of the much smaller size of the fore footprint (see Chapter 9). For the latter examples, a quadrupedal locomotion with overlap of the pes on the manus may be applied (Aramayo et al., 2015; Melchor et al., 2015).

Ichnogenus *Venatoripes* Frenguelli, 1950.

Fig. 8.32C.

Type ichnospecies: *Venatoripes riojanus* Frenguelli, 1950.

Diagnosis: Plantigrade quadruped; pes and manus digit imprints are not distinct but included in the mass of the autopodium; outline of pes print is ellipsoidal, and manus print is oval in contour; pes print located behind the hand print and partially overlapping it (translated from the Spanish from Frenguelli, 1950).

Discussion: Frenguelli (1950) described *Venatoripes riojanus* from the Miocene Vinchina Formation, La Rioja Province, Argentina. At first, he considered that the hosting sediments were of Permian age but later, through studies carried out in the Yeso ravine, Bonaparte (1965) proposed a Miocene–Pliocene age and attributed the ichnogenus to xenarthran mammals. Bonaparte (1965) also documented associated footprints assigned to macrauchenid mammals, and bird footprints (compared with rheiforms and charadriiforms). *Venatoripes* was also tentatively recorded from the overlying Late Miocene–Late Pliocene Toro Negro Formation (Krapovickas et al., 2009). The material originally described by Frenguelli (1950) was included under the Triassic ichnogenus *Tetrapodichnus* by Haubold (1971a), which is currently considered a nomen dubium by Klein and Lucas (2021). *Venatoripes* was erected on poorly preserved material and is considered a nomen dubium.

4.4.6 Insectivora

Historically, the Order Insectivora was a "wastebasket" taxon for families of uncertain affinities and comprise six orders, including: Order Macroscelidea (elephant shrews or sengis), Order Afrosoricida (tenrecs and golden moles), Order Tubulidentata (the aardvark), Order Scandentia (treeshrews) and the Order Dermoptera (colugos).

Only one ichnogenus attributed to Insectivora has been proposed by Sarjeant and Langston (1994).

Ichnogenus *Schyromorphipus* (Sarjeant and Langston, 1994).

Fig. 8.33.

Type ichnospecies: *Schyromorphipus oxypages* (Sarjeant and Langston, 1994).

Diagnosis. Small, plantigrade mammalian footprints; all digits with claws. Pes somewhat larger than manus. Manual Digits II to V are thick and short, diverging from one another only at small angles, their total interdigital span not exceeding 35 degrees. The pollex is set behind

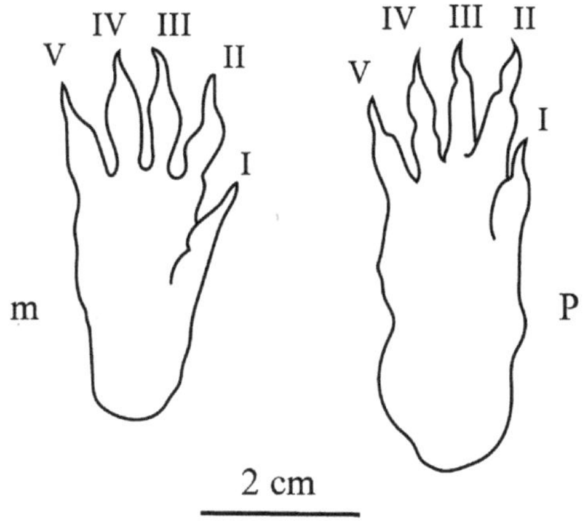

FIGURE 8.33 Sketch of outline of *Schyromorphipus oxypages. Modified from Sarjeant and Langston (1994).*

them. It is markedly shorter than Digits II to V, diverging from the axis of the manus at an angle of around 25 degrees. Palm proportionately long. Digits II to V of the pes are similar to those of the manus; the hallux is much shorter than the other digits but diverges at a lower angle (<15°) than does the pollex. Heel long (Sarjeant and Langston, 1994).

Discussion: Sarjeant and Langston (1994) pointed out that *Schyromorphipus oxypages* was made by an insectivore and compared it to European hedgehog footprints. The late Eocene fauna of Texas includes the record of three insectivores: *Leptictis, Apternodus*, and *Centetodon*. Sarjeant and Langston (1994) concluded that *Leptictis* may be responsible for producing *Schyromorphipus* because of similar size, whereas the remaining genera were much smaller.

5. Final considerations/thoughts

The tetrapod ichnotaxa during the Paleogene/Neogene interval show high ichnodiversity and various footprint morphologies. This ichnodiversity, however, does not necessarily indicate the diversity of vertebrate trackmakers. Although the morphology and size of the footprints are the main features for identification or establishment of new ichnotaxa, it is also necessary to consider some significant or accessory features that are used in the identification of ichnotaxa, and it differs between the vertebrate footprint morphotypes. Based on the current ichnotaxonomic revision of Paleogene/Neogene tetrapod footprints, about 16% of ichnogenera and 30% of ichnospecies are considered not valid or combined with other ichnotaxa (Table 8.6). The major decrease in the number of ichnospecies corresponds to the avian footprints, so that about 50% of the ichnospecies are considered not valid.

TABLE 8.6 Number of valid ichnogenera and ichnospecies of Paleogene/Neogene tetrapods revised in this study.

		Number of ichnotaxa		Number of valid ichnotaxa (based on this study)	
Main trackmakers		ichnogenera	ichnospecies	ichnogenera	ichnospecies
Amphibian		2	2	2	2
Reptilian		7	8	7	8
Avian		37	82	25	39
Mammals	Artiodactyl	20	45	17	37
	Perissodactyla and Paenungulata	16	28	14	27
	Carnivora and Creodonta	25	54	22	37
	Anagalida	5	7	5	7
	Xenarthra	3	3	3	3
	Insectivora	1	1	1	1
	Sum	116	230	97	161

Lockley et al. (2021) considered most used significant and accessory criteria as bird footprints ichnotaxobases, and it is necessary to make a similar evaluation for each main group of tetrapod footprints. The morphological features and geometric data of the tetrapod tracks, however, are two main ichnotaxobases. It seems that finding a unique and comprehensive framework for ichnotaxonomical identification of the tetrapod footprints is difficult. However, hopefully, we can agree on a set of main ichnotaxobases to be used for the major groups of trackmakers. We emphasize that morphological features of the footprints that are related to the track-maker anatomy are more significant than the preservation features (Marchetti et al., 2019a). Among these, the size of footprints is not the main ichnotaxobase (Bertling et al., 2006). Usually, quantitative study and statistical methods such as discriminant analysis of the morphospace of a specific ichnogenus facilitate identification and morphological variation analysis (e.g., Demathieu and Demathieu 2003) and/or identification of the producer. Hsieh and Uchman (2022) analyzed similarity among hoofed mammal tracks across the Cenozoic and concluded that Fourier analyses can be a useful tool for visualization of shape variation among mammal footprints, and perhaps other types of tetrapod ichnofossils or modern traces, which would be useful in testing hypotheses about their morphological similarities and differences across time and geographic location. However, ichnotaxonomy should not be based merely on any statistical analysis but on the identification of clear morphological contrasts.

Reconstruction of the paleobiogeographic distribution of the Paleogene/Neogene tetrapod tracks around the World is possible after the present compilation. Tables 8.7 and 8.8 show geographical distributions of the Paleogene/Neogene tetrapod ichnogenera that were named and reported for first time and include several countries: Argentina, Canada, Hungary, India, Iran, Mexico, Romania, Spain, Turkmenistan, Ukraine, and the USA (Figs. 8.34 and 8.35). Among these, the number of reports from the USA is

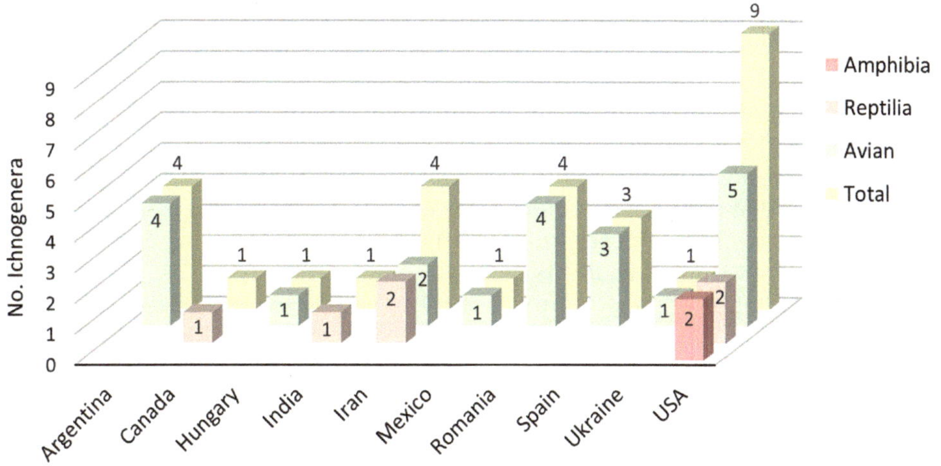

FIGURE 8.34 Geographical distribution and frequency of Paleogene/Neogene ichnogenera that are related to Amphibia, Reptilia and Aves type ichnogenera.

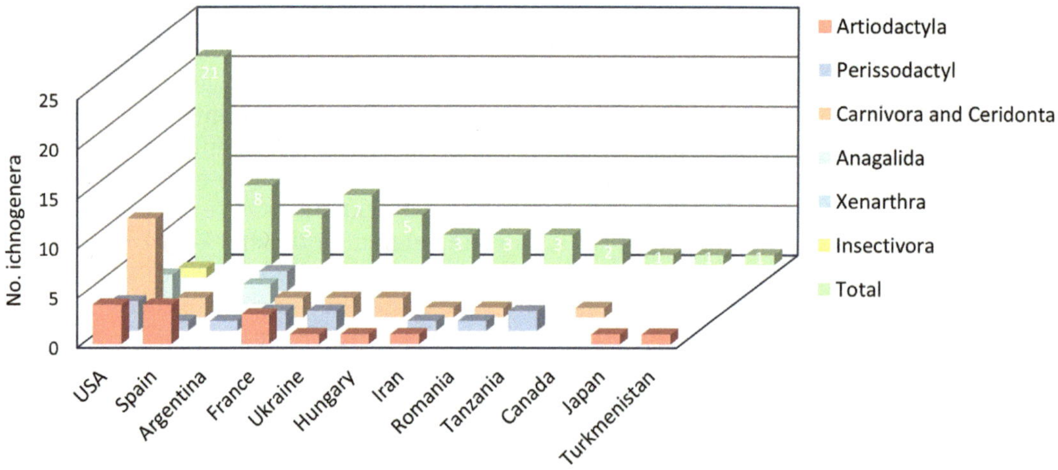

FIGURE 8.35 Geographical distribution and frequency of Paleogene/Neogene ichnogenera that are related to different groups of mammals.

considerable, and Argentina is next. The frequency of tetrapod footprints in each country is related to two main reasons, first the presence of outcrops of Paleogene/Neogene sedimentary rocks and second, the existence of vertebrate ichnologists working in those countries. The existence of some Paleogene/Neogene fossil tracksites in some countries has been known for a long time (e.g., Kordos et al., 2021), and recently, in some other countries, the study of Paleogene/Neogene vertebrate tracks is progressing (e.g., Abbassi, 2022; Xing et al., 2014).

The age distribution of the type ichnospecies of the Paleogene/Neogene tetrapod ichnotaxa is considerable, so that most reports belong to the Eocene and Miocene (Tables 8.7 and 8.8; Fig. 8.36). The geographical location of the reports of a given tetrapod ichnotaxon is thought-provoking, also, so that it helps us to understand the paleobiogeographical distribution of the ichnotaxon. For example, Abbassi et al. (2024) compared the geographical and age distributions of the proboscidean footprints (*Proboscipeda*) in the world (Fig. 8.37A). They showed that the oldest proboscidean footprints are from Eocene and most reports were from the Pleistocene strata. Lucas et al. (2023) evaluated the geographical distribution of *Gruipeda* around the world and with different ages (Fig. 8.37B). It shows that oldest *Gruipeda* appeared in the Triassic and was reported from the Mesozoic to Cenozoic and is most abundant in the Miocene. On the contrary, *Gruipeda* was been reported mostly from Argentina, the USA, and Spain. As far as countries are concerned, most reports are from the USA. Such evaluation is also required for other cases.

TABLE 8.7 Age and location of the Paleogene/Neogene ichnogenera assigned to Amphibia, Reptile and Aves (revised in this study).

Track-maker	Ichnogenus	Location	Age	Author
Amphibian	*Ambystomichnus*	USA	Paleocene	Peabody (1954)
	Ranipes	USA	Eocene	Lockley and Milner (2014)
Reptilian	*Albertasuchipes*	Canada	Paleocene	McCrea et al. (2004)
	Anticusuchipes	USA	Eocene	Mustoe (2019)
	Borealosuchipus	USA	Paleocene	Erickson (2005)
	Gandopodichnus	Iran	Miocene	Abbassi (2022)
	Indosuchipes	India	Oligocene	Rajkumar et al. (2015)
	Merthykhuwaripus	Iran	Miocene	Abbassi (2022)
Avian	*Anatipeda*	Romania	Miocene	Panin and Avram (1962)
	Ardeipeda	Romania	Miocene	Panin and Avram (1962)
	Avipeda	Ukraine	Miocene	Vyalov (1965)
	Charadriipeda	Romania	Miocene	Panin and Avram (1962)
	Gruipeda	Romania	Miocene	Panin and Avram (1962)
	Iranipeda	Iran	Miocene	Vyalov (1989)
	Leptoptilostipus	Spain	Eocene	Payros et al. (2000)
	Ornithoformipes	USA	Eocene	Patterson and Lockley (2004)
	Ornithotarnocia	Hungary	Miocene	Kordos (1985).
	Pavoformipes	Mexico	Miocene	Lockley and Delgado (2007)
	Persiavipes	Iran	Miocene	Abbassi and Dashtban (2021)
	Presbyorniformipes	USA	Eocene	Yang et al. (1995)
	Quadridigitus	USA	Eocene	Lockley et al. (2021)
	Reyesichnus	Argentina	Tertiary	Alonso et al. (1980)
	Rionegrina	Argentina	Miocene	Melchor et al. (2023)
	Rivavipes	USA	Eocene	Mustoe et al. (2012)
	Roepichnus	Spain	Miocene	Doyle et al. (2000)
	Uhangrichnus	USA	Eocene	Yang et al. (1995)
	Uvaichnites	Spain	Miocene	Díaz-Martínez et al. (2012)

TABLE 8.8 Age and location of the mammalian Paleogene/Neogene ichnogenera (revised in this study).

Track-maker		Ichnogenus	Location	Age	Author
Mammalia	Artiodactyl	*Anoplotheriipus*	France	Eocene	Ellenberger (1980)
		Bifidipes	France	Oligocene	Demathieu et al. (1984)
		Bijugopeda	USA	Miocene	Sarjeant and Reynolds (1999)
		Bijugopeda	USA	Miocene	Sarjeant and Reynolds (1999)
		Bothriodontipus	Spain	Oligocene	Casanovas and Santafé-Llopis (1982)
		Camelipeda	Turkmenistan	Pliocene	Vyalov (1983)
		Cervipus	Japan	Plio-Pleistocene	Matsukawa et al. (2007)
		Corymbipes	USA	Eocene	Sarjeant and Langston (1994)
		Diplartiopus	France	Eocene	Ellenberger (1980)
		Dizygopodium	USA	Miocene	Sarjeant and Reynolds (1999)
		Entelodontipus	Spain	Oligocene	Casanovas and Santafé (1982)
		Fustinianapodus	Spain	Miocene	Díaz-Martínez et al. (2020)
		Megapecoripeda	Hungary	Miocene	Kordos (1985)
		Pazhanipeda	Iran	Miocene	Abbassi and Kazemi (2023)
		Pecoripeda	Ukraine	Miocene	Vyalov (1965)
	Perissodactyl	*Apoxypus*	USA	Eocene	Sarjeant and Langston (1994)
		Dicerotinichnus	Tanzania	Pliocene	Guièrin and Demathieu (1993)
		Hippipeda	Ukraine	Miocene	Vyalov (1966)
		Llastaya	Argentina	Pliocene	Vera and Krapovickas (2022)
		Lophiopus	France	Eocene	Ellenberger (1980)
		Macrauchenichnus	Argentina	Mio-Pliocene	Angulo and Casamiquela (1982)
		Moropopus	Iran	Eocene	Abbassi et al. (2017)
		Oplidactylapes	USA	Eocene	Mustoe and Hopkins (2013)
		Palaeotheriipus	France	Eocene	Ellenberger (1980)
		Palimmecopus	USA	Eocene	Sarjeant and Langston (1994)
		Plagiolophustipus	Spain	Oligocene	Santamaria et al. (1989)
		Proboscipeda	Romania	Miocene	Panin and Avram (1962)
		Rhinoceripeda	Ukraine	Miocene	Vyalov (1965)
		Thrinaxopus	USA	Eocene	Sarjeant and Langston (1994)

TABLE 8.8 Age and location of the mammalian Paleogene/Neogene ichnogenera (revised in this study).—cont'd

Track-maker	Ichnogenus	Location	Age	Author
Carnivora and Creodonta	*Axiciapes*	USA	Eocene	Sarjeant and Langston (1994)
	Bestiopeda	Ukraine	Miocene	Vyalov (1965)
	Canipeda	Romania	Miocene	Panin and Avram (1962)
	Carnotipus	Spain	Miocene	López et al. (1992)
	Carnivoripeda	Hungary	Miocene	Kordos (1987)
	Creodontipus	Spain	Oligocene	Santamaria et al. (1989)
	Dischidodactylus	USA	Eocene	Sarjeant and Wilson (1988)
	Falcatipes	USA	Eocene	Sarjeant and Langston (1994)
	Felipeda	Ukraine	Miocene	Vyalov (1966)
	Hyaenodontipus	France	Eocene	Ellenberger (1980)
	Hirpexipes	USA	Miocene	Sarjeant et al. (2002)
	Khafipus	Iran	Oligocene	Abbassi et al. (2015)
	Mustelipeda	Hungary	Miocene	Kordos (1985)
	Phacelopus	USA	Eocene	Sarjeant and Langston (1994)
	Platykopus	USA	Miocene	Sarjeant et al. (2002)
	Pycnodactylopus	USA	Miocene	Sarjeant et al. (2002)
	Quiritipes	USA	Eocene	Sarjeant et al. (2002)
	Sarjeantipes	Canada	Paleocene	McCrea et al. (2004)
	Sarcotherichnus	France	Oligoene	Demathieu et al. (1984)
	Tetrastoibopus	USA	Eocene	Sarjeant and Langston (1994)
	Zanclonychopus	USA	Eocene	Sarjeant and Langston (1994)
Anagalida	*Musaltipes*	USA	Miocene	Lockley and Milner (2014)
	Porcellusignum	Argentina	Mio-Pliocene	Angulo and Casamiquela (1982)
	Ptyariopus	USA	Eocene	Sarjeant and Langston (1994)
	Tricorynopus	USA	Eocene	Sarjeant and Langston (1994)
	Tacheria	Argentina	Oligocene-Miocene	Krapovickas and Nasif (2011)
Xenarthra	*Falsatorichnum*	Argentina	Mio-Pliocene	Angulo and Casamiquela (1982)
	Megatherichnum	Argentina	Miocene-Pliocene	Casamiquela (1974)
	Venatoripes	Argentina	Mio-Pliocene	Frenguelli (1950)
Insectivora	*Schyromorphipus*	USA	Eocene	Sarjeant and Langston (1994)

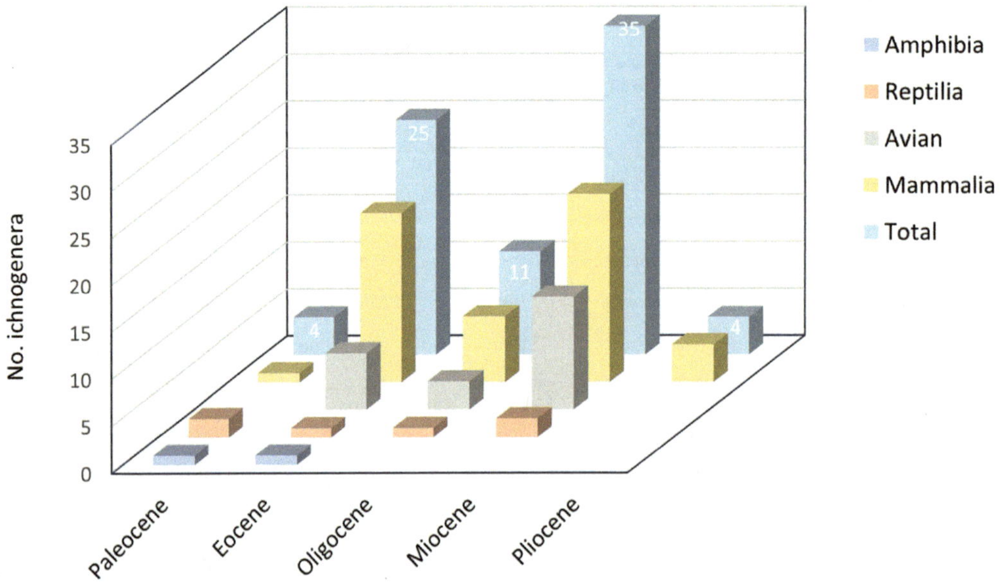

FIGURE 8.36 Age distribution of ichnogenera of the continental tetrapod land trackmakers during the Paleogene/Neogene.

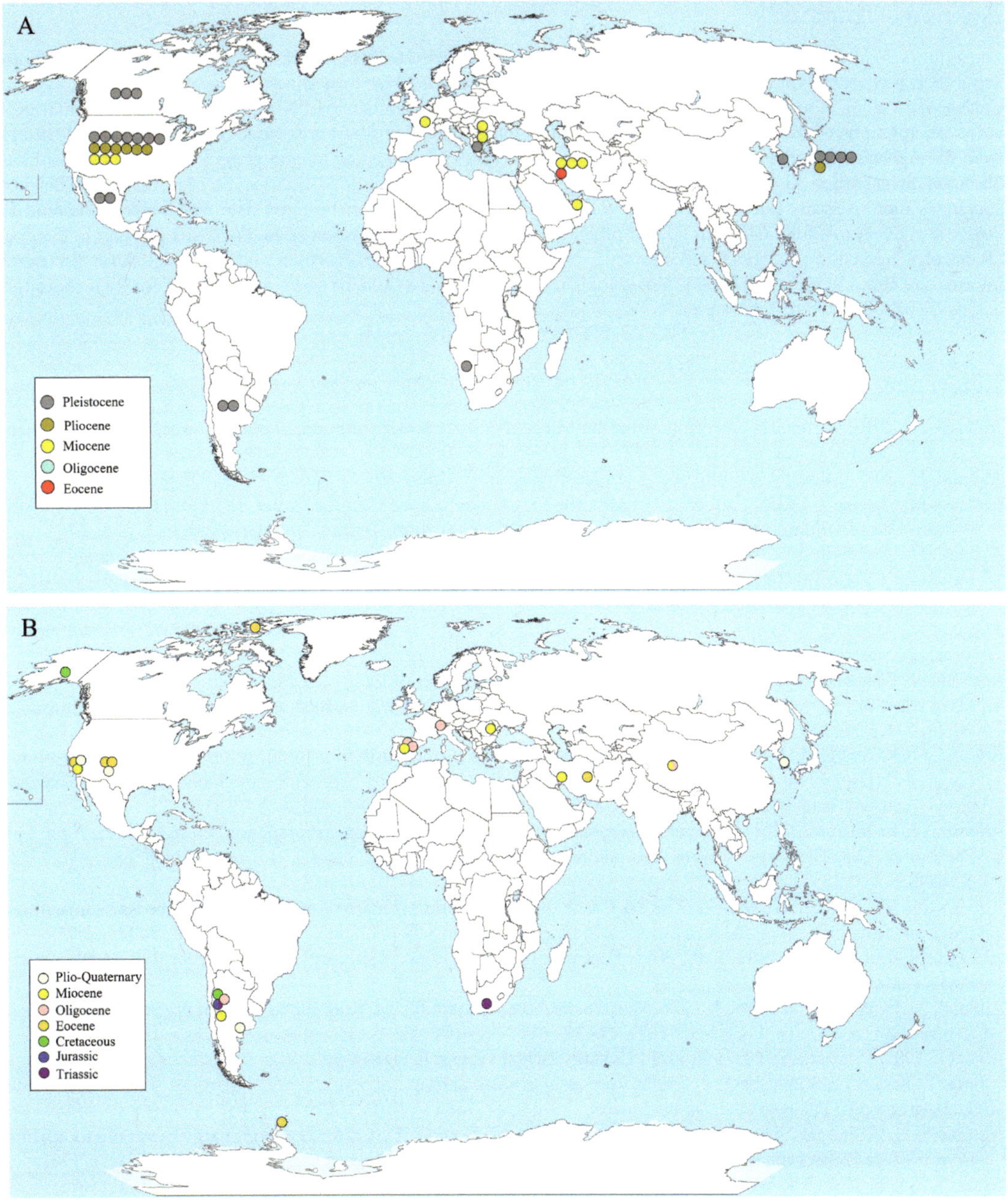

FIGURE 8.37 Location and age of the *Proboscipeda* (A) and *Gruipeda* (B) in the world. *(A) From Abbassi et al. (2024), and (B) Modified from Lucas et al. (2023). Map lines delineate study areas and do not necessarily depict accepted national boundaries.*

Acknowledgments

The authors thank the volume conveners the for invitation to contribute this chapter. Sara Mercado and Scott P. Kottkamp (San Bernardino County Museum, California, USA) are thanked for providing photographs of the holotype of *Culcitapeda*. Valeria Aquino (Cátedra de Paleontología General de la Universidad Nacional de Salta, Argentina) provided photographs of the type material of *Reyesichnus*. Participation of RNM was funded by project PICT2019-114 from the Agencia Nacional de Promoción Científica y Tecnológica of Argentina and project G22 from Universidad Nacional de La Pampa. I.D.M. is supported by a Ramón y Cajal fellowship (RYC-2022) and by the Ministry of Science and Innovation of Spain. We thank Dr. László Kordos for his useful information and data. We thank Dr. Leonid E. Popov and Dr. Mansoureh Ghobadi Pour for their help in translation of Russian texts. We thank Dr. László Kordos, Dr. Nicolae Panin and Dr. Titus Brustur for their helps. We also thanks Dr. Puértolas and Pérez-García for the useful information about Reptilia phylogeny. We thank Dr. Spencer G. Lucas and an anonymous referee for their thorough reviews which improved this chapter.

References

Abbassi, N., 2010. Vertebrate footprints from the Miocene Upper Red Formation, Shokorchi area, Zanjan province, NW Iran. Ichnos 17 (2), 115–126.

Abbassi, N., 2022. Miocene Wildlife of Zanjan Northwest Iran. University of Zanjan Press, 304 pp.

Abbassi, N., Amini, A., 2008. [Ichnology of Miocene vertebrate footprints in the upper Red Formation, eyvan-e-key section, east of Tehran, Iran]. Geological survey of Iran publication, Tehran, Iran. Geosciences 17, 54–67.

Abbassi, N., Dashtban, H., 2021. Vertebrate footprints from the Agha Jari Formation (late miocene–pliocene), Zagros mountains, and a review of the cenozoic vertebrate ichnites in the Persian Gulf region. Bull. Geosci. 96, 159–179. https://doi.org/10.3140/bull.geosci.1809.

Abbassi, N., Alinasiri, S., Lucas, S.G., 2017. New localities of late Eocene vertebrate footprints from the tarom mountains, northwestern Iran. Hist. Biol. 29, 987–1006. https://doi.org/10.1080/08912963.2016.1267162.

Abbassi, N., Gasparik, M., Kordos, L., Esmaeili, F., 2023. Evaluation of *Iranipeda abeli* (Avian track) from the Zagros Mountains, SW Iran: re-examination of the original material and overview of primary sources. Hist. Biol. https://doi.org/10.1080/08912963.2023.2173071.

Abbassi, N., Karim, K.H., Mohialdeen, I.M.J., Sharbazheri, K.M., 2021. Vertebrate footprints and a mammal mud-bath trace fossil (laspichnia) from the mukdadiya formation (late miocene–pliocene), Chamchamal area, Kurdistan region, Northeast Iraq. Ichnos 28 (1), 72–83. https://doi.org/10.1080/10420940.2020.1784156.

Abbassi, N., Kazemi, A., 2023. Artiodactyl footprint *Pazhanipeda kiyani* ichnog. ichnosp. nov. From Miocene Agha Jari Formation, Gotvand area, Zagros mountains, SW Iran. Hist. Biol. https://doi.org/10.1080/08912963.2023.2276145.

Abbassi, N., Mashhadizadeh Maleki, S., 2020. Carnivore and artiodactyl mammal footprints from the Karaj formation (Eocene), Taleghan valley, Alborz mountains, north Iran. Iran J. Sci. Technol. Trans. A Sci 44, 1407–1417.

Abbassi, N., Lucas, S.G., Zaare, G.R., 2015. First report of Oligocene vertebrate footprints from Iran. Palaeogeogr. Palaeoclimatol. Palaeoecol. 440, 78–89.

Abbassi, N., Shakeri, S., Rabbani, J., 2024. Miocene proboscidean footprints from the upper Red Formation, Avaj area, Central Iran. Sci. Quart. J. Geosci. 34 (1), 19–34.

Abel, O., 1935. Vorzeitliche Lebensspuren. Gustav Fischer Verlag, Jena, 644 pp.

Adatte, T., Keller, G., Stinnesbeck, W., 2002. Late cretaceous to early paleocene climate and sea-level fluctuations: the Tunisian record. Palaeogeogr. Palaeoclimatol. Palaeoecol. 178 (3–4), 165–196.

Agnolín, F.L., Chimento, N.R., Campo, D.H., Magnussen, M., Boh, D., De Cianni, F., 2019. Large carnivore footprints from the late Pleistocene of Argentina. Ichnos 26, 119–126. https://doi.org/10.1080/10420940.2018.1479962.

Alekseev, M.N., 1987. The Border between the Neogene and Quaternary Systems in the USSR. Nauka, Russia.

Alexander, R.M., 2003. Principles of Animal Locomotion. Princeton university press, Princeton, 371 pp.

Alibhai, S.K., Jewell, Z.C., Law, P.R., 2008. A footprint technique to identify white rhino *Ceratotherium simum* at individual and species levels. Endanger. Species Res. 4, 205–218.

Allen, J.R.L., 1997. Subfossil mammalian tracks (Flandrian) in the Severn Estuary, SW Britain: mechanics of formation, preservation and distribution. Philos. Trans. R. Soc. Lond. B Biol. Sci. 352, 481–518.

Alonso, R.N., Carbajal, E., Raskovsky, M., 1980. Hallazgo de icnitas (Aves, Charadriiformes) en el Terciario de la Puna Argentina, 2° Congreso Argentino de Paleontología y Bioestratigrafía y 1° Congreso Latinoamericano de Paleontología, Actas. Asociación Paleontológica Argentina, Buenos Aires, pp. 75–83, 1978.

Alonso, R.N., Marquillas, R.A., 1986. Nueva localidad con huellas de dinosaurios y primer hallazgo de huellas de aves en la Formación Yacoraite (Maastrichtiano) del Norte Argentino. In: Actas del IV Congreso Argentino de Paleontología y Bioestratigrafía. Editorial Inca, Mendoza, vol. 2, pp. 33–41.

Anderson, D.E., Miesner, M.D., Jones, M.L., 2013. Veterinary Techniques in Llamas and Alpacas. John Wiley & Sons, Hoboken, 348pp.

Angulo, R.J., Casamiquela, R.M., 1982. Estudio estratigráfico de las unidades aflorantes en los acantilados de la costa norte del Golfo de San Matías (Río Negro y extremo austral de Buenos Aires) entre los meridianos 62 30 y 64 30 W. Mundo Ameghiniano 2, 20–73.

Anton, M., López, G., Santamaria, R., 2004. Carnivore trackways from the Miocene site of Salinas de Añana (Alava, Spain). Ichnos 11 (3–4), 371–384.

Aramayo, S.A., 2007. Neogene vertebrate palaeoichnology of the north Atlantic coast of the Río Negro province, Argentina. Arq. Mus. Nac. 65 (4), 573–584.

Aramayo, S.A., Manera de Bianco, T., 1987. Hallazgo de una icnofauna continental (Pleistoceno tardío) en la localidad de Pehuen Co, Provincia de Buenos Aires, Argentina. Parte I: Edentata, Litopterna, Proboscidea. Parte II: Carnivora, Artiodactyla y Aves. In: 4° Congreso Latinoamericano de Geología, Actas, Santa Cruz de la Sierra. Asociación Boliviana de Paleontología, pp. 516–547.

Aramayo, S.A., Manera de Bianco, T., Bastianelli, N.V., Melchor, R.N., 2015. Pehuen Co: updated taxonomic review of a late Pleistocene ichnological site in Argentina. Palaeogeogr. Palaeoclimatol. Palaeoecol. 439, 144–165. https://doi.org/10.1016/j.palaeo.2015.07.006.

Asher, R.J., Gunnell, G.F., Seiffert, E.R., Pattinson, D., Tabuce, R., Hautier, L., Sallam, H.M., 2017. Dental eruption and growth in Hyracoidea (Mammalia, Afrotheria). J. Vertebr. Paleontol. 37, e1317638. https://doi.org/10.1080/02724634.2017.1317638.

Astibia, H., del Valle de Lersundi, J., Murelaga, X., 1994. Icnitas de artiodáctilos (Mammalia) del Paleógeno de Olcoz (Depresión del Ebro, Navarra). Estud. Geol. 50, 119–126.

Astibia, H., Pereda-Suberbiola, X., Payros, A., Murelaga, X., Berreteaga, A., Baceta, J.I., Badiola, A., 2007. Bird and mammal footprints from the tertiary of Navarre (western Pyrenees). Ichnos 14 (3–4), 75–184.

Ataabadi, M.M., Khazaee, A.R., 2004. New Eocene mammal and bird footprints from Birjand area, eastern Iran. Ichnos 11, 363–370.

Avanzini, M., Renesto, S., 2002. A review of *Rhynchosauroides tirolicus* Abel, 1926 ichnospecies (Middle Triassic: Anisian-Ladinian) and some inferences on *Rhynchosauroides* trackmaker. Riv. Ital. Paleontol. Stratigr. 108, 51–66.

Avanzini, M., García-Ramos, J.C., Lires, J., Menegon, M., Pinuela, L., Fernández, L.A., 2005. Turtle tracks from the Late Jurassic of Asturias, Spain. Acta Palaeontol. Pol. 50 (4), 743–755.

Avanzini, M., Piñuela, L., García-Ramos, J.C., 2010a. First report of a late jurassic lizard-like footprint (Asturias, Spain). J. Iber. Geol. 36 (2), 175–180.

Avanzini, M., Piñuela, L., Ruiz-Omeñaca, I., Garcia-Ramos, J.C., 2010b. The crocodile track *Hatcherichnus* from the upper jurassic of Asturia (Spain). Bull New Malique Mus. Nat. Hist. Sci. 51, 89–92.

Azuma, Y., Arakawa, Y., Tomida, Y., Currie, P.J., 2002. Early Cretaceous bird tracks from the Tetori group, Fukui Prefecture, Japan. Mem. Fukui Prefect. Dinosaur Mus. 1, 1–8.

Bai, B., Meng, J., Janis, C.M., Zhang, Z.-Q., Wang, Y.Q., 2020. Perissodactyl diversities and responses to climate changes as reflected by dental homogeneity during the Cenozoic in Asia. Ecol. Evol. 10, 6333–6355.

Barbosa, F.H.D.S., Porpino, K.D.O., Rothschild, B.M., Cabral, U.G., Bergqvist, L.P., 2017. Arthritic lesions and congenital fusion in foot bones of *Panochthus* sp. (Xenarthra, Cingulata). An. Acad. Bras. Cienc. 91, 1–6.

Barnet, J.S., Littler, K., Kroon, D., Leng, M.J., Westerhold, T., Röhl, U., Zachos, J.C., 2018. A new high-resolution chronology for the late Maastrichtian warming event: establishing robust temporal links with the onset of Deccan volcanism. Geology 46 (2), 147–150.

Basan, P.B., 1978. Introduction. In: Basan, P.B. (Ed.), Trace Fossil Concepts. Society of Economic Paleontologists and Mineralogists Short Course, Oklahoma City, pp. 1–10. Number 5.

Baskin, J.A., 1998. Mustelidae. In: Janis, C.M., Scott, K.M., Jacobs, L.L. (Eds.), Evolution of Tertiary Mammals of North America. Cambridge University press, Cambridge, pp. 152–173.

Bayer, F., 1885. Über die Extremitäten einer jungen Hatteria. Sitzungsberichte der Akademie der Wissenschaften mathematisch-naturwissenschaftliche Klasse 90, 237–245.

Bellani, G.G., 2019. Felines of the World: Discoveries in Taxonomic Classification and History. Elsevier Inc., New York, 486 pp.

Belvedere, M., Bennett, M.R., Marty, D., Budka, M., Reynolds, S.C., Bakirov, R., 2018. Stat-tracks and mediotypes: powerful tools for modern ichnology based on 3D models. PeerJ 6, e4247.

Bennett, M.R., Bustos, D., Belvedere, M., Martinez, P., Reynolds, S.C., Urbanet, T., 2019. Soft-sediment deformation below mammoth tracks at White Sands National Monument (New Mexico) with implications for biomechanical inferences from tracks. Palaeogeogr. Palaeoclimatol. Palaeoecol. 527, 25–38. https://doi.org/10.1016/j.palaeo.2019.04.023.

Benton, M.J., 1990. Phylogeny of the major tetrapod groups: morphological data and divergence dates. J. Mol. Evol. 30, 409–424.

Benton, M.J., 2005. Vertebrate Paleontology, 3th Edit. Blackwell Publishing Ltd, Oxford. 472 pp.

Berman, S.L., 1985. Convergent evolution in the hindlimb of bipedal rodents. Zoolog. Syst. Evol. Res. 23 (1), 59–77.

Bernor, R.L., Kaya, F., Kaakinen, A., Saarinen, J., Fortelius, M., 2021. Old world hipparion evolution, biogeography, climatology and ecology. Earth Sci. Rev. 221, 103784.

Bertling, M., Braddy, S.J., Bromley, R.G., Demathieu, G.R., Genise, J., Mikuláš, R., Nielsen, J.K., Nielsen, K.S.S., Rindsberg, A.K., Schlirf, M., Uchman, A., 2006. Names for trace fossils: a uniform approach. Lethaia 39, 265–286.

Bertling, M., Buatois, L.A., Knaust, D., Laing, B., Mángano, M.G., Meyer, N., Mikuláš, R., Minter, N.J., Neumann, C., Rindsberg, A.K., Uchman, A., Wisshak, M., 2022. Names for trace fossils 2.0: theory and practice in ichnotaxonomy. Lethaia 55, 1–19.

Biron, P.E., Dutuit, J.-M., 1981. Figurations sedimentaires et traces d'activite au sol dans le Trias de la formation d'Argana et de l'Ourika (Maroc). Bull. Mus. Natl. Hist. Nat. 4, 399–427.

Bonaparte, J.F., 1965. Nuevas icnitas de la Quebrada del Yeso (La Rioja) y reconsideración de la edad de los afloramientos. Acta Geol. Lilloana 7, 5–16.

Botfalvai, G., Magyar, J., Watah, V., Szarvas, I., Szolyák, P., 2023. Large-sized pentadactyl carnivore footprints from the early Miocene fossil track site at Ipolytarnóc (Hungary): 3D data presentation and ichnotaxonomical revision. Hist. Biol. 35 (9), 1709–1725.

Braddy, S.J., Morrissey, L.B., Yates, A.M., 2003. Amphibian swimming traces from the Lower Permian of southern New Mexico. Palaeontology 46 (4), 671–683.

Brand, L.R., 1996. Variations in salamander trackways resulting from substrate differences. J. Paleontol. 70 (6), 1004–1010.

Bravo Cuevas, V.M., Jimenez Hidalgo, E., Cabral Perdomo, M.A., Contreras-López, M., 2019. The first Mexican record of peccary footprints (Artiodactyla, Tayassuidae) from the late Cenozoic of Puebla: ichnotaxonomy and palaeobiological considerations. Hist. Biol. 31 (9), 1135–1144.

Brochu, C.A., 2001. Progress and future directions in archosaur phylogenetics. J. Paleontol. 75 (6), 1185–1201.

Bromley, R.G., 1981. Concepts in ichnotaxonomy illustrated by small round holes in shells. Acta Geol. Hisp. 16, 55–64.

Budras, K.D., Sack, W.O., Rock, S., 2009. Anatomy of the Horse: An Illustrated Text. Schlütersche, Hannover, 135pp.

Brown, R., Ferguson, J., Lawrence, M., Lees, D., 1987. Tracks & Signs of the Birds of Britain and Europe. Christopher Helm Ltd, Kent, p. 225.

Buck, P.V., Ghilardi, A.M., Aureliano, T., Fernandes, M.A., 2022. Lacertoid tracks from the Botucatu Formation (Lower Cretaceous) with different locomotor behaviors: a new trackmaker with novel paleoecological implications. J. South Am. Earth Sci. 116, 103825.

Buckley, L.G., McCrea, R.T., Lockley, M.G., 2016. Analyzing and resolving Cretaceous avian ichnotaxonomy using multivariate statistical analyses. Approaches and results. In: Falkingham, P.L., Marty, D., Richter, A. (Eds.), Dinosaur Tracks. The Next Steps. Indiana University Press, Bloomington, pp. 259–309.

Buffetaut, E., 1979. *Sokotosuchusian wilsoni* and the evolution of the dyrosaurid crocodiles. Niger. Field Monogr. 1, 31–41.

Buffetaut, E., 2004. Footprints of giant birds from the upper Eocene of the Paris basin: an ichnological enigma. Ichnos 11, 357–362.

Burrow, C.J., Turner, S., 2012. Fossil fish taphonomy and the contribution of microfossils in documenting Devonian vertebrate history. Earth Life Global Biodiv. Extinct. Interv. Biogeogr. Perturb. Through Time 189–223.

Buynevich, I.V., 2015. Recent vertebrate tracks in sandy substrates and their paleoenvironmental implications: examples from coastal Lithuania. Baltica 28 (1), 29–40.

Caldeira, K.G., Rampino, M.R., 1990. Deccan volcanism, greenhouse warming, and the Cretaceous/Tertiary boundary. Geol. Soc. Am. Spec. Pap 247, 117–124.

Camens, A.B., Worthy, T.H., 2019. Pliocene avian footprints from the lake Eyre basin, south Australia. J. Vertebr. Paleontol. 39 (4), e1676764.

Campos-Medina, J., Moreno, K., Rojas, J., Gómez, G., García, J.L., Buldrini, K.E., Lüthgens, C., Rodríguez, E., Álvarez, R., Abarzúa, A.M., 2022. The oldest record of the Lamini Tribe and Proboscidea order in the southwestern margin of the Andes Mountain range: late Pleistocene mammalian footprints at the Pelluco fossil forest sanctuary. J. South Am. Earth Sci. 118, 103940.

Casamiquela, R.M., 1961. Sobre la presencia de un mamífero en el primer elenco (icnológico) de vertebrados del Jurásico de la Patagonia. Physis 22 (63), 225–233.

Casamiquela, R.M., 1974. El bipedismo de los megaterioideos estudio de pisadas fósiles en la Formación Rio Negro típica. Ameghiniana 11 (3), 249–282.

Casamiquela, R., Chong, G., 1975. Icnitas (Mammalia: equidae?) en rocas del Plio- pleistoceno de la costa, Provincia de Antofagasta. In: 1er Congreso Argentino de Paleontología y Bioestratigrafía, vol. 2. Universidad Nacional de Tucumán, Tucumán, pp. 621–632.

Casanovas-Cladellas, M.L., Santafé-Llopis, J.V., 1982. Icnofauna Oligocena de Agramunt (Lérida, España). Acta Geol. Hisp. 17, 113–119.

Castanera, D., Pascual-Arribas, C., Canudo, J.I., Puértolas-Pascual, E., 2021. A new look at *Crocodylopodus meijidei*: implications for crocodylomorph locomotion. J. Vertebr. Paleontol. 41 (5), e2020803.

Castelló, J.R., 2018. Canids of the World: Wolves, Wild Dogs, Foxes, Jackals, Coyotes, and Their Relatives. Princeton University Press, Princeton, 332pp.

Čerňanský, A., Augé, M.L., Phelizon, A., 2020. Dawn of lacertids (Squamata, Lacertidae): new finds from the upper Paleocene and the lower Eocene. J. Vertebr. Paleontol. 40 (1), e1768539.

Cisneros, J.C., Day, M.O., Groenewald, J., Rubidge, B.S., 2020. Small footprints expand middle Permian amphibian diversity in the South African Karoo. Palaios 35 (1), 1–11.

Citton, P., Caratelli, M., Cónsole-Gonella, C., De Valais, S., Montes, R.M., Archuby, F.M., Díaz-Martínez, I., 2022. New vertebrate tracks from the Neogene Río Negro Formation (Río Negro province, Argentina). Ameghiniana 59 (6), 407–417.

Clack, J.A., 2009. The fin to limb transition: new data, interpretations, and hypotheses from paleontology and developmental biology. Annu. Rev. Earth Planet Sci. 37, 163–179.

Clack, J.A., 2012. Gaining Ground: The Origin and Evolution of Tetrapods. Indiana University Press.

Clifford, A.B., 2010. The evolution of the unguligrade manus in artiodactyls. J. Vertbr. Paleont 30, 1827–1839. https://doi.org/10.1080/02724634.2010.521216.

Coates, M.I., 2009. Beyond the age of fishes. Nature 458 (7237), 413–414.

Coates, M.I., Ruta, M., 2007. Skeletal Changes in the Transition from Fins to Limbs. Univ. of Chicago Press, Chicago, pp. 15–38.

Coates, M.I., Ruta, M., Friedman, M., 2008. Ever since Owen: changing perspectives on the early evolution of tetrapods. Annu. Rev. Ecol. Evol. Syst. 39, 571–592.

Coombs, M.C., 1983. Large mammalian clawed herbivores: a comparative study. Trans. Am. Phil. Soc. 73, 1–96.

Cope, E.D., 1882. The classification of the ungulate Mammalia. Proc. Am. Phil. Soc. 20, 438–447.

Coria, R.A., Currie, P.J., Eberth, D., Garrido, A., 2002. Bird footprints from the Anacleto Formation (Late Cretaceous), Neuquén, Argentina. Ameghiniana 39, 453–463.

Costeur, L., Balme, C., Legal, S., 2009. Early Oligocene mammal tracks from southeastern France. Ichnos 16, 257–267.

Covacevich, V.C., Lamperein, C.R., 1970. Hallazgo de icnitas en Peninsula Fildes, Isla Rey Jorge, Archipiélago Shetland del Sur, Antarctica. Series Cientifica del Instituto Antártico Chileno 1, 55–74.

Craigie, E.H., 1969. Bensley's Practical Anatomy of the Rabbit: An Elementary Laboratory Text-Book in Mammalian Anatomy (Revised and Edited, Reprinted 1969). University of Toronto press, Toronto, 406pp.

Crumly, C.R., Sánchez-Villagra, M.R., 2004. Patterns of variation in the phalangeal formulae of land tortoises (Testudinidae): developmental constraint, size, and phylogenetic history. J. Exp. Zool. B Mol. Dev. Evol. 302 (2), 134–146.

Cucina, A., Di Francesco, A., Saletti, R., Pittalà, M.G.G., Zilberstein, G., Zilberstein, S., Cunsolo, V., 2022. Metaproteomic analysis of two mammoth's trunks by EVA technology and high-resolution mass spectrometry for an indirect picture of their habitat and the characterization of the collagen type I, alpha-1 and alpha-2 sequence. Amino Acids 54 (6), 935–954.

Currie, P.J., 1981. Bird footprints from the Gething Formation (Aptian, Lower Cretaceous) of northeastern British Columbia, Canada. J. Vertebr. Paleontol. 1, 257–264.

Cuvier, G., 1807. Sur les différentes especes de crocodiles vivans et sur leurs caracteres distinctifs. Ann. Natl. Mus. Hist. Nat. Paris. 10, 8–86.

Dal Sasso, C., Pasini, G., Fleury, G., Maganuco, S., 2017. *Razanandrongobe sakalavae*, a gigantic mesoeucrocodylian from the Middle Jurassic of Madagascar, is the oldest known notosuchian. PeerJ 5, e3481.

Darlim, G., Lee, M.S., Walter, J., Rabi, M., 2022. The impact of molecular data on the phylogenetic position of the putative oldest crown crocodilian and the age of the clade. Biol. Lett. 18 (2), 20210603.

Davoodi, L., Mirzaie Ataabadi, M., Omrani, M., 2016. Paleoichnology of tertiary vertebrates from Taleghan area, NE Qazvin. In: Paper Presented at the 34th National and 2nd International Geosciences Congress Tehran.

de Andrade, R.C.L.P., Sayão, J.M., 2014. Paleohistology and lifestyle inferences of a dyrosaurid (Archosauria: Crocodylomorpha) from Paraíba basin (northeastern Brazil). PLoS One 9 (7), e102189.

de Gibert, J.M., Sáez, A., 2009. Paleohydrological significance of trace fossil distribution in Oligocene fluvial-fan-to-lacustrine systems of the Ebro Basin, Spain. Palaeogeogr. Palaeoclimatol. Palaeoecol. 272 (3–4), 162–175.

de Queiroz, K., Gauthier, J., 1990. Phylogeny as a central principle in taxonomy: phylogenetic definitions of taxon names. Syst. Zool. 39 (4), 307–322.

de Valais, S., Melchor, R.N., 2008. Ichnotaxonomy of bird-like footprints: an example from the Late Triassic-Early Jurassic of northwest Argentina. J. Vertebr. Paleontol. 28, 145–159.

De Valais, S., 2009. Ichnotaxonomic revision of *Ameghinichnus*, a mammalian ichnogenus from the Middle Jurassic La Matilde Formation, Santa Cruz province, Argentina. Zootaxa 2203 (1), 1–21.

de Valais, S., Cónsole-Gonella, C., 2019. An updated review of the avian footprint record from the Yacoraite Formation (Maastrichtian-Danian), northwestern Argentina. Ichnos 26, 224–241.

Demathieu, G., Demathieu, P., 2003. Concerning the erection of ichnogenera and ichnospecies in vertebrate ichnotaxonomy. Ichnos 9 (3–4), 117–121.

Demathieu, G., Ginsburg, L., Guerin, C., Truc, G., 1984. Étude paléontologique, ichnologique et paléoécologique du gisement oligocène de Saignon (bassin d'Apt,Vaucluse). Bull. Mus. Natl. Hist. Nat. 4e série, Section C 2, 153–183.

Denison, R.H., 1938. The broad-skulled Pseudocreodi. Ann. N. Y. Acad. Sci. 37, 163–256.

Desnoyers, J., 1859a. Note sur des empreintes de pas d'animaux dans le gypse des environs de Paris, particulièrement de la vallée de Montmorency. Comptes Rendus De L'académie Des Sciences De Paris 49, 67–73.

Desnoyers, J., 1859b. Sur des empreintes de pas d'animaux dans le Gypse des environs de Paris, et particulièrement de la vallée de Montmorency. Bull. Soc. Geol. Fr. 16, 936–944.

Díaz-Martínez, I., Murelaga, X., Astibia, H., 2018. Huellas cenozoicas de aves y mamíferos de los Pirineos occidentales. In: Registro fósil de los Pirineos occidentales: bienes de interés paleontológico y geológico. Proyección Social. Servicio Central de Publicaciones= Argitalpen Zerbitzu Nagusia, pp. 161–168.

Díaz-Martínez, I., Fernández, S.G., Hernández, J.M., Murelaga, X., Pérez-Lorente, F., 2011. Lower Miocene bird and mammal footprints from La Rioja (Ebro Basin, Spain). Ameghiniana 48, 139–153.

Díaz-Martínez, I., Hernández, J.M., García Fernández, S., Murelaga, X., Pérez-Lorente, F., 2012. *Uvaichnites riojana*: a new crane-like bird ichnotaxon from the lower Miocene of La Rioja (Ebro Basin, Spain). Proc. Geol. Assoc. 123, 464–470.

Díaz-Martínez, I., Suarez-Hernando, O., Larrasoaña, J.C., Martínez-García, B.M., Baceta, J.I., Murelaga, X., 2020. Multi-aged social behaviour based on artiodactyl tracks in an early Miocene palustrine wetland (Ebro Basin, Spain). Sci. Rep. 10, 1099.

Díaz-Martínez, I., Suarez-Hernando, O., Martínez-García, B., Hernández, J., Fernández, S.G., Pérez-Lorente, F., Murelaga, X., 2015. Early Miocene shorebird-like footprints from the Ebro basin, La Rioja, Spain: paleoecological and paleoenvironmental significance. Palaios 30, 424–431.

Díaz-Martínez, I., Suarez-Hernando, O., Martínez-García, B.M., Larrasoaña, J.C., Murelaga, X., 2016. First bird footprints from the lower Miocene Lerín Formation, Ebro Basin, Spain. Palaeontologia Electronica 19.1.7A, 1–15.

Diedrich, C.G., 2011a. An overview of the ichnological and ethological studies in the cave bear den in Urşilor Cave (western Carpathians, Romania). Ichnos 18 (1), 9–26.

Diedrich, C.G., 2011b. Late Pleistocene steppe lion *Panthera Leo spelaea* (Goldfuss, 1810) footprints and bone records from open air sites in northern Germany—evidence of hyena-lion antagonism and scavenging in Europe. Quat. Sci. Rev. 30 (15–16), 1883–1906.

Domning, D.P., 2009. Sirenian evolution. In: Perrin, W.F., Würsig, B., Thewissen, J.G.M. (Eds.), Encyclopedia of Marine Mammals, second ed. Acad. Press, San Diego, pp. 1016–1019.

Domning, D.P., Zalmput, I.S., Gingerich, P.D., 2010. Sirenia. In: Werdelin, L., William, J.S. (Eds.), Cenozoic Mammals of Africa. Univ of California Press, pp. 147–160.

Domning, D.P., 2001. The earliest known fully quadrupedal sirenian. Nature 413 (6856), 625–627.

Doyle, P., Wood, J.L., George, G.T., 2000. The shorebird ichnofacies: an example from the Miocene of southern Spain. Geol. Mag. 137, 517–536.

Duellman, W.E., Trueb, L., 1994. Biology of Amphibians. Johns Hopkins Univ. Press, Baltimore, 670 pp.

Durbin, J., Funk, S.M., Hawkins, F., Hills, D.M., Jenkins, P.D., Moncrieff, C.B., Ralainasolo, F.B., 2010. Investigations into the status of a new taxon of *Salanoia* (Mammalia: carnivora: Eupleridae) from the marshes of lac Alaotra, Madagascar. Syst. Biodivers. 8 (3), 341–355.

Elbroch, M., 2003. Mammal Tracks and Sign. A Guide to North American Species. Stackpole Books, Mechanicsburg (PA), 792pp.

Elbroch, M., Marks, E., 2001. Bird Tracks & Sign - A Guide to North American Species. Stackpole Books, Pennsylvania, 456 pp.

Elbroch, M., McFarland, C., 2019. Mammal Tracks & Sign: A Guide to North American Species. Rowman & Littlefield.

Ellenberger, P., 1980. Sur les empreintes de pas des gros mammifères de l'Eocène supérieur de Garrigues-ste-Eulalie (Gard). Palaeovertebrata 9 (ext), 37–78.

Erickson, B.R., 2005. Crocodile and arthropod tracks from the late Paleocene Wannagan Creek fauna of North Dakota, USA. Ichnos 12 (4), 303–308.

Evans, S.E., Klembara, J., 2005. A choristoderan reptile (Reptilia: Diapsida) from the lower Miocene of northwest bohemia (Czech Republic). J. Vertebr. Paleontol. 25, 171–184.

Falkingham, P.L., 2014. Interpreting ecology and behaviour from the vertebrate fossil track record. J. Zool. 292 (4), 222–228.

Falkingham, P.L., Margetts, L., Smith, I.M., Manning, P.L., 2009. Reinterpretation of palmate and semi-palmate (webbed) fossil tracks; insights from finite element modelling. Palaeogeogr. Palaeoclimatol. Palaeoecol. 271 (1–2), 69–76.

Falkingham, P.L., Milán, J., Manning, P.L., 2010. A crocodilian trace fossil from the Lance Formation (upper Cretaceous) of Wyoming. Bull New Malique Mus. Nat. Hist. Sci. 51, 171–174.

Fariña, R.A., Vizcaíno, S.F., De Iuliis, G., 2012. Megafauna: Giant Beasts of Pleistocene South America, 1-436 pp.

Farina, M.E., Krapovickas, V., Fernández Piana, L., Vera, R.B., Ordoñez, M.Á., 2021. Flamingo-like footprints and the problem of addressing biological diversity in the past. Hist. Biol. 33, 912–926.

Farlow, J.O., Robinson, N.J., Kumagai, C.J., Paladino, F.V., Falkingham, P.L., Elsey, R.M., Martin, A.J., 2018a. Trackways of the American crocodile (*Crocodylus acutus*) in northwestern Costa Rica: implications for crocodylian ichnology. Ichnos 25 (1), 30–65.

Farlow, J.O., Robinson, N.J., Turner, M.L., Black, J., Gatesy, S.M., 2018b. Footfall pattern of a bottom-walking crocodile (*Crocodylus acutus*). Palaios 33 (9), 406–413.

Feldman, A., Sabath, N., Pyron, R.A., Mayrose, I., Meiri, S., 2016. Body sizes and diversification rates of lizards, snakes, amphisbaenians and the tuatara. Global Ecol. Biogeogr. 25 (2), 187–197.

Feola, S.F., 2022. Icnología y Sedimentología del Yacimiento Paleoicnológico de Pehuen Co (Pleistoceno Tardío). Unpublisehd Doctoral Thesis, Departamento de Geología, Universidad Nacional del Sur, Bahía Blanca, 410 pp.

Ferreira, G.S., Bronzati, M., Langer, M.C., Sterli, J., 2018. Phylogeny, biogeography and diversification patterns of side-necked turtles (Testudines: Pleurodira). R. Soc. Open Sci. 5 (3), 171773.

Fiorillo, A.R., Hasiotis, S.T., Kobayashi, Y., Breithaupt, B.H., McCarthye, P.J., 2011. Bird tracks from the Upper Cretaceous Cantwell Formation of Denali National Park, Alaska, USA: a new perspective on ancient northern polar vertebrate biodiversity. J. Syst. Palaeontol. 9, 33–49.

Flynn, J.J., Finarelli, J.A., Spaulding, M., 2010. Phylogeny of the Carnivora and Carnivoramorpha, and the use of the fossil record to enhance understanding of evolutionary transformations. In: Goswami, A., Friscia, A. (Eds.), Carnivoran Evolution. Cambridge University Press, Cambridge, pp. 25–63.

Fornós, J.J., Bromley, R.G., Clemmensen, L.B., Rodríguez-Perea, A., 2002. Tracks and trackways of *Myotragus balearicus* bate (Artiodactyla, Caprinae) in Pleistocene aeolianites from Mallorca (Balearic Islands, western Mediterranean). Palaeogeogr. Palaeoclimatol. Palaeoecol. 180, 277–313.

Foster, J.R., 2001. Salamander tracks (*Ambystomichnus*?) from the cathedral bluffs tongue of the wasatch formation (Eocene), northeastern Green River basin, Wyoming. J. Paleontol. 75 (4), 901–904.

Foster, J.R., Lockley, M.G., 1997. Probable crocodilian tracks and traces from the Morrison Formation (Upper Jurassic) of eastern Utah. Ichnos 5 (2), 121–129.

François, L.M., Goddéris, Y., 1998. Isotopic constraints on the Cenozoic evolution of the carbon cycle. Chem. Geol. 145 (3–4), 177–212.

Frenguelli, J., 1950. Ichnites en el Paleozoico Superior del oeste argentino. Rev. Asoc. Geol. Argent. 5 (3), 136–149.

Fuentes Vidarte, C., 1996. Primeras huellas de aves en el Weald de Soria (Espana). Nuevo icnogénero, *Archaeornithipus* y nueva icnoespecie *A. meijidei.* Estudios Geológicos 52, 63–75.

Fuentes Vidarte, C., Meijide Calvo, M., 2001. Primeras huellas de cocodrilo en el Weald de Cameros (Soria, España). Nueva familia: Crocodylopodidae, nuevo icnogenero: *Crocodylopodus*, nueva icnoespecie: *C. meijidei.* In: Actas de las I Jornadas Internacionales sobre paleontología de dinosaurios y su entorno. Colectivo Arqueológico-Paleontológico de Salas, Salas de los Infantes, pp. 329–335.

Fuentes Vidarte, C., Meijide Calvo, M., Meijide Fuentes, F., Meijide Fuentes, M., 2003. Rastro de un tetrápodo de pequeño tamaño en el Weald de Cameros (Sierra de Oncala, Soria, España) nov. icnogen. nov. icnosp.: *Emydhipus cameroi.* In: Dinosaurios y otros reptiles mesozoicos en España. Instituto de Estudios Riojanos, Logroño, pp. 119–128.

Gardner, J.D., Nydam, R.L., 2013. Mesozoic and Cenozoic lissamphibian and squamate assemblages of Laurasia—introduction to the special issue. Paleobiodivers. Paleoenviron. 93, 391–395.

Gatesy, S.M., 2003. Direct and indirect track features: what sediment did a dinosaur touch? Ichnos 10, 91–98.

Gatesy, S.M., Falkingham, P.L., 2020. Hitchcock's leptodactyli, penetrative tracks, and dinosaur footprint diversity. J. Vertebr. Paleontol. 40, e1781142.

Gaubert, P., Veron, G., 2003. Exhaustive sample set among Viverridae reveals the sister-group of felids: the linsangs as a case of extreme morphological convergence within Feliformia. Proc. Royal Soc. B. 270 (1532), 2523–2530.

Gauthier, J.A., 1994. The diversification of the amniotes. Short Cour. Paleontol. 7, 129–159.

Gauthier, J.A., Kluge, A.G., Rowe, T., 1988. Amniote phylogeny and the importance of fossils. Cladistics 4, 105–209.

Geisler, J.H., Theodor, J.M., Uhen, M.D., Foss, S.E., 2007. Phylogenetic relationships of cetaceans to terrestrial artiodactyls. In: Prothero, D.R., Foss, S.E. (Eds.), The Evolution of Artiodactyls. Johns Hopkins Univ. Press, Baltimore, pp. 19–31.

Gilmore, C.W., 1928a. Fossil Footprints from the Fort Union (Paleocene) of Montana, vol. 74. U.S. National Museum Proceedings, pp. 1–4.

Gilmore, C.W., 1928b. Fossil footprints from the Grand Canyon: third contribution. Smithsonian Misc. Collect. 80 (8), 1–16.

Goldberg, J., Fabrezi, M., 2008. Development and variation of the anuran webbed feet (Amphibia, Anura). Zool. J. Linn. Soc. 152 (1), 39–58.

Goswami, A., 2010. Introduction to carnivora. In: Goswami, A., Friscia, A. (Eds.), Carnivoran Evolution. Cambridge University Press, Cambridge, pp. 1–24.

Goswami, A., Noirault, E., Coombs, E.J., Clavel, J., Fabre, A.C., Halliday, T.J., et al., 2022. Attenuated evolution of mammals through the Cenozoic. Science 378 (6618), 377–383.

Greer, A.E., 1991. Limb reduction in squamates: identification of the lineages and discussion of the trends. J. Herpetol. 25, 166–173.

Groves, C.P., Grubb, P., 2011. Ungulate Taxonomy. Johns Hopkins University Press, Baltimore, 336pp.

Guérin, C., Demathieu, G., 1993. Empreintes et pistes de rhinocerotidae (Mammalia, Perissodactyla) du gisement pliocéne terminal de Laetoli (Tanzanie). Geobios 26, 497–513.

Gunnell, G.F., 1998. Creodonta. In: Janis, C.M., Scott M., K., Jacobs, L.L. (Eds.), Evolution of Tertiary Mammals of North America, Volume 1: Terrestrial Carnivores, Ungulates, and Ungulatelike Mammals. Cambridge University Press, Cambridge, pp. 91–105.

Gunnell, G.F., Gingerich, P.D., 1991. Systematics and evolution of late Paleocene and early Eocene Oxyaenidae (Mammalia, Creodonta) in the Clarks Fork basin, Wyoming. Contrib. Mus. Paleontol. Univ. Mich. 28, 141–180.

Hammond, J.A., Hauton, C., Bennett, K.A., Hall, A.J., 2012. Phocid seal leptin: tertiary structure and hydrophobic receptor binding site preservation during distinct leptin gene evolution. PLoS One 7 (4), e35395.

Haubold, H., 1971a. Die Tetrapodenfährten des Buntsandsteins: Paläontologische Abhandlungen A, vol. 4, pp. 395–548.

Haubold, H., 1971b. Ichnia Amphibiorum et Reptiliorum fossilium. Encycl. Paleoherpetol. (18), 1–124.

Heissig, K., 1989. The rhinocerotidae. In: Prothero, D.R., Schoch, R.M. (Eds.), The Evolution of Perissodactyls. Oxford University Press, Oxford, pp. 399–417.

Helm, C.W., Bateman, M.D., Carr, A.S., Cawthra, H.C., De Vynck, J.C., Dixon, M.G., Lockley, M.G., Stear, W., Venter, J.A., 2023a. Pleistocene fossil snake traces on South Africa's Cape south coast. Ichnos 1–17.

Helm, C.W., Carr, A.S., Cawthra, H.C., De Vynck, J.C., Dixon, M.G., Lockley, M.G., Stear, W., Venter, J.A., 2023b. Large Pleistocene tortoise tracks on the Cape south coast of South Africa. Quat. Res. 112, 93–110.

Herrera-Flores, J.A., Stubbs, T.L., Benton, M.J., 2017. Macroevolutionary patterns in Rhynchocephalia: is the tuatara (*Sphenodon punctatus*) a living fossil? Palaeontology 60 (3), 319–328.

Herrero, C., Herrero, E., Martin-Chivelet, J., Pérez-Lorente, F., 2022. Vertebrate ichnofauna from Sierra de las Cabras tracksite (Late Miocene, Jumilla, SE Spain). Mammalian Ichnofauna. J. Iberian Geol. 48, 241–279. https://doi.org/10.1007/s41513-022-00192-5.

Hildebrand, M., 1985. Walking and running. In: Hildebrand, M., Bramble, M., Liem, K., Wake, D. (Eds.), Functional Vertebrate Morphology. Harvard Uni. Press, Harvard, pp. 38–57.

Hill, R.V., Mccartney, J.A., Roberts, E., Bouaré, M., Sissoko, F., O'leary, M.A., 2008. Dyrosaurid (Crocodyliformes: Mesoeucrocodylia) fossils from the Upper Cretaceous and Paleogene of Mali: implications for phylogeny and survivorship across the K/T boundary. Am. Mus. Novit. 3631, 1–19.

Hime, P.M., Lemmon, A.R., Lemmon, E.C.M., Prendini, E., Brown, J.M., Thomson, R.C., Kratovil, J.D., Noonan, B.P., Pyron, R.A., Peloso, P.L.V., Weisrock, D.W., 2021. Phylogenomics reveals ancient gene tree discordance in the amphibian tree of life. Syst. Biol. 70, 49–66.

Hitchcock, E., 1845. An Attempt to Name, Classify, and Describe the Animals that Made the Fossil Footmarks of New England. Proceedings of the 6th Meeting American Association of Geologists and Naturalists, New Haven, pp. 23–55.

Hooker, J.J., Dashzeveg, D., 2004. The origin of chalicotheres (Perissodactyla, Mammalia). Palaeontology 47, 1363–1386.

Houck, K.J., Lockley, M.G., Avanzini, M., 2009. A survey of tetrapod tracksites preserved in pyroclastic sediments, with special reference to footprints of hominids, other mammals and birds. Ichnos 16, 76–97. https://doi.org/10.1080/10420940802470870.

Howard, J.D., 1978. Sedimentology and trace fossils. In: Basan, P.B. (Ed.), Trace Fossil Concepts. SEPM Society for Sedimentary Geology, pp. 13–45. No. 5.

Howell, A.B., 1944. Speed in Animals, Their Specialization for Running and Leaping. Hafner Publishing Company, New York, 268 pp.

Hsieh, S., Uchman, A., 2022. Persistent similarities among hoofed mammal tracks across the Cenozoic revealed through Fourier outline shape analysis. J. Vertebr. Paleontol. 42 (1), e2119148.

Hughes, J.J., Berv, J.S., Chester, S.G., Sargis, E.J., Field, D.J., 2021. Ecological selectivity and the evolution of mammalian substrate preference across the K–Pg boundary. Ecol. Evol. 11 (21), 14540–14554.

Hunt, R., 1998. Amphicyonidae. In: Janis, C.M., Scott, K.M., Jacobs, L.L. (Eds.), Evolution of Tertiary Mammals of North America. Cambridge University Press, New York, pp. 196–227.

Hunt, A.P., Lucas, S.G., 2007. Cenozoic vertebrate trace fossils of North America: ichnofaunas, ichnofacies and biochronology. Bull. N. M.Mus. Nat. Hist. Sci. 42, 17–41.

Hutchinson, J.R., 2021. The evolutionary biomechanics of locomotor function in giant land animals. J. Exp. Biol. 224, jeb217463.

ICZN [International Commission on Zoological Nomenclature], 1999. International Code of Zoological Nomenclature, fourth. International Trust for Zoological Nomenclature, London, pp. 1–306.

Illiger, C.D., 1811. Prodromus Systematis Mammalium et Avium Additis Terminis Zoographicis Uttriusque Classis. Classis, Eorumque Versione Germanica. Salfeld, Berlin, 301 pp.

Jaekel, M., Wake, D.B., 2007. Developmental processes underlying the evolution of a derived foot morphology in salamanders. Proc. Natl. Acad. Sci. USA 104 (51), 20437–20442.

Janis, C.M., Bernor, R.L., 2019. The evolution of equid monodactyly: a review including a new hypothesis. Front. Ecol. Evol. 7, 119.

Jennings, A.P., Veron, G., 2015. Predicted distributions, niche comparisons, and conservation status of the spotted linsang (*Prionodon pardicolor*) and banded linsang (*Prionodon linsang*). Mammal Res. 60, 107–116.

Johnson, W.E., Eizirik, E., Pecon-Slattery, J., Murphy, W.J., Antunes, A., Teeling, E., O'Brien, S.J., 2006. The late Miocene radiation of modern Felidae: a genetic assessment. Science 311 (5757), 73–77.

Jones, M.E., Anderson, C.L., Hipsley, C.A., Müller, J., Evans, S.E., Schoch, R.R., 2013. Integration of molecules and new fossils supports a Triassic origin for *Lepidosauria* (lizards, snakes, and tuatara). BMC Evol. Biol. 13, 1–21.

Kälin, J.A., Peyer, B., 1936. *Hispanochampsa mülleri* nov. gen. sp., ein neuer Crocodilide aus dem unteren Oligocaen von Tárrega. (Catalonien) Abhandlungen der Schweizerische Paläontologischen Gesellschaft 58, 1–40.

Karl, H.V., Valdiserri, D., Tichy, G., 2012. Nuevo material del ichnotaxon de quelonio «Emydhipus cameroi» Fuentes Vidarte et al., 2003 del Berriasiense (Cretácico Inferior) del Norte de Alemania, vol. 9. Studia Palaeocheloniologica Salmanticensia, pp. 61–72.

Kaup, J.J., 1835. Über Thierfährten bei Hildburghausen. Neues Jahrbuch für Mineralogie. Geologie und Paläontologie 1835, 327–328.

Keith-Diagne, L.W., Barlas, M.E., Reid, J.P., Hodgson, A.J., Marsh, H., 2022. Diving and foraging behaviors. In: Ethology and Behavioral Ecology of Sirenia. Springer International Publishing, Cham, pp. 67–100.

Kellner, A.W., Pinheiro, A.E., Campos, D.A., 2014. A new sebecid from the Paleogene of Brazil and the crocodyliform radiation after the K–Pg boundary. PLoS One 9 (1), e81386.

Kim, H.M., 1969. A study of several sole marks in the Haman Formation. J. Geol. Soc. Korea 5, 243–258.

Kim, J.Y., Kim, S.H., Kim, K.S., Lockley, M., 2006. The oldest record of webbed bird and pterosaur tracks from South Korea (Cretaceous Haman Formation, Changseon and Sinsu Islands): more evidence of high avian diversity in East Asia. Cretac. Res. 27, 56–69.

Kim, J.Y., Kim, M.K., Oh, M.S., Lee, C.Z., 2013. A new semi-palmate bird track, *Gyeongsangornipes lockleyi* ichnogen. et ichnosp. nov., and *Koreanaornis* from the Early Cretaceous Jindong Formation of Goseong County, southern coast of Korea. Ichnos 20, 72–80.

Kim, K.S., Lockley, M.G., Lim, J.D., Piñuela, L., Xing, L., Moon, H.W., 2017. First report of lacertiform (lizard) tracks from the Cretaceous of Asia. Cretac. Res. 69, 62–70.

Kim, K.S., Lockley, M.G., Lim, J.D., Kim, D.H., 2019. The oldest known anuran (frog) trackways from the Jinju Formation, Lower Cretaceous, Korea. Cretac. Res. 96, 142–148.

Kim, J.Y., Lockley, M.G., Seo, S.J., Kim, K.S., Kim, S.H., Baek, K.S., 2012. A paradise of mesozoic birds: the world's richest and most diverse Cretaceous bird track assemblage from the Early Cretaceous Haman Formation of the Gajin tracksite, Jinju, Korea. Ichnos 19, 28–42.

King, A.S., McLelland, J., 1984. Birds Their Structure & Function, 2d ed. Bailliere Tindall, London. 334 pp.

King, M.R., Botterill, S.E., Gingras, M.K., Pemberton, S.G., 2020. *Rhizocorallium* and turtle tracks: a late Cretaceous proximal distributary channel trace-fossil assemblage, central Utah. Ichnos 27 (4), 406–427.

Klein, H., Lucas, S., 2021. The Triassic tetrapod footprint record. Bull. N. M.Mus. Nat. Hist. Sci. 83, 1–194.

Kordos, L., 1985. Footprints in the lower Miocene sandstone of ipolytarnóc. Geologica Hungarica ser. Palaeontologica 46, 359–415.

Kordos, L., Prakfalvi, P., 1990. Újabb adatok az európai neogén lábnyomos rétegek ismeretéhez [A contribution to the knowledge of Neogene beds with footprints in Europe]. In: A Magyar Állami Földtani Intézet Évi Jelentése 1988, pp. 200–212.

Kordos, L., Meszaros, I., Szarvas, I., 2021. Tracking a "prehistoric Pompeii", rhinoland and crocodilia: new discoveries and interpretations of Ipolytarnóc (N Hungary) lower Miocene track site. Geoconserv. Res. 4 (2), 621–634.

Korth, W.W., 1994. The Tertiary Record of Rodents in North America, vol. 12. Springer Science & Business Media.

Krapovickas, V., Nasif, N.L., 2011. Large caviomorph rodent footprints of the late Oligocene Vinchina Formation, Argentina. Palaeontol. Electron. 14, 1–13.

Krapovickas, V., Vizcaíno, S., 2016. The Cenozoic radiation of mammals. In: The Trace-Fossil Record of Major Evolutionary Events: Volume 2: Mesozoic and Cenozoic. Springer, pp. 371–410.

Krapovickas, V., Ciccioli, P.L., Mángano, M.G., Marsicano, C.A., Limarino, C.O., 2009. Paleobiology and paleoecology of an arid–semiarid Miocene South American ichnofauna in anastomosed fluvial deposits. Palaeogeogr. Palaeoclimatol. Palaeoecol. 124, 129–152.

Krapovickas, V., Munoz, N.A., Alonso-Muruaga, P.J., Ciccioli, P.L., Marenssi, S., Vizcaíno, S.F., 2018. *Tracheria troyana* (Krapovickas and Nasif, 2011): redefinition, environmental distribution, and heritage conservation. Ichnos 25 (2–3), 221–231.

Krapovickas, V., Vera, R.B., Farina, M.E., Piana, L.R.F., Koelewijn, A., 2023. A new vertebrate ichnological association sheds light on the small metatherian record of the Middle Miocene in South America. J. S. Am. Earth Sci. 129, 104529.

Kubo, T., 2010. Extant lizard tracks: variation and implications for paleoichnology. Ichnos 17 (3), 187–196.

Kulchytzky, Y.O., Kulchytzky, A.Y., 1980. Finding of the maxilla of [a] hollow-horned and the traces of vertebrates in the Miocene molasses of the Precarpathian and Transcarpathian Foredeeps. Paleontol. Sb. (L'vov) 17, 69–74.

Kumagai, C.J., Farlow, J.O., 2010. Observations on traces of the American crocodile (*Crocodylus acutus*) from northwestern Costa Rica. Bull. N. M.Mus. Nat. Hist. Sci. 51, 41–49.

Kuhn, O., 1958. Die Fährtender vorzeitlichen Amphibien und Reptilien. Meisenbach, Bamberg, 64 pp.

Kupriyanova, N.S., 2009. Current views of the origin and diversification of tetrapods. Mol. Biol. 43, 819–833.

Lambrecht, K., 1938. *Urmiornis abeli* n. sp., eine Pliozäne Vogel-Fährte aus Persien. Palaeobiologica 6, 242–245.

Lande, R., 1978. Evolutionary mechanisms of limb loss in tetrapods. Evolution 73–92.

Laporte, L.F., Behrensmeyer, A.K., 1980. Tracks and substrate reworking by terrestrial vertebrates in Quaternary sediments of Kenya. J. Sediment. Res. 50, 1337–1346.

Laurin, M., 2002. Tetrapod phylogeny, amphibian origins, and the definition of the name Tetrapoda. Syst. Biol. 51 (2), 364–369.

Laurin, M., Reisz, R.R., 1995. A reevaluation of early amniote phylogeny. Zool. J. Linn. Soc. 113 (2), 165–223.

Le Conte, J., 1891. Evolution: Its Nature, its Evidences, and its Relation to Religious Thought (2nd ed., rev.). D Appleton & Company. https://doi.org/10.1037/12974-000.

Lee, Y.N., Kong, D.Y., Jung, S.H., 2020. The first possible choristoderan trackway from the Lower Cretaceous Daegu Formation of South Korea and its implications on choristoderan locomotion. Sci. Rep. 10, 14442.

Leonardi, G., 1987. Glossary and Manual of Tetrapod Footprint Palaeoichnology. Ministério das Minas e Energia. Departamento Nacional da Produção Mineral, Brasilia, 75 pp.

Leonardi, G., 1981. *Brasilichnium elusivum* gen. n., sp. n.: pistas de tetrapóde mesozóico guardadas nas colecções do Museu Nacional do Rio de Janeiro. An Acad. Bras Ciências 53 (4), 793–805.

Li, J., Zhen, S., 1994. New materials of bird ichnites from Fildes Peninsula, King George Island of Antarctica and their biogeographic significance. In: Shen, Y. (Ed.), Stratigraphy and Palaeontology of Fildes Peninsula, King George Island, Antarctica. Science Press, Beijing, pp. 239–250.

Lichtig, A.J., Lucas, S.G., Klein, H., Lovelace, D.M., 2018. Triassic turtle tracks and the origin of turtles. Hist. Biol. 30 (8), 1112–1122.

Linares Montes, M., Luzón, A., Cuenca-Bescós, G., Canudo, J.I., Castanera, D., 2022. New mammal and bird tracks from the Lower Oligocene of the Ebro Basin (NE Spain): implications for the Palaeogene ichnological record. Hist. Biol. 35 (9), 1616–1636.

Liu, J., 2004. A nearly complete skeleton of *Ikechosaurus pijiagouensis* sp. nov. (Reptilia: Choristodera) from the Jiufotang Formation (Lower Cretaceous) of Liaoning, China. Vertebr. Palasiat. 42, 120–129.

Lockley, M.G., 2007. A tale of two ichnologies: the different goals and potentials of invertebrate and vertebrate (tetrapod) ichnotaxonomy and how they relate to ichnofacies analysis. Ichnos 14, 39–57.

Lockley, M.G., 2009. New perspectives on morphological variation in tridactyl footprints: clues to widespread convergence in developmental dynamics. Geol.l Quart. 53, 415–432.

Lockley, M.G., 2010. A solution to the *Mehiella* mystery: tracking, naming, identifying and measuring the first crocodilian trackway reported from the Cretaceous (Dakota Group, Colorado. Bull. N. M.Mus. Nat. Hist. Sci. 51, 157–174.

Lockley, M.G., Delgado, C.R., 2007. Tracking an ancient Turkey: a preliminary report on a new Miocene ichnofauna from near Durango, Mexico. Bull. N. M.Mus. Nat. Hist. Sci. 42, 67–72.

Lockley, M.G., Foster, J., 2003. Late Cretaceous mammal tracks from Northnorth America. Ichnos 10, 269–276.

Lockley, M.G., Foster, J.R., 2010. An assemblage of probable crocodylian traces and associated dinosaur tracks from the lower Morrison Formation (Upper Jurassic) of eastern Utah. Bull. N. M.Mus. Nat. Hist. Sci. 51, 93–98.

Lockley, M.G., Harris, J., 2010. On the trail of early birds: a review of the fossil footprint record of avian morphological evolution and behavior. In: Ulrich, P.K., Willett, J.H. (Eds.), Trends in Ornithological Research. Nova Science Publishers, pp. 1–63.

Lockley, M.G., Hunt, A.P., 1995. Ceratopsid tracks and associated ichnofauna from the Laramie Formation (Upper Cretaceous: Maastrichtian) of Colorado. J. Vertebr. Paleontol. 15 (3), 592–614.

Lockley, M.G., Meyer, C., 2004. Crocodylomorph trackways from the Jurassic to Early Cretaceous of North America and Europe: implications for ichnotaxonomy. Ichnos 11 (1–2), 167–178.

Lockley, M.G., Milner, A.R., 2014. The ichnotaxonomy of hopping vertebrate trackways from the Cenozoic. Bull. N. M.Mus. Nat. Hist. Sci. 62, 489–500.

Lockley, M.G., Cawthra, H.C., De Vynck, J.C., Helm, C.W., McCrea, R.T., Nel, R., 2019a. New fossil sea turtle trackway morphotypes from the Pleistocene of South Africa highlight role of ichnology in turtle paleobiology. Quat. Res. 92 (3), 626–640.

Lockley, M., Chin, K., Houck, K., Matsukawa, M., Kukihara, R., 2009. New interpretations of *Ignotornis*, the first-reported Mesozoic avian footprints: implications for the paleoecology and behavior of an enigmatic Cretaceous bird. Cretac. Res. 30, 1041–1061.

Lockley, M.G., Culver, T.S., Wegweiser, M., 2007. An ichnofauna of hopping rodent and arthropod trackways from the Miocene of Colorado. Bull. N. M.Mus. Nat. Hist. Sci. 42, 59–66.

Lockley, M.G., Goodell, Z., Evaskovich, J., Krall, A., Schumacher, B.A., Romilio, A., 2022. Small bird and mammal tracks from a mid-Cenozoic volcanic province in Southern Colorado: implications for palaeobiology. Hist. Biol. 34, 130–140.

Lockley, M.G., Houck, K., Yang, S.-Y., Matsukawa, M., Lim, S.-K., 2006. Dinosaur-dominated footprint assemblages from the Cretaceous Jindong Formation, Hallyo Haesang National Park area, Goseong county, South Korea: evidence and implications. Cretac. Res. 27, 70–101.

Lockley, M., Kim, K.S., Lim, J.D., Romilio, A., 2021. Bird tracks from the Green River Formation (Eocene) of Utah: ichnotaxonomy, diversity, community structure and convergence. Hist. Biol. 33, 2085–2102.

Lockley, M.G., Li, R., Matsukawa, M., Li, J., 2010b. Tracking Chinese crocodylians: *Kuaangyuanpus, Laiyangpus*, and implications for naming crocodylian and crocodylian-like tracks and associated inchnofacies. Bull. N. M.Mus. Nat. Hist. Sci. 51, 99–108.

Lockley, M.G., Lim, J.D., Kim, K.S., Huh, M., Hwang, K.G., 2012. Tracking Korea's early birds: a review of Cretaceous avian ichnology and its implications for evolution and behavior. Ichnos 19 (1–2), 17–27.

Lockley, M.G., Lim, J.D., Park, H.D., Romilio, A., Yoo, J.S., Choi, J.W., Kim, K.S., Choi, Y., Kang, S.H., Kim, D.H., Kim, T.H., 2020. First reports of *Crocodylopodus* from east Asia: implications for the paleoecology of the Lower Cretaceous. Cretac. Res. 111, 104441.

Lockley, M.G., Lucas, S.G., Milán, J., Harris, J.D., Avanzini, M., Foster, J.R., Spielmann, J.A., 2010a. The fossil record of crocodylian tracks and traces: an overview. Bull. N. M.Mus. Nat. Hist. Sci. 51, 1–13.

Lockley, M.G., Ritts, B.D., Leonardi, G., 1999. Mammal track assemblages from the early tertiary of China, Peru, Europe and North America. Palaios 14 (4), 398–404.

Lockley, M.G., Yang, S.-Y., Matsukawa, M., Fleming, F., Lim, S.-K., 1992. The track record of Mesozoic birds: evidence and implications. Philos. Trans. R. Soc. Lond. B Biol. Sci. 336, 113–134.

López, G., Mayoral, E., Muñoz, A., Pérez, A., Santamaría, R., 1992. Nuevas incitas de mamíferos carnívoros en el Mioceno Inferior del sector occidental de la depresión del Ebro (Logroño, La Rioja). Rev. Soc. Geol. Esp 5, 31–51.

Lovette, I.J., Fitzpatrick, J.W., 2016. Cornell Lab of Ornithology's Handbook of Bird Biology, third ed. Wiley-Blackwell, Chichester, West Sussex. 736 pp.

Lucas, S.G., 2005. Tetrapod ichnofacies and ichnotaxonomy: quo vadis? Ichnos 12, 157–162.

Lucas, S.G., 2007. Cenozoic vertebrate footprint ichnotaxa named by O. S. Vyalov in 1965 and 1966. Bull. N. M.Mus. Nat. Hist. Sci. 42, 113–148.

Lucas, S.G., Allen, B.D., Morgan, G.S., Myers, R.G., Love, D.W., Bustos, D., 2007. Mammoth footprints from the upper Pleistocene of the Tularosa basin, Doña Ana county, New Mexico. Bull. N. M.Mus. Nat. Hist. Sci. 42, 149–154.

Lucas, S.G., Hunt, A.P., 2007. Ichnotaxonomy of camel footprints. Bull. N. M.Mus. Nat. Hist. Sci. 42, 155–168.

Lucas, S., Schultz, G., 2007. Miocene vertebrate footprints from the Texas Panhandle. Cenozoic vertebrate tracks and traces. Bull. N. M.Mus. Nat. Hist. Sci. 42, 177–183.

Lucas, A.M., Stettenheim, P.R., 1972. Avian Anatomy: Integument. U.S. Department of Agriculture, Washington, DC, 750 pp.

Lucas, S.G., Williamson, T.E., 1993. Eocene vertebrates and late Laramide stratigraphy of New Mexico. Bull. N. M.Mus. Nat. Hist. Sci. 2, 145–158.

Lucas, S.G., Abbassi, N., Francischini, H., Denzien-Dias, P.A.U.L.A., Knight, P., 2023. Bird-dominated Pliocene track-site from the Santa Fe group, Rio Rancho, New Mexico. In: Lucas, S.G., et al. (Eds.), Fossil Record 9. New Mexico Museum of Natural History and Science Bulletin, vol. 94, pp. 399–410.

Lull, R.S., 1904. Fossil footprints of the Jura-Trias of North America. Memoirs Boston Soc. Nat. Hist. 5, 461–557.

MacFadden, B.J., 2005. Fossil horses–evidence for evolution. Science 307, 1728–1730.

Mansilla, H.G., De Valais, S., Stinnesbeck, W., Varela, N.A., Leppe, M.A., 2012. New avian tracks from the lower to middle Eocene at fossil Hill, King George Island, Antarctica. Antarct. Sci. 24, 500–506.

Magwood, J.P.A., 1992. Ichnotaxonomy: a burrow by any other name?. In: Maples, C.G., West, R.R., Fossils, T. (Eds.), Short Courses in Paleontology, vol. 5, pp. 15–33.

Maidwell, F.T., 1911. Notes on footprints from the Keuper of Runcorn Hill. Proc. Geol. Soc. Lond. 11, 140–152.

Marchetti, L., Petti, F.M., Zoboli, D., Pillola, G.L., 2018. Vertebrate and invertebrate trace fossils in the late Pennsylvanian (Carboniferous) fluvio-lacustrine San Giorgio basin (South-West Sardinia): remarks on the oldest continental ichnoassociation of Italy. Ichnos 25 (2–3), 94–105.

Marchetti, L., Belvedere, M., Voigt, S., Klein, H., Castanera, D., Díaz-Martínez, I., et al., 2019a. Defining the morphological quality of fossil footprints. Problems and principles of preservation in tetrapod ichnology with examples from the Palaeozoic to the present. Earth Sci. Rev. 193, 109–145.

Marchetti, L., Voigt, S., Lucas, S.G., Francischini, H., Dentzien-Dias, P., Sacchi, R., et al., 2019b. Tetrapod ichnotaxonomy in eolian paleoenvironments (Coconino and De Chelly formations, Arizona) and late Cisuralian (Permian) sauropsid radiation. Earth Sci. Rev. 190, 148–170.

Marsh, O.C., 1984. Description of tertiary artiodactyls. Am. J. Sci. 48, 259–274.

Marshall, S.K., Superina, M., Spainhower, K.B., Butcher, M.T., 2021. Forelimb myology of armadillos (Xenarthra: Cingulata, Chlamyphoridae): anatomical correlates with fossorial ability. J. Anat. 238, 551–575.

Marsicano, C.A., Wilson, J.A., Smith, R.M., 2014. A temnospondyl trackway from the Early Mesozoic of Western Gondwana and its implications for basal tetrapod locomotion. PLoS One 9 (8), e103255.

Martin, J.E., 2014. A sebecosuchian in a middle Eocene karst with comments on the dorsal shield in Crocodylomorpha. Acta Palaeontol. Pol. 60 (3), 673–680.

Martinez, B., Farke, A.A., 2018. Trackways of *Platykopus ilycalcator* from the White Narrows Formation (Pliocene), Nevada. In: Thomson, T.J., Harris, J., Milner, A.R.C., Kirkland, J. (Eds.), Western Association of Vertebrate Paleontology Annual Meeting. Western Assoc. Vert. Pal., p. 21. Program with Abstracts. PaleoBios 35 Supplement. ucmp_paleobios_38376.

Mateus, O., Milan, J., 2010. First record of crocodyle and pterosaur tracks in the Upper Jurassic of Portugal. Bull. N. M.Mus. Nat. Hist. Sci. 51, 83–88.

Mateus, O., Puértolas-Pascual, E., Callapez, P.M., 2019. A new eusuchian crocodylomorph from the Cenomanian (Late Cretaceous) of Portugal reveals novel implications on the origin of Crocodylia. Zool. J. Linn. Soc. 186 (2), 501–528.

Matsukawa, M., Baba, K., Lockley, M.G., 2007. A mammalian ichnofauna from Plio-Pleistocene terrestrial deposits of West Tokyo, Japan. Bull. N. M.Mus. Nat. Hist. Sci. 42, 185–199.

Matsumoto, R., Evans, S.E., 2010. Choristoderes and the freshwater assemblages of Laurasia. J. Iber. Geol. 36, 253–274.

Matsumoto, R., Buffetaut, E., Escuillie, F., Hervet, S., Evans, S.E., 2013. New material of the choristodere *Lazarussuchus* (Diapsida, Choristodera) from the Paleocene of France. J. Vertebr. Paleontol. 33 (2), 319–339.

Mautner, A.K., Latimer, A.E., Fritz, U., Scheyer, T.M., 2017. An updated description of the osteology of the pancake tortoise *Malacochersus tornieri* (Testudines: Testudinidae) with special focus on intraspecific variation. J. Morphol. 278, 321–333.

McCrea, R.T., Sarjeant, W.A.S., 2001. New Ichnotaxa of Bird and Mammal Footprints from the Lower Cretaceous (Albian) Gates Formation of Alberta. Mesozoic Vertebrate Life, pp. 453–478.

McCrea, R.T., Buckley, L.G., Plint, A.G., Currie, P.J., Haggart, J.W., Helm, C.W., Pemberton, S.G., 2014. A review of vertebrate track-bearing formations from the Mesozoic and earliest Cenozoic of western Canada with a description of a new theropod ichnospecies and reassignment of an avian ichnogenus. Bull. N. M.Mus. Nat. Hist. Sci. 62, 5–93.

McCrea, R.T., Buckley, L.G., Plint, A.G., Lockley, M.G., Matthews, N.A., Noble, T.A., Xing, L., Krawetz, J.R., 2015. Vertebrate ichnites from the Boulder Creek Formation (Lower Cretaceous: Middle to? Upper Albian) of northeastern British Columbia, with a description of a new avian ichnotaxon, *Paxavipes babcockensis* ichnogen. et isp. nov. Cretac. Res. 55, 1–18.

McCrea, R.T., Pemberton, S.G., Currie, P.J., 2004. New ichnotaxa of mammal and reptile tracks from the Upper Paleocene of Alberta. Ichnos 11, 323–339.

McDonald, G.H., White, R., Lockley, M.G., Mustoe, G., 2007. An indexed bibliography of Cenozoic vertebrate tracks. Bull. N. M.Mus. Nat. Hist. Sci. 42, 42–275.

McKenna, M.C., 1975. Toward a phylogenetic classification of the Mammalia. In: Luckett, P., Szalay, F.S. (Eds.), Phylogeny of the Primates. Plenum Press, New York, pp. 21–46.

McKee, E.D., 1947. Experiments on the development of tracks in fine cross-bedded sand. J. Sediment. Res. 17, 23–28.

McNeil, P., Hills, L.V., Tolman, M.S., Kooyman, B., 2007. Significance of latest Pleistocene tracks, trackways, and trample grounds from southern Alberta, Canada. Bull. N. M.Mus. Nat. Hist. Sci. 42, 209–223.

Mehl, M.G., 1931. Additions to the vertebrate record of the Dakota sandstone. Am. J. Sci. 21, 441–452.

Melchor, R.N., 2015. Application of vertebrate trace fossils to palaeoenvironmental analysis. Palaeogeogr. Palaeoclimatol. Palaeoecol. 439, 79–96.

Melchor, R.N., Genise, J.F., 2004. Invertebrate vs. vertebrate ichnology. . . or one ichnology? In: Buatois, L.A., Mángano, M.G. (Eds.), Ichnia 2004 First International Congress on Ichnology Abstract Book. Museo Paleontológico Egidio Feruglio, Trelew, pp. 55–56.

Melchor, R.N., Sarjeant, W.A.S., 2004. Small amphibian and reptile footprints from the Permian Carapacha basin, Argentina. Ichnos 11 (1–2), 57–78.

Melchor, R.N., Bromley, R., Bedatou, E., 2010. *Spongeliomorpha* in nonmarine settings: an ichnotaxonomic approach. Earth Environ. Sci. Trans. R. Soc. Edinb. 100, 429–436.

Melchor, R.N., Cardonatto, M.C., Visconti, G., 2012. Palaeonvironmental and palaeoecological significance of flamingo-like footprints in shallow-lacustrine rocks: an example from the Oligocene–Miocene Vinchina Formation, Argentina. Palaeogeogr. Palaeoclimatol. Palaeoecol. 315, 181–198.

Melchor, R.N., Perez, M., Umazano, A.M., 2013. Continental trace fossils from the Miocene-Pliocene Río Negro Formation (Río Negro and Buenos Aires provinces), Argentina. In: Melchor, R.N., Umazano, A.M. (Eds.), SLIC2013 Post-symposium Fieldtrip Guidebook. INCITAP - Universidad Nacional de La Pampa, Santa Rosa, Argentina, pp. 31–45.

Melchor, R.N., Feola, S.F., de Bianco, T.M., 2019. Canid paleoichnology: taxonomic review and producers of *Canipeda* from the Late Pleistocene of Argentina. Ichnos 26 (2), 85–107.

Melchor, R.N., Feola, S.F., Cardonatto, M.C., Espinoza, N., Rojas-Manriquez, M.A., Herazo, L., 2023. First terror bird footprints reveal functionally didactyl posture. Sci. Rep. 13, 16474.

Melchor, R.N., Perez, M., Cardonatto, M.C., Umazano, A.M., 2015. Late Miocene ground sloth footprints and their paleoenvironment: *Megatherichnum oportoi* revisited. Palaeogeogr. Palaeoclimatol. Palaeoecol. 439, 126–143.

Melchor, R.N., Uchman, A., Steel, R., 2020. Avian diversity and behavior in an Eocene coastal plain, Svalbard: the ichnological evidence. Ichnos 27, 334–343.

Meunier, S., 1906. Remarquables traces de pas sur un banc de gypse. Le Naturaliste 28, 19–21.

Mikuláš, R., 2010. *Characichnus* isp., possible crocodile swim traces from the Miocene of the Eské Bud Jovice basin, Czech Republic. Bull. N. M.Mus. Nat. Hist. Sci. 51, 179–181.

Milàn, J., Bromley, R.G., 2007. The impact of sediment consistency on track and undertrack morphology: experiments with emu tracks in layered cement. Ichnos 15, 19–27.

Mirzaie Ataabadi, M., Khazaee, A.R., 2004. New Eocene mammal and bird footprints from Birjand area, eastern Iran. Ichnos 11, 363–370.

Modesto, S.P., Anderson, J.S., 2004. The phylogenetic definition of Reptilia. Syst. Biol. 53 (5), 815–821.

Montgelard, C., Catzeflis, F.M., Douzery, E., 1997. Phylogenetic relationships of artiodactyls and cetaceans as deduced from the comparison of cytochrome b and 12S rRNA mitochondrial sequences. Mol. Biol. Evol. 14 (5), 550–559.

Montgelard, C., Ducrocq, S., Douzery, E., 1998. What is a Suiforme (Artiodactyla)?: contribution of Cranioskeletal and mitochondrial DNA data. Mol. Phylogenet. Evol. 9 (3), 528–532.

Morgan, G.S., Williamson, T.E., 2000. Middle Miocene (late Barstovian) vertebrates from the Benavidez Ranch local fauna, Albuquerque basin, New Mexico. New Mexico Museum Nat. Hist. Sci. Bullet. 16, 195–207.

Mudelsee, M., Bickert, T., Lear, C.H., Lohmann, G., 2014. Cenozoic climate changes: a review based on time series analysis of marine benthic $\delta^{18}O$ records. Rev. Geophys. 52 (3), 333–374.

Muizon, C., Lange-Badré, B., 2007. Carnivorous dental adaptations in tribosphenic mammals and phylogenetic reconstruction. Lethaia 30, 353–366.

Mujal, E., Schoch, R.R., 2020. Middle Triassic (Ladinian) amphibian tracks from the Lower Keuper succession of southern Germany: implications for temnospondyl locomotion and track preservation. Palaeogeogr. Palaeoclimatol. Palaeoecol. 543, 109625.

Murelaga, X., Baceta, J.I., Astibia, H., Badiola, A., Pereda Suberbiola, X., 2000. Icnitas de perisodáctilos en el Oligoceno de Navarra: posición estratigráfica y sistemática. Geogaceta 27, 15–17.

Murie, O.J., 1982. A Field Guide to Animal Tracks. Houghton Mifflin, Boston, 375 pp.

Mustoe, G.E., 2019. Lower Eocene footprints from northwest Washington, USA. Part 1: reptile tracks. Geosciences 9 (7), 321.

Mustoe, G.E., Hopkins, D.Q., 2013. Mammal and bird tracks from the Eocene Puget group, northwest Washington, USA. Ichnos 20, 36–42.

Mustoe, G.E., Tucker, D.S., Kemplin, K.L., 2012. Giant Eocene bird footprints from northwestnorthwest Washington, USA. Palaeontology 55, 1293–1305.

Nagashima, H., Kuraku, S., Uchida, K., Kawashima-Ohya, Y., Narita, Y., Kuratani, S., 2012. Body plan of turtles: an anatomical, developmental and evolutionary perspective. Anat. Sci. Int. 87, 1–13.

Nesbitt, S.J., 2011. The early evolution of archosaurs: relationships and the origin of major clades. Bull. Am. Mus. Nat. Hist. 352, 1–292.

Neto de Carvalho, C.N., Muñiz, F., Belaústegui, Z., Belo, J., Gómez, P., Toscano, A., Cachão, M., Cunha, P.P., Ruiz, F., Cáceres, L.M., Rodriguez-Vidal, J., 2020. Paleoecological implications of large-sized wild boar tracks recorded during the Last Interglacial (MIS 5) at Huelva (SW Spain). Palaios 35 (12), 512–523. https://doi.org/10.2110/palo.2020.058.

Neto de Carvalho, C.N., Muñiz, F., Cáceres, L.M., Belaústegui, Z., Rodríguez-Vidal, J., Belo, J., Moreira, N., Cachão, M., Cunha, P.P., Figueiredo, S., Galán, J.M., 2022. Aurochs roamed along the SW coast of Andalusia (Spain) during late Pleistocene. Sci. Rep. 12 (1), 9911.

O'Leary, M.A., Bloch, J.I., Flynn, J.J., Gaudin, T.J., Giallombardo, A., Giannini, N.P., et al., 2013. The placental mammal ancestor and the post–K-Pg radiation of placentals. Science 339 (6120), 662–667.

Oliva, C., Arregui, C., 2018. Mammalian Ichnopathology: a case study of Holartic Ungulates (Gomphotheriidae, Equidae, Camelidae) of the Late Pleistocene of South America. Ichnotaxomic implications. Boletin de la Soc. Geol. Mex. 70, 417–447.

Olsen, K., Padian, P.E., 1986. Earliest records of *Batrachopus* from the southwestern United States, and a revision of some Early Mesozoic crocodylomorph ichnogenera. In: Padian, K. (Ed.), The Beginning of the Age of Dinosaurs. Cambridge University Press, Cambridge, pp. 259–273.

Osborn, H.F., 1929. The Titanotheres of Ancient Wyoming, Dakota, and Nebraska. Department of the Interior, US Geological Survey.

Padian, K., Olsen, P.E., 1984. Footprints of the Komodo monitor and the trackways of fossil reptiles. Copeia 1984, 662–671.

Panin, N., 1965. Coexistence de traces de pas de vertébrés et des mêcanoglyphes dans la Molasse Miocéne des Carpates Orientales. Revue roumaine de géologie, géophysique et géographie. Ser. Geol. 9, 141–163.

Panin, N., Avram, E., 1962. Noe urme de vertebrate in Miocenul Subcarpaţilor romineşti. Stud. Cercet. Geol. 7, 455–484.

Panin, N., Lăzărescu, V., Grujinski, C., 1966. Un nou sector ichno-fosilifer in terenurile Miocene din Carpatii Orientali. Buletinul Institutului de petrol. Gaze Şi Geologie 14, 9–21.

Park, W.M., Lockley, M.G., Kim, J.Y., Kim, K.S., 2018. Anuran (frog) trackways from the Cretaceous of Korea. Cretac. Res. 86, 135–148.

Patterson, J., Lockley, M., 2004. A probable diatryma track from the Eocene of Washington: an intriguing case of controversy and skepticism. Ichnos 11, 341–347.

Payros, A., Astibia, H., Cearreta, A., Pereda-Suberbiola, X., Murelaga, X., Badiola, A., 2000. The upper Eocene South Pyrenean coastal deposits (Liedena Sandstone, Navarre): sedimentary facies, benthic foraminifera and avian ichnology. Facies 42, 107–131.

Peabody, F.E., 1954. Trackways of an ambystomid salamander from the Paleocene of Montana. J. Paleontol. 28 (1), 79–83.

Pérez-García, A., 2017. The Iberian fossil record of turtles: an update. J. Iber. Geol. 43, 155–191.

Pérez-García, A., 2020. Surviving the Cretaceous-Paleogene mass extinction event: a terrestrial stem turtle in the Cenozoic of Laurasia. Sci. Rep. 10 (1), 1489.

Pérez-García, A., 2022. Palaeoecological analysis and diversity of turtles and other reptiles. Diversity 14 (2), 88.

Pérez-Lorente, F., Serrano, F., Rodriguez, T., Mancheno, M.A., Romero, M., 1999. Pisadas fósiles de mamíferos en el Mioceno Superior de La Hoya de la Sima (Jumilla, Murcia. Espana). Rev. Esp. Palaontol. 14, 257–267.

Peterson, O.A., 1907. Preliminary notes on some American chalicotheres. Am. Nat. 41, 733–752.

Petti, F.M., Bernardi, M., Ashley-Ross, M.A., Berra, F., Tessarollo, A., Avanzini, M., 2014. Transition between terrestrial-submerged walking and swimming revealed by Early Permian amphibian trackways and a new proposal for the nomenclature of compound trace fossils. Palaeogeogr. Palaeoclimatol. Palaeoecol. 410, 278–289.

Pickford, M., Moya Sola, S., Mein, P., 1997. A revised phylogeny of Hyracoidea (Mammalia) based on new specimens of Pliohyracidae from Africa and Europe. Neues Jahrbuch Geol. Palaontol. Abhand. 205, 265–288.

Piras, P., Buscalioni, A.D., 2006. *Diplocynodon muelleri* comb. nov., an Oligocene diplocynodontine alligatoroid from Catalonia (Ebro Basin, Lleida province, Spain). J. Vertebr. Paleontol. 26 (3), 608–620.

Pocock, R.I., 1915. On the feet and glands and other external characters of the Paradoxurinae Genera *Paradoxurus, Arctictis, Arctogalidia*, and *Nandinia*. Proc. Zool. Soc. Lond. 28, 87–412.

Pocock, R.I., June 1916. On the external characters of the mongooses (Mungotidae). Proc. Zool. Soc. Lond. 86, 349–374.

Pough, F.H., Andrews, R.M., Cadle, J.E., Crump, M.L., Savitzky, A.H., Wells, K.D., 1998. Herpetology. Prentice-Hall, Upper Saddle River, New Jersey, 577 pp.

Prothero, D.R., 2013. Rhinoceros Giants: The Paleobiology of Indricotheres. Indiana University Press, Indiana, 160 pp.

Prothero, D.R., 2016. The Princeton Field Guide to Prehistoric Mammals, vol. 112. Princeton University Press, Princeton, 240 pp.

Prothero, D.R., 2017. Evolution: What the Fossils Say and Why it Matters. Columbia University Press, New York, 408 pp.

Prothero, D.R., Guerin, C., Manning, E., 1989. The history of the Rhinocerotoidea. In: Prothero, D.R., Schocho, R.M. (Eds.), The Evolution of Perissodactyls. Oxford University Press, Oxford, pp. 321–340.

Prothero, D.R., Domning, D., Fordyce, R.E., Foss, S., Janis, C., Lucas, S., Marriott, K.L., Metais, G., Naish, D., Padian, K., Rössner, G., 2022. On the Unnecessary and Misleading Taxon "Cetartiodactyla". Journal of Mammalian Evolution, pp. 1–5.

Puértolas-Pascual, E., Young, M.T., Brochu, C.A., 2020. Introducing the first European symposium on the evolution of Crocodylomorpha. Zool. J. Linn. Soc. 189 (2), 419–427.

Pyron, R.A., Burbrink, F.T., Wiens, J.J., 2013. A phylogeny and revised classification of Squamata, including 4161 species of lizards and snakes. BMC Evol. Biol. 13, 1–54.

Rădan, S., Brustur, T., 1993. Urme de paşi de păsări în Oligocenul superior din bazinul Petroşani (Carpaţii Meridionali). Stud. Cercet. Geol. 38, 71–80.

Rage, J.-C., Roček, Z., 1989. Redescription of *Triadobatrachus massinoti* (Piveteau, 1936) an anuran amphibian from the early Triassic. Palaeontogr. Am. 206, 1–16.

Radinsky, L.B., 1969. The early evolution of the Perissodactyla. Evolution 23, 308–328.

Rajkumar, H.S., Mustoe, G.E., Khaidem, K.S., Soibam, I., 2015. Crocodylian tracks from lower Oligocene flysch deposits of the Barail group, Manipur, India. Ichnos 22 (2), 122–131.

Remeika, P., 1999. Identification, stratigraphy, and age of Neogene vertebrate footprints from the Vallecito-fish Creek basin, Anza-Borrego Desert state Park, California. San Bernardino County Museum Assoc. Quart. 46 (2), 37–46.

Remeika, P., 2001. The Fish Creek Canyon Ichnofauna; a Pliocene (Blancan) Vertebrate Footprint Assemblage from Anza-Borrego Desert State Park. Technical Report - National Park Service, Geological Resource Division, California, pp. 55–75.

Remeika, P., 2006. Fossil footprints of Anza-Borrego. In: Jefferson, G.T., Lindsay, L. (Eds.), Fossil Treasures of the Anza-Borrego Desert: The Last Seven Million Years: El Cajon. Sunbelt Publ, pp. 311–327.

Reynolds, S.H., 1897. The Vertebrate Skeleton. Cambridge University Press, Cambridge, 562 pp.

Reynolds, R.E., 2012. Ichnites in the Bouse Formation, Amboy, San Bernardino County, California. In: Reynolds, R.E. (Ed.), Searching for the Pliocene: Southern Exposures. California State University Desert Studies Center, pp. 136–139.

Reynolds, R.E., Milner, A.R.C., 2007. Preliminary description of mammal trackways from Middle Miocene (late Barstovian NALMA) Enterprise Reservoir sediments in southwestern Utah. In: Lucas, S.G., Spielmann, J.A.,

Lockley, M.G. (Eds.), Cenozoic Vertebrate Tracks and Traces, vol. 42, pp. 261–266. Bull. New Mexico Mus Nat Hist Sci.

Riff, D., 2007. Anatomia apendicular de *Stratiotosuchus maxhechti* (Baurusuchidae, Cretáceo Superior do Brasil) e análise filogenética dos Mesoeucrocodylia. Museu Nacional da Universidade Federal do Rio de Janeiro, Brasil, 406 pp.

Rio, J.P., Mannion, P.D., 2021. Phylogenetic analysis of a new morphological dataset elucidates the evolutionary history of Crocodylia and resolves the long-standing gharial problem. PeerJ 9, e12094.

Robinson, S.F., 1991. Bird and frog tracks from the Late Cretaceous Blackhawk Formation in east-central Utah. In: Chidsey, T.C. (Ed.), Geology of East-Central Utah, vol. 19. Utah Geol Assoc Publ, pp. 325–334.

Rodríguez-de la Rosa, R.A., Guzmán-Gutiérrez, J.R., 2012. Huellas de aves y mamíferos del Neógeno de Jalisco, México. Palaontol. Mexic. 62, 147–158.

Rojas-Manriquez, M.A., Feola, S., Melchor, R.N., 2024. Trace fossil zonation in interdune lakes: example from the Neogene of eastern Patagonia, Argentina. Ichnos 30 (4), 303–337. https://doi.org/10.1080/10420940.2024.2309487.

Roscito, J.G., Nunes, P.M., Rodrigues, M.T., 2014. Digit evolution in gymnophthalmid lizards. Int. J. Dev. Biol. 58, 895–908. https://doi.org/10.1387/ijdb.140255jg.

Rose, K.D., 2006. The Beginning of the Age of Mammals. JHU Press, Baltimore, 429 pp.

Rühle von Lilienstern, H., 1939. Fährten und Spüren im Chirotheriem– Sandstein von Südthüringen. Fortschritte der Geologie und Paläontologie 12, 293–387.

Russell, A.P., Bauer, A.M., 2008. The appendicular locomotor apparatus of *Sphenodon* and normal-limbed squamates. In: Gans, C., Gaunt, A.S., Adler, K. (Eds.), Biology of Reptilia, Volume 21: The Skull and Appendicular Locomotor Apparatus of Lepidosauria, Contributions to Herpetology, Ithaca, New York. Society for the Study of Reptiles & Amphibians, pp. 1–465.

Rybczynski, N., Dawson, M.R., Tedford, R.H., 2009. A semi-aquatic Arctic mammalian carnivore from the Miocene epoch and origin of Pinnipedia. Nature 458, 1021–1024.

Sánchez-Villagra, M.R., Aguilera, O., Horovitz, I., 2003. The anatomy of the world's largest extinct rodent. Science 301, 1708–1710.

Sánchez-Villagra, M.R., Mitgutsch, C., Nagashima, H., Kuratani, S., 2007. Autopodial development in the sea turtles *Chelonia mydas* and *Caretta caretta*. Zool. Sci. 24 (3), 257–263.

Sánchez-Villagra, M., Ziermann, J., Olsson, L., 2008. Limb chondrogenesis in *Graptemys nigrinoda* (Emydidae), with comments on the primary axis and the digital arch in turtles. Amphib-reptil. 29 (1), 85–92.

San Mauro, D., 2010. A multilocus timescale for the origin of extant amphibians. Mol. Phylogenet. Evol. 56 (2), 554–561.

Santamaria, R., López, G., Casanovas-Cladellas, M.L., 1989–1990. Nuevos yacimientos con icnitas de mamíferos del Oligoceno de los alrededores de Agramunt (Lleida, España). Paleontol. Evol. 23, 141–152.

Sarjeant, W.A.S., Langston Jr., W., 1994. Vertebrate footprints and invertebrate traces from the Chadronian (late Eocene) of Trans-Pecos Texas. Texas Memorial Museum Bullet. 36, 1–86.

Sarjeant, W.A.S., Reynolds, R.E., 1999. Camel and horse footprints from the Miocene of California. San Bernardino County Museum Assoc. Quart. 46, 31–32.

Sarjeant, W.A.S., Reynolds, R.E., 2001. Bird footprints from the Miocene of California. In: Reynolds, R.E. (Ed.), 2001 Desert Symposium. The Changing Face of the East Mojave Desert. California State University, Fullerton, pp. 21–40.

Sarjeant, W.A., Wilson, J.A., 1988. Late Eocene (Duchesnean) mammal footprints from the skyline channels of Trans-Pecos Texas. Tex. J. Sci. 40 (4), 439–446.

Sarjeant, W.A.S., Reynolds, R.E., Kissell-Jones, M.M., 2002. Fossil creodont and carnivore footprints from California, Nevada, and Wyoming. In: Reynolds, R.E. (Ed.), Between the Basins: Exploring the Western Mojave and Southern Basin and Range Province. California State University, Desert Studies Consortium, pp. 37–50. Abstracts from the 2002 Desert Symposium.

Schaller, N.U., D'Août, K., Villa, R., Herkner, B., Aerts, P., 2011. Toe function and dynamic pressure distribution in ostrich locomotion. J. Exp. Biol. 214, 1123–1130.

Schoch, R.R., 2014. Amphibian Evolution: The Life of Early Land Vertebrates. John Wiley & Sons.

Scrivner, P.J., Bottjer, D.J., 1986. Neogene avian and mammalian tracks from Death Valley National Monument, California: their context, classification and preservation. Palaeogeogr. Paleoclimatol. Palaeoecol. 57, 285–331.

Serrano-Brañas, C.I., Espinosa-Chávez, B., Ventura, J.F., Barrera-Guevara, D., Torres-Rodríguez, E., Vega, F.J., 2022. New insights on the avian trace fossil record from NE Mexico: evidences on the diversity of latest Maastrichtian web-footed bird tracks. J. South Am. Earth Sci. 113, 103686.

Sheil, C.A., Portik, D., 2008. Formation and ossification of limb elements in *Trachemys scripta* and a discussion of autopodial elements in turtles. Zool. Sci. 25 (6), 622–641.

Shoshani, J., 1996. Skeletal and other basic anatomical features of elephants. In: Shoshani, J., Tassy, P. (Eds.), The Proboscidea: Evolution and Palaeoecology of Elephants and Their Relatives. Oxford University Press, Oxford, pp. 9–20.

Shoshani, J., Tassy, P., 2005. Advances in proboscidean taxonomy & classification, anatomy & physiology, and ecology & behavior. Quat. Int. 126, 5–20.

Shubin, N.H., Wake, D.B., 2003. Morphological variation, development, and evolution of the limb skeleton of salamanders. Amphib. Biol. 5, 1782–1808.

Simpson, G.G., 1931. A new insectivore from the Oligocene, Ulan Gochu horizon, of Mongolia. American Museum Novitates 505, 56.9,33A (1181:51.7).

Soibelzon, L.H., Schubert, B.W., 2011. The largest known bear, *Arctotherium angustidens*, from the early Pleistocene Pampean region of Argentina: with a discussion of size and diet trends in bears. J. Paleontol. 85 (1), 69–75.

Solounias, N., Danowitz, M., Stachtiaris, E., Khurana, A., Araim, M., Sayegh, M., Natale, J., 2018. The evolution and anatomy of the horse manus with an emphasis on digit reduction. R. Soc. Open Sci. 5 (1), 171782.

Sorkin, B., 2008. A biomechanical constraint on body mass in terrestrial mammalian predators. Lethaia 41, 333–347.

Spaulding, M., Flynn, J.J., 2012. Phylogeny of the Carnivoramorpha: the impact of postcranial characters. J. Syst. Palaeontol. 10 (4), 653–677.

Sterli, J., 2015. A review of the fossil record of Gondwanan turtles of the clade Meiolaniformes. Peabody Mus. Nat. Hist. 56 (1), 21–45.

Steyer, J.S., 2011. A Review of "The Rise of Amphibians: 365 Million Years of Evolution" by Robert L. Carroll, 2009. The John Hopkins University Press, Baltimore, 392 pp.

Sues, H.-D., 2019. Archosauromorpha: Choristodera. In: Sues, H.D. (Ed.), The Rise of Reptiles: 320 Million Years of Evolution. Johns Hopkins University Press, Baltimore, pp. 167–168.

Sulimski, A., 1969. Paleocene genus *Pseudictops* Matthew, Granger, and Simpson, 1929, (Mammalia) and its revision. Palaeontol. Pol. 1969 (19–1968), 101–129.

Tambussi, C.P., Degrange, F., 2012. South American and Antarctic Continental Cenozoic Birds: Paleobiogeographic Affinities and Disparities. Springer Science & Business Media, Dordrecht, 113 pp.

Tamura, K., Yonei-Tamura, S., Yano, T., Yokoyama, H., Ide, H., 2008. The autopod: its formation during limb development. Dev. Growth Differ. 50, S177–S187.

Tarquini, J., Toledo, N., Morgan, C.C., Soibelzon, L.H., 2017. The forelimb of †*Cyonasua* sp. (Procyonidae, Carnivora): ecomorphological interpretation in the context of carnivorans. Earth Environ. Sci. Trans. R Soc. Edinb. 106 (4), 325–335.

Taylor, M.E., 1988. Foot structure and phylogeny in the Viverridae (Carnivora). J. Zool. 216 (1), 131–139.

Theodor, J.M., Erfurt, J., Grégoire, M., 2007. The earliest artiodactyls: diacodexeidae, dichobunidae, homacodontidae, leptochoeridae and raoellidae. In: Prothero, D.R., Foss, S.E. (Eds.), Evolution of Artiodactyls. Johns Hopkins University Press, pp. 32–58.

Thenius, E., 1967. Säugetierfährten aus dem Rohrbacher Konglomerat (Pliozän) von Niederösterreich. Ann. Nathist Mus Wien 363–379.

Tineo, D.E., Pérez, L.M., Bona, P., Comerio, M.A., Noriega, J.I., Poiré, D.G., 2018. The oldest record of *Aramayoichnus rheae* from the Neogene of northwestern Argentina. Ameghiniana 55, 109–116.

Tito, G., De Iuliis, G., 2003. Morphofunctional aspects and palaeobiology of the manus in the giant ground sloth *Eremotherium* Spillmann 1948. Senckenberg. Biol. 83, 79–94.

Tsagkogeorga, G., McGowen, M.R., Davies, K.T.J., Jarman, S., Polanowski, A., Bertelsen, M.F., Rossiter, S.J., 2015. A phylogenomic analysis of the role and timing of molecular adaptation in the aquatic transition of cetartiodactyl mammals. R. Soc. Open Sci. 2, 150156.

Turazzini, G.F., Gómez, R.O., 2023. Comparative osteology of paradoxical frogs (Hylidae: Pseudae) with comments on diagnostic features, evolutionary trends and potential aquatic adaptations. Zool. Anz. 303, 47–70.

Turkalo, A., Barnes, R., 2013. *Loxodonta cyclotis*. Forest elephant. In: Kingdon, J., Happold, D., Hoffmann, M., Butynski, T., Happold, M., Kalina, J. (Eds.), Mammals of Africa, Volume I. Bloomsbury, London, pp. 195–200.

Valencia Fermin, G.L., 2020. Trace Fossils and Fluvial Architecture of the Miocene Vinchina Formation, Vinchina Basin, Northwest Argentina. Doctoral dissertation. University of Saskatchewan.

Valentini, M., Conti, M.A., Mariotti, N., 2007. Lacertoid footprints of the Upper Permian Arenaria di Val Gardena Formation (Northern Italy). Ichnos 14 (3–4), 193–218.

Vallon, L.H., Rindsberg, A.K., Bromley, R.G., 2016. An updated classification of animal behaviour preserved in substrates. Geodin. Acta 28 (1–2), 5–20.

Vaughan, T.A., Ryan, J.M., Czaplewski, N.J., 2013. Mammalogy. Jones & Bartlett Publishers, Burlington, 755 pp.

Vellekoop, J., Woelders, L., van Helmond, N.A., Galeotti, S., Smit, J., Slomp, C.P., Brinkhuis, H., Claeys, P., Speijer, R.P., 2018. Shelf hypoxia in response to global warming after the Cretaceous-Paleogene boundary impact. Geology 46 (8), 683–686.

Vera, R.B., Krapovickas, V., 2022. Paleobiology and paleoecology of ungulates from the Miocene of South America based on an ichnological analysis. Palaeogeogr. Palaeoclimatol. Palaeoecol. 606, 111247.

Vyalov, O.S., 1965. Sztratigrafija Neogenovüh Molassez Predkarpatszkovo Progiva [Stratigraphy of the Neogene Molasse of the PreCarpathian Basin]. Naukova Dumka, Kiev, 178 pp.

Vyalov, O.S., 1966. Sledy Zhiznedeyatelnosti Organizmov I Ikh Paleontologicheskoe Znacheniye [Traces of Living Organisms and Their Paleontological Significance]. Naukova Dumka, Akademiya Nauk Ukrainskoy SSR Institut Geologii I Geokhimii Goryuchikh Iskopayemykh, Kiev, 219 pp.

Vyalov, O.S., 1983. The tracks of prey beasts and birds from the Pliocene of the western Turkmenia. Paleontologicheskii Sbornik 20, 74–79.

Vyalov, O.S., 1985. Szledu Pozvonoesnuh Iz Nızsnego Miocena Vengru Paleont Szboi, vol. 22, pp. 71–78.

Vyalov, O.S., 1986. Artiodactyl traces from the lower Miocene of Hungary. Paleontologicheskii Sbornik 23, 63–69.

Vyalov, O.S., 1989. Pliocene bird tracks from Iran assigned to the genus *Urmiornis*. Paleontol. J. 23, 119–121.

Vilas, L., Ariasi, C., Herrero, C., Coruñai, F., Herrero, E., 2009. La geología en el museo municipal de Jumilla (Murcia, España). Bol. R. Soc. Esp. Hist. Nat. Secc. Geol. 103, 65–79.

Vincelette, A.R., Renders, E., Scott, K.M., Falkingham, P.L., Janis, C.M., 2023. *Hipparion* tracks and horses' toes: the evolution of the equid single hoof. R. Soc. Open Sci. 10 (6), 230358.

Viranta, S., 1996. European Miocene Amphicyonidae: taxonomy, systematics and ecology. Acta Zool. Fennica 204, 1–61.

Vislobokova, I.A., 2013. On the origin of Cetartiodactyla: comparison of data on evolutionary morphology and molecular biology. Paleontol. J. 47, 321–334.

Wagner, G.P., Griffith, O.W., Bergmann, P.J., Bello-Hellegouarch, G., Kohlsdorf, T., Bhullar, A., Siler, C.D., 2018. Are there general laws for digit evolution in squamates? The loss and re-evolution of digits in a clade of fossorial lizards (Brachymeles, Scincinae). J. Morphol. 279 (8), 1104–1119.

Warner, S.E., Pickering, P., Panagiotopoulou, O., Pfau, T., Ren, L., Hutchinson, J.R., 2013. Size-related changes in foot impact mechanics in hoofed mammals. PLoS One 8, e54784.

Weems, R.E., 2018. An early Pleistocene (early Irvingtonian) footprint fauna from the Bacons Castle formation, Westmoreland county, Virginia. Bull. N. M.Mus. Nat. Hist. Sci. 79, 731–748.

Weems, R.E., 2021. Additions to early Pleistocene (early Irvingtonian) Stratford Hall local ichnofauna from the Bacons Castle Formation, westmoreland county, Virgina (USA). Bull. N. M.Mus. Nat. Hist. Sci. 82, 487–496.

Weems, R.E., Bachman, J., Rayburn, W.T., McCloud, J.R., 2021. A vertebrate ichnofauna from the late Pleistocene (Sangamon interglacial) Tabb Formation at Stratford Hall Plantation, Westmoreland county, Virginia USA. Bull. N. M.Mus. Nat. Hist. Sci. 505–512.

Werdelin, L., Solounias, N., 1991. The Hyaenidae: taxonomy, systematics and evolution. Fossils Strata 30, 1–104.

Wiens, J.J., Brandley, M.C., Reeder, T.W., 2006. Why does a trait evolve multiple times within a clade? Repeated evolution of snakeline body form in squamate reptiles. Evolution 60 (1), 123–141.

Witton, M.P., 2017. Recreating an Age of Reptiles. The Crowood Press, Ramsbury, 112 pp.

Wolff, J.O., Sherman, P.W., 2008. Rodent societies as model systems. In: Wolff, J.O., Sherman, P.W. (Eds.), Rodent Societies: An Ecological and Evolutionary Perspective. University of Chicago Press, Chicago, pp. 3–7.

Xing, L., Belvedere, M., Buckley, L., Falk, A.R., Lockley, M.G., Klein, H., Abbassi, N., Zhang, X., Tang, Y., 2014. First record of bird tracks from Paleogene of China (Guangdong province). Palaeogeogr. Palaeoclimatol. Palaeoecol. 414, 415–425.

Yang, S.-Y., Lockley, M.G., Greben, R., Erickson, B.R., Lim, S.-K., 1995. Flamingo and duck-like bird tracks from the Late Cretaceous and early Tertiary: evidence and implications. Ichnos 4, 21–34.

Young, C.C., 1943. Note on some fossil footprints in China Bull. New Mexico Mus. Nat. Hist. Sci. 13, 151–154.
Young, C.C., 1960. Fossil footprints in China. Vertebr. Palasiat. 4 (2), 53–66.
Zapfe, H., 1979. *Chalicotherium grande* (Blainv.) aus der miozänen Spaltenfüllung von Neudorf an der March (Devinská Nová Ves), Tschechoslowakei. Neue Denkschr Naturh Mus Wien 2, 1–282.
Zardoya, R., Meyer, A., 2001. On the origin of and phylogenetic relationships among living amphibians. Proc. Natl. Acad. Sci. USA 98 (13), 7380–7383.
Zonneveld, J.P., Zaim, Y., Rizal, Y., Ciochon, R.L., Bettis III, E.A., Aswan, Gunnell, G.F., 2011. Oligocene shorebird footprints, Kandi, Ombilin basin, Sumatra. Ichnos 18, 221–227.
Zub, K., Szafrańska, P.A., Konarzewski, M., Redman, P., Speakman, J.R., 2009. Trade-offs between activity and thermoregulation in a small carnivore, the least weasel *Mustela nivalis*. Proc. R. Soc. B: Biol. Sci. 276 (1663), 1921–1927.

Further reading

Hildebrand, M., Bramble, M., Liem, K., Wake, D., 1985. Functional Vertebrate Morphology. Harvard University Press, Harvard, 430 pp.
Lockley, M.G., Xing, L., Xu, X., 2019b. The "lost" holotype of *Laiyangpus liui* (Lower Cretaceous, Shandong Province, China) is found: implications for trackmaker identification, ichnotaxonomy and interpretation of turtle tracks. Cretac. Res. 95, 260–267.

Index

'*Note:* Page numbers followed by "f" indicate figures and "t" indicate tables.'

B

T

V

W

X

Y

Z